Silver Nanomaterials for Agri-Food Applications

Nanobiotechnology for Plant Protection

Silver Nanomaterials for Agri-Food Applications

Edited by

Kamel A. Abd-Elsalam
Research Professor, Plant Pathology Research Institute,
Agricultural Research Center (ARC), Giza, Egypt

ELSEVIER

Elsevier
Radarweg 29, PO Box 211, 1000 AE Amsterdam, Netherlands
The Boulevard, Langford Lane, Kidlington, Oxford OX5 1GB, United Kingdom
50 Hampshire Street, 5th Floor, Cambridge, MA 02139, United States

Copyright © 2021 Elsevier Inc. All rights reserved.

No part of this publication may be reproduced or transmitted in any form or by any means, electronic or mechanical, including photocopying, recording, or any information storage and retrieval system, without permission in writing from the publisher. Details on how to seek permission, further information about the Publisher's permissions policies and our arrangements with organizations such as the Copyright Clearance Center and the Copyright Licensing Agency, can be found at our website: www.elsevier.com/permissions.

This book and the individual contributions contained in it are protected under copyright by the Publisher (other than as may be noted herein).

Notices
Knowledge and best practice in this field are constantly changing. As new research and experience broaden our understanding, changes in research methods, professional practices, or medical treatment may become necessary.

Practitioners and researchers must always rely on their own experience and knowledge in evaluating and using any information, methods, compounds, or experiments described herein. In using such information or methods they should be mindful of their own safety and the safety of others, including parties for whom they have a professional responsibility.

To the fullest extent of the law, neither the Publisher nor the authors, contributors, or editors, assume any liability for any injury and/or damage to persons or property as a matter of products liability, negligence or otherwise, or from any use or operation of any methods, products, instructions, or ideas contained in the material herein.

Library of Congress Cataloging-in-Publication Data
A catalog record for this book is available from the Library of Congress

British Library Cataloguing-in-Publication Data
A catalogue record for this book is available from the British Library

ISBN: 978-0-12-823528-7

For information on all Elsevier publications
visit our website at https://www.elsevier.com/books-and-journals

Publisher: Matthew Deans
Acquisitions Editor: Simon Holt
Editorial Project Manager: Rafael G. Trombaco
Production Project Manager: Selvaraj Raviraj
Cover designer: Greg Harris

Typeset by SPi Global, India

Contents

Contributors xxiii
Preface xxxi
Series preface xxxiii

CHAPTER 1 Silver-based nanomaterials for sustainable applications in agroecology: A note from the editor 1
Kamel A. Abd-Elsalam

- 1.1 Introduction 1
- 1.2 AgNP-based nanosystem applications 2
  - 1.2.1 Agri-food applications 3
  - 1.2.2 Veterinary applications 4
  - 1.2.3 Environmental applications 6
- 1.3 Challenges 7
- 1.4 Biosafety and regulations 7
- 1.5 Future trends 9
- 1.6 Conclusion 10
- References 11

PART 1 Antimicrobials

CHAPTER 2 Silver-based nanostructures as antifungal agents: Mechanisms and applications 17
Santwana Padhi and Anindita Behera

- 2.1 Introduction 17
- 2.2 Mechanism of action of antifungal agents 18
- 2.3 Mechanistic approach of silver nanoparticles as antifungal agents 22
- 2.4 Comparison of antifungal activity of biosynthesized silver nanoparticles with physically or chemically synthesized nanoparticles 25
- 2.5 Application of silver nanoparticles in human fungal infections 26
- 2.6 Application of silver nanoparticles in plant fungal infections 29
- 2.7 Conclusion 32
- References 32
- Further reading 38

CHAPTER 3		**Antimicrobial properties of surface-functionalized silver nanoparticles** 39
		Parteek Prasher and Mousmee Sharma
	3.1	Introduction 39
	3.2	Synthetic paradigm for colloidal stabilized AgNPs 40
	3.3	Biogenic silver nanoparticles as antimicrobial agents 41
	3.4	Aqueous phase transfer of colloidal stabilized AgNPs 42
	3.5	Oligodynamic antimicrobial mechanism of AgNPs 44
	3.6	Biocidal activity of AgNPs against MDR microbes 46
	3.7	Synergism of AgNPs with antibiotics 46
	3.8	Conclusion 46
		References 56
CHAPTER 4		**Silver-based nanoantimicrobials: Mechanisms, ecosafety, and future perspectives** 67
		Parinaz Ghadam, Parisa Mohammadi, and Ahya Abdi Ali
	4.1	Introduction 67
	4.2	AgNPs as nano-antimicrobials (antibacterial, antifungal, and antiviral) 68
		4.2.1 Antibacterial activity of AgNPs 68
		4.2.2 Antifungal activity of AgNPs 72
		4.2.3 Antiviral activity of AgNPs 73
	4.3	Toxicity 74
		4.3.1 Reducing nanoparticles toxicity 76
	4.4	Applications 78
		4.4.1 AgNPs and packaging food 78
		4.4.2 Nanocomposite fabrics 79
		4.4.3 AgNP as a pesticide 80
		4.4.4 Air decontamination 80
		4.4.5 Water sanitization 80
		4.4.6 Treatment of multidrug-resistant (MDR) microorganisms 82
		4.4.7 Antifungal activity and mycotoxin control 82
		4.4.8 Conservation and restoration of monuments 84
	4.5	Perspective and future trends 85
	4.6	Conclusion 86
		References 86

PART 2 Food applications

CHAPTER 5 Silver nanoparticles as nanomaterial-based nanosensors in agri-food sector 103
Mythili Ravichandran, Paulkumar Kanniah, and Murugan Kasi

5.1	Introduction 103
5.2	Nanosensors 104
5.3	Significance of silver as nanosensors 105
5.4	Contaminant exposure prevention and mitigation applications 105
5.5	Applications of silver nanosensors in agriculture 107
5.6	Silver nanosensors: An alarm for heavy metal contamination 107
5.7	Functionalization of AgNPs for sensing other metal ions 109
5.8	Mechanism of heavy metal detection using silver nanosensors 111
5.9	Silver nanosensors for pesticide detection 112
5.10	Food contaminant detection 114
5.11	Conclusion and future perspectives 116
	Acknowledgments 117
	References 117

CHAPTER 6 Silver-based nanomaterials for food packaging applications 125
Shiji Mathew and E.K. Radhakrishnan

- 6.1 Introduction 125
- 6.2 Food packaging and its importance in food safety 126
- 6.3 Nanotechnology in food packaging 127
- 6.4 Benefits of silver as nanofiller in food packagings 128
- 6.5 Mechanisms of antimicrobial action of AgNPs 129
- 6.6 Preparation of silver-based food packaging nanomaterials 131
- 6.7 Silver-based food packaging nanomaterials 131
  - 6.7.1 AgNP-based nondegradable nanocomposite film 132
  - 6.7.2 AgNP-based bio-nanocomposite film 133
  - 6.7.3 AgNP-based edible films/nanocoatings 133
  - 6.7.4 AgNP-based polymer blend nanocomposites 134
- 6.8 Practical application of silver-based nanocomposites on food systems 136
- 6.9 Safety assessments 138

6.10	Conclusion	139
	References	139

CHAPTER 7 Emerging silver nanomaterials for smart food packaging in combating food-borne pathogens..... 147

Divya Sachdev, Akanksha Joshi, Neetu Kumra Taneja, and Renu Pasricha

7.1	Introduction	147
7.2	Major foodborne pathogens responsible for food spoilage and FBDs	148
7.3	Sustainable food packaging (SFP)	150
	7.3.1 Demands and challenges	150
	7.3.2 Nanotechnological aspects of food packaging	152
	7.3.3 How nanomaterials improve barrier and strength properties	155
7.4	Silver nanomaterials (AgNMs) for food packaging	156
	7.4.1 Types of silver nanomaterial-based antimicrobial packaging materials	156
	7.4.2 Silver nanomaterial-based sensors for maintenance of food quality	168
	7.4.3 Commercial silver nonmaterial-based packaging materials in the marketplace	169
7.5	Mechanism of action	173
	7.5.1 Factors affecting antimicrobial potential of AgNPs	175
7.6	Safety aspects and regulations	175
7.7	Conclusion and future prospects	177
	References	178

CHAPTER 8 Novel silver-based nanomaterials for control of mycobiota and biocide analytical regulations in agri-food sector 187

Elena Piecková, Farah K. Ahmed, Renáta Lehotská, and Mária Globanová

8.1	Introduction	187
8.2	Silver nanoparticle biocides	188
8.3	Mycogenic-mediated silver nano-fungicides	190
8.4	Ag-based nanohybrid fungicides	192
8.5	Toxic effects of AgNP-based biocides under discussion	193
8.6	Mechanisms of antifungal activities in phytopathogens	195
8.7	Regulation of phytopathogenic mycobiota in agriculture	197
8.8	Regulation of the contaminant mycobiota in food processing	199

8.9	Sanitary procedures in agrotech and food production	201
8.10	Biocide (disinfectant) effectivity/efficacy tests in vitro	202
8.11	Future perspectives: Proteomic-based fungicides	207
8.12	Conclusion	208
	Conflict of interest	209
	References	209
	Further reading	216

PART 3 Plant science

CHAPTER 9 Silver nanoparticle applications in wood, wood-based panels, and textiles 219
Mohamed Z.M. Salem

9.1	Introduction	219
9.2	Importance of silver nanoparticles in practice	220
	9.2.1 Improving the biological durability properties of wood and wood products with AgNPs	220
	9.2.2 Effect of AgNPs on physical and mechanical characteristics of impregnated wood and wood products	222
	9.2.3 AgNP/wood as a water purification filter	225
	9.2.4 AgNPs as additives for fibers	225
9.3	Conclusions	229
	References	229

CHAPTER 10 Immobilization efficiency and modulating abilities of silver nanoparticles on biochemical and nutritional parameters in plants: Possible mechanisms 235
Luqmon Azeez, Amadu K. Salau, and Simiat M. Ogunbode

10.1	Introduction	235
10.2	Silver nanoparticles in agriculture	237
10.3	Immobilization efficiency of AgNPs	240
	10.3.1 Phytoremediation and AgNPs	241
	10.3.2 Chemical immobilization and adsorption with AgNPs	243
	10.3.3 Catalytic degradation and pollutant-sensing properties of AgNPs	245
10.4	Silver nanoparticles as modulators of plant biochemical and nutritional qualities	245
	10.4.1 Mode of entry	246

		10.4.2 As physiology and phytohormone modulators 248
		10.4.3 As regulators of biochemical, and nutritional parameters and for disease-suppression 249
	10.5	Conclusion 252
		References 252

CHAPTER 11 Controlled-release and positive effects of silver nanoparticles: An overview 265
Ambreen Ahmed and Shabana Wagi

	11.1	Introduction 265
	11.2	Synthesis of nanoparticles 266
		11.2.1 Destructive strategy 266
		11.2.2 Constructive strategy 266
	11.3	Green synthesis of silver nanoparticles 266
		11.3.1 Green synthesis of silver nanoparticles using bacteria 267
		11.3.2 Green synthesis of silver nanoparticles using plants 268
	11.4	Controlled release of silver nanoparticles 269
		11.4.1 Controlled release of silver nanoparticles as nanofertilizers 269
		11.4.2 Controlled release of silver nanoparticles as antioxidant 271
		11.4.3 Controlled release of nanoparticles as antimicrobial agents 272
		11.4.4 Controlled release of silver nanoparticles as nanopesticides 274
	11.5	Positive effects of silver nanoparticles 276
	11.6	Concerns regarding AgNPs 276
	11.7	Conclusion 277
		References 277

CHAPTER 12 Comparison of the effect of silver nanoparticles and other nanoparticle types on the process of barley malting 281
Dmitry V. Karpenko

	12.1	Introduction 281
	12.2	Method for determining the germination of barley and characteristics of barley batches used in experiments 282
	12.3	Characteristics of the nanopreparations used in the studies 284
		12.3.1 Multiwalled carbon nanotubes (MWCNTs) 284
		12.3.2 Nanopreparation of titanium dioxide (TiO_2) 284

		12.3.3 Nanopreparation of copper oxide (CuO)......................284
		12.3.4 Nanopreparation of nickel oxide (NiO) 285
		12.3.5 Silver nanopreparation (Ag)...285
	12.4	Effect of silver nanoparticles on the germination of brewing barley of varying quality...285
	12.5	Effect of multiwalled carbon nanotubes (MWCNTs) on the germination of brewing barley of varying quality 289
	12.6	Effect of titanium dioxide nanoparticles (TiO_2 NPs) on the germination of brewing barley of varying quality 290
	12.7	Effect of copper oxide nanoparticles (CuO NPs) on the germination of brewing barley of varying quality 291
	12.8	Effect of nickel oxide nanoparticles (NiO NPs) on the germination of brewing barley of varying quality 293
	12.9	Conclusion...294
		References 294

CHAPTER 13 Adverse effects of silver nanoparticles on crop plants and beneficial microbes 301
Faisal Zulfiqar and Muhammad Ashraf

	13.1	Introduction 301
	13.2	Effect of silver nanoparticles (AgNPs) on different growth processes..302
		13.2.1 Seed germination..302
		13.2.2 Plant morphological characteristics 303
		13.2.3 Effect on physiological characteristics........................303
		13.2.4 Effect on biochemical characteristics..........................308
	13.3	In situ negative effects of AgNPs on beneficial microbes...310
		13.3.1 Negative impacts on the abundance of beneficial microbes 310
		13.3.2 Negative impacts on the structure of beneficial microbe communities...310
		13.3.3 Negative impacts on the activity of beneficial microbes 311
	13.4	Conclusion and future perspectives 312
		References 313

CHAPTER 14 Silver nanoparticles phytotoxicity mechanisms.....317
Renata Biba, Petra Peharec Štefanic, Petra Cvjetko, Mirta Tkalec, and Biljana Balen

	14.1	Introduction 317
	14.2	AgNP stability in biological media for plant growth...............318

14.3	AgNP coating-dependent effects	330
14.4	Cytotoxic and genotoxic effects of AgNPs	334
14.5	AgNP impact on gene and protein expression	340
14.6	Conclusion and future perspectives	347
	Acknowledgments	347
	References	347

CHAPTER 15 Complex physicochemical transformations of silver nanoparticles and their effects on agroecosystems 357

Parteek Prasher, Mousmee Sharma, and Amit Kumar Sharma

15.1	Introduction	357
15.2	Sources of AgNPs for environmental release	358
	15.2.1 Natural sources	358
	15.2.2 Nanosilver impregnated textiles	358
	15.2.3 Nanosilver loaded medical bandages	360
	15.2.4 Nanopesticides and nanofertilizers	361
15.3	Physicochemical transformations of AgNPs	361
	15.3.1 Speciation	361
	15.3.2 Aggregation	362
	15.3.3 Oxidative dissolution	363
	15.3.4 Sulfidation for toxicity mitigation	365
	15.3.5 Chlorination	366
	15.3.6 Photochemical transformation	367
15.4	Effect on agroecosystems	368
15.5	Conclusion	368
	Acknowledgements	369
	References	373

PART 4 Plant protection applications

CHAPTER 16 Detection and mineralization of pesticides using silver nanoparticles 383

Shubhankar Dube and Deepak Rawtani

16.1	Introduction	383
16.2	Detection of pesticides	385
	16.2.1 Chemical-based assays	385
	16.2.2 Enzyme-based assays	390
	16.2.3 Surface-enhanced Raman spectroscopy based sensing	392
	16.2.4 Aptamer-based assays	396

	16.3	Mineralization of pesticides...397
	16.4	Conclusion..401
		References...401

CHAPTER 17 Pesticide degradation by silver-based nanomaterials 407
Manviri Rani, Jyoti Yadav, and Uma Shanker

	17.1	Introduction..407
		17.1.1 Banned pesticides in India 413
	17.2	Degradation of pesticides with silver nanoparticles..413
	17.3	Photocatalytic efficiency of silver nanoparticles 414
	17.4	Environmental concern of pesticides....................................416
	17.5	Utilization of green-synthesized Ag-based compounds for pesticide degradation..418
	17.6	Future scope and perspectives 420
	17.7	Conclusion..421
		References...421

CHAPTER 18 Application of silver nanoparticles as a chemical sensor for detection of pesticides and metal ions in environmental samples 429
Tushar Kant, Nohar Singh Dahariya, Vikas Kumar Jain, Balram Ambade, and Kamlesh Shrivas

	18.1	Introduction..429
	18.2	Synthesis of silver nanoparticles (AgNPs) 431
		18.2.1 Wet-chemical method..431
		18.2.2 Physical approach..433
		18.2.3 Hydrothermal approach..433
		18.2.4 Electrochemical approach 434
		18.2.5 Biological-assisted method 434
		18.2.6 Sonochemical synthesis 434
	18.3	Characterization techniques..435
	18.4	Working principle of a chemical sensor for the detection of pesticides and metal ions 435
		18.4.1 Colorimetric-based chemical sensor 435
		18.4.2 Fluorescence based-chemical sensor..........................436
		18.4.3 Electrochemical-based chemical sensor.....................437
		18.4.4 Surface-enhanced Raman spectroscopy (SERS) based chemical sensor 437

18.5	Silver nanoparticles as a chemical sensor for detection of pesticides and metal ions in environmental samples	438
	18.5.1 Determination of metals ions using AgNPs based chemical sensors	438
	18.5.2 Determination of pesticides using AgNPs based chemical sensors	442
18.6	Conclusions	444
	Acknowledgments	445
	References	445

CHAPTER 19 Applications of silver nanomaterial in agricultural pest control 453
Sharmin Yousuf Rikta and P. Rajiv

19.1	Introduction	453
19.2	Conventional pest control strategies	454
19.3	Nanotechnology in pest control	454
	19.3.1 Ag NPs and Ag-nanocomposites for insects' control	457
	19.3.2 Ag NPs and Ag-nanocomposites for control fungal plant diseases	457
	19.3.3 Ag NPs and Ag-nanocomposites for control bacterial plant diseases	461
	19.3.4 Ag NPs and Ag-nanocomposites for control viral plant diseases	463
	19.3.5 Anti-plant pathogenic mechanisms	463
19.4	Conclusion	465
	References	465

CHAPTER 20 Silver nanoparticles for insect control: Bioassays and mechanisms 471
Usha Rani Pathipati and Prasanna Laxmi Kanuparthi

20.1	Introduction	471
20.2	Nanosilver in agricultural insect control	473
20.3	The effects of AgNPs on insect vector management	476
20.4	AgNPs on postharvest insect management	482
20.5	Mode of action of AgNPs and their effect on insect biochemistry and physiology	484
20.6	Conclusion	488
	Acknowledgments	488
	References	488

CHAPTER 21 Silver-based nanomaterials for plant diseases management: Today and future perspectives 495
Heba I. Mohamed, Kamel A. Abd-Elsalam, Asmaa M.M. Tmam, and Mahmoud R. Sofy

- 21.1 Introduction 495
  - 21.1.1 Antimicrobial effect of silver nanoparticles 497
  - 21.1.2 Mechanism of antimicrobial effect of silver nanoparticles 497
  - 21.1.3 Antibacterial effect of silver nanoparticles 498
  - 21.1.4 Mechanism of antibacterial effect of silver nanoparticles 500
- 21.2 Nonoxidative mechanism 501
  - 21.2.1 Cell wall and cell membrane attachment in bacteria 501
  - 21.2.2 Damaging the intracellular structure in bacteria 501
- 21.3 Oxidative-stress mechanism 502
  - 21.3.1 Antifungal effect of silver nanoparticles 502
- 21.4 Mechanism of antifungal effect of silver nanoparticles 504
  - 21.4.1 Effect of silver nanoparticles on plant-parasitic nematodes 506
  - 21.4.2 Mechanism of nematocidal effect of silver nanoparticles 507
  - 21.4.3 Effect of silver nanoparticles on antiviral 507
  - 21.4.4 Mechanism of antifungal effect of silver nanoparticles 508
  - 21.4.5 Postharvest diseases 509
  - 21.4.6 Pathogens detection and diagnosis 511
- 21.5 Silver nanoparticles for bacterial detection 512
  - 21.5.1 Mycotoxin detection 513
  - 21.5.2 Mycotoxin management 513
- 21.6 Conclusion and future perspectives 514
  - References 515

CHAPTER 22 Nematicidal activity of silver nanomaterials against plant-parasitic nematodes 527
Benay Tuncsoy

- 22.1 Introduction 527
- 22.2 Types of plant-parasitic nematodes 528
  - 22.2.1 Root-knot nematodes (Meloidogyne spp.) 528
  - 22.2.2 Cyst nematodes (*Heterodera* and *Globodera* spp.) 529
  - 22.2.3 *Ditylenchus* species 529

22.3	Management of plant-parasitic nematodes	530
22.4	Nanotechnology on agriculture	532
	22.4.1 Types of nanoparticles	533
	22.4.2 Metal nanoparticles	533
22.5	Silver nanoparticles	533
	22.5.1 Silver nanoparticles as nanopesticides	534
	22.5.2 Silver nanoparticles as nanofertilizers	536
	22.5.3 Antioxidant and antimicrobial properties of silver nanoparticles	537
22.6	Effects of silver nanoparticles on plant-parasitic nematodes	539
22.7	Conclusion	541
	References	541

CHAPTER 23 Silver nanoparticles applications and ecotoxicology for controlling mycotoxins 549

Velaphi C. Thipe, Caroline S.A. Lima, Kamila M. Nogueira, Jorge G.S. Batista, Aryel H. Ferreira, Kattesh V. Katti, and Ademar B. Lugão

23.1	Introduction	549
23.2	Silver nanoparticles in agriculture as antifungicides	552
	23.2.1 Antifungal activity of commercial AgNPs	552
	23.2.2 AgNPs commercial applications in controlling mycotoxins	554
	23.2.3 Mycotoxin detection using AgNPs	556
	23.2.4 Green synthetic AgNP against mycotoxigenic fungi	556
23.3	Toxicology profile of AgNPs	560
	23.3.1 Ecotoxicology and phytotoxicology of AgNPs	560
23.4	Life-cycle analysis of AgNPs for controlling mycotoxins in the agricultural sector	565
23.5	Prospective ecotoxicological framework	569
23.6	Conclusion	570
	Acknowledgments	571
	References	571

PART 5 Water treatment and purification

CHAPTER 24 Comparing the biosorption of ZnO and Ag nanomaterials by consortia of protozoan and bacterial species 579

Anza-vhudziki Mboyi, Ilunga Kamika, and Maggy N.B. Momba

24.1	Introduction	579
24.2	Materials and methods	580
	24.2.1 Wastewater solution and adsorbent (NMs) preparation	580
	24.2.2 Nanomaterial characterization and solution preparation	581
	24.2.3 Consortia of protozoan isolates and bacterial isolates	581
	24.2.4 Batch biosorption kinetics studies	582
	24.2.5 Uptake of nanomaterials and percentage sorption	582
	24.2.6 pH effect on the biosorption	582
	24.2.7 Concentration effect on the biosorption	583
	24.2.8 Passive accumulation on cellular structure	583
	24.2.9 Statistical analysis	583
24.3	Results	583
	24.3.1 Characterization of NMs	583
	24.3.2 Kinetic modeling of nZnO and nAg biosorption by protozoan and bacterial consortia	584
	24.3.3 Reaction kinetics	586
	24.3.4 The removal of NMs by the microbial consortia	586
	24.3.5 The removal of COD by the microbial consortia	588
	24.3.6 Effects of passive accumulation of contaminants on the cellular structure of the organisms	590
24.4	Discussion	590
24.5	Conclusion	594
	References	595

CHAPTER 25 Silver-doped metal ferrites for wastewater treatment 599

Nimra Nadeem, Muhammad Zahid, Muhammad Asif Hanif, Ijaz Ahmad Bhatti, Imran Shahid, Zulfiqar Ahmad Rehan, Tajamal Hussain, and Qamar Abbas

25.1	Introduction	599
25.2	Chapter objectives	601
25.3	Disinfection of wastewater	601
25.4	Silver nanoparticles	601
25.5	Ferrite nanoparticles	603
25.6	Properties of ferrite NPs	603
25.7	Ag-doped metal ferrite nanoparticles (Ag-MFNPs)	604
	25.7.1 Structural properties	605

25.7.2 Magnetic properties 605
25.7.3 Optical properties 607
25.8 The formation mechanism of Ag-MFNPs using hydrothermal method 607
25.9 Antibacterial activity test of Ag-MFNPs 609
25.10 Applications of Ag-MFNPs in wastewater treatment 610
    25.10.1 Photodegradation of dyes using Ag-MFNPs 610
    25.10.2 An antibacterial activity using Ag-doped BFO nanocomposite 615
    25.10.3 The ROS antibacterial activity mechanism of Ag-doped catalysts 615
    25.10.4 Role of ROS in degradation 617
25.11 Conclusion 617
References 617

CHAPTER 26 Silver-doped ternary compounds for wastewater remediation 623
Noor Tahir, Muhammad Zahid, Haq Nawaz Bhatti, Asim Mansha, Khalid Mahmood Zia, Ghulam Mustafa, Muhammad Tahir Soomro, and Umair Yaqub Qazi

26.1 Introduction 623
26.2 Objectives 624
26.3 Semiconductor photocatalytic ternary compounds 625
    26.3.1 Metal tungstates 625
    26.3.2 Metal vanadates 625
    26.3.3 Ternary metal ferrites 626
26.4 Possible mechanisms of silver nanoparticles doping on ternary compounds 626
    26.4.1 Schottky barrier and surface plasmon response 626
    26.4.2 Dye removal mechanism for wastewater treatment using silver-doped ternary compounds 628
    26.4.3 Antimicrobial mechanism 629
26.5 Influence of various synthesis approaches on doping behavior and morphology of silver-doped compounds 630
26.6 Silver-doped ternary compounds for wastewater remediation 631
    26.6.1 Silver-doped tungstates for wastewater treatment 632
    26.6.2 Silver-doped molybdates for wastewater treatment 633
    26.6.3 Silver-doped ferrites for wastewater treatment 634
    26.6.4 Silver-doped vanadates for wastewater treatment 636

	26.6.5 Silver-doped Titanates for wastewater remediation	637
	26.6.6 Silver-doped bismuth oxyhalides for wastewater remediation	638
	26.6.7 Silver-doped Tantalates for wastewater treatment	639
	26.6.8 Silver-doped niobates for wastewater treatment	639
	26.6.9 Other silver-doped ternary compounds for wastewater treatment	640
26.7	Effects of silver doping on ternary compounds	641
	26.7.1 Effect of silver doping on the morphology of ternary compounds	641
	26.7.2 Effect of silver doping on physical properties of ternary compounds	641
	26.7.3 Effect of silver doping on enhancement of visible light response and photocatalytic activity	643
	26.7.4 Effect of doping content of silver	643
26.8	Other applications of silver-doped ternary compounds	644
	Conclusion	645
	References	646

PART 6 Silver nanoparticles in veterinary science

CHAPTER 27 Potential of silver nanoparticles for veterinary applications in livestock performance and health 657

Moyosore Joseph Adegbeye, Mona M.M.Y. Elghandour, P. Ravi Kanth Reddy, Othman Alqaisi, Sandra Oloketuyi, Abdelfattah Z.M. Salem, and Emmanuel K. Asaniyan

27.1	Introduction	657
27.2	Brief on silver nanoparticle synthesis	658
27.3	Potential routes of administration	659
27.4	Potential for nanoveterinary application of silver nanoparticles	660
27.5	Endoparasites (helminths)	661
27.6	Ectoparasites (ticks)	662
27.7	Potential application in poultry and hatcheries	663
27.8	Immunity	665
27.9	Wound and burn healing	667
27.10	Antimicrobial activity and synergy of silver nanoparticles	668
27.11	Infectious diseases	669
	27.11.1 Mastitis	669
	27.11.2 Tuberculosis	670

27.12	Contaminated/infected water	670
27.13	Biosecurity/disinfection	671
27.14	Mechanism of action	672
	27.14.1 Nutrient deliveries for fetuses/neonates	674
	27.14.2 Potential side effects	674
27.15	Conclusion	675
	References	675

CHAPTER 28 Silver nanoparticles in poultry health: Applications and toxicokinetic effects ... 685

Vinay Kumar, Neha Sharma, Sivarama Krishna Lakkaboyana, and Subhrangsu Sundar Maitra

28.1	Introduction	685
28.2	Silver nanoparticles application in poultry	686
	28.2.1 As antibacterial agents	686
	28.2.2 In poultry feed	687
	28.2.3 In poultry production	688
28.3	Nanoparticles uptake by cells	689
28.4	Silver nanoparticles distribution	689
	28.4.1 Distribution through placenta	689
	28.4.2 Distribution through blood barrier	690
28.5	Silver nanoparticles toxicity	691
	28.5.1 Nanoparticles size and toxicity	691
	28.5.2 Nanoparticles toxicity mechanisms	693
	28.5.3 ROS-mediated toxicity	694
	28.5.4 Reproductive toxicity of silver nanoparticles	695
28.6	Conclusion	696
	References	696

CHAPTER 29 Nanosilver-based strategy to control zoonotic viral pathogens ... 705

Yasemin Budama-Kilinc, Burak Ozdemir, Tolga Zorlu, Bahar Gok, Ozan Baris Kurtur, and Zafer Ceylan

29.1	Introduction	705
29.2	Production methods of silver nanoparticles	706
	29.2.1 Chemical synthesis	706
	29.2.2 Green synthesis	708
29.3	Biological activities of silver nanoparticles	709
	29.3.1 Antibacterial effects	709
	29.3.2 Antifungal effects	710
	29.3.3 Antiviral effects	711

29.4	Viral applications of silver nanoparticles		711
	29.4.1	Paramyxoviruses	711
	29.4.2	Coronaviruses	713
	29.4.3	Filoviruses	714
	29.4.4	West Nile virus	714
	29.4.5	Chikungunya virus	715
	29.4.6	Crimean-Congo hemorrhagic fever	716
	29.4.7	Zika virus	716
29.5	Conclusion and future perspectives		716
	References		717
Index			723

Contributors

Qamar Abbas
Institute for Chemistry and Technology of Materials, Graz University of Technology, Graz, Austria

Kamel A. Abd-Elsalam
Plant Pathology Research Institute, Agricultural Research Center (ARC), Giza, Egypt

Ahya Abdi Ali
Department of Microbiology, Faculty of Biological Sciences; Research Center for Applied Microbiology and Microbial Biotechnology, Alzahra University, Tehran, Iran

Moyosore Joseph Adegbeye
Department of Agriculture, College of Agriculture and Natural Sciences, Joseph Ayo Babalola University, Ilesha, Osun State, Nigeria

Ambreen Ahmed
Department of Botany, University of the Punjab, Lahore, Pakistan

Farah K. Ahmed
Biotechnology English Program, Faculty of Agriculture, Cairo University, Giza, Egypt

Othman Alqaisi
Animal and Veterinary Sciences Department, College of Agricultural & Marine Sciences, Sultan Qaboos University, Al-Khoud, Oman

Balram Ambade
Department of Chemistry, National Institute of Technology, Jamshedpur, Jharkhand, India

Emmanuel K. Asaniyan
Department of Animal Production and Health, Olusegun Agagu University of Science and Technology, Okitipupa, Ondo State, Nigeria

Muhammad Ashraf
University of Agriculture Faisalabad, Faisalabad; International Center for Chemical and Biological Sciences (ICCBS), University of Karachi, Karachi, Pakistan

Luqmon Azeez
Department of Pure and Applied Chemistry, Osun State University, Osogbo, Nigeria

Biljana Balen
Division of Molecular Biology, Department of Biology, Faculty of Science, University of Zagreb, Croatia

Contributors

Jorge G.S. Batista
Laboratório de Ecotoxicologia, Centro de Química e Meio Ambiente, Instituto de Pesquisas Energéticas e Nucleares (IPEN), Comissão Nacional de Energia Nuclear, IPEN/CNEN-SP, São Paulo, Brazil

Anindita Behera
School of Pharmaceutical Sciences, Siksha 'O' Anusandhan Deemed to be University, Bhubaneswar, Odisha, India

Haq Nawaz Bhatti
Department of Chemistry, University of Agriculture, Faisalabad, Pakistan

Ijaz Ahmad Bhatti
Department of Chemistry, University of Agriculture, Faisalabad, Pakistan

Renata Biba
Division of Molecular Biology, Department of Biology, Faculty of Science, University of Zagreb, Croatia

Yasemin Budama-Kilinc
Department of Bioengineering, Yildiz Technical University, Istanbul, Turkey

Zafer Ceylan
Van Yuzuncu Yıl University, Faculty of Tourism, Department of Gastronomy and Culinary Arts, Van, Turkey

Petra Cvjetko
Division of Molecular Biology, Department of Biology, Faculty of Science, University of Zagreb, Croatia

Nohar Singh Dahariya
Department of Chemistry, Government Brijlal College, Pallari, Chhattisgarh, India

Shubhankar Dube
School of Pharmacy, National Forensic Sciences University, Gandhinagar, Gujarat, India

Mona M.M.Y. Elghandour
Facultad de Medicina Veterinaria y Zootecnia, Universidad Autónoma Del Estado de México, Mexico

Aryel H. Ferreira
Laboratório de Ecotoxicologia, Centro de Química e Meio Ambiente, Instituto de Pesquisas Energéticas e Nucleares (IPEN), Comissão Nacional de Energia Nuclear, IPEN/CNEN-SP, São Paulo, Brazil

Parinaz Ghadam
Department of Biotechnology, Faculty of Biological Sciences, Alzahra University, Tehran, Iran

Mária Globanová
Slovak Medical University in Bratislava, Bratislava, Slovakia

Bahar Gok
Yildiz Technical University, Graduate School of Natural and Applied Science, Department of Bioengineering, Istanbul, Turkey

Muhammad Asif Hanif
Department of Chemistry, University of Agriculture, Faisalabad, Pakistan

Tajamal Hussain
Institute of Chemistry, University of the Punjab, Lahore, Pakistan

Vikas Kumar Jain
Department of Chemistry, Government Engineering College, Raipur, Chhattisgarh, India

Akanksha Joshi
National Institute of Food Technology Entrepreneurship and Management-Kundli, Sonipat, Haryana, India

Ilunga Kamika
Institute for Nanotechnology and Water Sustainability; School of Science; College of Science, Engineering and Technology; University of South Africa, Florida Campus, Johannesburg, South Africa

Paulkumar Kanniah
Department of Biotechnology, Manonmaniam Sundaranar University, Tirunelveli, Tamil Nadu, India

Tushar Kant
School of Studies in Chemistry, Pt. Ravishankar Shukla University, Raipur, Chhattisgarh, India

Prasanna Laxmi Kanuparthi
Biology and Biotechnology Division, Indian Institute of Chemical Technology, Hyderabad, India

Dmitry V. Karpenko
Department of Technology of Fermentation and Winemaking, Federal State Budgetary Educational Institution of Higher Education "Moscow State University of Food Production", Moscow, The Russian Federation

Murugan Kasi
Department of Biotechnology, Manonmaniam Sundaranar University, Tirunelveli, Tamil Nadu, India

Kattesh V. Katti
Institute of Green Nanotechnology, Department of Radiology, University of Missouri, Columbia, Columbia, MO, United States

Vinay Kumar
Department of Biotechnology, Indian Institute of Technology Roorkee, Uttarakhand, India

Ozan Baris Kurtur
Yildiz Technical University, Graduate School of Natural and Applied Science, Department of Bioengineering, Istanbul, Turkey

Sivarama Krishna Lakkaboyana
Department of Chemical Technology, Faculty of Science, Chulalongkorn University, Bangkok, Thailand

Renáta Lehotská
Slovak Medical University in Bratislava, Bratislava, Slovakia

Caroline S.A. Lima
Laboratório de Ecotoxicologia, Centro de Química e Meio Ambiente, Instituto de Pesquisas Energéticas e Nucleares (IPEN), Comissão Nacional de Energia Nuclear, IPEN/CNEN-SP, São Paulo, Brazil

Ademar B. Lugão
Laboratório de Ecotoxicologia, Centro de Química e Meio Ambiente, Instituto de Pesquisas Energéticas e Nucleares (IPEN), Comissão Nacional de Energia Nuclear, IPEN/CNEN-SP, São Paulo, Brazil

Subhrangsu Sundar Maitra
School of Biotechnology, Jawaharlal Nehru University, New Delhi, India

Asim Mansha
Department of Chemistry, Govt. College University Faisalabad, Faisalabad, Pakistan

Shiji Mathew
School of Biosciences, Mahatma Gandhi University, Kottayam, Kerala, India

Anza-vhudziki Mboyi
Department of Environmental, Water and Earth Sciences, Faculty of Science, Tshwane University of Technology, Pretoria, South Africa

Heba I. Mohamed
Biological and Geological Sciences Department, Faculty of Education, Ain Shams University, Cairo, Egypt

Parisa Mohammadi
Department of Microbiology, Faculty of Biological Sciences; Research Center for Applied Microbiology and Microbial Biotechnology, Alzahra University, Tehran, Iran

Maggy N.B. Momba
Department of Environmental, Water and Earth Sciences, Faculty of Science, Tshwane University of Technology, Pretoria, South Africa

Ghulam Mustafa
Department of Chemistry, University of Okara, Okara, Pakistan

Nimra Nadeem
Department of Chemistry, University of Agriculture, Faisalabad, Pakistan

Kamila M. Nogueira
Laboratório de Ecotoxicologia, Centro de Química e Meio Ambiente, Instituto de Pesquisas Energéticas e Nucleares (IPEN), Comissão Nacional de Energia Nuclear, IPEN/CNEN-SP, São Paulo, Brazil

Simiat M. Ogunbode
Biochemistry and Nutrition Unit, Department of Chemical Sciences, Fountain University, Osogbo, Nigeria

Sandra Oloketuyi
Laboratory for Environmental and Life Sciences, University of Nova Gorcia, Nova Gorcia, Slovenia

Burak Ozdemir
Yildiz Technical University, Graduate School of Natural and Applied Science, Department of Bioengineering, Istanbul, Turkey

Santwana Padhi
KIIT Technology Business Incubator, KIIT Deemed to be University, Bhubaneswar, Odisha, India

Renu Pasricha
New York University, Abu Dhabi, Dubai

Usha Rani Pathipati
Biology and Biotechnology Division, Indian Institute of Chemical Technology, Hyderabad, India

Elena Piecková
Slovak Medical University in Bratislava, Bratislava, Slovakia

Parteek Prasher
Department of Chemistry, UGC Sponsored Centre for Advanced Studies, Guru Nanak Dev University, Amritsar; Department of Chemistry, University of Petroleum & Energy Studies, Energy Acres, Dehradun, India

Umair Yaqub Qazi
Department of Chemistry, University of Hafar Al Batin, Hafar Al Batin, Saudi Arabia

E.K. Radhakrishnan
School of Biosciences, Mahatma Gandhi University, Kottayam, Kerala, India

P. Rajiv
Department of Biotechnology, Karpagam Academy of Higher Education, Coimbatore, Tamil Nadu, India

Manviri Rani
Department of Chemistry, Malaviya National Institute of Technology, Jaipur, Rajasthan, India

Mythili Ravichandran
Department of Microbiology, K.S. Rangasamy College of Arts and Science, Tiruchengode, Tamil Nadu, India

Deepak Rawtani
School of Pharmacy, National Forensic Sciences University, Gandhinagar, Gujarat, India

P. Ravi Kanth Reddy
Department of Livestock Farm Complex, College of Veterinary Science, Sri Venkateswara Veterinary University, Proddatur, Andhra Pradesh, India

Zulfiqar Ahmad Rehan
Department of Polymer Engineering, National Textile University Faisalabad, Faisalabad, Pakistan

Sharmin Yousuf Rikta
Department of Environmental Sciences, Jahangirnagar University, Dhaka, Bangladesh

Divya Sachdev
National Institute of Food Technology Entrepreneurship and Management-Kundli, Sonipat, Haryana, India

Amadu K. Salau
Biochemistry and Nutrition Unit, Department of Chemical Sciences, Fountain University, Osogbo, Nigeria

Abdelfattah Z.M. Salem
Facultad de Medicina Veterinaria y Zootecnia, Universidad Autónoma Del Estado de México, Mexico

Mohamed Z.M. Salem
Forestry and Wood Technology Department, Faculty of Agriculture (El-Shatby), Alexandria University, Alexandria, Egypt

Imran Shahid
Environmental Science Centre, Qatar University, Doha, Qatar

Uma Shanker
Department of Chemistry, Dr. B.R. Ambedkar National Institute of Technology, Jalandhar, Punjab, India

Amit Kumar Sharma
Department of Chemistry, University of Petroleum & Energy Studies, Energy Acres, Dehradun, India

Mousmee Sharma
Department of Chemistry, UGC Sponsored Centre for Advanced Studies, Guru Nanak Dev University, Amritsar; Department of Chemistry, Uttaranchal University, Dehradun, India

Neha Sharma
School of Biotechnology, Jawaharlal Nehru University, New Delhi, India

Kamlesh Shrivas
School of Studies in Chemistry, Pt. Ravishankar Shukla University, Raipur, Chhattisgarh, India

Mahmoud R. Sofy
Botany and Microbiology Department, Faculty of Science, Al-Azhar University, Cairo, Egypt

Muhammad Tahir Soomro
Center of Excellence in Environmental Studies (CEES), King Abdulaziz University, Jeddah, Saudi Arabia

Petra Peharec Štefanić
Division of Molecular Biology, Department of Biology, Faculty of Science, University of Zagreb, Croatia

Noor Tahir
Department of Chemistry, University of Agriculture, Faisalabad, Pakistan

Neetu Kumra Taneja
National Institute of Food Technology Entrepreneurship and Management-Kundli, Sonipat, Haryana, India

Velaphi C. Thipe
Laboratório de Ecotoxicologia, Centro de Química e Meio Ambiente, Instituto de Pesquisas Energéticas e Nucleares (IPEN), Comissão Nacional de Energia Nuclear, IPEN/CNEN-SP, São Paulo, Brazil

Mirta Tkalec
Division of Botany, Department of Biology, Faculty of Science, University of Zagreb, Croatia

Asmaa M.M. Tmam
Biological and Geological Sciences Department, Faculty of Education, Ain Shams University, Cairo, Egypt

Benay Tuncsoy
Adana Alparslan Turkes Science and Technology University, Faculty of Engineering, Department of Bioengineering, Adana, Turkey

Shabana Wagi
Department of Botany, University of the Punjab, Lahore, Pakistan

Jyoti Yadav
Department of Chemistry, Malaviya National Institute of Technology, Jaipur, Rajasthan, India

Muhammad Zahid
Department of Chemistry, University of Agriculture, Faisalabad, Pakistan

Khalid Mahmood Zia
Department of Chemistry, Govt. College University Faisalabad, Faisalabad, Pakistan

Tolga Zorlu
Department of Physical and Inorganic Chemistry and EMaS, Universitat Rovira I Virgili, Tarragona, Spain

Faisal Zulfiqar
Department of Horticultural Sciences, Faculty of Agriculture and Environment, The Islamia University of Bahwalpur, Punjab, Pakistan

Preface

This book represents the fourth volume of a series titled *Nanobiotechnology for Plant Protection*; the original book series approved by Elsevier. The present volume contains 29 chapters prepared by outstanding authors from Algeria, Austria, Bangladesh, Brazil, Croatia, Egypt, India, Iran, KSA, Nigeria, Oman, Mexico, Pakistan, Russia, Slovakia, Slovenia, Spain, Thailand, Turkey, Pakistan, Qatar, United Arab of Emirates, and United States of America. A combination of 29 chapters written by professionals and experts represents an outstanding knowledge of silver-based nanostructures for environmental and agricultural applications. The recent volume is divided into six main parts. **Part 1** (four chapters) highlights silver nanoantimicrobials' effect and drug synergism against drug-resistant pathogens, Ag nanoparticles as an antifungal agent, silver-based nanoantimicrobials: mechanisms, eco-safety, and future perspectives. The increasing importance of silver nanomaterials in food applications is reflected in the four chapters of **Part 2**. Part 2 gives an overview of silver nanomaterial-based nanosensors in Agri-food application, silver-based nanomaterials for food packaging applications, emerging silver nanomaterials for smart food packaging in combating food-borne diseases, nanosilver disinfectant control in agriculture and food processing. **Part 3** includes some chapters which describe the use of silver nanostructures in plant science applications such as silver nanoparticles application in wood and wood-based panels, immobilization efficiency and modulating abilities of silver nanoparticles on biochemical and nutritional parameters in plants: possible mechanism, controlled-released, and positive effects of silver nanofertilizers, comparison of the effect of silver nanoparticles and other nanoparticle's types on the process of barley malting, negative effects of silver nanoparticles on crop plants and beneficial microbes, in additions to Ag nanoparticles phytotoxicity mechanisms. **Part 4** of this volume is devoted to one of the most innovative topics of nanobiotechnology which concerns the plant protection applications. This part summarizes current approaches to use silver nanomaterials in the detection and mineralization of pesticides using silver nanoparticles, pesticides degradation by silver-based nanomaterials, applications of silver nanomaterial in agricultural pest control, nematocidal activity of silver nanomaterials against plant-parasitic nematodes, silver nanostructures: purification, detection, and management mycotoxins. **Part 5** (three chapters) which describes silver nanomaterial applications in photocatalytic removal of dyes and pesticides from wastewater, comparing the biosorption of nAg and $nZnO$ nanomaterials by consortia of protozoan and bacterial species in the wastewater, silver-doped ternary and metal ferrites for wastewater treatment and remediation. The **final part** is divided into four chapters summarizes current applications of silver nanomaterials in veterinary like silver nanoparticles as new weapon in veterinary science, application of sliver nanoparticles as feed additives in livestock production, silver nanoparticles toxicity to poultry, in addition to the use of nanosilver-based strategy for control zoonotic viral pathogens.

This volume is multidisciplinary and will be of great benefit to students, teachers, and researchers, agri-food and agroecosystem scientists working in biology, plant biotechnology, science materials, physics, technology, chemistry, microbiology, plant physiology, nanotechnology, and other interesting groups, such as the manufacturing sector. This would also be a valuable tool for environmental researchers researching technology to develop their understanding of silver-based nanomaterials toxicity. Several new topics will be included covering the application of nanotechnology in plant protection, plant disease management, controlling postharvest diseases, mycotoxins reduction, and pesticide sensing and degradations.

We hope to offer a reliable, informative, and creative perspective in this area, not only to advanced readers but also to technical decision-makers and those with minimal expertise in the context. The intended audience for this book includes scholars, undergraduates, postgraduate students, etc., from various fields of science and technology. We are thankful to the authors to provide this collection of high-quality manuscripts. We thank all the authors who have written the chapters of the book and edited the book with their suggestions and useful experiences. Writing this book might not have been possible without their participation and dedication. The service provided by Elsevier Publisher is also highly commendable, as they provided a high level of expertise, reliability, and tolerance throughout the project. We express our appreciation to the Elsevier staff, in particular Simon Holt, Rafael Trombaco, Editorial Project Manager Intern, Nirmala Arumugam, and Narmatha Mohan, for their great support and efforts. We also thank all the reviewers who have expanded their valuable time commenting on each chapter. We also thank the members of our family for their sustained support and assistance.

Kamel A. Abd-Elsalam
Agricultural Research Center, Giza, Egypt

Series preface

The field application of engineered nanomaterials (ENMs) has not been well investigated in the plant promotion and protection in agro-environment yet, and lots of most effective components have been taken into consideration theoretically or with prototypes, which make it hard to evaluate the utility of ENMs for plant promotion and protection. Nanotechnology is now invading the food enterprise and forming super potential. Nanotechnology applications in the food industry involve the following: encapsulation and delivery of materials in targeted sites, developing the flavor, introducing nano-antimicrobials agents into food, improvement of shelf life, sensing contamination, improved food preservative, monitoring, tracing, and logo protection. The list of environmental problems that the world faces may be huge, but a few strategies for fixing them are small. Scientists throughout the globe are developing nanomaterials that could use selected nanomaterials to capture poisonous pollutants from water and degrade solid waste into useful products. The market intake of nanomaterials is rushing, and the Freedonia Group predicts that nanostructures will grow to $100 billion through 2025. Nanotechnology research and development has been growing on a sharp slope across all scientific disciplines and industries. Based on this background, the scientific series entitled *Nanobiotechnology for Plant Protection* was inspired by the desire of the Editor, Kamel A. Abd-Elsalam, to put together detailed, up-to-date and applicable studies on the field of nanobiotechnology applications in agro-ecosystems, to foster awareness and extend our view of future perspectives.

The main appeal of the present book series is its specific focus on plant protection in agri-food, and environment which is one of the most topical nexus areas in the many challenges faced by humanity today. The discovery and highlighting of new book inputs, based on nanobiotechnology, that can be used at lower application rates will be critical to eco-agriculture sustainability. The research carried out in the concerned fields is scattered and not in a single place. The book series will cover the applications in Agri-food and environment sectors which is the new topic of research in the field of nanobiotechnology. This book series will be a comprehensive account of the literature on specific nanomaterials and their application in agriculture, food, and environment. The audience will be able to gather information from a single book series. Students, teachers, researchers from colleges, universities, research institutes, as well as industry will benefit from this book series. Four specific features that make the current book series one of a kind. First, the book series has a very specific editorial focus, and researchers can locate nanotechnology information precisely without looking into various full-text. Second, and more importantly, the series offers a crucial evaluation of the content material along with nanomaterials, technologies, applications, methods and equipment and safety and regulatory aspects in agri-food and environmental sciences. Third, the series offers the reader a concise precision of the content material, it will offer nanoscientists with clarity and deep information. Finally, presenting researchers with insights on new discoveries. The present book series gives the researchers a sense of what to do, what they would need to do, and how to do it properly—by searching for others who have done it. The fourth book entitled, *Silver Nanomaterials for Agri-Food Applications*, shows that this book gathered the know-how, observations and effective applications of zinc-based nanomaterials in the environmental and agri-food systems. The expected readership for this book series would be researchers in the field of environmental science, food, and agriculture science. Some readers may also be chemists, material scientists, government regulatory agencies, agro and food industry players, and academicians. A few readers from industrial personnel would also be interested in it. This book series is useful to a wide audience of food, agriculture, and environmental sciences research including undergraduates, graduate students, postgraduates, etc. In addition, agricultural producers could benefit from the applied knowledge that will be highlighted in the book, which, otherwise, would be buried in different journals. Both primary and secondary audiences are seeking the up to date knowledge of the nanotechnology application in environmental science, agriculture, and food science. It is a trending area, and lots of new studies get published every week. The readers need some good summaries to help them to learn the latest key findings, which could be review articles and/or books. This book series will help to put these pockets of knowledge together, and make it more easily accessible globally.

Kamel A. Abd-Elsalam
Agricultural Research Center, Giza, Egypt

CHAPTER 1

Silver-based nanomaterials for sustainable applications in agroecology: A note from the editor

Kamel A. Abd-Elsalam
Plant Pathology Research Institute, Agricultural Research Center (ARC), Giza, Egypt

1.1 Introduction

The next-generation technologies for the production of metal nanoparticles are varied in terms of internal pore space, surface porosity, and surface chemical adsorption properties and exhibit improved absorption capabilities, containment, and mode of delivery of plant tissue phytonutrients and various agrochemicals (Hendrickson et al., 2017). Silver nanoparticles (AgNPs) produced by biosynthetic methods have significant applications in agriculture and agro-industries, in particular as antimicrobial agents in the agricultural sector to combat bacterial and pathogenic fungi that can cause plant, waterborne, and foodborne diseases (Castillo-Henríquez et al., 2020). More than 23% of all nanotechnology-based devices, diagnostic, and therapeutic applications implanted with silver nanoparticles, which are well recognized for their antifungal, antibacterial, and antiviral properties, are used in textile fabrics. (Verma and Maheshwari, 2019). Many studies based on the use of nanosilver to process seed or as foliar- and plant-growth sprayers have been evaluated by researchers, including as a means to achieve maximum postharvest time for cut flowers (Andżelica-Byczyńska, 2017). AgNPs can be used in biodegradable and non-biodegradable polymers, which improve food protection and extend the shelf life of foods through the use of antimicrobial packaging (Simbine et al., 2019). Plants are the basic elements of the agroecosystem and are the primary food source for humans. Therefore it is important to assess the possible environmental risks due to AgNPs' effects on the growth and development of plants and thus on food and human health (Yan and Chen, 2019).

The existing literature does not include books focusing specifically on silver nanomaterial applications in the agri-food sector; hence this book is a unique contribution to the field. Part 1 largely deals with the antimicrobial effects of silver-based nanostructures. Part 2 focuses on food applications. Part 3 covers appropriate technological developments in plant science applications. Part 4 comprises seven chapters on the use of silver nanomaterials in plant protection. Part 5 presents silver nanomaterials applications in photocatalytic removal of dyes and pesticides from wastewater in addition to wastewater treatment and remediation. Part 6 presents examples of nanobiotechnological uses in veterinary medicine. Finally, the need for more research to gain a better understanding of AgNPs' cytotoxic mechanisms and how to extend their use in agroecology is addressed.

1.2 AgNP-based nanosystem applications

Silver nanoparticles' formulation processes distinguish it from other nanoparticle formulations and are among the most popular nanomaterials. For example, AgNPs have been used in 278 products, in 34 countries, and in 15 sectors, with a range of features and application areas (e.g., medicine, cosmetics, clothing, electronics, energy, food packaging, and coatings) (StatNano, 2019). Silver nanoparticles have also received much attention for use in products such as in electronics, pharmaceuticals, medical, veterinary medicine, industry, renewable energy, agri-food, and more (Fig. 1.1).

Pesticides, food security, aquaculture reservoirs, farm production, biomedical applications, and antimicrobial agents are important factors in water treatment (Ijaz et al., 2020; Simbine et al., 2019). Nanocomposite preparations have adjustable electrical conductivity, enhanced thermal conductivity and tensile strength, superior rigidity and durability, and resistance to erosion. These features make nanocomposites suitable for rendering satellite parts, aircraft spare parts, industrial components, electronic microchips, and so on (Yusuf, 2020). In several subfields of nanomedicine, AgNPs are emerging as a next-generation application, and the benefits of using them are widely recognized in biomedical and industrial sectors (Lee and Jun, 2019). AgNP-based nanosystems have demonstrated a high level of versatility in the biomedical field and remarkable potential for the production of new antimicrobial agents. AgNPs have been used for drug-delivery formulations, detection and diagnosis platforms, and biomaterial and medical device coatings (Burduşel et al., 2018).

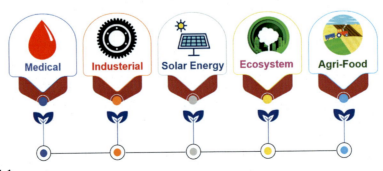

FIG. 1.1

Potential applications of silver-based nanomaterials in diverse industries.

1.2.1 Agri-food applications

By utilizing nanoparticles, nanotechnology can play an important role in improving agricultural productivity and maintaining sustainable agroecosystems. Its applications include, for example, nano-scale pesticides, nano-scale fertilizers, nano-scale herbicides, and agrochemical nanocarrier encapsulated systems for plant promotion and protection. These factors may contribute to higher crop production and inhibition of pathogens in plants, cost-effective elimination of unwanted weeds and insects, and energy-from-waste production. Also, the application of silver nanomaterials in food science enhances food safety, extends the life of postharvest products, improves flavor, and preserves nutritional value.

Several areas have been identified that show promise for the use of AgNPs, for example, as additives in food and as agrochemicals. Agribusinesses comprise multiple manufacturing sectors, including water-soluble aggregate nanofertilizers from 100 nm to 250 nm, nanopesticides that include nanoherbicides, nanocoatings, and an elegant framework for plant nutrients (Fouda et al., 2020; Khan et al., 2020). Changes resulting from AgNP-induced multiple cellular alterations are critical for preventing the production of antifungal resistance (Bocate et al., 2019). AgNPs have beneficial effects on plants' physiology and decrease oxidative damage due to salt stress. AgNPs have been used for increasing plant growth, increasing the basic nutrient supply by the uptake of water, and lowering salinity stress. However, more attention needs to be given to the effects of the dosage and duration of AgNPs, specific to the type of nanoparticles and the plant species (Khan et al., 2020). AgNPs improve the roots and root growth of several plants when applied at low concentrations. AgNPs also enhance antioxidant enzyme activity, which restricts the production of reactive oxygen species (ROS) in plant cells. Lower AgNP doses often help to boost chlorophyll output and improve chlorophyll fluorescence parameters (Sami et al., 2020). Plant-mediated and microbe-biosynthesized metal nanoparticles demonstrate effective antimicrobial action against pathogens and pesticides (Ali et al., 2020). An aqueous extract of the agriculturally beneficial fungus *Pythium oligandrum* is used in mycosynthesized AgNPs. The use of a broad-spectrum antibacterial agent against *Clavibacter michiganensis* subsp. *michiganensis* (Cmm) demonstrated a range of specific bactericidal activity at a low level of 0088 mg/L; no negative effect on in vitro germination rates and on plant physiology was observed during the in vivo experiment (Noshad et al., 2020). AgNPs can bind to coat protein virus particles on *Tomato mosaic virus* (ToMV) and *Potato virus Y* (PVY) (El-Dougdoug et al., 2018). Silver nanoparticles can be used with a high level of success against beetles and pesticides, which are significant threats to crops and stores. AgNPs can supply an inexpensive and secure source against *Tribolium castaneum* (Alif Alisha and Thangapandiyan, 2019). Silver nanostructures can be used as carriers for smart delivery of nutrients, anti-caking agents, antimicrobial agents, and fillers to improve mechanical strength and durability of packaging content, while nanosensing can be used to enhance food quality and safety (Ezhilarasi et al., 2013). Biopolymers can be adjusted to obtain nanocomposite functionality using silver nanoparticles.

AgNP-based polymer materials may play an ever-increasing role in food packaging. However, lack of information about the safety and toxicity of silver nanoparticles (Kraśniewska et al., 2020) may be a barrier for their use in packaging materials. On the other hand, adding AgNPs to packaging for foodstuffs enhances its antimicrobial properties, which protects against the growth of microorganisms, thereby increasing product safety (Simbine et al., 2019). A study on the impact of AgNPs on the shelf life of fruit showed that the shelf life increased more with nanomaterials than with other chemical and physical means used to reduce postharvest losses (Ijaz et al., 2020). Chitosan is a suitable biopolymer matrix for AgNP capping with excellent bactericidal and fungicidal capabilities and the capacity to reduce cytotoxic effects and eliminate various plant pathogens (Jena et al., 2012). The negative effects arising from the use of AgNPs in agri-food commodities include exposure pathways, mode of use, and method used for final disposal and can include damage to DNA, gene disruption, and metabolic changes.

Different nanobiotechnological methods used for detection of mycotoxins have been established. Ideally, various metal and magnetic nanoparticles are used in the fabrication of sensors to detect mycotoxin (Ingle et al., 2020). To detect mycotoxins, a sensor panel was developed by immobilizing carboxylated AgNPs and bovine serum albumin sequentially, combined with amino groups on a glass chip (Jiang et al., 2020). Citrinin, aflatoxin B1 and ochratoxin A mycotoxins were found consistently in red yeast rice. Its high selectivity, high performance, reasonable stability, and precision offer a promising outlook for monitoring and assessing mycotoxins in foods. Numerous studies have demonstrated the feasibility of using engineered nanoparticles as antifungals, especially against fungal pathogens that produce mycotoxins in human and animal foods (Alghuthaymi et al., 2020: Asghar et al., 2020). Various nanoparticles, such as FeNPs, CuNPs, and AgNPs, were developed effectively with *Syzygium cumini* leaf extracts. AgNPs show greater antibiotic activity and decreased development of aflatoxins compared with CuNPs and FeNPs (Asghar et al., 2020). Ag-Chit-NCs < 10 nm produced by metal vapor synthesis showed significant antifungal activity against *P. expansum*. Ag-Chit-NCs are a feasible and effective supplement for reducing mycotoxins in animal feed (Alghuthaymi et al., 2020). Fig. 1.2 shows a schematic diagram of silver nanomaterials and their applications in various sectors.

1.2.2 Veterinary applications

AgNPs are being seriously considered for biomedical applications such as diagnostics; delivery of care and medication; medical equipment coatings; wound dressings; textiles used in medicine; contraceptives; antimicrobial agents; antiviral, antifungal, and anti-inflammatory drugs; and angiogenesis inhibitors (Ge et al., 2014; Zhang et al., 2016; Bai et al., 2018). Nanoparticles hold promise in veterinary medicine

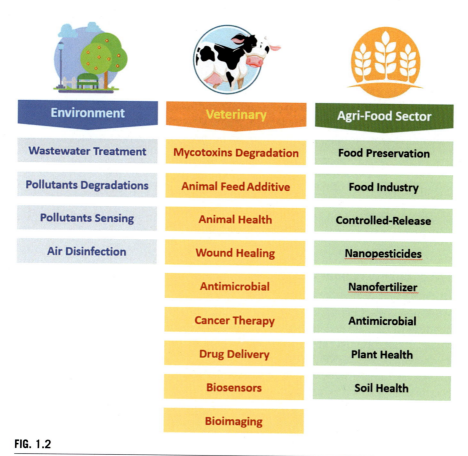

FIG. 1.2

AgNP-based applications promoted for applications in the environmental, veterinary, and agri-food sectors.

and animal production systems and have been used in innovative ways, for example, in sensors, imaging devices, drug delivery, and tissue engineering (Narducci, 2007; Bai et al., 2018). This volume provides a succinct overview of the developments in veterinary medicine through the use of nanotechnology tools. Dosage forms such as nanoparticulates can offer higher selectivity and lower drug toxicity, guarantee better and more efficient pest therapy, and improve safety for consumers of animal products (Carvalho et al., 2020), compared with traditional forms.

Nanoparticles are used to help resolve issues related to antibiotic thresholds, identification of pathogenic bacteria, and administration of new and enhanced vaccines. Candidates and medicines, diagnosis, therapies, food additives, nutrient provision, biocidal agents, reproductive and other medicinal items were applied to protect animal health (Bai et al., 2018; Carvalho et al., 2020; Hassan et al., 2020).

State-of-the-art strategies have been used to adapt nanocarriers to drug delivery approaches in animal models. Also, key challenges related to research, scale up studies, large-scale development, assessment techniques, and regulatory aspects of nanomedicine have been addressed (Carvalho et al., 2020). The potential advantages for using nanocarriers in veterinary medicine include vectorization of drugs, reduction of dosage released by therapeutic agent, reduction of severe adverse effects, and treatment for chronic pain. For example, pharmaceutical products are greatly needed in marine shrimp farming, and this need can be met through nanotechnology that will effectively treat and prevent infections and disease (Govindaraju et al., 2019). AgNPs can also be used as a diagnostic tool for detecting diseases in shrimp. Because the absorption spectrum of nanoparticles varies based on their size and shape, AgNPs with various features can be combined with pathogen-detection biomolecules (such as antibodies) (Yen et al., 2015). In this volume, the latest research on the positive and negative health effects of AgNPs is summarized and analyzed.

1.2.3 Environmental applications

To detect and degrade pollutants, biogenic silver nanoparticles obtained through a fast and cost-effective method were used as part of an environmental remediation project (Ragam and Mathew, 2020). Due to their high surface-to-volume ratio, chemical stability, and enhanced catalytic activity properties, plant extract-mediated green synthesized AgNPs may be used for water surveillance, purification, water storage, and industrial wastewater treatment. Nanocomposite cellulose fabrics are possible alternatives as effective and environmentally friendly materials used to monitor water contamination and treatment. Moreover, environmentally sustainable advantages can be obtained with the use of properly constructed recycled materials and cost-efficient recycling (Fiorati et al., 2020). It was found that once visible light was irradiated, green synthesized nanocrystal Ag_2O successfully decolorized or mineralized poisonous organic pollutants. This finding verified that sensitive, highly economical, and stable photocatalysts of green synthesized Ag_2Os are useful for degrading harmful organic compounds and visible light irradiation dyes (Shume et al., 2020). An eco-friendly technique for controlling shape and size based on NP synthesis is used in the effluent treatment of color degradation in diverse industries, such as textiles, paper, plastic, paints, and various chemical industries (Chand et al., 2020). Through a surface shift, functionally modifiable AgNPs can significantly increase testing output through responsive and precise identification of accelerated signal transductions and enhanced signal strength of different contaminants and chemical residues. However, future research is necessary to integrate colorimetric tests of alternative technologies, such as an imaging method for pesticide quantification in situ (Sulaiman et al., 2020). Their characteristics, such as durability, tunable monoclay, versatility, and low expense, make such technologies suitable for atmospheric and water quality monitoring such as quantitative chemical analysis of materials that sense atmosphere and water contaminants, with many benefits such as high mineral absorption and direct surface alterations (Prosposito et al., 2020; Singh et al., 2021).

1.3 Challenges

AgNPs are recognized for their advantages in physicochemical properties, compared with their bulk counterparts. However, the use of nanoparticles in agriculture is raising many concerns, including their synthesis and penetration mechanisms and the associated risks. One of the most important challenges for laboratories in resource-limited facilities is the difficultly and expense of obtaining and synthesizing AgNPs. It is estimated that the preparation of one kg of AgNPs costs approximately $4 million, while one kg of raw Ag costs around $14,000 (Ahmed et al., 2016; Mahawar and Prasanna, 2018). The current challenge associated with the development of food packaging with AgNPs is one of cost. It has been noted that the incorporation of AgNPs in packaging systems can increase the expense of packaging. It was recommended that the total packaging cost should be 10% of the product cost. Therefore, to assess the practicality of implementing AgNPs in food packaging, a proper cost-benefit analysis is needed (Dainelli et al., 2008). Although the synthesis and application of nanomaterials in agriculture may be enhanced significantly by green synthesis techniques, there are several uncertainties with regard to nanoparticle synthesis, their uses, their uptake by plants, and their incorporation into plant cells. AgNPs are highly vulnerable to oxygen, because the silver ion attaches to oxygen and forms a powerful bond called the ionic bond. The formation of this ionic bond changes nanoparticle composition, thereby modifying nanoparticles' physicochemical characteristics. Antibacterial properties of AgNPs often decrease because they are dependent on silver ions (Lok et al., 2007). The propensity to aggregate these particles is another weakness of AgNPs. In addition, nanomaterials can enter the food chain of animals and humans as a result of a bioaccumulation phenomenon in plants. Consequently, rigorous research and regulations regarding the safety of nanomaterial usage are essential. The initial interaction damaged by an increase of nanomaterials in the environment is the plant–soil micro symbiosis, which can reduce the growth of plants considerably. As nanomaterials enter plants' tissues through their root, stem, and leaf openings, these materials can induce stress mechanisms that can reach toxic levels, leading to plant degradation and alteration of entire crops or even ecosystems (Millán-Chiu et al., 2020). AgNPs can harm certain nitrogen-fixing and denitrifying microbes in vitro. In vitro studies of AgNPs in soil, sediment, or loam environments have shown that these functional processes are distorted by nitrogen-cycling microbial communities and nitrogen transformation rates (McGee, 2020). To fill in the knowledge gap on the side effect profile of long-term AgNP exposure, it will be important to determine the cumulative effects of silver in infected trees and fields over time. In veterinary medicine, there is a need for future research on ecotoxicology to determine possible adverse impacts on non-target species and site contamination resulting from treatments on a nanoparticle basis in the vicinity of livestock and farmers.

1.4 Biosafety and regulations

Because of their widespread use and distribution, AgNPs have permeated different environmental media. Although much research is being done on both the benefits and the negative impacts of AgNPs, studies are needed to determine the degree of toxicity and additional effects of AgNPs on species at the cellular level. These nanoparticles are seen as potential hazards due to their ability to induce cytotoxic mechanisms such as generation of ROS, DNA damage, and increased pro-inflammatory progression in cells. These nanoparticles can cause sublethal adverse effects because of the specific physicochemical properties of the AgNPs, especially of ROS if the concentration or development of ROS is not regulated properly (Calderón-Jiménez et al., 2017). The increased use of AgNPs presents major concerns about, for example, habitat degradation, decreased food quality and yield, and negative effects on human health (Yan and Chen, 2019). Studies are also needed on the AgNP toxicity mechanisms that pass through the food chain from plants to other populations leading to the destruction of healthy ecosystems (Tripathi et al., 2017). Different phytotoxicity studies of AgNPs show that the interactions between plants and AgNPs are highly complicated and depend not only on the intrinsic presence of AgNPs but also on plant organisms, developmental phases, and various tissues and syntheses. The aggregation of these nanoparticles and their cytotoxicity are key factors in the use of AgNPs. Surface chemistry, size, size distribution, shape, plant species, particle morphology, applications methods, particle reactivity in solution, growing conditions, efficiency of ion release, cell type, and the type of reducing agents used for AgNP synthesis are all important factors in determining their biological activity. Although the toxicity of AgNPs is still uncertain, the levels of toxicity differ greatly depending on some factors like size, shape, exposure time, concentrations and capping agents, as well as the exposure pathway. The degradation rate of the AgNPs is also influenced by numerous factors such as internal factors including nanoparticles' particle size, composition, concentrations, surface morphology, and adsorption (Fig. 1.3).

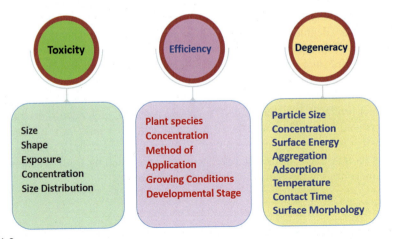

FIG. 1.3

Factors effecting AgNPs' toxicity, efficiency, and degradation.

The transfer of AgNPs through the food chain continues to be a problem because of the toxic effects on human health. To help determine the safety of AgNPs prior to their comprehensive use, a careful study was conducted on the toxicological properties of AgNPs released from food packaging (Istiqola and Syafiuddin, 2020). It has been noted that the utilization of AgNPs in packaging food technology is still in its early stages. Intensive research therefore is needed to analyze and evaluate the potential toxic effects of AgNPs in food commodities.

Future studies could provide a deeper understanding of this topic by using analytical techniques to do toxicological research. The extensive use of AgNPs in food packaging technologies has triggered concern about possible risks linked to the release of AgNPs from packaging into food. Food safety authorities in the European Union and the United States have released regulations on the use of AgNPs for food packaging. The European Food Safety Authority (EFSA) does not allow the use of AgNPs in products that come in contact with food (e.g., embalming fluids and dietary supplements) without authorization (Bumbudsanpharoke and Ko, 2015). The Korean Ministry of Food and Drug Safety (MFDS) has a variety of programs that analyze nanomaterial safety in food and food packaging technologies (Hwang et al., 2012). However, in agroecosystems, marine, and land-based organisms, there is still a strong focus on processing, internalization, and bioaccumulation of nanoparticles. Nanoecotoxicological research based on validated criteria needs to be structured according to regulatory frameworks. As every nanomaterial has its individual properties, toxicity is likely to be calculated individually (Mahler et al., 2012). To guarantee product quality, health and safety, and environmental legislation, regulatory agencies must establish the particular requirements of commercial products.

1.5 Future trends

It is predicted that by 2050 the global population will expand to 9.6 billion or more, increasing the pressure to feed the world's ever-growing population on a finite amount of land. In this situation, conventional fertilizers are not sufficient because of their high cost due to high energy requirements and their adverse environmental effects. Nanoscale carriers and biosensors have brought tremendous advantages, such as engineered xylem vessels, to the agricultural sector. Many nanotechnological inventions are contributing to the modernization of the agriculture and horticulture sectors, including silver nanoparticles. However, their use presents adverse risks related to their environmental effects, effects that are not yet completely understood (Fig. 1.4). There is an immediate need to research aspects of nanotechnology that are still untouched in order to expand our understanding of them. For example, there is a lack of literature on the functioning of azadirachtin, thymol, and curcumin in conjunction with different NPs; and minimal information is available on the use of biological or green or biogenous silver nanoparticles synthesized for quality postharvested plants. By using photocatalytic AgNPs, many pollutants can be eliminated from wastewater. By linking them to suitable organisms, photocatalytic AgNPs can

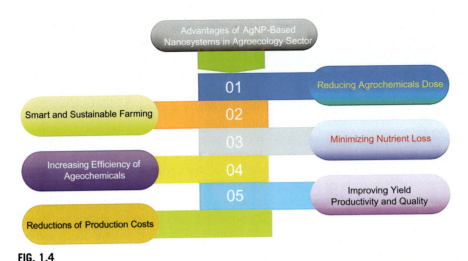

FIG. 1.4

Benefits of silver-based nanomaterials in agricultural and food sectors.

be improved. A well-established research program that clearly explains the optimal doses of nanoparticles to be used under different environmental conditions for different crops is required. Further studies are needed to clarify the effects of AgNPs on soil health or aquatic microorganisms, however, they may pose a threat to beneficial microorganism in the environment. Future studies could explore the environmental and health benefits derived from the use of AgNPs and guide the environmental tradeoffs in new AgNP-based consumer products.

1.6 Conclusion

The application of AgNPs and their efficacy in the agricultural sector are important in maintaining sustainable agroecosystems. Prior to their mass production and use, studies must evaluate AgNPs' effects on biotic and abiotic environmental factors and on human health. AgNPs may be useful in various areas, such as plant protection, agricultural precision, nutrient use efficiency, and nanoherbicides and nanocarrier agrochemical systems, among others. Potential applications in farming have been developed. The research included in this chapter deals with the synthesis and application of functionalized silver nanoparticles used as colorimetric sensing materials for the major polluting substances in water and agroecosystems such as heavy metals and pesticides. Also, AgNP-based bionanomaterial photocatalysts are promising for the control and treatment of water contamination. The delivery of drugs via various metallic nanoparticles for veterinary infections is promising in both the short term and the long term. Depending on plant types, concentration and form of NP, treatment time, and plant development stage, among other factors, the nanoparticle effects may be positive or negative. AgNPs entering soils are also underestimated and need to be explored to better understand their effects on beneficial and harmful microbes. A thorough study to decode engineered nanomaterials and determine their mechanistic use and agroecological toxicity is therefore needed. Volume parts that follow discuss the toxicity, legislation, and potential prospects for the use of silver-based nanostructures in agroecology.

References

Ahmed, S., Annu, Ikram, S., Yudha, S.S., 2016. Biosynthesis of gold nanoparticles: a green approach. J. Photochem. Photobiol. B 161, 141–153.

Alghuthaymi, M.A., Abd-Elsalam, K.A., Shami, A., Said-Galive, E., Shtykova, E.V., Naumkin, A.V., 2020. Silver/chitosan nanocomposites: preparation and characterization and their fungicidal activity against dairy cattle toxicosis penicillium expansum. J. Fungi 6 (2), 51.

Ali, M.A., Ahmed, T., Wu, W., Hossain, A., Hafeez, R., Islam Masum, M.M., Wang, Y., An, Q., Sun, G., Li, B., 2020. Advancements in plant and microbe-based synthesis of metallic nanoparticles and their antimicrobial activity against plant pathogens. Nanomaterials (Basel) 10 (6), 1146.

Alif Alisha, A.S., Thangapandiyan, S., 2019. Comparative bioassay of silver nanoparticles and malathion on infestation of red flour beetle, *Tribolium castaneum*. J. Basic Appl. Zool. 80 (1), 55.

Asghar, M.A., Zahir, E., Asghar, M.A., Iqbal, J., Rehman, A.A., 2020. Facile, one-pot biosynthesis and characterization of iron, copper and silver nanoparticles using *Syzygium cumini* leaf extract: as an effective antimicrobial and aflatoxin B1 adsorption agents. PLoS One 15 (7), e0234964.

Bai, D.P., Lin, X.Y., Huang, Y.F., Zhang, X.F., 2018. Theranostics aspects of various nanoparticles in veterinary medicine. Int. J. Mol. Sci. 19 (11), 3299.

Bocate, K.P., Reis, G.F., de Souza, P.C., Junior, A.G.O., Durán, N., Nakazato, G., Furlaneto, M.C., de Almeida, R.S., Panagio, L.A., 2019. Antifungal activity of silver nanoparticles and simvastatin against toxigenic species of Aspergillus. Int. J. Food Microbiol. 291, 79–86.

Bumbudsanpharoke, N., Ko, S., 2015. Nano-food packaging: an overview of market, migration research, and safety regulations. J. Food Sci. 80 (5), R910–R923.

Burduşel, A.C., Gherasim, O., Grumezescu, A.M., Mogoantă, L., Ficai, A., Andronescu, E., 2018. Biomedical applications of silver nanoparticles: an up-to-date overview. Nanomaterials 8 (9), 681.

Byczyńska, A., 2017. Nano-silver as a potential biostimulant for plant. Review. World Sci. News 86 (3), 180–192.

Calderón-Jiménez, B., Johnson, M.E., Montoro Bustos, A.R., Murphy, K.E., Winchester, M.R., Baudrit, J.R.V., 2017. Silver nanoparticles: technological advances, societal impacts, and metrological challenges. Front. Chem. 5, 1–26.

Carvalho, S.G., Araujo, V.H.S., dos Santos, A.M., Duarte, J.L., Silvestre, A.L.P., Fonseca-Santos, B., Villanova, J.C.O., Gremião, M.P.D., Chorilli, M., 2020. Advances and challenges in nanocarriers and nanomedicines for veterinary application. Int. J. Pharm. 580, 119214.

Castillo-Henríquez, L., Alfaro-Aguilar, K., Ugalde-Álvarez, J., Vega-Fernández, L., Montes de Oca-Vásquez, G., Vega-Baudrit, J.R., 2020. Green synthesis of gold and silver nanoparticles from plant extracts and their possible applications as antimicrobial agents in the agricultural area. Nanomaterials 10 (9), 1763.

Chand, K., Cao, D., Fouad, D.E., Shah, A.H., Dayo, A.Q., Zhu, K., Lakhan, M.N., Mehdi, G., Dong, S., 2020. Green synthesis, characterization and photocatalytic application of silver nanoparticles synthesized by various plant extracts. Arab. J. Chem. 13 (11), 8248–8261.

Dainelli, D., Gontard, N., Spyropoulos, D., Zondervan-van den Beuken, E., Tobback, P., 2008. Active and intelligent food packaging: legal aspects and safety concerns. Trends Food Sci. Technol. 19, S103–S112.

El-Dougdoug, N.K., Bondok, A.M., El-Dougdoug, K.A., 2018. Evaluation of silver nanoparticles as antiviral agent against ToMV and PVY in tomato plants. Middle East J. Appl. Sci. 8, 100–111.

Ezhilarasi, P.N., Karthik, P., Chhanwal, N., Anandharamakrishnan, C., 2013. Nanoencapsulation techniques for food bioactive components: a review. Food Bioprocess Technol. 6, 628–647.

Fiorati, A., Bellingeri, A., Punta, C., Corsi, I., Venditti, I., 2020. Silver nanoparticles for water pollution monitoring and treatments: ecosafety challenge and cellulose-based hybrids solution. Polymers 12 (8), 1635.

Fouda, M.M., Abdelsalam, N.R., El-Naggar, M.E., Zaitoun, A.F., Salim, B.M., Bin-Jumah, M., Allam, A.A., Abo-Marzoka, S.A., Kandil, E.E., 2020. Impact of high throughput green synthesized silver nanoparticles on agronomic traits of onion. Int. J. Biol. Macromol. 149, 1304–1317.

Ge, L., Li, Q., Wang, M., Ouyang, J., Li, X., Xing, M.M., 2014. Nanosilver particles in medical applications: synthesis, performance, and toxicity. Int. J. Nanomedicine 9, 2399–2407.

Govindaraju, K., Itroutwar, P.D., Veeramani, V., Kumar, T.A., Tamilselvan, S., 2019. Application of nanotechnology in diagnosis and disease management of white spot syndrome virus (WSSV) in aquaculture. J. Clust. Sci. 31, 1–9.

Hassan, A.A., Mansour, M.K., El Hamaky, A.M., El Ahl, R.M.S., Oraby, N.H., 2020. Nanomaterials and nanocomposite applications in veterinary medicine. In: Multifunctional Hybrid Nanomaterials for Sustainable Agri-Food and Ecosystems. Elsevier, pp. 583–638.

Hendrickson, C., Huffstutler, G., Bunderson, L., 2017. Emerging applications and future roles of nanotechnologies in agriculture. Agric. Res. Technol. 11, 1.

Hwang, M., Lee, E.J., Kweon, S.Y., Park, M.S., Jeong, J.Y., Um, J.H., Kim, S.A., Han, B.S., Lee, K.H., Yoon, H.J., 2012. Risk assessment principle for engineered nanotechnology in food and drug. Toxicol. Res. 28 (2), 73–79.

Ijaz, M., Zafar, M., Afsheen, S., Iqbal, T., 2020. A review on Ag-nanostructures for enhancement in shelf time of fruits. J. Inorg. Organomet. Polym. Mater. 30, 1475–1482.

Ingle, A.P., Gupta, I., Jogee, P., Rai, M., 2020. Role of nanotechnology in the detection of mycotoxins: a smart approach. In: Nanomycotoxicology. Academic Press, pp. 11–33.

Istiqola, A., Syafiuddin, A., 2020. A review of silver nanoparticles in food packaging technologies: regulation, methods, properties, migration, and future challenges. J. China Chem. Soc. 67, 1942–1956.

Jena, P., Mohanty, S., Mallicki, R., Jacob, B., Sonawane, A., 2012. Toxicity and antibacterial assessment of chitosan coated silver nanoparticles on human pathogens and macrophage cells. Int. J. Nanomedicine 7, 1805–1818.

Jiang, F., Li, P., Zong, C., Yang, H., 2020. Surface-plasmon-coupled chemiluminescence amplification of silver nanoparticles modified immunosensor for high-throughput ultrasensitive detection of multiple mycotoxins. Anal. Chim. Acta 1114, 58–65.

Khan, I., Raza, M.A., Awan, S.A., Shah, G.A., Rizwan, M., Ali, B., Tariq, R., Hassan, M.J., Alyemeni, M.N., Brestic, M., Zhang, X., 2020. Amelioration of salt induced toxicity in pearl millet by seed priming with silver nanoparticles (AgNPs): the oxidative damage, antioxidant enzymes and ions uptake are major determinants of salt tolerant capacity. Plant Physiol. Biochem. 156, 221–232.

References

Kraśniewska, K., Galus, S., Gniewosz, M., 2020. Biopolymers-based materials containing silver nanoparticles as active packaging for food applications—a review. Int. J. Mol. Sci. 21 (3), 698.

Lee, S.H., Jun, B.H., 2019. Silver nanoparticles: synthesis and application for nanomedicine. Int. J. Mol. Sci. 20 (4), 865.

Lok, C.N., Ho, C.M., Chen, R., He, Q.Y., Yu, W.Y., Sun, H., Tam, P.K.H., Chiu, J.F., Che, C.M., 2007. Silver nanoparticles: partial oxidation and antibacterial activities. J. Biol. Inorg. Chem. 12, 527–534.

Mahawar, H., Prasanna, R., 2018. Prospecting the interactions of nanoparticles with beneficial microorganisms for developing green technologies for agriculture. Environ. Nanotechnol. Monit. Manag. 10, 477–485.

Mahler, G.J., Esch, M.B., Tako, E., Southard, T.L., Archer, S.D., Glahn, R.P., et al., 2012. Oral exposure to polystyrene nanoparticles affects iron absorption. Nat. Nanotechnol. 7, 264–271.

McGee, C.F., 2020. The effects of silver nanoparticles on the microbial nitrogen cycle: a review of the known risks. Environ. Sci. Pollut. Res. 27, 1–13.

Millán-Chiu, B.E., del Pilar Rodriguez-Torres, M., Loske, A.M., 2020. Nanotoxicology in plants. In: Green Nanoparticles. Springer, Cham, pp. 43–76.

Narducci, D., 2007. An introduction to nanotechnologies: what's in it for us? Vet. Res. Commun. 1, 131–137.

Noshad, A., Iqbal, M., Hetherington, C., Wahab, H., 2020. Biogenic AgNPs—a nano weapon against bacterial canker of tomato (BCT). Adv. Agric. 2020, 9630785.

Prosposito, P., Burratti, L., Venditti, I., 2020. Silver nanoparticles as colorimetric sensors for water pollutants. Chem. Aust. 8 (2), 26.

Ragam, P.N., Mathew, B., 2020. Unmodified silver nanoparticles for dual detection of dithiocarbamate fungicide and rapid degradation of water pollutants. Int. J. Environ. Sci. Technol. 17 (3), 1739–1752.

Sami, F., Siddiqui, H., Hayat, S., 2020. Impact of silver nanoparticles on plant physiology: a critical review. In: Sustainable Agriculture Reviews 41. Springer, Cham, pp. 111–127.

Shume, W.M., Murthy, H.C., Zereffa, E.A., 2020. A review on synthesis and characterization of Ag_2O nanoparticles for photocatalytic applications. J. Chem. 2020. https://doi.org/10.1155/2020/5039479, 5039479.

Simbine, E.O., Rodrigues, L.D.C., Lapa-Guimaraes, J., Kamimura, E.S., Corassin, C.H., Oliveira, C.A.F.D., 2019. Application of silver nanoparticles in food packages: a review. Food Sci. Technol. 39 (4), 793–802.

Singh, R., Mehra, R., Walia, A., Gupta, S., Chawla, P., Kumar, H., Thakur, A., Kaushik, R., Kumar, N., 2021. Colorimetric sensing approaches based on silver nanoparticles aggregation for determination of toxic metal ions in water sample: a review. Int. J. Environ. Anal. Chem., 1–16. https://doi.org/10.1080/03067319.2021.1873315.

StatNano, 2019. Nanotechnology Products Database (NPD) (WWW Document) http://product.statnano.com/. (Accessed 16 September 2019).

Sulaiman, I.C., Chieng, B.W., Osman, M.J., Ong, K.K., Rashid, J.I.A., Yunus, W.W., Noor, S.A.M., Kasim, N.A.M., Halim, N.A., Mohamad, A., 2020. A review on colorimetric methods for determination of organophosphate pesticides using gold and silver nanoparticles. Microchim. Acta 187 (2), 1–22.

Tripathi, D.K., Tripathi, A., Singh, S., Singh, Y., Vishwakarma, K., Yadav, G., Sharma, S., Singh, V.K., Mishra, R.K., Upadhyay, R.G., Dubey, N.K., 2017. Uptake, accumulation and toxicity of silver nanoparticle in autotrophic plants, and heterotrophic microbes: a concentric review. Front. Microbiol. 8, 7.

Verma, P., Maheshwari, S.K., 2019. Applications of silver nanoparticles in diverse sectors. Int. J. Nano Dimens. 10 (1), 18–36.

Yan, A., Chen, Z., 2019. Impacts of silver nanoparticles on plants: a focus on the phytotoxicity and underlying mechanism. Int. J. Mol. Sci. 20 (5), 1003.

Yen, C.W., de Puig, H., Tam, J.O., Gómez-Márquez, J., Bosch, I., Hamad-Schifferli, K., Gehrke, L., 2015. Multicolored silver nanoparticles for multiplexed disease diagnostics: distinguishing dengue, yellow fever, and Ebola viruses. Lab Chip 15 (7), 1638–1641.

Yusuf, M., 2020. Silver Nanoparticles: Synthesis and Applications. Handbook of Ecomaterials. Springer, p. 2343.

Zhang, X.F., Liu, Z.G., Shen, W., Gurunathan, S., 2016. Silver nanoparticles: synthesis, characterization, properties, applications, and therapeutic approaches. Int. J. Mol. Sci. 17, 1534.

PART 1

Antimicrobials

CHAPTER 2

Silver-based nanostructures as antifungal agents: Mechanisms and applications

Santwana Padhi[a] and Anindita Behera[b]

[a]KIIT Technology Business Incubator, KIIT Deemed to be University, Bhubaneswar, Odisha, India
[b]School of Pharmaceutical Sciences, Siksha 'O' Anusandhan Deemed to be University, Bhubaneswar, Odisha, India

2.1 Introduction

Fungal diseases often appear in the form of topical infections in humans, predominantly affecting skin and mucous membranes but when invasive become systemic infections in internal organs. Limited diagnostic techniques and the lack of efficacious antifungal agents have led researchers to search for novel leading molecules or to repurposing the existing ones. Fungal infections are often seen in individuals with immunity disorders that can hasten diseases and postoperative complications (Brown et al., 2012). Moreover, plants with compromised immunity can become infected with fungi, which affects the quality and quantity of crop yield. When vegetative parts of a plant are affected, the fungal infection can appear either superficially or postpenetration inside the plant's system (Jain et al., 2019). Most fungal infections are seen in rice, wheat, maize, potatoes, soybeans, and tomatoes, which affects the agricultural economics of developing countries (Almeida et al., 2019).

Several antifungal drugs are hydrophobic, which can culminate in restricted water solubility, poor bioavailability, and constrained formulation strategies (Lewis, 2011). In such scenarios, rational designs of therapeutic drug delivery systems have the potential to boost delivery and overcome adverse conditions (Padhi et al., 2015).

In last two decades, nanotechnology has established a new paradigm in the efficient delivery of numerous drugs for an array of diseases (Khuroo et al., 2014). The entrapment of nano drugs often leads to high therapeutic efficacy with minimal adverse effects (Zhang et al., 2010; Padhi et al., 2020). Among different delivery systems, nanoparticles (NPs) have proven to enhance and maintain the therapeutic efficacy of drugs for a predetermined time (Padhi et al., 2018; Zazo et al., 2016). NPs are defined as having dimensions between approximately 1 and 100 nm with specific physicochemical, electrical, thermal, and optical characteristics (Behera et al., 2020; Dakal et al., 2016). NPs possess unique properties and thus can offer higher quality delivery systems in health care than bulk material can. Metal NPs can be divided into two categories, noble or magnetic based on the material from which they are derived; and are mostly prepared in the form of oxides. Gold and silver NPs are known as noble metal nanoparticles and have numerous applications because their formulation steps are not complicated and involve nontoxic excipients (Behera et al., 2020). Engineered NPs of gold and silver possess unique properties, such as a high surface-to-volume ratio and high chemical activity with improved antimicrobial properties (Gitipour et al., 2016). AgNPs are used widely as active antimicrobial agents, and their therapeutic efficacy is impacted by their physicochemical attributes, such as size, shape, concentration, and colloidal state (Bhattacharya and Mukherjee, 2008; Nateghi and Hajimirzababa, 2014; Pal et al., 2007; Raza et al., 2016).

The particle size of AgNPs plays an important role in exhibiting antimicrobial activity. AgNPs less than 50 nm show appreciable antimicrobial activity. AgNPs in smaller sizes are noticeably stable and biocompatible and offer high antimicrobial activity (Yacamán et al., 2001). Small size, high surface-to-volume ratio, and crystallographic structure are important factors responsible for increased permeability through cell membranes and causing cell membrane damage and cell death (Li et al., 2013b; Pal et al., 2007). Similarly, the shape of AgNPs affects the physicochemical properties of formulated AgNPs and determines interactions with fungi, viruses, and bacteria as well. Extensive research has been done on shape-dependent antimicrobial activity (Galdiero et al., 2011; Wu et al., 2014; Raza et al., 2016). It has been suggested that AgNPs of different shapes, but with the same surface area, show a difference in antimicrobial activity, as they vary with effective surface area and sites for active interactions (Dakal et al., 2016). The concentration of AgNPs affects their efficacy as antimicrobial agents, depending on the type of microorganism with which they interact. Studies have revealed that different types of cell structures and cell membranes require different concentrations of AgNPs, as the thickness and potential of cell membranes differ from species to species. Gram-positive and gram-negative bacteria require different concentrations of AgNPs, as the thickness of their cell membrane depends on the absence or presence of the peptidoglycan layer (Dakal et al., 2016). The colloidal AgNPs were found to be more effective as antimicrobial agents than AgNPs alone, as they act as a catalyst and cause destabilization of the enzymes required for oxygen utilization by pathogenic bacteria, viruses, yeasts, and fungi (Lkhagvajav et al., 2011; Dehnavi et al., 2012; Kumar et al., 2014; Suganya et al., 2015). The green synthesis method can be adopted for synthesis of AgNPs with controlled size, shape, stability, and enhanced antimicrobial activity (Dehnavi et al., 2012).

This chapter presents an in-depth discussion of the mode of action of antifungal agents along with the mechanisms by which AgNPs confer their antimicrobial efficacy. The application of AgNPs as antifungal agents against fungal infections in plants and humans is also discussed.

2.2 Mechanism of action of antifungal agents

Antifungal drug molecules can exert their action as either a fungicidal or a fungistatic agent. Fungicidal agents kill fungi, whereas fungistatic agents inhibit

2.2 Mechanism of action of antifungal agents

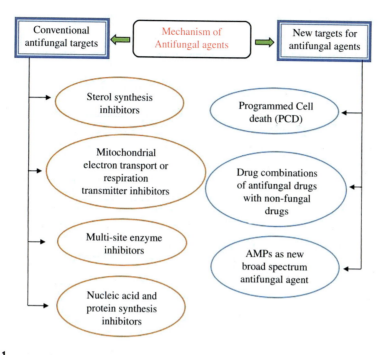

FIG. 2.1

Mechanisms of antifungal agents. (Conventional antifungal agents show biological effects by inhibiting various metabolic processes of fungi, such as sterol synthesis, mitochondrial electron transport or respiration transmitters, multisite enzymes, and nucleic acid and protein synthesis; whereas new targets for antifungal agents are programmed cell death, a combination of antifungal drugs with nonfungal drugs to show a synergistic effect or prevent fungal resistance, and AMPs as new broad-spectrum antifungal agents.)

the growth of fungi. As fungi are eukaryotes, they share commonalities in the metabolic processes and biological similarities with the host. Hence antifungal agents target a few specific processes occurring in the fungi to kill or stop their growth. The targets of the conventional antifungal agents are presented below (see also in Fig. 2.1).

- sterol synthesis inhibitors
- mitochondrial electron transport or respiration transmitter inhibitors
- multisite enzyme inhibitors
- nucleic acid and protein synthesis inhibitors

In addition, the following novel targets have emerged and offer better antifungal activity to prevent fungal resistance:

- programmed cell death
- drug combinations of fungal drugs with nonfungal drugs
- antimicrobial peptides (AMPs) as new broad-spectrum antifungal agents

One of the vital targets for the antifungal drugs is ergosterol, a cell membrane sterol. Ergosterol (5,7-diene oxysterol) is an abundant sterol in fungal cell membranes responsible for the maintenance of integrity, permeability, and fluidity of fungi (Douglas and Konopka, 2014). Membrane integrity is required for cell survival, as it protects the cell from the external environment and regulates cellular physiology. If cell membrane fluidity is lost, disruption of the cell wall and the plasma membrane follows, which affects the biosynthesis of the cell wall and results in a weak cell. In animal or human fungal infections, ergosterol biosynthesis is targeted by some inhibitors such as the azole group of molecules. Azoles are of two types, based on the number of nitrogen atoms present in the ring (presence of either two or three nitrogen atoms). Azoles block the biosynthetic pathway of ergosterol by accumulating a toxic derivative of sterol. The accumulation of toxic methylated sterol (i.e., 14-α-methyl-ergosta-8-ene-3,6-diol) blocks the production of ergosterol and thereby stops cell growth. The fluidity of the cell membrane is maintained when ergosterol is inserted between the phospholipids, as reduction of the ergosterol content directly affects the integrity and functionality of cell membranes (de Oliveira Pereira et al., 2013; Nett and Andes, 2016).

Another class of drugs that was introduced before azoles are polyenes. Amphotericin B was the first drug to be used in the polyenes class of antifungal drugs. Amphotericin B enters the cell cytoplasm by binding to the ergosterol layer by forming pores in the cytoplasmic membrane. The formation of pores leads to a loss of cell membrane integrity, which leads to leakage of intracellular ions, causing an unstable osmotic gradient (Teixeira-Santos et al., 2015). It was shown that formation of a pore in a cell membrane did not lead to cell death; rather, cell apoptosis was reported due to the arrest of the G_2/M stages of the cell cycle, oxidative stress, and abnormal morphology of the mitochondria (Guirao-Abad et al., 2017).

Apart from ergosterol, chitin, glucan, mannan, and glycoproteins form the major constituents of fungal cell walls. These components protect the cells from environmental stress and osmotic pressure along with maintaining their integrity and shape. Echinocandin is the latest antifungal agent added to the category. Echinocandins are semisynthetic lipopeptides that impair cell wall biosynthesis by inhibiting the β-1,3-glucan synthase enzyme, responsible for β-1,3-glucan biosynthesis. A decrease in the concentration of β-1,3-glucan leads to loss of integrity of the cell wall. Echinocandins are noncompetitive β-1,3-glucan synthase inhibitors and are not cytotoxic to host cells. A few publications reported that caspase-dependent apoptosis was triggered by echinocandin (e.g., Shirazi and Kontoyiannis, 2015).

In the case of phytopathogenic fungal infections, morpholines and demethylation inhibitors are the major class of molecules that act by blocking the biosynthesis of ergosterol. Morpholines inhibit two enzymes, namely sterol reductase and sterol isomerase, which play a pivotal role during ergosterol biosynthesis (Mazu et al., 2016). The demethylation inhibitors stop the C14-demethylation step during ergosterol biosynthesis. The chemical nuclei showing the demethylation inhibition are pyrimidines and azoles (Campoy and Adrio, 2017). Respiration inhibitors located at various sites of intracellular mitochondria are the second type of antifungal targets.

The respiration inhibitors are succinate dehydrogenase inhibitors and Q_o (quinone outside) inhibitors. Succinate dehydrogenase inhibitors inhibit fungal respiration by binding to the ubiquinone-binding site in the complex II of mitochondria (Avenot and Michailides, 2010), whereas Q_o inhibitors inhibit energy production from mitochondrial respiration by binding to the quinone outside the (Q_o) site of complex III in mitochondria (Fernández-Ortuño et al., 2008). Multisite enzyme inhibitors inhibit several thiols containing enzymes responsible for respiration. Nucleic acid metabolism and protein synthesis inhibitors inhibit DNA and RNA synthesis, affecting cell division and cellular metabolism. Benzimidazoles, phenylamides, and dicarboximides inhibit microtubule assembly during cell division, thereby causing cell death (Davidse, 1986).

The previously mentioned antifungal agents trigger apoptotic cell death along with their mode of action. But the physiology of apoptosis in fungi is different from mammalian apoptosis because the extrinsic pathway is absent in fungi. The intrinsic pathway in fungi is characterized by an increase in permeability of the outer mitochondrial membrane. This permeability is dependent on calcium and other voltage-gated ion channels. Thus cytochrome c and apoptosis-inducing factor (AIF) along with other intra-mitochondrial constituents are released into the cytoplasm. Cytochrome c binds to apoptotic protease-activating factors in the cytosol. These processes alter the mitochondrial membrane potential, and cysteine-dependent caspases are activated in the early stages of apoptosis. The reactive oxygen species (ROS) are accumulated in an early signaling pathway for apoptosis (Carmona-Gutierrez et al., 2010; Mazzoni and Falcone, 2008; Perrone et al., 2008). The early stages of apoptosis are triggered by a loss of mitochondrial membrane potential, accumulation of ROS, and externalization of phosphatidylserine; whereas later stages of apoptosis are accompanied by nuclear fission, DNA damage, cytochrome c and ATP release, and increased intracellular calcium ion levels (Madeo et al., 2004; Kyrylkova et al., 2012).

Using a combination of drugs is another approach to increasing drug efficacy and avoiding the fungi's resistance to drugs. However, a combination of two antifungal drugs was found to be costly and had serious adverse effects (Campitelli et al., 2017). Combination antifungal therapy can also be synergistic or antagonistic. Combinatorial therapy involving an antifungal agent with a nonfungal agent was evaluated. The majority of the combinations showed a synergistic action on resistant strains of fungi (Liu et al., 2014). The synergistic effect showed increased permeability of cell membranes, reduction in the outflow of antifungal drugs, interference with intracellular ionic homeostasis, inhibition of enzymes and proteins required for fungal survival, and inhibition of biofilm formation (Liu et al., 2014).

Antimicrobial peptides (AMPs) are another interesting class of antifungal drug targets and are an important part of the immune system in all forms of life. AMPs contain amino acid residues used for cell transfer. They contain cationic or hydrophobic residues, depending on the amino acid's composition, size, and shape. They act by disrupting the cell membranes of fungi with electrostatic and/or hydrophobic interactions, leading to leakage of intracellular contents. Development of resistance during the use of AMPs is low, as they target the cell membrane and reformation of the cell membrane is inefficient. AMPs act through various mechanisms, so the success rate is very high, without development of resistance. AMPs' mechanism involves extracellular binding, nonlytic ATP release, and efflux of ion channels. Therefore AMPs are known to be a potent and suitable alternative to conventional antifungal agents (Hayes et al., 2014; Wang et al., 2016b).

2.3 Mechanistic approach of silver nanoparticles as antifungal agents

Silver nanoparticles are known to be broad-spectrum antimicrobial agents against a wide range of microbes, such as bacteria, viruses, fungi, and yeasts. The extensive research carried out on the antibacterial activity of AgNPs has led to an understanding of their different underlying mechanisms and their efficacy. The mechanism and effects of AgNPs' antifungal activities are less studied than are their antibacterial activities. The antifungal activity of AgNPs is strongly related to their size, shape, and surface modification. To better describe the mechanism and mode of AgNPs' antifungal activity, selected reports focusing on such activity are discussed next and are illustrated in Fig. 2.2.

- AgNPs' positive charge helps to attach NPs to the surface of cells, that is, the cell membrane. An electrostatic attraction occurs between the AgNPs and the negatively charged fungal cell membrane. After adhesion of the NPs, the proteins of the cell wall undergo an irreversible alteration leading to disruption of the cell wall. This sequence affects the integrity and permeability of the cell membrane and leads to an increase in its permeability, which helps regulate the transport of substances through the cell membrane. An increase in membrane leakage of intracellular contents such as ions, proteins are responsible for cell death (Ghosh et al., 2012; Schreurs and Rosenberg, 1982; Kim et al., 2011).
- Most reports indicate the superior ability of AgNPs to disrupt the potential of the fungal cell membrane and to form pores, leading to leakage of intracellular ions, organelles, and proteins. These events trigger the apoptosis process and promote ultrastructural changes in fungi (Vazquez-Muñoz et al., 2014; Hwang et al., 2012; Kim et al., 2009). Lara et al. (2015) investigated the mechanism of AgNPs with *Candida albicans*, a prevalent human fungal pathogen, using a high-resolution TEM (i.e., STEM-HAADF, an atomic resolution analytical microscope). The cells of *C. albicans* were exposed to AgNPs for 24 h and enlargement of the cell was found with changes in the cell membrane and cell wall. The thickness of the cell membrane was increased, and there was no appearance of electron density. The cell walls of fungi comprised chitin, glucans, and mannoproteins. NPs' interactions disrupted the outer cell wall, leading to the formation of pores through which the AgNPs entered the cells. Antibiofilm formation was also reported, which might be due to disruption of

2.3 Mechanistic approach of silver nanoparticles as antifungal agents

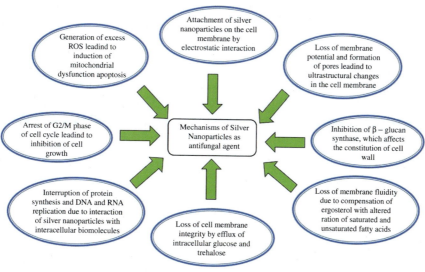

FIG. 2.2

Different mechanisms involved in antifungal activity of silver nanoparticles.

the cell wall that then affected the survival of the fungus and the filamentous structure of the fungus (hyphae). Hence it can be stated that positively charged AgNPs inhibited the growth of the *C. albicans* and subsequently prevent biofilm formation by the hyphae.

- Gutiérrez et al. (2018) reported that AgNPs retard fungal growth by inhibiting β-glucan synthase, the enzyme responsible for glucan synthesis for cell walls. The observed fungicidal effect was due to modification in the integrity of the cell wall, leading to loss of mechanical resistance and cell destruction by osmotic pressure variations.
- In a study on *C. albicans* (Radhakrishnan et al., 2018), it was found that, if ergosterol content is reduced, a cell can compensate for the loss by utilizing other membrane lipids to maintain the optimum membrane fluidity. In the case of this study, saturated and unsaturated fatty acids compensated to maintain membrane fluidity. Through the use of gas chromatogaphy-mass spectrometry (GC-MS), the fungal cell was assessed for saturated and unsaturated fatty acids levels in the presence of AgNPs. There was a significant increase in the percentage of saturated fatty acids with a simultaneous decrease in levels of unsaturated fatty acids (palmitoleic acid, oleic acid, linoleic acid, and linolenic acid). These changes contributed to a net decrease in membrane fluidity following treatment with AgNPs.
- Carbohydrate components of a cell's membrane such as glucose and trehalose protect the biological membrane from denaturation or inactivation due to stress conditions like dehydration, heat, cold, oxidation, and toxic agents. When the

fungal cell was exposed to AgNPs, it was found that the release of glucose and trehalose increased due to disruption of the cell membrane, which hampered the integrity of the cell membrane and produced pores, thereby increasing the permeability of intracellular elements such as ions, proteins, and other constituting intracellular material (Alvarez-Peral et al., 2002). These reactions resulted in the loss of electrical potential of the cell membrane and affected the growth and survival of the cell (Elbein et al., 2003; Kim et al., 2009).

- When AgNPs enter a cell due to interactions among intracellular structures and biomolecules (e.g., proteins, lipids, and DNA), the AgNPs interact with the ribosomes and cause their deactivation, leading to inhibition of protein synthesis by translation (Rai et al., 2014). Bhattacharya and Mukherjee (2008) reported the inhibition of carbohydrate metabolism due to inhibition of enzymes through interaction with AgNPs. AgNPs inhibit phosphomannose isomerase, an enzyme involved in the glycolytic cycle, leading to inhibition of sugar metabolism in fungal cells. Also, AgNPs interrupt denaturation of DNA, thereby hampering the process of cell division. AgNPs complex with fungal nucleic acid and interact with the nucleosides without interfering with a phosphate group. Ag^+ ion intercalates between purine and pyrimidine base pairs and breaks the hydrogen bonds between the base pairs of DNA strands, hence destroying the double-helix structure of DNA and interfering with the transcription process (Klueh et al., 2000).

- The antifungal activity of AgNPs is also associated with induction of mitochondrial dysfunctional apoptosis, which can result from excess oxidative stress due to ROS generation, such as hydrogen peroxide (H_2O_2), superoxide anion ($O_2^{\bullet-}$), hydroxyl radical ($OH^\bullet$), hypochlorous acid (HOCl), and singlet oxygen ($^1O_2$) (Kim et al., 2011). An excess of generated ROS damages the membrane of mitochondria and leads to necrosis and cell death. Polyunsaturated fatty acids on oxidation produce lipid hydroperoxides in the initial stages of ROS generation. Increases in ROS levels also increase the oxidation of lipids, proteins, and DNA. Free radicals interact with lipid molecules present in the cell membrane and produce lipid peroxidation products (Huang et al., 2010; Howden and Faux, 1996). The mechanism of ROS-induced cell damage can be studied through a combination of techniques such as staining with fluorogenic dye like $DCFH_2$-DA (2,7-dichlorodihydrofluorescein diacetate), followed by confocal laser scanning microscopy. AgNP-treated cells, when compared with a control group (untreated cells), show more fluorescence due to diffusion of the dye into the cell. Once the dye enters the cell, the intracellular esterase breaks down the diacetate group, and the ROS oxidize the nonfluorescent dye into the highly fluorescent 2,7-dichlolorofluorescein compound. Furthermore, confocal laser scanning microscopy shows increased green fluorescence in the AgNP-treated cells, as compared with the control group (untreated with NPs) or the positive control group (treated with H_2O_2) (Radhakrishnan et al., 2018).

- The role of AgNPs in the fungal life cycle was studied in the absence and presence of NPs on fungal cell cultures (Kim et al., 2009). The study was conducted by measuring DNA content employing flow cytometry and using propidium iodide (PI) as the staining dye [PI is a dye that stains the bases of DNA and RNA molecules (Tas and Westerneng, 1981)]. The results showed that AgNPs arrested the cell cycle at the G_2/M phase.

2.4 Comparison of antifungal activity of biosynthesized silver nanoparticles with physically or chemically synthesized nanoparticles

Biosynthesis of metallic NPs is more advantageous than synthesis by physical or chemical methods. For example, through the use of microorganisms or plant extracts, biological synthesis of AgNPs is more cost-effective, involves the use of nontoxic reducing and stabilizing agents and thus is superior in terms safety and toxicity, produces NPs that are more stable and more uniform in size and shape, and have more enhanced properties; moreover, the biological and therapeutic activity of biosynthesized NPs is greater. Kumari et al. (2019) did extensive studies on the efficacy of biosynthesized and chemically synthesized AgNPs and noted the following outcomes:

- Biosynthesized AgNPs (bAgNPs) showed increased antifungal activity, compared with chemically synthesized AgNPs (cAgNPs).
- Metal-based bAgNPs with biological reducing and stabilizing agents showed a synergistic effect inhibiting different physiological processes of fungus-like DNA damage and membrane disruption. The AgNPs could easily enter the fungal cell and bind with intracellular biomolecules. Fungal cells exposed to bAgNPs faced a challenge in the ability to develop resistance and survive.
- Comparison and assessment of ROS generation and its effect on the fungal cell were done using the nitroblue tetrazolium (NBT) reduction assay method. It was found that bAgNPs generated more ROS than cAgNPs. The study suggested that oxidative stress damages the DNA, cell membrane, and proteins leading to retardation of growth and ultimately cell death.
- bAgNPs were found to be responsible for downregulating the expression of enzymes responsible for managing oxidative stress. bAgNPs destroyed the intracellular redox homeostasis of the fungus, while cAgNPs failed to do the same.
- The genes responsible for maintaining cell wall integrity, fluidity, osmotic balance, spore viability, and cell division were down-regulated by bAgNPs, but they were noted to be upregulated by cAgNPs. bAgNPs caused a complete loss of membrane fluidity, cell wall integrity, and germination ability, leading to cell death; whereas the cAgNP-treated fungi sustained the stress and osmotic imbalance by increasing the expression of the gene. Hence it can be said that cAgNPs imparted membrane injury, not cellular death.

Based on the observed effects, it can be said that bAgNPs are more efficacious than cAgNPs. However, the biosynthesis of AgNPs requires the selection of suitable biological reducing and stabilizing agents and proper reaction conditions, such as pH, temperature, and concentration of reducing and stabilizing agents. In bAgNPs, surface modification is not required, as they are accompanied by other biomolecules as biological reducing and stabilizing agents, which enhances bAgNPs' physicochemical properties.

2.5 Application of silver nanoparticles in human fungal infections

Fungal infections in humans can range in severity from mild to life-threatening. Some fungal infections are cutaneous or superficial, whereas some cause severe systemic infections that damage vital organs such as the liver, kidney, lungs, and nervous system. The most prevalent fungi infecting humans belong to the *Candida* and *Aspergillus* species, which can cause systemic pathogenic diseases. Other strains of fungi causing infections are *Fusarium* spp., *Cryptococcus* spp., *Zygomycete*, and yeasts (Pfaller and Diekema, 2004; Badiee and Hashemizadeh, 2014).

Among the *Candida* spp., *Candida albicans* is the most infectious and causes most of the systemic infections in immune-compromised or hospitalized patients (Pfaller and Diekema, 2010). *Aspergillus* spp. differ based on geographical regions, and most pathogenic species are reported to be *A. fumigatus, A. flavus, A. niger*, or *A. terreus. Aspergillus* spp. generally infects the respiratory system of patients suffering from asthma or cystic fibrosis (Lin et al., 2001). *Cryptococcus neoformans* causes asymptomatic lung infections, and in severe cases may lead to meningitis or meningoencephalitis in patients with AIDS. *Cryptococcus* spp. can also infect soft tissues, skin, eyes, and bones or joints (McCarthy et al., 2006). Zygomycosis is most prevalent in patients with diabetes mellitus and hematological malignancy, and the most pathogenic species of this variety are *Rhizopus, Mucor*, and *Rhizomucor* (Kontoyiannis et al., 2005). The *Fusarium* species can cause both superficial and systemic infections (Nucci and Anaissie, 2007).

Robles-Martínez et al. (2020) synthesized AgNPs entrapping aqueous and ethanolic extracts of *Mentha piperita* and evaluated its antifungal activity against *Candida albicans*. It was noted that the AgNPs entrapping the ethanolic extract exhibited antimycotic activity superior to that of the aqueous extract. With a 10% solution of aqueous extract, AgNPs showed 10.66% inhibition of growth; whereas with an 8% solution of ethanolic extract, AgNPs showed complete inhibition of growth. The inhibitory capability for the growth of *C. albicans* with that of ethanolic extract entrapped in AgNPs was increased by 89% as compared to only ethanolic extract. The inhibition of growth was achieved due to the presence of secondary metabolites in the ethanolic extract, that is, anthocyanins and flavonoids that damaged the cell membrane of the fungi.

2.5 Application of silver nanoparticles in human fungal infections

Using green synthesis, Dutta et al. (2020) prepared AgNPs from *Citrus limetta* peel extracts, which was conducted to determine the antibacterial and antifungal activity of the AgNPs. The antifungal activity was conducted on four species of *Candida*: *C. albicans*, *C. glabrata*, *C. tropicalis*, and *C. Parapsilosis*. The formulation was subjected for evaluation of protein release by internalization of AgNPs. The micromorphological study by electron microscopy showed a change in the morphology of the cell membrane with leakage of intracellular material and deposition on the cell surface.

Guerra et al. (2020) studied the selective comparative antifungal activity of AgNPs on pathogenic *Candida tropicalis* and probiotic *Saccharomyces boulardii*. The comparative study was done against fluconazole and amphotericin B. The polygonal AgNPs selectively inhibited 90% of the growth of *C. tropicalis*, whereas 50% of the cells of *S. boulardii* were observed to be viable. It was concluded that polygonal AgNPs were selective inhibitors of growth against pathogenic fungal cells.

Khan et al. (2020) studied the antibacterial, antifungal, and cytotoxic activity of biologically synthesized AgNPs against gram-positive and gram-negative bacteria, *Aspergillus*, *Fusarium* spp., and mammalian cell lines. Using a biogenic method, Khan and colleagues found that AgNPs were synthesized from *Bacillus* sp. MB353 (PRJNA357966). The AgNPs showed very good antibacterial activity against *E. coli*, *Bacillus subtilis*, *Staphylococcus aureus*, *Klebsiella pneumoniae*, *Enterococcus faecium*, *Enterococcus faecalis*, and *Streptomyces laurentii*. Similarly, growth of *Aspergillus niger*, *Aspergillus fumigatus*, and *Fusarium soleni* was inhibited by the biogenic AgNPs. A cytotoxicity study was performed on mammalian cell lines (i.e., rhabdomyosarcomas and fibroblasts) that involved ROS generation and changes in the intracellular calcium level. Tutaj et al. (2016) studied the antifungal effect of amphotericin B conjugated with AgNPs against *Candida albicans*, *Aspergillus niger*, and *Fusarium culmorum*. Synergistic antifungal activity was reported owing to the antifungal activity of amphotericin B and silver ions released from the NPs.

Wang et al. (2016a) biosynthesized AgNPs from the fungus *Arthroderma fulvum* and reported antifungal activity on ten pathogenic strains of *Candida*, *Aspergillus*, and *Fusarium*, with itraconazole and fluconazole used as controls.

Parthiban et al. (2019) biosynthesized AgNPs from the leaves of *Annona Reticulata* and studied the larvicidal and antimicrobial effects on human pathogens. The study evaluated the efficacy of AgNPs in killing the larvae of *Aedes aegypti* and its antibacterial and antifungal efficacy toward the pathogens causing infections in humans. The synthesized AgNPs were crystalline with a face-centered cubic structure, and the minimum inhibitory concentration (MIC) was evaluated in six human pathogenic bacteria and fungi by multiwell plates (96 wells) through a serial dilution method. For the *Candida* species, the MIC was found to be 62.5 µg/mL.

Nayak et al. (2018) biosynthesized AgNPs from the fungus *Penicillium italicum*, isolated from wasp nest soil, and performed antimicrobial activity on different multidrug-resistant strains of pathogenic bacteria and the fungus *Candida albicans*. The antifungal activity was determined on clinical isolates of multidrug-resistant isolates of the pathogen, and the zone of inhibition was found to be 14 mm.

Owaid et al. (2015) biosynthesized AgNPs from the edible mushroom *Pleurotus cornucopiae* var. *citrinopileatus*. The antimycotic activity was reported against the *Candida* species. The anticandida activity was found to be significant by in vitro inhibition of growth of the pathogenic fungus.

Moazeni et al. (2012) synthesized AgNPs from three dermatophytic strains, *Trichophyton rubrum*, *T. mentagrophytes*, and *Microsporum canis* by extracellular synthesis. The three varieties of AgNPs varied in size (nm) and shape (from spherical to cylindrical). The antimycotic activity of AgNPs synthesized from *T. mentagrophytes* showed better activity against *Candida* species (4 µg/mL), compared with that of the positive control group, fluconazole (16 µg/mL). Nasrollahi et al. (2011) chemically synthesized AgNPs, and their efficacy against *Candida* species was found to be better than fluconazole and amphotericin B. Panáček et al. (2009) synthesized AgNPs by using modified Tollens' reagent and studied its efficacy against *Candida albicans* with different strains of *Candida* spp. collected from blood of patients suffering from Candida sepsis. The fungistatic and fungicidal activities were determined by using the polymer PVP; the surfactants polyoxyethylene lauryl ether, Tween 80 and sodium dodecyl sulfate (SDS)-stabilized AgNPs; and ionic silver ion. The SDS-stabilized AgNPs exhibited higher fungicidal activity, and the polymer- and surfactant-stabilized AgNPs showed better fungistatic activity when compared with an ionic form of silver.

Prucek et al. (2011) synthesized two magnetic binary nanocomposites (i.e., silver@magnetite (Ag@Fe_3O_4) and maghemite@silver (γ-Fe_2O_3@Ag)) by in situ chemical reduction of silver in the presence of maltose and a magnetic phase. The synthesized nanocomposites were evaluated for antibacterial and antifungal activity. Four strains of *Candida* (*C. albicans* I & II, *C. tropicalis*, and *C. parapsilosis*) were selected for the antifungal activity. The MIC for both nanocomposites were compared with AgNPs (26 nm) and found to range from 1.9 to 33.3 mg/L. The nanocomposites retained the antimicrobial property of the AgNPs and the magnetic property of iron oxide.

Wang et al. (2020) synthesized trimethyl chitosan nitrate (TMCN)-stabilized AgNPs and studied the fungicidal and antibiofilm activity of *Candida* spp. in zebrafish embryo. The MICs were as follows: for *C. albicans*, 0.46 mM; for *C. tropicalis*, 0.06 mM; and for *C. glabrata*, 0.46 mM. The antibiofilm activity was studied in zebrafish embryo infected with the *Candida* species. Li et al. (2013a) studied antifungal activity of graphene oxide-AgNPs against *C. albicans* and *C. tropicalis*. Graphene oxide alone did not demonstrate notable antifungal activity but graphene oxide-AgNP nanocomposites showed increased and sustained antifungal activity against the selected *Candida* spp., causing nosocomial infection and local fungal infection.

Malassezia furfur is a lipid-dependent fungus present in human skin microflora that causes various skin disorders. Mussin et al. (2019) formulated an antimicrobial carbopol hydrogel containing AgNPs with ketoconazole. The broad-spectrum antimicrobial properties and the AgNPs' fungicidal actions sustained the release of ketoconazole against *Malassezia* and other superficial pathogens.

Pereira et al. (2014) investigated the antifungal activity of chemically and biosynthesized AgNPs against *Trichophyton rubrum*. The chemically synthesized Ag NPs, which was done by chemical reduction method, were coated with polyvinylpyrrolidone; whereas the extracellular biosynthesis was achieved with *P. chrysogenum* MUM 03.22 and *A. oryzae* MUM 97.19. The antifungal activity was compared with the standard antifungal drugs terbinafine, itraconazole, and fluconazole. Biosynthesized AgNPs showed antifungal activity against *Trichophyton rubrum* more than fluconazole did, but less than the chemically synthesized AgNPs, terbinafine and itraconazole.

Ishida et al. (2013) synthesized AgNPs by microbial synthesis by using an aqueous extract of filamentous *Fusarium oxysporum*, and the antimycotic activity was tested on *Candida* and *Cryptococcus neoformans*. Treatment with AgNPs caused cell wall disruption and leakage of intracellular material, thereby inhibiting cell growth. The MIC for both strains was found to be $\leq 1.68\,\mu g/mL$. Dananjaya et al. (2017) prepared chitosan NP and chitosan AgNP nanocomposites and conducted the antifungal activity against *Fusarium oxysporum*. Both the chitosan NPs and the chitosan AgNPs inhibited the growth of the fungus and also caused changes in the morphological structure of the fungus. Reduction in the growth of *Fusarium oxysporum* in the infected zebrafish confirmed the antifungal activity.

Marimuthu et al. (2011) synthesized biogenic AgNPs from aqueous extract of red seaweed *Gelidiella acerosa* and studied the antifungal activity against *Humicola insolens*, *Fusarium dimerum*, *Mucor indicus*, and *Trichoderma reesei*. The synthesized AgNPs showed promising antifungal activity against the standard drug clotrimazole.

2.6 Application of silver nanoparticles in plant fungal infections

Plants, crops, and grains are highly prone to pathogenic fungal infections during postharvest storage, and it is important to note that there is great diversity among plant pathogenic fungi. These biotic stresses can be handled by the use of either fungistatic agents or fungicidal agents. Biogenic and chemically synthesized AgNPs have been studied extensively to explore their effectiveness as fungistatic and fungicidal agents. With that in mind, the following discussion supports the assessment that AgNPs are more effective antifungal agents than are their chemical counterparts.

Alternaria alternata causes leaf spot and blight disease in tomato and potato plants. El-Gazzar and Ismail (2020) prepared metal NPs by using a chemical and biogenic method. Titanium dioxide and silver NPs were prepared by biogenic synthesis using *Aspergillus versicolor* and selenium (Se) NPs through a chemical reduction method. The antifungal activity of all the metal NPs was tested at three different concentrations, 25, 50, and 100 ppm. The biosynthesized titanium dioxide NPs showed the maximum antifungal effect at 100 ppm, and Se NPs showed similar activity as titanium dioxide; whereas AgNPs showed a moderate effect as an antifungal agent.

Kumari et al. (2019) worked on a similar platform, directing the antifungal effect of biosynthesized AgNPs against *Alternaria solani*, which is known to be responsible for early blight disease in potatoes. AgNPs were synthesized from a cell-free extract of *Trichoderma viride*, and the NPs were able to demonstrate 100% reduction in spore count after 3 days of application and 73.3% reduction in biomass after 7 days of application. The AgNPs showed higher antimicrobial activities than the chemical counterpart due to the coating of the AgNPs with a secondary metabolite of *Trichoderma viride* and the small spherical size (2–5 nm).

Ismail et al. (2016) performed similar research by using biosynthesized AgNPs from *Trichoderma viride* and SeNPs prepared using the glutathione method. The AgNPs showed maximum inhibition at 25 μg/mL, compared with the same activity by Kocide fungicide at 600 μg/mL. Selenium NPs completely inhibited fungal growth at 800 μg/mL. The AgNPs can be used as an alternative to chemical pesticides, and the SeNPs can be used as an antioxidant to improve the immunity of the potato plant against *Alternaria solani*.

Kora et al. (2020) biosynthesized AgNPs from rice leaf extract and studied the antifungal activity against *Rhizoctonia solani*, which is known to be responsible for sheath blight disease in rice. Biosynthesized AgNPs inhibited the mycelial growth from 81.7% to 96.7% at 10 μg/mL. AgNPs inhibited the incidence of disease at 20 μg/mL.

Cereal grains often are spoiled during storage, which may be because of mycotoxin contamination due to the *Fusarium* species. Tarazona et al. (2019) synthesized AgNPs using a chemical reduction method and studied the antifungal effect on different strains of *Fusarium*. AgNPs inhibited the growth and influenced the spore viability and lag period of *F. graminearum*, *F. culmorum*, *F. sporotrichioides*, *F. langsethiae*, *F. poae*, *F. verticillioides*, *F. proliferatum*, and *F. oxysporum*.

Roseline et al. (2019) biosynthesized AgNPs from agar seaweeds (*Gracilaria corticata* and *G. edulis*) and carrageenan seaweeds (*Hypnea musciformis* and *Spyridia hypnoides*). The research group conducted antifungal activity against *Ustilaginoidea virens* causing false smut in rice and found promising results for the use of biosynthesized AgNPs as nanopesticides.

El-Saadony et al. (2019) biosynthesized AgNPs from *Bacillus pseudomycoides* isolated from soil and studied the antifungal effect of the biogenic AgNPs against ten plant pathogenic fungi—*Aspergillus niger*, *A. flavus*, *A. tereus*, *Penicillium notatum*, *Rhizoctonia solina*, *Fusarium solani*, *F. oxysporum*, *Trichoderma viride*, *Verticillium dahlia*, and *Pythium spinosum*—and the MIC was found to range from 70 to 90 μg/mL.

Sclerotinia sclerotiorum is known to cause white mold disease, and the resistant fungal infection effects crop yields. Guilger et al. (2017) biosynthesized AgNPs from *Trichoderma harzianum* and studied the cytotoxicity and genotoxicity against *S. sclerotiorum*. The AgNPs showed an inhibitory effect on sclerotia germination and mycelial growth.

El-Moslamy et al. (2017) biosynthesized AgNPs by using *Trichoderma harzianum* SYA. F4 strain. The physicochemical properties and the biomass of the AgNPs were noted to increase three times by applying the Taguchi experimental design. The antifungal activity was achieved at 100 µg/mL against *Alternaria alternate*, *Helminthosporium* sp., *Botrytis* sp., and *Phytopthora arenaria*.

Elgorban et al. (2016a) biosynthesized AgNPs from *Aspergillus versicolor* and evaluated the antifungal activity on phytopathogenic fungi. Phytopatogenic fungi, white mold (*Sclerotinia sclerotiorum*), and gray mold (*Botrytis cinerea*) in strawberry plants were evaluated, and the antifungal activity was found to be significantly dependent on the concentration of the AgNPs. Elgorban et al. (2016b) reported the antifungal activity on six variants of *Rhizoctonia solani* anastomosis groups (AGs) with different concentrations of chemically synthesized AgNPs. The AgNPs significantly reduced the growth of *R. solani* on cotton plants.

Balakumaran et al. (2015) biosynthesized the AgNPs by using the endophytic fungi, *Guignardia mangiferae*, isolated from leaves of citrus plants. The fabricated AgNPs were highly stable and spherical with sizes of approximately 5 to 30 nm. The antifungal activity was conducted by well diffusion assay on *A. niger*, *Fusarium* and *Colletotrichum* spp., *C. lunata*, and *Rhizoctonia solani*. The AgNPs showed efficient antifungal and cytotoxic activity.

Abd El-Aziz et al. (2015) biosynthesized AgNPs by using *Fusarium solani* and evaluated the antifungal activity of grain-borne fungi (in wheat, barley, and corn), that is, *F. oxysporum*, *F. moniliform*, *F. solani*, *F. verticillioides*, *F. semitectum*, *A. flavus*, *A. terreus*, *A. niger*, *A. ficuum*, *P. citrinum*, *P. islandicum*, *P. chrysogenum*, *Rhizopus stolonifer*, *A. alternate*, and *A. chlamydospora*. The grains of wheat, barley, and corn were soaked with a 2% solution of AgNPs, and the appearance of fungi on the grain was recorded. Silver nanoparticles were found to be effective in the control of the grain-borne fungi.

Ifuku et al. (2015) prepared a nanocomposite film using chitin nanofibers with UV radiation-reduced silver ions (9:1 and 9:5 ratio) and an acrylic polymer. The film was used for evaluating antifungal activity against spores of *Alternaria alternata*, *A. brassicae*, *A. brassicicola*, *Bipolaris oryzae*, *Botrytis cinerea*, and *Penicillium digitatum*. The nanocomposite film prohibited spore germination of all the tested fungi at 24-h applications and was sustained for 7 days. The antifungal activity was due mainly to the antifungal activity of AgNPs, rather than to chitin.

Al-Othman et al. (2014) synthesized biogenic AgNPs using *Aspergillus terreus* on growth and aflatoxin production by *A. flavus*. The AgNPs decreased the growth of *A. flavus* by impairing the functions of the cell, which caused degeneration in the hyphae of the fungus. AgNPs caused a reduction in spore number. The NPs also showed a greater reduction in aflatoxin production, compared to the control group.

Gajbhiye et al. (2009) prepared biogenic AgNPs using *Aspergillus alternate*, and the antifungal activity was evaluated against *Phoma glomerata*, *P. herbarum*, *Fusarium semitectum*, *Trichoderma* sp., and *Candida albicans* with fluconazole. The antifungal activity of fluconazole was increased in the presence of AgNPs.

Karu et al. (2020) biosynthesized AgNPs from leaf extracts of *Solenostemon monostachyus*, and the NPs showed strong antifungal activity against *Candida albicans* and *Aspergillus niger*.

Woo et al. (2009) reported the antifungal activity of AgNPs against the *Raffaelea* species against oak wilt disease. The AgNPs effectively inhibited the growth and development of fungal pathogens by damaging the cellular structure.

2.7 Conclusion

Conventional drug delivery systems are at times ineffective in regulating fungal infections, which has led to gradual development of novel drug delivery systems that boost the therapeutic effectiveness of specific antifungal drugs. Research on these novel systems has shown positive results for reducing fungal infections in plant species. Yet, the widespread use of these delivery systems faces enormous challenges related to their ability to scale up and become commercially viable. Therefore there is a need for new formulations comprising a range of antifungal drugs that target diseases caused by fungi. Strong focus needs to be placed on the release characteristics of moieties entrapped at the target site to achieve optimal therapeutic efficacy. Also, combinatorial drug therapy systems need to be developed in order to achieve a synergistic effect. In addition, fungal genes need to be explored in order to identify specific antifungal drug targets and to develop delivery systems to combat the related infections.

References

Abd El-Aziz, A., Al-Othman, M., Mahmoud, M., Metwaly, H., 2015. Biosynthesis of silver nanoparticles using *Fusarium solani* and its impact on grain borne fungi. Dig. J. Nanomater. Biostruct. 10, 655–662.

Almeida, F., Rodrigues, M., Coelho, C., 2019. The still underestimated problem of fungal diseases worldwide. Front. Microbiol. 10, 214.

Al-Othman, M., Abd El-Aziz, A., Mahmoud, M., Eifan, S., El-Shikh, M., Majrashi, M., 2014. Application of silver nanoparticles as antifungal and antiaflatoxin B1 produced by *Aspergillus flavus*. Dig. J. Nanomater. Biostruct. 9, 151–157.

Alvarez-Peral, F., Zaragoza, O., Pedreno, Y., Argüelles, J., 2002. Protective role of trehalose during severe oxidative stress caused by hydrogen peroxide and the adaptive oxidative stress response in *Candida albicans*. Microbiology 148, 2599–2606.

Avenot, H., Michailides, T., 2010. Progress in understanding molecular mechanisms and evolution of resistance to succinate dehydrogenase inhibiting (SDHI) fungicides in phytopathogenic fungi. Crop Prot. 29, 643–651.

Badiee, P., Hashemizadeh, Z., 2014. Opportunistic invasive fungal infections: diagnosis & clinical management, Indian. J. Med. Res. 139 (2), 195–204.

Balakumaran, M., Ramachandran, R., Kalaichelvan, P., 2015. Exploitation of endophytic fungus, *Guignardia mangiferae* for extracellular synthesis of silver nanoparticles and their in vitro biological activities. Microbiol. Res. 178, 9–17.

Behera, A., Mittu, B., Padhi, S., Patra, N., Singh, J., 2020. Bimetallic nanoparticle: green synthesis, applications and future perspectives. In: Multifunctional Hybrid Nanomaterials for Sustainable Agri-Food and Ecosystems. Elsevier Publications, Amsterdam, pp. 639–681.

Bhattacharya, R., Mukherjee, P., 2008. Biological properties of "naked" metal nanoparticles. Adv. Drug Deliv. Rev. 60, 1289–1306.

Brown, G., Denning, D., Gow, N., Levitz, S., Netea, M., White, T., 2012. Hidden killers: human fungal infections. Sci. Transl. Med. 4. 165rv13.

Campitelli, M., Zeineddine, N., Samaha, G., Maslak, S., 2017. Combination antifungal therapy: a review of current data. J. Clin. Med. Res. 9 (6), 451–456.

Campoy, S., Adrio, J., 2017. Antifungals. Biochem. Pharmacol. 133, 86–96.

Carmona-Gutierrez, D., Eisenberg, T., Büttner, S., Meisinger, C., Kroemer, G., Madeo, F., 2010. Apoptosis in yeast: triggers, pathways, subroutines. Cell Death Differ. 17, 763–773.

Dakal, T., Kumar, A., Majumdar, R., Yadav, V., 2016. Mechanistic basis of antimicrobial actions of silver nanoparticles. Front. Microbiol. 7, 1831.

Dananjaya, S., Erandani, W., Kim, C., Nikapitiya, C., Lee, J., De Zoysa, M., 2017. Comparative study on antifungal activities of chitosan nanoparticles and chitosan silver nano composites against *Fusarium oxysporum* species complex. Int. J. Biol. Macromol. 105, 478–488.

Davidse, L., 1986. Benzimidazole fungicides: mechanism of action and biological impact. Annu. Rev. Phytopathol. 24, 43–65.

de Oliveira Pereira, F., Mendes, J., de Oliveira Lima, E., 2013. Investigation on mechanism of antifungal activity of eugenol against *Trichophyton rubrum*. Med. Mycol. 51, 507–513.

Dehnavi, A.S., Raisi, A., Aroujalian, A., 2012. Control size and stability of colloidal silver nanoparticles with antibacterial activity prepared by a green synthesis method. Synth. React. Inorg. Met.-Org. Nano-Met. Chem. 43, 543–551.

Douglas, L., Konopka, J., 2014. Fungal membrane organization: the Eisosome concept. Annu. Rev. Microbiol. 68, 377–393.

Dutta, T., Ghosh, N., Das, M., Adhikary, R., Mandal, V., Chattopadhyay, A., 2020. Green synthesis of antibacterial and antifungal silver nanoparticles using *Citrus limetta* peel extract: experimental and theoretical studies. J. Environ. Chem. Eng. 8, 104019.

Elbein, A., Pan, Y., Pastuszak, I., Carroll, D., 2003. New insights on trehalose: a multifunctional molecule. Glycobiology 13, 17R–27R.

El-Gazzar, N., Ismail, A., 2020. The potential use of titanium, silver and selenium nanoparticles in controlling leaf blight of tomato caused by *Alternaria alternata*. Biocatal. Agric. Biotechnol. 27, 101708.

Elgorban, A., Aref, S., Seham, S., Elhindi, K., Bahkali, A., Sayed, S., Manal, M., 2016a. Extracellular synthesis of silver nanoparticles using *Aspergillus versicolor* and evaluation of their activity on plant pathogenic fungi. Mycosphere 7 (6), 844–852.

Elgorban, A., El-Samawaty, A., Yassin, M., Sayed, S., Adil, S., Elhindi, K., Bakri, M., Khan, M., 2016b. Antifungal silver nanoparticles: synthesis, characterization and biological evaluation. Biotechnol. Biotechnol. Equip. 30, 56–62.

El-Moslamy, S., Elkady, M., Rezk, A., Abdel-Fattah, Y., 2017. Applying Taguchi design and large-scale strategy for mycosynthesis of nano-silver from endophytic *Trichoderma harzianum* SYA.F4 and its application against phytopathogens. Sci. Rep. 7, 45297.

El-Saadony, M., El-Wafai, N., El-Fattah, H., Mahgoub, S., 2019. Biosynthesis, optimization and characterization of silver nanoparticles using a soil isolate of *Bacillus pseudomycoides* MT32 and their antifungal activity against some pathogenic fungi. Adv. Anim. Vet. Sci. 7, 238–249.

Fernández-Ortuño, D., Torés, J., de Vicente, A., Pérez-García, A., 2008. Mechanisms of resistance to QoI fungicides in phytopathogenic fungi. Int. Microbiol. 11, 1–9.

Gajbhiye, M., Kesharwani, J., Ingle, A., Gade, A., Rai, M., 2009. Fungus-mediated synthesis of silver nanoparticles and their activity against pathogenic fungi in combination with fluconazole. Nanomedicine 5, 382–386.

Galdiero, S., Falanga, A., Vitiello, M., Cantisani, M., Marra, V., Galdiero, M., 2011. Silver nanoparticles as potential antiviral agents. Molecules 16, 8894–8918.

Ghosh, S., Patil, S., Ahire, M., Kitture, R., Kale, S., Pardesi, K., Cameotra, S., Bellare, J., Dhavale, D., Jabgunde, A., Chopade, B., 2012. Synthesis of silver nanoparticles using *Dioscorea bulbifera* tuber extract and evaluation of its synergistic potential in combination with antimicrobial agents. Int. J. Nanomedicine 7, 483–496.

Gitipour, A., Thiel, S., Scheckel, K., Tolaymat, T., 2016. Anaerobic toxicity of cationic silver nanoparticles. Sci. Total Environ. 557–558, 363–368.

Guerra, J., Sandoval, G., Avalos-Borja, M., Pestryakov, A., Garibo, D., Susarrey-Arce, A., Bogdanchikova, N., 2020. Selective antifungal activity of silver nanoparticles: a comparative study between *Candida tropicalis and Saccharomyces boulardii*. Colloids Interface Sci. Commun. 37, 100280.

Guilger, M., Pasquoto-Stigliani, T., Bilesky-Jose, N., Grillo, R., Abhilash, P., Fraceto, L., Lima, R., 2017. Biogenic silver nanoparticles based on *Trichoderma harzianum*: synthesis, characterization, toxicity evaluation and biological activity. Sci. Rep. 7, 44421.

Guirao-Abad, J., Sanchez-Fresneda, R., Alburquerque, B., Hernandez, J., Arguelles, J., 2017. ROS formation is a differential contributory factor to the fungicidal action of amphotericin B and micafungin in Candida albicans. Int. J. Med. Microbiol. 307, 241–248.

Gutiérrez, J., Caballero, S., Díaz, L., Guerrero, M., Ruiz, J., Ortiz, C., 2018. High antifungal activity against Candida species of monometallic and bimetallic nanoparticles synthesized in nanoreactors. ACS Biomater. Sci. Eng. 4 (2), 647–653.

Hayes, B., Anderson, M., Traven, A., van der Weerden, N., Bleackley, M., 2014. Activation of stress signalling pathways enhances tolerance of fungi to chemical fungicides and antifungal proteins. Cell. Mol. Life Sci. 71, 2651–2666.

Howden, P., Faux, S., 1996. Fibre-induced lipid peroxidation leads to DNA adduct formation in *Salmonella typhimurium* TA104 and rat lung fibroblasts. Carcinogenesis 17, 413–419.

Huang, C., Aronstam, R., Chen, D., Huang, Y., 2010. Oxidative stress, calcium homeostasis, and altered gene expression in human lung epithelial cells exposed to ZnO nanoparticles. Toxicol. In Vitro 24, 45–55.

Hwang, I., Lee, J., Hwang, J., Kim, K., Lee, D., 2012. Silver nanoparticles induce apoptotic cell death in *Candida albicans* through the increase of hydroxyl radicals. FEBS J. 279, 1327–1338.

Ifuku, S., Tsukiyama, Y., Yukawa, T., Egusa, M., Kaminaka, H., Izawa, H., Morimoto, M., Saimoto, H., 2015. Facile preparation of silver nanoparticles immobilized on chitin nanofiber surfaces to endow antifungal activities. Carbohydr. Polym. 117, 813–817.

Ishida, K., Cipriano, T., Rocha, G., Weissmüller, G., Gomes, F., Miranda, K., Rozental, S., 2013. Silver nanoparticle production by the fungus *Fusarium oxysporum*: nanoparticle characterisation and analysis of antifungal activity against pathogenic yeasts. Mem. Inst. Oswaldo Cruz 109, 220–228.

Ismail, A., Sidkey, N., Arafa, R., Fathy, R., El-Batal, A., 2016. Evaluation of in vitro antifungal activity of silver and selenium nanoparticles against *Alternaria solani* caused early blight disease on potato. Br. Biotechnol. J. 12, 1–11.

Jain, A., Sarsaiya, S., Wu, Q., Lu, Y., Shi, J., 2019. A review of plant leaf fungal diseases and its environment speciation. Bioengineered 10, 409–424.

Karu, E., Magaji, B., Shehu, Z., Abdulsalam, H., 2020. Green synthesis of silver nanoparticles from *Solenostemon monostachyus* leaf extract and in vitro antibacterial and antifungal evaluation. EJ-Chem. 1 (4), 1–5.

Khan, T., Yasmin, A., Townley, H., 2020. An evaluation of the activity of biologically synthesized silver nanoparticles against bacteria, fungi and mammalian cell lines. Colloids Surf. B: Biointerfaces 194, 111156.

Khuroo, T., Verma, D., Talegaonkar, S., Padhi, S., Panda, A., Iqbal, Z., 2014. Topotecan–tamoxifen duple PLGA polymeric nanoparticles: investigation of *in vitro, in vivo* and cellular uptake potential. Int. J. Pharm. 473 (1–2), 384–394.

Kim, K., Sung, W., Suh, B., Moon, S., Choi, J., Kim, J., Lee, D., 2009. Antifungal activity and mode of action of silver nano-particles on *Candida albicans*. Biometals 22, 235–242.

Kim, S., Lee, H., Ryu, D., Choi, S., Lee, D., 2011. Antibacterial activity of silver-nanoparticles against *Staphylococcus aureus* and *Escherichia coli*. Korean J. Microbiol. Biotechnol. 39, 77–85.

Klueh, U., Wagner, V., Kelly, S., Johnson, A., Bryers, J., 2000. Efficacy of silver-coated fabric to prevent bacterial colonization and subsequent device-based biofilm formation. J. Biomed. Mater. Res. 53, 621–631.

Kontoyiannis, D., Lionakis, M., Lewis, R., Chamilos, G., Healy, M., Perego, C., Safdar, A., Kantarjian, H., Champlin, R., Walsh, T., Raad, I., 2005. Zygomycosis in a tertiary-care cancer center in the era of Aspergillus-active antifungal therapy: a case-control observational study of 27 recent cases. J. Infect. Dis. 191, 1350–1360.

Kora, A., Mounika, J., Jagadeeshwar, R., 2020. Rice leaf extract synthesized silver nanoparticles: an in vitro fungicidal evaluation against *Rhizoctonia solani*, the causative agent of sheath blight disease in rice. Fungal Biol. 124, 671–681.

Kumar, S., Singh, M., Halder, D., Mitra, A., 2014. Mechanistic study of antibacterial activity of biologically synthesized silver nanocolloids. Colloids Surf. A Physicochem. Eng. Asp. 449, 82–86.

Kumari, M., Giri, V., Pandey, S., Kumar, M., Katiyar, R., Nautiyal, C., Mishra, A., 2019. An insight into the mechanism of antifungal activity of biogenic nanoparticles than their chemical counterparts. Pestic. Biochem. Physiol. 157, 45–52.

Kyrylkova, K., Kyryachenko, S., Leid, M., Kioussi, C., 2012. Detection of apoptosis by TUNEL assay. Methods Mol. Biol. 887, 41–47.

Lara, H., Romero-Urbina, D., Pierce, C., Lopez-Ribot, J., Arellano-Jiménez, M., Jose-Yacaman, M., 2015. Effect of silver nanoparticles on *Candida albicans* biofilms: an ultrastructural study. J. Nanobiotechnol. 13, 91–102.

Lewis, R., 2011. Current concepts in antifungal pharmacology. Mayo Clin. Proc. 86, 805–817.

Li, J., Rong, K., Zhao, H., Li, F., Lu, Z., Chen, R., 2013a. Highly selective antibacterial activities of silver nanoparticles against *Bacillus subtilis*. J. Nanosci. Nanotechnol. 13, 6806–6813.

Li, C., Wang, X., Chen, F., Zhang, C., Zhi, X., Wang, K., Cui, D., 2013b. The antifungal activity of graphene oxide silver nanocomposites. Biomaterials 34, 3882–3890.

Lin, S., Schranz, J., Teutsch, S., 2001. Aspergillosis case-fatality rate: systematic review of the literature. Clin. Infect. Dis. 32, 358–366.

Liu, S., Hou, Y., Chen, X., Gao, Y., Li, H., Sun, S., 2014. Combination of fluconazole with non-antifungal agents: a promising approach to cope with resistant *Candida albicans* infections and insight into new antifungal agent discovery. Int. J. Antimicrob. Agents 43, 395–402.

Lkhagvajav, N., Yasab, I.C., Elikc, E., Koizhaiganova, M., Saria, O., 2011. Antimicrobial activity of colloidal silver nanoparticles prepared by sol gel method. Dig. J. Nanomater. Biostruct. 6, 149–154.

Madeo, F., Herker, E., Wissing, S., Jungwirth, H., Eisenberg, T., Frohlich, K., 2004. Apoptosis in yeast. Curr. Opin. Microbiol. 7, 655–660.

Marimuthu, V., Palanisamy, S., Sesurajan, S., Sellappa, S., 2011. Biogenic silver nanoparticles by *Gelidiella acerosa* extract and their antifungal effects. Avicenna J. Med. Biotechnol. 3, 143–148.

Mazu, T., Bricker, B., Flores-Rozas, H., Ablordeppey, S., 2016. The mechanistic targets of antifungal agents: an overview. Mini-Rev. Med. Chem. 16, 555–578.

Mazzoni, C., Falcone, C., 2008. Caspase-dependent apoptosis in yeast. Biochim. Biophys. Acta 1783, 1320–1327.

McCarthy, K., Morgan, J., Wannemuehler, K., Mirza, S., Gould, S., Mhlongo, N., Moeng, P., Maloba, B., Crewe-Brown, H., Brandt, M., Hajjeh, R., 2006. Population-based surveillance for cryptococcosis in an antiretroviral-naive South African province with a high HIV seroprevalence. AIDS 20, 2199–2206.

Moazeni, M., Rashidi, N., Shahverdi, A., Noorbakhsh, F., Rezaie, S., 2012. Extracellular production of silver nanoparticles by using three common species of dermatophytes: *Trichophyton rubrum, Trichophyton mentagrophytes* and *Microsporum canis*. Iran. Biomed. J. 16, 52–58.

Mussin, J., Roldán, M., Rojas, F., Sosa, M., Pellegri, N., Giusiano, G., 2019. Antifungal activity of silver nanoparticles in combination with ketoconazole against *Malassezia furfur*. AMB Express 9, 131.

Nasrollahi, A., Pourshamsian, K., Mansourkiaee, P., 2011. Antifungal activity of silver nanoparticles on some of fungi. Int. J. Nano Dimens. 1 (3), 233–239.

Nateghi, M., Hajimirzababa, H., 2014. Effect of silver nanoparticles morphologies on antimicrobial properties of cotton fabrics. J. Text. Inst. 105, 806–813.

Nayak, B., Nanda, A., Prabhakar, V., 2018. Biogenic synthesis of silver nanoparticle from wasp nest soil fungus, *Penicillium italicum* and its analysis against multi drug resistance pathogens. Biocatal. Agric. Biotechnol. 16, 412–418.

Nett, J., Andes, D., 2016. Antifungal agents: spectrum of activity, pharmacology, and clinical indications. Infect. Dis. Clin. North. Am 30, 51–83.

Nucci, M., Anaissie, E., 2007. Fusarium infections in immunocompromised patients. Clin. Microbiol. Rev. 20, 695–704.

Owaid, M., Raman, J., Lakshmanan, H., Al-Saeedi, S., Sabaratnam, V., Abed, I., 2015. Mycosynthesis of silver nanoparticles by *Pleurotus cornucopiae* var. *citrinopileatus* and its inhibitory effects against *Candida* sp. Mater. Lett. 153, 186–190.

Padhi, S., Mirza, M., Verma, D., Khuroo, T., Panda, A., Talegaonkar, S., Khar, R., Iqbal, Z., 2015. Revisiting the nanoformulation design approach for effective delivery of topotecan in its stable form: an appraisal of its *in vitro* behaviour and tumor amelioration potential. Drug Deliv. 23 (8), 2827–2837.

Padhi, S., Kapoor, R., Verma, D., Panda, A., Iqbal, Z., 2018. Formulation and optimization of topotecan nanoparticles: in vitro characterization, cytotoxicity, cellular uptake and pharmacokinetic outcomes. J. Photochem. Photobiol. B 183, 222–232.

Padhi, S., Nayak, A., Behera, A., 2020. Type II diabetes mellitus: a review on recent drug ased therapeutics. Biomed. Pharmacother. 131, 110708.

Pal, S., Tak, Y., Song, J., 2007. Does the antibacterial activity of silver nanoparticles depend on the shape of the nanoparticle? A study of the gram-negative bacterium Escherichia coli. Appl. Environ. Microbiol. 73, 1712–1720.

Panáček, A., Kolář, M., Večeřová, R., Prucek, R., Soukupová, J., Kryštof, V., Hamal, P., Zbořil, R., Kvítek, L., 2009. Antifungal activity of silver nanoparticles against *Candida* spp. Biomaterials 30, 6333–6340.

Parthiban, E., Manivannan, N., Ramanibai, R., Mathivanan, N., 2019. Green synthesis of silver-nanoparticles from *Annona reticulata* leaves aqueous extract and its mosquito larvicidal and anti-microbial activity on human pathogens. Biotechnol. Rep. 21, e00297.

Pereira, L., Dias, N., Carvalho, J., Fernandes, S., Santos, C., Lima, N., 2014. Synthesis, characterization and antifungal activity of chemically and fungal-produced silver nanoparticles against *Trichophyton rubrum*. J. Appl. Microbiol. 117, 1601–1613.

Perrone, G., Tan, S., Dawes, I., 2008. Reactive oxygen species and yeast apoptosis. Biochim. Biophys. Acta 1783, 1354–1368.

Pfaller, M., Diekema, D., 2004. Rare and emerging opportunistic fungal pathogens: concern for resistance beyond *Candida albicans* and *Aspergillus fumigatus*. J. Clin. Microbiol. 42, 4419–4431.

Pfaller, M., Diekema, D., 2010. Epidemiology of invasive mycoses in North America. Crit. Rev. Microbiol. 36, 1–53.

Prucek, R., Tuček, J., Kilianová, M., Panáček, A., Kvítek, L., Filip, J., Kolář, M., Tománková, K., Zbořil, R., 2011. The targeted antibacterial and antifungal properties of magnetic nanocomposite of iron oxide and silver nanoparticles. Biomaterials 32, 4704–4713.

Radhakrishnan, V., Mudiam, M., Kumar, M., Dwivedi, S., Singh, S., Prasad, T., 2018. Silver nanoparticles induced alterations in multiple cellular targets, which are critical for drug susceptibilities and pathogenicity in fungal pathogen (*Candida albicans*). Int. J. Nanomedicine 13, 2647–2663.

Rai, M., Kon, K., Ingle, A., Duran, N., Galdiero, S., Galdiero, M., 2014. Broad-spectrum bioactivities of silver nanoparticles: the emerging trends and future prospects. Appl. Microbiol. Biotechnol. 98, 1951–1961.

Raza, M., Kanwal, Z., Rauf, A., Sabri, A., Riaz, S., Naseem, S., 2016. Size- and shape-dependent antibacterial studies of silver nanoparticles synthesized by wet chemical routes. Nanomaterials 6, 74.

Robles-Martínez, M., Patiño-Herrera, R., Pérez-Vázquez, F., Montejano-Carrizales, J., González, J., Pérez, E., 2020. *Mentha piperita* as a natural support for silver nanoparticles: a new anti-*Candida albicans* treatment. Colloids Interface Sci. Commun. 35, 100253.

Roseline, T., Murugan, M., Sudhakar, M., Arunkumar, K., 2019. Nanopesticidal potential of silver nanocomposites synthesized from the aqueous extracts of red seaweeds. Environ. Technol. Innov. 13, 82–93.

Schreurs, W., Rosenberg, H., 1982. Effect of silver ions on transport and retention of phosphate by *Escherichia coli*. J. Bacteriol. 152, 7–13.

Shirazi, F., Kontoyiannis, D., 2015. Micafungin triggers caspase-dependent apoptosis in *Candida albicans* and *Candida parapsilosis* biofilms, including caspofungin nonsusceptible isolates. Virulence 6, 385–394.

Suganya, K., Govindaraju, K., Kumar, V., Dhas, T., Karthick, V., Singaravelu, G., Elanchezhiyan, M., 2015. Size controlled biogenic silver nanoparticles as antibacterial agent against isolates from HIV infected patients. Spectrochim. Acta A Mol. Biomol. Spectrosc. 144, 266–272.

Tarazona, A., Gómez, J., Mateo, E., Jiménez, M., Mateo, F., 2019. Antifungal effect of engineered silver nanoparticles on phytopathogenic and toxigenic *Fusarium* spp. and their impact on mycotoxin accumulation. Int. J. Food Microbiol. 306, 108259.

Tas, J., Westerneng, G., 1981. Fundamental aspects of the interaction of propidium diiodide with nuclei acids studied in a model system of polyacrylamide films. J. Histochem. Cytochem. 29, 929–936.

Teixeira-Santos, R., Ricardo, E., Guerreiro, S., Costa-de-Oliveira, S., Rodrigues, A., Pina-Vaz, C., 2015. New insights regarding yeast survival following exposure to liposomal amphotericin B. Antimicrob. Agents Chemother. 59, 6181–6187.

Tutaj, K., Szlazak, R., Szalapata, K., Starzyk, J., Luchowski, R., Grudzinski, W., Osinska-Jaroszuk, M., Jarosz-Wilkolazka, A., Szuster-Ciesielska, A., Gruszecki, W., 2016. Amphotericin B-silver hybrid nanoparticles: synthesis, properties and antifungal activity. Nanomedicine 12, 1095–1103.

Vazquez-Muñoz, R., Avalos-Borja, M., Castro-Longoria, E., 2014. Ultrastructural analysis of *Candida albicans* when exposed to silver nanoparticles. PLoS One 9, e108876.

Wang, L., He, D., Gao, S., Wang, D., Xue, B., Yokoyama, K., 2016a. Biosynthesis of silver nanoparticles by the fungus *Arthroderma fulvum* and its antifungal activity against genera of *Candida, Aspergillus* and *Fusarium*. Int. J. Nanomedicine 11, 1899–1906.

Wang, S., Zeng, X., Yang, Q., Qiao, S., 2016b. Antimicrobial peptides as potential alternatives to antibiotics in food animal industry. Int. J. Mol. Sci. 17 (5), 603–614.

Wang, S., Chen, C., Lee, C., Chen, X., Chang, T., Cheng, Y., Young, J., Lu, J., 2020. Fungicidal and anti-biofilm activities of trimethylchitosan-stabilized silver nanoparticles against *Candida* species in zebrafish embryos. Int. J. Biol. Macromol. 143, 724–731.

Woo, K., Kim, K., Lamsal, K., Kim, Y., Kim, S., Jung, M., Sim, S., Kim, H., Chang, S., Kim, J., Lee, Y., 2009. An in vitro study of the antifungal effect of silver nanoparticles on oak wilt pathogen *Raffaelea* sp. J. Microbiol. Biotechnol. 19, 760–764.

Wu, D., Fan, W., Kishen, A., Gutmann, J., Fan, B., 2014. Evaluation of the antibacterial efficacy of silver nanoparticles against *Enterococcus faecalis* biofilm. J. Endod. 40, 285–290.

Yacamán, M., Ascencio, J., Liu, H., Gardea-Torresdey, J., 2001. Structure shape and stability of nanometric sized particles. J. Vacuum Sci. Technol. B Microelectron. Nanometer. Struct. 19, 1091–1103.

Zazo, H., Colino, C., Lanao, J., 2016. Current applications of nanoparticles in infectious diseases. J. Control. Release 224, 86–102.

Zhang, L., Pornpattananangkul, D., Hu, C., Huang, C., 2010. Development of nanoparticles for antimicrobial drug delivery. Curr. Med. Chem. 17, 585–594.

Further reading

Behera, A., Padhi, S., 2020. Passive and active targeting strategies for the delivery of the camptothecin anticancer drug: a review. Environ. Chem. Lett. 18 (5), 1557–1567.

Kumari, M., Pandey, S., Bhattacharya, A., Mishra, A., Nautiyal, C., 2017. Protective role of biosynthesized silver nanoparticles against early blight disease in *Solanum lycopersicum*. Plant Physiol. Biochem. 121, 216–225.

CHAPTER 3

Antimicrobial properties of surface-functionalized silver nanoparticles

Parteek Prasher[a,b] and Mousmee Sharma[a,c]

[a]Department of Chemistry, UGC Sponsored Centre for Advanced Studies, Guru Nanak Dev University, Amritsar, India
[b]Department of Chemistry, University of Petroleum & Energy Studies, Energy Acres, Dehradun, India
[c]Department of Chemistry, Uttaranchal University, Dehradun, India

3.1 Introduction

Increasing antimicrobial resistance to conventional antimicrobial chemotherapies led to a paradigm shift in the design and development of new antibiotic drugs and novel targeted therapies, such as multidrug-resistant efflux pumps (Nikaido, 2009). A World Health Organization (WHO) report noted that 80% of bacterial strains exhibit resistance against commercial antibiotics (WHO, 2014). The situation became more critical when the administration of synergistic antibiotics failed to deliver positive results and when higher doses exhibited adverse side effects (Aslam et al., 2018), all of which indicated overall failure. Therefore comprehensive research is needed on the design and development of new antibiotics to counter multidrug-resistant organisms or to upgrade existing antibiotics (Karaiskos et al., 2019), which places a financial burden on low-income areas (Malik and Bhattacharyya, 2019). Surface functionalization of silver nanoparticles (AgNPs) presents a robust candidate for next-generation antibiotics (Gupta et al., 2019). Direct application of AgNPs has shown to be physiologically deleterious; however, surface functionalization of biologically benevolent molecules or biomolecules shows enhanced in vivo acceptability with minimal toxicity to living systems (Baptista et al., 2018). Besides, conjugation of AgNPs with biomolecules of interest promotes targeted engineering to obtain optimum therapeutic results (Mathur et al., 2018). Interestingly, the oligodynamic effect demonstrated by AgNPs improves upon synergistic administration with conventional antibiotics, and vice versa (Masri et al., 2018). Such approaches are gradually phasing out the use of conventional antibiotics and promoting the use of nano-silver-based systems (Patra and Baek, 2017). Finally, the procedures for synthesizing surface-functionalized AgNPs are not complicated and produce highly efficient results, such as plant-based green synthesis, microbial systems, biomolecules, and biological macromolecules (Roy et al., 2019).

3.2 Synthetic paradigm for colloidal stabilized AgNPs

Synthesis of AgNPs by physical methods includes techniques such as microemulsions (Richard et al., 2017), the sol-gel method (Goncalves, 2018), hydrolysis, and thermolysis (Majidi et al., 2016). It also includes sonochemical reactions (Vinoth et al., 2017), hydrothermal reactions (Darr et al., 2017), flow injection analysis (Passos et al., 2015), electrospray synthesis (Sridhar and Ramakrishna, 2013), evaporation-condensation process (Lee and Jun, 2019), and laser ablation (Sportelli et al., 2018). These techniques however have drawbacks related to the agglomeration of AgNPs in the colloidal phase in the absence of appropriate stabilizing agents (Siegel et al., 2012). Chemical synthesis of AgNPs occurs in organic solvents and reducing agents where a stabilizer or capping ligand ensures the colloidal stability of the nanoparticles (Kister et al., 2018). Chemical reducing agents for AgNP synthesis typically include hydrazine (El-Faham et al., 2014), hydrazine hydrate (Gurusamy et al., 2017), ascorbate (Malassis et al., 2016), ethylene glycol (Li et al., 2014), N-dimethylformamide (Santos and Marzan, 2009), citrate (Soliwoda et al., 2017), and $NaBH_4$ (Song et al., 2009). The weaker reducing agents exhibit a slower rate of reduction and hence promote better control over the morphology of the resulting AgNPs (Zhang et al., 2018a). Similarly, capping agents for obtaining the colloidal stabilized AgNPs include cellulosic natural polymers (Ogundare and van Zyl, 2018), oleylamine (Chen et al., 2007), gluconic acid (Meshram et al., 2013), and chitosan (Cinteza et al., 2018). Polymers (Batista et al., 2018) such as poly N-vinyl-2-pyrrolidone (PVP) (Jovanovik et al., 2012), polymethacrylic acid (PMAA), polymethylmethacrylate (PMMA), and polyethylene glycol (PEG) (Luo et al., 2005) stabilize the AgNPs electrostatically (Solomon and Umoren, 2015) or sterically (Li et al., 2013). Upon anchoring to the AgNPs' surface, anionic motifs such as –COOH, polyoxoanions, halides, and citrate impart a negative charge on particle surfaces and hence promote their electrostatic stabilization (Ajitha et al., 2016). Conversely, a branched polyethyleneimine (PEI) offers an amine-functionalized positive charge to the nanoparticle with similar stabilizing effect (Tan et al., 2007). In addition, the tagging of AgNPs with bulky organic polymers and alkylammonium cation promotes steric stabilization of particles by restricting their aggregation (Al-Thabaiti et al., 2015). The use of a capping agent enables engineering of the physicochemical parameters of AgNPs, including shape, size, and optical properties (Phan and Nguyen, 2017). Similarly, physical parameters (i.e., temperature, concentration, pH, nature of functionality, nature of reducing agents, and molar ratio of surfactant over capping agent) regulate the growth and nucleation of AgNPs (Patel et al., 2017). Conversely, during green synthesis of AgNPs, the macromolecular ingredients present in plant extracts (e.g., in tannins, alkaloids, saponins, flavonoids, terpenoids, and microbial cultures such as peptidoglycans, oligopeptides, polysaccharides, and coenzymes) play the dual role of reducing and capping agents (Ahmad et al., 2019). Interestingly, some microbes possess the ability to synthesize AgNPs intracellularly where the intracellular components mediate capping and reduction of precursor metal (Luo et al., 2018). For example, in the *Verticillium* species of fungus, the AgNPs' synthesis takes place on the surface of the cell wall (Mukherjee et al., 2001), where the negatively charged carboxylate groups of the membrane-bound enzymes electrostatically trap the Ag^+ ions on the surface of cells (Zhao et al., 2018). Similarly, several bacterial strains synthesize AgNPs intracellularly by bioreduction of Ag^+ ions through an intracellular electron donor in the presence of essential microbial enzymes (Heitzschold et al., 2019).

Likewise, in *B. licheniformis* and *aspergillus terreus*, the electron donor NADH converts Ag^+ ions to zero-valent silver (Ali et al., 2019). The surface-decorated AgNPs procured from these methods provide a suitable profile for further tethering of the desired surface functionalities.

3.3 Biogenic silver nanoparticles as antimicrobial agents

The green synthesis of silver nanoparticles occurs in the presence of a plant extract and microbial exopolysaccharides that act as a capping agent to stabilize the nanoparticles in the colloidal phase. Estevez et al. (2020) reported the synthesis of biogenic AgNPs with a 45-nm diameter from a *Phanerochaete chrysosporium* extract. The AgNPs demonstrated antimicrobial activity against *Escherichia coli* and *Candida albicans* by significantly lowering the cellular fatty acid content and attenuating the formation of biofilm. The AgNPs possessed a negative surface charge and showed marked stability in a pH ranging from 3 to 9. Rad and Pohel (2020) developed AgNPs from the aerial extract of *Pulicaria vulgaris* that displayed marked antioxidant activity and antimicrobial potency against the pathogenic strains of *Escherichia coli*, *Staphylococcus aureus*, *Candida glabrata*, and *Candida albicans* with IC50 values ranging from $40 \mu g\,mL^{-1}$ to $60 \mu g\,mL^{-1}$. The AgNPs reportedly exhibited a synergistic antifungal effect with fluconazole and amoxicillin. Saaho et al. (2020) yielded biogenic silver nanoparticles from cyanobacteria *Chroococcus minutus* with antibacterial properties against *Escherichia coli* and *Streptococcus pyogenes* that cause infections mainly in the upper respiratory tract. Bardania et al. (2020) prepared biogenic AgNPs by using ethanolic extract of *Teucrium polium*. The nanoparticles were embedded in polyethylene glycol or polylactic acid and resulted in absorbable wound dressings. The nanoparticles mainly exhibited antibacterial properties against *Staphylococcus aureus* and *Pseudomonas aeruginosa*. Aty et al. (2020) reported the statistically controlled synthesis of biogenic silver nanoparticles by *Penicillium chrysogenum* MF318506. These nanoparticles inhibited *Staphylococcus aureus*, *Escherichia coli*, *Candida albicans*, *Candida tropicalis*, *Aspergillus niger*, and *Fusarium solani*. Interestingly, the biogenic AgNPs showed antiangiogenic properties against HCT-116 and MCF-7 cancer cell lines due to their cellular uptake via macropinocytosis. Aygun et al. (2020) synthesized AgNPs from *Ganoderma lucidum* extract and evaluated their biological properties. The synthesized AgNPs showed diameters ranging from 15 nm to 22 nm and a face-centered cubic structure. These nanoparticles revealed a robust inhibitory potential against *Staphylococcus aureus*, *Escherichia coli*, *Bacillus cereus*, and *Candida albicans* with IC50 ranging from $16\,\mu g\,mL^{-1}$ to $28\,\mu g\,mL^{-1}$. The DNA cleaving potential and antioxidant activity further supported the biological profile of nanoparticles. Bolbanabad et al. (2020) synthesized biogenic AgNPs from *Yarrowia lipolytica* and studied their antibacterial activity against *Enterococcus faecalis*, *Proteu vulgaris*, *Streptococcus pyrogenes*, and *Pseudomonas aeruginosa*. The nanoparticles reportedly showed substantial antimicrobial activity superior to chloramphenicol, as validated by the MIC50 values. Pawar and Patil (2020) reported the green synthesis of AgNPs from a tuber extract of *Eulophia herbacea*. The AgNPs possessed a diameter of $11.70 \pm 2.43\,nm$ with face-centered cubic geometry. The nanoparticles showed antibacterial activity against the pathogenic strains of *Pseudomonas aeruginosa* and *Bacillus subtilis* and a significant synergism with the antibiotics streptomycin and chloramphenicol. Interestingly, the nanoparticles effectively removed the hazardous organic dyes congo red and methylene blue from their aqueous solution, which further validated their catalytic potency. Gol et al. (2020) reported green synthesis of spherical AgNPs with diameters ranging from 10nm to 20nm from a *Camellia sinensis* extract. The antibacterial activity of the nanoparticles against methicillin-resistant *Staphylococcus aureus* showed promise as a next-generation antibiotic against drug-resistant bacteria. The preceding studies validate future use of biogenic AgNPs as future pharmaceuticals with broad-spectrum activity against pathogenic strains.

3.4 Aqueous phase transfer of colloidal stabilized AgNPs

The current synthetic protocols for AgNPs include physical, chemical, and green methods (Prasher et al., 2018a) that yield colloidal dispersion of nanoparticles stabilized with the presence of an appropriate ligand on their surface (Ravindran et al., 2013). However, the presence of hydrophobic functionalities on AgNPs restrains in vivo biological applications, as most of the physiological processes take place in aqueous media (Fratoddi, 2018). Hence the hydrophilicity of surface ligands plays a strategic role in deciding the biological acceptance of AgNPs (Giron et al., 2016). Typical approaches for yielding aqueous phase-stabilized AgNPs include ligand-addition, ligand-exchange, and encapsulation (Fig. 3.1). The ligand-addition approach includes attaching organic phase-stabilized AgNPs to a biological inorganic hydrophilic head group such as silica, ZnS, and silanes that ensure enhanced colloidal stability in an aqueous phase and minimum nonspecific interactions (Palma et al., 2007). Also, this approach provides supplementary functionalities and chemical groups for further functionalization. Conversely, the ligand-exchange method substitutes hydrophobic ligands already present on the surface of AgNPs with strongly bonding hydrophilic ligands containing high affinity and polar functional head groups such as $-NH_2$, PH_3, $-COOH$, $-H$ that transfer the nanoparticles from the organic phase to the aqueous phase (Kumar et al., 2003). Encapsulation of hydrophobic ligand-tagged AgNPs with amphiphilic polymers or block co-polymers offers another effective strategy for aqueous phase transfer of functionalized AgNPs (Rinaldi et al., 2019). The polar head groups decorated on the encapsulating amphiphilic polymer enhance aqueous

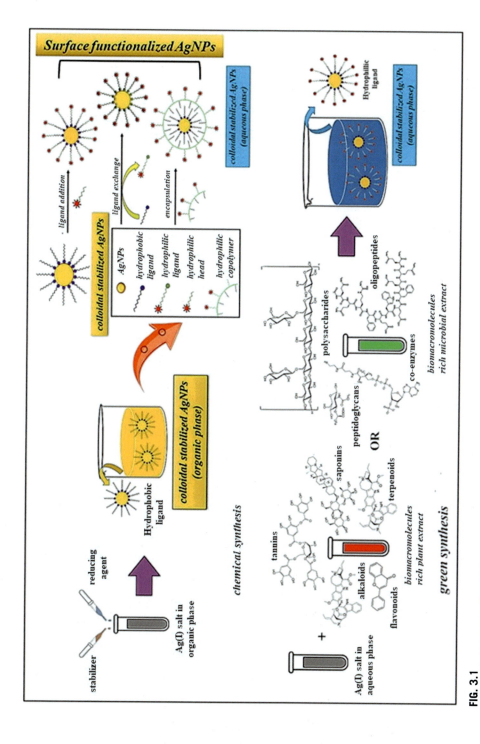

FIG. 3.1

Synthetic approaches for AgNPs.

phase stabilization of AgNPs. The phase transfer strategies improve the nanoparticle bioavailability, resulting in low cytotoxicity and enhanced physiological compatibility (Kumar et al., 2005). They also facilitate further bioconjugation of the AgNP system to the biomolecules of interest for designing diagnostic probes and high-efficiency theranostics.

3.5 Oligodynamic antimicrobial mechanism of AgNPs

Among the contemporary biocidal metal-based nanomaterials, AgNPs present the most significant oligodynamic effect (Prasher et al., 2018b) and offer broad-spectrum antimicrobial properties (Burdusel et al., 2018). The production of reactive oxygen species (ROS) and the intracellular release of Ag^+ ions contribute to biocidal activity (Lee et al., 2019). The AgNPs attach electrostatically with negatively charged microbial lipopolysaccharides (Slavin et al., 2017). The interactions also appear with negatively charged teichoic acid linked to peptidoglycans present in the cell membrane of gram-positive bacteria (Qing et al., 2018). It has been reported that Ag^+ cations also interact with membrane sulfhydryl groups and form stable thiol linkages that subsequently deactivate the associated peptide (Dakal et al., 2016). Similarly, AgNPs and Ag^+ ions distort the 3D structure of transmembrane proteins that mediate ionic transport across membranes and interactions with disulfide bonds. These factors block the active sites, thereby deforming the structure of macromolecules (Liu et al., 2017). Polysaccharide-capped AgNPs disrupt the optimal functioning of essential membrane-bound enzymes, such as lanosterol 14 α-demethylase that mediates the conversion of lanosterol to ergosterol, which is an essential ingredient for sustaining membrane integrity (Prasher et al., 2018c). Also, the small, biomolecule-stabilized AgNPs trigger oxidative stress in microbial cultures, leading to clumping and aggregation (Singh and Prasher, 2018). These factors enhance the membrane permeability that impairs cellular transit processes (Yan et al., 2018), and it also leaks essential cellular contents, including ions, enzymes, proteins, and electrolytes, reducing sugars and ATP reserves (Prabhu and Poulose, 2012). Upon entering cells through membrane porins, AgNPs interact with essential components of cells, including the vital organelles, essential proteins, enzymes, and DNA (Li et al., 1997). AgNPs' impact on ribosomes results in their gradual denaturation and interference with optimal peptide synthesis (Lok et al., 2006). The AgNPs induce adverse effects on the metabolism of microbes by altering the catalytic activity of enzymes involved in glycolysis, eventually starving the microbial cell of its energy (Lee et al., 2016). Moreover, AgNPs interact with DNA by disrupting the H-bonds between purine and pyrimidine base pairs, thereby instigating the denaturation of DNA by breaking the double-helical structure (Nallanthighal et al., 2017). AgNPs also hamper DNA replication and induces mutations in DNA repair genes, eventually causing a loss of microbial cell division (Ribeiro et al., 2018). Exposure to AgNPs produces cellular oxidative stress due to intracellular generation of superoxide anions and hydroxyl radicals (Onodera et al., 2015).

3.5 Oligodynamic antimicrobial mechanism of agnps

AgNPs' interactions with mitochondrial enzymes produce ROS by uncoupling the mitochondrial electron transport chain from oxidative phosphorylation (Belluco et al., 2016). These adverse intracellular events instigated by exposure to AgNPs eventually cause microbial necrotic-like death. AgNPs also modify microbial signal transduction pathways by disturbing the phosphorylation and dephosphorylation balance required for growth and metabolism of host microbes (Mao et al., 2018). Conversely, intracellular AgNPs dephosphorylate tyrosine residues on essential peptide substrates in host microbes, thereby marring their growth (Zhang et al., 2018b). Elevated stress conditions prompt bacterial colonies to develop thick, slimy glycocalyx sheaths, referred to as biofilms (Chibber et al., 2017). The biofilms contain essential nutrients that sustain microbial colonies during the unfavorable environmental conditions. The biofilms reportedly provide resistance against several antibiotics (de Campos et al., 2016). Surface-functionalized AgNPs inhibit the production of microbial biofilm and increase microbial susceptibility to antibiotics anchored to the surface of a nanoparticle (Habash et al., 2014). Fig. 3.2 illustrates the antimicrobial mechanism of AgNPs.

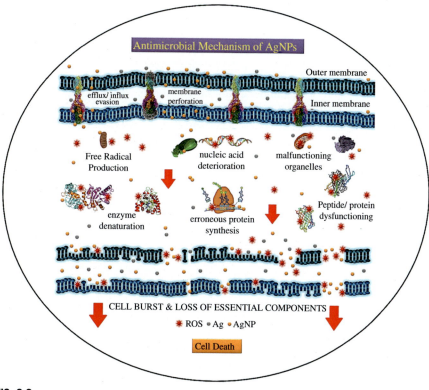

FIG. 3.2

Antimicrobial mechanism of AgNPs.

3.6 Biocidal activity of AgNPs against MDR microbes

Antimicrobial resistance strained the effectiveness of multiple antibiotics and necessitated the development of a next-generation strategy to treat microbes that are resistant to drugs. One strategy is the use of surface-functionalized AgNPs (Kovács et al., 2016). Surface-functionalized AgNPs' mechanism for inhibiting MDR microbes focuses MDR microbes on evading energy-driven efflux pumps (Du et al., 2015). It has been reported that AgNPs deactivate microbial MDR AcrAB-TolC efflux pumps by destabilizing AcrB transporter proteins (Mishra et al., 2018). Directly binding AgNPs to the active site of the efflux pumps prompts an inhibitory effect by obstructing the efflux kinetics coupled with disintegration of the proton motive force (PMF) required for optimal functioning of the efflux system (Dibrov et al., 2002). Table 3.1 shows recent reports on AgNPs' inhibition of MDR microbes.

3.7 Synergism of AgNPs with antibiotics

Conventional antibiotics have failed to alleviate drug-resistant microbes (Llor and Bjerrum, 2014). Combinatorial therapy in addition to polypharmacology and drug repurposing offer an improved therapeutic antibiotic profile (Alanis, 2005); however, the need for novel therapeutic systems continues (Wise, 2002). Polysaccharide-macromolecule-capped AgNPs serve as excellent drug delivery systems for intracellular transport of chemotherapeutics (Gallon et al., 2019; Krychowiak et al., 2018). Capping of AgNPs with antibiotics prompts an effective intracellular delivery of drugs (Esmaeillou et al., 2017; Kaur and Kumar, 2019; Habash et al., 2017). It also manifests synergistic inhibition of the targeted microbes (Adil et al., 2019; Cotton et al., 2019) due to the biocidal effect of AgNPs combined with antimicrobial efficacy in delivery of the cargo therapeutic (Tiwari et al., 2017). Table 3.2 presents recent reports on the synergistic effects of commercial antibiotics and AgNPs on targeted microorganisms.

3.8 Conclusion

Because of their oligodynamic effect and biocidal potency on both drug-resistant and nonresistant microbes, surface-functionalized AgNPs are plausible next-generation antibiotics. Bio-conjugation of AgNPs with desired biomacromolecules and AgNPs' synergistic effects with conventional antibiotics suggest that AgNPs can be used in future combination therapies. However, complications related to the secretion of AgNP-based formulations and their intracellular assimilation into tissues and organs restrict their current use in antibiotic therapies. Therefore, to achieve clinical success, researchers must consider these issues in antibiotic developments based on the use of AgNPs.

Table 3.1 Recent investigations of AgNPs targeting multidrug-resistant microbes.

AgNPs morphology	Target drug-resistant organism	Mechanism	Reference
Shape: Spherical Size: Diameter ranging from 48 nm to 125 nm with an average diameter of 53.2 nm	Rifampin resistant *E. coli* Strain-I (Donor), Streptomycin resistant *E. coli* Strain-II (Recipient), Streptomycin resistant *E. coli* AB1157 (Mutant)	Bactericidal effect occurs due to reactive oxygen species (ROS) generation and chromosomal aberration.	Kotakadi et al. (2015)
Shape: Spherical Size: Diameter ranging from 15.11 nm to 16.54 nm	Methicillin-resistant *S. aureus* Strains: CCARM and ATCC	The Ag^+ released from AgNPs strongly binds to essential microbial enzymes and DNA. AgNPs penetrate the cell wall, which increases cell permeability followed by cell death.	Jang et al. (2015)
Shape: Spherical Size: Diameter ranging from 10 nm to 25 ± 1.5 nm	Multidrug-resistant *B. cereus*, *B. subtilis*, *Micrococcus luteus*, *S. aureus*, *E. cloacae*, *Serratia marcescens*, *Shigella dysentery*	The Ag^+ released from AgNPs directly into the bacterial growth media creates metal stress in the bacteria, influencing them to recognize and uptake cationic polyamine-capped AgNPs that express biomimetic activity, restricting microbial growth.	Saha et al. (2016)
Shape: Spherical Size: Average diameter 40 ± 10 nm	Multidrug-resistant *P. aeruginosa*, *B. subtilis*	Bioconjugation of AgNPs with siRNA evades multidrug resistance.	Sun et al. (2016)
Shape: Spherical Size: Diameter ranging from 30 nm to 40 nm	Multidrug-resistant *E. coli*, *B. subtilis*	Causes alteration of bacterial cell division mechanism, including bacterial cytoskeletal proteins FtsA and FtsZ.	Sanyasi et al. (2016)
Shape: Spherical Size: Diameter ranging from 120 nm to 400 nm	Multidrug-resistant strains of *E. coli*, *B. subtilis*, *B. cereus*, *K. pneumonia*, *S. aureus*, *P. aeruginosa*, *Salmonella enteritidis*, *Acinetobacter radioresistens*, *Enterobacter xiangfanfensis*, *S. epidermidis*, *S. haemolyticus*, *Ponibacterium acnes*	Structural alterations in the bacterial cell membrane affect permeability, leading to the demise of the bacterial cell.	Jha et al. (2017)
Shape: Spherical Size: Diameter <60 nm	Multidrug-resistant *Acinetobacter baumannii* ATCC 19606	Adherence to the bacterial cell wall occurs causing the formation of pits, which subsequently leads to permeability loss and bacterial death.	Chang et al. (2017)

Continued

Table 3.1 Recent investigations of AgNPs targeting multidrug-resistant microbes—cont'd

AgNPs morphology	Target drug-resistant organism	Mechanism	Reference
Shape: Spherical Size: Average diameter 100 nm	Carbapenem-resistant RS-307 strain of *Acinetobacter baumannii*	Bactericidal effect occurs due to ROS generation.	Tiwari et al. (2017)
Shape: Spherical, distinct Size: Average diameter 11 nm	Drug-resistant strains of *S. aureus* and *P. aeruginosa*	Causes generation of ROS, malondialdehyde, leakage of proteins and sugars in bacterial cells, and lower lactate dehydrogenase activity, lower adenosine triphosphate levels, downregulated expression of glutathione, upregulation of glutathione S-transferase, and downregulation of both superoxide dismutase and catalase.	Yuan et al. (2017)
Shape: Spherical Size: Diameter ranging from 20.12 nm to 29.48 nm	Drug-resistant strains of *A. baumannii* (ATCC BAA-1605), *E. coli* (ATCC BAA-2523), *K. pneumoniae* (ATCC BAA-2524), *P. aeruginosa* (ATCC BAA-2108), *M. luteus* (ATCC 10240); methicillin-resistant strains of *S. aureus* (MRSA, ATCC 43330), *E. faecalis* (ATCC 51299), *S. typhimurium* (ATCC 14028), *S. sonnei* (ATCC25931), *S. epidermidis* (ATCC 1228), *S. bovis* (ATCC 49147)	Instigates oxidative stress.	Mahmoud et al. (2017)
Shape: Spherical Size: Diameter ranging from 10 nm to 15 nm	Drug-resistant strains of *A. baumannii*	Causes rapid bactericidal activity by altering the structural integrity and physicochemical changes in microbial cell wall.	Neethu et al. (2018)

Shape: Spherical, monodispersed Size: Average diameter 28.94 nm	Drug-resistant strains of S. typhimurium (98), E. coli (1610), S. aureus (96), B. cereus (430), Shigella flexneri (1457), Vibrio cholera (3904), Vibrio parahaemolyticus (451), P. aeruginosa (1688), B. subtilis (6939), E. aerogenes (13048)	Causes release of Ag$^+$ ions by nanoparticles inside microbial cells.	Manukumar et al. (2017)
Shape: Spherical, monodispersed Size: Average diameter 2.3 nm	Drug-resistant strains of E. coli (BTCB03), K. pneumonia (BTCB04), Acinetobacter sp. (BTCB05), P. aeruginosa (BTCB01), S. aureus (BTCB02)	Bactericidal mode of action is cell wall disruption.	Iqtedar et al. (2019)
Shape: Spherical, monodispersed Size: Average diameter ranging from 5 nm to 40 nm & 1 nm to 25 nm	Drug-resistant strains of E. coli, P. aeruginosa, methicillin-sensitive and resistant S. aureus	Causes ROS-mediated destruction of cell membrane, inhibition of growth, and biofilms.	Ahmed et al. (2018)
Shape: Spherical Size: Average diameter 25 nm	Drug-resistant isolates of Salmonella enterica serovar Typhi, isolates MCASMZU 1 to 13	The interaction of Ag$^+$ ions with microbes causes the denaturation of proteins, rupture of the plasma membrane, reduction of intracellular ATP, and finally cell death.	Balakrishnan et al. (2017)
Shape: Spherical Size: Average diameter ranges from 30 nm to 40 nm	Drug-resistant E. cloacae ATCC 13047	Evades AcrAB-TolC efflux pump.	Mishra et al. (2018)

Continued

Table 3.1 Recent investigations of AgNPs targeting multidrug-resistant microbes—cont'd

AgNPs morphology	Target drug-resistant organism	Mechanism	Reference
Shape: Spherical Size: Average diameter 10 nm	Drug-resistant strains of P. aeruginosa (ATCC 27853), E. coli, S. typhi (ATCC 14028), K. pneumonia (ATCC 13883)	Causes pore formation in the microbial membrane bilayer mediated through a hydrophobic collapse mechanism.	Pal et al. (2019)
Shape: Spherical Size: Average diameter ranges from 12 nm to 20 nm	Multidrug-resistant engineered E. coli QH4	Has synergistic bactericidal effect with commercial antibiotics.	Rahman et al. (2019)
Shape: Spherical, well dispersed Size: Average diameter ranges from 8.33 nm to 26.17 nm	Drug-resistant strains of S. aureus, P. aeruginosa, L. paracasei	Causes disruption of cell membrane permeability, leakage of protein and DNA, and formation of bactericidal ROS.	Gomaa (2019)
Shape: Spherical Size: Average diameter ranging from 24 nm to 46 nm	Drug-resistant strains of S. aureus, E. coli, K. pneumoniae, P. aeruginosa	Causes synergistic bactericidal effect with commercial antibiotics.	Arul et al. (2017)

Table 3.2 Synergistic effects of antibiotics in presence of AgNPs (2015–2019).

AgNPs morphology	Synergistic antibiotic	Target microorganism	Reference
Shape: Spherical Size: 50 nm to 100 nm in diameter	Tetracycline, neomycin, penicillin	Gram-negative: S. typhimurium (DT104)	McShan et al. (2015)
Shape: Spherical Size: 10 nm to 90 nm in diameter	Ampicillin, polymyxin, gentamycin, chloramphenicol, penicillin, amikacin, tetracycline, cephalothin, amoxyclav, cefpirome, clotrimazole	Gram-positive: S. aureus, B. cereus Gram-negative: E. coli, P. aeruginosa	Padalia et al. (2015)
Shape: Spherical Size: 50 nm in diameter	Gentamycin, cefotaxime, meropenem	Gram-negative: K. Pneumoniae, E. coli	Gurunathan (2015)
Shape: Spherical Size: 5.8 ± 2.4 nm in diameter	Streptomycin	Gram-positive: S. aureus (25923, 49834) Gram-negative: E. coli (25922), P. aeruginosa (27853)	Rastogi et al. (2015)
Shape: Spherical Size: 2.78 ± 1.47 nm in diameter	Ampicillin, azithromycin, kanamycin, netilmicin	Gram-negative: E. coli (MREC33)	Manna et al. (2015)
Shape: Spherical Size: 97 nm in diameter	Lincomycin, oleandomycin, novobiocin, vancomycin, penicillin G, rifampicin	Gram-positive: B. anthracis, B. cereus Gram-negative: S. enterica, E. coli, V. parahaemolyticus	Singh et al. (2015)
Shape: Spherical Size: 23.7 nm in diameter	Levofloxacin	Gram-positive: S. aureus, B. subtilis Gram-negative: E. coli, P. aeruginosa	Ibrahim (2015)
Shape: Spherical Size: 10 nm to 30 nm in diameter	Penicillins	Gram-positive: S. aureus	Marta et al. (2015)

Continued

Table 3.2 Synergistic effects of antibiotics in presence of AgNPs (2015–2019)—cont'd

AgNPs morphology	Synergistic antibiotic	Target microorganism	Reference
Shape: Spherical, crystallized in a face-centered cubic structure Size: 20 nm to 30 nm in diameter	Amoxycillin, ampicillin, streptomycin, amikacin, kanamycin, tetracycline, vancomycin, cefepime, cefetaxime	Gram-positive: *B. cereus*, *S. epidermidis*, *B. subtilis*, *S. aureus* Gram-negative: *S. marcescens*, *S. typhimurium*, *K. pneumoniae*, *E. coli*	Jyoti et al. (2016)
Shape: Spherical, narrow size distribution Size: 26 nm in diameter	Amoxycillin, ampicillin, piperacillin, penicillin, clindamycin, vancomycin, ciprofloxacin, ofloxacin, gentamycin, amikacin	Gram-positive: *S. aureus* (CCM 4223) Gram-negative: *E. coli* (CCM 4225), *P. aeruginosa* (CCM 3955)	Panacek et al. (2016a)
Shape: Spherical Size: 29.8 ± 6.4 nm in diameter	Enoxacin, kanamycin, neomycin, tetracycline	Gram-negative: *S. typhimurium* (DT 104)	Deng et al. (2016)
Shape: Spherical, narrow size distribution Size: 28 nm in diameter	Cefotaxime, ceftazidime, meropenem, ciprofloxacin, gentamicin	Gram-negative: multiresistant, β-lactamase, carbapenemase-producing *Enterobacteriaceae*	Panacek et al. (2016b)
Shape: Spherical, narrow size distribution Size: 5 nm to 15 nm in diameter (fewer particles possess 20 nm and 30 nm diameter)	Antibiotics (streptomycin, tetracycline, kanamycin, and rifampicin) and fungicides (amphotericin B, fluconazole and ketoconazole)	Gram-positive: *B. cereus*, *S. aureus* Gram-negative: *K. pneumoniae*, *E. coli*, *P. brassicacearum*, *A. hydrophila* Fungal strains: *Candida albicans*, *Fusarium oxysporum*, *Aspergillus flavus*	Aziz et al. (2016)
Shape: Spherical Size: 5 nm to 12 nm in diameter, with average size of 8.4 nm	Polymixin B, rifampicin, tigecycline	Gram-negative: *A. baumannii*	Wan et al. (2016)

Table 3.2 Synergistic effects of antibiotics in presence of AgNPs (2015–2019)—cont'd

AgNPs morphology	Synergistic antibiotic	Target microorganism	Reference
Shape: Spherical Size: Two types of AgNPs with diameters 8 nm and 28 nm	Amoxycillin, gentamycin, penicillin G, colistin	Gram-negative: S. enterica, E. coli, A. pleuropneumoniae, P. multocida Gram-positive: S. aureus, S. uberis	Smekalova et al. (2016)
Shape: Spherical Size: 15 nm to 20 nm in diameter	Amoxicillin, azithromycin, clarithromycin, linezolid, vancomycin	Gram-positive: Methicillin-resistant S. aureus	Akram et al. (2016)
Shape: Spherical Size: 5 nm to 15 nm in diameter	Amoxycillin, ampicillin, erythromycin, kanamycin, tetracycline	Gram-positive: B. cereus, S. epidermidis Gram-negative: S. aureus, E. coli, K. pneumoniae, P. aeruginosa, S. typhimurium, V. vulnificus	Prema et al. (2017)
Shape: Spherical Size: 10 nm to 100 nm in diameter	Cephalexin	Gram-negative: S. aureus	Salarian et al. (2017)
Shape: Spherical Size: 5 nm to 40 nm in diameter	Vancomycin, streptomycin, tetracycline, gentamycin, amoxycillin, ciprofloxacin, erythromycin	Gram-negative: S. aureus, E. coli	Saratale et al. (2017)
Shape: Spherical Size: 20 nm to 170 nm in diameter	Ampicillin	Gram-negative: S. aureus, E. coli	Tippayawat et al. (2017)
Shape: Spherical Size: average size 20 nm	Gentamycin, chloramphenicol	Gram-positive: E. faecalis	Katva et al. (2017)

Continued

Table 3.2 Synergistic effects of antibiotics in presence of AgNPs (2015–2019)—cont'd

AgNPs morphology	Synergistic antibiotic	Target microorganism	Reference
Shape: Spherical Size: 35 nm to 60 nm in diameter	Ampicillin, ciprofloxacin	Gram-negative: S. aureus (VN3), V. cholera (VN1)	Naik et al. (2017)
Shape: Spherical Size: 10 nm in diameter	Gentamycin	Gram-negative: S. aureus	Zhou et al. (2017)
Shape: Spherical Size: Different sizes AgNPs (7.22 ± 3.07 nm, 17.0 ± 5.78 nm and 62.12 ± 31.5 nm)	Ampicillin, chloramphenicol, streptomycin, tetracycline	Gram-positive: B. subtilis (MTCC 441) Gram-negative: S. aureus (MTCC 737), E. coli (MTCC 1687), K. Pneumoniae (MTCC 4030)	Phanjom and Ahmed (2017)
Shape: Spherical Size: 20 nm in diameter	Ceftriaxone	Gram-positive: B. cereus Gram-negative: S. aureus, K. pneumoniae, P. aeruginosa	Shanmuganathan et al. (2018)
Shape: Crystalline Size: 8 nm to 48 nm in diameter	Kanamycin, ampicillin, tetracycline, streptomycin	Gram-positive: B. subtilis Gram-negative: S. aureus, K. pneumoniae, P. aeruginosa, E. coli, P. mirabilis	Buszewski et al. (2018)
Shape: Irregular Size: 10 nm to 40 nm in diameter	Erythromycin, novobiocin, lincomycin, penicillin, vancomycin, oleandomycin	Gram-negative: P. aeruginosa, E. coli, S. enterica	Singh et al. (2018)
Shape: Spherical Size: < 140 nm in diameter	Ciprofloxacin, fluconazole	Gram-positive: S. pyogenes Gram-negative: P. aeruginosa, Salmonella	Ghiuta et al. (2018)
Shape: Spherical Size: Average radius of 73.91 ± 15.33 nm	Ampicillin	Gram-negative: E. coli	Gonzalez et al. (2018)

Table 3.2 Synergistic effects of antibiotics in presence of AgNPs (2015–2019)—cont'd

AgNPs morphology	Synergistic antibiotic	Target microorganism	Reference
Shape: Spherical, polydispersed Size: Average radius of 5nm to 50nm and 5nm to 20nm	Kanamycin, tetracycline, fluconazole, ketoconazole, amphotericin B	Gram-positive: S. pyogenes Gram-negative: E. coli, A. baumanii, E. aerogenes, P. aeruginosa, S. typhimurium	Wypij et al. (2018a)
Shape: Spherical Size: 8.57 ± 1.17nm in diameter	Ampicillin, amikacin	Gram-positive: E. faecium, S. aureus Gram-negative: E. coli, A. baumanii, E. cloacae, P. aeruginosa, K. pneumonia	Carrizales et al. (2018)
Shape: Spherical, polydispersed Size: 5nm to 20nm in diameter	Ampicillin, kanamycin, tetracycline, amphotericin B, fluconazole, ketoconazole	Gram-positive: S. aureus Gram-negative: E. coli, P. aeruginosa	Wypij et al. (2018b)
Shape: Spherical, dispersed Size: 10nm in diameter	Vancomycin, gentamycin	Gram-positive: S. mutans Gram-negative: E. fergusonii	Gurunathan (2019)
Shape: Spherical, polydispersed Size: 10nm to 35nm in diameter	Amoxicillin, ampicillin, trimethoprim, nitrofurantoin	Gram-positive: S. epidemidis Gram-negative: P. aeruginosa	Qaralleh et al. (2019)
Shape: Spherical, polydispersed Size: < 200nm	Polymyxin B	Gram-negative: P. aeruginosa	Salman et al. (2019)
Shape: Spherical, polydispersed Size: 3nm to 17nm in diameter	Tetracycline, ampicillin	Gram-negative: E. coli (K12)	Anush et al. (2019)

Continued

Table 3.2 Synergistic effects of antibiotics in presence of AgNPs (2015–2019)—cont'd

AgNPs morphology	Synergistic antibiotic	Target microorganism	Reference
Shape: Spherical, polydispersed Size: 70 nm to 91 nm in diameter	Vancomycin	Gram-positive: S. aureus Gram-negative: E. coli	Kaur et al. (2019)
Shape: Spherical Size: 70 nm to 80 nm in diameter	Violacein	Gram-positive: S. aureus (ATCC 29213) Gram-negative: E. coli (ATCC 25922)	Nakazato et al. (2019)
Shape: Spherical Size: 5 nm to 30 nm in diameter	Streptomycin	Gram-positive: B. cereus Gram-negative: E. coli, P. tolaasii	Teimoori et al. (2019)
Shape: Spherical Size: 86.36 ± 0.22 nm in diameter	Tetracycline	Gram-positive: S. aureus, B. subtilis	Rafinska et al. (2019)

References

Adil, M., Khan, T., Aasim, M., Khan, A.A., Ashraf, M., 2019. Evaluation of the antibacterial potential of silver nanoparticles synthesized through the interaction of antibiotic and aqueous callus extract of *Fagonia indica*. AMB Express 9, 75.

Ahmad, S., Munir, S., Zeb, N., Ullah, A., Khan, B., Ali, J., Bilal, M., Omer, M., Alamzeb, M., Salman, S.M., Ali, S., 2019. Green nanotechnology: a review on green synthesis of silver nanoparticles—an ecofriendly approach. Int. J. Nanomedicine 14, 5087–5107.

Ahmed, B., Hashmi, A., Khan, M.S., Musarrat, J., 2018. ROS mediated destruction of cell membrane, growth and biofilms of human bacterial pathogens by stable metallic AgNPs functionalized from bell pepper extract and quercetin. Adv. Powder Technol. 29, 1601–1616.

Ajitha, B., Reddy, Y.A.K., Reddy, P.S., Jeon, H.-J., Ahn, C.W., 2016. Role of capping agents in controlling silver nanoparticles size, antibacterial activity and potential application as optical hydrogen peroxide sensor. RSC Adv. 6, 36171–36179.

Akram, F.E., El-Tayeb, T., Abou-Aisha, K., El-Azizi, M., 2016. A combination of silver nanoparticles and visible blue light enhances the antibacterial efficacy of ineffective antibiotics against methicillin-resistant *Staphylococcus aureus* (MRSA). Ann. Clin. Microbiol. Antimicrob. 15, 48.

Alanis, A.J., 2005. Resistance to antibiotics: are we in the post-antibiotic era? Arch. Med. Res. 36, 697–705.

Ali, J., Ali, N., Wang, L., Waseem, H., Pan, G., 2019. Revisiting the mechanistic pathways for bacterial mediated synthesis of noble metal nanoparticles. J. Microbiol. Methods 159, 18–25.

Al-Thabaiti, S.A., Obaid, A.Y., Hussain, S., Khan, Z., 2015. Shape-directing role of cetyltrimethylammonium bromide on the morphology of extracellular synthesis of silver nanoparticles. Arab. J. Chem. 8, 538–544.

Anush, K., Shushanik, K., Sussana, T., Ashkhen, H., 2019. Antibacterial effect of silver and iron oxide nanoparticles in combination with antibiotics on *E. coli* K12. BioNanoScience 9, 587–596.

Arul, D., Balasubramanian, G., Natarajan, T., Perumal, P., 2017. Antibacterial efficacy of silver nanoparticles and ethyl acetate's metabolites of the potent halophilic (marine) bacterium, *Bacillus cereus* A30 on multidrug resistant bacteria. Pathog. Glob. Health 111, 367–382.

Aslam, B., Wang, W., Arshad, M.I., Khurshid, M., Muammil, S., Rasool, M.H., Nisar, M.A., Alvi, R.F., Aslam, M.A., Qamar, M.U., Salamat, M.K.F., Baloch, Z., 2018. Antibiotic resistance: a rundown of a global crisis. Infect. Drug Resist. 11, 1645–1658.

Aty, A.A.A.E., Mohamed, A.A., Zohair, M.M., Soliman, A.A.F., 2020. Statistically controlled biogenesis of silver nano-size by *Penicillium chrysogenum* MF318506 for biomedical application. Biocatal. Agric. Biotechnol. 25, 101592.

Aygun, A., Ozdemir, S., Gulcan, M., Cellat, K., Sen, F., 2020. Synthesis and characterization of Reishi mushroom-mediated green synthesis of silver nanoparticles for the biochemical applications. J. Pharm. Biomed. Anal. 178, 112970.

Aziz, N., Pandey, R., Barman, I., Prasad, R., 2016. Leveraging the attributes of Mucor hiemalis-derived silver nanoparticles for a synergistic broad-spectrum antimicrobial platform. Front. Microbiol. 7, 1984.

Balakrishnan, S., Sivaji, I., Kandasamy, S., Duraisamy, S., Kumar, N.S., Gurusubramanian, G., 2017. Biosynthesis of silver nanoparticles using *Myristica fragrans* seed (nutmeg) extract and its antibacterial activity against multidrug-resistant (MDR) *Salmonella enterica serovar Typhi* isolates. Environ. Sci. Pollut. Res. 24, 14758–14769.

Baptista, P.V., McCusker, M.P., Carvalho, A., Ferreira, D.A., Mohan, N.M., Martins, M., Fernandes, A.R., 2018. Nano-strategies to fight multidrug resistant bacteria—"a battle of the titans". Front. Microbiol. 9, 1441.

Bardania, H., Mahmoudi, R., Bagheri, H., et al., 2020. Facile preparation of a novel biogenic silver-loaded nanofilm with intrinsic anti-bacterial and oxidant scavenging activities for wound healing. Sci. Rep. 10, 6129.

Batista, C.C.S., Albuquerque, L.J.C., de Araujo, I., Albuquerque, B.L., da Silva, F.D., Giacomelli, F.C., 2018. Antimicrobial activity of nano-sized silver colloids stabilized by nitrogen-containing polymers: the key influence of the polymer capping. RSC Adv. 8, 10873–10882.

Belluco, S., Losasso, C., Patuzzi, I., Rigo, L., Conficoni, D., Gallocchio, F., Cibin, V., Catellani, P., Segato, S., Ricci, A., 2016. Silver as antibacterial toward *Listeria monocytogenes*. Front. Microbiol. 7, 307.

Bolbanabad, E.M., Ahengroph, M., Darvishi, F., 2020. Development and evaluation of different strategies for the clean synthesis of silver nanoparticles using *Yarrowia lipolytica* and their antibacterial activity. Process Biochem. 94, 319–328. https://doi.org/10.1016/j.procbio.2020.03.024.

Burdusel, A.-C., Gherasim, O., Grumezescu, A.M., Mogoanta, L., Ficai, A., Andronescu, E., 2018. Biomedical applications of silver nanoparticles: an up-to-date overview. Nanomaterials (Basel) 8, 681.

Buszewski, B., Railean-Plugaru, V., Pomastowski, P., Rafinska, K., Szultka-Mlynska, M., 2018. Antimicrobial activity of biosilver nanoparticles produced by a novel *Streptacidiphilus durhamensis* strain. J. Microbiol. Immunol. Infect. 51, 45–54.

Carrizales, M.L., Velasco, K.I., Castillo, C., Flores, A., Magana, M., Castanon, G.A.M., Gutierrez, F.M., 2018. *In vitro* synergism of silver nanoparticles with antibiotics as an alternative treatment in multiresistant uropathogens. Antibiotics 7, 50.

Chang, T.Y., Chen, C.C., Cheng, K.M., Chin, C.Y., Chen, Y.H., Chen, X.A., Sun, J.R., Young, J.J., Chiueh, T.S., 2017. Trimethyl chitosan-capped silver nanoparticles with positive surface charge: their catalytic activity and antibacterial spectrum including multidrug-resistant strains of *Acinetobacter baumannii*. Colloid Surf. B Biointerfaces 155, 61–70.

Chen, M., Feng, Y.-G., Wang, X., Li, T.-C., Zhang, J.-Y., Qian, D.-J., 2007. Silver nanoparticles capped by oleylamine: formation, growth, and self-organization. Langmuir 23, 5296–5304.

Chibber, S., Gondil, V.S., Sharma, S., Kumar, M., Wangoo, N., Sharma, R.K., 2017. A novel approach for combating *Klebsiella pneumoniae* biofilm using histidine functionalized silver nanoparticles. Front. Microbiol. 8, 1804.

Cinteza, L.O., Scomoroscenco, C., Voicu, S.N., Nistor, C.L., Nitu, S.G., Trica, B., Jecu, M.-L., Petcu, C., 2018. Chitosan-stabilized Ag nanoparticles with superior biocompatibility and their synergistic antibacterial effect in mixtures with essential oils. Nanomaterials (Basel) 8, 826.

Cotton, G.C., Gee, C., Jude, A., Duncan, W.J., Abdelmoneim, D., Coates, D.E., 2019. Efficacy and safety of alpha lipoic acid-capped silver nanoparticles for oral applications. RSC Adv. 9, 6973–6985.

Dakal, T.C., Kumar, A., Majumdar, R.S., Yadav, V., 2016. Mechanistic basis of antimicrobial actions of silver nanoparticles. Front. Microbiol. 7, 1831.

Darr, J.A., Zhang, J., Makwana, N.M., Weng, X., 2017. Continuous hydrothermal synthesis of inorganic nanoparticles: applications and future directions. Chem. Rev. 117, 11125–11238.

de Campos, P.A., Royer, S., Batistão, D.W., Araújo, B.F., Queiroz, L.L., 2016. Multidrug resistance related to biofilm formation in *Acinetobacter baumannii* and *Klebsiella pneumoniae* clinical strains from different pulsotypes. Curr. Microbiol. 72, 617–723.

Deng, H., Mc Shan, D., Zhang, Y., Sinha, S.S., Arslan, Z., Ray, P.C., Yu, H., 2016. Mechanistic study of the synergistic antibacterial activity of combined silver nanoparticles and common antibiotics. Environ. Sci. Technol. 50, 8840–8848.

Dibrov, P., Dzioba, J., Gosink, K.K., Häse, C.C., Ha, C.C., 2002. Chemiosmotic mechanism of antimicrobial activity of Ag^+ in *Vibrio cholerae*. Antimicrob. Agents Chemother. 46, 2668–2670.

Du, D., Wang, Z., James, N.R., Voss, J.E., Klimont, E., Ohene, T., Venter, H., Chiu, W., Luisi, B.F., 2015. Structure of the AcrAB-TolC multidrug efflux pump. Nature 509, 512–515.

El-Faham, A., Elzatahry, A.A., Al-Othman, Z.A., Elsayed, E.A., 2014. Facile method for the synthesis of silver nanoparticles using 3-hydrazino-isatin derivatives in aqueous methanol and their antibacterial activity. Int. J. Nanomedicine 9, 1167–1174.

Esmaeillou, M., Zarrini, G., Rezaee, M.A., Mojarrad, J.S., Bahadori, A., 2017. Vancomycin capped with silver nanoparticles as an antibacterial agent against multi-drug resistance bacteria. Adv. Pharm. Bull. 7, 479–483.

Estevez, M.B., Raffaelli, S., Mitchell, S.G., Faccio, R., Albores, S., 2020. Biofilm eradication using biogenic silver nanoparticles. Molecules 25, 2023.

Fratoddi, I., 2018. Hydrophobic and hydrophilic au and Ag nanoparticles. Breakthroughs and perspectives. Nanomaterials (Basel) 8, 11.

Gallon, S.M.N., Alpaslan, E., Wang, M., Larese-Casasnova, P., Londono, M.E., Atehortua, L., Pavon, J.J., Webster, T.J., 2019. Characterization and study of the antibacterial mechanisms of silver nanoparticles prepared with microalgal exopolysaccharides. Mater. Sci. Eng. C 99, 685–695.

Ghiuta, I., Cristea, D., Croitoru, C., Kost, J., Wenkert, R., Vyrides, I., Anayiotos, A., Munteanu, D., 2018. Characterization and antimicrobial activity of silver nanoparticles, biosynthesized using *Bacillus* species. Appl. Surf. Sci. 438, 66–73.

Giron, J.V.M., Vico, R.V., Maggio, B., Zelaya, E., Rubert, A., Benitez, E., Carro, P., Salvarezza, R.C., Vela, M.E., 2016. Role of the capping agent in the interaction of hydrophilic Ag nanoparticles with DMPC as a model biomembrane. Environ. Sci. Nano 3, 462–472.

Gol, F., Aygun, A., Seyrankaya, A., Gur, T., Yenikaya, C., Sen, F., 2020. Green synthesis and characterization of *Camellia sinensis* mediated silver nanoparticles for antibacterial ceramic applications. Mater. Chem. Phys. 250, 123037.

Gomaa, E.Z., 2019. Synergistic antibacterial efficiency of bacteriocin and silver nanoparticles produced by probiotic *Lactobacillus paracasei* against multidrug resistant bacteria. Int. J. Pept. Res. Ther. 25, 1113–1125.

Goncalves, M.C., 2018. Sol-gel silica nanoparticles in medicine: a natural choice. Design, synthesis and products. Molecules 23, 2021.

Gonzalez, D.A., Ulate, D., Alvarado, R., Miranda, O.A., 2018. Chitosan-silver nanoparticles as an approach to control bacterial proliferation, spores and antibiotic-resistant bacteria. Biomed. Phys. Eng. Express 4, 035011.

Gupta, A., Mumtaz, S., Li, C.-H., Hussain, I., Rotello, V.M., 2019. Combatting antibiotic-resistant bacteria using nanomaterials. Chem. Soc. Rev. 48, 415–427.

Gurunathan, S., 2015. Biologically synthesized silver nanoparticles enhances antibiotic activity against *gram-negative* bacteria. J. Ind. Eng. Chem. 29, 217–226.

Gurunathan, S., 2019. Rapid biological synthesis of silver nanoparticles and their enhanced antibacterial effects against *Escherichia fergusonii* and *Streptococcus mutans*. Arab. J. Chem. 12, 168–180.

Gurusamy, V., Krishnamoorthy, R., Gopal, B., Veeraragavan, V., Neelamegam, P., 2017. Systematic investigation on hydrazine hydrate assisted reduction of silver nanoparticles and its antibacterial properties. Inorg. Nano-Metal Chem. 47, 761–767.

Habash, M.B., Park, A.J., Vis, E.C., Harris, R.J., Khursigara, C.M., 2014. Synergy of silver nanoparticles and aztreonam against *Pseudomonas aeruginosa* PAO1 biofilms. Antimicrob. Agents Chemother. 58, 5818–5830.

Habash, M.B., Goodyear, M.C., Park, A.J., Surette, M.D., Vis, E.C., Harris, R.J., Khursigara, C.M., 2017. Potentiation of tobramycin by silver nanoparticles against *Pseudomonas aeruginosa* biofilms. Antimicrob. Agents Chemother. 61, e00415-17.

Heitzschold, S., Walter, A., Davis, C., Taylor, A.A., Sepunaru, L., 2019. Does nitrate reductase play a role in silver nanoparticle synthesis? Evidence for NADPH as the sole reducing agent. ACS Sustain. Chem. Eng. 7, 8070–8076.

Ibrahim, H.M.M., 2015. Green synthesis and characterization of silver nanoparticles using banana peel extract and their antimicrobial activity against representative microorganisms. J. Radiat. Res. Appl. Sci. 8, 265–275.

Iqtedar, M., Aslam, M., Akhtyar, M., Shehzaad, A., Abdullah, R., Kaleem, A., 2019. Extracellular biosynthesis, characterization, optimization of silver nanoparticles (AgNPs) using *Bacillus mojavensis* BTCB15 and its antimicrobial activity against multidrug resistant pathogens. Prep. Biochem. Biotechnol. 49, 136–142.

Jang, H., Lim, S.H., Choi, J.S., Park, Y., 2015. Antibacterial properties of cetyltrimethylammonium bromide-stabilized green silver nanoparticles against *methicillin-resistant Staphylococcus aureus*. Arch. Pharm. Res. 38, 1906–1912.

Jha, D., Thiruveedula, P.K., Pathak, R., Kumar, B., Gautam, H.K., Agnihotri, S., Sharma, A.K., Kumar, P., 2017. Multifunctional biosynthesized silver nanoparticles exhibiting excellent antimicrobial potential against multi-drug resistant microbes along with remarkable anticancerous properties. Mater. Sci. Eng. C 80, 659–669.

Jovanovik, Z., Rodasavljevic, A., Siljegovic, M., Bibic, N., Miskovic-Stankovic, V., Kacarevic-Popovic, Z., 2012. Structural and optical characteristics of silver/poly(N-vinyl-2-pyrrolidone) nanosystems synthesized by γ-irradiation. Radiat. Phys. Chem. 81, 1720–1728.

Jyoti, K., Baunthiyal, M., Singh, A., 2016. Characterization of silver nanoparticles synthesized using *Urtica dioica* Linn. leaves and their synergistic effects with antibiotics. J. Radiat. Res. Appl. Sci. 9, 217–227.

Karaiskos, I., Lagou, S., Pontikis, K., Rapti, V., Poulakou, G., 2019. The "old" and the "new" antibiotics for MDR *gram-negative* pathogens: for whom, when, and how. Front. Public Health 7, 151.

Katva, S., Das, S., Moti, H.S., Jyoti, A., Kaushik, S., 2017. Antibacterial synergy of silver nanoparticles with gentamicin and chloramphenicol against *Enterococcus faecalis*. Pharmacogn. Mag. 13, S828–S833.

Kaur, A., Kumar, R., 2019. Enhanced bactericidal efficacy of polymer stabilized silver nanoparticles in conjugation with different classes of antibiotics. RSC Adv. 9, 1095–1105.

Kaur, A., Preet, S., Kumar, V., Kumar, R., 2019. Synergetic effect of vancomycin loaded silver nanoparticles for enhanced antibacterial activity. Colloids Surf. B: Biointerfaces 176, 62–69.

Kister, T., Monego, D., Mulvaney, P., Cooper, A.-W., Kraus, T., 2018. Colloidal stability of apolar nanoparticles: the role of particle size and ligand shell structure. ACS Nano 12, 5969–5977.

Kotakadi, V.S., Gaddam, S.A., Venkata, S.K., Gopal, D.V.R.S., 2015. New generation of bactericidal silver nanoparticles against different antibiotic resistant *Escherichia coli* strains. Appl. Nanosci. 5, 847–855.

Kovács, D., Szoke, K., Igaz, N., Spengler, G., Molnár, J., Tóth, T., Madarasz, D., Razga, Z., Konya, Z., Boros, I.M., Kiricsi, M., 2016. Silver nanoparticles modulate ABC transporter activity and enhance chemotherapy in multidrug resistant cancer. Nanomedicine 12, 601–610.

Krychowiak, M., Kawiak, A., Narajczyk, M., Borowik, A., Krolicka, A., 2018. Silver nanoparticles combined with naphthoquinones as an effective synergistic strategy against *Staphylococcus aureus*. Front. Pharmacol. 9, 816.

Kumar, A., Joshi, H., Pasricha, R., Mandale, A.B., Sastry, M., 2003. Phase transfer of silver nanoparticles from aqueous to organic solutions using fatty amine molecules. J. Colloid Interface Sci. 264, 396–401.

Kumar, A., Swami, A., Sastry, M., 2005. Process for Phase Transfer of Hydrophobic Nanoparticles. US20050165120A1.

Lee, S.H., Jun, B.-H., 2019. Silver nanoparticles: synthesis and application for nanomedicine. Int. J. Mol. Sci. 20, 865.

Lee, M.J., Lee, S.J., Yun, S.J., Yang, J.-Y., Kang, H., Kim, K., Choi, I.-H., Park, S., 2016. Silver nanoparticles affect glucose metabolism in hepatoma cells through production of reactive oxygen species. Int. J. Nanomedicine 11, 55–68.

Lee, B., Lee, M.J., Yun, S.J., Kim, K., Choi, I.-H., Park, S., 2019. Silver nanoparticles induce reactive oxygen species-mediated cell cycle delay and synergistic cytotoxicity with 3-bromopyruvate in *Candida albicans*, but not in *Saccharomyces cerevisiae*. Int. J. Nanomedicine 14, 4801–4816.

Li, X.Z., Nikaido, H., Williams, K.E., 1997. Silver-resistant mutants of *Escherichia coli* display active efflux of Ag$^+$ and are deficient in porins. J. Bacteriol. 179, 6127–6132.

Li, C.-C., Chang, S.-J., Su, F.-J., Lin, S.-W., Chou, Y.-C., 2013. Effects of capping agents on the dispersion of silver nanoparticles. Colloids Surf. A Physicochem. Eng. Asp. 419, 209–215.

Li, J., Zhu, J., Liu, X., 2014. Ultrafine silver nanoparticles obtained from ethylene glycol at room temperature: catalyzed by tungstate ions. Dalton Trans. 43, 132–137.

Liu, W., Worms, I.A.M., Herlin-Boime, N., Truffier-Boutry, D., Michaud-Soret, I., Mintz, E., Vidaud, C., Rollin-Genetet, F., 2017. Interaction of silver nanoparticles with metallothionein and ceruloplasmin: impact on metal substitution by Ag (I), corona formation and enzymatic activity. Nanoscale 9, 6581–6594.

Llor, C., Bjerrum, L., 2014. Antimicrobial resistance: risk associated with antibiotic overuse and initiatives to reduce the problem. Ther. Adv. Drug Saf. 5, 229–241.

Lok, C.N., Ho, C.M., Chen, R., He, Q.Y., Yu, W.Y., Sun, H., Tam, P.K., Chiu, J.F., Che, C.M., 2006. Proteomic analysis of the mode of antibacterial action of silver nanoparticles. J. Proteome Res. 5, 916–924.

Luo, C., Zhang, Y., Zeng, X., Zeng, Y., Wang, Y., 2005. The role of poly(ethylene glycol) in the formation of silver nanoparticles. J. Colloid Interface Sci. 288, 444–448.

Luo, K., Jung, S., Park, K.-H., Kim, Y.-R., 2018. Microbial biosynthesis of silver nanoparticles in different culture media. J. Agric. Food Chem. 66, 957–962.

Mahmoud, W., Elazzazy, A.M., Danial, E.N., 2017. In vitro evaluation of antioxidant, biochemical and antimicrobial properties of biosynthesized silver nanoparticles against multidrug-resistant bacterial pathogens. Biotechnol. Biotechnol. Equip. 31, 373–379.

Majidi, S., Sehrig, F.Z., Farkhani, S.M., Goloujeh, M.S., Akbazadeh, A., 2016. Current methods for synthesis of magnetic nanoparticles. Artif. Cells Nanomed. Biotechnol. 44, 722–734.

Malassis, L., Dreyfus, R., Murphy, R.J., Hough, L.A., Donnio, B., Murray, C.B., 2016. One-step green synthesis of gold and silver nanoparticles with ascorbic acid and their versatile surface post-functionalization. RSC Adv. 6, 33092–33100.

Malik, B., Bhattacharyya, S., 2019. Antibiotic drug-resistance as a complex system driven by socio-economic growth and antibiotic misuse. Sci. Rep. 9, 9788.

Manna, D.K., Mandal, A.K., Sen, I.K., Maji, P.K., Chakraborti, S., Chakraborti, R., Islam, S.S., 2015. Antibacterial and DNA degradation potential of silver nanoparticles synthesized via green route. Int. J. Biol. Macromol. 80, 455–459.

Manukumar, H.M., Umesha, S., Kumar, H.N.N., 2017. Promising biocidal activity of thymol loaded chitosan silver nanoparticles (T-C@AgNPs) as anti-infective agents against perilous pathogens. Int. J. Biol. Macromol. 102, 1257–1265.

Mao, B.H., Chen, Z.-Y., Wang, Y.-J., Yan, S.-J., 2018. Silver nanoparticles have lethal and sublethal adverse effects on development and longevity by inducing ROS-mediated stress responses. Sci. Rep. 8, 2445.

Marta, B., Potara, M., Iliut, M., Jakab, E., Radu, T., Imre-Lukaki, F., Katona, G., Popescu, O., Astilean, S., 2015. Designing chitosan-silver nanoparticles-graphene oxide nanohybrids with enhanced antibacterial activity against *Staphylococcus aureus*. Colloids Surf. A Physicochem. Eng. Asp. 487, 113–120.

Masri, A., Anwar, A., Ahmed, D., Siddiqui, R.B., Raza Shah, M., Khan, N.A., 2018. Silver nanoparticle conjugation-enhanced antibacterial efficacy of clinically approved drugs cephradine and vildagliptin. Antibiotics 7, E100.

Mathur, P., Jha, S., Ramteke, S., Jain, N.K., 2018. Pharmaceutical aspects of silver nanoparticles. Artif. Cells Nanomed. Biotechnol. 46, 115–126.

McShan, D., Zhang, Y., Deng, H., Ray, P.C., Yu, H., 2015. Synergistic antibacterial effect of silver nanoparticles combined with ineffective antibiotics on drug resistant *Salmonella typhimurium* DT104. J. Environ. Sci. Health C 33, 369–384.

Meshram, S.M., Bonde, S.R., Gupta, I.R., Gade, A.K., Rai, M.K., 2013. Green synthesis of silver nanoparticles using white sugar. IET Nanobiotechnol. 7, 28–32.

Mishra, M., Kumar, S., Majhi, R.K., Goswami, L., Goswami, C., Mohapatra, H., 2018. Antibacterial efficacy of polysaccharide capped silver nanoparticles is not compromised by AcrAB-TolC efflux pump. Front. Microbiol. 9, 823.

Mukherjee, P., Ahmad, A., Mandal, D., Senapati, S., Sainkar, S.R., Khan, M.I., Parishcha, R., Ajaykumar, P.V., Alam, M., Kumar, R., Sastry, M., 2001. Fungus-mediated synthesis of silver nanoparticles and their immobilization in the mycelial matrix: a novel biological approach to nanoparticle synthesis. Nano Lett. 1, 515–519.

Naik, M.M., Prabhu, M.S., Samant, S.N., Naik, P.M., Shirodhkar, S., 2017. Synergistic action of silver nanoparticles synthesized from silver resistant estuarine *Pseudomonas aeruginosa* strain SN5 with antibiotics against antibiotic resistant bacterial human pathogens. Thalassas 33, 73–80.

Nakazato, G., Goncalves, M.C., Neves, M.S., Kobayashi, R.K.T., Brocchi, M., Duran, N., 2019. Violacein@biogenic Ag system: synergistic antibacterial activity against *Staphylococcus aureus*. Biotechnol. Lett. 41, 1433–1437.

Nallanthighal, S., Chan, C., Murray, T.M., Mosier, A.P., Cady, N.C., Reliene, R., 2017. Differential effects of silver nanoparticles on DNA damage and DNA repair gene expression in Ogg1-deficient and wild type mice. Nanotoxicology 11, 996–1011.

Neethu, S., Midhun, S.J., Radhakrishnan, E.K., Jyothis, M., 2018. Green synthesized silver nanoparticles by marine endophytic fungus *Penicillium polonicum* and its antibacterial efficacy against biofilm forming, multidrug-resistant *Acinetobacter baumanii*. Microb. Pathog. 116, 263–272.

Nikaido, H., 2009. Multidrug resistance in bacteria. Annu. Rev. Biochem. 78, 119–146.

Ogundare, S.A., van Zyl, W.E., 2018. Nanocrystalline cellulose as reducing- and stabilizing agent in the synthesis of silver nanoparticles: application as a surface-enhanced raman scattering (SERS) substrate. Surf. Interface 13, 1–10.

Onodera, A., Nishiumi, F., Kakiguchi, K., Tanaka, A., Tanabe, N., Honma, A., Yayama, K., Yoshioka, Y., Nakahira, K., Yonemura, S., Yanagihara, I., Tsutsumi, Y., Kawai, Y., 2015. Short-term changes in intracellular ROS localisation after the silver nanoparticles exposure depending on particle size. Toxicol. Rep. 2, 574–579.

Padalia, H., Moteria, P., Chanda, S., 2015. Green synthesis of silver nanoparticles from marigold flower and its synergistic antimicrobial potential. Arab. J. Chem. 8, 732–741.

Pal, I., Bhattacharyya, D., Kar, R.K., Zarena, D., Bhunia, A., Atreya, H.S., 2019. A peptide-nanoparticle system with improved efficacy against multidrug resistant bacteria. Sci. Rep. 9, 4485.

Palma, R.D., Peeters, S., van Bael, M.J., van del Rul, H., Bonroy, K., Laureyn, W., Mullens, J., Borghs, G., Maes, G., 2007. Silane ligand exchange to make hydrophobic superparamagnetic nanoparticles water-dispersible. Chem. Mater. 19, 1821–1831.

Panacek, A., Smekalova, M., Kilianova, M., Prucek, R., Bogdanova, K., Vecerova, R., Kolar, M., Havrdova, M., Plaza, G.A., Chojniak, J., Zboril, R., Kvitek, L., 2016a. Strong and nonspecific synergistic antibacterial efficiency of antibiotics combined with silver nanoparticles at very low concentrations showing no cytotoxic effect. Molecules 21, 26.

Panacek, A., Smekalova, M., Vecerova, R., Bogdanova, K., Vecerova, R., Kilianova, M., Froning, J.P., Zboril, R., Kvitek, L., Prucek, R., 2016b. Silver nanoparticles strongly enhance and restore bactericidal activity of inactive antibiotics against multiresistant *Enterobacteriaceae*. Colloid Surf. B Biointerfaces 142, 392–399.

Passos, M.L.C., Costa, D., Lima, J.L.F.C., Saraiva, M.L.M.F.S., 2015. Sequential injection technique as a tool for the automatic synthesis of silver nanoparticles in a greener way. Talanta 133, 45–51.

Patel, K., Bharatiya, B., Mukherjee, T., Soni, T., Shukla, A., Suhagia, B.N., 2017. Role of stabilizing agents in the formation of stable silver nanoparticles in aqueous solution: characterization and stability study. J. Dispers. Sci. Technol. 38, 626–631.

Patra, J.K., Baek, K.-H., 2017. Antibacterial activity and synergistic antibacterial potential of biosynthesized silver nanoparticles against foodborne pathogenic bacteria along with its anticandidal and antioxidant effects. Front. Microbiol. 8, 167.

Pawar, J.S., Patil, R.H., 2020. Green synthesis of silver nanoparticles using *Eulophia herbacea* (Lindl.) tuber extract and evaluation of its biological and catalytic activity. SN Appl. Sci. 2, 52.

Phan, C.M., Nguyen, H.M., 2017. Role of capping agent in wet synthesis of nanoparticles. J. Phys. Chem. A 121, 3213–3219.

Phanjom, P., Ahmed, G., 2017. Effect of different physicochemical conditions on the synthesis of silver nanoparticles using fungal cell filtrate of *Aspergillus oryzae* (MTCC no. 1846) and their antibacterial effect. Adv. Nat. Sci. Nanosci. Nanotechnol. 8, 045016.

Prabhu, S., Poulose, E.K., 2012. Silver nanoparticles: mechanism of antimicrobial action, synthesis, medical applications, and toxicity effects. Int. Nano. Lett. 2, 32.

Prasher, P., Singh, M., Mudila, H., 2018a. Green synthesis of silver nanoparticles and their antifungal properties. BioNanoScience 8, 254–263.

Prasher, P., Singh, M., Mudila, H., 2018b. Oligodynamic effect of silver nanoparticles: a review. BioNanoScience 8, 951–962.

Prasher, P., Singh, M., Mudila, H., 2018c. Silver nanoparticles as antimicrobial therapeutics: current perspectives and future challenges. 3 Biotech 8, 411.

Prema, P., Thangapandiyan, S., Immanuel, G., 2017. CMC stabilized nano silver synthesis, characterization and its antibacterial and synergistic effect with broad-spectrum antibiotics. Carbohydr. Polym. 158, 141–148.

Qaralleh, H., Khleifat, K.M., Al-Limoun, M.O., Alzedaneen, F.Y., Al-Tawarah, N., 2019. Antibacterial and synergistic effect of biosynthesized silver nanoparticles using the fungi *Tritirachium oryzae* W5H with essential oil of *Centaurea damascena* to enhance conventional antibiotics activity. Adv. Nat. Sci. Nanosci. Nanotechnol. 10, 2.

Qing, Y., Cheng, L., Li, R., Liu, G., Zhang, Y., Tang, X., Wang, J., Le, H., Qin, Y., 2018. Potential antibacterial mechanism of silver nanoparticles and the optimization of orthopedic implants by advanced modification technologies. Int. J. Nanomedicine 13, 3311–3327.

Rad, M.S., Pohel, P., 2020. Synthesis of biogenic silver nanoparticles (AgCl-NPs) using a *Pulicaria vulgaris* Gaertn. aerial part extract and their applications as antibacterial, antifungal and antioxidant agents. Nanomaterials (Basel) 10, 638.

Rafinska, K., Pomastowski, P., Buzsewski, B., 2019. Study of *Bacillus subtilis* response to different forms of silver. Sci. Total Environ. 661, 120–129.

Rahman, A.U., Khan, A.U., Yuan, Q., Wei, Y., Ahmad, A., Ullah, S., Khan, Z.U.H., Shams, S., Tariq, M., Ahmad, W., 2019. Tuber extract of *Arisaema flavum* eco-benignly and effectively synthesize silver nanoparticles: photocatalytic and antibacterial response against multidrug resistant engineered *E. coli* QH4. J. Photochem. Photobiol. B 193, 31–38.

Rastogi, L., Kora, A.J., Sashidhar, R.B., 2015. Antibacterial effects of gum kondagogu reduced/stabilized silver nanaoparticles in combination with various antibiotics: a mechanistic approach. Appl. Nanosci. 5, 535–543.

Ravindran, A., Chandran, P., Khan, S.S., 2013. Biofunctionalized silver nanoparticles: advances and prospects. Colloid Surf. B: Biointerfaces 105, 342–352.

Ribeiro, A.P.C., Anbu, S., Alegria, E.C.B.A., Fernandes, A.R., Baptista, P.V., Mende, R., Matias, A.S., Mendes, M., da Silva, M.F.C., Pombeiro, A.J.L., 2018. Evaluation of cell toxicity and DNA and protein binding of green synthesized silver nanoparticles. Biomed. Pharmacother. 101, 137–144.

Richard, B., Lemyre, J.-L., Ritcey, A.M., 2017. Nanoparticle size control in microemulsion synthesis. Langmuir 33, 4748–4757.

Rinaldi, F., Del Favero, E., Moeller, J., Henieh, P.N., Passeri, D., Rossi, M., Angeloni, L., Venditti, I., Marianecci, C., Carafa, M., Fratodi, I., 2019. Hydrophilic silver nanoparticles loaded into niosomes: physical–chemical characterization in view of biological applications. Nanomaterials (Basel) 9, 1177.

Roy, A., Bulut, O., Some, S., Mandal, A.K., Yilmaz, M.D., 2019. Green synthesis of silver nanoparticles: biomolecule-nanoparticle organizations targeting antimicrobial activity. RSC Adv. 9, 2673–2702.

Saaho, C.R., Maharana, S., Mandhata, C.P., Bishoyi, A.K., Paidesetty, S.K., Padhy, R.N., 2020. Biogenic silver nanoparticle synthesis with cyanobacterium *Chroococcus minutus* isolated from Baliharachandi Sea-mouth, Odisha, and in vitro antibacterial activity. Saudi J. Biol. Sci. https://doi.org/10.1016/j.sjbs.2020.03.020.

Saha, S., Gupta, B., Gupta, K., Chaudhuri, M.G., 2016. Production of putrescine-capped stable silver nanoparticle: its characterization and antibacterial activity against multidrug-resistant bacterial strains. Appl. Nanosci. 6, 1137–1147.

Salarian, A.A., Mollamahale, Y.B., Hami, Z., Rad, M.S.R., 2017. Cephalexin nanoparticles: synthesis, cytotoxicity and their synergistic antibacterial study in combination with silver nanoparticles. Mater. Chem. Phys. 198, 125–130.

Salman, M., Rizwana, R., Khan, H., Munir, I., Humayun, M., Iqbal, A., Rehman, A., Amin, K., Ahmed, G., Khan, M., Khan, A., Amin, F.U., 2019. Synergistic effect of silver nanoparticles and polymyxin B against biofilm produced by *Pseudomonas aeruginosa* isolates of pus samples in vitro. Artif. Cells Nanomed. Biotechnol. 47, 2465–2472.

Santos, I.P., Marzan, L.M.L., 2009. N,N-dimethylformamide as a reaction medium for metal nanoparticle synthesis. Adv. Funct. Mater. 19, 679–688.

Sanyasi, S., Majhi, R.K., Kumar, S., Mishra, M., Ghosh, A., Suar, M., Satyam, P.V., Mohapatra, H., Goswami, C., Goswami, L., 2016. Polysaccharide-capped silver nanoparticles inhibit biofilm formation and eliminate multi-drug-resistant bacteria by disrupting bacterial cytoskeleton with reduced cytotoxicity towards mammalian cells. Sci. Rep. 6, 24929.

Saratale, G.D., Saratale, R.G., Benelli, G., Kumar, G., Pugazhendi, A., Kim, D.-S., Shin, H.-S., 2017. Anti-diabetic potential of silver nanoparticles synthesized with *Argyreia nervosa* leaf extract high synergistic antibacterial activity with standard antibiotics against foodborne bacteria. J. Clust. Sci. 28, 1709–1727.

Shanmuganathan, R., Ali, D.M., Prabhakar, D., Muthukumar, H., Thajuddin, N., Kumar, S.S., Pughazhendi, A., 2018. An enhancement of antimicrobial efficacy of biogenic and ceftriaxone-conjugated silver nanoparticles: green approach. Environ. Sci. Pollut. Res. 25, 10362–10370.

Siegel, J., Kvitek, O., Ulbrich, P., Kolska, Z., Slepicka, P., Svorcik, V., 2012. Progressive approach for metal nanoparticle synthesis. Mater. Lett. 89, 47–50.

Singh, M., Prasher, P., 2018. Ultrafine silver nanoparticles: synthesis and biocidal studies. BioNanoScience 8, 735–741.

Singh, P., Kim, Y.J., Singh, H., Wang, C., Hwang, K.H., El-Agamy Farh, M., Yang, D.C., 2015. Biosynthesis, characterization and antimicrobial applications of silver nanoparticles. Int. J. Nanomedicine 10, 2567–2577.

Singh, H., Du, J., Singh, P., Yi, T.H., 2018. Extracellular synthesis of silver nanoparticles by Pseudomonas sp. THG-LS1.4 and their antimicrobial application. J. Pharm. Anal. 8, 258–264.

Slavin, Y.N., Asnis, J., Hafeli, U.O., Bach, H., 2017. Metal nanoparticles: understanding the mechanisms behind antibacterial activity. J. Nanobiotechnol. 15, 65.

Smekalova, M., Aragon, V., Panacek, A., Prucek, R., Zboril, R., 2016. Enhanced antibacterial effect of antibiotics in combination with silver nanoparticles against animal pathogens. Vet. J. 209, 174–179.

Soliwoda, K.R., Tomaszewska, E., Socha, E., Krzyczmonik, P., Ignaczak, A., Orlowski, P., Krzyzowska, M., Celichowsky, G., Grobelny, J., 2017. The role of tannic acid and sodium citrate in the synthesis of silver nanoparticles. J. Nanopart. Res. 19, 273.

Solomon, M.M., Umoren, S.A., 2015. Performance assessment of poly (methacrylic acid)/silver nanoparticles composite as corrosion inhibitor for aluminium in acidic environment. J. Adhes. Sci. Technol. 29, 2311–2333.

Song, K.C., Lee, S.M., Park, T.S., Lee, B.S., 2009. Preparation of colloidal silver nanoparticles by chemical reduction method. Korean J. Chem. Eng. 26, 153–155.

Sportelli, M.C., Izzi, M., Volpe, A., Clemente, M., Picca, R.A., Ancona, A., Lugara, P.M., Palazzo, G., Cioffi, N., 2018. The pros and cons of the use of laser ablation synthesis for the production of silver nano-antimicrobials. Antibiotics 7, 67.

Sridhar, R., Ramakrishna, S., 2013. Electrosprayed nanoparticles for drug delivery and pharmaceutical applications. Biomatter 3, 24281.

Sun, D., Zhang, W., Li, N., Zhao, Z., Mou, Z., Yang, E., Wang, W., 2016. Silver nanoparticles-quercetin conjugation to siRNA against drug-resistant *Bacillus subtilis* for effective gene silencing: in vitro and in vivo. Mater. Sci. Eng. C 63, 522–534.

Tan, S., Erol, M., Attygalle, A., Du, H., Sukhishvili, S., 2007. Synthesis of positively charged silver nanoparticles via photoreduction of $AgNO_3$ in branched polyethyleneimine/HEPES solutions. Langmuir 23, 9836–9843.

Teimoori, B.B., Pourianfar, H.R., Akhlaghi, M., Tanhaeian, A., Rezayi, M., 2019. Biosynthesis and antibiotic activity of silver nanoparticles using different sources: glass industrial sewage-adapted *Bacillus sp.* and herbaceous *Amaranthus sp.* Biotechnol. Appl. Biochem. 66, 900–910.

Tippayawat, P., Sapa, V., Srijampa, S., Boueroy, P., Chompoosor, A., 2017. D-maltose coated silver nanoparticles and their synergistic effect in combination with ampicillin. Monatsh. Chem. 148, 1197–1203.

Tiwari, V., Tiwari, M., Solanki, V., 2017. Polyvinylpyrrolidone-capped silver nanoparticle inhibits infection of carbapenem-resistant strain of *Acinetobacter baumannii* in the human pulmonary epithelial cell. Front. Immunol. 8, 973.

Vinoth, V., Wu, J.J., Asiri, A.M., Anandan, S., 2017. Sonochemical synthesis of silver nanoparticles anchored reduced graphene oxide nanosheets for selective and sensitive detection of glutathione. Ultrason. Sonochem. 39, 363–373.

Wan, G., Ruan, L., Yin, Y., Yang, T., Ge, M., Cheng, X., 2016. Effects of silver nanoparticles in combination with antibiotics on the resistant bacteria *Acinetobacter baumannii*. Int. J. Nanomedicine 11, 3789–3800.

WHO, 2014. Antimicrobial Resistance Global Report on Surveillance. WHO.
Wise, R., 2002. Antimicrobial resistance: priorities for action. J. Antimicrob. Chemother. 49, 585–586.
Wypij, M., Swiecimska, M., Czarnecka, J., Dahm, H., Rai, M., Golinska, P., 2018a. Antimicrobial and cytotoxic activity of silver nanoparticles synthesized from two *haloalkaliphilic actinobacterial* strains alone and in combination with antibiotics. J. Appl. Microbiol. 124, 1411–1424.
Wypij, M., Czarnecka, J., Swiecimska, M., Dahm, H., Rai, M., Golinska, P., 2018b. Synthesis, characterization and evaluation of antimicrobial and cytotoxic activities of biogenic silver nanoparticles synthesized from *Streptomyces xinghaiensis* OF1 strain. World J. Microbiol. Biotechnol. 34, 23.
Yan, X., He, B., Liu, L., Qu, G., Shi, J., Hu, L., Jiang, G., 2018. Antibacterial mechanism of silver nanoparticles in *Pseudomonas aeruginosa*: proteomics approach. Metallomics 10, 557–564.
Yuan, Y.-G., Peng, Q.-L., Gurunathan, S., 2017. Effects of silver nanoparticles on multiple drug-resistant strains of *Staphylococcus aureus* and *Pseudomonas aeruginosa* from mastitis-infected goats: an alternative approach for antimicrobial therapy. Int. J. Mol. Sci. 18, 569.
Zhang, Z., Shen, W., Xue, J., Liu, Y., Yan, P., Liu, J., Tang, J., 2018a. Recent advances in synthetic methods and applications of silver nanostructures. Nanoscale Res. Lett. 13, 54.
Zhang, L., Wu, L., Si, Y., Shu, K., 2018b. Size-dependent cytotoxicity of silver nanoparticles to *Azotobacter vinelandii*: growth inhibition, cell injury, oxidative stress and internalization. PLoS One 13, e0209020.
Zhao, X., Zhou, L., Riaz, R.M.S., Yan, L., Jiang, C., Shao, D., Zhu, J., Shi, J., Huang, Q., Yang, H., Jin, M., 2018. Fungal silver nanoparticles: synthesis, application and challenges. Crit. Rev. Biotechnol. 38, 817–835.
Zhou, W., Jia, Z., Xiong, P., Yan, J., Li, Y., Li, M., Cheng, Y., Zheng, Y., 2017. Bioinspired and biomimetic AgNPs/gentamicin-embedded silk fibroin coatings for robust antibacterial and osteogenetic applications. ACS Appl. Mater. Interfaces 9, 25830–25846.

CHAPTER 4

Silver-based nanoantimicrobials: Mechanisms, ecosafety, and future perspectives

Parinaz Ghadam[a], Parisa Mohammadi[b,c], and Ahya Abdi Ali[b,c]
[a]*Department of Biotechnology, Faculty of Biological Sciences, Alzahra University, Tehran, Iran*
[b]*Department of Microbiology, Faculty of Biological Sciences, Alzahra University, Tehran, Iran*
[c]*Research Center for Applied Microbiology and Microbial Biotechnology, Alzahra University, Tehran, Iran*

4.1 Introduction

Nanomaterials are substances with a size of fewer than one hundred nanometers in at least one dimension. They are classified into four groups: zero-dimensional (nanoparticles), one-dimensional (nanotubes), two-dimensional (nanoplates), and three-dimensional (nanocomposites) (Alagarasi, 2013).

The characteristics of nanoparticles, e.g., electrical, optical, thermal, mechanical, and chemical reactivity properties, vary with bulk types because of their ratio of surface area to volume increases (Parikh et al., 2008).

The use of AgNPs is highly beneficial due to their conductivity, chemical resistance, antibacterial, antiviral, antifungal, antiangiogenic, and antiinflammatory properties (Deepak et al., 2011). AgNPs are classified into Ag^0 (Janata, 2003), Ag_2S, $AgO/Ag_2O/Ag_2O_3/Ag_3O_4/Ag_4O_4$ (Levard et al., 2012), AgCl (Levard et al., 2013), Ag_3PO_4 (Rahimi et al., 2017), and AgBr and AgI (Holade et al., 2016). The net charge in solution and pH influence the distribution of ion around the nanoparticles (Paseban et al., 2019; Sabbah et al., 2016).

In ancient times, Romans, Greeks, Egyptians, and other nations used silver for their food and drink protection (Alexander, 2009). Silver has also long been employed to prevent burnt-skin infection (Klasen, 2000a, b). In the first half of the twentieth century, before the introduction of antibiotics, silver was primarily used as a colloidal aqueous dispersion for oral application and the treatment of infections. Herodotus has already described silver's antibacterial activity (Alexander, 2009), and this has been linked to silver ions since the 19th century (Lansdown, 2006, 2010). These days, with several resistant bacterial strains to antibiotics, silver in the form of ionic or nanoparticles is used as a disinfectant in many cases as a coating. Applying

Ag as a microbiocidal agent involves oxidation Ag^0 to the Ag^+, which is a slow process under normal circumstances and results in small silver concentrations. Silver salts have also been used for disinfection since the 1st century BC (Alexander, 2009). A common example of that is silver nitrate used as early as 69 BC in medicine (Alexander, 2009). Since the Middle Ages, when the "lunar caustic" (Latin: Lapis infernalis) was used in solid shape to treat inflammations and warts (Klasen, 2000b). Silver nitrate's extreme solubility results in a large concentration of local silver and often kills bacteria and affects the underlying tissue. In interaction with water, metallic silver, AgNPs, and sparsely soluble silver salts expel silver ions (Chernousova and Epple, 2013).

There are two nanoparticle synthesis strategies: top-down and bottom-up. In the first strategy, the bulk materials are transited to nanoparticles using physical and chemical processes, while the atoms and molecules aggregate and create nanoparticles in the second strategy (Prema, 2011). The synthesis of nanoparticles is also divided into two categories: green and nongreen. The green method consists of applying biomaterials to synthesize nanoparticles.

The purity, monodispersity, and yield of chemical and physical synthesis of nanoparticles are better than their biosynthesis (Ghorbani et al., 2011; Ingale and Chaudhari, 2013), but they have certain drawbacks because some of the used materials in chemical synthesis are toxic and flammable (Sahayaraj and Rajesh, 2011), and physical methods consume high energy due to their need for extreme temperature and pressure (Ghorbani et al., 2011). Biological synthesis of nanoparticles is economical, and environmental-friendly (Narayanan and Sakthivel, 2010). Moreover, synthesis and capping of chemically and physically synthesized nanoparticles are carried out in a two-step process, but they occur in one step in the biological method because of both producer and capping agents' presence in the biomass (Fig. 4.1) (Abbasi et al., 2017; Deepak et al., 2011; Feizi et al., 2018; Jain et al., 2011; Mousavi et al., 2020; Singh et al., 2015; Somee et al., 2018; Thakkar et al., 2010).

In this chapter, we try to explain antimicrobial mechanisms of AgNPs toxicity and their application in packaging food, fabrics, air/water decontamination, and monument protection.

4.2 AgNPs as nano-antimicrobials (antibacterial, antifungal, and antiviral)

4.2.1 Antibacterial activity of AgNPs

AgNPs are well-known for their excellent antimicrobial effects against numerous organisms, including bacteria, fungi, and viruses. Yet, the pathways responsible for AgNP's bactericidal activity have remained unclear (Aletayeb et al., 2020; Feizi et al., 2018; Liao et al., 2019).

AgNPs toxicity depends on the size, pH, and concentration, exposure period to pathogens, the form of AgNPs, and their capping agents. Organic groups such as carbonyl and protein within the bacterial plasma membrane are electron donors;

4.2 AgNPs as nano-antimicrobials (antibacterial, antifungal, and antiviral)

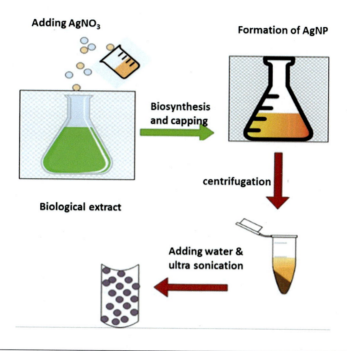

FIG. 4.1

The scheme of AgNP biosynthesis.

therefore, they are unable to produce silver ions from silver atoms; nevertheless, Ag^+ ions are produced, confirming the existence of a chemical agent (Ma et al., 2011). Some researchers have shown that contact between the positive charges on some AgNPs and the negative charge on the microorganism membrane is essential to microorganism growth inhibition (Dibrov et al., 2002; Siddiqi et al., 2018).

Direct contact of AgNPs with large surfaces on the cell wall of bacteria can result in damage to the membrane, causing leakage of cell contents and death of cells (Barros et al., 2018; Sondi and Salopek-Sondi, 2004). Specifically, AgNPs smaller than 10 nm are more toxic to bacteria. The AgNPs can associate with biomolecules such as proteins, lipids, and DNA by penetrating the cytoplasm. In some instances, AgNPs can interact with the respiratory system enzymes, producing reactive oxygen species (ROS) such as hydrogen peroxide (H_2O_2), hydroxyl (OH^-), and superoxide (O^{2-}) radicals that make oxidative stress and damage biomolecules (Gurunathan et al., 2018). Earlier research showed that the building of the AgNPs (12 nm) on the *E. coli* cell wall ends in the creation of pits (Sondi and Salopek-Sondi, 2004), which cause a loss of the cohesion of the outer membrane, leading to the leakage of molecules and membrane proteins from lipopolysaccharides (LPS), and eventual death. Morones et al. (2005) reported anchoring the AgNPs (1–10 nm) to the *E. coli* cell wall and disrupting its normal function such as respiration and permeability. The nanoparticles also enter the cytoplasm and interact with the protein and DNA leading to cell death (Morones et al., 2005). Gahlawat et al. have recently shown that

AgNPs (10nm) bind to the *Vibrio cholera* cell membrane; thus, they interrupt the cell's permeability and metabolic pathways and induce cell death (Gahlawat et al., 2016). On the contrary, AgNPs' cytotoxic impact on bacteria could be attributed to chemically oxidize of Ag to give Ag^+ in aqueous solutions (Behra et al., 2013; Tjong et al., 1982; Tjong and Yeager, 1981). In this respect, AgNPs are often oxidized to yield Ag^+ ions in aerated solutions (Chernousova and Epple, 2013). Xiu et al. demonstrated that the Ag^+ produced by AgNPs are responsible for the antibacterial activity in aerobic conditions (Xiu et al., 2011). Small AgNPs (5nm) can release more Ag^+ ions under aerobic conditions than larger AgNPs (11nm), due to their higher surface to volume ratio. On the contrary, low Ag^+ are produced in an anaerobic environment, and AgNPs are nontoxic to bacteria. The antimicrobial action of Ag^+ ions is also related to their association with thiol groups. Ag^+ may, therefore, react to thiol groups of cell wall-bound enzymes and proteins that interact with the bacteria's respiratory chain and break the bacterial cell wall. Also, those ions released from AgNPs can penetrate the cytoplasm; as a result, they degrade DNA (Hsueh et al., 2015) or react with thiol groups of cytoplasmic proteins. As a consequence, DNA lacks the capacity to replicate; therefore, the proteins necessary for ATP output are inactivated. Smaller silver particles can usually get more easily into the cytoplasm than larger particles (Khalandi et al., 2017). In addition, Ivask et al. showed that positively charged ions released from AgNPs tend to interfere with the normal function of the *E. coli* bacterial electron transport chain, thus facilitating the formation of ROS (Ivask et al., 2014). Generation of ROS is usually responsible for bacterial death as it facilitates lipid peroxidation but inhibits the production of ATP and replication of DNA (Quinteros et al., 2016).

All these results can be described as an initial attachment of AgNPs or Ag^+ ion to the bacterial cell wall, accompanied by ROS and free radical production, DNA damage, and protein denaturation (Liao et al., 2019).

Fig. 4.2 shows a schematic representation of bactericidal effects or the degradation of these two effects due to AgNPs mediated membrane destruction and Ag^+ release from the nanoparticles.

4.2.1.1 Antibiofilm activity of AgNPs

The biofilm of bacteria has a serious medical problem. Due to the growing ineffectiveness of traditional antibiotics, various alternative approaches are considered to counter bacterial biofilms and one of them is the usage of the NPs (Feizi et al., 2018; Mousavi et al., 2020; Somee et al., 2018). Recently, AgNPs have received greater attention for their antimicrobial effects and potential clinical applications. The strong antibiofilm effect of AgNPs is, nonetheless, undeniable. Several studies have shown inhibition of the formation of in vitro biofilms by a range of bacterial species at specific NP concentrations (Markowska et al., 2013). AgNPs (with an average diameter of 25.5 ± 4 nm) have been shown to effectively prevent the development of *Pseudomonas aeruginosa* biofilms and destroy bacteria in the biofilm structures, indicating that they may be used to prevent and treat biofilm-related infections (Martinez-Gutierrez et al., 2013; Somee et al., 2018). Another recent research

4.2 AgNPs as nano-antimicrobials (antibacterial, antifungal, and antiviral)

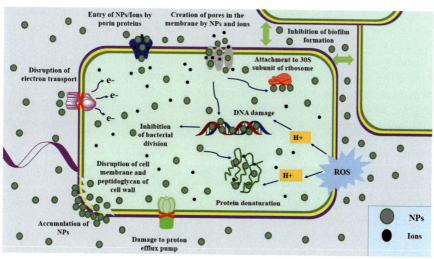

FIG. 4.2

Different mechanisms of AgNPs action in bacterial cells, combating MDR bacteria and preventing the formation of biofilms.

has revealed that AgNPs with an average diameter of 12.6 ± 5.7 nm are also successful against biofilm *Mycobacterium* spp. Compared to other antimicrobial strategies, prevention of microbial adhesion and development of biofilms including silver ions or silver impregnation is more prolonged. Therefore, the impregnation of medical instruments with AgNPs prevents both the external and internal surfaces of the products and the continual regeneration of silver ions (Darouiche, 1999; Wilcox et al., 1998). Consequently, AgNPs coating or impregnated on the medical system will appear as a new generation of antibiofilm substances (Chen and Schluesener, 2008). While research has been conducted on AgNPs biosafety, several research studies indicated the effectiveness of AgNPs in reducing or preventing the production of biofilms on catheter materials in vitro (Samuel and Guggenbichler, 2004) as well as in animal models (Hsu et al., 2010; Roe et al., 2008). Studies are also scant on individual models. The toxicity of NPs of eukaryotic cells is, however, a major problem and remains uncharacterized. A mechanism for mitigating this potential disadvantage can be based on the accumulation of concentrated AgNPs at the geographic site of an infection. Although changes in the way AgNPs are integrated into medical devices could increase their efficacy and decrease any side effects, considerable efforts are still required to complete this technology (Martinez-Gutierrez et al., 2013). Some samples of antimicrobial catheters, implants, and wound dressings containing AgNPs are available for clinical use, but sophisticated antibiofilm strategies are still underdeveloped, and studies focused mainly on in vitro research studies, and only two products are currently used in clinical trials: Arikace TM (Phase III) and fluidosom (Phase II) for the treatment of respiratory tract infections associated with cystic fibrosis (Liu et al., 2019).

It is necessary to bear in mind the need for rigorous in vivo trials to assess the efficacy of these emerging developments within the biomedical background. Specificity may be essential in clinical applications, especially where it is required to make a distinction between pathogenic bacteria and microbiota and host tissue (Fulaz et al., 2019). AgNPs' antibiofilm behavior against the studied strains indicates a less substantial impact on gram-positive bacteria's biofilm development than the gram-negative bacteria's biofilm. This discovery can also be derived from structural variations in the composition of these bacteria cell wall, while planktonic bacteria, which have been documented as the belonging of both groups, exhibit equal mortality to AgNPs (Fayaz et al., 2010). Another possible mechanism reducing the formation of biofilms by AgNPs (Ag^0 metallic and Ag^+) is the interference or inhibition of external polymeric substances (EPS) control. With AgNP, EPS is assumed to provide physical defense for bacteria. When bacterial biofilms were removed from loosely bound EPS, they were more sensitive to AgNPs.

Given the dynamic physicochemical structure of the EPS, a multitude of various associations influence the transmission of NPs inside the biofilm matrix (Fulaz et al., 2019). For example, the assembly of EPS in *P. aeruginosa* biofilm is regulated by a gene cluster of 15 genes, and a potential protein inhibition by AgNPs impairs EPS synthesis. A reactive binding of Ag^+ to the cysteine residues found in proteins may also clarify the process for deactivating proteins (Martinez-Gutierrez et al., 2013); hence, AgNPs will be embedded in the material used in the manufacture of medical devices to prevent the adherence, colonization, and formation of microbial biofilms on these devices' surfaces. Achieving this technology in the therapeutic and preventive approaches calls for more attempts (Martinez-Gutierrez et al., 2013).

4.2.2 Antifungal activity of AgNPs

Several factors can be considered as the reason for pathogenicity of most fungal diseases such as the formation of the germ tube, tissue damage due to the extracellular enzymes such as proteinase, phospholipase, lipase, and hemolysin, which facilitate the adherence and penetration of fungi, and alteration of hyphal morphogenesis. It was reported that the AgNPs can inhibit the production of different kinds of extracellular hydrolytic enzymes as well as germ tube formation (Haji Esmaeil Hajjar et al., 2015; Hamid et al., 2018; Jalal et al., 2019). Also, AgNPs influence the biosynthesis of organic acid in some fungi via reducing the amount of citric and oxalic acid, in particular, and alter the profile of extracellular enzyme, although total enzymatic activity is increased (Pietrzak et al., 2015). Furthermore, AgNPs can disturb the integrity of cell wall and cell regulatory mechanisms, resulting in pore formation and cell collapse. Consequently, AgNPs interfere with ion transport, change the osmotic balance, and embed the protein into the plasma membrane, resulting in the macromolecules release, leading to cell death (Yun et al., 2015). It was reported that in pathogenic cells exposed to the AgNPs, a significantly large number of ROS was generated; subsequently, more H_2O_2 was converted by superoxide dismutase. H_2O_2 produces more OH radicals, which is very reactive. OH radicals can indiscriminately interact with DNA, protein, and fatty acids. In addition, it was shown that in eukaryotic cells such as fungi, AgNPs can disrupt mitochondria, decrease membrane integrity, and release cytochromes and hydrogenase. Cytochrome C, which is an apoptosis-inducing factor and its release to the cytosol, causes the degradation and depolymerizations of mitochondrial membranes, resulting in the fragmentation of mitochondria (Eisenberg et al., 2007; Halliwell and Aruoma, 1991; Perrone et al., 2008; Wong et al., 2017). Besides, in the presence of ROS, cytochrome C, which is bonded to the mitochondrial membrane, is detached and extruded to the cytosol, and it can result in the apoptosis process and cell death after some steps (Hafez and Kabeil, 2013). At this stage of apoptosis, not only DNA is going to fragment, but also cells' nucleus shrinks and subsequently, the growth of fungi is inhibited with a lower amount of sterol (Gupta and Chauhan, 2016). Furthermore, it was reported that fungal DNA loses its ability to replicate after treatment with Ag ions (Elgorban et al., 2016; Kim et al., 2012a).

Moreover, genes coding for protein and enzyme functions involved in energy reactions are affected by AgNPs (Gogoi et al., 2006). AgNPs have also been shown to induce an important inhibitory effect of budding process and also mycelial development and sporulation phase (Lee et al., 2010; Pinto et al., 2013). The studies indicate that AgNPs can inhibit secondary metabolite synthesis in several filamentous fungi that produce mycotoxins. Such effects are correlated with growth suppression and thus involve a concentration of sublethal to lethal NPs.

Furthermore, microscopic observations demonstrated that the phenotype of fungi was changed to hyphae shrinkage. Subsequently, AgNPs caused malformation and hypertrophy, leading to spores' damages and decrease in spore numbers. In addition, AgNPs antimicrobial activity may also be linked to altered cell wall, cytoplasm, and membrane permeability (Manjumeena et al., 2014). SEM observation by metal nanoparticles of the treated fungal cells shows the disruptions and fissures of the cell walls or membranes (Gajbhiye et al., 2009).

4.2.3 Antiviral activity of AgNPs

Viruses constitute one of the main causes of sickness and death worldwide. Several various infections were eradicated due to vaccination programs such as small pox in 1979 or greatly the disease burden, as in the case of paralytic poliomyelitis (Hull et al., 1994). There are no vaccines for the majority of the most common infectious diseases.

On the other hand, the utilization of vaccine adjuvants and follow-up evaluations are very critical in the vaccination process. For example, silver NPs can be deployed as an adjuvant in mucosal vaccines based on inactivated influenza virus, stimulating the remarkable antigen-specific formation of IgA with low toxicity by stimulating neogenesis of the bronchus-associated lymphoid tissue (BALT) (Nasrollahzadeh et al., 2020; Sanchez-Guzman et al., 2019).

Because of remarkable capacity of viruses to adjust to their current host, migrate to a new host, and evolve methods to evade antiviral action, evolving and re-emerging viruses can be treated as a continuing danger to human health. Due to the emergence of the evolving infectious diseases triggered by multiple pathogenic viruses and the growth of antiviral tolerance to conventional antiviral medications, pharmaceutical firms and several researchers are seeking new antiviral agents like AgNPs which the antiviral properties of it have been reported recently. AgNPs are recorded to associate with glycoproteins on the viral surface and have links to the host cells and then introduce their virus behavior by contact with the viral genome (Galdiero et al., 2014). Furthermore, AgNPs are stated to be virucidal against not only the human immunodeficiency virus (HIV) by interacting with the subunit of gp 120 glycoprotein (Elechiguerra et al., 2005), but also hepatitis B virus by double-strand DNA or viral particle binding (Lu et al., 2008), and a respiratory syncytial virus is caused by viral attachment intervention (Sun et al., 2008).

AgNPs affect HSV-1 by stopping virus attachment and virus entry or by inhibiting virus dissemination (Baram-Pinto et al., 2009). They also have an effect on monkeypox virus by blocking and breaching the host cell nucleus (Rogers et al., 2008). Due to its strong antiviral activity, silver nanoscale materials have attracted interests. The main antiviral function, theoretically correlated with silver nanoscale materials, is the direct contact between the surface proteins of the virus and silver nanoscale materials to prevent the entry of the viruses into cells (Alizadeh and Khodavandi, 2020; Gaikwad et al., 2013; Galdiero et al., 2014; Haggag et al., 2019; Lv et al., 2014).

Remarkably, AgNP-based antiviral agents are highly successful against many viruses such as bovine herpesvirus 1 (El-Mohamady et al., 2018), Coxsackievirus B3 (Shaheen et al., 2019), Chikungunya, human parainfluenza type 3 (Gaikwad and Sasane, 2013), HIV type 1 (Trefry and Wooley, 2013), and influenza A virus (Park et al., 2018). AgNPs can also block viral receptors and/or interact via the viral genome, and particulate viruses inactivated outside the host cells (Khandelwal et al., 2014)

4.3 Toxicity

AgNPs are widely used as antimicrobials in different fields, as dental restorative products including endodontic retro filling resin, dental implants, bedding, washing facilities, purification of the wash, toothpaste, perfume, bottles for breastfeeding, toys, and lifting humidifiers (Chernousova and Epple, 2013). Given the positive advantages of utilizing AgNPs in various systems, the possible safety risks correlated with human use are such nanomaterials are neither well known nor well established. Nanoparticles are a complex object with many characteristics that extend beyond its chemical composition, including particle size, distribution, particle morphology, crystallinity, surface functionality, and charging (Grainger and Castner, 2008; Krug and Wick, 2011; Nel et al., 2009; Oberdörster et al., 2005; Seaton et al., 2010; Stark, 2011). In the case of cytotoxicity, this can rely on the form of nanoparticles used, i.e., their morphology (shape, size), chemical purity, type of solvent correlated with choosing the preparation phase of nanocarriers, method of functionalization and capping agents, as well as their stability and aggregation resistance (Gorjanc et al., 2010). Problems resulting from the usage of nanoparticles contribute to their reduced mobility, their tendency to agglomerate, the possibility of producing metal ions or modifying the composition by oxidation of their surface. This means that smaller AgNPs that are normalized to the same amount of silver are more toxic than larger AgNPs or silver microparticles (Kim et al., 2012b; Liu et al., 2010; Martínez-Castañon et al., 2008; Maurer and Meyer, 2016; Park et al., 2011). Even more complex are assumptions regarding the interaction between particle dynamics and biological impact. Very few research have been conducted in that regard (Pal et al., 2007). Besides releasing silver ions, the kinetics of cellular absorption may play a significant part. Therefore, the agglomeration of biological media should be the bioavailability effects (Kittler et al., 2010; Limbach et al., 2005; Stark, 2011; Teeguarden et al., 2007). Using biological approaches also results in the eventual functionalization of the surface, as the used natural materials have an attraction to the surface of metal nanoparticles, which are reducing agents, stabilizers, or other materials. In addition, cell absorption of AgNPs is depending on the scale, shape, and surface charge. The multiple in vivo experiments have shown that intravenous injection, oral administration (van der Zande et al., 2012), intranasal instillation, and intraperitoneal injection (Daniel et al., 2010) could have been treated and detected in several organs, including the lungs, kidneys, liver, spleen, and brain (Song et al., 2016; Tang et al., 2009). Lung toxicity, including inflammatory reactions and histopathological modifications in animal model trials (Schulte et al., 2019; Song et al., 2013), pulmonary function alterations (Song et al., 2013), or allergic reactions, are recorded to be responsible while further studies could not establish hematological consequences, systemic changes, and improvements in pulmonary function following exposure to AgNPs (Ji et al., 2007; Sung et al., 2011). Possible inhalation therapy for respiratory infections and allergic irritation of the airways has been studied in fine mist form for AgNPs (Davidson et al., 2015).

Additionally, AgNPs inserted into the bloodstream may induce blood-brain barrier disruption, astrocyte swelling, cytoskeletal degradation, pre- and post-synaptic protein modification, mitochondrial dysfunction, and neuronal degeneration (Sharma et al., 2009a, b; Xu et al., 2013, 2015a, b).

AgNPs are correlated with genotoxicity due to DNA and chromosomal disruption, which can induce mutagenicity. Researchers have noted improvements in gene expression affecting xenobiotic metabolism, the production and integrity of motor neurons, molecular intracellular patterns that regulate numerous cellular processes, including anatomy, adhesion, motility, and apoptosis, potentially correlated with neurotoxic and immunotoxic effects of nanoparticles (Dong et al., 2013; Lee et al., 2010). A study that involved mice fed with 13 nm AgNPs demonstrated increases in gene expression, apoptosis, and liver cell inflammation. It has motivated further work into the properties of AgNPs (Hansen and Baun, 2012).

Despite the decade-long history of adding colloidal silver, there is a large volume of possibly toxicologically relevant data with little reported harmful effects, which is therefore probably not appropriate for excessive silver anxiety (Nowack et al., 2011).

Endocytosis and micropinocytosis include eukaryotic, nonphagocytic cells (Luther et al., 2011), and AgNPs (Chernousova and Epple, 2013). Like most nanoparticles, AgNPs are quickly coated with a protein layer, labeled "corona" when transferred into a biological medium (Grainger and Castner, 2008; Hellstrand et al., 2009; Monopoli et al., 2011a, b; Mueller et al., 2010; Rezwan et al., 2004; Treuel et al., 2012). This corona has a direct effect on the interaction between cells (Limbach et al., 2005; Walczyk et al., 2010). Beyond containing sparsely soluble silver salts (AgCl, Ag_2S), silver ions have a detrimental impact, leading to associations with thiol and amino groups of proteins, nucleic acids and cell membranes (Brett, 2006; Choi et al., 2008; Feng et al., 2000; Lansdown, 2010; Powers et al., 2011; Silver et al., 1999). Observations have also demonstrated the development of ROS (Foldbjerg et al., 2011; Liu et al., 2010; Piao et al., 2011; Tiwari et al., 2011; Valodkar et al., 2011), which has been overlooked in many cases. It seems that ROS structure depends on the type of involving cell in the presence of silver (Greulich et al., 2011; Luther et al., 2011; Ma et al., 2011). There are also various possibilities for silver to affect biological systems, and a general inference regarding the sources of the toxicity of silver is not feasible.

The concern is whether increased sensitivity to silver (e.g., in cosmetics) results in intolerance between bacteria and whether small quantities of silver will influence microbiota in the skin or bowels. Because of the coselection process, nanosilver has a strong likelihood of promoting not only silver resistance but also antibiotic resistance. Coselection happens as bacteria affected by one antimicrobial agent discover a gene immune to it by swapping DNA with antimicrobial-resistant bacteria (Baker-Austin et al., 2006). Up till now, environmental concerns and limited silver-resistance data were available. There are at least two issues in terms of biosafety or from an environmental point of view: 1. Do they destroy the bacteria required to break down the pollution? 2. What occurs if they make their way into the rivers and soils as the wastewater sludge is added to the field as fertilizer? (Colman et al., 2013).

4.3.1 Reducing nanoparticles toxicity

Several methods avoid or constrain the harmful consequences of AgNPs. Studies have shown that modifying the size and shape of particles and methods to change their composition will contribute to the development of nanoparticles with the required properties and without toxic effects (Ungor et al., 2019). A typical technique for defensive coating is to use suitable surface modifications to influence the properties of metallic nanoparticles, including reducing their toxic effects. Equally essential is the functionalization of the surface of the nanoparticle by inserting the right functional groups (Nayak et al., 2016).

Compared with spherical- and wire-shaped nanoparticles, the platelet-shaped AgNPs revealed greater toxicity to the fish gill epithelial cell lines (RT-W1). They experienced fewer surface defects than the latter (George et al., 2012).

The primary aim of the usage of coating compounds is to enhance the durability of NPs by avoiding the release of ions from within, avoiding the oxidation of the surface of the nanoparticle, and inhibiting nanoparticles aggregation and eventual agglomeration.

Chitosan loaded with doxorubicin was updated by examined nanoparticles against safe lung cells with cell survival above 80%. The findings of the stability check show that the existence of rubber particles on the surface of the nanoparticle not only acts as a reduction agent but also gives the nanoparticles colloidal stability (Zeng et al., 2014).

Recently, engineered nanoparticle-chitosan (CHI-AgNPs) together with antibacterial, antibiofilm, and anticancer medicines are possible candidate products for biomedical usage. To combine antibacterial activity and toxicity, formulated AgNPs were established in the study of Cinteza et al. Chitosan was used as a capping agent for stabilization and for improving biocompatibility. Biocompatibility of AgNPs and antimicrobial actions against specific bacteria have been evaluated in vitro. The findings showed that chitosan-stabilized AgNPs were nontoxic to regular fibroblasts relative to bare nanoparticles, even at large concentrations, though substantial antibacterial activity was noted. Their findings indicate that the synergistic activity of polymer-stabilized AgNPs and essential oils, with limited adverse effects, may provide a significant utility against a broad form of microorganism. These results show that formulations containing chitosan-stabilized AgNPs and thyme or clove oils can be used as antibacterial materials for medical use because of their enhanced bactericidal properties and increased biocompatibility (Cinteza et al., 2018).

Repeated exposure to AgNPs at subinhibitory concentrations may trigger bacterial resistance to potential diseases. Treatment with a combination of antibiotics and AgNPs has been proposed to address this problem, contributing to dose reduction and its toxicity. In vitro experiments found that AgNPs with different antibiotics showed the synergistic antibacterial activities against both multidrug resistance and virulence of *Enterococcus faecium, Staphylococcus aureus, Klebsiella pneumoniae, Acinetobacter baumannii, P. aeruginosa,* and *Enterobacter* spp. (Golińska et al., 2016; Habash et al., 2017; Panáček et al., 2016; Singh et al., 2018; Wypij et al., 2018). In another study, the synergistic antibacterial activity of AgNPs with polymyxin B was seen in *A. baumannii*-infected mouse model with a survival rate of 60% compared to controls treated with antibiotics or AgNPs alone (Mulani et al., 2019).

AgNPs can also improve traditional antibacterial and antibiofilm activities. Studies identify synergistic action between AgNPs and ampicillin, kanamycin, streptomycin, or vancomycin against *E. coli* and *P. aeruginosa* (Wolska et al., 2012).

Given the usage of AgNPs as a potential therapeutic agent, a literature survey reveals a lack of information from in vivo research used to test AgNPs' toxicity, effectiveness, pharmacokinetic, and immune-modulatory reaction. More work by well-defined experiments and clinical trials contributes to the usage of AgNPs in wound dressings or medical devices (Mulani et al., 2019).

The use of AgNPs as an important antimicrobial agent, particularly after long-term usage, does not induce microbial resistance. Through laboratory usage of therapeutic doses of AgNPs, only very small amounts (less than 0.5 μg/g of an organ) were observed through mice's organs, indicating that AgNPs are healthy at such low concentrations (Wong and Liu, 2010).

For example, the widespread use of AgNPs as a disinfectant ingredient may facilitate the spread of bacterial strains that are silver-resistant. Nevertheless, repeated exposure to AgNPs by bacteria has not contributed to the creation of resistant mutant cells. Being biocides, AgNPs appear to concentrate on several locations on or inside bacterial cells and therefore have broad-spectrum action. In addition, this property of AgNPs means that they can resolve the current mechanisms of microbial drug resistance, including reduced absorption and enhanced efflux from the microbial cell, and biofilm formation (Pelgrift and Friedman, 2013).

So far, there are just a few findings on the human health effects of AgNPs, but several experiments have reported biodistribution and toxicity in rats and mice in vivo. Another recent research reported the dose-dependent effect of AgNPs on the bioactivity of bone-forming cells and the likelihood of human mesenchymal stem cells and human osteoblasts picking up nanoparticles (Pauksch et al., 2014).

The potential side effects of nanoparticles have certainly not been provided adequate consideration; therefore, comprehensive research is desperately needed before therapeutic usage of AgNPs is accepted (Markowska et al., 2013). Given the effectiveness of nanocoatings and AgNPs to avoid the development of biofilms on catheters, a key drawback remains, namely, the capacity of the substance to adsorb still in the same concentration of the medication and the ability to monitor its release, which in most situations results in an unregulated elution of the product within the first hours of injection (Crisante et al., 2009).

There is a strong antifungal synergistic effect of AgNPs when it was coupled with some azoles as shown in several studies. There are some obstacles in the use of antifungals, including those of the emergence of drug-resistant microorganisms and the toxicity of these chemicals that are not friendly and they restrict their application. Two main strategies are available to limit the use of these chemicals. The first is to substitute chemicals with adequate amounts of AgNPs alone, which several studies have noted a significant antimicrobial effect on various pathogens (Khatami et al., 2016; Nejad et al., 2016); the second one is to use a mixture of AgNPs and chemical compounds to help improve the effectiveness of antibiotics (Dar et al., 2013). To minimize the concentration of nanoparticles, alleviate microbial resistance to traditional antibiotics, and contribute to more effective antimicrobial and antimycotoxin action of metal nanoparticles in the treatment of diseases in human beings and animals, AgNP combination therapies with several other conventional antibiotics are greatly required (Hassan et al., 2016).

It is also shown that the application of citrate-coated AgNPs at a defined nonlethal dose could lead to a rise of more than two folds in inactivation of carcinogenic mycotoxin biosynthesis, aflatoxin B1 in the filamentous fungus, and *Aspergillus parasiticus*, without restricting fungal growth (Mitra et al., 2017).

4.4 Applications
4.4.1 AgNPs and packaging food

AgNPs are highly useful nanoparticles and are known as novel substances embedded in a polymer matrix to produce innovative nanocomposite structures in food packaging. Despite their intense antimicrobial action, they are effective antimicrobial agents against a large variety of micro-organisms, including extremely pathogenic gram-positive and gram-negative bacteria and fungi contributing to prolonging food shelf-life. Also, their correlation with reduced impact on the sensory properties of food is among the advantages of utilizing AgNPs of edible packaging. Essentially, films and coatings generated using biopolymers can maintain the integrity and improve the longevity and shelf-life of food items by (a) controlling the movement of moisture, gasses, and fatty acids between food and the outside world, (b) protecting against microbial contamination, and (c) preventing the loss of attractive compounds such as volatile flavors. Therefore, several studies have suggested strengthening the biopolymer matrix by nanoparticles, since they have become a promising alternative to boost the specific functional properties of packaging materials (Kraśniewska et al., 2020; Kraśniewska et al., 2019; Mihindukulasuriya and Lim, 2014; Palza, 2015; Zambrano-Zaragoza et al., 2018). Natural packaging products based on biopolymers are becoming increasingly popular due to their environmental safety, nontoxicity, biodegradability, and biocompatible properties. In addition, biopolymer may be used as a carrier/container of antimicrobials, enzymes, color and flavoring chemicals, vitamins or other nutrients, thereby improving the sensory and nutritious properties of the packaged drugs (Mihindukulasuriya and Lim, 2014; Vodnar et al., 2015).

In addition to the contact between AgNPs and polymer matrices, certain other considerations such as particle size, form, and silver concentration should be regarded to fully exploit the antimicrobial activity of polymer nanocomposites (Chaloupka et al., 2010; Liao et al., 2019).

The polymer synthesis path guarantees a reasonable dispersion of nanoparticles in a polymer matrix, which consequently affects the final structural stability and homogeneity of the nanocomposite film and maintains high antimicrobial properties in nanocomposite film (Islam and Yeum, 2013; Venkatesan et al., 2018; Youssef et al., 2015).

Thus, the preparation of these films is divided into two methods: first, in situ, where the polymer matrix is used as a reaction medium to form AgNPs and works as a stabilizing agent for them; and second, ex situ, where the polymer matrix is primarily used as dispersion and stabilizing medium for AgNPs separately presynthesized (Palza, 2015).

Throughout this situation, polymer matrices function as a stabilizer to avoid agglomeration of nanoparticles, and thus, the nanomaterial exhibits significant antimicrobial action against different bacterial strains (Palza, 2015).

The spread of AgNPs in the polymer matrix as a discontinuous phase decreases water vapor of the investigated film, which affected fewer water molecules through the films. One possible limitation in the use of nanoparticles in food packages is their migration to the food, which could lead to potential toxicity problems (Panea et al., 2014). More studies are also required to decide the effective rates for the inclusion of AgNPs in food packages (Simbine et al., 2019).

4.4.2 Nanocomposite fabrics

Nanocomposite fabrics as known to make inevitable contributions to cytotoxicity and genotoxicity through direct interaction of AgNPs with the human body. In washing machines, mechanical vibration coupled with the high-temperature condition can detach AgNPs from the fabrics. The absence of chemical interactions between the AgNPs and cotton fibers allow AgNPs to bind fragile fabrics. Consequently, during washing cycles, such AgNPs are removed from the fabrics. The composite fabrics' antibacterial activity counteracts all bacterial strains, although decreased by more than 40% after 20 washing cycles. Researchers have made several attempts to enhance the adherence of AgNPs to fabric fibers by plasma deposition, fabric material use, graft polymerization, etc. In recent years, significant attention has been given to the production of antimicrobial composite fabrics because of its attractive applications in healthcare and medicine (Liao et al., 2019; Rivero et al., 2015).

4.4.3 AgNP as a pesticide

Modern nanotechnology has the potential to tackle the various problems of conventional farming. Careless use of pesticides increases pathogen and pest resistance, reduces soil biodiversity, kills useful soil microbes, causes biomagnification of pesticides, decreases pollinators, and destroys the natural habitat of farmers friends such as birds (Duhan et al., 2017; Tilman et al., 2002). The use of antimicrobial properties of AgNPs for plant disease control is of importance (Mishra et al., 2014). Biosynthesized AgNPs were used to rid plants of toxic microorganisms. Specific AgNP concentrations were used, and tests revealed that AgNPs could be used to greatly improve seed germination capacity, mean germination time, seed germination index, seed vigor index, fresh seedling weight, and dry weight. AgNPs are concerned with reducing the burden of pests from crops as pesticides. Antifungal activity of zinc oxide (35–45 nm), silver (20–80 nm), and titanium dioxide (85–100 nm) nanoparticles was tested against *Macrophomina phaseolina*, a large pulse and oilseed pathogen born in soil. The higher antifungal activity in AgNPs was observed at concentrations lower than zinc oxide and titanium dioxide nanoparticles (Shyla et al., 2014). Table 4.1 indicates that AgNPs are active against plant pathogens in a wide variety of microbiocide activities.

4.4.4 Air decontamination

Tran and Le (2018) observed that antimicrobial activity against validated bacterial bioaerosols increased when Ag/CNTs (carbon nanotube) is accumulated on the surface of an air filter media. It can be contrasted with CNT or AgNP deposition alone, while the filter pressure decreased and the filtration capacity of bioaerosols was similar to those of CNT deposition alone (Tran and Le, 2018).

4.4.5 Water sanitization

In several developing countries, drinking water contamination and the consequent spread of waterborne pathogens are the leading cause of death (Pradeep, 2009). In addition to the use of chemically produced nanosilver for water bacterial disinfection, several guidelines supporting the use of biologically created nanosilver

Table 4.1 Application of AgNP as a pesticide.

Application	Pathogens	References
Bactericidal action against plant pathogens and some pathogenic bacteria	*Bacillus cereus, Escherichia coli, Bacillus subtilis, Streptococcus thermophiles, Lactobacillus strains, Corynebacterium* sp., *Klebsiella pneumoniae, Pseudomonas aeruginosa, Salmonella paratyphi, Streptococcus pyogenes, Staphylococcus aureus, Aeromonas hydrophila, Pseudomonas fluorescens,* and *Flavobacterium branchiophilum*	Duhan et al. (2017), Kumar et al. (2017), and Avila-Quezada and Espino-Solis (2019)
Fungicidal action against plant pathogenic fungi	*Bipolaris sorokiniana, Magnaporthe grisea, Alternaria solani, Pythium spinosum, Pythium aphanidermatum, Cylindrocarpon destructans, Cladosporium cucumerinum, Glomerella cingulata, Didymella bryoniae, Stemphylium lycopersici, Monosporascus cannonballus, Fusarium graminearum, Aspergillus niger, Fusarium oxysporum, Curvularia lunata, Rhizopus arrhizus, Alternaria alternata, Penicillium digitatum, Alternaria citri, Candida* spp., *Aspergillus* spp., and *Fusarium* spp.	Duhan et al. (2017), Ibrahim et al. (2020), and Kumar et al. (2017)
Larvicide, herbicide, insecticide and nematicide action against plant pathogen	*Aedes aegypti, Meloidogyne incognita, Eichhornia crassipes,* and *Tribolium castaneum*	Ahmed Hammad and Bahig Ahmed (2018), Duhan et al. (2017), Namasivayam et al. (2014), Waris et al. (2020), and As and Thangapandiyan (2019)

for water virus disinfection (De Gusseme et al., 2011). This demonstrated that the chem-AgNPs formed on these copolymer beads by a chemical reduction process were stable underwater washing, and their stability was attributable to the interaction of the chem-AgNPs on the copolymer beads via the carboxylic group. The chem-AgNPs were also successfully built on the macroporous methacrylic acid copolymer beads to disinfect water (Gangadharan et al., 2010). Also, bio-AgNPs were inserted into polymer membranes made of polyvinylidene fluoride (PVDF) to examine the continuous disinfection of viruses in water (De Gusseme et al., 2011). A new class of hybrid ultrafiltration membranes of polyethersulfone bending with changed halloysite nanotubes charged with chem-AgNPs has recently been documented for water purification (Zhang et al., 2012). The application of silver-based NPs is crucial in preventing outbreaks of waterborne diseases linked to inadequate treatment of drinking water. Such AgNPs-porous ceramic composites have been tested extensively for drinking water purification (Yakub and Soboyejo, 2012). The silver-based NPs may be inserted into the core materials and polymeric membranes to decontaminate the water-infecting bacteria and viruses.

4.4.6 Treatment of multidrug-resistant (MDR) microorganisms

The unnecessary and indiscriminate use of antibiotics together with the decreasing research and development of new antibiotics contributed to a rapid rise in antibiotic resistance. The assembly of novel antibiotics against various life-threatening bacterial pathogens is therefore highly urgent (Griffin et al., 2010; Hall and Barlow, 2004; Hancock and Sahl, 2006; Rossiter et al., 2017).

Nanoparticles have a tremendous potential to overcome this problem by acting as a functional solution to managing multiple infections induced by MDR microorganisms, in particular. AgNPs have antimicrobial activity against a wide variety of microorganisms, as well as resistant microbes, fungi, and viruses (Aletayeb et al., 2020; Blecher et al., 2011; Elechiguerra et al., 2005; Feizi et al., 2018; Huh and Kwon, 2011; Mousavi et al., 2020; Somee et al., 2018).

It is important to notice that AgNPs display greater antimicrobial activity than antibiotics such as gentamicin or vancomycin against *P. aeroginosa* and methicillin-resistant *S. aureus* (Saeb et al., 2014). Lara et al. have demonstrated AgNPs' potential bactericidal effect against MDR *P. aeroginosa*, ampicillin-resistant *E. coli* O157:H7 and erythromycin-resistant *S. pyogenes* (Baptista et al., 2018). Synergistic effect of imipenem and AgNPs against planktonic cells and biofilm was demonstrated a promising candidate for a replacement treatment of infections caused by MDR *A. baumannii* strains. Further investigation should be applied to make a certain of the effectiveness of this nanoparticle in vivo experiments, and to assay probable cytotoxicity of this material (Hendiani et al., 2015).

4.4.7 Antifungal activity and mycotoxin control

Mycotoxin, secondary fungal metabolites, are detrimental to animals or plants secreted by both molds and toxic mushroom. At preharvest and postharvest stages, as well as in processed food products, they produce harmful mycotoxicological risks. Although mycotoxins can provide adequate benefits in the form of biological adaptation, competition with the other microorganisms, and consumer safety, together with many of derivatives are used in medicine, according to Food and Agriculture Organization (FAO); mycotoxins are in charge for contamination of around 25% of the world's food crops (Al-Othman et al., 2014). Lethal amatoxins present in some Amanita mushrooms and ergot alkaloids cause severe epidemics of ergotism in people consuming rye or similarly infected cereals with ergot fungus sclerotia, *Claviceps purpurea* (Collins et al., 2006). Many mycotoxins include aflatoxins, which are hazardous to hepatic toxins and highly carcinogenic metabolites generated by some *Aspergillus* species. In addition, ochratoxins, patulins, trichothecenes, and fumonisins have a dramatic impact on human or livestock food (Veselý and Vesela, 1995).

The genera of *Aspergillus, Fusarium, Alternaria,* and *Penicillium* are the most common toxigenic fungi. It is recognized that mycotoxins such as patulin, PR toxin, roquefortine, and mycophenolic compounds have a high potential to cause animal mycotoxicosis (Xu et al., 2017). The other mycotoxins are deoxynivalenol, nivalenol, citrin, and zearalenone. Although many mycotoxins showed acute toxic effects, especially if consumed at large doses, others would have a detrimental impact after a long time. There are now three approaches to overcome the problem of mycotoxin: fungal growth suppression, adsorption of mycotoxins, and minimization of the impact of mycotoxins on the body (Gacem et al., 2020). Thermo-resistant fungi are also capable of not only developing toxic secondary metabolites but also a wide range of extracellular enzymes. Furthermore, ascospores of many heat-resistant fungi, even in combination with alterations of pH and other extreme conditions, are resistant to conservators. Loading the thermo-resistant fungi after harvest, during fruit and related products processing and storage, along with the heat treatment conditions, affects the level of spoilage of such foods. Moreover, pectinolytic enzymes released by heat-resistant fungi are hydrolytic or they break down the texture, leading to canned fruit softening (Jogee and Rai, 2020). AgNPs were reported to have impacts against pathogenic and toxigenic fungal strains as well as mycotoxin production (Table 4.2).

Table 4.2 Some fungal growth inhibition and aflatoxin reduction reports as affected by AgNPs.

Organisms and mycotoxins	References
Aspergillus flavus, Aflatoxin B1, and *A. nomius*	Ismaiel and Tharwat (2014), Zhao et al. (2017), Gómez et al. (2019), and Bocate et al. (2019)
A. niger and *A. melleus*	Pinto et al. (2013), Yu et al. (2013), Gómez et al. (2019), and Bocate et al. (2019)
A. parasiticus and Aflatoxin B1	Mousavi and Pourtalebi (2015), Gómez et al. (2019), and Bocate et al. (2019)
A. carbonarius, A. ochraceus, A. steynii, A. westerdijkiae, and *Penicillium verrucosum* (OTA)	Gómez et al. (2019) and Bocate et al. (2019)
Aflatoxin B1	Al-Othman et al. (2014) and Deabes et al. (2018)
Aflatoxin B1 and Ochratoxin A	El-Desouky and Ammar (2016) and Khalil et al. (2019)
Penicillium vulpinum and patulin	Ismaiel and Tharwat (2014)
Alternaria brassicicola, A. flavus, A. niger, Fusarium culmorum, F. oxysporum, P. igitatum, Sclerotinia, F.graminearum, F. sporotrichioides, F. langsethiae, F. poae, F. verticillioides, and *F. proliferatum*	Venat et al. (2018), Khalil et al. (2019), and Tarazona et al. (2019)
A. brassicicola	Gupta and Chauhan (2016)
Tricophyton mentagrophytes and *Candida species*	Kim et al. (2009), Takamiya et al. (2021), and Shah et al. (2021)

AgNPs in biological stabilization of footwear materials were used to prevent the growth of pathogenic fungi and yeast cells (Falkiewicz-Dulik and Macura, 2008). AgNPs antifungal impact can be directly determined by their size and form. The spherical AgNPs synthesized by green method from various plant sources show specific antifungal activities observed against several fungal species (Feizi et al., 2018) and indicates that AgNPs may enable the development of O_2 from fungal mycelia, and hence, there is a process involving O_2 to understand the reduction in the output of aflatoxin from fungal mycelia *A. flavus*. The spherical AgNPs with a mean size of 29 nm were more effective than the biosynthesized AgNPs using specific plant extracts (Mohammadlou et al., 2017).

4.4.8 Conservation and restoration of monuments

Biodeterioration is the consequence of various microbial community interactions, which are noticeable in the predominant form of biofilms on heritage materials. The microorganisms are implicated in stone deterioration since they secrete enzymes and organic acids throughout their physiological activities, which are extremely hazardous to the materials of heritage. Nowadays, major studies have centered on novel approaches that include applying the combination of consolidants, water repellents with antimicrobial nanoparticles. Moreover, the use of nanoparticles has a significant impact in the area of cultural heritage and building to increase the stability and better performance of building/construction materials and its safety. Following conservation treatment, these products also had effective results as a preservative therapy against recolonization (La Russa et al., 2014). It should be noted that the study of the efficacy, stability and durability of such novel nanomaterials is important to avoid unsuitable treatment that alters the aesthetic, physical and chemical properties of stone materials, resulting in new disorders. The risks to human health and the environmental consequences of using the current nanomaterials should be taken into consideration when planning nanomaterial-based treatments. Not only AgNPs has been studied as antimicrobial coatings for stone heritage, but also there are reports that the NPs can be functionalized by condensing a silane precursor such as tetraethyl orthosilicate on the surface. It was demonstrated to be efficient in reducing cell viability by the used doses, and it was not inhibited the antimicrobial activity of NPs (Egger et al., 2009). Aflori and his team constructed antibacterial and antifungal coatings based on 2 hybrid nanocomposites based on silsesquioxane with methacrylate units modified with AgNPs, which showed an antibacterial/antifungal activity against *E. coli* and *C. albicans* (Aflori et al., 2013). Along with this, AgNPs was green synthesized (Carrillo-González et al., 2016) and their effectiveness in controlling in vitro bacteria and fungi, as well as various types of substrates, such as stucco, basalt, and calcite, which was widely used in cultural heritage, was investigated and proposed AgNPs were realized as promising antimicrobials for the conservation and preservation of cultural heritage that it highly depends on the particular doses. In addition, owing to their thicker cell wall, gram-positive cells are usually more tolerant of NPs than gram-negative bacteria

(Feng et al., 2000). Due to their high antimicrobial activity, silver nanoprisms were preferred over any other nanostructures (Pal et al., 2007). There are some major criteria for the essential features of materials used in the restoration of cultural heritage. The efficacy/strength of cleaning should be controllable by the application time to adjust the uses of the formulation in various artwork locations; and the used materials and protocols should minimize any negative impact on the environment (https://www.nanowerk.com/spotlight/spotid=52761.php).

4.5 Perspective and future trends

Scientific attempts are required to overcome the lack of knowledge on toxicological and environmental issues through deeper insights into the detrimental biological properties of nanomaterials. Consequently, many scientists have tried to evaluate the impact of the appropriate dose and exposure time on human and animal health as well as its probably mutagenic effect. More investigation should be carried out to gain insights about nanomaterials to assess the efficient doses to be taken by human beings and animals. Also, antifungal NPs that can be used as antimicrobials or drugs would be produced on a large scale. Moreover, green synthesis of NPs provides additional benefits including the formation of nanobiocomposites *via* microorganisms, plant and animal sources, which induce less toxicity (Abd-Elsalam et al., 2020). This kind of green nanotechnology compounds will bring a new period of fungicidal and bactericidal compounds that will eradicate major agricultural problems faced by human beings today. Moreover, it will avoid potential starvation, food waste, and ecological degradation (Thipe et al., 2020).

To integrate all information into the insights of pathogenesis and pathophysiology of fungal disease and the development of novel therapeutic approaches in human beings and animals, a multidisciplinary and system biology strategy is required (Chaud et al., 2020).

Adding metal nanomaterials to animal feeds with certain current antibiotics induces a decrease in the concentration of utilized nanoparticles, based on their synergistic impact and results in eliminating the microbial resistance to conventional medicines as well as small nontoxic levels of nanomaterials to animal feeds.

More experiments are required to determine the usage of more efficient synergistic nanomaterial additives as natural herbs to provide nontoxic doses to animal feed. Molecular biology must be used to determine the efficacy of nanomaterial therapies for human and animal feed and to examine different genes for pre- and post-treatment mycotoxin or fungus (Hassan et al., 2020).

In the field of cultural heritage, investigation of the behavior and stability of used nanoparticles when exposed to environmental factors such as relative humidity, exposure period, and concentration of CO_2 is very important. Also, a comprehensive awareness of the petrophysical properties of stone materials and other substrates and their stability with nanoparticles, pre- and post-treatments are crucial factors to take into account to evaluate the efficiency of treatments.

4.6 Conclusion

Nanoparticles are dynamic materials with many features that go beyond its chemical composition including particle size, propagation, the shape of particles, crystallinity, the flexibility of the surface and charging. AgNPs are renowned for their excellent antimicrobial activity on various species, including bacteria, fungi, and viruses. Direct interaction of AgNPs with large areas on the bacterial cell wall may result in damaging to the membrane, allowing cell contents to leak and cell death. Furthermore, this property of AgNPs means that they can overcome the existing processes of microbial drug resistance, including decreased absorption and increased microbial cell efflux, and the formation of biofilms. Additionally, antimicrobial activity of AgNPs can also be associated with the altered cell wall, cytoplasm, and permeability of membranes. All these findings can be represented as the initial attachment of AgNPs or Ag^+ ions to the bacterial cell wall, followed by ROS and free radical growth, DNA damage and denaturation of proteins. There are many studies focused on the antipathogenic fungal behavior of AgNPs and the use of AgNPs in biologically stabilizing footwear products to inhibit the growth of fungi and yeast cells. Additionally, it has been shown that AgNPs can damage mitochondria in eukaryotic cells such as fungi, decrease membrane integrity, and release cytochromes and hydrogenase. While research on AgNPs biosafety was carried out, several investigations indicated the effectiveness of AgNPs in reducing or preventing the production of biofilms on in vitro catheter materials as well as in animal models. In vitro studies find AgNPs demonstrating the synergistic antibacterial behavior of specific antibiotics. Finally, the use of AgNPs, as an antimicrobial agent, in food processing, clothing, decontamination of air and water, protection of monuments, etc., makes it popular for other studies.

References

Abbasi, Z., Feizi, S., Taghipour, E., Ghadam, P., 2017. Green synthesis of silver nanoparticles using aqueous extract of dried Juglans regia green husk and examination of its biological properties. Green Process. Synth., 6.

Abd-Elsalam, K.A., El-Naggar, M.A., Ghannouchi, A., Bouqellah, N.A., 2020. Nanomaterials and ozonation: safe strategies for mycotoxin management. In: Nanomycotoxicology. Elsevier, pp. 285–308.

Aflori, M., Simionescu, B., Bordianu, I.-E., Sacarescu, L., Varganici, C.-D., Doroftei, F., Nicolescu, A., Olaru, M., 2013. Silsesquioxane-based hybrid nanocomposites with methacrylate units containing titania and/or silver nanoparticles as antibacterial/antifungal coatings for monumental stones. Mater. Sci. Eng. B 178, 1339–1346.

Ahmed Hammad, N.E.D., Bahig Ahmed, E.D., 2018. Effectiveness of silver nanoparticles against root-knot nematode, Meloidogyne incognita infecting tomato under greenhouse conditions. J. Agric. Sci. 10, 148–156.

Alagarasi, A., 2013. Chapter-Introduction to Nanomaterials. Indian Inst. Technol. Madras, pp. 1–24.

Aletayeb, P., Ghadam, P., Mohammadi, P., 2020. Green synthesis of $AgCl/Ag_3PO_4$ nanoparticle using cyanobacteria and assessment of its antibacterial, colorimetric detection of heavy metals and antioxidant properties. IET Nanobiotechnol. 14 (8), 707–713.

Alexander, J.W., 2009. History of the medical use of silver. Surg. Infect. (Larchmt) 10, 289–292.

Alizadeh, F., Khodavandi, A., 2020. Systematic review and meta-analysis of the efficacy of nanoscale materials against coronaviruses–possible potential antiviral agents for SARS-CoV-2. IEEE Trans. Nanobiosci. 19 (3), 485–497.

Al-Othman, M.R., Abd El-Aziz, A.R.M., Mahmoud, M.A., Eifan, S.A., El-Shikh, M.S., Majrashi, M., 2014. Application of silver nanoparticles as antifungal and antiaflatoxin B1 produced by *Aspergillus flavus*. Dig. J. Nanomater. Biostruct. 9, 151–157.

As, A.A., Thangapandiyan, S., 2019. Comparative bioassay of silver nanoparticles and malathion on infestation of red flour beetle, *Tribolium castaneum*. J. Basic Appl. Zool. 80, 55.

Avila-Quezada, G.D., Espino-Solis, G.P., 2019. Silver nanoparticles offer effective control of pathogenic bacteria in a wide range of food products. In: Pathogenic Bacteria. IntechOpen.

Baker-Austin, C., Wright, M.S., Stepanauskas, R., McArthur, J.V., 2006. Co-selection of antibiotic and metal resistance. Trends Microbiol. 14, 176–182.

Baptista, P.V., McCusker, M.P., Carvalho, A., Ferreira, D.A., Mohan, N.M., Martins, M., Fernandes, A.R., 2018. Nano-strategies to fight multidrug-resistant bacteria—"A Battle of the Titans". Front. Microbiol. 9, 1441.

Baram-Pinto, D., Shukla, S., Perkas, N., Gedanken, A., Sarid, R., 2009. Inhibition of herpes simplex virus type 1 infection by silver nanoparticles capped with mercaptoethane sulfonate. Bioconjug. Chem. 20, 1497–1502.

Barros, C.H.N., Fulaz, S., Stanisic, D., Tasic, L., 2018. Biogenic nanosilver against multidrug-resistant bacteria (MDRB). Antibiotics 7, 69.

Behra, R., Sigg, L., Clift, M.J.D., Herzog, F., Minghetti, M., Johnston, B., Petri-Fink, A., Rothen-Rutishauser, B., 2013. Bioavailability of silver nanoparticles and ions: from a chemical and biochemical perspective. J. R. Soc. Interface 10, 20130396.

Blecher, K., Nasir, A., Friedman, A., 2011. The growing role of nanotechnology in combating infectious disease. Virulence 2, 395–401.

Bocate, K.P., Reis, G.F., de Souza, P.C., Junior, A.G.O., Durán, N., Nakazato, G., Furlaneto, M.C., de Almeida, R.S., Panagio, L.A., 2019. Antifungal activity of silver nanoparticles and simvastatin against toxigenic species of *Aspergillus*. Int. J. Food Microbiol. 291, 79–86.

Brett, D.W., 2006. A discussion of silver as an antimicrobial agent: alleviating the confusion. Ostomy. Wound. Manage. 52, 34–41.

Carrillo-González, R., Martínez-Gómez, M.A., del González-Chávez, M.C.A., Hernández, J.C.M., 2016. Inhibition of microorganisms involved in deterioration of an archaeological site by silver nanoparticles produced by a green synthesis method. Sci. Total Environ. 565, 872–881.

Chaloupka, K., Malam, Y., Seifalian, A.M., 2010. Nanosilver as a new generation of nanoproduct in biomedical applications. Trends Biotechnol. 28, 580–588.

Chaud, M.V., Rios, A.C., dos Santos, C.A., de Barros, C.T., de Souza, J.F., Alves, T.F.R., 2020. Nanostructure self-assembly for direct nose-to-brain drug delivery: a novel approach for cryptococcal meningitis. In: Nanomycotoxicology. Elsevier, pp. 449–480.

Chen, X., Schluesener, H.J., 2008. Nanosilver: a nanoproduct in medical application. Toxicol. Lett. 176, 1–12.

Chernousova, S., Epple, M., 2013. Silver as antibacterial agent: ion, nanoparticle, and metal. Angew. Chem. Int. Ed. 52, 1636–1653.

Choi, O., Deng, K.K., Kim, N.-J., Ross Jr., L., Surampalli, R.Y., Hu, Z., 2008. The inhibitory effects of silver nanoparticles, silver ions, and silver chloride colloids on microbial growth. Water Res. 42, 3066–3074.

Cinteza, L.O., Scomoroscenco, C., Voicu, S.N., Nistor, C.L., Nitu, S.G., Trica, B., Jecu, M.-L., Petcu, C., 2018. Chitosan-stabilized Ag nanoparticles with superior biocompatibility and their synergistic antibacterial effect in mixtures with essential oils. Nanomaterials 8, 826.

Collins, T.F.X., Sprando, R.L., Black, T.N., Olejnik, N., Eppley, R.M., Alam, H.Z., Rorie, J., Ruggles, D.I., 2006. Effects of zearalenone on in utero development in rats. Food Chem. Toxicol. 44, 1455–1465.

Colman, B.P., Arnaout, C.L., Anciaux, S., Gunsch, C.K., Hochella Jr., M.F., Kim, B., Lowry, G.V., McGill, B.M., Reinsch, B.C., Richardson, C.J., 2013. Low concentrations of silver nanoparticles in biosolids cause adverse ecosystem responses under realistic field scenario. PLoS One 8, e57189.

Crisante, F., Francolini, I., Bellusci, M., Martinelli, A., D'Ilario, L., Piozzi, A., 2009. Antibiotic delivery polyurethanes containing albumin and polyallylamine nanoparticles. Eur. J. Pharm. Sci. 36, 555–564.

Daniel, S.C.G.K., Tharmaraj, V., Sironmani, T.A., Pitchumani, K., 2010. Toxicity and immunological activity of silver nanoparticles. Appl. Clay Sci. 48, 547–551.

Dar, M.A., Ingle, A., Rai, M., 2013. Enhanced antimicrobial activity of silver nanoparticles synthesized by *Cryphonectria* sp. evaluated singly and in combination with antibiotics. Nanomed. Nanotechnol. Biol. Med. 9, 105–110.

Darouiche, R.O., 1999. Anti-infective efficacy of silver-coated medical prostheses. Clin. Infect. Dis. 29, 1371–1377.

Davidson, R.A., Anderson, D.S., Van Winkle, L.S., Pinkerton, K.E., Guo, T., 2015. Evolution of silver nanoparticles in the rat lung investigated by X-ray absorption spectroscopy. J. Phys. Chem. A 119, 281–289.

De Gusseme, B., Hennebel, T., Christiaens, E., Saveyn, H., Verbeken, K., Fitts, J.P., Boon, N., Verstraete, W., 2011. Virus disinfection in water by biogenic silver immobilized in polyvinylidene fluoride membranes. Water Res. 45, 1856–1864.

Deabes, M.M., Khalil, W.K.B., Attallah, A.G., El-Desouky, T.A., Naguib, K.M., 2018. Impact of silver nanoparticles on gene expression in *Aspergillus flavus* producer aflatoxin B1. Open Access Maced. J. Med. Sci. 6, 600.

Deepak, V., Kalishwaralal, K., Pandian, S.R.K., Gurunathan, S., 2011. An insight into the bacterial biogenesis of silver nanoparticles, industrial production and scale-up. In: Metal Nanoparticles in Microbiology. Springer, pp. 17–35.

Dibrov, P., Dzioba, J., Gosink, K.K., Häse, C.C., 2002. Chemiosmotic mechanism of antimicrobial activity of Ag^+ in *Vibrio cholerae*. Antimicrob. Agents Chemother. 46, 2668–2670.

Dong, M.S., Choi, J.-Y., Sung, J.H., Kim, J.S., Song, K.S., Ryu, H.R., Lee, J.H., Bang, I.S., An, K., Park, H.M., 2013. Gene expression profiling of kidneys from Sprague-Dawley rats following 12-week inhalation exposure to silver nanoparticles. Toxicol. Mech. Methods 23, 437–448.

Duhan, J.S., Kumar, R., Kumar, N., Kaur, P., Nehra, K., Duhan, S., 2017. Nanotechnology: the new perspective in precision agriculture. Biotechnol. Rep. 15, 11–23.

Egger, S., Lehmann, R.P., Height, M.J., Loessner, M.J., Schuppler, M., 2009. Antimicrobial properties of a novel silver-silica nanocomposite material. Appl. Environ. Microbiol. 75, 2973–2976.

Eisenberg, T., Büttner, S., Kroemer, G., Madeo, F., 2007. The mitochondrial pathway in yeast apoptosis. Apoptosis 12, 1011–1023.

El-Desouky, T.A., Ammar, H.A.M., 2016. Honey mediated silver nanoparticles and their inhibitory effect on aflatoxins and ochratoxin A. J. Appl. Pharm. Sci. 6, 83–90.

Elechiguerra, J.L., Burt, J.L., Morones, J.R., Camacho-Bragado, A., Gao, X., Lara, H.H., Yacaman, M.J., 2005. Interaction of silver nanoparticles with HIV-1. J. Nanobiotechnol. 3, 1–10.

Elgorban, A.M., El-Samawaty, A.E.-R.M., Yassin, M.A., Sayed, S.R., Adil, S.F., Elhindi, K.M., Bakri, M., Khan, M., 2016. Antifungal silver nanoparticles: synthesis, characterization and biological evaluation. Biotechnol. Biotechnol. Equip. 30, 56–62.

El-Mohamady, R.S., Ghattas, T.A., Zawrah, M.F., Abd El-Hafeiz, Y.G.M., 2018. Inhibitory effect of silver nanoparticles on bovine herpesvirus-1. Int. J. Vet. Sci. Med. 6, 296–300.

Falkiewicz-Dulik, M., Macura, A.B., 2008. Nanosilver as substance biostabilising footwear materials in the foot mycosis prophylaxis. Mikol. Lek. 15, 145–150.

Fayaz, A.M., Balaji, K., Girilal, M., Yadav, R., Kalaichelvan, P.T., Venketesan, R., 2010. Biogenic synthesis of silver nanoparticles and their synergistic effect with antibiotics: a study against gram-positive and gram-negative bacteria. Nanomedicine Nanotechnology. Biol. Med. 6, 103–109.

Feizi, S., Taghipour, E., Ghadam, P., Mohammadi, P., 2018. Antifungal, antibacterial, antibiofilm and colorimetric sensing of toxic metals activities of eco-friendly, economical synthesized Ag/AgCl nanoparticles using Malva sylvestris leaf extracts. Microb. Pathog. 125, 33–42.

Feng, Q.L., Wu, J., Chen, G.Q., Cui, F.Z., Kim, T.N., Kim, J.O., 2000. A mechanistic study of the antibacterial effect of silver ions on *Escherichia coli* and *Staphylococcus aureus*. J. Biomed. Mater. Res. 52, 662–668.

Foldbjerg, R., Dang, D.A., Autrup, H., 2011. Cytotoxicity and genotoxicity of silver nanoparticles in the human lung cancer cell line, A549. Arch. Toxicol. 85, 743–750.

Fulaz, S., Vitale, S., Quinn, L., Casey, E., 2019. Nanoparticle-biofilm interactions: the role of the EPS matrix. Trends Microbiol. 27, 915–926.

Gacem, M.A., Gacem, H., Telli, A., Khelil, A.O.E.H., 2020. Mycotoxins: decontamination and nanocontrol methods. In: Nanomycotoxicology. Elsevier, pp. 189–216.

Gahlawat, G., Shikha, S., Chaddha, B.S., Chaudhuri, S.R., Mayilraj, S., Choudhury, A.R., 2016. Microbial glycolipoprotein-capped silver nanoparticles as emerging antibacterial agents against cholera. Microb. Cell Fact. 15, 25.

Gaikwad, R.W., Sasane, V.V., 2013. Assessment of groundwater quality in and around Lonar lake and possible water treatment. Int. J. Environ. Sci. 3.

Gaikwad, S., Ingle, A., Gade, A., Rai, M., Falanga, A., Incoronato, N., Russo, L., Galdiero, S., Galdiero, M., 2013. Antiviral activity of mycosynthesized silver nanoparticles against herpes simplex virus and human parainfluenza virus type 3. Int. J. Nanomed. 8, 4303.

Gajbhiye, M., Kesharwani, J., Ingle, A., Gade, A., Rai, M., 2009. Fungus-mediated synthesis of silver nanoparticles and their activity against pathogenic fungi in combination with fluconazole. Nanomed. Nanotechnol. Biol. Med. 5, 382–386.

Galdiero, S., Falanga, A., Cantisani, M., Ingle, A., Galdiero, M., Rai, M., 2014. Silver nanoparticles as novel antibacterial and antiviral agents. In: Handbook of Nanobiomedical Research: Fundamentals, Applications and Recent Developments: Volume 1. Materials for Nanomedicine. World Scientific, pp. 565–594.

Gangadharan, D., Harshvardan, K., Gnanasekar, G., Dixit, D., Popat, K.M., Anand, P.S., 2010. Polymeric microspheres containing silver nanoparticles as a bactericidal agent for water disinfection. Water Res. 44, 5481–5487.

George, S., Lin, S., Ji, Z., Thomas, C.R., Li, L., Mecklenburg, M., Meng, H., Wang, X., Zhang, H., Xia, T., 2012. Surface defects on plate-shaped silver nanoparticles contribute to its hazard potential in a fish gill cell line and zebrafish embryos. ACS Nano 6, 3745–3759.

Ghorbani, H.R., Safekordi, A.A., Attar, H., Sorkhabadi, S.M., 2011. Biological and non-biological methods for silver nanoparticles synthesis. Chem. Biochem. Eng. Q. 25, 317–326.

Gogoi, S.K., Gopinath, P., Paul, A., Ramesh, A., Ghosh, S.S., Chattopadhyay, A., 2006. Green fluorescent protein-expressing *Escherichia coli* as a model system for investigating the antimicrobial activities of silver nanoparticles. Langmuir 22, 9322–9328.

Golińska, P., Wypij, M., Rathod, D., Tikar, S., Dahm, H., Rai, M., 2016. Synthesis of silver nanoparticles from two acidophilic strains of *Pilimelia columellifera* subsp. *pallida* and their antibacterial activities. J. Basic Microbiol. 56, 541–556.

Gómez, J.V., Tarazona, A., Mateo, F., Jiménez, M., Mateo, E.M., 2019. Potential impact of engineered silver nanoparticles in the control of aflatoxins, ochratoxin A and the main aflatoxigenic and ochratoxigenic species affecting foods. Food Control 101, 58–68.

Gorjanc, M., Bukošek, V., Gorenšek, M., Mozetič, M., 2010. CF4 plasma and silver functionalized cotton. Text. Res. J. 80, 2204–2213.

Grainger, D.W., Castner, D.G., 2008. Nanobiomaterials and nanoanalysis: opportunities for improving science to benefit biomedical technologies. Adv. Mater. 20, 867–877.

Greulich, C., Diendorf, J., Gessmann, J., Simon, T., Habijan, T., Eggeler, G., Schildhauer, T.A., Epple, M., Köller, M., 2011. Cell type-specific responses of peripheral blood mononuclear cells to silver nanoparticles. Acta Biomater. 7, 3505–3514.

Griffin, M.O., Fricovsky, E., Ceballos, G., Villarreal, F., 2010. Tetracyclines: a pleitropic family of compounds with promising therapeutic properties. Review of the literature. Am. J. Physiol. Physiol. 299, C539–C548.

Gupta, D., Chauhan, P., 2016. Fungicidal activity of silver nanoparticles against *Alternaria brassicicola*. In: AIP Conference Proceedings. AIP Publishing LLC, p. 20031.

Gurunathan, S., Choi, Y.-J., Kim, J.-H., 2018. Antibacterial efficacy of silver nanoparticles on endometritis caused by *Prevotella melaninogenica* and *Arcanobacterum pyogenes* in dairy cattle. Int. J. Mol. Sci. 19, 1210.

Habash, M.B., Goodyear, M.C., Park, A.J., Surette, M.D., Vis, E.C., Harris, R.J., Khursigara, C.M., 2017. Potentiation of tobramycin by silver nanoparticles against *Pseudomonas aeruginosa* biofilms. Antimicrob. Agents Chemother. 61.

Hafez, E.E., Kabeil, S.S., 2013. Antimicrobial activity of nano-silver particles produced by microalgae. J. Pure Appl. Microbiol. 7, 35–42.

Haggag, E.G., Elshamy, A.M., Rabeh, M.A., Gabr, N.M., Salem, M., Youssif, K.A., Samir, A., Muhsinah, A.B., Alsayari, A., Abdelmohsen, U.R., 2019. Antiviral potential of green synthesized silver nanoparticles of *Lampranthus coccineus* and *Malephora lutea*. Int. J. Nanomed. 14, 6217.

Haji Esmaeil Hajjar, F., Jebali, A., Hekmatimoghaddam, S., 2015. The inhibition of *Candida albicans* secreted aspartyl proteinase by triangular gold nanoparticles. Nanomed. J. 2, 54–59.

Hall, B.G., Barlow, M., 2004. Evolution of the serine β-lactamases: past, present and future. Drug Resist. Updat. 7, 111–123.

Halliwell, B., Aruoma, O.I., 1991. DNA damage by oxygen-derived species Its mechanism and measurement in mammalian systems. FEBS Lett. 281, 9–19.

Hamid, S., Zainab, S., Faryal, R., Ali, N., Sharafat, I., 2018. Inhibition of secreted aspartyl proteinase activity in biofilms of *Candida* species by mycogenic silver nanoparticles. Artif. Cells Nanomed. Biotechnol. 46, 551–557.

Hancock, R.E.W., Sahl, H.-G., 2006. Antimicrobial and host-defense peptides as new anti-infective therapeutic strategies. Nat. Biotechnol. 24, 1551–1557.

Hansen, S.F., Baun, A., 2012. When enough is enough. Nat. Nanotechnol. 7, 409–411.

Hassan, A.A., Mansour, M.K., Oraby, N.H., Mohamed, A.M., 2016. The efficiency of using silver nanoparticles singly and in combination with traditional antimicrobial agents in control of some fungal and bacterial affection of buffaloes. Int. J. Curr. Res. 8 (4), 29758–29770.

Hassan, A.A., Sayed-Elahl, R.M., Oraby, N.H., El-Hamaky, A.M.A., 2020. Metal nanoparticles for management of mycotoxigenic fungi and mycotoxicosis diseases of animals and poultry. In: Nanomycotoxicology. Elsevier, pp. 251–269.

Hellstrand, E., Lynch, I., Andersson, A., Drakenberg, T., Dahlbäck, B., Dawson, K.A., Linse, S., Cedervall, T., 2009. Complete high-density lipoproteins in nanoparticle corona. FEBS J. 276, 3372–3381.

Hendiani, S., Abdi-Ali, A., Mohammadi, P., Kharrazi, S., 2015. Synthesis of silver nanoparticles and its synergistic effects in combination with imipenem and two biocides against biofilm-producing *Acinetobacter baumannii*. Nanomed. J. 2, 291–298.

Holade, Y., Hickey, D.P., Minteer, S.D., 2016. Halide-regulated growth of electrocatalytic metal nanoparticles directly onto a carbon paper electrode. J. Mater. Chem. A 4, 17154–17162.

Hsu, S., Tseng, H.-J., Lin, Y.-C., 2010. The biocompatibility and antibacterial properties of waterborne polyurethane-silver nanocomposites. Biomaterials 31, 6796–6808.

Hsueh, Y.-H., Lin, K.-S., Ke, W.-J., Hsieh, C.-T., Chiang, C.-L., Tzou, D.-Y., Liu, S.-T., 2015. The antimicrobial properties of silver nanoparticles in *Bacillus subtilis* are mediated by released Ag^+ ions. PLoS One 10, e0144306.

Huh, A.J., Kwon, Y.J., 2011. "Nanoantibiotics": a new paradigm for treating infectious diseases using nanomaterials in the antibiotics resistant era. J. Control. Release 156, 128–145.

Hull, H.F., Ward, N.A., Hull, B.P., Milstien, J.B., de Quadros, C., 1994. Paralytic poliomyelitis: seasoned strategies, disappearing disease. Lancet 343, 1331–1337.

Ibrahim, E., Zhang, M., Zhang, Y., Hossain, A., Qiu, W., Chen, Y., Wang, Y., Wu, W., Sun, G., Li, B., 2020. Green-synthesization of silver nanoparticles using endophytic bacteria isolated from garlic and its antifungal activity against wheat fusarium head blight pathogen *Fusarium graminearum*. Nanomaterials 10, 219.

Ingale, A.G., Chaudhari, A.N., 2013. Biogenic synthesis of nanoparticles and potential applications: an eco-friendly approach. J. Nanomed. Nanotechol. 4, 1–7.

Islam, M.S., Yeum, J.H., 2013. Electrospun pullulan/poly(vinyl alcohol)/silver hybrid nanofibers: preparation and property characterization for antibacterial activity. Colloids Surf. A Physicochem. Eng. Asp. 436, 279–286.

Ismaiel, A.A., Tharwat, N.A., 2014. Antifungal activity of silver ion on ultrastructure and production of aflatoxin B1 and patulin by two mycotoxigenic strains, *Aspergillus flavus* OC1 and *Penicillium vulpinum* CM1. J. Mycol. Med. 24, 193–204.

Ivask, A., Kurvet, I., Kasemets, K., Blinova, I., Aruoja, V., Suppi, S., Vija, H., Käkinen, A., Titma, T., Heinlaan, M., 2014. Size-dependent toxicity of silver nanoparticles to bacteria, yeast, algae, crustaceans and mammalian cells in vitro. PLoS One 9, e102108.

Jain, N., Bhargava, A., Majumdar, S., Tarafdar, J.C., Panwar, J., 2011. Extracellular biosynthesis and characterization of silver nanoparticles using *Aspergillus flavus* NJP08: a mechanism perspective. Nanoscale 3, 635–641.

Jalal, M., Ansari, M.A., Alzohairy, M.A., Ali, S.G., Khan, H.M., Almatroudi, A., Siddiqui, M.I., 2019. Anticandidal activity of biosynthesized silver nanoparticles: effect on growth, cell morphology, and key virulence attributes of *Candida* species. Int. J. Nanomed. 14, 4667.

Janata, E., 2003. Structure of the trimer silver cluster Ag_3^{2+}. J. Phys. Chem. B 107, 7334–7336.

Ji, J.H., Jung, J.H., Kim, S.S., Yoon, J.-U., Park, J.D., Choi, B.S., Chung, Y.H., Kwon, I.H., Jeong, J., Han, B.S., 2007. Twenty-eight-day inhalation toxicity study of silver nanoparticles in Sprague-Dawley rats. Inhal. Toxicol. 19, 857–871.

Jogee, P., Rai, M., 2020. Application of nanoparticles in inhibition of mycotoxin-producing fungi. In: Nanomycotoxicology. Elsevier, pp. 239–250.

Khalandi, B., Asadi, N., Milani, M., Davaran, S., Abadi, A.J.N., Abasi, E., Akbarzadeh, A., 2017. A review on potential role of silver nanoparticles and possible mechanisms of their actions on bacteria. Drug Res. (Stuttg.) 11, 70–76.

Khalil, N.M., Abd El-Ghany, M.N., Rodríguez-Couto, S., 2019. Antifungal and anti-mycotoxin efficacy of biogenic silver nanoparticles produced by *Fusarium chlamydosporum* and *Penicillium chrysogenum* at non-cytotoxic doses. Chemosphere 218, 477–486.

Khandelwal, N., Kaur, G., Kumar, N., Tiwari, A., 2014. Application of silver nanoparticles in viral inhibition: a new hope for antivirals. Dig. J. Nanomater. Biostruct. 9.

Khatami, M., Mehnipor, R., Poor, M.H.S., Jouzani, G.S., 2016. Facile biosynthesis of silver nanoparticles using *Descurainia sophia* and evaluation of their antibacterial and antifungal properties. J. Clust. Sci. 27, 1601–1612.

Kim, K.-J., Sung, W.S., Suh, B.K., Moon, S.-K., Choi, J.-S., Kim, J.G., Lee, D.G., 2009. Antifungal activity and mode of action of silver nano-particles on *Candida albicans*. Biometals 22, 235–242.

Kim, S.W., Jung, J.H., Lamsal, K., Kim, Y.S., Min, J.S., Lee, Y.S., 2012a. Antifungal effects of silver nanoparticles (AgNPs) against various plant pathogenic fungi. Mycobiology 40, 53–58.

Kim, T., Kim, M., Park, H., Shin, U.S., Gong, M., Kim, H., 2012b. Size-dependent cellular toxicity of silver nanoparticles. J. Biomed. Mater. Res. A 100, 1033–1043.

Kittler, S., Greulich, C., Gebauer, J.S., Diendorf, J., Treuel, L., Ruiz, L., Gonzalez-Calbet, J.M., Vallet-Regi, M., Zellner, R., Köller, M., 2010. The influence of proteins on the dispersability and cell-biological activity of silver nanoparticles. J. Mater. Chem. 20, 512–518.

Klasen, H.J., 2000a. Historical review of the use of silver in the treatment of burns. I. Early uses. Burns 26, 117–130.

Klasen, H.J., 2000b. A historical review of the use of silver in the treatment of burns. II. Renewed interest for silver. Burns 26, 131–138.

Kraśniewska, K., Pobiega, K., Gniewosz, M., 2019. Pullulan-biopolymer with potential for use as food packaging. Int. J. Food Eng. 15.

Kraśniewska, K., Galus, S., Gniewosz, M., 2020. Biopolymers-based materials containing silver nanoparticles as active packaging for food applications—a review. Int. J. Mol. Sci. 21, 698.

Krug, H.F., Wick, P., 2011. Nanotoxicology: an interdisciplinary challenge. Angew. Chem. Int. Ed. 50, 1260–1278.

Kumar, P., Chopra, S., Maninder Singh, A.S., 2017. Green synthesis of silver nanoparticles for plant disease diagnosis. Int. J. Curr. Res. 9, 48283–48288.

La Russa, M.F., Macchia, A., Ruffolo, S.A., De Leo, F., Barberio, M., Barone, P., Crisci, G.M., Urzì, C., 2014. Testing the antibacterial activity of doped TiO_2 for preventing biodeterioration of cultural heritage building materials. Int. Biodeterior. Biodegradat. 96, 87–96.

Lansdown, A.B.G., 2006. Silver in health care: antimicrobial effects and safety in use. In: Biofunctional Textiles and the Skin. Karger Publishers, pp. 17–34.

Lansdown, A.B.G., 2010. Silver in Healthcare: Its Antimicrobial Efficacy and Safety in Use. Royal Society of Chemistry.

Lee, H.-Y., Choi, Y.-J., Jung, E.-J., Yin, H.-Q., Kwon, J.-T., Kim, J.-E., Im, H.-T., Cho, M.-H., Kim, J.-H., Kim, H.-Y., 2010. Genomics-based screening of differentially expressed genes in the brains of mice exposed to silver nanoparticles via inhalation. J. Nanopart. Res. 12, 1567–1578.

Levard, C., Hotze, E.M., Lowry, G.V., Brown Jr., G.E., 2012. Environmental transformations of silver nanoparticles: impact on stability and toxicity. Environ. Sci. Technol. 46, 6900–6914.

Levard, C., Mitra, S., Yang, T., Jew, A.D., Badireddy, A.R., Lowry, G.V., Brown Jr., G.E., 2013. Effect of chloride on the dissolution rate of silver nanoparticles and toxicity to *E. coli*. Environ. Sci. Technol. 47, 5738–5745.

Liao, C., Li, Y., Tjong, S.C., 2019. Bactericidal and cytotoxic properties of silver nanoparticles. Int. J. Mol. Sci. 20, 449.

Limbach, L.K., Li, Y., Grass, R.N., Brunner, T.J., Hintermann, M.A., Muller, M., Gunther, D., Stark, W.J., 2005. Oxide nanoparticle uptake in human lung fibroblasts: effects of particle size, agglomeration, and diffusion at low concentrations. Environ. Sci. Technol. 39, 9370–9376.

Liu, W., Wu, Y., Wang, C., Li, H.C., Wang, T., Liao, C.Y., Cui, L., Zhou, Q.F., Yan, B., Jiang, G.B., 2010. Impact of silver nanoparticles on human cells: effect of particle size. Nanotoxicology 4, 319–330.

Liu, Y., Bhattarai, P., Dai, Z., Chen, X., 2019. Photothermal therapy and photoacoustic imaging via nanotheranostics in fighting cancer. Chem. Soc. Rev. 48, 2053–2108.

Lu, X., Zhang, H., Ni, Y., Zhang, Q., Chen, J., 2008. Porous nanosheet-based ZnO microspheres for the construction of direct electrochemical biosensors. Biosens. Bioelectron. 24, 93–98.

Luther, E.M., Koehler, Y., Diendorf, J., Epple, M., Dringen, R., 2011. Accumulation of silver nanoparticles by cultured primary brain astrocytes. Nanotechnology 22, 375101.

Lv, X., Wang, P., Bai, R., Cong, Y., Suo, S., Ren, X., Chen, C., 2014. Inhibitory effect of silver nanomaterials on transmissible virus-induced host cell infections. Biomaterials 35, 4195–4203.

Ma, J., Zhang, J., Xiong, Z., Yong, Y., Zhao, X.S., 2011. Preparation, characterization and antibacterial properties of silver-modified graphene oxide. J. Mater. Chem. 21, 3350–3352.

Manjumeena, R., Duraibabu, D., Sudha, J., Kalaichelvan, P.T., 2014. Biogenic nanosilver incorporated reverse osmosis membrane for antibacterial and antifungal activities against selected pathogenic strains: an enhanced eco-friendly water disinfection approach. J. Environ. Sci. Health A 49, 1125–1133.

Markowska, K., Grudniak, A.M., Wolska, K.I., 2013. Silver nanoparticles as an alternative strategy against bacterial biofilms. Acta Biochim. Pol. 60.

Martínez-Castañon, G.-A., Nino-Martinez, N., Martinez-Gutierrez, F., Martinez-Mendoza, J.R., Ruiz, F., 2008. Synthesis and antibacterial activity of silver nanoparticles with different sizes. J. Nanopart. Res. 10, 1343–1348.

Martinez-Gutierrez, F., Boegli, L., Agostinho, A., Sánchez, E.M., Bach, H., Ruiz, F., James, G., 2013. Anti-biofilm activity of silver nanoparticles against different microorganisms. Biofouling 29, 651–660.

Maurer, L.L., Meyer, J.N., 2016. A systematic review of evidence for silver nanoparticle-induced mitochondrial toxicity. Environ. Sci. Nano 3, 311–322.

Mihindukulasuriya, S.D.F., Lim, L.-T., 2014. Nanotechnology development in food packaging: a review. Trends Food Sci. Technol. 40, 149–167.

Mishra, S., Singh, B.R., Singh, A., Keswani, C., Naqvi, A.H., Singh, H.B., 2014. Biofabricated silver nanoparticles act as a strong fungicide against *Bipolaris sorokiniana* causing spot blotch disease in wheat. PLoS One 9, e97881.

Mitra, C., Gummadidala, P.M., Afshinnia, K., Merrifield, R.C., Baalousha, M., Lead, J.R., Chanda, A., 2017. Citrate-coated silver nanoparticles growth-independently inhibit aflatoxin synthesis in *Aspergillus parasiticus*. Environ. Sci. Technol. 51, 8085–8093.

Mohammadlou, M., Jafarizadeh-Malmiri, H., Maghsoudi, H., 2017. Hydrothermal green synthesis of silver nanoparticles using Pelargonium/Geranium leaf extract and evaluation of their antifungal activity. Green Process. Synth. 6, 31–42.

Monopoli, M.P., Bombelli, F.B., Dawson, K.A., 2011a. Nanoparticle coronas take shape. Nat. Nanotechnol. 6, 11–12.

Monopoli, M.P., Walczyk, D., Campbell, A., Elia, G., Lynch, I., Baldelli Bombelli, F., Dawson, K.A., 2011b. Physical-chemical aspects of protein corona: relevance to in vitro and in vivo biological impacts of nanoparticles. J. Am. Chem. Soc. 133, 2525–2534.

Morones, J.R., Elechiguerra, J.L., Camacho, A., Holt, K., Kouri, J.B., Ramírez, J.T., Yacaman, M.J., 2005. The bactericidal effect of silver nanoparticles. Nanotechnology 16, 2346.

Mousavi, S.A.A., Pourtalebi, S., 2015. Inhibitory effects of silver nanoparticles on growth and aflatoxin B1 production by *Aspergillus parasiticus*. Iran. J. Med. Sci. 40, 501.

Mousavi, S.S., Ghadam, P., Mohammadi, P., 2020. Screening of soil fungi in order to biosynthesize AgNPs and evaluation of antibacterial and antibiofilm activities. Bull. Mater. Sci. 43, 1–8.

Mueller, B., Zacharias, M., Rezwan, K., 2010. Bovine serum albumin and lysozyme adsorption on calcium phosphate particles. Adv. Eng. Mater. 12, B53–B61.

Mulani, M.S., Kamble, E.E., Kumkar, S.N., Tawre, M.S., Pardesi, K.R., 2019. Emerging strategies to combat ESKAPE pathogens in the era of antimicrobial resistance: a review. Front. Microbiol. 10, 539.

Namasivayam, K.R.S., Aruna, A., Gokila, 2014. Evaluation of silver nanoparticles-chitosan encapsulated synthetic herbicide paraquate (AgNp-CS-PQ) preparation for the controlled release and improved herbicidal activity against *Eichhornia crassipes*. Res. J. Biotechnol. 9, 19–27.

Narayanan, K.B., Sakthivel, N., 2010. Biological synthesis of metal nanoparticles by microbes. Adv. Colloid Interface Sci. 156, 1–13.

Nasrollahzadeh, M., Sajjadi, M., Soufi, G.J., Iravani, S., Varma, R.S., 2020. Nanomaterials and nanotechnology-associated innovations against viral infections with a focus on coronaviruses. Nanomaterials 10, 1072.

Nayak, D., Minz, A.P., Ashe, S., Rauta, P.R., Kumari, M., Chopra, P., Nayak, B., 2016. Synergistic combination of antioxidants, silver nanoparticles and chitosan in a nanoparticle-based formulation: characterization and cytotoxic effect on MCF-7 breast cancer cell lines. J. Colloid Interface Sci. 470, 142–152.

Nejad, M.S., Bonjar, G.H.S., Khatami, M., Amini, A., Aghighi, S., 2016. In vitro and in vivo antifungal properties of silver nanoparticles against *Rhizoctonia solani*, a common agent of rice sheath blight disease. IET Nanobiotechnol. 11, 236–240.

Nel, A.E., Mädler, L., Velegol, D., Xia, T., Hoek, E.M.V., Somasundaran, P., Klaessig, F., Castranova, V., Thompson, M., 2009. Understanding biophysicochemical interactions at the nano-bio interface. Nat. Mater. 8, 543–557.

Nowack, B., Krug, H.F., Height, M., 2011. 120 Years of Nanosilver History: Implications for Policymakers.

Oberdörster, G., Oberdörster, E., Oberdörster, J., 2005. Nanotoxicology: an emerging discipline evolving from studies of ultrafine particles. Environ. Health Perspect. 113, 823–839.

Pal, S., Tak, Y.K., Song, J.M., 2007. Does the antibacterial activity of silver nanoparticles depend on the shape of the nanoparticle? A study of the gram-negative bacterium *Escherichia coli*. Appl. Environ. Microbiol. 73, 1712–1720.

Palza, H., 2015. Antimicrobial polymers with metal nanoparticles. Int. J. Mol. Sci. 16, 2099–2116.

Panáček, A., Smékalová, M., Večeřová, R., Bogdanová, K., Röderová, M., Kolář, M., Kilianová, M., Hradilová, Š., Froning, J.P., Havrdová, M., 2016. Silver nanoparticles strongly enhance and restore bactericidal activity of inactive antibiotics against multiresistant *Enterobacteriaceae*. Colloids Surf. B Biointerfaces 142, 392–399.

Panea, B., Ripoll, G., González, J., Fernández-Cuello, Á., Albertí, P., 2014. Effect of nanocomposite packaging containing different proportions of ZnO and Ag on chicken breast meat quality. J. Food Eng. 123, 104–112.

Parikh, R.Y., Singh, S., Prasad, B.L.V., Patole, M.S., Sastry, M., Shouche, Y.S., 2008. Extracellular synthesis of crystalline silver nanoparticles and molecular evidence of silver resistance from *Morganella* sp.: towards understanding biochemical synthesis mechanism. ChemBioChem 9, 1415–1422.

Park, J.H., Rivière, I., Gonen, M., Wang, X., Sénéchal, B., Curran, K.J., Sauter, C., Wang, Y., Santomasso, B., Mead, E., 2018. Long-term follow-up of CD19 CAR therapy in acute lymphoblastic leukemia. N. Engl. J. Med. 378, 449–459.

Park, M.V.D.Z., Neigh, A.M., Vermeulen, J.P., de la Fonteyne, L.J.J., Verharen, H.W., Briedé, J.J., van Loveren, H., de Jong, W.H., 2011. The effect of particle size on the cytotoxicity, inflammation, developmental toxicity and genotoxicity of silver nanoparticles. Biomaterials 32, 9810–9817.

Paseban, N., Ghadam, P., Pourhosseini, P.S., 2019. The fluorescence behavior and stability of AgNPs synthesized by Juglans Regia green husk aqueous extract. Int. J. Nanosci. Nanotechnol. 15, 117–126.

Pauksch, L., Hartmann, S., Rohnke, M., Szalay, G., Alt, V., Schnettler, R., Lips, K.S., 2014. Biocompatibility of silver nanoparticles and silver ions in primary human mesenchymal stem cells and osteoblasts. Acta Biomater. 10, 439–449.

Pelgrift, R.Y., Friedman, A.J., 2013. Nanotechnology as a therapeutic tool to combat microbial resistance. Adv. Drug Deliv. Rev. 65, 1803–1815.

Perrone, G.G., Tan, S.-X., Dawes, I.W., 2008. Reactive oxygen species and yeast apoptosis. Biochim. Biophys. Acta (BBA)-Mol. Cell Res. 1783, 1354–1368.

Piao, M.J., Kim, K.C., Choi, J.-Y., Choi, J., Hyun, J.W., 2011. Silver nanoparticles down-regulate Nrf2-mediated 8-oxoguanine DNA glycosylase 1 through inactivation of extracellular regulated kinase and protein kinase B in human Chang liver cells. Toxicol. Lett. 207, 143–148.

Pietrzak, K., Twarużek, M., Czyżowska, A., Kosicki, R., Gutarowska, B., 2015. Influence of silver nanoparticles on metabolism and toxicity of moulds. Acta Biochim. Pol. 62.

Pinto, R.J.B., Almeida, A., Fernandes, S.C.M., Freire, C.S.R., Silvestre, A.J.D., Neto, C.P., Trindade, T., 2013. Antifungal activity of transparent nanocomposite thin films of pullulan and silver against *Aspergillus niger*. Colloids Surf. B Biointerfaces 103, 143–148.

Powers, C.M., Badireddy, A.R., Ryde, I.T., Seidler, F.J., Slotkin, T.A., 2011. Silver nanoparticles compromise neurodevelopment in PC12 cells: critical contributions of silver ion, particle size, coating, and composition. Environ. Health Perspect. 119, 37–44.

Pradeep, T., 2009. Noble metal nanoparticles for water purification: a critical review. Thin Solid Films 517, 6441–6478.

Prema, P., 2011. Chemical mediated synthesis of silver nanoparticles and its potential antibacterial application. Prog. Mol. Environ. Bioeng. Anal. Model. Technol. Appl. 6, 151–166.

Quinteros, M.A., Aristizábal, V.C., Dalmasso, P.R., Paraje, M.G., Páez, P.L., 2016. Oxidative stress generation of silver nanoparticles in three bacterial genera and its relationship with the antimicrobial activity. Toxicol. Vitr. 36, 216–223.

Rahimi, M., Hosseini, M.R., BaNhshi, M., Baghbanan, A., 2017. Biosynthesis of silver-phosphate nanoparticles using the extracellular polymeric substance of *Sporosarcina pasteurii*. Int. J. Biotechnol. Bioeng. 11, 120–123.

Rezwan, K., Meier, L.P., Rezwan, M., Vörös, J., Textor, M., Gauckler, L.J., 2004. Bovine serum albumin adsorption onto colloidal Al_2O_3 particles: a new model based on zeta potential and UV-Vis measurements. Langmuir 20, 10055–10061.

Rivero, P.J., Urrutia, A., Goicoechea, J., Arregui, F.J., 2015. Nanomaterials for functional textiles and fibers. Nanoscale Res. Lett. 10, 501.

Roe, D., Karandikar, B., Bonn-Savage, N., Gibbins, B., Roullet, J.-B., 2008. Antimicrobial surface functionalization of plastic catheters by silver nanoparticles. J. Antimicrob. Chemother. 61, 869–876.

Rogers, J.V., Parkinson, C.V., Choi, Y.W., Speshock, J.L., Hussain, S.M., 2008. A preliminary assessment of silver nanoparticle inhibition of monkeypox virus plaque formation. Nanoscale Res. Lett. 3, 129.

Rossiter, S.E., Fletcher, M.H., Wuest, W.M., 2017. Natural products as platforms to overcome antibiotic resistance. Chem. Rev. 117, 12415–12474.

Sabbah, M., Esposito, M., Pierro, P.D., Giosafatto, C.V., Mariniello, L., Porta, R., 2016. Insight into zeta potential measurements in biopolymer film preparation. J. Biotechnol. Biomat 6, e126.

Saeb, A., Alshammari, A.S., Al-Brahim, H., Al-Rubeaan, K.A., 2014. Production of silver nanoparticles with strong and stable antimicrobial activity against highly pathogenic and multidrug-resistant bacteria. Sci. World J. 2014.

Sahayaraj, K., Rajesh, S., 2011. Bionanoparticles: synthesis and antimicrobial applications. Sci. Against Microb. Pathog. Commun. Curr. Res. Technol. Adv. 23, 228–244.

Samuel, U., Guggenbichler, J.P., 2004. Prevention of catheter-related infections: the potential of a new nano-silver impregnated catheter. Int. J. Antimicrob. Agents 23, 75–78.

Sanchez-Guzman, D., Le Guen, P., Villeret, B., Sola, N., Le Borgne, R., Guyard, A., Kemmel, A., Crestani, B., Sallenave, J.M., Garcia-Verdugo, I., 2019. Silver nanoparticle-adjuvanted vaccine protects against lethal influenza infection through inducing BALT and IgA-mediated mucosal immunity. Biomaterials 217, 119308.

Schulte, P.A., Leso, V., Niang, M., Iavicoli, I., 2019. Current state of knowledge on the health effects of engineered nanomaterials in workers: a systematic review of human studies and epidemiological investigations. Scand. J. Work. Environ. Health 45, 217.

Seaton, A., Tran, L., Aitken, R., Donaldson, K., 2010. Nanoparticles, human health hazard and regulation. J. R. Soc. Interface 7, S119–S129.

Shah, Z., Gul, T., Khan, S.A., Shaheen, K., Anwar, Y., Suo, H., Ismail, M., Alghamdi, K.M., Salman, S.M., 2021. Synthesis of high surface area AgNPs from *Dodonaea viscosa* plant for the removal of pathogenic microbes and persistent organic pollutants. Mater. Sci. Eng. B 263, 114770.

Shaheen, M.N.F., El-hadedy, D.E., Ali, Z.I., 2019. Medical and microbial applications of controlled shape of silver nanoparticles prepared by ionizing radiation. Bionanoscience 9, 414–422.

Sharma, H.S., Ali, S.F., Hussain, S.M., Schlager, J.J., Sharma, A., 2009a. Influence of engineered nanoparticles from metals on the blood-brain barrier permeability, cerebral blood flow, brain edema and neurotoxicity. An experimental study in the rat and mice using biochemical and morphological approaches. J. Nanosci. Nanotechnol. 9, 5055–5072.

Sharma, H.S., Ali, S.F., Tian, Z.R., Hussain, S.M., Schlager, J.J., Sjöquist, P.-O., Sharma, A., Muresanu, D.F., 2009b. Chronic treatment with nanoparticles exacerbate hyperthermia-induced blood-brain barrier breakdown, cognitive dysfunction and brain pathology in the rat. Neuroprotective effects of nanowired-antioxidant compound H-290/51. J. Nanosci. Nanotechnol. 9, 5073–5090.

Shyla, K.K., Natarajan, N., Nakkeeran, S., 2014. Antifungal activity of zinc oxide, silver and titanium dioxide nanoparticles against *Macrophomina phaseolina*. J. Mycol. Plant Pathol. 44, 268–273.

Siddiqi, K.S., Rahman, A., Husen, A., 2018. Properties of zinc oxide nanoparticles and their activity against microbes. Nanoscale Res. Lett. 13, 1–13.

Silver, S., Gupta, A., Matsui, K., Lo, J.-F., 1999. Resistance to Ag (I) cations in bacteria: environments, genes and proteins. Met. Based. Drugs 6, 315–320.

Simbine, E.O., da Rodrigues, L.C., Lapa-Guimaraes, J., Kamimura, E.S., Corassin, C.H., de Oliveira, C.A.F., 2019. Application of silver nanoparticles in food packages: a review. Food Sci. Technol. 39, 793–802.

Singh, P., Kim, Y.J., Singh, H., Wang, C., Hwang, K.H., Farh, M.E.-A., Yang, D.C., 2015. Biosynthesis, characterization, and antimicrobial applications of silver nanoparticles. Int. J. Nanomed. 10, 2567.

Singh, R., Upadhyay, A.K., Chandra, P., Singh, D.P., 2018. Sodium chloride incites reactive oxygen species in green algae *Chlorococcum humicola* and *Chlorella vulgaris*: implication on lipid synthesis, mineral nutrients and antioxidant system. Bioresour. Technol. 270, 489–497.

Somee, L.R., Ghadam, P., Abdi-Ali, A., Fallah, S., Panahi, G., 2018. Biosynthesised AgCl NPs using *Bacillus* sp. 1/11 and evaluation of their cytotoxic activity and antibacterial and antibiofilm effects on multi-drug resistant bacteria. IET Nanobiotechnol. 12, 764–772.

Sondi, I., Salopek-Sondi, B., 2004. Silver nanoparticles as antimicrobial agent: a case study on *E. coli* as a model for Gram-negative bacteria. J. Colloid Interface Sci. 275, 177–182.

Song, B., Zhang, Y., Liu, J., Feng, X., Zhou, T., Shao, L., 2016. Is neurotoxicity of metallic nanoparticles the cascades of oxidative stress? Nanoscale Res. Lett. 11, 291.

Song, K.S., Sung, J.H., Ji, J.H., Lee, J.H., Lee, J.S., Ryu, H.R., Lee, J.K., Chung, Y.H., Park, H.M., Shin, B.S., 2013. Recovery from silver-nanoparticle-exposure-induced lung inflammation and lung function changes in Sprague Dawley rats. Nanotoxicology 7, 169–180.

Stark, W.J., 2011. Nanopartikel in biologischen Systemen. Angew. Chem. 123, 1276–1293.

Sun, L., Singh, A.K., Vig, K., Pillai, S.R., Singh, S.R., 2008. Silver nanoparticles inhibit replication of respiratory syncytial virus. J. Biomed. Nanotechnol. 4, 149–158.

Sung, J.H., Ji, J.H., Song, K.S., Lee, J.H., Choi, K.H., Lee, S.H., Yu, I.J., 2011. Acute inhalation toxicity of silver nanoparticles. Toxicol. Ind. Health 27, 149–154.

Takamiya, A.S., Monteiro, D.R., Gorup, L.F., Silva, E.A., de Camargo, E.R., Gomes-Filho, J.E., de Oliveira, S.H.P., Barbosa, D.B., 2021. Biocompatible silver nanoparticles incorporated in acrylic resin for dental application inhibit *Candida albicans* biofilm. Mater. Sci. Eng. C 118, 111341.

Tang, J., Xiong, L., Wang, S., Wang, J., Liu, L., Li, J., Yuan, F., Xi, T., 2009. Distribution, translocation and accumulation of silver nanoparticles in rats. J. Nanosci. Nanotechnol. 9, 4924–4932.

Tarazona, A., Gómez, J.V., Mateo, E.M., Jiménez, M., Mateo, F., 2019. Antifungal effect of engineered silver nanoparticles on phytopathogenic and toxigenic *Fusarium* spp. and their impact on mycotoxin accumulation. Int. J. Food Microbiol. 306, 108259.

Teeguarden, J.G., Hinderliter, P.M., Orr, G., Thrall, B.D., Pounds, J.G., 2007. Particokinetics in vitro: dosimetry considerations for in vitro nanoparticle toxicity assessments. Toxicol. Sci. 95, 300–312.

Thakkar, K.N., Mhatre, S.S., Parikh, R.Y., 2010. Biological synthesis of metallic nanoparticles. Nanomed. Nanotechnol. Biol. Med. 6, 257–262.

Thipe, V.C., Bloebaum, P., Khoobchandani, M., Karikachery, A.R., Katti, K.K., Katti, K.V., 2020. Green nanotechnology: nanoformulations against toxigenic fungi to limit mycotoxin production. In: Nanomycotoxicology. Elsevier, pp. 155–188.

Tilman, D., Cassman, K.G., Matson, P.A., Naylor, R., Polasky, S., 2002. Agricultural sustainability and intensive production practices. Nature 418, 671–677.

Tiwari, D.K., Jin, T., Behari, J., 2011. Dose-dependent in-vivo toxicity assessment of silver nanoparticle in Wistar rats. Toxicol. Mech. Methods 21, 13–24.

Tjong, S.C., Yeager, E., 1981. ESCA and SIMS studies of the passive film on iron. J. Electrochem. Soc. 128, 2251.

Tjong, S.C., Hoffman, R.W., Yeager, E.B., 1982. Electron and ion spectroscopic studies of the passive film on iron-chromium alloys. J. Electrochem. Soc. 129, 1662.

Tran, Q.H., Le, A.-T., 2018. Corrigendum: silver nanoparticles: synthesis, properties, toxicology, applications and perspectives (Adv. Nat. Sci: Nanosci. Nanotechnol. 4 033001). Adv. Nat. Sci. Nanosci. Nanotechnol. 9, 49501.

Trefry, J.C., Wooley, D.P., 2013. Silver nanoparticles inhibit vaccinia virus infection by preventing viral entry through a macropinocytosis-dependent mechanism. J. Biomed. Nanotechnol. 9, 1624–1635.

Treuel, L., Malissek, M., Grass, S., Diendorf, J., Mahl, D., Meyer-Zaika, W., Epple, M., 2012. Quantifying the influence of polymer coatings on the serum albumin corona formation around silver and gold nanoparticles. J. Nanopart. Res. 14, 1102.

Ungor, D., Dékány, I., Csapó, E., 2019. Reduction of tetrachloroaurate (III) ions with bioligands: role of the thiol and amine functional groups on the structure and optical features of gold nanohybrid systems. Nanomaterials 9, 1229.

Valodkar, M., Jadeja, R.N., Thounaojam, M.C., Devkar, R.V., Thakore, S., 2011. In vitro toxicity study of plant latex capped silver nanoparticles in human lung carcinoma cells. Mater. Sci. Eng. C 31, 1723–1728.

van der Zande, M., Vandebriel, R.J., Van Doren, E., Kramer, E., Herrera Rivera, Z., Serrano-Rojero, C.S., Gremmer, E.R., Mast, J., Peters, R.J.B., Hollman, P.C.H., 2012. Distribution, elimination, and toxicity of silver nanoparticles and silver ions in rats after 28-day oral exposure. ACS Nano 6, 7427–7442.

Venat, O., Iacomi, B., Peticilă, A.G., 2018. In vitro studies of antifungal activity of colloidal silver against important plants pathogens. Not. Bot. Horti Agrobot. Cluj-Napoca 46, 533–537.

Venkatesan, J., Singh, S.K., Anil, S., Kim, S.-K., Shim, M.S., 2018. Preparation, characterization and biological applications of biosynthesized silver nanoparticles with chitosan-fucoidan coating. Molecules 23, 1429.

Veselý, D., Vesela, D., 1995. Embryotoxic effects of a combination of zearalenone and vomitoxin (4-dioxynivalenole) on the chick embryo. Vet. Med. (Praha) 40, 279–281.

Vodnar, D.C., Pop, O.L., Dulf, F.V., Socaciu, C., 2015. Antimicrobial efficiency of edible films in food industry. Not. Bot. Horti Agrobot. Cluj-Napoca 43, 302–312.

Walczyk, D., Bombelli, F.B., Monopoli, M.P., Lynch, I., Dawson, K.A., 2010. What the cell "sees" in bionanoscience. J. Am. Chem. Soc. 132, 5761–5768.

Waris, M., Nasir, S., Abbas, S., Azeem, M., Ahmad, B., Khan, N.A., Hussain, B., Al-Ghanim, K.A., Al-Misned, F., Mulahim, N., 2020. Evaluation of larvicidal efficacy of *Ricinus communis* (Castor) and synthesized green silver nanoparticles against *Aedes aegypti* L. Saudi J. Biol. Sci.

Wilcox, M., Kite, P., Dobbins, B., 1998. Antimicrobial intravascular catheters-which surface to coat? J. Hosp. Infect. 38, 322–324.

Wolska, K.I., Grzes, K., KuREK, A., 2012. Synergy between novel antimicrobials and conventional antibiotics or bacteriocins. Pol. J. Microbiol. 61, 95–104.

Wong, H.-S., Dighe, P.A., Mezera, V., Monternier, P.-A., Brand, M.D., 2017. Production of superoxide and hydrogen peroxide from specific mitochondrial sites under different bioenergetic conditions. J. Biol. Chem. 292, 16804–16809.

Wong, K.K.Y., Liu, X., 2010. Silver nanoparticles-the real "silver bullet" in clinical medicine? Medchemcomm 1, 125–131.

Wypij, M., Czarnecka, J., Świecimska, M., Dahm, H., Rai, M., Golinska, P., 2018. Synthesis, characterization and evaluation of antimicrobial and cytotoxic activities of biogenic silver nanoparticles synthesized from *Streptomyces xinghaiensis* OF1 strain. World J. Microbiol. Biotechnol. 34, 23.

Xiu, Z.-M., Ma, J., Alvarez, P.J.J., 2011. Differential effect of common ligands and molecular oxygen on antimicrobial activity of silver nanoparticles versus silver ions. Environ. Sci. Technol. 45, 9003–9008.

Xu, F., Piett, C., Farkas, S., Qazzaz, M., Syed, N.I., 2013. Silver nanoparticles (AgNPs) cause degeneration of cytoskeleton and disrupt synaptic machinery of cultured cortical neurons. Mol. Brain 6, 1–15.

Xu, H., Hao, S., Gan, F., Wang, H., Xu, J., Liu, D., Huang, K., 2017. In vitro immune toxicity of ochratoxin A in porcine alveolar macrophages: a role for the ROS-relative TLR4/MyD88 signaling pathway. Chem. Biol. Interact. 272, 107–116.

Xu, L., Dan, M., Shao, A., Cheng, X., Zhang, C., Yokel, R.A., Takemura, T., Hanagata, N., Niwa, M., Watanabe, D., 2015a. Silver nanoparticles induce tight junction disruption and astrocyte neurotoxicity in a rat blood-brain barrier primary triple coculture model. Int. J. Nanomed. 10, 6105.

Xu, L., Shao, A., Zhao, Y., Wang, Z., Zhang, C., Sun, Y., Deng, J., Chou, L.L., 2015b. Neurotoxicity of silver nanoparticles in rat brain after intragastric exposure. J. Nanosci. Nanotechnol. 15, 4215–4223.

Yakub, I., Soboyejo, W.O., 2012. Adhesion of E. coli to silver-or copper-coated porous clay ceramic surfaces. J. Appl. Phys. 111, 124324.

Youssef, A.M., Abou-Yousef, H., El-Sayed, S.M., Kamel, S., 2015. Mechanical and antibacterial properties of novel high-performance chitosan/nanocomposite films. Int. J. Biol. Macromol. 76, 25–32.

Yu, K.-P., Huang, Y.-T., Yang, S.-C., 2013. The antifungal efficacy of nano-metals supported TiO_2 and ozone on the resistant *Aspergillus niger* spore. J. Hazard. Mater. 261, 155–162.

Yun, J., Lee, H., Ko, H.J., Woo, E.-R., Lee, D.G., 2015. Fungicidal effect of isoquercitrin via inducing membrane disturbance. Biochim. Biophys. Acta (BBA)-Biomembr. 1848, 695–701.

Zambrano-Zaragoza, M.L., González-Reza, R., Mendoza-Muñoz, N., Miranda-Linares, V., Bernal-Couoh, T.F., Mendoza-Elvira, S., Quintanar-Guerrero, D., 2018. Nanosystems in edible coatings: a novel strategy for food preservation. Int. J. Mol. Sci. 19, 705.

Zeng, Y., Zhang, D., Wu, M., Liu, Y., Zhang, X., Li, L., Li, Z., Han, X., Wei, X., Liu, X., 2014. Lipid-AuNPs@ PDA nanohybrid for MRI/CT imaging and photothermal therapy of hepatocellular carcinoma. ACS Appl. Mater. Interfaces 6, 14266–14277.

Zhang, J., Zhang, Y., Chen, Y., Du, L., Zhang, B., Zhang, H., Liu, J., Wang, K., 2012. Preparation and characterization of novel polyethersulfone hybrid ultrafiltration membranes bending with modified halloysite nanotubes loaded with silver nanoparticles. Ind. Eng. Chem. Res. 51, 3081–3090.

Zhao, J., Wang, L., Xu, D., Lu, Z., 2017. Involvement of ROS in nanosilver-caused suppression of aflatoxin production from *Aspergillus flavus*. RSC Adv. 7, 23021–23026.

PART 2

Food applications

CHAPTER 5

Silver nanoparticles as nanomaterial-based nanosensors in agri-food sector

Mythili Ravichandran[a], Paulkumar Kanniah[b], and Murugan Kasi[b]

[a]Department of Microbiology, K.S. Rangasamy College of Arts and Science, Tiruchengode, Tamil Nadu, India
[b]Department of Biotechnology, Manonmaniam Sundaranar University, Tirunelveli, Tamil Nadu, India

5.1 Introduction

Significant attention is being paid to the development of nanotechnology due to its notable applications in various fields, especially in the food and agriculture sector (Srivastava et al., 2018), but also in areas such as remediation of pollutants, medicine, solar cell technology, and pharmaceuticals (Somu and Paul, 2018; Rashad and Shalan, 2013; Otsuka et al., 2003). In general, the nanoparticles are useful in chemical and biological applications based on their unique characteristics including surface-to-volume ratio, optical, and catalytic properties (Abdel-Karim et al., 2020; Nasrollahzadeh et al., 2015).

Since the antiquities, humans have used nanomaterials in their daily lives. However, it was not until the end of the 20th century that the use of nanomaterials became popular due to the potency of their antimicrobial and photocatalytic properties (Sampathkumar et al., 2020). They are now playing prominent role as catalysts for antimicrobial, anticancer, antidiabetic, antioxidant, antiinflammatory, antilarval, and anticorrosion agents (Rajakumar and Rahuman, 2011; Bindhu and Umadevi, 2015; Jang et al., 2016; Govindappa et al., 2018; Solomon et al., 2018; Hulikere and Joshi, 2019; Sampathkumar et al., 2020; Mustafa and Andreescu, 2020). Most countries are now investing more in innovative nanoparticle-based technologies in order to acquire newly designed products and to develop strategies to deal with challenges in the agri-food industry. Nevertheless, more research is needed to find innovative technologies appropriate for rural economies (Sastry et al., 2010, 2011).

The agriculture and food sector is a critical factor in nations' economic development and growth. These sectors essentially strive to provide sustainable, high-quality, hygienic, and safe agri-food products for human and animal consumption (Sekhon, 2014; Srivastava et al., 2018). Chemical and biological contaminants used in spraying crops (e.g., pesticides, herbicides, synthetic fertilizers, heavy metals, and harmful bacterial and fungal pathogens) penetrate food crops and thus their products (Aktar et al., 2009). These contaminants can creep into food products at all stages, from production, processing, storage, and marketing (Sastry et al., 2010). Therefore, to provide contaminant-free crops and food products, it is necessary to introduce novel strategies with nanotechnology-enabled protocols.

Nanoparticles offer extensive opportunities to upgrade the quality, hygiene, and security of agri-food products through sensing, antimicrobial agents, safe packaging materials, and innocuous preservatives (Kumar et al., 2017; Nile et al., 2020). In this case, sensing is critical in determining the presence of chemicals, biological contaminants, humidity, and odors in agri-food products with accuracy and expeditious monitoring (Kumar et al., 2017; Srivastava et al., 2018; Nile et al., 2020). Therefore, to ensure disease-free, healthy products in the agriculture and food sector, they need to undergone nanotechnology-based sensing processes before human consumption. Nanoparticles, including silver, copper, and gold, express outstanding features as sensors in agri-food applications (Xiang et al., 2011; Shafiq et al., 2020). Among them, silver is a particularly excellent nanosensor due to its munificent optical properties (Bulbul et al., 2015). This chapter focuses on the importance of silver nanosensors in detecting pesticides, heavy metals, herbicides, bacterial and fungal pathogens, synthetic preservatives, and odors in agri-food sectors.

5.2 Nanosensors

Detection of contaminants in products prior to their use is a security measure with extensive applications in the food, pharmaceutical, environmental, and agricultural sectors. Screening and characterization of chemical and biological compounds in samples and specimens offer precise signaling and are essential tasks in sensor technology (Zhang et al., 2020). Nanomaterials are small devices with the unique ability to sense components rapidly with high specificity, sensitivity, and selectivity. Nanosensors are generally categorized as four types based on their properties: optical nanosensors, electrochemical nanosensors, mechanical nanosensors, and electromagnetic nanosensors (Abdel-Karim et al., 2020). Generally, nanosensors bind with specific ligands based on the targeted application, and the nanoparticle-conjugated ligands bind with specific receptors (chemicals or biological substances) (Liao et al., 2014; Munawar et al., 2019; Nosheen et al., 2019). Surface-enhanced Raman scattering (SERS), which is based on electromagnetic and chemical theories, is a sensing process that detects heavy metals, pesticides, and biological contaminants; and involves two significant procedures: Raman spectroscopy and localized surface plasmon resonance (LSPR). Transition metals such as silver, copper, and gold are generally used as substrates for SERS studies due to their enhanced functional plasmon properties in the near-infrared and visible regions (Sur, 2013; Sur and Chowdhury, 2013; Nosheen et al., 2019).

5.3 Significance of silver as nanosensors

The versatile properties of AgNPs are responsible for their various uses, including medicine, agriculture, pharmaceuticals, food processing, electronics, and in catalytic and other sensors (Alshehri et al., 2012; De Moura et al., 2012; Kirubaharan et al., 2012; Mishra and Singh, 2015; Khan et al., 2016; Burdusel et al., 2018; Jouyban and Rahimpour, 2020). Compared with other transition metal nanoparticles, AgNPs comprise inherent optical properties such as an excellent extinction coefficient, strong bandgap energy, and high absorbance of scattered plasmon (Caro et al., 2010; Vilela et al., 2012; Kailasa et al., 2018; Prosposito et al., 2020). According to the synthesis process, the production of AgNPs is more facile, rapid, and cost-effective, compared with other transition metals. Nanoparticles synthesized by plant resources have recently received considerable attention due to their nontoxic, fast, uncomplicated, and eco-friendly nature (Chung et al., 2016). It is noteworthy that the synthesis of nanoparticles by chemical reducing agents requires suitable environmental parameters, including high temperature and pressure; and also that the production of unsolicited compounds along with nanoparticles may disturb the purity and functional properties of nanoparticles (Bar et al., 2009). The microorganism-mediated synthesis process is also an excellent alternative to chemical and physical processes, whereas avoiding cross-contamination and maintaining a single, pure colony are challenging (Saxena et al., 2012). The flawless biocompatibility feature of AgNPs makes them a promising candidate for effective binding capacity with biological moieties such as antibodies, enzymes, proteins, and DNA (Caro et al., 2010).

Another important characteristic of AgNPs is stability, which needs additional care to avoid aggregation, crystallinity, and surface assembly. Their size and shape also significantly impact the stability of nanoparticles (Prosposito et al., 2020). These characteristics are required to maintain stability and result in nanoparticles that are appropriate for sensing applications. The fabrication of nanoparticles and their interactions with specific functional groups (e.g., hydroxyls, amines, esters, thiols, and carboxylate ions) change the surface property, which results in enhancement of nanoparticle-analyte affinity. Functional groups tend to hinder nanoparticle aggregation and also impact the formation of shape- and size-controlled nanoparticles. According to colorimetric sensing, the color change and the SPR absorbance peak are vital for determining accurate, qualitative detection of toxic agri-food analytes. After the addition of nanoparticles with analytes, the changes in color and SPR peak enable the detection of contaminated analytes. By optimizing physicochemical parameters such as temperature, pH, and nanoparticle concentration, it is possible to standardize the contaminants and analytes' sensing behavior of nanoparticles (Prosposito et al., 2020).

5.4 Contaminant exposure prevention and mitigation applications

The agriculture and food sector is the backbone of socioeconomic development (Sekhon, 2014). In developing and undeveloped countries, agricultural productivity can help to meet the essential needs of impoverished people. As a result, countries are seeking to use novel strategies to improve agriculture in order to provide quality and hygienic food products. Toxic contaminants discharged by various industries (e.g., heavy metals such as arsenic, mercury, copper, zinc, lead, and chromium), domestic waste, and sewage affect the quality of agricultural products (Chopra et al., 2009; Arunakumara et al., 2013) (Fig. 5.1). Pesticides are the prominent chemicals used on agricultural land to protect crops from hazardous pests (Agostini et al., 2020). After harvest, the presence of heavy metals and trace amounts of pesticides in crop products can create hazards for the health of humans and other animals (Zhou et al., 2015; Qaswar et al., 2020). The food industry also uses chemical-based preservatives to, for example, extend the shelf life and prevent changes in the color of food products (Singh and Gandhi, 2015). Most additives are safe for most humans. However, in some cases, they can harm human health and the environment (Zengin et al., 2011; Amchova et al., 2015). It is therefore essential to trace the level of toxic contaminants in agri-food products before they enter the market (Fig. 5.1). The monitoring techniques and equipment are expensive and time consuming. Consequently, devices are needed that will enable straightforward, portable, and rapid determination of the level of contaminants in food samples.

Because of intense research and development, innovative, portable sensor devices that that can rapidly identify contaminants in agri-food samples are widely available (Srivastava et al., 2018). These sensors are precautionary devices used to monitor and report products' purity in the agriculture and food industries. Nanoparticle-based

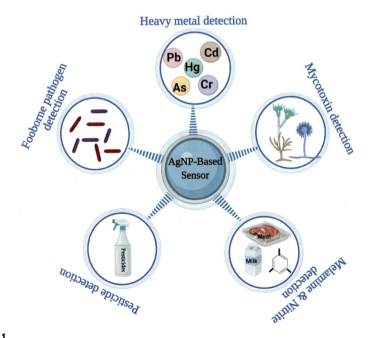

FIG. 5.1

Various applications of AgNPs-based sensors.

sensors functioning as nanodevices have pushed efforts to accurately monitor agri-food sector contaminants. Also, compared with other analytical equipment, nanodevices' lower production costs and analysis time make them attractive tools (Painuli et al., 2018). Because of their biocompatibility and ability to rapidly trace analytes with accuracy, AgNPs are well-known in sensor technologies. These salient features make AgNP nanosensors as excellent monitors in the food and agriculture sectors (Nile et al., 2020). The stability and optoelectronic properties of AgNPs also give them a promising role as nano-based sensing technology. Moreover, AgNPs' potent antimicrobial activity enables them to function as nanoshields against bacterial and fungal pathogens.

5.5 Applications of silver nanosensors in agriculture

Heavy metal contamination of agricultural lands is a huge concern and needs to be rectified in order to have food products that are safe for consumption. Toxic heavy metals emitted by industrial plants very often wind up in agricultural fields, damaging soil quality and plant growth. Crops cultivated and harvested in such areas, if consumed, can lead to diseases and disorders in humans. Nanoparticle-based sensor technology is therefore a popular means for detecting the presence of heavy metals in the environment. As a result of rainwater and industrial wastes, heavy metals—such as mercury (Hg), cadmium (Cd), arsenic (As), lead (Pb), chromium (Cr), copper (Cu), nickel (Ni), cobalt (Co), zinc (Zn), manganese (Mn), and ferric (Fe)—are available in the environment and can contaminate agricultural land.

5.6 Silver nanosensors: An alarm for heavy metal contamination

AgNPs synthesized via the green method rule in environmental nanotechnology because of their remarkable use in heavy metal sensors (Fig. 5.2). Hg^{2+} ions were sensed selectively by AgNPs synthesized while reacting with the surfactant plant *Acanthe phylum bracteatum* (Farhadi et al., 2012). From among various metal ions (Zn, Ni, Cd, Cu, Co, and Mn), the AgNPs selectively sensed only the Hg^{2+} ions. After the addition of a 100-μM concentration of Hg, the yellowish color of the AgNPs solution became colorless, revealing the binding of Hg^{2+} ions with the AgNPs (Farhadi et al., 2012). Similarly, when AgNPs were synthesized by *Hibiscus sabdariffa* leaf extract in a 300-μM/μL solution of Hg (pH 7), the yellowish-brown solution changed and became colorless. Whereas other metal ions (i.e., Cd^{2+}, Cr^{2+}, Cu^{2+}, Fe^{3+}, Zn^{2+}, Mg^{2+}, and Pb^{2+}) did not exhibit color changes, indicating selective Hg^{2+} ion detection. AgNPs synthesized at a pH of 3 showed the disappearance of the pink color in all metal ions, and there was no selective sensing of Hg^{2+} ions. In addition, AgNPs synthesized under alkaline conditions (pH 10) exhibited a change of color at 2000-μM/μL solution of Hg. In contrast, there was no selective sensing of Hg^{2+} at minimal-level concentrations.

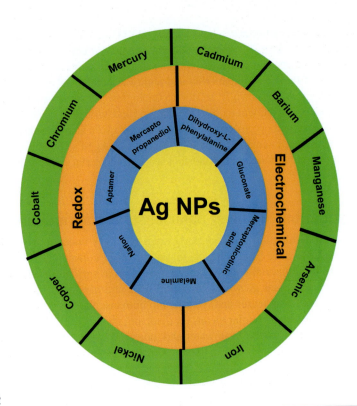

FIG. 5.2

Involvement of chemicals and biomolecules with AgNPs for detection of heavy metal ions.

The AgNPs synthesized (pH 3) by *Hibiscus sabdariffa* stem extracts were also actively involved in the selective sensing of Hg^{2+} ions. AgNPs synthesized without heated stem extract sensed the Hg only at a concentration of 200 μM/μL, but AgNPs synthesized using the heated (60°C) stem extract detected the Hg at 300 μM/μL (Kumar et al., 2014). Ahmed et al. (2015) experimented with sensing Hg^{2+} ions selectively in a sequence of the metal ions Co^{3+}, Mn^{2+}, Fe^{3+}, Zn^{2+}, Cu^{2+}, Cd^{2+}, Pb^{2+}, Ca^{2+}, Ni^{2+}, and Cr^{3+}. The SPR peak of the AgNPs vanished entirely with the addition of Hg^{2+} ions in a 55-μM concentration. Furthermore, Ahmed and colleagues noted that AgNPs produced from spinach leaf extract sensed Hg ions present in a sewage water sample. In addition, Sengan et al. (2018) documented the disappearance of the yellow color in AgNPs synthesized by *Durio zibethinus* leaf extract after the addition of Hg^{2+}. It has been used to a selective finding of Hg^{2+} among the divalent and trivalent ions including Zn, Pb, Ni, Mn, Mg, Cd, Mg, Ca, Fe, Cr, and Co. The SPR absorption band of AgNPs disappeared at an Hg^{2+} ion concentration of 33 μM.

Ghosh et al. (2018) reported selective finding of Hg^{2+} ions, even at the lower volume of 2 μM using AgNPs fabricated by *Allium sativum* extract in a phosphate buffer solution. In the aqueous medium, the AgNPs also sensed Fe^{3+} ions; and in the phosphate buffer medium, they selectively found only the Hg^{2+}, where the discharge of Ag^+ ions from the surface of AgNPs in the phosphate buffer medium strongly quenching Fe^{3+} ions, with no effect on Hg^{2+}. As a result, the Ag^+ ions from AgNPs strongly sensed Hg^{2+} ions in phosphate buffer, which is used to detect Hg^{2+} ions effectively.

Alzahrani (2020) stated that AgNPs synthesized by the peel extract of onion successfully sensed Hg^{2+} ions. Various concentrations of Hg^{2+} ions were used (0 to 1400 µL), and the color of the AgNPs vanished at 1400 µL. Alzahrani also reported detection of Hg^{2+} ions using fabricated AgNPs in ground and tap water samples. In all concentrations (300, 600, and 900 µM) of Hg^{2+} ions, there was a recovery of Hg^{2+} in more than 90% of the ground and tap water samples using AgNPs.

5.7 Functionalization of AgNPs for sensing other metal ions

When biomolecules and metal oxides attach to the surface of AgNPs, sensing of various metal ions can be achieved. Cheon and Park (2016) documented that sensing Cu^{2+} ions using 3,4-dihydroxy-L-phenylalanine (DOPA) stabilized AgNPs. The 20 nm (average size) and spherical-shaped AgNPs' detection limit was 8.1×10^{-5} M of Cu^{2+} ions. Cheon and Park (2016) also analyzed the presence of Cu^{2+} ions by the addition of Cu^{2+} ions in water samples obtained from different places including laboratory (detection limit—1.5×10^{-4} M) and market (detection limit—9.3×10^{-5} M). Another straightforward approach for sensing Cu^{2+} ions using AgNPs coated with carbon dots (AgNPs-CDs) was reported by Beiraghi and Najibi-Gehraz (2017). The color of the AgNPs-CDs changed from orange to dark brown after the addition of a Cu^{2+} substrate copper(II) nitrate trihydrate solution, revealing the selective detection of Cu^{2+} ions; whereas the other metal ions did not exhibit a color change after being mixed with an AgNPs-CD solution. Increasing the concentration of copper(II) nitrate trihydrate from 0.3 to 8 µM increased the SPR band from 430 to 525 nm. The detection limit of Cu^{2+} ions was found to be 0.037 µM, which was very low compared with WHO's recommendation of a Cu^{2+} concentration (31 µM) in drinking water (Beiraghi and Najibi-Gehraz, 2017).

Similarly, Cheng et al. (2019) stated that reduced graphene oxide-AgNPs (rGO-AgNPs) are excellent materials for detecting Cu^{2+} ions. In their study, the rGO-AgNPs acted as electrochemical nanosensors, and the detection limit was calculated at 10^{-15} M Cu^{2+} ions. Prior to this, triethylenetetramine (TETA) had been used to functionalize rGOs and was used again in this study as a reducing agent in the hydrothermal process to fabricate AgNPs size 10–20 nm on the surface of rGOs. Cheng and colleagues also used rGO-AgNPs as nanoprobes for selective detection of Hg^{2+} ions with a detection limit of 10^{-29} M.

In contrast, Maiti et al. (2016) documented that without functionalization and modification processes, AgNPs sensitized with *Citrullus lanatus* fruit extract selectively detected Cu^{2+} ions. The SPR peak was attained at 406 nm for AgNPs, while the addition of different concentrations of copper(II) nitrate solutions (0 to 200 µM) reduced the intensity of the SPR peak of AgNPs, indicating a complex formation of AgNPs and Cu^{2+} ions. On the other hand, from 20 to 200 μM, the SPR peak centered at 770 nm; the increased peak intensity also revealed the binding of Cu^{2+} ions with AgNPs. However, it was shown that bare AgNPs were not suitable for sensing Hg^{2+} ions. For that, Maiti and colleagues modified the surface of AgNPs with 3-mercapto-1,2-propanediol (MP). After the addition of a mercury (II) nitrate solution to the MP-AgNPs solution, the color of the solution changed from yellow to black, indicating binding of Hg^{2+} ions with MP-AgNPs. The study also showed that a rise of SPR peaks to 606 and 404 nm (MP-AgNPs) revealed a complex formation between MP-AgNPs and Hg^{2+} ions.

The accumulation of Co^{2+} ions in crops is responsible for reducing the minerals K, Zn, and Mn. The stress Co ions place on plants may induce oxidative stress and substantially reduce plant growth (Gopal et al., 2003; Lwalaba et al., 2020). Using AgNPs nanosensor technology offers a solution for detecting contamination of Co^{2+} ions. For example, AgNPs functionalized with glutathione (GSH) and stabilized with cetyl tri ammonium bromide (CTAB) can be considered as prominent method for detecting Co^{2+} ions. GSH-AgNPs with a spherical shape can sense not only Co^{2+} ions but also As^{3+}, Cd^{2+}, Pb^{2+}, and Ni^{2+} metal ions; in contrast, rod-shaped GSH-AgNPs selectively detected Co^{2+} ions (Sung et al., 2013).

Like Co^{2+} ions, the extreme accumulation of Pb^{2+} ions was shown to impact physiological and biological activities, including metabolism, membrane penetrability, and ATP formation of plants (Pourrut et al., 2011). The release of Pb^{2+} ions from various sources (e.g., plastics, insecticides, gasoline, paints, and sewage water) drastically affects crops and humans (Feleafel and Mirdad, 2013). Therefore, prior to human consumption, it is essential to detect Pb^{2+} ion contamination in crops. To help solve this problem, sensitive, selective, and rapid sensor technology is needed to detect toxic Pb^{2+} ions effectively. Ibuprofen-capped AgNPs (IP/AgNPs) coated on glassy carbon electrode (GCE)-based nanosensors developed by Tagar et al. (2011) successfully sensed Pb^{2+} ions. The IP/AgNPs/GCE selectively sensed a detection limit of 0.01 ppb Pb^{2+} ions. In this study, Tagar and colleagues also examined Pb^{2+} ion contamination in various water samples, including river, tap, and drinking water. Choudhury and Misra (2018) reported that AgNPs stabilized by gluconate (Glu) served as an excellent candidate for sensing Pb^{2+} ions.

The Glu-AgNPs detected Pb^{2+} ions effectively than other monovalent metal ions (Na, K), divalent ions (Ba, Ca, Ni, Hg, Cd, Zn, and Cu), and trivalent ions (Al, Fe). The reported detection limit is 0.2029 μM (Choudhury and Misra, 2018). A combinatory nanocomposite using AgNPs combined with polyamide nanofibers and cellulose nanowhiskers was effectively involved in sensing Pb^{2+} ions, where the developed PAM/CNW/AgNPs sensor detected contamination of Pb^{2+} ions below the level of 10 nmol/L (Teodoro et al., 2019).

Cd^{2+} ions in agricultural land affect plant growth by inhibiting mineral uptake, photosynthesis, and chlorosis. Cd^{2+} ions discharged from plastics, pigments, and power stations can destroy soil and crops (Benavides et al., 2005). Kumar and Anthony (2014) reported the AgNPs functionalized with *N*-(2-hydroxy benzyl)-valine (V/AgNPs) and *N*-(2-hydroxy benzyl)-isoleucine (I/AgNPs) selectively detected

Cd^{2+} ions. Further, they stated the occurrence of two SPR peaks in functionalized AgNPs around 400–410, and 500 nm revealed the presence of AgNPs and Cd^{2+} ions. The addition of the stabilizing agents trisodium citrate (SC) and ethylenediaminetetraacetate (EDTA) with V/AgNPs and I/AgNPs was shown to enhance Cd^{2+} ions' sensing capability (Kumar and Anthony, 2014). In contrast, Aravind et al. (2018a) explored biofabrication of AgNPs with cloves of *Allium sativum* to detect toxic Cd^{2+} ions. They reported that, without functionalization, modification, and stabilization, AgNPs sense Cd^{2+} ions with a detection limit of 0.277 μM. A report by Pawar et al. (2019) stated that AgNPs' memristive switching property enabled precise sensing of Cd^{2+} ions, even at a minimum level of 10 ppm in the frequency of 0.16 Hz.

In heavy metal sensing, Cr^{4+} ions have been detected by a AgNPs/Nafion/GCE composite. Nafion is a well-known cationic exchange ionomer matrix used in electrochemical sensor technology. Nafion acts as a stabilizer for AgNPs synthesis and as a substrate to bind AgNPs on GCE/Nafion electrodes. The reported minimum detection limit is 0.67 ppb of Cr^{4+} ions was achieved through the use of modified AgNPs-coated electrodes (Xing et al., 2011). Through naked eye visualization, Cr^{3+} ions were detected easily using AgNPs modified with melamine (M) and 6-mercaptonicotinic acid (MCNA). The color of the AgNPs/M/MCNA complex changed from yellow to reddish-brown after the addition of chromium(III) nitrate, revealing selective detection of Cr^{3+} ions, with a detection limit of 64.51 nM (Modi et al., 2014). Interestingly, the AgNPs/M/MCNA complex sensed the Ba^{2+} ions with a detection limit of 80.21 nM.

In contrast, Aravind et al. (2018a,b) reported that AgNPs synthesized with the *Lycopersicon esculentum* fruit extract without any functionalization sensed Cr^{3+} ions. Compared with the previously mentioned reports, the report by Aravind and colleagues is more straightforward and sensitive and shows that supplementary stabilizing and functionalizing agents are not required. The AgNPs-modified platinum electrode selectively sensed Cr^{3+} ions, with a minimum detection limit of 0.804 μM.

Because of the aptamer's high affinity to its target, it was functionalized with AgNPs and was involved in the effective detection of As^{3+} ions. The minimum level of detection of As^{3+} ions by AgNPs was 6 μg/L. The AgNPs effectively sensed the presence of As^{3+} ions in tap, river, and well-water samples (Divsar et al., 2015). Boruah et al. (2019) synthesized polyethylene glycol (PEG)-functionalized AgNPs for effective sensing of As^{3+} ions. After the addition of As^{3+} ions to the PEG-AgNPs solution, the color of the solution changed from yellow to bluish, revealing the binding of As^{3+} ions with PEG-AgNPs. The detection limit was found to be 1 ppb of As^{3+} ions.

5.8 Mechanism of heavy metal detection using silver nanosensors

Numerous reports exemplify bare AgNPs and functionalized AgNPs' mechanisms for sensing heavy metals and pesticides. The redox and electrochemical reactions are the primary processes that occur between AgNPs and heavy metal ions.

MP-functionalized AgNPs bind with Hg^{2+} ions through the O atom present in the MP. The aggregation of AgNPs and Hg^{2+} ions is revealed through a change in color from yellow to blackish (Maiti et al., 2016). In addition, the AgNPs without functionalization successfully bind with Hg^{2+} ions through a redox reaction. It was shown that *Acanthe phylum bracteatum* plant stabilizing agents are replaced by Hg^{2+} ions and are deposited on the surface of AgNPs (Farhadi et al., 2012). Another mechanism demonstrated by Kumar et al. (2014) and Sengan et al. (2018) the difference in the standard reduction potentials of Ag (+0.80 V) and Hg (+0.92 V) resulted in the Hg^{2+} ions being effective oxidizing agents. The oxidation potential of AgNPs may reduce the Hg^{2+} ions to Hg^0 and result in an amalgamation reaction between the reduced Hg and Ag (Kumar et al., 2014; Sengan et al., 2018).

Interestingly, AgNPs functionalized with the biological molecule DOPA specifically bind with Pb^{2+} ions through the bond formation between oxygen and nitrogen atoms. The oxygen and nitrogen atoms are derived from the DOPA present on the surface of AgNPs (Cheon and Park, 2016). AgNPs stabilized with gluconate bind with Pb^{2+} ions through the carboxylate ions of gluconate. Binding of Pb^{2+} ions with gluconate results in the aggregation of AgNPs, with a change in color from yellow to pinkish-red (Choudhury and Misra, 2018). Other AgNPs-functionalized molecules (e.g., aptamer, nafion, melamine, and mercaptonicotinic acid) are able to sense noxious heavy metal ions (Divsar et al., 2015; Xing et al., 2011; Modi et al., 2014) (Fig. 5.2).

5.9 Silver nanosensors for pesticide detection

Plant diseases affect food security and the economy worldwide. Proper management that efficiently detects, diagnoses, and ultimately treats plant diseases is needed. Such management can help ensure healthier plants and better crop yield, assess and reduce the spread of diseases, analyze plant metabolic flux, and reduce waste in agricultural settings. In general, nanoparticles are potent nanocides against insects and pests (Kannan et al., 2020). They provide a greater surface area to fertilizers and pesticides (Shang et al., 2019). Moreover, they offer controlled, site-targeted release of active agro compounds, which enhances crop production (Shang et al., 2019) and helps to overcome the uncertainties in the food and agriculture sector.

Currently, AgNPs-based nanomaterials are being used in agriculture (Fig. 5.3). Despite its specificity, it monitors plant signaling pathways and regulatory metabolism. In addition, it protects and inspects plant health through potent sensing ability of early and on-site detection (Kashyap et al., 2019). AgNPs are an emerging nanodiagnostic tool for the detection of plant pathogens. Single-stranded DNA sensors functionalized with silver and DNA hybridization based on a SERS substrate were used to detect *P. lateralis* and *P. ramorum* analytes against plant pathogens (Li et al., 2020).

The overuse of pesticides has contaminated the world's food and environment (Li et al., 2020), thus there is great concern about food safety and other health risks. These pesticides include organochlorines, organophosphates, carbamates, and

5.9 Silver nanosensors for pesticide detection

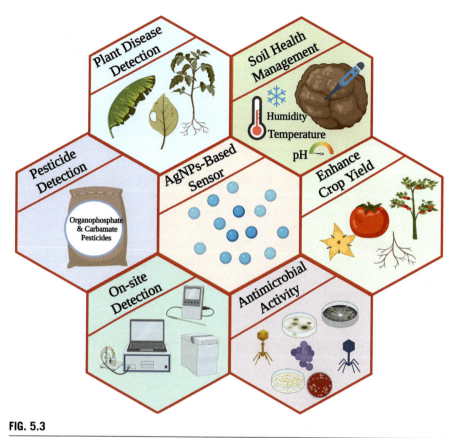

FIG. 5.3

Applications of AgNPs-based sensors in agriculture.

pyrethroids, which can cause neurological disorders and even death. To help alleviate this crisis, immediate and appropriate measures are required, with a particular focus on the development of sustainable agriculture. The concept of "controlled-loss-fertilizers" meant to decrease nonpoint source pollution is currently of interest (Jiang et al., 2006).

AgNPs have been improved over time and are attractive candidates for pesticide detection. Le et al. (2019) noted that silver-incorporated chitosan nanocomposites (Ag/CS) combined with antracol fungicides showed enhanced antifungal activity. The stem extract of *Gossypium hirsutum* was used to synthesize AgNPs and showed positive antibacterial activity against prominent bacterial pathogens such as *Xanthomonas campestris* pv. *campestris* and *Xanthomonas axonopodis* pv. *malvacearum* (Vanti et al., 2019). He et al. (2015) reported that the silver nanoparticle-based chemiluminescent (CL) sensor array used to detect carbamate and organophosphate pesticides including carbofuran, chlorpyrifos, carbaryl, dipterex, and dimethoate were distinguished at 24 μg/mL with 95% accuracy. However, the sensors' working mechanism is based mainly on optical and electrochemical transduction of the triple-channel properties of the luinol-functionalized silver nanoparticle (L/AgNP), which is conjugated with a hydrogen peroxide-based CL detection system to engender CL fingerprinting.

A similar approach was demonstrated by detecting organophosphorus (OPs) pesticides using a colorimetric sensor during the modulated synthesis of AgNPs by enzyme-based assays (Kumar et al., 2015). AgNPs fabricated with graphitic carbon nitride (g-C_3N_4/AgNPs) were also used for pesticide analysis. The g-C_3N_4/AgNPs is a dual-signal sensor accompanied to detect parathion-methyl, an organophosphorous pesticide (Li et al., 2020). An electrochemical-based sensor conjugated with AgNPs-decorated polyaniline-nanocrystalline zeolite, employed to detect lindane is broadly used to control disease vectors in agriculture. The linearity ranged from 10 nM to 900 μM, with a detection limit of 5 nM (Kaur et al., 2015). In addition, the AgNPs-based sensors interconnected with sulphurazon-ethyl herbicides, which demonstrated improved and selective detection of different concentrations of herbicidal contaminants, inducing color variations, ranging from yellow to orange to purple (Dubas and Pimpan, 2008).

5.10 Food contaminant detection

The quality and type of foods we eat are important to our health. However, our food can be contaminated and those contaminants come from many sources—for example, overuse of chemical fertilizers, residue from pesticides, foodborne pathogens, microbial toxins, preservatives, adulterants, and processing and packaging methods. Such contaminants negatively impact human and environmental health (Kuswandi et al., 2017; Madhavan et al., 2019).

AgNPs-fabricated sensors that can detect foodborne contaminants are opening new doors in food technology to determine the interactions and influence of food components at a nano level (Sharif et al., 2018). These food contaminant sensors are used due to their high efficiency, specificity, and sensitivity to food quality and security. AgNPs-based chemical sensors are employed to analyze pesticides in food samples. They have been used to do a quantitative analysis of the pesticide diazinon based on π-conjugated pyrimidine nitrogen and sulfur moieties and noncovalent interactions of diazinon with AgNPs (Shrivas et al., 2019). Madhavan et al. (2019) reported that AgNPs-based sensors functionalized with cysteine and histidine sensed milk spoilage based on the lactic acid concentration. Nitrite is an essential food preservatives; it is also a carcinogen. A hyperbranched polyethyleneimine-protected silver nanocluster (hbPEI-Ag NC) sensor is used to analyze nitrite (Chen et al., 2016). In another study, Ping et al. (2012) have used label-free AgNPs probe for the detection of common adulterant melamine in the milk sample. The assay was based on the aggregation behavior of AgNPs, which induces a color change from yellow to red, and the sensitivity was 2.32 μM (Ping et al., 2012).

In general, food is the primary source of health management. However, it is contaminated with pathogens, it will cause severe food-borne infections. The detection of pathogens in food plays a significant role in disease prevention (Rubab et al., 2018), and several methods have been used to detect them. AgNPs-based biosensors are attractive in on-site applications. Wang et al. (2017) demostrated a multifunctional bioassay-derived conjugated polyelectrolytes (CPs)-silver nanostructure pair that was used to detect, prevent, and kill bacteria, which acts as the new dual-platform for multifunctional and antibacterial sensing. Besides, the detection was based on the plasmon-enhanced fluorescent response of the paired silver nanostructure (Wang et al., 2017). Abbaspour et al. (2015) developed a bioassay that used a dual-based-aptamer sandwich immunosensor, conjugated with AgNPs, to detect *Staphylococcus aureus*. AgNPs acted as a reporter for this sensing mechanism. The capture probe that is streptavidin-coated magnetic beads immobilized with anti-*S. aureus* aptamer. In the presence of the target organism (*S. aureus*), an Apt/*S. aureus*/apt-AgNPs sandwich composite formed above the surface of the magnetic beads. The electrochemical signal amplification was carried out by anodic stripping voltammetry from the AgNPs. This method is quick, precise, and offers multiplexed detection.

Heavy metal contaminants in food, including arsenic, chromium, cadmium, lead, and mercury, act as causative agents for many neurological, reproductive, cardiovascular, and developmental disorders (Kim et al., 2012). They are highly toxic to public health, even at trace levels; therefore the development of sensitive methods, such as nanosensors, is highly advantageous. Accordingly, many Au- and AgNPs-based nanosensors have been developed for sensing these metal ions (Cao et al., 2001). To achieve the highest sensitivity, a catalytic and molecular beacon (CAMB) sensor integrated with silver nanoclusters (AgNCs) has been used as an amplified sensing array to detect Pb^{2+} ions. However, AgNCs are used to induce fluorescence response to enhance recognition of CAMB sensors (Gong et al., 2015). Based on a similar scheme, a colorimetric sensor based on AgNPs with methionine was used to effectively detect Hg^{2+} ions (Balasurya et al., 2020).

Nanomaterial-based apta sensing is an emerging sensing field. Significant progress has been achieved with nanosilver-assisted aptasensors, including analysis of a broad range of molecules with particular reference to mycotoxin analysis and monitoring (Rhouati et al., 2017). Mycotoxins are secondary metabolites of molds. They act as the source of harmful effects on living entities, especially humans, animals, and plants. Mycotoxins such as aflatoxins, ochratoxins, trichothecenes, zearalenones, fumonisins, tremorgenic toxins, and ergot alkaloids are significant mycotoxins in the agriculture sector (Hussein and Brasel, 2001). In a study conducted by Karimi et al. (2017), single-stranded DNA aptamer-functionalized AgNPs with the support of carbon fiber microelectrode (CFME) selectively detected OTA, with a detection limit of 0.05 nM. According to work conducted by Chen et al. (2014), a fluorescent DNA-scaffolded AgNC aptasensor was designed to detect ochratoxin (OTA) in wheat samples. It exhibited a sensitivity limit as low as 20 pg/mL.

In a recent study, AgNPs were utilized as SERS signal inducers combined with gold NPs. In SERS, molecules were placed near the surface, and usually gold and

AgNPs were employed at 10^{-5} and 10^{-8}, respectively (Wilson and Willets, 2013). In a study by Zhao et al. (2015), mycotoxins such as ochratoxin (OTA) and aflatoxin B1 (AFB1) were detected in cornmeal, which was achieved through SERS labels embedded with Ag-Au core-shell (CS) NPs through a galvanic replacement-free deposition. The SERS-engineered Ag-Au CS NPs produced a strong plasmonic coupling to achieve powerful SERS signal amplification. The specificity pattern was susceptible as low as 0.006 and 0.03 µg/L for OTA and AFB1, respectively. In another study conducted by Li et al. (2016), a SERS sensor fabricated with heterogeneous gold nanostar (AuNS) core-silver (AgNPs) satellites was employed for ultrasensitive analysis of AFB1 in a peanut milk sample. The research was carried out with the aid of surface Ag satellites, amplified by SERS signals. AFB1 sensitivity was detected at a concentration ranging from 1 to 1000 pg/mL, with a detection limit of 0.48 pg/mL.

5.11 Conclusion and future perspectives

Despite criticism related to the use of silver-based nanosensors, there is great expectation for their use in a variety of ways, including in the agri-food sector, as has been reported in various studies. The growing interest in silver nanosensors' role in the agricultural and food sectors is due mainly to their diverse capabilities and their ability to make an innovative paradigm shift from conventional agriculture to energy-efficient, smart, and sustainable agriculture. For example, nanosensing devices that incorporate AgNPs are a practical means for quickly detecting agricultural- and food-related contaminants and damage from pesticides, thus protecting plants and plant products. As compared with other nanomaterial-based nanosensors, silver nanosensors have shown greater potential in scientific and industrial fields due to their useful functionalities, such as identifying and measuring analytes; on-site monitoring and analysis; multiplexed, continuous detection with high sensitivity; and real-time results. The remarkable achievements derived from incorporating AgNPs into sensors include the scope with which they can be used within a wide range of sectors, their rapid response, their precise detection and sensitivity, and their cost-effectiveness and economic viability.

In this chapter, we explored existing literature on AgNPs and broadened the understanding of the vast potential for smart use of AgNPs. Nevertheless, despite much progress, the use of AGNP nanosensors in agri-food and other sectors still faces challenges, and our understanding of them and their applications in all sectors is still in its infancy. In the agriculture and food sector specifically, in-depth studies are needed to establish and consider the impact of AgNPs on nanotechnology in phytopathology designed to protect and improve plants. More generally, because their underlying mechanisms remain in large part undetermined, future studies could focus on developing stabilization criteria and optimizing parameters to further enhance the analytical and practical advancements of AgNPs-based biosensors, all of which would significantly improve their marketability.

Acknowledgments

The authors thank BioRender for the drawings in this chapter.

References

Abbaspour, A., Norouz-Sarvestani, F., Noori, A., Soltani, N., 2015. Aptamer-conjugated silver nanoparticles for electrochemical dual-aptamer-based sandwich detection of *Staphylococcus aureus*. Biosens. Bioelectron. 68, 149–155.

Abdel-Karim, R., Reda, Y., Abdel-Fattah, A., 2020. Nanostructured materials-based nanosensors. J. Electrochem. Soc. 167 (3), 037554.

Agostini, M.G., Roesler, I., Bonetto, C., Ronco, A.E., Bilenca, D., 2020. Pesticides in the real world: the consequences of GMO-based intensive agriculture on native amphibians. Biol. Conserv. 241, 108355.

Ahmed, K.B.A., Senthilnathan, R., Megarajan, S., Anbazhagan, V., 2015. Sunlight mediated synthesis of silver nanoparticles using redox phytoprotein and their application in catalysis and colorimetric mercury sensing. J. Photochem. Photobiol. B 151, 39–45.

Aktar, W., Sengupta, D., Chowdhury, A., 2009. Impact of pesticides use in agriculture: their benefits and hazards. Interdiscip. Toxicol. 2 (1), 1–12.

Alshehri, A.H., Jakubowska, M., Młożniak, A., Horaczek, M., Rudka, D., Free, C., Carey, J.D., 2012. Enhanced electrical conductivity of silver nanoparticles for high frequency electronic applications. ACS Appl. Mater. Interfaces 4 (12), 7007–7010.

Alzahrani, E., 2020. Colorimetric detection based on localized surface plasmon resonance optical characteristics for sensing of mercury using green-synthesized silver nanoparticles. J. Anal. Methods Chem. 2020. https://doi.org/10.1155/2020/6026312. Article ID 6026312.

Amchova, P., Kotolova, H., Ruda-Kucerova, J., 2015. Health safety issues of synthetic food colorants. Regul. Toxicol. Pharmacol. 73 (3), 914–922.

Aravind, A., Sebastian, M., Mathew, B., 2018a. Green silver nanoparticles as a multifunctional sensor for toxic Cd (II) ions. New J. Chem. 42 (18), 15022–15031.

Aravind, A., Sebastian, M., Mathew, B., 2018b. Green synthesized unmodified silver nanoparticles as a multi-sensor for Cr (III) ions. Environ. Sci. Water Res. Technol. 4 (10), 1531–1542.

Arunakumara, K.K.I.U., Walpola, B.C., Yoon, M.H., 2013. Current status of heavy metal contamination in Asia's rice lands. Rev. Environ. Sci. Biotechnol. 12 (4), 355–377.

Balasurya, S., Syed, A., Thomas, A.M., Marraiki, N., Elgorban, A.M., Raju, L.L., Das, A., Khan, S.S., 2020. Rapid colorimetric detection of mercury using silver nanoparticles in the presence of methionine. Spectrochim. Acta A Mol. Biomol. Spectrosc. 228, 117712.

Bar, H., Bhui, D.K., Sahoo, G.P., Sarkar, P., Pyne, S., Misra, A., 2009. Green synthesis of silver nanoparticles using seed extract of *Jatropha curcas*. Colloids Surf. A Physicochem. Eng. Asp. 348 (1–3), 212–216.

Beiraghi, A., Najibi-Gehraz, S.A., 2017. Carbon dots-modified silver nanoparticles as a new colorimetric sensor for selective determination of cupric ions. Sensors Actuators B Chem. 253, 342–351.

Benavides, M.P., Gallego, S.M., Tomaro, M.L., 2005. Cadmium toxicity in plants. Braz. J. Plant Physiol. 17 (1), 21–34.

Bindhu, M.R., Umadevi, M., 2015. Antibacterial and catalytic activities of green synthesized silver nanoparticles. Spectrochim. Acta A Mol. Biomol. Spectrosc. 135, 373–378.

Boruah, B.S., Daimari, N.K., Biswas, R., 2019. Functionalized silver nanoparticles as an effective medium towards trace determination of arsenic (III) in aqueous solution. Results Phys. 12, 2061–2065.

Bulbul, G., Hayat, A., Andreescu, S., 2015. Portable nanoparticle-based sensors for food safety assessment. Sensors 15 (12), 30736–30758.

Burdusel, A.C., Gherasim, O., Grumezescu, A.M., Mogoanta, L., Ficai, A., Andronescu, E., 2018. Biomedical applications of silver nanoparticles: an up-to-date overview. Nanomaterials 8 (9), 681.

Cao, Y., Jin, R., Mirkin, C.A., 2001. DNA-modified core-shell Ag/Au nanoparticles. J. Am. Chem. Soc. 123 (32), 7961–7962.

Caro, C., Castillo, P.M., Klippstein, R., Pozo, D., Zaderenko, A.P., 2010. Silver nanoparticles: sensing and imaging applications. In: Silver Nanoparticles. InTech, Croatia, pp. 201–224 (Republished).

Chen, J., Zhang, X., Cai, S., Wu, D., Chen, M., Wang, S., Zhang, J., 2014. A fluorescent aptasensor based on DNA-scaffolded silver-nanocluster for ochratoxin A detection. Biosens. Bioelectron. 57, 226–231.

Chen, C., Yuan, Z., Chang, H.T., Lu, F., Li, Z., Lu, C., 2016. Silver nanoclusters as fluorescent nanosensors for selective and sensitive nitrite detection. Anal. Methods 8 (12), 2628–2633.

Cheng, Y., Li, H., Fang, C., Ai, L., Chen, J., Su, J., Zhang, Q., Fu, Q., 2019. Facile synthesis of reduced graphene oxide/silver nanoparticles composites and their application for detecting heavy metal ions. J. Alloys Compd. 787, 683–693.

Cheon, J.Y., Park, W.H., 2016. Green synthesis of silver nanoparticles stabilized with mussel-inspired protein and colorimetric sensing of lead (II) and copper (II) ions. Int. J. Mol. Sci. 17 (12), 2006.

Chopra, A.K., Pathak, C., Prasad, G., 2009. Scenario of heavy metal contamination in agricultural soil and its management. J. Appl. Nat. Sci. 1 (1), 99–108.

Choudhury, R., Misra, T.K., 2018. Gluconate stabilized silver nanoparticles as a colorimetric sensor for Pb^{2+}. Colloids Surf. A Physicochem. Eng. Asp. 545, 179–187.

Chung, I.M., Park, I., Seung-Hyun, K., Thiruvengadam, M., Rajakumar, G., 2016. Plant-mediated synthesis of silver nanoparticles: their characteristic properties and therapeutic applications. Nanoscale Res. Lett. 11 (1), 40.

De Moura, M.R., Mattoso, L.H., Zucolotto, V., 2012. Development of cellulose-based bactericidal nanocomposites containing silver nanoparticles and their use as active food packaging. J. Food Eng. 109 (3), 520–524.

Divsar, F., Habibzadeh, K., Shariati, S., Shahriarinour, M., 2015. Aptamer conjugated silver nanoparticles for the colorimetric detection of arsenic ions using response surface methodology. Anal. Methods 7 (11), 4568–4576.

Dubas, S.T., Pimpan, V., 2008. Green synthesis of silver nanoparticles for ammonia sensing. Talanta 76 (1), 29–33.

Farhadi, K., Forough, M., Molaei, R., Hajizadeh, S., Rafipour, A., 2012. Highly selective Hg^{2+} colorimetric sensor using green synthesized and unmodified silver nanoparticles. Sensors Actuators B Chem. 161 (1), 880–885.

Feleafel, M.N., Mirdad, Z.M., 2013. Hazard and effects of pollution by lead on vegetable crops. J. Agric. Environ. Ethics 26 (3), 547–567.

Ghosh, S., Maji, S., Mondal, A., 2018. Study of selective sensing of Hg^{2+} ions by green synthesized silver nanoparticles suppressing the effect of Fe^{3+} ions. Colloids Surf. A Physicochem. Eng. Asp. 555, 324–331.

Gong, L., Kuai, H., Ren, S., Zhao, X.H., Huan, S.Y., Zhang, X.B., Tan, W., 2015. Ag nanocluster-based label-free catalytic and molecular beacons for amplified biosensing. Chem. Commun. 51 (60), 12095–12098.

Gopal, R., Dube, B.K., Sinha, P., Chatterjee, C., 2003. Cobalt toxicity effects on growth and metabolism of tomato. Commun. Soil Sci. Plant Anal. 34 (5–6), 619–628.

Govindappa, M., Hemashekhar, B., Arthikala, M.K., Rai, V.R., Ramachandra, Y.L., 2018. Characterization, antibacterial, antioxidant, antidiabetic, anti-inflammatory and antityrosinase activity of green synthesized silver nanoparticles using *Calophyllum tomentosum* leaves extract. Results Phys. 9, 400–408.

He, Y., Xu, B., Li, W., Yu, H., 2015. Silver nanoparticle-based chemiluminescent sensor array for pesticide discrimination. J. Agric. Food Chem. 63 (11), 2930–2934.

Hulikere, M.M., Joshi, C.G., 2019. Characterization, antioxidant and antimicrobial activity of silver nanoparticles synthesized using marine endophytic fungus-*Cladosporium cladosporioides*. Process Biochem. 82, 199–204.

Hussein, H.S., Brasel, J.M., 2001. Toxicity, metabolism, and impact of mycotoxins on humans and animals. Toxicology 167 (2), 101–134.

Jang, S.J., Yang, I.J., Tettey, C.O., Kim, K.M., Shin, H.M., 2016. In-vitro anticancer activity of green synthesized silver nanoparticles on MCF-7 human breast cancer cells. Mater. Sci. Eng. C 68, 430–435.

Jiang, J., Cai, D.Q., Yu, Z.L., Wu, Y.J., 2006. Loss-Control Fertilizer Made BY Active Clay, Flocculant and Sorbent. Chinese Patent Specification ZL200610040631, 1.

Jouyban, A., Rahimpour, E., 2020. Optical sensors based on silver nanoparticles for determination of pharmaceuticals: an overview of advances in the last decade. Talanta 217, 121071.

Kailasa, S.K., Koduru, J.R., Desai, M.L., Park, T.J., Singhal, R.K., Basu, H., 2018. Recent progress on surface chemistry of plasmonic metal nanoparticles for colorimetric assay of drugs in pharmaceutical and biological samples. Trends Anal. Chem. 105, 106–120.

Kannan, M., Elango, K., Tamilnayagan, T., Preetha, S., Kasivelu, G., 2020. Impact of nanomaterials on beneficial insects in agricultural ecosystems. In: Nanotechnology for Food, Agriculture, and Environment. Springer, Cham, pp. 379–393.

Karimi, A., Hayat, A., Andreescu, S., 2017. Biomolecular detection at ssdna-conjugated nanoparticles by nano-impact electrochemistry. Biosens. Bioelectron. 87, 501–507.

Kashyap, P.L., Kumar, S., Jasrotia, P., Singh, D.P., Singh, G.P., 2019. Nanosensors for plant disease diagnosis: current understanding and future perspectives. In: Nanoscience for Sustainable Agriculture. Springer, Cham, pp. 189–205.

Kaur, B., Srivastava, R., Satpati, B., 2015. Silver nanoparticle decorated polyaniline–zeolite nanocomposite material based non-enzymatic electrochemical sensor for nanomolar detection of lindane. RSC Adv. 5 (71), 57657–57665.

Khan, F.U., Chen, Y., Khan, N.U., Khan, Z.U.H., Khan, A.U., Ahmad, A., Tahir, K., Wang, L., Khan, M.R., Wan, P., 2016. Antioxidant and catalytic applications of silver nanoparticles using *Dimocarpus longan* seed extract as a reducing and stabilizing agent. J. Photochem. Photobiol. B Biol. 164, 344–351.

Kim, H.N., Ren, W.X., Kim, J.S., Yoon, J., 2012. Fluorescent and colorimetric sensors for detection of lead, cadmium, and mercury ions. Chem. Soc. Rev. 41 (8), 3210–3244.

Kirubaharan, C.J., Kalpana, D., Lee, Y.S., Kim, A.R., Yoo, D.J., Nahm, K.S., Kumar, G.G., 2012. Biomediated silver nanoparticles for the highly selective copper (II) ion sensor applications. Ind. Eng. Chem. Res. 51 (21), 7441–7446.

Kumar, V.V., Anthony, S.P., 2014. Silver nanoparticles based selective colorimetric sensor for Cd^{2+}, Hg^{2+} and Pb^{2+} ions: tuning sensitivity and selectivity using co-stabilizing agents. Sensors Actuators B Chem. 191, 31–36.

Kumar, V.V., Anbarasan, S., Christena, L.R., SaiSubramanian, N., Anthony, S.P., 2014. Biofunctionalized silver nanoparticles for selective colorimetric sensing of toxic metal ions and antimicrobial studies. Spectrochim. Acta A Mol. Biomol. Spectrosc. 129, 35–42.

Kumar, D.N., Rajeshwari, A., Alex, S.A., Sahu, M., Raichur, A.M., Chandrasekaran, N., Mukherjee, A., 2015. Developing acetylcholinesterase-based inhibition assay by modulated synthesis of silver nanoparticles: applications for sensing of organophosphorus pesticides. RSC Adv. 5 (76), 61998–62006.

Kumar, V., Guleria, P., Mehta, S.K., 2017. Nanosensors for food quality and safety assessment. Environ. Chem. Lett. 15 (2), 165–177.

Kuswandi, B., Futra, D., Heng, L.Y., 2017. Nanosensors for the detection of food contaminants. In: Nanotechnology Applications in Food. Academic Press, United States, pp. 307–333.

Le, V.T., Bach, L.G., Pham, T.T., Le, N.T.T., Ngoc, U.T.P., Tran, D.H.N., Nguyen, D.H., 2019. Synthesis and antifungal activity of chitosan-silver nanocomposite synergize fungicide against *Phytophthora capsici*. J. Macromol. Sci. A 56 (6), 522–528.

Li, A., Tang, L., Song, D., Song, S., Ma, W., Xu, L., Kuamg, H., Wu, X., Liu, L., Chen, X., Xu, C., 2016. A SERS-active sensor based on heterogeneous gold nanostar core–silver nanoparticle satellite assemblies for ultrasensitive detection of aflatoxinB1. Nanoscale 8 (4), 1873–1878.

Li, Y., Wan, M., Yan, G., Qiu, P., Wang, X., 2020. A dual-signal sensor for the analysis of parathion-methyl using silver nanoparticles modified with graphitic carbon nitride. J. Pharm. Anal. https://doi.org/10.1016/j.jpha.2020.04.007.

Liao, L., Wang, S., Xiao, J., Bian, X., Zhang, Y., Scanlon, M.D., Hu, X., Tang, Y., Liu, B., Girault, H.H., 2014. A nanoporous molybdenum carbide nanowire as an electrocatalyst for hydrogen evolution reaction. Energy Environ Sci. 7 (1), 387–392.

Lwalaba, J.L.W., Louis, L.T., Zvobgo, G., Richmond, M.E.A., Fu, L., Naz, S., Mwamba, M., Mundende, R.P.M., Zhang, G., 2020. Physiological and molecular mechanisms of cobalt and copper interaction in causing phyto-toxicity to two barley genotypes differing in Co tolerance. Ecotoxicol. Environ. Saf. 187, 109866.

Madhavan, A.A., Qotainy, R., Nair, R., 2019. Synthesis of functionalized silver nanoparticles and its application as chemical sensor. IEEE, Dubai, United Arab Emirates, pp. 1–4.

Maiti, S., Barman, G., Laha, J.K., 2016. Detection of heavy metals (Cu^{+2}, Hg^{+2}) by biosynthesized silver nanoparticles. Appl. Nanosci. 6 (4), 529–538.

Mishra, S., Singh, H.B., 2015. Biosynthesized silver nanoparticles as a nanoweapon against phytopathogens: exploring their scope and potential in agriculture. Appl. Microbiol. Biotechnol. 99 (3), 1097–1107.

Modi, R.P., Mehta, V.N., Kailasa, S.K., 2014. Bifunctionalization of silver nanoparticles with 6-mercaptonicotinic acid and melamine for simultaneous colorimetric sensing of Cr^{3+} and Ba^{2+} ions. Sensors Actuators B Chem. 195, 562–571.

Munawar, A., Ong, Y., Schirhagl, R., Tahir, M.A., Khan, W.S., Bajwa, S.Z., 2019. Nanosensors for diagnosis with optical, electric and mechanical transducers. RSC Adv. 9 (12), 6793–6803.

Mustafa, F., Andreescu, S., 2020. Nanotechnology-based approaches for food sensing and packaging applications. RSC Adv. 10 (33), 19309–19336.

Nasrollahzadeh, M., Sajadi, S.M., Babaei, F., Maham, M., 2015. *Euphorbia helioscopia* Linn as a green source for synthesis of silver nanoparticles and their optical and catalytic properties. J. Colloid Interface Sci. 450, 374–380.

Nile, S.H., Baskar, V., Selvaraj, D., Nile, A., Xiao, J., Kai, G., 2020. Nanotechnologies in food science: applications, recent trends, and future perspectives. Nanomicro Lett. 12 (1), 45.

Nosheen, E., Shah, A., Iftikhar, F.J., Aftab, S., Bakirhan, N.K., Ozkan, S.A., 2019. Optical nanosensors for pharmaceutical detection. In: New Developments in Nanosensors for Pharmaceutical Analysis. Academic Press, New York, NY, USA, pp. 119–140.

Otsuka, H., Nagasaki, Y., Kataoka, K., 2003. PEGylated nanoparticles for biological and pharmaceutical applications. Adv. Drug Deliv. Rev. 55 (3), 403–419.

Painuli, R., Joshi, P., Kumar, D., 2018. Cost-effective synthesis of bifunctional silver nanoparticles for simultaneous colorimetric detection of Al (III) and disinfection. Sensors Actuators B Chem. 272, 79–90.

Pawar, A.V., Kanapally, S.S., Kadam, K.D., Patil, S.L., Dongle, V.S., Jadhav, S.A., Kim, S., Dongale, T.D., 2019. MemSens: a new detection method for heavy metals based on silver nanoparticle assisted memristive switching principle. J. Mater. Sci. Mater. Electron. 30 (12), 11383–11394.

Ping, H., Zhang, M., Li, H., Li, S., Chen, Q., Sun, C., Zhang, T., 2012. Visual detection of melamine in raw milk by label-free silver nanoparticles. Food Control 23 (1), 191–197.

Pourrut, B., Shahid, M., Dumat, C., Winterton, P., Pinelli, E., 2011. Lead uptake, toxicity, and detoxification in plants. In: Reviews of Environmental Contamination and Toxicology. vol. 213. Springer, New York, NY, pp. 113–136.

Prosposito, P., Burratti, L., Venditti, I., 2020. Silver nanoparticles as colorimetric sensors for water pollutants. Chemosensors 8 (2), 26.

Qaswar, M., Yiren, L., Jing, H., Kaillou, L., Mudasir, M., Zhenzhen, L., Hongqian, H., Xianjin, L., Ahmed, W., Dongchu, L., Humin, Z., 2020. Soil nutrients and heavy metal availability under long-term combined application of swine manure and synthetic fertilizers in acidic paddy soil. J. Soils Sediments 20 (4), 2093–2106.

Rajakumar, G., Rahuman, A.A., 2011. Larvicidal activity of synthesized silver nanoparticles using *Eclipta prostrata* leaf extract against filariasis and malaria vectors. Acta Trop. 118 (3), 196–203.

Rashad, M.M., Shalan, A.E., 2013. Surfactant-assisted hydrothermal synthesis of titania nanoparticles for solar cell applications. J. Mater. Sci. Mater. Electron. 24 (9), 3189–3194.

Rhouati, A., Bulbul, G., Latif, U., Hayat, A., Li, Z.H., Marty, J.L., 2017. Nano-aptasensing in mycotoxin analysis: recent updates and progress. Toxins 9 (11), 349.

Rubab, M., Shahbaz, H.M., Olaimat, A.N., Oh, D.H., 2018. Biosensors for rapid and sensitive detection of *Staphylococcus aureus* in food. Biosens. Bioelectron. 105, 49–57.

Sampathkumar, K., Tan, K.X., Loo, S.C.J., 2020. Developing nano-delivery systems for agriculture and food applications with nature-derived polymers. iScience 23 (5), 101055.

Sastry, R.K., Rashmi, H.B., Rao, N.H., Ilyas, S.M., 2010. Integrating nanotechnology into agri-food systems research in India: a conceptual framework. Technol. Forecast Soc. Change 77 (4), 639–648.

Sastry, R.K., Rashmi, H.B., Rao, N.H., 2011. Nanotechnology for enhancing food security in India. Food Policy 36 (3), 391–400.

Saxena, A., Tripathi, R.M., Zafar, F., Singh, P., 2012. Green synthesis of silver nanoparticles using aqueous solution of *Ficus benghalensis* leaf extract and characterization of their antibacterial activity. Mater. Lett. 67 (1), 91–94.

Sekhon, B.S., 2014. Nanotechnology in agri-food production: an overview. Nanotechnol. Sci. Appl. 7, 31.

Sengan, M., Veeramuthu, D., Veerappan, A., 2018. Photosynthesis of silver nanoparticles using *Durio zibethinus* aqueous extract and its application in catalytic reduction of nitroaromatics, degradation of hazardous dyes and selective colorimetric sensing of mercury ions. Mater. Res. Bull. 100, 386–393.

Shafiq, M., Anjum, S., Hano, C., Anjum, I., Abbasi, B.H., 2020. An overview of the applications of nanomaterials and nanodevices in the food industry. Foods 9 (2), 148.

Shang, Y., Hasan, M., Ahammed, G.J., Li, M., Yin, H., Zhou, J., 2019. Applications of nanotechnology in plant growth and crop protection: a review. Molecules 24 (14), 2558.

Sharif, M.K., Awan, K.A., Butt, M.S., Sharif, H.R., 2018. Nanotechnology: a pioneering rebellion for food diligence. In: Impact of Nanoscience in the Food Industry. Academic Press, United States, pp. 29–56.

Shrivas, K., Sahu, S., Sahu, B., Kurrey, R., Patle, T.K., Kant, T., Karbhal, I., Satnami, M.L., Deb, M.K., Ghosh, K.K., 2019. Silver nanoparticles for selective detection of phosphorus pesticide containing π-conjugated pyrimidine nitrogen and sulfur moieties through non-covalent interactions. J. Mol. Liq. 275, 297–303.

Singh, P., Gandhi, N., 2015. Milk preservatives and adulterants: processing, regulatory and safety issues. Food Rev. Int. 31 (3), 236–261.

Solomon, M.M., Gerengi, H., Umoren, S.A., Essien, N.B., Essien, U.B., Kaya, E., 2018. Gum Arabic-silver nanoparticles composite as a green anticorrosive formulation for steel corrosion in strong acid media. Carbohydr. Polym. 181, 43–55.

Somu, P., Paul, S., 2018. Casein based biogenic-synthesized zinc oxide nanoparticles simultaneously decontaminate heavy metals, dyes, and pathogenic microbes: a rational strategy for wastewater treatment. J. Chem. Technol. Biotechnol. 93 (10), 2962–2976.

Srivastava, A.K., Dev, A., Karmakar, S., 2018. Nanosensors and nanobiosensors in food and agriculture. Environ. Chem. Lett. 16 (1), 161–182.

Sung, H.K., Oh, S.Y., Park, C., Kim, Y., 2013. Colorimetric detection of Co^{2+} ion using silver nanoparticles with spherical, plate, and rod shapes. Langmuir 29 (28), 8978–8982.

Sur, U.K., 2013. Surface-enhanced Raman scattering (SERS) spectroscopy: a versatile spectroscopic and analytical technique used in nanoscience and nanotechnology. Adv. Nano Res. 1 (2), 111.

Sur, U.K., Chowdhury, J., 2013. Surface-enhanced Raman scattering: overview of a versatile technique used in electrochemistry and nanoscience. Curr. Sci. 105 (7), 923–939.

Tagar, Z.A., Memon, N., Agheem, M.H., Junejo, Y., Hassan, S.S., Kalwar, N.H., Khattak, M.I., 2011. Selective, simple and economical lead sensor based on ibuprofen derived silver nanoparticles. Sensors Actuators B Chem. 157 (2), 430–437.

Teodoro, K.B., Shimizu, F.M., Scagion, V.P., Correa, D.S., 2019. Ternary nanocomposites based on cellulose nanowhiskers, silver nanoparticles and electrospun nanofibers: use in an electronic tongue for heavy metal detection. Sensors Actuators B Chem. 290, 387–395.

Vanti, G.L., Nargund, V.B., Vanarchi, R., Kurjogi, M., Mulla, S.I., Tubaki, S., Patil, R.R., 2019. Synthesis of *Gossypium hirsutum*-derived silver nanoparticles and their antibacterial efficacy against plant pathogens. Appl. Organomet. Chem. 33 (1), e4630.

Vilela, D., González, M.C., Escarpa, A., 2012. Sensing colorimetric approaches based on gold and silver nanoparticles aggregation: chemical creativity behind the assay. A review. Anal. Chim. Acta 751, 24–43.

Wang, X., Cui, Q., Yao, C., Li, S., Zhang, P., Sun, H., Lv, F., Liu, L., Li, L., Wang, S., 2017. Conjugated polyelectrolyte–silver nanostructure pair for detection and killing of bacteria. Adv. Mater. Technol. 2 (7), 1700033.

Wilson, A.J., Willets, K.A., 2013. Surface-enhanced Raman scattering imaging using noble metal nanoparticles. Wiley Interdiscip. Rev. Nanomed. Nanobiotechnol. 5 (2), 180–189.

Xiang, L., Zhao, C., Wang, J., 2011. Nanomaterials-based electrochemical sensors and biosensors for pesticide detection. Sens. Lett. 9 (3), 1184–1189.

Xing, S., Xu, H., Chen, J., Shi, G., Jin, L., 2011. Nafion stabilized silver nanoparticles modified electrode and its application to Cr (VI) detection. J. Electroanal. Chem. 652 (1–2), 60–65.

Zengin, N., Yüzbaşıoğlu, D., Unal, F., Yılmaz, S., Aksoy, H., 2011. The evaluation of the genotoxicity of two food preservatives: sodium benzoate and potassium benzoate. Food Chem. Toxicol. 49 (4), 763–769.

Zhang, R., Belwal, T., Li, L., Lin, X., Xu, Y., Luo, Z., 2020. Nanomaterial-based biosensors for sensing key foodborne pathogens: advances from recent decades. Compr. Rev. Food Sci. Food Saf. 19 (4), 1465–1487.

Zhao, Y., Yang, Y., Luo, Y., Yang, X., Li, M., Song, Q., 2015. Double detection of mycotoxins based on SERS labels embedded Ag@ Au core–shell nanoparticles. ACS Appl. Mater. Interfaces 7 (39), 21780–21786.

Zhou, S., Liu, J., Xu, M., Lv, J., Sun, N., 2015. Accumulation, availability, and uptake of heavy metals in a red soil after 22-year fertilization and cropping. Environ. Sci. Pollut. Res. 22 (19), 15154–15163.

CHAPTER 6

Silver-based nanomaterials for food packaging applications

Shiji Mathew and E.K. Radhakrishnan
School of Biosciences, Mahatma Gandhi University, Kottayam, Kerala, India

6.1 Introduction

One of the major issue faced by the global food industry is the microbial contamination of food products and the resulting food wastage and foodborne illnesses (Carbone et al., 2016). Microbial contamination of food can occur due to poor practices during harvesting, handling, storage, and transport. A better solution for enabling sustainable food production and enhanced market development is to tackle the source of food wastage directly rather than increasing food production (Morris et al., 2017). So, one of the main challenges with the food quality assurance sector is the development of methodologies to provide safer and healthier food, free of microbiological hazards (Rossi et al., 2017).

Food packaging plays an eminent role in containment, preservation, and protection of food from physical, chemical, and environmental factors until it reaches the consumers. However, the conventionally used packaging materials do not possess inherent antimicrobial activity to protect food from unexpected microbial spoilage. In this context, antimicrobial food packaging was developed as an emergency requirement, which could prevent/delay microbial spoilage, extend product shelf-life, and improve food quality (Carbone et al., 2016). The concept of antimicrobial packaging was first accomplished by the incorporation of biocidal agents such as organic acids, salts, antibiotics, bacteriocins, enzymes, nitrites and sulphites, natural plant extracts, or polymers, which could be directly released into food or the space surrounding the food (Ahmed et al., 2017; Baldevraj and Jagadish, 2011; Vermeiren et al., 2002). Later, many inorganic metal or metal oxide antimicrobial nanoparticles were introduced into packaging systems. These nanoparticles gained more advantages as they were found to successfully withstand exacting process conditions when compared to organic biocidal agents (Metak and Ajaal, 2013).

In recent times, advanced nanotechnological interventions in material science have allowed the direct reinforcement of nanoparticles into polymeric matrices through the development of polymer nanocomposites. When compared to conventional food packaging systems, polymer nanocomposite has evolved as a promising alternative food packaging material, which exhibits better antimicrobial, improved physiochemical and biodegradable properties (Mathew and Radhakrishnan, 2019). Among the varied nanoparticles used, silver nanoparticles (AgNPs) have gained much attraction and appreciation as a promising nanofiller in polymer nanocomposites. The remarkable and broad-spectrum antimicrobial activity of AgNPs serves it to act as an important component in food packaging nanocomposites, which inhibits spoilage and pathogenic microorganisms, thereby aiding in preserving foods for longer periods (Liao et al., 2019; Mihindukulasuriya and Lim, 2014). This chapter presents the state of art related to the various silver-based nanomaterials used for food packaging application. This section mainly discusses the mechanisms of antimicrobial action of AgNPs, the techniques employed for the development of silver-based nanocomposites and its practical applications in food packaging. Finally, this chapter also discusses the possible migration of silver nanoparticles from the nanocomposite packages and the related toxicity concerns that may arise from its wide usage.

6.2 Food packaging and its importance in food safety

Packaging of food serves an inevitable role in preserving the quality of food from its manufacturing step until its consumption (Ghaani et al., 2016). The functions performed by a conventional food package include (1) protection of food against physical, chemical, and external barriers; (2) acting as handy containers for food, which comes in attractive shapes and sizes; and (3) communication with the consumer about the ingredients, date of manufacturing, nutritive value, etc., and enables the effective distribution of food and facilitates the end-use convenience (Kotzekidou, 2016; Vanderroost et al., 2014). Due to its importance, the packaging industry is estimated to contribute 2% of Gross National Product (GNP) in developed countries and is the third largest industry in the world (Mihindukulasuriya and Lim, 2014). Depending on the food type, the common packaging materials include paper, cardboard paper, plastics, glass, and a combination of different metals.

Since the 20th century, petroleum-based polymers have been widely used for food packaging purposes relatively with other packaging materials. This is because of their unique features such as ease in availability, high durability and ductility, good physiochemical properties, lightweight, cost effectiveness, and ease of processing (Arora and Padua, 2010; Rhim et al., 2013). One of the most alarming drawbacks of these petroleum-based polymers are their nonrenewable nature, which renders them nonbiodegradable. These plastics hence get accumulated in the disposal area and impose huge threat to the ecosystem (Ahmad et al., 2015). In addition, the high permeability to gases/vapors and absence of inherent antimicrobial property adds to the decline in their commercial value and usage.

An ideal packaging should possess the essential basic features such as physical and mechanical, thermal, gas/water vapor barrier functions along with antibacterial and optical properties to protect the food during transportation and storage. The advances in material science and nanotechnology, as well as changing demands of modern society and industrial production, have led to a consistent refinement in the food packaging industry (Mihindukulasuriya and Lim, 2014). Today, with nanotechnological interventions, food packaging has evolved from simple containment and preservation systems to aspects of safety, food freshness, material reduction, marketing appraisal, convenience, ecofriendly, and tamper-proofing (Bouwmeester et al., 2014).

6.3 Nanotechnology in food packaging

Nanotechnology is an interdisciplinary research area, which deals with the portrayal, synthesis, and manipulation of particles, structures, materials, or gadgets of size ranging from 1 to 100 nm (Jeevanandam et al., 2018). Due to the small size range, morphology, and distribution, nanoparticles exhibit new and improved physicochemical properties distinct from their bulk counterparts. Nanotechnology has emerged as a promising tool for the development of pioneering products, and every aspect of human life is benefited with these (Matteucci et al., 2018). Nanotechnology and nanoscience have influenced almost all fields including electronics, optics, catalysis, mechanics, human health, and recently the food industry (Duncan, 2011). In the food industry, nanotechnological interventions have beneficially influenced three main areas such as food processing, food packaging, and food preservation.

Nanotechnology has created a breakthrough in composite materials by the invention of nanocomposites, which enables more benefits to the product for varied applications (Youssef, 2013). The nanocomposite is defined as a two-phase material where the continuous phase consists of a dispersed phase at nanometer (10^{-9} m) level. Polymer nanocomposites are composed of polymers or copolymers that are reinforced with nano range particles in small quantities (Hernández-Muñoz et al., 2019). Reinforcement of nanofillers into polymer matrices in trace amounts makes it cost-effective (Farhoodi, 2016) and modifies the nanocomposite's Young's modulus and hardness (Guo and Mei, 2014). Moreover, polymer nanocomposites also offer temperature and flame resistance, enhanced elasticity and durability, UV-visible light barrier properties, enhanced processability due to its lower viscosity, and delivery of active materials into the biological system in an ecofriendly manner (Mathew and Radhakrishnan, 2019).

Nanotechnology has successfully addressed the critical issues and shortcomings of conventional food packaging by the reinforcement of nanomaterials in various forms to develop polymer nanocomposites with (1) improved physicochemical and gas/vapor barrier properties, known as improved packaging; (2) active releasing/absorbing agent or antibacterial properties, termed active packaging; and (3) indicators/sensors to provide the traceability and feedback about the quality of food, defined as the intelligent/smart packaging (Lim, 2011; Silvestre et al., 2011). More recently, nanomaterials are also used as a thin layer of coating over food surfaces to the extent its shelf-life, called as edible nanocoatings (Yousuf et al., 2018). Hence, the use of nanomaterials in food packaging articles can be a promising solution for extending food shelf-life by tackling microbial growth and avoiding contamination.

Antimicrobial packaging is a type of active packaging, which release antimicrobial agents to the food matrix or to the package headspace. With the incorporation of antimicrobial nanoparticles into polymeric matrices, development of a nanocomposite material is accomplished, which possess excellent biocidal activity along with improved mechanical and barrier properties. Such active antimicrobial packaging materials can aid in extending the shelf-life of food, as well as ensure the quality and safety of food. To impart antimicrobial effect to the packaging material, the antimicrobial agents can either be dispersed directly into packaging materials or incorporated as sachets in packs or applied as coatings on the surface of the packaging material or the food surface. Direct incorporation of antimicrobial substances into the packaging material is found to be the most effective method for imparting high antimicrobial properties. Many metallic nanoparticles such as silver (Ag) (Gautam and Ram, 2010) and metal oxide nanoparticles such as zinc oxide (ZnO), copper oxide (CuO) (Rao et al., 2015), and titanium dioxide (TiO_2) have been studied as food contact materials. Besides these inorganic nanoparticles, nanosized clays such as smectite, hectorite, laponite, kaolinite, mica, and montmorillonite are also widely incorporated in food contact and packaging materials (Youssef, 2013). The use of such nanoparticles in polymer films can also improve polymeric mechanical properties.

6.4 Benefits of silver as nanofiller in food packagings

Since the time immemorial, silver (Ag) has been used as an effective antimicrobial agent for protecting food and beverages against spoilage. For instance, in ancient times, silver vessels were used to store wine and water, and silver spoons were placed in milk containers for preservation (Duncan, 2011). Silver nanoparticles (AgNPs) have gained much popularity among the other nanoparticles and are the most commonly used component for developing active antimicrobial food packaging systems (Rai et al., 2009). This is due to the varied and unique characteristics of AgNPs which are detailed below.

The high surface area to volume ratio of AgNPs permits larger area for contact with microbes, thereby resulting in its excellent broad-spectrum antimicrobial activity, which makes it a promising agent in attacking spoilage organisms and thereby extend food shelf-life. Moreover, AgNPs can be easily synthesized and incorporated into different synthetic nondegradable and natural biodegradable polymers to develop desirable antimicrobial food packages (Almeida et al., 2015). Interestingly, AgNPs are highly stable nanoparticles, which can effectively exhibit its antimicrobial activity for longer periods when embedded in polymeric matrices. Another important advantage is that only less quantity of silver nanoparticles is needed to be incorporated into packages to impart better and long-term antimicrobial activity, which is not possible with other traditional antimicrobial agents. Another important feature of AgNPs is its low volatility and low toxicity to eukaryotic cells, which makes them appropriate in food packaging application (Nile et al., 2020). Also, it is well documented that AgNPs generated from biological routes are less toxic and hence safe for usage in

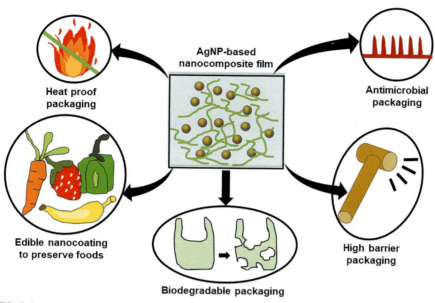

FIG. 6.1

Food packaging related applications of AgNP-based nanocomposite films.

food packaging than those synthesized from other routes (Shankar et al., 2015). Also, recently, AgNPs have emerged as an important component in edible packagings with much public acceptance, owing to their less impact on the sensory attributes of food. The strong odor and flavor of essential oils and plant extracts normally used in edible packages are found to interfere with the organoleptic properties of the food, which thereby limits their use (Dhall, 2013). Fig. 6.1 depicts the various applications of AgNP-based nanocomposite materials in packaging field.

6.5 Mechanisms of antimicrobial action of AgNPs

AgNPs are found to be highly efficient in inhibiting a wide range of multidrug-resistant microorganisms and are hence considered as the next-generation antibiotics. Although AgNPs have been proved to be active against more than 650 microorganisms, still the actual mechanisms behind their mode of action are not understood (Malarkodi et al., 2013). However, the use of sophisticated microscopic techniques, such as TEM, FE-SEM, AFM, and spectroscopic techniques, such as XRD, UV-vis spectroscopy, DLS, and fluorescence spectroscopy, has provided better mechanistic insights into the antimicrobial effects of AgNPs. The antimicrobial activity of AgNPs is generally considered to be associated with four mechanisms, which include (1) attachment of AgNPs to microbial cell wall and membrane surface,

(2) penetration of AgNPs inside the cell, leading to damage to internal structures and biomolecules, (3) generation and release of ROS (reactive oxygen species) and free radicals, leading to oxidative stress and damage, and (4) modulation of signal transduction pathways. Moreover, in contrast to traditional antibiotics, nanoparticles are found to act by multiple mechanisms, which act either singly or simultaneously. This is considered to be the reason for increased antimicrobial activity of AgNPs (Rafique et al., 2017).

Silver, being a Lewis acid usually interacts with Lewis base. The major components of bacterial cell membrane, DNA and protein, are phosphorous and sulfur containing biomolecules. Therefore, AgNPs have a greater affinity to bind to these biomolecules. When exposed to bacteria, AgNPs can initially accumulate on the surface of the cell wall and cause changes like cytoplasm shrinkage, detachment of membrane, formation of pits, and disrupted cell membrane (Chopade et al., 2012; Sondi and Salopek-Sondi, 2004). The AgNPs can also bind to membrane proteins like transport and respiratory chain proteins, which in turn increase the membrane permeability and interfere with ion transport chain, respiratory chain, and the cell division. Further penetration of AgNPs into the bacteria and their interaction with lipids, proteins, and DNA results in the inhibition of transcription, translation, protein synthesis, and cell functions (Rinna et al., 2015). Antibacterial action of AgNPs is also mediated by the generation of increased oxidative stress, due to the release of ROS and free radicals such as hydrogen peroxide, hydroxyl radical, superoxide anions, singlet oxygen, and hypochlorous acid. A schematic diagram depicting the various mechanisms of antimicrobial action of AgNPs is shown in Fig. 6.2.

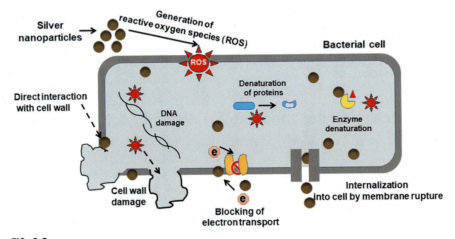

FIG. 6.2

Mechanism of antimicrobial action of silver nanoparticles (AgNPs).

Modified image from Qayyum, S., Khan, A.U., 2016. Nanoparticles vs. biofilms: a battle against another paradigm of antibiotic resistance. MedChemComm 7, 1479–1498.

6.6 Preparation of silver-based food packaging nanomaterials

The preparative routes of silver nanocomposites require proper dispersion of AgNPs throughout the polymer matrix to achieve the desired properties. The techniques for nanocomposite preparation depend upon where AgNPs are synthesized (Kraśniewska et al., 2020). AgNPs can be incorporated in polymer composites mainly by two strategies; ex situ or "top-down" method where the polymer just acts as a dispersion medium for the preformed nanoparticles and in situ or "bottom-down" method in which the polymer matrix acts as the reaction medium for the generation of nanoparticles (Gautam and Ram, 2010; Palza, 2015).

Ex situ method is also termed direct compounding in which nanomaterials and polymeric solution are prepared separately and then blended together using different ways such as by direct mixing, ball milling, melt blending, solvent casting, electrospinning, and self-assembly. Ex situ synthesis is a widely accepted technique, but possess the demerit of nanoparticle aggregation, which hinders its uniform distribution in the polymer matrices (Ucankus et al., 2018; Vasile, 2018). An ex situ method of AgNPs inclusion in antibacterial nanocomposite film was reported by Bahrami et al. (2019). Here, HPMC/tragacanth/beeswax polymer matrix was reinforced with commercially purchased AgNPs, and the film was prepared by solvent casting method.

On the other hand, in situ synthesis method involves the simultaneous process of generation of nanoparticles and polymerization. Here, the precursors for nanomaterial synthesis (metal salts and reducing agents) and monomers for the polymer are combined, and the polymer nanocomposite preparation is allowed to occur. In some cases, the polymer itself may act as a reducing agent for AgNPs synthesis. The main methods involve in situ polymerization, spin coating, and casting. Recently, in situ method was adopted for the development of highly antibacterial PVA/rice starch water/AgNP blend nanocomposite film. Here, the polymer blends and the reducing agent ($AgNO_3$) for AgNP generation was added in a single step, and the so-formed nanocomposite solution was solvent cast to form films for packaging application (Mathew et al., 2019a). In this study, the rice starch water acted both as a polymer and reducing agent for the generation of AgNPs. Fig. 6.3 illustrates the general schemes involved in the development of AgNP-based nanocomposite.

6.7 Silver-based food packaging nanomaterials

Until now, numerous studies have reported the development of AgNP-based food packaging materials. Besides imparting antibacterial ability, the inclusion of AgNPs in nanocomposites has also demonstrated to improve the material's mechanical (Jo et al., 2018), gas/vapor barrier properties (Wang et al., 2011), thermal resistance (Faghihi and Hajibeygi, 2013), and also biodegradability (Deka et al., 2010). Herein, we discuss the applications and features of some of nonbiodegradable, biodegradable, and edible films or nanocoatings reinforced with AgNPs used for packaging application.

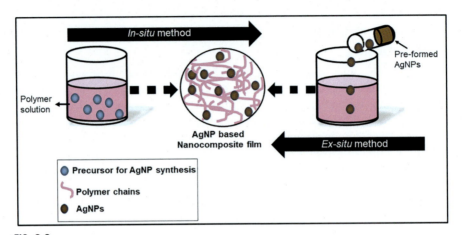

FIG. 6.3

General schemes for preparing AgNP-based nanocomposites.

6.7.1 AgNP-based nondegradable nanocomposite film

The antibacterial effectiveness and successful incorporation of AgNPs into various petroleum-based polymer matrices such as low-density polyethylene (LDPE), polyacrylic acid (PAA), polyethylene (PE), polypropylene (PP), polystyrene (PS), poly(vinyl) chloride (PVC), polyethylene terephthalate (PET), and polyamides (PA) are available. Deng et al. (2020) reported the development of AgNP-based LDPE composite film, which showed its suitability as an antibacterial food packaging material along with high gas/moisture barrier ability and improved mechanical properties. PVC/AgNP nanocomposites for food packaging application were prepared by an in situ approach, and its antibacterial and antifungal characteristics were evaluated (Khalil and Rabie, 2017). Mallakpour and Naghdi (2016) reported the preparation of poly(amide-imide) (PAI) matrix embedded with PVA capped AgNPs. This nanocomposite exhibited better thermal stability between 250°C and 420°C when compared to their control films, which was pointed out because of the incorporation of AgNPs in the matrix. In another approach, antimicrobial PET films were fabricated from postconsumption discarded PET waste by secondary cycling and then functionalized with chitosan assisted AgNPs (Singh et al., 2017). In a similar study conducted by Oliani et al. (2015), AgNPs (1%) were included as biocidal agents for the generation of antibacterial nanocomposites of polypropylene (PP), which showed activity against *Escherichia coli* and *Staphylococcus aureus*. In another study, polystyrene (PS) was used as a matrix for the ex situ incorporation of AgNPs for the fabrication of antibacterial, thermally stable, and mechanically stable nanocomposite packaging (Krystosiak et al., 2017).

All these nanocomposite films were found to be potentially beneficial and functional than the pristine polymeric films without AgNPs.

6.7.2 AgNP-based bio-nanocomposite film

Biodegradable polymers or biopolymers are derived from natural sources and can be degraded naturally by the metabolic action of soil microorganisms. Owing to their striking features and sustainable development, biopolymers have gained much attraction and have efficiently replaced traditional packages (Arora and Padua, 2010). Biopolymer-based food packaging is cost-effective, highly biodegradable, and environment friendly. On the contrary, the biggest shortcoming includes its weak mechanical and barrier properties compared to conventional plastics. This limits the use of pristine biopolymeric materials as packaging materials to a certain extent. Several studies have reported that the inclusion of AgNPs, particularly in biopolymers, could not only impart antimicrobial activity but also improve their mechanical properties (Chang et al., 2010; Kadam et al., 2019; Roy et al., 2019).

Biopolymers can be of two types: natural and synthetic. Commonly available natural biopolymers include cellulose, agar, carrageenan, gelatin, chitosan, alginate, and starch from various sources. AgNPs have been already used as antibacterial nanofillers in many of these natural polymers. Kanmani and Rhim (2014b) have developed and evaluated the antibacterial efficiency of gelatin-based nanocomposite containing AgNP and organoclay Cloisite 30B against foodborne pathogens. Also, in another study, Kanmani and Rhim (2014a) have reported the antimicrobial activity of AgNPs incorporated agar films against *Listeria monocytogenes* and *E. coli* O157:H7. Alginate-based nanocomposites were developed by Shankar et al. (2016) using Ag in from various sources such as metallic Ag, laser-ablated AgNPs, citrate reduced AgNPs, silver nitrate, and silver zeolite. These nanocomposites were evaluated for the effect of different AgNPs on their antibacterial and physicochemical properties. In another study, tamarind nut powder reduced AgNPs were used as nanofillers in cellulose matrix for the in situ development of nanocomposites, and their antibacterial characteristics against five bacteria were studied (Mamatha et al., 2019).

The synthetic biopolymers include poly(vinyl) alcohol (PVA), polylactic acid (PLA), polyhydroxy butyrate (PHB), etc. PLA-nanosilver-clay nanocomposites were found to show antibacterial activity against *Salmonella* spp. (Busolo et al., 2010). PVA-based nanocomposite with AgNP and cellulose nanocrystals displayed antibacterial activity against *E. coli* and *S. aureus* (Fortunati et al., 2013a, b). Antimicrobial PVA-AgNP nanocomposites were prepared (Mathew et al., 2019b) in a complete green synthetic method in which ginger-mediated AgNPs were generated in situ within the PVA matrix under the influence of solar radiation. Castro-Mayorga et al. (2018) used the bacteria *Cupriavidus nectar* for the simultaneous synthesis of AgNP embedded in PHB films suitable for packaging, which showed remarkable antibacterial performance against foodborne pathogens, *Salmonella enterica* and *L. monocytogenes*.

6.7.3 AgNP-based edible films/nanocoatings

Silver nanoparticles have been already applied as a promising component in varied coatings to protect many food-contact surfaces such as storage containers, dishes, mugs, cutting boards, utensils, and even on fridge surfaces (Deshmukh et al., 2019). Recent research studies include the investigation of the application of AgNPs as direct nanocoatings on food surface or on other packaging materials.

AgNPs and cellulose nanofibrils were deposited as a coating on the surfaces of kraft and greaseproof base papers to develop antibacterial packaging papers. These papers were found effective against *E. coli* and *S. aureus* and also showed improved physicochemical properties (Amini et al., 2016). In a recent study, AgNPs were synthesized from *Artemisia scoparia* plant extract and mixed with calcium alginate coating material. This, when applied on strawberries and loquats, acted as an active and edible coating, which increased the fruit's shelf-life and protected from microbial spoilage (Hanif et al., 2019). Jung et al. reported the application of starch reduced AgNPs as antimicrobial coatings on Whatman No. 1 filter paper for its potential use as a food packaging material. This starch-AgNP coated paper was found to be biodegradable, antibacterial with good oil resistant property (Jung et al., 2018b). In a study, starch-assisted AgNPs were prepared and incorporated ex situ into chitosan solution and then coated on to cellulose papers. This food packaging paper was found to show improved tensile and burst strengths together with better water and oil resistance and successful in inhibiting the growth of *S. aureus, E. coli,* and *Penicillium expansum* (Jung et al., 2018a).

6.7.4 AgNP-based polymer blend nanocomposites

An alternative to improve the weak barrier and mechanical properties of pristine polymers is to blend it with another polymer. AgNPs containing polymer blends have been extensively used in food packaging applications and also for developing antibacterial nanocomposites with highly improved physicochemical properties. A recent study reported the development of green, antibacterial nanocomposite blend made up of konjac glucomannan, poly ε-caprolactone (PCL) incorporated with AgNPs (Lin et al., 2020). This blended material also exhibited improved mechanical and thermal properties and was hydrophobic, which shows promises as effective packaging material. Lee et al. (2019) developed AgNP incorporated pullulan/pectin biopolymer blend, which showed better physicochemical properties and antibacterial activity against foodborne pathogens. In another study, agar/lignin composite films incorporated with AgNPs were developed for packaging purpose (Shankar and Rhim, 2017). Similarly, agar/banana powder blend films incorporated with green synthesized AgNPs (reduced by banana powder) displayed potential antibacterial activity against *L. monocytogenes* and *E. coli* (Orsuwan et al., 2016). Mathew et al. (2019a) reported the use of boiled rice water starch for the simultaneous generation of AgNP and blending with PVA for the development of active and biodegradable packaging material. In all these cases, the nanocomposite blend showed excellent mechanical, gas/vapor barrier properties compared to the native control polymers. Table 6.1 summarizes the recently developed AgNP-based food packaging nanomaterials with their preparation routes, size, and concentration of AgNPs used, and antimicrobial activity is also shown.

Table 6.1 Antimicrobial effect of AgNP-based food packaging nanomaterials.

Polymer matrix	Material developed	AgNP/AgNO$_3$ concentration	Synthesis method	Antibacterial effects	Reference
Cellulose	Packaging film	0.005, 0.01, 0.02, 0.04, and 0.08 g of AgNPs	In situ	S. aureus, E. coli	Chen et al. (2020)
LDPE	Packaging film	1.5, 3.75, 7.5, 15, 30, 60, and 75 μg/mL of AgNPs	Ex situ	S. aureus, E. faecalis, E. coli, S. Typhimurium, and Penicillium expansum	da Brito et al. (2019)
LDPE	Packaging film	0.5, 1, 2 wt% AgNPs	Ex situ	E. coli	Olmos et al. (2018)
PAA	Packaging film	5 mM AgNO$_3$	In situ	C. albicans, S. aureus	Mofidfar et al. (2019)
PLA & oligomeric lactic acid (OLA)	Blend film material	0.4, 0.8, 2.4, 4 (wt%) AgNPs	Ex situ	S. aureus, E. coli	Sonseca et al. (2019)
PVA & rice starch water	Biodegradable blend film	1 mM AgNO$_3$	In situ	S. aureus, S. Typhimurium	Mathew et al. (2019a)
Chitosan	Packaging film	0.1%, 0.2%, 0.3% (w/v) AgNPs	Ex situ	S. aureus, B subtilis, P. aeruginosa, and E. coli	Kadam et al. (2019)
Gum tragacanth/HPMC	Edible film	2%, 4%, 8% (w/v) AgNPs	Ex situ	B cereus, S. aureus, Strep. pneumoniae L. monocytogenes, E. coli, S. typhimurium, P. aeruginosa, and K. pneumoniae	Bahrami et al. (2019)
PLA	Packaging film	0.34 (wt%) AgNPs	Ex situ	E. coli, S. aureus	Kostic et al. (2019)
Methyl cellulose	Bio-nanocomposite	2 mL AgNO$_3$	Ex situ	E. coli, S. aureus	Nunes et al. (2018)
Cellulose acetate/triethyl citrate	Bio-nanocomposite	40 mM AgNO$_3$	In situ	E. coli, S. aureus, P. aeruginosa, S. arizonae, Aspergillus flavus, and Aspergillus niger	Dairi et al. (2019)
Carrageenan	Food coating	20 μg AgNPs	Ex situ	E. coli, S. aureus	Vishnuvarthanan and Rajeswari (2019)
Agar & Lignin	Packaging film	0.5, 1, 1.5, 2 wt% AgNO$_3$	In situ	E. coli, L. monocytogenes	Shankar and Rhim (2017)
Agar & banana powder	Blend film	1 mM AgNO$_3$	In situ	E. coli, L. monocytogenes	Orsuwan et al. (2016)
Banana powder	Packaging film	0.5, 1, 2 mM AgNO$_3$	In situ	E. coli, L. monocytogenes	Orsuwan et al. (2017)
PVA	Packaging film	1 mM AgNO$_3$	In situ	S. aureus, S. Typhimurium	Mathew et al. (2019b)

6.8 Practical application of silver-based nanocomposites on food systems

Numerous studies have already reported the laboratory scale application of AgNP containing nanocomposites as food packaging materials, which have resulted in extending the food shelf-life and providing protection against food pathogens because of its antibacterial efficiency. For this purpose, AgNP-based nanomaterials can be fabricated in different forms such as thin flexible packaging films, hydrogels, edible films, or coatings (Panea et al., 2014). Usually, AgNP-based nanomaterials are commonly used to pack minimally processed foods that require long period of transport and storage. Currently, various AgNP-based nanocomposites and coatings are already tested with a variety of fresh produce (fruits and vegetables) as well as with highly perishable food kinds of stuff (meat and dairy products). Table 6.2 gives some examples of silver nanoparticle-based nanocomposite materials applied in different food systems.

Table 6.2 AgNP-based nanomaterials used to test food systems.

Nanocomposite	Types of food tested	Effect on packed food	References
Chitosan/laurel essential oil/lignin mediated AgNPs / liposome coating	Pork	Antibacterial effect against *S. aureus, E. coli*	Wu et al. (2019)
Gellan gum/AgNPs sensory hydrogel	Chicken breast and silver carp fish	Increased food shelf-life	Zhai et al. (2019)
Chitosan/PVA/AgNPs	Meat	Antibacterial effect against *E. coli* and *L. monocytogenes*	Pandey et al. (2020)
Cellulose/AgNPs	Kiwi and melon juices Beef Fresh cut melon	Inhibited bacteria and yeast *Pseudomonas* spp. *Enterobacteriaceae* Yeast, mold, psychrotrophic bacteria	Lloret et al. (2012) Fernández et al. (2010)
Carboxy methyl cellulose/guar gum-AgNPs nanocoatings	Kinnow mandarin	Increased fruit shelf-life, Reduced growth of psychrotrophic aerobic yeast and mold	Shah et al. (2015)
Chitosan/AgNPs	Strawberries	Protected from soft rot by *Botrytis cinerea*	Moussa et al. (2013)
Ethyl vinyl alcohol/AgNPs	Cheese, lettuce, apples, chicken and pork	Antibacterial effect against *Salmonella* spp. and *L. monocytogenes*	Martínez-Abad et al. (2012)
LDPE/AgNPs	Barberry	Inhibited Total aerobic bacteria	Valipoor Motlagh et al. (2013)
PVC/AgNPs	Beef	Inhibited growth of *E. coli, S. aureus*	Mahdi et al. (2012)

6.8 Practical application of silver-based nanocomposites on food systems

Table 6.2 AgNP-based nanomaterials used to test food systems—cont'd

Nanocomposite	Types of food tested	Effect on packed food	References
Pullulan/AgNPs	Turkey meat	Antibacterial effect against L. monocytogenes, S. aureus	Khalaf et al. (2013)
Chitosan, HPMC/AgNPs/lysozyme hydrosol coating	Meat	B. cereus, Micrococcus flavus, E. coli, and Pseudomonas fluorescens	Zimoch-Korzycka and Jarmoluk (2015)
LDPE/AgNPs + TiO_2 NPs	Rice	Inhibited growth of Aspergillus flavus	Li et al. (2017)
LDPE/AgNPs + ZnO NPs	Orange juice	Inhibited yeasts, molds	Emamifar et al. (2010)
LDPE/kaolin/AgNPs/TiO_2 NPs	Strawberry	Inhibited Botrytis cinerea	Yang et al. (2010)
Agar/Ag-MMT NPs hydrogel coating	Fior di latte cheese	Pseudomonas and coliforms	Incoronato et al. (2011)
Sodium alginate/AgNPs coating	Fior di latte cheese	Controlled the growth of spoilage microbes	Gammariello et al. (2011) Mastromatteo et al. (2015)
Calcium alginate/Ag-MMT nanoparticles	Fresh cut carrots	Mesophilic & psychrotrophic bacteria, Enterobacteriaceae spp., Pseudomonas spp., yeasts, and molds	Costa et al. (2012)
Polyethylene/AgNPs/TiO_2 NPs	Apples, carrots, orange juice, cheese, milk powder	Antibacterial effect against Lactobacillus, Penicillium	Metak and Ajaal (2013)
Polylactide/cinnamon oil/AgNPs/Cu NPs	Chicken	Antibacterial effect against S. ser. Typhimurium, Campylobacter jejuni, and L. monocytogenes	Ahmed et al. (2018)
PVA/montmorillonite clay/ginger mediated AgNPs	Chicken sausage	Inhibited growth of S. ser. Typhimurium, S. aureus	Mathew et al. (2019c)
Chitosan /gelatin /AgNPs	Red grapes	Increased shelf-life	Kumar et al. (2018)
Alginate/AgNP coating	Shiitake mushrooms	inhibited the growth of mesophilic, psychrophilic, pseudomonas, yeast, and molds	Jiang et al. (2013)
Gum Arabic/AgNPs composite coating	Bell pepper	Led to extension of fruit shelf-life	Hedayati and Niakousari (2015)
Agar/AgNP coating	Thornless lime and apple	E. coli, S. aureus, C. violaceum	Rai et al. (2014)
AgNPs colloidal solution coating	Chicken sausages	Reduced the growth of Lactic acid bacilli	Marchiore et al. (2017)

6.9 Safety assessments

Although the use of AgNPs in food packaging is a well-accepted and promising technology, the concerns regarding the release of nanosilver from such packages into food and its associated toxicity is highly arising. The biocidal activity of AgNPs incorporated in packaging materials is found to be attributed to the release of silver ions into the surrounding space. So, it is important to conduct migration studies to understand the possibility and quantity of nanoparticles released from the packagings into food (Huang et al., 2018). Migration tests are performed in distinguished food simulants such as ethanol (10, 20, and 50% v/v), acetic acid (3% v/v), and vegetable oils at specified temperatures. The experiment is conducted by immersing 6 dm^2 of the test material in 1 kg of food simulant, and the AgNP release amount as set by European Food Safety Authority should not exceed 0.05 mg/kg (EFSA, 2011).

According to the AgNPs migration study conducted by Echegoyen and Nerín (2013), the release of AgNPs occurs either due to dissolution of silver on oxidation or as a result of the detachment of silver from the nanocomposite. Migration study was conducted by Marchiore et al. (2017) to evaluate the release of silver from starch edible nanocoatings into packed sausages. By ICP-MS, the actual concentration of AgNPs in the developed coating was found to be 23.47 μg AgNPs/mL, but after simple washing and cooking, the concentration on the sausage surface was found to be reduced to 5.3 ng AgNPs/g sausage. This indicated the most of the AgNPs got removed during these steps, and only very less amount of AgNPs was retained on the sausage's surface. Moreover, the authors also added that the AgNPs did not migrate from the sausage surface to its interior. Silver release studies of PLA films with Ag/alginate microbeads (Kostic et al., 2019) were performed on distilled water and ethanol for 10 days at different temperatures. Here also, the Ag$^+$ release was within the limit of 0.05 mg/kg. In another study, Nair et al. (2017) performed ICP-MS studies to measure the migration of AgNPs from glucomannan cassava starch nanocomposite films to bread samples while it was stored for 14 days. It was found that the amount of silver migrated was less (4.3–4.5 μg/mL) than the permitted level.

Several other studies have also specified that the amount of silver migrated from nanocomposites is below the maximum migration limit as stated by European Union legislation (Cushen et al., 2013, 2014). Research has also suggested that several factors such as a method of nanocomposite preparation, initial silver concentration, temperature, time, and type of contact medium determine the rate of AgNPs migrated into food stimulants (Vasile, 2018). Nevertheless, a huge gap is evident in the scientific knowledge regarding the safety and toxicity of AgNPs in the environment and living beings. Hence, further research is highly recommended to determine their accurate level in foods and to assess their possible risks on consumer health (Fahmy et al., 2020; de Morais et al., 2019).

6.10 Conclusion

Silver nanoparticles, due to their remarkable antimicrobial activity, have earned an eminent role as nanofillers in antimicrobial food packaging nanocomposites. The reinforcement of silver nanoparticles in nanocomposites not only make them active against microbes but also provide other essential functionalities such as improvement in mechanical, water barrier properties, thermal resistance, and biodegradability. Until now, silver nanoparticles have been included for the development of numerous composites composed of natural and synthetic polymers, and have also been applied on food systems to evaluate its role in maintaining the food fresh and extension of its shelf-life. Although the inclusion of AgNPs as antimicrobial agents in food packaging is an established technology, concerns regarding the migration of Ag^+ ions into foods and drinks and its related toxicity persists. So, extensive research is critical for assessing the optimum levels of silver nanoparticles that retain in food and their ecotoxicological effects.

References

Ahmad, M., Hani, N.M., Nirmal, N.P., Fazial, F.F., Mohtar, N.F., Romli, S.R., 2015. Optical and thermo-mechanical properties of composite films based on fish gelatin/rice flour fabricated by casting technique. Prog. Org. Coat. 84, 115–127.

Ahmed, I., Lin, H., Zou, L., Brody, A.L., Li, Z., Qazi, I.M., Pavase, T.R., Lv, L., 2017. A comprehensive review on the application of active packaging technologies to muscle foods. Food Control 82, 163–178.

Ahmed, J., Arfat, Y.A., Bher, A., Mulla, M., Jacob, H., Auras, R., 2018. Active chicken meat packaging based on polylactide films and bimetallic Ag-Cu nanoparticles and essential oil: Active chicken meat packaging…. J. Food Sci.

Almeida, A.C.S., Franco, E.A.N., Peixoto, F.M., Pessanha, K.L.F., Melo, N.R., 2015. Aplicação de nanotecnologia em embalagens de alimentos. Polímeros 25, 89–97.

Amini, E., Azadfallah, M., Layeghi, M., Talaei-Hassanloui, R., 2016. Silver-nanoparticle-impregnated cellulose nanofiber coating for packaging paper. Cellulose 23, 557–570.

Arora, A., Padua, G.W., 2010. Review: nanocomposites in food packaging. J. Food Sci. 75, R43–R49.

Bahrami, A., Rezaei Mokarram, R., Sowti Khiabani, M., Ghanbarzadeh, B., Salehi, R., 2019. Physico-mechanical and antimicrobial properties of tragacanth/hydroxypropyl methylcellulose/beeswax edible films reinforced with silver nanoparticles. Int. J. Biol. Macromol. 129, 1103–1112.

Baldevraj, R.S.M., Jagadish, R.S., 2011. Incorporation of chemical antimicrobial agents into polymeric films for food packaging. In: Multifunctional and Nanoreinforced Polymers for Food Packaging. Elsevier, pp. 368–420.

Bouwmeester, H., Brandhoff, P., Marvin, H.J.P., Weigel, S., Peters, R.J.B., 2014. State of the safety assessment and current use of nanomaterials in food and food production. Trends Food Sci. Technol. 40, 200–210.

Busolo, M.A., Fernandez, P., Ocio, M.J., Lagaron, J.M., 2010. Novel silver-based nanoclay as an antimicrobial in polylactic acid food packaging coatings. Food Addit. Contam. A 27, 1617–1626.

Carbone, M., Donia, D.T., Sabbatella, G., Antiochia, R., 2016. Silver nanoparticles in polymeric matrices for fresh food packaging. J. King Saud Univ. Sci. 28, 273–279.

Castro-Mayorga, J.L., Freitas, F., Reis, M.A.M., Prieto, M.A., Lagaron, J.M., 2018. Biosynthesis of silver nanoparticles and polyhydroxybutyrate nanocomposites of interest in antimicrobial applications. Int. J. Biol. Macromol. 108, 426–435.

Chang, P.R., Jian, R., Yu, J., Ma, X., 2010. Fabrication and characterisation of chitosan nanoparticles/plasticised-starch composites. Food Chem. 120, 736–740.

Chen, Q.-Y., Xiao, S.-L., Shi, S.Q., Cai, L.-P., 2020. A one-pot synthesis and characterization of antibacterial silver nanoparticle-cellulose film. Polymers 12, 440.

Chopade, B., Ghosh, P., Ahire, K., Jabgunde, K., Pardesi, C., Bellare, D., 2012. Synthesis of silver nanoparticles using *Dioscorea bulbifera* tuber extract and evaluation of its synergistic potential in combination with antimicrobial agents. Int. J. Nanomed. 483.

Costa, C., Conte, A., Buonocore, G.G., Lavorgna, M., Del Nobile, M.A., 2012. Calcium-alginate coating loaded with silver-montmorillonite nanoparticles to prolong the shelf-life of fresh-cut carrots. Food Res. Int. 48, 164–169.

Cushen, M., Kerry, J., Morris, M., Cruz-Romero, M., Cummins, E., 2013. Migration and exposure assessment of silver from a PVC nanocomposite. Food Chem. 139, 389–397.

Cushen, M., Kerry, J., Morris, M., Cruz-Romero, M., Cummins, E., 2014. Evaluation and simulation of silver and copper nanoparticle migration from polyethylene nanocomposites to food and an associated exposure assessment. J. Agric. Food Chem. 62, 1403–1411.

da Brito, S.C., Bresolin, J.D., Sivieri, K., Ferreira, M.D., 2019. Low-density polyethylene films incorporated with silver nanoparticles to promote antimicrobial efficiency in food packaging. Food Sci. Technol. Int. 108201321989420.

Dairi, N., Ferfera-Harrar, H., Ramos, M., Garrigós, M.C., 2019. Cellulose acetate/AgNPs-organoclay and/or thymol nano-biocomposite films with combined antimicrobial/antioxidant properties for active food packaging use. Int. J. Biol. Macromol. 121, 508–523.

de Morais, L.O., Macedo, E.V., Granjeiro, J.M., Delgado, I.F., 2019. Critical evaluation of migration studies of silver nanoparticles present in food packaging: a systematic review. Crit. Rev. Food Sci. Nutr., 1–20.

Deka, H., Karak, N., Kalita, R.D., Buragohain, A.K., 2010. Bio-based thermostable, biodegradable and biocompatible hyperbranched polyurethane/Ag nanocomposites with antimicrobial activity. Polym. Degrad. Stab. 95, 1509–1517.

Deng, J., Ding, Q.M., Li, W., Wang, J.H., Liu, D.M., Zeng, X.X., Liu, X.Y., Ma, L., Deng, Y., Su, W., Ye, B., 2020. Preparation of nano-silver-containing polyethylene composite film and Ag ion migration into food-simulants. J. Nanosci. Nanotechnol. 20, 1613–1621.

Deshmukh, S.P., Patil, S.M., Mullani, S.B., Delekar, S.D., 2019. Silver nanoparticles as an effective disinfectant: a review. Mater. Sci. Eng. C 97, 954–965.

Dhall, R.K., 2013. Advances in edible coatings for fresh fruits and vegetables: a review. Crit. Rev. Food Sci. Nutr. 53, 435–450.

Duncan, T.V., 2011. Applications of nanotechnology in food packaging and food safety: barrier materials, antimicrobials and sensors. J. Colloid Interface Sci. 363, 1–24.

Echegoyen, Y., Nerín, C., 2013. Nanoparticle release from nano-silver antimicrobial food containers. Food Chem. Toxicol. 62, 16–22.

EFSA, 2011. Scientific opinion on the safety evaluation of the substance, silver zeolite A (silver zinc sodium ammonium aluminosilicate), silver content 2–5%, for use in food contact materials. EFSA J. 9, 1999.

Emamifar, A., Kadivar, M., Shahedi, M., Soleimanian-Zad, S., 2010. Evaluation of nanocomposite packaging containing Ag and ZnO on shelf life of fresh orange juice. Innov. Food Sci. Emerg. Technol. 11, 742–748.

Faghihi, K., Hajibeygi, M., 2013. Synthesis and properties of polyimide/silver nanocomposite containing dibenzalacetone moiety in the main chain. J. Saudi Chem. Soc. 17, 419–423.

Fahmy, H.M., Salah Eldin, R.E., Abu Serea, E.S., Gomaa, N.M., AboElmagd, G.M., Salem, S.A., Elsayed, Z.A., Edrees, A., Shams-Eldin, E., Shalan, A.E., 2020. Advances in nanotechnology and antibacterial properties of biodegradable food packaging materials. RSC Adv. 10, 20467–20484.

Farhoodi, M., 2016. Nanocomposite materials for food packaging applications: characterization and safety evaluation. Food Eng. Rev. 8, 35–51.

Fernández, A., Picouet, P., Lloret, E., 2010. Cellulose-silver nanoparticle hybrid materials to control spoilage-related microflora in absorbent pads located in trays of fresh-cut melon. Int. J. Food Microbiol. 142, 222–228.

Fortunati, E., Peltzer, M., Armentano, I., Jiménez, A., Kenny, J.M., 2013a. Combined effects of cellulose nanocrystals and silver nanoparticles on the barrier and migration properties of PLA nano-biocomposites. J. Food Eng. 118, 117–124.

Fortunati, E., Puglia, D., Luzi, F., Santulli, C., Kenny, J.M., Torre, L., 2013b. Binary PVA bionanocomposites containing cellulose nanocrystals extracted from different natural sources: Part I. Carbohydr. Polym. 97, 825–836.

Gammariello, D., Conte, A., Buonocore, G.G., Del Nobile, M.A., 2011. Bio-based nanocomposite coating to preserve quality of Fior di latte cheese. J. Dairy Sci. 94, 5298–5304.

Gautam, A., Ram, S., 2010. Preparation and thermomechanical properties of Ag-PVA nanocomposite films. Mater. Chem. Phys. 119, 266–271.

Ghaani, M., Cozzolino, C.A., Castelli, G., Farris, S., 2016. An overview of the intelligent packaging technologies in the food sector. Trends Food Sci. Technol. 51, 1–11.

Guo, X., Mei, N., 2014. Assessment of the toxic potential of graphene family nanomaterials. J. Food Drug Anal. 22, 105–115.

Hanif, J., Khalid, N., Khan, R.S., Bhatti, M.F., Hayat, M.Q., Ismail, M., Andleeb, S., Mansoor, Q., Khan, F., Amin, F., Hanif, R., Hashmi, M.U., Janjua, H.A., 2019. Formulation of active packaging system using *Artemisia scoparia* for enhancing shelf life of fresh fruits. Mater. Sci. Eng. C 100, 82–93.

Hedayati, S., Niakousari, M., 2015. Effect of coatings of silver nanoparticles and gum arabic on physicochemical and microbial properties of green bell pepper (*C apsicum annuum*): effect of nanosilver and gum arabic coatings on bell pepper. J. Food Process. Preserv. 39, 2001–2007.

Hernández-Muñoz, P., Cerisuelo, J.P., Domínguez, I., López-Carballo, G., Catalá, R., Gavara, R., 2019. Nanotechnology in food packaging. In: Nanomaterials for Food Applications. Elsevier, pp. 205–232.

Huang, Y., Mei, L., Chen, X., Wang, Q., 2018. Recent developments in food packaging based on nanomaterials. Nanomaterials 8, 830.

Incoronato, A.L., Conte, A., Buonocore, G.G., Del Nobile, M.A., 2011. Agar hydrogel with silver nanoparticles to prolong the shelf life of Fior di Latte cheese. J. Dairy Sci. 94, 1697–1704.

Jeevanandam, J., Barhoum, A., Chan, Y.S., Dufresne, A., Danquah, M.K., 2018. Review on nanoparticles and nanostructured materials: history, sources, toxicity and regulations. Beilstein J. Nanotechnol. 9, 1050–1074.

Jiang, T., Feng, L., Wang, Y., 2013. Effect of alginate/nano-Ag coating on microbial and physicochemical characteristics of shiitake mushroom (*Lentinus edodes*) during cold storage. Food Chem. 141, 954–960.

Jo, Y., Garcia, C.V., Ko, S., Lee, W., Shin, G.H., Choi, J.C., Park, S.-J., Kim, J.T., 2018. Characterization and antibacterial properties of nanosilver-applied polyethylene and polypropylene composite films for food packaging applications. Food Biosci. 23, 83–90.

Jung, J., Kasi, G., Seo, J., 2018a. Development of functional antimicrobial papers using chitosan/starch-silver nanoparticles. Int. J. Biol. Macromol. 112, 530–536.

Jung, J., Raghavendra, G.M., Kim, D., Seo, J., 2018b. One-step synthesis of starch-silver nanoparticle solution and its application to antibacterial paper coating. Int. J. Biol. Macromol. 107, 2285–2290.

Kadam, D., Momin, B., Palamthodi, S., Lele, S.S., 2019. Physicochemical and functional properties of chitosan-based nano-composite films incorporated with biogenic silver nanoparticles. Carbohydr. Polym. 211, 124–132.

Kanmani, P., Rhim, J.-W., 2014a. Physical, mechanical and antimicrobial properties of gelatin based active nanocomposite films containing AgNPs and nanoclay. Food Hydrocoll. 35, 644–652.

Kanmani, P., Rhim, J.-W., 2014b. Antimicrobial and physical-mechanical properties of agar-based films incorporated with grapefruit seed extract. Carbohydr. Polym. 102, 708–716.

Khalaf, H.H., Sharoba, A.M., Sharoba, A.M., El-Tanahi, H.H., Morsy, M., 2013. Stability of antimicrobial activity of pullulan edible films incorporated with nanoparticles and essential oils and their impact on turkey deli meat quality. J. Food Dairy Sci. 4, 557–573.

Khalil, A.M., Rabie, S.T., 2017. Antimicrobial behavior and photostability of polyvinyl chloride/1-vinylimidazole nanocomposites loaded with silver or copper nanoparticles. J. Vinyl Addit. Technol. 23, E25–E33.

Kostic, D., Vukasinovic-Sekulic, M., Armentano, I., Torre, L., Obradovic, B., 2019. Multifunctional ternary composite films based on PLA and Ag/alginate microbeads: physical characterization and silver release kinetics. Mater. Sci. Eng. C 98, 1159–1168.

Kotzekidou, P., 2016. Factors influencing microbial safety of ready-to-eat foods. In: Food Hygiene and Toxicology in Ready-to-Eat Foods. Elsevier, pp. 33–50.

Kraśniewska, K., Galus, S., Gniewosz, M., 2020. Biopolymers-based materials containing silver nanoparticles as active packaging for food applications—a review. Int. J. Mol. Sci. 21, 698.

Krystosiak, P., Tomaszewski, W., Megiel, E., 2017. High-density polystyrene-grafted silver nanoparticles and their use in the preparation of nanocomposites with antibacterial properties. J. Colloid Interface Sci. 498, 9–21.

Kumar, S., Shukla, A., Baul, P.P., Mitra, A., Halder, D., 2018. Biodegradable hybrid nanocomposites of chitosan/gelatin and silver nanoparticles for active food packaging applications. Food Packag. Shelf Life 16, 178–184.

Lee, J.H., Jeong, D., Kanmani, P., 2019. Study on physical and mechanical properties of the biopolymer/silver based active nanocomposite films with antimicrobial activity. Carbohydr. Polym. 224, 115159.

Li, L., Zhao, C., Zhang, Y., Yao, J., Yang, W., Hu, Q., Wang, C., Cao, C., 2017. Effect of stable antimicrobial nano-silver packaging on inhibiting mildew and in storage of rice. Food Chem. 215, 477–482.

Liao, C., Li, Y., Tjong, S., 2019. Bactericidal and cytotoxic properties of silver nanoparticles. Int. J. Mol. Sci. 20, 449.

Lim, L.-T., 2011. Active and intelligent packaging materials. In: Comprehensive Biotechnology. Elsevier, pp. 629–644.

Lin, W., Ni, Y., Pang, J., 2020. Size effect-inspired fabrication of konjac glucomannan/polycaprolactone fiber films for antibacterial food packaging. Int. J. Biol. Macromol. 149, 853–860.

Lloret, E., Picouet, P., Fernández, A., 2012. Matrix effects on the antimicrobial capacity of silver based nanocomposite absorbing materials. LWT—Food Sci. Technol. 49, 333–338.

Mahdi, S.S., Vadood, R., Nourdahr, R., 2012. Study on the antimicrobial effect of nanosilver tray packaging of minced beef at refrigerator temperature. Glob. Vet. 9, 284–289.

Malarkodi, C., Rajeshkumar, S., Vanaja, M., Paulkumar, K., Gnanajobitha, G., Annadurai, G., 2013. Eco-friendly synthesis and characterization of gold nanoparticles using *Klebsiella pneumoniae*. J. Nanostruct. Chem. 3, 30.

Mallakpour, S., Naghdi, M., 2016. Evaluation of nanostructure, optical absorption, and thermal behavior of poly(vinyl alcohol)/poly(*N*-vinyl-2-pyrrolidone) based nanocomposite films containing coated SiO_2 nanoparticles with citric acid and L-(+)-ascorbic acid. Polym. Compos.

Mamatha, G., Varada Rajulu, A., Madhukar, K., 2019. Development and analysis of cellulose nanocomposite films with *in situ* generated silver nanoparticles using tamarind nut powder as a reducing agent. Int. J. Polym. Anal. Charact. 24, 219–226.

Marchiore, N.G., Manso, I.J., Kaufmann, K.C., Lemes, G.F., de Pizolli, A.P.O., Droval, A.A., Bracht, L., Gonçalves, O.H., Leimann, F.V., 2017. Migration evaluation of silver nanoparticles from antimicrobial edible coating to sausages. LWT—Food Sci. Technol. 76, 203–208.

Martínez-Abad, A., Lagaron, J.M., Ocio, M.J., 2012. Development and characterization of silver-based antimicrobial ethylene-vinyl alcohol copolymer (EVOH) films for food-packaging applications. J. Agric. Food Chem. 60, 5350–5359.

Mastromatteo, M., Conte, A., Lucera, A., Saccotelli, M.A., Buonocore, G.G., Zambrini, A.V., Del Nobile, M.A., 2015. Packaging solutions to prolong the shelf life of Fiordilatte cheese: bio-based nanocomposite coating and modified atmosphere packaging. LWT—Food Sci. Technol. 60, 230–237.

Mathew, S., Radhakrishnan, E.K., 2019. Polymer nanocomposites: alternative to reduce environmental impact of non-biodegradable food packaging materials. In: Ahmed, S., Chaudhry, S.A. (Eds.), Composites for Environmental Engineering. Wiley, pp. 99–133.

Mathew, S., Jayakumar, A., Kumar, V.P., Mathew, J., Radhakrishnan, E.K., 2019a. One-step synthesis of eco-friendly boiled rice starch blended polyvinyl alcohol bionanocomposite films decorated with in situ generated silver nanoparticles for food packaging purpose. Int. J. Biol. Macromol. 139, 475–485.

Mathew, S., Mathew, J., Radhakrishnan, E.K., 2019b. Polyvinyl alcohol/silver nanocomposite films fabricated under the influence of solar radiation as effective antimicrobial food packaging material. J. Polym. Res. 26.

Mathew, S., Mathew, J., Radhakrishnan, E.K., 2019c. Biodegradable and active nanocomposite pouches reinforced with silver nanoparticles for improved packaging of chicken sausages. Food Packag. Shelf Life 19, 155–166.

Matteucci, F., Giannantonio, R., Calabi, F., Agostiano, A., Gigli, G., Rossi, M., 2018. Deployment and exploitation of nanotechnology nanomaterials and nanomedicine. In: Presented at the EMERGING TECHNOLOGIES: MICRO TO NANO (ETMN-2017): Proceedings of the 3rd International Conference on Emerging Technologies: Micro to Nano, Solapur, India, p. 020001.

Metak, A.M., Ajaal, T.T., 2013. Investigation on polymer based nano-silver as food packaging materials. Int. J. Biol. Food Vet. Agric. Eng. 7, 772–777.

Mihindukulasuriya, S.D.F., Lim, L.-T., 2014. Nanotechnology development in food packaging: a review. Trends Food Sci. Technol. 40, 149–167.

Mofidfar, M., Kim, E.S., Larkin, E.L., Long, L., Jennings, W.D., Ahadian, S., Ghannoum, M.A., Wnek, G.E., 2019. Antimicrobial activity of silver containing crosslinked poly(acrylic acid) fibers. Micromachines 10, 829.

Morris, M.A., Padmanabhan, S.C., Cruz-Romero, M.C., Cummins, E., Kerry, J.P., 2017. Development of active, nanoparticle, antimicrobial technologies for muscle-based packaging applications. Meat Sci. 132, 163–178.

Moussa, S.H., Tayel, A.A., Alsohim, A.S., Abdallah, R.R., 2013. Botryticidal activity of nano-sized silver-chitosan composite and its application for the control of gray mold in strawberry: botryticidal activity of nano composite…. J. Food Sci. 78, M1589–M1594.

Nair, S.B., Alummoottil, J.N., Moothandasserry, S.S., 2017. Chitosan-konjac glucomannan-cassava starch-nanosilver composite films with moisture resistant and antimicrobial properties for food-packaging applications: nanocomposite films with moisture resistant and antimicrobial properties. Starch—Stärke 69, 1600210.

Nile, S.H., Baskar, V., Selvaraj, D., Nile, A., Xiao, J., Kai, G., 2020. Nanotechnologies in food science: applications, recent trends, and future perspectives. Nano-Micro Lett. 12.

Nunes, M.R., de Souza Maguerroski Castilho, M., de Lima Veeck, A.P., da Rosa, C.G., Noronha, C.M., Maciel, M.V.O.B., Barreto, P.M., 2018. Antioxidant and antimicrobial methylcellulose films containing *Lippia alba* extract and silver nanoparticles. Carbohydr. Polym. 192, 37–43.

Oliani, W.L., Parra, D.F., Lima, L.F.C.P., Lincopan, N., Lugao, A.B., 2015. Development of a nanocomposite of polypropylene with biocide action from silver nanoparticles. J. Appl. Polym. Sci. 132.

Olmos, D., Pontes-Quero, G., Corral, A., González-Gaitano, G., González-Benito, J., 2018. Preparation and characterization of antimicrobial films based on LDPE/Ag nanoparticles with potential uses in food and health industries. Nanomaterials 8, 60.

Orsuwan, A., Shankar, S., Wang, L.-F., Sothornvit, R., Rhim, J.-W., 2016. Preparation of antimicrobial agar/banana powder blend films reinforced with silver nanoparticles. Food Hydrocoll. 60, 476–485.

Orsuwan, A., Shankar, S., Wang, L.-F., Sothornvit, R., Rhim, J.-W., 2017. One-step preparation of banana powder/silver nanoparticles composite films. J. Food Sci. Technol. 54, 497–506.

Palza, H., 2015. Antimicrobial polymers with metal nanoparticles. Int. J. Mol. Sci. 16, 2099–2116.

Pandey, V.K., Upadhyay, S.N., Niranjan, K., Mishra, P.K., 2020. Antimicrobial biodegradable chitosan-based composite nano-layers for food packaging. Int. J. Biol. Macromol. 157, 212–219.

Panea, B., Ripoll, G., González, J., Fernández-Cuello, Á., Albertí, P., 2014. Effect of nanocomposite packaging containing different proportions of ZnO and Ag on chicken breast meat quality. J. Food Eng. 123, 104–112.

Rafique, M., Sadaf, I., Rafique, M.S., Tahir, M.B., 2017. A review on green synthesis of silver nanoparticles and their applications. Artif. Cells Nanomed. Biotechnol. 45, 1272–1291.

Rai, M., Yadav, A., Gade, A., 2009. Silver nanoparticles as a new generation of antimicrobials. Biotechnol. Adv. 27, 76–83.

Rai, M., Gudadhe, J.A., Yadav, A., Durán, N., Marcato, P.D., Gade, A., 2014. Preparation of an agar-silver nanoparticles (A-AgNp) film for increasing the shelf-life of fruits. IET Nanobiotechnol. 8, 190–195.

Rao, J.K., Raizada, A., Ganguly, D., Mankad, M.M., Satayanarayana, S.V., Madhu, G.M., 2015. Investigation of structural and electrical properties of novel CuO-PVA nanocomposite films. J. Mater. Sci. 50, 7064–7074.

Rhim, J.-W., Park, H.-M., Ha, C.-S., 2013. Bio-nanocomposites for food packaging applications. Prog. Polym. Sci. 38, 1629–1652.

Rinna, A., Magdolenova, Z., Hudecova, A., Kruszewski, M., Refsnes, M., Dusinska, M., 2015. Effect of silver nanoparticles on mitogen-activated protein kinases activation: role of reactive oxygen species and implication in DNA damage. Mutagenesis 30, 59–66.

Rossi, M., Passeri, D., Sinibaldi, A., Angjellari, M., Tamburri, E., Sorbo, A., Carata, E., Dini, L., 2017. Nanotechnology for food packaging and food quality assessment. In: Advances in Food and Nutrition Research. Elsevier, pp. 149–204.

Roy, S., Shankar, S., Rhim, J.-W., 2019. Melanin-mediated synthesis of silver nanoparticle and its use for the preparation of carrageenan-based antibacterial films. Food Hydrocoll. 88, 237–246.

Shah, S., Jahangir, M., Qaisar, M., Khan, S., Mahmood, T., Saeed, M., Farid, A., Liaquat, M., 2015. Storage stability of kinnow fruit (*Citrus reticulata*) as affected by CMC and guar gum-based silver nanoparticle coatings. Molecules 20, 22645–22661.

Shankar, S., Rhim, J.-W., 2017. Preparation and characterization of agar/lignin/silver nanoparticles composite films with ultraviolet light barrier and antibacterial properties. Food Hydrocoll. 71, 76–84.

Shankar, S., Prasad, R.G.S.V., Selvakannan, P.R., Jaiswal, L., Laxman, R.S., 2015. Green synthesis of silver nanoribbons from waste X-ray films using alkaline protease. Mater. Express 5, 165–170.

Shankar, S., Wang, L.-F., Rhim, J.-W., 2016. Preparations and characterization of alginate/silver composite films: effect of types of silver particles. Carbohydr. Polym. 146, 208–216.

Silvestre, C., Duraccio, D., Cimmino, S., 2011. Food packaging based on polymer nanomaterials. Prog. Polym. Sci. 36, 1766–1782.

Singh, A., Kumari, K., Kundu, P.P., 2017. Extrusion and evaluation of chitosan assisted AgNPs immobilized film derived from waste polyethylene terephthalate for food packaging applications. J. Packag. Technol. Res. 1, 165–180.

Sondi, I., Salopek-Sondi, B., 2004. Silver nanoparticles as antimicrobial agent: a case study on *E. coli* as a model for Gram-negative bacteria. J. Colloid Interface Sci. 275, 177–182.

Sonseca, A., Madani, S., Rodríguez, G., Hevilla, V., Echeverría, C., Fernández-García, M., Muñoz-Bonilla, A., Charef, N., López, D., 2019. Multifunctional PLA blends containing chitosan mediated silver nanoparticles: thermal, mechanical, antibacterial, and degradation properties. Nanomaterials 10, 22.

Ucankus, G., Ercan, M., Uzunoglu, D., Culha, M., 2018. Methods for preparation of nanocomposites in environmental remediation. In: New Polymer Nanocomposites for Environmental Remediation. Elsevier, pp. 1–28.

Valipoor Motlagh, N., Hamed Mosavian, M.T., Mortazavi, S.A., 2013. Effect of polyethylene packaging modified with silver particles on the microbial, sensory and appearance of dried barberry: AG-LDPE PACKAGES FOR WELL PRESERVING BARBERRY QUALITY. Packag. Technol. Sci. 26, 39–49.

Vanderroost, M., Ragaert, P., Devlieghere, F., De Meulenaer, B., 2014. Intelligent food packaging: the next generation. Trends Food Sci. Technol. 39, 47–62.

Vasile, C., 2018. Polymeric nanocomposites and nanocoatings for food packaging: a review. Materials 11, 1834.

Vermeiren, L., Devlieghere, F., Debevere, J., 2002. Effectiveness of some recent antimicrobial packaging concepts. Food Addit. Contam. 19, 163–171.

Vishnuvarthanan, M., Rajeswari, N., 2019. Preparation and characterization of carrageenan/silver nanoparticles/laponite nanocomposite coating on oxygen plasma surface modified polypropylene for food packaging. J. Food Sci. Technol. 56, 2545–2552.

Wang, J., Wang, X., Xu, C., Zhang, M., Shang, X., 2011. Preparation of graphene/poly(vinyl alcohol) nanocomposites with enhanced mechanical properties and water resistance. Polym. Int. 60, 816–822.

Wu, Z., Zhou, W., Pang, C., Deng, W., Xu, C., Wang, X., 2019. Multifunctional chitosan-based coating with liposomes containing laurel essential oils and nanosilver for pork preservation. Food Chem. 295, 16–25.

Yang, F.M., Li, H.M., Li, F., Xin, Z.H., Zhao, L.Y., Zheng, Y.H., Hu, Q.H., 2010. Effect of nano-packing on preservation quality of fresh strawberry (*Fragaria ananassa* Duch. cv Fengxiang) during storage at 4°C. J. Food Sci. 75, C236–C240.

Youssef, A.M., 2013. Polymer nanocomposites as a new trend for packaging applications. Polym. Plast. Technol. Eng. 52, 635–660.

Yousuf, B., Qadri, O.S., Srivastava, A.K., 2018. Recent developments in shelf-life extension of fresh-cut fruits and vegetables by application of different edible coatings: a review. LWT 89, 198–209.

Zhai, X., Li, Z., Shi, J., Huang, X., Sun, Z., Zhang, D., Zou, X., Sun, Y., Zhang, J., Holmes, M., Gong, Y., Povey, M., Wang, S., 2019. A colorimetric hydrogen sulfide sensor based on gellan gum-silver nanoparticles bionanocomposite for monitoring of meat spoilage in intelligent packaging. Food Chem. 290, 135–143.

Zimoch-Korzycka, A., Jarmoluk, A., 2015. The use of chitosan, lysozyme, and the nano-silver as antimicrobial ingredients of edible protective hydrosols applied into the surface of meat. J. Food Sci. Technol. 52, 5996–6002.

CHAPTER 7

Emerging silver nanomaterials for smart food packaging in combating food-borne pathogens

Divya Sachdev[a], Akanksha Joshi[a], Neetu Kumra Taneja[a], and Renu Pasricha[b]

[a]National Institute of Food Technology Entrepreneurship and Management-Kundli, Sonipat, Haryana, India
[b]New York University, Abu Dhabi, Dubai

7.1 Introduction

Although worldwide concerns for sufficient quantities of food at affordable prices continue, we have now moved into an era when food nutrition and safety are paramount, which requires strategic approaches that promote good health and sustain life. One approach, coined *preventive strategy*, focuses on all phases of the food supply chain, from agricultural production to consumption of the products. This strategy includes packaging of food products in a way that reduces adverse physical, chemical, and microbial effects and helps to prevent the outbreak of foodborne diseases (FBDs) that have resulted in numerous international health and economic emergencies requiring shutdowns that negatively affected global trade (Mehlhorn, 2015). In particular, every year developed and developing countries incur huge losses in food products due to inefficient and unhygienic packaging and storage problems that lead to microbial contamination. Most microbes (i.e., bacteria, fungus, virus, protozoa, and helminth worms) cause food spoilage, making it unfit for consumption. If consumed, almost all these microbes can cause a variety of illnesses, with symptoms ranging from abdominal pain and temporary flatulence to severe diarrhea, vomiting, food poisoning, and sometimes death. Therefore packaging is a crucial defense against microbial agents that preserves the food during postharvest processing and storage while ensuring the quality of the products and extension of the shelf life.

In the food supply chain, multidrug resistance to pathogenic microbes poses a very serious threat to food safety and public health. This threat is a consequence of the evolutionary adaptation of microbes that has led to their survival even in extreme conditions. The rampant use of antimicrobials in veterinary medicines, as feed additives in the poultry, aquaculture, and agriculture (as biocides in crop production) has led to antimicrobial-resistant (AMR) pathogens in the food chain (e.g., in products such as meat, eggs, and milk), exposing humans to highly infectious and resilient microbial strains (Levy and Bonnie, 2004).

The need of the hour is to mitigate AMR threats via multifaceted strategies in order to combat the serious threats of food infections, illnesses, and deaths due to the spread of multidrug-resistant (MDR) organisms in the food chain (Bengtsson-Palme, 2017). Nanotechnology can open up new horizons to combat food-borne pathogens incorporating it either as an antimicrobial material in food packaging inclusive of enhanced barrier strength or as an antimicrobial detector in the form of nanosensor. Nanotechnology has attracted the attention of researchers to combat food-borne pathogens and multidrug-resistant microbial biofilms (microbial consortia embedded in exopolysaccharide matrix) to resolve the issue of food-borne pathogenesis. AgNPs are good candidates for this purpose (Carbone et al., 2016). Thus, this chapter covers the role of synthesized AgNPs in achieving sustainable packaging materials via greener synthesis of materials and other synthetic methodologies. We also discuss AgNPs' influence on three major components of food packaging: improved properties of packaging (e.g., enhanced mechanical and barrier properties), intelligent/smart packaging (e.g., real-time feedback on quality of packaged foods), and active packaging (e.g., monitoring the internal environment of a package to prevent biofouling and increasing shelf life). AgNPs' antimicrobial mechanisms, factors related to their safety, and their market potential will be comprehensively discussed in the chapter.

7.2 Major foodborne pathogens responsible for food spoilage and FBDs

As mentioned earlier, the severity of FBDs varies, with symptoms ranging from diarrhea, vomiting, mild gastrointestinal problems where the patient readily recovers within 1–7 days to severe symptoms ranging from gastroenteritis, dysentery, haemorrhagic colitis, and meningitis which in extreme cases may lead to death. Thirty-one main foodborne pathogens have been identified: 21 bacteria, 5 parasites, and 5 viruses (Adley and Ryan, 2016). Major bacterial foodborne pathogens belong to the genera *Salmonella, Listeria, Escherichia, Clostridium,* and *Campylobacter* and to toxin-producing fungal species such as *Aspergillus* sp., *Candida* sp. (Kumar et al., 2017b). However, most cases of foodborne diseases are caused by viral infections due to the presence of noroviruses, rotaviruses, and hepatitis A and E viruses in contaminated food (O'Shea et al., 2019).

It has been observed that the microbes capable of biofilm formation are the dominant causative agent for FBDs. These microbes are usually drug-resistant and are capable of adhering to food contact surfaces or to exposed food surfaces and often have fatal outcomes. Some of the major biofilm-forming pathogens include *Listeria monocytogenes, Salmonella enterica, E. coli* O157:H7, and *Yersinia enterocolitica* (Bridier et al., 2015). Fig. 7.1A highlights common foodborne pathogens responsible for pathogenesis via products from major sectors of the food industry (Koutsoumanis et al., 2014).

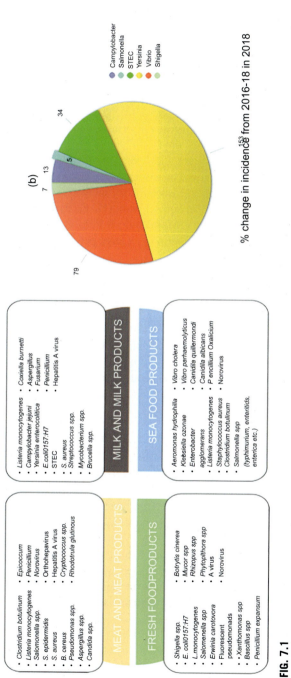

FIG. 7.1

(A). Different pathogens prevalent in major sectors of the food industry (Koutsoumanis et al., 2014). (B) Incidents of foodborne pathogens, especially as a result of *Yersinia* and *Vibrio* infections, showed maximum hype, as compared with the rate of infections in 2016, 2018, and 2019 (Tack et al., 2020).

According to a Centers for Disease Control and Prevention (CDC) report, a significant increase in bacterial infections, specifically those of *Yersinia* (153%), *Vibrio* (79%), *E. coli* (Shiga toxin-producing *E. coli*, or STEC) (34%), and *Campylobacter* (13%), appeared in 2019, as compared with data from 2016 through 2018, as shown in Fig. 7.1B. The report suggests that the increase may be attributed to advanced testing methods for disease-causing pathogens (including the detection of non-culture species); nonetheless, increased exposure to these pathogens cannot be ignored (Tack et al., 2020).

Although scientific developments already have led to better diagnostic techniques and more efficient clinical treatments, the field continues to evolve and novel strategies to control the spread of FBDs are still being developed. For example, to help ensure healthy practices and safeguard consumers' health, industries are employing recent developments in food safety measures. Also, methods such as clean-in-place (CIP) and hazard analysis and critical control point (HACCP) include modern treatments such as nonthermal treatments (Morales-de la Peña et al., 2019) and innovations in packaging (Risch, 2009). Furthermore, a large percentage of novel practices focus on sustainable packaging options.

7.3 Sustainable food packaging (SFP)

To reduce the ecological footprint and optimize resources, there is a pressing need to transition from packaging that generates a high load of waste to recyclable packaging that reduces waste and improves the quality of life (Guillard et al., 2018). Packaging that helps to prolong the shelf life of food products (e.g., during transportation and storage) while supporting their marketability is also important. Because packaging is an integral and costly part of the food industry, achieving sustainability relies mainly on economical, societal, and environmental factors. Influences via economical performances are targeted mainly on manufacturers, who should bear only a fraction of the cost of the product in packaging henceforth most manufacturers compromise the quality of packaging. In terms of societal factors, packaging needs to be customer-friendly and trustworthy and needs to match customers' attitudes and behaviors regarding waste disposal. Likewise, the environmental factor relates to the level of packagings' overall footprint, which should be kept low by reducing, recycling, and reusing materials.

SFP requires a holistic approach that will enable the food industry to adapt according to lifestyle changes. In addition to their increasing demand for microbiologically safer foods, customers are inclined to purchase packaged foods that are easy to handle, ready to consume and that maintain integrity (quality and freshness) from farm to fork.

7.3.1 Demands and challenges

The packaging industry is an important component in the food industry. While using packaging that will decrease the carbon footprint and minimize the amount of discarded materials, the industry needs to avoid overexploitation of available resources in order be sustainable. To achieve the goal of sustainable development, it is essential to reduce global food waste at retail and consumer levels and decrease postharvest food loss. Food waste is also a rising impediment to sustainable foods and requires both responsible production of food products and responsible consumption of those products.

Before reformatting and introducing eco-friendly strategies for SFP, certain conditions need to be fulfilled, as shown here (Robertson, 2009):

- *Legality:* Regulations on the use and packaging of materials need to be checked.
- *Effectiveness:* The packaging used needs to provide social and economic benefits. It should add real value by reducing the manufacturing cost, while maintaining the quality of products throughout the supply chain.
- *Efficiency:* Packaging needs to utilize the materials and energy efficiently throughout a product's life cycle and to maintain its quality, while improving and following safety standards and perceptions.
- *Cyclical:* Recovery rates need to be optimized, and the packaging material must be recycled continuously, either by natural or industrial processes, thereby reducing the ecological load. The amount of recycled content in packaging needs to be a factor.
- *Safety:* The packaging material should be nontoxic to both consumers and the environment. It should also be derived from nonpolluting sources and follow current standards.
- *Customer satisfaction and convenience:* Packaging products need to evolve to meet changing consumer needs, demonstrate food traceability throughout the supply chain, respond to consumers' complaints, demonstrate social responsibility, and respond to competitive pressure.
- *Minimization of negative impacts:* The packing industry needs to eliminate toxic and hazardous substances and comply with environmental regulations.
- *Disposal:* Packaging needs to be designed in such a way that before disposal it can be compressed to minimize its volume in a landfill.
- *Design for litter reduction:* Antilitter information needs to be included on the packaging.

Despite rigorous research in the development of SFP, currently available SFP materials have encountered barriers to dynamic marketability. To overcome these barriers and enter the global packaging market successfully, certain factors need to be considered (Guillard et al., 2018; Robertson, 2009):

- *Efficiency-to-cost ratio:* Sustainable packaging (SP) materials generally have a higher production cost, compared with their conventional counterparts, which makes manufacturers reluctant to adopt them.
- *Ecological impact:* Biobased and biodegradable materials are highly relevant in the production of SP materials, whereas conventional materials are less eco-friendly, as most of these materials are nonbiodegradable, depending on the raw material from which they were developed. Therefore thorough assessments are needed.

- *Sustainability assessments:* Overall sustainability assessments of proposed materials are another challenge; however there is no clearcut definition of sustainability, which makes it difficult to establish appropriate criteria for such assessments.
- *Recyclability:* SP material needs to be recyclable, either by natural or industrial methods. This scenario remains in doubt for most conventional materials.
- *Technical issues:* Common technical issues that impede the development and use of SPs relate to variations in raw materials; a very narrow processing window, which limits a standardized manufacturing protocol; and a lack of tools to tailor the packaging according to food needs, which hinders their acceptance among manufacturers.

These criteria for packaging designs will ensure the quality, safety, and textural stability and conduciveness of stored foods. Moreover, their commercial value will be enhanced. In essence, the current demand for engineered, sustainable packaging materials will minimize the ecological footprint of the food industry by reducing food waste due to the accumulation of nonbiodegradable packaging, consequently promoting sustainable food production and consumption, all of which will help maintain natural resources.

Polymer bio-nanocomposites are seen as the best alternative to conventional packaging materials, as they are nontoxic, ecologically compatible, and renewable. The major drawbacks associated with biopolymer based materials include poor mechanical strength, poor chemical and gaseous barriers, a high water vapor transmission rate (WVTR), and no antimicrobial activity. However, these advanced packaging materials with active technologies will combat and decrease the growth and accumulation of foodborne pathogens that causes biofouling and ultimately FBDs. The application of nanotechnology can help combat the major challenges to achieving SFP and enable successful competition in the global food market by promoting cost-effectiveness and societal and environmental sustainability. These technologies will also reduce loss by controlling food wastage, enable longer storage periods while maintaining low susceptibility to food pathognens, and avoid toxic chemicals used in preservatives, among other benefits. The use of nanomaterials in packaging material can also enhance polymeric properties such as barriers and mechanical- and heat-resistant properties (Bumbudsanpharoke and Ko, 2015). Both metallic and nonmetallic nanoparticles (NPs) have been established that can enhance antimicrobial effects and be utilized in novel packaging materials.

7.3.2 Nanotechnological aspects of food packaging

Nanotechnology can provide innovative food packing by direct incorporation of nanomaterials into food products and food packaging materials and during food processing. Fig. 7.2A provides a brief overview of advancements in food packaging. Nanoenabled packaging can protect the color, flavor, taste, texture, and consistency of foodstuff via improved barrier, mechanical and antimicrobial properties (Sharma et al., 2017).

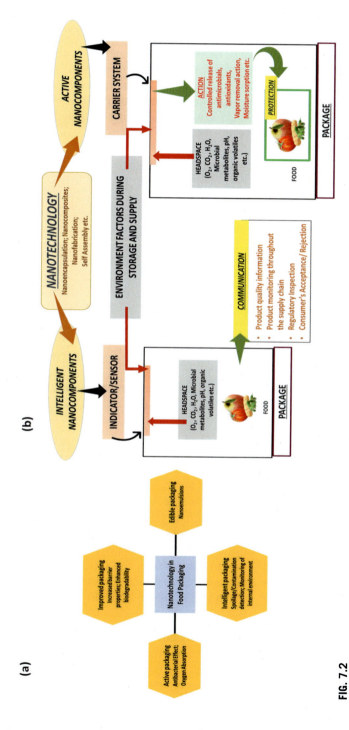

FIG. 7.2

(A) Prospects for nanotechnological applications in food packaging. (B) Two uses for nanotechnology in the food industry include smart packaging systems that enhance communication with consumers indicating the suitability of the product and nanocomponents whose mechanisms prevent food from spoiling.

Nanomaterials and nanocomposites can positively affect food packaging with following major approaches (Grumezescu and Holban, 2018):

- *Improved packaging:* Biodegradable nanomaterials and nanocomposites can be combined with polymeric packaging material, which will improve the material's mechanical and barrier properties (e.g., elasticity, barrier against oxygen and carbon dioxide, diffusion of flavor compounds, and stability under varied temperature-moisture conditions). As a result, biodegradable nanomaterials and nanocomposites protect food products from undesirable mechanical, chemical, and microbial effects.
- *Active packaging:* This technology involves the incorporation of nanomaterials (as active constituents) into food packaging materials where the carrier component interacts with intrinsic and extrinsic factors, which stimulates actions and ultimately enhances the shelf life, quality, and safety of food products (Gokularaman et al., 2017).
- *Intelligent packaging:* This type of packaging provides information to consumers such as the quality, nutritional value, and instructions for preparing packaged foods. This information is important to consumers and thus the products' marketability. It is also a guard against food fraud. Additionally, when nanosensors are embedded, they can be used as labels that indicate to customers information such as food spoilage or contamination due to the presence of toxins, pesticides, and microbes or due to environmental changes inside a package due to temperature, moisture, and gases caused by microbial growth.
Intelligent packaging can also contribute to the hazard analysis critical control point (HACCP) and the quality analysis critical control point (QACCP) projects designed to ensure safe foods and identify health hazards on site.
Fig. 7.2B shows how NPs function in smart packaging systems as reactive particles to augment product quality (active packaging) and to communicate with customers about the condition of packaged products (intelligent system). These technologies can be used together and thus work in synergy to improve the packaging sector. Nanotechnology-based techniques that can detect foodborne pathogens at various stages of food production and provide quick results are already being used in the marketplace.
- *Edible packaging:* The wide use of edible nanolaminate coatings helps to preserve and prolong the shelf life of food products. This type of packaging acts as a barrier against mass transfer incidents. It is a broad-based packaging technique that utilizes AgNPs, which is discussed at length in a later section in this chapter.

Nanomaterials can be categorized as inorganic and organic and are applied in food packaging systems. The inorganic nanoagents Ag, CuO, ZnO, TiO_2, MgO, and Fe_2O_3 have shown to strongly inhibit foodborne pathogens (Huang et al., 2018). Among the plethora of metallic NPs, AgNPs have attracted much attention because of their unique properties, particularly their high antimicrobial action even in solid-state samples. Although silver has been known for its bacteriostatic property since antiquity and could probably be discussed ad infinitum, our focus in this chapter is on silver NPs/nanocomposites as emerging "smart food packaging" materials used to combat food pathogens and thereby enhance the shelf life of foods.

7.3.3 How nanomaterials improve barrier and strength properties

When packaging materials (polymers) are reinforced with nanomaterials, the gas and moisture barrier properties are enhanced, which prolongs the shelf life of foods. Polymer matrices filled with nanomaterials are used to produce stiff nanofillers for packaging. Depending mostly on the amount and diversity of nanomaterials used, their dimensions, morphology, and agglomeration state, nanomaterial dispersion is expected to improve the mechanical properties of the polymer matrix (Rai et al., 2019). Improvements in mechanical properties allow the use of thinner packages and thus fewer materials, which reduces the cost for raw materials and recycling, giving such products a commercial advantage.

Polymers generally have low stiffness and strength, and metal-based nanomaterials like silver act as nanofillers that reinforce polymer matrices. NPs with large surface areas favor filler-matrix interactions and change the performance of the resulting material due primarily to the following conditions (de Azeredo, 2009):

- The uniform dispersion of NPs leads to a large matrix-filler interface, which changes the molecular mobility and relaxation behavior, resulting in better thermal and mechanical properties of materials.
- A high surface-area-to-volume ratio (with nano dimensions) provides better reinforcement and improves the properties of packaging materials.
- The nonaggregated network of NPs within the matrix allows NPs to infiltrate the matrix interphase network, which improves the functionality of nanocomposite materials.

Antibacterial packaging systems have used metal nanomaterials as fillers. A study by De Moura et al. (2012) reported that the elasticity modulus increased along with the size of the NPs. Nanofillers have an impact on packaging materials' water vapor barrier capacity. Other reports have suggested that when AgNPs are incorporated into pectin films, the mechanical properties of the films slightly increase, which may be attributed to the hydrogen bonding between the AgNPs and the pectin polymers.

In addition, the incorporation of nanofillers can lead to structural changes in the polymer matrix and a less dense structure, which will facilitate interactions between the components, rather than via hydrogen bonding with water molecules. Upon the addition of AgNPs, whether there is an improvement or weakness in a material's strength also depends on the nature of the matrix, homogeneity of the components, processing conditions, and the components' compatibility (Kraśniewska et al., 2020).

7.4 Silver nanomaterials (AgNMs) for food packaging

As early as 4000 BCE (and before the realization that microbes were causing infections), silver was used in the treatment of numerous medical conditions (Alexander, 2009). Silver has also been used as an antimicrobial agent in the storage of food and beverages. There are reports that ancient societies stored wine and water in silver vessels and also that early settlers in United States placed silver dollars and silver spoons at the bottom of milk containers to prolong its self life. Silver also has long been used as a disinfectant. In the early 20th century (before the introduction of antibiotics in the early 1940s), silver became the most obvious choice as an antimicrobial agent. In 2009, the FDA modified the regulations for permitting the use of silver nitrate as a disinfectant in water bottles (Duncan, 2011). With the present era, microbes are increasingly developing resistance to antibiotics and are colonizing to form biofilms. These biofilms are most prevalent on the surfaces of food materials, food processing equipment, and surface pipelines. Bacterial biofilms that are resistant to conventional antibiotics have been a major cause of FBD outbreaks and further contamination of food products. As a consequence, food processing units, food-based manufacturers, and agri-producers urgently need to minimize the emerging threat of biofilms.

Because AgNMs are effective antimicrobial agents with the ability to enhance the quality and shelf life of food and help in the safe distribution food worldwide, there is an urgent need to understand both their potential and their challenges. Studies have looked at AgNPs and hybrid silver nanocomposites (AgNCs) because of their broad-spectrum antimicrobial activity. Silver, due to its low volatility and high-temperature stability at the nanoscale, has been claimed to be effective against 100 types of bacteria (Mustafa and Andreescu, 2020). Their efficacies and applications are very much dependent on their size, polymer matrices, stability, and shape. Table 7.1 depicts the broad-spectrum antimicrobial application of AgNPs and hybrid AgNCs for food packaging. AgNPs used in food packaging can be administered through a variety of polymeric matrices that can be either biodegradable (chitosan, cellulose, etc.) or nonbiodegradable (polyethylene, polyvinyl alcohol, etc.) (Carbone et al., 2016).

7.4.1 Types of silver nanomaterial-based antimicrobial packaging materials

Diverse packaging materials embedded with AgNPs or hybrid AgNCs have been developed for the purpose of increasing the shelf life of food. Polymers are considered as a suitable medium for synthesis and stabilization of AgNPs. Although a great deal of research has been done on the types of AgNP-based packaging materials, the majority of them can be classified as biodegradable or nonbiodegradable.

7.4.1.1 Nonbiodegradable AgNMs for food packaging

AgNP-functionalized, nonbiodegradable packaging materials gained ground due to their effective antimicrobial actions, while retaining the thermal, mechanical, and physical features of traditional packaging. Multiple strategies are used to incorporate

Table 7.1 Diverse silver-nanomaterials and their potential applications in food packaging.

Inorganic nanoparticles alone/combination	Packaging material (biodegradable/nonbiodegradable)	Size/concentration of NPs	Tested food/film	Property enhanced antimicrobial/sensor/barrier properties	Commercial/application	Remarks	Reference
Chemically synthesized AgNPs	C + PEG	12 nm	Edible film	Antimicrobial	Porous/sustain release of silver ions application	Superior inhibition	Vimala et al. (2010)
	C + bacterial cellulose nanocrystals	AgNPs 1.0% (w/w)	As a film	Antimicrobial	As an application	Improved physical/mechanical property of film	Salari et al. (2018)
Chemically synthesized AgNPs	C + starch	25 nm	As a film	Antimicrobial and improves barrier properties for oxygen Diminishes moisture barrier properties	As an application	Improved tensile strength	Yoksan and Chirachanchai (2010)
Chemically synthesized AgNPs	Cellulose	5–35 nm	Adsorbent pads for melons	Antimicrobial	Lessened the ripening effect	Retarded the senescence of the melon cuts	De Moura et al. (2012) and Fernández et al. (2010a)
	Hydroxypropyl methyl cellulose (HPMC)	41 nm	Films		Increased the juicier appearance up to 10 days; bactericidal effect		
Green synthesized AgNPs	C (biodegradable)	61.5–76.89 nm	Minced beef stored for 10 days at 4°C	Antimicrobial	Proposed sustainable packaging	Improved shelf life and acceptance	Badawy et al. (2019)

Continued

Table 7.1 Diverse silver-nanomaterials and their potential applications in food packaging—cont'd

Inorganic nanoparticles alone/combination	Packaging material (biodegradable/nonbiodegradable)	Size/concentration of NPs	Tested food/film	Property enhanced antimicrobial/sensor/barrier properties	Commercial/application	Remarks	Reference
UV radiation and heat synthesized AgNPs	Cellulose	1% adsorbed on cellulose fibers	Beef meat	Antimicrobial	Experimental application	Reduction by silver adsorbent pads under modified atmospheric packaging (MAP)	Fernández et al. (2010b)
Chemically synthesized AgNPs	Bacterial cellulose	10–15 nm	Tomatoes	Antimicrobial	Can be commercialized	Tomatoes were prevented from spoilage for 30 days	Adepu and Khandelwal (2018)
UV synthesized AgNPs	LDPE	19–27 nm	Packaging	Antimicrobial	Potential for active packaging	Improved mechanical properties of films	Azlin-Hasim et al. (2016)
AgNPs	LDPE (nonbiodegradable)	1 wt% Ag particles	Barberry	Antimicrobial	Application	Potential of retaining the red color and brightness of barberry retained for 3 weeks	Motlagh et al. (2013)
AgNPs chemically synthesized	LDPE (nonbiodegradable)	10 nm	Chicken breast fillet 4°C for 12 days	Antimicrobial and significant increase in shelf life	Application especially for nonacidic foods	AgNPs/LDPE films retard the lipid oxidation	Azlin-Hasim et al. (2015) and Cushen et al. (2014)

AgNPs purchased	PVC	40–50 nm	Minced beef 3 ± 1°C for 14 days	Antimicrobial Microbial count significantly decreased due to AgNPs	Application	The shelf life improved from 2 to 10 days	Mahdi et al. (2012)
AgNPs + calcium chloride	Sodium alginate		Fior de Latte cheese 8 ± 1°C	Antimicrobial coating	Application	Shelf life increased to 10 days	Mastromatteo et al. (2015)
AgNPs	Alginate	1–2%	Shiitake mushroom 4 ± 1°C	Antimicrobial action preserved the physicochemical and sensory quality of mushrooms	Commercial applicability	Shelf life increased for 16 days Lower microbial counts, including mesophilic, psychrophilic bacteria, yeasts, and molds	Jiang et al. (2013)
Chemically synthesized silver nanoclusters (AgNCs)	Zein films (biodegradable)	Less than 10 nm	Films for packaging	Antimicrobial	Commercial applicability	The film has low environmental impact and low volatility	Mei et al. (2017)
AgNPs	Ethylene vinyl alcohol copolymer (EVOH)	1–20 µm	Chicken, pork, cheese, lettuce, apples, peels, eggshells	Antimicrobial	Commercials applicability	2 log reduction of bacterial count	MartínezAbad et al. (2012)

Continued

Table 7.1 Diverse silver-nanomaterials and their potential applications in food packaging—cont'd

Inorganic nanoparticles alone/combination	Packaging material (biodegradable/nonbiodegradable)	Size/concentration of NPs	Tested food/film	Property enhanced antimicrobial/sensor/barrier properties	Commercial/application	Remarks	Reference
AgNPs	Alphasan material (polymer used in refrigerator)	Varied percentage	Meat, cheese, vegetables	Antimicrobial (high sensory quality, no suppression of antibacterial effect of silver due to proteins of meat and cheese)	Commercial applicability	Silver inner liners have an additional safety barrier that contributes to the overall hygienic concept within a refrigerator.	Kampmann et al. (2008)
Nanosilver	BPA free plastic	—	—	—	Manufacturer-Kinetic Go Green (commercial product)	Food containers in market Food and beverage storage	Bumbudsanpharoke and Ko (2015)
					Manufacturer-Always Fresh (food lasts for weeks)	Release all-natural minerals that act like magnets to remove these damaging gases	
Green synthesized AgNPs	C + gelatin	AgNPs 0.1% (w/w)	Red grapes	Antimicrobial	Extends shelf life for 14 days	Composite film	Kadam et al. (2019) and Kumar et al. (2018)
	C alone	8 nm	Film	Antimicrobial	pH-dependent release		

Green synthesized AgNPs	C (biodegradable)	80 nm	Tomato, ladyfinger (perishable)	Antimicrobial property	Edible coating (packaging) application	Can be directly used Improved the shelf life at room temperature Cost effectivity to be enhanced	Ragunathan et al. (2015)
Green synthesized AgNPs	Polyethylene matrix	20 nm	Nuts, hazelnuts, almonds, and pistachios	Antimicrobial	Increases the shelf life to 20 months	Effects of AgNPs from packaging material to the food and effect on health needs to be investigated	Darroudi et al. (2010) and Tavakoli et al. (2017)
AgNPs	Agar hydrogel	3 concentrations of silver montmorillonite used	Fior di Latte cheese $10 \pm 1°C$	Antimicrobial	Commercials applicability Shelf life for active packaging increased	Suitability for diffusion of active systems (silver) as food contact materials, needs to be undertaken	Incoronato et al. (2011)
AgNPs	Polyvinylpyrrolidone	20 nm	Asparagus stored from 2 to 10°C	Improved barrier properties and antimicrobial	Application Shelf life improved to 10 days at 2°C	Slowed down the weight loss as well as loss of ascorbic acid and total chlorophyll	An et al. (2008)
AgNPs	Agar, zein, and poly(e-caprolactone)	40 nm	Film for active packaging	Antimicrobial	Application	Agar-based nanocomposites were the most active against the test microorganisms	Incoronato et al. (2010)

Continued

Table 7.1 Diverse silver-nanomaterials and their potential applications in food packaging—cont'd

Inorganic nanoparticles alone/combination	Packaging material (biodegradable/nonbiodegradable)	Size/concentration of NPs	Tested food/film	Property enhanced antimicrobial/sensor/barrier properties	Commercial/application	Remarks	Reference
AgNPs	Polyethylene oxide (PEO)	90 nm	Apple juice	Antimicrobial Inhibiting a thermal resistant food spoilage microorganism, in acidic beverages	Commercial applicability	Inhibiting the growth of an *Alicyclobacillus acidoterrestris* stra	Del Nobile et al. (2001)
Green synthesized AgNPs	Sodium alginate (biodegradable)	5–40 nm	Pear carrot	Antimicrobial	Experimental application	Shelf life and sensory acceptability improved till 7 days of preservation Give good tensile strength, flexibility, resistance to tearing	Costa et al. (2012)
Green synthesized AgNPs	PVA (polyvinyl alcohol)-montmorillonite K10 clay (biodegradable)	40 μl of $AgNO_3$ solution in film	Chicken sausages	Antimicrobial	Application	Superior mechanical properties, water resistivity, and light barrier ability	Mathew et al. (2019)
Green synthesized AgNPs	Polyethylene glycol (biodegradable)	4.2 ± 1.3 nm	Film	Antimicrobial	Sustainable packaging	Effective antimicrobial action with improved shelf life and no cytotoxicity	Rolim et al. (2019)

Nanoparticle	Polymer	Size	Food/Product	Function	Industrial packaging films	Effect	Reference
AgNPs	LLDPE + cyclo olefin copolymer (nonbiodegradable)	50 nm	Film	Antimicrobial		Exhibited excellent antifungal activity with improved water vapor barrier properties	Sanchez-Valdes (2014)
Green synthesized AgNPs	Agar (biodegradable)	15–20 nm	Film	Antimicrobial	Sustainable packaging	Improved UV barrier properties and mechanical strength	Shankar and Rhim (2015)
AgNPs	Carboxy methyl cellulose (biodegradable)	8 nm	Films	Antimicrobial	Sustainable packaging	Enhanced water vapor paper	Siqueira et al. (2014)
AgNPs (green synthesis)	Sodium alginate (biodegradable)	5–40 nm	Carrot and pear	Antimicrobial	Sustainable packaging	Increased shelf life of fruits and vegetables by 10 days	Mohammed Fayaz et al. (2009)
AgNPs + ZnO	LDPE (nonbiodegradable)	10%	Chicken breast cooked 4°C for 21 days	Antimicrobial (delayed spoilage and lipid oxidation)	Experimental application	Migration of NPs in accordance to permissible limits	Panea et al. (2014)
Commercial AgNPs + ZnO	C + rice starch	–	Peach	Antimicrobial	Experimental application	Increases	Kaur et al. (2017)
AgNPs + ZnO + essential oils	Pullalan	110 nm (both)	Turkey deli meat	Antimicrobial	Application	Edible coating extended shelf life	Khalaf et al. (2013) and Morsy et al. (2014)
			Poultry products	Antimicrobials		Safety of fresh refrigerated poultry and meat	

Continued

Table 7.1 Diverse silver-nanomaterials and their potential applications in food packaging—cont'd

Inorganic nanoparticles alone/ combination	Packaging material (biodegradable/ nonbiodegradable)	Size/ concentration of NPs	Tested food/film	Property enhanced antimicrobial/ sensor/barrier properties	Commercial/ application	Remarks	Reference
Ag-CuNPs cinnamon essential oil	Polylactide (PLA) and polyethylene glycol (PEG)	<100nm	Chicken meat	Antimicrobial and improves barrier properties	Film with thermal stability has potential for commercialization for active food packaging	Composite films with UV protection and prevent meat spoilage	Ahmed et al. (2018)
AgNPs + CuO + ZnO	LDPE (nonbiodegradable)	Ag 35 nm	Ultra-filtrated cheese 4 ± 0.5°C for 28 days	No synergism of all the inorganic nanoparticles	For active packaging	Only CuO showed antimicrobial and increase the harshness of food processing	Beigmohammadi et al. (2016)
TiO_2 + AgNPs	Polylactic acid (PLA)	10nm	Yunnan cottage cheese 5 ± 1°C for 25 days	Antimicrobial Inhibition over the microbial, yeast, and molds	Potential for commercialization	Migration of NPs in accordance with the EFSA	Li et al. (2017)
Bergamot essential oils + TiO_2 + AgNPs	Poly lactic acid (PLA)	<150nm	Mangoes room temperature, 15 days	Antimicrobial Effectiveness against total bacteria count	Potential for commercialization Maintain the quality and extend the postharvest life for 15 days.	Improved barrier properties, delayed the loss of firmness, retard the harmful color changes, microbial count, and total acidity	Chi et al. (2019)

Nanocomposite	Polymer matrix	Size	Food/Application	Function	Potential for commercialization	Observations	Reference
Calcium alginate + Ag-montmorillonite NPs	Montmorillonite (active coating)	10–40 nm	Carrots 4 ± 1°C	Antimicrobial and prevent dehydration		Combined use of active packaging and the active coating maintains the preservation up to 70 days	Costa et al. (2012)
Ag + ZnO NPs	LDPE (nonbiodegradable)	70 nm	Orange juice	Antimicrobial and increased the shelf life to 28 days	Application	The sensorial parameters were not affected.	Emamifar et al. (2010, 2011)
Ag + TiO$_2$	LDPE polyethylene	300 nm	Strawberry 4 ± 1°C	Maintain the sensory, physicochemical, and physiological quality of strawberry fruits	Application	Extend the shelf life	Yang et al. (2010)
Pectin + AgNPs (green synthesis) + laponite	Polypropylene (biodegradable)	–	Film for active packaging	Antimicrobial with enhanced oxygen barrier	Biodegradable active packaging	Improved mechanical and barrier properties	Vishnuvarthanan and Rajeswari (2019)
Ceric ammonium nitrate + AgNPs (green synthesis)	Cellulose (biodegradable)	25 nm	Film	Antimicrobial	Biodegradable active packaging	No harsh treatments used	Tankhiwale and Bajpai (2009)
Nanosilica-carbon-silver ternary hybrid	Commercial biopolymer	<100 nm	Chicken thigh	Antimicrobial	Biodegradable biostatic packaging	3-D printed packaging films	Biswas et al. (2019)

AgNPs into a nonbiodegradable host matrix (generally polyethylene, polypropylene, polystyrene, polyvinylchloride, or ethylene-vinyl chloride).

Sa'nchez-Valdes et al. (2009) deposited AgNPs on a five-layer barrier film structure of linear low-density polyethylene (LLDPE) via three different methods (laminating, casting, and spraying). The AgNPs developed were found to be more homogeneous with an average diameter of 25 nm and less aggregation after ultrasonic irradiation. The casting method and weight percentage affected the antimicrobial potential and barrier properties of the multilayered film. Laminated films with 0.6 wt% AgNP showed minimal oxygen permeability ($33 \pm 0.3\,cm^3\,m^2\,d^1$), compared with films developed through other methods, although this was 10% more than the reference film (without AgNPs). The WVTR was minimal ($3.2 \pm 0.4\,g/m^2$/day), compared with the reference film and other nanocomposite films, which was attributed to the tiny pores on the surface of films developed through casting and spraying methods, which enhanced the rate of diffusion of gases, water vapors, and Ag^+ ions and led to the films' enhanced antimicrobial action. All these films exhibited antimicrobial activity against *Psuedomonas oleovorans* and *Aspergillus niger*, but the effect was much more enhanced against the bacterial species than the fungal species.

Jokar et al. (2012) incorporated AgNPs (spherical, average size of 5.5 ± 1.1 nm) produced by chemical reduction using polyethylene glycol (PEG) into low-density polyethylene (LDPE) using the melt blending method. The mechanical properties of AgNPs incorporated blend were not significantly different from silver-free LDPE containing PEG film, although the nanocomposite films demonstrated potential antimicrobial action against *E. coli*, *S. aureus*, and *C. albicans* when tested by the agar diffusion method, and the dynamic shake flask test against *S. aureus* showed the highest susceptibility. Sanchez-Valdes (2014) employed the sonochemical deposition method to deposit AgNPs on LLDPE/cycloolefin copolymer blends. Ethylene glycol and ammonium hydroxide acted as reducing agents to produce spherical AgNPs < 50 nm in size. Fungicidal activity against *Aspergillus niger* was found to increase with AgNP concentrations from 0.02 mol/L (84.5% inhibition) to 0.08 mol/L (95.4%), but no significant difference was observed at the concentration of 0.10 mol/L (95.6%). WVTR was found to decrease, which may have been due to hindrance created by impervious AgNPs in the polymer matrix. Also, Becaro et al. (2015) developed LDPE films with AgNPs embedded into different carriers (named as PEN with 5% SiO_2 and PEC with 15% TiO_2). PEC demonstrated marginally superior antimicrobial action presumably due to the synergic action of highly oxidative TiO_2.

7.4.1.2 Biodegradable AgNMs for food packaging

Bio-based polymer matrices have attracted the attention of researchers who are seeking to develop sustainable and biodegradable packaging materials. The high demand for biopolymer-based antimicrobial packaging poses great potential for novel developments due to their low cost, renewability, nontoxicity, and environmental suitability. Various biopolymers that have gained the FDA approval are generally regarded as safe, for example, zein, polylactic acid (PLA), cellulose, chitosan, and pullulan. These natural polymers exhibit good properties as packaging materials with cohesive structures and the ability to create a thin protective layer on food surfaces; however, their poor mechanical properties pose problems for their use. Nevertheless, these polymers can be used in active and smart packaging materials wherein the active components are incorporated into polymer-based materials. Extensive research is being conducted on strategies for implanting antimicrobial agents into biodegradable polymers, targeted as sustainable and green alternatives.

Blends with metallic NPs such as TiO_2, Zn, Cu, and Ag are being employed, and among metallic NPs, AgNPs are of prime interest. Various blends and nanocomposites utilizing AgNPs have shown remarkable antimicrobial properties. Mei et al. (2017) explored a novel idea by incorporating silver nanoclusters (AgNCs) into the polymeric matrix of zein. The ultra-small size of these AgNCs, about 2 nm, render them superior activity (due to high surface-area to volume ratio), as compared with AgNPs. The AgNCs were prepared by UV-A irradiation in a reaction mixture, that is, $AgNO_3$ and poly methacrylic acid (PMAA). PMAA acted as the stabilizer, and the resultant AgNCs were embedded into zein films through a dry cast process. Agar diffusion assay and microbial growth rate measurements were performed to assess the antimicrobial potential of the AgNCs against *E. coli* O157:H7 and were compared to $AgNO_3$ and AgNPs (60 and 100 nm, respectively). It was observed that AgNCs exhibited a potential zone of activity (ZOI) of 1.95 mm in contrast with either $AgNO_3$ or AgNPs and a minimum inhibitory concentration (MIC) of 1.09 μg/cm^2 within the zein films. AgNCs were found to exhibit low cytotoxicity against human colon cancer cells of cell line HCT116 with a cell viability of 80% and IC_{50} of 34.68 μg/mL.

Rhim et al. (2014) synthesized nonagglomerated spherical AgNPs by laser ablation of a silver plate. Agar/AgNP composite films prepared by the solvent casting method, which enhanced thermal stability, followed by the addition of metallic silver. Improved water barrier properties (WBPs) were observed, as compared with the agar films. The WBP increased slightly along with the increasing AgNP content, which was attributed to the tortuous path created by the impermeable NPs, although this increase was not statistically significant. However, reduced mechanical strength and stiffness was observed in the composite. The film exhibited good antimicrobial potential against both Gram-positive and Gram-negative bacteria (*Listeria monocytogenes* and *E. coli* O157:H7), which was tested through the viable cell count method.

Carboxymethylcellulose (CMC) was explored as a host matrix for AgNPs by Siqueira et al. (2014) using the casting method for composite film preparation. AgNPs were prepared by chemical reduction using sodium borohydride ($NaBH_4$) and polyvinyl alcohol (PVA) as reducing and stabilizing agents. The nanocomposite film showed decreased water vapor permeability (WVP) and high thermal stability. The susceptibility of Gram-positive (*Enterococcus faecalis*) and Gram-negative (*E. coli*) bacteria to the films was tested, and the MIC value was found to be 0.1 μg/cm^3 for both strains, indicating high bacteriostatic potential.

Similarly, different studies involved the use of hydroxypropyl methylcellulose (HPMC) (De Moura et al., 2012), silica (Biswas et al., 2019), and PVA (Tao et al., 2017), and numerous other polymeric matrices of biological origin. Recently, Poly(3-hydroxybutyrate-*co*-3-hydroxyvalerate), a biodegradable material known as PHBV, was incorporated with 0.04% AgNPs, which led to a decrease in oxygen permeability to about 56% while retaining optical and thermal stability. Moreover, the film produced was antimicrobial against foodborne pathogens (*Salmonella enterica* and *Listeria monocytogenes*) (Fahmy et al., 2020).

7.4.2 Silver nanomaterial-based sensors for maintenance of food quality

Sensors incorporated into food packaging provide not only a platform for protecting food but also a way to communicate with customers via information that appears on packages about the quality and safety of packaged food products. Currently, sensing systems in packages involve the detection of microorganisms, spoiled gases, and changes in pH during spoilage. Nanobased metal oxides, quantum dots, and magnetic NPs for bar codes have been employed to detect spoilage and contaminants; however, very few examples have involved the use of AgNPs. Silver nanocolloids have exhibited properties that can be used for diagnostic purposes in the form of optical chemosensors and biosensors.

Heli et al. (2016) utilized AgNPs' plasmic properties while embedding them in bacterial cellulose paper for visual detection of corrosive vapors emitted from spoiled fish and meat. The spoiled meat released ammonia gas that consequently led to a change in the color of the cellulose paper from amber to gray or taupe. Sachdev et al. (2016) also implemented the plasmic behavior of AgNPs as a visual sensor for selectively detecting spoilage in onions. The spoiled onions emitted volatile sulfur gases that showed distinct color changes of the AgNPs from yellow to orange to pink and then to transparent in 10 days (as the spoilage level increased), while no color changes were observed in healthy onions. Such visual sensors can be used in the form of paper-based sensors in packaging that detect spoilage and help minimize food wastage.

Zhai et al. (2019) also developed sensors based on gellan gum AgNP for intelligent packaging wherein colorimetric sensors selectively and sensitively detected H_2S vapors for monitoring the spoilage of meat. As the H_2S concentration varied from 0 to 85 pM, changes in surface plasmon resonance of the AgNPs yielded solutions that changed from bright yellow to light yellow to pink and transparent. In addition, silver nanoparticles (AgNPs@gold nanorods) acting as a sensing material over agarose hydrogels have shown characteristic visual changes for monitoring volatile biogenic amines during food spoilage. As a result, color changes due to the longitudinal LSPR peak led to multiple changes in color from peach to brick-red, according to the concentration of the biogenic amine and the degree of spoilage (Lin et al., 2016).

Furthermore, AgNPs have been used in various nutritional analyses of food before packaging. AgNPs have also been tested as electrochemical sensors for robust quantification of antioxidants in the food matrix at the postharvest level before and after packaging. An aptasensor in which the primary aptamer is immobilized on a magnetic bead and the second aptamer is conjugated to AgNPs acts as a sandwiched electrochemical sensor for the detection of *S. aureus* (Abbaspour et al., 2015). The results suggest excellent sensitivity and discriminatory power originate because of the use of two aptamers. Similarly, AgNPs have shown the capacity to determine the antioxidant activity of polyphenols such as sinapic acid, gallic acid, caffeic acid, ascorbic acid, and quercetin in rapeseed (Szydłowska-Czerniak et al., 2012). AgNPs have a variety of uses in diagnostic procedures, especially for detecting the antioxidant capacity of rapeseed oil, as gas sensors (for measuring freshness of meat, chicken, and beef). Also, RFID tags for monitoring the supply chain are framed by metallic inks such as AgNPs that will behave as ethylene and moisture sensors. The experimental applicability of AgNPs as sensitive diagnostics for detecting spoilage of food products will pave the road to future possibilities and capabilities for smart packaging (Kuswandi et al., 2011). Fig. 7.3 illustrates TTI-based Ag hybrid nanorods on the package of perishable goods like milk, where the deterioration process is indicated by a change in color from pink to red (Zhang et al., 2013). Intelligent packaging will enable simultaneous detection and bacteria inactivation, leading to better distribution of products in the marketplace.

7.4.3 Commercial silver nonmaterial-based packaging materials in the marketplace

Having a larger surface area per unit mass with a wide scope of pathogen inactivation, engineered AgNPs are widely accepted as antibacterial agents, including in the food packaging industry. The antimicrobial properties of AgNPs depend mostly on particle size and uniformity of distribution in packaging materials. AgNPs are added to food containers and food packaging materials globally. A brief overview of commercial products containing AgNPs is provided in Table 7.2. The products claim

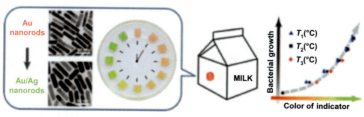

FIG. 7.3

Microbial growth governed by time and temperature eventually depicted by color change as a result of silver growth on gold nanorods directed by the growing microbial population. The clock corresponds to the color changes in Ag/Au nanorod-hydrogel cubes, indicating the changes over time (Zhang et al., 2013). © 2013, American Chemical Society.

Table 7.2 Commercially used AgNPs incorporated products (Bumbudsanpharoke and Ko, 2005).

Nanomaterial alone/combination/purpose	Polymer	Commercial product name	Product image
AgNPs Store fresh food for longer time	Polypropylene (PP)	Sharper Image Company (USA)	
AgNPs Keeps food fresh/remove bad smell/growth/prevent from dirt and fungus	PP silicon	Dai Dong Tien Corporation (Vietnam)	
AgNPs Prevent bad odor/sterilize food container	PP, Co polyester	Window Nano-silver Airtight Container (South Korea)	
AgNPs is present in silicon seal Prevents from spoilage and keep fresh (keep bacteria away)	N/A	New Life Co. Ltd. (South Korea)	

AgNPs Container to reduce bacteria and fungus growth to 99%	Co-polyester (TRITAN™)	Dong Yang Chemical Co. Ltd. Incense Nano-silver food container (South Korea)	
AgNPs Shown antibacterial/antifungal properties	PE (polyethylene)	Fine Polymer, Inc.(South Korea)	
AgNPs These feeding bottles are antibacterial help protect babies with weak immunity from germs. The bottles do not need any sterilization by boiling	PES, PP	Baby dream Co. Ltd. (South Korea)	
AgNPs combined with zeolites Antimicrobial film (bacteria enzymes and molds)	PP, PS	Sinanen Zeomic Co. Ltd. (Japan)	

Continued

Table 7.2 Commercially used AgNPs incorporated products (Bumbudsanpharoke and Ko, 2005)—cont'd

Nanomaterial alone/combination/purpose	Polymer	Commercial product name	Product image
AgNPs for films, storage bags and containers Keep the food safe longer FDA approved	PP, PE	Anson Nano-Biotechnology (Zhuhai) Co. Ltd. (China)	
AgNPs Antibacterial, keep costly foods fresher and longer. Bacteria do not develop any further resistance against silver ops	PP, silicon	Cixi Mingx in Plastic & Rubber Factory (China)	
AgNPs based baby bottles, food grade (antibacterial)	PP, silicon	Shenzhen Ibecare Commodity Limited Company (China)	
AgNPs	BPA free plastic	Manufacturer-Kinetic GoGreen (commercial product) Food containers in market Food and beverage storage	

to have nanosilver that ensures long-term freshness in food and reduced growth of bacteria (Bumbudsanpharoke and Ko, 2015).

7.5 Mechanism of action

A complete understanding of AgNPs' antimicrobial actions is yet to be recognized; however, a few mechanisms are shown here:

- adhesion, accumulation, and coagulation on the bacterial cell wall and membrane along with progressive leakage of lipopolysaccharides and membrane proteins (Dakal et al., 2016);
- formation of "pits" and penetration into the cytosol, leading to leakage of cellular content (Sondi and Salopek-sondi, 2004);
- inducing oxidative stress by generation of free radicals and reactive oxygen species (ROS), which causes DNA damage, protein inactivation, and lipid peroxidation and triggers cytotoxicity (Kim et al., 2007b); and
- modulating microbial signal transduction pathways critical to cellular homeostasis and survival (Shrivastava et al., 2007).

Fig. 7.4 shows that exposure of bacterial cells to AgNPs onsets multiple, simultaneous antimicrobial activities against the target microbe, which executes as a combined effect of the identified probable mechanism (Ogunsona et al., 2020).

The antimicrobial effects of AgNPs are closely related to their electrostatic interactions. At the biological pH, bacterial cell membranes exhibit a net negative charge (due to lipopolysaccharides), which commences interactions between cell membranes and Ag^+ ions (Dong et al., 2019; Morones et al., 2005). Both these studies suggest that this step may initiate a degenerative cascade of events that ultimately executes AgNPs' biocidal mechanism. The downstream processes may involve conformational changes within cell membranes as the slightly acidic Ag^+ ions spontaneously react with soft phosphorus (DNA) and sulfur entities (sulfur-containing proteins on the exterior and interior of cells).

Further, the biocidal mechanism of AgNPs is enhanced through membrane disintegration and pits formation (Sondi and Salopek-sondi, 2004). The membrane destruction is synergistically catalyzed by increased permeability induced by AgNPs. Only NPs interacting with cell surfaces penetrate a membrane and enter a cell's cytosol, where they react with cell organelles and impede the basic metabolism (e.g., respiration, replication, and cell division) of the target. Das et al. (2017) and Su et al. (2009) demonstrated that ROS burst in the cells of *E. coli*. AgNP invasion is also linked to altered protein expressions in bacterial cells.

In addition to this, access to AgNPs within bacterial cytoplasm triggers various biochemical alterations including upregulation of proteins involved in oxidative stress and ROS generation, while genes involved in metabolic processes of amino compounds and macromolecular substances were downregulated (Liao et al., 2019). The study suggests that excessive formation of ROS has two major implications. First, it promotes lipid peroxidation, membrane permeability augmentation, and

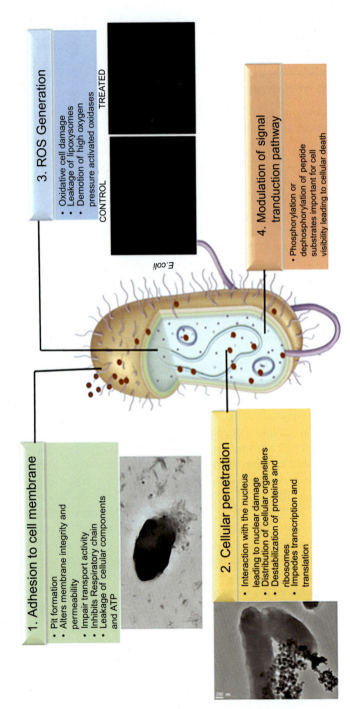

FIG. 7.4

Diagrammatic representation of antimicrobial action of AgNPs. Adhesion of the nanoparticles onto the cell membrane (1) of the pathogen initiates a signaling cascade followed by cellular penetrations (2) © 2019 Dong et al. (3) 2,7-di chloroflourescin-di acetate (DCFH$_2$-DA) is used to map ROS activity. AgNP-treated bacterial cells uptake DCFH$_2$-DA (DCF+), which results in fluorescence; the intensity of DCF fluorescence is proportional to the amount of ROS generated. Control cells do not possess fluorescence (Das et al., 2017).

oxidative damage to proteins and nucleic acid. Second, as ROS generation requires a high amount of oxygen, it depletes the cellular oxygen pressure and as a consequence reduces the high-oxygen pressure that regulates oxidases such as CcoP1, CcoO1, and Cyt-bo3; and activates low-oxygen-pressure-regulated oxidases. ROS significantly enrich proteins participating in oxidative phosphorylation, ribosomal metabolism, RNA degradation, DNA replication, synthesis and degradation of ketone bodies, and fatty acid degradation.

Another effect exhibited after the injection of AgNPs is modulation of the signal transduction pathway. This alteration is executed by phosphorylation and dephosphorylation processes. Shrivastava et al. (2007) reported modulation by tyrosine phosphorylation of putative peptide substrates critical for cell viability and division, which led to a complete shutdown of the metabolic factory.

The previously mentioned mechanisms cumulatively lead to bacterial cell death (Fig. 7.4). Nevertheless, the extent of antimicrobial activity depends on numerous physical and environmental factors.

7.5.1 Factors affecting antimicrobial potential of AgNPs

Various factors determine the antimicrobial efficacy of AgNPs incorporated into a polymeric matrix, for example, concentration, shape, size, and distribution of the NPs in the base matrix. The external environment to which AgNPs are exposed also influences their actions. AgNPs are sensitive to the presence of oxygen and are efficient when partially oxidized. This fact was underscored in a study by Levard et al. (2012), which showed a loss of antimicrobial action of nano-Ags under anaerobic conditions even at higher concentrations, as Ag^+ ions are not released in the absence of oxygen. Similar results were reported by Lok et al. (2007) who emphasized that partially oxidized nano-Ags may act as a carrier of chemisorbed Ag^+ in quantities sufficient to elicit antimicrobial action. As reported in many studies, the shape and size of AgNPs are highly associated with the efficiency of these NPs against microbes. The size of AgNPs is inversely related to their antimicrobial potential as a result of an enhanced surface-to-volume ratio, which allows better interactions between AgNPs and the target microbes. Cheon et al. (2019) showed that spherical AgNPs exhibit higher potential, compared with rod-shaped and triangular-shaped particles. Several studies show that increasing concentrations of AgNPs enhances the antimicrobial impact (Kim et al., 2007a,b; López-Carballo et al. 2013; Tavakoli et al. 2017). Pal et al. (2007) noted that antibacterial activity is dependent on the initial bacterial concentration and that the lower the initial concentration, the higher the antimicrobial activity. Their results also showed that the presence of cationic surfactants led to an agglomeration of AgNPs, probably due to charge neutralization.

7.6 Safety aspects and regulations

The food sector is greatly influenced by the emerging field of nanotechnology due to its great potential and consistent support from the world's scientific communities and governments. Extensive research is being conducted to establish the safety and sustainability of this budding technology; however, elaborative regulations are yet to be established. Major groups that are actively pursuing these goals include the European Unions' scientific committees and agencies, the Organization for Economic Co-operation and Development (OECD), the International Organization for Standardization (ISO), and the US Food and Drug Administration (FDA).

Europe has been successful in implementing nanotechnology guidelines in its legislation through the European Food Safety Authority (EFSA); however, the 2016 EFSA regulations did note the need for additional data in order to establish the safety of nanosilver.

The United States has adopted a decentralized approach to the regulation of nanotechnology. The Environmental Protection Agency, Food and Drug Administration, Occupational Safety and Health Administration, and Consumer Product Safety Commission coordinate nanotechnological regulations (Amenta et al., 2015).

The Food Safety and Standards Authority of India (FSSAI) and the Government of India are highly optimistic about the advancement of nanotechnology in India. This fact was made evident through the 2001 Nano Science and Technology Initiative (NSTI) followed by a coordinated program launched by FSSAI and the Department of Sciences & Technology in 2007 under the name "Nano Mission." Although at the early stage of development, nanotechnology has great potential to provide novel opportunities and well-defined standards to the agriculture and food sector in India. However, better regulation of government policies and guidelines is needed to ensure the use of nano-based materials in that sector, which would encourage commercial use of nanotechnology in products such as nanofertilizers and nano-pesticides and simultaneously ensure their quality and safety for human consumption and the environment.

The nanotechnological revolution faces a plethora of challenges in the food packaging industry. The prime challenge is to ascertain the safety of nanomaterials both biologically and ecologically. There is a need to fill the knowledge gap in regard to safety, and the key to resolving this issues is a critical review and assessment of potential nanotechnology-based advancements by scientific, academic, and industrial communities.

The documented studies reporting the suspected toxicology of AgNPs has been diverted and hence the big picture regarding the toxicity of NPs is not yet clear. As expected, the fate, transformation, and ecological and biological interactions of nanomaterials are modulated by intrinsic factors (size, shape, nature of capping agent, concentration, etc.) as well as by extrinsic factors (environmental pH, ionic strength, presence of other ions, etc.) (Tortella et al., 2020). Few studies indicate the toxic effects of AgNPs, and the underlying mechanism is still unclear. Khan et al. (2019) tested the effects of AgNPs on the blood-brain barrier established via co-culturing of the human brain microvascular endothelial cell line (hCMEC/D3) and primary astrocytes. Interestingly, the group discovered that the effect of AgNPs is time-dependent; that is, during the initial hours of encounter, high inflammation and activation of prooxidative pathways occur and are diminished during the later stages. Yu et al. (2019) showed that cellulose nanofibril/AgNP composite films did not demonstrate significant variations in cell viability assays. Furthermore, extensive research has focused on the anticancer potential of AgNPs against diverse kinds of cancer cells; few studies have received a positive response in terms of nontoxicity of AgNPs (Murugesan et al., 2019; Priya et al., 2020; Sivakumar et al., 2020).

Another major concern is leaching of AgNPs from packaging material into food, which is also affected by similar extrinsic and intrinsic factors, such as concentration, pH, temperature, and humidity (Carbone et al., 2016). Most recent studies have underlined the fact that no significant migration of AgNPs has been observed in food contact materials (Choi et al., 2018; Marchiore et al., 2017; Metak et al., 2015).

7.7 Conclusion and future prospects

All novel and emerging technologies are accompanied by opportunities as well as limitations. To be accepted by society and the industrial sector, the budding technologies must be cleared of doubt and criticism. Many studies have shown the inherent potential of using nanotechnology in food packaging; however, the findings underscore the need for rigorous testing of the risks related to the safety and quality, freshness, toxicity, transportation, and degradation of foods. Furthermore, human and environmental safety needs to be ascertained before large-scale applications of AgNPs in food supplements as well as in nanopackaging materials. A detailed evaluation of AgNMs for potential threats is vital to their application. Multidisciplinary approaches must be followed to define and quantify the characteristics of AgNMs and their biotransformation in food. Conventional techniques are of little use due to their poor efficiency, low-throughput formats, and inability to analyze variable and complex food matrices. Additionally, for efficient deployment of AgNPs, their production cost and price to consumers, their biocompatibility, and their ability to scale up are required for optimal application. Additional measures and technological validations are needed to ensure long-term stability as well as potency during storage under in situ conditions. Studies on silver nanomaterials' migration from packaging into food and the environment are also needed. Appropriate regulatory standards for the use of AgNMs as well as proper labeling are required to address legal and other issues and to eventually attain acceptance from consumers and industries, including their willingness to cover the cost of the materials. Although the safety, toxicological effects, and ecological impact of AgNMs are still of prime concern, nanopackaging systems that include silver-based active packaging materials have opened up a plethora of new opportunities for sustainable, eco-friendly alternatives that will help create new jobs in the food packaging industry. Finally, interdisciplinary studies in the scientific community and among governments are key to developing the vast potential of this emerging technology.

References

Abbaspour, A., Norouz-Sarvestani, F., Noori, A., Soltani, N., 2015. Aptamer-conjugated silver nanoparticles for electrochemical dual-aptamer-based sandwich detection of staphylococcus aureus. Biosens. Bioelectron. 68, 149–155.

Adepu, S., Khandelwal, M., 2018. Broad-spectrum antimicrobial activity of bacterial cellulose silver nanocomposites with sustained release. J. Mater. Sci. 53, 1596–1609.

Adley, C.C., Ryan, M.P., 2016. The nature and extent of foodborne disease. Antimicrob. Food Packag., 1–10.

Ahmed, J., Arfat, Y.A., Bher, A., Mulla, M., Jacob, H., Auras, R., 2018. Active chicken meat packaging based on polylactide films and bimetallic ag–cu nanoparticles and essential oil. J. Food Sci. 83, 1299–1310.

Alexander, J.W., 2009. History of the medical use of silver. Surg. Infect. (Larchmt) 10 (3), 289–292.

Amenta, V., Aschberger, K., Arena, M., Bouwmeester, H., Botelho Moniz, F., Brandhoff, P., Gottardo, S., Marvin, H.J.P., Mech, A., Quiros Pesudo, L., Rauscher, H., Schoonjans, R., Vettori, M.V., Weigel, S., Peters, R.J., 2015. Regulatory aspects of nanotechnology in the Agri/feed/food sector in EU and non-EU countries. Regul. Toxicol. Pharmacol. 73, 463–476.

An, J., Zhang, M., Wang, S., Tang, J., 2008. Physical, chemical and microbiological changes in stored green asparagus spears as affected by coating of silver nanoparticles-PVP. LWT - Food Sci. Technol. 41, 1100–1107.

Azlin-Hasim, S., Cruz-Romero, M.C., Morris, M.A., Cummins, E., Kerry, J.P., 2015. Effects of a combination of antimicrobial silver low density polyethylene nanocomposite films and modified atmosphere packaging on the shelf life of chicken breast fillets. Food Packag. Shelf Life 4, 26–35.

Azlin-Hasim, S., Cruz-Romero, M.C., Cummins, E., Kerry, J.P., Morris, M.A., 2016. The potential use of a layer-by-layer strategy to develop LDPE antimicrobial films coated with silver nanoparticles for packaging applications. J. Colloid Interface Sci. 461, 239–248.

Badawy, M.E.I., Lotfy, T.M.R., Shawir, S.M.S., 2019. Preparation and antibacterial activity of chitosan-silver nanoparticles for application in preservation of minced meat. Bull. Natl. Res. Cent. 43.

Becaro, A.A., Puti, F.C., Correa, D.S., Paris, E.C., Marconcini, J.M., Ferreira, M.D., 2015. Polyethylene films containing silver nanoparticles for applications in food packaging: characterization of physico-chemical and anti-microbial properties. J. Nanosci. Nanotechnol. 15, 2148–2156.

Beigmohammadi, F., Peighambardoust, S.H., Hesari, J., Azadmard-Damirchi, S., Peighambardoust, S.J., Khosrowshahi, N.K., 2016. Antibacterial properties of LDPE nanocomposite films in packaging of UF cheese. LWT - Food Sci. Technol. 65, 106–111.

Bengtsson-Palme, J., 2017. Antibiotic resistance in the food supply chain: where can sequencing and metagenomics aid risk assessment? Curr. Opin. Food Sci. 14, 66–71.

Biswas, M.C., Tiimob, B.J., Abdela, W., Jeelani, S., Rangari, V.K., 2019. Nano silica-carbon-silver ternary hybrid induced antimicrobial composite films for food packaging application. Food Packag. Shelf Life 19, 104–113.

Bridier, A., Sanchez-Vizuete, P., Guilbaud, M., Piard, J.C., Naïtali, M., Briandet, R., 2015. Biofilm-associated persistence of food-borne pathogens. Food Microbiol. 45, 167–178.

Bumbudsanpharoke, N., Ko, S., 2015. Nano-food packaging: an overview of market, migration research, and safety regulations. J. Food Sci. 80, R910–R923.

Carbone, M., Donia, D.T., Sabbatella, G., Antiochia, R., 2016. Silver nanoparticles in polymeric matrices for fresh food packaging. J. King Saud Univ. Sci. 28, 273–279.

Cheon, J.Y., Kim, S.J., Rhee, Y.H., Kwon, O.H., Park, W.H., 2019. Shape-dependent antimicrobial activities of silver nanoparticles. Int. J. Nanomedicine 14, 2773–2780.

Chi, H., Song, S., Luo, M., Zhang, C., Li, W., Li, L., Qin, Y., 2019. Effect of PLA nanocomposite films containing bergamot essential oil, TiO2 nanoparticles, and Ag nanoparticles on shelf life of mangoes. Sci. Hortic. (Amsterdam) 249, 192–198.

Choi, J.I., Chae, S.J., Kim, J.M., Choi, J.C., Park, S.J., Choi, H.J., Bae, H., Park, H.J., 2018. Potential silver nanoparticles migration from commercially available polymeric baby products into food simulants. Food Addit. Contam. Part A Chem. Anal. Control. Expo. Risk Assess. 35, 996–1005.

Costa, C., Conte, A., Buonocore, G.G., Lavorgna, M., Del Nobile, M.A., 2012. Calcium-alginate coating loaded with silver-montmorillonite nanoparticles to prolong the shelf-life of fresh-cut carrots. Food Res. Int. 48, 164–169.

Cushen, M., Kerry, J., Morris, M., Cruz-Romero, M., Cummins, E., 2014. Evaluation and simulation of silver and copper nanoparticle migration from polyethylene nanocomposites to food and an associated exposure assessment. J. Agric. Food Chem. 62, 1403–1411.

Dakal, T.C., Kumar, A., Majumdar, R.S., Yadav, V., 2016. Mechanistic basis of antimicrobial actions of silver nanoparticles. Front. Microbiol. 7, 1–17.

Darroudi, M., Ahmad, M.B., Abdullah, A.H., Ibrahim, N.A., Shameli, K., 2010. Effect of accelerator in green synthesis of silver nanoparticles. Int. J. Mol. Sci. 11, 3898–3905.

Das, B., Dash, S.K., Mandal, D., Ghosh, T., Chattopadhyay, S., Tripathy, S., Das, S., Dey, S.K., Das, D., Roy, S., 2017. Green synthesized silver nanoparticles destroy multidrug resistant bacteria via reactive oxygen species mediated membrane damage. Arab. J. Chem. 10, 862–876.

de Azeredo, H.M.C., 2009. Nanocomposites for food packaging applications. Food Res. Int. 42, 1240–1253.

De Moura, M.R., Mattoso, L.H.C., Zucolotto, V., 2012. Development of cellulose-based bactericidal nanocomposites containing silver nanoparticles and their use as active food packaging. J. Food Eng. 109, 520–524.

Del Nobile, M.A., Cannarsi, M., Altieri, C., Sinigaglia, M., Favia, P., Iacoviello, G., D'agostino, R., 2001. E: food engineering and physical properties effect of Ag-containing nano-composite active packaging system on survival of Alicyclobacillus acidoterrestris. J. Food Sci. 69.

Dong, Y., Zhu, H., Shen, Y., Zhang, W., Zhang, L., 2019. Antibacterial activity of silver nanoparticles of different particle size against Vibrio Natriegens. PLoS One 14, 1–12.

Duncan, T.V., 2011. Applications of nanotechnology in food packaging and food safety: barrier materials, antimicrobials and sensors. J. Colloid Interface Sci. 363, 1–24.

Emamifar, A., Kadivar, M., Shahedi, M., Soleimanian-Zad, S., 2010. Evaluation of nanocomposite packaging containing Ag and ZnO on shelf life of fresh orange juice. Innov. Food Sci. Emerg. Technol. 11, 742–748.

Emamifar, A., Kadivar, M., Shahedi, M., Soleimanian-Zad, S., 2011. Effect of nanocomposite packaging containing Ag and ZnO on inactivation of lactobacillus plantarum in orange juice. Food Control 22, 408–413.

Fahmy, H.M., Salah Eldin, R.E., Abu Serea, E.S., Gomaa, N.M., AboElmagd, G.M., Salem, S.A., Elsayed, Z.A., Edrees, A., Shams-Eldin, E., Shalan, A.E., 2020. Advances in nanotechnology and antibacterial properties of biodegradable food packaging materials. RSC Adv. 10, 20467–20484.

Fan, L., Zhang, H., Gao, M., Zhang, M., Liu, P., Liu, X., 2019. Cellulose nanocrystals/silver nanoparticles: in-situ preparation and application in PVA films. Holzforschung 74 (5), 523–528.

Fernández, A., Picouet, P., Lloret, E., 2010a. Cellulose-silver nanoparticle hybrid materials to control spoilage-related microflora in absorbent pads located in trays of fresh-cut melon. Int. J. Food Microbiol. 142, 222–228.

Fernández, A., Picouet, P., Lloret, E., 2010b. Reduction of the spoilage-related microflora in absorbent pads by silver nanotechnology during modified atmosphere packaging of beef meat. J. Food Prot. 73, 2263–2269.

Geethalakshmi, R., Sarada, D.V.L., 2010. Synthesis of plant-mediated silver nanoparticles using Trianthema decandra extract and evaluation of their anti microbial activities. J. Eng. Sci. Technol. 2 (5), 970–975.

Gokularaman, S., Stalin Cruz, A., Pragalyaashree, M.M., Nishadh, A., 2017. Nanotechnology approach in food packaging—review. J. Pharm. Sci. Res. 9, 1743–1749.

Grumezescu, A.M., Holban, A.M., 2018. Impact of Nanoscience in the Food Industry, Impact of Nanoscience in the Food Industry. Elsevier Inc.

Guillard, V., Gaucel, S., Fornaciari, C., Angellier-Coussy, H., Buche, P., Gontard, N., 2018. The next generation of sustainable food packaging to preserve our environment in a circular economy context. Front. Nutr. 5, 1–13.

Heli, B., Morales-Narváez, E., Golmohammadi, H., Ajji, A., Merkoçi, A., 2016. Modulation of population density and size of silver nanoparticles embedded in bacterial cellulose: via ammonia exposure: visual detection of volatile compounds in a piece of plasmonic nanopaper. Nanoscale 8, 7984–7991.

Huang, Y., Mei, L., Chen, X., Wang, Q., 2018. Recent developments in food packaging based on nanomaterials. Nano 8, 1–29.

Incoronato, A.L., Buonocore, G.G., Conte, A., Lavorgna, M., Del Nobile, M.A., 2010. Active systems based on silver-montmorillonite nanoparticles embedded into bio-based polymer matrices for packaging applications. J. Food Prot. 73, 2256–2262.

Incoronato, A.L., Conte, A., Buonocore, G.G., Del Nobile, M.A., 2011. Agar hydrogel with silver nanoparticles to prolong the shelf life of Fior di latte cheese. J. Dairy Sci. 94, 1697–1704.

Jenaa, J., et al., 2013. Biosynthesis and Characterization of silver nanoparticles using microalga Chlorococcum humicola and its antibacterial activity. Int. J. Nanomater. Biostruct. 3 (1), 1–8.

Jiang, T., Feng, L., Wang, Y., 2013. Effect of alginate/nano-ag coating on microbial and physicochemical characteristics of shiitake mushroom (Lentinus edodes) during cold storage. Food Chem. 141, 954–960.

Jokar, M., Abdul Rahman, R., Ibrahim, N.A., Abdullah, L.C., Tan, C.P., 2012. Melt production and antimicrobial efficiency of low-density polyethylene (LDPE)-silver nanocomposite film. Food Bioprocess Technol. 5, 719–728.

Kadam, D., Momin, B., Palamthodi, S., Lele, S.S., 2019. Physicochemical and functional properties of chitosan-based nano-composite films incorporated with biogenic silver nanoparticles. Carbohydr. Polym. 211, 124–132.

Kampmann, Y., De Clerck, E., Kohn, S., Patchala, D.K., Langerock, R., Kreyenschmidt, J., 2008. Study on the antimicrobial effect of silver-containing inner liners in refrigerators. J. Appl. Microbiol. 104, 1808–1814.

Kanmani, P., Lim, S.T., 2013. Synthesis and structural characterization of silver nanoparticles using bacterial exopolysaccharide and its antimicrobial activity against food and multidrug resistant pathogens. Process Biochem. 48, 1099–1106.

Kaur, M., Kalia, A., Thakur, A., 2017. Effect of biodegradable chitosan–rice-starch nanocomposite films on post-harvest quality of stored peach fruit. Starch/Starke 69.

Khalaf, H., Sharoba, A., El-Tanahi, H., Morsy, M., 2013. Stability of antimicrobial activity of pullulan edible films incorporated with nanoparticles and essential oils and their impact on Turkey deli meat quality. J. Food Dairy Sci. 4, 557–573.

Khan, A.M., Korzeniowska, B., Gorshkov, V., Tahir, M., Schrøder, H., Skytte, L., Rasmussen, K.L., Khandige, S., Møller-Jensen, J., Kjeldsen, F., 2019. Silver nanoparticle-induced expression of proteins related to oxidative stress and neurodegeneration in an in vitro human blood-brain barrier model. Nanotoxicology 13, 221–239.

Kim, J.S., Kuk, E., Yu, K.N., Kim, J.H., Park, S.J., Lee, H.J., Kim, S.H., Park, Y.K., Park, Y.H., Hwang, C.Y., Kim, Y.K., Lee, Y.S., Jeong, D.H., Cho, M.H., 2007a. Antimicrobial effects of silver nanoparticles. Nanomedicine nanotechnology. Biol. Med. 3, 95–101.

Kim, J.S., Kuk, E., Yu, N., Kim, J., Park, S.J., Lee, J., Kim, H., Park, Y.K., Park, H., Hwang, C., Kim, Y., Lee, Y., Jeong, D.H., Cho, M., 2007b. Antimicrobial effects of silver nanoparticles. Biol. Med. 3, 95–101.

Koutsoumanis, K.P., Lianou, A., Sofos, J.N., 2014. Food safety: emerging pathogens. Encycl. Agric. Food Syst. 3, 250–272.

Kraśniewska, K., Galus, S., Gniewosz, M., 2020. Biopolymers-based materials containing silver nanoparticles as active packaging for food applications—a review. Int. J. Mol. Sci. 21.

Kumar, M., Bala, R., Gondil, V.S., Pandey, S.K., Chhibber, S., 2017a. Combating food pathogens using sodium benzoate functionalized silver nanoparticles: synthesis, characterization and antimicrobial evaluation. J. Mater. Sci. 52 (14), 8568–8575.

Kumar, P., Mahato, D.K., Kamle, M., Mohanta, T.K., Kang, S.G., 2017b. Aflatoxins: a global concern for food safety, human health and their management. Front. Microbiol. 7, 1–10.

Kumar, S., Shukla, A., Baul, P.P., Mitra, A., Halder, D., 2018. Biodegradable hybrid nanocomposites of chitosan/gelatin and silver nanoparticles for active food packaging applications. Food Packag. Shelf Life 16, 178–184.

Kuswandi, B., Wicaksono, Y.A.J., Abdullah, A., Heng, L.Y., Ahmad, M., 2011. Smart packaging: sensors for monitoring of food quality and safety. Sens. & Instrumen. Food Qual. 5, 137–146.

Levard, C., Hotze, E.M., Lowry, G.V., Brown, G.E., 2012. Environmental transformations of silver nanoparticles: impact on stability and toxicity. Environ. Sci. Technol. 46, 6900–6914.

Levy, S.B., Bonnie, M., 2004. Antibacterial resistance worldwide: causes, challenges and responses. Nat. Med. 10, S122–S129.

Li, W., Li, L., Zhang, H., Yuan, M., Qin, Y., 2017. Evaluation of PLA nanocomposite films on physicochemical and microbiological properties of refrigerated cottage cheese. J. Food Process. Preserv. 42, 1–9.

Liao, S., Zhang, Y., Pan, X., Zhu, F., Jiang, C., Cheng, Z., Wu, G., Chen, L., 2019. Antibacterial activity and mechanism of silver nanoparticles against multidrug-resistant *Pseudomonas aeruginosa*. Int. J. Nanomed. 14, 1469–1487.

Lin, T., Wu, Y., Li, Z., Song, Z., Guo, L., Fu, F., 2016. Visual monitoring of food spoilage based on hydrolysis-induced silver metallization of au nanorods. Anal. Chem. 88, 11022–11027.

Lok, C.N., Ho, C.M., Chen, R., He, Q.Y., Yu, W.Y., Sun, H., Tam, P.K.H., Chiu, J.F., Che, C.M., 2007. Silver nanoparticles: partial oxidation and antibacterial activities. J. Biol. Inorg. Chem. 12, 527–534.

López-Carballo, G., Higueras, L., Gavara, R., Hernández-Muñoz, P., 2013. Silver ions release from antibacterial chitosan films containing in situ generated silver nanoparticles. J. Agric. Food Chem. 61, 260–267.

Mahdi, S.S., Vadood, R., Nourdahr, R., 2012. Study on the antimicrobial effect of nanosilver tray packaging of minced beef at refrigerator temperature. Glob. Vet. 9, 284–289.

Marchiore, N.G., Manso, I.J., Kaufmann, K.C., Lemes, G.F., Pizolli, A.P.O., Droval, A.A., Bracht, L., Gonçalves, O.H., Leimann, F.V., 2017. Migration evaluation of silver nanoparticles from antimicrobial edible coating to sausages. LWT - Food Sci. Technol. 76, 203–208.

MartínezAbad, A., Lagaron, J.M., Ocio, M.J., 2012. Development and characterization of silver-based antimicrobial ethylene–vinyl alcohol copolymer (EVOH) films for food-packaging applications. J. Agric. Food Chem. 60, 5350–5359.

Mastromatteo, M., Conte, A., Lucera, A., Saccotelli, M.A., Buonocore, G.G., Zambrini, A.V., Del Nobile, M.A., 2015. Packaging solutions to prolong the shelf life of Fiordilatte cheese: bio-based nanocomposite coating and modified atmosphere packaging. LWT - Food Sci. Technol. 60, 230–237.

Mathew, S., Snigdha, S., Mathew, J., Radhakrishnan, E.K., 2019. Biodegradable and active nanocomposite pouches reinforced with silver nanoparticles for improved packaging of chicken sausages. Food Packag. Shelf Life 19, 155–166.

Mehlhorn, H., 2015. Food-borne disease burden epidemiology reference group. Encycl. Parasitol. 1–1.

Mei, L., Teng, Z., Zhu, G., Liu, Y., Zhang, F., Zhang, J., Li, Y., Guan, Y., Luo, Y., Chen, X., Wang, Q., 2017. Silver nanocluster-embedded Zein films as antimicrobial coating materials for food packaging. ACS Appl. Mater. Interfaces 9, 35297–35304.

Metak, A.M., Nabhani, F., Connolly, S.N., 2015. LWT - food science and technology migration of engineered nanoparticles from packaging into food products. LWT - Food Sci. Technol. 64, S156–S162.

Mohammed Fayaz, A., Balaji, K., Girilal, M., Kalaichelvan, P.T., Venkatesan, R., 2009. Mycobased synthesis of silver nanoparticles and their incorporation into sodium alginate films for vegetable and fruit preservation. J. Agric. Food Chem. 57, 6246–6252.

Morales-de la Peña, M., Welti-Chanes, J., Martín-Belloso, O., 2019. Novel technologies to improve food safety and quality. Curr. Opin. Food Sci. 30, 1–7.

Morones, J.R., Elechiguerra, J.L., Camacho, A., Holt, K., Kouri, J.B., Ram, J.T., Yacaman, M.J., 2005. The bactericidal effect of silver nanoparticles. Nanotechnology 16, 2346–2353.

Morsy, M.K., Khalaf, H.H., Sharoba, A.M., El-Tanahi, H.H., Cutter, C.N., 2014. Incorporation of essential oils and nanoparticles in pullulan films to control foodborne pathogens on meat and poultry products. J. Food Sci. 79.

Motlagh, N.V., Mosavian, M.T.H., Mortazavi, S.A., 2012. Effect of polyethylene packaging modified with silver particles on the microbial, sensory and appearance of dried barberry. Packag. Technol. Sci. 26 (1), 39–49.

Murugesan, K., Koroth, J., Srinivasan, P.P., Singh, A., Mukundan, S., Karki, S.S., Choudhary, B., Gupta, C.M., 2019. Effects of green synthesised silver nanoparticles (ST06-AgNPs) using curcumin derivative (ST06) on human cervical cancer cells (HeLa) in vitro and EAC tumor bearing mice models. Int. J. Nanomed. 14, 5257–5270.

Mustafa, F., Andreescu, S., 2020. Nanotechnology-based approaches for food sensing and packaging applications. RSC Adv. 10, 19309–19336. https://doi.org/10.1039/d0ra01084g.

O'Shea, H., Blacklaws, B.A., Collins, P.J., McKillen, J., Fitzgerald, R., 2019. Viruses Associated with Foodborne Infections, Reference Module in Life Sciences. Elsevier Ltd.

Ogunsona, E.O., Muthuraj, R., Ojogbo, E., Valerio, O., Mekonnen, T.H., 2020. Engineered nanomaterials for antimicrobial applications: a review. Appl. Mater. Today 18, 100473.

Pal, S., Tak, Y.K., Song, J.M., 2007. Does the antibacterial activity of silver nanoparticles depend on the shape of the nanoparticle? A study of the Gram-negative bacterium Escherichia coli. Appl. Environ. Microbiol. 73, 1712–1720.

Panea, B., Ripoll, G., González, J., Fernández-Cuello, Á., Albertí, P., 2014. Effect of nanocomposite packaging containing different proportions of ZnO and Ag on chicken breast meat quality. J. Food Eng. 123, 104–112.

Patra, J.K., Baek, K., Perera, C.O., 2017. Antibacterial activity and synergistic antibacterial potential of biosynthesized silver nanoparticles against foodborne pathogenic bacteria along with its anticandidal and antioxidant effects. Front. Microbiol. 8 (167), 1–14.

Priya, K., Vijayakumar, M., Janani, B., 2020. Chitosan-mediated synthesis of biogenic silver nanoparticles (AgNPs), nanoparticle characterisation and in vitro assessment of anticancer activity in human hepatocellular carcinoma HepG2 cells. Int. J. Biol. Macromol. 149, 844–852.

Ragunathan, R., Kumar, R.R., Tamilenthi, A., Jeteena, J., 2015. Green synthesis of chitosan silver nanocomposites, its medical and edible coating on fruits and vegetables. Int. J. Biol. Pharm. Res. 6, 129–136.

Rai, M., Ingle, A.P., Gupta, I., Pandit, R., Paralikar, P., Gade, A., Chaud, M.V., dos Santos, C.A., 2019. Smart nanopackaging for the enhancement of food shelf life. Environ. Chem. Lett. 17, 277–290.

Rhim, J.W., Wang, L.F., Lee, Y., Hong, S.I., 2014. Preparation and characterization of bio-nanocomposite films of agar and silver nanoparticles: laser ablation method. Carbohydr. Polym. 103, 456–465.

Risch, S.J., 2009. Food packaging history and innovations. J. Agric. Food Chem. 57, 8089–8092.

Robertson, G.L., 2009. Sustainable Food Packaging, Handbook of Waste Management and Co-Product Recovery in Food Processing. Woodhead Publishing Limited.

Rolim, W.R., Pelegrino, M.T., de Araújo Lima, B., Ferraz, L.S., Costa, F.N., Bernardes, J.S., Rodigues, T., Brocchi, M., Seabra, A.B., 2019. Green tea extract mediated biogenic synthesis of silver nanoparticles: characterization, cytotoxicity evaluation and antibacterial activity. Appl. Surf. Sci. 463, 66–74.

Roy, S., Shankar, S., Rhim, J.W., 2019. Melanin-mediated synthesis of silver nanoparticle and its use for the preparation of carrageenan-based antibacterial films. Food Hydrocoll. 88, 237–246.

Sa'nchez-Valdes, S., Ortega-Ortiz, H., Valle, L.F.R., Medellı'n-Rodrı'guez, F.J., Guedea-Miranda, R., 2009. Mechanical and antimicrobial properties of multilayer films with a polyethylene/silver nanocomposite layer. J. Appl. Polym. Sci. 111, 953–962.

Sachdev, D., Kumar, V., Maheshwari, P.H., Pasricha, R., Deepthi, Baghel, N., 2016. Silver based nanomaterial, as a selective colorimetric sensor for visual detection of post harvest spoilage in onion. Sensors Actuators B Chem. 228, 471–479.

Salari, M., Sowti Khiabani, M., Rezaei Mokarram, R., Ghanbarzadeh, B., Samadi Kafil, H., 2018. Development and evaluation of chitosan based active nanocomposite films containing bacterial cellulose nanocrystals and silver nanoparticles. Food Hydrocoll. 84, 414–423.

Sanchez-Valdes, S., 2014. Sonochemical deposition of silver nanoparticles on linear low density polyethylene/cyclo olefin copolymer blend films. Polym. Bull. 71, 1611–1624.

Shankar, S., Rhim, J.W., 2015. Amino acid mediated synthesis of silver nanoparticles and preparation of antimicrobial agar/silver nanoparticles composite films. Carbohydr. Polym. 130, 353–363.

Shankar, S., Rhim, J.W., 2016. Tocopherol-mediated synthesis of silver nanoparticles and preparation of antimicrobial PBAT/silver nanoparticles composite films. LWT - Food Sci. Technol. 72, 149–156.

Sharma, C., Dhiman, R., Rokana, N., Panwar, H., 2017. Nanotechnology: an untapped resource for food packaging. Front. Microbiol. 8.

Shrivastava, S., Bera, T., Roy, A., Singh, G., Ramachandrarao, P., Dash, D., 2007. Characterization of enhanced antibacterial effects of novel silver nanoparticles. Nanotechnology 18.

Siqueira, M.C., Coelho, G.F., De Moura, M.R., Bresolin, J.D., Hubinger, S.Z., Marconcini, J.M., Mattoso, L.H.C., 2014. Evaluation of antimicrobial activity of silver nanoparticles for carboxymethylcellulose film applications in food packaging. J. Nanosci. Nanotechnol. 14, 5512–5517.

Sivakumar, M., Surendar, S., Jayakumar, M., Seedevi, P., Sivasankar, P., Ravikumar, M., Anbazhagan, M., Murugan, T., Siddiqui, S.S., Loganathan, S., 2020. Parthenium hysterophorus mediated synthesis of silver nanoparticles and its evaluation of antibacterial and antineoplastic activity to combat liver cancer cells. J. Clust. Sci., 1–11.

Sondi, I., Salopek-sondi, B., 2004. Silver nanoparticles as antimicrobial agent: a case study on E. coli as a model for Gram-negative bacteria. 275, 177–182.

Su, H.L., Chou, C.C., Hung, D.J., Lin, S.H., Pao, I.C., Lin, J.H., Huang, F.L., Dong, R.X., Lin, J.J., 2009. The disruption of bacterial membrane integrity through ROS generation induced by nanohybrids of silver and clay. Biomaterials 30, 5979–5987.

Szydłowska-Czerniak, A., Tułodziecka, A., Szłyk, E., 2012. A silver nanoparticle-based method for determination of antioxidant capacity of rapeseed and its products. Analyst 137, 3750–3759.

Tack, D.M., Marder, E.P., Griffin, P.M., Cieslak, P.R., Dunn, J., Hurd, S., Scallan, E., Lathrop, S., Muse, A., Ryan, P., Smith, K., Tobin-D'Angelo, M., Vugia, D.J., Holt, K.G., Wolpert, B.J., Tauxe, R., Geissler, A.L., 2020. Preliminary incidence and trends of infections with pathogens transmitted commonly through food—foodborne diseases active surveillance network, 10 U.S. sites, 2015–2018. Am. J. Transplant. 19, 1859–1863.

Tankhiwale, R., Bajpai, S.K., 2009. Graft copolymerization onto cellulose-based filter paper and its further development as silver nanoparticles loaded antibacterial food-packaging material. Colloids Surf. B Biointerfaces 69 (2), 164–168.

Tao, G., Cai, R., Wang, Y., Song, K., Guo, P., Zhao, P., Zuo, H., He, H., 2017. Biosynthesis and characterization of AgNPs-silk/PVA film for potential packaging application. Materials (Basel) 10.

Tavakoli, H., Rastegar, H., Taherian, M., Samadi, M., Rostami, H., 2017. The effect of nanosilver packaging in increasing the shelf life of nuts: an in vitro model. Ital. J. Food Saf. 6, 156–161.

Tortella, G.R., Rubilar, O., Durán, N., Diez, M.C., Martínez, M., Parada, J., Seabra, A.B., 2020. Silver nanoparticles: toxicity in model organisms as an overview of its hazard for human health and the environment. J. Hazard. Mater. 390, 121974.

Vimala, K., Mohan, Y.M., Sivudu, K.S., Varaprasad, K., Ravindra, S., Reddy, N.N., Padma, Y., Sreedhar, B., MohanaRaju, K., 2010. Fabrication of porous chitosan films impregnated with silver nanoparticles: a facile approach for superior antibacterial application. Colloids Surf. B Biointerfaces 76, 248–258.

Vishnuvarthanan, M., Rajeswari, N., 2019. Food packaging: pectin–laponite–Ag nanoparticle bionanocomposite coated on polypropylene shows low O2 transmission, low Ag migration and high antimicrobial activity. Environ. Chem. Lett. 17, 439–445.

Yang, F.M., Li, H.M., Li, F., Xin, Z.H., Zhao, L.Y., Zheng, Y.H., Hu, Q.H., 2010. Effect of nano-packing on preservation quality of fresh strawberry (fragaria ananassa duch. cv fengxiang) during storage at 4°c. J. Food Sci. 75.

Yoksan, R., Chirachanchai, S., 2010. Silver nanoparticle-loaded chitosan-starch based films: fabrication and evaluation of tensile, barrier and antimicrobial properties. Mater. Sci. Eng. C 30, 891–897.

Yu, Z., Wang, W., Kong, F., Lin, M., Mustapha, A., 2019. Cellulose nano fibril/silver nanoparticle composite as an active food packaging system and its toxicity to human colon cells. Int. J. Biol. Macromol. 129, 887–894.

Zhai, X., Li, Z., Shi, J., Huang, X., Sun, Z., Zhang, D., Zou, X., Sun, Y., Zhang, J., Holmes, M., Gong, Y., Povey, M., Wang, S., 2019. A colorimetric hydrogen sulfide sensor based on gellan gum-silver nanoparticles bionanocomposite for monitoring of meat spoilage in intelligent packaging. Food Chem. 290, 135–143.

Zhang, Y., Yang, D., Kong, Y., Wang, X., Pandoli, O., Gao, G., 2010. Synergetic antibacterial effects of silver nanoparticles@Aloe vera prepared via a green method. Nano Biomed. Eng. 2, 252–257.

Zhang, C., Yin, A.X., Jiang, R., Rong, J., Dong, L., Zhao, T., Sun, L.D., Wang, J., Chen, X., Yan, C.H., 2013. Time-temperature indicator for perishable products based on kinetically programmable Ag overgrowth on au nanorods. ACS Nano 7, 4561–4568.

CHAPTER 8

Novel silver-based nanomaterials for control of mycobiota and biocide analytical regulations in agri-food sector

Elena Piecková[a], Farah K. Ahmed[b], Renáta Lehotská[a], and Mária Globanová[a]

[a]Slovak Medical University in Bratislava, Bratislava, Slovakia
[b]Biotechnology English Program, Faculty of Agriculture, Cairo University, Giza, Egypt

8.1 Introduction

The rising threat of resistance to pathogenic fungicide in plants is a major challenge in the agriculture sector. Therefore innovative and cost-effective antimicrobial agents, such as bactericidal and fungicidal agents, need to be investigated (Ventola, 2015). Nanomaterials are already being used to produce new antimicrobial agents (e.g., graphene oxide, titanium oxide, carbon-based nanomaterials, copper nanoparticles, and silver nanoparticles) that can control plant diseases (Besinis et al., 2014). The next-generation pesticides are defined as nanobiofungicides that will provide greater benefits such as improved effectiveness and resilience and reduced use of active substances under field conditions. Emerging strategies against specific pathogenic and toxic fungi could be used as environmentally friendly antimicrobials for biohybrid nanocide substances (Abd-Elsalam et al., 2019). One potential application of silver is in the management of plant diseases. Silver exhibits several modes of inhibition against microorganisms (Soliman et al., 2020); thus, compared to synthetic fungicides, silver can be used with relative safety to control different plant pathogens (Villamizar-Gallardo et al., 2016). Plant and microorganism biosynthesis of nanoparticles is feasible, secure, affordable, and less time consuming, providing satisfactory results without the use of dangerous chemicals (Ibrahim et al., 2020; Othman et al., 2019). Mycosilver nanoparticles and their formulations can be an alternative and more cost-effective source for the prevention of fungal resistance and sustainable farming, compared with chemical fungicides. Biosynthesized silver nanoparticles (AgNPs) are also seen as potent antifungal agents against phytopathogens (Akther and Hemalatha, 2019). One strategy that shows promise for overcoming fungicide resistance and reducing environmental risks posed by xenobiotics involves the synergetic combination of novel antifungal agents and traditional drugs (Malandrakis et al., 2020a,b). This chapter evaluates the inhibitory properties of biofabricated AgNPs, mycogenic-mediated silver nano-fungicides, and Ag-based nanohybrid fungicides against various plant-pathogenic fungi and assesses silver compounds' effectiveness in the suppression of plant-pathogenic fungi. Toxic effects and antifungal mechanisms to control mycobiota contamination in food processing are also discussed.

8.2 Silver nanoparticle biocides

Noble metallic nanoparticles' antimicrobial activities have attracted scientific and pragmatic attention for more than 20 years. Among these noble metallic nanoparticles, AgNPs have especially exhibited large-scale biocidal effects, mostly against bacteria, though, there are also reports of their antiviral and antifungal (especially levurocidal) capabilities (Abou-El-Sherbini et al., 2018). In the agri-food industry, AgNPs (particle size up to 100 nm) are applied predominantly in animal husbandry and food processing, including water supply systems. They were found to be effective against the ascomycete *Aspergillus* fungi, *Stachybotrys chartarum* (Marambio-Jones and Hoek, 2010; Deshmukh et al., 2019), and if used as feed supplements, they showed the capability to reduce mycotoxin (aflatoxin) load in contaminated feed in animals (Gholami-Ahangaran and Zia-Jahromi, 2013).

AgNPs (size 6–12 nm, 100–250 mg/L) have been applied successfully as a general disinfectant in the cultivation of grass explants. When applied at higher concentrations, the NPs caused plant tissue necrosis, but in certain cases, 4–16 mg/L worked well (Parzymies et al., 2019). According to another source, even concentrations as high as 400 mg/L did not damage an araucaria plant, while showing high antibacterial effectiveness (Sarmast et al., 2011). Biocides always penetrate more into buds, kernels, and hard plants. If AgNPs are applied to young, quickly developing plants, they inhibit cell division in root tips (e.g., in onion) (Kumari et al., 2009). Other nanoparticles, such as nano-TiO_2, have shown to affect the growth of host plants and reduce growth and transpiration in leaves of corn because of inhibited water flow due to nano-binding in plant cell walls (Asli and Neumann, 2009). Nano-magnetite in water has shown to accumulate in leaves and roots of cucumbers and lettuce (Zhu et al., 2008; Larue et al., 2014), and in another case, seed germination was diminished by nano-hydrogen particles (Lee et al., 2010). It was also found that some plants (e.g., brassicas) may be hyperaccumulators of nanoparticles from the environment (McGrath and Zhao, 2003). The NPs' harm to macroorganisms may be a result of the production of reactive oxygen radicals damaging cell organelles or to their binding to enzymes, thereby blocking their activity and causing osmotic stress (Klaine et al., 2008; Bhatt and Tripathi, 2011).

AgNPs are highly effective against the formation of microbial biofilms, and they also work like antiadhesive agents and disinfectants. The antibiofouling performance depends on the Ag particle's morphology and the leaching ability of silver and its synthesis of round particles or nanowires. The nanowires exhibit higher antibiofilm activity due to the enhanced speed of the flow around the surfaces prone to film adhesion (Bhatt and Tripathi, 2011; Hu et al., 2017a). Such a method has been successfully applied in purification and disinfection management of drinking water (Ben-Sasson et al., 2014). A yeast *Candida albicans* is the causative agent of the most common hospital-acquired fungal infection due to its strong ability to form biofilms on medical devices. For example, to prevent biofilm formations, biogenic AgNPs are applied onto the predominantly plastic surfaces of catheters. The highest efficiency was reached by applying 50 μg/mL concentration of AgNPs. However, the significant level of Ag toxicity exhibited from the long-term use of such devices in animal studies indicates continuing problems (Ansari et al., 2015; Hamid et al., 2017). Also, the overuse of AgNPs, especially in the health-care sector, means that introducing microbial resistance to silver must be carefully considered.

In indoor environments, especially those with specific hygienic standards (e.g., hospitals, health-care and social service institutions, schools, open-space offices, pharma and food processing), effective air purification is needed to reduce pathogenic airborne particles containing viruses, bacteria, fungi, and other pathogenic matter. The main inhalatory infectious diseases include SARS, MERS, COVID-19, anthrax, and aspergillosis (including its farmer's lungs' form in handlers of organic materials like hay, straw, cotton, and flax that are heavily contaminated with gliotoxin and its producer *Aspergillus fumigatus* antigens). The bioaerosols are deposited on the filters of air conditioning systems and multiply under elevated moisture. The quality of the filters may be improved significantly by the application of AgNPs. Several technical solutions were developed to employ AgNPs to enhance air quality. For example, an atomizer generating NPs proved to be highly effective as an antibacterial and antisporal device (Yoon et al., 2008). Also, silver-coated silica NPs were layered on air filters and significantly eradicated viral loads in the air (Joe et al., 2016). Additionally, in an air-liquid interface model of human airway epithelium, aerosolized AgNPs proved not to be cytotoxic (Herzog et al., 2013). Nevertheless, investigations of the effects of humans' long-term inhalation of NPs are needed to establish a safe ratio between their antimicrobial effectivity and toxicological safety.

AgNPs are a very promising additive in food processing and packaging in terms of prolonging the shelf life of food and animal feed and to ensure hygienic conditions. NPs act as effective antimicrobials not only in eliminating direct viral, bacterial, or fungal contamination of food but also in preventing the production of microbial toxins in food matrices, including mycotoxins. The migration of Ag ions from packages to food must also be considered risky from a consumer health point of view (Cushen et al., 2012). The European Food Safety Authority (EFSA) recommended 0.05 mg/L in water (liquid) and 0.05 mg/kg in solid food as the upper limits of silver migration from packages to foodstuffs (ES Committee, 2011). AgNPs are large used in two types of packaging: improved packaging, with nanomaterial embedded into polymers to enhance their gas barrier properties; and active packaging, where nanoparticles directly interact with food to protect it from microbial growth and multiplication (e.g., AgNP film coatings or direct incorporation into the matrice) (Duncan, 2011; Vasile et al., 2017).

AgNP active packing of vegetables that was low in cost and noticeably environmentally friendly was investigated. The study revealed that the shelf life of vegetables was prolonged and no nutritional value was lost (Singh and Sahareen, 2017). Another study was conducted with Ag/TiO_2 nanocomposites in high-density polyethylene packaging of bread. Its microbial stability was improved as well (Mihaly Cozmuta et al., 2015). Polyethylene packages enriched with a nanocomposite of Ag/TiO_2/Fe kept the organoleptic properties of orange juice almost identical to that of the fresh juice for 10 days of storage. Silver and iron also proved to be more antimicrobial than TiO_2 (Peter et al., 2014). PVC nanocomposite films with AgNPs prolonged the shelf life and diminished lipid oxidation in packed chicken breast meat (Azlin-Hasim et al., 2016).

During a study on ion migration from packages into foods, AgNPs revealed less oxidation and agglomeration from plastic bottles and containers, depending on the type of polymer and storage conditions (Ramos et al., 2016). Silver and copper NPs impregnated into guar gum nanocomposites showed good antimicrobial properties, but their health and safety aspects must be investigated for a longer period of time (Arfat et al., 2017). Migration of nanoparticles from plastic materials was shown to be significantly higher after being heated in a microwave, compared with conventional heating, and under an acid condition, but the release depended on the size and aggregation of NPs (Echegoyen and Nerín, 2013).

8.3 Mycogenic-mediated silver nano-fungicides

Due to their strong metal resistance and ability to internalize and bioaccumulate metals, Kingdom Mycota organisms are generally used as decreasing and stabilizing agents. Also, fungi can be cultivated easily on a broad scale (nanofactories) and can produce nanoparticles with a controlled size and morphology (Guilger-Casagrande and de Lima, 2019; Khan et al., 2017). Fungi have benefits over other microorganisms because they produce large quantities of proteins, including enzymes, both of which can be used to synthesize nanoparticles easily and sustainably (Alghuthaymi et al., 2015). For example, *Trichoderma harzianum* is used for silver nanoparticle synthesis. Under glasshouse conditions, the generated nanoparticles were tested against tomato *Fusarium* wilt. In comparison with the control plants and *Trichoderma* formulation-treated plants, nanoparticle-treated plants demonstrated a promoter effect on all tested seedlings' quality parameters (germination, plant height, root length, total yield, and root dry weight) (Pandey et al., 2015). Green and biogenic synthesis of AgNPs was demonstrated using the *Curvularia pallescens* fungus. *Cladosporium fulvum*, which is the main cause of a severe plant disease known as tomato leaf mold, was tested for the antifungal activity of the as-prepared AgNPs. The synthesized AgNPs led to outstanding fungicidal activity against various fungal pathogens, including *C. fulvum* (Elgorban et al., 2017). The silver nanoparticle was biosynthesized extracellularly by employing the biocontrol agent *T. longibrachiatum*. Mycosynthesized AgNPs from the presented fungus not only enhanced the seed germination percentage and quality index but also reduced the incidence of seedborne diseases caused by *F. oxysporum* on faba bean, tomato, and barley.

The smaller dimensions of AgNPs used in the work by Elamawi and coworkers were shown to be the most efficient against various plant pathogenic fungi (Elamawi et al., 2018). AgNPs were produced using *T. harzianum* cultivated with AgNP/TS, without need of enzymatic stimulation (AgNP/T) by the cell walls of *Sclerotinia sclerotiorum*. Mycogenic nanoparticles were evaluated in vitro for the regulation of *S. sclerotiorum*. Both AgNP-TS and AgNP-T were able to monitor the growth of *S. sclerotiorum*, inhibiting mycelial growth and preventing the development of new sclerotia of these pathogenic fungus.

The most effective regulation of mycelial growth was accomplished by using AgNP/TS, which could have been due to the lower hydrodynamic diameter of the nanoparticles (Guilger-Casagrande et al., 2017, 2019). A total of 15 isolates of *Trichoderma* spp. were screened for the biosynthesis of AgNPs by Tomah and colleagues. Maximum synthesis of AgNPs occurred using a cell-free aqueous filtrate of gliotoxin containing *T. virens* HZA14. AgNPs' antifungal activity against soilborne pathogens was also investigated. There was a high percentage of inhibition of a *Sclerotiorum* pathogen and of its hyphal growth, sclerotial formation, and myceliogenic sclerotia germination (Tomah et al., 2020). Consequently, further research on the potential of biogenic nanoparticles combined with *Trichoderma* spp. is required, with further focus on biocontrol agents' possible threats to public and environmental health (Fraceto et al., 2018). Novel techniques have been developed for microencapsulation of fungi and for the synthesis of bionanoparticles and mycogenic nanoparticles to improve the biological control of pathogens and contribute to sustainable farming practices (Fig. 8.1).

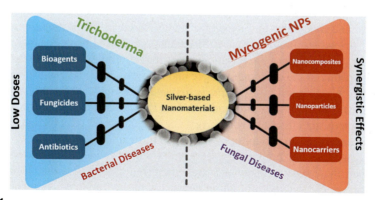

FIG. 8.1

Current applications of silver-based nanomaterials in the management of plant diseases.

8.4 Ag-based nanohybrid fungicides

The antimicrobial activity of inorganic NPs in combination with conventional fungicides has been studied in terms of plant defense mechanisms. CuNPs, AgNPs, and ZnO NPs have shown considerable antifungal activity against diverse pathogenic fungi, either alone or in combination with traditional antifungals, particularly thiophanate-methyl (TM), tebuconazole, mancozeb, fluazinam, fludioxonil, carbendazim, thiram, and propineb (Huang et al., 2018, 2020; Jamdagni et al., 2018; Malandrakis et al., 2020a,b). Gajbhiye et al., (2009) observed the synergistic effect of fluconazole and AgNPs regarding their antifungal activity against the plant pathogenic fungi, *Phoma glomerata, P. herbarum, Fusarium semitectum,* and *Trichoderma* sp., using the disc diffusion method. The formulation, which was highly inhibitory to *P. glomerata* and *Trichoderma* sp., involved the combination of fluconazole with AgNPs (Higa et al., 2013). The efficiency of AgNPs was expanded by combining them with the antifungal agent miconazole, which, with the aid of increasing levels of reactive oxygen species (ROS), inhibited ergosterol biosynthesis and biofilm formation (Kumar and Poornachandra, 2015). Studies have shown that, due to the existence of bio-AgNPs, the antifungal activity of fluconazole and itraconazole against the examined pathogenic fungi improved (e.g., Singh et al., 2013). The synergistic effect of ethanolic and acetone extracts of *Cydonia oblonga* leaves and AgNPs was also assessed and showed that the growth of *Asperillus niger* was prevented (Alizadeh et al., 2014). A new combination of amphotericin B (AmB) blended with AgNPs was evaluated against *A. niger* and *F. culmorum*. Interactions between AmB and AgNPs created high antifungal activity (Tutaj et al., 2016). The in vitro and in vivo antifungal effect of mycosynthesized AgNPs on *Cephalosporium maydis* has been evaluated alone and in combination with one or two different fungicides. Under greenhouse conditions, parallel findings were observed where AgNPs induced a strong reduction in the severity of *C. maydis* disease: 57.7% of contaminated plants alone, 75.5% combined with MAXIM XL, and 90% combined with Vitavax (Elgazzar and Rabie, 2018). Conventional zineb fungicides (Zi) were combined with functionalized nanocomposites of Ag-chitosan, which enhanced stability and synergized with *Neoscytalidium dimidiatum* fungicide to enhance antifungal efficacy. Silver-incorporated chitosan nanocomposites (Ag@CS) showed a low-concentration of antifungal properties in both AgNPs and CS participating in the compound (Ngoc and Nguyen, 2018). For antifungal action against rice blast caused by *Pyricularia oryzae* fungus isolated from blast-infected leaves, Ag@CS mixed with Trihexad 700 WP (Tri) was introduced. Results showed that the synergistic effect of Tri and Ag@CS NPs may be a possible candidate for the use of antibiotics in agriculture with high antifungal activity (Pham et al., 2018). Le with coworkers first prepared Ag@CS with CS used as a reducing and stabilizing agent; then these nanocompounds were synergized against *Phytophthora capsici* (which causes *Phytophthora* blight in pepper) with the fungicide Antracol (An) as Ag@CS/An. In vitro findings have shown that Ag@CS/An has substantially greater antifungal potential than each part alone (Le et al., 2019).

Ligustrum lucidum leaf extract, a major synergistic antifungal, was synthesized by plant-mediated AgNPs where AgNPs were combined with epoxiconazole at 8:2 and 9:1 ratios. The findings of this study included a new fungistat that provided not only comprehensive control of plant fungi but also reduced the use of chemical pesticides and prevented the generation of drug-resistant phytopathogens (Huang et al., 2020).

Compared with individual treatments, the combination of AgNPs with TM resulted in a significantly improved fungitoxic effect, regardless of the resistant phenotype (BEN-R/S), both in vitro and when applied to apple fruit, thus, AgNPs could be used effective against *Monilia fructicola*. When used in conjunction with traditional fungicides, the combination may provide the means to counteract benzimidazole resistance while reducing the environmental effect of synthetic fungicides due to the reduced dosage required for the control of pathogens (Malandrakis et al., 2020b). Therefore, it can be said that based on the dosage of fungicides and the counterresistance of plant pathogenic fungi such as *Botrytis cinerea* (Fig. 8.2), a combination of nanomaterials and traditional fungicides offers a resistance management strategy that can lead to a healthy, sustainable environment.

Nanotechnology has also allowed the active agents of fungicides to be encapsulated into polymer nanoparticles to build a nano-delivery system of fungicides that can more efficiently transport them to the target fungal pathogens. To gain a deeper understanding of the synergistic process and extend the scope of application of polymer NPs with other fungicides, more research is required.

8.5 Toxic effects of AgNP-based biocides under discussion

As the use of AgNP antimicrobial applications increases, health policy makers are becoming concerned about their toxicity and negative environmental and health impacts, which is understandable as NPs have a certain degree of toxicity due to their physicochemical properties. They are absorbed by nontarget organisms such as aquatic species. Even at a concentration of 1–5 μg/L, reactions are noticeable, perhaps because of transformation of the NPs' surface area, phase, aggregation state, and rate of sulfidation in the environment (Nowack et al., 2011), which may lead to problems when they are discharged into the environment. Their extensive use in common disinfection practices may both increase microbes' resistance and, thus, decrease the use of AgNP-based biocides. For example, it has been reported that Argyria disease, a rare condition in which the color of skin becomes blue-gray, is due to the toxic effects of AgNPs. Argyria disease appears at a silver threshold limit of 0.9 g per body weight over a lifetime (Pulit-Prociak and Banach, 2016). A safe drinking water limit has been set at 100 μg/L of nanosilver or the dissolved form of silver. According to recent studies (AshaRani et al., 2009; Gliga et al., 2014), the nano form of silver seems to possess higher toxicity than the dissolved form. The nano form revealed cytotoxic, genotoxic, and antiproliferative parameters in human cell lines (e.g., lung cells).

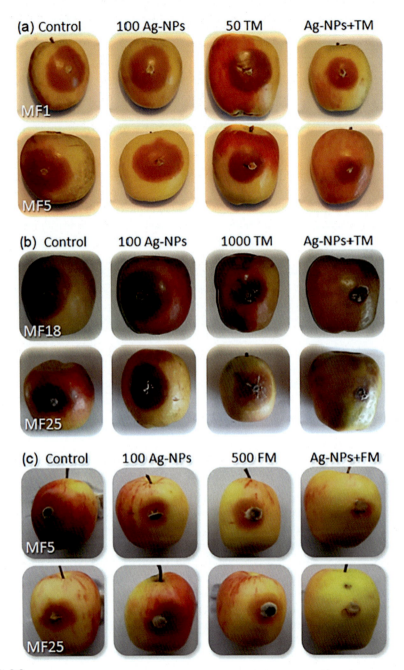

FIG. 8.2

Synergistic activity of Ag-NPs (100 μg/mL) in combination with (A, B) thiophanate-methyl (50) and (C) fluazinam (500 μg/mL) on apple fruit against selected *Botrytis cinerea* isolates sensitive (MF1, MF5) and resistant (MF18, MF25) to thiophanate-methyl (*TM*, thiophanate-methyl; *FM*, fluazinam).

Data from Malandrakis, A.A., Kavroulakis, N., Chrysikopoulos, C.V., 2020b. Use of silver nanoparticles to counter fungicide resistance in Monilinia fructicola. Sci. Total Environ. 747, 141287, with permission from Elsevier.

8.6 Mechanisms of antifungal activities in phytopathogens

It is difficult to clarify a biocides' exact mechanism of antifungal activity, including fungicides and disinfectants, which may function two ways: fungistatically by inhibiting fungus growth and fungicidally through lethal activity on microbes.

Lethal activity is related especially to changes in membrane permeability or even decomposition of the cytoplasmatic membrane, when nonspecific cytosolic compounds are inevitably released from the cell. The blocking of a key enzyme of cell energetic metabolism, the ATPase, which leads to the affected cell's death, is another crucial fungicidal mechanism (Ritota and Manzi, 2020).

The most common antifungal compound, which is usually applied as a disinfectant in the agri-food sector, is 50%–70% alcohol. Ethanol interacts with microbial cell membranes, resulting in elevated permeability, leakage of inner cell solutes, and finally cell death. The greatest effect results from concentrations of 70% and higher. Ethanol vapor has been shown to easily inactivate spores of *Penicillium chrysogenum*, *P. digitatum*, and *P. italicum* (Rogawansamy et al., 2015).

The antifungal mechanisms of etheric oils have been studied only partially so far. These lipophilic compounds with many low-molecular components can easily penetrate cell membranes and damage the organization of cells. At any stage, they can stop the biosynthesis of ergosterol, the main sterol present in fungal cell membranes. For example, inhibition of demethylase lowers ergosterol production of lanosterol and causes an accumulation of methylated sterols. Ergosterol serves as a bioregulator of membrane fluidity, asymmetry, and integrity and also stimulates fungal growth. Azole fungicides possess a similar mechanism of activity to etheric oils—that is, they affect ergosterol synthesis negatively (Kathiravan et al., 2012; Hu et al., 2017a,b).

Antifungals, including etheric oils, cause mitochondrial dysfunction that leads to metabolic stagnation observed as delayed growth. Such an effect was studied on *Aspergillus flavus* (Hu et al., 2017a,b). The researchers observed depression of aflatoxin B1 production due to a particular gene's expression. Etheric oils may inhibit the production of aflatoxin B1 by blocking certain key enzymes in fungal carbohydrate catabolism (Tian et al., 2011). Other teams studied the effects of chitosan with different molecular masses on *A. niger*. These polysaccharides can enter a fungal cell and inhibit its proliferation and spore production, cause thinning of the cell wall, and destroy the cell's nucleus, all of which are lethal for the cell (Li et al., 2008). Dibrov et al. (2002) stated that metallic NPs' reduction of bacterial and fungal contamination resulted from the release of metal ions (e.g., Ag ions), which can break the microorganisms' membrane structures and, therefore, penetrate them. A general overview of known antifungal targets is shown in Table 8.1.

Knowledge of biocide mechanisms of action and an understanding of quantitative structure-activity relationships provide a platform for emerging biocides with enhanced activity and environmental acceptability. These mechanisms also add light to the antifungal mechanisms of resistance of particular classes of chemicals (Table 8.2), as fungi strive to survive biocidal exposure by decreasing the toxic concentration in their cells.

Table 8.1 Antifungal target sites and mechanisms of action.

Target site	Mechanism of action	Biocide group
Cell wall	Cross-linking of cell proteins and chitin Cell agglutination	Glutaraldehyde AgNPs
Cytoplasmatic membrane	Induced leakage of intracellular material and protoplast lysis; loss of structural organization and integrity; disruption to physiological functions Alteration of membrane properties and activation of an efflux pump system; membrane perturbation Inhibition of the proton motive force (delta pH component); induced leakage of intracellular material Inhibition of respiration and energy transfer; inhibition of ATP synthesis Interaction with ergosterol and destabilization of cell membrane functions; inhibition of cytochrome P450 in ergosterol biosynthetic pathway Inhibition of the electron transport system	Chlorhexidin, quaternary ammonium compounds (QACs), ethanol Organic acids Esters Quinone outside inhibitors (QoIs) Demythylation inhibitors (DMI) groups Benzylcarbamate
DNA/RNA	Interference with NA synthesis	Pyrimidine analogue flucytosine, AgNPs
Proteins	Interaction with alkylating and oxidizing agents; binding to key functional groups of fungal enzymes Inhibition of cell division; binding to proteins of tubulin; cytoskeleton formation	Heavy metals, AgNPs (–SH groups) Benzimidazole

Based on Fernandez Acero, F.J., Carbú, M., El-Akhal, M.R., Garrido, C., González-Rodríguez, V.E., Cantoral, J.M., 2011. Development of proteomics-based fungicides: new strategies for environmentally friendly control of fungal plant diseases. Int. J. Mol. Sci. 12(1), 795–816; Abou-El-Sherbini, K.S., Amer, M.H., Abdel-Aziz, M.S., Hamzawy, E.M., Sharmoukh, W., Elnagar, M.M., 2018. Encapsulation of biosynthesized nanosilver in silica composites for sustainable antimicrobial functionality. Global Chall. 2(10), 1800048.

Fungi have certain intrinsic biocide resistance mechanisms, and other mechanisms can be acquired. For example, secondary resistance can result from mutations in one or several target sites' genes. Acquisition of biocide resistance was observed, for example, in *Candida* spp. and *A. fumigatus* (Warnock et al., 1999; Kontoyiannis and Lewis, 2002). Other yet unrecognized mechanisms may also be activated to achieve fungicide resistance in cells.

Table 8.2 Mechanisms of fungal resistance to biocides.

Target site	Mechanism of resistance	Biocide group
Cell wall	Entry of the drug is prevented. The drug is pumped out of the cell by an efflux pump.	DMI groups and many other antifungals
Cytoplasma	The drug target is altered; the drug cannot bind to the target. The target enzyme is overproduced; the drug does not completely inhibit the biochemical reaction. Certain fungal enzymes' capability to convert an inactive drug to its active form is inhibited. The cell secretes drug-degrading enzymes or chelating substances to the extracellular side. Synthesis of an alternative enzyme replaces the drug target (mutations).	Qols, benzimidazole DMI groups Flucytosine Formaldehyde, bisphenol triclosan, diphenyl ether, biphenyl, phenols, heavy metals Strobilurin, carboxamides

Based on Fernandez Acero, F.J., Carbú, M., El-Akhal, M.R., Garrido, C., González-Rodríguez, V.E., Cantoral, J.M., 2011. Development of proteomics-based fungicides: new strategies for environmentally friendly control of fungal plant diseases. Int. J. Mol. Sci. 12(1), 795–816.

8.7 Regulation of phytopathogenic mycobiota in agriculture

Phytopathogenic fungi may infect all parts of the plant body, and plants may suffer from a fungal attack in all their developmental stages. Fungal spores, sexual and asexual, can survive unfavorable conditions and after a period of dormancy germinate into a plant via a germ tube produced during germination. The growing fungus breaks through the plant cell wall through an appressorium producing a battery of enzymes—for example, cuticle degradation enzymes (Fernández-Acero et al., 2008, 2010) and toxins (phytoalexins or even trichothecenes produced by *Giberella* spp. (Idnurm and Howlett, 2001)—to defend the invaders against the plant's defense system. A thin invasion hypha penetrates plant tissue and is transformed into a haustorium responsible for degrading a plant's digestive processes. The mycelium then continues to destroy the host, producing more asexual reproductive fruiting bodies to generate more spores; therefore, the disease quickly spreads in the plant as a secondary infection (Agrios, 2005). All chemicals involved in the infection cycle of phytopathogenic fungi are encoded by specific genes. These genes and their products are categorized according to their pathogenicity and virulence qualities. Pathogenicity refers to the potential of a pathogen to cause disease. Virulence refers to the capacity to regulate the intensity of disease and is the degree of pathogenicity caused by the genes and their products (Fernandez-Acero et al., 2007). To date, dozens of genomes of pathogenic fungi were reported in scientific and applied databases, including *Aspergillus nidulans, A. terreus, B. cinerea, Colletotrichum graminicola, Fusarium graminearum, F. oxysporum, F. verticillioides, Mycosphaerella graminicola, Nectria haematococca, Neurospora crassa, Puccinia triticina, Rhizopus oryzae, Sclerotinia sclerotiorum,* and *Verticillium dahliae* (see http://cogeme.ex.ac.uk). Experimentally verified pathogenicity and virulence factors from fungal and bacterial pathogens infecting animal, plant, fungal, and insect hosts are listed in the pathogen-host interactions database (www.phi-base.org). A proteomics perspective of pathogenic fungi, including phytopathogens, has been presented; see for example, the Swiss Expert Protein Analysis System (ExPASy) proteomics server (https://www.expasy.org). Genomic and proteomic studies have completed the functional analysis of the gene products and cellular pathways and unveiled the developmental mechanisms of particular diseases, with the result that we can combat them effectively. In this context, the use of natural products or related compounds as specific enzyme inhibitors offers a promising path toward specific target-oriented treatments with minimal environmental impacts.

Fungal infections as major threats to agricultural plants depress production, resulting in high economic losses, which forced the development of new fungicides. Although the chemicals involved were first studied from an antifungal point of view, their input to the environment was evaluated as well. Although they can't cure the problem, fungicides can help protect plants from diseases, eradicate phytopathogenic propagules, and enhance plant production. They should exhibit the following characteristics (Gadd and White, 1989; Gupta, 2018):

- low plant and animal toxicity,
- high fungal toxicity,
- effective penetration into fungal spores and development of mycelium that target their (selective) toxic action, and
- ability to cover plants with a protective film resistant to environmental conditions.

Though synthetic fungicides are highly effective, their extensive application has led to severe consequences such as environmental accumulation and contamination, resistance of phytopathogens, residual toxicity, teratogenicity, and carcinogenicity. However, an appropriate compromise should be sought because without the use of fungicides, the agriculture and food sector may be badly damaged (e.g., plant disease epidemics, food shortages, and enormous economic, finally resulting in human losses) (Yoon et al., 2013; Berger et al., 2017). Regulation (EC) No. 396/2005 (Regulation (EC), 2005) set the maximum residual concentration of particular fungicides tolerated in food and animal feed. An updated register of allowed fungicides, synthetic and natural, is released regularly by the EC authorities.

Today, a broad spectrum of antifungal compounds are being used to control the spread of fungal pathogens. However, during the early part of the 20th century, fungicides comprised organic matter such as salt, sulfur, lime, copper, arsenic, and mercury. Around 1940, farmers began using synthetic fungicides such as the highly toxic hexachlorobenzene (unfortunately, it is still being used in some developing countries). For example, azoles were used a great deal during the mid-20th century.

Their chemical structure is based on unsaturated aromatic molecules with at least one atom of nitrogen. They are still the most widely used antimycotics and are highly effective, have a broad spectrum of use, and are low in cost (Berger et al., 2017). Their use, however, comes with risks. In their 2017 study, Berger and colleagues hypothesized that as fungicides penetrate the human body, some people may acquire secondary azole resistance (opportunistic) even to other fungal pathogens, for example, *Aspergillus fumigatus*, which causes human aspergillosis. According to their study, such an azole resistance occurs in humans after the consumption of food containing residual azole fungicides, after interactions with contaminated insects, or by inhalation of azole plant sprayings.

Biopesticides and botanic fungicides are being studied as alternatives to synthetic fungicides. They are proposed to be biologically degradable, much less (or not at all) damaging to the environment, highly effective and selective, and safe for humans and animals. Biopesticides are generally characterized as natural materials obtained from microorganisms (e.g., bacteria, protozoa, and nematodes), viruses, fungi, animals, and plant extracts. Botanical agro-fungicides have high potential to inspire and affect today's agrochemical industry. Their development can minimize the negative ecologic consequences of synthetic compounds' (Yoon et al., 2013). Cinnamaldehyde, one of the promising biofungicides, is a yellow liquid with high viscosity, a cinnamon smell, and strong antifungal activity. Its mechanism of activity remains unknown, but due to its low toxicity and practical properties, cinnamaldehyde is ideal for agriculture (Choi et al., 2004). In some countries, products containing chrysopanol and parietin isolated from *Rumex crispus* L. roots have been applied as ecological agrofungicides for a decade. These particular anthraquinones can also be produced by certain fungi (*Alternaria* spp., *Aspergillus* spp., and *Penicillium* spp.). The antifungal and antimycotic effects of an etheric oil obtained from the roots and leaves of *Curcuma longa* L. against *Aspergillus flavus*-infected maize were studied. The oil proved to be antifungal, so *Curcuma* may be a promising fungicide in bio-agriculture (Hu et al., 2017a,b).

8.8 Regulation of the contaminant mycobiota in food processing

The level and degree of microbial food contamination depend on several factors: temperature, humidity, water activity, microbiota present, physical damage, and harvest and storage conditions (European Food Safety Authority (EFSA), 2007). Physical, chemical, and microbial methods may be applied to control the growth of microscopic fungi in food and feedstuff.

Physical procedures include thermal processing (e.g., pasteurization), reduction of water activity (salting or drying), cooling, air processing, and packing in a modified atmosphere.

Microbial methods rely on the antimycotic activity of lactic acid produced by bacteria (Fernandez et al., 2017). *Propionibacterium* and *Lactobacillus* strains showed the highest antifungal effect against the common cheese spoiling fungi *Penicillium chrysogenum, Mucor racemosus, Aspergillus versicolor*, and *Cladosporium herbarum*.

Chemical methods imply the application of conservants, as they are highly potent, low in cost, and easy to acquire, but their concentrations must be limited due to possible negative health effects. According to Regulation EC (European Community) ES (European Standard) No 1333/2008 on food additives, food conservants are defined as "the chemicals prolonging the life of food by protecting them against the microbial damage and/or preventing the proliferation of pathogenic microorganisms" (Regulation (EC), 2008). Commission (EU) Declaration No. 1129/2011 changed and completed Annex II of the aforementioned legislation ES 1333/2008 by listing the officially proven food additives, including conservants such as sorbic acid and sorbates (E 200–203) for unripened and ripening cheeses, nisin (E 234) only for mascarpone and ripening cheeses and acetic acid (E 260) and lactic acid (E 270) only for mozzarella (Commission regulation (EU), 2011). The use of matured lysosyme (E 1105), propionic acid, propionates (E 280–283), occasionally natamycin (E 235), hexamethylenetetramine (E 239), and nitrates (E 251–2252) are legal exclusively in ripening cheeses (Ritota and Manzi, 2020). Natamycin is a polyene antibiotic (also called polyene antimycotic) and can be applied only on the surface of cheeses (Jalilzadeh et al., 2015). Regarding the conservant effects of mycobiota, a study showed that certain fungi can convert sorbates into metabolized *trans*-1,3-pentadiene or transpiperylene giving a plastic or kerosine aroma to the cheeses, which is undesirable and indicative of food spoilage (Sensidoni et al., 1994).

Although using chemical conservants is the easiest way to prolong the shelf life of food, consumers want more natural products, thus food producers had to rethink this strategy and use natural alternative conservants. In 1980, the inhibition of food-spoiling microbiota and production of their toxins by common spices was studied. The extraction of eugenol from cloves and thymol from tymian (0.4 mg/mL) completely inhibited the growth of *Aspergillus flavus* and *A. versicolor*. Anetol from anise seeds at 2 mg/mL suppressed the growth of all fungi tested. Twenty-four other spices did not inhibit the proliferation of fungi, but acted against their toxin production (Hitokoto et al., 1980). In 1978, Hitokoto et al. found cinnamon to be a very strong inhibitor of toxic aspergilli.

A certain degree of antifungal activity has been exhibited by many other natural compounds: L-glutamic acid; gama-aminobutyric acid; laminarin (polysaccharide isolated from the alga *Laminaria digitata*); curcuma; phenolic compounds (flavons, flavonoid glycosides, cumarines and their derivatives, anthraquinones, resorcinols); etheric oils from thyme, tymian, basalicum, bergamot, and tee plant; and animal products such as lactoferrin, lysozyme, and chitosan. Some curcuminoids (demethoxycurcumin, curcumin, bismethoxycurcumin) can inhibit mycelium prolongation and spore germination of microscopic filamentous fungi (Yoon et al., 2013). Despite its specific taste, aroma, and strong color, curcuma may play an important role in cheese conservation. In some Oriental countries, curcuma has been widely used for centuries as a spice and to protect food against microbial spoilage.

8.9 Sanitary procedures in agrotech and food production

Fungi can colonize practically any material unless the humidity and nutrition sources are limited. The growth and sporulation of fungi depend on the effect of the material's water activity as well as thermal conditions. Fungal reproduction is based mostly on spore formation and propagation. All fungal spores are resistant to unfavorable environmental conditions. Thus inhibition of sporulation may be a component in effective antifungal activity, although, to our knowledge, this fact is not mentioned in the existing literature. Indoor and outdoor fungi are similar in composition but not in quantity. Both characteristics change with the seasons and have locally specific patterns. The most common fungi are saprophytes inhabiting decaying plant material, that is *Cladosporium* spp., *Alternaria* spp., *Epicoccum* spp., and *Aureobasidium* spp. Air-borne species of aspergilli and penicillia might be present in lower concentrations outdoors but are usually more dominant in indoor air. About 80 fungal species have been recognized as being important indoor allergens and enter mostly via ventilation and air conditioning systems. Visible fungal growth can be eliminated by cleaning and using disinfectants based on alcohol, chlorine, hyperoxide etc. (Rogawansamy et al., 2015).

To eliminate pathogens and prevent recontamination, a primary sanitation procedure in food plants is disinfection of water, machines, tools and other types of equipment, and all surfaces. For example, chlorine products are used extensively in the food industry, as they can kill pathogens in solutions effectively and are also inexpensive. During a disinfection process, reactions occur between natural organic matter and inorganic ions and side effects can take place that strongly affect employees' health. Disinfected water is a primary source of volatile trihalomethanes and nonvolatile halogen acetic acids, which are carcinogenic and toxic to humans. Disinfection of water is of crucial value, especially in the production of cheese, where salty solutions in disinfected water are in direct contact with the cheese for hours. The same salty solution is reutilized in a factory for several weeks and in the meanwhile is enriched with precursors from the processed food (lipids, proteins, saccharides) that can react with free chlorine anions. The byproducts of disinfection can be absorbed by the cheese, and the longer the cheese is in contact with the salty water, the stronger the effect. Milk, as the raw material in dairy factories, represents another source of the adverse effects of disinfection on final products due to chlorine disinfectants that are applied directly or in disinfected water to clean storage cisterns, tanks, and pipings, as noted by Cardador et al. (2017). Their study dealt with the occurrence of adverse side effects in cheeses due to the use of disinfection products. Trihalomethanes (trichloromethane, bromodichloromethane) and halogen acids (dichloroacetic, trichloroacetic, and bromchloroacetic acid) were detected at levels that fell within the tolerable daily intake of 10–40 µg/kg per day, according to WHO standards. Halogen acids are hydrophilic and nonvolatile, so, they can persist longer in a cheese matrice. Their concentration in cheese grows in the course of its ripening, while the content of water declines simultaneously (Collado et al., 2007).

The spectrum of effective sanitary and disinfection products applicable in today's food industry is relatively broad; thus, the safety of personnel who handle foods in the process of cleaning and disinfection needs to be considered. Many such commercial products are potentially toxic and caustic dermally or when inhaled. At the same time, their residua contaminate underground water and the environment in general.

Particular fungal isolates, even of the same species, can express different levels of sensitivity to disinfectants. The inhibitory activity of different concentrations of acetic acid and 70% ethanol on *Penicillium chrysogenum* and *Aspergillus fumigatus* were tested. It was found that acetic acid (4.0%–4.2%) inhibited growth and sporulation of the *Penicillium* fungus but not of the *Aspergillus* fungus (Parzymies et al., 2019). Notably, due to its antimicrobial activity (rather limited) and acidic pH, vinegar is commonly used as a natural and inexpensive disinfectant for wet cleaning of surfaces.

8.10 Biocide (disinfectant) effectivity/efficacy tests in vitro

The European Committee for Standardization (CEN) is the EU institution responsible for setting the legislation on control and standardization of materials, products, and processes. Regarding the effectivity (the term used for the activity of biocidal products in the field) and efficacy (distinguishes the activity of biocides under laboratory conditions) tests on antimicrobial activity of biocides for control as well as commercial purposes, CEN adopted several European norms (EN):

- EN 1650: Quantitative suspension test to evaluate fungicidal or levurocidal activity of chemical disinfectants and antiseptics used in food processing, industrial, domestic, and institutional areas (EN 1650:2019, 2019);
- EN 1275: Chemical disinfectants and antiseptics; quantitative suspension test to evaluate basic fungicidal or levurocidal activity of chemical disinfectants; procedure and conditions test (EN 1275:2005, 2005) (this norm is applicable for testing active substances (antifungal biocides) and products under development that are planned to be used in the food processing, industrial, domestic, institutional, medical, and veterinarian sectors); and
- EN 12353: Chemical disinfectants and antiseptics; deposition of microbial strains utilized in the evaluation of bactericidal (including *Legionella*), mycobactericidal, sporicidal, fungicidal, and virucidal (including bacteriophages) activity (EN 12353:2013, 2013).

In Slovakia, the last two ENs norms shown above were adopted to prepare the Slovak Technical Norm (STN) EN 1650 (last edition 2020) Chemical Disinfectants and Antiseptics (STN EN 1650:2020, 2020) stipulation. The norm specifies testing procedures and minimal requirements for the quality of chemical disinfection and antiseptic products that form homogenous and physically stable solutions while diluted in hard water or in ready-to-use preparations already diluted with water. The test set the maximum concentration as 80%, as after the addition of the test organism, the product will be further diluted. The norm cannot be applied in areas and situations where medical disinfection is indicated and on products used on living tissues, excluding hand hygiene products.

The general purpose of STN EN 1650 is to eliminate the number of living microorganisms at the minimum level of 4-log (99.99%) if the product is diluted with hard water or in the case of ready-to-use products. This dilution-neutralization method is the test of choise in general practice of testing efficacy of disinfectants. In this method, a sample of the disinfectant diluted with hard water is added to a suspension of filamentous fungi diluted in a solvent; it is then kept at a particular testing temperature for a particular amount of time in water; the test is performed under simulated clean conditions (with the addition of the interfering compound at a low concentration to simulate real-life conditions) or under dirty conditions (with the addition of the interfering compound at a higher concentration), according to the practical application of the product and to conditions specified by the EN. Before the test, during its course, and afterward, samples are taken and inoculated into the media and counts of vital microorganisms (*Aspergillus brasiliensis* or *Candida albicans* from the culture collections recommended) are compared. All the chemicals used in the test should be anhydrous, commercially available, and suitable for microbiological purposes. Other conditions must comply with recommendations for the production of the disinfectant being tested. The entire process must be run aseptically and all solutions must be sterile. The results of the test are interpreted in terms of whether the general fungicidal activity and the biocide concentration fulfilled the desired criterion, that is, 99.99% reduction of microbial vitality.

We used the STN EN 1650 method to test the antifungal effects of different commonly used disinfectants in food processing, especially in the dairy and fruit canning industry. Namely, hydroperoxides in different concentrations of applied solutions (0.5%, 1%, and 2%), halogens (CL_2 and I_2), chlorinated compounds (chlorhexidine and chloramine) and quaternary ammonium tensides (QACs) were evaluated for their ability to suppress the growth of the following: the dermatophyte *Trichophyton mentagrophytes* (may be present on the skin or hair of workers), the cultured yeast *Geotrichum candidum* (on the other hand, airy mold also forms stubborn biofilms, e.g., in milk tanks and pipe linings from which this strain was isolated), *Penicillium roqueforti* used in cheese production, and the heat-resistant fungi *Byssochlamys nivea, Dichotomomyces cejpii, Eupenicillium baarnense, Neoartorya fischeri, Talaromyces avellaneus, T. flavus*, and *T. trachyspermus* causing spoilage of canned fruit juices and fruit compotes up to 120 min (with cultivation readings of fungal growth after 5, 10, 15, 30, 60, and 120 min of antifungal treatment) (Fig. 8.3).

As expected, peroxide showed the best antifungal efficacy and stopped the growth of all fungi in the experiment after a maximum of 30 min at 1% concentration, while the lower and higher concentrations tested were less effective against both yeasts and molds, including dermatophytes. QACs, chlorinated compounds, and halogens were not effective against the fungi in the experiment, as the fungi survived even after 120 min of activity. Thus biocides' antifungal effects were shown to be ineffective within a reasonable time frame, even biocides that were traditionally believed to be effective.

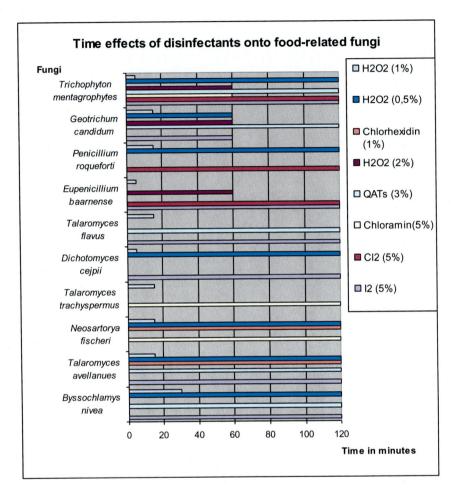

FIG. 8.3

Time course of antifungal treatment of foodborne fungi with biocides. *Legend*: The bars represent the last time point of fungal growth after particular antifungal testing action.

The dilution-neutralization method we also applied to evaluate the efficacy of several antifungals (0.5%, 1%, and 2% H_2O_2; chloramine; iodine; glutaraldehyde; and formaldehyde at concentrations commonly recommended by their producers) against *Purpureocillium lilacinum* (synonym, *Paecilomyces lilacinus*) that had caused spoilage of washing solutions used by a manufacturer of ewe cheese (Fig. 8.4). The fungus, which is originally airborne, is frequently found in soil and storage places, especially with cereals, and is able to cause human mycoses. The biocides were left to act up to 3 h, with cultivation activity checked at 10, 30, 60, 120, and 180 min intervals (Fig. 8.5).

FIG. 8.4

Spoiled cheese washing solution (L) and a causative agent culture of *Purpureocillium lilacinum* on malt extract agar after 7 days of cultivation at 25°C (R).

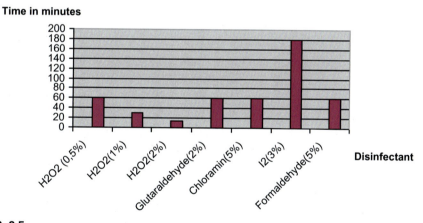

FIG. 8.5

The fungicidal effect of biocides against the airborne mold *Purpureocillium lilacinum*. *Legend*: The bars represent the last time point of fungal growth after particular antifungal testing action.

As in the previously mentioned study, hydroperoxide proved to be the most effective agent against mold, with fungicidal effects observed at 10 and 30 min intervals in 1% and 2% concentrated solutions. The iodine count registered low to none as an antifungal biocide, although it is still used as a disinfectant.

The American Society for Testing and Materials (ASTM) established similar norms for antifungal testing applicable in the United States:

ASTM E2315: Standard handbook for evaluating antimicrobial activity via gradually timed devitalization; with the dermatophyte *Trichophyton interdigitale* as the testing fungus (ASTM E2315, 2020), and
ASTM E2197: Standard quantitative method to test disc carriers and evaluate bactericidal, virucidal, fungicidal, mycobactericidal, and sporicidal effects of chemical compounds (ASTM E2197, 2020).

For disinfectants applied on surfaces, other international norms on efficiency testing are designated:

- EN 13697: Chemical disinfectants and antiseptics; quantitative nonporous surface test to evaluate bactericidal and/or fungicidal activity of chemical disinfectants used in food processing, other industries, and domestic and institutional sectors; testing method and recommendations without mechanical action (EN 13697:2015, 2015);
- AFNOR NFT 72-281 (French norm): Method of disinfection of surfaces interfacing with air; evaluation of bactericidal, fungicidal, levurocidal, mycobactericidal, tuberculocidal, sporicidal, and virucidal activity, including bacteriophages (AFNOR NFT 72-281, 2020; Woodall, 2020); and
- AOAC 955.17: Fungicidal activity of disinfectants on nonporous surfaces (AOAC 955.17, 2020).

The US Office of Chemical Safety and Pollution Prevention (OCSPP) developed several documents dealing with the evaluation of biocides:

- OCSPP 810.2000: General considerations on antimicrobial pesticides testing in the area of public health (OSCPP 810.2000, 2020);
- OCSPP 810.2100: Sterilates, sporicides, and decontaminants (OSCPP 810.2100, 2020); and
- OCSPP 810.2200: Disinfectants to be used on environmental surfaces (OSCPP 810.2200, 2020).

If a biocide will be used as a pesticide in agriculture, its efficiency is tested according to norm 961.02 set by the Association of Official Analytical Chemists (AOAC) for germicidal spray products. The test was originally developed in 1961 and is the standard method for evaluating the biocidal activity of liquid products on hard, nonporous surfaces. Although the method is mandatory, products must be registered by the US Environmental Protection Agency (EPA). This method is suitable for testing the products without solution and for application on surfaces by spraying. The method primarily tests bactericidal activity but also collects data on fungicidal (*Trichophyton interdigitale*) and levurocidal agents (AOAC 961.02, 2020).

The Organization for Economic Cooperation and Development (OECD) also publishes directives regarding testing of chemicals and provides a platform for evaluating their potential effects on health and the environment. The directives are accepted internationally as standard edicts on testing the safety of chemicals in industrial and academic research. These recommendations are continuously updated according to the latest scientific and technical information and procedures. Under this framework, OECD member countries and allied ones accept the results of sound testing procedures and laboratory practices due to their positive implications regarding human health and the environment. The OECD has described quantitative evaluation of microbicidal activity on nonporous surfaces. The standard testing fungal organism is *Aspergillus niger*, but other culture collections and strains are accepted as well, according to ISO 111331 (2009) standards (Da Silva et al., 2018; AFNOR NFT 72-281, 2020; OECD, 2020).

8.11 Future perspectives: Proteomic-based fungicides

Classic methods to control fungal diseases and spoilage are based far too much on the use of chemical products. These strategies produce serious adverse effects related to environmental pollution and the development of multidrug resistance. As a result, biosynthetic fungicides are established as a new area in fungicide development (Pinedo et al., 2008). Based on an in-depth study of fungal biology, the use of alien or modified natural compounds provided a potential species-specific method of controlling fungal pathogens by inhibition of the proteins involved in the infection cycle (Xu et al., 2007a,b). Such compounds are biodegradable, of high specificity, and are poorly integrated into the food chain.

Proteomic techniques can separate and characterize complex sets of proteins, and the majority of currently known drug targets are proteins. With proteomic techniques, it is possible to select drug targets of a specific compound or search for other therapeutic objectives, in pharmacy and the agrochemistry as well (Fernandez-Acero et al., 2007). Math algorithms may be another useful tool for drug discovery through proteomic analysis. The concept of "drug target-likeness" of a protein as an independent set of characteristics of well-operating targets is already established. It is possible to model and determine whether a derived protein sequence fits the activity of the drug target (Rudolph et al., 2003).

Bioinformatics try to identify the shared features of disease targets by the assumption that

- structurally similar chemicals (drugs) may require similar target structures, and
- a good therapeutic (effective) index can be achieved by regulating proteins that are specific to diseases, do not cause severe toxicity, are robust (less affected by individual genomic characteristics), and are effective (not in a complex homeostasis network).

Good targets therefore should display shared characteristics in terms of structure (sequence), function, network compatibility (connectivity), and single nucleotide polymorphism. These characteristics are not easily applicable in plant fungicide design, although math algorithms to predict the role of a protein as a fungicide are being improved (Fernandez Acero et al., 2011). The use of small molecules to selectively perturb gene function is referred to as "chemical genetics"; if applied on a genome-wide scale, it is referred to as "chemogenomics." The application of chemogenomics to protein targets is called "chemoproteomics" or TRAP (target related affinity profiling), that is, the use of biology to inform chemistry (Rudolph et al., 2003).

A very promising technology in targeted design is the use of peptide sequences able to modify protein activities. Those aptamers can disrupt fungal development and strengthen plant immunity by interacting with specific proteins obtained from proteome mining. Peptide aptamers are molecules that contain 8–20 amino acids defined by their ability to bind to specific proteins and inhibit or activate them. Functional peptide aptamers have properties similar to antibodies (Crawford et al., 2003). They can even be used as antibodies in several applications (e.g., nitrocellulose immunoblots). Peptide aptamer libraries are constructed from yeast hybrid libraries, yeast and bacterial expression libraries, and retroviral expression libraries in mammalian cells (Fernandez Acero et al., 2011).

Another novel chemical proteomic tool is activity-based protein profiling (ABPP) (Richau and Van Der Hoorn, 2010), which uses a small-molecule fluorescent probe reacting irreversibly with the site of the catalytic subunits in an activity-dependent manner. The activities can be quantified both in vitro and in vivo (Gu et al., 2010). This technology of suppressing protein activities, known as targeted chemical genetics, has the potential to clarify the role of specific enzymes in biological or pathogenic processes. The inhibition of a specific gene product or its protein-targeted mutants (PTMs) by knockout mutants results in a loss of fungal pathogenicity, if the selected protein is a pathogenicity factor.

RNA silencing has been studied in a few phytopathogenic fungi and may be another promising way of controlling plant diseases (Nakayashiki et al., 2006). This method solves the problem of the low efficiency of homologous recombination found in knockout strategies in microorganisms, as well as problems in multinucleate fungal cells. It allows gene inactivation in a specific stage or tissue, as the gene is not eliminated permanently. The technology is especially useful in unravelling fungal pathology processes (Nakayashiki, 2005).

8.12 Conclusion

Plant pathogenic fungi, which are among the most devastating causes of plant diseases from growth to distribution and storage, result in significant financial losses for producers, and the human public health risks due to food shortage, too. Another important group of fungi with negative economic, technological, and public health impacts are fungi that cause spoilage of food and animal feedstuff during production and distribution. The indiscriminate use of chemical biocides (agrochemicals, disinfectants) has a detrimental effect on the environment. It is necessary to develop new strategies to attack pathogenic fungi without causing irreversible damage to the environment. Genomics and proteomics hold promise to identify pathogenicity and virulence factors of fungi and their mechanisms of resistance in order to minimize their adverse effects. Nanotechnology is well known for playing a significant role in the effective management of phytopathogens, nutrient utilization, controlled release of pesticides, and fertilizers. Nano-fungicides are expected to play an important role in future plant disease management as eco-friendly alternatives for conventional synthetic fungicides Fungitoxicity of various types of nanoparticles dramatically increase when applied to spores rather than fungal hyphae, indicating the great potential of these compounds to be used as protective antifungal agents. Moreover, to commercialize nanotechnological products obtained via biogenic synthesis, it is necessary to establish protocols for standardizing the preparation of these biocontrol agents, as well as methods for scaling up production processes. There is tremendous potential for the development and commercialization of novel products for the biological control of pests and pathogens based on the fungi of genus *Trichoderma* spp., especially considering the use of these products in sustainable agriculture. At the present time, precisely controlling the effectivity and efficacy of antifungal biocides by using different normalized methods is an immediate and easy way to eliminate the overuse of fungicides in the agricultural, horticultural, and food processing industries.

Conflict of interest

This publication resulted from the project, "Centre of Excellence of Environmental Health," ITMS project No. 24240120033, financially supported by the EU Structural Fund on Regional Development, Research and Development operation program.

References

Abd-Elsalam, K.A., Al-Dhabaan, F.A., Alghuthaymi, M., Njobeh, P.B., Almoammar, H., 2019. Nanobiofungicides: present concept and future perspectives in fungal control. In: Nano-Biopesticides Today and Future Perspectives. Academic Press, USA, pp. 315–351.

Abou-El-Sherbini, K.S., Amer, M.H., Abdel-Aziz, M.S., Hamzawy, E.M., Sharmoukh, W., Elnagar, M.M., 2018. Encapsulation of biosynthesized nanosilver in silica composites for sustainable antimicrobial functionality. Global Chall. 2 (10), 1800048.

AFNOR NFT 72-281, 2020. Procédes de désinfection des surfaces par voie aérienne—Détermination de l´activité bactéricide, fongicide, levuricide, mycobactéricide, tuberculocide, sporicide et virucide incluant bacériophages. http://www.boutiqe.afnor.org. (Online June 1, 2020).

Agrios, G.N., 2005. Plant Pathology. Academic Press, San Diego, p. 952.

Akther, T., Hemalatha, S., 2019. Mycosilver nanoparticles: synthesis, characterization and its efficacy against plant pathogenic fungi. BioNanoScience 9 (2), 296–301.

Alghuthaymi, M.A., Almoammar, H., Rai, M., Said-Galiev, E., Abd-Elsalam, K.A., 2015. Myconanoparticles: synthesis and their role in phytopathogens management. Biotechnol. Biotechnol. Equip. 29, 221–236.

Alizadeh, H., Rahnema, M., Semnani, S.N., Ajalli, M., 2014. Synergistic antifungal effects of quince leaf's extracts and silver nanoparticles on *Aspergillus niger*. J. Appl. Biol. Sci. 8, 10–13.

Ansari, M.A., Khan, H.M., Khan, A.A., Cameotra, S.S., Alzohairy, M.A., 2015. Anti-biofilm efficacy of silver nanoparticles against MRSA and MRSE isolated from wounds in a tertiary care hospital. Indian J. Med. Microbiol. 33 (1), 101.

AOAC 955.17, 2020. Fungicidal Activity of Disinfectants. http://www.microchemlab.com. (Online June 1, 2020).

AOAC 961.02, 2020. Germicidal Spray Test Modified for Fungi. http://www.microchemlab.com. (Online June 1, 2020).

Arfat, Y.A., Ejaz, M., Jacob, H., Ahmed, J., 2017. Deciphering the potential of guar gum/Ag-Cu nanocomposite films as an active food packaging material. Carbohydr. Polym. 157, 65–71.

AshaRani, P.V., Low Kah Mun, G., Hande, M.P., Valiyaveettil, S., 2009. Cytotoxicity and genotoxicity of silver nanoparticles in human cells. ACS Nano 3 (2), 279–290.

Asli, S., Neumann, P.M., 2009. Colloidal suspensions of clay or titanium dioxide nanoparticles can inhibit leaf growth and transpiration via physical effects on root water transport. Plant Cell Environ. 32 (5), 577–584.

ASTM E2197, 2020. Standard Quantitative Disc Carrier Test Method for Determining Bactericidal, Virucidal, Fungicidal, Mycobactericidal and Sporocidal Activities of Chemicals. http://www.microchemlab.com. (Online June 1, 2020).

ASTM E2315, 2020. Standard Guide for Assessment of Antimicrobial Activity Using a Time-Kill Procedure. http://www.microchemlab.com. (Online June 1, 2020).

Azlin-Hasim, S., Cruz-Romero, M.C., Morris, M.A., Padmanabhan, S.C., Cummins, E., Kerry, J.P., 2016. The potential application of antimicrobial silver polyvinyl chloride nanocomposite films to extend the shelf-life of chicken breast fillets. Food Bioproc. Tech. 9 (10), 1661–1673.

Ben-Sasson, M., Zodrow, K.R., Genggeng, Q., Kang, Y., Giannelis, E.P., Elimelech, M., 2014. Surface functionalization of thin-film composite membranes with copper nanoparticles for antimicrobial surface properties. Environ. Sci. Technol. 48, 384–393. https://doi.org/10.1021/es404232s.

Berger, S., El Chazli, Y., Babu, A.F., Coste, A.T., 2017. Azole resistance in *Aspergillus fumigatus*: a consequence of antifungal use in agriculture? Front. Microbiol. 8, 1024.

Besinis, A., De Peralta, T., Handy, R.D., 2014. The antibacterial effects of silver, titanium dioxide and silica dioxide nanoparticles compared to the dental disinfectant chlorhexidine on *Streptococcus mutans* using a suite of bioassays. Nanotoxicology 8 (1), 1–16.

Bhatt, I., Tripathi, B.N., 2011. Interaction of engineered nanoparticles with various components of the environment and possible strategies for their risk assessment. Chemosphere 82 (3), 308–317.

Cardador, M.J., Gallego, M., Prados, F., Fernández-Salguero, J., 2017. Origin of disinfection by-products in cheese. Food Addit. Contamin: Part A 34 (6), 928–938.

Choi, G.J., Lee, S.W., Jang, K.S., Kim, J.S., Cho, K.Y., Kim, J.C., 2004. Effects of chrysophanol, parietin, and nepodin of *Rumex crispus* on barley and cucumber powdery mildews. Crop Prot. 23 (12), 1215–1221.

Collado, I.G., Sánchez, A.J.M., Hanson, J.R., 2007. Fungal terpene metabolites: biosynthetic relationships and the control of the phytopathogenic fungus *Botrytis cinerea*. Nat. Prod. Rep. 24 (4), 674–686.

Commission regulation (EU), 2011. Commission regulation (EU) No. 1129/2011 establishing an Union list of food additives. Off. J. Eur. Communities 295, 1–160.

Crawford, M., Woodman, R., Ferrigno, P.K., 2003. Peptide aptamers: tools for biology and drug discovery. Brief. Funct. Genomics 2 (1), 72–79.

Cushen, M., Kerry, J., Morris, M., Cruz-Romero, M., Cummins, E., 2012. Nanotechnologies in the food industry—recent developments, risks and regulation. Trends Food Sci. Technol. 24 (1), 30–46.

Da Silva, N., Taniwaki, M.H., Junqueira, V.C., Silveira, N., Okazaki, M.M., Gomes, R.A.R., 2018. Microbiological Examination Methods of Food and Water: A Laboratory Manual. CRC Press.

Deshmukh, S.P., Patil, S.M., Mullani, S.B., Delekar, S.D., 2019. Silver nanoparticles as an effective disinfectant: a review. Mater. Sci. Eng. C 97, 954–965.

Dibrov, P., Dzioba, J., Gosink, K.K., Häse, C.C., 2002. Chemiosmotic mechanism of antimicrobial activity of Ag^+ in *Vibrio cholerae*. Antimicrob. Agents Chemother. 46 (8), 2668–2670.

Duncan, T.V., 2011. Applications of nanotechnology in food packaging and food safety: barrier materials, antimicrobials and sensors. J. Colloid Interface Sci. 363 (1), 1–24.

Echegoyen, Y., Nerín, C., 2013. Nanoparticle release from nano-silver antimicrobial food containers. Food Chem. Toxicol. 62, 16–22.

Elamawi, R.M., Al-Harbi, R.E., Hendi, A.A., 2018. Biosynthesis and characterization of silver nanoparticles using *Trichoderma longibrachiatum* and their effect on phytopathogenic fungi. Egypt. J. Biol. Pest Control 28 (1), 28.

Elgazzar, N.S., Rabie, G.H., 2018. Application of silver nanoparticles on *Cephalosporium maydis in vitro* and *in vivo*. Egypt. J. Microbiol. 53 (1), 69–81.

Elgorban, A.M., El-Samawaty, A.E.R.M., Abd-Elkader, O.H., Yassin, M.A., Sayed, S.R., Khan, M., Adil, S.F., 2017. Bioengineered silver nanoparticles using *Curvularia pallescens* and its fungicidal activity against *Cladosporium fulvum*. Saudi J. Biol. Sci. 24 (7), 1522–1528.

EN 12353:2013, 2013. Chemical Disinfectants and Antiseptics—Prservation of Test Organisms Used for the Determination of Bactericidal (Including *Legionella*), Mycobactericidal, Sporocidal, Fungicidal and Virucidal (Including Bacteriphages) Acitivity.

EN 1275:2005, 2005. Chemical Disinfectants and Antiseptics—Quantitative Suspension Test for the Evaluation of Basic Fungicidal or Basic Yeasticidal Activity of Chemical Disinfectants and Antiseptics—Test Method and Requirements.

EN 13697:2015, 2015. Chemical Disinfectants and Antiseptics—Quantitative Non-Porous Surface Test for Evaluation of Bactericidal and/or Fungicidal Acitivity of Chemical Disinfectants Used in Food, Industrial, Domestic and Institutional Areas—Test Method and Requirements Without Mechanical Action.

EN 1650:2019, 2019. Chemical Disinfectants and Antiseptics. Quantitative Suspension Test for the Evaluation of Fungicidal or Yeasticidal Activity of Chemical Disinfectants and Antispetics Used in Food, Industrial, Domestic and Institutional Areas. Test Method and Requirements.

ES Committee, 2011. Guidance on the risk assessment of the application of nanoscience and nanotechnologies in the food and feed chain. EFSA J. 9 (2), 134.

European Food Safety Authority (EFSA), 2007. Opinion of the scientific panel on contaminants in the food chain [CONTAM] related to the potential increase of consumer health risk by a possible increase of the existing maximum levels for aflatoxins in almonds, hazelnuts and pistachios and derived products. EFSA J. 5 (3), 446.

Fernandez Acero, F.J., Carbú, M., El-Akhal, M.R., Garrido, C., González-Rodríguez, V.E., Cantoral, J.M., 2011. Development of proteomics-based fungicides: new strategies for environmentally friendly control of fungal plant diseases. Int. J. Mol. Sci. 12 (1), 795–816.

Fernandez, B., Vimont, A., Desfossés-Foucault, É., Daga, M., Arora, G., Fliss, I., 2017. Antifungal activity of lactic and propionic acid bacteria and their potential as protective culture in cottage cheese. Food Control 78, 350–356.

Fernandez-Acero, F.J., Carbú, M., Garrido, C., Vallejo, I., Cantoral, J.M., 2007. Proteomic advances in phytopathogenic fungi. Curr. Proteomics 4 (2), 79–88.

Fernández-Acero, F.J., Colby, T., Harzen, A., Wieneke, U., Garrido, C., Carbú, M., Vallejo, I., Schmidt, J., Cantoral, J.M., 2008. Analysis of *Botrytis cinerea* secretome. J. Plant Pathol. 90 (S2), 197.

Fernández-Acero, F.J., Colby, T., Harzen, A., Carbú, M., Wieneke, U., Cantoral, J.M., Schmidt, J., 2010. 2-DE proteomic approach to the *Botrytis cinerea* secretome induced with different carbon sources and plant-based elicitors. Proteomics 10 (12), 2270–2280.

Fraceto, L.F., Maruyama, C.R., Guilger, M., Mishra, S., Keswani, C., Singh, H.B., de Lima, R., 2018. *Trichoderma harzianum*-based novel formulations: potential applications for management of Next-Gen agricultural challenges. J. Chem. Technol. Biotechnol. 93 (8), 2056–2063.

Gadd, G.M., White, C., 1989. Heavy metal and radionuclide accumulation and toxicity in fungi and yeasts. In: Special Publications of the Society for General Microbiology.

Gajbhiye, M., Kesharwani, J., Ingle, A., Gade, A.K., 2009. Fungus-mediated synthesis of silver nanoparticles and their activity against pathogenic fungi in combination with fluconazole. Nanomed. Nanotechnol. Biol. Med. 5 (4), 382–386. https://doi.org/10.1016/j.nano.2009.06.005.

Gholami-Ahangaran, M., Zia-Jahromi, N., 2013. NPsilver effects on growth parameters in experimental aflatoxicosis in broiler chickens. Toxicol. Ind. Health 29 (2), 121–125.

Gliga, A.R., Skoglund, S., Wallinder, I.O., Fadeel, B., Karlsson, H.L., 2014. Size-dependent cytotoxicity of silver nanoparticles in human lung cells: the role of cellular uptake, agglomeration and Ag release. Part. Fibre Toxicol. 11 (1), 1–17.

Gu, C., Kolodziejek, I., Misas-Villamil, J., Shindo, T., Colby, T., Verdoes, M., Richau, K.H., Schmidt, J., Overkleeft, H.S., van der Hoorn, R.A., 2010. Proteasome activity profiling: a simple, robust and versatile method revealing subunit-selective inhibitors and cytoplasmic, defense-induced proteasome activities. Plant J. 62 (1), 160–170.

Guilger-Casagrande, M., de Lima, R., 2019. Synthesis of silver nanoparticles mediated by fungi: a review. Front. Bioeng. Biotechnol. 7, 1–18.

Guilger-Casagrande, M., Pasquoto-Stigliani, T., Bilesky-Jose, N., Grillo, R., Abhilash, P.C., Fraceto, L.F., De Lima, R., 2017. Biogenic silver nanoparticles based on *Trichoderma harzianum*: synthesis, characterization, toxicity evaluation and biological activity. Sci. Rep. 7, 44421.

Guilger-Casagrande, M., Germano-Costa, T., Pasquoto-Stigliani, T., Fraceto, L.F., de Lima, R., 2019. Biosynthesis of silver nanoparticles employing *Trichoderma harzianum* with enzymatic stimulation for the control of *Sclerotinia sclerotiorum*. Sci. Rep. 9 (1), 1–9.

Gupta, P.K., 2018. Toxicity of fungicides. In: Veterinary Toxicology. Academic Press, pp. 569–580.

Hamid, S., Zainab, S., Faryal, R., Ali, N., 2017. Deterrence in metabolic and biofilms forming activity of *Candida* species by mycogenic silver nanoparticles. J. Appl. Biomed. 15 (4), 249–255.

Herzog, F., Clift, M.J., Piccapietra, F., Behra, R., Schmid, O., Petri-Fink, A., Rothen-Rutishauser, B., 2013. Exposure of silver-nanoparticles and silver-ions to lung cells in vitro at the air-liquid interface. Part. Fibre Toxicol. 10 (1), 11.

Higa, L.H., Schilrreff, P., Perez, A.P., Morilla, M.J., Romero, E.L., 2013. The intervention of nanotechnology against epithelial fungal diseases. J. Biomater. Tissue Eng. 3, 70–88.

Hitokoto, H., Morozumi, S., Wauke, T., Sakai, S., Ueno, I., 1978. Inhibitory effects of spices on toxigenic fungi. J. Appl. Environ. Microbiol. 37, 692–695.

Hitokoto, H., Morozumi, S., Wauke, T., Sakai, S., Kurata, H., 1980. Inhibitory effects of spices on growth and toxin production of toxigenic fungi. Appl. Environ. Microbiol. 39 (4), 818–822.

Hu, M., Zhong, K., Liang, Y., Ehrman, S.H., Mi, B., 2017a. Effects of particle morphology on the antibiofouling performance of silver embedded polysulfone membranes and rate of silver leaching. Ind. Eng. Chem. Res. 56 (8), 2240–2246.

Hu, Y., Zhang, J., Kong, W., Zhao, G., Yang, M., 2017b. Mechanisms of antifungal and anti-aflatoxigenic properties of essential oil derived from turmeric (*Curcuma longa* L.) on *Aspergillus flavus*. Food Chem. 220, 1–8.

Huang, W., Wang, C., Duan, H., Bi, Y., Wu, D., Du, J., Yu, H., 2018. Synergistic antifungal effect of biosynthesized silver nanoparticles combined with fungicides. Int. J. Agric. Biol. 20, 1225–1229.

Huang, W., Yan, M., Duan, H., Bi, Y., Cheng, X., Yu, H., 2020. Synergistic antifungal activity of green synthesized silver nanoparticles and epoxiconazole against *Setosphaeria turcica*. J. Nanomater. 2020.

Ibrahim, E., Zhang, M., Zhang, Y., Hossain, A., Qiu, W., Chen, Y., Wang, Y., Wu, W., Sun, G., Li, B., 2020. Green-synthesization of silver nanoparticles using endophytic bacteria isolated from garlic and its antifungal activity against wheat *Fusarium* head blight pathogen *Fusarium graminearum*. Nanomaterials 10 (2), 219.

Idnurm, A., Howlett, B.J., 2001. Pathogenicity genes of phytopathogenic fungi. Mol. Plant Pathol. 2 (4), 241–255.

Jalilzadeh, A., Tunçtürk, Y., Hesari, J., 2015. Extension shelf life of cheese: a review. Int. J. Dairy Sci. 10 (2), 44–60.

Jamdagni, P., Rana, J.S., Khatri, P., 2018. Comparative study of antifungal effect of green and chemically synthesised silver nanoparticles in combination with carbendazim, mancozeb, and thiram. IET Nanobiotechnol. 12 (8), 1102–1107.

Joe, Y.H., Park, D.H., Hwang, J., 2016. Evaluation of Ag nanoparticle coated air filter against aerosolized virus: anti-viral efficiency with dust loading. J. Hazard. Mater. 301, 547–553.

Kathiravan, M.K., Salake, A.B., Chothe, A.S., Dudhe, P.B., Watode, R.P., Mukta, M.S., Gadhwe, S., 2012. The biology and chemistry of antifungal agents: a review. Bioorg. Med. Chem. 20 (19), 5678–5698.

Khan, N.T., Khan, M.J., Jameel, J., Jameel, N., Rheman, S.U.A., 2017. An overview: biological organisms that serves as nanofactories for metallic nanoparticles synthesis and fungi being the most appropriate. Bioceram. Dev. Appl. 7, 101.

Klaine, S.J., Alvarez, P.J., Batley, G.E., Fernandes, T.F., Handy, R.D., Lyon, D.Y., Mahendra, S., McLaughlin, M.J., Lead, J.R., 2008. Nanomaterials in the environment: behavior, fate, bioavailability, and effects. Environ. Toxicol. Chem. Int. J. 27 (9), 1825–1851.

Kontoyiannis, D.P., Lewis, R.E., 2002. Antifungal drug resistance of pathogenic fungi. Lancet 359 (9312), 1135–1144.

Kumar, C.G., Poornachandra, Y., 2015. Biodirected synthesis of miconazole-conjugated bacterial silver nanoparticles and their application as antifungal agents and drug delivery vehicles. Colloids Surf. B Biointerfaces 125, 110–119.

Kumari, M., Mukherjee, A., Chandrasekaran, N., 2009. Genotoxicity of silver nanoparticles on *Allium cepa*. Sci. Total Environ. 407, 5243–5246. https://doi.org/10.1016/j.scitotenv.2009.06.024.

Larue, C., Castillo-Michel, H., Sobanska, S., Cécillon, L., Bureau, S., Barthès, V., Ouerdane, L., Carrière, M., Sarret, G., 2014. Foliar exposure of the crop *Lactuca sativa* to silver nanoparticles: evidence for internalization and changes in Ag speciation. J. Hazard. Mater. 264, 98–106.

Le, V.T., Bach, L.G., Pham, T.T., Le, N.T.T., Ngoc, U.T.P., Tran, D.H.N., Nguyen, D.H., 2019. Synthesis and antifungal activity of chitosan-silver nanocomposite synergize fungicide against *Phytophthora capsici*. J. Macromol. Sci. A 56 (6), 522–528.

Lee, C.W., Mahendra, S., Zodrow, K., Li, D., Tsai, Y.C., Braam, J., Alvarez, P.J., 2010. Developmental phytotoxicity of metal oxide nanoparticles to *Arabidopsis thaliana*. Environ. Toxicol. Chem. Int. J. 29 (3), 669–675.

Li, X.F., Feng, X.Q., Yang, S., Wang, T.P., Su, Z.X., 2008. Effects of molecular weight and concentration of chitosan on antifungal activity against *Aspergillus niger*. Iran. Polym. J. 17, 843–852.

Malandrakis, A.A., Kavroulakis, N., Chrysikopoulos, C.V., 2020a. Synergy between Cu-NPs and fungicides against *Botrytis cinerea*. Sci. Total Environ. 703, 135557.

Malandrakis, A.A., Kavroulakis, N., Chrysikopoulos, C.V., 2020b. Use of silver nanoparticles to counter fungicide-resistance in *Monilinia fructicola*. Sci. Total Environ. 747, 141287.

Marambio-Jones, C., Hoek, E.M., 2010. A review of the antibacterial effects of silver nanomaterials and potential implications for human health and the environment. J. Nanopart. Res. 12 (5), 1531–1551.

McGrath, S.P., Zhao, F.J., 2003. Phytoextraction of metals and metalloids from contaminated soils. Curr. Opin. Biotechnol. 14 (3), 277–282.

Mihaly Cozmuta, A., Peter, A., Mihaly Cozmuta, L., Nicula, C., Crisan, L., Baia, L., Turila, A., 2015. Active packaging system based on Ag/TiO_2 nanocomposite used for extending the shelf life of bread. Chemical and microbiological investigations. Packag. Technol. Sci. 28 (4), 271–284.

Nakayashiki, H., 2005. RNA silencing in fungi: mechanisms and applications. FEBS Lett. 579 (26), 5950–5957.

Nakayashiki, H., Kadotani, N., Mayama, S., 2006. Evolution and diversification of RNA silencing proteins in fungi. J. Mol. Evol. 63 (1), 127–135.

Ngoc, U.T.P., Nguyen, D.H., 2018. Synergistic antifungal effect of fungicide and chitosan-silver nanoparticles on *Neoscytalidium dimidiatum*. Green Processes Synth. 7 (2), 132–138.

Nowack, B., Krug, H.F., Height, M., 2011. 120 Years of Nanosilver History: Implications for Policymakers.

OECD, 2020. Environmental Health and Safety Publications. Guidance Document on Quantitative Methods for Evaluating the Activity of Microbicides Used on Hard Non-Porous Surfaces. Series on Testing and Assessment. No. 187 Series on Biocides. No. 6.2013. http://www.oecd.org. (Online June 1, 2020).

OSCPP 810.2000, 2020. Antimicrobial Efficacy Test Guidelines. General Considerations for Testing Public Health Antimicrobial Pesticides, Guide for Efficacy Testing. http://epa.gov. (Online June 1, 2020).

OSCPP 810.2100, 2020. EPA Produect Performace Test Guideline. Sterilants – Efficacy Data Recommendations. http://nepis.epa.gov. 2020. (Online June 1, 2020).

OSCPP 810.2200, 2020. Disinfectants for Use on Environmental Surfaces, Guide for Efficacy Testing. http://epa.gov. (Online June 1, 2020).

Othman, A.M., Elsayed, M.A., Al-Balakocy, N.G., Hassan, M.M., Elshafei, A.M., 2019. Biosynthesis and characterization of silver nanoparticles induced by fungal proteins and its application in different biological activities. J. Genet. Eng. Biotechnol. 17 (1), 8.

Pandey, S., Shahid, M., Srivastava, M., Singh, A., Kumar, V., Trivedi, S., Srivastava, Y.K., 2015. Biosynthesis of silver nanoparticles by *Trichoderma harzianum* and its effect on the germination, growth and yield of tomato plant. J. Pure Appl. Microbiol. 9 (4), 3335–3342.

Parzymies, M., Pudelska, K., Poniewozik, M., 2019. The use of nano-silver for disinfection of *Pennisetum alopecuroides* plant material for tissue culture. Acta Sci. Pol. 18 (3), 127–135.

Peter, A., Mihaly-Cozmuta, L., Mihaly-Cozmuta, A., Nicula, C., Indrea, E., Barbu-Tudoran, L., 2014. Testing the preservation activity of $AgTiO_2$-Fe and TiO_2 composites included in the polyethylene during orange juice storage. J. Food Process Eng. 37 (6), 596–608.

Pham, D.C., Nguyen, T.H., Ngoc, U.T.P., Le, N.T.T., Tran, T.V., Nguyen, D.H., 2018. Preparation, characterization and antifungal properties of chitosan-silver nanoparticles synergize fungicide against *Pyricularia oryzae*. J. Nanosci. Nanotechnol. 18 (8), 5299–5305.

Pinedo, C., Wang, C.M., Pradier, J.M., Dalmais, B., Choquer, M., Le Pêcheur, P., Morgant, G., Collado, I.G., Cane, D.E., Viaud, M., 2008. Sesquiterpene synthase from the botrydial biosynthetic gene cluster of the phytopathogen *Botrytis cinerea*. ACS Chem. Biol. 3 (12), 791–801.

Pulit-Prociak, J., Banach, M., 2016. Silver nanoparticles—a material of the future...? Open Chem. 14 (1), 76–91.

Ramos, K., Gómez-Gómez, M.M., Cámara, C., Ramos, L., 2016. Silver speciation and characterization of nanoparticles released from plastic food containers by single particle ICPMS. Talanta 151, 83–90.

Regulation (EC), 2005. Regulation (EC) No. 396/2005 on maximum residue levels of pesticides in or on food and feed of plant and animal origin. Off. J. Eur. Communities 70, 1–16.

Regulation (EC), 2008. Regulation (EC) No. 1333/2008 on food additives. Off. J. Eur. Communities 34, 16–33.

Richau, K.H., van der Hoorn, R.A.L., 2010. Studies on plant-pathogen interactions using activity-based proteomics. Curr. Proteomics 7 (4), 328–336.

Ritota, M., Manzi, P., 2020. Natural preservatives from plant in cheese making. Animals 10 (4), 749.

Rogawansamy, S., Gaskin, S., Taylor, M., Pisaniello, D., 2015. An evaluation of antifungal agents for the treatment of fungal contamination in indoor air environments. Int. J. Environ. Res. Public Health 12 (6), 6319–6332.

Rudolph, C., Schreier, P.H., Uhrig, J.F., 2003. Peptide-mediated broad-spectrum plant resistance to tospoviruses. Proc. Natl. Acad. Sci. U. S. A. 100 (8), 4429–4434.

Sarmast, M., Salehi, H., Khosh-Khui, M., 2011. NPsilver treatment is effective in reducing bacterial contaminations of *Araucaria excelsa* R. Br. var. *glauca* explants. Acta Biol. Hung. 62 (4), 477–484.

Sensidoni, A., Rondinini, G., Peressini, D., Maifreni, M., Bortolomeazzi, R., 1994. Presence of an off-flavour associated with the use of sorbates in cheese and margarine. Ital. J. Food Sci. 6, 237–242.

Singh, M., Sahareen, T., 2017. Investigation of cellulosic packets impregnated with silver nanoparticles for enhancing shelf-life of vegetables. Food Sci. Technol. 86, 116–122.

Singh, M., Kumar, M., Kalaivani, R., Manikandan, S., Kumaraguru, A.K., 2013. Metallic silver nanoparticle, a therapeutic agent in combination with antifungal drug against human fungal pathogen. Bioprocess Biosyst. Eng. 36, 407–415.

Soliman, A.M., Abdel-Latif, W., Shehata, I.H., Fouda, A., Abdo, A.M., Ahmed, Y.M., 2020. Green approach to overcome the resistance pattern of *Candida* spp. using biosynthesized silver nanoparticles fabricated by *Penicillium chrysogenum* F9. Biol. Trace Elem. Res., 1–12.

STN EN 1650:2020, 2020. Chemical Disinfectants and Antiseptics—Quantitative Suspension Test for the Evaluation of Fungicidal or Yeasticidal Activity of Chemical Disinfectants and Antiseptics Used in Food, Industrial, Domestic and Institutional Areas—Test Method and Requirements (Phase 2, Step 1).

Tian, J., Ban, X., Zeng, H., He, J., Huang, B., Wang, Y., 2011. Chemical composition and antifungal activity of essential oil from *Cicuta virosa* L. var. latisecta Celak. Int. J. Food Microbiol. 145 (2–3), 464–470.

Tomah, A.A., Alamer, I.S.A., Li, B., Zhang, J.Z., 2020. Mycosynthesis of silver nanoparticles using screened *Trichoderma* isolates and their antifungal activity against *Sclerotinia sclerotiorum*. Nanomaterials 10 (10), 1955.

Tutaj, K., Szlazak, R., Szalapata, K., Starzyk, J., Luchowski, R., Grudzinski, W., OsinskaJaroszuk, M., Jarosz-Wilkolazka, A., Szuster-Ciesielska, A., Gruszecki, W.I., 2016. Amphotericin B-silver hybrid nanoparticles, synthesis, properties and antifungal growth. Nanomedicine 124, 1095–1103.

Vasile, C., Râpă, M., Ştefan, M., Stan, M., Macavei, S., Darie-Niţă, R.N., Barbu-Tudoran, L., Vodnar, D.C., Popa, E.E., Ştefan, R., Borodi, G., 2017. New PLA/ZnO:Cu/Ag bionanocomposites for food packaging. Express Polym. Lett. 11 (7).

Ventola, C.L., 2015. The antibiotic resistance crisis: part 1: causes and threats. Pharmacol. Ther. 40 (4), 277.

Villamizar-Gallardo, R., Cruz, J.F.O., Ortíz-Rodriguez, O.O., 2016. Fungicidal effect of silver nanoparticles on toxigenic fungi in cocoa. Pesq. Agrop. Brasileira 51 (12), 1929–1936.

Warnock, D.W., Arthington-Skaggs, B.A., Li, R.K., 1999. Antifungal drug susceptibility testing and resistance in Aspergillus. Drug Resist. Updat. 2 (5), 326–334.

Woodall, C., 2020. Testing strategies and international standards for disinfectants. In: Walker, J. (Ed.), Decontamination in Hospitals and Healthcare. Woodhead Publ, Sawston, pp. 371–376.

Xu, H., Fang, Y., Yao, L., Chen, Y., Chen, X., 2007a. Does drug-target have a likeness? Methods Inf. Med. 46 (03), 360–366.

Xu, H., Xu, H., Lin, M., Wang, W., Li, Z., Huang, J., Chen, Y., Chen, X., 2007b. Learning the drug target-likeness of a protein. Proteomics 7 (23), 4255–4263.

Yoon, K.Y., Byeon, J.H., Park, J.H., Ji, J.H., Bae, G.N., Hwang, J., 2008. Antimicrobial characteristics of silver aerosol nanoparticles against *Bacillus subtilis* bioaerosols. Environ. Eng. Sci. 25 (2), 289–294.

Yoon, M.Y., Cha, B., Kim, J.C., 2013. Recent trends in studies on botanical fungicides in agriculture. Plant Pathol. J. 29 (1), 1.

Zhu, H., Han, J., Xiao, J.Q., Jin, Y., 2008. Uptake, translocation, and accumulation of manufactured iron oxide nanoparticles by pumpkin plants. J. Monitor. 10 (6), 713–717.

Further reading

Beroza, P., Villar, H.O., Wick, M.M., Martin, G.R., 2002. Chemoproteomics as a basis for post-genomic drug discovery. Drug Discov. Today 7 (15), 807–814.

Liu, Y., Rosenfield, E., Hu, M., Mi, B., 2013. Direct observation of bacterial deposition on and detachment from nanocomposite membranes embedded with silver nanoparticles. Water Res. 47 (9), 2949–2958.

PART 3

Plant science

CHAPTER 9

Silver nanoparticle applications in wood, wood-based panels, and textiles

Mohamed Z.M. Salem

Forestry and Wood Technology Department, Faculty of Agriculture (El-Shatby), Alexandria University, Alexandria, Egypt

9.1 Introduction

Numerous wood preservatives (e.g., pesticides with toxic chemicals, waterborne preservatives such as chromated copper arsenate [CCA], and wood biopesticides) have evolved to provide long-term protection for wood and wood products against conditions such as moisture and fungal and insect attacks, among others (Grace, 1998; Salem et al., 2019; Ashmawy et al., 2020; El-Hefny et al., 2020). From an environmental perspective, woods treated with highly toxic chemicals pose public health risks; for example, CCA is a carcinogen as well as a skin irritant (Vasselinovitch et al., 1979).

In recent years, the scientific community has focused more on finding methods for treating wood, wood-based panels, and textiles that are safer for humans and other animals. Nanoparticles (NPs) have received much attention because of their unique properties such as their small size and various functionalities (Li et al., 2017; Abdelsalam et al., 2018). Silver nanoparticles (AgNPs), which are inorganic materials (mainly as silver hydrosols), have numerous uses in various fields and can produce an antimicrobial product when applied to certain substances (Durán et al., 2005; Guzman et al., 2012; Lakshmanan and Chakraborty, 2017; Pakdel et al., 2017). Silver ions have been used as antimicrobial agents for wood coatings (Liu et al., 2008; Gao, 2013). AgNPs and other nanomaterials have shown antibacterial activity against *Pseudomonas aeruginosa* (Polo et al., 2011), as well as the ability to form a thin water film on wood surfaces to prevent biological growth (Polo et al., 2011; Gladis et al., 2010; Clausen et al., 2011). AgNPs synthesized with the leaf extract of *Brassica rapa* showed potential antifungal activity against degradation of wood by *Gloeophyllum abietinum*, *G. trabeum*, *Phanerochaete sordida*, and *Chaetomium globosum*) (Narayanan and Park, 2014).

Several methods can be used to synthesize AgNPs, such as chemical reduction of silver ions, bio or green synthesis to induce thermal decomposition, and reduction of solutions (Ibrahim et al., 2017; Abdelsalam et al., 2018). It was shown that substrates can be imparted through electrical conductivity with AgNPs (Chou et al., 2005; Tien et al., 2011; Gao et al., 2015). On the other hand, the water absorption properties of wood, medium-density fiberboard (MDF), particleboard (PB), and oriented strand board (OSB) improve with the impregnation of AgNPs (Mantanis and Papadopoulos, 2010a,b; Soltani et al., 2013) by reducing the pore size and space available in the cell wall, which is used for the absorption of water molecules, where a rough hydrophobic surface is created (Papadopoulos et al., 2019). The application of AgNPs has enhanced the protection of wood (Liu et al., 2002a,b,c); also AgNPs have improved the physical and mechanical properties of wood, such as resistance against fire and densification of wood, including PB and kiln-dried wood (Rassam et al., 2012; Taghiyari, 2012; Tarmian et al., 2012).

Several spectroscopic analyses have been used to identify and characterize synthesized AgNPs, such as TEM, SEM, XRD, and EDX (Gao et al., 2015; Abdelsalam et al., 2018). For example, EDX spectroscopy showed the presence of oxygen, silver, gold, and carbon on an AgNP-treated wood surface (Gao et al., 2015). Additionally, the amount of AgNPs that adsorbed on different cellulose nanofibrils were characterized through the use of XRD, Fourier-transform infrared spectroscopy (FITR), and X-ray photoelectron spectroscopy; and the results showed that the adsorption increased in this order: lignocellulose nanofibers (20% lignin) > pure cellulose nanofibers > lignocellulose nanofibers (31% lignin) > holocellulose nanofibers (Kwon et al., 2020). Therefore this chapter aims to summarize AgNPs' ability to improve the properties of wood, wood products, and textiles when used as a coating or to permeate materials.

9.2 Importance of silver nanoparticles in practice

AgNPs are used to improve the properties of wood and wood products through biological resistance against stained or wood deteriorates fungi. AgNPs, also, improve the physical and chemical properties of wood, as well as in the manufacture of composite panels and modified woods.

9.2.1 Improving the biological durability properties of wood and wood products with AgNPs

The durability of certain tropical woods (e.g., *Cedrela odorata*, *Enterolobium cyclocarpum*, *Gmelina arborea*, *Tectona grandis*, and *Vochysia ferruginea*) coated with AgNPs against *Trametes versicolor* and *Lenzites acuta* fungi increased, compared with uncoated tropical wood (Moya et al., 2017). Also, *Acacia mangium*, *Cedrela odorata*, and *Vochysia guatemalensis* woods impregnated with AgNPs (50 ppm) under pressure showed a mass loss of less than 5% in wood inoculated with *T. versicolor*, compared with 20% found in the control treatment (Moya et al., 2014).

Growth of the stained fungi *Ophiostoma flexuosum, O. polonicum, O. tetropii,* and *O. ips* was inhibited when AgNPs (100 ppm) were incorporated into the growth media (Velmurugan et al., 2009). The addition of different agents, such as fungicides, to a solution of AgNPs led to decreased wood decay (Liu et al., 2001, 2002a,b,c). The cellulose enzyme activity of the fungi *Postia placenta, Tephrophana palustris,* and *Gloeophyllum trabeum* was inhibited by the application of silver ions in an aqueous solution of AgNPs (Highley, 1975; Dorau et al., 2004).

Poplar wood treated with AgNPs with a low concentration (400 ppm) was not able to inhibit *T. versicolor* (Rezaei et al., 2011), while *Paulownia fortune* wood impregnated with AgNPs (400 ppm) showed resistance to decay caused by *T. versicolor* (Akhtari and Arefkhani, 2013). Suspension of AgNP-treated PB increased resistance against *T. versicolor* (Taghiyari et al., 2014). Ionic Ag^+-based biocides with a 1% Ag solution did not show good activity against brown-rot fungi when colonized with southern yellow pine wood (Dorau et al., 2004).

Due to their slow oxidation reaction and small particle size (<100 nm), Ags have a high surface area available for oxidation. For example, AgNPs, unlike organic biocides, are nonvolatile, nondegradable over time, odorless, and exhibit long-term efficacy (Clausen, 2007). AgNPs impregnated into wood at 400 ppm upon heat treatment showed potential resistance to *T. versicolor* (Moradi et al., 2013; Reza et al., 2013). AgNP-impregnated wood samples showed moderate activity against *T. versicolor* with wood from *Fagus sylvatica* and *Poria placenta* with *Pinus sylvestris* wood (Pařil et al., 2017).

Research has shown that the biological effect of AgNPs is more than Ag^+ as a result of the formation of reactive oxygen species (ROS), which damages pathogen membranes (Ivask et al., 2013). An AgNP aqueous solution (400 ppm) showed effectiveness against *T. versicolor* with a weight loss of 2.1% in the treated wood. Furthermore, *Populus* spp. wood treated with AgNPs (5–20 ppm) showed good antifungal activity against *T. versicolor* with the weight loss ranging from 8.49% to 8.94%, compared with 24.79% in the untreated wood. A slight degradation in the cell wall was found in the vessels and fibers of wood treated with AgNPs, compared with untreated wood exposed to *T. versicolor* for 16 weeks (Casado-Sanz et al., 2019).

Another study showed that, even with AgNP-impregnated *P. sylvestris* sapwood and *F. sylvatica* with high resistance to leaching, the antifungal activity against *Coriolus versicolor* and *Coniophora puteana* was not promising (Bak and Németh, 2018). Also, building materials such as wooden flooring treated with AgNPs showed only moderate activity against *Aspergillus brasiliensis* and *Penicillium funiculosum* (Huang et al., 2015), while the combination of AgNPs with CuNP- or ZnO NP-treated wood showed strong insecticidal activity against termites (Green and Arango, 2007).

Wood specimens of teak and beech coated with AgNPs (50–150 μg/mL) have exhibited good durability against the growth of the wood-decay fungi *Ganoderma*, with an inhibition of fungal mycelia of 90.9%, which explains the use of thin films of AgNPs as protective coating agents against fungi (Nair et al., 2013).

P. sylvestris wood impregnated with a soybean oil polymer that was autoxidized and contained AgNPs (Agsbox) was characterized by FTIR. Agsbox increased decay resistance against *Coniophora puteana*, with low efficacy against *Hylotrupes bajulus* larvae, a wood-destroying insect (Can et al., 2019).

Wood products such as PB and MDF treated with coatings of AgNP/acrylic showed efficiency against *Escherichia coli* and *Staphylococcus aureus* (Iždinský et al., 2018), compared with acrylic coatings (without NPs) that did not protect the PB and MDF top surfaces against bacterial attacks. Transparent wood coatings with AgNPs (<50 ppm) did not protect a wooden façade against microorganisms when used for outdoor applications (Künniger et al., 2014a); also, their protective effects against blue stain, mold, and algae were insufficient when exposed to outdoor weathering conditions for 1 year (Künniger et al., 2014b).

Commercial acrylic latexes mixed with different concentrations of AgNPs and applied to *P. resinosa* sapwood inhibited the growth of *Sclerophoma pityophila* and *Eppicoccum nigrum* at low concentration and *Aureobasidium pullulans* at high concentration (Boivin et al., 2019). Also, the incorporation of AgNPs into the latexes did not affect the coatings' optical properties, such as color or gloss. At a high concentration, AgNPs did not show much inhibition to *A. pullulans,* while at a low concentration, inhibition occurred, suggesting that AgNPs are well dispersed at high concentrations.

9.2.2 Effect of AgNPs on physical and mechanical characteristics of impregnated wood and wood products

Wood surface coatings or impregnation treatments with nanoparticles, including AgNPs, have a great effect on a wood's properties (Papadopoulos and Taghiyari, 2019; Taghiyari et al., 2020), as shown in Figs. 9.1 and 9.2.

A superhydrophobic and electrically conductive coating with a high ability to repel motor oil was obtained by coating *P. ussuriensis* wood with AgNPs modified by fluoroalkyl silane, resulting in a coating that can be used for self-cleaning and electronic biomedical devices (Gao et al., 2015).

In one study, wood treated with AgNPs along with (heptadecafluoro-1,1,2,2-tetradecyl)trimethoxysilane ($CF_3(CF_2)7CH_2CH_2Si(OCH_3)$) exhibited a high ability to repel liquids. Subsequently, droplets of water and oil appeared as spheres. The superhydrophobic coating on wood surfaces with a contact angle (CA) greater than 150° has the potential to act as a barrier and protect wood from oil pollution (Gao et al., 2015).

Improvement in the sorption behavior of AgNP-treated wood has been shown, due to a reduction in the total sorption of water vapor (Mantanis and Papadopoulos, 2010a,b; Sahin and Mantanis, 2011). Also, wood durability was enhanced against a series of tropical white decay fungi (Moya et al., 2014).

Gas and liquid permeability in PB was significantly decreased when produced at industrial scale using AgNPs (Taghiyari, 2011). For example, an AgNP suspension added to the mat (at 100 and 150 mL/kg levels) of oven-dried wood particles

9.2 Importance of silver nanoparticles in practice

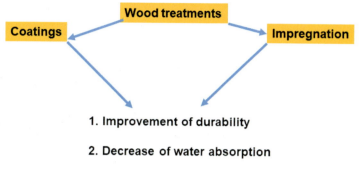

FIG. 9.1

Classification of wood surface impregnation treatments.

Data from Papadopoulos, A.N., Taghiyari, H.R., 2019. Innovative wood surface treatments based on nanotechnology: a review. Coatings 9(12), 866, with permission of MDPI. This article is an open access article distributed under the terms and conditions of the Creative Commons Attribution (CC BY) license (http://creativecommons.org/licenses/by/4.0/).

significantly decreased the permeability of all nano-treated composite panels, compared with the control treatment. These decreases resulted from the high thermal conductivity coefficient of AgNPs, which caused better heat transfer to the mat. It was recommended that a 100 mL suspension be used to prevent resin depolymerization on the surface of the composite mat. Moreover, the hot-press time of the PBs decreased by 10.9% when an AgNP suspension (100 mL/kg wood particle) was used.

Conversely, AgNP-treated wood composite panels with accelerated heat transfer can influence the permeability of wood. Heat treatments were reported to significantly affect fluctuations in the permeability of different woods (Taghiyari, 2013). It was shown that during the manufacturing of boards, the water absorption and amount of swelling in PBs significantly decreased with the addition of AgNPs (Taghiyari et al., 2011; Taghiyari and Farajpour Bibalan, 2013). Overall, mineral nanomaterials have demonstrated the ability to improve heat transference in wood and wood-based products such as PB and MDF, thereby increasing their physical and mechanical properties (Sadeghi and Rastgo, 2012; Haghighi Poshtiri et al., 2013; Taghiyari et al., 2013).

Another study reported that fire-retardant-treated plywood combined with high temperature can degrade the strength of a product (Winandy et al., 2002). When impregnated with AgNPs, *F. orientalis*, *P. nigra*, *Platanus orientalis*, *Alnus* spp., and

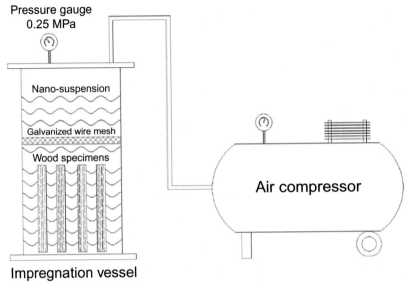

FIG. 9.2

Laboratory scale pressure vessel for impregnating specimens cut to size for different tests.

Data from Taghiyari, H.R., Bayani, S., Militz, H., Papadopoulos, A.N., 2020. Heat treatment of pine wood: possible effect of impregnation with silver nanosuspension: a review. Forests 11(4), 466, with permission of MDPI. This article is an open access article distributed under the terms and conditions of the Creative Commons Attribution (CC BY) license (http://creativecommons.org/licenses/by/4.0/).

A. alba showed an improvement in some fire-retarding properties, such as glowing time and heat-transfer properties (Taghiyari, 2012).

Heat treatment of AgNP-impregnated wood specimens shows more potential to significantly impact the mechanical properties of woods than their physical properties (Reza et al., 2013). An aqueous suspension of AgNPs (400 ppm) added in different amounts (100, 150, and 200 mL/kg based on the rate of dry wood fibers) for MDF-matrix production showed an increase in hardness under a 6 min hot-press time. However, another study showed that increasing mat hot-press time and the subsequent overheating decreased the hardness of specimens (Taghiyari and Norton, 2014). It has also been shown that AgNP-impregnated wood improved almost equally the thermal conductivity of both the outer part and the inner part of the wood (Taghiyari et al., 2012; Taghiyari and Samandarpour, 2015; Bayani et al., 2019).

AgNP-treated spruce wood showed a spring-back effect in a 0.04% suspension when compressed at 150°C for 4 h (Rassam et al., 2012). Additionally, significant increases in the modulus of rupture (MOR), the modulus of elasticity (MOE), and the impact resistance were found with percentages of 53%, 41.2%, and 175.7%, respectively, compared with the control treatments (wood compressed at 150°C for 4 h) (Rassam et al., 2012). On the other hand, heat-impregnated beech wood specimens resulted in a decrease in the mechanical properties of the wood (Bayani et al., 2019).

Another study reported no negative effects on MOR, MOE, and compression strength of the grain properties due to impregnation of AgNPs (Nosal and Reinprecht, 2018).

Prepared coatings of waterborne polyurethane incorporated with nanocellulose crystalline and AgNPs and used to treat larch wood specimens exhibited different antibacterial activities against *E. coli* in terms of inhibition zones around the treated wood. As shown in Fig. 9.3, the activity increased gradually as the Ag increased from 5% to 10% (Cheng et al., 2020).

The use of liquid wood granules (pellets) coated with a thin film of AgNPs obtained by injection molding as bioengineering or biomedical applications is under discussion (Motas et al., 2020). Ag and AgNP compounds were reported to change the color of treated wood, where the color of Ags was dark brown (Hazer and Kalaycı, 2017; Pařil et al., 2017; Lakshmanan et al., 2018); also, the original wood color was noticeably changed with impregnation of Agsbox, with positive values being reddish and negative values being greenish and bluish (Sivrikaya and Can, 2016; Can et al., 2019). Table 9.1 shows several advantages of using nanomaterials as wood preservatives.

9.2.3 AgNP/wood as a water purification filter

Che et al. (2019) showed that the advanced properties of AgNP/wood composites make them suitable materials in water purification filters, where more than 98.5% of methylene blue was removed through catalytic degradation and physical adsorption. The study also showed that after passing through an AgNP/wood filter, the inactivation and removal of *E. coli* and *S. aureus* extended by 6 and 5.2 orders of magnitude, respectively.

Recently, cellulose nanofibers produced from bleached birch wood pulp prepared by the kraft process were used to fabricate aerogel membranes whose surface was further decorated with AgNPs. The prepared membranes showed excellent discoloration of wastewater due to cationic and anionic dyes, which suggested potential industrial applications (Zhang et al., 2020).

Basswood blocks immersed in a silver nitrate solution were used to create an AgNP/basswood membrane for filtration and sterilization of river water. The membrane was shown to kill *E. coli*, *S. aureus*, *Bacillus subtilis*, *P. aeruginosa*, and *Candida albicans* (Chen et al., 2019).

9.2.4 AgNPs as additives for fibers

In the production process, AgNPs can be incorporated into textiles for individual fibers or through chemical modification of the final textile product (Filipowska et al., 2011). Cotton fabrics coated with AgNPs have been obtained by aqueous KOH and $AgNO_3$ treatment, which is suitable for subsequent surface hydrophobization (Shateri Khalil-Abad and Yazdanshenas, 2010; Xue et al., 2012). AgNP-coated cotton fabric functionalized from seaweed extract showed significant antibacterial activity against *S. aureus* and *E. coli* (Rajaboopathi and Thambidurai, 2018), compared with uncoated cotton.

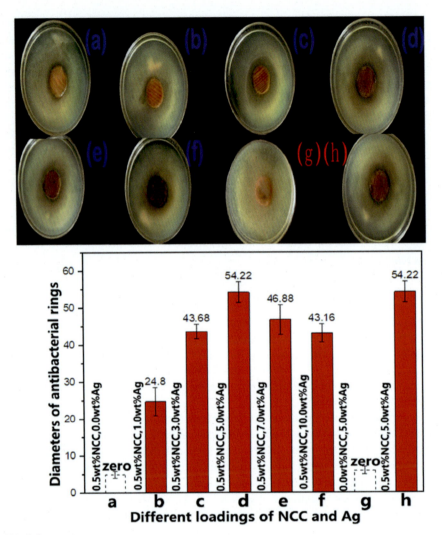

FIG. 9.3

Antimicrobial experiment of wood loaded with crystalline from different nanocellulose sources (NCCs) and silver (Ag).

Data from Cheng, L., Ren, S., Lu, X., 2020. Application of eco-friendly waterborne polyurethane composite coating incorporated with nano cellulose crystalline and silver nano particles on wood antibacterial board. Polymers 12(2), 407, with permission of MDPI. This article is an open access article distributed under the terms and conditions of the Creative Commons Attribution (CC BY) license (http://creativecommons.org/licenses/by/4.0/).

Table 9.1 The advantages of nanomaterials in wood preservation versus conventional treatments.

Property	Advantage	Disadvantage
Hydrophobicity	Nanocomposite coatings create rough hydrophobic surfaces without affecting the softness and abrasion resistance of wood. The impregnation of nanomaterials reduces the pore size and space available within the cell wall, which is used for the absorption of water molecules.	Although nanotechnology is widely integrated into wood treatments, criticisms and discussions about their potential health and environmental risks are increasing. Even if incomplete knowledge or a lack of information about nanomaterials delays a study, addressing the potential impact of nano-based treatments using tools such as a life-cycle assessment will help avoid mistakes made in the past when new technological innovations were released prior to an impact assessment.
UV protection	Surface modifications with nano-sized inorganic fillers have found many applications as they are nontoxic and stable under exposure to UV radiation. Their large surface-to-volume ratio makes them effective for improving the photoresistance property of wood.	
Fire performance	Wood is composed of cellulose, hemicellulose, and lignin, which can enduce thermal degradation if the wood contacts a source of ignition. Although the fire performance characteristics of wood are improved with the use different fire retardants, use of these chemicals has resulted in low efficiency, leaching, and high environmental and health risks. The use of nanoparticles alone or in combination with other fire-retardant chemicals can reduce the ignitability of wood and limit the leaching of fire-retardant chemicals.	
Antimicrobial	Nanoparticles increase the decay resistance of wood by reducing the wood's uptake of moisture either by preventing the absorption of moisture or by blocking the flow path of liquid water.	
Mechanical properties	The mechanical properties of wood depend on environmental agents, its isotropic properties, and sometimes on the type of treatment used to improve the wood's characteristics. Impregnating wood with nanoparticles enhances its hardness by filling the cavities of the wood.	

Data from Papadopoulos, A.N., Bikiaris, D.N., Mitropoulos, A.C., Kyzas, G.Z., 2019. Nanomaterials and chemical modification technologies for enhanced wood properties: a review. Nanomaterials 9, 607, with permission from MDPI. This article is an open access article distributed under the terms and conditions of the Creative Commons Attribution (CC BY) license (http://creativecommons.org/licenses/by/4.0/).

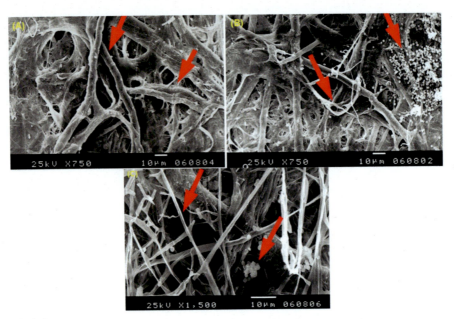

FIG. 9.4

SEM examination of flax paper sheets produced with or without AgNP additives and inoculated with *A. flavus*.

Reproduced from Abo Elgat, W.A.A., Taha, A.S., Böhm, M., Vejmelková, E., Mohamed, W.S., Fares, Y.G., Salem, M.Z.M., 2020. Evaluation of the mechanical, physical, and anti-fungal properties of flax laboratory paper sheets with the nanoparticles treatment. Materials 13(2), 363, with permission of MDPI. This article is an open-access article distributed under the terms and conditions of the Creative Commons Attribution (CC BY) license (http://creativecommons.org/licenses/by/4.0/).

Fibers of viscose rayon decorated with NPs of Ag/Ag_2CO_3 and Ag/Ag_3PO_4 showed excellent photocatalytic activity. Additionally, protection against pathogens and UV radiation was achieved, with colors ranging from pale yellow (Ag/Ag_2CO_3/viscose fibers) to pale purple/violet (Ag/Ag_3PO_4/viscose fibers) (Rehan et al., 2019).

Engineered AgNPs used for fabrics are being utilized to kill bacteria that cause bad odors, especially in socks and sports clothing, as well as for use in biomedical applications due to their antibacterial and antiviral properties (Bhatacharyya and Joshi, 2011; Boroumand et al., 2015).

The combination of chitosan + AgNP (Fig. 9.4A), chitosan + Ag (Fig. 9.4B), and Paraloid B-72 + AgNP (Fig. 9.4C) added at 1%, 3%, and 3% to flax pulp, respectively, exhibited good inhibition against *A. flavus*; while pulp treated with Paraloid B-72 + Ag NP (1%) showed some activity against *Stemphylium solani* (Abo Elgat et al., 2020). The combined treatment of chitosan + AgNP (3%) showed the highest percentage of brightness (57.50%) in paper sheets produced from flax (Abo Elgat et al., 2020). The AgNPs' mode of action causes the release of Ag ions that penetrate the cell membranes of bacteria and thus retard DNA reproduction and apoptosis of the bacterial unit (Xue et al., 2012; Kang et al., 2016), which leads to the destruction of the bacterial species (El-Shishtawy et al., 2011).

An AgNP polystyrene-*block*-polyacrylic copolymer solution coated onto textile fabrics showed the ability to enhance the permanency of antibacterial activity up to five cycles of washing against *E. coli* and up to 20 washings against *S. aureus* (Budama et al., 2013). AgNPs synthesized by fungi and incorporated onto cotton fabrics showed promising activity against *S. aureus* (Durán et al., 2007). AgNPs deposited by the ultrasound irradiation method onto the surface of certain textiles, including nylon, cotton, and polyester, exhibited good bioactivity against *E. coli* and *S. aureus* (Perelshtein et al., 2008). Other reports have shown that AgNPs in the form of colloidal solutions on cellulose-based and synthetic fabrics have good antimicrobial activity even after many cycles of laundering (Lee et al., 2003).

9.3 Conclusions

Solid wood and wood-based composites, such as particleboards, oriented strand boards, and medium-density fiberboards, are susceptible to attacks by different microorganisms and insects. Synthesized AgNPs have shown to increase the protection of wood and wood products against rot fungi (*T. versicolor*) and to resist against mold and mildew. Also, studies indicate that the addition of AgNPs improves the physical properties of wood, such as heat transfer and fire retardation. Finally, AgNP-impregnated wood and wood products are considered as the next-generation, eco-friendly bioactive agents to protect wood and wood products.

References

Abdelsalam, N.R., Abdel-Megeed, A., Ali, H.M., Salem, M.Z.M., Al-Hayali, M.F., Elshikh, M.S., 2018. Genotoxicity effects of silver nanoparticles on wheat (*Triticum aestivum* L.) root tip cells. Ecotoxicol. Environ. Saf. 155, 76–85.

Abo Elgat, W.A.A., Taha, A.S., Böhm, M., Vejmelková, E., Mohamed, W.S., Fares, Y.G., Salem, M.Z.M., 2020. Evaluation of the mechanical, physical, and anti-fungal properties of flax laboratory paper sheets with the nanoparticles treatment. Materials 13 (2), 363.

Akhtari, M., Arefkhani, M., 2013. Study of microscopy properties of wood impregnated with nanoparticles during exposed to white-rot fungus. Agric. Sci. Dev. 2, 116–119.

Ashmawy, N.A., Salem, M.Z.M., El Shanhorey, N., Al-Huqail, A.A., Ali, H.M., Behiry, S.I., 2020. Eco-friendly wood-biofungicidal and antibacterial activities of various *Coccoloba uvifera* L. leaf extracts: HPLC analysis of phenolic and flavonoid compounds. BioResources 15 (2), 4165–4187.

Bak, M., Németh, R., 2018. Effect of different nanoparticle treatments on the decay resistance of wood. BioResources 13 (4), 7886–7899.

Bayani, S., Taghiyari, H.R., Papadopoulos, A.N., 2019. Physical and mechanical properties of thermally-modified beech wood impregnated with silver nano-suspension and their relationship with the crystallinity of cellulose. Polymers 11 (10), 1538.

Bhatacharyya, M., Joshi, A., 2011. Nanotechnology—a new route to high-performance functional textiles. Text. Prog. 43 (3), 155–233.

Boivin, G., Ritcey, A.M., Landry, V., 2019. The effect of silver nanoparticles on the black-stain resistance of acrylic resin for translucent wood coating application. BioResources 14 (3), 6353–6369.

Boroumand, M.A., Namvar, F., Moniri, M., Tahir, P., Azizi, S., Mohamad, R., 2015. Nanoparticles biosynthesized by fungi and yeast: a review of their preparation, properties, and medical applications. Molecules 20, 16540–16565.

Budama, L., Çakır, B.A., Topel, Ö., Hoda, N., 2013. A new strategy for producing antibacterial textile surfaces using silver nanoparticles. Chem. Eng. J. 228, 489–495.

Can, A., Palanti, S., Sivrikaya, H., Hazer, B., Stefanı, F., 2019. Physical, biological and chemical characterisation of wood treated with silver nanoparticles. Cellul. 26 (8), 5075–5084.

Casado-Sanz, M.M., Silva-Castro, I., Ponce-Herrero, L., Martín-Ramos, P., Martín-Gil, J., Acuña-Rello, L., 2019. White-rot fungi control on *Populus* spp. Wood by pressure treatments with silver nanoparticles, chitosan oligomers and propolis. Forests 10 (10), 885.

Che, W., Xiao, Z., Wang, Z., Li, J., Wang, H., Wang, Y., Xie, Y., 2019. Wood-based mesoporous filter decorated with silver nanoparticles for water purification. ACS Sustain. Chem. Eng. 7 (5), 5134–5141.

Chen, Q., Fei, P., Hu, Y., 2019. Hierarchical mesopore wood filter membranes decorated with silver nanoparticles for straight-forward water purification. Cellul. 26 (13–14), 8037–8046.

Cheng, L., Ren, S., Lu, X., 2020. Application of eco-friendly waterborne polyurethane composite coating incorporated with nano cellulose crystalline and silver nano particles on wood antibacterial board. Polymers 12 (2), 407.

Chou, K.-S., Huang, K.-C., Lee, H.-H., 2005. Fabrication and sintering effect on the morphologies and conductivity of nano-Ag particle films by the spin coating method. Nanotechnology 16, 779.

Clausen, C.A., 2007. Nanotechnology: Implications for the Wood Preservation Industry (Document No. IRG/WP 07-3041). International Research Group on Wood Protection, Stockholm, Sweden. 5 pp.

Clausen, C.A., Kartal, S.N., Arango, R.A., Green, F., 2011. The role of particle size of particulate nano-zinc oxide wood preservatives on termite mortality and leach resistance. Nanoscale Res. Lett. 6, 427.

Dorau, B., Arango, R., Green, F.I.I.I., 2004. An investigation into the potential of ionic silver as a wood preservative. In: Woodframe Housing Durability and Disaster Issues, 4–6 October. Forest Products Society, Las Vegas, NV, pp. 133–145.

Durán, N., Marcato, P.D., Alves, O.L., De Souza, G.I., Esposito, E., 2005. Mechanistic aspects of biosynthesis of silver nanoparticles by several *Fusarium oxysporum* strains. J. Nanobiotechnol. 3 (1), 8.

Durán, N., Marcato, P., De Souza, G.I.H., Alves, O.L., Esposito, E., 2007. Antibacterial effect of silver nanoparticles produced by fungal process on textile fabrics and their effluent treatment. J. Biomed. Nanotechnol. 3, 203–208.

El-Hefny, M., Salem, M.Z.M., Behiry, S.I., Ali, H.M., 2020. The potential antibacterial and antifungal activities of wood treated with *Withania somnifera* fruit extract, and the phenolic, caffeine, and flavonoid composition of the extract according to HPLC. Processes 8 (1), 113. https://doi.org/10.3390/pr8010113.

El-Shishtawy, R.M., Asiri, A.M., Abdelwahed, N.A., Al-Otaibi, M.M., 2011. In situ production of silver nanoparticle on cotton fabric and its antimicrobial evaluation. Cell 18 (1), 75–82.

Filipowska, B., Rybicki, E., Walawska, A., Matyjas-Zgondek, E., 2011. New method for the antibacterial and antifungal modification of silver finished textiles. Fibres Text. East. Eur. 19 (4), 124–128.

Gao, F., 2013. Preparation Method of Novel Nano Modified Antibiosis Aqueous Woodenware Paint. (International Patent Application No. CN103073964A).

Gao, L., Lu, Y., Li, J., Sun, Q., 2015. Superhydrophobic conductive wood with oil repellency obtained by coating with silver nanoparticles modified by fluoroalkyl silane. Holzforschung 70 (1), 63–68.

Gladis, F., Eggert, A., Karsten, U., Schumann, R., 2010. Prevention of biofilm growth on man-made surfaces: evaluation of antialgal activity of two biocides and photocatalytic nanoparticles. Biofouling 26, 89–101.

Grace, J.K., 1998. Resistance of pine treated with chromated copper arsenate to the Formosan subterranean termite. For. Prod. J. 48 (3), 79–82.

Green, F., Arango, R.A., 2007. Wood Protection by Commercial Silver Formulations Against Eastern Subterranean Termites (IRG/WP 07-30422). International Research Group on Wood Protection, Stockholm, Sweden.

Guzman, M., Dille, J., Godet, S., 2012. Synthesis and antibacterial activity of silver nanoparticles against gram-positive and gram-negative bacteria. Nanomed. Nanotechnol. Biol. Med. 8 (1), 37–45.

Haghighi Poshtiri, A., Taghiyari, H.R., Karimi, A.N., 2013. The optimum level of nano-wollastonite consumption as fire-retardant in poplar wood (*Populus nigra*). Int. J. Nano Dimens. 4 (2), 141–151.

Hazer, B., Kalaycı, Ö.A., 2017. High fluorescence emission silver nano particles coated with poly (styrene-g-soybean oil) graft copolymers: antibacterial activity and polymerization kinetics. Mater. Sci. Eng. C 74, 259–269.

Highley, T.L., 1975. Inhibition of Celluloses of Wood-Decay Fungi (Res Pap FPL-247). USDA For Serv Forest Prod Lab, Madison, WI. 9 pp.

Huang, H.-L., Lin, C.-C., Hsu, K., 2015. Comparison of resistance improvement to fungal growth on green and conventional building materials by nano-metal impregnation. Build. Environ. 93 (Part 2), 119–127.

Ibrahim, N.A., Eid, B.M., Abdel-Aziz, M.S., 2017. Effect of plasma superficial treatments on antibacterial functionalization and coloration of cellulosic fabrics. Appl. Surf. Sci. 392, 1126–1133.

Ivask, A., ElBadawy, A., Kaweeteerawat, C., Boren, D., Fischer, H., Ji, Z., Chang, C.H., Liu, R., Tolaymat, T., Telesca, D., et al., 2013. Toxicity mechanisms in *Escherichia coli* vary for silver nanoparticles and differ from ionic silver. ACS Nano 8, 374–386.

Iždinský, J., Reinprecht, L., Nosál, E., 2018. Antibacterial efficiency of silver and zinc-oxide nanoparticles in acrylate coating for surface treatment of wooden composites. Wood Res. 63 (3), 365–372.

Kang, C.K., Kim, S.S., Kim, S., Lee, J., Lee, J.H., Roh, C., Lee, J., 2016. Antibacterial cotton fibers treated with silver nanoparticles and quaternary ammonium salts. Carbohydr. Polym. 151, 1012–1018.

Künniger, T., Heeb, M., Arnold, M., 2014a. Antimicrobial efficacy of silver nanoparticles in transparent wood coatings. Eur. J. Wood Wood Prod. 72 (2), 285–288.

Künniger, T., Gerecke, A.C., Ulrich, A., Huch, A., Vonbank, R., Heeb, M., Wichser, A., Haag, R., Kunz, P., Faller, M., 2014b. Release and environmental impact of silver nanoparticles and conventional organic biocides from coated wooden façades. Environ. Pollut. 184, 464–471.

Kwon, G.J., Han, S.Y., Park, C.W., Park, J.S., Lee, E.A., Kim, N.H., Alle, M., Bandi, R., Lee, S.H., 2020. Adsorption characteristics of Ag nanoparticles on cellulose nanofibrils with different chemical compositions. Polymers 12 (1), 164.

Lakshmanan, A., Chakraborty, S., 2017. Coating of silver nanoparticles on jute fibre by in situ synthesis. Cell 24 (3), 1563–1577.

Lakshmanan, G., Sathiyaseelan, A., Kalaichelvan, P.T., Murugesan, K., 2018. Plant-mediated synthesis of silver nanoparticles using fruit extract of *Cleome viscosa* L.: assessment of their antibacterial and anticancer activity. Karbala Int. J. Mod. Sci. 4 (1), 61–68.

Lee, H.J., Yeo, S.Y., Jeong, S.H., 2003. Antibacterial effect of nanosized silver colloidal solution on textile fabrics. J. Mater. Sci. 38, 2199–2204.

Li, Z., Meng, J., Wang, W., Wang, Z., Li, M., Chen, T., Liu, C.J., 2017. The room temperature electron reduction for the preparation of silver nanoparticles on cotton with high antimicrobial activity. Carbohydr. Polym. 161, 270–276.

Liu, Y., Laks, P., Heiden, P., 2001. Use of nanoparticles for controlled release of biocides in solid wood. J. Appl. Polym. Sci. 79 (3), 458–465.

Liu, Y., Laks, P., Heiden, P., 2002a. Controlled release of biocides in solid wood: I. Efficacy against brown rot wood decay fungus (*Gloeophyllum trabeum*). J. Appl. Polym. Sci. 86 (3), 596–607.

Liu, Y., Laks, P., Heiden, P., 2002b. Controlled release of biocides in solid wood: II. Efficacy against *Trametes versicolor* and *Gloeophyllum trabeum* wood decay fungi. J. Appl. Polym. Sci. 86 (3), 608–614.

Liu, Y., Laks, P., Heiden, P., 2002c. Controlled release of biocides in solid wood: III. Preparation and characterization of surfactant-free nanoparticles. J. Appl. Polym. Sci. 86 (3), 615–621.

Liu, J., Jin, K., Huang, J., 2008. Nanometre Ag Antibacterial Aqueous Wood-Ware Paint and Its Preparation Method. International Patent Application No. CN100387666C, 14 May 2008.

Mantanis, G., Papadopoulos, A.N., 2010a. Reducing the thickness swelling of wood based panels by applying a nanotechnology compound. Eur. J. Wood Wood Prod. 68, 237–239.

Mantanis, G., Papadopoulos, A.N., 2010b. The sorption of water vapour of wood treated with a nanotechnology compound. Wood Sci. Technol. 44, 515–522.

Moradi, B.M., Ghorban, M.K., Taghiyari, H.R., Mirshokraie, S.A., 2013. Effects of silver nanoparticles and fungal degradation on density and chemical composition of heattreated poplar wood (*Populus euroamerica*). Eur. J. Wood Wood Prod. 71 (4), 491–495.

Motas, J.G., Quadrini, F., Nedelcu, D., 2020. Silver nano-coating of liquid wood for nanocomposite manufacturing. Procedia Manuf. 47, 974–979.

Moya, R., Berrocal, A., Rodriguez-Zuniga, A., Vega-Baudrit, J., Noguera, S.C., 2014. Effect of silver nanoparticles on white-rot wood decay and some physical properties of three tropical wood species. Wood Fiber Sci. 46 (4), 527–538.

Moya, R., Zuniga, A., Berrocal, A., Vega, J., 2017. Effect of silver nanoparticles synthesized with NPsAg-ethylene glycol on brown decay and white decay fungi of nine tropical woods. J. Nanosci. Nanotechnol. 17, 5233–5240.

Nair, M., Shankar, S.M., Rajesh, E.M., Anulakshmi, A., Anulakshmi, A., 2013. Synthesis of silver nanoparticles by biological method and its application in protection of wood from Ganoderma species. J. NanoSci. NanoEng. Appl. 3 (1), 37–43.

Narayanan, K.B., Park, H.H., 2014. Antifungal activity of silver nanoparticles synthesized using turnip leaf extract (*Brassica rapa* L.) against wood rotting pathogens. Eur. J. Plant Pathol. 140 (2), 185–192.

Nosal, E., Reinprecht, L., 2018. Preparation and application of silver and zinc oxide nanoparticles in wood industry: the review. Acta Fac. Xylolog. Zvolen Publ. Slov. 60 (2), 5–23.

Pakdel, E., Daoud, W.A., Afrin, T., Sun, L., Wang, X., 2017. Enhanced antimicrobial coating on cotton and its impact on UV protection and physical characteristics. Cell 24 (9), 4003–4015.

Papadopoulos, A.N., Taghiyari, H.R., 2019. Innovative wood surface treatments based on nanotechnology: a review. Coatings 9 (12), 866.

Papadopoulos, A.N., Bikiaris, D.N., Mitropoulos, A.C., Kyzas, G.Z., 2019. Nanomaterials and chemical modification technologies for enhanced wood properties: a review. Nanomaterials 2019 (9), 607.

Pařil, P., Baar, J., Čermák, P., Rademacher, P., Prucek, R., Sivera, M., Panáček, A., 2017. Antifungal effects of copper and silver nanoparticles against white and brown-rot fungi. J. Mater. Sci. 52 (5), 2720–2729.

Perelshtein, I., Applerot, G., Perkas, N., Guibert, G., Mikhailov, S., Gedanken, A., 2008. Sonochemical coating of silver nanoparticles on textile fabrics (nylone, polyester and cotton) and their antibacterial activity. Nanotechnology 19, 1–6.

Polo, A., Diamanti, M.V., Bjarnsholt, T., Hoiby, N., Villa, F., Pedeferri, M.P., Cappitelli, F., 2011. Effects of photo activated titanium dioxide nano powders and coating on plank-tonic and biofilm growth of *Pseudomonas aeruginosa*. Photochem. Photobiol. 87, 1387–1394.

Rajaboopathi, S., Thambidurai, S., 2018. Evaluation of UPF and antibacterial activity of cotton fabric coated with colloidal seaweed extract functionalized silver nanoparticles. J. Photochem. Photobiol. B Biol. 183, 75–87.

Rassam, G., Ghofrani, M., Taghiyari, H.R., Jamnani, B., Khajeh, M.A., 2012. Mechanical performance and dimensional stability of nano-silver impregnated densified spruce wood. Eur. J. Wood Wood Prod. 70 (5), 595–600.

Rehan, M., Barhoum, A., Khattab, T.A., Gätjen, L., Wilken, R., 2019. Colored, photocatalytic, antimicrobial and UV-protected viscose fibers decorated with Ag/Ag_2CO_3 and Ag/Ag_3PO_4 nanoparticles. Cell 26 (9), 5437–5453.

Reza, H.T., Enayati, A., Gholamiyan, H., 2013. Effects of nanosilver impregnation on brittleness, physical and mechanical properties of heat-treated hardwoods. Wood Sci. Technol. 47 (3), 467–480.

Rezaei, V.T., Usefi, A., Soltani, M., 2011. Wood protection by nano silver against white rot. In: Proceedings of 5th Symposium on Advanced in science and technology, Mashhad.

Sadeghi, B., Rastgo, S., 2012. Study of the shape controlling silver nanoplates by reduction process. Int. J. Bio-Inorg. Hybrid Nanomater. 1 (1), 33–36.

Sahin, H.T., Mantanis, G.I., 2011. Nano-based surface treatment effects on swelling, water sorption and hardness of wood. Maderas-Cienc. Tecnol. 13, 41–48.

Salem, M.Z.M., Hamed, S.A.E.K.M., Mansour, M.M.A., 2019. Assessment of efficacy and effectiveness of some extracted bio-chemicals as bio-fungicides on wood. Drv. Ind. 70 (4), 337–350.

Shateri Khalil-Abad, M., Yazdanshenas, M.E., 2010. Superhydrophobic antibacterial cotton textiles. J. Colloid Interface Sci. 351, 293–298.

Sivrikaya, H., Can, A., 2016. Effect of weathering on wood treated with tall oil combined with some additives. Maderas-Cienc. Tecnol. 18, 723–732.

Soltani, M., Najafi, A., Bakar, E., 2013. Water repellent effect and dimensional stability of beech wood impregnated with nano-zinc oxide. BioResources 8, 6280–6287.

Taghiyari, H.R., 2011. Effects of nano-silver on gas and liquid permeability of particleboard. Dig. J. Nanomater. BioRes. 6 (4), 1517–1525.

Taghiyari, H.R., 2012. Fire-retarding properties of nanosilver in solid woods. Wood Sci. Technol. 46 (5), 939–952.

Taghiyari, H.R., 2013. Effects of heat-treatment on permeability of untreated and nanosilver-impregnated native hardwoods. Maderas-Cienc. Tecnol. 15 (2), 183–194.

Taghiyari, H.R., Farajpour Bibalan, O., 2013. Effect of copper nanoparticles on permeability, physical, and mechanical properties of particleboard. Eur. J. Wood Prod. 71 (1), 69–77.

Taghiyari, H.R., Norton, J., 2014. Effect of silver nanoparticles on hardness in medium-density fiberboard (MDF). iForest 8, 677–680.

Taghiyari, H.R., Samandarpour, A., 2015. Effects of nanosilver-impregnation and heat treatment on coating pull-off adhesion strength on solid wood. Drv. Ind. 66, 321–327.

Taghiyari, H.R., Rangavar, H., Farajpour Bibalan, O., 2011. Nano-silver in particleboard. BioResources 6 (4), 4067–4075.

Taghiyari, H.R., Enayati, A., Gholamiyan, H., 2012. Effects of nano-silver impregnation on brittleness, physical and mechanical properties of heat-treated hardwoods. Wood Sci. Technol. 47, 467–480.

Taghiyari, H.R., Moradiyan, A., Farazi, A., 2013. Effect of nanosilver on the rate of heat transfer to the core of the medium density fiberboard mat. Int. J Bio-Inorg. Hybrid Nanomater. 2 (1), 303–308.

Taghiyari, H.R., Schmidt, O., Bari, E., Tahir, M.D.P., Karimi, A., Nouri, P., Jahangiri, A., 2014. Effect of Silver Nano-Particles on the Rate of Heat Transfer to the Core of the Medium-Density Fiberboard Mat. (IRG/WP 14-40653).

Taghiyari, H.R., Bayani, S., Militz, H., Papadopoulos, A.N., 2020. Heat treatment of pine wood: possible effect of impregnation with silver nanosuspension: a review. Forests 11 (4), 466.

Tarmian, A., Sepehr, A., Gholamiyan, H., 2012. The use of nano-silver particles to determine the role of the reverse temperature gradient in moisture flow in wood during low-intensity convective drying. Spec. Top. Rev. Porous Media 3 (2), 149–156.

Tien, H.-W., Huang, Y.-L., Yang, S.-Y., Wang, J.-Y., Ma, C.-C.M., 2011. The production of graphene nanosheets decorated with silver nanoparticles for use in transparent, conductive films. Carbon 49, 1550–1560.

Vasselinovitch, S., et al., 1979. Neoplastic Response of Mouse Tissues during Perinatal Age Periods and Its Significance in Chemical Carcinogensis. Perinatal Carcinogenesis. National Cancer Institute Monograph, vol. 51. AnalyticalChemistry/fsAC001f.htm.

Velmurugan, N., Kumar, G.G., Han, S.S., Nahm, K.S., Lee, Y.S., 2009. Synthesis and characterization of potential fungicidal silver nano-sized particles and chitosan membrane containing silver particles. Iran. Polym. J. 18 (5), 383–392.

Winandy, J.E., Lebow, P.K., Murphy, J.F., 2002. Predicting current serviceability and residual service life of plywood roof sheathing using kinetics-based models. In: The 9th Durability of Building Materials and Components Conference, Brisbane, Australia, p. 7.

Xue, C.-H., Chen, J., Yin, W., Jia, S.T., Ma, J.Z., 2012. Superhydrophobic conductive textiles with antibacterial property by coating fibers with silver nanoparticles. Appl. Surf. Sci. 258, 2468–2472.

Zhang, W., Wang, X., Zhang, Y., Seppälä, J., Xu, W., Willför, S., Xu, C., 2020. Robust shape-memory nanocellulose-based aerogels decorated with silver nanoparticles for fast continuous catalytic discoloration of organic dyes. Sep. Purif. Technol. 242, 116523.

CHAPTER 10

Immobilization efficiency and modulating abilities of silver nanoparticles on biochemical and nutritional parameters in plants: Possible mechanisms

Luqmon Azeez[a], Amadu K. Salau[b], and Simiat M. Ogunbode[b]

[a]*Department of Pure and Applied Chemistry, Osun State University, Osogbo, Nigeria*
[b]*Biochemistry and Nutrition Unit, Department of Chemical Sciences, Fountain University, Osogbo, Nigeria*

10.1 Introduction

Sustainable agriculture, as the focal goal of every development, is intertwined with soil health that adequately provides nutrients to plants when needed and as required. Sustainable agriculture is the new world order now as the world is faced with possible population explosion alongside overwhelming demands for functional foods (Belal and El-Ramady, 2016; Chhipa and Joshi, 2016; Rizwan et al., 2019; Zulfiqar et al., 2019; Zhao et al., 2020). This is more compounded in the developing countries as there is a significant drop in available arable lands to cope with their exponential population growth coupled with snowballing environmental pollution arising from various hazardous practices such as water pollution (Mukherjee et al., 2016; White and Gardea-Torresdey, 2018; Azeez et al., 2018, 2020b). Conventional agricultural practices involving soil tillage, fertilizer applications, pesticide usage, and burning of crop debris have led to soil degradation in addition to pollution of other environmental matrices due to run-off of unused agrochemicals resulting in eutrophication (Farooq and Siddique, 2015; Shalaby et al., 2016; Chen et al., 2018; Das et al., 2018a; Chatterjee et al., 2020). Although the use of chemical fertilizers and pesticides has been effective in improving crop yield and controlling pathogens, problems such as ecological disturbance, low efficiency, the formation of precipitates, and nonremediating potentials warrant the search for alternatives (Prasad et al., 2017; Elmer and White, 2018; Roberto et al., 2020). Therefore, there is a need for a paradigm shift from conventional to sustainable agriculture where the entirety of crop improvement, pest/weed management, and remediation of pollutants can be achieved with less or no burden on the environment. This necessitates the call for an inclusion of nanoparticles (NPs) as nanofertilizers, nanopesticides, and soil conditioners due to their exceptional and unequalled morphological, chemical, and biological characteristics (Elmer and White, 2018; Ma et al., 2018; Adisa et al., 2019; Elemike et al., 2019; Kopittke et al., 2019; Sanzari et al., 2019; Zulfiqar et al., 2019; Azeez et al., 2020b).

Nanoparticles are particles of ≤100 nm in size having unique properties controlled by shape, size, concentration, surface area, and surface chemistry. There are metal, metal-oxide, carbon-based, and cellulose nanomaterials that have found usefulness in various spheres of human lives such as for cancer diagnosis, as biosensors in agriculture, for food safety, in pharmaceutical, electronic, and textile industries (Alam et al., 2015; Servin et al., 2015; Lateef et al., 2015a,b, 2016a,b,c,d; Chhipa and Joshi, 2016; El-Argawy et al., 2017; Uddin et al., 2017; Buledi et al., 2020; Elemike et al., 2020). Many studies have reported their distinctive antioxidant, antibacterial, antifungal, antiviral, antithrombolytic, and anticancer properties that aid their functions as growth modulators, hormone mimics, antioxidant regulators, terminators of pathogenic organisms, and inhibitors of abiotic stresses in plants (Kumar et al., 2014; Lateef et al., 2015a,b, 2016a,b,c,d; Ammar and El-Desouky, 2016; Azeez et al., 2017; Iqbal et al., 2017; Yang et al., 2017; Ejaz et al., 2018; Nakamura et al., 2019; Rajwade et al., 2020). Moreover, they contain vastly reactive sites and possess catalytic properties required to achieve complete remediation and immobilization of pollutants in the soil for optimum growth of plants (Fazal et al., 2016; Azeez et al., 2018, 2019a,b, 2020a,b; Singh et al., 2018; Tewari, 2019).

In comparison with conventional agrochemicals and other immobilizing agents, NPs are more superior because they can boost plant nutrition and improve tolerance indices as against conventional agrochemicals that could hardly guarantee pollutant-free soil (Cai et al., 2017; Das et al., 2018b; Ma et al., 2018; Zheng et al., 2020). Additionally, NPs have intrinsic abilities to function as immobilizing agents for the removal of soil pollutants in comparison to other immobilizing agents that can barely improve biochemical and nutritional attributes of plants (Jasim et al., 2017; Gorczyca et al., 2018; Das et al., 2019; Nile et al., 2020). Combination of these purposes confers a greater superiority on NPs. Furthermore, target-specificity, sustainable stability, controllable release, and non-tissue-damaging are foremost characteristics for the consideration of NPs in agriculture (Khan et al., 2017; Adisa et al., 2019).

Nanoparticles of essential minerals are good sources of vital nutrients needed for plant quality, while NPs from nonessential minerals could also serve as vehicles for the supply of both micro- and macronutrients (Tripathi et al., 2017; Achari and Kowshik, 2018). Nanoparticles are promoters of phytohormones and suppressors of phytopathogens. Nanoparticles can regulate these hormones via induction of protein-coding gene expression to develop physiological indicators through various reactions with root exudates prompting their absorption (Raliya et al., 2015; Azeez et al., 2017; Siddiqi and Husen, 2017; Sheykhbaglou et al., 2018; Khan et al., 2019a,c). One of the incontrovertible facts about NPs is their ability to subdue diseases and control insects and other pests. Many studies have reported fungicidal, antimicrobial, nematocidal, pesticidal, insecticidal, and antiviral properties of NPs, especially AgNPs against resistant, pathogenic, and nonpathogenic microorganisms infesting plants (Dhand et al., 2016; Pacheco and Buzea, 2017, 2018; Goncalves et al., 2019; Elemike et al., 2019, 2020; Roberto et al., 2020).

Contrarily, some studies have reported negative influences of NPs on plants ranging from phytotoxicity, reduction in nutrient accumulation, and health concerns when consumed by human beings (Iannone et al., 2016; Chung et al., 2019; Kumara et al., 2019; Lalau et al., 2020). It is pertinent to note that stimulatory and phytotoxic potentials of NPs depend on the type/method of synthesis, synthetic routes, mode of application, concentration of NPs, plant part used, experimental design/conditions, duration of exposure to NPs, plant developmental stages, and plant species (Raliya et al., 2015; Cox et al., 2016; Tripathi et al., 2017; Yang et al., 2017; Ahmed et al., 2018; Ejaz et al., 2018; Feregrino-Perez et al., 2018; Nair, 2018; Chung et al., 2019; Yan and Chen, 2019).

Delivery of NPs to plants takes different routes whether apoplastically (extracellular routes) or symplastically (intracellular routes). Foliar and root applications are frequently used but recently, petiole and trunk injection for delivery of nanoparticles have been reported too. For foliar application, NPs are transported from cuticles through stomata to xylem while for soil application, NPs travel through rhizodermis, root hairs to phloem (Siddiqi and Husen, 2017; Achari and Kowshik, 2018; Ahmed et al., 2018; Hu et al., 2018; Verma et al., 2018; Ahmadov et al., 2020; Su et al., 2019; Usman et al., 2020; Zhao et al., 2020).

Silver nanoparticles (AgNPs) have been mostly studied of all NPs and have enjoyed inclusion into numerous medical, pharmaceutical, food packaging, agricultural, and electronic products, as well as for remediation purposes largely due to their physicochemical, morphological, antimicrobial, antioxidant, antifungal, antiviral, and anticancer properties. This is not without reported toxicity from its application, but green-synthesized AgNPs are less toxic, ecofriendlier, and less hazardous (Ammar and El-Desouky, 2016; Park and Ahn, 2016; Azeez et al., 2017, 2018, 2019a,b; Gupta et al., 2018; Singh et al., 2018; Barkataki and Singh, 2019).

This chapter overviews the immobilization efficiency and stimulatory potentials of AgNPs on biochemical and nutritional characteristics in plants with an outline of the possible mechanisms involved.

10.2 Silver nanoparticles in agriculture

Silver nanoparticles (AgNPs) are zero-valent silver particles that have enjoyed good patronage in agriculture due to their unparalleled properties, functioning as nanofertilizers, nanopesticides, growth modulators, and immobilizing agents (Fig. 10.1) (Yang et al., 2017; Zuverza-Mena et al., 2017; Achari and Kowshik, 2018). They are equally used for preservation of food, disinfecting water, purifying wastewater, wound-healing, cancer diagnosis, and as antimicrobials in consumer products (Dhand et al., 2016;

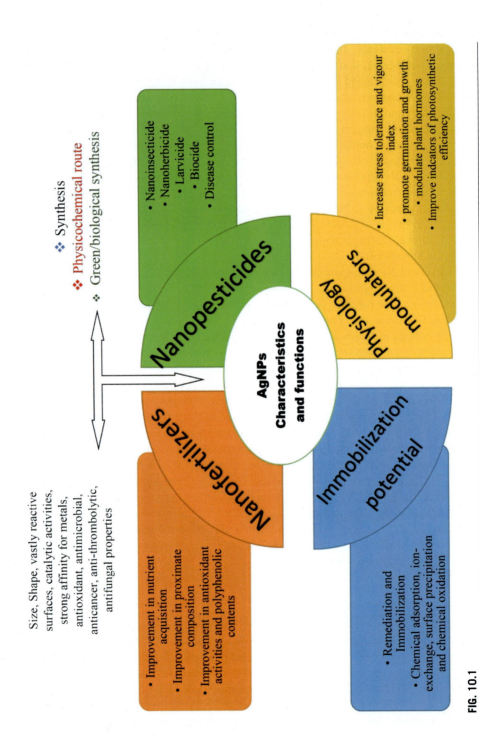

FIG. 10.1

Potentials of AgNPs in agriculture.

Najafpoor et al., 2020). These functions stem from their antioxidant, antimicrobial, antiviral, anticancer, antithrombolytic, and antifungal properties (Ammar and El-Desouky, 2016; Lateef et al., 2017; Nakamura et al., 2019). Silver nanoparticles can be synthesized by chemical and biological/green methods, while green synthesis is a better approach because it is ecofriendly, less toxic, and produce AgNPs with uniform properties. Numerous extracts from pods, peels, shells of cocoa, cola, banana, pomegranate, coconut, orange, watermelon, cashew, mango, cobweb, *Bacillus safensis* as well as leaf extracts of *Jatropha curcus*, *Morinda tinctorial*, *Gmelina arborea*, *Stemona tuberosa Lour*, *Ficus benjamina*, and Benjamina among others have been used to biologically synthesize AgNPs with some having higher antioxidant activities than standard ascorbic acid, vitamin E, and greater radical scavenging abilities (Vanaja et al., 2014; Ibrahim, 2015; Lateef et al., 2015a,b, 2016a,b,c,d, 2017; Saha et al., 2017; Bonigala et al., 2018; Su et al., 2019).

Survival of plants is of utmost obligation as they are producers of food, absorbers of carbon dioxide, and sometimes as sinks for pollutants. However, they are continuously exposed to both biotic and abiotic stresses like environmental factors (drought, water deficit, heat, and salinity), pests and pollutants (heavy metals, dyes, and organic pollutants) that hinder their functions. These are the major causes of deficiencies in morphological, physiological, and biochemical parameters in plants that drastically reduce their yield, nutritional quality, and may eventually lead to death (Servin et al., 2015; Elmer and White, 2016; Al-Sherbini et al., 2019; Shabnam et al., 2019; Hamid et al., 2019, 2020).

Plants are equipped with enzymic antioxidants (superoxide dismutase, catalase, glutathione peroxidase, glutathione reductase, and ascorbate peroxidase) and non-enzymic antioxidants (polyphenols, carotenoids, ascorbic acid, and α-tocopherol) to quench/scavenge the continuous generation of reactive oxygen species (ROS) via exposure to stresses. When these lines of defence are overwhelmed, plants are susceptible to possible attack with manifestation in growth redundancy, delayed fruiting, reduced water absorption, discoloration of leaves, and probable death. This will adversely affect food production (Adisa et al., 2019; Khan et al., 2017, 2018, 2019b,c; Kopittke et al., 2019; Tombuloglu et al., 2019a,b,c; Liu et al., 2020; Ren et al., 2020; Usman et al., 2020; Zhao et al., 2020).

Silver nanoparticles, like other NPs, can protect plants against the onslaught of ROS generated during exposure to stresses by modulating the stimulation of antioxidant compounds and also providing antioxidant activities to mop-up radicals, thus improving plant health. This is aside adsorption of pollutants and control of pathogenic organisms that endanger plants' survival that AgNPs are reported to perform (Syu et al., 2014; Mehta et al., 2016; Jasim et al., 2017; Azeez et al., 2017, 2019a,b; Elemike et al., 2019, 2020).

Silver nanoparticles (AgNPs) are irrefutably the unsurpassed NPs for disease-suppression and control of pathogenic microorganisms in soil, plants, and clinical isolates. They have been in use for ages and have continually been found worthy in many products. Silver nanoparticles are target-specific and were found to be effective in halting *Artemisia absinthium*, *Aspergillus terreus*, *A. niger*, *Alternaria alternate*,

Bacillus subtilis, *Bipolaris sorokiniana*, *Cladosporium* spp., *Citrobacter freundii*, *Colletotrichum gloeosporioides*, *Erwinia cacticida*, *Escherichia coli*, *Fusarium* spp., *Klebsiella pneumoniae*, *Phytophthora parasitica*, *Podosphaera xanthii*, *Pseudomonas aeruginosa*, *P. fluorescens*, *Sclerotium cepivorum*, *Staphylococcus aureus*, *Trichoderma viride*, and xanthomonas species. Nematocidal and larvicidal properties of AgNPs led to inhibition of larvae development in *Anopheles gambiae*, *Culex*, and total extermination of *Meloidogyne* spp. (Ali et al., 2015; Ibrahim, 2015; Lallawmawma et al., 2015; Lateef et al., 2015a,b; Mishra et al., 2014; Prasad et al., 2017; Azeez et al., 2019a; Barkataki and Singh, 2019; Chhipa, 2019; Kumara et al., 2019; Sanzari et al., 2019; Liu et al., 2020; Rajwade et al., 2020).

Silver nanoparticles have been employed to remediate and adsorb polycyclic aromatic hydrocarbons, heavy metals, dyes, and many other pollutants with excellent performances better than results obtained using other NPs. Silver nanocomposites and adsorbents like hen feather, starch, and activated carbon modified with AgNPs have been reported for removal of recalcitrant pollutants (Abbasi et al., 2014; Ali et al., 2017b; Al-Qahtani, 2017; Saha et al., 2017; Pacheco and Buzea, 2017, 2018; Werkneh and Rene, 2019; Chatterjee et al., 2020).

Safety assessment of AgNPs for cytogenotoxicity on *Allium cepa* as well as biochemical and histological indices on chicken reported no deleterious effects (Yekeen et al., 2018; Ogunbode et al., 2019).

10.3 Immobilization efficiency of AgNPs

Resulting from anthropogenic activities of human beings without due regard for ecosystems, wastes from households and industries are discarded onto arable agricultural lands, which deposit chemically active species (CAS) such as heavy metals, reactive dyes, and other organic pollutants. Plants are continually faced with challenges of combating adverse toxicity and oxidative damage caused by CAS with consequent structural defects, growth retardation, and disturbance to hormonal signaling pathways (Bethi et al., 2016; Al-Sherbini et al., 2019; Song et al., 2019; Roberto et al., 2020).

Heavy metals such as Cd, Cr, Pb, and As are extremely toxic and do not have any known biological values. They alter physiological and morphological appearances of plants ranging from stunted growth, discoloration of leaves, impairment of photosynthetic pigment, and inhibition of nutrient uptake (Fig. 10.2). They are difficult to degrade and, hence, must be immobilized or remediated to prevent the death of plants and entry into the food chain (Ali et al., 2017a; Chen et al., 2018; Das et al., 2019; Shabnam et al., 2019). Dyes are ubiquitously used in laboratories, textile, paper, leather, and photographic industries. They cause unpleasant difficulty to plants, disrupting hormonal balance and damaging cell activities. Consequences of organic pollutants such as polycyclic aromatic hydrocarbons, volatile organic compounds, and other agrochemicals are notably deadly to plants inducing oxidative stress triggered by imbalances in physiological and morphological changes

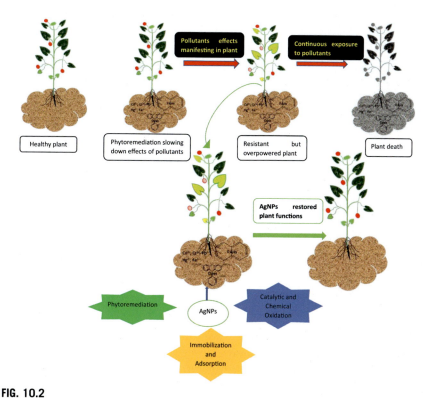

FIG. 10.2

AgNPs-assisted phytoremediation of pollutants.

(Chen et al., 2015; Chhipa and Joshi, 2016; Guerra et al., 2018; Ma et al., 2018; Ogunkunle et al., 2018). Plants are inherently built to respond appropriately to detoxify induced toxicities by CAS, but under high concentrations, phytotoxicity can significantly affect basic functions of plants (Khan et al., 2017; Ali et al., 2019; Ren et al., 2020).

Different management methods, viz., phytoremediation, chemical immobilization, adsorption, catalytic degradation, and chemical oxidation, are usually employed for reducing mobility and concentrations of toxic pollutants in soil (Ali et al., 2017b; Seshadri et al., 2017; Desoky et al., 2018; Rizwan et al., 2018).

10.3.1 Phytoremediation and AgNPs

Phytoremediation is an inexpensive ecofriendly system of remediating soil contaminated with toxic pollutants and has been proven to be a good management system. This achieves removal, volatilization, degradation, extraction, and precipitation of pollutants via phytostabilization, phytoextraction, phytodegradation, phytovolatilization, and rhizodegradation methods (Fig. 10.3) (Kang, 2014; Liu et al., 2015,

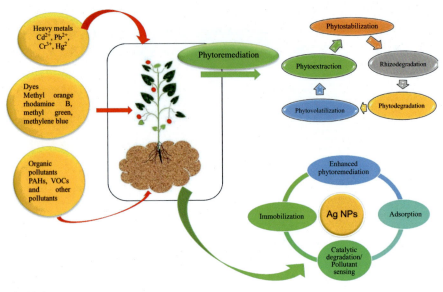

FIG. 10.3

Applications of AgNPs for remediation.

2018; Shtangeeva et al., 2018; Agoun-Bahar et al., 2019; Mahgoub, 2019; Sivaram et al., 2019; Werkneh and Rene, 2019; Chatterjee et al., 2020).

i. Phytostabilization also known as phytoimmobilization describes the immobilization/adsorption of pollutants with probable complexation in hyperaccumulating plants to reduce their bioavailability, thwart their migration into other plant parts, and precipitate the products into rhizosphere (Fernandez-Luqueno et al., 2017; Birolli et al., 2018).
ii. Phytoextraction, which is also called phytosequestration or phytoabsorption, indicates the uptake of pollutants by plants and storing them in other harvestable parts (Song et al., 2019; Roberto et al., 2020).
iii. Phytodegradation refers to the transformation of pollutants during metabolic processes in plants into smaller innocuous compounds, which are afterwards incorporated to support plant growth (Song et al., 2019; Roberto et al., 2020).
iv. Phytovolatilization defines the ability of plants to volatilize pollutants into the atmosphere through absorption, translocation, and transpiration (Song et al., 2019; Roberto et al., 2020).
v. Rhizodegradation is referred to as phytostimulation of microorganisms and microbial activities in rhizosphere that exudes enzymes, carbohydrates, and amino acids in plant roots to metabolically break down pollutants (Song et al., 2019; Roberto et al., 2020).

Phytoremediation procedures have been effectively applied to clean-up toxic contaminants in soil with worthy results using some plants (Birolli et al., 2018;

Shtangeeva et al., 2018; Agoun-Bahar et al., 2019; Mahgoub, 2019; Sivaram et al., 2019). However, the formation of toxic intermediates and secondary metabolites are largely overlooked during this process in addition to being plant-specific because not all plants possess the capacity to withstand bioaccumulation of pollutants. Moreover, there could be an eventual transfer of life-threatening pollutants into the food chain and the worst concern is the resultant phytotoxicity induced in plants having been exposed to pollutants since the concentrations of pollutants are not stagnant (Fig. 10.2) (Liang et al., 2017; Chen, 2018).

Considering the damaging consequences on plants and the possibility of pollutant entry into food chain arising from phytoremediation; ecofriendly NPs such as AgNPs can enhance phytoremediation potentials of plants with little or no effects on them (Fazal et al., 2016; Fernandez-Luqueno et al., 2017; Das et al., 2019).

Nanoparticle-assisted phytoremediation has been reported to augment plant hormones, promote growth, enlarge roots, improve tolerance, and facilitate greater remediation of pollutants (Achari and Kowshik, 2018; Ma et al., 2018; Adisa et al., 2019; Das et al., 2019). This happens in two ways either acting directly on plants potential to remediate depending on the application procedure or indirectly by significantly reducing the concentrations of pollutants to lessen the burden of removal on plants, which would be very effective in metabolizing the remnants (Fernandez-Luqueno et al., 2017; Ali et al., 2019; Scott et al., 2019; Werkneh and Rene, 2019).

Silver nanoparticles reportedly improved phytoremediation potentials of *Moringa oleifera* from 20% to 50% for immobilization of Cd, and Pb and was found to enlarge roots of maize to photostimulated rhizosphere extracting Cd, Pb, and Ni. Similar results were obtained for other NPs for the enhancement of plants' potential to remove heavy metals, organic, and pharmaceutical pollutants (Jin et al., 2016; Liang et al., 2017; Liu et al., 2018; Vıtkova et al., 2018; Azeez et al., 2019a).

10.3.2 Chemical immobilization and adsorption with AgNPs

Another management system for remediating soil pollutants is the application of chemical immobilizing agents/adsorbents that considerably decrease their mobility to prevent plant death and eventual entry into the food chain. Some of the chemical immobilizing agents reported in the literature are manure, compost, hydroxyapatite, biochar, clay, chitosan, activated carbon, zeolite, and NPs (Chen et al., 2015, 2018; Ali et al., 2017a,b; Guerra et al., 2018; Noori et al., 2020). These are good soil nourishers, conditioners, and excellent adsorbents of pollutants. They prevent translocation of pollutants into plant parts through sequestration, complexation, stabilization, precipitation, and coprecipitation (Seshadri et al., 2017; Desoky et al., 2018; Rizwan et al., 2018; Hamid et al., 2019, 2020). Unfortunately, some of these agents do not improve biochemical, physiological, and morphological attributes of plants. Remediation/immobilization with metal NPs proffer the best choice as they remediate, chelate, catalyze, adsorb, and immobilize pollutants with about 80% efficiencies reported in some studies, as well as a considerable decrease in pollutants' bioaccessibility and availability to plants (Fig. 10.2) (Singh et al., 2018; Song et al., 2019; Lv et al., 2019).

Applications of NPs to immobilize/remediate pollutants are more effective in performance, faster in action, easily dispersed in soil, cost-effective, and controllable due to their well-developed surface chemistry, massive reactive sites, modifiable shapes, and large surface area to volume ratio. Nanoparticles with small size are quite helpful for reducing the availability of heavy metals owing to the provision of large surface area for binding heavy metals (Bethi et al., 2016; Khan et al., 2017, 2019c; Boente et al., 2018; Buledi et al., 2020; Chatterjee et al., 2020; Usman et al., 2018, 2020). Another advantage of NPs for immobilization/adsorption is their reasonable use either for in situ or ex situ management. They are equally reusable, highly selective for profiling, and sensing pollutants with a strong affinity (Belal and El-Ramady, 2016; Tombuloglu et al., 2019a,b,c). The ability of NPs to function as immobilizing agents is principally dependent on pH as it controls the surface charges. When the pH point of zero charge (pH_{pzc}) is below the soil pH, the surface of NPs would be negatively charged and hence cationic pollutants (heavy metals, cationic dyes) are easily remediated due to electrostatic attractions (Azeez et al., 2018; Al-Sherbini et al., 2019).

As for AgNPs, they show high adsorption capacity over a wide range of pH from acidic to basic, and are thus able to uptake both cationic and anionic pollutants. This is also dependent on large surface area and pristine morphology to trap pollutants via chemisorption or physisorption processes (Elemike et al., 2019; Azeez et al., 2020a,b).

Silver nanoparticles (AgNPs) are extensively researched affordable and excellent adsorbents for dyes, organic pollutants, and immobilizing agents for heavy metals in soil. Green synthesized AgNPs have been found to perform better with higher adsorption capacity than chemically synthesized AgNPs (Cai et al., 2017; Scott et al., 2019; Werkneh and Rene, 2019). This is connected to their suitably appropriate properties such as highly reactive sites, exceptional sorption capacity, strong affinity, great catalytic potential, and modifiable surface charges. They can directly adsorb pollutants to reduce toxicity, which is essential to the survival of plants, improve soil activities to ensure precipitation of their nontoxic metabolites and enhance plant tolerance toward their phytotoxicity (Figs. 10.2 and 10.3) (Khan et al., 2019a; Kumara et al., 2019; Lv et al., 2019; Ren et al., 2020; Roberto et al., 2020).

The use of AgNPs has been previously documented as adsorbents for dye, for desulphurization of oil and as immobilizing agents for Cd, Pb, Hg, Fe, and Cr. Silver nanoparticles achieved over 70% immobilization of Cd and Pb contaminated soil and prevented over 60% translocation to other parts in *Moringa oleifera*. In addition to these abilities, AgNPs attenuated toxic effects of heavy metals on plants by decreasing their absorption via the roots. Similarly, AgNPs as adsorbents were able to remediate approximately 90% of rhodamine B, methyl orange, Hg^{2+}, Fe^{3+}, Cr^{2+}, Cd^{2+}, Cu^{2+}, and Pb^{2+} in wastewater and soil. Silver nanoparticles equally remediated over 80% of phenanthrene, pyrene, and anthracene (Abbasi et al., 2014; Al-Qahtani, 2017; Azeez et al., 2018, 2019b, 2020a,b; Das et al., 2019; El-Tawil et al., 2019).

10.3.3 Catalytic degradation and pollutant-sensing properties of AgNPs

The catalytic action of AgNPs on pollutants is an innovative environmental remediation method to degrade or convert pollutants to more tolerant and useful forms. This is preferred to other conventional methods and involves the transfer of electrons from biomolecules on NPs surface to pollutants (Lateef et al., 2015a,b; Groiss et al., 2017). It is a redox process that involves donation or removal of electrons to degrade or stabilize pollutants, especially dyes. Silver nanoparticles were used for catalytic conversion of toxic Fe^{3+} to essential Fe^{2+}, which is more accessible, required for nutrition and promoted antioxidant activity in the plant due to inherent ferric reducing ability (Azeez et al., 2017, 2019a,b; Lateef et al., 2016a,b,c,d). Numerous reports have described catalytic degradation and decoloration ability of dyes by AgNPs with exceptional results achieving about 96% decoloration of methyl orange, methylene blue, cresol green, rhodamine B, methyl red, methyl green, and p-nitrophenol. This involves breakage of chromophoric functional groups responsible for color formation to degrade dyes (Vanaja et al., 2014; Saha et al., 2017; Bonigala et al., 2018; Veisi et al., 2018).

The use of AgNPs as sensors for profiling pollutants to selectively degrade them is another unique property; it leaves no toxic by-product. Silver nanoparticles are used as sensing probe to detect, chelate, and precipitate pollutants even at nanolevel, owing to their high surface reactivity, large surface area to volume ratio, and associated sensitivity from size-dependent characteristics. They have been successfully deployed to detect Hg^{2+}, Cd^{2+}, Pb^{2+}, and Cu^{2+} at nanolevel (Kumar et al., 2014; Alam et al., 2015; Uddin et al., 2017; Nakkala et al., 2018; Veisi et al., 2018; Roberto et al., 2020).

10.4 Silver nanoparticles as modulators of plant biochemical and nutritional qualities

The primary purposes of cultivating crops are to serve as food sources for nutrition and protection. Most arable lands have been intensely cropped over many years necessitating the applications of agrochemicals to boost harvestable and marketable yields of crops since soils are greatly deficient of required nutrients to support plant growth (Verma et al., 2018; Azeez et al., 2019a). Many studies have linked an increase in plant nutrition to agrochemical use, regrettably, the deficiency in synchronization between the applied agrochemicals and uptake by plants contributes to contamination of other ecological matrixes. This intensifies calls for sustainable agriculture using nanotechnology (Elmer and White, 2016; White and Gardea-Torresdey, 2018).

Numerous NPs have been reported to positively promote physiological indices, nutritional composition, pharmacological uses, and for repressing disease-causing organisms in plants. Although toxicity arising from their use have been sources of worry, this is well covered using green synthesis (Jasim et al., 2017; Gupta et al., 2018; Noori et al., 2020).

Both stimulatory and phytotoxicity of NPs are dependent on size, shape, dissolution, mode of synthesis, concentrations, functionalization, and exposure duration. These functions are reliant on plant physiology (Dang et al., 2020). In some studies, higher concentrations have been reportedly toxic, while others reported stimulation of physiological and biochemical parameters. Also, low concentrations have been found worthier as stimulators with least toxicity (Khan et al., 2017, 2019c; Chung et al., 2019; Lalau et al., 2020). As for the size and shape of NPs, they are major conditions to be considered for use in plants. Smaller sized NPs are speedily internalized after passing through cuticles (<5 nm NPs), stomata, and root hairs. These routes of passage ordinarily allow small-sized NPs to pass through due to their small spaces, while NPs larger than 20 nm are restricted or stopped in roots, and NPs with 10 nm are not allowed via stomata (Cox et al., 2016; Shalaby et al., 2016; El-Argawy et al., 2017; Siddiqi and Husen, 2017). Rod-shaped NPs rapidly dissolve and ionize faster, thus are more toxic. Release of ion during dissolution is mostly responsible for the toxicity of NPs (Hernandez-Hernandez et al., 2017; Chaudhuri and Malodia, 2017; Cvjetko et al., 2017). Due to plant internal environment and soil composition, zero-valent NPs are favored for better distribution and greater influence on plants as they are not restrained within the cells and not retained in the soil. Different plants react with different nanoparticle compositions. Some immediately interact, initiating molecules in root exudates for NPs absorption while some plants delay NPs absorption due to their transformed nature (Reddy et al., 2016; Line et al., 2017; Tripathi et al., 2017; Ahmed et al., 2018; Elmer and White, 2018; Barkataki and Singh, 2019).

10.4.1 Mode of entry

Nanoparticles as nanofertilizers and nanopesticides can be applied mainly in two ways: foliar and soil (Fig. 10.4). Foliar route application entails the deposition of NPs on the aerial paths via cuticle to stomata to other parts in plants. The fate of foliar NPs is influenced by an interplay between environmental factors, such as temperature, relative humidity, and sunlight, and morphological factors, such as leaf morphology, its chemical composition, hydathodes on leaf tip, presence of leaf wax, exudates, and trichomes (Raliya et al., 2015; Ahmed et al., 2018; Achari and Kowshik, 2018; Zulfiqar et al., 2019).

Nanoparticles applied to the soil undergo different transformation (physical, chemical, and biological) with soil components and root rhizosphere. These transformations can either diminish their translocation or enhance it, but usually, interaction with natural organic matter contributes to their effectiveness as nanofertilizers (Cornelis et al., 2013, 2014; Cornelis, 2015; Wagner et al., 2014; Zulfiqar et al., 2019). They are absorbed via physiologically active roots and adhere firmly through van der Waal forces, and hydrophobic and electrostatic interactions. This is followed by wrapping and docking with root surface and subsequent entry into plants. They are translocated into different parts of plants by forming complexes with root exudates and the carrier protein (Reddy et al., 2016; Zuverza-Mena et al., 2017; Pacheco and Buzea, 2017, 2018; Khan et al., 2019a,b,c).

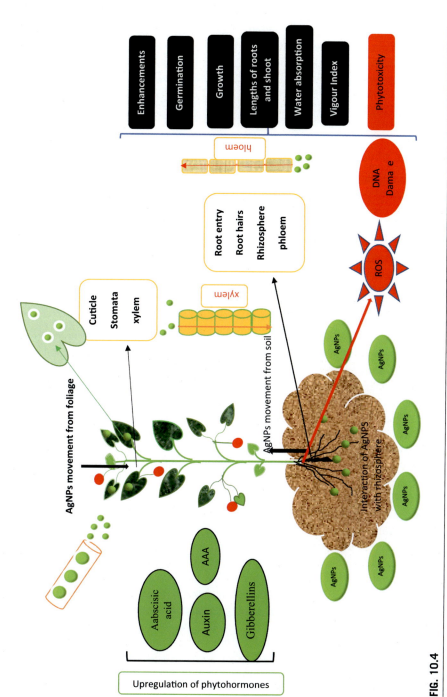

FIG. 10.4

Mode of entry of AgNPs and their influence on physiology and phytohormones in plants.

Nanoparticles enter into plant walls by forming new pores, binding to transporter proteins, endocytosis, aquaporins, and via ion channels, while endocytosis is identified as the major mechanism responsible for NPs uptake by plants. Nanoparticles travel intercellularly via apoplastic and symplastic translocation pathways downwardly and upwardly in xylem and phloem tissues (Nair and Chung, 2014; Line et al., 2017; Barkataki and Singh, 2019; Lv et al., 2019; Ahmadov et al., 2020; Usman et al., 2020).

10.4.2 As physiology and phytohormone modulators

Physiological indicators are indices to access healthy and diseased plants. They give an immediate assessment of the plant environment, depict normal/abnormal metabolic activities and stress/stimulators currently being experienced (Iqbal et al., 2017; Sheykhbaglou et al., 2018; Noori et al., 2020).

There are diverse parameters in use for determining plant physiology comprising percentage germination, growth, biomass, number of leaves, root and shoot lengths, relative water content, and vigor indices. These parameters define the outlook of a plant; when it is healthy, they are in right proportion, but when stressed or diseased, they are grossly below the required (Sharma et al., 2012a,b; Shams et al., 2013; Xiong et al., 2018; Dang et al., 2020).

Enhancement of physiological indicators by NPs is associated with modulation of phytohormones, regulation of noncoding microRNA required for transcriptional genes expression and protein-coding. Phytohormones are organic molecules produced during metabolic activities in the plant to control and modulate physiological responses needed for growth parameters. These are indole acetic acid, ethylene, auxins, gibberellins, abscisic acid, jasmonates, brassinosteroids, and cytokinins. MicroRNA is vital in metabolic processes in plants and is believed to be evolutionarily responsible for regulating molecules needed for physiological development in plants (Rezvani et al., 2012; Nair and Chung, 2014; Nair, 2018; Syu et al., 2014; Vinkovic et al., 2017; Yang et al., 2017; Ma et al., 2018; Azeez et al., 2019b; Khan et al., 2019a,b,c). Silver nanoparticles have been found to positively accelerate seed germination, growth, and root and shoot lengths by altering phytohormones responsible for cell division, proliferation, and signaling pathways. Formation of new pores and enlargement as one of the mechanisms of NPs uptake by plants might be accountable for longer roots and shoots under AgNPs application (Chaudhuri and Malodia, 2017; Hernandez-Hernandez et al., 2017; Khan et al., 2017; Vinkovic et al., 2017; Ma et al., 2018; Goncalves et al., 2019; Khan et al., 2019a,b,c). Studies conducted on physiological parameters of *Oryza sativa*, *Pisum sativum*, *Corchorus olitorius*, *Moringa oleifera*, *Cucumis sativus*, *Crocus sativus*, *Arabidopsis thaliana*, *Brassica napus*, *Amaranthus caudatus*, *Brassica juncea*, *Eupatorium fustulosum*, lettuce, and barley with AgNPs indicated their positive contribution to influencing better physiology. Silver nanoparticles achieved this by upregulating signaling pathways and activating genes necessary for cell proliferation (Sharma et al., 2012a,b; Yin et al., 2012; Shams et al., 2013; Nair and Chung,

2014; Syu et al., 2014; Zuverza-Mena et al., 2016; Tripathi et al., 2017; Ejaz et al., 2018; Sheykhbaglou et al., 2018; Xiong et al., 2018; Azeez et al., 2017, 2019a,b).

Compromised water level can significantly hinder seed viability, tolerance status, and vigor index. Studies using AgNPs reported improved relative water content and vigor index of *Moringa oleifera*, *Brassica juncea*, and *Corchorus olitorius* under an influence of heavy metals that dwarfed these parameters. This is related to the ability of AgNPs to regulate physiological processes, promote water absorption, and enhance cell proliferation enzymes (Sharma et al., 2012a,b; Cvjetko et al., 2017; Alsaeedi et al., 2019; Azeez et al., 2019a,b; Khan et al., 2019a,b,c; Liu et al., 2020).

10.4.3 As regulators of biochemical, and nutritional parameters and for disease-suppression

The significance of fertilizer application in agriculture is to enhance the supply of essential nutrients required by plants for their activities. This is without drawbacks as has been noted severally in literature, especially eutrophication of water from lack of synchronization and formation of organoprecipitates. Using nanofertilizers on plants provides a controllable release of nutrients to halt environmental contamination, ensures effective use of nutrients in multiple folds, and protects against harsh environmental conditions (Alsaeedi et al., 2019; Rizwan et al., 2019; Usman et al., 2018, 2020; Zhang et al., 2020; Zhao et al., 2020).

Nanoparticles can be designed as suppliers or carriers of nutrients or carriers for effectiveness to improve nutritional quality and pharmacological uses of plants. This is connected to morphological characteristics of NPs such as small size and large surface area, which offer surface binding of fertilizer to facilitate more absorption and subsequent delivery into plants (Khan et al., 2017; Yang et al., 2017; Ma et al., 2018; Sheykhbaglou et al., 2018). Previous studies have widely reported NPs influence on nutritional quality improving carbohydrate, sugar, protein, fatty acids, essential oils, metabolizable energy, and macro- and micronutrients (Al-Oubaidi and Mohammed-Ameen, 2014; Nair and Chung, 2014; Morales-Díaz et al., 2017; Belluco et al., 2018; Khan et al., 2017, 2018, 2019a,b,c; Azeez et al., 2020b). Silver nanoparticles were reported to improve carbohydrate, protein, and polyphenols in *Bacopa monnieri*, mustards, beans, and corns. They do this by ensuring efficient delivery of mineral elements to plants and thereby improving plants' nutrition (Arora et al., 2012; Salama, 2012; Krishnaraj et al., 2012; Rani et al., 2016; Zuverza-Mena et al., 2016).

Exposure of plants to NPs is expected to alter antioxidant activities for scavenging of free radicals, induce polyphenol production, and suppress disease-causing microorganisms. Application of NPs has proved effective in regulating plant functions to further understand mechanistic responses in biochemical parameters upon exposure. Sometimes, the effects of NPs on the physiology of plants may not be showing but may translate to enhanced biochemical functions in higher antioxidant activities, increased protection against ROS, reduced oxidative stress levels, and better protection against insects (Groiss et al., 2017; Das et al., 2018a).

Normally, ROS are produced during regular metabolism in plants but under duress from abiotic and biotic stresses. Their production increases forcing plants to activate their defence mechanism comprising enzymic and nonenzymic antioxidant molecules to halt their continuous production (Fig. 10.5). These defence molecules play significant roles in quenching ROS to protect plant functions. However, when plant defence system is stunned, highly reactive radicals such as hydroxyl ion, superoxide ion, hydrogen peroxide, and lipid peroxidation ion are excessively produced prompting adverse oxidative degradation of lipids, DNA, proteins, and membrane. Increase in malondialdehyde and hydrogen peroxide levels in plants is typically considered as an indicator of phytotoxicity due to ROS (Fazal et al., 2016; Achari and Kowshik, 2018; Ma et al., 2018; Hu et al., 2018; Adisa et al., 2019; Usman et al., 2020).

Nanoparticles are significant in protecting plants against oxidative stress from ROS by activating and upregulating antioxidant enzymes. Additionally, they have been reported to decrease levels of lipid peroxidation and protein oxidation. Moreover, some NPs are inherently antioxidant compounds leading to induction of phytochemicals, notably polyphenols that are secreted for defence (Thuesombat et al., 2014; Rico et al., 2011, 2015a,b; Iannone et al., 2016; Kumari et al., 2017; Gupta et al., 2018; Zhang et al., 2020; Zhao et al., 2020). Phytochemicals are antioxidant compounds produced for adaptation in the ecosystem and defence against

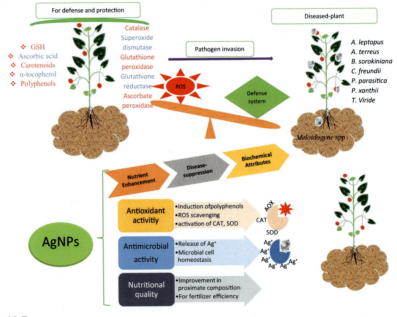

FIG. 10.5

Mechanism of AgNPs regulation of nutritional quality, antioxidant activities, and disease-suppression.

pathogenic organisms together with environmental stresses. Owing to their antioxidant activities, they can scavenge radicals and hinder the oxidation of proteins and lipids coupled with their metal chelation ability. These compounds are reputable for their medicinal values such anti-inflammatory, antioxidant, and anticancer activities that have been courted for the wellbeing of human beings (Elmer and White, 2016; Azeez et al., 2017; Siddiqi and Husen, 2017; Yang et al., 2017; Khan et al., 2017, 2019a,b,c; Marslin et al., 2017).

Nanoparticles induce a boost in phytochemical contents by simultaneously inciting ROS generation and triggering phytochemical production to cope with the generation. Since most NPs also possess antioxidant activities, they quickly quench ROS while phytochemical levels remain high. Several studies have reported that AgNPs enhanced the antioxidant status of plants (Arora et al., 2012; Krishnaraj et al., 2012; Salama, 2012; Sharma et al., 2012a,b; Kaveh et al., 2013; Fazal et al., 2016; Rani et al., 2016; Azeez et al., 2017; Khan et al., 2017, 2018, 2019a).

Nanoparticles have similarly been reported to improve photosynthesis for higher CO_2 capture and production of food. This is related to an increase in pigment synthesis, light absorbance, photophosphorylation; regulation of rubisco enzyme and carbonic anhydrase; inhibition of ethylene perception; promotion of plant hormones; and improvement in sensing of signaling molecules (Rico et al., 2011, 2015a,b; Tripathi et al., 2017; Khan et al., 2019a,b,c; Tombuloglu et al., 2019a,b,c).

Silver nanoparticles likewise have been reported to boost photosynthetic pigment formation in *Trigonella foenum-graecum*, *Daucus carota*, *Vigna radiata*, *Moringa oleifera*, *Corchorus olitorius*, *Oryza sativa*, and *Arabidopsis* (Sharma et al., 2012a,b; Yin et al., 2012; Nair and Chung, 2014; Thuesombat et al., 2014; Park and Ahn, 2016; Jasim et al., 2017; Kumari et al., 2017; Gupta et al., 2018; Azeez et al., 2019a,b).

Due to their inherent antimicrobial, antifungal, and antibacterial properties, NPs including AgNPs can effectively subdue pathogenic microorganisms in soil and on plants (Fig. 10.5). Silver nanoparticles have demonstrated abilities to prevent and control many plant pathogens achieving about 90% success. This is believed to be largely connected to NPs possessing antimicrobial, antifungal, and antibacterial properties in microbial cells exhibiting toxicity resulting in cell lysis. Release of Ag^+ from AgNPs is also thought to play a major role. Ionic silver (Ag^+) is known to be toxic to microorganisms creating oxidative burst that damages pathogen cells via disruption of electron transport chain and microbial cellular homeostasis. This ultimately compromises cell integrity and consequently cause pathogen death (Marslin et al., 2017). Another mechanism is via induction of polyphenols in plants, which generate ROS toxic to pathogens but support plants with antioxidant activities to cope with ROS (Aguilar-Mendez et al., 2011; Mishra et al., 2014; Servin et al., 2015; Fraceto et al., 2016; Elmer and White, 2016; El-Argawy et al., 2017; Ma et al., 2018; Adisa et al., 2019; Elemike et al., 2020; Roberto et al., 2020).

Silver nanoparticles have been investigated to suppress *Antigonon leptopus*, *Artemisia absinthium*, *Aspergillus terreus*, *A. niger*, *Alternaria alternate*, *Bacillus subtilis*, *Bipolaris sorokiniana*, *Cladosporium* spp., *Cassia auriculate*, *Citrobacter freundii*, *Coffea arabica*, *Colletotrichum gloeosporioides*, *Eclipta prostrata, Erwinia cacticida, Escherichia coli, Fusarium* spp., *Jasminum nervosum, Klebsiella pneumoniae, Pantoea ananatis, Penicillium expansum, Phytophthora parasitica, Podosphaera xanthii, Pseudomonas aeruginosa, Pseudomonas fluorescens, Sclerotium cepivorum, Staphylococcus aureus, Trichoderma Viride*, and *Xanthomonas* species. Nematocidal and larvicidal properties of AgNPs led to inhibition of larvae development of *Anopheles gambiae* and total extermination of *Meloidogyne* spp. (Ali et al., 2015; Lallawmawma et al., 2015; Lateef et al., 2015a,b; Mishra et al., 2014; Prasad et al., 2017; Azeez et al., 2019a; Barkataki and Singh, 2019; Chhipa, 2019; Kumara et al., 2019; Sanzari et al., 2019; Liu et al., 2020; Rajwade et al., 2020).

10.5 Conclusion

Agriculture is the bedrock of sustenance as it provides food for consumption and raw materials for industries. The use of traditional agrochemicals is a burden to the environment coupled with the inability to prevent toxicities of pollutants. Nanoparticles have found worthiness in agriculture as promoters of growth, modulators of nutrition, regulators of antioxidant activities, and protectors against pathogens. These are due to their intrinsic properties, easy dispersal in soil, easy translocation within plants, and the ability to improve fertilizer use efficiency. The selection of suitable, less/nontoxic, highly degradable with excellent remediation, and better disease-suppressing potentials brings to fore green synthesized silver nanoparticles possessing these abilities. The stimulatory influence of AgNPs is multifaceted so also are concerns arising from their phytotoxic influence on plants. The size, shape, mode of entry, concentration, dissolution, mode of synthesis, reactivity, and plant species are connecting factors that determine stimulatory or phytotoxic actions of AgNPs. Extensively researched ecofriendly AgNPs possess attributes to act as immobilizing agents, nanofertlizers, and nanopesticides for pollutants removal, physiology regulation, enhancement of biochemical properties, and controlling pathogens.

References

Abbasi, M., Saeed, F., Rafique, U., 2014. Preparation of silver nanoparticles from synthetic and natural sources: remediation model for PAHs. IOP Conf. Ser. Mater. Sci. Eng. 60, 1.

Achari, G.A., Kowshik, M., 2018. Recent developments on nanotechnology in agriculture: plant mineral nutrition, health, and interactions with soil microflora. J. Agric. Food Chem. 66, 8647–8661.

Adisa, I.O., Pullagurala, V.L.R., Peralta-Videa, J.R., Dimkpa, C.O., Elmer, W.H., Gardea-Torresdey, J.L., White, J.C., 2019. Recent advances in nano-enabled fertilizers and pesticides: a critical review of mechanisms of action. Environ. Sci. Nano. 6, 2002–2030.

Agoun-Bahar, S., Djebbar, R., Achour, T.N., Abrous-Belbachir, O., 2019. Soil-to-plant transfer of naphthalene and its effects on seedlings pea (*Pisum sativum* L.) grown on contaminated soil. Environ. Technol. 40 (28), 3713–3723.

Aguilar-Mendez, M.A., Martın-Martınez, E.S., Ortega-Arroyo, L., Cobian-Portillo, G., Sanchez Espındola, E., 2011. Synthesis and characterization of silver nanoparticles: effect on phytopathogen *Colletotrichum gloesporioides*. J. Nanopart. Res. 13, 2525–2532.

Ahmadov, I.S., Ramazanov, M.A., Gasimov, E.K., Rzayev, F.H., Veliyeva, S.B., 2020. The migration study of nanoparticles from soil to the leaves of plants. Biointerface Res. Appl. Chem. 10 (5), 6101–6111.

Ahmed, B., Khan, M.S., Saquib, Q., Al-Shaeri, M., Musarrat, J., 2018. Chapter 2: Interplay between engineered nanomaterials (ENMs) and edible plants: a current perspective. In: Faisal, M., et al. (Eds.), Phytotoxicity of Nanoparticles. Springer International Publishing AG, Part of Springer Nature, pp. 63–102.

Alam, A., Ravindran, A., Chandran, P., Khan, S.S., 2015. Highly selective colorimetric detection and estimation of Hg^{2+} at nanomolar concentration by silver nanoparticles in the presence of glutathione. Spectrochim. Acta A Mol. Biomol. Spectrosc. 137, 503–508.

Ali, M., Kim, B., Belfield, K.D., Norman, D., Brennan, M., Ali, G.S., 2015. Inhibition of *Phytophthora parasitica* and *P. capsici* by silver nanoparticles synthesized using aqueous extract of *Artemisia absinthium*. Phytopathology 105 (9), 1183–1190.

Ali, A., Guo, D., Zhang, Y., Sun, X., Jiang, S., Guo, Z., Huang, H., Liang, W., Li, R., Zhang, Z., 2017a. Using bamboo biochar with compost for the stabilization and phytotoxicity reduction of heavy metals in mine-contaminated soils of China. Nat. Sci. Rep. 7 (1), 2690.

Ali, I., Peng, C.S., Naz, I., Khan, Z.M., Sultan, M., Islam, T., Abbasi, I.A., 2017b. Phytogenic magnetic nanoparticles for wastewater treatment: a review. RSC Adv. 7 (64), 40158–40178.

Ali, S., Rizwan, S., Hussain, A., Zia ur Rehman, M., Ali, B., Yousaf, B., Wijaya, L., Alyemeni, M.N., Ahmad, P., 2019. Silicon nanoparticles enhanced the growth and reduced the cadmium accumulation in grains of wheat (*Triticum aestivum* L.). Plant Physiol. Biochemist. 140, 1–8.

Al-Oubaidi, H.K.M., Mohammed-Ameen, A.S., 2014. The effect of ($AgNO_3$) NPs on increasing of secondary metabolites of *Calendula officinalis* L. in vitro. Int. J. Pharm. Pract. 5, 267–272.

Al-Qahtani, K.M., 2017. Cadmium removal from aqueous solution by green synthesis zero valent silver nanoparticles with *Benjamina* leaves extract. Egypt. J. Aquat. Res. 43, 269–274.

Alsaeedi, A., El-Ramady, H., Alshaal, T., El-Garawany, M., Elhawat, N., Al-Otaibi, A., 2019. Silica nanoparticles boost growth and productivity of cucumber under water deficit and salinity stresses by balancing nutrients uptake. Plant Physiol. Biochem. 139, 1–10.

Al-Sherbini, A.A., Ghannam, H.E.A., El-Ghanam, G.M.A., El-Ella, A.A., Youssef, A.M., 2019. Utilization of chitosan/Ag bionanocomposites as eco-friendly photocatalytic reactor for bactericidal effect and heavy metals removal. Heliyon 5, e01980.

Ammar, H.A., El-Desouky, T.A., 2016. Green synthesis of nanosilver particles by *Aspergillus terreus* HA_1N and *Penicillium expansum* HA_2N and its antifungal activity against mycotoxigenic fungi. J. Appl. Microbiol. 121 (1), 89–100.

Arora, S., Sharma, P., Kumar, S., Nayan, R., Khanna, P., Zaidi, M., 2012. Gold-nanoparticle induced enhancement in growth and seed yield of *Brassica juncea*. Plant Growth Regul. 66, 303–310.

Azeez, L., Lateef, A., Adebisi, S.A., 2017. Silver nanoparticles (AgNPs) biosynthesized using pod extract of *Cola nitida* enhances antioxidant activity and phytochemical composition of *Amaranthus caudatus* Linn. Appl. Nanosci. 7 (1–2), 59–66.

Azeez, L., Lateef, A., Adebisi, S.A., Oyedeji, A.O., Adetoro, R.O., 2018. Novel biosynthesized silver nanoparticles from cobweb as adsorbent for Rhodamine B: equilibrium isotherm, kinetic and thermodynamic studies. Appl. Water Sci. 8, 32.

Azeez, L., Adejumo, A.L., Lateef, A., Adebisi, S.A., Adetoro, R.O., Adewuyi, S., Tijani, K.O., Olaoye, S., 2019a. Zero-valent silver nanoparticles attenuate Cd and Pb toxicities on *Moringa oleifera* via immobilization and induction of phytochemicals. Plant Physiol. Biochem. 139, 283–292.

Azeez, L., Lateef, A., Wahab, A.A., Rufai, M.A., Salau, A.K., Ajayi, E.I.O., Ajayi, M., Adegbite, M.K., Adebisi, B., 2019b. Phytomodulatory effects of silver nanoparticles on *Corchorus olitorius*: its antiphytopathogenic and hepatoprotective potentials. Plant Physiol. Biochem. 136, 109–117.

Azeez, L., Lateef, A., Adejumo, A.L., Adeleke, T.A., Adetoro, R.O., Mustapha, Z., 2020a. Adsorption behaviour of Rhodamine B on hen feather and corn starch functionalized with green synthesized silver nanoparticles (AgNPs) mediated with cocoa pod extract. Chem. Afr. 3, 237–250.

Azeez, L., Adejumo, A.L., Ogunbode, S.M., Lateef, A., 2020b. Influence of calcium nanoparticles (CaNPs) on nutritional qualities, radical scavenging attributes of *Moringa oleifera* and risk assessments on human health. Food Meas. 14, 2185–2195.

Barkataki, M.P., Singh, T., 2019. Plant-nanoparticle interactions: mechanisms, effects, and approaches. Compr. Anal. Chem. 87, 55–83.

Belal, E., El-Ramady, H., 2016. Chapter 10: Nanoparticles in water, soils and agriculture. In: Ranjan, S., et al. (Eds.), Nanoscience in Food and Agriculture 2, Sustainable Agriculture Reviews 21. Springer International Publishing, Switzerland, pp. 311–358.

Belluco, S., Gallocchio, F., Losasso, C., Ricci, A.J., 2018. State of art of nanotechnology applications in the meat chain: a qualitative synthesis. Crit. Rev. Food Sci. Nutr. 58, 1084–1096.

Bethi, B., Sonawane, S.H., Bhanvase, B.A., Gumfekar, S.P., 2016. Nanomaterials-based advanced oxidation processes for wastewater treatment: a review. Chem. Eng. Process. Process Intensif. 109, 178–189.

Birolli, W.G., Santos, D.A., Alvarenga, N., Garcia, A.C.F.S., Romao, L.P.C., Porto, A.L.M., 2018. Biodegradation of anthracene and several PAHs by the marine-derived fungus *Cladosporium sp.* CBMAI 1237. Mar. Pollut. Bull. 129 (2), 525–533.

Boente, C., Sierra, C., Martínez-Blanco, D., Menéndez-Aguado, J.M., Gallego, J.R., 2018. Nanoscale zero-valent iron-assisted soil washing for the removal of potentially toxic elements. J. Hazard. Mater. 350, 55–65.

Bonigala, B., Kasukurthi, B., Konduri, V.V., Mangamuri, U.K., Gorrepati, R., Poda, S., 2018. Green synthesis of silver and gold nanoparticles using *Stemona tuberosa* Lour and screening for their catalytic activity in the degradation of toxic chemicals. Environ Sci Pollut. Res. Int. 25 (32), 32540–32548.

Buledi, J.A., Amin, S., Haider, S.I., Bhanger, M.I., Solangi, A.R., 2020. A review on detection of heavy metals from aqueous media using nanomaterial-based sensors. Environ. Sci. Pollut. Res. https://doi.org/10.1007/s11356-020-07865-7.

Cai, Z., Sun, Y., Liu, W., Pan, F., Sun, P., Fu, J., 2017. An overview of nanomaterials applied for removing dyes from wastewater. Environ. Sci. Pollut. Res. 24, 15882–15904.

Chatterjee, A., Kwatra, N., Abraham, J., 2020. Chapter 8. Nanoparticles fabrication by plant extracts. In: Phytonanotechnology. Elsevier Inc, pp. 143–157.

Chaudhuri, S.K., Malodia, L., 2017. Biosynthesis of zinc oxide nanoparticles using leaf extract of *Calotropis gigantea*: characterization and its evaluation on tree seedling growth in nursery stage. Appl. Nanosci. 7, 501–512.

Chen, H., 2018. Metal based nanoparticles in agricultural system: behavior, transport, and interaction with plants. Chem. Speciat. Bioavailab. 30 (1), 123–134.

Chen, M., Xu, P., Zeng, G., Yang, C., Huang, D., Zhang, J., 2015. Bioremediation of soils contaminated with polycyclic aromatic hydrocarbons, petroleum, pesticides, chlorophenols and heavy metals by composting: applications, microbes and future research needs. Biotechnol. Adv. 33, 745–755.

Chen, T., Lei, M., Wan, X., Yang, J., Zhou, X., 2018. Arsenic hyperaccumulator *Pteris vittata* L. and its application to the field. In: Luo, Y., et al. (Eds.), Twenty Years of Research and Development on Soil Pollution and Remediation in China, pp. 465–476.

Chhipa, H., 2019. Chapter 6: Applications of nanotechnology in agriculture. Methods Microbiol. 46, 115–142.

Chhipa, H., Joshi, P., 2016. Chapter 9: Nanofertilisers, nanopesticides and nanosensors in agriculture. In: Ranjan, S., et al. (Eds.), Nanoscience in Food and Agriculture 1. Sustainable Agriculture Reviews 20. Springer International Publishing, Switzerland, pp. 247–282.

Chung, I.-M., Rekha, K., Venkidasamy, B., Thiruvengadam, B., 2019. Effect of copper oxide nanoparticles on the physiology, bioactive molecules, and transcriptional changes in Brassica rapa ssp seedlings. Water Air Soil Pollut. 230, 48.

Cornelis, G., 2015. Fate descriptors for engineered nanoparticles: the good, the bad, and the ugly. Environ. Sci. Nano 2, 19–26.

Cornelis, G., Pang, L., Doolette, C., Kirby, J.K., McLaughlin, M.J., 2013. Transport of silver nanoparticles in saturated columns of natural soils. Sci. Total Environ. 463–464, 120–130.

Cornelis, G., Hund-Rinke, K., Kuhlbusch, T., van den Brink, N., Nickel, C., 2014. Fate and bioavailability of engineered nanoparticles in soils: a review. Crit. Rev. Environ. Sci. Technol. 44, 2720–2764.

Cox, A., Venkatachalam, P., Sahi, S., Sharma, N., 2016. Silver and titanium dioxide nanoparticle toxicity in plants: a review of current research. Plant Physiol. Biochem. 107, 147–163.

Cvjetko, P., Milošić, A., Domijan, A.M., Vinković Vrček, I., Tolić, S., Peharec Štefanić, P., Letofsky-Papst, I., Tkalec, M., Balen, B., 2017. Toxicity of silver ions and differently coated silver nanoparticles in *Allium cepa* roots. Ecotoxicol. Environ. Saf. 137, 18–28.

Dang, F., Wang, Q., Cai, W., Zhou, D., Xing, B., 2020. Uptake kinetics of silver nanoparticles by plant: relative importance of particles and dissolved ions. Nanotoxicology 14 (5), 654–666.

Das, P., Barua, S., Sarkar, S., Karak, N., Bhattacharyya, P., Raza, N., Kim, K.-H., Bhattacharya, S.S., 2018a. Plant extract-mediated green silver nanoparticles: efficacy as soil conditioner and plant growth promoter. J. Hazard. Mater. 346, 62–72.

Das, S., Chakraborty, J., Chatterjee, S., Kumar, H., 2018b. Prospects of biosynthesized nanomaterials for the remediation of organic and inorganic environmental contaminants. Environ. Sci. Nano 5 (12), 2784–2808.

Das, A., Kamle, M., Bharti, A., Kumar, P., 2019. Nanotechnology and its applications in environmental remediation: an overview. Vegetos 32, 227–237.

Desoky, E.M., Merwad, A.M., Rady, M.M., 2018. Natural biostimulants improve saline soil characteristics and salt stressed-sorghum performance. Commun. Soil Sci. Plant Anal. 49 (8), 967–983.

Dhand, V., Soumya, L., Bharadwaj, S., Chakra, S., Bhatt, D., Sreedhar, B., 2016. Green synthesis of silver nanoparticles using *Coffea arabica* seed extract and its antibacterial activity. Mater. Sci. Eng. C 58, 36–43.

Ejaz, M., Raja, N.I., Ahmad, M.S., Hussain, M., Iqbal, M., 2018. Effect of silver nanoparticles and silver nitrate on growth of rice under biotic stress. IET Nanobiotechnol. 12, 299.

El-Argawy, E., Rahhal, M.M.H., El-Korany, A., Elshabrawy, E.M., Eltahan, R.M., 2017. Efficacy of some nanoparticles to control damping-off and root rot of sugar beet in El-Behiera governorate. Asian J. Plant Pathol. 11, 35–47.

Elemike, E.E., Uzoh, I.M., Onwudiwe, D.C., Babalola, O.O., 2019. The role of nanotechnology in the fortification of plant nutrients and improvement of crop production. Appl. Sci. 9 (3), 499.

Elemike, E.E., Ibe, K.A., Mbonu, J.I., Onwudiwe, D.C., 2020. Chapter 5: Phytonanotechnology and synthesis of silver nanoparticles. In: Phytonanotechnology. Elsevier Inc, pp. 71–96.

Elmer, W.H., White, J.C., 2016. The use of metallic oxide nanoparticles to enhance growth of tomatoes and eggplants in disease-infested soil or soilless medium. Environ. Sci. Nano 3 (5), 1072–1079.

Elmer, W., White, J.C., 2018. The future of nanotechnology in plant pathology. Annu. Rev. Phytopathol. 56, 111–133.

El-Tawil, R.S., El-Wakeel, S.T., Abdel-Ghany, A.E., Abuzeid, H.A.M., Selim, K.A., Hashem, A.M., 2019. Silver/quartz nanocomposite as an adsorbent for removal of mercury (II) ions from aqueous solutions. Heliyon 5, e02415.

Farooq, M., Siddique, K.H.M., 2015. Conservation agriculture. Concepts, brief history, and impacts on agricultural systems. In: Farooq, M., Siddique, K. (Eds.), Conservation Agriculture. Springer, Cham.

Fazal, H., Abbasi, B.H., Ahmad, N., Ali, M., 2016. Elicitation of medicinally important antioxidant secondary metabolites with silver and gold nanoparticles in callus cultures of *Prunella vulgaris* L. Appl. Biochem. Biotechnol. 180 (6), 1076–1092.

Feregrino-Perez, A.A., Magaña-López, E., Guzmán, C., Esquivel, K., 2018. A general overview of the benefits and possible negative effects of the nanotechnology in horticulture. Sci. Hortic. 238, 126–137.

Fernandez-Luqueno, F., Lopez-Valdez, F., Sarabia-Castillo, C.R., Garcı'a-Mayagoitia, S., Perez-Rıos, S.R., 2017. Bioremediation of polycyclic aromatic hydrocarbons-polluted soils at laboratory and field scale: a review of the literature on plants and microorganisms. In: Naser, A., Saravjeet, G., Narendra, T. (Eds.), Enhancing Cleanup of Environmental Pollutants. Volume 1: Biological Approaches. Springer, Cham, pp. 43–64.

Fraceto, L.F., Grillo, R., de Medeiros, G.A., Scognamiglio, V., Rea, G., Bartolucci, C., 2016. Nanotechnology in agriculture: which innovation potential does it have? Front. Environ. Sci. 4, 20.

Goncalves, S.P.C., Delite, F.S., Coa, F., Neto, L.L.R., da Silva, G.H., Bortolozzo, L.S., Ferreira, A.G., de Medeiros, A.M.Z., Strauss, M., Martinez, D.S.T., 2019. Chapter 7: Biotransformation of nanomaterials in the soil environment: nanoecotoxicology and nanosafety implications. In: Nanomaterials Applications for Environmental Matrices. Elsevier, pp. 265–304.

Gorczyca, A., Przemieniecki, S.W., Kurowski, T., Ocwieja, M., 2018. Early plant growth and bacterial community in rhizoplane of wheat and flax exposed to silver and titanium dioxide nanoparticles. Environ. Sci. Pollut. Res. 25, 33820–33826.

Groiss, S., Selvaraj, R., Varadavenkatesan, T., Vinayagam, R., 2017. Structural characterization, antibacterial and catalytic effect of iron oxide nanoparticles synthesised using the leaf extract of *Cynometra ramiflora*. J. Mol. Struct. 1128, 572–578.

Guerra, F.D., Attia, M.F., Whitehead, D.C., Alexis, F., 2018. Nanotechnology for environmental remediation: materials and applications. Molecules 23, 1760.

Gupta, S.D., Agarwal, A., Pradhan, S., 2018. Phytostimulatory effect of silver nanoparticles (AgNPs) on rice seedling growth: an insight from antioxidative enzyme activities and gene expression patterns. Ecotoxicol. Environ. Saf. 161, 624–633.

Hamid, Y., Tang, L., Sohail, M.I., Cao, X., Hussain, B., Aziz, M.Z., Usman, M., He, Z.-l., Yang, X., 2019. An explanation of soil amendments to reduce cadmium phytoavailability and transfer to food chain. Sci. Total Environ. 660, 80–96.

Hamid, Y., Tang, L., Hussain, B., Usman, M., Gurajala, H.K., Rashid, M.S., He, Z., Yang, X., 2020. Efficiency of lime, biochar, Fe containing biochar and composite amendments for Cd and Pb in a co-contaminated alluvial soil. Environ. Pollut. 257, 113609.

Hernandez-Hernandez, H., Benavides-Mendoza, A., Ortega-Ortiz, H., Hernandez-Fuentes, A.D., Juarez-Maldonado, A., 2017. Cu nanoparticles in chitosan-PVA hydrogels as promoters of growth, productivity and fruit quality in tomato. Emirates J. Food Agric. 29 (8), 573–580.

Hu, J., Jiang, J., Wang, N., 2018. Control of citrus huanglongbing *via* trunk injection of plant defense activators and antibiotics. Phytopathology 108, 186–195.

Iannone, M.F., Groppa, M.D., de Sousa, M.E., van Raap, M.B.F., Benavides, M.P., 2016. Impact of magnetite iron oxide nanoparticles on wheat (*Triticum aestivum* L.) development: evaluation of oxidative damage. Environ. Exp. Bot. 131, 77–88.

Ibrahim, H.M., 2015. Green synthesis and characterization of silver nanoparticles using banana peel extract and their antimicrobial activity against representative microorganisms. J. Radiat. Res. Appl. Sci. 8, 265–275.

Iqbal, M., Raja, N.I., Hussain, M., Ejaz, M., Yasmeen, F., 2017. Effect of silver nanoparticles on growth of wheat under heat stress. Iran. J. Sci. Technol. Trans. A Sci. 43, 1–9.

Jasim, B., Thomas, R., Mathew, J., Radhakrishnan, E., 2017. Plant growth and diosgenin enhancement effect of silver nanoparticles in fenugreek (*Trigonella foenum-graecum* L.). Saudi Pharm. J. 25, 443–447.

Jin, Y., Liu, W., Li, X., Shen, S., Liang, S., Liu, C., Shan, L., 2016. Nano-hydroxyapatite immobilized lead and enhanced plant growth of ryegrass in a contaminated soil. Ecol. Eng. 95, 25–29.

Kang, J.W., 2014. Removing environmental organic pollutants with bioremediation and phytoremediation. Biotechnol. Lett. 36, 1129–1139.

Kaveh, R., Li, Y.-S., Ranjbar, S., Tehrani, R., Brueck, C.L., Van Aken, B., 2013. Changes in *Arabidopsis thaliana* gene expression in response to silver nanoparticles and silver ions. Environ. Sci. Technol. 47, 10637–10644.

Khan, M.N., Mobin, M., Abbas, Z.K., AlMutairi, K.A., Siddiqui, Z.H., 2017. Role of nanomaterials in plants under challenging environments. Plant Physiol. Biochem. 110, 194–209.

Khan, T., Abbasi, B.H., Zeb, A., Shad, A.G., 2018. Carbohydrate-induced biomass accumulation and elicitation of secondary metabolites in callus cultures of *Fagonia indica*. Ind. Crop. Prod. 126, 168–176.

Khan, M.R., Adam, V., Rizvi, T.F., Zhang, B., Ahamad, F., Josko, I., Zhu, Y., Yang, M., Mao, C., 2019a. Nanoparticle–plant interactions: two-way traffic. Small 2019, 1901794.

Khan, T., Khan, T., Hano, C., Abbasi, B.H., 2019b. Effects of chitosan and salicylic acid on the production of pharmacologically attractive secondary metabolites in callus cultures of *Fagonia indica*. Ind. Crop. Prod. 129, 525–535.

Khan, M.A., Khan, T., Mashwani, Z.-R., Riaz, M.S., Ullah, N., Ali, H., Nadhman, A., 2019c. Chapter 2: Plant cell nanomaterials interaction: growth, physiology and secondary metabolism. In: Comprehensive Analytical Chemistry. vol. 84, pp. 23–54.

Kopittke, P.M., Lombi, E., Wang, P., Schjoerring, J.K., Husted, S., 2019. Nanomaterials as fertilizers for improving plant mineral nutrition and environmental outcomes. Environ. Sci. Nano 6, 3513–3524.

Krishnaraj, C., Jagan, E.G., Ramachandran, R., Abirami, S.M., Mohan, N., Kalaichelvan, P.T., 2012. Effect of biologically synthesized silver nanoparticles on *Bacopa monnieri* (Linn.) Wettst. Plant growth metabolism. Process Biochem. 47, 651–658.

Kumar, V., Anbarasan, S., Christena, L.R., SaiSubramanian, N., Anthony, S.P., 2014. Bio functionalized silver nanoparticles for selective colorimetric sensing of toxic metal ions and antimicrobial studies. Spectrochim. Acta A Mol. Biomol. Spectrosc. 129, 35–42.

Kumara, V., Lakkaboyanaa, S.K., Sharma, N., Abdelaal, A.S., Maitra, S.S., Pant, D., 2019. Chapter 4: Engineered nanomaterials uptake, bioaccumulation and toxicity mechanisms in plants. In: Comprehensive Analytical Chemistry. vol. 87, pp. 111–131.

Kumari, R., Singh, J.S., Singh, D.P., 2017. Biogenic synthesis and spatial distribution of silver nanoparticles in the legume mungbean plant (Vigna radiata L.). Plant Physiol. Biochemist 110, 158–166.

Lalau, C.M., Simioni, C., Vicentini, D.S., Ouriques, L.C., Mohedano, R.A., Puerari, R.C., Matias, W.G., 2020. Toxicological effects of AgNPs on duckweed (*Landoltia punctata*). Sci. Total Environ. 710, 136318.

Lallawmawma, H., Sathishkumar, G., Sarathbabu, S., Ghatak, S., Sivaramakrishnan, S., Gurusubramanian, G., Kumar, N.S., 2015. Synthesis of silver and gold nanoparticles using *Jasminum nervosum* leaf extract and its larvicidal activity against filarial and arboviral vector *Culex quinquefasciatus* say (Diptera: Culicidae). Environ. Sci. Pollut. Res. Int. 22 (22), 17753–17768.

Lateef, A., Ojo, S.A., Akinwale, A.S., Azeez, L., Gueguim-Kana, E.B., Beukes, L.S., 2015a. Biogenic synthesis of silver nanoparticles using cell-free extract of *Bacillus safensis* LAU 13: antimicrobial, free radical scavenging and larvicidal activities. Biologia 70 (10), 1295–1306.

Lateef, A., Azeez, M.A., Asafa, T.B., Yekeen, T.A., Akinboro, A., Oladipo, I.C., Ajetomobi, F.E., Gueguim-Kana, E.B., Beukes, L.S., 2015b. *Cola nitida*-mediated biogenic synthesis of silver nanoparticles using seed and seed shell extracts and evaluation of antimicrobial activities. BioNanoScience 5, 196–205.

Lateef, A., Azeez, M.A., Asafa, T.B., Yekeen, T.A., Akinboro, A., Oladipo, I.C., Azeez, L., Ojo, S.A., Gueguim-Kana, E.B., Beukes, L.S., 2016a. Cocoa pod husk extract mediated tivities. J. Nanostruct. Chem. 6 (2), 159–169.

Lateef, A., Azeez, M.A., Asafa, T.B., Yekeen, T.A., Akinboro, A., Oladipo, I.C., Azeez, L., Ajibade, S.E., Ojo, S.A., Gueguim-Kana, E.B., Beukes, L.S., 2016b. Biogenic synthesis of silver nanoparticles using a pod extract of *Cola nitida*: antibacterial and antioxidant activities and application as a paint additive. J. Taibah Uni. Sci. 10, 551–562.

Lateef, A., Akande, M.A., Azeez, M.A., Ojo, S.A., Folarin, B.I., Gueguim-Kana, E.B., Beukes, L.S., 2016c. Phytosynthesis of silver nanoparticles (AgNPs) using miracle fruit plant (*Synsepalum dulcificum*) for antimicrobial, catalytic, anti-coagulant and thrombolytic applications. Nanotechnol. Rev. 5 (6), 507–520.

Lateef, A., Akande, M.A., Ojo, S.A., Folarin, B.I., Gueguim-Kana, E.B., Beukes, L.S., 2016d. Paper wasp nest-mediated biosynthesis of silver nanoparticles for antimicrobial, catalytic, anti-coagulant and thrombolytic applications. 3 Biotech 6, 140.

Lateef, A., Ojo, S.A., Elegbede, J.A., Azeez, M.A., Yekeen, T.A., Akinboro, A., 2017. Evaluation of some biosynthesized silver nanoparticles for biomedical applications: hydrogen peroxide scavenging, anticoagulant and thrombolytic activities. J. Clust. Sci. 28, 1379–1392.

Liang, J., Yang, Z., Tang, L., Zeng, G., Yu, M., Li, X., Wu, H., Qian, Y., Li, X., Luo, Y., 2017. Changes in heavy metal mobility and availability from contaminated wetland soil remediated with combined biochar-compost. Chemosphere 181, 281–288.

Line, C., Camille, L.A., Flahaut, E., 2017. Carbon nanotubes: impacts and behaviour in the terrestrial ecosystem—a review. Carbon 123, 767–785.

Liu, W., Tian, S., Zhao, X., Xie, W., Gong, Y., Zhao, D., 2015. Application of stabilized nanoparticles for in situ remediation of metal-contaminated soil and groundwater: a critical review. Curr. Pollut. Rep. 1, 280–291.

Liu, L., Li, W., Song, W., Guo, M., 2018. Remediation techniques for heavy metal contaminated soils: principles and applicability. Sci. Total Environ. 633, 206–219.

Liu, W., Worms, I., Slaveykova, V.I., 2020. Interaction of silver nanoparticles with antioxidant enzymes. Environ. Sci. Nano. https://doi.org/10.1039/c9en01284b.

Lv, J., Christie, P., Zhang, S., 2019. Uptake, translocation, and transformation of metal-basednanoparticles in plants: recent advances and methodological challenges. Environ. Sci. Nano 6 (1), 41–59.

Ma, C., White, J.C., Zhao, J., Zhao, Q., Xing, B., 2018. Uptake of engineered Nanoparticlesby food crops: characterization, mechanisms, and implications. Annu. Rev. Food Sci. Technol. 9, 129–153.

Mahgoub, H.B., 2019. Nanoparticles used for extraction of polycyclic aromatic hydrocarbons. J. Chem. 2019. https://doi.org/10.1155/2019/4816849.

Marslin, G., Sheeba, C.J., Franklin, G., 2017. Nanoparticles alter secondary metabolism in plants via ROS burst. Front. Plant Sci. 8, 832.

Mehta, C., Srivastava, R., Arora, S., Sharma, A., 2016. Impact assessment of silver nanoparticles on plant growth and soil bacterial diversity. 3 Biotech 6, 254.

Mishra, S., Singh, B.R., Singh, A., Keswani, C., Naqvi, A.H., Singh, H.B., 2014. Biofabricated silver nanoparticles act as a strong fungicide against *Bipolaris sorokiniana* causing spot blotch disease in wheat. PLoS One 9 (5), e97881.

Morales-Díaz, A.B., Ortega-Ortíz, H., Juárez-Maldonado, A., Cadenas-Pliego, G., González Morales, S., Benavides-Mendoza, A., 2017. Application of nanoelements in plant nutrition and its impact in ecosystems. Adv. Nat. Sci. Nanosci. Nanotechnol. 8, 013001.

Mukherjee, A., Peralta-Videa, J.R., Gardea-Torresdey, J., White, J.C., 2016. Chapter 20: Effects and uptake of nanoparticles in plants. In: Xing, B., Vecitis, C.D., Senesi, N. (Eds.), Engineered Nanoparticles and the Environment: Biophysicochemical Processes and Toxicity, first ed. John Wiley & Sons, Inc, pp. 386–408.

Nair, R., 2018. Chapter 17: Plant response strategies to engineered metal oxide nanoparticles: a review. In: Faisal, M., et al. (Eds.), Phytotoxicity of Nanoparticles. Springer International Publishing AG, part of Springer Nature, pp. 377–393.

Nair, P.M.G., Chung, I.M., 2014. Physiological and molecular level effects of silver nanoparticles exposure in rice (*Oryza sativa* L.) seedlings. Chemosphere 112, 105–113.

Najafpoor, A., Norouzian-Ostad, R., Alidadi, H., Rohani-Bastami, T., Davoudi, M., Barjasteh Askari, F., Zanganeh, J., 2020. Effect of magnetic nanoparticles and silver-loaded magnetic nanoparticles on advanced wastewater treatment and disinfection. J. Mol. Liq. 303, 112640.

Nakamura, S., Sato, M., Sato, Y., Ando, N., Takayama, T., Fujita, M., Ishihara, M., 2019. Synthesis and application of silver nanoparticles (Ag NPs) for the prevention of infection in healthcare workers. Int. J. Mol. Sci. 20, 3620.

Nakkala, J.R., Mata, R., Raja, K., Chandra, V.K., Sadras, S.R., 2018. Green synthesized silver nanoparticles: catalytic dye degradation, *in vitro* anticancer activity and *in vivo* toxicity in rats. Mater. Sci. Eng. C 91, 372–381.

Nile, S.H., Baskar, V., Selvaraj, D., Nile, A., Xiao, J., Kai, J., 2020. Nanotechnologies in food science: applications, recent trends and future perspectives. Nano-Micro Lett. 12, 45.

Noori, A., Ngo, A., Gutierrez, P., Theberge, S., White, J.C., 2020. Silver nanoparticle detection and accumulation in tomato (*Lycopersicon esculentum*). J. Nanopart. Res. 22, 131.

Ogunbode, S.M., Salau, A.K., Azeez, L., Osineye, S.O., Olaogun, O.O., Adebisi, J.O., Akinlade, H.O., Isa, F.O., 2019. Carcass characteristics and selected tissue enzymes activities of birds injected with cocoa pod extract-mediated silver nanaoparticles (CPHE-AgNPs) at day old. Sci. Focus 23 (2), 81–90.

Ogunkunle, C.O., Jimoh, M.A., Asogwa, N.T., Viswanathan, K., Vishwakarma, V., Fatoba, P.O., 2018. Effects of manufactured nano-copper on copper uptake, bioaccumulation and enzyme activities in cowpea grown on soil substrate. Ecotoxicol. Environ. Saf. 155, 86–93.

Pacheco, I., Buzea, C., 2017. Nanoparticle interaction with plants. In: Ghorbanpour, M., Manika, K., Varma, A. (Eds.), Nanoscience and Plant–Soil Systems. Soil Biology. 48. Springer, Cham, pp. 323–355.

Pacheco, I., Buzea, C., 2018. Chapter 1: Nanoparticle uptake by plants: beneficial or detrimental? In: Faisal, M., et al. (Eds.), Phytotoxicity of Nanoparticles. Springer International Publishing AG, part of Springer Nature, pp. 1–61.

Park, S., Ahn, Y.-J., 2016. Multi-walled carbon nanotubes and silver nanoparticles differentially affect seed germination, chlorophyll content, and hydrogen peroxide accumulation in carrot (*Daucus carota* L.). Biocatal. Agric. Biotechnol. 8, 257–262.

Prasad, R., Bhattacharyya, A., Nguyen, Q.D., 2017. Nanotechnology in sustainable agriculture: recent developments, challenges, and perspectives. Front. Microbiol. 8, 1014 (1–13).

Rajwade, J.M., Chikte, R.G., Paknikar, K.M., 2020. Nanomaterials: new weapons in a crusade against phytopathogens. Appl. Microbiol. Biotechnol. 104, 1437–1461.

Raliya, R., Nair, R., Chavalmane, S., Wang, W.N., Biswas, P., 2015. Mechanistic evaluation of translocation and physiological impact of titanium dioxide and zinc oxide nanoparticles on the tomato (*Solanum lycopersicum* L.) plant. Metallomics 7, 1584–1594.

Rani, P.U., Yasur, J., Loke, K.S., Dutta, D., 2016. Effect of synthetic and biosynthesized silver nanoparticles on growth, physiology and oxidative stress of water hyacinth: *Eichhornia crassipes* (Mart) Solms. Acta Physiol. Plant. 38, 1–9.

Reddy, P.V.L., Hernandez-Viezcas, J.A., Peralta-Videa, J.R., Gardea-Torresday, J.L., 2016. Lessons learned: are engineered nanomaterials toxic to terrestrial plants? Sci. Total Environ. 568, 470–479.

Ren, Y., Wang, W., He, J., Zhang, L., Wei, Y., Yang, M., 2020. Nitric oxide alleviates salt stress in seed germination and early seedling growth of pakchoi (*Brassica chinensis* L.) by enhancing physiological and biochemical parameters. Ecotoxicol. Environ. Saf. 187, 109785.

Rezvani, N., Sorooshzadeh, A., Farhadi, N., 2012. Effects of nano-silver on growth of saffron in flooding stress. World Acad. Sci. Eng. Technol. 6, 517–522.

Rico, C.M., Majumdar, S., Duarte-Gardea, M., Peralta-Videa, J.R., Gardea-Torresdey, J.L., 2011. Interaction of nanoparticles with edible plants and their possible implications in the food chain. J. Agric. Food Chem. 59, 3485–3498.

Rico, C.M., Peralta-Videa, J.R., Garadea-Torresdey, J.L., 2015a. Chemistry, biochemistry of nanoparticles, and their role in antioxidant defense system in plants. In: Siddiqui, H.M., et al. (Eds.), Nanotechnology and Plant Sciences. Springer International Publishing, Switzerland, pp. 1–17.

Rico, C.M., Barrios, A.C., Tan, W., Rubenecia, R., Lee, S.C., Varela-Ramirez, A., Peralta-Videa, J.R., Gradea-Torresdey, J.L., 2015b. Physiological and biochemical response of soil grown barley (*Hordeum vulgare* L.) to cerium oxide nanoparticles. Environ. Sci. Pollut. Res. 22, 10551–10558.

Rizwan, M., Ali, S., Abbas, T., Adrees, M., Zia-ur-Rehman, M., Ibrahim, M., Abbas, F., Qayyum, M.F., Nawaz, R., 2018. Residual effects of biochar on growth, photosynthesis and cadmium uptake in rice (*Oryza sativa* L.) under Cd stress with different water conditions. J. Environ. Manag. 206, 676–683.

Rizwan, M., Ali, S., Ali, B., Adrees, M., Arshad, M., Hussain, A., Zia ur Rehman, M., Waris, A.A., 2019. Zinc and iron oxide nanoparticles improved the plant growth and reduced the oxidative stress cadmium concentration in wheat. Chemosphere 214, 269–277.

Roberto, S.-C.S., Andrea, P.-M., Andres, G.-O., Norma, F.-P., Hermes, P.-H., Gabriela, M.-P., Fabiána, F.-L., 2020. Chapter 9: Phytonanotechnology and environmental remediation. In: Phytonanotechnology. Elsevier Inc, pp. 159–185.

Saha, J., Begum, A., Mukherjee, A., Kumar, S., 2017. A novel green synthesis of silver nanoparticles and their catalytic action in reduction of methylene blue dye. Sustain. Environ. Res. 27, 245–250.

Salama, H.M., 2012. Effects of silver nanoparticles in some crop plants, common bean (*Phaseolus vulgaris* L.) and corn (*Zea mays* L.). Int. Res. J. Biotechnol. 3 (10), 190–197.

Sanzari, I., Leone, A., Ambrosone, A., 2019. Nanotechnology in plant science: to make a long story short. Front. Bioeng. Biotechnol. 7, 120. https://doi.org/10.3389/fbioe.2019.00120.

Scott, T., Zhao, H.L., Deng, W., Feng, X.H., Li, Y., 2019. Photocatalytic degradation of phenol in water under simulated sunlight by an ultrathin MgO coated Ag/TiO_2 nanocomposite. Chemosphere 216, 1–8.

Servin, A., Elmer, W., Mukherjee, A., De la Torre-Roche, R., Hamdi, H., White, J.C., Dimkpa, C., 2015. A review of the use of engineered nanomaterials to suppress plant disease and enhance crop yield. J. Nanopart. Res. 17 (2), 92.

Seshadri, B., Bolan, N.S., Choppala, G., Kunhikrishnan, A., Sanderson, P., Wang, H., Currie, L.D., Tsang, D.C.W., Ok, Y.S., Kim, G., 2017. Potential value of phosphate compounds in enhancing immobilization and reducing bioavailability of mixed heavy metal contaminants in shooting range soil. Chemosphere 184, 197–206.

Shabnam, N., Kim, M., Kim, H., 2019. Iron (III) oxide nanoparticles alleviate arsenic induced stunting in *Vigna radiata*. Ecotoxicol. Environ. Saf. 183, 109496.

Shalaby, T.A., Bayoumi, Y., Abdalla, N., Taha, H., Alshaal, T., Shehata, S., Amer, M., Domokos Szabolcsy, E., El-Ramady, H., 2016. Nanoparticles, soils, plants and sustainable agriculture. In: Ranjan, S. (Ed.), Nanoscience in Food and Agriculture 1, Sustainable Agriculture Reviews 20. Springer International Publishing, Switzerland, pp. 283–312.

Shams, G., Ranjbar, M., Amiri, A., 2013. Effect of silver nanoparticles on concentration of silver heavy element and growth indexes in cucumber (*Cucumis sativus* L. negeen). J. Nanopart. Res. 15, 1630.

Sharma, P., Bhatt, D., Zaidi, M.G., Saradhi, P.P., Khanna, P.K., Arora, S., 2012a. Silver nanoparticle mediated enhancement in growth and antioxidant status of *Brassica juncea*. Appl. Biochem. Biotechnol. 167, 2225–2233.

Sharma, P., Jha, A.B., Dubey, R.S., Pessarakli, M., 2012b. Reactive oxygen species, oxidative damage, and antioxidative defense mechanism in plants under stressful conditions. J. Bot., 217037.

Sheykhbaglou, R., Sedghi, M., Fathi-Achachlouie, B., 2018. The effect of ferrous nano-oxide particles on physiological traits and nutritional compounds of soybean (*Glycine max* L.) seed. Ann. Braz. Acad. Sci. 90 (1), 485–494.

Shtangeeva, I., Perämäki, P., Niemelä, M., et al., 2018. Potential of wheat (*Triticum aestivum* L.) and pea (*Pisum sativum*) for remediation of soils contaminated with bromides and PAHs. Int. J. Phytorem. 20 (6), 560–566.

Siddiqi, K.S., Husen, A., 2017. Plant response to engineered metal oxide nanoparticles. Nanoscale Res. Lett. 12 (1), 92.

Singh, J., Dutta, T., Kim, K.H., Rawat, M., Samddar, P., Kumar, P., 2018. Green' synthesis of metals and their oxide nanoparticles: applications for environmental remediation. J. Nanobiotechnol. 16, 84.

Sivaram, A.K., Logeshwaran, P., Lockington, R., Naidu, R., Megharaj, 2019. Phytoremediation efficacy assessment of polycyclic aromatic hydrocarbons contaminated soils using garden pea (*Pisum sativum*) and earthworms (*Eisenia fetida*). Chemosphere 229, 227–235.

Song, B., Xu, P., Chen, M., Tang, W., Zeng, G., Gong, J., Zhang, P., Ye, S., 2019. Using nanomaterials to facilitate the phytoremediation of contaminated soil. Crit. Rev. Environ. Sci. Technol. 49, 791–824.

Su, Y., Ashworth, V., Kim, C., Adeleye, A.S., Rolshausen, P., Roper, C., White, J., Jassby, D., 2019. Delivery, uptake, fate, and transport of engineered nanoparticles in plants: a critical review and data analysis. Environ. Sci. Nano 6, 2311–2331.

Syu, Y.-Y., Hung, J.-H., Chen, J.-C., Chuang, H.-W., 2014. Impacts of size and shape of silver nanoparticles on Arabidopsis plant growth and gene expression. Plant Physiol. Biochem. 83, 57–64.

Tewari, B.B., 2019. Critical reviews on engineered nanoparticles in environmental remediation. CJAST 36 (4), 1–21.

Thuesombat, P., Hannongbua, S., Akasit, S., Chadchawan, S., 2014. Effect of silver nanoparticles on rice (*Oryza sativa* L. cv. KDML 105) seed germination and seedling growth. Ecotoxicol. Environ. Saf. 104, 302–309.

Tombuloglu, H., Slimani, Y., Tombuloglu, G., Demir-Korkmaz, A., Baykal, A., Almessiere, M., Ercan, I., 2019a. Impact of superparamagnetic iron oxide nanoparticles (SPIONs) and ionic iron on physiology of summer squash (*Cucurbita pepo*): a comparative study. Plant Physiol. Biochem. 139, 56–65.

Tombuloglu, H., Slimani, Y., Tombuloglu, G., Almessiere, A., Demir-Korkmaz, H.S.A., AlShammari, T.M., Baykal, A., Ercan, I., Hakeem, K.R., 2019b. Impact of calcium and magnesium substituted strontium nano-hexaferrite on mineral uptake, magnetic character, and physiology of barley (*Hordeum vulgare* L.). Ecotoxicol. Environ. Saf. 186, 109751.

Tombuloglu, H., Slimani, Y., Tombuloglu, G., Almessiere, M., Baykal, A., Ercan, I., Sozeri, H., 2019c. Tracking of $NiFe_2O_4$ nanoparticles in barley (*Hordeum vulgare* L.) and their impact on plant growth, biomass, pigmentation, catalase activity, and mineral uptake. Environ. Nanotechnol. Monit. Manag. 11, 100223.

Tripathi, D.K., Singh, S., Singh, S., Pandey, R., Singh, V.P., Sharma, N.C., Prasad, S.M., Dubey, N.K., Chauhan, D.K., 2017. An overview on manufactured nanoparticles in plants: uptake, translocation, accumulation and phytotoxicity. Plant Physiol. Biochem. 110, 2–12.

Uddin, I., Ahmad, K., Khan, A.A., Kazmi, M.A., 2017. Synthesis of silver nanoparticles using *Matricaria recutita* (Babunah) plant extract and its study as mercury ions sensor. Sens. Biosens. Res. 16, 62–67.

Usman, M., Byrne, J.M., Chaudhary, A., Orsetti, S., Hanna, K., Ruby, C., Kappler, A., Haderlein, S.B., 2018. Magnetite and green rust: synthesis, properties, and environmental applications of mixed-valent iron minerals. Chem. Rev. 118, 3251–3304.

Usman, M., Farooq, M., Wakeel, A., Nawaz, A., Cheema, S.A., Rehman, H., Ashraf, I., Sanaullah, M., 2020. Nanotechnology in agriculture: current status, challenges and future opportunities. Sci. Total Environ. 721, 137778.

Vanaja, M., Paulkumar, K., Baburaja, M., Rajeshkumar, S., Gnanajobitha, G., Malarkodi, C., Sivakavinesan, M., Annadurai, G., 2014. Degradation of methylene blue using biologically synthesized silver nanoparticles. Bioinorg. Chem. Appl. https://doi.org/10.1155/2014/742346.

Veisi, H., Azizi, S., Mohammadi, P., 2018. Green synthesis of the silver nanoparticles mediated by *Thymbra spicata* extract and its application as a heterogeneous and recyclable nanocatalyst for catalytic reduction of a variety of dyes in water. J. Clean. Prod. 170, 1536–1543.

Verma, S.K., Das, A.K., Patel, M.K., Shah, A., Kumar, V., Gantait, S., 2018. Engineered nanomaterials for plant growth and development: a perspective analysis. Sci. Total Environ. 630, 1413–1435.

Vinkovic, T., Novák, O., Strnad, M., Goessler, W., Jurašin, D.D., Paradikovic, N., Vrcek, I.V., 2017. Cytokinin response in pepper plants (*Capsicum annuum* L.) exposed to silver nanoparticles. Environ. Res. 156, 10–18.

Vítkova, M., Puschenreiter, M., Komarek, M., 2018. Effect of nano zero-valent iron application on As, Cd, Pb, and Zn availability in the rhizosphere of metal(loid) contaminated soils. Chemosphere 200, 217–226.

Wagner, S., Gondikas, A., Neubauer, E., Hofmann, T., von der Kammer, F., 2014. Spot the difference: engineered and natural nanoparticles in the environment—release, behaviour, and fate. Angew. Chem. Int. Ed. 53, 12398–12419.

Werkneh, A.A., Rene, E.R., 2019. Applications of nanotechnology and biotechnology for sustainable water and wastewater treatment. In: Bui, X.T., Chiemchaisri, C., Fujioka, T., Varjani, S. (Eds.), Water and Wastewater Treatment Technologies. Energy, Environment, and Sustainability. Springer, Singapore, pp. 405–430.

White, J.C., Gardea-Torresdey, J.L., 2018. Achieving food security through the very small. Nat. Nanotechnol. 13, 621–629.

Xiong, J.-L., Li, J., Wang, H.-C., Zhang, C.-L., Naeem, M.S., 2018. Fullerol improves seed germination, biomass accumulation, photosynthesis and antioxidant system in *Brassica napus* L. under water stress. Plant Physiol. Biochemist 129, 130–140.

Yan, A., Chen, Z., 2019. Impacts of silver nanoparticles on plants: a focus on the phytotoxicity and underlying mechanism. Int. J. Mol. Sci. 20, 1003. https://doi.org/10.3390/ijms20051003.

Yang, J., Cao, W., Rui, Y., 2017. Interactions between nanoparticles and plants: phytotoxicity and defense mechanisms. J. Plant Inter. 12 (1), 158–169.

Yekeen, T.A., Azeez, M.A., Akinboro, A., Lateef, A., Asafa, T.B., Oladipo, I.C., Oladokun, S.O., Ajibola, A.A., 2018. Safety evaluation of green synthesized Cola nitida pod, seed and seed shell extracts-mediated silver nanoparticles (AgNPs) using *Allium cepa* assay. J. Taibah Univ. Sci. 11, 895–909.

Yin, L., Colman, B.P., McGill, B.M., Wright, J.P., Bernhardt, E.S., 2012. Effects of silver nanoparticle exposure on germination and early growth of eleven wetland plants. PLoS One 7, e47674.

Zhang, P., Guo, Z., Zhang, Z., Fu, H., White, J.C., Lynch, I., 2020. Nanomaterial transformation in the soil–plant system: implications for food safety and application in agriculture. Small. https://doi.org/10.1002/smll.202000705.000705.

Zhao, L., Lu, L., Wang, A., Zhang, A., Huang, M., Wu, H., Xing, B., Wang, Z., Ji, R., 2020. Nanobiotechnology in agriculture: use of nanomaterials to promote plant growth and stress tolerance. J. Agric. Food Chem. 68 (7), 1935–1947.

Zheng, M., Huang, Z., Ji, H., Qiu, F., Zhao, D., Bredar, A.R.C., Farnum, B.H., 2020. Simultaneous control of soil erosion and arsenic leaching at disturbed land using polyacrylamide modified magnetite nanoparticles. Sci. Total Environ. 702, 134997.

Zulfiqar, F., Navarro, M., Ashraf, M., Akram, N.A., Munné-Bosch, S., 2019. Nanofertilizer use for sustainable agriculture: advantages and limitations. Plant Sci. 289, 110270.

Zuverza-Mena, N., Armendariz, R., Peralta-Videa, J.R., Gardea-Torresdey, J.L., 2016. Effects of silver nanoparticles on radish sprouts: root growth reduction and modifications in the nutritional value. Front. Plant Sci. 7, 90.

Zuverza-Mena, N., Martínez-Fernandez, D., Du, W., Hernandez-Viezcas, J.A., Bonilla-Bird, N., Lopez-Moreno, M.L., Komarek, M., Peralta-Vide, J.R., Gardea-Torresdey, J.L., 2017. Exposure of engineered nanomaterials to plants: insights into the physiological and biochemical responses—a review. Plant Physiol. Biochem. 110, 236–264.

CHAPTER 11

Controlled-release and positive effects of silver nanoparticles: An overview

Ambreen Ahmed and Shabana Wagi
Department of Botany, University of the Punjab, Lahore, Pakistan

11.1 Introduction

Nanotechnology is the recent discipline of science that deals with nanosize particles ranging in size from 1 to 100 nm. Due to their unique size, dimensions, and optical properties, nanoparticles exhibit unique characteristics and properties, which are different from the molecules or atoms of relevant metal or its bulk matter in many ways (Yadav et al., 2015; Wagi and Ahmed, 2020). Green nanotechnology is a branch of nanotechnology that deals with the ability of living organisms to manipulate metals by converting them to nanosize atoms and transform them into nanoparticles. The biomolecules in these living organisms enhance the efficiency of these green nanoparticles by affecting their optical properties, in addition to the modifications in their physical and chemical properties. Green synthesis of nanoparticles makes it possible to obtain nanoparticles with desired size and composition by controlling various factors involved in the process. Among these factors, the primary role is played by the organism used for this purpose. Other factors that need to be monitored include pH, reaction time, temperature, the path followed and basic compounds/solvents selected (Patra and Baek, 2014; Wagi and Ahmed, 2019b). Manipulation of these factors under highly controlled conditions along with in-depth knowledge of all the key components involved enables the researcher to synthesize nanoparticles with desired characteristics, thereby making possible their successful application. It is primarily important to characterize the synthesized green nanoparticles before their application to fully evaluate their potential for a successful outcome.

Nanoparticles synthesized using green technology offer a wide range of variability concerning their size, shape, nature, dimensions, and applicability owing to the wide diversity of bacteria and plants and depend upon the type of organism selected. Among the various properties, high surface to volume ratio is one of the unique properties of AgNPs for their widespread application (Guzel and Erdal, 2018). With a rapid increase in the human population, continuous food supply is one of the major concerns for every country's economy. This is the high time to find out alternative strategies to improve plant growth and crop yield and replace the hazardous, toxic, and expensive chemical fertilizers with the nontoxic, ecofriendly, and cost-effective nanofertilizers, which will not only help to minimize environmental pollution including soil, water, and air pollution caused by chemical fertilizers but will also be helpful economically by reducing the expenditure involved in the purchase of synthetic fertilizers, pesticides, etc. (Wagi and Ahmed, 2019a; Abbas and Ahmed, 2020). Green synthesis methods allow the exploitation of indigenous resources available in the form of indigenous bacteria and plants making the nanofertilizers more compatible with the local crops as well as local environment, leading to a higher rate of successful application with minimum cost involved.

11.2 Synthesis of nanoparticles

There are two strategies which can be followed for the synthesis of nanoparticles, i.e., destructive and constructive strategy.

11.2.1 Destructive strategy

In the destructive strategy, the bulk matter is converted to nanoparticles by applying physical methods such as milling, grinding, sputting, laser ablation method, nanosphere lithography (NSL), etc. Thus, in this strategy, physical methods are used to break down the bulk matter to particles of nanosize. This strategy is also named as top-down or top-to-bottom approach (Patra and Baek, 2014; Parveen et al., 2015; Ealia and Saravanakumar, 2017; Guzel and Erdal, 2018; Gour and Jain, 2019; Castillo-Henríquez et al., 2020). This method is expensive, highly time-consuming, and energy-requiring.

11.2.2 Constructive strategy

In the constructive strategy, metal or metal oxides are converted to nanosize particles by following either the chemical method or green method. Chemical methods include chemical reduction, electrolysis, sol–gel process, microemulsion, etc., whereas green method involved the use of living organism including microbes and plants (Das et al., 2017). This strategy is also named as bottom-up or bottom-to-top approach (Patra and Baek, 2014; Parveen et al., 2015; Ealia and Saravanakumar, 2017; Guzel and Erdal, 2018; Gour and Jain, 2019; Castillo-Henríquez et al., 2020).

There are also chances of toxicity in the nanoparticles synthesized via chemical or physical methods depending upon the chemicals utilized so toxicity testing is required for these *AgNPs* before their use as nanofertilizers.

11.3 Green synthesis of silver nanoparticles

Green synthesis of silver nanoparticles follows the constructive strategy for nanoparticle synthesis (Fig. 11.1). Green nanoparticles work more efficiently in the living system than the chemically or physically synthesized nanoparticles (Patra and Baek, 2014; Gour and Jain, 2019).

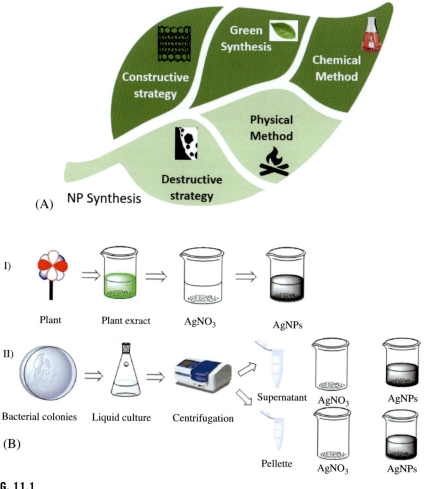

FIG. 11.1

(A) Various strategies for nanoparticle synthesis and (B) various strategies (I and II) for nanoparticle synthesis.

11.3.1 Green synthesis of silver nanoparticles using bacteria

Bacterial synthesis of silver nanoparticles involves the reduction of silver ions to zerovalent silver atoms leading to the formation of AgNPs. Some of the biomolecules in the bacterial environment act as capping agents and provide stability to the silver nanoparticles by binding to their surface during their synthesis. Bacteria generally synthesize many inorganic and organic substances including various enzymes, proteins, and secondary metabolites, etc., both extracellularly and intracellularly. These substances are involved in the manipulation of silver metal ions for the development of silver nanoparticles (Das et al., 2017). The synthesis of these intracellular and extracellular substances follows different patterns in different bacteria. Wide range of diversity is also found in the type and quantity of these substances in nature (Patra and Baek, 2014). This factor plays a major role in the variable synthesis of silver nanoparticles by using different bacteria regarding their type, size, shape, and quantity. During the synthesis of AgNPs using bacteria, the enzymes in the cell wall of bacteria interact with Ag ions and cause reduction of the positively charged silver ions, thereby manipulating them to AgNPs. Various bacterial proteins act as capping agent, which help in the greater stability of these AgNPs (Wagi and Ahmed, 2019b; Gour and Jain, 2019). Ambient temperature is required for the green synthesis of AgNPs using bacteria, which makes the process very fast, easy, and economic. Bacterially synthesized AgNPs have shown close similarity to the chemically synthesized AgNPs regarding their properties (Das et al., 2017).

11.3.2 Green synthesis of silver nanoparticles using plants

Plants are the group of living organisms exhibiting the highest level of diversity both in the form, structures, and production of variable substances with a wide range of secondary metabolites (Fig. 11.2). Most of the plants are capable of reducing metal ions by either using various tissues or organs or at their surfaces (Patra and Baek, 2014). Plant metabolites cause the reduction of silver ions resulting in the formation of silver nanoparticles. Plant biomolecules, especially proteins, terpenoids, flavonoids, amino acids, vitamins, sugars, etc., play a significant role in binding to the silver ions and cause their reduction (Gour and Jain, 2019). Reduction of silver ions to silver nanoparticles is followed by the growth and termination phase, which

FIG. 11.2

Green synthesis of AgNPs using bacteria.

results in the formation of stable nanoparticles with most suitable conformation (Makarov et al., 2014). Plant biomolecules play a key role in the formation of stable AgNPs (Makarov et al., 2014; Parveen et al., 2015; Gour and Jain, 2019).

11.4 Controlled release of silver nanoparticles

As the world population is increasing tremendously, we need more resources and consequent increment in the agricultural sector to meet the growing world food demand. Green revolution in agriculture has occurred during 1960–1970 in which crop yield was substantially enhanced using macrofertilizers, breeding techniques, and plant protective agents. Modern agricultural practices heavily rely on chemical fertilizers, which provide nourishment to the plants and, hence, ensure better plant growth and development. Upto 40%–60% of world food production depends on utilization of chemical fertilizers (Marchiol, 2018). Chemical fertilizers are responsible for changing the environment by affecting its quality and directly contribute to environmental pollution. Chemical fertilizers are not only toxic for the ecosystem in the longer run, but at the same time, these are quite expensive and are a great burden to the country's economy. Nanofertilizers offer an intelligent economical and ecofriendly alternative over chemical fertilizers, with the efficient and target-based delivery system and efficient potential for improved nutrient uptake. Silver nanoparticles have unique physicochemical properties, size, and shape, which allow them to be used in a new paradigm in agricultural fertilization with improved bioavailability of these NPs to the soil. Besides the use of nanofertilizers, nanodimensional adsorbent also allows the slow release of nanofertilizers as compared to the chemical or mineral fertilizers and also protect the plants from abiotic stress. The nanofertilizers could either be utilized in combination with bacteria (biofertilizers) or separately when utilized in combination, and their efficiency is recorded to be improved (Wagi and Ahmed, 2019b). Nowadays, various nanoparticles (Cu, Ag, Zn, Mn, Mo, and Fe) and several bioformulations (urea, phosphate, and validamycin) of different fertilizers and pesticides are generally being applied for better crop yield. Nanofertilizers have the potential to enhance nutrient efficiency due to improved nutrient uptake with the advent of surface area to volume ratio of nanoparticles, which enhanced nutrient–surface interface. Hence, the use of nanofertilizers not only sustainably improve crop yield but also minimize the risk of environmental pollution.

11.4.1 Controlled release of silver nanoparticles as nanofertilizers

AgNPs were transported into the plant body via two main channels. One is short distance transport using intracellular spaces, while the second most important pathway is long-distance transport, in which entry point is vascular tissue. AgNPs also utilize cell wall and cell membrane to get entry into the cell. Initially, only the passage of small AgNPs is allowed, which later develops larger holes into the cell wall and ultimately allow the macrosize AgNPs into the cell. Xylem tissue in the root acts as a vehicle for the transfer of AgNPs to the shoots (Yan and Chen, 2019). There are also some clues about the mechanism of endocytosis for the entry of AgNPs into the plant cell, which is still under investigation (Dykman and Shchyogolev Sergei, 2018). After entry inside, AgNPs use either the apoplast or simplest pathway for further translocation inside the plant (Goswami and Mathur, 2019).

There are also several materials utilized as bioabsorbent for the slow release of AgNPs to the plants. In this case, mineral fertilizers were usually loaded with nanomaterials such as zeolite, which is an interlinked cage structure and ensures slow availability of both micro-and macronutrients to the crops. Zeolite has been used since the 1980s. Chitosan is also an important biocompatible and biodegradable material utilized to load nitrogen, phosphorous, and potassium (Marchiol, 2019). Water-based nanoformulations are quite smart alternatives, which increase biological efficiency. Nanocapsules, nanoemulsions, nanospheres, nanomicelles, and nanosuspensions are improved formulations with properties such as greater stability, bioavailability, dispersion, target adhesion, permeability, and controlled release (Zhao et al., 2017). Controlled release of nanofertilizers is essential to alleviate the increasing environmental problems due to excessive use of chemical fertilizers, which are extensively utilized in the farms to improve crop yield. The deliberate release of fertilizers coated with nanoparticles also pointedly reduced leaching and denitrification of chemical fertilizers, especially nitrogen fertilizers, which are prone to quick degradation and waste. Moreover, the controlled release of silver nanoparticles ensures significantly controlled liberation of nutrients through diffusion and is an excellent strategy for supplying plant nutrients at the right time and right stage of growth. Nanofertilizers formulated using nanomaterial coatings such as polymer films in addition to the encapsulation techniques using sodium alginate and nanoemulsions, significantly support porous plant roots for enhanced transport of nanomaterials as compared to the conventional fertilizers. Interactions of micropore and channels create nanopores, which in combination with stomatal pores also facilitate nanofertilizer uptake (Iqbal, 2019). High surface area to volume ratio of AgNPs supports their penetration in the seeds followed by radicle and roots, hence improve their bioavailability (Abobatta, 2018). Nanoforms are responsible for the active delivery of the nanoparticles to the plants and consequently improve plant nutrient value, seed germination, and biomass and provide a sustainable alternative to their counter bulk part (chemical fertilizers). With the advent of nanotools, smart and site-specific delivery of nanofertilizers and nanopesticides are now possible, which will not only reduce the cost but will also save resources. Even minute concentrations of silver nanoparticles are quite effective in improving overall plant growth development, immunity, and also the rhizosphere microbial community with the beneficial microbes that further improve plant growth and development (Fatima et al., 2020; Wagi and Ahmed, unpublished data). NPs may also act as "magic bullets" able to replenish nutrients, supporting beneficial genes and triggering organic compounds, which are targeted to specific plant areas or structures to enhance their productivity. Thus, NPs represent smart nanodelivery systems for agricultural administration, specifically affecting crop nutrition (Castillo-Henríquez et al., 2020).

11.4.2 Controlled release of silver nanoparticles as antioxidant

Antioxidants are intracellular or extracellular compounds that act as a defensive system against oxidants and prevent cell damage. Antioxidant potential of AgNPs is well known with effective scavenging activity that results in improved cell viability and mitigation of mitochondrial and DNA damage (Taha et al., 2019; Abd El-Maksoud et al., 2019; Bedlovičová et al., 2020). AgNPs have the potential to increase the activity of antioxidant enzymes such as peroxidase, catalase, which affect early growth of plant (Soliman et al., 2020). These antioxidants protect the cells from toxic elements, resulting in the induction of oxidative stress (Alkhalaf et al., 2020).

A state of nonequilibrium between cell oxidants and antioxidants due to the accumulation of an excessive number of oxidants in a cell is called oxidative stress. This state of nonequilibrium develops due to excessive accumulation of reactive oxygen species (ROS) and reactive nitrogen species (RNS). When these species are retained in the cell due to failure of cell defensive mechanisms for a longer period of time, it results in cell damage leading to cell death. The ROS exhibit radicals such as hydroxyl (OH), superoxide (O^{2-}), and nonradicals such as hydrogen peroxide and an organic peroxide. Superoxide formation results due to the reduction of oxygen by electrons. ROS and RNS cause double bond polyunsaturated fatty acids oxidation, changing membrane permeability, disturb Ca^{2+} ions homeostasis, inactivate proteins, modify enzymatic activity, and damage DNA, leading to mutation, translational error, and inhibition of photosynthesis. These free radicals also have some important functions in a cell when produced in a controlled amount such as phagocytosis and protect cell against pathogenic microorganisms.

AgNPs have the potential to produce phenolic compounds and flavonoids under stress conditions, which facilitate plant to cope with oxidative damage caused by ROS and ultimately improve plant growth and development. AgNPs also influence antioxidant enzymes such as catalase (CAT), superoxide dismutase (SOD), peroxidase (POD), glutathione reductase (GR), glutathione peroxidase (GPX), and increase their concentration, but it varies from species to species and also depends on the genotype of plants. AgNP scan improve plant growth and enhance plant stress tolerance by improving plant physiology and plant growth. AgNPs were found to upregulate genes that are involved in recirculation of nitrogen, hence improve plant photosynthetic efficiency resulting in improved production of photoassimilates and consequently improve plant growth. AgNPs also reduce proline content in plants, which may be accumulated due to salt stress (Khan et al., 2020). Reactive oxygen species (ROS) act as signaling molecules for stress tolerance. However, excessive release of ROS can collapse the antioxidant defense system, leading to the damage of DNA, lipids, and proteins (Fig. 11.3). Antioxidant enzymes are essential modulator of AgNPs, which induce oxidative stress. Catalase and superoxide dismutase are prominent for maintaining the level of ROS in organisms and are used as a bioindicator of ROS production. When AgNPs penetrate the cell membrane, it binds to the NADH dehydrogenase and generates a high amount of ROS, which disrupt the ATP and interrupt the respiratory chain.

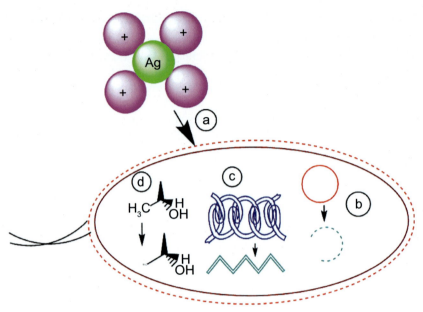

FIG. 11.3

Mechanism of action of Ag NPs as antimicrobial agent:(A) degradation of bacterial cell wall, (B) degradation of plasmid, (C) relaxation in the conformity of DNA, and (D) degradation of a bacterial cell.

11.4.3 Controlled release of nanoparticles as antimicrobial agents

Several nanoformulations are being efficiently utilized as nanofungicide, nanoinsecticides, nanopesticides, and nanoherbicides (Pho et al., 2020). Cell wall of gram-positive bacteria is a thick layer made up of peptidoglycan. In gram-negative bacteria, cell wall is composed of a double layer of lipopolysaccharide. The antimicrobial activity of AgNPs depends on the size, shape, diameter, and surface changes of bacterial cells and it acts in a dose-dependent manner. The shape of the silver nanoparticles is quite important and contribute toward their biocidal properties regarding their antibacterial effect (Docea et al., 2020; Hamad et al., 2020).

The interaction between bacterial cell wall (negatively charged) and AgNPs (positively/less negatively charged) with antibacterial activity may result in the bacterial cell wall disruption. This electrostatic interaction facilitates adhesion of AgNPs on bacterial cell wall and produces changes in bacterial cell wall morphology. This disruption of bacterial cell wall results in the disruption of membrane permeability, leakage of cell sap, as well as it halts the bacterial respiratory functions.

AgNPs adhesion on bacterial cell wall forms irregular pits on cell wall, which facilitates the entry of AgNPs into the periplasmic spaces. AgNPs interact with bacterial proteins of the outer cell membrane and develop complexes with oxygen, phosphorous, and produce irreversible changes in cell wall. Protein thiol group specifically interacts with AgNPs, which is an integral site of Ag$^+$ binding, and it results in the inactivation of thiol group-containing protein molecules. Iodonitrotetrazolium chloride (INT) is a colorless compound and is reduced by bacterial respiratory chain dehydrogenases to a dark red H$_2$O insoluble iodonitrotetrazolium formazan (INF) under normal conditions (show absorbance at 490 nm). AgNPs destruct cell membrane and peptidoglycan, enter the cell, and inhibit the respiratory chain dehydrogenases, leading to bacterial cell death. Ag$^+$ converts DNA from a relaxed state to condensed state, and hence, DNA ultimately loses its replication power. Ag$^+$ interacts with thiol group of proteins leading to enzyme inactivation.

AgNPs and Ag$^+$ both produce conformational changes as well as structural changes in the bacterial DNA, leading to denaturation of DNA. Silver has a high affinity for sulfur and amine groups of bacterial proteins located on bacterial outer cell membrane, which facilitates in the uptake of AgNPs by cell. Ag$^+$ ions attract toward DNA and interact with nucleoside part of the nucleotide, which leads to disruption of complementary base pairing and breaking of hydrogen bonds. The DNA damage is also mediated by oxidative damage caused by AgNPs. Moreover, antioxidant balance also plays a vital role in antibacterial effect of AgNPs and silver ions. Thiol/thiolate is a functional group of amino acid cysteine, which is a highly conserved residue, and plays a significant role in biochemical reactions due to its high affinity to metal-binding proteins. Nucleophilic role in catalytic reactions and ability to form disulfide bond, which is vital in the folding of the three-dimensional structure of proteins (Roy et al., 2019). AgNPs bind to sulfur and a phosphorous group of bacterial DNA that results in aggregation and damage of DNA, resulting in the interruption of transcription and translation.

AgNPs halt the respiratory chain reaction of bacteria by interacting with sulfhydryl group, which ultimately leads to the lipid peroxidation and oxidative damage of DNA and proteins and ultimately cell death. AgNPs cause dephosphorylation of phosphotyrosines and hinder signal transduction. As a result of the action of AgNPs on bacteria, the redox homeostasis is disturbed, and oxidative stress response is promoted, and consequently, increment in the level of superoxide dismutase, peroxidases, and catalases was observed. AgNPs inhibit the activity of catalase and peroxidase, which enhance reactive oxygen species (i.e., H$_2$O$_2$ and peroxidase) accumulation in the cell and delay their removal from the cell, hence hamper DNA, ribosomes, and other macromolecules synthesis (Liao et al., 2019). They have the potential to convert into silver sulfide, which is a stable silver form, and would not allow the release of silver ions into the environment, hence cannot kill the bacterial cells (Seltenrich, 2013).

The antibacterial potential of *AgNPs* is directly proportional to their surface area and size. The *AgNPs* with smaller particle have higher antibacterial activity since a reduction in the particle size would increase the surface area of the synthesized particles facilitating greater contact, which allows the dissolution of materials and facilitate the release of Ag$^+$ that can exhibit bactericidal impact on the pathogenic bacterial cells, hence improved antibacterial efficacy. Moreover, NP with different shapes has distinctive effects on bacterial cells (León-Silva et al., 2016).

11.4.4 Controlled release of silver nanoparticles as nanopesticides

As extensive use of pesticides resulted in the gradual accumulation of these chemicals in the soil as an off-target waste, ultimately leading to the genetic modification of pests against that specific pesticides. The pesticide drift and volatile loss also resulted in environmental pollution. Urgent attention for the protection of resources is thus required to alleviate the hazardous impact of chemical pesticides. Nanoparticles provide one of the most suitable answers to this problem owing to their smaller size, unique physiochemical properties, smart delivery system, and efficient solubility, which makes them most reactive, less volatile, less persistent, and, hence reduce environmental pollution and provide an alternative ecofriendly approach (Gahukar and Das, 2020; Pradhan and Mailapalli, 2020).

Nanoencapsulation of pesticides is quite important for the target-specific, active, and controlled release of the pesticides into the crops. Nanofertilizers allow slow release of nutrients via allosperse delivery system of pesticides, which is an efficient technique (Chhipa, 2017). AgNPs are effective against a wide variety of phytopathogens such as *Botrytis cinerea*, *Biploaris sorokinniana*, *Colletotrichum gloeosporioides*, *Fusarium culmorum*, *Phythium ultimum*, *Phoma*, *Megnaporthe grisea*, *Scalerotinia sclerotiorum*, *Trichoderma* sp., *Sphaerotheca pannasa*, and *Rhizoctonia solani* (Goswami and Mathur, 2019).

Several organic and inorganic nanoparticles have been successfully utilized against a wide range of noxious pests, arthropods, and fungi and have the potential to successfully inhibit their growth at various growth stages (Saranya et al., 2020). Smart delivery system and utilization of various stimuli-based nanoformulations further improve the concept of utilization of nanoformulations as pesticides for efficient eradication of pests (Liang et al., 2020).

Nanomaterials triggered by enzymes are responsible for enzyme stimulated release of nanopesticides in a site-specific manner, and hence can be utilized as an excellent pest control agent, specifically designed according to the nature of enzymes present in the salivary glands and midgut of insects such asprotease and carbohydrases. Some were designed for soil-borne pathogens as they exhibited stimulation in response to several soil enzymes like alkaline phosphatase, dehydrogenase, catalase, and urease in soil. Some nanoformulations acting as nanopesticides are also designed specifically for phytopathogenic fungi that release pectinase and cellulose and are responsible for degrading plant cell wall. The pesticides either encapsulated with nanoparticles or with alginate beads when released in an enzyme-rich environment quickly release and trigger the activity of specific pathogens in an efficient manner (Camara et al., 2019). The mechanism of action of nanopesticides is not fully understood, but when in the vicinity of the pathogen, they influence cell ROS content as well as induce changes in nutrient availability and toxin secreting capabilities of pathogens. NPs also halt reproductive capabilities of the pathogens leading to its death. Nanoparticles are a powerful tool to combat harmful pest of crops to the advent of a smart delivery system, active nature, and these nanopesticides are most promising, sustainable, and ecofriendly substitute to the conventional pesticides (de Oliveira et al., 2020).

11.4 Controlled release of silver nanoparticles

One of the most suitable ways to mitigate the toxicity of the pesticides is the utilization of nanocarriers as a source for the smart delivery of these active compounds. These nanoengineered pesticides can be one of the most important smart tools for efficient use in future (Jampílek and Kráľová, 2017). For the control release of nanoparticles as nanopesticides, fabrication of nanoencapsulated pesticides is recommended as it has improved slow release of pesticides and amplified porosity, specificity, stability, and solubility. These nanopesticides are ecofriendly with less vulnerability to living organisms and can remarkably reduce the application of pesticides for crop protection gradually, hence revolutionizing the global food industry and minimize the detrimental influences of synthetic pesticides on the ecosystem (Shafi et al., 2020).

The process of release of silver nanoparticles is an oxidation reaction utilizing the mutual effect of oxygen and protons. Surface modification of silver nanoparticles using various chemicals such as citrate sodium sulfate (Na_2S) and mercaptoundecanoic acid (MUA) also facilitate its controlled release (Fig.11.4). Furthermore, the reverse binding of Ag, control of oxidation of silver nanoparticles to silver ions by utilizing different antioxidant agents (CAT and POX), binding with free thiol group such as glutathione and cysteine, in addition to sodium citrate, inhibits the release of Ag^+ ions and also improves controlled release of silver nanoparticles (Liu et al., 2010).

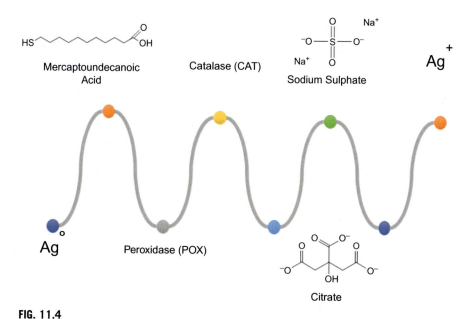

FIG. 11.4

Different chemical agents utilized for the controlled release of silver nanoparticles.

11.5 Positive effects of silver nanoparticles

The impact of AgNPs on plant growth depends upon their size, shape, and concentration, and it also varies according to the species as well as stage of plant growth and development. AgNPs with low input can enable farmers to get better yield. AgNPs have been used as nanofertilizers, nanopesticides, and nanoherbicides efficiently. Application of AgNPs improved *Triticum aestivum* biomass and decreased shoot weight. AgNPs, when applied on *Ricinus communis*, resulted in increased reactive oxygen species (ROS), which improved root length upto a certain level and also resulted in the activation of antioxidant defense mechanism such as peroxidase, superoxide dismutase, and phenolic compounds, which proved to be potent repellent of pathogen, hence improved plants defense system (Goswami and Mathur, 2019).

Due to unique physiochemical properties, nanosize dimension, tunable pore size, and high surface area to volume ratio, AgNPs are considered to be quite beneficial for improving the nutritional values of the plants. Sublethal concentrations of AgNPs have phytostimulatory effect on plants growth and development with improved evapotranspiration, enhanced root length, and biomass production. Root growth improvement was reported in *Crocus sativa* L. due to blockage in ethylene production by nanofertilizers (Dykman and Shchyogolev Sergei, 2018).

AgNPs improved chlorophyll and ascorbic acid content in growing seedlings. Very low toxic effects of AgNPs were evaluated in *Lactuca sativa* L. Germination percentage in *Oryza sativa* L. is concentration-dependent, and improvement in germination under low concentration was recorded (Dykman and Shchyogolev Sergei, 2018). The actual mechanism involved in plant growth promotion by nanoparticles is still unclear, but they are responsible for improving plant growth regulators, and hence improve plant growth. Controlled use of AgNPs improves plant growth and development through improvement in seed germination and seedling growth (Wagi and Ahmed, 2019a, b; Abbas and Ahmed, 2020). Silver nanoparticles does not produce damaging effects on environment and living organisms, but when toxic chemicals were utilized for the synthesis of silver nanoparticles, these toxic chemicals gets incorporated into the silver nanoparticles and produce negative impacts on the environment and living organisms. This toxic effect depends upon the structure and solubility of these silver nanoparticles (Shafi et al., 2020).

Utilization of AgNPs as nanofertilizers, nanopesticides, and nanoherbicides is a smart, efficient, and more ecofriendly approach that may reduce the dose and frequency of application as compared to the conventional fertilizers, pesticides, and herbicides (Castillo-Henríquez et al., 2020). Antioxidant activity of AgNPs offers the utilization of AgNPs in food industry as a food preservative, in food packaging, food safety, and also improvement of shelf life of food products (Zorraquín-Peña et al., 2020).

11.6 Concerns regarding AgNPs

Although AgNPs have a broad spectrum of applications when they are released into the environment, they are subjected to transformation and may result in aggregation, stabilization, dissolution, and oxidation and may also affect the microbial profile of the soil. The change in the microbial profile is due to a change in soil metabolomics. AgNPs pose substantial influence in agriculture, but there is a halt in the marketability of nanofertilizers due to uncertainty in the interaction of NPs with the environment. Also, the potential risk of using nanoparticles on human health still needs to be investigated (Zulfiqar et al., 2019).

11.7 Conclusion

In conclusion, green AgNPs utilized in agriculture have shown improvement in plant growth and development efficiently and effectively in the form of nanofertilizers and nanopesticides as compared to mineral fertilizers in an ecofriendly and sustainable manner, offering great opportunities to improve plant nutrition and stress tolerance to get better yields. However, since green nanotechnology is passing through the earlier phase of its development, an in-depth research is needed to declare it as an independent discipline with extensive application of nanoparticles in various fields. Before the approval of all different nanoformulations, a complete toxicological and pharmacological profile analysis is prerequisite.

References

Abbas, G., Ahmed, A., 2020. Biosynthesis of silver nanoparticles and their impact on the growth of *Triticum aestivum* L. MSc dissertation, University of the Punjab.

Abd El-Maksoud, E.M., Lebda, M.A., Hashem, A.E., Taha, N.M., Kamel, M.A., 2019. Ginkgo biloba mitigates silver nanoparticles-induced hepatotoxicity in Wistar rats via improvement of mitochondrial biogenesis and antioxidant status. Environ. Sci. Pollut. Res. 26 (25), 25844–25854.

Abobatta, W.F., 2018. Nanotechnology application in agriculture. Acta. Sci. Agric. 2 (6) (ISSN: 2581-365X).

Alkhalaf, M.I., Hussein, R.H., Hamza, A., 2020. Green synthesis of silver nanoparticles by *Nigella sativa* extract alleviates diabetic neuropathy through anti-inflammatory and antioxidant effects. Saudi J. Biol. Sci. 27 (9), 2410–2419.

Bedlovičová, Z., Strapáč, I., Baláž, M., Salayová, A., 2020. A brief overview on antioxidant activity determination of silver nanoparticles. Molecules 25 (14), 3191.

Camara, M.C., Campos, E.V.R., Monteiro, R.A., Santo Pereira, A.D.E., de Freitas Proença, P.L., Fraceto, L.F., 2019. Development of stimuli-responsive nano-based pesticides: emerging opportunities for agriculture. J. Nanobiotechnol. 17 (1), 100.

Castillo-Henríquez, L., Alfaro-Aguilar, K., Ugalde-Álvarez, J., Vega-Fernández, L., Montes de Oca-Vásquez, G., Vega-Baudrit, J.R., 2020. Green synthesis of gold and silver nanoparticles from plant extracts and their possible applications as antimicrobial agents in the agricultural area. Nanomaterials 10 (9), 1763.

Chhipa, H., 2017. Nanofertilizers and nanopesticides for agriculture. Environ. Chem. Lett. 15 (1), 15–22.

Das, R.K., Pachapur, V.L., Lonappan, L., Naghdi, M., Pulicharla, R., Maiti, S., Cledon, M., Dalila, L.M.A., Sarma, S.J., Brar, S.K., 2017. Biological synthesis of metallic nanoparticles: plants, animals and microbial aspects. Nanotechnol. Environ. Eng. 2, 18. https://doi.org/10.1007/s41204-017-0029-4.

de Oliveira, C.R.S., Mulinari, J., Júnior, F.W.R., da Silva Júnior, A.H., 2020. Nano-delivery systems of pesticides active agents for agriculture applications—an overview. In: International Agrobusiness Congress (CIAGRO)—Science, Technology and Innovation: From the Field to the Meal., https://doi.org/10.31692/ICIAGRO.2020.0051.

Docea, A.O., Calina, D., Buga, A.M., Zlatian, O., Paoliello, M.M.B., Mogosanu, G.D., Vasile, B.Ş., 2020. The effect of silver nanoparticles on antioxidant/pro-oxidant balance in a murine model. Int. J. Mol. Sci. 21 (4), 1233.

Dykman, L.A., Shchyogolev Sergei, Y., 2018. The effect of gold and silver nanoparticles on plant growth and development. In: Metal Nanoparticles. Nova Science Publisher Incharge. ISBN: 978-1-53614-115-3.

Ealia, S.A.M., Saravanakumar, M.P., 2017. IOP Conference Series: Materials Science and Engineering. 263.032019.

Fatima, F., Hashim, A., Anees, S., 2020. Efficacy of nanoparticles as nanofertilizer production: a review. Environ. Sci. Pollut. Res., 1–12.

Gahukar, R.T., Das, R.K., 2020. Plant-derived nanopesticides for agricultural pest control: challenges and prospects. Nanotechnol. Environ. Eng. 5 (1), 3.

Goswami, P., Mathur, J., 2019. Positive and negative effects of nanoparticles on plants and their applications in agriculture. Plant Sci. Today 6 (2), 232–242.

Gour, A., Jain, N.K., 2019. Advances in green synthesis of nanoparticles. Artif. Cells Nanomed. Biotechnol. 47 (1), 844–851.

Guzel, R., Erdal, G., 2018. Synthesis of Silver Nanoparticles. Intech Open. https://doi.org/10.5772/intechopen.75363.

Hamad, A., Khashan, K.S., Hadi, A., 2020. Silver nanoparticles and silver ions as potential antibacterial agents. J. Inorg. Organomet. Polym. Mater., 1–18.

Iqbal, M.A., 2019. Nano-fertilizers for sustainable crop production under changing climate: a global perspective. In: Sustainable Crop Production. Intech Open, https://doi.org/10.5772/intechopen.89089.

Jampílek, J., Kráľová, K., 2017. Nanopesticides: preparation, targeting, and controlled release. In: New pesticides and soil sensors. Academic Press, pp. 81–127.

Khan, I., Raza, M.A., Awan, S.A., Shah, G.A., Rizwan, M., Ali, B., Zhang, X., 2020. Amelioration of salt induced toxicity in pearl millet by seed priming with silver nanoparticles (*AgNPs*): the oxidative damage, antioxidant enzymes and ions uptake are major determinants of salt tolerant capacity. Plant Physiol. Biochem. 156, 221–232.

León-Silva, S., Fernández-Luqueño, F., López-Valdez, F., 2016. Silver nanoparticles (AgNP) in the environment: a review of potential risks on human and environmental health. Water Air Soil Pollut. 227 (9), 306.

Liang, Y., Gao, Y., Wang, W., Dong, H., Tang, R., Yang, J., Cao, Y., 2020. Fabrication of smart stimuli-responsive mesoporous organosilica nano-vehicles for targeted pesticide delivery. J. Hazard. Mater. 389, 122075.

Liao, S., Zhang, Y., Pan, X., Zhu, F., Jiang, C., Liu, Q., Chen, L., 2019. Antibacterial activity and mechanism of silver nanoparticles against multidrug-resistant *Pseudomonas aeruginosa*. Intl. J. Nanomedicine 14, 1469.

Liu, J., Sonshine, D.A., Shervani, S., Hurt, R.H., 2010. Controlled release of biologically active silver from nanosilver surfaces. ACS Nano 4 (11), 6903–6913.

Makarov, V.V., Love, A.J., Sinitsyna, O.V., Makarov, S.S., Yaminsky, I.V., Taliansky, M.E., Kalinina, N.O., 2014. Green nanotechnology: synthesis of metal nanoparticles using plants. Acta Nat. 6 (1), 35–44 (20).

Marchiol, L., 2018. Nanotechnology in agriculture: new opportunities and perspectives. In: New Visions in Plant Science. vol. 121.

Marchiol, L., 2019. Nanofertilisers. An outlook of crop nutrition in the fourth agricultural revolution. Int. J. Agric. 14, 1367.

Parveen, K., Ledwani, L., Banse, V., 2015. Green synthesis of nanoparticles: their advantagess and disadvantages. AIP Conf. Proc. 1724, 020048 (2016) https://doi.org/10.1063/1.4945168.

Patra, A.J.K., Baek, K.-H., 2014. Green nanotechnology: factors affecting synthesis and characterization techniques. J. Nanomater. https://doi.org/10.1155/2014/417305.

Pho, Q.H., Losic, D., Ostrikov, K., Tran, N.N., Hessel, V., 2020. Perspectives on plasma-assisted synthesis of N-doped nanoparticles as nanopesticides for pest control in crops. React. Chem. Eng. 5 (8), 1374–1396.

Pradhan, S., Mailapalli, D.R., 2020. Nanopesticides for pest control. In: Sustainable Agriculture Reviews. vol. 40. Springer, Cham, pp. 43–74.

Roy, A., Bulut, O., Some, S., Mandal, A.K., Yilmaz, M.D., 2019. Green synthesis of silver nanoparticles: biomolecule-nanoparticle organizations targeting antimicrobial activity. RSC Adv. 9 (5), 2673–2702.

Saranya, S., Selvi, A., Babujanarthanam, R., Rajasekar, A., Madhavan, J., 2020. Insecticidal activity of nanoparticles and mechanism of action. In: Model Organisms to Study Biological Activities and Toxicity of Nanoparticles. Springer, Singapore, pp. 243–266.

Seltenrich, N., 2013. Nanosilver: weighing the risks and benefits. Environ. Health Perspect. 121 (7), a220–a225.

Shafi, A., Qadir, J., Sabir, S., Khan, M.Z., Rahman, M.M., 2020. Nanoagriculture: aholistic approach for sustainable development of agriculture. In: Kharissova, O.V. (Ed.), Handbook of Nanomaterials and Nanocomposites for Energy and Environmental Applications. Springer Nature Switzerland AG. https://doi.org/10.1007/978-3-030-11155-7_48-1.

Soliman, M., Qari, S.H., Abu-Elsaoud, A., El-Esawi, M., Alhaithloul, H., Elkelish, A., 2020. Rapid green synthesis of silver nanoparticles from blue gum augment growth and performance of maize, fenugreek, and onion by modulating plants cellular antioxidant machinery and genes expression. Acta Physiol. Plant. 42 (9), 1–16.

Taha, Z.K., Hawar, S.N., Sulaiman, G.M., 2019. Extracellular biosynthesis of silver nanoparticles from Penicillium italicum and its antioxidant, antimicrobial and cytotoxicity activities. Biotechnol. Lett. 41 (8–9), 899–914.

Wagi, S., Ahmed, A., 2019a. Bacillus spp.: potent microfactories of bacterial IAA. PeerJ 7, e7258.

Wagi, S., Ahmed, A., 2019b. Green production of *AgNPs* and their phytostimulatory impact. Green Process. Synth. 8 (1), 885–894.

Wagi, S., Ahmed, A., 2020. Bacterial nanobiotic potential. Green Process. Synth. 9, 203–211.

Yadav, A., Theivasanthi, T., Paul, P.K., Upadhyay, K.C., 2015. Extracellular biosynthesis of silver nanoparticles from plant growth promoting rhizobacteria *Pseudomonas* sp. Int. J. Curr. Microbiol. Appl. Sci. 4 (8), 1057–1068.

Yan, A., Chen, Z., 2019. Impacts of silver nanoparticles on plants: a focus on the phytotoxicity and underlyingmechanism. Int. J. Mol. Sci. 20 (5), 1003. https://doi.org/10.3390/ijms20051003.

Zhao, X., Cui, H., Wang, Y., Sun, C., Cui, B., Zeng, Z., 2017. Development strategies and prospects of nano-based smart pesticide formulation. J. Agric. Food Chem. 66 (26), 6504–6512.

Zorraquín-Peña, I., Cueva, C., Bartolomé, B., Moreno-Arribas, M., 2020. Silver nanoparticles against foodborne bacteria. Effects at intestinal level and health limitations. Microorganisms 8 (1), 132.

Zulfiqar, F., Navarro, M., Ashraf, M., Akram, N.A., Munné-Bosch, S., 2019. Nanofertilizer use for sustainable agriculture: advantages and limitations. Plant Sci., 110270.

CHAPTER 12

Comparison of the effect of silver nanoparticles and other nanoparticle types on the process of barley malting

Dmitry V. Karpenko

Department of Technology of Fermentation and Winemaking, Federal State Budgetary Educational Institution of Higher Education "Moscow State University of Food Production", Moscow, The Russian Federation

12.1 Introduction

Obtaining, studying, and applying nanoparticles and nanopreparations became one of the leading trends in science and industry in the first several decades of the 21st century. Several publications (Gopal et al., 2013; Huk et al., 2014; Stanković et al., 2012; Yang and Park, 2019) have shown that a change in geometric dimensions can significantly alter a variety of properties of matter. Thus, nanoparticles often exhibit distinct properties and behaviors in comparison to atoms of the same substance, or macrosized arrays. As a rule, the practical or recommended applications of nanoparticles, nanopreparations, and nanomaterials are based on these differences. Such materials are used in the manufacture of food (Amini et al., 2014; Gmoshinskii et al., 2018; He and Hwang, 2016; Oehlke et al., 2014; Peters et al., 2014) and nonfood products (Cominetti et al., 2015; Furberg et al., 2018; Mohajerani et al., 2019; Rotta et al., 2019; Sharma and Sirotiak, 2019; Timoshina et al., 2018; Vinyas et al., 2019; Zhang et al., 2020), in medicine (He et al., 2015; Matsumoto et al., 2019; Pérez-Herrero and Fernández-Medarde, 2015; Vallet-Regi and Tamanoi, 2018), veterinary medicine (Shkil et al., 2016), and agriculture (Kah et al., 2018; Mohd Yusof et al., 2019; White and Gardea-Torresdey, 2018), and as part of packaging materials (Gmoshinskii et al., 2018; Gressler et al., 2018; Kavoosi et al., 2014; Rokbani et al., 2019) and cosmetics (Djearamane et al., 2019; Sharma et al., 2019).

The ability of nanoparticles to overcome various natural barriers (Colombo et al., 2019) means that they can move from the material/object into which they were intentionally introduced to solve a particular problem, into objects and media in which the presence of nanoscale particles was not assumed (Deng et al., 2020; Weiner et al., 2018). Moreover, the "loss" of nanoparticles at all stages of their production and use in the production of certain products is inevitable (Haliullin et al., 2015). As a result, an additional quantity of nanoscale particles enters the environment and accumulates in the soil (Garner and Keller, 2014; Keller and Lazareva, 2014) and natural waters of various origins (Moise, 2019; Sanchís et al., 2018; Xing et al., 2016). From there, nanoparticles can enter the tissues of cultivated agricultural plants, from which they can move into the technological flow of food production, in other words, into production environments. At the same time, in the scientific literature of the last decade, an increasing amount of experimental data indicating the potential negative impact of nanoparticles on objects of different origins and level of organization has been reported (Akcetim and Erdem, 2019; Bind and Kumar, 2019; Ozgur, 2019; Rajput et al., 2019; Saxena and Harish, 2019; Senoh et al., 2020; Sharma et al., 2019; Shen et al., 2019; Zheng et al., 2019), including plants (Cekic et al., 2017) and, in particular, cereals (El-Shazoly and Amro, 2019; Chen et al., 2019).

Thus, it seems expedient to form an objective idea of the influence of various types of nanoparticles, including silver NPs, on biological objects of various levels of an organization, including plants, because a variety of food products are plant-based. In particular, the process of barley germination is fundamental in malt production, and thus in the production of beer. Sources have reported on the possible negative effect of nanoparticles on barley characteristics (Rajput et al., 2019). Moreover, the results of our previous studies have confirmed that nanoparticles of different types, depending on specific conditions and primarily on their content in the medium, can significantly and negatively impact the technologically important characteristics of some types of raw materials and brewing processes: brewer's yeast (Karpenko et al., 2011; Karpenko and Gernet, 2017; Karpenko et al., 2015a, 2017b), amylases (Karpenko et al., 2016a, 2017a), the proteases (Karpenko et al., 2015b, 2016b, 2017c, 2018b) and cellulases (Karpenko and Pozdnjakova, 2019) of barley malt and microbial enzyme preparations, the mashing process (Karpenko and Uvarov, 2012), and beer wort fermentation (Karpenko et al., 2012). The array of data that our research group has accumulated confirms the perspective (Garner and Keller, 2014) that the intensity of the negative impact on a substance under similar conditions significantly depends on the type (nature) of the nanoparticles to which it has been exposed.

For this reason, recent research efforts at the Moscow State University of Food Production have begun to explore the effect of the presence of various nanoparticles in the steep water used to increase the moisture content of barley grains before malting (Karpenko et al., 2018a; Uvarov and Karpenko, 2012). Some of the obtained experimental data are given in this chapter.

12.2 Method for determining the germination of barley and characteristics of barley batches used in experiments

Our studies were carried out in the context of an in-depth investigation of the influence of nanoparticles on the processes and objects of brewing. Because of this, the basic characteristics of the barley used as an object of exposure to nanoparticles were evaluated specifically from the point of view of their suitability for use in malt and the brewing industry. Foremost, the germination of barley grains was assessed, using the generally accepted procedure in Russian brewing and the corresponding index of germination energy (except for experiments to determine the effect of silver NPs, the description of these procedures is given in the corresponding section). Accordingly, 500 grains (for one repetition: a total of three repetitions were used for the analysis of each sample unless otherwise indicated) were submerged in tap water and held for 4 h at a temperature of 22–25 °C. Then, the water was removed, and the grains were allowed to stand for 15–18 h at the same temperature. Afterward, the grains were again held under a layer of water for 4 h. Two days (48 h) after the start of the germination process, the grain layer was stirred. After 72 h of germination, the number of germinated and hatched grains was counted; the obtained value was expressed as a percentage of the total number of grains used for germination, thus determining the index of germination energy. Each option was analyzed in triplicate; the average values, calculated on this basis, are given.

In malt production, only the germinative power of barley is standardized (percentage of grains germinated in 120 h), which should be at least 95% for class I barley and at least 90% for class II barley. Therefore, we rated the barley with germinative energy of 90%–99% as brewing barley with high germinative energy and thus of high quality.

In addition to the high-quality barley, it was decided to use barley with medium (31.7%–45.0%) and low (0.1%–4.6%) germinative energies. Although such barley is not suitable for the production of brewing malt, nonetheless it seemed reasonable to assess the effect of the presence of nanoparticles in technological media (steep water) on barley grains with different physiological and, likely, biochemical characteristics. In part, we made this assumption based on our previous experiments, namely, that the effect of nanoparticles on biological objects of various levels of the organization is largely due to their effect on the protein molecules in enzymes and, accordingly, on a change in the enzymes' catalytic activities. Therefore, it seemed worthwhile to establish the effect of nanoparticles on plant enzymatic systems with different initial activities.

In addition to the germination energy, the number of other indices of barley batches used in our experiments were determined. These indices are given in Tables 12.1 and 12.2.

Table 12.1 Characteristics of barley batches with different germinative energies used in a series of experiments with MWNTs and TiO_2 nanopreparation.

Barley	Indices				
	Extract (%)	Coarseness of grain (%)	Grains uniformity (%)	W (%)	Thousand-kernels weight (g)
With low germinability	85.75 ± 1.30	94.0 ± 0.5	94.0 ± 0.5	14.74 ± 0.12	40.45 ± 0.21
With high germinability	83.12 ± 1.21	96.0 ± 0.5	96.0 ± 0.5	15.16 ± 0.14	48.86 ± 0.19

Table 12.2 Characteristics of barley batches with different germinative energies used in a series of experiments with CuO and NiO nanopreparations.

Barley	Indices				
	Extract (%)	Coarseness of grain (%)	Grains uniformity (%)	W (%)	Thousand-kernels weight (g)
With low germinability	65.96±1.16	96.2±0.5	96.2±0.5	12.16±0.08	40.12±0.17
With medium germinability	76.51±1.31	89.0±0.5	89.0±0.5	15.13±0.11	39.91±0.20
With high germinability	84.09±1.02	96.7±0.5	96.7±0.5	12.21±0.13	43.95±0.27

12.3 Characteristics of the nanopreparations used in the studies

12.3.1 Multiwalled carbon nanotubes (MWCNTs)

The MWCNT "DEALTOM" preparation (Mnogosloynye uglerodnye nanotrubki (MUNT), 2020) was used for the experimental work. Multiwalled carbon nanotubes without purification (directly after the reactor) are a powder consisting of nanotubes of essentially two sizes, $D_{out} = 49.3 \pm 0.45$ nm and 72.0 ± 0.45 nm; $D_{in} = 13.3 \pm 0.45$ nm; $L = 5$ μm; and specific surface area, $S = 97.55 \pm 0.02$ m^2/g. $C_{ash} = 3.52 \pm 0.2$ wt%—the content of the residue that is not combustible up to 1000°C (ash-content): catalyst residue. C_{non}–CNT = 0.65 ± 0.08 wt%—the content of unstructured carbon forms that burn before the start of CNT combustion. Dealtom MWCNTs are intercalated by ~10 nm Ni nanoparticles, which creates magnetically soft ferromagnetic MWCNTs with a low residual magnetization.

12.3.2 Nanopreparation of titanium dioxide (TiO$_2$)

A nanopreparation of titanium dioxide manufactured by "Plazmotherm," of the Russian Federation, was used in the experiments. This preparation is produced by using plasma synthesis technology. The resulting average particle size is 50–70 nm. The specific surface area $S = 22$–30 m^2/g. The color is white. The shape of the particles is spherical. The chemical composition of such particles is (wt%) TiO$_2$ 99.8%; Cl$_2$ < 0.2% (Plazmoterm TiO2, anatase/rutile, 65–190 nm, 2020).

12.3.3 Nanopreparation of copper oxide (CuO)

This nanopreparation is also produced by "Plazmotherm," of the Russian Federation, using plasma synthesis technology. The average particle size is 60–110 nm. The specific surface area $S = 6$–12 m^2/g. The color is brown. The shape of the particles is spherical. The phase composition is a mixture of the oxides CuO and Cu_2O. The chemical composition of the particles is (wt%) CuO 99.9% (Plazmoterm CuO, 30–110 nm, 2020).

12.3.4 Nanopreparation of nickel oxide (NiO)

A nanopreparation of nickel oxide manufactured by "Plazmotherm," of the Russian Federation, was used in the experiments. This preparation is produced by using plasma synthesis technology. The average particle size is 40–130 nm. The specific surface area $S = 10$–$20\,m^2/g$. The color is black. The shape of the particles is spherical. The phase composition is divalent nickel oxide. The chemical composition of the particles is (wt%) nickel oxide 99.6% (Plazmoterm NiO, 40–130 nm, 2020).

12.3.5 Silver nanopreparation (Ag)

This nanopreparation was carried out according to the veterinary medicinal preparation of Argovit (Арговит), produced by LLC RPC "Vector-Vita" of the Russian Federation. The active ingredient is clustered colloidal silver, with auxiliary substances of medical polyvinylpyrrolidone and distilled water. Each cubic centimeter (1 cm^3) contains 13 mg of clustered silver and 187 mg of polyvinylpyrrolidone. In appearance, this preparation is a dark brown liquid with a greenish-grey tint (Argovit klasternoe serebro [Argovit Silver Cluster], 2020).

12.4 Effect of silver nanoparticles on the germination of brewing barley of varying quality

Initially, low-quality barley was used as the object of the Ag NPs. Air–water steeping was employed, with the duration of water pauses being 3 h and that of the air pauses, 7.5 h, for a total process duration of 24 h. Potassium permanganate was added to the initial steep water at the amount of 46 mg/kg of grain. Solutions of the Argovit preparation were added to the final steep water of the experimental variants so that the silver nanoparticle concentrations were 32.0 and 1.6 $\mu g/cm^3$. No Ag NP preparation was added to the steep water of the control sample.

After removing the final steep water, germination was carried out for 4 days at a temperature of 19–23 °C until the acrospires reached a length of 2/3–1 grain length. During the germination process, the grain mass of the samples was moistened two times per day by spraying with tap water, followed by agitation. At the end of the process, the number of germinated grains in each sample was counted. The data obtained are shown in Fig. 12.1.

It can be seen that the presence of silver nanoparticles in the final steep water at any of the concentrations considered led to a decrease in the number of germinated grains: at 1.6 $\mu g/cm^3$ Ag NPs, by 22%; at 32 $\mu g/cm^3$ Ag NPs, by 36.5%. The magnitude of the decrease in germination energy was significant. Also, it is reasonably

286 CHAPTER 12 Comparison of the effect of silver nanoparticles

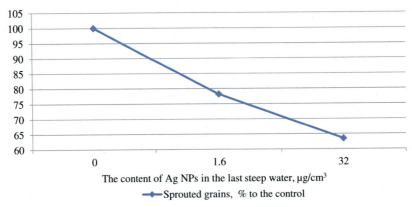

FIG. 12.1

Dependence of the germination of low-quality barley grains on the content of Ag NPs in the last steep water, % to the control.

assumed that any nanoparticles that had been originally introduced via the steep water could remain in the grain mass and adversely affect the processes that occur during the subsequent stages of brewing production with the participation of enzymes of plant origin (mashing, lautering) and brewer's yeast (main fermentation, lagering, and maturation of beer). To clarify the validity of this assumption, we decided to determine the number of indices of green malt of each of the three examined variants. In the brewing industry, these indices are usually determined in the malt extracts. Hence, aqueous extracts were prepared from the samples of green malt, and the following indices were determined from these extracts:

- viscosity, mm^2/s;
- dry matter (DM) concentration, %; and
- amylolytic activity, which was evaluated by the amount of starch (mg) hydrolyzed by malt extract amylases in a 1% soluble starch solution at $T=30\,°C$ for 10 min.

The results are presented in Fig. 12.2. For descriptive reasons, they are expressed as a percentage of the values in the control variant, which are taken as 100%. It can be seen that the negative effect of the Ag NPs present in the final steep water depended on the NP concentration, with the intensity of this effect varying with the malt extract indices. Thus, the greatest decrease, greater than 50% relative to that of the control, was recorded for the activity of the extract amylases. The dry matter content in the green malt extract was the least changed: under the influence of the presence of the nanoparticles in the final steep water, this value decreased by only 3%. The viscosity of the extract in the experimental samples exceeded that of the control index by 6%–35%, with its greatest increase recorded in the variant obtained with an Ag NP content of $1.6\,mg/cm^3$. This can be explained by the fact that with an increase in the content in the medium (in the final steep water, for example), the nanoparticles can

12.4 Effect of silver nanoparticles

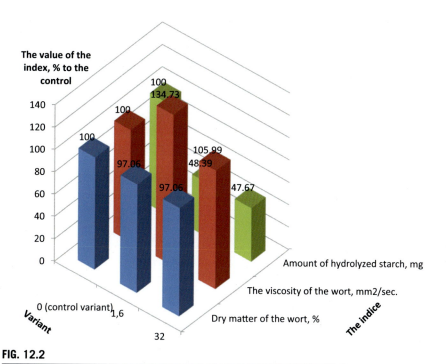

FIG. 12.2

Extracts indices of green malt variants, obtained from the barley with low germinability in the presence of Ag NPs in the last steep water at a concentration of 0 μg/cm^3 (control), 1.6 μg/cm^3, 32.0 μg/cm^3, % to the control.

begin to "conglutinate," forming aggregates that cause them to "lose" their nanoscale size and, consequently, the properties that are inherent to being nanoparticles of the same material. Nevertheless, on the whole, it can be concluded that the presence of Ag NPs in the final steep water significantly and negatively affected both the process of barley germination and the technologically important characteristics of the extract (draw) obtained from the green malt. As a result, it was considered expedient to determine the effect of Ag NPs on the germination process of high-quality barley. The associated experiment was carried out under the conditions described above; its results are presented in Figs. 12.3 and 12.4.

From Fig. 12.3, it can be seen that the presence of Ag NPs in the final steep water had a less significant, but still negative, effect when using the high-quality barley with the high initial germination energy (92%) as the object of nanoparticle influence. Likely, the more active metabolic/enzymatic systems in the high-quality barley are more resistant to the effects of the Ag NPs. As in the previous case, we sought to determine the indices of the extracts obtained from the control and experimental samples of the green malt (Fig. 12.4), although we expanded the list to include determinations of the concentrations of reducing substances (RS, %) and free amino acid nitrogen contents (FAN, mg/100 cm^3).

288 CHAPTER 12 Comparison of the effect of silver nanoparticles

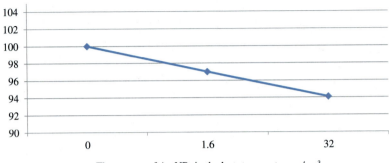

FIG. 12.3
Dependence of the germination of high-quality barley grains on the content of Ag NPs in the last steep water, % to the control.

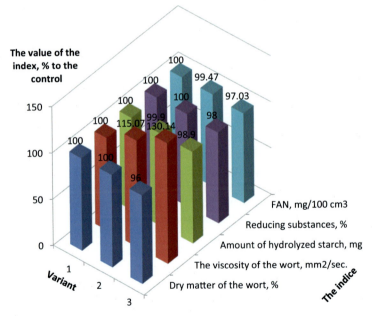

FIG. 12.4
Extracts indices of green malt variants, obtained from the barley with high germinability in the presence of Ag NPs in the last steep water at a concentration of 0 μg/cm^3 (control), 1.6 μg/cm^3, 32.0 μg/cm^3, % to the control.

It can be seen that the sample parameters of the malt extracts obtained in the presence of Ag NPs during the final steep water were inferior to the similar values recorded for the control variant. Notably, in most cases, the decrease was insignificant, except for the wort viscosity, which increased in proportion to the increase in the content of silver nanoparticles.

In general, it can be concluded that high-quality barley is more resistant to the negative effects of Ag NPs than grain with low germinative energy. However, from the point of view of brewing, and even in this case, the presence of silver nanoparticles in the steep water negatively affected the technological characteristics of the barley malt and the wort obtained from it. Unfortunately, the Ag NP preparation that we had at our disposal did not allow us to increase the content of the nanoparticles in the steep water; thus, we proceeded to study the effect of other types of NPs on the malting process.

12.5 Effect of multiwalled carbon nanotubes (MWCNTs) on the germination of brewing barley of varying quality

A series of experiments with powdered nanoparticle preparations, including MWCNTs, were carried out under conditions similar to those described above, except that the nanoparticles were introduced into the initial steep water, and not the final. Besides, the content of nanoparticles in the steep water was significantly increased in comparison to that of the Ag NPs. We made this change for two reasons:

- according to published data, the negative effect of silver nanoparticles on natural objects of various properties was significantly more pronounced than that of other nanoparticles and
- the preparations required for the experimental volumes of the solutions (suspensions) of the powdered nanopreparations called for concentration of less than $0.1 \, mg/cm^3$, which would have led to a significant increase in the error in the specified content of the nanoparticles.

Following the addition of the MWCNTs to the initial steep water, the effects on the germination of the low-quality barley (the characteristics of the grain batch are given in Table 12.1) were assessed and are presented in Fig. 12.5A.

Despite expectations, there was no negative effect of nanoparticles on the germinative energy of the low-quality barley over the entire concentration range of MWCNTs considered ($0.1–2.0 \, mg/cm^3$ in the initial steep water). Moreover, the presence of nanoparticles at low concentrations under the experimental conditions actually led to an increase in the values of the controlled index: at $0.25 \, mg/cm^3$, by 21%; at $0.1 \, mg/cm^3$, by 57%. The results of the influence of the MWCNTs on the germination of the high-quality barley is shown in Fig. 12.5B.

In this case, the presence of the nanoparticles did not give rise to a negative effect on the germinative energy of the grains. However, the positive effect was extremely weak; the only variant in which the increase in the controlled index significantly exceeded

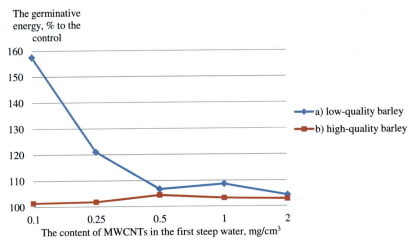

FIG. 12.5

Effect of multiwalled carbon nanotubes (MWCNTs) content in the first steep water on the germinative energy of (A) low-quality barley and (B) high-quality barley.

the error of determination (on average, ±2.5%) was that in which the concentration of the MWCNTs in the initial steep water was $0.5\,mg/cm^3$, amounting to almost 4.5%. A comparison of the data in Fig. 12.5A and B allows us to conclude that in this case, the influence of the nanoparticles was less significant in this experiment when the object of the NPs influence was the grain with the higher initial germinability.

12.6 Effect of titanium dioxide nanoparticles (TiO₂ NPs) on the germination of brewing barley of varying quality

The experiments were carried out under the same conditions as in the previous series. The values of the germinative energy of low-quality barley in the samples subjected to titanium dioxide nanoparticles in the initial steep water are shown in Fig. 12.6A.

The dependence between the content of TiO_2 NPs in the steep water and the germination energy is very close to that recorded for the MWCNTs. None of the concentrations of titanium dioxide nanoparticles used in the experiments showed a negative effect on the process of barley germination. In contrast, the presence of TiO_2 NPs in the lower concentration range used in the experimental variants led to a significant increase in the germination energy in comparison to the control: by 26% at $0.25\,mg/cm^3$ and by 50% at $0.1\,mg/cm^3$. The germinative energy increase in the variants with high concentrations of nanoparticles ($0.5–2.0\,mg/cm^3$) was less significant, although these values exceeded the average error of determination of the method used. The results for determining the dependence of the germinative energy of the high-quality barley on the presence of TiO_2 NPs in the initial steep water are shown in Fig. 12.6B.

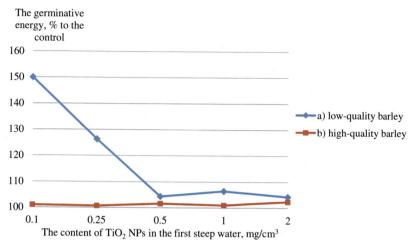

FIG. 12.6

Effect of TiO$_2$ NPs content in the first steep water on the germinative energy of (A) low-quality barley and (B) high-quality barley.

In all experimental variants, the values of the germinative energy were found to have slightly exceeded the analogous value in the control variant, while remaining within the error of the determination method. It can be concluded that under the conditions of this series of experiments, the presence of TiO$_2$ NPs did not affect the controlled parameter. Moreover, analysis of the experimental data allowed us to conclude, again, that grain with a low initial germination rate is more sensitive to the effect of nanoparticles, albeit positive in this particular case.

12.7 Effect of copper oxide nanoparticles (CuO NPs) on the germination of brewing barley of varying quality

Experiments on the effect of the presence of CuO NPs in the first steep water on the germinative energy of barley were carried out under the same conditions as the experiments of the previous series, except that three different batches of low-, medium-, and high-quality barley were used as the objects of nanoparticle influence. The characteristics of these batches are given in Table 12.2. For the low-quality barley, the experimental data are presented in Fig. 12.7A.

These data do not allow a smooth dependence of the germinative energy on the content of CuO NPs in the steep water to be constructed. Thus, with the NP content equal to 0.1 mg/cm^3, the value of the germinative energy in the experimental variant was consistent with the value of the same index in the control. However, with an increase in the content of CuO NPs to 0.25 mg/cm^3, the germinative energy in the experimental variant increased by nearly a factor of two (increased by 94%). Nevertheless, a further increase in the content of CuO NPs led to a sharp decline in

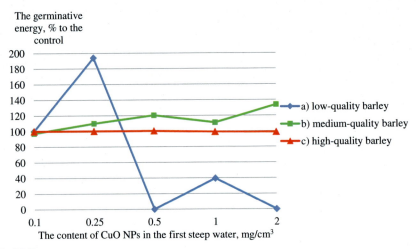

FIG. 12.7

Effect of CuO NPs content in the first steep water on the germinative energy of (A) low-quality barley; (B) medium-quality barley; and (C) high-quality barley.

the barley germinative energy, until a complete absence of the growth of the grain vegetative organs at 0.5 and 2.0 mg/cm^3. In general, it can be concluded that the presence of copper oxide nanoparticles in the first steep water in concentrations exceeding 0.25 mg/cm^3 can have significantly and negatively affected the metabolic processes in barley grain with a low germination rate.

The opposite conclusion can be made regarding barley of medium quality subjected to CuO NPs (Fig. 12.7B). In this case, a decrease in the germinative energy, only slightly exceeding the error of the determination method, was recorded only when the content of the copper oxide nanoparticles in the first steep water was 0.1 mg/cm^3. This decrease was 2.7% in comparison with the control. In all other experimental variants, the presence of CuO NPs in the steep water led to an increase in the germination energy, more or less proportional to the increase in the content of nanoparticles (except for the variant with the NP concentration of 1.0 mg/cm^3). The largest increase in germinative energy was 33%, which was observed at the maximum (2.0 mg/cm^3) of the examined range of copper oxide nanoparticles in the first steep water.

In the final series of experiments of the research phase under discussion, the object of the CuO NPs influence was high-quality barley. In this case (Fig. 12.7C), the presence of nanoparticles did not significantly affect the germinative energy, the values of which in the experimental variants ranged from −1.4 to +0.2% in comparison to the control. However, it can be assumed that a further increase in the content of CuO NPs would lead to a significant decrease in barley germination.

It can be argued that the effect (positive or negative) of copper oxide nanoparticles under the experimental conditions was increasingly pronounced as barley with a lower initial germination rate was used. This confirmed the conclusion made on the basis of the results of the series of experiments described above.

12.8 Effect of nickel oxide nanoparticles (NiO NPs) on the germination of brewing barley of varying quality

In the last stage of our work, we studied the effect of NiO NPs on the germinative energy of barley. The conditions were similar to those previously applied. When using low-quality barley as the object of nanoparticle influence, the results presented in Fig. 12.8A were obtained.

It can be seen that when the content of NiO NPs in the first steep water was 0.1 mg/cm^3, the germinative energy increased by 98% in comparison to the control. In the experimental variants with the NP contents of 0.25 and 0.50 mg/cm^3, the values of the controlled index were the same as in the control. However, a further increase in the concentration of NiO NPs in the first steep water led to significant inhibition of the process of barley germination, with the formation of vegetative organs completely ceasing at the nanoparticle content of 2.0 mg/cm^3. As a result, it was confirmed that the presence of nanoscale particles in the steep water significantly affects the development of grain with initially reduced physiological characteristics. Such an effect can be either positive or negative, depending on the nature of the nanoparticles and their concentration.

The influence of NiO NPs on the germinative energy of medium-quality barley under the experimental conditions was multidirectional (Fig. 12.8B). At concentrations of nanoparticles in the steep water equal to 0.25 and 1.0 mg/cm^3, a minor decline (2.8%–5.0% in comparison to the control) of the germinative energy was observed. In contrast, at other NiO NP concentrations, this index increased from 5.8% (at 0.1 mg/cm^3) to 25.8% (at 0.5 mg/cm^3). It is possible that this is due to the interaction of components of the germinating grain metabolic systems and various

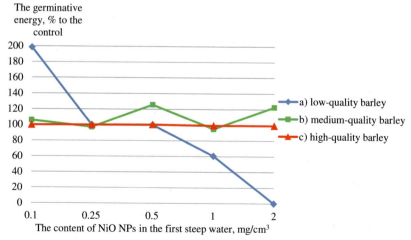

FIG. 12.8

Effect of NiO NPs content in the first steep water on the germinative energy of (A) low-quality barley; (B) medium-quality barley; and (C) high-quality barley.

amounts of nanoparticles or to the aggregation processes of the nanoparticles themselves, which varies based on the concentration of NPs in the medium.

The data obtained in the study of the influence of NiO NPs on the germination of high-quality barley (Fig. 12.8C) allowed us to conclude that in this case, the germinative energy marginally depends on the concentration of NPs in the first steep water. The values of the controlled index in the experimental variants varied in the range from −1.4% to +0.3% in comparison with the value in the control variant, which lies within the error of the determination method. Nevertheless, with the increase in the concentration of NiO NPs in the first steep water (up to 1.0–2.0 mg/cm^3), the germination energy tended to decrease. This suggests that even higher concentrations of nanoparticles of this type would have a significant and negative effect on the physiological processes in barley grain.

12.9 Conclusion

Based on the entire array of data presented, it can be concluded that the influence of various types of nanoparticles on the process of brewing barley germination becomes increasingly significant as the quality of the grain is reduced with respect to its initial germinability. Grains with good physiological characteristics are less "sensitive" to the presence of NPs in the steep water. In addition, the nature of the nanoparticles to which the grains are exposed plays a significant role. According to the intensity of the negative effect on the germinative energy of at least low-quality grains, the studied nanoparticles can be arranged in the following order: Ag ≫ CuO > NiO > TiO$_2$ ~ MWCNT, with Ag NPs standing apart. Under similar conditions, a comparable decrease in the germinative energy is due to a content of silver nanoparticles that is 1–2 orders of magnitude lower than that of the other NPs.

Nanoparticles of natural origin have been present in the biosphere of the Earth for billions of years. The intentional production of engineered nanoparticles increases the content of existing types of nanoscale objects, and besides, new varieties that previously were not formed in nature now appear in the environment. In all likelihood, the nanoparticles derived from anthropogenic sources are neither good nor evil and should be used in the applications and process in which it is impossible to do without them. However, their widespread, uncontrolled distribution in the Earth's ecosphere can cause problems, the elimination of which will require significant effort and money, and is likely to lead to victims of their pollution and toxicity. The story of science fiction writer Robert Sheckley "Watchbird" involuntarily recalls...

References

Akcetim, M., Erdem, A., 2019. Ecotoxic effects of cerium oxide nanoparticles on bacteria. Cumhuriyet Sci. J. 40, 544–553.

Amini, S.M., Gilaki, M., Karchani, M., 2014. Safety of nanotechnology in food industries. Electron. Physician 6 (4), 962–968.

Argovit klasternoe serebro [Argovit Silver Cluster], 2020. http://xn- -80aecv2aks.xn- -p1ai/instruction.html. (Accessed 16 November 2019 (In Russian).

Bind, V., Kumar, A., 2019. Current issue on nanoparticle toxicity to aquatic organism. MOJ Toxicol. 5 (2), 66–67.

Cekic, F.O., Ekinci, S., İnal, M.S., Unal, D., 2017. Silver nanoparticles induced genotoxicity and oxidative stress in tomato plants. Turk. J. Biol. 41, 700–707.

Chen, Y., Wu, N., Mao, H., Zhou, J., Su, Y., Zhang, Z., Zhang, H., Yuan, S., 2019. Different toxicities of nanoscale titanium dioxide particles in the roots and leaves of wheat seedlings. RSC Adv. 9, 19243–19252.

Colombo, G., Cortinovis, C., Moschini, E., Bellitto, N., Perego, M.C., Albonico, M., Astori, E., Dalle-Donne, I., Bertero, A., Gedanken, A., Perelsthein, I., Mantecca, P., Caloni, F., 2019. Cytotoxic and proinflammatory responses induced by ZnO nanoparticles in in vitro intestinal barrier. J. Appl. Toxicol. 39 (8), 1155–1163.

Cominetti, A., Pellegrino, A., Longo, L., Po, R., Tacca, A., Carbonera, C., Salvalaggio, M., Baldrighi, M., Meille, S., 2015. Polymer solar cells based on poly(3-hexylthiophene) and fullerene: pyrene acceptor systems. Mater. Chem. Phys. 159, 46–55.

Deng, J., Ding, Q.M., Li, W., Wang, J.H., Liu, D.M., Zeng, X.X., Liu, X.Y., Ma, L., Deng, Y., Su, W., Ye, B., 2020. Preparation of nano-silver-containing polyethylene composite film and Ag ion migration into food-simulants. J. Nanosci. Nanotechnol. 20 (3), 1613–1621.

Djearamane, S., Lim, Y., Ling Shing, W., Lee, P.F., 2019. Cellular accumulation and cytotoxic effects of zinc oxide nanoparticles in microalga *Haematococcus pluvialis*. Peer J. 7, e582.

El-Shazoly, R.M., Amro, A., 2019. Comparative physiological and biochemial effects of CuO NPs and bulk CuO phytotoxicity onto the maize (*Zea mays*) seedlings. Global Nest J. 20 (3), 276–289.

Furberg, A., Arvidsson, R., Molander, S., 2018. Dissipation of tungsten and environmental release of nanoparticles from tire studs: a Swedish case study. J. Clean. Prod. 207, 920–928.

Garner, K.L., Keller, A.A., 2014. Emerging patterns for engineered nanomaterials in the environment: a review of fate and toxicity studies. J. Nanopart. Res. 16 (8), 2503.

Gmoshinskii, I., Shipelin, V., Khotimchenko, S., 2018. Nanomaterials in food products and their package: comparative analysis of risks and advantages. J. Health Risk Anal. 4, 134–142.

Gopal, J., Hasan, N., Manikandan, M., Wu, H.-F., 2013. Bacterial toxicity/compatibility of platinum nanospheres, nanocuboids and nanoflowers. Sci. Rep. 3, 1260.

Gressler, S., Part, F., Gazsó, A., Huber-Humer, M., 2018. Nanotechnological Applications for Food Contact Materials. ITA NanoTrust-Dossiers, https://doi.org/10.1553/ita-nt-049en.

Haliullin, T.O., Zalyalov, R.R., Shvedova, A.A., Tkachov, A.G., Fathutdinova, L.M., 2015. Gigienicheskaya otsenka proizvodstva mnogosloynykh uglerodnykh nanotrubok [Hygienic evaluation of multilayer carbon nanotubes]. Med. Tr. Prom. Ekol. 7, 37–42 (In Russian).

He, X., Hwang, H.-M., 2016. Nanotechnology in food science: functionality, applicability, and safety assessment. J. Food Drug Anal. 24 (4), 671–681.

He, X., Aker, W., Huang, M.-J., Watts, J., Hwang, H.-M., 2015. Metal oxide nanomaterials in nanomedicine: applications in photodynamic therapy and potential toxicity. Curr. Top. Med. Chem. 15 (18), 1887–1900.

Huk, A., Izak-Nau, E., Reidy, B., Boyles, M., Duschl, A., Lynch, I., Dušinska, M., 2014. Is the toxic potential of nanosilver dependent on its size? Part. Fibre Toxicol. 11, 65.

Kah, M., Kookana, R., Gogos, A., Bucheli, T., 2018. A critical evaluation of nanopesticides and nanofertilizers against their conventional analogues. Nat. Nanotechnol. 13 (8), 677–684.

Karpenko, D.V., Gernet, M.V., 2017. Effect of nanopreparations on development of the populations of Saccharomyces brewer's yeasts. Microbiology 86 (5), 596–601.

Karpenko, D.V., Pozdnjakova, I.J., 2019. Vliyanie nanopreparatov na tsellyuloliticheskuyu aktivnost' mikrobnykh fermentnykh preparatov [Effect of nanopreparations on the cellulolytic activity of microbial enzyme preparations]. In: Pivo i napitki: bezalkogol'nye, alkogol'nye, soki, vino [Beer and Beverages: Non-Alcoholic, Alcoholic, Juices, Wine]. vol. 1, pp. 22–25 (In Russian).

Karpenko, D.V., Uvarov, Y.A., 2012. Vliyanie nanochastits serebra na rezul'taty zatiraniya v pivovarenii [Effect that silver nanoparticles have on mashing in the course of brewing]. In: Pivo i napitki: bezalkogol'nye, alkogol'nye, soki, vino [Beer and Beverages: Non-Alcoholic, Alcoholic, Juices, Wine]. vol. 2, pp. 6–7 (In Russian).

Karpenko, D.V., Uvarov, Y.A., Shaburova, L.N., 2011. Vliyanie nanochastits serebra na pivnye drozhzhi [Effect of silver nanoparticles on brewer's yeast]. In: Pivo i napitki: bezalkogol'nye, alkogol'nye, soki, vino [Beer and Beverages: Non-Alcoholic, Alcoholic, Juices, Wine]. vol. 6, pp. 32–33 (In Russian).

Karpenko, D.V., Uvarov, J.A., Marinin, A.I., Olishevskij, V.V., 2012. Vlijanie nanochastic metallov na sbrazhivanie pivnogo susla [The influence of metal nanoparticles on the fermentation of brewing wort]. In: Pivo i napitki: bezalkogol'nye, alkogol'nye, soki, vino [Beer and Beverages: Non-Alcoholic, Alcoholic, Juices, Wine]. vol. 1, pp. 16–17 (In Russian).

Karpenko, D.V., Rajnina, E.O., Himacheva, N.A., 2015a. Vlijanie nanochastic cinka na pivnye drozhzhi [The influence of zinc nanoparticles on brewing yeasts]. In: Pivo i napitki: bezalkogol'nye, alkogol'nye, soki, vino [Beer and Beverages: Non-Alcoholic, Alcoholic, Juices, Wine]. vol. 6, pp. 22–24 (In Russian).

Karpenko, D.V., Vartanova, U.V., Borisova, N.O., Shirokova, M.A., 2015b. Vlijanie nanocinka na aktivnost' proteoliticheskih fermentov [The nanozink influence on the activity of proteolytic enzymes]. In: Pivo i napitki: bezalkogol'nye, alkogol'nye, soki, vino [Beer and Beverages: Non-Alcoholic, Alcoholic, Juices, Wine]. vol. 5, pp. 16–19 (In Russian).

Karpenko, D.V., Drozdov, S.M., Evseeva, A.A., 2016a. Vlijanie nanopreparatov na aktivnost' amilaz [The effect of nanopreparations on the activity of amylases]. In: Pivo i napitki: bezalkogol'nye, alkogol'nye, soki [Beer and Beverages: Non-Alcoholic, Alcoholic, Juices, Wine]. vol. 5, pp. 28–31 (In Russian).

Karpenko, D.V., Zhitkov, V.V., Karjazin, S.A., 2016b. Vlijanie nanopreparatov na aktivnost' proteaz [The effect of nanopreparations on the activity of proteases]. In: Pivo i napitki: bezalkogol'nye, alkogol'nye, soki, vino [Beer and Beverages: Non-Alcoholic, Alcoholic, Juices, Wine]. vol. 4, pp. 46–49 (In Russian).

Karpenko, D.V., Kashankov, V.O., Savina, M.V., 2017a. Vlijanie nanopreparatov na aktivnost' amilaz svetlogo jachmennogo soloda [The influence of nanopreparations on the activity of light barley malt amylases]. In: Pivo i napitki: bezalkogol'nye, alkogol'nye, soki, vino [Beer and Beverages: Non-Alcoholic, Alcoholic, Juices, Wine]. vol. 6, pp. 18–21 (In Russian).

Karpenko, D.V., Komolova, A.O., Homenko, A.A., 2017b. Vlijanie nanopreparatov na razvitie drozhzhevyh populjacij [The influence of nanopreparations on the development of yeast populations]. In: Pivo i napitki: bezalkogol'nye, alkogol'nye, soki, vino vino [Beer and Beverages: Non-Alcoholic, Alcoholic, Juices, Wine]. vol. 1, pp. 14–17 (In Russian).

Karpenko, D.V., Zhitkov, V.V., Kanaev, S.A., Syrova, E.M., 2017c. Vlijanie nanopreparatov na aktivnost' proteaz svetlogo jachmennogo soloda [The influence of nanopreparations on the activity of light barley malt proteases]. In: Pivo i napitki: bezalkogol'nye, alkogol'nye, soki, vino [Beer and Beverages: Non-Alcoholic, Alcoholic, Juices, Wine]. vol. 4, pp. 14–17 (In Russian).

Karpenko, D.V., Smirnova, A.A., Vasil'chenko, D.N., 2018a. Vliyanie nanochastits na prorastaemost' pivovarennogo yachmenya [The influence of nanoparticles on germination of brewing barley]. In: Pivo i napitki: bezalkogol'nye, alkogol'nye, soki, vino [Beer and Beverages: Non-Alcoholic, Alcoholic, Juices, Wine]. vol. 4, pp. 14–18 (In Russian).

Karpenko, D.V., Zhitkov, V.V., Orlov, I.A., Sloveckij, S.A., 2018b. Vliyanie nanopreparatov oksida medi i nikelya na proteoliticheskuyu aktivnost' svetlogo yachmennogo soloda [The influence of copper and nickel oxide Nanopreparations on the Proteolytic activity of light barley malt]. In: Pivo i napitki: bezalkogol'nye, alkogol'nye, soki, vino [Beer and Beverages: Non-Alcoholic, Alcoholic, Juices, Wine]. vol. 3, pp. 15–19 (In Russian).

Kavoosi, G., Dadfar, S.M.M., Ali Dadfar, S.M., Ahmadi, F., Niakousari, M., 2014. Investigation of gelatin/multi-walled carbon nanotube nanocomposite films as packaging materials. Food Sci. Nutr. 2 (1), 65–73.

Keller, A.A., Lazareva, A., 2014. Predicted releases of engineered nanomaterials: from global to regional to local. Environ. Sci. Technol. Lett. 1 (1), 65–70.

Matsumoto, K., Saitoh, H., Doan, T.L.H., Shiro, A., Nakai, K., Komatsu, A., Tsujimoto, M., Yasuda, R., Kawachi, T., Tajima, T., Tamanoi, F., 2019. Destruction of tumor mass by gadolinium-loaded nanoparticles irradiated with monochromatic X-rays: implications for the auger therapy. Sci. Rep. 9 (1), 13275.

Mnogosloynye uglerodnye nanotrubki (MUNT), 2020. DEALTOM [Multi-wall carbon nanotubes (MWCNT), DEALTOM]. http://dealtom.ru/content/options. (Accessed 16 November 2019 (In Russian).

Mohajerani, A., Burnett, L., Smith, J.V., Kurmus, H., Milas, J., Arulrajah, A., Horpibulsuk, S., Kadir, A.A., 2019. Nanoparticles in construction materials and other applications, and implications of nanoparticle use. Materials 12 (19), 3052.

Mohd Yusof, H., Mohamad, R., Zaidan, U., 2019. Microbial synthesis of zinc oxide nanoparticles and their potential application as an antimicrobial agent and a feed supplement in animal industry: a review. J. Anim. Sci. Biotechnol. 10, 57.

Moise, M., 2019. FTIR study of the binary effect of titanium dioxide nanoparticles ($nTiO_2$) and copper ($Cu^{2}+$) on the biochemical constituents of liver tissues of catfish (*Clarias gariepinus*). Toxicol. Rep. 6, 1061–1070.

Oehlke, K., Adamiuk, M., Behsnilian, D., Gräf, V., Mayer-Miebach, E., Walz, E., Greiner, R., 2014. Potential bioavailability enhancement of bioactive compounds using food-grade engineered nanomaterials: a review of the existing evidence. Food Funct. 5, 1341–1359.

Ozgur, M.E., 2019. The protective effects of vitamin C and Trolox on kinematic and oxidative stress indices in rainbow trout sperm cells against flower-like ZnO nanoparticles. Aquacult. Res. 50 (10), 2838–2845.

Pérez-Herrero, E., Fernández-Medarde, A., 2015. Advanced targeted therapies in cancer: drug nanocarriers, the future of chemotherapy. Eur. J. Pharm. Biopharm. 93, 52–79.

Peters, R., van Bemmel, G., Herrera-Rivera, Z., Helsper, J., Marvin, H.J.P., Weigel, S., Tromp, P., Oomen, A., Rietveld, A., Bouwmeester, H., 2014. Characterization of titanium dioxide nanoparticles in food products: analytical methods to define nanoparticles. J. Agric. Food Chem. 62 (27), 6285–6293.

Anon., 2020. Plazmoterm CuO, 30–110 nm. http://plasmotherm.ru/catalog/oxides/item_29.html. (Accessed 16 November 2019 (In Russian).

Anon., 2020. Plazmoterm NiO, 40–130 nm. http://plasmotherm.ru/catalog/oxides/item_30.html. (Accessed 16 November 2019 (In Russian).

Anon., 2020. Plazmoterm TiO_2, anatase/rutile, 65–190 nm. http://plasmotherm.ru/catalog/oxides/item_23.html. (Accessed 16 November 2019 (In Russian).

Rajput, V.D., Minkina, T., Fedorenko, A., Mandzhieva, S., Sushkova, S., Lysenko, V., Duplii, N., Azarov, A., Chokheli, V., 2019. Destructive effect of copper oxide nanoparticles on ultrastructure of chloroplast, plastoglobules and starch grains in spring barley (*Hordeum sativum*). Int. J. Agric. Biol. 21 (1), 171–174.

Rokbani, H., Daigle, F., Ajji, A., 2019. Long- and short-term antibacterial properties of low-density polyethylene-based films coated with zinc oxide nanoparticles for potential use in food packaging. J. Plast. Film Sheeting 35 (2), 117–134.

Rotta, M., Motta, M., Pessoa, A.L., Carvalho, C.L., Ortiz, W., Zadorosny, R., 2019. Solution blow spinning control of morphology and production rate of complex superconducting $YBa_2Cu_3O_{7-x}$ nanowires. J. Mater. Sci. Mater. Electron. 30 (9), 9045–9050.

Sanchís, S.J., Milacic, R., Zuliani, T., Vidmar, J., Abad, E., Farré, M., Barcelo, D., 2018. Occurrence of C 60 and related fullerenes in the Sava River under different hydrologic conditions. Sci. Total Environ. 643, 1108–1116.

Saxena, P., Harish, 2019. Toxicity assessment of ZnO nanoparticles to freshwater microalgae Coelastrella terrestris. Environ. Sci. Pollut. Res. 26 (26), 26991–27001. https://doi.org/10.1007/s11356-019-05844-1.

Senoh, H., Kano, H., Suzuki, M., Fukushima, S., Oshima, Y., Kobayashi, T., Morimoto, Y., Izumi, H., Ota, Y., Takehara, H., Numano, T., Kawabe, M., Gamo, M., Takeshita, J., 2020. Inter-laboratory comparison of pulmonary lesions induced by intratracheal instillation of NiO nanoparticle in rats: histopathological examination results. J. Occup. Health 62 (1), e12117.

Sharma, A., Sirotiak, T.L., 2019. TiO2 nanoparticles in portland cement: a life cycle inventory analysis (LCI). In: Proceedings of the 54th Annual ASC Conference, April 18–21, 2018, Minneapolis, MN, pp. 642–650.

Sharma, S., Sharma, R.K., Gaur, K., Torres, J., Loza-Rosas, S., Torres, A., Saxena, M., Julin, M., Tinoco, A., 2019. Fueling a hot debate on the application of TiO_2 nanoparticles in sunscreen. Materials 12 (14), 2317. https://doi.org/10.3390/ma12142317.

Shen, J., Yang, D., Zhou, X., Wang, Y., Tang, S., Yin, H., Wang, J., Chen, R., Chen, J., 2019. Role of autophagy in zinc oxide nanoparticles-induced apoptosis of mouse LEYDIG cells. Int. J. Mol. Sci. 20 (16), 4042.

Shkil, N.A., Burmistrov, V.A., Shkil, N.N., Yushkov, Y.G., Sokolov, M.Y., Saichenko, V.I., 2016. Use of argovit product with silver content in veterinary. Int. Res. J. 4 (46), 68–70. Part 5.

Stanković, A., Dimitrijević, S., Uskoković, D., 2012. Influence of size scale and morphology on antibacterial properties of ZnO powders hydrothemally synthesized using different surface stabilizing agents. Colloids Surf. B. Biointerfaces 102, 21–28.

Timoshina, Y.A., Trofimov, A.V., Miftakhov, I.S., Voznesenskii, E.F., 2018. Modification of textile materials with nanoparticles using low-pressure high-frequency plasma. Nanotechnol. Russ. 13 (11–12), 561–564.

Uvarov, Y.A., Karpenko, D.V., 2012. Vozdeystvie nanochastits serebra na prorastanie yachmenya i kachestvo svezheprorosshego soloda [The impact of silver nanoparticles on the germination of barley and green malt quality]. In: Pivo i napitki: bezalkogol'nye, alkogol'nye, soki, vino [Beer and Beverages: Non-Alcoholic, Alcoholic, Juices, Wine]. vol. 3, pp. 32–33 (In Russian).

Vallet-Regi, M., Tamanoi, F., 2018. Overview of studies regarding mesoporous silica nanomaterials and their biomedical application. Enzyme 43, 1–10.

Vinyas, M., Atul, S.J., Harursampath, D., Loja, M., Nguyen Thoi, T., 2019. A comprehensive review on analysis of nanocomposites: from manufacturing to properties characterization. Mater. Res. Express 6 (9), 092002.

Weiner, R.G., Sharma, A., Xu, H., Gray, P., Duncan, T., 2018. Assessment of mass transfer from poly(ethylene) nanocomposites containing noble-metal nanoparticles: a systematic study of embedded particle stability. ACS Appl. Nano Mater. 1 (9), 5188–5196.

White, J., Gardea-Torresdey, J., 2018. Achieving food security through the very small. Nat. Nanotechnol. 13, 627–629.

Xing, B., Vecitis, C.D., Senesi, N., 2016. Engineered Nanoparticles and the Environment: Biophysicochemical Processes and Toxicity. John Wiley & Sons, Somerset, NJ, USA.

Yang, G., Park, S.J., 2019. Deformation of single crystals, polycrystalline materials, and thin films: a review. Materials 12 (12), 2003.

Zhang, R.-K., Wang, D., Wu, Y.-J., Hu, Y.-H., Chen, J.-Y., He, J.-C., Wang, J.-X., 2020. A cataluminescence sensor based on NiO nanoparticles for sensitive detection of acetaldehyde. Molecules 25 (5), 1097.

Zheng, X., Zhang, Y., Chen, W., Wang, W., Xu, H., Shao, X., Yang, M., Xu, Z., Zhu, L., 2019. Effect of increased influent COD on relieving the toxicity of CeO_2 NPs on aerobic granular sludge. Int. J. Environ. Res. Public Health 16, 3609.

CHAPTER 13

Adverse effects of silver nanoparticles on crop plants and beneficial microbes

Faisal Zulfiqar[a] and Muhammad Ashraf[b,c]

[a]Department of Horticultural Sciences, Faculty of Agriculture and Environment, The Islamia University of Bahwalpur, Punjab, Pakistan
[b]University of Agriculture Faisalabad, Faisalabad, Pakistan
[c]International Center for Chemical and Biological Sciences (ICCBS), University of Karachi, Karachi, Pakistan

13.1 Introduction

Consistent increase in human population is exerting direct pressure on crop production and hence overall food security. Consequently, there will be an increase in demand by 2050 to fulfil food requirements of predicted 10 billion population (FAO, 2019). Additionally, climate change may increase pressure, which will limit crop yields. Under these pressures, the scientific community is searching to find novel solutions for sustaining crop production. One putative solution is the use of nanotechnology for enhancing crop production as fertilizers or as pest and disease control tools. Nanomaterials are classified based on their size, shape, structure, and origin. Thus, the efficacy of nanomaterials for agriculture use depends on many factors including coating material, concentration, size, stability, and the tested plant or microbial species (Yin et al., 2012; Cox et al., 2016). The adjustable characteristics of nanomaterials along with better performance ability than their bulk counterpart are the reasons behind the popularity of nanomaterials. Silver nanoparticles (AgNPs) are a component in the field of nanomaterials and are being used predominantly in the agriculture sector (Yan and Chen, 2019) because they possess growth-stimulating, fungicidal and fruit ripening ability (Zulfiqar et al., 2019). Recent reports have shown that AgNPs can penetrate, shift, and accumulate in plants (Torrent et al., 2020). While, studies demonstrating the effects of the AgNPs on plants are still scarce because most studies are carried out during the initial stage of plant development. Thus, the question remained unanswered regarding the role of AgNPs on the development stages.

For example, several research reports have demonstrated the negative impacts of AgNPs on seed germination (Kumari et al., 2017; Qian et al., 2013), morphological (Cao et al., 2017; Nair and Chung, 2015), physiological (Song et al., 2013;

Vishwakarma et al., 2017; Falco et al., 2020), and biochemical characteristics (Dimkpa et al., 2013; Olchowik et al., 2017; Zuverza-Mena et al., 2016). Similarly, reports on the adverse impacts of AgNPs on the soil biological processes and beneficial microbial communities have also urged to search further the limitations of nanoparticles' use in the agriculture and horticulture sectors (Grun et al., 2017; Rahmatpour et al., 2017; Kumar et al., 2014; Liu et al., 2017; McGee et al., 2017; Grün and Emmerling, 2018). Additionally, the application method of AgNPs also has an impact on the phytotoxicity responses of plants. For example, Wu et al. (2020) reported higher root-induced phytotoxicity of AgNPs in lettuce plants than foliar application. In this regard, understanding of the basic interactions among AgNPs, plants and microbes is fundamental because plants and soil are the basis of food chains and vital in all ecosystems. Thus, we have attempted to discuss here how far silver nanoparticles could hamper crop growth either by adversely affecting rhizosphere microbes and other factors responsible for regulating growth and development of crops.

13.2 Effect of silver nanoparticles (AgNPs) on different growth processes

13.2.1 Seed germination

The success of annual crops in terms of optimum growth first depends on reasonably good seed germination, which in turn depends on healthy noncontaminated seed, good quality water enriched with all essential inorganic nutrients, and optimum temperature regimes. Seed germination of most crops is believed to be hampered due to the supply of silver nanoparticles (Abd-Alla et al., 2016; Cox et al., 2017; Qian et al., 2013; Lee et al., 2012). Amooaghaie et al. (2015) tested the comparative impact of AgNPs and $AgNO_3$ applications on seed germination of *Brassica nigra* at physiological and molecular levels and observed that both these compounds had an injurious impact on seed germination, soluble and reducing sugar contents, and lipase activity in both germinating seeds and their seedlings. Similar adverse effects of AgNPs on seed germination have been observed in different crops, e.g., mung bean (Kumari et al., 2017) and zucchini (*Cucurbita pepo*) (Stampoulis et al., 2009). Both these reports show that the effect of the AgNPs was found to be dependent on concentration and particle size. Likewise, in rice, AgNPs different particle sizes such as 20, 30, 60, 70, 120, and 150 nm at varying levels (0.1, 1, 10, 1000 mg/L) were tested and a marked reduction in seed germination and seedling growth was observed at higher AgNPs concentrations and lower particle sizes (Thuesombat et al., 2014). Furthermore, Ag^+ ions released from AgNPs inhibited germination more than that of AgNPs themselves (Speranza et al., 2013). More recently, Scherer et al. (2019) have reported that AgNPs penetrate the root cells of *Allium cepa* and they reduce germination index. Since such studies use varied size and different coatings along with variation in concentration and application method, it is hard to evaluate the exact reason for the variation in responses, which can be further linked with variation in sensitivity of test plant species to AgNPs.

13.2.2 Plant morphological characteristics

Silver nanoparticles not only affect seed germination, but they also significantly affect later growth and development of plants (Table 13.1). For example, while investigating the effect of high levels of AgNPs on Arabidopsis, Wang et al. (2013) reported that higher concentrations of AgNPs reduced morphological characteristics such as root and shoot lengths of the plants compared to those of the untreated controls. Nair and Chung (2015) and Vishwakarma et al. (2017) demonstrated that exposure to AgNPs caused a significant decline in plant biomass, leading to shorter shoots relative to those of untreated plants. Similar findings have been observed by Qian et al. (2013). AgNPs are reported to be more phytotoxic and cause a greater biomass loss compared to ionic silver by disturbing metabolic machinery of exposed plants (Stampoulis et al., 2009; El-Temsah and Joner, 2010). The effect of spherical AgNPs of 20.4 nm size at different concentrations on maize (*Zea mays* L.) growth and rhizospheric microbial communities has been tested in comparison with bulk Ag (Cao et al., 2017). The results revealed that AgNPs at the concentration of 2.5 mg/kg reduced the plant biomass and dissolved organic carbon content in rhizospheric soils compared to the bulk Ag. The accumulation of AgNPs in plant tissues was also reported, resulting in increased antioxidant enzymes' activity. Hence, the application of AgNPs at high concentrations impaired crop growth together with soil fertility (Cao et al., 2017). Thus, AgNPs have a profound negative effect on the growth and development of plants at the vegetative and later growth stages, analogous to what has been reported in the preceding section.

13.2.3 Effect on physiological characteristics

When the growth of different plants is impaired due to application of AgNPs, this impairment is attributable to dysfunctioning of key physiological processes (Sami et al., 2020). Tripathi et al. (2017) demonstrated that AgNPs negatively affected growth and photosynthesis in *Pisum sativum*. Studies by Speranza et al. (2013) using kiwi fruits reported pollen lethality in an AgNPs-dose increasing trend, disruption of pollen tube elongation, specific change at the ultrastructural level, and changes in calcium content. Photosynthetic pigments such as chlorophyll content were also negatively affected by the application of AgNPs (Qian et al., 2013; Song et al., 2013; Nair and Chung, 2014, 2015; Vishwakarma et al., 2017).

In a recent study, Falco et al. (2020) tested the phytotoxicity of AgNPs of different size (20, 51, and 73 nm) and the bulk silver particle on the photochemical efficacy of photosystem II and leaf gas exchange of broad bean. The authors reported an adverse impact of AgNPs as the small diameter triggers a decrease in the photochemical efficiency of photosystem II (PSII) along with an increase in the nonphotochemical quenching. Additionally, the stomatal conductance (gs) and CO_2 assimilation were also negatively affected. Moreover, AgNPs cause overproduction of ROS. The authors concluded that the adverse effect of AgNPs is diameter dependent. Torrent et al. (2020) also reported a phytotoxic effect of AgNPs, especially on transpiration and stomatal conductance. Similarly, in another study, Li et al. (2018)

Table 13.1 Phytotoxic effects of AgNPs supplementation in different plant species.

Test plant species	Exposure time	Size	Concentration	Experiment setup
Arabidopsis (*Arabidopsis thaliana*)	1-month	10 and 12 nm	12.5 mg/kg	Pot experiment
Lettuce (*Lactuca sativa*)	9 days	60 and 100 nm	Commercial solutions of sodium citrate (2 mM) stabilized silver nanoparticles (citrate-AgNPs); polyvinylpyrrolidone coated AgNPs (PVP-AgNPs) stabilized with sodium citrate (2 mM); and polyethylene glycol coated AgNPs (PEG-AgNPs)	Polypropylene containers
Lettuce (*Lactuca sativa*)	15 days	20 nm	0.1, 0.5, and 1 mg/L solutions	Petri dishes
Seagrass (*Cymodocea nodosa*)	8 days	12.30–212.64 nm	0.0002, 0.002, 0.02, and 0.2 mg/L	Glass-made aquaria
Pea (*Pisum sativum*)	15 days	~20 nm	1000 and 3000 µM	Pot experiment
Broad bean (*Vicia faba*)	One time injection	25, 50, and 75 nm	100 mg/L	Pot experiment

13.2 Effect of silver nanoparticles (AgNPs) on different growth processes

Exposure methodology	Growth stage	Negative effects	References
Addition to the soil at the beginning of growth	Seedling to vegetative	Parental plant exposure inhibited pod quality and seed growth; Delayed flowering in both parents and offspring; transgenerational effect on floral organs via downregulation of related genes	Ke et al. (2020)
Supplied in the stock solution	Seedling stage	Negatively affected transpiration and stomatal conductance	Torrent et al. (2020)
Root or foliar applications	Germination	Penetrated cells; translocated in different tissues; induced oxidative stress; high penetration and accumulation via root application than foliar	Wu et al. (2020)
Provided in solution form	Seedling	Induced oxidative stress; impaired cytoskeleton, endoplasmic reticulum, cell ultrastructure, leaf growth and actin filament damage	Mylona et al. (2020)
Provided with Hoagland's nutrient solution	Seedling	Caused a decline in growth, photosynthetic pigments and chlorophyll fluorescence; accumulated in roots and shoots; induced oxidative stress	Tripathi et al. (2017)
1.5 mL of solution was injected at the beginning of the central vein of each leaf	Seedling	Decreased photochemical efficiency of photosystem II (PSII) and increased nonphotochemical quenching; negatively affected stomatal conductance and CO_2 assimilation; induced an overproduction of ROS leading to oxidative stress condition	Falco et al. (2020)

Continued

Table 13.1 Phytotoxic effects of AgNPs supplementation in different plant species—cont'd

Test plant species	Exposure time	Size	Concentration	Experiment setup
Pearl millet (*Pennisetum glaucum*)	24 h	30 nm	2, 4, and 6 mM	Pot experiment
Mustard (*Brassica* spp.)	–	47 nm	1 and 3 mM	Pot experiment
Red clover (*Trifolium pratense*)	2 weeks	–	20–100 mg/L	Seedling trays
Faba bean (*Vicia faba*)	14 days	5–50 nm	100, 200, 300, 400, 500, 600, 700, 800, and 900 mg/kg soil	Pot experiment
Barley (*Hordeum vulgare*)	1 week	25 nm	0.1, 0.5, and 1 mM	Petri dishes
Wheat (*Triticum aestivum*)	4 months	3.1–8.7 nm	20, 200, and 2000 mg/kg	Pot experiment
Chrysanthemum (*Chrysanthemum × grandiflorum*), Gerbra (*Gerbera × jamesonii*), Cape Primrose (*Streptocarpus × hybridus*)	3 weeks	20 nm	10 and 30 and 50 ppm (0.01; 0.03, and 0.05 mg/mL)	In vitro

13.2 Effect of silver nanoparticles (AgNPs) on different growth processes

Exposure methodology	Growth stage	Negative effects	References
Applied in pots with Hoagland's nutrient solution	Seedling	Induced oxidative stress; tissue accumulation; decreased growth; altered quantum yield (Fv/Fm), photochemical (qP), and nonphotochemical quenching (NPQ); blocked electron transport chain (ETC); inhibited photosynthesis	Khan et al. (2019)
Provided with Hoagland's nutrient solution	Seedling	Declined growth; accumulated in plant tissues; induce oxidative stress; DNA degradation; cell death	Vishwakarma et al. (2017)
Provided with Hoagland's nutrient solution	Seedling	Decreased photosynthetic pigments; induced oxidative stress; damaged genetic material; accumulated in tissues	Mo et al. (2020)
Added in pot soil	Seedling	Decreased germination; lowered dry mass, root and shoot length; lowered leaf area; accumulated in roots; decreased nitrogen fixation; induced autophagy	Abd-Alla et al. (2016)
Provided with Hoagland's nutrient solution	Seedling	Decreased germination; reduced root length; induced leaf chlorosis, dense aggregation of nuclear chromatin materials and degeneration of mitochondria	Fayez et al. (2017)
Mixed with soil in powder form		At higher concentration, lower biomass, shorter plant height, and lower grain weight. Accumulation in vegetal tissues including grains; decreased the content of micronutrients (Fe, Cu, and Zn) in AgNP-treated grains; decreased grain quality via lowering the contents of arginine and histidine	Yang et al. (2018)
Mixed in MS media	Explants	Regenerated less adventitious roots and decreased root area; decreased shoot regeneration	Tymoszuk and Miler (2019)

demonstrated the effect of AgNPs either alone or in combination with Diclofop-methyl (DM), a common postemergence herbicide on the growth and photosynthesis and antioxidant machinery of *Arabidopsis thaliana*. They observed a reduction in the growth and photosynthesis and an elevation in the oxidative stress in response to AgNPs individual application, while combined application of AgNPs with DM has an antagonistic effect on the studied characteristics. Additionally, Ag^+ released from solution and accumulation in plant tissues was also observed while DM decreased the release and accumulation of Ag^+ of AgNPs. Ke et al. (2018) studied the effect of both AgNPs (9 and 12 nm) and Ag ions on *A. thaliana* in hydroponic and soil conditions. The authors observed a decrease in photosynthesis after exposure to AgNPs that ultimately reduced biomass. In red clover, AgNPs exposure, especially at high doses, showed an elevated level of lipid peroxidation and variation in antioxidant activities. Further the microstructure and altered RADP fingerprinting evidenced the damage to hereditary material. The authors also observed accumulation of Ag particles in roots thus concluded a toxic effect of AgNPs, especially at higher doses (Mo et al., 2020) (Fig. 13.1).

13.2.4 Effect on biochemical characteristics

Silver nanoparticles are also known to alter biochemical processes in different plant species. For example, Qian et al. (2013) reported that AgNPs had negative effects on *A. thaliana* at the molecular, physiological, and ultrastructural levels. The silver

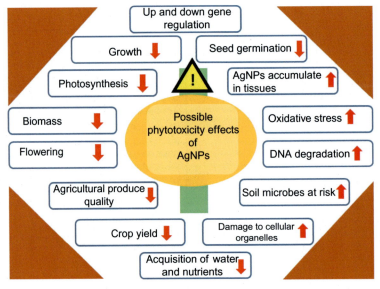

FIG. 13.1

Some examples of the possible ecotoxicity outcomes associated with the application of AgNPs.

nanoparticles were also found accumulated significantly in the leaves, thereby reducing the chlorophyll content, disturbing the thylakoid membrane structure and the normal plant growth. In another study, AgNPs applied as 20 and 50 mg/mL for 15 days to mung bean plants caused the generation of reactive oxygen species (ROS) and increased lipid and H_2O_2 peroxidation, ultimately causing cellular damage (Nair and Chung, 2015a). Olchowik et al. (2017) compared the effect of copper and AgNPs on English Oak (*Quercus robur* L.) seedlings and observed that only AgNPs caused significant changes in the chloroplast ultrastructure. Silver nanoparticles induce a disturbance in photosynthesis led to an excess of electrons, which resulted in the damage of chloroplast structure along with excess ROS (Qian et al., 2013). Furthermore, some changes at a genetic level were also observed, i.e., in the transcription of antioxidant effectors related genes involved in the regulation of aquaporins (Qian et al., 2013), glutathione GSH-biosynthesis, and sulfur assimilation (Nair and Chung, 2014). Furthermore, it has been reported that AgNPs trapped in the middle of plasmodesmata can block symplastic transport, resulting in disturbance of the intercellular transport, leading to the inhibition of plant growth (Geisler-Lee et al., 2013). Mirzajani et al. (2013) reported that AgNPs do not enter the root cells of *Oryza sativa* at low concentration, while at higher levels large-sized AgNPs tend to enter more in vegetal tissues (Torrent et al., 2020). From these findings, it is obvious that the entry of AgNPs is a crucial cause of a toxic effect. Sun et al. (2017) showed growth reduction after exposure to AgNPs through down-regulation of auxin receptors related genes resulting in decreased levels of auxins. Exposure to AgNPs is reported to decrease nutrient content in plants (Zuverza-Mena et al., 2016). Geisler-Lee et al. (2013) found a pronounced effect of AgNPs on the Ag content accumulation in plant tissues compared to that of $AgNO_3$ applications. Similarly, transcription of genes become more severely affected after exposure to AgNPs than to $AgNO_3$ (Qian et al., 2013; Nair and Chung, 2014). However, the negative effects of AgNPs are mostly related to small size particles as compared to larger ones in many studies (Geisler-Lee et al., 2013; Thuesombat et al., 2014; Zaka et al., 2016; Cvjetko et al., 2017). This could be due to extended dissolution of smaller particles, because of their larger specific surface area as well as greater absorption by organisms (Johnston et al., 2010). In another experiment by Mazumdar and Ahmed (2011), a supply of AgNPs (25 nm) damaged the cell wall and vacuole of roots of *Oryza sativa*, hence causing a toxic effect on rice. AgNPs application has been shown to enhance the synthesis of ROS, depicting a state of stress and ultimately enhanced synthesis of stress tolerating molecules such as antioxidant enzymes as an adaptive mechanism against AgNPs induced stress (Dimkpa et al., 2013). Zuverza-Mena et al. (2016) observed a significant change in several macromolecules such as proteins, lipids, lignin, cellulose, and pectin in *Raphanus sativus* when its plants were exposed to 2 nm AgNP (500 mg/L concentration). In a recent study, the photoperiod and vernalization were down-regulated following the exposure to AgNPs, resulting in decreased transcription of flowering key genes (*AP1*, *LFY*, *FT*, and *SOC1*), and a delayed flowering time of 5 days than control (Ke et al., 2018). In a more recent study, the exposure of AgNPs on the developmental stage of *A. thaliana* decreased petal and pollen viability and hence decreased production. Furthermore, transcription of flowering and floral organ development genes was downregulated by approximately 10%–40% in response to AgNPs (Ke et al., 2020). Regarding the accumulation and translocation of AgNPs, Torrent et al. (2020) studied the effect of different size, coating, and concentrations. They reported that AgNPs concentration in lettuce roots was influenced by the concentration and size, and not by the coating, while AgNPs' translocation to shoots was more pronounced for large size and neutrally charged NPs at higher concentrations.

Overall, some investigations conducted under simplified regimes have shown that AgNPs may have negative effects on plants. The individual biochemical attributes influenced by AgNPs could be used as bioindicators to get a better insight into the subcellular effects on plants.

13.3 In situ negative effects of AgNPs on beneficial microbes

Because of the abundant use of AgNPs in agriculture and horticulture domains, it is crucial to bring under consideration the negative impacts of AgNPs to the beneficial microbes in order to bring sustainability in the agriculture production sectors (Ottoni et al., 2020).

13.3.1 Negative impacts on the abundance of beneficial microbes

A decrease in the abundance of total beneficial microbe community after direct exposure of AgNPs to the soil is reported in several research reports (Hänsch and Emmerling, 2010; He et al., 2016). For example, Samarajeewa et al. (2017a, b) observed the effect of polyvinylpyrrolidone coated with AgNPs of 30 nm size with a concentration level of 49 and 1815 mg/kg on different biological characteristics such as enzyme activity, respiration, nitrification, microbial respiration, and molecular and physiological profiles. The authors demonstrated that in all the studies, biological characteristics were affected at all applied concentrations of the silver nanoparticles. The negative impacts of the AgNPs are mostly reported to be time-oriented. For instance, Grün et al. (2018) reported no effects on the abundance of nitrifiers during the start of the experiment (almost 1 month), but at a later stage, they showed negative effects. The authors suggested that this could have been due to the long-term release of Ag^+ from AgNPs.

13.3.2 Negative impacts on the structure of beneficial microbe communities

Various reports on changes in the structure of beneficial microbe communities have shown negative impacts of AgNPs, especially on bacterial communities (Kumar et al., 2014; Liu et al., 2017; McGee et al., 2017). These changes increased with increasing the toxicity with time (Grün and Emmerling, 2018) and in a dose-dependent way

(Samarajeewa et al., 2017a, b; Wang et al., 2017). However, fungal communities showed less change in their structure (Sillen et al., 2015). Similarly, in a comparative study, Kumar et al., 2014 observed pronounced negative effects of AgNPs on bacterial communities than on fungal community structure. Some reports have shown negative impacts on both bacterial and fungal community structures (McGee et al., 2017; Sillen et al., 2015). Doolette et al. (2016) examined the effect of AgNPs, sulphidized-silver NPs and silver ion on soil microbial communities by metagenomic sequencing. They observed that toxic levels of both NPs regarding operational taxonomic units (OTUs) were same than observed due to silver ions. However, sulfidized-silver NPs were relatively less toxic. Kim et al. (2018) studied the effect of nanocomposites of silver–graphene oxide at different concentrations (between 0.1 and 1 mg/g) and observed a significant decrease in the soil enzymatic activities up to 80%, together with a decrease in the nitrification process by 82%. These changes in microbe structure are hypothesized to be linked with a resistance of microbes against AgNPs. Kumar et al. (2014), while studying the effect of AgNPs of various size on the bacterial communities, observed a pronounced impact of NPs compared to microparticles, while no effect of particle size was observed on the bacterial structure (Doolette et al., 2016).

13.3.3 Negative impacts on the activity of beneficial microbes

Soil functionality is directly linked with activity rates of microbes. Different processes such as dehydrogenases, carbon respiration, and urease activities are related to the activities of microbes and are evaluated in various studies related to AgNPs. Several studies have shown that exposure to AgNPs decreases the activities of these processes (Peyrot et al., 2014; Schlich et al., 2016). Microbial activities in response to AgNPs applied at different concentrations (ranged from 0 to 50 mg/kg) in the calcareous soil having different salinity and texture levels were observed by Rahmatpour et al. (2017). They suggested that changes in the biological activities were related to AgNPs dose and soil texture type and hence demonstrated that at higher concentrations of AgNPs along with low clay soils, the negative effects on the microbial community/biological activities were pronounced. However, Hänsch and Emmerling (2010) did not note any negative effect on the soil respiration until 0.32 mg/kg of AgNPs exposure. The effect of AgNPs on a microbial community is reported to be temperature-dependent. For instance, Batista et al. (2017) tested litter decomposition mainly in their experiment and reported that the effect of AgNPs on microbial communities was more than that by temperature. This fact highlights the significance in considering the temperature range for future studies regarding the AgNPs effect on soil microbial communities. Asadishad et al. (2018) compared the effect of metallic NPs at different concentrations on soil enzymes and microbial community composition. They reported that the impact of individual NPs is dependent on the particle shape, chemical surface, concentration, and dissolution behavior. Among the evaluated NPs, AgNPs of 50 nm at the concentration of 100 mg/kg soil decreased soil enzyme activities (Asadishad et al., 2018). Yuan et al. (2018) reported enhanced stress in response to engineered AgNPs to aquatic plant species (*Egeria densa* and

Juncus effuses). The authors observed increased MDA depicting membrane damage and oxidative stress in response to AgNPs exposure (Yuan et al., 2018). The effect of AgNPs on soil microbes depends on the nature of NPs. For instance, it has been demonstrated that the subsidized based AgNPs was less injurious to soil microbial life than pristine based AgNPs (Doolette et al., 2016).

Negative impacts of AgNPs on the microbial community were more pronounced at the temperature of 10 and 23 °C (Batista et al., 2017). Grun et al. (2017) evaluated the toxic effect of AgNPs of 30 and 70 nm applied at different concentrations (from 20 to 600 mg/L) on the ecologically important soil protozoan strains (*Acanthamoeba castellanii* ATCC 30234 strain and *Acanthamoeba* strain) for 24 and 96 h. The authors reported a decrease in the adherence ability and metabolic activity for both strains in a dose-dependent manner for 96 h. Moreover, the authors compared the AgNPs with silver ions and observed similar toxic effects on protozoan organisms. In general, AgNPs application to the soil had an obvious negative effect on microbial activities, especially at higher concentrations.

13.4 Conclusion and future perspectives

The use of AgNPs in the agriculture sector is still at the infancy stage. However, while summarizing the foregoing discussion, it is now amply clear that AgNPs pose injurious impacts on plants and microbial community. Excessive interstudies variability and associated contradictory results are the main challenges for transferring existing knowledge from laboratory to field or real world. It is obvious from the surveyed literature that there is a continuous demand for the setting of realistic procedures to examine the toxicity of AgNPs. Existing literature discussed in this chapter reveals that AgNPs toxicity toward plants or microbial organisms varies considerably, so it is mandatory to further evaluate different plants and microbial species, especially at a molecular level. The lethal effects mostly relate to the use of nonoptimal higher concentrations. Although higher accumulation of AgNPs is mostly reported in roots, translocation and accumulation in shoots and other edible plant parts may cause health issues in consumers. Most phytotoxicity reports are pot-based growth room or glasshouse studies and it is a need of the hour to evaluate field-based results of AgNPs application. With a few exceptions, most studies focused seedling stage response to AgNPs supply, but the response of plants at the later growth stages should also be evaluated. A major challenge in confirming the accurate hazardous effects of AgNPs is due to the variability in the physicochemical characteristics of AgNPs that greatly depend on size, shape, coating material, concentration, composition, nature of testing media, surface charge, and test organism. Toxicity in soil beneficial microbes may have ecological concerns and hence demands a thorough investigation in the scenario of an increasing trend of NPs use. Thus, further experimental findings related to phytotoxicity and microbial health can serve as vital guidance and hence a reference in formulating long-term policies and regulations. Therefore, research on the influence of AgNPs on different plants and microbial health is of urgent need.

References

Abd-Alla, M.H., Nafady, N.A., Khalaf, D.M., 2016. Assessment of silver nanoparticles contamination on faba bean-*Rhizobium leguminosarum* bv. *viciae-Glomus aggregatum* symbiosis: implications for induction of autophagy process in root nodule. Agric. Ecosyst. Environ. 218, 163–177.

Amooaghaie, R., Tabatabaei, F., Ahadi, A.M., 2015. Role of hematin and sodium nitroprusside in regulating *Brassica nigra* seed germination under nanosilver and silver nitrate stresses. Ecotoxicol. Environ. Saf. 113, 259–270.

Asadishad, B., Chahal, S., Akbari, A., et al., 2018. Amendment of agricultural soil with metal nanoparticles: effects on soil enzyme activity and microbial community composition. Environ. Sci. Technol. 52, 1908–1918.

Batista, D., Pascoal, C., Cássio, F., 2017. Temperature modulates AgNP impacts on microbial decomposer activity. Sci. Total Environ. 601, 1324–1332.

Cao, J., Feng, Y., He, S., et al., 2017. Silver nanoparticles deteriorate the mutual interaction between maize (*Zea mays* L.) and arbuscular mycorrhizal fungi: a soil microcosm study. Appl. Soil Ecol. 119, 307–316.

Cox, A., Venkatachalam, P., Sahi, S., Sharma, N., 2017. Silver and titanium dioxide nanoparticle toxicity in plants: a review of current research. Plant Physiol. Biochem. 107, 147–163.

Dimkpa, C.O., McLean, J.E., Martineau, N., Britt, D.W., Haverkamp, R., Anderson, A.J., 2013. Silver nanoparticles disrupt wheat (*Triticum aestivum* L.) growth in a sand matrix. Environ. Sci. Technol. 47, 1082–1090.

Doolette, C.L., Gupta, V.V., Lu, Y., Payne, J.L., Batstone, D.J., Kirby, J.K., McLaughlin, M.J., 2016. Quantifying the sensitivity of soil microbial communities to silver sulfide nanoparticles using metagenome sequencing. PLoS One 11 (8).

El-Temsah, Y.S., Joner, E.J., 2010. Impact of Fe and ag nanoparticles on seed germination and differences in bioavailability during exposure in aqueous suspension and soil. Environ. Toxicol. 27, 42–49.

Falco, W.F., Scherer, M.D., Oliveira, S.L., Wender, H., Colbeck, I., Lawson, T., Caires, A.R., 2020. Phytotoxicity of silver nanoparticles on Vicia faba: evaluation of particle size effects on photosynthetic performance and leaf gas exchange. Sci. Total Environ. 701, 134816.

Fayez, K.A., El-Deeb, B.A., Mostafa, N.Y., 2017. Toxicity of biosynthetic silver nanoparticles on the growth, cell ultrastructure and physiological activities of barley plant. Acta Physiol. Plant. 39, 155.

Geisler-Lee, J., Wang, Q., Yao, Y., Zhang, W., Geisler, M., Li, K., Huang, Y., Chen, Y., Kolmakov, A., Ma, X., 2013. Phytotoxicity, accumulation and transport of silver nanoparticles by Arabidopsis thaliana. Nanotoxicology 7, 323–337.

Grün, A.-L., Emmerling, C., 2018. Long-term effects of environmentally relevant concentrations of silver nanoparticles on major soil bacterial phyla of a loamy soil. Environ. Sci. Europe 30 (1).

Grun, A.-L., Scheid, P., Hauroder, B., et al., 2017. Assessment of € the effect of silver nanoparticles on the relevant soil protozoan genus Acanthamoeba. J. Plant Nutr. Soil Sci. 180, 602–613.

Grün, A.-L., Straskraba, S., Schulz, S., Schloter, M., Emmerling, C., 2018. Long-term effects of environmentally relevant concentrations of silver nanoparticles on microbial biomass, enzyme activity, and functional genes involved in the nitrogen cycle of loamy soil. J. Environ. Sci. 69, 12–22.

Hänsch, M., Emmerling, C., 2010. Effects of silver nanoparticles on 1047 the microbiota and enzyme activity in soil. J. Plant Nutr. Soil Sci. 173, 554–558.

He, S., Feng, Y., Ni, J., Sun, Y., Xue, L., Feng, Y., Yu, Y., Lin, X., Yang, L., 2016. Different responses of soil microbial metabolic activity to silver and iron oxide nanoparticles. Chemosphere 147, 195–202.

Ke, M., Qu, Q., Peijnenburg, W.J.G.M., Li, X., Zhang, M., Zhang, Z., Qian, H., 2018. Phytotoxic effects of silver nanoparticles and silver ions to Arabidopsis thaliana as revealed by analysis of molecular responses and of metabolic pathways. Sci. Total Environ. 644, 1070–1079.

Ke, M., Li, Y., Qu, Q., Ye, Y., Peijnenburg, W.J.G.M., Zhang, Z., Qian, H., 2020. Offspring toxicity of silver nanoparticles to *Arabidopsis thaliana* flowering and floral development. J. Hazard. Mater. 386, 121975.

Khan, I., Raza, M.A., Khalid, M.H.B., Awan, S.A., Raja, N.I., Zhang, X., Huang, L., 2019. Physiological and biochemical responses of pearl millet (*Pennisetum glaucum* L.) seedlings exposed to silver nitrate (AgNO3) and silver nanoparticles (AgNPs). Int. J. Environ. Res. Public Health 16, 2261.

Kumar, N., Palmer, G.R., Shah, V., Walker, V.K., 2014. The effect of silver nanoparticles on seasonal change in arctic tundra bacterial and fungal assemblages. PLoS One 9, e99953.

Kumari, R., Singh, J.S., Singh, D.P., 2017. Biogenic synthesis and spatial distribution of silver nanoparticles in the legume mungbean plant (*Vigna radiata* L.). Plant Physiol. Biochem. 110, 158–166.

Lee, W.M., Kwak, J.I., An, Y.J., 2012. Effect of silver nanoparticles in crop plants *Phaseolus radiatus* and *Sorghum bicolor*: media effect on phytotoxicity. Chemosphere 86, 491–499.

Li, X., Ke, M., Zhang, M., Peijnenburg, W.J.G.M., Fan, X., Xu, J., Zhang, Z., Lu, T., Fu, Z., Qian, H., 2018. The interactive effects of diclofop-methyl and silver nanoparticles on *Arabidopsis thaliana*: growth, photosynthesis and antioxidant system. Environ. Pollut. 232, 212–219.

Liu, G., Zhang, M., Jin, Y., Fan, X., Xu, J., Zhu, Y., Fu, Z., Pan, X., Qian, H., 2017. The effects of low concentrations of silver nanoparticles on wheat growth, seed quality, and soil microbial communities. Water Air Soil Pollut. 228 (9), 1–12.

Mazumdar, H., Ahmed, G.U., 2011. Phytotoxicity effect of silver nanoparticles on *Oryza sativa*. Int. J. Chem. Tech. Res. 3, 1494–1500.

McGee, C.F., Storey, S., Clipson, N., Doyle, E., 2017. Soil microbial community responses to contamination with silver, aluminium oxide and silicon dioxide nanoparticles. Ecotoxicology 26, 449–458.

Mirzajani, F., Askari, H., Hamzelou, S., Farzaneh, M., Ghassempour, A., 2013. Effect of silver nanoparticles on *Oryza sativa* L. and its rhizosphere bacteria. Ecotoxicol. Environ. Saf. 88, 48–54.

Mo, F., Li, H., Li, Y., Cui, W., Wang, M., Li, Z., Wang, H., 2020. Toxicity of ag+ on microstructure, biochemical activities and genic material of *Trifolium pratense* L. seedlings with special reference to phytoremediation. Ecotoxicol. Environ. Saf. 195, 110499.

Mylona, Z., Panteris, E., Moustakas, M., Kevrekidis, T., Malea, P., 2020. Physiological, structural and ultrastructural impacts of silver nanoparticles on the seagrass *Cymodocea nodosa*. Chemosphere 248, 126066.

Olchowik, J., Bzdyk, R., Studnicki, M., Bederska-Błaszczyk, M., Urban, A., Aleksandrowicz-Trzcińska, M., 2017. The effect of silver and copper nanoparticles on the condition of English oak (*Quercus robur* L.) seedlings in a container nursery experiment. Forests 8, 310.

Ottoni, C.A., Neto, M.L., Léo, P., Ortolan, B.D., Barbieri, E., De Souza, A.O., 2020. Environmental impact of biogenic silver nanoparticles in soil and aquatic organisms. Chemosphere 239, 124698.

Peyrot, C., Wilkinson, K.J., Desrosiers, M., Sauvé, S., 2014. Effects of silver nanoparticles on soil enzyme activities with and without added organic matter. Environ. Toxicol. Chem. 33, 115–125.

Qian, H., Peng, X., Han, X., Ren, J., Sun, L., Fu, Z., 2013. Comparison of the toxicity of silver nanoparticles and silver ions on the growth of terrestrial plant model *Arabidopsis thaliana*. J. Environ. Sci. 25, 1947–1956.

Rahmatpour, S., Shirvani, M., Mosaddeghi, M.R., Nourbakhsh, F., Bazarganipour, M., 2017. Dose–response effects of silver nanoparticles and silver nitrate on microbial and enzyme activities in calcareous soils. Geoderma 285, 313–322.

Samarajeewa, A.D., Velicogna, J.R., Princz, J.I., Subasinghe, R.M., Scroggins, R.P., Beaudette, L.A., 2017a. Effect of silver nanoparticles on soil microbial growth, activity and community diversity in a sandy loam soil. Environ. Pollut. 220, 504–513.

Samarajeewa, A.D., Velicogna, J.R., Princz, J.I., Subasinghe, R.M., Scroggins, R.P., Beaudette, L.A., 2017b. Effect of silver nano-particles on soil microbial growth, activity and community diversity in a sandy loam soil. Environ. Pollut. 220, 504–513.

Sami, F., Siddiqui, H., Hayat, S., 2020. Impact of silver nanoparticles on plant physiology: a critical review. In: Sustainable Agriculture Reviews. vol. 41. Springer, Cham, pp. 111–127.

Scherer, M.D., Sposito, J.C., Falco, W.F., Grisolia, A.B., Andrade, L.H., Lima, S.M., Oliveira, S.L., 2019. Cytotoxic and genotoxic effects of silver nanoparticles on meristematic cells of *Allium cepa* roots: a close analysis of particle size dependence. Sci. Total Environ. 660, 459–467.

Schlich, K., Beule, L., Hund-Rinke, K., 2016. Single versus repeated applications of CuO and ag nanomaterials and their effect on soil microflora. Environ. Pollut. 215, 322–330.

Sillen, W.M.A., Thijs, S., Abbamondi, G.R., Janssen, J., Weyens, N., White, J.C., Vangronsveld, J., 2015. Effects of silver nanoparticles on soil microorganisms and maize biomass are linked in the rhizosphere. Soil Biol. Biochem. 91, 14–22.

Speranza, A., Crinelli, R., Scoccianti, V., Taddei, A.R., Iacobucci, M., Bhattacharya, P., et al., 2013. In vitro toxicity of silver nanoparticles to kiwifruit pollen exhibits peculiar traits beyond the cause of silver ion release. Environ. Pollut. 179, 258–267.

Stampoulis, D., Sinha, S.K., White, J.C., 2009. Assay-dependent phytotoxicity of nanoparticles to plants. Environ. Sci. Technol. 43, 9473–9479.

Sun, J., Wang, L., Li, S., Yin, L., Huang, J., Chen, C., 2017. Toxicity of silver nanoparticles to Arabidopsis: inhibition of root gravitropism by interfering with auxin pathway. Environ. Toxicol. Chem. 36, 2773–2780.

Thuesombat, P., Hannongbua, S., Akasit, S., Chadchawan, S., 2014. Effect of silver nanoparticles on rice (*Oryza sativa* L. cv. KDML 105) seed germination and seedling growth. Ecotoxicol. Environ. Saf. 104, 302–309.

Torrent, L., Iglesias, M., Marguí, E., Hidalgo, M., Verdaguer, D., Llorens, L., Vogel-Mikuš, K., 2020. Uptake, translocation and ligand of silver in Lactuca sativa exposed to silver nanoparticles of different size, coatings and concentration. J. Hazard. Mater. 384, 121201.

Tripathi, D.K., Singh, S., Singh, S., Srivastava, P.K., Singh, V.P., Singh, S., Chauhan, D.K., 2017. Nitric oxide alleviates silver nanoparticles (AgNps)-induced phytotoxicity in Pisum sativum seedlings. Plant Physiol. Biochem. 110, 167–177.

Tymoszuk, A., Miler, N., 2019. Silver and gold nanoparticles impact on in vitro adventitious organogenesis in chrysanthemum, gerbera and cape primrose. Sci. Hortic. 257, 108766.

Vishwakarma, K., Shweta, Upadhyay, N., Singh, J., Liu, S., Singh, V.P., Prasad, S.M., Chauhan, D.K., Tripathi, D.K., Sharma, S., 2017. Differential phytotoxic impact of plant mediated silver nanoparticles (AgNPs) and silver nitrate ($AgNO_3$) on Brassica sp. Front. Plant Sci. 8.

Wang, J., Koo, Y., Alexander, A., Yang, Y., Westerhof, S., Zhang, Q., et al., 2013. Phytostimulation of poplars and Arabidopsis exposed to silver nanoparticles and ag+ at sublethal concentrations. Environ. Sci. Technol. 47, 5442–5449.

Wang, J., Shu, K., Zhang, L., Si, Y., 2017. Effects of silver nanoparticles on soil microbial communities and bacterial nitrification in suburban vegetable soils. Pedosphere 27, 482–490.

Wu, J., Wang, G., Vijver, M.G., Bosker, T., Peijnenburg, W.J., 2020. Foliar versus root exposure of AgNPs to lettuce: Phytotoxicity, antioxidant responses and internal translocation. Environ. Pollut. 261, 114117.

Yan, A., Chen, Z., 2019. Impacts of silver nanoparticles on plants: a focus on the phytotoxicity and underlying mechanism. Int. J. Mol. Sci. 20, 1003.

Yang, J., Jiang, F., Ma, C., Rui, Y., Rui, M., Adeel, M., Xing, B., 2018. Alteration of crop yield and quality of wheat upon exposure to silver nanoparticles in a life cycle study. J. Agric. Food Chem. 66, 2589–2597.

Yin, L., Colman, B.P., McGill, B.M., Wright, J.P., Bernhardt, E.S., 2012. Effects of silver nanoparticle exposure on germination and early growth of eleven wetland plants. PLoS One 7, e47674.

Yuan, L., Richardson, C.J., Ho, M., Willis, C.W., Colman, B.P., Wiesner, M.R., 2018. Stress responses of aquatic plants to silver nanoparticles. Environ. Sci. Technol. 52, 2558–2565.

Zulfiqar, F., Navarro, M., Ashraf, M., Akram, N.A., Munné-Bosch, S., 2019. Nanofertilizer use for sustainable agriculture: advantages and limitations. Plant Sci. 110270.

CHAPTER 14

Silver nanoparticles phytotoxicity mechanisms

Renata Biba[a], Petra Peharec Štefanić[a], Petra Cvjetko[a], Mirta Tkalec[b], and Biljana Balen[a]

[a]*Division of Molecular Biology, Department of Biology, Faculty of Science, University of Zagreb, Croatia*
[b]*Division of Botany, Department of Biology, Faculty of Science, University of Zagreb, Croatia*

14.1 Introduction

The production and utilization of nanoparticles (NPs) is a rapidly growing field of industry since their small size, with dimensions between 1 and 100 nm, results in improved chemical and physical characteristics in comparison to the bulk material. Among different types of NPs, silver nanoparticles (AgNPs) are of particular interest because of their well-known antimicrobial properties, which make them very interesting for potential application in agriculture and food production due to many potential benefits (Prasad et al., 2017). However, increasing usage and unregulated release of AgNPs into aquatic and terrestrial systems bring concerns about their impact on the environment and living organisms, among which plants, as sessile organisms, are particularly affected. Moreover, plants play an important role in the accumulation of many environmentally released substances and their biodistribution into food chains (Rico et al., 2011). Uptake of AgNPs by crop plants has been proven in many studies (reviewed in Yan and Chen, 2019), resulting with their entrance into the food chain (Ma et al., 2010), with possible detrimental impacts not only to food production but also to human health (Colman et al., 2013). Therefore, there are an increasing number of publications that investigate AgNP potential phytotoxicity. However, the interaction of the AgNPs with the medium used for plant cultivation is often overlooked although it can lead to various modifications of AgNP properties influencing their transport, fate, and possible toxicity (Reidy et al., 2013). It is mandatory to establish AgNP behavior patterns in different relevant plant exposure media before the assessment of their potential toxic or beneficial effects. Moreover, different coating agents are applied to increase the AgNP stability, but they also affect AgNP behavior in a biological environment (Domazet Jurašin et al., 2016) and many of the AgNP-induced effects on plants have recently been found to be coating dependent (Biba et al., 2020, 2021; Su et al., 2020). Considering AgNP-induced phytotoxic effects, it is well known that they increase the formation of reactive oxygen species (ROS) (Tkalec et al., 2019), which as a consequence can generate cytotoxic and genotoxic effects in plants (Yan and Chen, 2019). Moreover, they can interfere with important biological macromolecules, such as DNA and proteins, thus inducing their destabilization and damage (Yan and Chen, 2019; Tripathi et al., 2017). In this chapter, recently published studies on AgNP stability in different biological media, applied for plant exposure, have been reviewed. Moreover, we present here a discussion about AgNP roles in promoting cytotoxic and genotoxic effects, changes in gene and protein expression, to reveal the AgNP-triggered molecular response in plants and assess potential phytotoxic mechanisms of AgNPs.

14.2 AgNP stability in biological media for plant growth

Stability of AgNPs and susceptibility to transformation depend on parameters that are directly related to AgNP synthesis, i.e., their surface chemistry, mostly size, charge, chemical functionality, and hydrophilicity (Bae et al., 2011). However, recently, it was emphasized that stabilization mechanism, aging, and environment also significantly influence the stabilization of AgNPs (De Leersnyder et al., 2019). AgNP synthesis (Fig. 14.1) usually starts with a solution of silver salt (for example, $AgNO_3$) as a precursor, and a reducing agent, most commonly sodium borohydride that accelerates fast nucleation or ascorbic acid used to achieve slower growth (Heinz et al., 2017). Formed nanoparticles are unstable due to the van der Waals attraction between them, which can lead to aggregation and subsequent precipitation of nanoparticles. To prevent interactions between AgNPs, different coating agents (also known as capping and stabilizing agents) are used during the synthesis as their attachment onto the surface of nanoparticles produce repulsive forces preventing their further growth and agglomeration (Li et al., 2013).

However, coating agents also determine particle shape and size and influence their solubility, reactivity, and overall stability (Tolaymat et al., 2010). PubMed literature search showed 12 different coatings used in assessment of AgNPs toward plants (Fig. 14.2), polyvinylpyrrolidone (PVP), and citrate coating being the most used ones. PVP is a stable nonionic polymer that provides excellent steric stabilization of NP (Koczkur et al., 2015). It is widely used in investigations of AgNP toxic effects in plants (Vannini et al., 2014; Jiang et al., 2017; Falco et al., 2020; Noori et al., 2019). Other commonly used polymers are polyvinyl alcohol (PVA) (Mirzajani et al., 2013; Mirzajani et al., 2014; Lalau et al., 2020) and polyethylene glycol (PEG) (Wang et al., 2013; Abdelsalam et al., 2018). Steric stabilization can also be achieved using gum arabic (GA), a mixture of polysaccharides and glycoproteins (Yin et al., 2011; Kong et al., 2014). Electrostatic stabilization of nanoparticles includes developing a surface charge, usually by physical adsorption of charged species onto the surface (Yu and Xie, 2012). Citrate is a common stabilizing molecule that provides a negative charge of AgNPs, and it has been used in many AgNP toxicology studies on plants (Cvjetko et al., 2018; Peharec Štefanić et al., 2018; Gubbins et al., 2011; Abdel-Aziz and Rizwan, 2019). Positively charged AgNP surface is gained using various cationic

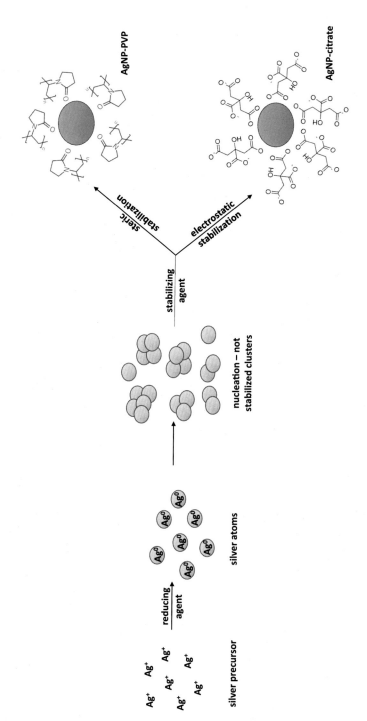

FIG. 14.1

AgNP synthesis and stabilization using steric (PVP) and electrostatic (citrate) stabilization.

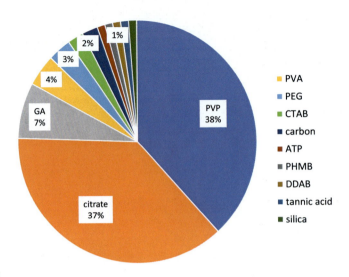

FIG. 14.2

Proportional representation of coatings used for AgNP stabilization in plant research.

surfactants, such as cetyltrimethylammonium bromide (CTAB) (Cvjetko et al., 2017; Biba et al., 2020), didecyldimethylammonium bromide (DDAB) (Barabanov et al., 2018) or polyhexamethylene biguanide (PHMB) (Gusev et al., 2016).

Even though AgNP effects on plants were extensively studied in the last decade, only a minority of reports included the AgNP stability analyses in media used for cultivation of plants (Table 14.1). When added in the media, AgNPs may interact with the variety of dissolved or particulate, inorganic, or organic compounds (present in the exposure medium or released by living cells), which influence their aggregation dynamics and thus colloidal stability, which as a consequence might influence the toxicity of AgNPs on living cells (Bundschuh et al., 2018). The extent of AgNP interaction with the medium depends on the composition of the medium, coating used for AgNP stabilization (Akter et al., 2018), and period of exposure time for a certain experiment. Therefore, it is important to control the AgNP stability in toxicology tests, particularly since stable laboratory-produced AgNPs may become unstable after exposure to different biological media. Special physicochemical properties of AgNPs, defined by their size on the nanoscale, can be changed, even lost after their transformation (Barrena et al., 2009), thus inducing different intracellular responses in the treated plants.

Different techniques are employed to study AgNP stability depending on the exposure medium (Table 14.1). UV/vis spectrometry is often used as a characterization technique that provides information on AgNP stability in media. Silver nanoparticles efficiently absorb and scatter light because the conduction electrons in the metal collectively oscillate when excited by light at specific wavelengths—a phenomenon known as surface plasmon resonance (SPR), which highly depend on the size and shape of AgNPs (Englebienne et al., 2003). Smaller spherical nanoparticles (10–50nm) have peaks near 400nm, while for larger ones (100nm) increased scattering

Table 14.1 Studies which included analyses of AgNP stability in different media for plant exposure.

Exposure medium	Type of the medium	Plant species and tissue	AgNP type	Methodology employed for stability analyses	Major findings	References
Water	Ultrapure	Tobacco (*Nicotiana tabacum*) adult plants	Commercial, citrate-coated	UV-VIS spectrometry, DLS measurement, ζ potential, ICP-MS	Loss of negatively charged citrate coatings, dissociation of Ag$^+$ ions	Peharec Štefanič et al. (2019)
	Moderately hard	Maize (*Zea mays* L.), cabbage (*Brassica oleracea* var. *capitata* L.)	Laboratory synthesized, citrate-coated	DLS measurement	Fairly stable AgNPs	Pokhrel and Dubey (2013)
	Deionised	Broad beans (*Vicia faba*)	Commercial, uncoated	DLS measurement, ζ potential	Moderate AgNP agglomeration and stable charge	Patlolla et al. (2012)
	Ultrapure	Common grass (*Lolium multiflorum*) seeds and seedlings	Laboratory synthesized, GA-coated	DLS measurement, ζ potential, TEM imaging	No measurable change in AgNP size and charge	Yin et al. (2011)
	Distilled	Lettuce (*Lactuca sativa*), cucumber (*Cucumis sativus*)	Laboratory synthesized, uncoated	DLS measurement, ζ potential, TEM imaging	No significant changes AgNP size distribution, morphology and shape	Barrena et al. (2009)
Liquid culture medium	½ strength Murashige and Skoog (MS)	Tobacco (*N. tabacum*) seeds and seedlings	Commercial, citrate-coated	UV-VIS spectrometry, DLS measurement, ζ potential, TEM imaging	Strong, rapid AgNP agglomeration; a slight shift toward less negative charge values	Biba et al. (2021)
	Plant culture medium prepared with O$_2$-free water	Wheat (*Triticum aestivum* L.) seedlings and root exudates	Commercial, PVP-coated	ICP-MS	Oxic conditions induce higher AgNP dissolution compared to anoxic	Dang et al. (2020)
	1/15 strength Hewitt (in the absence or presence of Fe^{2+}-EDTA)	Rice (*Oryza sativa*)	Laboratory synthesized, PVP-coated	UV-VIS-NIR spectrometry	Stable AgNPs with minimal dissolution	Yang et al. (2020)
	Hewitt medium (with Cl$^−$ ions) Hoagland medium (without Cl$^−$ ions)	Rice (*O. sativa*)	Laboratory synthesized, PVP-coated	UV-VIS spectrometry, DLS measurement, ζ potential, LC-ICP-MS	Cl$^−$ ions increased AgNP stability in Hoagland compared to Hewitt medium	Yang et al. (2019)
	Full strength MS	*Arabidopsis thaliana*	Commercial, citrate-coated	ζ potential	Decrease of ζ potential	Li et al. (2018a, b)

Continued

Table 14.1 Studies which included analyses of AgNP stability in different media for plant exposure—cont'd

Exposure medium	Type of the medium	Plant species and tissue	AgNP type	Methodology employed for stability analyses	Major findings	References
	Full strength MS	A. thaliana	Commercial, citrate-coated	DLS measurement, ζ potential, SEM imaging, ICP-MS	Extensive AgNP aggregation and dissolution	Ke et al. (2018)
	½ Hutner's solution	Duckweed (Landoltia punctata)	Laboratory synthesized, PVP-coated	DLS measurement, ζ potential, ICP-MS	Some AgNP aggregation and dissolution	Stegemeier et al. (2017)
	10% Hoagland	Duckweed (Spirodela polyrhiza)	Laboratory synthesized, PVP-coated	DLS measurement, TEM imaging, ICP-MS, X-ray diffraction	Stable AgNPs with slight increase of size distribution and Ag$^+$ release	Jiang et al. (2017)
	10% Hoagland	Duckweed (S. polyrhiza)	Laboratory synthesized, GA- and PVP-coated	TEM imaging, X-ray diffraction	Slight change of AgNP size	Jiang et al. (2014)
	¼ Hoagland	A. thaliana	Commercial, uncoated	TEM imaging, ICP-MS	AgNP aggregation and dissolution, difference in shape	Nair and Chung (2014a)
	High strength Hoagland's E+	Duckweed (S. punctuta)	Commercial, uncoated	DLS measurement, ICP-OES	Significant AgNP agglomeration With increasing AgNP concentration; changes in ζ potential; negligible AgNP dissolution	Thwala et al. (2013)
	¼ Hoagland	A. thaliana	Laboratory synthesized, citrate-coated	DLS measurement, ζ potential, TEM imaging, ICP-MS	Significant AgNP aggregation, almost linear increase of hydrodynamic radius, release of Ag$^+$ ions	Geisler-Lee et al. (2013)
	¼ Hoagland	Poplar (Populus deltoides × nigra, DN-34) and A. thaliana cuttings	Commercial, carbon-coated vs laboratory synthesized, PEG-coated	TEM imaging, ICP-MS	Aggregated carbon-coated AgNPs of nonuniform size; not significantly aggregated AgNP-PEG of uniform size	Wang et al. (2013)
	10% Hoagland	Duckweed (S. polyrhiza)	Laboratory synthesized, GA-coated	TEM imaging	Slight change in AgNP size	Jiang et al. (2012)
	Nutrient solution prepared according to OECD 221 guidelines	Duckweed (Lemna minor)	Laboratory synthesized, citrate-coated	DLS measurement, ζ potential, TEM imaging	Substantial reduction in ζ potential, increase in hydrodynamic diameter and presence of large star shaped agglomerates	Gubbins et al. (2011)

Table 14.1 Studies which included analyses of AgNP stability in different media for plant exposure—cont'd

Exposure medium	Type of the medium	Plant species and tissue	AgNP type	Methodology employed for stability analyses	Major findings	References
Solid culture medium	½ strength MS	Tobacco (N. tabacum) seeds and seedlings	Commercial, citrate-coated	UV-VIS spectrometry	Little AgNP agglomeration	Biba et al. (2021)
	½ strength MS	Tobacco (N. tabacum) seeds and seedlings	Laboratory synthesized, PVP- and CTAB-coated	UV-VIS spectrometry	AgNP-PVP were found to be more unstable compared to AgNP-CTAB	Biba et al. (2020)
	Full strength MS	Tobacco (N. tabacum) seedlings	Commercial, citrate-coated	UV-VIS spectrometry, dark-field microscopy	Stable AgNPs with a minor decrease in the absorption maximum, slight peak shift toward lower wavelengths	Peharec Štefanić et al. (2018)
Soil	Immature Pallic soil, Templeton loamy silt (pH = 5.1) granular silt loam (pH = 6.0)	Carrot [Daucus carota L. ssp. sativus (Hoffm.) Arcang.], radish [Raphanus raphanistrum L. var. sativus (L.) G. Beck], leek [Allium ampeloprasum L.], lettuce, parsley [Petroselinum crispum (Mill.) Nyman ex A.W. Hill], rocket, beetroot [Beta vulgaris ssp. vulgaris L.], silverbeet [Beta vulgaris L. ssp. maritima (L.) Arcang.], and spinach [Spinacia oleracea L.]	Laboratory synthesized, citrate-coated	ICP-OES	AgNPs were transformed into Ag$^+$ during the course of the plant growth experiments	Saleeb et al. (2019)
	Clay, loam with high OM, loam with high carbonate, sandy loam	Tall fescue (Festuca arundinacea)	Commercial, SiO$_2$- and PVP-coated	ICP-MS	Lower Ag concentrations in the solutions of clay soil and loam with high carbonate content than in the solutions of sandy soil and loam soil with high OM content, regardless silver source	Layet et al. (2019)
	Sand matrix	Wheat (Triticum aestivum L.)	Commercial, uncoated	DLS measurement, ζ potential, AFM imaging, ICP-MS	AgNP agglomeration, highly negative ζ potential indicates AgNP stabilization in the sand matrix	Dimkpa et al. (2013)
	OECD artificial soil	Mung bean (P. radiatus) seedlings, Sorghum (S. bicolor) seedlings	Commercial, citrate-coated	ICP-AES	Lower rates of AgNPs dissolution due to reduction of surface area attributed to greater soil aggregation	Lee et al. (2012)

broadened peaks and shifts them toward longer wavelengths. As the nanoparticles destabilize, the original extinction peak decrease in intensity (due to the depletion of stable AgNPs) and often gets broader or a secondary peak appears at longer wavelengths (due to the formation of aggregates) (Eccles et al., 2010). Dynamic light scattering (DLS) analysis allows the measurement of the AgNP size distribution profiles (Carvalho et al., 2018) and in combination with zeta (ζ) potential, which gives AgNP surface charge, can provide information about the AgNP tendency to agglomerate (Carvalho et al., 2018). Analysis with transmission electron microscopy (TEM) can also give insight into changes of AgNP size, shape, and morphology after their exposure in biological media (Peretyazhko et al., 2014). To determine the dissociation of Ag^+ ions from AgNPs, the exposure medium is usually centrifuged using ultrafiltration units with a small pore size (3 kDa) and collected Ag^+ ions, which pass through the membrane, are analyzed with either inductively coupled plasma mass spectrometry (ICP-MS) or inductively coupled plasma atomic emission spectroscopy (ICP-AES) (Wojcieszek et al., 2020).

Water and liquid culture media are frequently used as exposure media in laboratory studies of AgNP impact on plants. Deionised, demineralized, and ultrapure water are in general an excellent environment to keep and preserve AgNPs stable for a longer time (Capjak et al., 2018; Cañamares et al., 2008), and it is known from the literature that the rate of AgNP dissolution is much lower in water than in the culture media (Sharma et al., 2014; Zook et al., 2011) (Fig. 14.3). In the study by Peharec Štefanić et al. (2019), in which ultrapure water was used as a medium for exposure of tobacco plants to commercial citrate-coated AgNPs, UV–vis spectrometry revealed that nanoparticles remained quite stable during the 7 days, although a graduate decrease of absorbance after the 2nd day indicated some instability. Additional analyses revealed a decrease of hydrodynamic diameter and less negative ζ potential with prolonged exposure, which suggested a decrease in AgNP size due to the loss of negatively charged citrate coatings, which favored dissociation of Ag^+ ions. Silver concentration determined after centrifugal ultrafiltration showed that Ag dissociation in solution increased during the 7 days, but did not exceed 1%. In the study of AgNP effects on *Lolium multiflorum*, the similar results were reported; in the aqueous solution, AgNP-GA were found to be well dispersed with a negative ζ potential and no measurable change in particle size at the end of experiment was noticed (Yin et al., 2011). Only moderate AgNP agglomeration was found for uncoated AgNPs in deionised water used for exposure of common beans (*Vicia faba*) (Patlolla et al., 2012) as well as for AgNP-citrate in demineralized water for exposure of two crop plants, maize, and cabbage (Pokhrel and Dubey, 2013). In the study in which germination of lettuce (*Lactuca sativa*) and cucumber (*Cucumis sativus*) was investigated after exposure to uncoated AgNPs, DLS and TEM analyses showed that there were no significant changes in their size distribution and morphology during experiments (Barrena et al., 2009).

However, water does not represent relevant plant growth media, which are usually very complex ionic mixtures with different ionic strengths and pH values, which can affect AgNP stability and induce their transformation (Fig. 14.3). Variation in

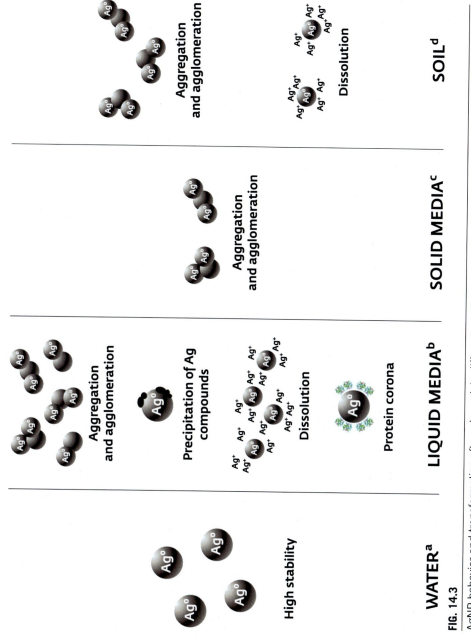

FIG. 14.3

AgNP behavior and transformation after release into different media applied in plant exposure experiments. (A) In water (deionised, demineralized or ultrapure) the highest AgNP stability has been recorded. (B) Hewitt, Hoagland, Hutner, and Murashige and Skoog liquid nutrient media of different strengths have been applied in which the highest AgNP aggregation, agglomeration and dissolution rate has been observed. (C) Only full and ½ strength Murashige and Skoog medium has been tested and some AgNP aggregation and agglomeration were detected, although in a lower rate compared to liquid media and soil. (D) In soil (immature Pallic soil, Templeton loamy silt, granular silt loam, clay, and OECD artificial soil) media AgNPs exhibited aggregation, agglomeration, and dissolution, which, however, was not as high as it was recorded in liquid media.

only one physical (change in pH) or chemical (change in electrolyte concentration or composition) parameter can influence the AgNP aggregation; laboratory synthesized AgNP-citrate was found to be more stable in alkaline and neutral than acidic pH, which was indicated by smaller ζ average values, more negative ζ potentials and smaller changes in the SPR peaks (Bélteky et al., 2019). Moreover, NaCl addition resulted in aggregation of citrate-coated AgNPs even to micrometer size range, as determined by the increased average hydrodynamic diameter of AgNPs and decreased SPR values. Such drastic changes in AgNP size were ascribed to the presence of Na^+ ions, which can shield the negatively charged surface groups, decreasing their repulsion and allows the formation of larger aggregates (Bélteky et al., 2019). Stability analysis of AgNP-PVP in two hydroponic media commonly used for rice growth, modified Hewitt solution (with Cl^- ions) and Hoagland solution (without Cl^- ions), revealed that the presence of Cl^- ions may increase the negative charge of AgNPs and subsequently stabilize AgNPs (Yang et al., 2019). Namely, compared to Hoagland medium where decreased SPR intensity as well as partially neutralized negative charge and significantly increased hydrodynamic diameter of AgNPs was recorded, in Hewitt solution, only minor changes in UV–vis spectrum, ζ potential, and hydrodynamic diameter was observed. However, the impact of Cl^- ions can also be concentration-dependent; high concentrations enhanced the AgNP aggregation due to the bridging by AgCl, while low concentrations stabilized AgNPs by formation of negatively charged surface precipitates (Bundschuh et al., 2018). Several other papers also demonstrated that the composition of the exposure medium has a strong impact on stability of AgNPs, causing their aggregation, dissociation of Ag^+ ions, changes in their ζ potential, and loss of their surface coating. Li et al. (2018b) recorded a significant decrease of ζ potential of the commercial, citrate-coated AgNP after exposure to the full strength (Murashige and Skoog, 1962, MS) medium. Substantial aggregation of citrate-coated AgNPs was also observed in nutrient solutions used for treatments of *Lemna minor* (Gubbins et al., 2011) and *Arabidopsis thaliana* (Ke et al., 2018), due to the high ionic strength of media. The hydrodynamic diameter of AgNP-PVP in ½ strength Hutner's medium revealed AgNP aggregation, although the particle size was found to be relatively constant after 2 days in the growth medium (Stegemeier et al., 2017). In the high strength Hoagland's E^+ medium, significant agglomeration of commercial uncoated AgNP was detected. Similarly, after exposure to the ¼ Hoagland medium, citrate-coated AgNPs (Geisler-Lee et al., 2013), as well as uncoated AgNPs (Nair and Chung, 2014a), exhibited aggregation and Ag^+ release. However, dissolution of uncoated AgNPs, observed in the ¼ Hoagland medium after 1 day of exposure, significantly decreased after 14 days (Nair and Chung, 2014a), probably due to the precipitation of Ag^+ ions with electrolytes in the medium or adsorption with other particles in the solution (Lowry et al., 2012). On the other hand, exposure of AgNPs in very diluted 10% Hoagland solution only slightly changed the size and dissociation of Ag^+ ions of GA- and PVP-coated AgNPs (Jiang et al., 2012, 2014, 2017). UV–vis measurements of the PVP-coated AgNPs in diluted ¹⁄₁₅ strength Hewitt nutrient solution, in the absence or presence of Fe^{2+}-EDTA also showed AgNP stability and minimal dissolution after different incubation times (Yang et al., 2020). Stability of the AgNPs in liquid medium was also found to be coating-dependent; namely, in the study in which poplar and *Arabidopsis* cuttings were exposed to AgNPs in ¼ strength Hoagland nutrient medium, carbon-coated AgNPs were found to be of the nonuniform size and prone to aggregation after 15 days in the exposure medium, while AgNP-PEG exhibited more size uniformity and did not reveal significant aggregation even after a 1-month incubation period (Wang et al., 2013). Since the complexity of the nutrient medium and the concentration of its components have been directly correlated with the stability of AgNPs in future experiments diluted culture media with the minimum nutrients possible for plant growth should be used. Also, as experiments with plants usually last for a certain period to monitor possible effects, it is necessary to analyze AgNP stability not only at the beginning but also for as long as the experiment lasts as AgNPs can change in time due to interaction with media components.

Several studies, which analyzed the effects of AgNPs on plant growth and physiology, were performed on the solid nutrient media which are more adequate for cultivation of terrestrial plants. However, these media are less convenient in terms of AgNP stability analyses compared to liquid ones due to the lack of adequate techniques (Coutris et al., 2012; Anjum et al., 2013). Namely, DLS analysis of ζ potential and size is only applicable for liquid medium (colloid solutions), while the measurement of dissolution rates of nanoparticles is almost impossible in the solid medium because AgNPs become encapsulated in agar added to solidify nutrient medium and they cannot be separated from the Ag^+ ions. In the study of Biba et al. (2021), the stability of the citrate-coated AgNPs in the liquid and solid medium of the same composition (½ strength MS medium), used for treatment of tobacco seedlings, was compared. In the liquid medium, a stronger reduction of the absorption maximum compared to solid medium at zero minutes was recorded accompanied by a peak shift toward lower wavelengths compared to the absorption maximum obtained in ultrapure water. Also, over a period of 5 h, the SPR intensity has exponentially decreased, which suggested rapid AgNP agglomeration. Additional DLS and TEM analysis corroborated these findings (Biba et al., 2021). On the contrary, in the agar-solidified medium, a reduction of the AgNP absorption maximum and a small peak shift toward lower wavelengths were recorded immediately after medium solidification in comparison to the absorption maximum measured in ultrapure water. The intensity of the SPR peak and its position remained relatively constant during the 5-days period, which suggested that AgNP agglomeration was negligible. These findings indicated encapsulation of AgNPs, which inhibited their aggregation and/or dissociation of Ag^+ ions. High stability of AgNP-citrate in the solid full-strength MS nutrient medium was also reported by Peharec Štefanić et al. (2018), who suggested that this was probably a result of AgNP stabilization with Phytagel, a natural polymer used for solidification of nutrient media. It is known that some polymer ligands can control access of O_2 to the AgNP surface, which would decrease their oxidative dissolution rate (Gunsolus et al., 2015). Biba et al. (2020) reported that stability of the AgNPs in the solid ½ strength MS nutrient medium is coating-dependent; AgNP-PVP were found to be unstable as they exhibited an almost immediate shift toward higher wavelengths in comparison to the SPR absorbance maximum in water, which suggested their aggregation. Moreover, measurement during 5 days of exposure in the medium revealed an SPR at lower wavelengths with a more intense absorbance peak indicating a size reduction, probably due to dissolution of AgNPs. In the same medium, the SPR peak intensity of AgNP-CTAB stayed relatively constant during the 5-days period, indicating almost no aggregation and/or dissolution of the AgNP-CTAB.

The least number of studies investigating AgNP impact on plants were performed in the soil as a medium for exposure. Namely, soil is an extremely complex mixture of organic matter (OM), minerals, gases, liquids, and organisms in different amounts depending on soil, which may affect AgNP behavior and toxicity (Anjum et al., 2013). Analysis of AgNP stability is done in soil solution obtained by soil extraction, and mostly, silver dissolution is measured because there are no adequate techniques for stability analysis in solids. In general, it seems that AgNP phytotoxicity in soil is lower compared to different culture media. Lee et al. (2012) found lower phytotoxic effects after exposure of mung bean and *Sorghum* seedlings to citrate-coated AgNPs in artificial soil compared to agar medium, which was ascribed to the lower rates of AgNPs dissolution, probably due to greater aggregation of AgNPs with soil particles. In the study of Dimkpa et al. (2013), in which sand matrix (low in OM) was used as an exposure medium for wheat plants, DLS and ζ potential measurement indicated agglomeration of the uncoated AgNPs. Saleeb et al. (2019) provided strong evidence that citrate-coated AgNPs transformed into Ag^+ ions in different soil types used for plant growth during the experiment. The interaction of nanoparticles with dissolved OM present in the soil particularly contribute to AgNP transformation, e.g., the colloidal stability of uncoated AgNPs increased through sorption of dissolved organic matter, and the release of Ag^+ ions from uncoated and citrate-coated AgNPs decreased due to formation of organic matter coatings (Klitzke et al., 2015). Moreover, AgNP mobility may be altered by the electrostatic interaction with different soil types; soil with positive charge may prevent mobility of AgNPs which bear a negatively charged coating (such as AgNP-citrate), while soil with negative charge may induce mobility of negatively charged citrate-coated AgNPs (Tolaymat et al., 2010; Yu et al., 2013). Layet et al. (2019) measured lower Ag concentrations in the solutions of clay soil and loam with high carbonate content than in the solutions of sandy soil and loam soil with high OM content, regardless of the form of Ag using AgNPs with inorganic (SiO_2) and organic (PVP) coatings as well as dissolved Ag ($AgNO_3$) as a silver source.

Studies have shown that in contact with living cells and/or plant root exudates AgNPs may get in touch with various small organic molecules (such as sugars, amino acids, or phenolic compounds) that can influence their surface charge upon adsorption and affect their colloidal stability. For example, it was found that root exudates may increase the AgNP dissolution and/or reduction of Ag^+ ions into AgNPs (Guo et al., 2019; Stegemeier et al., 2015). Bélteky et al. (2019) found that in the presence of glucose and glutamine, biomolecule interactions with the nanoparticle surface could cause a shift in the characteristic absorbance peak of AgNPs, although no significant changes in aggregation grade were recorded, probably due to the well-known biomolecular corona formation (Bélteky et al., 2019). Since AgNP toxicity is related to aggregation grade, the impact of the corona effect on the biological activity of nanoparticles should be evaluated and emphasized. In the study in which rice plants were hydroponically exposed to AgNP-PVP, it was found that due to the presence of iron plaque on the rice roots, AgNPs underwent oxidative dissolution, sulfidation, and chlorination, which all increased the uptake and translocation of silver in the rice plant (Yang et al., 2020). Moreover, AgNP dissolution can be additionally increased by O_2 released from aquatic plants (Soda et al., 2007) or by ROS generated from plant roots (Huang et al., 2019a). In the recent study, Dang et al. (2020) reported that under anoxic conditions environmentally relevant concentrations of root exudates exhibited a minimal effect on the dissociation of Ag^+ ions from PVP-coated AgNPs. However, under oxidizing conditions, substantially higher AgNP dissolution was recorded. Similar results were obtained when the experiment was conducted in the presence of wheat seedlings instead of root exudates in an anoxic and oxidizing environment, thus indicating that dissolved O_2 is a key factor responsible for the enhanced Ag^+ dissociation from AgNPs (Dang et al., 2020; Zhang et al., 2018).

After AgNP uptake by plant cells, it remains unclear how the complex internal plant cell environment affects AgNP stability and behavior. In the studies with other types of nanoparticles, it was found that the presence of organic molecules, originating from plant and algal cells, induced NP agglomeration (Adeleye and Keller, 2016; Schwabe et al., 2013). Therefore, it is reasonable to expect for AgNPs to respond similarly. Su et al. (2019) suggested that for example, different compounds in the plant sap might induce AgNP aggregation due to the high ionic strength, while the presence of organic molecules could impact AgNP dissolution rates. A recent study by Su et al. (2020) shed some light on this matter. Namely, AgNPs stabilized with different coatings (citrate, PVP, and GA) were examined concerning aggregation and dissolution when exposed to synthetic citrus sap. Hydrodynamic diameters of all three types of AgNPs significantly increased after exposure, which indicated their aggregation. Interestingly, the extent of aggregation was found to be coating-dependent; namely, citrate failed to stabilize AgNPs, PVP was successful to a moderate extent, while GA was found to be the best stabilizing agent for AgNPs in this synthetic sap. Zeta potentials were found to be less negative and of similar values for all tested AgNPs and authors concluded that high ionic strength of citrus sap limited the contribution of electrostatic repulsion to the AgNP stability. Moreover, a very low rate of AgNP dissolution was recorded, which was ascribed to the inhibitory effect of organic compounds in the sap. However, since different plants have different sap composition (Gourieroux et al., 2016; Hijaz and Killiny, 2014; López-Portillo et al., 2014), different results considering the behavior and fate of AgNPs might be expected in various plant species. Therefore, more studies on the impact of plant internal environment on AgNP stability should be conducted in the future.

14.3 AgNP coating-dependent effects

It is known that coating, surface charge, and particle size greatly determine the effects of NPs (Choi and Hu, 2008; Elechiguerra et al., 2005; El Badawy et al., 2011), but the result can also be affected by the plant species used and specifications of the experimental details, such as dosage, duration, and medium (Tkalec et al., 2019). Even though there are more than 90 research papers that question potential phytotoxic effects of AgNPs stabilized with various surface coatings, only a small number of experiments implement differently coated AgNPs in the same experimental setup which enables comparing (Table 14.2). Most of these studies have found that differently coated AgNPs induce differential plant response. However, it is not always completely clear whether the observed differences depend solemnly on the coating characteristics of AgNPs or if the size of the AgNPs applied also plays a role, as AgNPs of different size were investigated. Yin et al. (2012) have examined germination and early growth of 11 wetland plants exposed to AgNP-PVP, and AgNP-GA in two experimental setups: directly through simple culture exposure and in greenhouse experiment through soil. Their results showed that phytotoxic effects greatly differ between Ag sources, treated plants and medium used in the experiment. Effects of AgNP-GA were far more deleterious compared to AgNP-PVP, both in culture medium and in soil. However, this outcome could also be a result of a difference in AgNP sizes (21 nm AgNP-PVP and 6 nm AgNP-GA) since smaller NPs can penetrate the cell wall much easier than the bigger ones (Tripathi et al., 2017). Similar observations were made in an experiment with *Spirodela polyrhiza* treated with AgNP-PVP (20 nm), AgNP-GA (6 nm) and μm-sized Ag (Jiang et al., 2014). Both types of AgNPs induced formation of reactive oxygen species (ROS) and altered antioxidant status of the plant, with the effects far more pronounced in the AgNP-GA treatment. Wang et al. (2013) have observed a difference in response of *A. thaliana* and poplar tree to AgNP-carbon (20 nm) and AgNP-PEG (5 and 10 nm) where carbon-coated AgNPs showed stimulatory effects on evapotranspiration and growth of both plants at sublethal concentration, and AgNP-PEG caused mostly negative trends in root elongation and fresh weight. Difference between coating effects was also investigated in research by Cvjetko et al. (2017), who treated *Allium cepa* roots with three types of AgNPs (citrate, PVP-, and CTAB-coated). Increased ROS and oxidative damage were significant in all treatments at higher concentrations but were most pronounced after AgNP-CTAB treatment which also caused a reduction of root growth and mitotic index. Those effects were attributed to the smaller size of AgNP-CTAB compared to other AgNPs used, as well as a positive surface charge. Positively charged NPs can interact with the negatively charged cell membrane and in that way affect silver uptake in the plant (Tripathi et al., 2017). On the other side, citrate-coated AgNPs showed the weakest impact on *A. cepa* roots. Their bigger size and overall negative charge disabled their uptake, reducing overall phytotoxic effects.

Only a few investigations on plants reported the effect of AgNPs with different coatings but of similar size on plants. Obtained differences in response were mostly related to the impact of medium as AgNPs can undergo different chemical and morphological changes in exposure medium, from oxidation and Ag^+ release to aggregation (Shen et al., 2015). However, in most of these studies, there was no detailed analysis of AgNP stability. In bryophyte *Physcomitrella patens*, effects of AgNPs without surface coating and two different coatings (citrate and PVP) were investigated in two stages of gametophyte development, protonema and as a leafy gametophyte (Liang et al., 2018). Results showed that uncoated AgNPs have had the most detrimental effect on protonemal growth, whereas citrate-coated AgNPs most significantly damaged the leafy gametophyte. In both cases, AgNP-PVP effects were least harmful to *P. patens*. However, higher dissolution of AgNP-citrate was measured in medium during course of experiment and although it was not correlated with toxic effect observed in the protonema stage, authors suggested that different factors including media composition, light, and temperature could contribute to aggregation, transformation and/or sedimentation of AgNPs. Investigating nanotoxic response of *L. minor* to AgNPs, Pereira et al. (2018) observed that PVP-coated AgNPs (90 nm) strongly affected the growth while citrate-ones (80 nm) altered activities of antioxidative enzymes. Differential responses were related to differences in their stability in Steinberg medium used for plant cultivation, e.g., AgNP-PVP were found to be more aggregated. Layet et al. (2019) reported that Ag flux and uptake in roots and leaves of *Festuca arundinacea* grown in different soil types did not differ after treatment with 50 nm AgNP-PVP or AgNP-SiO_2 of 50 nm, but soil properties controlled silver phytoavailability.

Several studies indicated that the properties of the coating molecule itself could influence AgNP effects on plants. A study by Torrent et al. (2020) showed that the levels of Ag uptake in *L. sativa* depend mainly on AgNP size and concentration, but not on coating. However, translocation of AgNPs to shoots was more pronounced for AgNP-PEG, which have a neutral surface charge compared to negatively charged AgNP-PVP and AgNP-citrate. In another study on the aquatic plant *Wolffia globosa*, Zou et al. (2017) have chosen citrate and adenosine triphosphate (ATP) as coating agents. The effects of AgNP-ATP included loss of Hill reaction, reduction of chlorophyll content and sugar depletion, and an increase in superoxide dismutase (SOD) and peroxidase (POD) activities. On the contrary, plants treated with AgNP-citrate maintained Hill reaction activity and increased accumulation of soluble sugars, with no effect on antioxidative enzyme activities. The authors ascribed the difference in physiological response to different metabolic role of caping molecule released from AgNP; citrate serves as a substrate of tricarboxylic acid cycle and pentose phosphate pathways, while ATP acts as an exogenous metabolic energy source. A great difference between 40 nm PVP- and CTAB-coated AgNPs was recorded in a study by Biba et al. (2020). *Nicotiana tabacum* seedling growth was mildly affected by AgNP-PVP and severely damaged by AgNP-CTAB, even at the lowest tested concentrations. Treatment with CTAB alone showed severe detrimental effects on tobacco seedlings, suggesting that AgNP-CTAB phytotoxic effects result from the surface coating rather than nanoparticles. In another study, Barabanov et al. (2018) showed that AgNPs with PVP and dimethyldidodecylammonium bromide (DDAB) coatings promoted growth of *Pisum sativum* at lower concentrations, but when applied at

Table 14.2 Studies that compared the effects of differently coated AgNPs in plants.

Plant species	AgNPs compared	AgNP concentration	Exposure time	Exposure medium	Type of analysis	Findings	References
Carex spp./Eupatorium fistulosum	AgNP-PVP (21.0 ± 17.0 nm) and AgNP-GA (6.0 ± 1.7 nm)	1, 10 and 40 mg/L	20 days	Pure culture and soil experiment	Seed germination and growth	AgNP-GA > AgNP-PVP ≈ Ag$^+$	Yin et al. (2012)
Arabidopsis thaliana/Populus deltoides x nigra	AgNP-carbon (27.3 ± 5.6 nm) and AgNP-PEG (5.1 ± 0.3 and 10.4 ± 1.7 nm)	0.01–100 mg/L	11 days/14 days	¼ strength Hoagland solution	Evapotranspiration, plant growth	Ag$^+$ > AgNP-carbon > AgNP-PEG	Wang et al. (2013)
Spirodela polyrhiza	AgNP-PVP (20.0 ± 7.8 nm), AgNP-GA (6.5 ± 2.3 nm), and micron-Ag particles	0.5, 1, 5 and 10 mg/L	72 h	10% Hoagland solution	ROS accumulation, MDA and GSH content, activity of antioxidative enzymes, ultrastructure	AgNP-GA > AgNP-PVP > micron-Ag	Jiang et al. (2014)
Wolffia globosa	AgNP-citrate (18.8 nm) and AgNP-ATP (20.1 nm)	0.1, 1, and 10 mg/L	3 days	1/10 strength modified Hoagland solution	Silver and nutrient uptake, photosynthesis, activity of antioxidative enzymes	AgNP-ATP ≈ Ag$^+$ > AgNP-citrate	Zou et al. (2017)
Allium cepa	AgNP-citrate (61.2 ± 33.9 nm), AgNP-PVP (9.4 ± 1.3 nm), and AgNP-CTAB (5.6 ± 2.1 nm)	25, 50, 75, and 100 μM	72 h	Ultrapure Milli-Q water	Silver uptake, root growth and dry weight, mitotic index, oxidative damage, activity of antioxidative enzymes	Ag$^+$ > AgNP-CTAB > AgNP-PVP > AgNP-citrate	Cvjetko et al. (2017)

Plant	Nanoparticles	Concentration	Duration	Medium	Parameters measured	Results	Reference
Pisum sativum	AgNP-PVP (50 nm) and AgNP-DDAB (20 nm)	0.0005%–0.05% (w/v)	96 h	Ultrapure Milli-Q water	Seed germination, root length	AgNP-DDAB > Ag$^+$ > AgNP-PVP	Barabanov et al. (2018)
Physcomitrella patens	AgNP-citrate (21.5 ± 4.2 nm), AgNP-PVP (29.0 ± 6.3 nm), and bare AgNPs (37.4 ± 13.4 nm)	2, 4, 6, and 8 μg/mL	10 days (protonema) 40 days (leafy gametophyte)	Solid BCDAT medium (protonema) Solid NFM medium (leafy gametophyte)	Silver and nutrient uptake, morphology, chlorophyll content	Bare AgNP > AgNP-citrate > AgNP-PVP (protonema) AgNP-citrate > AgNP-PVP ≈ bare AgNP (leafy gametophyte)	Liang et al. (2018)
Lemna minor	AgNP-citrate (80.78 ± 7.46 nm) and AgNP-PVP (91.81 ± 7.07 nm)	0.05–2 mg/L	7 and 14 days	Steinberg medium	Growth rate, chlorosis, fronds per colony, activity of antioxidative enzyme	Ag$^+$ > AgNP-PVP > AgNP-citrate	Pereira et al. (2018)
Festuca arundinacea	AgNP-PVP (50 nm) and AgNP-SiO$_2$ (50 nm)	0.0015 and 0.15 mg/kg	8 days	Clay, loam (high OM), sandy loam, loam (high carbonate)	Silver flux and uptake in roots and leaves	AgNP-PVP ≈ AgNP-SiO$_2$ ≈ Ag$^+$	Layet et al. (2019)
Nicotiana tabacum	AgNP-PVP (41.12 ± 16.25 nm) and AgNP-CTAB (40.30 ± 12.16 nm)	25, 50, and 100 μM	3 weeks	Solid ½ strength MS medium	Silver uptake, seed germination and growth	AgNP-CTAB > Ag$^+$ > Ag-PVP	Biba et al. (2020)
Lactuca sativa	AgNP-citrate (60 and 100 nm), AgNP-PVP (75 and 100 nm), and AgNP-PEG (100 nm)	1 mg/L	9 days	Hoagland solution	Silver uptake and translocation	AgNP-citrate ≈ AgNP-PVP ≈ AgNP-PEG ≈ Ag$^+$	Torrent et al. (2020)

higher concentrations, AgNP-DDAB significantly inhibited root length. Since the treatment with DDAB alone showed a similar inhibitory effect, it was concluded that phytotoxicity of AgNP-DDAB most likely originates from the coating molecule itself. CTAB and DDAB are both cationic surfactants, and although CTAB is commonly used in synthesis of various nanomaterials (Li et al., 2018a; Singh and Singh, 2019), it exerts high toxicity to cells even at submicromolar doses (Alkilany et al., 2009). On the other hand, treatments with polymers like PEG and PVP did not cause significant toxic effects in plants investigated so far (Wang et al., 2013; Vannini et al., 2013, 2014; Biba et al., 2020). The occurrence of free surfactant molecules in nanoparticle suspension could be a consequence of inadequate purification of the stock solution upon synthesis or later desorption of the surfactant from nanoparticle surface (Alkilany and Murphy, 2010). Therefore, greater care should be taken during preparation of nanoparticles for testing and a coating control should be implemented in future AgNP phytotoxic investigations.

14.4 Cytotoxic and genotoxic effects of AgNPs

AgNPs are known to exert toxic effects, although the exact mechanism on how they affect plant growth is not well understood. Observed effects were mostly related to the silver ions released from AgNPs and/or ROS, but recent studies conducted on plants have unequivocally found that nanoparticles themselves show cyto- and genotoxic potential (Cvjetko et al., 2017; Scherer et al., 2019). Toxic effects are related to AgNPs uptake, distribution, and translocation within the plant (Chichiriccò and Poma, 2015) (Fig. 14.4). Phytotoxic markers at a higher level of biological organization are more easily recognized as they are usually monitored as changes in germination rate, fresh and dry weight, root length, chlorosis symptoms, and they can implicate the cell toxicity, but when investigating AgNP toxicity at the cellular level, the variety of models used make comparison difficult (Casillas-Figueroa et al., 2020).

Cytotoxicity is related to any unfavorable effect, which perturbs normal cell homeostasis and, in consequence, leads to disruption in maintaining the cell biochemistry and physiology. AgNP-induced cytotoxic effects indicate their ability to adversely damage different cell structures, affect the rate of cell division, induce genomic instability or cause irreparable cell death (Yan and Chen, 2019). The cytotoxicity of AgNPs in plants has been directly correlated with their uptake and accumulation within the cells. In *A. cepa* roots (Cvjetko et al., 2017) as well as in the roots of tobacco seedlings (Peharec Štefanić et al., 2018) and adult plants (Cvjetko et al., 2018) TEM observation, along with energy-dispersive X-ray spectroscopy (EDX), showed that AgNPs were deposited in the cell wall, intermembrane space and within the cell, inducing cytotoxic effects. Scherer et al. (2019) have demonstrated that *A. cepa* roots, exposed to nano- and microsized Ag particles, incorporated only nanosized ones. They elicited cytotoxic effects, which was particularly pronounced for AgNPs of smaller sizes. Genotoxicity in the broad sense of the word

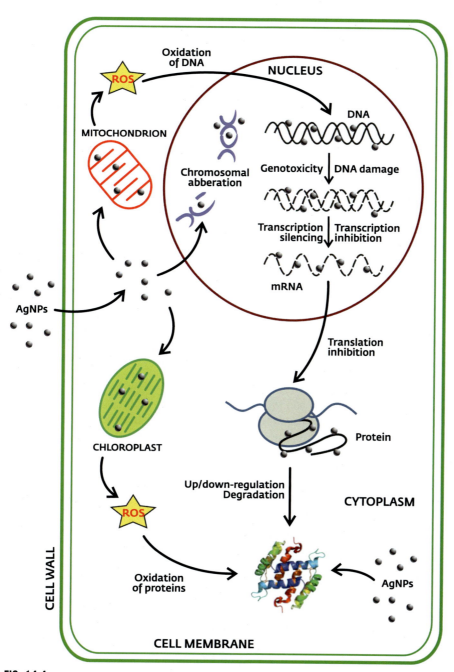

FIG. 14.4

Possible damage which AgNPs may induce after entering the plant cell through direct (by binding to important biomolecules, DNA, mRNA and/or proteins) or indirect (by the formation of ROS after interaction with mitochondria and chloroplasts) effects.

describes any effects of DNA-damaging agents. AgNPs can damage DNA directly, since due to their small size they diffuse inside the nucleus through its pores; or indirectly, through nuclear membrane damage. Moreover, due to the degradation of the nuclear membrane during mitosis, the interaction of NPs with genomic DNA is truly facilitated (Huang et al., 2019b). Although in bacterium *Staphylococcus aureus* TEM analyses visualized AgNP deposition in nucleoid with AgNPs directly bound on DNA fiber (Grigor'eva et al., 2013), to our knowledge, AgNPs deposition in plant nucleus has not yet been reported. Therefore, for now, it can be assumed that in plants AgNPs have an indirect effect on DNA, probably through ROS formation.

In plant cells, the cell wall is the first barrier and primary interaction site for AgNPs internalization. The differences in the complex architecture and porosity of cell wall have been debated as a reason for easier accumulation of NPs in dicots vs monocots (Schwab et al., 2016). However, a damaged cell wall was found in root cells upon exposure of rice seedlings to AgNPs (Mirzajani et al., 2013; Mazumdar and Ahmed, 2011), which can also facilitate NP intake. Several reports demonstrated that AgNPs can permeate cell membranes and enter the cell (Tkalec et al., 2019). Cytotoxic impact of PVP-coated AgNPs on kiwifruit during the germination of pollen has been correlated with AgNP uptake and translocation through the cell membrane (Speranza et al., 2013). Namely, TEM analysis revealed some irregular structure, which probably resulted from cell membrane invagination in vegetative cell; on the contrary, such structures were not identified in $AgNO_3$-treatment (Speranza et al., 2013). This finding supports the idea of endocytosis as a process for AgNP uptake within the plant cell.

Studies have revealed that AgNPs can impact many plant organs and cell organelles, among which roots and root cell vacuoles seem to be particularly affected, which is somewhat expected since roots are usually in direct contact with AgNPs during experiments, while vacuoles are the major storage place within the plant cell. In AgNP-treated *Lolium multiflorum*, damaged epidermis and root cap along with collapsed cortical cells with enlarged vacuole were found (Yin et al., 2011). Moreover, oxidative stress was reported in root cells, which led to apoptosis. On the macroscopic scale, the roots were growing opposite from gravity, probably because of the disturbed auxin transport toward the root tip (Yin et al., 2011). Studies of Mirzajani et al. (2013) and Mazumdar and Ahmed (2011) revealed disrupted vacuoles in root cells of rice seedlings, while reduced vacuole and cell size, as well as cell turgidity, were reported in maize and cabbage (Pokhrel and Dubey, 2013) upon exposure to AgNPs. Moreover, Vannini et al. (2013) reported highly vacuolated root tip cells in rocket seedlings treated with AgNPs. Beside vacuoles, changes were also observed in plastids and endoplasmic reticulum (ER) of root cells. Namely, a reduced number of amyloplasts and decreased size of the smooth ER in the root cap columella cells of AgNP-exposed rocket seedlings were reported by Vannini et al. (2013). Absence of starch grains in amyloplasts of root cap columella and in the plastids of meristematic cells, accompanied with the extensive ER swelling, was reported for wheat seedlings treated with AgNPs (Vannini et al., 2014). Although usually not in the direct contact with AgNPs, leaf organelles, primarily chloroplasts, were also found to be affected by AgNP-exposure (Tkalec et al., 2019). The translocation of AgNPs to the shoots is mostly very low (Tkalec et al., 2019) and cytotoxic effects have been rarely correlated with AgNP presence in leaf cells. Nevertheless, in *A. thaliana* treated with AgNPs, decreased chlorophyll content has been related to the NP presence in the leaves, where modified thylakoid membranes were visualized by TEM (Qian et al., 2013). Larue et al. (2014) reported AgNP penetration through leaf stomata in *L. sativa* after foliar application and confirmed internalization of AgNPs in the cell wall of epidermal cells as well as inside the cell, between guard cell and in the substomatal chamber, but no AgNP-related phytotoxic effects were observed. Contrarily, foliar application of AgNPs resulted in changes in chloroplast shape from lenticular to round in Scots pine (Aleksandrowicz-Trzcińska et al., 2019) and an increase in size of starch granules in English oak (Olchowik et al., 2017). Peharec Štefanić et al. (2018) reported swollen and ruptured chloroplasts in leaf cells of tobacco seedlings exposed to AgNPs, although NP could not be detected. Disrupted chloroplasts with reduced grana thylakoids were also observed in leaves of AgNP-treated barley plants (Fayez et al., 2017); moreover, destruction of mitochondria and nucleus was also reported.

Besides structural changes, it was found that AgNP presence within the cell may provoke the most severe cytotoxic effect involved in cell metabolism, oxidative stress, that eventually may lead to cell death. The induction of oxidative stress, a mechanism underlying the AgNP-toxicity in plants, is described in several reports in details (revised in Tkalec et al., 2019). Here we rather concentrate on visualization of ROS, mediators of oxidative stress in plant cells. Different histochemical staining procedures confirmed ROS formation in plant cells during AgNP-exposure. In *Solanum tuberosum*, plantlets exposed to AgNPs, overall ROS production in leaves was measured by assay with $2',7'$-dichloro-dihydrofluorescein diacetate (H_2DCFDA), and results have shown concentration-dependent induction of oxidative stress (Bagherzadeh Homaee and Ehsanpour, 2016). Moreover, leaves incubated with nitroblue tetrazolium, which specifically reacts with superoxide anion ($O_2^{.-}$), have formed dark blue formazan precipitation upon exposure to AgNPs, while quantification of H_2O_2 with $3,3'$-diaminobenzidine showed brownish colored leaves. As a consequence of higher ROS generation, disturbed membrane integrity facilitated ion leakage, resulting in cell death (Bagherzadeh Homaee and Ehsanpour, 2016). Nair and Chung (2014b) have performed staining with dihydroethidium (DHE) to detect $O_2^{.-}$ and with 30-(p-hydroxyphenyl) fluorescein to detect H_2O_2 in rice seedlings after exposure to AgNPs and found that fluorescence intensity increased with increasing concentration of AgNPs, thus confirming the generation of ROS in a dose-dependent manner. Additionally, cell death, as a consequence of oxidative stress, as indicated by increased fluorescence of propidium iodide (Nair and Chung, 2014b). PI is impermeable for undamaged cells, so when membrane damage occurs, the content of PI inside the cell increases. Decreased viability of cells of *Lemna gibba* plants exposed to AgNPs was observed using the fluorescein diacetate, a nonpolar ester that passes through cell membranes, while a strong increase in ROS formation in the same cells was detected by H_2DCFDA (Oukarroum et al., 2013).

Allium test has been highly used in the assessment of the cyto- and genotoxicity of AgNPs due to its simplicity and sensitivity. Originally introduced by Levan (1938) and modified by Fiskesjo (1979), the Allium test has been implemented for detection of chromosome aberrations of variety of toxic compounds, which arise either as results of an aneugenic or a clastogenic mode of action. The clastogenic effects, detectable as chromosome breaks or formation of chromosome bridges, are consequences of DNA breaks or disturbed DNA synthesis and replication, whereas aneugenic ones come from disturbances in mitotic spindle leading to chromosome laggards, C-metaphases, or multipolar anaphases. Debnath et al. (2018) have shown that AgNPs caused chromosome bridges accompanied by a decrease in mitotic index (MI), while Scherer et al. (2019) have shown that beside chromosome bridges, AgNPs induced C-metaphase, micronucleus and reduced MI, which was depended on the NP size. Kumari et al. (2009) and Sobieh et al. (2016) reported a significant decrease in MI upon treatment with AgNPs in a dose-response manner, with a majority of cells being accumulated in the prophase stage of mitosis. Moreover, Sobieh et al. (2016) have also found chromosome stickiness and spindle disturbance to be induced in a higher percentage in comparison to chromosome bridges and breaks, thus indicating that AgNPs exhibited cytotoxic potency by arresting the cells in particular cell cycle stages. Our results, derived from AgNP-treated roots of *A. cepa*, showed that exposure to AgNPs coated with either citrate or PVP had no significant effect on MI of root cells in any of the applied concentrations, while AgNP-CTAB severely affected root MI, showing more prominent effects at higher concentrations (Cvjetko et al., 2017). Casillas-Figueroa et al. (2020) compared Allium test results of Argovit™AgNPs with those of various AgNPs formulations from the literature and found no cytotoxic or genotoxic damage. Since the shape, size, and coating agent were quite similar in all studies, the authors suggested that the difference in Ag/coating agent ratio ([Ag]/[PVP] ratio in Argovit™AgNPs was 6:94 compared to 34:66 in NanoComposix AgNPs) could explain different toxic response. Under normal circumstances, cell division is regulated at several cell cycle checkpoints that ensure proper distribution of genetic material. Under adverse conditions, cell cycle arrest may occur at the transition from G1 to S phase to suppress DNA duplication or at the transition from G2 to M phase. In the study of Fouad and Hafez (2018), the expression of cyclin-dependent kinase 2 (*cdc2*), important not only for an aforementioned transition during interphase but also for a transition from metaphase to anaphase, was investigated in *A. cepa* root tips exposed to PVP-coated AgNPs. Obtained results revealed decreased expression of *cdc2*, accompanied by a decrease in MI. Moreover, enhanced spindle disturbances, clastogenic aberrations, and chromosome stickiness suggested strong AgNP genotoxicity (Fouad and Hafez, 2018). In general, decreased *cdc2* gene expression has been associated with arrest in certain phases of the cell cycle, such as prolonged S-phase duration or delayed entry into mitosis. Babu et al. (2008) have found in AgNP-treated *A. cepa* a high frequency of disturbed metaphase and anaphase altering the orientation of chromosomes, thus indicating impairment of mitotic spindle. In the *A. cepa* flower buds, Saha and Gupta (2017) investigated the impact of AgNPs in pollen mother cell during meiosis and have found a significant decrease in meiotic index accompanied with different types of aberration such as undifferentiated bivalents, stickiness, multipolar anaphase, fragmented and more condensed chromosome at metaphase stage, multipolar anaphase, micronuclei, abnormal conjugation, segregation, and precocious movements of chromosomes, including abnormal tetrads with micronuclei, tetrads with connected chromosome or triad formation, thus indicating that the process of meiosis can also be severely affected by AgNPs.

Beside *A. cepa*, chromosome aberrations and changes in MI have been investigated in other plant species upon exposure to AgNPs. The reports for several studies which have used *V. faba* root tip chromosomal aberration assay corroborate well with Allium test in assessment of dose–response relationship upon AgNPs exposure. Both tests are very similar in terms of technical performance and both detect the same endpoints in the cyto- and genotoxic effect assessment. Patlolla et al. (2012) have tested similar AgNPs concentrations in *V. faba* root tips as in *A. cepa* ones, and demonstrated that in both plant species AgNPs exposure significantly increased the number of chromosomal aberrations and micronuclei and decreased the MI. Abdel-Azeem and Elsayed (2013) also examined the effect of different sizes of AgNPs on *V. faba* and found that with decreasing NP size, the number of aberrant cells and chromosome stickiness increased, while a decrease in MI was correlated to root growth in a time- and concentration-dependent manner. Abdelsalam et al. (2018) analyzed cyto- and genotoxicity of AgNPs in *Triticum aestivum* and have found metaphasic plate distortion, i.e., incorrect orientation and movement during metaphase together with eroded chromosome, chromosomal gaps, c-metaphase, fragments, bridges, ring chromosomes and cap chromosomes after.

The Comet assay, also known as the single-cell gel electrophoresis (SCGE), is one of the most frequently used tests for assessing the genotoxic potential of different chemicals in higher plants (Gichner et al., 2003). It detects limited types of DNA lesions such as single or double-strand breaks as well as apurinic/apyrimidinic sites at the level of a single cell. This method can also be used to monitor DNA repair progress in a plant cell, but with some limitations, as the method itself is not indicative for the quality of repair process (Cvjetko et al., 2014). According to the literature, only a few studies have used Comet test to determine the genotoxic potential of AgNPs. In one of the earliest published papers, Comet test was used to determine the genotoxic effect of AgNPs in *N. tabacum* and *A. cepa* (Ghosh et al., 2012). As expected, in both species, DNA damage was more pronounced in the root than in the shoot and was negatively correlated with the applied AgNP concentration, which was ascribed to the possible AgNP agglomeration at higher concentrations. When differently coated AgNPs were applied on *A. cepa* roots, no DNA damage was observed with AgNP-citrate, while AgNP-CTAB exhibited significantly higher DNA damage compared to AgNP-PVP (Cvjetko et al., 2017). A similar result was obtained in roots of tobacco plants, in which citrate-coated AgNP did not induce DNA damage evaluated by Comet test (Cvjetko et al., 2018), which indicates that genotoxicity of AgNPs might also be dependent on the surface coating.

To date, DNA-based techniques have been very rarely applied in assessing genotoxic effect of AgNPs in plants. However, a few that have been reported rely on polymerase chain reaction (PCR) and have been already proven to be sensitive and accurate in identifying potentially genotoxic effect of other NPs. These techniques use genomic DNA from a single sample to create a series of amplified fragments, separated by size that creates a so-called DNA "fingerprint." Inter simple sequence repeat (ISSR) PCR method uses microsatellite repeats as primers that generate multilocus markers. This method was applied to detect genomic template stability (GTS) in *Solanum lycopersicum* exposed to AgNPs for 2 weeks (Cekic et al., 2017). It was shown that GTS decreased in a dose–response relationship. Moreover, analysis of ISSR bands has revealed numbers of newly appearing bands, a disappearance of existing ones as well as changes in the band intensity in comparison to control. Unique band profiling of the highest concentration suggested the genotoxic effect to be exerted in a dose–response manner (Cekic et al., 2017). Another DNA fingerprinting method, amplified fragment length polymorphism (AFLP), also focuses on microsatellite regions, but before amplification isolated genomic DNA is digested with a pair of restriction enzymes. After PCR reaction, differences in restriction fragment length, that created or abolished restriction endonuclease recognition sites, will manifest them as polymorphic marker. AFLP was applied to asses genotoxic effect of AgNPs on *T. aestivum* seedlings (Vannini et al., 2014), and even though the cytotoxic effect was confirmed in both roots and shoots, only 4.6% of all DNA bands analyzed in roots, and 3.7% examined in shoots were polymorphic. Therefore, AFLP analyses did not detect a significant genotoxic effect as a result of AgNP-exposure nor did the proteomic analysis reveal any changes in expression of enzymes involved in DNA repair mechanism. Moreover, the level of polymorphism corresponded to the level of genetic variability commonly found among different *T. aestivum* cultivars (Vannini et al., 2014).

14.5 AgNP impact on gene and protein expression

It has been shown that, among a wide range of plant responses, AgNPs triggers altered gene expression, which as a consequence results in changes in the expression of proteins (Hossain et al., 2016; Yan and Chen, 2019) (Fig. 14.4). Therefore, gene and protein expression analyses have been implemented as powerful methods to provide new insights into the molecular mechanisms of plant response to AgNPs. They can be performed at any one of several different levels at which gene expression is regulated: transcriptional, post-transcriptional, translational, and post-translational protein modification (Baginsky et al., 2010). Among methods used to investigate AgNP impact on gene expression, quantitative real-time PCR (qPCR) alone or combined with reverse transcription-PCR (qRT-PCR) are the most commonly applied. However, the throughput capability of the current technology of qRT-PCR is not suitable for transcriptome-wide gene expression analyses (Teo et al., 2016). On the other hand, proteomic analyses directly deal with the functional molecule providing a link between gene expression and cell metabolism (Hossain et al., 2016, 2020). In general, both gene and proteomic studies revealed that despite the similarity in gene and/or protein expression profile, there are a certain number of genes/proteins specific for AgNPs and not affected by Ag^+ ions, thus making AgNPs novel abiotic stress factors (Kohan-Baghkheirati and Geisler-Lee, 2015). However, in both approaches plant responses to AgNP-induced stress have been investigated in various plant species, grown in different exposure media, applying AgNPs of different sizes, stabilizations and concentrations, and employing different exposure times, so it is not surprising that results are not always unambiguous.

One of the most important mechanisms of AgNP-phytotoxicity is induction of oxidative stress, so it is not surprising that most gene and protein analyses revealed changes in expression of plant cell antioxidant machinery. Kohan-Baghkheirati and Geisler-Lee (2015) using publicly available microarray data showed that ROS associated genes were upregulated by AgNPs. The study of Nair and Chung (2014b) showed upregulation in expression of three SOD genes (cytosol Fe-SOD), mitochondrial Mn-SOD and cytosol Cu/Zn-SOD) in roots and shoots of rice seedlings upon exposure to different concentrations of AgNP-citrate, although the chloroplast Cu/Zn-SOD was significantly downregulated. Upregulation of SOD genes (both cytosol and mitochondrial isoforms) was also observed after exposure of *Arabidopsis* seedlings to AgNPs (Qian et al., 2013; Kaveh et al., 2013). SOD acts on the superoxide radical, the first form of ROS to be formed in stress conditions; therefore, SOD can be considered a primary defence against oxygen radicals. Furthermore, an increased relative transcript level of ascorbate peroxidase (APX) and glutathione peroxidase (GPX) genes was recorded in *A. thaliana*, although transcription level of APX genes was found to be dependent on the applied AgNP concentration since transcript levels of some APX genes were decreased after exposure to the highest applied concentration (Qian et al., 2013). Nair and Chung (2014b) also reported significant transcriptional upregulation of APX and POD genes, which in correlation with higher H_2O_2 content suggest protection from excess accumulation of H_2O_2 through the ascorbate–glutathione cycle, where APX acts as a key player. Moreover, the upregulation of genes for catalase (CAT) (Nair and Chung, 2014b; Thiruvengadam et al., 2015), an enzyme that catalyzes the conversion of H_2O_2 to H_2O and O_2 corroborate this conclusion. Furthermore, the upregulation of a gene for glutathione *S*-transferase (GST), an enzyme involved in metal detoxification, was recorded after treatment with AgNPs in turnip seedlings (Thiruvengadam et al., 2015) and *Arabidopsis* (Nair and Chung, 2014a). Majority of proteomic analyses also confirm the importance of antioxidant defence in response to AgNPs. Increased abundance of detoxification enzymes, including SOD, APX and CAT was reported for AgNP-treated seedlings of rice (Mirzajani et al., 2014), rocket (Vannini et al., 2013), and tobacco (Peharec Štefanić et al., 2018). Moreover, rice seedlings exhibited upregulation of proteins GST and NAC transcription factor involved in defence signaling pathways (Mirzajani et al., 2014). However, in roots of adult tobacco plants exposed to AgNPs, Fe-SOD, POD, salicylic acid-binding catalase, monodehydroascorbate reductase and quinone reductase were found to be downregulated, while only Mn-SOD was upregulated (Peharec Štefanić et al.,

2019). Decreased expression of SOD (Galazzi et al., 2019) and oxidation-reduction cascade related proteins, e.g., GDSL motif lipase 5, SKU5 similar 4, galactose oxidase and quinone reductase, was also found in soybean roots after treatment with AgNPs (Hossain et al., 2016), which indicates that in roots, which are in the direct contact with AgNPs, excessive metal concentration can result in reduced expression of antioxidant enzymes (Komatsu and Hossain, 2013).

In addition to enzymatic defence, exposure to AgNPs can also affect genes and proteins involved in biosynthesis of molecules of nonenzymatic defence against ROS, important for stress tolerance and maintenance of cellular redox homeostasis, such as sulfur-containing compounds, glutathione (GSH) and glucosinolates as well as phenolics and carotenoids (Schiavon et al., 2007; Waskiewicz et al., 2014). A significant increase in transcript levels of genes encoding for enzymes, which play a key role in sulfur assimilation (ATP sulfurylase, 3′-phosphoadenosine 5′-phosphosulfate reductase, sulfite reductase, and cysteine synthase) and in phytochelatin synthesis (phytochelatin synthase), was observed in *Arabidopsis* exposed to AgNPs (Nair and Chung, 2014a). Significant upregulation was also found for GSH synthetase, involved in the final step of GSH biosynthesis, as well as for GST and glutathione reductase (GR) genes (Nair and Chung, 2014a). Since it has been documented that AgNPs could inhibit the activity of GSH-synthesizing enzymes (Piao et al., 2011), by the upregulation of GSH biosynthesis genes plants may compensate decreased GSH production to be able to activate an efficient protective mechanism to cope with stressful environmental conditions. Dimkpa et al. (2013) analyzed the accumulation of transcripts encoding metallothionein (MT), a cysteine-rich protein involved in detoxification by metal ion sequestration, in root extracts of AgNP-exposed wheat plants and found that MT gene expression was highly induced. Moreover, proteomic analyses revealed upregulation of proteins involved in metabolism of sulfur-containing amino acids, methionine and cysteine. In rocket roots, AgNP-exposure induced accumulation of the vitamin-B12-independent methionine synthase isozyme, which is involved in the biosynthesis of the methionine as well as cysteine synthase, a key enzyme in cysteine biosynthesis (Vannini et al., 2013). Cysteine is a direct coupling step between sulfur and its incorporation into GSH, important in plant stress tolerance to ROS. Besides, a higher expression of the protein involved in the methionine synthesis was observed in *Arabidopsis* (Kumari et al., 2020). In the study by Hossain et al. (2016), upregulation of methionine gamma lyase protein was recorded in soybean roots upon exposure to AgNPs, indicating a shift in the metabolism of methionine in an alternative pathway, leading to the isoleucine biosynthesis, which is considered as an adaptive strategy of protecting from single or multiple stresses (Atkinson et al., 2013).

Among phenolic compounds, flavonoid pigment anthocyanin, which can act as a nonenzymatic antioxidant by scavenging excess ROS as well as a metal chelator (Carocho and Ferreira, 2013), was the most frequently analyzed after exposure to AgNPs. Thiruvengadam et al. (2015) reported that transcript levels of anthocyanin pigment 1, anthocyanin synthase and phenylalanine ammonia-lyase genes, all involved in anthocyanin biosynthesis, were gradually upregulated after exposure of turnip plants to AgNPs, which is in a good correlation with studies on AgNP-exposed

Arabidopsis plants, in which anthocyanin accumulation was significantly induced (Qian et al., 2013; Syu et al., 2014; Nair and Chung, 2014a). Moreover, gene expression analysis in *Arabidopsis* treated with AgNP-citrate showed upregulation of key genes of flavonoid and anthocyanin biosynthesis pathway (García-Sánchez et al., 2015). Also, AgNPs induced upregulation of genes involved in synthesis of lignin and lignans, another phenolic compound important for cell wall structure (Kohan-Baghkheirati and Geisler-Lee, 2015). As for the glucosinolates (GSL), sulfur-containing metabolites that are involved in biotic and abiotic stress resistance (Kim et al., 2013), studies in turnip seedlings exposed to AgNPs have shown elevated expression of genes related to glucosinolate biosynthesis (Thiruvengadam et al., 2015). Also, the same study showed higher transcript levels of β-cyclase and zeaxanthin epoxidase-1, genes involved in biosynthesis of antioxidant carotenoids, after exposure to 1.0 mg/L AgNPs, while the same genes were downregulated at higher AgNP concentrations (5.0 and 10 mg/L).

Several studies also investigated transcriptional responses of genes related to plant responses to pathogens and wounding after exposure to AgNPs. In *Arabidopsis* seedlings exposed to AgNP-PVP, downregulation of genes involved in systemic acquired resistance (SAR) was recorded (Kaveh et al., 2013). SAR is triggered upon infection by certain pathogens or by mechanical damage and results in thickening of the cell wall and other physiological responses that enhance general plant defence. Similarly, in the study of García-Sánchez et al. (2015), repression of pathogen-activated genes and salicylic acid-mediated pathways were found after *Arabidopsis* exposure to AgNPs. However, several defence-related genes were found to be upregulated in *Arabidopsis* after AgNP exposure, e.g., genes involved in salt stress (tumor necrosis factor receptor-associated factor-like protein); in defence against insects and pathogens (myrosinase-binding protein); in thalianol biosynthetic pathway (which is thought to be involved in plant defence system); and in the response to wounding (miraculin-like protein, MLP) (Kaveh et al., 2013). Upregulation of two MLP genes were also found in cucumber leaves after exposure to Ag_2S-NPs (Wang et al., 2017), while in turnip, expression of genes involved in response to biotic stress, pathogenesis-related (PR) gene 1 and lipoxygenase 2, increased with increasing concentration of AgNP (Thiruvengadam et al., 2015). AgNP-treatments also altered the expression of some PR and PR-like proteins, involved in response to abiotic and biotic stimuli, such as basic beta-1,3-glucanase, acidic chitinase, pathogen- and wound-inducible antifungal protein CBP20 precursor, osmotin and cysteine-rich secretory protein, although expression differed depending on the tissue examined (Peharec Štefanić et al., 2018, 2019). Similarly, AgNPs treatment changed the concentration of some chitinases and PRs in shoots of wheat seedling (Vannini et al., 2014), and increased disease resistance protein in leaves of *Arabidopsis* (Kumari et al., 2020), thus indicating that these proteins could be components of the plant defence response against AgNPs.

Besides the various defence proteins, AgNPs can also affect genes associated with hormone signaling. Sun et al. (2017) reported downregulation of genes involved in auxin signaling (TIR1/AFB family of F-box proteins) in root tips of *Arabidopsis* seedlings after exposure to PVP-coated AgNPs. On the other hand, auxin transport carrier-related genes of the PIN family as well as YUC8, which encode flavin monooxygenase that catalyzes conversion of tryptophan into indole-3-acetic acid, the most common naturally occurring hormone of the auxin class, were upregulated. Another gene involved in auxin signaling pathway, indoleacetic acid protein 8, which encodes an auxin-inducible AUX/IAA protein, was also found to be induced in AgNP-exposed *Arabidopsis* seedlings (Syu et al., 2014). Obtained results indicate that auxin perception might be inhibited by AgNPs aggregation in the cell membrane, which causes interference with the binding of auxin and its receptors (Sun et al., 2017). Abscisic acid (ABA), another important plant hormone that has an important role in plant environmental stress response, has also been analyzed upon *Arabidopsis* exposure to decahedral AgNPs; induced expression was found for gene encoding 9-cis-epoxycarotenoid dioxygenase in ABA bisynthetic pathway and dehydration-responsive gene encoding ABA-mediated dehydration responsive protein (Syu et al., 2014). Furthermore, in the same study, it was shown that the gene expression of enzymes involved in synthesis of ethylene (ET), whose signaling transduction plays a significant role in the plant response to biotic and abiotic stress, were slightly downregulated in response to AgNPs, suggesting that AgNPs might act as inhibitors of ET perception and could interfere with its biosynthesis (Syu et al., 2014). Moreover, in cucumber seedlings, six genes involved in ET signaling pathway were significantly upregulated after exposure to silver sulfide nanoparticles (Ag_2S-NPs), suggesting that NPs can potentially reduce plant growth through the direct interactions with the ET signaling pathway (Wang et al., 2017).

Transcriptional responses of aquaporin genes, which encode aquaporins that regulate inter- and intracellular water flow and maintain water homeostasis, were also studied in response to AgNP exposure. Qian et al. (2013) found that AgNPs could change the transcription of aquaporin genes in *Arabidopsis* seedlings. Downregulation of aquaporin genes was recorded in both roots and shoots of wheat seedlings exposed to Ag_2S-NPs (Wang et al., 2017), although the magnitude of expression was less in shoots than in roots. However, in cucumber, opposite level expressions were found for some aquaporin genes between shoots and roots, thus suggesting that the effect on aquaporin genes varies with both plant species and tissue (Wang et al., 2017).

Majority of the proteomic analyses showed that the plant proteins expressed differently after AgNP-imposed stress were those involved in the processes of primary metabolism (Tkalec et al., 2019). The studies of the AgNP effects on tobacco (Peharec Štefanić et al., 2018, 2019), soybean (Mustafa et al., 2015; Galazzi et al., 2019), and wheat (Vannini et al., 2014; Jhanzab et al., 2019) revealed a strong impact on photosynthesis protein expression. The proteome study of tobacco seedlings showed that AgNP upregulated expression of several proteins important for electron transport and ATP synthesis as well as carbon metabolism (Rubisco, Rubisco activase 2, and beta-carbonic anhydrase) (Peharec Štefanić et al., 2018), probably to enhance production of energy required to overbear stress imposed by AgNPs. Moreover, ferritin and ATP-dependent Clp protease proteolytic subunit, both involved in photosynthesis, were found to be upregulated after exposure of wheat seedlings to AgNPs (Jhanzab et al., 2019). However, in other studies, negative effects on photosynthesis-related protein expression were recorded upon AgNP-treatment. After exposure of soybean seedlings to AgNPs, Mustafa et al. (2015) reported decreased expression of several photosystem (PS) II proteins, while Galazzi et al. (2019) found downregulation of PSII-related oxygen-evolving enhancer protein 2 and small and large Rubisco subunits. Moreover, negative effect of AgNP-exposure on photosynthesis was also reported for wheat seedlings, where downregulation of HCF136, a protein essential for assembly of the PSII reaction centre, was recorded (Vannini et al., 2014). In the study of Peharec Štefanić et al. (2019), proteomic analysis of leaves after exposure of adult tobacco plants to AgNPs revealed downregulation of several proteins engaged in the photosynthesis light-dependent reactions, e.g., peripheral components of PSI (PsaD and PsaE), extrinsic Mn-binding PSII proteins (Psb27 and PsbP), ferredoxin-NADP reductase (FNR) and ATP synthase (Peharec Štefanić et al., 2019). Consistently with decreased primary electron transport processes, photosynthesis carbon reactions were also impaired. Namely, although partial Rubisco subunit RbcL of 47 kDa was upregulated after exposure to AgNPs, the native RbcL 53-kDa protein was downregulated, which indicated Rubisco degradation (Peharec Štefanić et al., 2019). Glycolysis, an important metabolic pathway responsible for conversion of glucose to pyruvate, was also found to be affected by AgNPs. The enhanced expression of several glycolytic proteins, i.e., plastidic aldolase, triosephosphate isomerase (TPI), glyceraldehyde-3-phosphate dehydrogenase (GAPDH) and 2,3-bisphosphoglycerate-independent phosphoglycerate mutase was recorded in tobacco seedlings after AgNP-exposure (Peharec Štefanić et al., 2018), while contrary, downregulation of GAPDH, TPI and plastidic aldolase was detected in leaves of adult tobacco plants (Peharec Štefanić et al., 2019). Moreover, GAPDH was also found to be a responsive protein in AgNP-exposed rice seedlings (Mirzajani et al., 2014). Study of Jhanzab et al. (2019) showed that some glycolysis-related proteins, i.e., GAPDH and carboxypeptidase, were downregulated in wheat seedlings exposed to AgNPs, while another group of glycolytic proteins, e.g., glyceraldehyde dehydrogenase, acetyl-CoA synthetase and glusose-6-phosphate 1-epimerase, were upregulated. Moreover, exposure of rocket seedlings to AgNP-PVP induced the accumulation of the glyoxalase 1, a ubiquitous cellular enzyme that participates in the detoxification of methylglyoxal, a cytotoxic byproduct of glycolysis that accumulates in cells in response to environmental stress (Vannini et al., 2013), while in the roots of AgNP-treated soybean seedlings, another important enzyme of the glyoxalase pathway, glyoxalase II 3 protein, was found to be time-dependently decreased (Mustafa et al., 2015). Changes in protein expression were also detected for enzymes involved in tricarboxylic acid cycle (TCA), which links glycolysis to the mitochondrial electron transport chain. The most affected proteins were malate dehydrogenase (MDH), which was upregulated in wheat roots (Vannini et al., 2013, 2014) and tobacco seedlings (Peharec Štefanić et al., 2018), but downregulated in roots of adult tobacco plants (Peharec Štefanić et al., 2019) upon exposure to AgNPs. Moreover, lower abundance in tobacco roots was also recorded for isocitrate dehydrogenase, another TCA protein, as well as for mitochondrial ATP synthase 24-kDa, which indicates decreased ATP production (Peharec Štefanić et al., 2019).

In several studies, AgNP-induced stress was found to result in activation of proteins responsible for protein synthesis and folding (Vannini et al., 2014; Mirzajani et al., 2014; Peharec Štefanić et al., 2018; Jhanzab et al., 2019), which is probably related to enhanced production of proteins important to cope with induced stress and to preserve cell homeostasis. However, in rocket roots, AgNP-exposure induced downregulation of two chaperones related to protein folding: resident protein of endoplasmic reticulum (ER), binding protein 1 (BiP1) and the heat shock protein 70-2, both involved in ER-associate degradation, while the expression of beta-glucosidase 23, a major component in the ER body, was upregulated (Vannini et al., 2013). A similar observation was recorded in roots of wheat seedlings, where a strong influence of AgNPs on the ER was indicated by reduced levels of three ER-resident proteins (Vannini et al., 2014). Furthermore, AgNP treatment decreased two vacuolar-type proton ATPase subunits, which have an important role in the trans-Golgi network. These results indicate that ER and Golgi might be the target organelles for the AgNPs action (Vannini et al., 2013). Proteins involved in protein degradation processes, particularly ubiquitin-proteasome-related proteins (Jhanzab et al., 2019; Hashimoto et al., 2020) and Kunitz family trypsin and protease inhibitor proteins (Mustafa et al., 2015) were also of enhanced expression after AgNP treatment, probably to degrade irreversibly damaged proteins since stress conditions result in protein misfolding and aggregation (Liu and Howell, 2010). Moreover, increased expression of proteins related to amino acid metabolism was recorded in soybean Mustafa et al. (2015) and tobacco seedlings (Peharec Štefanić et al., 2018) after AgNP-exposure.

Exposure to AgNP-treatments also had an impact on proteins involved in cell signaling. In wheat seedlings exposed to chemoblended AgNP, proteins related to cell signaling, i.e., F-box associated interaction domain, acetyltransferase-superfamily, small GTPase superfamily and carbon–nitrogen hydrolase, were found to be downregulated (Jhanzab et al., 2019). However, Mirzajani et al. (2014) reported upregulation of calmodulin in rice plants after AgNP-treatment. Calmodulin is one of the calcium-binding messenger proteins, which senses levels of Ca^{2+} ions in response to different biotic and abiotic stimuli and transmits signals to various calcium-sensitive enzymes, ion channels and other proteins to ensure duly response to unfavorable environmental conditions (Wilkins et al., 2016).

Proteomic studies also revealed the negative impact of AgNP-imposed stress on some important structural proteins. The abundance of expansin, a member of cell wall loosening proteins that are involved in the cell elongation, was decreased under AgNPs exposure and its downregulation was dependent on the AgNP-size (Mustafa et al., 2016). Additionally, Jhanzab et al. (2019) reported a decreased expression of cell wall proteins after exposure of wheat to AgNPs. Moreover, in AgNP-treated roots of soybean (Mustafa et al., 2015) and tobacco (Peharec Štefanić et al., 2019), annexins, proteins involved in cell organization, were found to be downregulated, which implies stagnation of root cell division and elongation under AgNP-imposed stress conditions.

14.6 Conclusion and future perspectives

Findings presented in this chapter suggest that AgNP phytotoxicity can be either direct or indirect and can be correlated with several possible mechanisms, which can impact the whole cell (induction of ROS generation and/or apoptotic cell death) as well as cellular substructures (disruption of cell-membrane integrity) and biologically important molecules (protein and DNA binding and damage). However, since AgNPs are prone to transformation in biological media and since it is still not completely elucidated to which degree their toxicity results from the nanoparticulate form and how much toxicity can be correlated with the dissolved Ag^+ ions, a careful design of experimental model studies for plant exposures to AgNPs is mandatory. Obtained results suggest that the method of AgNP synthesis and a choice of an appropriate surface coating are of particular importance. Moreover, AgNP behavior has to be comprehensively investigated, with the special emphasis to their stability in growth media, since changes in their chemical and physical properties have a direct impact on their uptake and possible detrimental effects in plants.

Acknowledgments

We gratefully acknowledge the funding for research provided by Croatian Science Foundation [grant number IP-2014-09-6488], European Union Seventh Framework Programme under Grant Agreement 312483-ESTEEM2 (Integrated Infrastructure Initiative–I3) and European Social Fund [grant number HR.3.2.01-0095]. The authors wish to thank Mr. Juraj Balen for technical assistance in graphical design of the figures.

References

Abdel-Azeem, E.A., Elsayed, B.A., 2013. Phytotoxicity of silver nanoparticles on *Vicia faba* seedlings. New York Sci. J. 6, 148–156.

Abdel-Aziz, H., Rizwan, M., 2019. Chemically synthesized silver nanoparticles induced physio-chemical and chloroplast ultrastructural changes in broad bean seedlings. Chemosphere 235, 1066–1072.

Abdelsalam, N.R., Abdel-Megeed, A., Ali, H.M., Salem, M.Z.M., Al-Hayali, M.F.A., Elshikh, M.S., 2018. Genotoxicity effects of silver nanoparticles on wheat (*Triticum aestivum* L.) root tip cells. Ecotoxicol. Environ. Saf. 155, 76–85.

Adeleye, A.S., Keller, A.A., 2016. Interactions between algal extracellular polymeric substances and commercial TiO_2 nanoparticles in aqueous media. Environ. Sci. Technol. 50, 12258–12265.

Akter, M., Sikder, M.T., Rahman, M.M., Ullah, A.K.M.A., Hossain, K.F.B., Banik, S., Hosokawa, T., Saito, T., Kurasaki, M., 2018. A systematic review on silver nanoparticles-induced cytotoxicity: physicochemical properties and perspectives. J. Adv. Res. 9, 1–16.

Aleksandrowicz-Trzcińska, M., Szaniawski, A., Studnicki, M., Bederska, M., Olchowik, J., Urban, A., 2019. The effect of silver and copper nanoparticles on the growth and mycorrhizal colonisation of scots pine (*Pinus sylvestris* L.) in a container nursery experiment. iForest—Biogeosci. For. 11, 690–697.

Alkilany, A., Murphy, C., 2010. Toxicity and cellular uptake of gold nanoparticles: what we have learned so far? J. Nanopart. Res. 12, 2313–2333.

Alkilany, A., Nagaria, P., Hexel, C., Shaw, T., Murphy, C., Wyatt, M., 2009. Cellular uptake and cytotoxicity of gold nanorods: molecular origin of cytotoxicity and surface effects. Small 5, 701–708.

Anjum, N.A., Gill, S.S., Duarte, A.C., Pereira, E., Ahmad, I., 2013. Silver nanoparticles in soil-plant systems. J. Nanopart. Res. 15, 1896. https://doi.org/10.1007/s11051-013-1896-7.

Atkinson, N.J., Lilley, C.J., Urwin, P.E., 2013. Identification of genes involved in the response to simultaneous biotic and abiotic stresses. Plant Physiol. 162, 2028–2041. https://doi.org/10.1104/pp.113.222372.

Babu, K., Deepa, A., Sabesan, G., Rai, S., 2008. Effect of nano-silver on cell division and mitotic chromosomes: a prefatory siren. Int. J. Nanotechnol. 2, 2–7.

Bae, E., Park, H.-J., Park, J., Yoon, J., Kim, Y., Choi, K., Yi, J., 2011. Effect of chemical stabilizers in silver nanoparticle suspensions on nanotoxicity. Bull. Kor. Chem. Soc. 32, 613.

Bagherzadeh Homaee, M., Ehsanpour, A.A., 2016. Silver nanoparticles and silver ions: oxidative stress responses and toxicity in potato (*Solanum tuberosum* L) grown in vitro. Hortic. Environ. Biotechnol. 57, 544–553.

Baginsky, S., Hennig, L., Zimmermann, P., Gruissem, W., 2010. Gene expression analysis, proteomics, and network discovery. Plant Physiol. 152, 402–410.

Barabanov, P.V., Gerasimov, A.V., Blinov, A.V., Kravtsov, A.A., Kravtsov, V.A., 2018. Influence of nanosilver on the efficiency of *Pisum sativum* crops germination. Ecotoxicol. Environ. Saf. 147, 715–719.

Barrena, R., Casals, E., Colon, J., Font, X., Sanchez, A., Puntes, V., 2009. Evaluation of the ecotoxicity of model nanoparticles. Chemosphere 75, 850–857.

Bélteky, P., Rónavári, A., Igaz, N., Szerencsés, B., Tóth, I.Y., Pfeiffer, I., Kiricsi, M., Kónya, Z., 2019. Silver nanoparticles: aggregation behavior in biorelevant conditions and its impact on biological activity. Int. J. Nanomedicine 14, 667–687.

Biba, R., Matić, D., Lyons, D., Peharec Štefanić, P., Cvjetko, P., Tkalec, M., Pavoković, D., Letofsky-Papst, I., Balen, B., 2020. Coating-dependent effects of silver nanoparticles on tobacco seed germination and early growth. Int. J. Mol. Sci. 21, 3441.

Biba, R., Tkalec, M., Cvjetko, P., Peharec Štefanić, P., Šikić, S., Pavoković, D., Balen, B., 2021. Silver nanoparticles affect germination and photosynthesis in tobacco seedlings. Acta Bot. Croat. 80 (1).

Bundschuh, M., Filser, J., Lüderwald, S., McKee, M.S., Metreveli, G., Schaumann, G.E., Schulz, R., Wagner, S., 2018. Nanoparticles in the environment: where do we come from, where do we go to? Environ. Sci. Eur. 30, 6.

Cañamares, M.V., Garcia-Ramos, J.V., Sanchez-Cortes, S., Castillejo, M., Oujja, M., 2008. Comparative SERS effectiveness of silver nanoparticles prepared by different methods: a study of the enhancement factor and the interfacial properties. J. Colloid Interface Sci. 326, 103–109.

Capjak, I., Zebić Avdičević, M., Dutour Sikirić, M., Domazet Jurašin, D., Hozic, A., Pajić, D., Dobrović, S., Goessler, W., Vinković Vrček, I., 2018. Behavior of silver nanoparticles in wastewater: systematic investigation on the combined effects of surfactants and electrolytes in the model systems. Environ. Sci. Water Res. Technol. 4, 2146–2159.

Carocho, M., Ferreira, I.C.F.R., 2013. A review on antioxidants, prooxidants and related controversy: natural and synthetic compounds, screening and analysis methodologies and future perspectives. Food Chem. Toxicol. 51, 15–25.

Carvalho, P.M., Felício, M.R., Santos, N.C., Gonçalves, S., Domingues, M.M., 2018. Application of light scattering techniques to nanoparticle characterization and development. Front. Chem. 6, 237.

Casillas-Figueroa, F., Arellano-García, M.E., Leyva-Aguilera, C., Ruíz-Ruíz, B., Luna Vázquez-Gómez, R., Radilla-Chávez, P., Chávez-Santoscoy, R.A., Pestryakov, A., Toledano-Magaña, Y., García-Ramos, J.C., Bogdanchikova, N., 2020. Argovit™ silver nanoparticles effects on *Allium cepa*: plant growth promotion without cyto genotoxic damage. Nanomaterials (Basel) 10 (7), 1386.

Cekic, F.O., Ekinci, S., İnal, M., Unal, D., 2017. Silver nanoparticles induced genotoxicity and oxidative stress in tomato plants. Turk. J. Biol. 41, 700–707.

Chichiriccò, G., Poma, A., 2015. Penetration and toxicity of nanomaterials in higher plants. Nanomaterials (Basel, Switzerland) 5, 851–873.

Choi, O., Hu, Z., 2008. Size dependent and reactive oxygen species related nanosilver toxicity to nitrifying bacteria. Environ. Sci. Technol. 42, 4583–4588.

Colman, B.P., Arnaout, C.L., Anciaux, S., Gunsch, C.K., Hochella, M.F.J., Kim, B., Lowry, G.V., McGill, B.M., Reinsch, B.C., Richardson, C.J., Unrine, J.M., Wright, J.P., Yin, L., Bernhardt, E.S., 2013. Low concentrations of silver nanoparticles in biosolids cause adverse ecosystem responses under realistic field scenario. PLoS One 8, e57189.

Coutris, C., Joner, E.J., Oughton, D.H., 2012. Aging and soil organic matter content affect the fate of silver nanoparticles in soil. Sci. Total Environ. 420, 327–333.

Cvjetko, P., Balen, B., Peharec Štefanić, P., Debogović, L., Pavlica, M., Klobučar, G.I.V., 2014. Dynamics of heat-shock induced DNA damage and repair in senescent tobacco plants. Biol. Plant. 58, 71–79. https://doi.org/10.1007/s10535-013-0362-9.

Cvjetko, P., Milošić, A., Domijan, A.-M., Vinković Vrček, I., Tolić, S., Peharec Štefanić, P., Letofsky-Papst, I., Tkalec, M., Balen, B., 2017. Toxicity of silver ions and differently coated silver nanoparticles in *Allium cepa* roots. Ecotoxicol. Environ. Saf. 137, 18–28.

Cvjetko, P., Zovko, M., Štefanić, P.P., Biba, R., Tkalec, M., Domijan, A.-M., Vrček, I.V., Letofsky-Papst, I., Šikić, S., Balen, B., 2018. Phytotoxic effects of silver nanoparticles in tobacco plants. Environ. Sci. Pollut. Res. 25, 5590–5602.

Dang, F., Wang, Q., Cai, W., Zhou, D., Xing, B., 2020. Uptake kinetics of silver nanoparticles by plant: relative importance of particles and dissolved ions. Nanotoxicology 14, 654–666.

De Leersnyder, I., De Gelder, L., Van Driessche, I., Vermeir, P., 2019. Revealing the importance of aging, environment, size and stabilization mechanisms on the stability of metal nanoparticles: a case study for silver nanoparticles in a minimally defined and complex undefined bacterial growth medium. Nanomaterials (Basel, Switzerland) 9, 1684.

Debnath, P., Mondal, A., Hajra, A., Das, C., Mondal, N.K., 2018. Cytogenetic effects of silver and gold nanoparticles on *Allium cepa* roots. J. Genet. Eng. Biotechnol. 16, 519–526.

Dimkpa, C.O., McLean, J.E., Martineau, N., Britt, D.W., Haverkamp, R., Anderson, A.J., 2013. Silver nanoparticles disrupt wheat (*Triticum aestivum* L.) growth in a sand matrix. Environ. Sci. Technol. 47, 1082–1090.

Domazet Jurašin, D., Ćurlin, M., Capjak, I., Crnković, T., Lovrić, M., Babič, M., Horák, D., Vinković Vrček, I., Gajović, S., 2016. Surface coating affects behavior of metallic nanoparticles in a biological environment. Beilstein J. Nanotechnol. 7, 246–262.

Eccles, J., Bangert, U., Bromfield, M., Christian, P., Harvey, A., Thomas, P., 2010. UV-vis plasmon studies of metal nanoparticles. J. Phys. Conf. Ser. 241, 012090.

El Badawy, A.M., Silva, R.G., Morris, B., Scheckel, K.G., Suidan, M.T., Tolaymat, T.M., 2011. Surface charge-dependent toxicity of silver nanoparticles. Environ. Sci. Technol. 45, 283–287.

Elechiguerra, J., Burt, J., Morones-Ramirez, J., Camacho, A., Gao, X., Lara, H., Yacaman, M., 2005. Interaction of silver nanoparticles with HIV-1. J. Nanobiotechnol. 3, 6.

Englebienne, P., Hoonacker, A., Verhas, M., 2003. Surface plasmon resonance: principles, methods and applications in biomedical sciences. J. Spectrosc. 17, 255–273.

Falco, W.F., Scherer, M.D., Oliveira, S.L., Wender, H., Colbeck, I., Lawson, T., Caires, A.R.L., 2020. Phytotoxicity of silver nanoparticles on *Vicia faba*: evaluation of particle size effects on photosynthetic performance and leaf gas exchange. Sci. Total Environ. 701, 134816.

Fayez, K., El-Deeb, B., Mostafa, N., 2017. Toxicity of biosynthetic silver nanoparticles on the growth, cell ultrastructure and physiological activities of barley plant. Acta Physiol. Plant. 39, 155.

Fiskesjo, G., 1979. Mercury and selenium in a modified Allium test. Hereditas 91, 169–178.

Fouad, A.S., Hafez, R.M., 2018. The effects of silver ions and silver nanoparticles on cell division and expression of cdc2 gene in *Allium cepa* root tips. Biol. Plant. 62, 166–172.

Galazzi, R.M., Lopes Júnior, C.A., de Lima, T.B., Gozzo, F.C., Arruda, M.A.Z., 2019. Evaluation of some effects on plant metabolism through proteins and enzymes in transgenic and non-transgenic soybeans after cultivation with silver nanoparticles. J. Proteomics 191, 88–106.

García-Sánchez, S., Bernales, I., Cristobal, S., 2015. Early response to nanoparticles in the *Arabidopsis* transcriptome compromises plant defense and root-hair development through salicylic acid signaling. BMC Genomics 16, 341.

Geisler-Lee, J., Wang, Q., Yao, Y., Zhang, W., Geisler, M., Li, K., Huang, Y., Chen, Y., Kolmakov, A., Ma, X., 2013. Phytotoxicity, accumulation and transport of silver nanoparticles by *Arabidopsis thaliana*. Nanotoxicology 7, 323–337.

Ghosh, M., Manivannan, J., Sinha, S., Chakraborty, A., Mallick, S.K., Bandyopadhyay, M., Mukherjee, A., 2012. In vitro and in vivo genotoxity of silver nanoparticles. Mutat. Res. 749, 60–69.

Gichner, T., Patková, Z., Kim, J.K., 2003. DNA damage measured by the comet assay in eight agronomic plants. Biol. Plant. 47, 185–188.

Gourieroux, A.M., Holzapfel, B.P., Scollary, G.R., McCully, M.E., Canny, M.J., Rogiers, S.Y., 2016. The amino acid distribution in rachis xylem sap and phloem exudate of *Vitis vinifera* "cabernet sauvignon" bunches. Plant Physiol. Biochem. 105, 45–54.

Grigor'eva, A., Saranina, I., Tikunova, N., Safonov, A., Timoshenko, N., Rebrov, A., Ryabchikova, E., 2013. Fine mechanisms of the interaction of silver nanoparticles with the cells of *Salmonella typhimurium* and *Staphylococcus aureus*. Biometals 26, 479–488.

Gubbins, E.J., Batty, L.C., Lead, J.R., 2011. Phytotoxicity of silver nanoparticles to *Lemna minor* L. Environ. Pollut. 159, 1551–1559.

Gunsolus, I.L., Mousavi, M.P.S., Hussein, K., Buhlmann, P., Haynes, C.L., 2015. Effects of humic and fulvic acids on silver nanoparticle stability, dissolution, and toxicity. Environ. Sci. Technol. 49, 8078–8086.

Guo, Y., Cichocki, N., Schattenberg, F., Geffers, R., Harms, H., Müller, S., 2019. AgNPs change microbial community structures of wastewater. Front. Microbiol. 9, 3211.

Gusev, A.A., Kudrinsky, A.A., Zakharova, O.V., Klimov, A.I., Zherebin, P.M., Lisichkin, G.V., Vasyukova, I.A., Denisov, A.N., Krutyakov, Y.A., 2016. Versatile synthesis of PHMB-stabilized silver nanoparticles and their significant stimulating effect on fodder beet (*Beta vulgaris* L.). Mater. Sci. Eng. C 62, 152–159.

Hashimoto, T., Mustafa, G., Nishiuchi, T., Komatsu, S., 2020. Comparative analysis of the effect of inorganic and organic chemicals with silver nanoparticles on soybean under flooding stress. Int. J. Mol. Sci. 21, 1300.

Heinz, H., Pramanik, C., Heinz, O., Ding, Y., Mishra, R.K., Marchon, D., Flatt, R.J., Estrela-Lopis, I., Llop, J., Moya, S., Ziolo, R.F., 2017. Nanoparticle decoration with surfactants: molecular interactions, assembly, and applications. Surf. Sci. Rep. 72, 1–58.

Hijaz, F., Killiny, N., 2014. Collection and chemical composition of phloem sap from *Citrus sinensis* L. Osbeck (sweet orange). PLoS One 9, e101830.

Hossain, Z., Mustafa, G., Sakata, K., Komatsu, S., 2016. Insights into the proteomic response of soybean towards Al_2O_3, ZnO, and ag nanoparticles stress. J. Hazard. Mater. 304, 291–305.

Hossain, Z., Yasmeen, F., Komatsu, S., 2020. Nanoparticles: synthesis, morphophysiological effects, and proteomic responses of crop plants. Int. J. Mol. Sci. 21, 3056.

Huang, H., Ullah, F., Zhou, D.-X., Yi, M., Zhao, Y., 2019a. Mechanisms of ROS regulation of plant development and stress responses. Front. Plant Sci. 10, 800.

Huang, R., Zhou, Y., Hu, S., Zhou, P.-K., 2019b. Targeting and non-targeting effects of nanomaterials on DNA: challenges and perspectives. Rev. Environ. Sci. Biotechnol. 18, 617–634.

Jhanzab, H.M., Razzaq, A., Bibi, Y., Yasmeen, F., Yamaguchi, H., Hitachi, K., Tsuchida, K., Komatsu, S., 2019. Proteomic analysis of the effect of inorganic and organic chemicals on silver nanoparticles in wheat. Int. J. Mol. Sci. 20, 825.

Jiang, H.-S., Li, M., Chang, F.-Y., Li, W., Yin, L.-Y., 2012. Physiological analysis of silver nanoparticles and $AgNO_3$ toxicity to *Spirodela polyrhiza*. Environ. Toxicol. Chem. 31, 1880–1886.

Jiang, H.S., Qiu, X.N., Li, G.B., Li, W., Yin, L.Y., 2014. Silver nanoparticles induced accumulation of reactive oxygen species and alteration of antioxidant systems in the aquatic plant *Spirodela polyrhiza*. Environ. Toxicol. Chem. 33, 1398–1405.

Jiang, H.S., Yin, L.Y., Ren, N.N., Zhao, S.T., Li, Z., Zhi, Y., Shao, H., Li, W., Gontero, B., 2017. Silver nanoparticles induced reactive oxygen species via photosynthetic energy transport imbalance in an aquatic plant. Nanotoxicology 11, 157–167.

Kaveh, R., Li, Y.-S., Ranjbar, S., Tehrani, R., Brueck, C.L., Van Aken, B., 2013. Changes in *Arabidopsis thaliana* gene expression in response to silver nanoparticles and silver ions. Environ. Sci. Technol. 47, 10637–10644.

Ke, M., Qu, Q., Peijnenburg, W.J.G.M., Li, X., Zhang, M., Zhang, Z., Lu, T., Pan, X., Qian, H., 2018. Phytotoxic effects of silver nanoparticles and silver ions to *Arabidopsis thaliana* as revealed by analysis of molecular responses and of metabolic pathways. Sci. Total Environ. 644, 1070–1079.

Kim, Y.B., Li, X., Kim, S.-J., Kim, H.H., Lee, J., Kim, H., Park, S.U., 2013. MYB transcription factors regulate glucosinolate biosynthesis in different organs of Chinese cabbage (*Brassica rapa* ssp. *pekinensis*). Molecules 18, 8682–8695.

Klitzke, S., Metreveli, G., Peters, A., Schaumann, G.E., Lang, F., 2015. The fate of silver nanoparticles in soil solution—sorption of solutes and aggregation. Sci. Total Environ. 535, 54–60.

Koczkur, K., Mourdikoudis, S., Lakshminarayana, P., Skrabalak, S., 2015. Polyvinylpyrrolidone (PVP) in nanoparticle synthesis. Dalton Trans. 44 (41), 17883–17905.

Kohan-Baghkheirati, E., Geisler-Lee, J., 2015. Gene expression, protein function and pathways of *Arabidopsis thaliana* responding to silver nanoparticles in comparison to silver ions, cold, salt, drought, and heat. Nanomaterials (Basel, Switzerland) 5, 436–467.

Komatsu, S., Hossain, Z., 2013. Organ-specific proteome analysis for identification of abiotic stress response mechanism in crop. Front. Plant Sci. 4, 71.

Kong, H., Yang, J., Zhang, Y., Fang, Y., Nishinari, K., Phillips, G.O., 2014. Synthesis and antioxidant properties of gum arabic-stabilized selenium nanoparticles. Int. J. Biol. Macromol. 65, 155–162.

Kumari, M., Mukherjee, A., Chandrasekaran, N., 2009. Genotoxicity of silver nanoparticles in *Allium cepa*. Sci. Total Environ. 407, 5243–5246.

Kumari, M., Pandey, S., Mishra, S.K., Giri, V.P., Agarwal, L., Dwivedi, S., Pandey, A.K., Nautiyal, C.S., Mishra, A., 2020. Omics-based mechanistic insight into the role of bioengineered nanoparticles for biotic stress amelioration by modulating plant metabolic pathways. Front. Bioeng. Biotechnol. 8, 242.

Lalau, C.M., Simioni, C., Vicentini, D.S., Ouriques, L.C., Mohedano, R.A., Puerari, R.C., Matias, W.G., 2020. Toxicological effects of AgNPs on duckweed (*Landoltia punctata*). Sci. Total Environ. 710, 136318.

Larue, C., Castillo-Michel, H., Sobanska, S., Cécillon, L., Bureau, S., Barthès, V., Ouerdane, L., Carrière, M., Sarret, G., 2014. Foliar exposure of the crop *Lactuca sativa* to silver nanoparticles: evidence for internalization and changes in Ag speciation. J. Hazard. Mater. 264, 98–106.

Layet, C., Santaella, C., Auffan, M., Chevassus-Rosset, C., Montes, M., Levard, C., Ortet, P., Barakat, M., Doelsch, E., 2019. Phytoavailability of silver at predicted environmental concentrations: does the initial ionic or nanoparticulate form matter? Environ. Sci. Nano 6, 127–135.

Lee, W.M., Kwak, J.I., An, Y.J., 2012. Effect of silver nanoparticles in crop plants *Phaseolus radiatus* and *Sorghum bicolor*: media effect on phytotoxicity. Chemosphere 86, 491–499.

Levan, A., 1938. The effect of colchicine on root mitoses in Allium. Hereditas 24, 471–486.

Li, C.-C., Chang, S.-J., Su, F.-J., Lin, S.-W., Chou, Y.-C., 2013. Effects of capping agents on the dispersion of silver nanoparticles. Colloids Surf. A Physicochem. Eng. Asp. 419, 209–215.

Li, Q., Huang, C., Liu, L., Hu, R., Qu, J., 2018a. Effect of surface coating of gold nanoparticles on cytotoxicity and cell cycle progression. Nanomaterials 8, 1063.

Li, X., Ke, M., Zhang, M., Peijnenburg, W.J.G.M., Fan, X., Xu, J., Zhang, Z., Lu, T., Fu, Z., Qian, H., 2018b. The interactive effects of diclofop-methyl and silver nanoparticles on *Arabidopsis thaliana*: growth, photosynthesis and antioxidant system. Environ. Pollut. 232, 212–219.

Liang, L., Tang, H., Deng, Z., Liu, Y., Chen, X., Wang, H., 2018. Ag nanoparticles inhibit the growth of the bryophyte, *Physcomitrella patens*. Ecotoxicol. Environ. Saf. 164, 739–748.

Liu, J.-X., Howell, S.H., 2010. Endoplasmic reticulum protein quality control and its relationship to environmental stress responses in plants. Plant Cell 22, 2930–2942.

López-Portillo, J., Ewers, F.W., Méndez-Alonzo, R., Paredes López, C.L., Angeles, G., Alarcón Jiménez, A.L., Lara-Domínguez, A.L., Torres Barrera, M.D.C., 2014. Dynamic control of osmolality and ionic composition of the xylem sap in two mangrove species. Am. J. Bot. 101, 1013–1022.

Lowry, G.V., Gregory, K.B., Apte, S.C., Lead, J.R., 2012. Transformations of nanomaterials in the environment. Environ. Sci. Technol. 46, 6893–6899.

Ma, X., Geisler-Lee, J., Deng, Y., Kolmakov, A., 2010. Interactions between engineered nanoparticles (ENPs) and plants: phytotoxicity, uptake and accumulation. Sci. Total Environ. 408, 3053–3061.

Mazumdar, H., Ahmed, G.U., 2011. Phytotoxicity effect of silver nanoparticles on *Oryza sativa*. Int. J. Chem. Technol. Res. 3, 1494–1500.

Mirzajani, F., Askari, H., Hamzelou, S., Farzaneh, M., Ghassempour, A., 2013. Effect of silver nanoparticles on *Oryza sativa* L. and its rhizosphere bacteria. Ecotoxicol. Environ. Saf. 88, 48–54.

Mirzajani, F., Askari, H., Hamzelou, S., Schober, Y., Römpp, A., Ghassempour, A., Spengler, B., 2014. Proteomics study of silver nanoparticles toxicity on *Oryza sativa* L. Ecotoxicol. Environ. Saf. 108, 335–339.

Murashige, T., Skoog, F., 1962. A revised medium for rapid growth and bio assays with tobacco tissue cultures. Physiol. Plant. 15, 473–497.

Mustafa, G., Sakata, K., Hossain, Z., Komatsu, S., 2015. Proteomic study on the effects of silver nanoparticles on soybean under flooding stress. J. Proteomics 122, 100–118.

Mustafa, G., Sakata, K., Komatsu, S., 2016. Proteomic analysis of soybean root exposed to varying sizes of silver nanoparticles under flooding stress. J. Proteomics 148, 113–125.

Nair, P.M.G., Chung, I.M., 2014a. Assessment of silver nanoparticle-induced physiological and molecular changes in *Arabidopsis thaliana*. Environ. Sci. Pollut. Res. 21, 8858–8869.

Nair, P.M.G., Chung, I.M., 2014b. Physiological and molecular level effects of silver nanoparticles exposure in rice (*Oryza sativa* L.) seedlings. Chemosphere 112, 105–113.

Noori, A., Donnelly, T., Colbert, J., Cai, W., Newman, L., White, J., 2019. Exposure of tomato (*Lycopersicon esculentum*) to silver nanoparticles and silver nitrate: physiological and molecular response. Int. J. Phytoremediation 22, 1–12.

Olchowik, J., Bzdyk, R., Studnicki, M., Bederska, M., Urban, A., Aleksandrowicz-Trzcińska, M., 2017. The effect of silver and copper nanoparticles on the condition of english oak (*Quercus robur* L.) seedlings in a container nursery experiment. Forests 8, 310.

Oukarroum, A., Barhoumi, L., Pirastru, L., Dewez, D., 2013. Silver nanoparticle toxicity effect on growth and cellular viability of the aquatic plant *Lemna gibba*. Environ. Toxicol. Chem. 32, 902–907.

Patlolla, A.K., Berry, A., May, L., Tchounwou, P.B., 2012. Genotoxicity of silver nanoparticles in *Vicia faba*: a pilot study on the environmental monitoring of nanoparticles. Int. J. Environ. Res. Public Health 9, 1649–1662.

Peharec Štefanić, P., Cvjetko, P., Biba, R., Domijan, A., Letofsky-Papst, I., Tkalec, M., Šikić, S., Cindric, M., Balen, B., 2018. Physiological, ultrastructural and proteomic responses of tobacco seedlings exposed to silver nanoparticles and silver nitrate. Chemosphere 209, 640–653.

Peharec Štefanić, P., Jarnević, M., Cvjetko, P., Biba, R., Šikić, S., Tkalec, M., Cindrić, M., Letofsky-Papst, I., Balen, B., 2019. Comparative proteomic study of phytotoxic effects of silver nanoparticles and silver ions on tobacco plants. Environ. Sci. Pollut. Res. Int. 26, 22529–22550.

Pereira, S.P.P., Jesus, F., Aguiar, S., de Oliveira, R., Fernandes, M., Ranville, J., Nogueira, A.J.A., 2018. Phytotoxicity of silver nanoparticles to *Lemna minor*: surface coating and exposure period-related effects. Sci. Total Environ. 618, 1389–1399.

Peretyazhko, T.S., Zhang, Q., Colvin, V.L., 2014. Size-controlled dissolution of silver nanoparticles at neutral and acidic pH conditions: kinetics and size changes. Environ. Sci. Technol. 48, 11954–11961.

Piao, M., Kang, K., Lee, I., Kim, H., Kim, S., Choi, J., Choi, J., Hyun, J., 2011. Silver nanoparticles induce oxidative cell damage in human liver cells through inhibition of reduced glutathione and induction of mitochondria-involved apoptosis. Toxicol. Lett. 201, 92–100.

Pokhrel, L.R., Dubey, B., 2013. Evaluation of developmental responses of two crop plants exposed to silver and zinc oxide nanoparticles. Sci. Total Environ. 452–453, 321–332.

Prasad, R., Bhattacharyya, A., Nguyen, Q.D., 2017. Nanotechnology in sustainable agriculture: recent developments, challenges, and perspectives. Front. Microbiol. 8, 1014.

Qian, H., Peng, X., Han, X., Ren, J., Sun, L., Fu, Z., 2013. Comparison of the toxicity of silver nanoparticles and silver ions on the growth of terrestrial plant model *Arabidopsis thaliana*. J. Environ. Sci. (China) 25, 1947–1955.

Reidy, B., Haase, A., Luch, A., Dawson, K.A., Lynch, I., 2013. Mechanisms of silver nanoparticle release, transformation and toxicity: a critical review of current knowledge and recommendations for future studies and applications. Materials (Basel, Switzerland) 6, 2295–2350.

Rico, C.M., Majumdar, S., Duarte-Gardea, M., Peralta-Videa, J.R., Gardea-Torresdey, J.L., 2011. Interaction of nanoparticles with edible plants and their possible implications in the food chain. J. Agric. Food Chem. 59, 3485–3498.

Saha, N., Gupta, D.S., 2017. Low-dose toxicity of biogenic silver nanoparticles fabricated by *Swertia chirata* on root tips and flower buds of *Allium cepa*. J. Hazard. Mater. 330, 18–28.

Saleeb, N., Gooneratne, R., Cavanagh, J.-A., Bunt, C., Hossain, A.K.M.M., Gaw, S., Robinson, B., 2019. The mobility of silver nanoparticles and silver ions in the soil-plant system. J. Environ. Qual. 48, 1835–1841.

Scherer, M.D., Sposito, J.C.V., Falco, W.F., Grisolia, A.B., Andrade, L.H.C., Lima, S.M., Machado, G., Nascimento, V.A., Gonçalves, D.A., Wender, H., Oliveira, S.L., Caires, A.R.L., 2019. Cytotoxic and genotoxic effects of silver nanoparticles on meristematic cells of *Allium cepa* roots: a close analysis of particle size dependence. Sci. Total Environ. 660, 459–467.

Schiavon, M., Wirtz, M., Borsa, P., Quaggiotti, S., Hell, R., Malagoli, M., 2007. Chromate differentially affects the expression of a high-affinity sulfate transporter and isoforms of components of the sulfate assimilatory pathway in *Zea mays* (L.). Plant Biol. (Stuttg.) 9, 662–671.

Schwab, F., Zhai, G., Kern, M., Turner, A., Schnoor, J.L., Wiesner, M.R., 2016. Barriers, pathways and processes for uptake, translocation and accumulation of nanomaterials in plants-critical review. Nanotoxicology 10, 257–278.

Schwabe, F., Schulin, R., Limbach, L.K., Stark, W., Bürge, D., Nowack, B., 2013. Influence of two types of organic matter on interaction of CeO_2 nanoparticles with plants in hydroponic culture. Chemosphere 91, 512–520.

Sharma, V.K., Siskova, K.M., Zboril, R., Gardea-Torresdey, J.L., 2014. Organic-coated silver nanoparticles in biological and environmental conditions: fate, stability and toxicity. Adv. Colloid Interface Sci. 204, 15–34.

Shen, M.-H., Zhou, X.-X., Yang, X.-Y., Chao, J.-B., Rui, L., Liu, J.-F., 2015. Exposure medium: key in identifying free Ag^+ as the exclusive species of silver nanoparticles with acute toxicity to *Daphnia magna*. Sci. Rep. 5, 9674.

Singh, Z., Singh, I., 2019. CTAB surfactant assisted and high pH nano-formulations of CuO nanoparticles pose greater cytotoxic and genotoxic effects. Sci. Rep. 9, 5880. https://doi.org/10.1038/s41598-019-42419-z.

Sobieh, S., Kheiralla, Z., Rushdy, A., Yakob, N., 2016. In vitro and in vivo genotoxicity and molecular response of silver nanoparticles on different biological model systems. Caryologia 69, 1–15.

Soda, S., Ike, M., Ogasawara, Y., Yoshinaka, M., Mishima, D., Fujita, M., 2007. Effects of light intensity and water temperature on oxygen release from roots into water lettuce rhizosphere. Water Res. 41, 487–491.

Speranza, A., Crinelli, R., Scoccianti, V., Taddei, A.R., Iacobucci, M., Bhattacharya, P., Ke, P.C., 2013. In vitro toxicity of silver nanoparticles to kiwifruit pollen exhibits peculiar traits beyond the cause of silver ion release. Environ. Pollut. 179, 258–267.

Stegemeier, J.P., Schwab, F., Colman, B.P., Webb, S.M., Newville, M., Lanzirotti, A., Winkler, C., Wiesner, M.R., Lowry, G.V., 2015. Speciation matters: bioavailability of silver and silver sulfide nanoparticles to alfalfa (*Medicago sativa*). Environ. Sci. Technol. 49, 8451–8460.

Stegemeier, J.P., Colman, B.P., Schwab, F., Wiesner, M.R., Lowry, G.V., 2017. Uptake and distribution of silver in the aquatic plant *Landoltia punctata* (duckweed) exposed to silver and silver sulfide nanoparticles. Environ. Sci. Technol. 51, 4936–4943.

Su, Y., Ashworth, V., Kim, C., Adeleye, A.S., Rolshausen, P., Roper, C., White, J., Jassby, D., 2019. Delivery, uptake, fate, and transport of engineered nanoparticles in plants: a critical review and data analysis. Environ. Sci. Nano 6, 2311–2331.

Su, Y., Ashworth, V.E.T.M., Geitner, N.K., Wiesner, M.R., Ginnan, N., Rolshausen, P., Roper, C., Jassby, D., 2020. Delivery, fate, and mobility of silver nanoparticles in citrus trees. ACS Nano 14, 2966–2981.

Sun, J., Wang, L., Li, S., Yin, L., Huang, J., Chen, C., 2017. Toxicity of silver nanoparticles to *Arabidopsis*: inhibition of root gravitropism by interfering with auxin pathway. Environ. Toxicol. Chem. 36, 2773–2780.

Syu, Y., Hung, J.-H., Chen, J.-C., Chuang, H., 2014. Impacts of size and shape of silver nanoparticles on *Arabidopsis* plant growth and gene expression. Plant Physiol. Biochem. 83, 57–64.

Teo, Z., Savas, P., Loi, S., 2016. Gene expression analysis: current methods. In: Lakhani, S.R., Fox, S.B. (Eds.), Molecular Pathology in Cancer Research. Springer, pp. 107–136.

Thiruvengadam, M., Gurunathan, S., Chung, I.-M., 2015. Physiological, metabolic, and transcriptional effects of biologically-synthesized silver nanoparticles in turnip (*Brassica rapa* ssp. rapa L.). Protoplasma 252, 1031–1046.

Thwala, M., Musee, N., Sikhwivhilu, L., Wepener, V., 2013. The oxidative toxicity of ag and ZnO nanoparticles towards the aquatic plant *Spirodela punctuta* and the role of testing media parameters. Environ. Sci. Process. Impacts 15 (10), 1830–1843.

Tkalec, M., Peharec Štefanić, P., Balen, B., 2019. Phytotoxicity of silver nanoparticles and defence mechanisms. Compr. Anal. Chem. 84, 145–198.

Tolaymat, T.M., El Badawy, A.M., Genaidy, A., Scheckel, K.G., Luxton, T.P., Suidan, M., 2010. An evidence-based environmental perspective of manufactured silver nanoparticle in syntheses and applications: a systematic review and critical appraisal of peer-reviewed scientific papers. Sci. Total Environ. 408, 999–1006.

Torrent, L., Iglesias, M., Marguí, E., Hidalgo, M., Verdaguer, D., Llorens, L., Kodre, A., Kavčič, A., Vogel-Mikuš, K., 2020. Uptake, translocation and ligand of silver in *Lactuca sativa* exposed to silver nanoparticles of different size, coatings and concentration. J. Hazard. Mater. 384, 121201.

Tripathi, D.K., Tripathi, A., Shweta, Singh, S., Singh, Y., Vishwakarma, K., Yadav, G., Sharma, S., Singh, V.K., Mishra, R.K., Upadhyay, R.G., Dubey, N.K., Lee, Y., Chauhan, D.K., 2017. Uptake, accumulation and toxicity of silver nanoparticle in autotrophic plants, and heterotrophic microbes: a concentric review. Front. Microbiol. 8, 7.

Vannini, C., Domingo, G., Onelli, E., Prinsi, B., Marsoni, M., Espen, L., Bracale, M., 2013. Morphological and proteomic responses of *Eruca sativa* exposed to silver nanoparticles or silver nitrate. PLoS One 8.

Vannini, C., Domingo, G., Onelli, E., De Mattia, F., Bruni, I., Marsoni, M., Bracale, M., 2014. Phytotoxic and genotoxic effects of silver nanoparticles exposure on germinating wheat seedlings. J. Plant Physiol. 171, 1142–1148.

Wang, J., Koo, Y., Alexander, A., Yang, Y., Westerhof, S., Zhang, Q., Schnoor, J.L., Colvin, V.L., Braam, J., Alvarez, P.J.J., 2013. Phytostimulation of poplars and arabidopsis exposed to silver nanoparticles and Ag^+ at sublethal concentrations. Environ. Sci. Technol. 47, 5442–5449. https://doi.org/10.1021/es4004334.

Wang, P., Lombi, E., Sun, S., Scheckel, K., Malysheva, A., McKenna, B., Menzies, N., Zhao, F.-J., 2017. Characterizing the uptake, accumulation and toxicity of silver sulfide nanoparticles in plants. Environ. Sci. Nano 4, 448–460.

Waskiewicz, A., Beszterda, M., Goliński, P., 2014. Chapter 7. Nonenzymatic antioxidants in plants. In: Ahmad, P. (Ed.), Oxidative Damage to Plants: Antioxidant Networks and Signaling. Academic Press, pp. 201–234.

Wilkins, K.A., Matthus, E., Swarbreck, S.M., Davies, J.M., 2016. Calcium-mediated abiotic stress signaling in roots. Front. Plant Sci. 7, 1296.

Wojcieszek, J., Jiménez-Lamana, J., Ruzik, L., Szpunar, J., Jarosz, M., 2020. To-do and not-to-do in model studies of the uptake, fate and metabolism of metal-containing nanoparticles in plants. Nanomaterials (Basel, Switzerland) 10, E1480.

Yan, A., Chen, Z., 2019. Impacts of silver nanoparticles on plants: a focus on the phytotoxicity and underlying mechanism. Int. J. Mol. Sci. 20, 1003.

Yang, Q., Shan, W., Hu, L., Zhao, Y., Hou, Y., Yin, Y., Liang, Y., Wang, F., Cai, Y., Liu, J., Jiang, G., 2019. Uptake and transformation of silver nanoparticles and ions by rice plants revealed by dual stable isotope tracing. Environ. Sci. Technol. 53, 625–633.

Yang, Q., Xu, W., Liu, G., Song, M., Tan, Z., Mao, Y., Yin, Y., Cai, Y., Liu, J., Jiang, G., 2020. Transformation and uptake of silver nanoparticles and silver ions in rice plant (*Oryza sativa* L.): the effect of iron plaque and dissolved iron. Environ. Sci Nano 7, 599–609.

Yin, L., Cheng, Y., Espinasse, B., Colman, B.P., Auffan, M., Wiesner, M., Rose, J., Liu, J., Bernhardt, E.S., 2011. More than the ions: the effects of silver nanoparticles on *Lolium multiflorum*. Environ. Sci. Technol. 45, 2360–2367.

Yin, L., Colman, B.P., McGill, B.M., Wright, J.P., Bernhardt, E.S., 2012. Effects of silver nanoparticle exposure on germination and early growth of eleven wetland plants. PLoS One 7.

Yu, W., Xie, H., 2012. A review on nanofluids: preparation, stability mechanisms, and applications. J. Nanomater. 2012, 435873.

Yu, S., Yin, Y., Liu, J., 2013. Silver nanoparticles in the environment. Environ. Sci. Process. Impacts 15, 78–92.

Zhang, W., Xiao, B., Fang, T., 2018. Chemical transformation of silver nanoparticles in aquatic environments: mechanism, morphology and toxicity. Chemosphere 191, 324–334.

Zook, J.M., Long, S.E., Cleveland, D., Geronimo, C.L.A., MacCuspie, R.I., 2011. Measuring silver nanoparticle dissolution in complex biological and environmental matrices using UV-visible absorbance. Anal. Bioanal. Chem. 401, 1993–2002.

Zou, X., Li, P., Lou, J., Zhang, H., 2017. Surface coating-modulated toxic responses to silver nanoparticles in *Wolffia globosa*. Aquat. Toxicol. 189, 150–158.

CHAPTER 15

Complex physicochemical transformations of silver nanoparticles and their effects on agroecosystems

Parteek Prasher[a,b], Mousmee Sharma[a,c], and Amit Kumar Sharma[b]

[a]Department of Chemistry, UGC Sponsored Centre for Advanced Studies, Guru Nanak Dev University, Amritsar, India
[b]Department of Chemistry, University of Petroleum & Energy Studies, Energy Acres, Dehradun, India
[c]Department of Chemistry, Uttaranchal University, Dehradun, India

15.1 Introduction

The unique surface properties of silver nanoparticles (AgNPs) manifest diverse applications in optoelectronics, storage batteries, and high-performance solar cells (Bhojanaa et al., 2020). Over the past decades, the biological properties of AgNPs revolutionized pharmaceutical sector by demonstrating state-of-the-art applications as theranostics, next-generation antibiotics, biosensors and diagnostic probes, and molecular medicine (Jouyban and Rahimpour, 2020). The introduction of nanotechnology in modern agriculture promoted the utilization of AgNPs as nanofertilizers for enhanced mineral uptake by plants, resulting in an optimal development (Lijuan et al., 2020). The biocidal and microbicidal potency of AgNPs prompts their application as pesticides and insecticides and for effectively managing the microborne diseases in agricultural plants (Rehmanullah et al., 2020). Similarly, the oligodynamic effect of AgNPs for killing microbes by multiple mechanisms validates their utility as wound dressings, and antiodour textiles (Prasher et al., 2018). These applications make AgNPs as one of the most extensively utilized nanoresource in the contemporary era. However, the escalating demand for this nanoresource prompted an enhanced environmental exposure, which includes complex transformations of AgNPs in the ecosystem before meeting their ultimate fate (Yu et al., 2020). It is pertinent to study the environmental transformations of AgNPs because their interactions with soil, microbiota, and autotrophs followed by accumulation in various trophic levels of food chain ultimately influence the agroecosystems (Reidy et al., 2013). In addition to agriculture sector, the application of AgNPs in textile industry poses a major concern in emanating environmental exposure owing to the erosion of nanosilver under the influence of surfactants and detergents, mechanical stress during laundering, and washing under variable temperature and pH (Riaz and Ashraf, 2020). The present chapter deals with the source, accumulation, transformation, and fate of AgNPs in agroecosystems.

15.2 Sources of AgNPs for environmental release
15.2.1 Natural sources

River estuarine waters and silver mining areas present the natural sources of nanosilver formed by the reducing agents such as humic acids and fulvic acid that possess diverse functional groups such as hydroxyls, phenolic, aldehydes, and quinines that prompt the reduction of ionic silver to nanosilver (Shao and Wang, 2020). The dissolved organic materials and polyaromatic hydrocarbons present in natural waters and environment facilitate the photochemical reduction of ionic silver to AgNPs in the presence of sunlight. An extended irradiation results in the precipitation of AgNPs. Reportedly, the dissolved oxygen concentration in natural water and reactive intermediates, including triplet natural organic matter, and superoxide does not produce a marked effect on the natural formation of AgNPs (Hou et al., 2013). However, the temperature, pH, and concentration of natural organic matter significantly influence the natural formation of AgNPs in water. The oxidative dissolution of silver objects including silver jewelry, utensils, and silver wires followed by reduction generates AgNPs naturally. Also, the photosynthetic plants tissues provide suitable medium and plant enzymes prompt the natural synthesis of AgNPs by photofabrication (Torresdey et al., 2003). The biomolecules present in plants including flavonoids, sterols, terpenes, saponins, polysaccharides, alkaloids, and polyphenols serve as reducing and capping agent for natural synthesis of AgNPs (Prasad, 2014). Besides, leaching from the natural bedrocks, disposal by photographic industry, in addition to mining processes, make a significant contribution to the silver contamination in surface water bodies (McGillicuddy et al., 2017). The wastewater serves as the chief source of AgNPs in the environment. The wastewater treatment plants retain AgNPs in the sewage sludge, which releases the AgNPs in agriculture land when used as a fertilizer (Kuhr et al., 2018). The surface run-off from the farmlands further transfers AgNPs to aquatic systems. Application of sewage sludge a landfills and subsequent leaching transfers AgNPs to aquatic ecosystem and groundwater or subsoil (Sagee et al., 2012).

15.2.2 Nanosilver impregnated textiles

The anthropogenic factors present a potential source of AgNPs in the environment. The unique properties of AgNPs present innumerable applications in many sectors as displayed in Fig. 15.1, which further potentiates their environmental disposal. Vance et al. (2015) described the widespread applications of AgNPs among other

15.2 Sources of AgNPs for environmental release

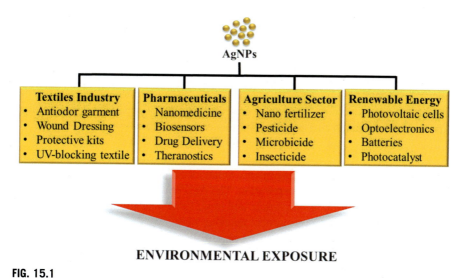

FIG. 15.1

Environmental exposure of AgNPs due to extensive application.

nanomaterials in consumer products, which raises imminent concerns regarding their environmental exposure. Benn and Westerhoff (2008) investigated the environmental release of AgNPs by commercial sock fabrics impregnated with nanoparticles to impart antibacterial property. The sock fabric contained up to 1360 μg-Ag/g-sock and leached up to 650 μg silver having size 10–500 nm, in 500 mL distilled water while washing. Typically, the AgNPs released by the fabrics while washing goes to the wastewater treatment plants becomes a part of the sewage sludge that eventually becomes a part of the agricultural fields when used as fertilizers. Mitrano et al. (2014) investigated the presence of AgNPs in wash water from the textiles impregnated with conventional silver and nanosilver. The textiles impregnated with conventional silver yielded more particulate sized silver particles and AgNPs compared to the textiles impregnated directly with nanosilver. The washing solution displayed the varying percentages of metallic Ag, AgS, and AgCl particles owing to the preliminary speciation of silver in the impregnated fabric. Also, the transformation of Ag^+ ions generated by the impregnated silver particles generated these species. Gagnon et al. (2019) investigated the release of silver nanomaterials in actual wearing conditions. The impregnated fabrics exhibited a marked 29% reduction in the silver content after three cycles of wearing. The wash water contained >90% of particulate silver in the size range >1–2 nm. Further wearing caused an increase in the number of small-sized nanoparticle in the size range of 50–100 nm. The physical stress caused fractures of the impregnated surface of fibers that resulted in release of the flakes, sheets of silver, in addition to the spherical particles. Further grinding of larger particles yielded smaller sheet-like morphology with irregular shapes. Kulthong et al. (2010) reported an interesting investigation for the determination of the release behavior of AgNPs in artificial sweat from antibacterial fabrics. The incubation of nanosilver impregnated textile into artificial sweat resulted in the varying release of AgNPs ranging from 0 to 322 mg/Kg of the fabric. The amount of release of nanosilver depended on the silver coating, fabric quality, and the formulation of artificial sweat, including the physical parameters such as its pH. Pasricha et al. (2012) studied the leaching of AgNPs from fabrics and treatment of the nanoparticle-containing effluent to mitigate the subsequent ecotoxicology. The rate of leaching reportedly depended upon the fiber type and the manufacturing process. Subsequently, the treatment of wastewater effluent containing nanosilver with bacterial strains resulted in the recovery of AgNPs via biosorption.

Yan et al. (2012) appraised the release behavior of nanosilver textiles in the stimulated fluids associated with perspiration. The AgNPs reportedly solubilized mostly in slightly acidic fluid, both in particle and ionic form. The slightly basic fluid contained the metallic silver in a particulate state with a diameter of 100–200 nm. In the salt perspiration solution, the partial aggregation of AgNPs occurred. Similarly, the release amount of AgNPs depended on the treatment time of the fabric with perspiration fluids. Stefaniak et al. (2014) evaluated the physiological factors deciding the release of nanosilver from textiles. Reportedly, the released silver was primarily present as ionic form in the artificial sweat. Interestingly, the simulated laundering experiment revealed a higher percentage of total cumulative amount of released silver in the finishing process textile as compared to the masterbatch process textile. These findings indicated that the textile treatment processes presented a higher exposure of nanosilver compared to artificial sweat and saliva. Kim et al. (2017) evaluated the nature of AgNPs released from the textiles into the laundry wash water and artificial sweat. Reportedly, the AgNPs release from textile depends markedly on the nature of the manufacturing process and the exposure environment and less on the total content of AgNPs in the impregnated textile. The highest total exposure of nanosilver impregnated textiles after 60 min of sweating appeared to be 0.81–2.02 mg Ag/Kg body weight for an average weighing human. These results suggested a minimal human exposure of AgNPs from impregnated textiles, hence raising least concerns for the toxicity of textiles released nanosilver toward human beings. Reed et al. (2016) studied the environmental impact of nanosilver containing textiles caused by the release of AgNPs during various stages of the life cycle. Reportedly, the impregnation methodology of nanosilver and their tethering to the fabric decided the release of AgNPs during washing. The release of nanosilver while washing maintained the antimicrobial potency of the fabric with >99% retention inactivity for 2 µg/g of residual nanosilver left on the fabric. However, the leaching of the released nanosilver and their subsequent loading in landfills resulted in enhanced ecotoxicology.

15.2.3 Nanosilver loaded medical bandages

Nanosilver impregnated wound dressings and plasters present applications for serious burns and wounds. Besides, the conjugation of AgNPs with antibiotics and drug molecules via physisorption, and ionic and covalent bonding to potentiate the effect

(Kalantari et al., 2020). Pourzahedi and Eckelman (2014) assessed the environmental life cycle of AgNPs released by nanosilver-loaded bandages. Despite low nanosilver loading, copious amounts of AgNPs released in the environment by the impregnated bandages. Incineration of hospital waste containing the nanosilver-bandages at 800 °C releases incinerated AgNPs that either remain in the slag or release in the atmosphere (0.05%–1%). The airborne incinerated AgNPs subsequently settle on land or water bodies where they further transform to sulfide and sulfhydryl compounds. Reportedly, for 50% dissociation and 6.8% deposition rate on water bodies, 0.002–0.04 mg silver releases form the bandage upon incineration. In addition, the incineration of nanosilver-bandages results in partial volatilization of AgNPs in furnace followed by subsequent condensation and surface reaction with other constituents present in flue gas.

15.2.4 Nanopesticides and nanofertilizers

AgNPs present myriad applications as nanopesticides by producing the toxic Ag^+ ions that disrupt the microbial membrane by binding to the essential cysteine-containing proteins present on the membrane (Adisa et al., 2019). However, the excessive applications of AgNP-based pesticides and fertilizers result in a hazardous effect on nitrogen-fixing bacteria, ammonifying bacteria, and chemolithotrophic bacteria, thereby depriving the soil of essential nutrients (Panyala et al., 2008). Similarly, the application of AgNP-based fertilizers results in an enhanced uptake and accumulation in the crop plants, hence resulting in the bioaccumulation and incorporation of AgNPs in various trophic levels in a food chain (Prasher et al., 2019). Eventually, due to the intricate association between the soil–plant systems, the biogeochemical changes in the soil directly influences plant growth and development. Hence, the buildup of engineered nanoparticles in soil by application of nanofertilizers and nanopesticides intensely influences the thriving and sustenance of crops (Raliya et al., 2018). The various properties of AgNPs including aging, rate of dissolution, dispersibility, agglomeration, surface area, and charge alter upon interacting with the soil components, hence affecting their stability in terrestrial ecosystems, their environmental transport, and subsequent availability and toxicity to the organisms (Anjum et al., 2013). The surface functionalization of AgNPs while applying in the agriculture fields mitigates the toxicity; however, the subsequent physicochemical transformations in the soil and while uptake by the plants results in the loss of surface functionality, hence causing a direct environmental exposure of AgNPs and Ag^+ ions causing imminent ecotoxicology (Panpatte and Jhala, 2019). Fig. 15.1 highlights the important sources of nanosilver released in the environment.

15.3 Physicochemical transformations of AgNPs

15.3.1 Speciation

The surface functionalization of AgNPs change degradation, sorption, and exchange with natural ligands by various chemical and biological processes. The natural organic and inorganic materials undergoing sorption and surface coating on the AgNPs influence the aggregation and dispersion behavior of AgNPs in the environment (Philippe and Schaumann, 2014). Reportedly, the humic acid interacts with citrate capped AgNPs via aromatic substituents at pH 7; however, at acidic pH, the interaction occurs via carboxylate groups. The higher aromatic/aliphatic functional group percentage in humic acid compared to fulvic acid results in high-affinity interactions of the former with citrate capped AgNPs, whereas the latter interacts with nonfunctionalized AgNPs. At pH 7, the high molecular weight fractions of humic acid tethered to the surface of citrate capped AgNPs, whereas the low molecular weight fractions of fulvic acid absorbed preferably on the surface of nonfunctionalized AgNPs at same pH (Lau et al., 2013). However, some studies suggest that both the humic acid and fulvic acid components of natural organic matter effectively displace surface citrate coating with fulvic acid interacting with the ring and polyvinyl functionality of polyvinylpyrrolidone coating on the surface of AgNPs (Khan et al., 2011a). Similarly, the natural exopolysaccharides play a role in AgNPs functionalization, which depends on the pH of medium and concentration of Cl^- ions. Change in pH from acidic to basic lowers the adsorption of exopolysaccharides on the surface of AgNPs due to reduction in the electrostatic interactions. Similarly, the increase in the salt concentration lowers the adsorption due to the less availability of the effective surface area of AgNPs for functionalization due to salt-induced aggregation (Sanchez-Cortes et al., 1998). Koser et al. (2017) predicted the speciation and toxicology of AgNPs in ecotoxicological media where a limited initial release of the Ag^+ ions controlled the concentration of dissolved silver in the media. After the initial release, the Cl^- ions, proteins and other components present in the media controlled the dissolution of silver via precipitation and complexation with these components. However, the further release of Ag^+ ions by AgNPs decreased with time in all the media except the one containing Cl^- ions. Similarly, the presence of surface stabilizers prevented the release of Ag^+ ions by AgNPs due to enhancement in the redox stability. Hashimoto et al. (2017) investigated the chemical speciation of AgNPs in aerobic and anaerobic condition to appraise their phase transformation in soil. One month of incubation of AgNPs in aerobic conditions in the soil resulted in the 88% persistence. However, the $AgNO_3$ transformed to Ag in the same conditions by associating with the minerals present in clay and humus. Similarly, 83% of AgNPs in anaerobic soil transformed to Ag_2S and 50% of $AgNO_3$ transformed to metallic Ag under same conditions. Both aerobic and anaerobic soils demonstrate trace amounts of Ag_2O and Ag_2CO_3. The redox potential of the soil played a critical role in the determination of various phase-transformation pathways for AgNPs in aerobic and anaerobic conditions.

15.3.2 Aggregation

Several factors determine the aggregation behavior of AgNPs in a medium such as surface coating, pH, nature of the electrolyte, dissolved organic matter, and dissolved oxygen. Sterically and electrostatically stabilized AgNPs display high stability compared to the nonfunctionalized and electrostatically stabilized AgNPs due to their additional steric repulsions that stabilize their colloidal solutions (El Badawy et al., 2010). Similarly, the pH of the parent solution significantly influences the zeta potential of AgNPs functionalized with natural organic matter such as exopolysaccharides and microbial extracellular proteins. A pH of the parent medium affects the colloidal stability of AgNPs via isoelectric addition of H^+ and OH^- ions that induce surface charges on nanoparticles (Khan et al., 2011b). The adsorption of these materials on the AgNPs surface provides essential steric repulsions necessary to stabilize the colloidal AgNPs. Interestingly, the presence of electrolytes neutralized the overall charge of the AgNPs resulting in their aggregation due to the loss of repulsive forces responsible for their colloidal stabilization (El Badawy et al., 2010). Low concentration of electrolytes potentiates this effect whereas the high electrolyte concentration completely shields the overall charge of AgNPs, thus increasing the aggregation kinetics significantly. Notably, the divalent electrolytes efficiently destabilize the colloidal AgNPs compared to the monovalent electrolytes (Li et al., 2012). The increase in the ionic strength lowers the thickness of the electric double layer, hence resulting in the aggregation of the AgNPs. Dissolved oxygen markedly contributes to the aggregation rate of AgNPs. Reportedly, the presence of dissolved oxygen increases the aggregation rate of colloidal citrate capped AgNPs with periodic fluctuations in their hydrodynamic radii that reversed with deoxygenation of the solution (Zhang et al., 2011). The presence of dissolved organic matter improved the electrostatic repulsions in colloidal AgNPs. Low molecular weight dissolved organic molecules improve the stability of AgNPs upon adsorption to their surface (Baalousha et al., 2013). Zehlike et al. (2019) reported the effect of particle size and dissolved organic matter on aggregation behavior of AgNPs in soil solution. The initial particle size and the concentration of particles markedly influenced the aggregation in the presence of dissolved organic matter. The AgNPs formed large aggregates in the presence of hydrophilic dissolved organic matter, whereas in the presence of hydrophobic dissolved organic matter, small-sized aggregates of AgNPs appear. In addition, the former demonstrated larger charge density compared to the latter and significantly influenced the aggregation characteristics of AgNPs in the presence of multivalent cations. Klitzke et al. (2015) reported the ultimate fate of AgNPs in soil solution. The dissolved organic matter present in the soil improved the colloidal stability of AgNPs while lowering the aggregation behavior and mediating the suppression of their oxidation to Ag^+ ions. The adsorption of AgNPs on soil particles depended on the surface charge. The nonfunctionalized AgNPs preferentially adsorbed the short-chain organic matter whose concentration influences the stability of AgNPs.

15.3.3 Oxidative dissolution

The presence of dissolved oxygen results in the surface oxidation of the AgNPs in the acidic medium. However, the presence of chlorides and sulfides thermodynamically disfavor the formation of Ag_2O, which precipitates out upon atmospheric exposure containing a higher concentration of oxygen compared to the H_2S and Cl_2.

The AgNPs oxidation becomes thermodynamically favorable with low particle size and ultrafine AgNPs with diameter < 2–3 nm due to lower oxidation potential exhibit high vulnerability toward oxidation. In addition, the high surface/volume ratio of smaller AgNPs enhance their reactivity toward oxidation. The oxidation of AgNPs results in the generation of highly toxic Ag^+ ions; however, the oxidation in the presence of chlorine and sulfide result in the formation of less harmful AgCl and Ag_2S precipitates (Lee et al., 2005). The presence of an excessive amount of chlorine results in the formation of $AgCl_2^-$ $AgCl_3^{2-}$ species that are further less toxic compared to the Ag^+ ions. Liu and Hurt (2010) investigated the Ag^+ ion release kinetics by citrate capped AgNPs. Reportedly, the ion release depended on the concentration of dissolved dioxygen and acidic pH. The rate of release of Ag^+ ions enhanced with temperature and lowered at basic pH. The addition of humic acid and fulvic acid lowered the rate of ion release. The kinetic and thermodynamic data suggested the vulnerability of AgNPs in the oxygenated environment; however, the replacement of dioxygen with noble gases such as argon led to the suppression in ion release by AgNPs. Further investigations validated that the rate of release of Ag^+ ions decelerates with thiol and citrate capping, in addition to the sulfide fabrication and scavenging of peroxides. Conversely, the peroxidation of AgNPs followed by a subsequent reduction in particle size accelerate the release rate. Further, the surface fabrication of AgNPs with labile polymeric coatings alter the Ag^+ release profile (Liu et al., 2010). Li et al. (2017) investigated the redox transformations of AgNPs in paddy soils impregnated with polyvinylpyrrolidone-coated AgNPs and $AgNO_3$. Reportedly, the soil organic matter and soil redox potential influence the AgNPs transformation. >94% AgNPs oxidize to Ag_2S after 28-day incubation, whereas the remaining oxidize to AgCl depending on the soil redox potential and initial concentration of AgNPs. The soil solids reportedly exhibited maximum retention of AgNPs compared to the liquid components of soil. Notably, the presence of organic matter in soil enhanced the retention of AgNPs and lowered the percentage of dissolved Ag salts. Conversely, the higher values of soil oxidation potential lowered the sulfidation oxidation of AgNPs and enhanced the release of dissolved Ag salts in the soil. The flooded paddy soil displayed a higher percentage of Ag_2S independent of the initial form of AgNPs impregnated to the soil. Mahdi et al. (2018) performed a saturated column experiment for tracking of AgNPs in soil. The transport and retention of AgNPs in soil depend on soil organic matter and cation exchange capacity with a high concentration of organic matter resulting in decreased content of AgNPs. The sandy soils containing the lowest percentage of the organic matter displayed the highest percentage of AgNPs. Notably, the soils with high cation exchange capacity exhibited a substantial decrease in the particle size of AgNPs. These investigations confirmed the transportation and leaching of AgNPs in deeper layers of soil. He et al. (2019) studied the effect of initial concentration, particle size, and surface coating for the transformation and retention of AgNPs in soil. Reportedly, the AgNPs displayed lower migration to ultisol owing to a high surface area and a large number of retention sites. The optimal transport of AgNPs occurred at high initial concentration and smaller size of particles. The presence of surface coating and functionalization prompted the transport of AgNPs by blocking the complexing sites at solid phase. Molleman and Hiemstra (2017) studied the effect of time, pH, and particle size on the oxidative dissolution of AgNPs. Acidification of solution containing pristine shaped AgNPs covered with partially oxidized silver resulted in undersaturation of the parent solution, thereby triggering oxidative dissolution, which eventually exposes the pristine structure of AgNPs due to the formation of Ag^+ ions. Radwan et al. (2019) further confirmed the influence of particle size and capping agent on the oxidative dissolution of AgNPs, where the variation in the capping agent, particle size, and surface area of AgNPs exhibited marked differences in the dissolution trends.

15.3.4 Sulfidation for toxicity mitigation

The nanosilver released by various factors discussed in Fig. 15.1 undergo several transformation processes in the environment by the reaction of metallic silver with organic ligands that eventually decides their physicochemical stability and toxicity. Sulfidation of AgNPs acts as a natural antidote to mitigate their toxic environmental effects. The metallic silver demonstrates strong affinity toward both organic and inorganic sulfur that eventually result in lowering of the toxicity posed by nanosilver due to its subsequent oxidation. The high stability of nanosilver sulfidation prevents the exposure of metallic silver toward the oxidants to generate the toxic, dissolved Ag^+ ions. Hence, sulfidation serves as one of the principal measures to attenuate AgNPs ecological toxicity. Levard et al. (2013a) described that even a lower sulfidation degree of 0.019 S/Ag mass ratio results in significant mitigation of AgNPs toxicity. The sulfidation of AgNPs mainly occur in anoxic conditions containing high levels of sulfide ions, such as in the wastewater treatment plants and sulfur-rich soils (Liu et al., 2011). Reportedly, at the elevated sulfide concentration, the heterogeneous direct oxysulfidation of AgNPs occurs. The reduction in pH from 11.7 to 7.0 improved the sulfidation of AgNPs. Interestingly, at high pH and in the presence of a sulfide donor, the extent of sulfidation depended on the amount of dissolved oxygen in the medium. However, the sulfidation completely diminished in the absence of oxygen at high pH. Levard et al. (2011) reported that the sulfide ions in hypoxic conditions react with Ag^{+2} ions resulting in their oxysulfidation to generate Ag_2S that due to their lower solubility precipitate out to form novel nanosilver materials. However, in the complete absence of oxygen, the direct sulfidation of AgNPs results in the formation of Ag_2S coating around the existing nanosilver particles. Similarly, the metal sulfides also act as a source of sulfide ions for the indirect oxysulfidation of AgNPs (Thalmann et al., 2014). The reaction of citrate stabilized AgNPs having an average diameter in the range 10–100 nm, with crystalline CuS and ZnS in the presence of oxygen at pH 7.5 resulted in the sulfidation of AgNPs. The extent of sulfidation increased with the concentration of Cu^{+2} and Zn^{+2} ions while following the pseudo-first-order kinetics. These investigations suggested the sulfidation of AgNPs in oxic conditions even in the absence of free sulfide ions. Reinsch et al. (2012) studied the rate of aggregation of AgNPs on their sulfidation rates. Notably, the larger aggregates of AgNPs demonstrate a slower rate of sulfidation. The degree of sulfidation depended on the ratio of the sulfide donor and the metallic silver. While comparing the sulfidation behavior of monodisperse and polydisperse AgNPs, the former fully transformed to Ag_2S, whereas the latter only displayed a partial conversion to sulfide in similar reaction conditions, hence suggesting retention in the toxicity for polydisperse AgNPs. The surface coating plays a substantial role in the environmental transformation of AgNPs. PVP-coated AgNPs, mainly with diameter 15 and 50 nm, display differential transformation to sulfide, nitrates, and oxides. The uncoated AgNPs with diameter 50 nm predominantly form complexes with humic acid (VandeVoort et al., 2014). The soils treated with metallic silver for 18 months retained 47% of Ag, whereas the 53% of metallic silver converted to Ag^+ ions, followed by their subsequent sulfidation to Ag_2S via an indirect oxysulfidation pathway. However, the oxidation and sulfidation of metallic Ag in a terrestrial environment depended on the soil dryness and concentration of oxygen (Lowry et al., 2012).

15.3.5 Chlorination

The lower solubility product of AgCl in water at RT prompts the lowering of direct environmental exposure of the toxic Ag^+ ions (Gherrou et al., 2002). The environmental chlorination of metallic silver takes place in the presence of hypochlorous, whereas the direct chlorination of Ag^+ ions generated by the oxidation of AgNPs results in the formation of AgCl (Impellitteri et al., 2009). The washing of AgNP impregnated fabrics with detergent and oxidant result in the transformation of 50% of nanosilver to AgCl. Reportedly, the chlorination of nanosilver occurs via an oxidation step resulting in the formation of AgCl, while an increase in the concentration of Cl^- ions results in the formation of negatively charged species $AgCl_x^{(x-1)-}$ (Chambers et al., 2014). Similarly, the presence of surface ligands, such as glutathione, results in the dissolution of chlorinated nanosilver to form silver–glutathione complexes. Garg et al. (2016) investigated the oxidative dissolution of AgNPs in the presence of chlorine in an air-saturated atmosphere with 2 mol of AgNPs oxidized for 1 mol of chlorine. The presence of dioxygen in air converted chlorine to OCl^- ion that prompts the oxidation of AgNPs to less toxic AgCl. The oxidation of AgNPs by chlorine bleach via this mechanism results in the attenuation of their disinfection properties due to the generation of less reactive AgCl. Levard et al. (2013b) investigated the effect of chloride on the rate of dissolution of environmentally fragile pristine AgNPs. The chloride ions possess a strong affinity toward the oxidized Ag^+ ions, and the dissolution kinetics of AgNPs depend on the ratio of Ag/Cl ions asserted by the thermodynamically driven speciation of Ag^+ in the presence of Cl^-. Further, the toxic effects of AgNPs depend on the levels of $AgCl_x^{(x-1)-}$ species, thereby suggesting the role of ions rather than the nanoparticles for imminent toxicity. Guo et al. (2019) studied the transformation of Ag^+ ions to nanosilver loaded AgCl microcubes in the root zone of plants. Notably, the natural organic matter and extracellular polymeric substances mediate the conversion of Ag^+ ions to silver nanoparticles. In addition, the plant roots and their exudates promote the transformation of Ag^+ ions to AgNPs in the presence of sunlight. The process takes place via the formation of AgCl microtubules via complex formation of Ag^+ ions with Cl^- ions released by root exudates. Further, the stabilization of AgCl microtubules occurs by biomolecules present in root exudates followed by their partial photoreduction to core–shell AgNPs. Similarly, the AgCl microcubes transform into core–shell nanostructures containing nanosilver cluster shell, and AgCl microcubes as a core, with the transformation following second-order kinetics. Zhang et al. (2015) investigated the Cl^- ion mediated shape transformation of AgNPs in environment. The presence of Cl^- ion resulted in the formation of triangular Ag–AgCl heterostructures that transform into symmetric hexapod, eventually forming small-sized AgNPs. The presence of environmental components, humic acid, bovine serum albumin, and surfactants markedly influences this transformation process.

15.3.6 Photochemical transformation

The toxicity of AgNPs arises from the generated Ag^+ ions resulting from the oxidation and dissolution of nanosilver formulations. However, the presence of natural organic matter and electron donors under optimal conditions promote the reduction of Ag^+ ions generated by AgNPs. Cheng et al. (2011) studied the sunlight mediated toxicity reduction of polymer-stabilized AgNPs. The AgNPs with particle size 6 and 25 nm, coated with gum arabic and polyvinylpyrrolidone irreversibly aggregated in the presence of sunlight. Notably, the UV content of incident sunlight prompted the aggregation behavior, followed by the nanoparticle destabilization caused by the oscillating dipole–dipole interactions. The aggregation and destabilization resulted in a marked reduction in the toxicity of AgNPs. Gorham et al. (2012) investigated the UV-mediated photochemical transformations of citrate fabricated AgNPs. The UV irradiation resulted in the surface oxidation of AgNPs, eventually leading to a reduction in the particle size and the adoption of core–shell structure. The UV light-induced loss of surface plasmon resonance resulted in photodegradation of AgNPs depending upon the particle morphology and solution phase condition. Yin et al. (2012) performed a photocatalytic reduction of Ag^+ ions to AgNPs in the presence of dissolved organic matter. The superoxide ions generated by photoirradiation of phenolic functional group in the dissolved organic matter accelerated the generation of AgNPs. The transformation depended on the pH, initial concentration of AgNPs, and nature of dissolved organic matter. Zhang et al. (2016) demonstrated light-mediated reduction of Ag^+ ions to AgNPs. In the presence of extracellular polysaccharides present on microbial membranes that contain several reducing functional groups. The reduction of Ag^+ ions by exopolysaccharides followed pseudo-first-order kinetics in the presence of UV–visible radiation that significantly accelerated the transformation. The excitation of AgNPs for their surface plasmon resonance thereby accelerates the electrons from the functional head groups on exopolysaccharides to the adjacent Ag^+ ions, resulting in their reduction. Similarly, the UV-irradiation prompt the generation of high affinity reducing species from exopolysaccharides that enable the reduction of Ag^+ ions to AgNPs. Yu et al. (2016) performed double isotope labeling experiments to investigate the transformation kinetics of AgNPs and Ag^+ ions. Reportedly, in a simple water solution containing AgNPs and Ag^+ ions, the oxidation of the former predominated. The presence of sunlight further enhanced the dissolution of AgNPs; however, irradiation for extended periods resulted in their aggregation thereby reducing the rate of reaction. Conversely, the presence of dissolved organic matter resulted in the reduction of Ag^+ ions that further improved at high pH. The presence of divalent cations of other metal atoms prompted aggregation of AgNPs without much affecting the reduction of Ag^+ ions. These investigations suggested the role of external conditions on the transformation of AgNPs and Ag^+ ions. Shi et al. (2013) investigated the effect of particle size on light-mediated reduction in toxicity of AgNPs. Reportedly, a negative correlation existed between the size of nanoparticle and its toxicity, with AgNPs having diameter 5–10 nm exhibiting maximum toxicity. The presence of visible light lowered the toxicity of small-sized AgNPs to 32% and by 11% for the nanoparticles with larger size. The mitigation of toxicity appeared mainly due to the reduction in the release of Ag^+ ions, reduction in particle growth, and restrain on the particle aggregation in the presence of light. Similarly, the light irradiation promoted the growth of small-sized AgNPs to obtain a marked increase in the particle size and aggregation behavior. These factors collaborated for minimizing the Ag^+ ion release by small-sized AgNPs, thereby mitigating the toxicity by latter (Fig. 15.2).

15.4 Effect on agroecosystems

The physicochemical transformation of AgNPs eventually affects the environment in one way or the other. The transformed AgNPs prove detrimental toward the plants and soil biogeochemistry and their phytotoxic effects disturb the whole ecosystem. The transformed AgNPs interact with various components of soil and microbiota, thereby inducing changes in the soil chemistry and microbial diversity. Customarily, the Ag^+ ions generated by the oxidized AgNPs result in the toxicity, and their further transformation to sulfides, chlorides may potentiate the toxic effect (Table 15.1).

15.5 Conclusion

The AgNPs find ubiquitous commercial applications that raises environmental concerns owing to the complex physicochemical transformations undergone by the nanosilver. The AgNPs in environment interacts with the various biotic components such as microbes, plants in addition to interacting with the abiotic components including soil, water, and metals. While undergoing these interactions, the AgNPs change to other forms such as ionic silver, halides, nitrates, and sulfides whose properties differ markedly compared to the parent AgNPs. The environment acts

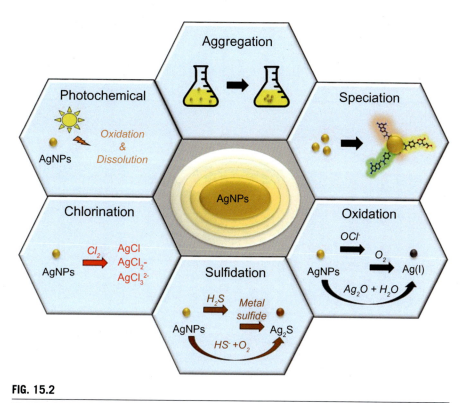

FIG. 15.2

Physicochemical transformations of AgNPs in the ecosystem.

as the sink for these species where they interact with the different components of the ecosystem. An agroecosystem consists of a complex intertwining of producers, wastewater, soil, microbiota, and crop residues. Principally, the surface run-off, sludge-based landfill, nanofertilizers, and nanopesticides serve as the main source of AgNPs and the transformed species in the agroecosystems. The interaction of nanosilver species with these components of an agroecosystem produces long-lasting effects that influence the soil quality and crop output. The utility of AgNPs in the agroecosystems must be, therefore, critically analyzed to appraise their effects on the ecosystem.

Acknowledgements

The authors thank Guru Nanak Dev University for providing the research infrastructure. SEED support from the University of Petroleum & Energy Studies is duly acknowledged.

Table 15.1 highlights the effect of transformed AgNPs and its products.

Type of particle/ion	Transformation	Plant	Effect	References
Effect on crop and plants				
AgNPs and Ag$^+$ ions	Oxidation dissolution	*Arabidopsis thaliana*	Inhibition of photosynthesis, deleterious effects on metabolic pathways	Ke et al. (2018)
AgNPs and AgNO$_3$	Oxidation	*Allium cepa*	Generation of oxidative stress in roots	Cvjetko et al. (2017)
AgNPs and Ag$^+$ ions	Oxidation	*Triticum aestivum*	Oxidative damage to chloroplast	Rastogi et al. (2019)
AgNPs	Speciation	*Triticum aestivum*	Disruption in the antioxidative defense system	Iqbal et al. (2019)
Ag$^+$ ions	Photochemical transformation	*Triticum aestivum* and *Helianthus*	Enhanced light sensitivity, perturbation in light harnessing properties	Saradhi et al. (2018)
AgNPs and AgNO$_3$	Oxidation	*Triticum aestivum*	Differential expression in superoxide dismutase (SOD)	Karimi et al. (2017)
AgNPs and Ag$^+$ ions	Oxidation dissolution	*Cucumis sativus* and *Triticum aestivum*	Enhanced phytotoxicity during germination and vegetative growth stage	Cui et al. (2014)
AgNPs and AgNO$_3$	Oxidation	*Brassica* sp.	Oxidative damage, DNA degradation	Vishwakarma et al. (2017)
AgNPs	Speciation	*Triticum aestivum*	Damaging influence on plant growth and microorganism communities in soil	Liu et al. (2017)
AgNPs and AgNO$_3$	Oxidation	*Solanum tuberosum*	Enhanced lipid peroxidation, Inhibition of ethylene perception	Homaee and Ehsanpour (2015)
AgNPs and NO$_3^-$	Oxidation	*Pisum sativum*	Alleviation in nanoparticle induced phytotoxicity	Tripathi et al. (2017)
AgNPs	Speciation	*Triticum aestivum*	Phytotoxic and genotoxic effects	Vannini et al. (2014)
Ag$_2$O	Reduction	*Zea maize* and *Brassica oleracea*	Developmental phytotoxicity	Pokhrel and Dubey (2013)
AgNPs	Speciation	*Lactuca sativa*	Penetrates in leaf tissues, phytotoxicity	Larue et al. (2014)
AgNPs	Speciation	*Oryza sativa*	Kills the rhizosphere bacteria	Mirzajani et al. (2013)
AgNPs	Oxidation	*Oryza sativa*	Leave cell deformation, disrupts seedling growth	Thuesombat et al. (2014)
AgNPs	Several transformations	*Oryza sativa*	Generation of redox stress, weakening of antioxidative defense	Nair and Chung (2014)
AgNPs	Several transformations	*Oryza sativa*	Photostimulatory effect on redox balance	Gupta et al. (2018)

Type of particle/ion	Transformation	Soil microbe	Effect	References
Effect on soil microbes				
AgNPs	Oxidation	*Rhodanobacter* sp.	Enhanced tolerance in sandy loam soil	Samarajeewa et al. (2017)
AgNPs	Oxidation dissolution	Bacteria: *Pseudomonas, Janthinobacterium* and Fungi: *Cordyceps*, and *Isaria* present in the soils of arctic tundra	Disruption in seasonal progression, inhibition of soil friendly bacteria	Kumar et al. (2014)
AgNPs	Several transformations	*Paracoccus denitrificans*	Inhibition of aerobic respiration, modification of enzyme activity	Hansch and Emmerling (2010)
AgNPs	Several transformations	*Pseudomonas putida* KT2440	Bacteriostatic effect on the beneficial soil microbe	Gajjar et al. (2009)
AgNPs	Several transformations	Loamy soil microbiome	Effects microbial biomass, disruption of genes associated with the nitrogen cycle	Grun et al. (2018)
AgNPs	Several transformations	Rhizospheric bacterial diversity	Toxic effect on soil microbiota	Pallavi et al. (2016)
Ag$^+$ ions	Oxidation dissolution	Not available	Reduction in denitrification	Throback et al. (2007)
AgNPs	Oxidative dissolution	*Bacillus thuringiensis* SBURR1, and *Bacillus amyloliquefaciens*	Disruption in bacterial membrane, pit formation in bacterial cell wall	Mirzajani et al. (2013)
AgNPs	Speciation and aggregation in presence of Ca^{+2}	*Pseudomonas chlororaphis* O6	Loss of bacterial culturability in loamy soil	Calder et al. (2012)

Continued

Table 15.1 highlights the effect of transformed AgNPs and its products—cont'd

Type of particle/Ion	Transformation	Soil microbe	Effect	References
AgNPs	Oxidative dissolution	Nitrogen cycling bacteria, Nitrosomonas europaea, Pseudomonas stutzeri, Azotobacter vinelandii	Effects on cellular and transcriptional activity	Yang et al. (2013)
Ag_2S, $AgNO_3$	Sulfidation, speciation	Arthrobacter globiformis	Toxicity effects	Schulz et al. (2018)
AgNPs	Speciation, oxidation	Pseudomonas chlororaphis	Impaired growth due to extracellular production of reactive oxygen species	Dimkpa et al. (2011)
AgNPs	Speciation	Acidobacteria, Bacteroidetes, and beta-Proteobacteria	Bactericidal property in loamy soil	Grun et al. (2018)
AgNPs and Ag^+ ions	Oxidation	Not available	Reduction in the number of nitrate and nitrite bacteria	Wu et al. (2019)
AgNPs	Speciation	Ammonia oxidizing Archaea and bacteria	Decrease in soil nitrification activity	Huang et al. (2018)

References

Adisa, I.O., Pullagurala, V.L.R., Peralta-Videa, J.R., Dimkpa, C.O., Elmer, W.H., Gardea-Torresdey, J.L., White, J.C., 2019. Recent advances in nano-enabled fertilizers and pesticides: a critical review of mechanism of action. Environ. Sci. Nano 6, 2002–2030.

Anjum, N.A., Gill, S.S., Duarte, A.C., Pereira, E., Ahmad, I., 2013. Silver nanoparticles in soil-plant systems. J. Nanopart. Res. 15, 1896.

Baalousha, M., Nur, Y., Romer, I., Tejamaya, M., Lead, J.R., 2013. Effect of monovalent and divalent cations, anions and fulvic acid on aggregation of citrate-coated silver nanoparticles. Sci. Total Environ. 454, 119–131.

Benn, T.M., Westerhoff, P., 2008. Nanoparticle silver released into water from commercially available sock fabrics. Environ. Sci. Technol. 42, 4133–4139.

Bhojanaa, K.B., Ramesh, M., Pandikumar, A., 2020. Complementary properties of silver nanoparticles on the photovoltaic performance of titania nanospheres based photoanode in dye-sensitized solar cells. Mater. Res. Bull. 122, 110672.

Calder, A.J., Dimkpa, C.O., McLean, J.E., Britt, D.W., Johnson, W., Anderson, A.J., 2012. Soil components mitigate the antimicrobial effects of silver nanoparticles towards a beneficial soil bacterium, *Pseudomonas chlororaphis O6*. Sci. Total Environ. 429, 215–222.

Chambers, B.A., Afrooz, A.R.M.N., Bae, S., Aich, N., Katz, L., Saleh, N.B., Kirisits, M.J., 2014. Effects of chloride and ionic strength on physical morphology, dissolution, and bacterial toxicity of silver nanoparticles. Environ. Sci. Technol. 48, 761–769.

Cheng, Y., Yin, L., Lin, S., Wiesner, M., Bernhardt, E., Liu, J., 2011. Toxicity reduction of polymer-stabilized silver nanoparticles by sunlight. J. Phys. Chem. C 115, 4425–4432.

Cui, D., Zhang, P., Ma, Y.-H., He, X., Li, Y.-Y., Zhao, Y.-C., Zhang, Z.-Y., 2014. Phytotoxicity of silver nanoparticles to cucumber (*Cucumis sativus*) and wheat (*Triticum aestivum*). J. Zheijang Univ. Sci. A 15, 662–670.

Cvjetko, P., Milosic, A., Domijan, A.-M., Vrcek, I.V., Tolic, S., Stefanic, P.P., Papst, I.L., Tkalek, M., Balen, B., 2017. Toxicity of silver ions and differently coated silver nanoparticles in Allium cepa roots. Ecotoxicol. Environ. Saf. 137, 18–28.

Dimkpa, C.O., Calder, A., Gajjar, P., Merugu, S., Huang, W., Britt, D.W., McLean, J.E., Johnson, W.P., Anderson, A.J., 2011. Interaction of silver nanoparticles with an environmentally beneficial bacterium, *Pseudomonas chlororaphis*. J. Hazard. Mater. 188, 428–435.

El Badawy, A.M., Luxton, T.P., Silva, R.G., Scheckel, K.G., Suidan, M.T., Tolaymat, T.M., 2010. Impact of environmental conditions (pH, ionic strength, and electrolyte type) on the surface charge and aggregation of silver nanoparticles suspensions. Environ. Sci. Technol. 44, 1260–1266.

Gagnon, V., Button, M., Boparai, H.K., Nearing, M., O'Carroll, D.M., Weber, K.P., 2019. Influence of realistic wearing on the morphology and release of silver nanomaterials from textiles. Environ. Sci. Nano 6, 411–424.

Gajjar, P., Pettee, B., Brit, D.W., Huang, W., Johnson, W.P., Anderson, A.J., 2009. Antimicrobial activities of commercial nanoparticles against an environmental soil microbe, *Pseudomonas putida* KT2440. J. Biol. Engg. 3, 9.

Garg, S., Rong, H., Miller, C.J., Waite, T.D., 2016. Oxidative dissolution of silver nanoparticles by chlorine: implications to silver nanoparticle fate and toxicity. Environ. Sci. Technol. 50, 3890–3896.

Gherrou, A., Kerdjoudj, H., Molinari, R., Drioli, E., 2002. Removal of silver and copper ions from acidic thiourea solutions with a supported liquid membrane containing D2EHPA as carrier. Sep. Purif. Technol. 28, 235–244.

Gorham, J.M., MacCuspie, R.I., Klein, K.L., Fairbrother, D.H., Holbrook, R.D., 2012. UV-induced photochemical transformations od citrate-capped silver nanoparticle suspensions. J. Nanopart. Res. 14, 1139.

Grun, A.-L., Straskraba, S., Schulz, S., Schloter, M., Emmerling, C., 2018. Long-term effects of environmentally relevant concentrations of silver nanoparticles on microbial biomass, enzyme activity, and functional genes involved in the nitrogen cycle of loamy soil. J. Environ. Sci. 69, 12–22.

Guo, H., Ma, C., Thistle, L., Huynh, M., Yu, C., Clasby, D., Chefetz, D., Polubesova, T., White, J.C., He, L., Xing, B., 2019. Transformation of Ag ions into Ag nanoparticle-loaded AgCl microcubes in the plant root zone. Environ. Sci. Nano 6, 1099–1110.

Gupta, S.D., Agarwal, A., Pradhan, S., 2018. Phytostimulatory effect of silver nanoparticles (AgNPs) on rice seedling growth: an insight from antioxidative enzyme activities and gene expression patterns. Ecotoxicol. Environ. Saf. 161, 624–633.

Hansch, M., Emmerling, C., 2010. Effects of silver nanoparticles on the microbiota and enzyme activity in soil. J. Plant Nutr. Soil Sci. 173, 554–558.

Hashimoto, Y., Takeuchi, S., Mitsunobu, S., Ok, Y.-S., 2017. Chemical speciation of silver (Ag) in soils under aerobic and anaerobic conditions: Ag nanoparticles vs. ionic Ag. J. Hazard. Mater. 322, 318–324.

He, J., Wang, D., Zhou, D., 2019. Transport and retention of silver nanoparticles in soil: Effect of input concentration, particle size and surface coating. Sci. Total Environ. 648, 102–108.

Homaee, M.B., Ehsanpour, A.A., 2015. Physiological and biochemical responses of potato (*Solanum tuberosum*) to silver nanoparticles and silver nitrate treatments under in vitro conditions. Indian J. Plant Physiol. 20, 353–359.

Hou, W.-C., Stuart, B., Howes, R., Zepp, R.G., 2013. Sunlight-driven reduction of silver ions by natural organic matter: formation and transformation of silver nanoparticles. Environ. Sci. Technol. 47, 7713–7721.

Huang, J., Cao, C., Li, R., Guan, W., 2018. Effects of silver nanoparticles on soil ammonia-oxidizing microorganisms under temperatures of 25 and 5°C. Pedosphere 28, 607–616.

Impellitteri, C.A., Tolaymat, T.M., Scheckel, K.G., 2009. The speciation of silver nanoparticles in antimicrobial fabric before and after exposure to a hypochlorite/detergent solution. J. Environ. Qual. 38, 1528–1530.

Iqbal, M., Raja, N.I., Mashwani, Z.U.R., Wattoo, F.H., Hussain, M., Ejaz, M., Saira, H., 2019. Assessment of AgNPs exposure on physiological and biochemical changes and antioxidative defence system in wheat (*triticum aestivum* L) under heat stress. IET Nanobiotechnol. 13, 230–236.

Jouyban, A., Rahimpour, E., 2020. Optical sensors based on silver nanoparticles for determination of pharmaceuticals: an overview of advances in the last decade. Talanta 217, 121071.

Kalantari, K., Motafavi, E., Afifi, A.M., Izadiyan, Z., Jahangirian, H., Rafiee-Moghaddam, R., Webster, T.J., 2020. Wound dressings functionalized with silver nanoparticles: promises and pitfalls. Nanoscale 12, 2268–2291.

Karimi, J., Mohsenzadeh, S., Niazi, A., Moghadam, A., 2017. Differential expression of mitochondrial *Manganese Superoxide Dismutase (SOD)* in *Triticum aestivum* exposed to silver nitrate and silver nanoparticles. Iran. J. Biotechnol. 15, 284–288.

Ke, M., Qu, Q., Peijnenburg, W.J.G.M., Li, X., Zhang, M., Zhang, Z., Lu, T., Pan, X., Qian, X., 2018. Phytotoxic effects of silver nanoparticles and silver ions to Arabidopsis thaliana as revealed by analysis of molecular responses and of metabolic pathways. Sci. Total Environ. 644, 1070–1079.

Khan, S.S., Mukherjee, A., Chandrasekaran, N., 2011a. Impact of exopolysaccharides on the stability of silver nanoparticles in water. Water Res. 45, 5184–5190.

Khan, S.S., Srivatsan, P., Vaishnavi, N., Mukherjee, A., Chandrasekaran, N., 2011b. Interaction of silver nanoparticles (SNPs) with bacterial extracellular proteins (ECPs) and its adsorption isotherms and kinetics. J. Hazard. Mater. 192, 299–306.

Kim, J.B., Kim, J.Y., Yoon, T.H., 2017. Determination of silver nanoparticle species released form textiles into artificial sweat and laundry wash for a risk assessment. Human Ecol. Risk Assess. 23, 741–750.

Klitzke, S., Metreveli, G., Peters, A., Schaumann, G.E., Lang, F., 2015. The fate of silver nanoparticles in soil solution-sorption of solutes. Sci. Total Environ. 535, 54–60.

Koser, J., Engelke, M., Hoppe, M., Nogowski, A., Filser, J., Thoming, J., 2017. Predictability of silver nanoparticle speciation and toxicity in ecotoxicological media. Environ. Sci. Nano 4, 1470–1483.

Kuhr, S., Schneider, S., Meisterjahn, B., Schlich, K., Hund-Rinke, K., Schlechtriem, C., 2018. Silver nanoparticles in sewage treatment plant effluents: chronic effects and accumulation of silver in the freshwater amphipod hyalella azteca. Environ. Sci. Europe 30, 7.

Kulthong, K., Srisung, S., Boonpavanitchakul, K., Kangwansupamonkon, W., Maniratanachote, R., 2010. Determination of silver nanoparticle release from antibacterial fabrics into artificial sweat. Part. Fibre Toxicol. 7, 8.

Kumar, N., Palmer, G.R., Shah, V., Walker, V.K., 2014. The effect of silver nanoparticles on seasonal change in Arctic tundra bacterial and fungal assemblages. PLoS One 9, e99953.

Larue, C., Michel, H.C., Sobanska, S., Cecillon, L., Bureau, S., Barthes, V., Ouerdane, L., Carriere, M., Sarret, G., 2014. Foliar exposure of the crop Lactuca sativa to silver nanoparticles: evidence for internalization and changes in Ag speciation. J. Hazard. Mater. 264, 98–106.

Lau, B.L.T., Hockaday, W.C., Ikuma, K., Furman, O., Decho, A.W., 2013. A preliminary assessment of the interactions between the capping agents of silver nanoparticles and environmental organics. Colloid Surf. A. 435, 22–27.

Lee, D.-Y., Fortin, C., Campbell, P.G.C., 2005. Contrasting effects of chloride on the toxicity of silver to two green algae, Pseudokirchneriella subcapitata and Chlamydomonas reinhardtii. Aquat. Toxicol. (Amst) 75, 127–135.

Levard, C., Reinsch, B.C., Michel, F.M., et al., 2011. Sulfidation processes of PVP-coated silver nanoparticles in aqueous solution: impact on dissolution rate. Environ. Sci. Technol. 45, 5260–5266.

Levard, C., Hotze, E.M., Colman, B.P., Truong, L., Yang, X.Y., Bone, A., Brown, G.E., Tanguay, R.L., Giulio, R.T.D., Bernhardt, E.S., Meyer, J.N., Wiesner, M.R., Lowry, G.V., 2013a. Sulfidation of silver nanoparticles: natural antidote to their toxicity. Environ. Sci. Technol. 47, 13440–13448.

Levard, C., Mitra, S., Yang, T., Jew, A.D., Badireddy, A.R., Lowry, G.V., Brown, G.E., 2013b. Effect of chloride on the dissolution rate of silver nanoparticles and toxicity to *E. coli*. Environ. Sci. Technol. 47, 5738–5745.

Li, X., Lenhart, J.J., Walker, H.W., 2012. Aggregation kinetics and dissolution of coated silver nanoparticles. Langmuir 28, 1095–1104.

Li, M., Wang, P., Dang, F., Zhou, D.-M., 2017. The transformation and fate of silver nanoparticles in paddy soils: effects of soil organic matter and redox conditions. Environ. Sci. Nano 4, 919–928.

Lijuan, Z., Lu, L., Wang, A., Zhang, H., Huang, M., Xing, B., Wang, Z., Ji, R., 2020. Nanobiotechnology in agriculture: use of nanomaterials to promote plant growth and stress tolerance. J. Agric. Food Chem. 68, 1935–1947.

Liu, J., Hurt, R.H., 2010. Ion release kinetics and particle persistence in aqueous nano-silver colloids. Environ. Sci. Technol. 44, 2169–2175.

Liu, J., Sonshine, D.A., Shervani, S., Hurt, R.H., 2010. Controlled release of biologically active silver from nanosilver surfaces. ACS Nano 4, 6903–6913.

Liu, J., Pennell, K.G., Hurt, R.H., 2011. Kinetics and mechanisms of nanosilver oxysulfidation. Environ. Sci. Technol. 45, 7345–7353.

Liu, G., Zhang, M., Jin, Y., Fan, X., Xu, J., Zhu, Y., Fu, Z., Pan, X., Qian, H., 2017. The effects of low concentrations of silver nanoparticles on wheat growth, seed quality, and soil microbial communities. Water Air Soil Pollut. 228, 348.

Lowry, G.V., Espinasse, B.P., Badireddy, A.R., Richardson, C.J., Reinsch, B.C., Bryant, L.D., Bone, A.J., Deonarine, A., Chae, S., Therezien, M., Colman, B.P., Hsu-Kim, H., Bernhardt, E.S., Matson, C.W., Wiesner, M.R., 2012. Long-term transformation and fate of manufactured Ag nanoparticles in a simulated large-scale freshwater emergent wetland. Environ. Sci. Technol. 46, 7027–7036.

Mahdi, K.N.M., Peters, R., Ploeg, M., Ritsema, C., Geissen, V., 2018. Tracking the transport of silver nanoparticles in soil: a saturated column experiment. Water Air Soil Pollut. 229, 334.

McGillicuddy, E., Murray, I., Kavanagh, S., Morrison, S., Fogarty, A., Cormican, M., Dockery, P., Prendergast, M., Rowan, N., Morris, D., 2017. Silver nanoparticles in the environment: sources, detection and ecotoxicology. Sci. Total Environ. 575, 21–246.

Mirzajani, F., Askari, H., Hamzelou, S., Farzaneh, M., Ghassempour, A., 2013. Effect of silver nanoparticles on Oryza sativa L. and its rhizosphere bacteria. Ecotoxicol. Environ. Saf. 88, 48–54.

Mitrano, D.M., Rimelle, E., Wichser, A., Erni, R., Height, M., Nowack, B., 2014. Presence of nanoparticles in wash water from conventional silver and nano silver textiles. ACS Nano 8, 7208–7219.

Molleman, B., Hiemstra, T., 2017. Time, pH, and size dependency of silver nanoparticle dissolution: the road to equilibrium. Environ. Sci. Nano 4, 1314–1327.

Nair, P.M.G., Chung, M., 2014. Physiological and molecular level effects of silver nanoparticles exposure in rice (*Oryza sativa L.*) seedlings. Chemosphere 112, 105–113.

Pallavi, A., Mehta, C.M., Srivastava, R., Arora, S., Sharma, A.K., 2016. Impact assessment of silver nanoparticles on plant growth and soil bacterial diversity. 3 Biotech 6, 254.

Panpatte, D., Jhala, Y. (Eds.), 2019. Nanotechnology for Agriculture: Crop Production and Protection. Springer Nature Singapore.

Panyala, N.R., Pena-Mendez, E.M., Havel, J., 2008. Silver or silver nanoparticles: a hazardous threat to the environment and human health. J. Appl. Biomed. 6, 117–129.

Pasricha, A., Jangra, S.L., Singh, N., Dilbaghi, N., Sood, K.N., Arora, K., Pasricha, R., 2012. Comparative study of leaching of silver nanoparticles from fabric and effective effluent treatment. J. Environ. Sci. 2, 852–859.

Philippe, A., Schaumann, G.E., 2014. Interactions of dissolved organic matter with natural and engineered inorganic colloids: a review. Environ. Sci. Technol. 48, 8946–8962.

Pokhrel, L.R., Dubey, B., 2013. Evaluation of developmental responses of two crop plants exposed to silver and zinc oxide nanoparticles. Sci. Total Environ. 452–453, 321–332.

Pourzahedi, L., Eckelman, M.J., 2014. Environmental life cycle assessment of nanosilver-enabled bandages. Environ. Sci. Technol. 49, 361–368.

Prasad, R., 2014. Synthesis of silver nanoparticles in photosynthetic plants. J. Nanopart. 2014, 963961.

Prasher, P., Singh, M., Mudila, H., 2018. Oligodynamic effects of silver nanoparticles: a review. BioNanoScience 8, 951–962.

Prasher, P., Sharma, M., Mudila, H., Khati, B., 2019. Uptake, accumulation, and toxicity of metal nanoparticles in autotrophs. In: Panpatte, D., Jhala, Y. (Eds.), Nanotechnology for Agriculture. Springer, Singapore.

Radwan, I.M., Gitipour, A., Potter, P.M., Dionysiou, D.D., Al-Abed, S.R., 2019. Dissolution of silver nanoparticles in colloid consumer products: effects of particle size and capping agent. J. Nanopart. Res. 21, 155.

Raliya, R., Saharan, V., Dimkpa, C., Biswas, P., 2018. Nanofertilizer for precision and sustainable agriculture: current state and future perspectives. J. Agric. Food Chem. 66, 6487–6503.

Rastogi, A., Zivcak, M., Tripathi, D.K., Yadav, S., Kalaji, H.M., Brestic, M., 2019. Phytotoxic effect of silver nanoparticles in Triticum aestivum: improper regulation of photosystem I activity as the reason for oxidative damage in the chloroplast. Photosynthetica 57, 209–216.

Reed, R.B., Zaikova, T., Barber, A., Simonich, M., Lankone, R., Marco, M., Hristovski, K., Herckes, P., Passantino, L., Fairbrother, D.H., Tanguay, R., Ranville, J.F., Huchison, J.E., Westerhoff, P.K., 2016. Potential environmental impacts and antimicrobial efficacy of silver- and nanosilver-containing textiles. Environ. Sci. Technol. 50, 4018–4026.

Rehmanullah, M.Z., Inayat, N., Majeed, A., 2020. Application of nanoparticles in agriculture as fertilizers and pesticides: challenges and opportunities. In: Rakshit, A., Singh, H., Singh, A., Singh, U., Fraceto, L. (Eds.), New Frontiers in Stress Management for Durable Agriculture. Springer, Singapore.

Reidy, B., Haase, A., Luch, A., Dawson, K.A., Lynch, I., 2013. Mechanisms of silver nanoparticle release, transformation and toxicity: a critical review of current knowledge and recommendations for future studies and applications. Materials (MDPI Basel) 6, 2295–2350.

Reinsch, B.C., Levard, C., Li, Z., Ma, R., Wise, A., Gregory, K.B., Brown, G.E., Lowry, G.V., 2012. Sulfidation of silver nanoparticles decreases Escherichia coli growth inhibition. Environ. Sci. Technol. 46, 6992–7000.

Riaz, S., Ashraf, M., 2020. Recent advances in development of antimicrobial textiles. In: Shahid, M., Adivarekar, R. (Eds.), Advances in Functional Finishing of Textiles. Textile Science and Clothing Technology. Springer, Singapore.

Sagee, O., Dror, I., Berkowitz, B., 2012. Transport of silver nanoparticles in soil. Chemosphere 88, 670–675.

Samarajeewa, A.D., Velicogna, J.R., Princz, J.I., Subasinghe, R.M., Scroggins, R.P., Beaudette, L.A., 2017. Effect of silver nano-particles on soil microbial growth, activity and community diversity in a sandy loam soil. Environ. Pollut. 220, 504–513.

Sanchez-Cortes, S., Francioso, O., Ciavatta, C., Garcia-Ramos, J.V., Gessa, C., 1998. pH-dependent adsorption of fractionated peat humic substances on different silver colloids studied by surface-enhanced Raman spectroscopy. J. Colloid Interface Sci. 198, 308–318.

Saradhi, P.P., Shabnam, N., Sharmila, P., Ganguli, A.K., Kim, H., 2018. Differential sensitivity of light-harnessing photosynthetic events in wheat and sunflower to exogenously applied ionic and nanoparticulate silver. Chemosphere 194, 340–351.

Schulz, C.L., Gray, J., Verweij, R.A., Fite, M.B., Puntes, V., Svendsen, C., Lahive, E., Matzke, M., 2018. Aging reduces the toxicity of pristine but not sulphidised silver nanoparticles to soil bacteria. Environ. Sci. Nano 5, 2618–2630.

Shao, Z., Wang, W.-X., 2020. Biodynamics of silver nanoparticles in an estuarine oyster revealed by ^{110m}AgNP tracing. Environ. Sci. Technol. 54, 965–974.

Shi, J., Xu, B., Sun, X., Ma, C., Yu, C., Zhang, H., 2013. Light induced toxicity reduction of silver nanoparticles to *Tetrahymena pyriformis*: effect of particle size. Aquat. Toxicol. 132–133, 53–60.

Stefaniak, A.B., Duling, M.G., Lawrence, R.B., Thomas, T.A., LeBouf, R.F., Wade, E.E., Virji, M.A., 2014. Dermal exposure potential from textiles that contain silver nanoparticles. Int. J. Occup. Environ. Health 20, 220–234.

Thalmann, B., Voegelin, A., Sinnet, B., Morgenroth, E., Kaegi, R., 2014. Sulfidation kinetics of silver nanoparticles reacted with metal sulfides. Environ. Sci. Technol. 48, 4885–4892.

Throback, I.N., Johansson, M., Rosenquist, M., Pell, M., Hansson, M., Hallin, S., 2007. Silver (Ag$^+$) reduces denitrification and induces enrichment of novel nirK genotypes in soil. FEMS Mirobiol. Lett. 270, 189–194.

Thuesombat, P., Hannongbua, S., Akasit, S., Chadchawan, S., 2014. Effect of silver nanoparticles on rice (*Oryza sativa L. cv. KDML 105*) seed germination and seedling growth. Ecotoxicol. Environ. Saf. 104, 302–309.

Torresdey, J.L.G., Gomez, E., Videa, J.R.P., Parsons, J.G., Troiani, H., Yacaman, M.J., 2003. Alfalfa sprouts: a natural source for the synthesis of silver nanoparticles. Langmuir 19, 1357–1361.

Tripathi, D.K., Singh, S., Singh, S., Srivastava, P.K., Singh, V.P., Singh, S., Prasad, S.M., Singh, P.K., Dubey, N.K., Pandey, A.C., Chauhan, D.K., 2017. Nitric oxide alleviates silver nanoparticles (AgNps)-induced phytotoxicity in Pisum sativum seedlings. Plant Physiol. Biochem. 110, 167–177.

Vance, M.E., Kuiken, T., Vejerano, E.P., McGinnis, S.P., Hochella, M.F., Rejeski, D., Hull, M.S., 2015. Nanotechnology in the real world: redeveloping the nanomaterial consumer products inventory. Beilstein J. Nanotechnol. 6, 1769–1780.

VandeVoort, A.R., Tappero, R., Arai, Y., 2014. Residence time effects on phase transformation of nanosilver in reduced soils. Environ. Sci. Pollut. R. 21, 7828–7837.

Vannini, C., Domingo, G., Onelli, E., Mattia, F.D., Bruni, I., Marsoni, M., Bracale, M., 2014. Phytotoxic and genotoxic effects of silver nanoparticles exposure on germinating wheat seedlings. J. Plant Physiol. 171, 1142–1148.

Vishwakarma, K., Shweta, Upadhyay, N., Singh, J., Liu, S., Singh, V.P., Prasad, S.M., Chauhan, D.K., Tripathi, D.K., Sharma, S., 2017. Differential phytotoxic impact of plant mediated silver nanoparticles (AgNPs) and silver nitrate (AgNO$_3$) on *Brassica sp.*. Front. Plant Sci. 8, 1501.

Wu, L.-L., Zhang, X., Shu, K.-H., Zhang, L., Si, Y.-B., 2019. Effects of silver nanoparticles and silver ions on soil nitrification microorganisms and ammoxidation. Huan Jing Ke Xue. 40, 2939–2947.

Yan, Y., Yang, H., Li, J., Lu, X., Wang, C., 2012. Release behavior of nano-silver textiles in simulated perspiration fluids. Text. Res. J. 82, 1422–1429.

Yang, Y., Wang, J., Xiu, Z., Alvarez, P.J.J., 2013. Impacts of silver nanoparticles on cellular and transcriptional activity of nitrogen-cycling bacteria. Environ. Toxicol. Chem. 32, 1488–1494.

Yin, Y., Liu, J., Jiang, G., 2012. Sunlight-induced reduction of Ionic Ag and Au to metallic nanoparticles by dissolved organic matter. ACS Nano 6, 7910–7919.

Yu, S., Yin, Y., Zhou, X., Dong, L., Liu, J., 2016. Transformation kinetics of silver nanoparticles and silver ions in aquatic environments revealed by double stable isotope labeling. Environ. Sci. Nano 3, 883–893.

Yu, S.-J., Yin, Y.-G., Liu, J.-F., 2020. Silver nanoparticles in the environment. Environ. Sci. Process. Impacts 15, 78–92.

Zehlike, L., Peters, A., Ellerbrock, R.H., Degenkolb, R.H., Klitzke, S., 2019. Aggregation of TiO2 and Ag nanoparticles in soil solution—effects of primary nanoparticle size and dissolved organic matter characteristics. Sci. Total Environ. 688, 288–298.

Zhang, W., Yao, Y., Li, K.G., Huang, Y., Chen, Y.S., 2011. Influence of dissolved oxygen on aggregation kinetics of citrate-coated silver nanoparticles. Environ. Pollut. 159, 3757–3762.

Zhang, L., Li, X., He, R., Wu, L., Zhang, L., Zeng, J., 2015. Chloride-induced shape transformation of silver nanoparticles in a water environment. Environ. Pollut. 204, 145–151.

Zhang, X., Yang, C.-Y., Yu, H.-Q., Sheng, G.-P., 2016. Light-induced reduction of silver ions to silver nanoparticles in aquatic environments by microbial extracellular polymeric substances (EPS). Water Res. 106, 242–248.

PART 4

Plant protection applications

CHAPTER 16

Detection and mineralization of pesticides using silver nanoparticles

Shubhankar Dube and Deepak Rawtani

School of Pharmacy, National Forensic Sciences University, Gandhinagar, Gujarat, India

16.1 Introduction

Pesticides are natural, synthetic, or semisynthetic substances which could have a single active ingredient or multiple active ingredients which primarily work to prevent, repel, mitigate, or kill pests. They could also be defoliants, plant regulators, desiccants, or nitrogen stabilizers. The active ingredients are generally mixed with inert ingredients which are not usually nontoxic but are instead utilized to improve pesticide's effectiveness and performance (EPA, n.d.).

Pesticides could be classified either based on their origin or by their use, as depicted in Fig. 16.1 (Rawtani et al., 2018). These classes of pesticides are not very specific to their targets and generally affect an extensive variety of pests. Insecticides, herbicides, rodenticides, and fungicides are the four commonly used classes of pesticides (Costa, 2008).

Pesticides have far-ranging effects on public health and ecosystems. Humans are exposed to pesticides through touch, ingestion via certain medium and inhalation of air-borne particulates. Exposure to pesticides can cause different outcomes depending on the class and amount of pesticide the said person is exposed to and the individual's health status. Pesticides can damage different systems of the body viz. neurological, respiratory, reproductive, gastrointestinal, respiratory, dermatological, carcinogenic, and endocrine (Nicolopoulou-Stamati et al., 2016). Exposure to such agents could be accidental, intentional, or part of an occupational hazard. Insecticides as a class are most toxic whereas herbicides are generally moderate to low acutely toxic except highly toxic paraquat. Acute toxicity varies among different fungicides, but it is usually low and rodenticides are not as toxic to humans compared with rats. According to studies, most human poisoning cases were due to organophosphates (a subclass of insecticides) and paraquat (a subclass of herbicides) (Mohanty et al., 2004; Nagami et al., 2005).

The two commonly used classes of insecticides, carbamates, and organophosphates are neurotoxic primarily because both of them inhibit the enzyme acetylcholinesterase (AChE), which plays a vital physiological role of hydrolysing the neurotransmitter acetylcholine. Inhibition of AChE causes an accumulation of

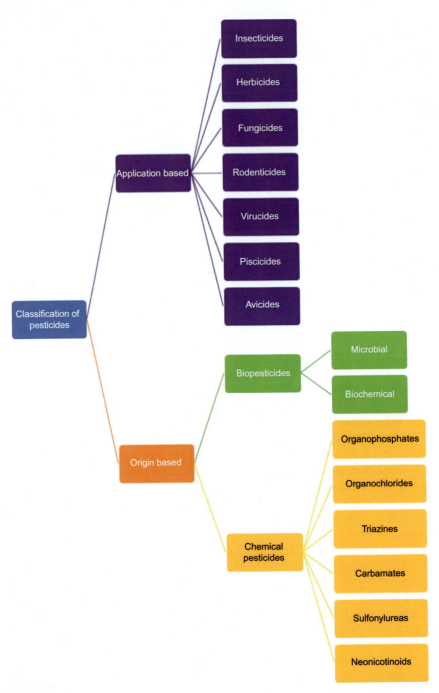

FIG. 16.1

A brief classification of pesticides based on their origin and use (Rawtani et al., 2018).

acetylcholine which overstimulates cholinergic receptors. This condition is known as cholinergic syndrome. The herbicide paraquat accumulates mainly in the lungs and kidney producing toxicity to cells by the generation of superoxide anions which cause lipid peroxidation (Costa, 2008).

Nanotechnology deals with the manipulation and application of peculiar properties of different materials at the nanoscale. Among the plethora of available nanomaterials, silver nanoparticles (AgNPs) are most studied along with gold nanoparticles (AuNPs). This is mainly owing to localized surface plasmon resonance (LSPR) and antimicrobial properties which confers them new properties like broad-spectrum antimicrobials (Franci et al., 2015; Stiufiuc et al., 2013), surface-enhanced Raman spectroscopy (SERS) (Konrad et al., 2013). AgNPs have unique electrical, optical, and thermal properties which render them optimal for a variety of industrial applications (Natsuki, 2015).

Many breakthroughs have been made in the field of agricultural nanotechnology which includes nanocides to control pests, Buckyball fertilizers, airtight packaging with nanoparticles, development of nanofibers, and nano-veterinary medicine (Mukhopadhyay, 2014). Many techniques have been devised to detect and degrade pesticides from a range of consumables. Degradation is usually achieved by converting them into harmless minerals. This chapter aims at highlighting the full range of approaches developed to detect and mineralize different pesticides using AgNPs. The proposals will be segregated into two sections, namely sensors and mineralization techniques.

16.2 Detection of pesticides

Surface modification of AgNPs is an essential step toward the preparation of sensors as it enhances the analytical potential of nanomaterial thus providing augmented selectivity and sensitivity toward the target pesticides (Aragay et al., 2012; Unser et al., 2015). There is a range of well-documented methods for synthesis of surface-modified nanoparticles which include microwave-assisted, chemical, microbial, and green synthesis methods (Chawla et al., 2018; Srivastava, 2012). This section will summarize a wide range of techniques to detect different pesticides from consumables.

16.2.1 Chemical-based assays

Chemical-based or nonenzymatic assays are more robust systems with better practical applicability. Generally, these assays are based on nanoparticles capped with capping agent changing from dispersed state to aggregated state on interaction with pesticide.

A study published in 2008 used p-Sulfonatocalix[n]arene-modified AgNPs (pSC_n-Ag NPs) to colorimetrically detect optunal in water samples (Xiong and Li, 2008). The study used pSC_4-Ag NPs that showed a significant change in color from yellow to red in the presence of the optunal. A colorimetric response was produced

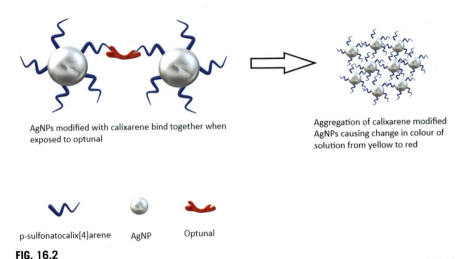

FIG. 16.2

Schematic diagram of the development of colorimetric response due to optunal induced aggregation of AgNPs.

due to aggregation of pSC$_4$-Ag NPs owing to interaction with the pesticide as explained in Fig. 16.2. This sensor exhibited a Limit of detection (LOD) of 10^{-7} M and provided real-time on-site detection of optunal (Xiong and Li, 2008). In another study, researchers used humic acid (HA) capped AgNPs to colorimetrically detect the herbicide sulfurazon-ethyl. HA-reduced silver nitrate while also acting as a capping agent (Dubas and Pimpan, 2008). When the herbicide is present, color of the solution changes from yellow to purple owing to the clumping of AgNPs (Dubas and Pimpan, 2008). The aggregation is caused due to the bonding of phenyl moieties of HA and hydrophobic groups of herbicides. The sensitivity of this sensor to the herbicide, however, was in the range of 100–500 ppm (Dubas and Pimpan, 2008). Another group of researchers proposed the use of plasmonic metal nanoparticles for colorimetric detection of organophosphorous pesticides (Dissanayake et al., 2019). The sulfur atoms of pesticides interacted with the AgNPs destabilizing them and thus causing wavelength shifts in LSPR. The AgNP solution showed a change in color from yellow to different colors depending on the pesticide it interacted with while being selective toward the organophosphorous pesticide class. The plasmonic AgNPs showed LOD of 9, 44, 18, and 11 ppm for the pesticides ethion, parathion, malathion, and fenthion, respectively (Dissanayake et al., 2019).

A study proposed use of 5-sulfo anthranilic acid dithiocarbamate functionalized silver nanoparticles (SAADTC-AgNPs) as a selective and sensitive colorimetric sensor for detection of tricyclazole fungicide (Rohit and Kailasa, 2014a). On interaction with the target, SAADTC-AgNPs changed the color of the solution from yellow to pink (Rohit and Kailasa, 2014a). This color change was produced due to the aggregation of SAADTC-AgNPs, and it showed good selectivity against interfering pesticides. The sensor showed an LOD of 1.8×10^{-7} M and it could be used for in situ analysis of the fungicide in complex food samples (Rohit and Kailasa, 2014a). Another group of researchers devised a colorimetric sensor using dopamine dithiocarbamate functionalized silver nanoparticles (DDTC-AgNPs) to detect mancozeb in fruit juice and water samples (Rohit et al., 2014). The phenolic group of DDTC interacts with amine groups of mancozeb to induce aggregation of DDTC-Ag NPs thus causing a color change from brownish orange to blue. This detection technique was highly selective with LOD of 21.1×10^{-6} M and provided results with minimal sample preparation (Rohit et al., 2014). A study used cyclen dithiocarbamate-functionalized silver nanoparticles (CN-DTC-AgNPs) to detect paraquat and thiram pesticides (Rohit and Kailasa, 2014b). The CN-DTC groups act as supramolecular structures which capture the two pesticides selectively to cause clumping of CN-DTC-AgNPs leading to change of color from yellow to pink for thiram and orange for paraquat as described in Fig. 16.3 (Rohit and Kailasa, 2014b). The color and optical changes were produced due to "host-guest" chemistry where the target pesticides mediated assembly of CN-DTC-AgNPs. The sensor had selectivity against similar pesticides and had fair sensitivity with LOD of 2.81×10^{-6} M for thiram and 7.21×10^{-6} M for paraquat (Rohit and Kailasa, 2014b).

Another study used 4-aminobenzenethiol (ABT) functionalized AgNPs as a colorimetric nanoprobe for detection of carbendazim in food and environmental water samples. The ABT-AgNPs developed strong ion pair and π-π interaction with the pesticide which causes aggregation of these nanoparticles, thus changing color of the solution from yellow to orange. The sensor had high selectivity as well as fair sensitivity with a low LOD of 1.04 μM (Patel et al., 2015). A study developed colorimetric sensor using bare AgNPs for sensing of diazinon in fruit and vegetables. The sulfur and pyrimidine nitrogen moieties of pesticide induced clumping of AgNPs

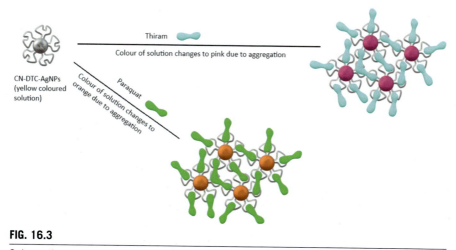

FIG. 16.3

Schematic representation of the use of CN-DTC-AgNPs as a colorimetric probe for selective detection of pesticides thiram and paraquat.

via noncovalent interactions to produce a color change from yellow to pinkish-red (Shrivas et al., 2019). The sensor had high selectivity and had a low detection limit of 7 ng/mL while also being cheap to fabricate (Shrivas et al., 2019). A study proposed use of AgNPs immobilized on polyethylene imine modified reduced graphene oxide (rGO-PEI-AgNPs) as a biocompatible interface for the sensing trifluralin in plasma samples (Jafari et al., 2019). Trifluralin interacted with rGO-PEI-AgNPs nanosensor producing an increase in peak current due to the increase in electron transfer. The sensor had LOD of 1 nM and showed great potential as a point-of-care diagnostic tool for trifluralin poisoning (Jafari et al., 2019). Another group of researchers developed an electrochemical sensor using dissolved organic matter (DOM) capped AgNPs to detect atrazine in water samples (Zahran et al., 2020). A composite was prepared using DOM, AgNPs and glassy carbon electrode (GCE) to detect the pesticide via its electrochemical oxidation and clumping of AgNPs. The technique showed high selectivity while also being stable enough to test real samples like water from streams (Zahran et al., 2020).

A study published in 2013 used a novel fluorescence-based sensor developed by researchers for the sensing of malathion. The sensor used chitosan-capped silver nanoparticles (Chi-AgNPs) as a fluorophore (Vasimalai and John, 2013). The color of Chi-AgNPs solution changed from yellow to brown on the introduction of malathion in the micromolar range due to clumping of AgNPs (Vasimalai and John, 2013). No change in absorption spectrum was observed below the micromolar range; however, the emission intensity decreased when malathion was introduced in picomolar concentration. Along with being highly selective, it had a very low LOD of 94 fM/L (Vasimalai and John, 2013). Another study developed a colorimetric sensor for cypermethrin using L-cysteine modified silver nanoparticles (L-cys-AgNPs) produced via green synthesis (Kodir et al., 2016). On interaction with pesticide, the color of solution changes from brownish-yellow to a clear solution. This was mainly due to COO– groups present on the L-cys-AgNPs which interact with positively charged pesticide along with NaCl to cause aggregation of the particles (Kodir et al., 2016). NaCl mainly increased aggregation rate as it had a bridging effect on the pesticide-Na^+ ion complex. This sensor was selective for the target while being sensitive enough to be used for quantitative analysis of cypermethrin of concentration 20–100 ppm (Kodir et al., 2016). Another research study developed a colorimetric sensor using 2-mercapto-5-nitrobenzimidazole capped silver nanoparticles (MNBZ-AgNPs) to detect glyphosate in water samples (Rawat et al., 2016). They used Mg^{2+} ions as a trigger that could enhance the sensitivity of MNBZ-AgNPs by capturing glyphosate at nanomolar concentration with high selectivity. The metal ion complexes with the nitro group and phosphate group of MNBZ-AgNPs and glyphosate, respectively, causing aggregation and thus producing color change from yellow to orange-red (Rawat et al., 2016). The sensor had LOD of 17.1 nM and was suitable for performing in-situ analysis of glyphosate (Rawat et al., 2016).

A year later, an approach was devised using unmodified AgNPs to detect prothioconazole in paddy water samples colorimetrically. Citrate-capped AgNPs under appropriate pH were negatively charged, and thus they interacted with the positively charged NH^{2+} groups of the pesticide leading to aggregation of nanoparticles. NaCl salt also was added in a controlled quantity to increase clumping of AgNPs. Aggregation of AgNPs produced color change from yellow to pink-orange (Ivrigh et al., 2017). The sensor was fairly selective with a very low detection limit of 1.7 ng/mL (Ivrigh et al., 2017). A fluorometric sensing method for dimethoate was developed by researchers who used oxidized multiwalled carbon nanotubes decorated with silver nanoparticles (AgNPs/oxMWCNTs) (Hsu et al., 2017). The composite showed remarkable peroxidase-like activity with hydrogen peroxide-amplex red (AR) system, where the oxidized form of AR fluoresced at 584 nm (Hsu et al., 2017). On the interaction of AgNPs of the composite with dimethoate, the catalytic activity was inhibited, and this event was exhibited by the diminishing fluorescence. This technique allowed detection in the range of 0.01–0.35 $\mu g\ mL^{-1}$ with a low LOD of 0.003 $\mu g\ mL^{-1}$ (Hsu et al., 2017). A study used citrate-capped AgNPs to colorimetrically detect triazophos in fruit and water samples. Here a triazophos-Cit-AgNP complex is formed by the carboxylic groups of citrate interacting with two ethoxy groups of triazophos which forms π-π interactions between the other complexes to induce clumping of AgNPs to change the color of the solution from yellow to prunosus (Ma et al., 2018). The sensor showed fair selectivity and with a low LOD of 5 nM (Ma et al., 2018). Another study in 2018 used green synthesized of CDs-capped AgNPs to devise a colorimetric sensor for phoxim in fruit and water samples. The CDs help to stabilize the AgNPs while the phoxim contains cyano and ethoxy groups which under acidic conditions bind to the $-NH_2$ and $-COOH$ moieties on the surface of CDs, thus forming coordination complexes (Zheng et al., 2018a). This results in the clumping of CDs-AgNPS, causing a color change from yellow to red. The sensor was fairly specific with sensitivity in the range of 0.1–100 μM (Zheng et al., 2018a).

A study proposed use of AgNPs conjugated with polystyrene-*block*-poly(2-vinylpyridine) (PS-*b*-P2VP) as a colorimetric nanoprobe for cartap in blood plasma and water samples. PS-*b*-P2VP tends to create micelles in toluene where AgNPs enter the core along with poly(2-vinylpyridine) (P2VP). In this process, the pesticide gets attracted toward the positive charges of P2VP inside the core leading to blue shift phenomenon. The sensor was highly selective and also fairly sensitive with LOD of 0.06 μg/L (Rahim et al., 2018). In 2019, sodium dodecyl sulfate capped silver nanoparticles (SDS-AgNPs) was used to colorimetrically detect ziram, maneb, and zineb in environmental samples (Ghoto et al., 2019). Dithiocarbamate (DTC) pesticides have disulfide groups that made it capable of adsorbing on the exterior of AgNPs while the sulfate ions of SDS had an electrostatic attraction toward N-H groups of DTCs thus causing aggregation of SDS-AgNPs. Change in color from yellow to grayish or dark brownish was observed selectively for DTCs and it showed LOD of 4.0, 149.3, and 9.1 ng/mL for zineb, ziram, and maneb, respectively (Ghoto et al., 2019). A group of researchers developed AgNP-based pH-assisted colorimetric sensor for concomitant detection and discrimination of Phosalone (PS) and Azinphosmethyl (AM) pesticides. The two pesticides showed different capabilities of inducing clumping of AgNPs as these organophosphates exhibit anionic, cationic, and neutral forms depending on the pH of the solution. At acidic pH, PS-induced clumping of AgNPs while AM produced the same effect at basic pH, thus becoming an efficient fingerprinting tool for agrichemical applications (Orouji et al., 2019). Another study used green synthesized AgNPs via *Syzygium aromaticum* extract to synthesize a colorimetric nanoprobe for the sensing of vinclozolin. The clove extract capped AgNPs aggregate when the fungicide is present thus producing color change from yellow to dark brown (Hussain et al., 2019). The sensor showed detection limit of 21 nM with high selectivity while also being facile and cost-effective (Hussain et al., 2019).

16.2.2 Enzyme-based assays

A large group of pesticides including carbamates and organophosphates are well-known inhibitors of cholinesterase enzymes which catalyze the hydrolysis of choline-based substrates. Butyryl cholinesterase (BChE) and acetylcholinesterase (AChE) are mainly found in vertebrates. Several biosensors make use of enzyme inhibiting effect of anticholinergic compounds; however, BChE inhibition-based sensors are yet to be explored (Štěpánková and Vorčáková, 2016). The spectral properties of nanoparticles, coupled with the catalytic behavior of these enzymes, allow colorimetric sensing of pesticides (Che Sulaiman et al., 2020).

In the same year, a few researchers used unmodified AgNPs to detect dipterex in different water bodies via its inhibiting activity against AChE. The enzyme catalyzes the generation of thiol group-containing compound, which causes aggregation of citrate-capped AgNPs, thus producing a color change from bright yellow to pink (Li et al., 2014). The pesticide irreversibly inhibits the activity of enzyme and thus prevents aggregation of AgNPs. This sensing method worked for all organophosphate pesticides while being fairly sensitive with LOD of 0.18 ng/mL. The study indicated the potential of this type of sensor to be extended for the use on vegetable and fruit samples as well (Li et al., 2014). In 2013, researchers developed a novel electrochemical sensor for the sensing of chlorpyrifos and carbaryl. It had AChE immobilized on the surface of AgNPs, carboxylic graphene (CGR) and Nafion (NF) hybrid-modified glassy carbon electrode (GCE) (Liu et al., 2014). The target pesticides are inhibitors of AChE. The biosensor showed good affinity toward acetylthiocholine chloride (ATCl), and its oxidation led to an amperometric response, which was altered with the presence of pesticides (Liu et al., 2014). This allowed the researchers to detect the pesticides carbaryl and chlorpyrifos in concentrations ranging between from 1.0×10^{-12} to 1×10^{-8} M and from 1.0×10^{-13} to 1×10^{-8} M, respectively (Liu et al., 2014).

Another study published a year later proposed use of a novel enzyme-based sensor which used AgNPs decorated with tetrathiacalix[4]arene for detection of carbamate and organophosphate pesticides in grape juice and peanut samples. The enzyme AChE was immobilized onto GCE modified with carbon black (CB) and AgNPs (Evtugyn et al., 2014). Hydrolysis of the substrate acetylthiocholine (ATCh) produced an oxidation current which was measured by the cyclic voltammetry apparatus. The sensor had excellent sensitivity with LOD of 0.05, 0.1, 0.1, and 10 nM for paraoxon, malaoxon, carbofuran, and aldicarb respectively. The study highlighted the protective effect of the macrocycles over the AgNPs, thus preventing them from direct anodic oxidation (Evtugyn et al., 2014). In 2017, a study proposed use of enzyme-based sensor using NF and AgNPs supported on amine-functionalized reduced graphene oxide (rGO-NH$_2$) for the sensing of methidathion, malathion and chlorpyrifos ethyl pesticides. The nanocomposite Ag@rGO-NH$_2$ with AChE immobilized onto it was coated on GCE to construct a working electrode. AChE hydrolyses the substrate ATCl to produce an oxidation peak current which was recorded using cyclic voltammetry. The biosensor was reasonably sensitive with LOD being 4.5, 9.5, and 14 ng/mL for malathion, methidathion, and chlorpyrifos ethyl, respectively (Guler et al., 2017). Another study conducted in 2016 fabricated enzyme-mediated sensor for the detection of trichlorfon and malathion pesticides. AChE catalyzes the hydrolysis of ATCh into thiocholine (TCh), which inhibits aggregation of AgNPs. AChE-immobilized glass slides were immersed in AgNP solution. The pesticides inhibit AChE and thus allow AgNPs to aggregate over the glass slide leading to an increase in absorbance intensity at 420 nm. This technique permitted very sensitive detection with LOD of 5.46 and 0.455 nM for trichlorfon and malathion, respectively (Kumar et al., 2016). A group of researchers developed a dual signal sensor using AgNPs modified with graphitic carbon nitride (g-C$_3$N$_4$) for sensing of parathion-methyl in water and vegetable samples (Li et al., 2020b). The sensor consisted of AgNPs which inhibited the blue-colored fluorescence of g-C$_3$N$_4$ by inner filter effect (IFE). The enzyme AChE hydrolysed ATCh to produce TCh which induces clumping of AgNPs, thus causing recovery of fluorescence of g-C$_3$N$_4$. However, when pesticide is present, the enzyme activity is inhibited, causing a reduction in fluorescence along with an increase in absorption peak of g-C$_3$N$_4$/AgNPs at 390 nm. The sensor had high selectivity with LOD 0.0324 μg/L (Li et al., 2020b).

In another study, rhodamine B coated silver nanoparticles (RB-AgNPs) were used to develop a highly sensitive dual-signal assay for carbaryl (Luo et al., 2017). AChE hydrolyses acetylthiocholine to generate TCh which causes clumping of AgNPs, thus changing the color of the solution from pale yellow to gray and unquenching of RB. The sensor devised was highly sensitive with detection range extending from 0.1 to 8.0 ng/L with very low LOD of 0.0023 ng/L. The study proposed that the technique was suitable for carbamate analysis of complex samples (Luo et al., 2017). In 2016, a carbon dots (CDs)-based detector for inhibitors of AChE by using IFE of silver nanoparticles. Carbaryl-spiked apple juice samples and tacrine, a well-known AChE inhibitor was used in this study. CDs generally produce fluorescence, but due to overlap of the absorption spectrum of AgNPs with excitation and emission spectra of CDs, IFE occurs, and fluorescence is turned off. The AChE in the solution hydrolyses ATCh producing thiocholine which aggregates AgNPs thus changing yellow-colored solution to gray. When carbaryl is present, it inhibits AChE activity causing IFE providing a yellow colored solution with no fluorescence as is depicted by Fig. 16.4. The sensor was highly sensitive with a low detection limit of 0.021 mU/mL by colorimetric measurement and 0.016 mU/mL fluorometric measurement (Zhao et al., 2016).

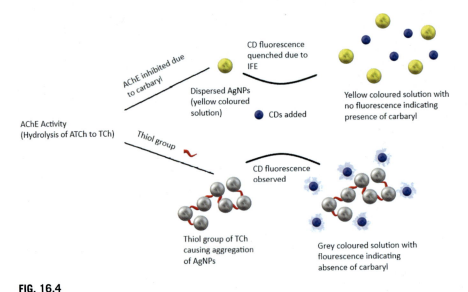

FIG. 16.4

Schematic representation of the working of CD-assisted dual readout sensor for carbaryl.

Another study described a colorimetric assay based on enzyme triggered deposition of AgNPs over AuNPs which detects methamidophos and malathion. The enzyme alkaline phosphatase (ALP) dephosphorylates p-aminophenyl phosphate (p-APP) to form p-aminophenol (p-AP) which reduces silver ions, thus depositing AgNPs over AuNPs forming core-shell Au@AgNPs nanoparticles (Lv et al., 2016). As explained in Fig. 16.5, under the uninhibited condition, ALP activity causes a gradual change of color of the solution from red to yellow and then to gray as the thickness of Ag shell increases while producing little to no change in color in the presence of OPPs as it inhibits the enzyme. The sensor showed fair sensitivity with LOD of 0.025 µg/L for methamidophos and 0.036 µg/L for malathion (Lv et al., 2016). Few researchers developed an enzyme-based double signal assay of parathion methyl (PM) by using reversible quenching of graphene quantum dots (GQDs) with AgNPs (Li et al., 2020a). In this approach, AChE hydrolyses ATCh to produce TCh which displaces AgNPs from GQD/AgNP platform thus inducing fluorescence which would otherwise be quenched due to inner filtration effect. When organophosphate pesticides are present, the enzyme activity is inhibited leading to diminished fluorescence, and a linear increase in absorbance of GQD/AgNP at 390-nm wavelength was recorded in accordance with parathion methyl (PM) concentration. (Li et al., 2020a). The sensor had fair selectivity with a low LOD of 0.017 µg/L (Li et al., 2020a).

16.2.3 Surface-enhanced Raman spectroscopy based sensing

SERS is a highly sensitive analytical technique which uses roughened noble metal surfaces or nanostructures to improve the Raman signals of the analyte. Since every type of molecule yielded sharp Raman spectral profiles, the technique could

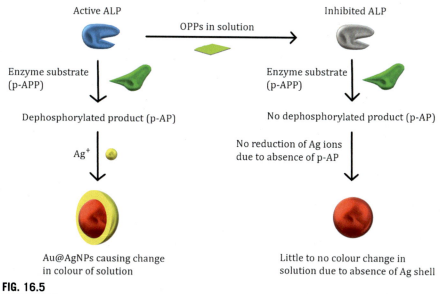

FIG. 16.5

Schematic representation of working of Au@AgNPs based colorimetric sensor for organophosphate pesticides.

potentially be used for a range of analytes including pesticides (Pang et al., 2016) and biomolecules (Zheng et al., 2018b).

In 2014, a few researchers devised a novel AgNP decorated filter paper which acted as a substrate for dynamic SERS (D-SERS) to detect paraoxon and thiram residues on fruit peels (Zhu et al., 2014). D-SERS allows collection of Raman signals during the transition of a substrate from wet to dry state (Qian et al., 2013). AgNPs were decorated over filter paper using the silver mirror reaction method. The use of filter as a substrate provided with the advantage of a better collection of trace quantity of analytes present on rough surfaces of fruits and other samples. Furthermore, the swabbing action did not change the size distribution of AgNPs, thus not interfering with the sensing and identification of pesticides (Zhu et al., 2014). The facile filter paper-based substrate is cost-effective and portable compared with conventional substrates. The test was fairly sensitive with LOD of 7.2 ng/cm^2 and 0.23 μg/cm^2 for thiram and paraoxon, respectively (Zhu et al., 2014). A year later, another study proposed development of a column substrate for SERS using AgNPs functionalized with polymethacrylate monoliths for sensing of trace quantities of phosmet on oranges, tea leaves, and apples. In this study, the researchers synthesized (glycidyl methacrylate)-co-(ethylene dimethacrylate) in a silica capillary and immobilized AgNPs onto the cross-section of this column. The substrate showed increase in Raman signal when the volume of Ag colloid was in the 50–100 μL range, beyond which it showed a substantial decrease in signal due to aggregation of AgNPs at the top of the column (Pan et al., 2015). The test showed good reproducibility with a low detection limit of 0.34 μg/L

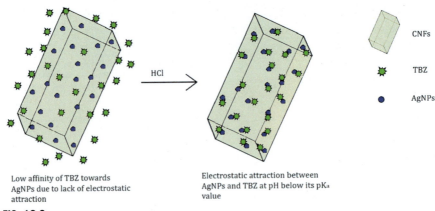

FIG. 16.6

Schematic representation of electrostatic attraction developed between TBZ and AgNPs due to pH change.

and was proposed to serve as an alternative protocol to qualify monitoring requirement of trace phosmet in fruits in China (Pan et al., 2015). A study published in 2016 developed a flexible SERS substrate using cellulose nanofibers (CNFs) coated with AgNPs to detect thiabendazole (TBZ) in apples (Liou et al., 2017). TBZ is a neutral molecule with hydrophobic nature, and it has a low electrostatic attraction towards the negative charge of AgNPs. Hence, the secondary amine groups of the pesticide were protonated by decreasing the pH below their respective pK_a levels, thereby inducing electrostatic attraction between AgNPs and TBZ, as depicted in Fig. 16.6 (Liou et al., 2017). The clumping of AgNPs due to a low pH environment was prevented by their adherence to CNFs. The LOD for the system was 5 ppm and was proposed to be used for sensing of other neutral molecules found in food (Liou et al., 2017). Another study developed silver-coated gold nanoparticles (Au@AgNPs) as SERS substrate to concomitantly detect multiclass pesticides like thiacloprid, oxamyl, and profenofos in peach (Yaseen et al., 2019). The substrate exhibits hot spots through the clumping of Au@AgNPs on interaction with pesticides. There were characteristic peaks of all the pesticides in the sample, thus making this technique a potential candidate in environmental monitoring and food safety (Yaseen et al., 2019). A study proposed the fabrication of flexible SERS substrate using AgNPs grown on nonwoven (NW) fabric surface based on mussel-inspired polydopamine (PDA) molecules to detect residues of carbaryl on fruit surfaces (Zhang et al., 2019). NW@PDA@AgNPs fabric contained a multitude of hot spots for SERS sensing, thus drastically increasing the sensitivity of sensor while also improving collection efficiency from irregular surfaces owing to its flexibility. The sensor exhibited a very low detection limit and had versatility in sensing of carbaryl in various samples (Zhang et al., 2019).

In the same year, another study used AgNPs assembled on cotton swabs as a substrate for SERS for sensing of carbaryl on cucumber. The cotton swabs were treated with (3-Aminopropyl) trimethoxysilane (APTMS), and then after activation of fibers, it was soaked in a concentrated solution of AgNPs (Qu et al., 2016). This SERS substrate showed enhanced performance compared with commercial substrates due to presence of effective "hot spots" among the assembled AgNPs. The cotton swabs showed LOD of 5×10^{-7} M and were a cheap and robust alternative to conventional SERS substrates while also being able to detect pesticides and pollutants on irregular surfaces (Qu et al., 2016). A study published in 2018 used flexible SERS substrate based on polydimethylsiloxane (PDMS) with silver nanoparticles@nanowire (AgNP@AgNW) network embedded into it to detect thiram on Crucian carp scales and spathiphyllum leaves (Wei et al., 2018). The AgNP@AgNW films were developed by mixing well-dispersed AgNPs to the dispersed solution of AgNW further coating this mixed dispersion over Teflon mould via a spin-coating method. PDMS was used as a support structure which held the AgNP@AgNW structure. The substrate allowed very sensitive detection of thiram with very low LOD of 10^{-10} M, excellent reusability and fair reproducibility with a robust and flexible structure (Wei et al., 2018). Another study proposed detection of atrazine in water samples via SERS using AgNPs-modified CDs (CD@Ag). Rhodamine 6G (R6G) adsorbed on CD@Ag was used as Raman reporter (Tang et al., 2019a). On interaction with target CD@Ags aggregated, and thus amplifying the SERS signal produced by R6G. The sensor had a detection limit of 10 nM and served as a cost-effective solution in monitoring natural water samples (Tang et al., 2019a). In the same year, a study proposed use of nonplanar SERS substrate using AgNP-coated glass beads to detect chlorpyrifos and imidacloprid. The substrate was produced by simple silver mirror reaction on beads and SERS measurements could be conducted on these beads glued to a glass slide providing fast label-free quantitation of pesticides (Tang et al., 2019b). The sensor showed LOD of 50 and 10 ng/mL for imidacloprid and chlorpyrifos in apple extract and could potentially be used in food and environment monitoring (Tang et al., 2019b). A study proposed use of AgNP assembled microbowl array prepared by sphere monolayer templated electrodeposition technique as a SERS substrate for quick detection of methyl parathion and thiram on vegetables (Zhu et al., 2020). The substrate provided 3-D distributed hotspots for improved SERS activity. The LOD for methyl parathion and thiram was 1.26 and 1.20 ppb, respectively (Zhu et al., 2020). A few researchers developed an eco-friendly and flexible SERS substrate using Plasmonic silver nanoparticle bacterial nanocellulose paper (AgNP-BNCP) for sensing of methomyl on various fruit peels (Parnsubsakul et al., 2020). The AgNPs provided superior Raman enhancement compared with other metals, and it provided plasmonic 3D hotspots on highly porous BNCP. This "paste-and-read" approach toward SERS sensing provided rapid in-situ analysis as it can be directly used when the substrate is in proximity with the contaminated site (Parnsubsakul et al., 2020). Another study developed flower like AgNPs as SERS substrate for detection of 2,4-dichlorophenoxyacetic acid (2,4-D), acetamiprid and methomyl in tea (Hassan et al., 2021). The coarse surface of AgNPs interacted with imidothioate group, ring structure and chlorine atom of methomyl, acetamiprid and 2,4-D, respectively, to produce strong SERS signal. This method showed great versatility in the inspection of several food samples (Hassan et al., 2021).

16.2.4 Aptamer-based assays

Aptamers are short single-stranded oligonucleotides which offer several advantages over antibodies such as low molecular weight, high specificity, facile modification, and synthesis. With the advent of aptasensors which used aptamer as a recognition module, a range of targets could be detected at trace concentrations and that included pesticides (Liu et al., 2019b).

A few researchers developed an aptasensor with an aptamer that selectively bonded with acetamiprid to produce a colorimetric signal via salt-induced clumping of AgNPs. The unfolded aptamer (ssDNA) binds to AgNPs via nitrogen atoms, thus enhancing its stability against NaCl salt-induced aggregation (Jokar et al., 2016). However, when acetamiprid is present, the aptamer is unable to provide stability to AgNPs due to structural changes. The salt thus bridges the AgNPs together, causing aggregation and thus producing a color change from yellow to gray. The study showed outstanding selectivity with LOD of 0.02 ppm (Jokar et al., 2016). Another study proposed electrochemical detection of acetamiprid using AgNP decorated nitrogen-doped graphene (NG) nanocomposites (Jiang et al., 2015). Ag-NG nanocomposite was deposited over GCE, and then aptamer was added over it as recognition module. Electron transfer was blocked by the formation of acetamiprid-aptamer complex and thus showed a remarkable increase in impedance (Jiang et al., 2015). With 6-Mercapto-1-hexanol (MCH) blocking the nonspecific binding of aptamer/Ag-NG/GCE, the sensor had high selectivity with very low LOD of 3.3×10^{-14} M (Jiang et al., 2015). In 2018, a study proposed use of aptamer-based sensor for the detection of malathion. The sensor used a cationic hexapeptide KKKRRR which binds to malathion-specific aptamer via electrostatic interaction, thus preventing clumping of AgNPs (Bala et al., 2018). However, when malathion is present, the aptamer preferentially binds to it, thus leaving enough peptide to clump the AgNPs producing a color change from yellow to orange. The aptasensor had excellent selectivity while having a tremendously low LOD of 0.5 pM (Bala et al., 2018). Another group developed an electrochemiluminescence (ECL) aptasensor using AgNPs to detect profenofos in vegetables (Liu et al., 2019a). The sensor used platinum electrode (PE) with luminol-hydrogen peroxide system built on it. AgNPs catalyzed the excitation of luminol, increasing the ECL intensity while shortening the response time (Liu et al., 2019a). Presence of analyte led to a diminishing of ECL thus providing measurable response. The sensor was highly selective and had LOD of 0.13 ng/mL (Liu et al., 2019a). A group of researchers developed an aptasensor for chlorpyrifos using peroxidase-mimic tyrosine capped silver NanoZymes (Ag-NanoZyme) (Weerathunge et al., 2019). The Ag-NanoZyme oxidizes 3,3′,5,5′-tetramethylbenzidine (TMB) to produce a blue-colored product. This activity is inhibited when the Ag-NanoZymes are exposed to chlorpyrifos-specific aptamers. These aptamers bind with target pesticide on exposure, thus causing to the recovery of peroxidase-like activity and color of solution changes to blue (Weerathunge et al., 2019). The sensor had fair selectivity with LOD being 11.3 ppm (Weerathunge et al., 2019). A study proposed the use of label-free electrochemical aptasensor, which used reduced graphene with silver nanoparticles (rGO-AgNPs) and prussian blue-gold (PB-AuNPs) nanocomposites to determine presence of acetamiprid in vegetable samples (Shi et al., 2020). GCE was modified with this nanocomposite where rGO-AgNPs and PB-AuNPs enhanced sensitivity of aptasensor by increasing surface area and amplifying the current signal. The sensor exhibited reasonable recovery rates in different samples while also being highly selective with very low LOD of 0.30 pM (Shi et al., 2020). Another group of researchers developed AgNPs/histidine-functionalized graphene quantum dots/graphene hybrid (Ag/His-GQD/G) for electrochemical sensing of acetamiprid in tea samples (Dan et al., 2020). Chitosan was used to immobilize Ag/His-GQD/G on GCE while MCH blocked nonspecific binding. Acetamiprid-specific aptamer binds to form a complex on the electrode surface, thus dampening electron transfer. The sensor exhibited high specificity with LOD of 4.0×10^{-17} M (Dan et al., 2020). A study used aptamer specific to malathion for developing an SERS sensor. AgNPs were modified with the cation spermine which binds to aptamer (Nie et al., 2018). Aptamer-malathion complex gets adsorbed onto AgNPs thus producing characteristic peak in the presence of NaCl as an aggregate reagent. This sensor was selective and can be improved by developing it to detect other pesticides (Nie et al., 2018).

Now that we have learnt about the various ways to detect pesticides as summarized in Table 16.1, the following section will elaborate on the approaches developed to mineralize the pesticides which can convert them to biologically inactive form.

16.3 Mineralization of pesticides

Mineralization is the formation of pesticide salts with noble metals. As shown in Figs. 16.7 and 16.8, the AgNPs are introduced in the sample contaminated with pesticides. This can be accomplished either by directly suspending these nanoparticles in water or by immobilizing them on a support and then dipping the support in the sample. After keeping the solution for a certain amount of incubation time, the salts are precipitated down, and the sample is now free of the active form of pesticide and is thus made safe for consumption. This section will give an outline of all the approaches devised for the mineralization of pesticides.

A study published in 2007 used AuNPs and AgNPs to adsorb and quantitatively remove malathion and chlorpyrifos which are commonly found in water bodies of developing countries (Nair and Pradeep, 2007). They used two approaches for adsorption, one being solution-based support free approach and the other being alumina used as a support for the nanoparticles. Both of the methods were able to remove pesticides quite efficiently; however, the alumina supported nanoparticles were able to altogether remove all the pesticides (Nair and Pradeep, 2007). A study published in 2014 used of AgNPs immobilized on cellulose acetate membrane (CAM) to mineralize the pesticides in water (Manimegalai et al., 2014). The model pesticides malathion and chlorpyrifos underwent mineralization and led to the formation of silver metal salts once exposed to AgNPs. In both cases, mineralization time was inversely proportional to the concentration of AgNPs (Manimegalai et al., 2014). This study claimed that the method of development of AgNPs is economical and that

Table 16.1 Comparative data on different sensing approaches for pesticides.

Materials used	Type of response	Linearity	LOD	Analyte	References
pSC$_n$-AgNPs	Colorimetric	5×10^{-6} to 10^{-3} M	10^{-7} M	Optunal	Xiong and Li (2008)
Plasmonic AgNPs	Colorimetric	—	9 ppm (ethion) 11 ppm (fenthion) 18 ppm (malathion) 44 ppm (parathion)	Ethion, fenthion, malathion, and parathion	Dissanayake et al. (2019)
HA-capped AgNPs	Colorimetric	100–500 ppm	—	Sulfurazon-ethyl	Dubas and Pimpan (2008)
SAADTC-AgNPs	Colorimetric	10–100 μM	1.8×10^{-7} M	Tricyclazole	Rohit and Kailasa (2014a)
CN-DTC-AgNPs	Colorimetric	10–20 μM (thiram) 50–250 μM (paraquat)	2.81×10^{-6} M (thiram) 7.21×10^{-6} M (paraquat)	Thiram and paraquat	Rohit and Kailasa (2014b)
ABT-AgNPs	Colorimetric	10–100 μM	1.04 μM	Carbendazim	Patel et al. (2015)
Bare AgNPs	Colorimetric	20–600 ng/mL	7 ng/mL	Diazinon	Shrivas et al. (2019)
rGO-PEI-AgNPs	Electrochemical	1 mM–1 nM	1 nm	Trifluralin	Jafari et al. (2019)
Chi-AgNPs	Fluorometric	1×10^{-9}–10×10^{-12} M	94 fM/L	Malathion	Vasimalai and John (2013)
Citrate-capped AgNPs	Colorimetric	0.25–37.5 ng mL^{-1}	0.18 ng/mL	Dipterex	Li et al. (2014)
Citrate-capped AgNPs	Colorimetric	0.1–280 μM	5 nM	Triazophos	Ma et al. (2018)
L-cys-AgNPs	Colorimetric	20 100 ppm	—	Cypermethrin	Kodir et al. (2016)
Clove extract capped AgNPs	Colorimetric		21 nM	Vinclozolin	Hussain et al. (2019)
Citrate-capped AgNPs	Colorimetric	0.01–0.4 μg mL^{-1}	1.7 ng/mL	Prothioconazole	Ivrigh et al. (2017)
AgNPs-CGR-NF/GCE	Electrochemical	1.0×10^{-13} to 1×10^{-8} M (chlorpyrifos) 1.0×10^{-12} to 1×10^{-8} M (carbaryl)	5.3×10^{-14} M (chlorpyrifos) 5.45×10^{-13} M (carbaryl)	Chlorpyrifos and carbaryl	Liu et al. (2014)
AgNPs decorated with tetrathiacalix[4]arene	Electrochemical	0.2 nM–0.2 μM (paraoxon) 0.4 nM–0.2 μM (malaoxon) 0.2 nM–2.0 μM (carbofuran), and 0.01–0.20 μM (aldicarb)	0.05 nM (paraoxon), 0.1 nM (malaoxon), 0.1 nM (carbofuran), and 10 nM (aldicarb)	Malathion, paraoxon, malaoxon, aldicarb, and carbofuran	Evtugyn et al. (2014)
Ag@rGO-NH$_2$	Electrochemical	0.0063–0.077 mg/mL (malathion) 0.012–0.105 mg/mL (methidathion) 0.021–0.122 mg/mL (chlorpyrifos ethyl)	4.5 ng/mL (malathion), 9.5 ng/mL (methidathion), and 14 ng/mL (chlorpyrifos ethyl)	Malathion, methidathion, and chlorpyrifos ethyl	Guler et al. (2017)
AChE based immobilization of AgNPs on glass	Colorimetric	1–100 nM (malathion) 5–100 nM (trichlorfon)	0.455 nM (malathion) and 5.46 nM (trichlorfon)	Malathion and trichlorfon	Kumar et al. (2016)

Material	Method	Linear range	LOD	Analyte	References
g-C$_3$N$_4$/AgNPs	Fluorometric and UV-vis	0.1–1.0 µg/L	0.056 µg/L	Parathion-methyl	Li et al. (2020b)
RB-AgNPs	Colorimetric and fluorometric	0.1–8.0 ng/L	0.0023 ng/L	Carbaryl	Luo et al. (2017)
CDs + AgNPs	Colorimetric and fluorometric	0.025–2 mU/mL (colorimetric) 0.025–2 mU/mL (fluorometric)	0.021 mU/mL (colorimetric) and 0.016 mU/mL (fluorometric)	Carbaryl	Zhao et al. (2016)
CD capped AgNPs	Colorimetric	0.1–100 µM	0.04 µM	Phoxim	Zheng et al. (2018a)
AgNPs/oxMWCNTs	Fluorometric	0.01–0.35 µg mL^{-1}	0.003 µg mL^{-1}	Dimethoate	Hsu et al. (2017)
Au@AgNPs	Colorimetric	0.05–500 µg L^{-1} (methamidophos) and 0.1–500 µg L^{-1} (malathion)	0.025 µg L^{-1} (methamidophos) and 0.036 µg L^{-1} (malathion)	Methamidophos and malathion	Lv et al. (2016)
GQDs/AgNPs	Fluorometric and UV-vis	0.1–6 µg/L	0.017 µg/L	Parathion methyl	Li et al. (2020a)
PS-b-P2VP-AgNPs	Colorimetric	0.036–0.36 µg/L	0.06 µg/L	Cartap	Rahim et al. (2018)
DDTC-AgNPs	Colorimetric	5×10^{-5} to 3×10^{-4} M	21.1×10^{-6} M	Mancozeb	Rohit et al. (2014)
MNBZ-AgNPs	Colorimetric	1.0×10^{-7} to 1.2×10^{-6} M	17.1 nM	Glyphosate	Rawat et al. (2016)
SDS-AgNPs	Colorimetric	195.7–733.9 ng/mL (ziram) 17.6–66.2 ng/mL (zineb) 16.9–63.6 ng/mL (maneb)	149.3 ng/mL (ziram), 4.0 ng/mL (zineb), and 9.1 ng/mL (maneb)	Ziram, Zineb, and Maneb	Ghoto et al. (2019)
AgNPs with aptamer and NaCl	Colorimetric	1–50 ppm	0.02 ppm	Acetamiprid	Jokar et al. (2016)
Aptamer/Ag-NG/GCE	Electrochemical	1×10^{-13} M to 5×10^{-9} M	3.3×10^{-14} M	Acetamiprid	Jiang et al. (2015)
Aptamer and AgNPs	Colorimetric	0.01–0.75 nM	0.5 pM	Malathion	Bala et al. (2018)
AgNPs with aptamer and luminol	Electro-chemiluminescent	0.5–100 µg/mL	0.13 ng/mL	Profenofos	Liu et al. (2019a)
Ag-NanoZyme	Colorimetric	35–210 ppm	11.3 ppm	Chlorpyrifos	Weerathunge et al. (2019)
PB-AuNPs/rGO-AgNPs/GCE with aptamer	Electrochemical	1 pM to 1 µM	0.30 pM	Acetamiprid	Shi et al. (2020)
Ag/His-GQD/G/GCE with aptamer	Electrochemical	1.0×10^{-16} to 5.0×10^{-12} M	4.0×10^{-17} M	Acetamiprid	Dan et al. (2020)

ABT, 4-aminobenzenethiol; AChE, acetylcholinesterase; CDs, carbon dots; CGR, carboxylic graphene; Chi, chitosan; CN-DTC, cyclen dithiocarbamate; DDTC, dopamine dithiocarbamate; G, graphene hybrid; g-C$_3$N$_4$, graphitic carbon nitride; GCE, glassy carbon electrode; GQDs, graphene quantum dots; HA, humic acid; His, histidine; L-cys, L-cysteine; MNBZ, 2-mercapto-5-nitrobenzimidazole; NF, Nafion; NG, nitrogen-doped graphene; oxMWCNTs, oxidized multiwalled carbon nanotubes; PEI, polyethylene imine; PS-b-P2VP, polystyrene-block-poly(2-vinylpyridine); pSCn, p-sulfonatocalix[n]arene; RB, rhodamine B; rGO, reduced graphene oxide; SAADTC, 5-sulfo anthranilic acid dithiocarbamate; SDS, sodium dodecyl sulphate.

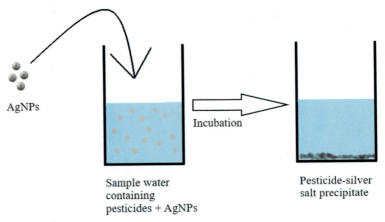

FIG. 16.7

A general description of the mineralization process.

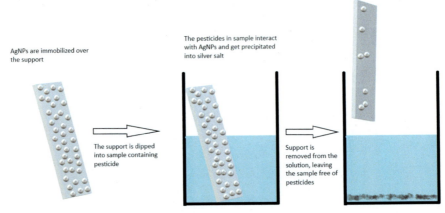

FIG. 16.8

Schematic representation of support-based mineralization process.

this technique is suitable for remediation of contaminated water bodies in rural areas (Manimegalai et al., 2014).

Another study proposed use of AgNPs incorporated on polyurethane foam (PUF) to mineralize pesticides present in water. They incorporated AgNPs on PUF by soaking it in the AgNPs solution overnight (Manimegalai et al., 2012). Malathion and chlorpyrifos were used here as representatives for commonly used pesticides. The complete mineralization time was about 100 min for 1 ppm solutions for both the pesticides. The study stated that pesticide type and concentration does not influence the mineralization time and that it could be used in rural areas where contamination by pesticides is prevalent (Manimegalai et al., 2012). In 2015, Hydrophilic fungal secretes of *Penicillium pinophilum* were used to synthesize mycogenic AgNPs, which were further used to degrade chlorpyrifos (Deka and Sinha, 2015). The mineralization of pesticides was studied under three different pH conditions and pH 6 was optimal. The study proposed use of this fungus as a source of hydrophilic secretes, which can provide reproducible batches of AgNPs (Deka and Sinha, 2015). Another approach was proposed in the same year where reduced graphene oxide-silver nanocomposites (RGO@Ag) was used to remove halocarbon pesticide, lindane. Here lindane was converted to different isomers of trichlorobenzenes (TCB) on the surface of RGO@Ag. The TCBs adsorb onto the composite and thus get easily removed from the contaminated water samples (Sen Gupta et al., 2015). The reaction was unique to RGO@Ag and was not shown by RGO or Ag on their own. The study shows that the composite could be reused by removing the bound TCBs using hexane treatment and further reloading of AgNPs onto regenerated RGO (Sen Gupta et al., 2015). In 2017, a study used green AgNPs to intensify the degradation of fipronil in soil and water. This study used leaf extracts of *Camellia siensis*, *Plantago major*, *Ipomoea carnea*, and *Brassica alba* to synthesize AgNPs (Romeh, 2018). The plant extracts acted as stabilizing and reducing agents in AgNP synthesis. The breakdown of fipronil varied according to the AgNP sources with *Brassica*-AgNPs reducing fipronil residues in water by 95.45% (Romeh, 2018). Other AgNPs also showed good activity and the degradation products fipronil desulfenyl and fipronil amide were found in water after being treated (Romeh, 2018).

Apart from these approaches, adsorption of pesticides onto high surface area nanostructures have also been studied on extensively and should also be considered as an efficacious way to remediate polluted sites.

16.4 Conclusion

This chapter outlines the diverse approaches devised to detect different pesticides on many samples; however, a lot literature revolves mainly around the commonly used pesticides, and not many papers mention the other less frequently used pesticides. It is worth mentioning that the mineralization approaches primarily are developed for contaminated water and not soil samples. Further strategies should be streamlined more on tackling the pesticide contamination in the soil to avoid further leaching into groundwater and other water bodies. Development of simultaneous detection and mineralization kits should also be considered as an avenue of research as it could help with the automation of pesticide monitoring and removal process.

References

Aragay, G., Pino, F., Merkoçi, A., 2012. Nanomaterials for sensing and destroying pesticides. Chem. Rev. 112, 5317–5338.

Bala, R., Mittal, S., Sharma, R.K., Wangoo, N., 2018. A supersensitive silver nanoprobe based aptasensor for low cost detection of malathion residues in water and food samples. Spectrochim. Acta A Mol. Biomol. Spectrosc. 196, 268–273.

Chawla, P., Kaushik, R., Shiva Swaraj, V.J., Kumar, N., 2018. Organophosphorus pesticides residues in food and their colorimetric detection. Environ. Nanotechnol. Monit. Manag. 10, 292–307.

Che Sulaiman, I.S., Chieng, B.W., Osman, M.J., Ong, K.K., Rashid, J.I.A., Wan Yunus, W.M.Z., Noor, S.A.M., Kasim, N.A.M., Halim, N.A., Mohamad, A., 2020. A review on colorimetric methods for determination of organophosphate pesticides using gold and silver nanoparticles. Microchim. Acta 187, 131.

Costa, L.G., 2008. Neurotoxicity of pesticides: a brief review. Front. Biosci. 13, 1240.

Dan, X., Ruiyi, L., Zaijun, L., Haiyan, Z., Zhiguo, G., Guangli, W., 2020. Facile strategy for synthesis of silver-graphene hybrid with controllable size and excellent dispersion for ultrasensitive electrochemical detection of acetamiprid. Appl. Surf. Sci. 512, 145628.

Deka, A.C., Sinha, S.K., 2015. Mycogenic silver nanoparticle biosynthesis and its pesticide degradation potentials. Int. J. Technol. Enhanc. Emerg. Eng. Res. 3.

Dissanayake, N.M., Arachchilage, J.S., Samuels, T.A., Obare, S.O., 2019. Highly sensitive plasmonic metal nanoparticle-based sensors for the detection of organophosphorus pesticides. Talanta 200, 218–227.

Dubas, S.T., Pimpan, V., 2008. Humic acid assisted synthesis of silver nanoparticles and its application to herbicide detection. Mater. Lett. 62, 2661–2663.

EPA, n.d. Pesticides. http://www.epa.gov/pesticides.

Evtugyn, G.A., Shamagsumova, R.V., Padnya, P.V., Stoikov, I.I., Antipin, I.S., 2014. Cholinesterase sensor based on glassy carbon electrode modified with Ag nanoparticles decorated with macrocyclic ligands. Talanta 127, 9–17.

Franci, G., Falanga, A., Galdiero, S., Palomba, L., Rai, M., Morelli, G., Galdiero, M., 2015. Silver nanoparticles as potential antibacterial agents. Molecules 20, 8856–8874.

Ghoto, S.A., Khuhawar, M.Y., Jahangir, T.M., 2019. Silver nanoparticles with sodium dodecyl sulfate as a colorimetric probe for the detection of dithiocarbamate pesticides in environmental samples. Anal. Sci. 35, 631–637.

Guler, M., Turkoglu, V., Basi, Z., 2017. Determination of malation, methidathion, and chlorpyrifos ethyl pesticides using acetylcholinesterase biosensor based on Nafion/Ag@rGO-NH$_2$ nanocomposites. Electrochim. Acta 240, 129–135.

Hassan, M.M., Zareef, M., Jiao, T., Liu, S., Xu, Y., Viswadevarayalu, A., Li, H., Chen, Q., 2021. Signal optimized rough silver nanoparticle for rapid SERS sensing of pesticide residues in tea. Food Chem. 338, 127796.

Hsu, C.-W., Lin, Z.-Y., Chan, T.-Y., Chiu, T.-C., Hu, C.-C., 2017. Oxidized multiwalled carbon nanotubes decorated with silver nanoparticles for fluorometric detection of dimethoate. Food Chem. 224, 353–358.

Hussain, M., Nafady, A., Sirajuddin, Avcı, A., Pehlivan, E., Nisar, J., Sherazi, S., Balouch, A., Shah, M., Almaghrabi, O., Ul-Haq, M., 2019. Biogenic silver nanoparticles for trace colorimetric sensing of enzyme disrupter fungicide vinclozolin. Nanomaterials 9, 1604.

Ivrigh, Z.J.-N., Fahimi-Kashani, N., Hormozi-Nezhad, M.R., 2017. Aggregation-based colorimetric sensor for determination of prothioconazole fungicide using colloidal silver nanoparticles (AgNPs). Spectrochim. Acta A. Mol. Biomol. Spectrosc. 187, 143–148.

Jafari, M., Hasanzadeh, M., Karimian, R., Shadjou, N., 2019. Sensitive detection of Trifluralin in untreated human plasma samples using reduced graphene oxide modified by polyethylene imine and silver nanoparticles: a new platform on the analysis of pesticides and chemical injuries. Microchem. J. 147, 741–748.

Jiang, D., Du, X., Liu, Q., Zhou, L., Dai, L., Qian, J., Wang, K., 2015. Silver nanoparticles anchored on nitrogen-doped graphene as a novel electrochemical biosensing platform with enhanced sensitivity for aptamer-based pesticide assay. Analyst 140, 6404–6411.

Jokar, M., Safaralizadeh, M.H., Hadizadeh, F., Rahmani, F., Kalani, M.R., 2016. Design and evaluation of an apta-nano-sensor to detect acetamiprid in vitro and in silico. J. Biomol. Struct. Dyn. 34, 2505–2517.

Kodir, A., Imawan, C., Permana, I.S., Handayani, W., 2016. Pesticide colorimetric sensor based on silver nanoparticles modified by L-cysteine. In: 2016 International Seminar on Sensors, Instrumentation, Measurement and Metrology (ISSIMM). Presented at the 2016 International Seminar on Sensors, Instrumentation, Measurement and Metrology (ISSIMM), IEEE, Malang, Indonesia, pp. 43–47.

Konrad, M.P., Doherty, A.P., Bell, S.E.J., 2013. Stable and uniform SERS signals from self-assembled two-dimensional interfacial arrays of optically coupled Ag nanoparticles. Anal. Chem. 85, 6783–6789.

Kumar, D.N., Alex, S.A., Chandrasekaran, N., Mukherjee, A., 2016. Acetylcholinesterase (AChE)-mediated immobilization of silver nanoparticles for the detection of organophosphorus pesticides. RSC Adv. 6, 64769–64777.

Li, Y., Chen, S., Lin, D., Chen, Z., Qiu, P., 2020a. A dual-mode nanoprobe for the determination of parathion methyl based on graphene quantum dots modified silver nanoparticles. Anal. Bioanal. Chem. 412, 5583–5591.

Li, Y., Wan, M., Yan, G., Qiu, P., Wang, X., 2020b. A dual-signal sensor for the analysis of parathion-methyl using silver nanoparticles modified with graphitic carbon nitride. J. Pharm. Anal. S2095177919311517.

Li, Z., Wang, Y., Ni, Y., Kokot, S., 2014. Unmodified silver nanoparticles for rapid analysis of the organophosphorus pesticide, dipterex, often found in different waters. Sens. Actuators B Chem. 193, 205–211.

Liou, P., Nayigiziki, F.X., Kong, F., Mustapha, A., Lin, M., 2017. Cellulose nanofibers coated with silver nanoparticles as a SERS platform for detection of pesticides in apples. Carbohydr. Polym. 157, 643–650.

Liu, H., Cheng, S., Shi, X., Zhang, H., Zhao, Q., Dong, H., Guo, Y., Sun, X., 2019a. Electrochemiluminescence aptasensor for profenofos detection based on silver nanoparticles enhanced luminol luminescence system. J. Electrochem. Soc. 166, B1562–B1566.

Liu, M., Khan, A., Wang, Z., Liu, Y., Yang, G., Deng, Y., He, N., 2019b. Aptasensors for pesticide detection. Biosens. Bioelectron. 130, 174–184.

Liu, Y., Wang, G., Li, C., Zhou, Q., Wang, M., Yang, L., 2014. A novel acetylcholinesterase biosensor based on carboxylic graphene coated with silver nanoparticles for pesticide detection. Mater. Sci. Eng. C 35, 253–258.

Luo, Q.-J., Li, Y.-X., Zhang, M.-Q., Qiu, P., Deng, Y.-H., 2017. A highly sensitive, dual-signal assay based on rhodamine B covered silver nanoparticles for carbamate pesticides. Chin. Chem. Lett. 28, 345–349.

Lv, B., Wei, M., Liu, Y., Liu, X., Wei, W., Liu, S., 2016. Ultrasensitive photometric and visual determination of organophosphorus pesticides based on the inhibition of enzyme-triggered formation of core-shell gold-silver nanoparticles. Microchim. Acta 183, 2941–2948.

Ma, S., He, J., Guo, M., Sun, X., Zheng, M., Wang, Y., 2018. Ultrasensitive colorimetric detection of triazophos based on the aggregation of silver nanoparticles. Colloids Surf. Physicochem. Eng. Asp. 538, 343–349.

Manimegalai, G., Shanthakumar, S., Sharma, C., 2014. Silver nanoparticles: synthesis and application in mineralization of pesticides using membrane support. Int. Nano Lett. 4, 105.

Manimegalai, G., Shanthakumar, S., Sharma, C., 2012. Pesticide mineralization in water using silver nanoparticles incorporated on polyurethane foam. Int. J. Sci. Res. IJSR 1, 91–94.

Mohanty, M.K., Kumar, V., Bastia, B.K., Arun, M., 2004. An analysis of poisoning deaths in Manipal. India. Vet. Hum. Toxicol. 46, 208–209.

Mukhopadhyay, S.S., 2014. Nanotechnology in agriculture: prospects and constraints. Nanotechnol. Sci. Appl. 63.

Nagami, H., Nishigaki, Y., Matsushima, S., Matsushita, T., Asanuma, S., Yajima, N., Usuda, M., Hirosawa, M., 2005. Hospital-based survey of pesticide poisoning in Japan, 1998–2002. Int. J. Occup. Environ. Health 11, 180–184.

Nair, A.S., Pradeep, T., 2007. Extraction of chlorpyrifos and malathion from water by metal nanoparticles. J. Nanosci. Nanotechnol. 7, 1871–1877.

Natsuki, J., 2015. A review of silver nanoparticles: synthesis methods, properties and applications. Int. J. Mater. Sci. Appl. 4, 325.

Nicolopoulou-Stamati, P., Maipas, S., Kotampasi, C., Stamatis, P., Hens, L., 2016. Chemical pesticides and human health: the urgent need for a new concept in agriculture. Front. Publ. Health, 4.

Nie, Y., Teng, Y., Li, P., Liu, W., Shi, Q., Zhang, Y., 2018. Label-free aptamer-based sensor for specific detection of malathion residues by surface-enhanced Raman scattering. Spectrochim. Acta. A Mol. Biomol. Spectrosc. 191, 271–276.

Orouji, A., Abbasi-Moayed, S., Hormozi-Nezhad, M.R., 2019. Development of a pH assisted AgNP-based colorimetric sensor array for simultaneous identification of phosalone and azinphosmethyl pesticides. Spectrochim. Acta A Mol. Biomol. Spectrosc. 219, 496–503.

Pan, Y., Guo, X., Zhu, J., Wang, X., Zhang, H., Kang, Y., Wu, T., Du, Y., 2015. A new SERS substrate based on silver nanoparticle functionalized polymethacrylate monoliths in a capillary, and it application to the trace determination of pesticides. Microchim. Acta 182, 1775–1782.

Pang, S., Yang, T., He, L., 2016. Review of surface enhanced Raman spectroscopic (SERS) detection of synthetic chemical pesticides. TrAC Trends Anal. Chem. 85, 73–82.

Parnsubsakul, A., Ngoensawat, U., Wutikhun, T., Sukmanee, T., Sapcharoenkun, C., Pienpinijtham, P., Ekgasit, S., 2020. Silver nanoparticle/bacterial nanocellulose paper composites for paste-and-read SERS detection of pesticides on fruit surfaces. Carbohydr. Polym. 235, 115956.

Patel, G.M., Rohit, J.V., Singhal, R.K., Kailasa, S.K., 2015. Recognition of carbendazim fungicide in environmental samples by using 4-aminobenzenethiol functionalized silver nanoparticles as a colorimetric sensor. Sens. Actuators B Chem. 206, 684–691.

Qian, K., Yang, L., Li, Z., Liu, J., 2013. A new-type dynamic SERS method for ultrasensitive detection. J. Raman Spectrosc. 44, 21–28.

Qu, L.-L., Geng, Y.-Y., Bao, Z.-N., Riaz, S., Li, H., 2016. Silver nanoparticles on cotton swabs for improved surface-enhanced Raman scattering, and its application to the detection of carbaryl. Microchim. Acta 183, 1307–1313.

Rahim, S., Khalid, S., Bhanger, M.I., Shah, M.R., Malik, M.I., 2018. Polystyrene-block-poly(2-vinylpyridine)-conjugated silver nanoparticles as colorimetric sensor for quantitative determination of Cartap in aqueous media and blood plasma. Sens. Actuators B Chem. 259, 878–887.

Rawat, K.A., Majithiya, R.P., Rohit, J.V., Basu, H., Singhal, R.K., Kailasa, S.K., 2016. Mg^{2+} ion as a tuner for colorimetric sensing of glyphosate with improved sensitivity via the aggregation of 2-mercapto-5-nitrobenzimidazole capped silver nanoparticles. RSC Adv. 6, 47741–47752.

Rawtani, D., Khatri, N., Tyagi, S., Pandey, G., 2018. Nanotechnology-based recent approaches for sensing and remediation of pesticides. J. Environ. Manage. 206, 749–762.

Rohit, J.V., Kailasa, S.K., 2014a. 5-Sulfo anthranilic acid dithiocarbamate functionalized silver nanoparticles as a colorimetric probe for the simple and selective detection of tricyclazole fungicide in rice samples. Anal. Methods 6, 5934–5941.

Rohit, J.V., Kailasa, S.K., 2014b. Cyclen dithiocarbamate-functionalized silver nanoparticles as a probe for colorimetric sensing of thiram and paraquat pesticides via host-guest chemistry. J. Nanopart. Res. 16, 2585.

Rohit, J.V., Solanki, J.N., Kailasa, S.K., 2014. Surface modification of silver nanoparticles with dopamine dithiocarbamate for selective colorimetric sensing of mancozeb in environmental samples. Sens. Actuators B Chem. 200, 219–226.

Romeh, A.A.A., 2018. Green silver nanoparticles for enhancing the phytoremediation of soil and water contaminated by fipronil and degradation products. Water Air Soil Pollut. 229, 147.

Sen Gupta, S., Chakraborty, I., Maliyekkal, S.M., Adit Mark, T., Pandey, D.K., Das, S.K., Pradeep, T., 2015. Simultaneous dehalogenation and removal of persistent halocarbon pesticides from water using graphene nanocomposites: a case study of lindane. ACS Sustain. Chem. Eng. 3, 1155–1163.

Shi, X., Sun, J., Yao, Y., Liu, H., Huang, J., Guo, Y., Sun, X., 2020. Novel electrochemical aptasensor with dual signal amplification strategy for detection of acetamiprid. Sci. Total Environ. 705, 135905.

Shrivas, K., Sahu, S., Sahu, B., Kurrey, R., Patle, T.K., Kant, T., Karbhal, I., Satnami, M.L., Deb, M.K., Ghosh, K.K., 2019. Silver nanoparticles for selective detection of phosphorus pesticide containing π-conjugated pyrimidine nitrogen and sulfur moieties through non-covalent interactions. J. Mol. Liq. 275, 297–303.

Srivastava, R., 2012. Synthesis and characterization techniques of nanomaterials. Int. J. Green Nanotechnol. 4, 17–27.

Štěpánková, Š., Vorčáková, K., 2016. Cholinesterase-based biosensors. J. Enzyme Inhib. Med. Chem. 31, 180–193.

Stiufiuc, R., Iacovita, C., Lucaciu, C.M., Stiufiuc, G., Dutu, A.G., Braescu, C., Leopold, N., 2013. SERS-active silver colloids prepared by reduction of silver nitrate with short-chain polyethylene glycol. Nanoscale Res. Lett. 8, 47.

Tang, J., Chen, W., Ju, H., 2019a. Sensitive surface-enhanced Raman scattering detection of atrazine based on aggregation of silver nanoparticles modified carbon dots. Talanta 201, 46–51.

Tang, J., Chen, W., Ju, H., 2019b. Rapid detection of pesticide residues using a silver nanoparticles coated glass bead as nonplanar substrate for SERS sensing. Sens. Actuators B Chem. 287, 576–583.

Unser, S., Bruzas, I., He, J., Sagle, L., 2015. Localized surface plasmon resonance biosensing: current challenges and approaches. Sensors 15, 15684–15716.

Vasimalai, N., John, S.A., 2013. Biopolymer capped silver nanoparticles as fluorophore for ultrasensitive and selective determination of malathion. Talanta 115, 24–31.

Weerathunge, P., Behera, B.K., Zihara, S., Singh, M., Prasad, S.N., Hashmi, S., Mariathomas, P.R.D., Bansal, V., Ramanathan, R., 2019. Dynamic interactions between peroxidase-mimic silver NanoZymes and chlorpyrifos-specific aptamers enable highly-specific pesticide sensing in river water. Anal. Chim. Acta 1083, 157–165.

Wei, W., Du, Y., Zhang, L., Yang, Y., Gao, Y., 2018. Improving SERS hot spots for on-site pesticide detection by combining silver nanoparticles with nanowires. J. Mater. Chem. C 6, 8793–8803.

Xiong, D., Li, H., 2008. Colorimetric detection of pesticides based on calixarene modified silver nanoparticles in water. Nanotechnology 19, 465502.

Yaseen, T., Pu, H., Sun, D.-W., 2019. Fabrication of silver-coated gold nanoparticles to simultaneously detect multi-class insecticide residues in peach with SERS technique. Talanta 196, 537–545.

Zahran, M., Khalifa, Z., Zahran, M.A.-H., Abdel Azzem, M., 2020. Dissolved organic matter-capped silver nanoparticles for electrochemical aggregation sensing of atrazine in aqueous systems. ACS Appl. Nano Mater. 3, 3868–3875.

Zhang, Z., Si, T., Liu, J., Zhou, G., 2019. In-situ grown silver nanoparticles on nonwoven fabrics based on mussel-inspired polydopamine for highly sensitive SERS carbaryl pesticides detection. Nanomaterials 9, 384.

Zhao, D., Chen, C., Sun, J., Yang, X., 2016. Carbon dots-assisted colorimetric and fluorometric dual-mode protocol for acetylcholinesterase activity and inhibitors screening based on the inner filter effect of silver nanoparticles. Analyst 141, 3280–3288.

Zheng, M., Wang, C., Wang, Y., Wei, W., Ma, S., Sun, X., He, J., 2018a. Green synthesis of carbon dots functionalized silver nanoparticles for the colorimetric detection of phoxim. Talanta 185, 309–315.

Zheng, X.S., Jahn, I.J., Weber, K., Cialla-May, D., Popp, J., 2018b. Label-free SERS in biological and biomedical applications: Recent progress, current challenges and opportunities. Spectrochim. Acta A Mol. Biomol. Spectrosc. 197, 56–77.

Zhu, C., Zhao, Q., Meng, G., Wang, X., Hu, X., Han, F., Lei, Y., 2020. Silver nanoparticle-assembled micro-bowl arrays for sensitive SERS detection of pesticide residue. Nanotechnology 31, 205303.

Zhu, Y., Li, M., Yu, D., Yang, L., 2014. A novel paper rag as 'D-SERS' substrate for detection of pesticide residues at various peels. Talanta 128, 117–124.

CHAPTER

Pesticide degradation by silver-based nanomaterials

17

Manviri Rani[a], Jyoti Yadav[a], and Uma Shanker[b]

[a]*Department of Chemistry, Malaviya National Institute of Technology, Jaipur, Rajasthan, India*
[b]*Department of Chemistry, Dr. B.R. Ambedkar National Institute of Technology, Jalandhar, Punjab, India*

17.1 Introduction

Due to swift industrialization, pollution is regularly increasing and is having a serious impact on the environment. In developing countries, water pollution is a life-threatening problem because of pesticides. The preservation of agriculture through killing unwanted organisms produces a cluster of chemicals of pesticides. Due to high persistence with toxic behavior and a contaminant nature, these pesticides must be regulated in public health and agriculture. Unnaturally fabricated pesticides highly bioaccrue in the atmosphere. Industrial wastes as well as agricultural fields and orchards treated with pesticides are the chief sources of pesticides (insecticides, herbicides, fungicides, rodenticides, etc.) in water. The continued use of such pesticides and discharged effluents into the surface and groundwater bodies are affecting water consumption. Moreover, pesticides blowing out over an area have negative effects on the health of living organisms. A report published by Cahill and his team concluded that only 10%–15% of applied pesticides approach the real target while the remaining 85%–90% is dispersed in the environment (Cahill et al., 2011). Pesticides are invariably employed in the food supply for the ever-increasing world population, despite less awareness among the public. The potential sources of pesticides are surface overflow from agricultural areas as well as the liberation of urban activities such as mining and industries (Neumann et al., 2003; Cahill et al., 2011).

Pesticides are lethal to human health and several involuntary species. However, some pesticides are assumed carcinogenic with endocrine disruptor potential (Kuroda et al., 1992; Dolara et al., 1994; Rehana et al., 1995; Bhatt et al., 2020). Highly persistent and recalcitrant pesticides are causing many health problems such as genetic disorders, leukemia, non-Hodgkin's lymphoma, and lung and pancreatic cancer.

In 2015, the International Agency for Research on Cancer (IARC) affirmed the carcinogenic behavior of the organophosphate insecticides tetrachlorvinphos, diazinon, parathion, malathion, and glyphosate (Nawab et al., 2003). Fig. 17.1 shows the

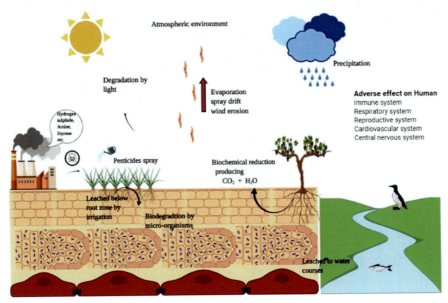

FIG. 17.1

Schematic representation of different stages of pesticide migration from the aerial atmosphere, soil, and water sources involved in the pesticide cycle.

conveyance of pesticides to humans via aerial, water, and soil modes. However, more alertness related to the risks in drinking water pollution by pesticides, their acceptable limits have been amended. This chapter will describe the role of various silver-based nanomaterials in the degradation of pesticides. It will also provide a detailed discussion of the environmental concerns of pesticides as well as the mechanism of pesticide degradation. (See Tables 17.1–17.3.)

Because of the harmful effects of pesticides, it is necessary to modify new techniques for their remediation, even when they are present at very low concentrations (ppb or ppm levels). For the elimination of toxic pesticides, several techniques are used by researchers such as hydrolysis, ultrasonic irradiation, oxygen plasma treatment, the Fenton process, ionizing radiation, and photocatalysis.

Higher cost and a longer time are serious issues in the replacement of traditional methods such as adsorption, biodegradation, and membrane separation. Nanotechnology is a developed technology that can have the peculiar characteristics of size-dependent physical and chemical properties of a material that are different from their bulk counterpart (Ju-Nam and Lead, 2008). Nanoparticles with at least one dimension in the nanometer range exhibit excellent properties due to their high surface-area-to-volume ratio, semiconducting nature, and variable bandgap energy. In the mineralization of organic and inorganic pollutants as well as halocarbons, several metallic nanoparticles such as zero-valent iron, copper, gold, and silver show unique catalytic activity. Apart from the use of zero-valent iron, gold, and silver

Table 17.1 Some important organic pollutant-related incidents in India.

Pesticide	Place	Year	Causes
Parathion	India	1958	Contaminated food due to leakage
HCH	India	1963	Contaminated rice
Endrin	India	1964	Contaminated food
HCH	India	1963	Contaminated rice
DDT	India	1965	Contaminated chutney
Diazinon	India	1968	Contaminated food
HCH	India	1976	Mixed with wheat
BHC	India	1989	Contaminated food
Endosulfan	India	1997	Contamination due to aerial spray
Phorate	India	2001	Spray drift from banana field
Endosulfan	India	2002	Contaminated wheat flour
Endosulfan	India	2011	Contaminated fruits (litches)
Monocrotophos	India	2013	Contaminated food
Phorate	India	2015	Spray of sugar cane field
Endosulfan	India	2017	Spray in crop field
Mancozeb	India	2017	Spray in cotton field

nanoparticles, several other metal oxides such as zinc oxide and titanium dioxide have been used extensively because of their ecofriendly nature, lower toxicity, and low-charge efficiency.

During the previous decades, the synthesis of silver metal nanoparticles and their environmental applications has become a hot theme for researchers because of their ability to remove contaminants and microorganisms, their bioimaging and catalysis behavior, and their latent implementation in electronics, optical and electric fields, and surface-enhanced Raman scattering (SERS) (Nickel et al., 2000; Leopold and Lend, 2003; McFarland and VanDuyne, 2003; Evanoff and Chumanov, 2004; Mallick et al., 2006). Numerous modified and advanced techniques have been studied for the degradation and reduction of various contaminants, including their byproducts, from the environment. When comparing developed oxidation processes and conventional treatment approaches, the former are preferred because of their advanced applicability and ability to develop the potential of multifunctional materials.

The distinct energy levels and controlled electron movement of a material improves its efficiency. Many significant possessions such as definite surface-to-volume ratio, high affinity (aqueous form) for sorption of metabolites, greater electron mobility, wider light absorption, and surface defects are enhanced by functional material. The fabricated nanoparticles were found to be stable and reusable that don't damage the capacity and can easily be recovered and recycled from the reaction (Alula et al., 2020). For the synthesis of visible-light photocatalysts, silver is a noble metal that has have unique properties such as nontoxic nature, high visible light photocatalytic activity, photo corrosion, size-dependent optical properties, and the ability to reduce the bandgap (surface plasmon resonance (SPR) effect).

Table 17.2 Physical and chemical assets of some pesticides commonly used in the metro-city environment.

Pesticide	Water solubility (mg L^{-1})	Octanol–water partition coefficient (Log Kow) at 25°C	Vapor pressure at 20–25°C (mmHg)	Half-life in soil (days)	Soil–water partition coefficient (Log Koc) (cm^3 g^{-1})	References
Acephate	818,000	0.13	2×10^{-6}	4.5–32	2–21.8	Vogue et al. (1994)
Arsenates	17,000			1–50	10^4–10^5	Vogue et al. (1994)
Bensulide	Partly miscible	4.12	8.0×10^{-7}	27.1–44.3	9911	
Benefin	0.1	5.29	6.5×10^{-5}	21–120	10,715	
Bendiocarb	40	NA	NA	5	570	Andreu and Picó (2004)
Benomyl	2	2.12	3.7×10^{-9}	67–240	1900	
Boric acid	50,000	0.18	1.6×10^{-6}			
Bromacil	700	2.11	3.07×10^{-7}	60	32	Vogue et al. (1994)
Carbaryl	120	2.36	1.17×10^{-6}	10	300	
Captan	5.1	2.80	9×10^{-8}	2.5	200	
Chlorothalonil	0.6	3.05	5.7×10^{-7}	21–180	10^3–10^5	Kookana and Simpson (2000)
Chlorpyrifos	0.4	4.96	2.02×10^{-7}	30–94	4981	Vogue et al. (1994)
Cypermethrin	0.004	6.60	1.7×10^{-9}	3–30	160,000	Kookana and Simpson (2000)
Chlorthal dimethyl	0.5	4.40	2.5×10^{-6}	11–289	5900	Vogue et al. (1994)
Chloroxuron	2.5	NA	NA	60	3000	Kookana and Simpson (2000)
2,4-D	890	2.81	1.40×10^{-7}	10–180	1–100	Vogue et al. (1994)
DCPA	0.5	4.40	2.5×10^{-6}	14–100	5000	
DDT	0.025	6.91	1.6×10^{-6}	1460–1095	5.2	Rani and Shanker (2018)
Dieldrin	0.11	5.4	5.9×10^{-5}	369	6.7	
Methoxychlor	0.1	5.08	4.2×10^{-5}	120	80,000	
Aldrin	0.19	6.5	1.2×10^{-4}	266	5.4	
Endosulfan	0.32	4.7	1.2 pa	50	4.0	
Dicamba	400,000	2.21	1.25×10^{-5}	3–180	1–100	Vogue et al. (1994)
Diuron	42	2.68	8.25×10^{-9}	10–90	10^2–10^4	Kookana and Simpson (2000)
Diazinon	60	3.81	1.2×10^{-2}	23–40	272	Vogue et al. (1994)
Dimethoate	39,800	0.78	1.1×10^{-3}	7	20	

Ethion	1.1	5.07	1.5×10^{-6}	150–280	10,000–15,435	Vogue et al. (1994)
Ethoprophos	750	3.59	3.8×10^{-4}	25	70	
Fenoxaprop ethyl	0.8	NA	0.19×10^{-5}	9	9490	
Fenarimol	14	3.6	2.25×10^{-7}	360	600	Andreu and Picó (2004)
Glyphosate	900,000	−3.40	9.8×10^{-8}	3–130	24,000	
HCH	0.005	3.78	NA	400–1000	1100–50,000	
Imidacloprid	610	0.57	1.5×10^{-9}	48–190	26.87	
Isazophos	69	133.72	NA	34	100	Vogue et al. (1994)
Isofenphos	24	4.12	3.00×10^{-6}	150	600	
Malathion	130	2.36	2.26×10^{-4}	1	1800	Andreu and Picó (2004)
Maneb	6	0.62	7.5×10^{-8}	70	2000	Vogue et al. (1994)
MCPA	866,000	3.25	5.90×10^{-6}	7–41	1–100	Kookana and Simpson (2000)
Mecoprop	660,000	3.13	1.2×10^{-5}	21	20	Vogue et al. (1994)
Neonicotinoids	340–4100	−0.13 to 0.91	NA	NA	NA	Klarich et al. (2017)
Oryzalin	2.5	3.73	9.75×10^{-9}	20	600	Vogue et al. (1994)
Oxadiazon	0.7	4.80	1.15×10^{-7}	60	3200	
Pendimethalin	0.275	5.20	9.4×10^{-6}	30–100	10^4–10^5	Kookana and Simpson (2000)
Phosalone	3	4.38	4.54×10^{-8}	21	1800	Vogue et al. (1994)
Propoxur	1800	1.52	9.68×10^{-6}	30	30	
Pyrethrins	0.001	5.9	2.03×10^{-5}	12–30	100,000	
Rotenone	0.2	4.10	6.9×10^{-10}	3	10,000	
Simazine	6.2	2.18	2.2×10^{-8}	1–90	20–2500	
Sulfur	Insoluble	NA	3.96×10^{-6}	NA	NA	
Sethoxydim	4390	1.65–4.51	1.6×10^{-7}	5	100	Vogue et al. (1994)
Triclopyr	2,100,000	0.42	1.26×10^{-6}	46	20	
Trichlorfon	120,000	0.51	7.8×10^{-6}	10	10	
Terbuthylazine	22	3.40	6.75×10^{-7}	30–90	500	Andreu and Picó (2004)
Thiophenate	3.5	1.40	7.13×10^{-8}	10	1830	Vogue et al. (1994)
Triadimefon	71.5	2.77	1.5×10^{-8}	26	300	

Table 17.3 Remediation of pesticides by silver nanoparticles into less-poisonous compounds with different technologies.

S. no.	Pesticide (mg L^{-1})	Nanoparticles	Brief summary	Reference
1.	Imidacloprid	Ag-reduced graphene oxide	100% degradation	Keshvardoostchokami et al. (2018)
2.	Tetrabromobisphenol-A	Ag-reduced graphene oxide-modified TiO$_2$	99.6% degradation in 80 min by sunlight irradiation mechanism	Zhou et al. (2019)
3.	Thiram and methamidophos	ZnO micron rod with Ag (ZMR/Ag arrays)	100% degradation in 30 min	Yao et al. (2020)
4.	Diazinon	N-TiO$_2$/Ag/Ti electrode	87.5% mineralization	Sheydaei et al. (2019)
5.	Thidiazuron (TDZ; 1-phenyl-3-(1,2,3-thiadiazol-5-yl) urea)	Ag/AgCl-activated carbon	91% degradation in 210 min via photocatalytic activity	Zinjarde et al. (2016)
6.	Chlorpyrifos	Graphene-oxide-silver nanocomposite (RGO@Ag)	95% degradation; pseudo-second-order rate equation	Pradeep et al. (2016)
7.	Reactive blue 19 (RB19) and reactive yellow 186 (RY186)	Ag nanoparticles with *Trigonella foenum-graecum* (TF) leaf extract	88% and 86%, respectively, in 180 min	Singh et al. (2019)
8.	Chlordane	Reduced graphene oxide-silver nanocomposites	Complete degradation in 11 min	Sarno et al. (2017)
9.	4-Nitrophenol	Ag/Cu/TiO$_2$	96% in 30 min	Rodríguez-González and Hernández-Gordillo (2019)
10.	4-Nitrophenol	Ag and zinc hydroxide layered hybrid material	100% degradation in 4 min	Quites et al. (2017)
11.	Diazinon	Ag-modified clinoptilolite zeolite	100% degradation in 120 min by photocatalytic degradation	Hallajiqomi et al. (2018)
12.	Lindane	ZnO/Ag nanostructures	99.5% degradation in 40 min.	Jung et al. (2018)

17.1.1 Banned pesticides in India

A list of toxic chemicals is published periodically by the World Health Organization (WHO), which set the lowest and highest hazardous forms of particular pesticides (upon serious risk to human health). On the basis of their toxic behavior these pesticides are classified into different categories. From the 2009 report, pesticides are classified as moderately hazardous, extremely hazardous, to some extent hazardous, and unlikely to present an intense hazard. In different countries, the index of pesticides used by growers for pest control and crop cultivation is the same and they are unaware of the toxic behavior of these hazardous pesticides. In India, a study showed that many growers stored pesticides in unsafe places, such as galleys (3.2%), restrooms (10.8%), and storerooms (29.6%) where they can be easily accessible to children and pose potential dangers. The fact that farmers were oblivious of the toxic nature of the compounds was quite evident as 21.2% of them did not use any protective measures while spraying and an additional 18% just washed their clothes after spraying (Singh et al., 2010). According to the International Labor Organization, 14% of all professional injuries are due to hazardous pesticides and a further 10% of agrochemical ingredients are deadly (Rodgers et al., 2009). The United Nations Environmental Program (UNEP) and WHO determined that fifty lacs cases of pesticide poisoning occur among agricultural labors each year, with about 20,000 mortalities (WHO, 1990).

Most techniques in the fabrication of silver nanoparticles are those in which the reduction of silver nitrate takes place through reductants such as ascorbic acid (Sondi et al., 2003; Velikov et al., 2003), borohydride (Creighton et al., 1979), or citrate (Lee and Meisel, 1982; Pillai and Kamat, 2004). Spherical and rod-shaped silver nanoparticles with a size of ~10 nm were obtained via reduction with citrate and borohydride (Dong et al., 2009). Separate use of citrate or borohydride as the reductant is not so much of effective in manipulation of the nucleation and growth processes at different pH, the temperature of the reactions and molar fraction of the reductant/silver precursor salts (Anusuya and Banu, 2016; Singh et al., 2013). A recent study showed that the factors that influence the biofabrication of silver nanoparticles are discovered systematically and their catalytic activity is also evaluated. Of the advanced oxidation processes and conventional treatment techniques, the former is preferred because of its advanced applicability and ability to develop potential multifunctional materials.

17.2 Degradation of pesticides with silver nanoparticles

For the complete removal of large-ranging pesticides, an economical, environmentally safe, and commercial method is still needed. Therefore, the development of a well-organized, biodegradable, profitable technique is required that could offer quick and high removal rates. Bioremediation, aerobic degradation (Murthy and Manonmani, 2007), adsorption (Dąbrowski, 2001; Kyriakopoulos and Doulia, 2006), coagulation (Jiang et al., 2006), fluid extraction (Kubátová et al., 2001), solid-phase extraction

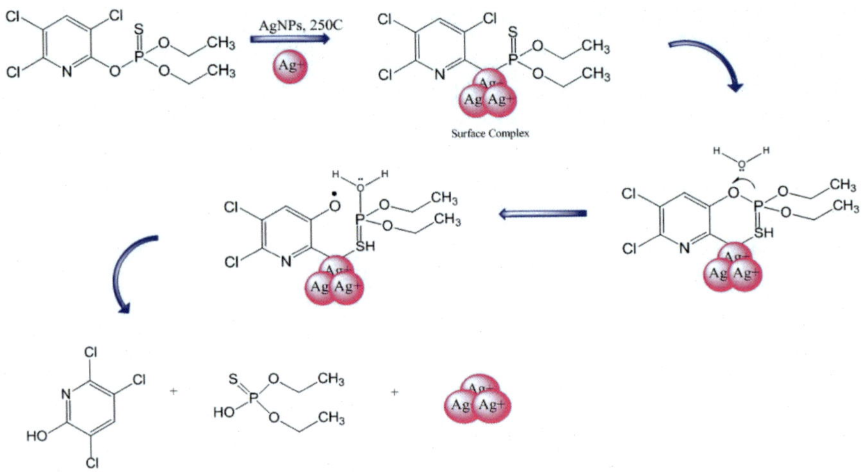

FIG. 17.2

Degradation of chloropyrifos (CP) on silver nanoparticles.

(Masselon et al., 1996; Danish et al., 2010), electrochemical and biological oxidation, advanced oxidation photocatalysis including homogeneous and heterogeneous processes (Ikehata and Gamal El-Din, 2005; Maldonado et al., 2006; Bandala et al., 2002; Martínez-Huitle et al., 2008), and nanofiltration membranes (Ahmad et al., 2006) are various approaches used for the degradation of pesticides. The most modest and operative technique applied for the elimination of toxic pollutants is adsorption (Foo and Hameed, 2010).

The complete degradation of chlorpyrifos to 3,5,6-trichloro-2-pyridinol (TCP) (3h) was achieved with Au and Ag metal supported on alumina (neutral). The pathways were shown by Fig. 17.2. In this process, the P–O bond is cleaved and a new Ag-S complex is produced, resulting in a stable aromatic species, TCP (Bootharaju and Pradeep, 2012).

17.3 Photocatalytic efficiency of silver nanoparticles

To get rid of recalcitrant material, photocatalytic oxidation is an ecofriendly process. It enhances the biodegradability of dangerous and nonbiodegradable contaminants. Photocatalysis can also be applied as an improving step to treat obdurate organic compounds (De Lasa and Serrano-Rosales, 2009). When ultraviolet (UV) or solar light irradiation falls on the surface of a semiconductor, it get photoexcited to give mobile electrons and positive surface charges. This phenomenon is named as photocatalysis. The excitation of positions and the quickening of electrons (oxidation and reduction) are two essential steps for toxin degradation (Roggo et al., 2013). A semiconductor photocatalyst has been modified with the development of nanotechnology that helps in the degradation of harmful pesticides. The photoexcitation of electrons from valence band (VB) to conduction band (CB) created a hole in VB on UV light irradiation. The deficiency of generated hole is recovered by absorbing water molecules to gives $OH^·$ radicals. Hydroxyl radicals are generated on the interface with water, which leads to the oxidation of any organic complex into water and CO_2 (Akpan and Hameed, 2009). To start the recombination process, the e- from CB has no hole, resulting in a superoxide radical anion ($^·O^{2-}$) from oxygen (TiO_2 surface). Further on, the reaction of the superoxide and water results in $HO^·$ (as explained in Fig. 17.3) (Akpan and Hameed, 2009; Linsebigler et al., 1995). The characteristics of such semiconductors (TiO_2, In_2O_3, Cu_2S, CdTe, and NiO) can be enhanced by doping some noble metal such as gold, platinum, or silver. Silver NPs doped with TiO_2 increase the photocatalytic activity of the semiconductor by decreasing the bandgap and avoiding e-/h recombination (Seery et al., 2007; Tian et al., 2014; Xu et al., 2012). The hybrid materials of TiO_2/AgNPs can be promoted for photodegradation by both UV-visible light based on three possible paths that subordinate the photocatalytic properties of TiO_2 with the antimicrobial and photocatalytic properties of silver nanoparticles (Zhang et al., 2014) and the interaction between titanium oxide and silver.

Zinc oxide is an operative semiconductor used for the degradation of hazardous pesticides. The photodegradation of pesticides in contaminated water was recurred by tandem zinc oxide/sodium oxide, where ZnO behaves as photosensitizer and sodium oxide as oxidant under day light irradiation (Navarro et al., 2009). The above example concluded that solar photocatalysis with zinc oxide (photosensitizer) is a fruitful method for the reduction of pesticides from the environment. In other studies, silver-doped zinc oxide nanoparticles had effective antibacterial and antifungal properties against variable microbes such as *Staphylococcus epidermidis*, which had sensitivity toward silver-doped zinc oxide nanoparticles (hydrothermal route). This process follows the footprints of Fig. 17.4.

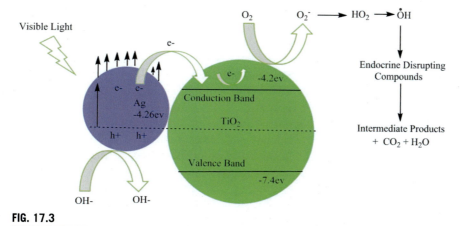

FIG. 17.3

A schematic diagram shows the electron-hole pair departure at the TiO_2 doped with Ag heterojunction boundary and represents a possible pathway for the degradation of organic contaminants under visible light irradiation.

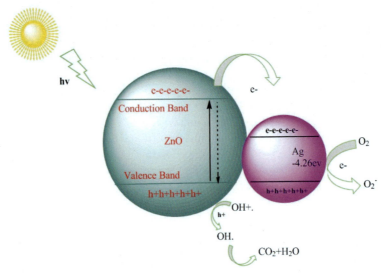

FIG. 17.4

Schematic representation of the proposed photocatalytic degradation mechanism showing the separation of photogenerated electron holes in an Ag-doped ZnO system under visible light.

17.4 Environmental concern of pesticides

These harmful pesticides are listed in the Federal Economics Poison Act because of their utilization on a large scale and adverse impact on the environment (Krishnaraj et al., 2010; Ahamed et al., 2011). Very low amounts ($ng\,mL^{-1}$) could be perilous due to their accumulating nature and high dwelling time (a few months to many years) (Chandran et al., 2006; Ankanna et al., 2010;). The signaling pathway of the nervous system (sodium-potassium gates and inhibition of cholinesterase) is impassable by the regular consumption of insecticides (Jain et al., 2009). It may cause Parkinson's disease, childhood cancer, dementia, genetic diseases, and amyotrophic lateral sclerosis (Kannan et al., 2011; Kaviya et al., 2011; Subramanian and Suja, 2012). The pesticide heptachlor, which initiates vomiting and nausea, is rapidly absorbed in the soil and is impervious to degradation (Vanaja and Annadurai, 2013; Kesharwani et al., 2009). Pyrethroids have been identified as endocrine-disrupting chemicals (EDCs) which are responsible for the male reproductive impairments. The interference in cellular signaling may causes vomiting, dizziness, headache and incoordination of muscles (Dubey et al., 2009). The major effects of pesticides are observed in miscarriage or birth-defects in humans, eggshells of birds and reproductive toxins in fish (Sukirtha et al., 2012). Size-dependent silver nanoparticles are effective in health problems. Variable modern strategies described in Fig. 17.5 are used for the fabrication of silver nanocomposites. A hybrid nanostructure of silver nanoparticles is helpful for antibacterial materials. The combination of silver nanoparticles with

17.4 Environmental concern of pesticides

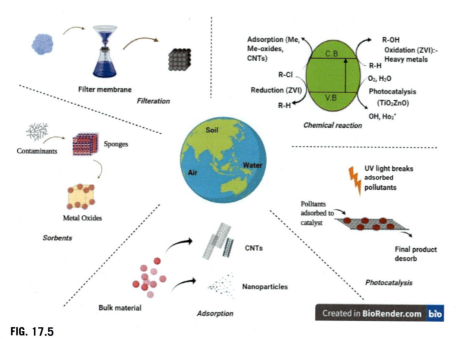

FIG. 17.5

Approaches for the environmental remediation of contaminants from soil, groundwater or surface water, and air.

carbon nanomaterials minimizes the toxicity of silver nanoparticles (Yildirimer et al., 2011). The combination of organic and inorganic nanocomposites outcome with silver-based nanomaterials might showed antibacterial activity. Silver-based nanocomposites with inorganic supports (Ag/TiO$_2$) (Li et al., 2011b) or polymer matrices (i.e., Ag/PVA) (Chen et al., 2007) creates an innovative material with the encapsulation of silver nanoparticles. Hence, these nanocomposites can be safer for the environment when applied for emerging novel products.

Increasing population and speedy industrialization have raised environmental pollution in the form of air, water, and soil contaminants to a high level. These toxins have high persistency in the environment and a negative impact on human health. Many biological, chemical, and physical techniques are applied for the purification of air, water, and soil by using nanomaterials of different forms. The excellent capacity of these materials because of the nanosize is efficient to remove impurities because of the high surface/volume ratio, adsorption capability, and efficient interaction (Yunus et al., 2012). Nanotechnology and nanomaterials have been used to abate the problem of air pollution in several different ways. Many of these synthesized nanoparticles may used for the conversion of harmful exhaust gases/vapors from vehicles and industries into harmless gases. Manganese oxide is a nanoparticle-based nanofiber catalyst that has been used to treat manufacturing discharges for volatile organic compounds. Nanogold particles have been proved very operative in catalyzing the renovation of largely toxic CO and CO$_2$.

17.5 Utilization of green-synthesized Ag-based compounds for pesticide degradation

In the remediation of environmental problems, green-synthesized nanomaterials are highly exploited. Recently, to make the green fabrication process efficient, researchers worldwide have used water, microorganisms, sunlight, and plant-based surfactant for nanomaterial synthesis. For the remediation of environmental difficulties, the green route is economical and efficacious for the fabrication of nanocatalysts. Growth of nanoparticles via greener approach was confirmed by observing the change in color of precursor on adding plant extract. The plant extract-based biological and chemical reduction process is adaptable. Silver nanoparticles have been enormously synthesized utilizing plant part such as *Azadirachta indica* and *Ocimum tenuiflorum*. Nanoparticles are fabricated by mixing a metal precursor solution and a plant leaf extract at diverse reaction conditions (such as phytochemical concentration, metal salt concentration, pH, temperature, and presence of different phytochemicals) (Mittal et al., 2013; Dwivedi and Gopal, 2010). The plant leaf extract contains phytochemicals that have the ability of metal ion reduction in a few minutes compared with microorganisms such as fungi and bacteria (large gestation time) (Jha et al., 2009).

To fabricate metal and metal oxides, plant leaf extracts are a benevolent and incomparable source. In the fabrication of nanoparticles, plant leaf extract behaves as a reducing as well as a stabilizing agent (Malik et al., 2014). Different plants have an inconstant number of phytochemicals (terpenoids, aldehydes, flavones, glucose, amides, and ketones) at different concentrations, which are affected by the composition of the plant leaf extract in nanoparticle fabrication, as illustrated in Fig. 17.6 (Li et al., 2011a,b; Mukunthan and Balaji, 2012). Flavonoids may hike the reducing capacity of metal ions due to the presence of various functional groups. Tautomeric transformations in flavonoids release reactive hydrogen atoms, which convert the enol form into the keto form. Biogenic silver nanoparticles are synthesized by *Ocimum basilicum* plant extract with an enol to keto transformation process (Ahmad et al., 2010). Glucose and fructose sugars are present in plant extracts, enhancing the fabrication of metallic nanoparticles with diverse shapes and sizes (Pillai and Kamat, 2004). Nanoparticles produced with plant leaf extract are connected with proteins that are proven by Fourier transform infrared spectroscopy (FTIR) analysis (Zayed et al., 2012). Amino acids with cysteine, arginine, lysine, and methionine help reduce metal ions in different ways and are capable of binding with silver ions (Gruen, 1975). FTIR analysis established that functional groups such as –C–O–, –C=C–, –C–O–C, and –C=O are the capping ligands of nanoparticles. The agglomeration and further growth of fabricated nanoparticles are controlled by capping ligands. According to Kesharwani and his coworkers, when light emerges on photographic films, it sensitizes the silver bromide emulsion (silver present in a metallic form) present in these films (Kesharwani et al., 2009). Then, these films are located in an aqueous solution of hydroquinone, which gives quinone on oxidation by the activity of the explained silver ion. Hydroquinone/plastohydroquinone/quinol (alcoholic compound) behave as main reducing agents in the chemistry of photography by reducing silver metal

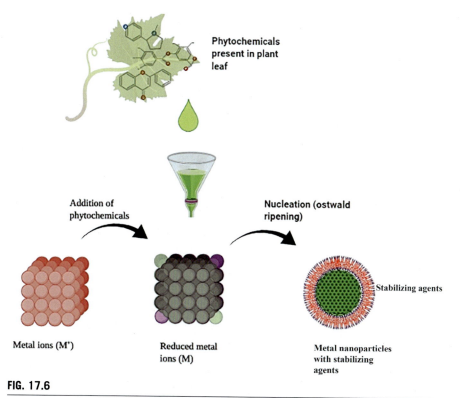

FIG. 17.6

Pathway of nanoparticle development by plant leaf extract (phytochemicals) with reduction process.

ions into metal nanoparticles through a noncyclic photophosphorylation process (Kesharwani et al., 2009). This proves that the heterocyclic compounds and biomolecules in plant extracts were answerable for the extracellular fabrication of metallic nanoparticles. To illustrate, geranium leaf extract with terpenoids as the phytochemical are actively accountable for the translation of silver ions into nanoparticles. The bioreduction of silver nitrate metal salts into metal nanoparticles is proficient with Eugenol (terpenoid component) of *Cinnamomum zeylanisum* (cinnamon) extracts. With the formation of silver nanoparticles, −OH groups originated from eugenol disappeared, which is documented with FTIR analysis.

However, the actual vital pathway for metal oxide nanoparticles synthesized by plant extracts is not completely applicable. Three stages are followed for the fabrication of metallic nanoparticles from plant extracts: (1) The activation phase with bioreduction and nucleation of metal ions, (2) the progressive phase with a spontaneous mixture of tiny particles with superior ones, and (3) the last phase defines the concluding shape of the nanoparticles (Li et al., 2007; Hoffmann et al., 1995). There are numerous applications in various contaminants such as pesticides, polyaromatic hydrocarbons (PAHs) (Shanker et al., 2016, 2017a,b; Jassal et al., 2016a,b,c), and dyes (alizarin red S, Eriochrome Black T, malachite green, methylene blue, and methyl orange). Sunlight and a natural surfactant (*Aegle marmelos*) are used for potassium metal hexacyanoferrates and transition metal oxide nanoparticle synthesis, respectively, with 90% dye photodegradation (Jassal et al., 2015a,b, 2016a,b,c). The complete degradation (100%) of dyes in 13 min has been observed by silver nanoparticles synthesized with yeast (*Saccharomyces cerevisiae*) or the *Morinda tinctoria* leaf extract, *Polygonum hydropiper* (Bonnia et al., 2016) 95% in 72 h (Vanaja et al., 2014) and 90% in 360 min (Roy et al., 2015) under solar light irradiation. Bimetallic iron with palladium nanoparticles results in the 98% degradation of methyl orange II dye (Luo et al., 2016).

17.6 Future scope and perspectives

Pesticides are critical challenges in the remediation of wastewater because of their persistence in the environment and harmful effects on human health. A small concentration of pesticides even in $ng\,L^{-1}$ can increase their existence for a longer time. In place of conventional methods, various kind of advanced treatment should be promoted for degradation of pesticides. Some concerning points to which scientists should pay significant attention for handling micropollutants are given below:

- The synthesized material should be profitable, effectual, and technically achievable, so we need to extend our research area toward this.
- Primary phase pesticides are not considered as hazardous as a secondary phase because the degradation of secondary phase pesticides is more problematic. The world market needs the exhaustive information about each and every pesticide including their persistence, degraded intermediate or bi-products and treatment techniques. Therefore, those data should be provided by the researcher in systematic ways.
- Very few studies are reported on the degradation of pesticides by functionalized nanomaterials at different operating parameters such as pH, existence of other ions, electrical conductivity and temperature. The probable insights of material behavior are proofed from these parameters for acute wastewater treatment.
- The reusability and recyclability of synthesized nanoparticles play an important role for developing sustainable treatment technique.
- To enhance collaboration with micropollutants, materials should be immobilized onto a solid substrate and be capable of material removal from aqueous solution.
- To ignore secondary contamination, functional material should be safe. Though numerous publishers have conveyed the use of toxic materials such as oxides of rare earth metals and metals such as vanadium, cadmium, lead, etc., used for the formation of many organic compounds, the possible secondary contamination has hardly been tested.

17.7 Conclusion

To fulfill the requirements and raise the living standards of a continuously growing population, growers use a large number of pesticides on their fields. Excessively used industrial and organic discharges including dyes, pesticides, and lower aromatic hydrocarbons with heavy metals are major intractable pollutants that contaminate water bodies. This chapter presents a detailed study on the adverse impact of persistent toxic contaminants on the environment and variable techniques for their removal or degradation with silver nanoparticles. Pesticides are hazardous pollutants, contaminating water, air, soil, turf, and vegetation due to their high persistence in the environment. Worldwide, organizations are making their best efforts for the remediation of the toxicity and persistence of these pesticides. In the recent era, many advanced technologies and approaches have been developed for the fabrication of silver nanoparticles. Silver nanoparticles can be fabricated by many chemical, physical, spectroscopic, biological, and green approaches. In chemical methods, mainly organic and inorganic solvents with chemical reducing agents such as tollens reagent, sodium borohydride, and citrate ascorbate are required for the development of silver nanoparticles. This results in microscopic-sized nanoparticles that cause serious issues related to respiration and are the reason behind fatal diseases. Similarly, physical methods require a lot of time, the tube furnace occupies a large, and they require a large amount of energy. In the biological method, optimal reaction conditions, cell growth, and enzyme activity are taken into consideration. This chapter briefly explains an overview of recent developments in versatile wastewater treatment nanotechnologies for instantaneous treatment of several pesticides long with coexisting contaminants by silver nanomaterials. Recently, due to the insertion of phytochemicals in a lattice of crystals, green-fabricated nanomaterials have been efficiently and conventionally manufactured. Researchers in laboratory settings have followed and reported the use of nanotechnology for environmental remediation. Nanoparticles act as reactive adsorbents and work in the degradation of pesticides by SN^2 reactions. Green-synthesized nanoparticles are free from secondary pollutants as well as being cost-effective, easy, and have low equipment requirements for potential large-scale production (simple method). A plant extract used in silver nanoparticle fabrication serves as a reducing and capping agent. This approach avoids the production of unwanted or harmful byproducts. An excellent solvent system or natural solvent is needed in the procedure.

References

Ahamed, A., Randhawa, M.A., Yusuf, M.J., Khalid, N., 2011. Effect of processing on pesticides residue in food crops—a review. J. Agric. Res. 49, 379–388.

Ahmad, M.T., Saha, S., Chaabane, D., Akretche, R., Maachi, D., 2006. Nanofiltration process applied to the tannery solutions. Desalination 200, 419–420.

Ahmad, N., Sharma, S., Alam, M.K., Singh, V., Shamsi, S., Mehta, B.R., Fatma, A., 2010. Rapid synthesis of silver nanoparticles using dried medicinal plant of basil. Colloids Surf. B. 81, 81–86.

Akpan, U.G., Hameed, B.H., 2009. Parameters affecting the photocatalytic degradation of dyes using TiO_2-based photocatalysts, a review. J. Hazard. Mater. 170, 520–529.

Alula, M.T., Lemmens, P., Madiba, M., Present, B., 2020. Synthesis of free-standing silver nanoparticles coated filter paper for recyclable catalytic reduction of 4-nitrophenol and organic dyes. Cellulose 27, 2279–2292.

Andreu, V., Picó, Y., 2004. Determination of pesticides and their degradation products in soil, critical review and comparison of methods. TrAC Trends Anal. Chem. 23, 772–789.

Ankanna, S., Prasad, T., Elumalai, E., Savithramma, N., 2010. Production of biogenic silver nanoparticles using Boswellia ovalifoliolata stem bark. Dig. J. Nanomater. Biostruct. 5, 369–372.

Anusuya, S., Banu, K.N., 2016. Silver-chitosan nanoparticles induced biochemical variations of chickpea (*Cicer arietinum* L.). Biocatal. Agric. 8, 39–44.

Bandala, E.R., Gelover, S., Leal, M.T., Arancibia-Bulnes, C., Jimenez, A., Estrada, C.A., 2002. Solar photocatalytic degradation of Aldrin. Catal. Today 76, 189–199.

Bhatt, J., Jangid, M., Kapoor, N., Ameta, R., Ameta, S.C., 2020. Detection and degradation of pesticides using nanomaterials. In: Biogenic Nano-Particles and Their Use in Agro-Ecosystems. Springer, pp. 431–455.

Bonnia, N., Kamaruddin, M., Nawawi, M., Ratim, S., Azlina, H., Ali, E., 2016. Green biosynthesis of silver nanoparticles using 'Polygonum Hydropiper' and study its catalytic degradation of methylene blue. Procedia Chem. 19, 594–602.

Bootharaju, M., Pradeep, T., 2012. Understanding the degradation pathway of the pesticide, chlorpyrifos by noble metal nanoparticles. Langmuir 28, 2671–2679.

Cahill, M.G., Caprioli, G., Stack, M., Vittori, S., James, K.J., 2011. Semi-automated liquid chromatography–mass spectrometry (LC–MS/MS) method for basic pesticides in wastewater effluents. Anal. Bioanal. Chem. 400, 587–594.

Chandran, S.P., Chaudhary, M., Pasricha, R., Ahmad, A., Sastry, M., 2006. Synthesis of gold nanotriangles and silver nanoparticles using Aloevera plant extract. Biotechnol. Prog. 22, 577–583.

Chen, M., Feng, Y.G., Wang, X., Li, T.C., Zhang, J.Y., Qian, D.J., 2007. Silver nanoparticles capped by oleylamine, formation, growth, and self-organization. Langmuir 23, 5296–5304.

Creighton, J.A., Blatchford, C.G., Albrecht, M.G., 1979. Plasma resonance enhancement of Raman scattering by pyridine adsorbed on silver or gold sol particles of size comparable to the excitation wavelength. J. Chem. Soc. Faraday Trans. 75, 790–798.

Dąbrowski, A., 2001. Adsorption—from theory to practice. Adv. Colloid Interf. Sci. 93, 135–224.

Danish, M., Sulaiman, O., Rafatullah, M., Hashim, R., Ahmad, A., 2010. Kinetics for the removal of paraquat dichloride from aqueous solution by activated date (*Phoenix dactylifera*) stone carbon. J. Dispers. Sci. Technol. 31, 248–259.

De Lasa, H., Serrano-Rosales, B., 2009. Advances in Chemical Engineering, Photocatalytic Technologies. vol. 36 Academic Press, pp. 111–145.

Dolara, P., Torricelli, F., Antonelli, N., 1994. Cytogenetic effects on human lymphocytes of a mixture of fifteen pesticides commonly used in Italy. Mutat. Res. 325, 47–51.

Dong, X., Ji, X., Wu, H., Zhao, L., Li, J., Yang, W., 2009. Shape control of silver nanoparticles by stepwise citrate reduction. J. Phys. Chem. C 113, 6573–6576.

Dubey, M., Bhadauria, S., Kushwah, B., 2009. Green synthesis of nanosilver particles from extract of Eucalyptus hybrida (safeda) leaf. Dig. J. Nanomater. Biostruct. 4, 537–543.

Dwivedi, A.D., Gopal, K., 2010. Biosynthesis of silver and gold nanoparticles using *Chenopodium album* leaf extract. Colloids Surf. A Physicochem. Eng. Asp. 369, 27–33.

Evanoff, D.D., Chumanov, G., 2004. Size-controlled synthesis of nanoparticles. 1."Silver-only" aqueous suspensions via hydrogen reduction. J. Phys. Chem. B 108, 13948–13956.

Foo, K.Y., Hameed, B.H., 2010. Insights into the modeling of adsorption isotherm systems. Chem. Eng. J. 156, 2–10.

Gruen, L.C., 1975. Interaction of amino acids with silver (I) ions. Biochim. Biophys. Acta, Protein Struct. 386, 270–274.

Hallajiqomi, M., Mehdipour, G.M., Varaminian, F., 2018. Degradation of diazinon from aqueous solution using silver-modified clinoptilolite zeolite in photocatalytic process. Adv. Environ. Technol. 4, 175–182.

Hoffmann, M.R., Martin, S.T., Choi, W., Bahnemann, D.W., 1995. Environmental applications of semiconductor photocatalysis. Chem. Rev. 95, 69–96.

Ikehata, K., Gamal El-Din, M., 2005. Aqueous pesticide degradation by ozonation and ozone-based advanced oxidation processes, a review (part I). Ozone Sci. Eng. 27, 83–114.

Jain, D., Daima, H.K., Kachhwaha, S., Kothari, S., 2009. Synthesis of plant-mediated silver nanoparticles using papaya fruit extract and evaluation of their anti microbial activities. Dig. J. Nanomater. Biostruct. 4, 557–563.

Jassal, V., Shanker, U., Kaith, B.S., Shankar, S., 2015a. Green synthesis of potassium zinc hexacyanoferrate nanocubes and their potential application in photocatalytic degradation of organic dyes. RSC Adv. 5, 26141–26149.

Jassal, V., Shanker, U., Shankar, S., 2015b. Synthesis characterization and applications of nano-structured metal hexacyanoferrates, a review. Int. J. Environ. Anal. Chem.Int. J. Environ. Anal. Chem. 2, 2380–2391. 1000128.

Jassal, V., Shanker, U., Gahlot, S., 2016a. Green synthesis of some iron oxide nanoparticles and their interaction with 2-amino, 3-amino and 4-aminopyridines. Mater. Today Proceed. 3, 1874–1882.

Jassal, V., Shanker, U., Gahlot, S., Kaith, B.S., Iqubal, M.A., Samuel, P., 2016b. Sapindus mukorossi mediated green synthesis of some manganese oxide nanoparticles interaction with aromatic amines. Appl. Phys. A Mater. Sci. Process. 122, 271.

Jassal, V., Shanker, U., Kaith, B.S., 2016c. Aegle marmelos mediated green synthesis of different nanostructured metal hexacyanoferrates, activity against photodegradation of harmful organic dyes. Scientifica.

Jha, A.K., Prasad, K., Kumar, V., Prasad, K., 2009. Biosynthesis of silver nanoparticles using Eclipta leaf. Biotechnol. Prog. 25, 1476–1479.

Jiang, J.Q., Wang, S., Panagoulopoulos, A., 2006. The exploration of potassium ferrate (VI) as a disinfectant/coagulant in water and wastewater treatment. Chemosphere 63, 212–219.

Ju-Nam, Y., Lead, J.R., 2008. Manufactured nanoparticles, an overview of their chemistry, interactions and potential environmental implications. Sci. Total Environ. 400, 396–414.

Jung, H.J., Koutavarapu, R., Lee, S., Kim, J.H., Choi, H.C., Choi, M.Y., 2018. Enhanced photocatalytic degradation of lindane using metal–semiconductor Zn@ ZnO and ZnO/Ag nanostructures. J. Envion. Sci. 74, 107–115.

Kannan, N., Mukunthan, K., Balaji, S., 2011. A comparative study of morphology, reactivity and stability of synthesized silver nanoparticles using Bacillus subtilis and *Catharanthus roseus* (L.) G. Don. Colloid Surf. B 86, 378–383.

Kaviya, S., Santhanalakshmi, J., Viswanathan, B., Muthumary, J., Srinivasan, K., 2011. Biosynthesis of silver nanoparticles using *Citrus sinensis* peel extract and its antibacterial activity. Spectrochim. Acta A 79, 594–598.

Kesharwani, J., Yoon, K.Y., Hwang, J., Rai, M., 2009. Phytofabrication of silver nanoparticles by leaf extract of *Datura metel*, hypothetical mechanism involved in synthesis. J. Biosci. 3, 39–44.

Keshvardoostchokami, M., Bigverdi, P., Zamani, A., Parizanganeh, A., Piri, F., 2018. Silver@ graphene oxide nanocomposite, synthesize and application in removal of imidacloprid from contaminated waters. Environ. Sci. Pollut. Res. 25, 6751–6761.

Klarich, K.L., Pflug, N.C., DeWald, E.M., Hladik, M.L., Kolpin, D.W., Cwiertny, D.M., 2017. Occurrence of neonicotinoid insecticides in finished drinking water and fate during drinking water treatment. Environ. Sci. Technol. Lett. 4, 168–173.

Kookana, R.S., Simpson, B.W., 2000. Pesticide fate in farming systems, research and monitoring. Commun. Soil Sci. Plant Anal. 31, 1641–1659.

Krishnaraj, C., Jagan, E., Rajasekar, S., Selvakumar, P., Kalaichelvan, P., Mohan, N., 2010. Synthesis of silver nanoparticles using *Acalypha indica* leaf extracts and its antibacterial activity against water borne pathogens. Colloid Surf. B 76, 50–56.

Kubátová, A., Lagadec, A.J., Miller, D.J., Hawthorne, S.B., 2001. Selective extraction of oxygenates from savory and peppermint using subcritical water. Flavour Fragr. J. 16, 64–73.

Kuroda, K., Yamaguchi, Y., Endo, G., 1992. Mitotic toxicity, sister chromatid exchange, and rec assay of pesticides. Arch. Environ. Contam. Toxicol. 23, 13–18.

Kyriakopoulos, G., Doulia, D., 2006. Adsorption of pesticides on carbonaceous and polymeric materials from aqueous solutions, a review. Sep. Purif. Rev. 35, 97–191.

Lee, P., Meisel, D., 1982. Adsorption and surface-enhanced Raman of dyes on silver and gold sols. J. Phys. Chem. 86, 3391–3395.

Leopold, N., Lend, B., 2003. A new method for fast preparation of highly surface-enhanced Raman scattering (SERS) active silver colloids at room temperature by reduction of silver nitrate with hydroxylamine hydrochloride. J. Phys. Chem. B 107, 5723–5727.

Li, S., Shen, Y., Xie, A., Yu, X., Qiu, L., Zhang, L., Zhang, Q., 2007. Green synthesis of silver nanoparticles using *Capsicum annuum* L. extract. Green Chem. 9, 852–858.

Li, H., Mehler, W.T., Lydy, M.J., You, J., 2011a. Occurrence and distribution of sediment-associated insecticides in urban waterways in the Pearl River Delta, China. Chemosphere 82, 1373–1379.

Li, M., Noriega-Trevino, M.E., Nino-Martinez, N., Marambio-Jones, C., Wang, J., Damoiseaux, R., Ruiz, F., Eric, M.V., 2011b. Synergistic bactericidal activity of Ag-TiO_2 nanoparticles in both light and dark conditions. Environ. Sci. Technol. 45, 8989–8995.

Linsebigler, A.L., Lu, G., Yates, J.T., 1995. Photocatalysis on TiO_2 surfaces, principles, mechanisms, and selected results. Chem. Rev. 95, 735–758.

Luo, F., Yang, D., Chen, Z., Megharaj, M., Naidu, R., 2016. One-step green synthesis of bimetallic Fe/Pd nanoparticles used to degrade Orange II. J. Hazard. Mater. 303, 145–153.

Maldonado, M., Malato, S., Pérez-Estrada, L., Gernjak, W., Oller, I., Doménech, X., Peral, J., 2006. Partial degradation of five pesticides and an industrial pollutant by ozonation in a pilot-plant scale reactor. J. Hazard. Mater. 138, 363–369.

Malik, P., Shankar, R., Malik, V., Sharma, N., Mukherjee, T.K., 2014. Green chemistry based benign routes for nanoparticle synthesis. J. Nanomater.

Mallick, K., Witcomb, M., Scurrell, M., 2006. Silver nanoparticle catalysed redox reaction, an electron relay effect. Mater. Chem. Phys. 97, 283–287.

Martínez-Huitle, C.A., DeBattisti, A., Ferro, S., Reyna, S., Cerro-López, M., Quiro, M.A., 2008. Removal of the pesticide methamidophos from aqueous solutions by electrooxidation using Pb/PbO$_2$, Ti/SnO$_2$, and Si/BDD electrodes. Environ. Sci. Technol. 42, 6929–6935.

Masselon, C., Krier, G., Muller, J.F., Nélieu, S., Einhorn, J., 1996. Laser desorption Fourier transform ion cyclotron resonance mass spectrometry of selected pesticides extracted on C18 silica solid-phase extraction membranes. Analyst 121, 1429–1433.

McFarland, A.D., VanDuyne, R.P., 2003. Single silver nanoparticles as real-time optical sensors with zeptomole sensitivity. Nano Lett. 3, 1057–1062.

Mittal, A.K., Chisti, Y., Banerjee, U.C., 2013. Synthesis of metallic nanoparticles using plant extracts. Biotechnol. Adv. 31, 346–356.

Mukunthan, K., Balaji, S., 2012. Cashew apple juice (*Anacardium occidentale* L.) speeds up the synthesis of silver nanoparticles. Int. J. Green Nanotechnol. 4, 71–79.

Murthy, H.R., Manonmani, H., 2007. Aerobic degradation of technical hexachlorocyclohexane by a defined microbial consortium. J. Hazard. Mater. 149, 18–25.

Navarro, S., Fenoll, J., Vela, N., Ruiz, E., Navarro, G., 2009. Photocatalytic degradation of eight pesticides in leaching water by use of ZnO under natural sunlight. J. Hazard. Mater. 172, 1303–1310.

Nawab, A., Aleem, A., Malik, A., 2003. Determination of organochlorine pesticides in agricultural soil with special reference to γ-HCH degradation by Pseudomonas strains. Bioresour. Technol. 88, 41–46.

Neumann, M., Liess, M., Schulz, R., 2003. A qualitative sampling method for monitoring water quality in temporary channels or point sources and its application to pesticide contamination. Chemosphere 51, 509–513.

Nickel, U., zu Castell, A., Pöppl, K., Schneider, S., 2000. A silver colloid produced by reduction with hydrazine as support for highly sensitive surface-enhanced Raman spectroscopy. Langmuir 16, 9087–9091.

Pillai, Z.S., Kamat, P.V., 2004. What factors control the size and shape of silver nanoparticles in the citrate ion reduction method? J. Phys. Chem. B 108, 945–951.

Pradeep, T., Koushik, D., Koushik, D., Gupta, S.S., Maliyekkal, S.M., 2016. Rapid dehalogenation of pesticides and organics at the interface of reduced graphene oxide–silver nanocomposite. J. Hazard. Mater. 308, 192–198.

Quites, F.J., Azevedo, C.K., Alves, E.P., Germino, J.C., Vinhas, R.C., 2017. Ag nanoparticles-based zinc hydroxide-layered hybrids as novel and efficient catalysts for reduction of 4-nitrophenol to 4-aminophenol. J. Braz. Chem. Soc. 28, 106–115.

Rani, M., Shanker, U., 2018. Degradation of traditional and new emerging pesticides in water by nanomaterials, recent trends and future recommendations. Int. J. Environ. Sci. Technol. 15, 1347–1380.

Rehana, Z., Malik, A., Ahmad, M., 1995. Mutagenic activity of the Ganges water with special reference to the pesticide pollution in the river between Kachla to Kannauj (UP), India. Mutat. Res. Genet. Toxicol. Environ. Mutagen. 343, 137–144.

Rodgers, G., Lee, E., Swepston, L., Van Daele, J., 2009. The International Labour Organization and the Quest for Social Justice, 1919–2009. International Labour Office (ILO), Geneva.

Rodríguez-González, V., Hernández-Gordillo, A., 2019. Silver-based photocatalysts, a special class. In: Nanophotocatalysis and Environmental Applications. Springer, pp. 221–239. Number of 221–39 pp.

Roggo, C., Coronado, E., Moreno-Forero, S.K., Harshman, K., van der Meer, J.R., Weber, J., 2013. Genome-wide transposon insertion scanning of environmental survival functions in the polycyclic aromatic hydrocarbon degrading bacterium Sphingomonas wittichii RW 1. Environ. Microbiol. 15, 2681–2695.

Roy, K., Sarkar, C., Ghosh, C., 2015. Photocatalytic activity of biogenic silver nanoparticles synthesized using yeast (*Saccharomyces cerevisiae*) extract. Appl. Sci. 5, 953–959.

Sarno, M., Casa, M., Cirillo, C., Ciambelli, P., 2017. Complete removal of persistent pesticide using reduced graphene oxide–silver nanocomposite. Chem. Eng. Trans. 60, 151–156.

Seery, M.K., George, R., Floris, P., Pillai, S.C., 2007. Silver doped titanium dioxide nanomaterials for enhanced visible light photocatalysis. J. Photochem. Photobiol. A Chem. 189, 258–263.

Shanker, U., Jassal, V., Rani, M., 2016. Towards green synthesis of nanoparticles: from bio-assisted sources to benign solvents. A review. Int. J. Environ. Anal. Chem. 96, 801–835.

Shanker, U., Jassal, V., Rani, M., 2017a. Green synthesis of iron hexacyanoferrate nanoparticles, potential candidate for the degradation of toxic PAHs. J. Environ. Chem. 5, 4108–4120.

Shanker, U., Rani, M., Jassal, V., 2017b. Degradation of hazardous organic dyes in water by nanomaterials. Environ. Chem. Lett. 15, 623–642.

Sheydaei, M., Karimi, M., Vahid, V., 2019. Continuous flow photoelectrocatalysis/reverse osmosis hybrid reactor for degradation of a pesticide using nano N-TiO$_2$/Ag/Ti electrode under visible light. J. Photochem. Photobiol. A Chem. 384, 112068.

Singh, A., Jain, D., Upadhyay, M., Khandelwal, N., Verma, H., 2010. Green synthesis of silver nanoparticles using *Argemone mexicana* leaf extract and evaluation of their antimicrobial activities. Dig. J. Nanomater. Biostruct. 5, 483–489.

Singh, A., Sharma, M.M., Batra, A., 2013. Synthesis of gold nanoparticles using chick pea leaf extract using green chemistry. J. Optoelectron. Adv. Mater. 5, 27–32.

Singh, J., Kumar, V., Jolly, S.S., Kim, K.H., Rawat, M., Kim, K.H., Rawat, M., Kukkar, D., Tsang, Y.F., 2019. Biogenic synthesis of silver nanoparticles and its photocatalytic applications for removal of organic pollutants in water. J. Ind. Eng. Chem. 80, 247–257.

Sondi, I., Goia, D.V., Matijević, E., 2003. Preparation of highly concentrated stable dispersions of uniform silver nanoparticles. J. Colloid Interface Sci. 260, 75–81.

Subramanian, V., Suja, S., 2012. Green synthesis of silver nanoparticles using *Coleus amboinicus* Lour, antioxitant activity and in vitro cytotoxicity against Ehrlich's Ascite carcinoma. J. Pharm. Res. 5, 1268–1272.

Sukirtha, R., Priyanka, K.M., Antony, J.J., Kamalakkannan, S., Thangam, R., Gunasekaran, P., Krishnan, M., Achiraman, S., 2012. Cytotoxic effect of green synthesized silver nanoparticles using *Melia azedarach* against in vitro HeLa cell lines and lymphoma mice model. Process Biochem. 47, 273–279.

Tian, B., Dong, R., Zhang, J., Bao, S., Yang, F., Zhang, J., 2014. Sandwich-structured AgCl@ Ag@ TiO$_2$ with excellent visible-light photocatalytic activity for organic pollutant degradation and E. coli K12 inactivation. Appl. Catal. B 158, 76–84.

Vanaja, M., Annadurai, G., 2013. Coleus aromaticus leaf extract mediated synthesis of silver nanoparticles and its bactericidal activity. Appl. Sci. 3, 217–223.

Vanaja, M., Paulkumar, K., Baburaja, M., Rajeshkumar, S., Gnanajobitha, G., Malarkodi, C., Sivakavinesan, M., Annadurai, G., 2014. Degradation of methylene blue using biologically synthesized silver nanoparticles. Bioinorg. Chem. Appl. 2014, 1–8.

Velikov, K.P., Zegers, G.E., van Blaaderen, A., 2003. Synthesis and characterization of large colloidal silver particles. Langmuir 19, 1384–1389.

Vogue, P.A., Kerle, E.A., Jenkins, J.J., 1994. Understanding pesticide persistence and mobility for groundwater and surface water protection. Report, Extension Service, Oregon State University.

WHO, 1990. Public Health Impact of Pesticides Used in Agriculture. World Health Organization.

Xu, J., Xiao, X., Ren, F., Wu, W., Dai, Z., 2012. Enhanced photocatalysis by coupling of anatase TiO_2 film to triangular Ag nanoparticle island. Nanoscale Res. Lett. 7, 1–6.

Yao, J., Quan, Y., Chen, L., Liu, Y., 2020. Detect, remove and re-use: sensing and degradation pesticides via 3D tilted ZMRs/Ag arrays. J. Hazard. Mater. 391, 122222.

Yildirimer, L., Thanh, N.T., Loizidou, M., Seifalian, A.M., 2011. Toxicology and clinical potential of nanoparticles. Nano Today 6, 585–607.

Yunus, I.S., Harwin, K.A., Adityawarman, D., Indarto, A., 2012. Nanotechnologies in water and air pollution treatment. Environ. Technol. Rev. 1, 136–148.

Zayed, M.F., Eisa, W.H., Shabaka, A., 2012. *Malva parviflora* extract assisted green synthesis of silver nanoparticles. Spectrochim. Acta A Mol. Biomol. Spectrosc. 98, 423–428.

Zhang, T., Wang, Q., Chen, C., 2014. Cytotoxic potential of silver nanoparticles. Yonsei Med. J. 55, 283–291.

Zhou, Q., Wang, M., Tong, Y., Wang, H., Zhou, X., Sheng, X., Sun, Y., Chen, C., 2019. Improved photoelectrocatalytic degradation of tetrabromobisphenol a with silver and reduced graphene oxide-modified TiO_2 nanotube arrays under simulated sunlight. Ecotoxicol. Environ. Saf. 182, 109472.

Zinjarde, S., Bapat, G., Labade, C., Chaudhari, A., 2016. Silica nanoparticle based techniques for extraction, detection, and degradation of pesticides. Adv. Colloid Interface Sci. 237, 1–14.

CHAPTER 18

Application of silver nanoparticles as a chemical sensor for detection of pesticides and metal ions in environmental samples

Tushar Kant[a], Nohar Singh Dahariya[b], Vikas Kumar Jain[c], Balram Ambade[d], and Kamlesh Shrivas[a]

[a]*School of Studies in Chemistry, Pt. Ravishankar Shukla University, Raipur, Chhattisgarh, India*
[b]*Department of Chemistry, Government Brijlal College, Pallari, Chhattisgarh, India*
[c]*Department of Chemistry, Government Engineering College, Raipur, Chhattisgarh, India*
[d]*Department of Chemistry, National Institute of Technology, Jamshedpur, Jharkhand, India*

18.1 Introduction

Pesticides are organic chemical substances, which are commonly used to destroy pests and weeds that damage the growth of crops in agriculture fields (Bala et al., 2016). It has been classified, based on target organisms and chemical properties, as an insecticide, fungicide, herbicide, algaecide, bactericide, rodenticide, ovicide, piscicide, avicide, nematicide, antimicrobial, and molluscicide (Bala et al., 2016; Obare et al., 2010; Zehani et al., 2014). The classification of different pesticides is summarized in Fig. 18.1. The use of these different types of pesticides on agriculture field increases the production of food material many folds. However, the application of these pesticides on agriculture land contaminates the soil and water, and finally, it enters in our body through skin contact, ingestion, breathing, and eye contact (Zehani et al., 2014). The harmful effects of these pesticides in the human body are headache, dizziness, nausea, excessive sweating, salivation, abdominal cramps, and unconsciousness (Zehani et al., 2014; Ly, 2008). Long-term exposure of these pesticides causes cardiac attack, and high and low blood pressure (Li et al., 2018a). Thus, the monitoring of these pesticides in water, soil, and vegetable samples is highly necessary. Several analytical techniques, such as gas chromatography (GC), high-performance liquid chromatography (HPLC), high-performance liquid

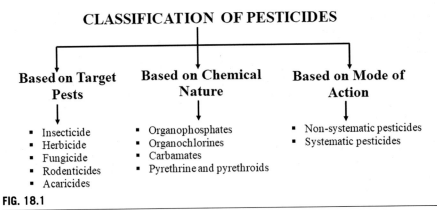

FIG. 18.1

Classification of pesticides.

chromatography-mass spectrometry (HPLC-MS), nuclear magnetic resonance (NMR), electrochemical, UV–vis, and fluorescence spectrometry, are reported for the determination of pesticides in different environmental samples (Li et al., 2018a; Liu et al., 2015; Shrivas et al., 2017, 2016a, b; Pogacnik and Franko, 2003). These techniques are expensive, tedious, time-consuming for solution preparation, and expensive chemicals for determination process (Pogacnik and Franko, 2003). Therefore, it is urged to develop a simple method for the determination of toxic chemical substances from different compartments of the environments.

Moreover, metal ions, such as mercury (Hg), arsenic (As), lead (Pb), cadmium (Cd), and chromium (Cr), have characteristics of their toxicity and non-biodegradable property (Liu et al., 2017; Li et al., 2018b; Yola et al., 2014). The wide distribution of these substances into the environmental samples, such as soil, water, and food ingredients, is the major concern as it serves harmful impact on human health, ranging from protein lose activity, chronic poisoning, Minamata disease, and even deaths (Gracia and Snodgrass, 2007; Liu et al., 2017; Sahu et al., 2020). From past decades, several quantitative techniques have been developed for the determination of these metal ions even at a nanomolar concentration (Abelsohn and Sanborn, 2010). Nowadays, low-cost technologies are highly needed for monitoring of environmental pollutant with the facile operation and in-field procedure (Sargazi and Kaykhaii, 2020). These toxic metals adversely affect all living organisms and the environment. Hence, there is need to develop a more effective and convenient method, which easily monitors the metal ions from several environmental samples for providing a good public health (Ghaedi et al., 2006; Chena et al., 2020).

Currently, research in the fields of nanomaterial-based colorimetric, fluorescence, electrochemical, and surface-enhanced Raman spectroscopy sensing approaches have been well established to determine pesticides and metal ions in environmental samples (Chena et al., 2020; Krishna et al., 2015; Yantasee et al., 2007; Christensen, 1995). Recently, nanoparticles (NPs) have been used as sensing materials due to their simple, low cost, and portability, facile operation, and excellent sensitivity for the detection of these chemical substances in environmental samples (Rafols et al., 2016; Tang et al., 2016; Zhou et al., 2016). On the contrary, designing nanoparticle-based sensors with suitable interface helps in the identification of trace level of pesticides and metal ions in a rapid and efficient route (Rafols et al., 2016; Du et al., 2013).

Moreover, AgNPs is considered as noble metal due to its distinctive physicochemical properties and wide range of their potential applications. Ag is found in different forms in the environment and living organisms, such as metallic, ionic, colloidal, and its complexes (Shenashen et al., 2013; Tian et al., 2007). It shows special optical properties such as localized surface plasmon resonance (LSPR), which is employed in surface-enhanced Raman scattering (SERS) and UV–visible (UV–vis) spectrophotometry. It has great potential applications in plasmonics, antibacterial applications, and sensing for quantitative as well as a qualitative determination of toxic substances from air, soil, water, and food samples (Yantasee et al., 2007). The colorimetric detection method is based on the alteration of AgNPs by aggregation after the addition of analyte molecules, and hence, it could be also visualized by naked eye (Du et al., 2013). In this chapter, the synthesis, characterization, and principles of sensing techniques and applications of AgNPs are demonstrated.

18.2 Synthesis of silver nanoparticles (AgNPs)

Recently, the understanding of physicochemical properties of NPs and their potential in monitoring the environmental pollutants has gained increasing interest for many researchers (Du et al., 2013). Hence, the synthesis of AgNPs and Ag-based nanocomposite has been investigated extensively (Du et al., 2013; Guo et al., 2014). Herein, AgNPs has an emphasis on the preparation of surface area with tunable and uniform particle shapes and sizes. Several methods, such a chemical reduction (Shenashen et al., 2013), electrochemical (Tian et al., 2007), solvothermal (Ghoto et al., 2019) hydrothermal (Rohit et al., 2014a), photochemical (Ravindran et al., 2012), sonochemical (Salkar et al., 1999), sputtering (Kruis et al., 2000), γ-radiation (Sylvestre et al., 2004), and laser ablation (Shen et al., 2011), have been reported for the synthesis of AgNPs with their controlled size, shape, and crystalline structure. The synthesis method of AgNPs is summarized in Fig. 18.2.

18.2.1 Wet-chemical method

This is simple and the most applicable route for the synthesis of stable AgNPs in aqueous or organic solvents (Shenashen et al., 2013; Salkar et al., 1999). This method is based on the chemical reduction of Ag^+ by the use of citrate, borohydride, and ascorbic acid for the formation of colloidal AgNPs with particle diameters of different sizes, as shown in Fig. 18.3. The reduction of Ag by wet-chemical approach generates wide range distribution of particles shapes and sizes. The chemical route synthesis of AgNPs showed evidence in controlling the shape and size of NPs. However, ecofriendly synthesis methods are still needed where the use of toxic chemicals are

432 CHAPTER 18 Application of silver nanoparticles

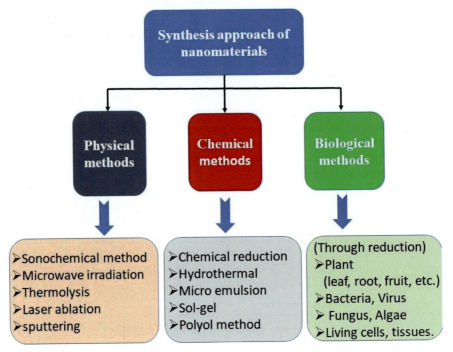

FIG. 18.2

Synthesis method used for the synthesis of AgNPs.

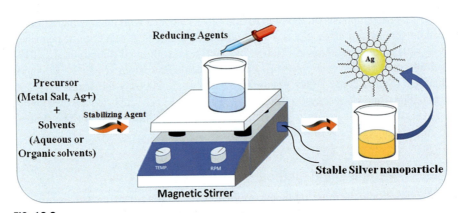

FIG. 18.3

Wet-chemical reduction method for the synthesis of stable AgNPs by reduction of Ag^+ using $NaBH_4$ as a reducing agent.

avoided during the synthesis of NPs (Shenashen et al., 2013). Therefore, it is highly needed to develop a green synthesis method for the preparation of AgNPs.

18.2.2 Physical approach

The synthesis of AgNPs through physical route usually involves the evaporation and condensation, which has been carried out using a furnace at atmospheric pressure. Herein, NPs are produced by evaporation and condensation in which metal precursor is vaporized into a carrier gas at constant temperature and pressure (Kruis et al., 2000; Sylvestre et al., 2004). The synthesis of AgNPs by laser ablation technique mainly depends on wavelength of laser impinging for target metal, laser pulses duration, and liquid medium. Among all, laser fluence is most important method for synthesizing NPs with uniform size. The presence of other analytes also affects the size of particles (Sylvestre et al., 2004). Moreover, the advantages of physical design for the preparation of NPs are that it is free from the use of solvent contamination and yields high purity materials compared to chemical methods. By contrast, the disadvantages of this method are use of high amount of energy and much time to achieve thermal stability.

18.2.3 Hydrothermal approach

This is a simple, economic, and ecofriendly method in which NPs have been synthesized through a hydrothermal or solvothermal process based on the use of hot solvent under high pressure shown in Fig. 18.4. Shen et al. (2011) developed a one-pot hydrothermal method to synthesize a reduced graphene oxide sheet (RGO)–AgNPs composite by the use of graphite oxide and Ag nitrate (NO_3) as a starting material. These methods include the advantages of preparation of controlled particle and size. The drawbacks of this method involve the small-scale process and require more time to get the thermal stability.

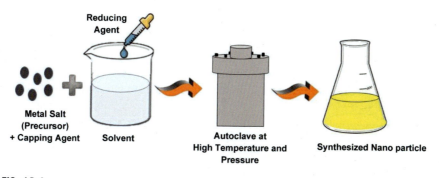

FIG. 18.4

Hydrothermal route for synthesis of AgNPs at high pressure and high temperature.

18.2.4 Electrochemical approach

Electrochemical deposition is a process that has been normally used to synthesize the metal nanoparticles. At the interface of an electrolyte solution, first metal ion forms coordinate bond with capping agent like poly(N-vinylpyrrolidone) (PVP) and poly(vinyl alcohol) (PVA), etc. After this complex formation electrochemical reduction of metal-capping agent occurs at the cathode/electrolyte interface and form a metal (M^0). The synthesis of AgNPs by electrochemical method has been reported by various researchers. Yin and his team demonstrated the synthesis of water-based AgNPs. In this technique, the PVP is used as a protecting agent for the AgNPs. The use of PVP promoted the synthesis of AgNPs by electroreduction of bulk silver ions (Yin et al., 2003). Rodríguez-Sánchez et al. (2000) synthesized 2–7 nm range of AgNPs by the dissolution of a metallic anode in the electrolyte. AgNPs capped with polyphenylpyrrole is synthesized at the liquid/liquid interface by electrochemical reduction process with a size of 3–20 nm. The method also involves use of ionic liquid (green electrolytes) for the synthesis of AgNPs, which is then applied for metal reduction reaction-related studies (Tsai et al., 2010). The electrochemical method is found feasible and economic in controlling the morphology of NPs by adjusting the specific electrolysis parameters. This technique improves the homogeneity of Ag particles. Moreover, the limitation of this method is obtaining the small-scale products, which require the consumption of more amount of energy.

18.2.5 Biological-assisted method

From a decade, microorganisms (bacteria, fungus, or plant extracts) are used as a greener approach for synthesis of several NPs, which is free from the use of toxic chemicals. In this synthesis procedure, extracts obtained from plants (leaf, fruit, bark, flower, etc.) and bio-organisms have been used as reducing and protecting agents. Hence, ecofriendly extracellular synthesis of NPs can be obtained with desired morphological properties (Shenashen et al., 2013; Huang et al., 2007; Iravani, 2011).

18.2.6 Sonochemical synthesis

A sonochemical method has been established for the synthesis of uniform Ag nanoplates using N,N-dimethylformamide (DMF) solvent. Here, Ag nanoplates have been synthesized by the reduction of $AgNO_3$ with DMF, and polyvinylpyrrolidone (PVP) is used as a stabilizing agent. In this method, plate-like structures are generated in the presence of PVP (Jiang et al., 2004). Darroudi et al. developed a sonochemical method for formulating silver nanoparticles (AgNPs) in water-based gelatin solutions. The outcome of the Ag(I) concentrations, reducing agent, ultrasonic time, and amplitude on the nanoparticle size have been studied. The extent of the AgNPs increases with increasing the ultrasonic time and decreases with the ultrasonic amplitude. The well-dispersed AgNPs with a spherical size of 3.5 nm are synthesized under ultrasonic procedure (Darroudi et al., 2012).

18.3 Characterization techniques

The characterization of nanomaterials is important to determine size, shape, structure, and coating material on the surface of functionalized material. The characterization of nanomaterials is done by using scanning electronic microscopy (SEM), transmission electron microscopy (TEM), scanning tunnelling microscopy (STM), X-ray diffraction (XRD) analysis, X-ray photoelectron spectroscopy (XPS), UV–vis spectroscopy, and Fourier transform infrared (FT-IR) spectroscopy. The basic principle involved in the electron microscopic techniques (SEM, TEM, and STEM) is that the electron beam strikes onto the surface of a sample and specific signals in the form of secondary electrons, backscattered electrons, and X-rays are produced due to the characteristics NPs. This whole process contains information about the surface topography and composition of samples. Spectroscopic techniques (UV–vis, FTIR, XRD, and XPS) based on the interaction of light with matter and wavelength or energy give information such as coating material on the surface of a material, crystalline structure, and composition. The optical spectroscopic techniques, such as XPS and XRD, are considered more attractive for the characterization of materials due to its rapid responsiveness, nondestructiveness, and has a better high resolution of chemical components in the sample. The use of characterization techniques is summarized in Table 18.1.

18.4 Working principle of a chemical sensor for the detection of pesticides and metal ions

18.4.1 Colorimetric-based chemical sensor

The method is based on colorimetric sensing approach due to the unique optical behavior of AgNPs. The noble metal NPs (Ag, Au, and Cu) exhibit a localized LSPR property, which is generated from the free movement of electrons on the

Table 18.1 The characteristic information of AgNPs by different spectroscopic and electron microscopic techniques.

Techniques	Characterization
SEM	Topography (texture of sample), morphology (shape and size of particles), elemental composition, and crystalline arrangements
TEM	Morphology (size and shape, and arrangement of particles), crystalline structure, and composition
EDX	Elemental composition of AgNPs
XRD	Lattice constant, d-spacing, crystal structure of samples, thickness or films, orientation of sample, and grain size
XPS	Polymer composite material and analysis of carbon fiber
UV–vis	Optical properties, size, concentration, and agglomeration state
FT-IR	Composition, concentration, and atomic structure

conduction band of metal surface in a specific vibrational mode with the response of UV–vis light. The vibrations of electron results the bright color of NPs, which can be tuned by changing the particle size and shapes. The LSPR surface property of AgNPs depends on the shape, size, dielectric environments, and supraparticle assemblies (Shenashen et al., 2013). This size-dependent property of noble NPs is described by Mie theory, which shows the redshift and broadening of the plasmon band due to the increase in the size of particles. The LSPR band also results in enhanced local electromagnetic fields onto the surface of NP, as it shows extremely high molar extinction coefficients and resonant Rayleigh scattering property (Lee and Jun, 2019). So far, various chemical sensors based on LSPR have been developed for detection of a variety of chemical substances from environmental samples (Thapa et al., 2015).

In colorimetric determination, the yellow color of AgNPs is shown due to the LSPR absorption band appeared at 400 nm, and it has been altered from yellow to red, pink, and grey after the addition of analyte molecules This change in color intensity is measure in UV–vis spectrophotometry for the analysis of chemical substances in a different type of unknown samples (Lee and Jun, 2019; Thapa et al., 2015; Kumar and Anthony, 2014). The colorimetric analysis of different analytes has been studied after the addition of analyte onto the solution of AgNPs in the ratio of 3:1 and determine the effect of these substances in NPs solution based on colorimetric sensing strategy. The colorimetric method for the determination of chemical substances is simple, rapid, and economic and can be applied at the sample source. Kumar and Anthony (2014) synthesized AgNPs-based colorimetric sensor using *N*-(2-hydroxybenzyl)-valine and *N*-(2-hydroxybenzyl)-isoleucine organic ligands as stabilizing and surface functionalizing agent. The method has been established for detection of Cd(II) ions in polluted groundwater samples. Balasurya et al. (2020) used AgNPs as a colorimetric sensor for detection of mercury. In this method, AgNPs is synthesized by Creighton's method where PVP is used as a protecting agent (Balasurya et al., 2020). The sensing mechanism has been applied in environmental samples with a recovery percentage of 101.6%.

18.4.2 Fluorescence based-chemical sensor

Fluorescence sensing plays a major role in detection of chemical substances in different types of samples using AgNPs as a colorimetric probe (Wu et al., 2018a, 2018b; Caro et al., 2010). Metallic NPs interact with fluorophores of sample and affects the emission intensity of fluorescence spectrum. The enhancement of signal intensity is known as metal enhanced fluorescence (MEF) (Gryczynski et al., 2002; Lakowicz, 2005; Cade et al., 2009). Among the different types of metal NPs, Ag greatly enhances the luminescence of fluorophores when it has been localized with the distance of 4–10 nm from the surface of AgNPs (Huang et al., 2020; Wu et al., 2018a, b; Zeng et al., 2019; Xia et al., 2017). Moreover, some factors also affect the enhancement of signals such as the distance of NPs to analytes, aggregation of AgNPs toward the emission band of an attached fluorophore (Cade et al., 2009; Bharadwaj et al., 2007).

AgNPs has also emerged as promising tools for direct determination of many environmental samples by conjugation with target analyte molecules and fluorophores (Bharadwaj et al., 2007).

18.4.3 Electrochemical-based chemical sensor

In recent years, electrochemical-based application of AgNPs is of great interest due to its better physicochemical properties. AgNPs provides excellent electron transfer rates and decreases over potential of oxidizing as well as reducing agents generated from enzymatic products (Nantaphol et al., 2015). With the addition of this technique, inkjet printing method is also used for the production of electronic devices due to its low cost and rapid response to the target analyte (Fig. 18.5). Kant et al. (2020) fabricated flexible printed paper electrode with Ag nanoink and had applied for electrochemical applications. This study is exploited to analyze NO_3^- ion in environmental water sample using cyclic voltammetry (CV) (Kant et al., 2020). Arduini et al. (2016) described electrochemical biosensors for the detection of pesticides in different food materials using nanomaterials. Herein, the pretreatment of the sample has also been illustrated for fast and accurate determination several analytes (Arduini et al., 2016).

18.4.4 Surface-enhanced Raman spectroscopy (SERS) based chemical sensor

SERS is a viable technique based on the measurement of radiation produced when the target analyte is adsorbed on the surface of AgNPs. Here, the intensity of scattering radiation is found to be enhanced due to the surface of nanostructured materials. Moram et al. (2018) studied the salt-induced Ag/Au aggregated NPs, which has been produced using a laser ablation method onto the SERS substrate (Moram et al., 2018). Tang et al. developed AgNPs modified carbon dots (CDs) for detection of atrazine. They synthesized AgNPs by thermal process after the

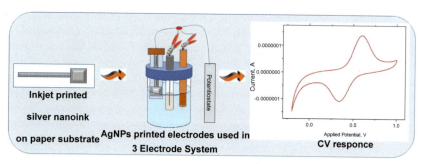

FIG. 18.5

Electrochemical application of AgNPs with CV response.

addition of CDs onto the solution of AgNO$_3$. The use of CDs@Ag greatly enhanced the signal intensity of atrazine pesticide in environmental water sample due to the aggregation of NPs (Tang et al., 2019a,b). Qu et al. (2016) synthesized AgNPs on cotton swabs for the detection of carbaryl based on SERS method. The sensing strategy of this method is demonstrated by the in situ detection of carbaryl fungicide on a cucumber sample with high reproducibility and accessibility. Ly et al. (2019) presented SERS and density functional theory (DFT) for the detection of fipronil pesticide using AgNPs. This method is applied for detection of fipronil within the range of 0.0001–0.1 mgL^{-1} with a correlation coefficient of 0.985 (R^2) (Ly et al., 2019).

18.5 Silver nanoparticles as a chemical sensor for detection of pesticides and metal ions in environmental samples

Some of metal ions such as Hg(II), As(III), Cr(VI), Pb(II), and Cd(II) showed adverse effects on human health as well as all living organisms. Therefore, it is highly necessary to detect trace level of these substances to minimize their harmful effect in the environment to safeguard the public health (El-Sayed et al., 2019).

18.5.1 Determination of metals ions using AgNPs based chemical sensors

18.5.1.1 Mercury (Hg)

Hg isa toxic heavy metal released from coal-burning power plants, gold mining, volcanic emissions, and also generated from solid waste incineration (Darbha et al., 2008). The inorganic Hg(II) is converted into methyl-mercury and it affects all living organisms. Hg damages the central nervous system, liver, and gastrointestinal tract, and hence, it is highly necessary to develop a more sensitive and selective to determine Hg levels from environmental samples (Darbha et al., 2008; Harano et al., 2007). There is a number of colorimetric based sensors have been developed for the quantitative determination of Hg(II). Farhadi et al. (2012) used green synthesis approach for the preparation of AgNPs and employed as a colorimetric sensor for determination of Hg(II). The detection method is based on the color change from yellow to colorless and blue shift of LSPR band in the visible region (Farhadi et al., 2012). Sharma et al. (2018) demonstrated the application of thiol terminated chitosan-AgNPs for the selective determination of Hg(II) from water samples. Here, the presence of Cu(II) and Fe(III) in water samples may interfere in determination of Hg(II), which is overcome by preparing thiol terminated chitosan for selective detection of target analyte (Sharma et al., 2018). Jiang et al. (2014) employed the use of AgNPs as a fluorescence signal amplifier for the selective and sensitive detection of Hg(II). The result shows the increment of fluorescence intensity when CdTe−CdS quantum dots bind with AgNPs in the presence of target analyte (Jiang et al., 2014).

In another study, Dai et al. (2015) studied autocatalytic sensor for the detection of Hg(II), which is based on oxidation of o-phenylenediamine to 2,3-diaminophenazine by Ag(I) ion followed by the formation AgNPs. Further, Hg(II) is adsorbed on the surface of AgNPs and inhibited the oxidation of o-phenylenediamine and a decrease in the fluorescence intensity of chemical substance is appeared (Dai et al., 2015). Eksin et al. (2019) fabricated pencil graphite electrode (PGE) and modified with AgNPs and folic acid (FA) for the selective detection of Hg(II) in CV. The prepared sensor is found to be selective for detection of target Hg(II) in the presence of many interfering metal ions in the sample solution (Eksin et al., 2019). Suherman et al. (2017) illustrated the use of AgNPs-glassy carbon electrode (GCE) for the linear sweep voltammetric determination of ultraconcentration of Hg(II) from water samples. The method is based on the measurement of alteration in Ag stripping peak for the introduction of different concentration of Hg(II) obtained from the galvanic displacement reaction (Suherman et al., 2017).

In SERS, Hg(II) detected by sodium 2-mercapto-ethanesulfonate-functionalized AgNPs where the sulfur group of NPs strongly binds with Hg(II), which cause the decrease of SERS signal. This is a simple and selective SERS method for the determination of Hg(II) in environmental samples (Chen et al., 2012). In another study by the Kang et al. (2014), AgNPs functionalized with dialkyne 1,4-diethynylbenzene is employed as a SERS probe for the selective determination of Hg(II). Here, the functional group of terminal ethynyl group linked with Hg(III) resulted the aggregation of NPs. Thus, hot spots are formed that enhances the signal intensity for the determination of target analyte in river water samples (Kang et al., 2014). Vasimalai et al. (2012) described the fluorescence-quenching method for the detection of Hg(II) in aqueous solution using mercapto-thiadiazole capped AgNPs. The result indicates that the decrease in emission intensity after the addition of Hg(II) ascribed to the aggregation of AgNPs (Vasimalai et al., 2012). Thus, the AgNPs is successfully demonstrated for the colorimetric determination of Hg(II) in environmental samples.

18.5.1.2 Arsenic (As)

As is a naturally occurring heavy metal found in soil and water. The concentration of As varies according to their geographic location, in anaerobic groundwater. It is usually found in the form of arsenite As (III) and arsenate (V), where As(III) is more toxic than As(V) (Divsar et al., 2015; Jena and Raj, 2008). Hence, the determination of As(III) or As(V) in drinking water is an important issue because of several health concerns such as skin cancer, and damage of liver and other body parts (Divsar et al., 2015). Currently, many methods have been established for determination of As(III) and As(V), such as chromatography, electrochemistry, and spectroscopic analyses, but these techniques demand much time for sample preparation and need significant maintenance (Wang et al., 2008). Hence, the methods based on spectroscopic approaches are increasing great interest.

Divsar et al. (2015) developed aptamer conjugated AgNPs for the selective determination of As(III) from water samples where the interaction between aptamer present on the surface of NPs specifically interacts with the target analyte followed by the decrease in the signal intensity at 403 nm, measured by UV–vis spectrophotometry. Boruah et al. (2019) demonstrated the use of polyethylene glycol functionalized AgNPs for the selective detection of As(III) from complex metal ion solution, and the detection is based on the aggregation of NPs followed by the color change from yellow to blue. Dar et al. (2014) used green synthesis approach for the preparation of AgNPs-graphene oxide (GO) composite material using beta-cyclodextrin and ascorbic acid as a capping and reducing agent, respectively. The surface of graphene is utilized for the reduction of Ag(I), where the anodic stripping analysis of As(III) is found to be enhanced many times compared to use of only GO. The cyclodextrin stabilized AgNPs/GO/GCE is used for selective detection of As(III) in ground and river water samples (Dar et al., 2014). Sonkoue et al. (2018) used electrochemical determination of As(III) using AgNPs as an electrochemical sensor. In this procedure, Au-electrode modified with AgNPs showed a well-defined reduction peak of As(III) in CV compared to bare-Au-electrode (Sonkoue et al., 2018). Li et al. (2011) proposed SERS method for detection of As(III) by the use of glutathione (GSH) capped AgNPs. The detection of As(III) indicated that as it interacts with GSH, the aggregation of AgNPs has been observed significantly. Li et al. (2011) illustrated the application of GSH and 4-mercaptopyridine (4-MPY) functionalized AgNPs for the selective and sensitive determination of As(III) in aqueous solution. The -SH functional group of GSH and 4-MPY binds with As(III) that resulted in the aggregation of NPs, and more hot spots are formed for SERS signal enhancement. The method can detect As(III) with LOD value of 0.73 ppb in a sample solution (Li et al., 2011).

18.5.1.3 Lead (Pb)

Pb is another toxic metal that is widely distributed in water, soil, and dust particles (Gracia et al., 2007; Lee and Huang, 2011). Pb exposure mainly affects the reproductive, nervous, immune, and cardiovascular systems and developmental processes in children (Lee et al., 2011; Poręba et al., 2011). From past decades, several quantitative techniques have been developed for the detection of Pb(II). Shrivas et al. (2019a) reported the synthesis of PVA capped AgNPs through the reduction of Ag(I) for the selective determination of Pb in contaminated water samples. The sensing principle depends on the yellow color of PVA-AgNPs, which is changed to red after the addition of Pb(II) followed by the redshift of LSPR band in visible region (Shrivas et al., 2019a). Qi et al. (2012) discussed the application of iminodiacetic acid (IDA) capped AgNPs for the selective and sensitive detection of Pb(II) from tap water samples. The sensing principle of Pb(II) depends on the color change of IDA capped AgNPs from yellow to green and redshift of LSPR band from 396 to 650 nm (Qi et al., 2012). Cheon and coworkers used polymerized 3,4-dihydroxy-L-phenylalanine (PDOPA) as a stabilizing agent and reducing agents for the synthesis of AgNPs in aqueous media. The PDOPA functionalized AgNPs is used as a sensing probe for detection of Pb(II) and Cu(II) in real water samples (Cheon and Park, 2016). Wang et al. (2015) employed nanoclusters of Ag/Au alloy as a fluorescent probe for analysis of Pb(II) in lake and drinking water samples. The nanoclusters are protected by bovine albumin serum (BSA), which selectively binds with Pb(II) followed by the aggregation-induced fluorescence quenching of sample solution (Wang et al., 2015). In another work, GCE with poly(1,8-diaminonaphthalene) and AgNPs coated nanocomposite is used for simultaneous determination of Pb(II), Cd(II), and Cu(II) in water samples using square wave anodic stripping voltammetry (ASV) with a scan rate of $5\,mV\,s^{-1}$ (Hassan et al., 2019). Singh et al. (2012) described the application of AgNPs for selective detection of Pb(II) at a very trace level using fluorescence technique.

18.5.1.4 Cadmium (Cd)

Cd is another most toxic heavy metal widely used in many industries for fabrication of metal alloys, batteries, and other products (Dong et al., 2017; Sung and Wu, 2014). Cd exposure to human beings cause acute poisoning and cancer in lungs, pancreas, and kidney (Poręba et al., 2011; Jin et al., 2015; Godt et al., 2006). According to World Health Organization (WHO), the maximum permissible unit of Cd in drinking water is $3\,\mu gL^{-1}$ (WHO, Geneva, 2004). Dong et al. (2017) used chalcon carboxylic acid coated AgNPs for the colorimetric determination of Cd(II), where the addition of Cd(II) into NPs solution caused the aggregation of Ag particle by changing the color from yellow to orange and red shift of absorption peaks from 396 to 522 nm. Jin et al. (2015) studied the colorimetric assay of Cd(II) through the 5-sulfosalicylic acid-functionalized AgNPs with excellent selectivity over the 16 metal ions. Makwana et al. (2016) synthesized an octamethoxy resorcin (4) arene tetrahydrazide (OMRTH)-functionalized AgNPs by one-pot preparation method. The principle for the detection of Cd(II) from sample solution is based on a decrease in the fluorescence intensity of OMRTH-AgNPs. The method is simple, rapid, and sensitive for the detection of Cd(II) from ground and industrial water samples (Makwana et al., 2016). Palisoc et al. (2018) showed the application of AgNPs modified graphene paste for the determination of Pb(II), Cd(II), and Cu(II) in ASV. Yao et al. (2019) employed the single-walled carbon nanohorns as an electrochemical sensor for the detection of Cd(II) and Pb(II) ions. The method is applied for the determination of Pb(II) and Cd(II) ions in honey and milk samples with detection limit of $0.2\,\mu gL^{-1}$ and $0.4\,\mu L^{-1}$ for Cd(II) and Pb(II) ions, respectively (Yao et al., 2019).

18.5.1.5 Chromium (Cr)

Cr (III) is an important trace element required in much biological process in human body and other organisms. However, the oxidation state of Cr(VI) has toxic property due to the excellent solubility even in biological systems, and hence, it disturbs the physiological process (Kotas and Stasicka, 2000; Nriagu and Pacyna, 1988). The main sources for contamination of water reservoirs are electroplating, water-cooling towers, chrome plating, pesticides, etc. (Sharma, 2003). The chronic exposure of Cr(VI) results the several disorders such as chronic ulcer, lung, and gastrointestinal tract (Basu et al., 1983). Hence, the determination of chromium from a variety of water samples is required to assure better public health.

Wu et al. (2013) prepared ascorbic acid (AA) capped AgNPs for the quantitative determination of Cr(VI) in environmental water samples. The sensing mechanism is based on cross-linkage of AA with Cr(III), resulting in the reduction of Cr(VI), which caused the aggregation of NPs, and then color change from yellow to red (Wu et al., 2013). Kailasa group used AgNPs modified with citrate ions and melamine for the selective determination of Cr(VI) and Hg(II) from water samples. Here, different ligands present on the AgNPs surface are responsible for binding with Cr(VI) and Hg(II) to cause color change as well as redshift of plasmon band in the visible region (Kailas et al., 2018). Ismail et al. (2018) synthesized AgNPs using aqueous extract of fruit of *Durenta erecta (D. erecta)*, which acts as a reducing agent and stabilizing agent. The AgNPs prepared through greener approach is applied for the analysis of Cr(VI), based on the measurement of LSPR band after the addition of target analyte in the sample (Ismail et al., 2018). Elavarasi et al. (2012) developed AgNPs as a fluorescence probe for Cr(VI) determination without the use of extrinsic fluorophore. The detection principle depends on the aggregation of AgNPs with the addition of Cr(VI) followed by the redshift of fluorescence peak from 420 to 684 nm (Elavarasi et al., 2012). Bu and coworkers employed sodium alginate (SA) as reducing and stabilizing agent for preparation of AgNPs. Herein, the detection method is based on the reduction of carbimazole with an increase in the concentration of Cr(VI) in sample solution. The reduction of carbimazole also causes the decrease of SERS signal intensity, which is directly proportional to the concentration of analyte (Bu et al., 2018). The electrochemical determination of Cr(VI) demonstrated using Nafion stabilized AgNPs deposited on GCE. The Nafion is used as a conducting matrix where AgNPs is tightly bounded. The AgNPs on the surface of GCE helped in the excellent catalytic reduction of Cr(VI). Thus, the method is found to be simple for the analysis of Cr(VI) in the presence of other metal ions in the water sample solution (Xing et al., 2011). Ye et al. (2012) established SERS based colorimetric method for detection of Cr(III) using citrate-coated AuNPs. The method indicates that Cr(III) linked with citrate molecules and causes the aggregation of AuNPs as well as enhancement of particle junctions (Ye et al., 2012).

18.5.2 Determination of pesticides using AgNPs based chemical sensors

18.5.2.1 Chlorinated pesticides

Among different kinds of pesticides, endrin is considered as organochlorine pesticide due to the presence of chlorine atom in its structure. Endrin is generally used to prevent the growth of insects on crops such as rice, wheat, cotton, sugarcane, and maize (Shrivas et al., 2016c). The pesticide used in agricultural fields and industries discharge their wastes into the ground and environmental water reservoirs, which adversely affect all living organisms. Furthermore, toxic effects of these pesticides cause headache, nausea, vomiting, and inhibition of neurotransmitter hormones (Cai et al., 2013). However, a variety of analytical methods such as GC, GC–MS, HPLC, and HPLC-MS have been developed for monitoring of pesticides in environmental samples (Deme et al., 2014; Deme and Upadhyayula, 2015). Recently, Shrivas et al. (2016a) fabricated sucrose-capped AgNPs for the detection of endrin in water and food samples based on stereoselective endorecognition phenomenon. The method is based on an aggregation of NPs after incorporation of endrin through the interaction of more electron density of oxygen atoms to the Ag atoms (Shrivas et al., 2016a). Moreover, electrochemical-based sensing of paraquat pesticide using AgNPs/water-soluble carboxylated pillar[5]arene (CP5) functionalized GO adapted glassy carbon electrode in differential pulse voltammetry is illustrated with high sensitivity. From this method, the limit of detection is found to be 1.0×10^{-8} M (S/N=3). Here, the better result is attributed to the synergic electrocatalytic effect of AgNPs and GO through the host–guest recognition of CP5 with paraquat (Sun et al., 2019). Further, Alak and Dinh (1988) studied the SERS based determination method for chlorinated pesticide including carbophenothion, methyl chloropyrifos, 1-hydroxychlordene chlordan, bromophos, dichloran, and linuron. For this, Ag-coated substrates are used as SERS-active media and pesticides have been detected in nanogram levels (Alak and Dinh, 1988).

18.5.2.2 Phosphorus pesticide

Diazinon is an organophosphorus pesticide commonly used to protect fruit, vegetable, nut, and field crops from insects. It enters into the body through ingestion, inhalation, skin contact, or breathing (Ly, 2008). The high exposure of diazinon pesticide to the human causes weakness, seizures, stiffness, muscle spasms, or rapid heart rate, and sometimes coma (Dahlgren et al., 2004). Different analytical techniques have been used for the determination of diazinon in environmental samples. Shrivas et al. (2019b) studied the detection of diazinon pesticide through the noncovalent interaction of π-conjugated pyrimidine nitrogen and sulfur moieties with bare-AgNPs. The procedure has been applied for diazinon determination in fruit and vegetable samples (Shrivas et al., 2019b). Triazophos is another class of organophosphorus pesticide used to increase productivity and to control many insects that harm to various crops. However, it is moderately toxic to human beings but highly toxic to fish and honeybees (Bala et al., 2017). Ma et al. (2018) studied detection of triazophos by AgNPs using ultrasensitive detection procedure. The interaction process has been studied via density functional theory (DFT) approach (Ma et al., 2018). Zheng et al. (2018) described detection of phoxim (an organophosphate pesticide) using a green synthesis of CD-AgNPs. This method is based on aggregation and red-shifting of plasmon band after the addition of phoxim onto the CDs-AgNPs sample (Zheng et al., 2018). He et al. (2015) demonstrated a simple AgNPs based chemiluminescent (CL) probe for the detection of organophosphate and carbamate pesticides. The use of CL-based sensing device utilizes the triple-channel characteristics of the luminol-functionalized AgNP. AgNPs based chemiluminescent is exploited for the simultaneous detection of five pesticides such as dipterex, dimethoate, carbaryl, carbofuran, and chlorpyrifos. Using this method, all the pesticides are determined in 20 environmental samples with a precision of 95%. The simple plan of this study is projected to promote the expansion of AgNPs based sensor arrays for simultaneous detection of different pesticides from water samples (He et al., 2015). Bala et al. (2017) applied peptide and aptamer-based nanoprobe for the detection of malathion pesticide in real food and water samples. The method provided LOD value of 1.94 pM for the concentration range of 0.01–0.75 nM (Bala et al., 2017). Tang et al. (2019b) developed SERS substrate using AgNPs for the rapid detection of chlorpyrifos and imidacloprid pesticide residues. The method obtained for the determination of chlorpyrifos and imidacloprid with values of LOD of 10 ngmL^{-1} and 50 ngmL^{-1}, respectively (Tang et al., 2019b).

18.5.2.3 Nitrogen containing pesticides

Tricyclazole (5-methyl-1,2,4-triazolo(3,4-b) benzothiazole) is known to be a fungicide employed for controlling the blast disease on rice and to enhance the quality and quantity of rice production. This is considered to be a pesticide by WHO, and hence, it is critically important to develop a more simple, rapid, and sensitive method to monitor tricyclazole in rice samples (Geneva, 2004). Rohit and Kailasa (2014b) described a simple colorimetric method by employing 5-sulfo anthranilic acid dithiocarbamate (SAADTC) functionalized AgNPs as an optical probe for the analysis of tricyclazole. The aggregation of AgNPs is induced in the presence of target analytes and color change for colorimetric determination in real samples (Rohit and Kailasa, 2014b). Ozcan et al. (2020) described the use of modified fumed silica (FS)@AgNPs for the electrochemical detection of carbendazim (CBZ). An electrochemical sensor was established using FS@Ag with a very low LOD value (9.4×10^{-10} M) is demonstrated for the detection of CBZ in sample solution (Ozcan et al., 2020).

Furthermore, atrazine (2-chloro-4-ethylamino-6-isopropylamino-s-triazine) is a different class of herbicide, which is manufactured or formulated by several agrochemical companies. Atrazine commonly used to control growth of broadleaf and grass weeds, which prevent the development of corn, sorghum, and sugar cane plants (Huang et al., 2003). There are several methods such as GC, HPLC, MS, electrochemiluminescence, voltammetric methods to monitor atrazine in food residues (Riemenschneider et al., 2017; Zacco et al., 2006). Patel et al. (2015) synthesized 4-aminobenzenethiol functionalized AgNPs as a colorimetric recognition probe for the selective detection of carbendazim fungicide, and the method is applied for the determination of pesticides in water and food samples. The study indicates that the carbendazim linked with AgNPs via strong ion-pair and π–π interactions results in the color alteration of AgNPs from yellow to orange. Moreover, electrochemical-based green synthesis approach has also been studied for the determination of atrazine using a boron-doped diamond electrode. The linear range for atrazine determination is obtained in the range of 0.05–40 M with R^2 of 0.999 (Svorc et al., 2013).

18.6 Conclusions

Different methods for the synthesis of several modified AgNPs are demonstrated for the selective determination of metal ions and pesticides from a different type of samples such as water, soil, and air. AgNPs are used as sensing probes in colorimetric, fluorescence, electrochemical SERS for the selective and sensitive detection of target chemical susbtances from sample solutions. The synthesis process of AgNPs should be simple and economic, which can be utilized by many laboratories for the quantitative determination of different metal ions and pesticides from several environmental samples. AgNPs shows unique and novel physicochemical properties for the facile determination of different toxic substances in comparison to other sophisticated analytical instruments.

Acknowledgments

We would like to thank the Science and Engineering Research Board (SERB), New Delhi, for awarding Kamlesh Shrivas as Extra Mural Research Project (File No.: EMR/2016/005813).

References

Abelsohn, A.R., Sanborn, M., 2010. Lead and children: clinical management for family physicians. Can. Fam. Physician 56, 531–535.

Alak, A.M., Dinh, T.V., 1988. Surface–enhanced Raman spectrometry of chlorinated pesticides. Anal. Chim. Acta 206, 333–337.

Arduini, F., Cinti, S., Scognamiglio, V., Moscone, D., 2016. Nanomaterials in electrochemical biosensors for pesticide detection: advances and challenges in food analysis. Microchim. Acta 183, 2063–2083.

Bala, R., Sharma, R.K., Wangoo, N., 2016. Development of gold nanoparticles–based aptasensor for the colorimetric detection of organophosphorus pesticide phorate. Anal. Bioanal. Chem. 408, 333–338.

Bala, R., Dhingra, S., Kumar, M., Bansal, K., Mittal, S., Sharma, R.K., Wangoo, N., 2017. Detection of organophosphorus pesticide—malathion in environmental samples using peptide and aptamer based nanoprobes. Chem. Eng. Trans. 311, 111–116.

Balasurya, S., Syed, A., Thomas, A.M., Marraiki, N., Elgorban, A.M., Raju, L.L., Das, A., Khan, S.S., 2020. Rapid colorimetric detection of mercury using silver nanoparticles in the presence of methionine. Spectrochim. Acta A 228, 117712.

Basu, D., Blackburn, K., Harris, B., Neal, M., Stoss, F., 1983. Health Assessment document for chromium (1983 final). U.S. Environmental Protection Agency, Washington, D.C. EPA/600/8–83/014F.

Bharadwaj, P., Anger, P., Novotny, L., 2007. Nanoplasmonic enhancement of single–molecule fluorescence. Nanotechnology 18, 044017.

Boruah, B.S., Daimari, N.K., Biswas, R., 2019. Functionalized silver nanoparticles as an effective medium towards trace determination of arsenic (III) in aqueous solution. Results Phys. 12, 2061–2065.

Bu, X., Zhang, Z., Zhang, L., Li, P., Wu, J., Zhang, H., Tian, Y., 2018. Highly sensitive SERS determination of chromium (VI) in water–based on carbimazole functionalized alginate–protected silver nanoparticles. Sens. Actuators B 273, 1519–1524.

Cade, N.I., Ritman-Meer, T., Kwakwa, K.A., Richards, D., 2009. The plasmonic engineering of metal nanoparticles for enhanced fluorescence and Raman scattering. Nanotechnology 20, 285201.

Cai, J., Zhu, F., Ruan, W., Liu, L., Lai, R., Zeng, F., Ouyang, G., 2013. Determination of organochlorine pesticides in textiles using solid–phase microextraction with gas chromatography–mass spectrometry. Microchem. J. 110, 280–284.

Caro, C., Castillo, P.M., Klippstein, R., Pozo, D., Zaderenko, A.P., 2010. Silver Nanoparticles: Sensing and Imaging Applications., https://doi.org/10.5772/8513.

Chen, Y., Wu, L., Chen, Y., Bi, N., Zheng, X., Qi, H., Qin, M., Liao, X., Zhang, H., Tian, Y., 2012. Determination of mercury(II) by surface–enhanced Raman scattering spectroscopy based on thiol–functionalized silver nanoparticles. Microchim. Acta 177, 341–348.

Chena, T., Jiang, L., Yuan, H.-Q., Zhang, Y., Sud, D., Bao, G.-M., 2020. A flexible paper–based chemosensor for colorimetric and ratiometric fluorescence detection of toxic oxalyl chloride. Sens. Actuators B 319, 128289.

Cheon, J.Y., Park, W.H., 2016. Green synthesis of silver nanoparticles stabilized with mussel–inspired protein and colorimetric sensing of Lead(II) and copper(II) ions. Int. J. Mol. Sci. 17.

Christensen, J.M., 1995. Human exposure to toxic metals: factors influencing interpretation of biomonitoring results. Sci. Total Environ. 166, 89–135.

Dahlgren, J.G., Takhar, H.S., Ruffalo, C.A., Zwass, M., 2004. Health effects of diazinon on a family. J. Toxicol. Clin. Toxicol. 42, 579–591.

Dai, H., Ni, P., Sun, Y., Hu, J., Jiang, S., Wang, Y., Li, Z., 2015. Label–free fluorescence detection of mercury ions based on the regulation of the Ag autocatalytic reaction. Analyst 140, 3616.

Dar, R.A., Khare, N.G., Cole, D.P., Karna, S.P., Srivastava, A.K., 2014. Green synthesis of a silver nanoparticles–graphene oxide composite and its application for As (III) detection. RSC Adv. 4, 14432–14440.

Darbha, G.K., Singh, A.K., Rai, U.S., Yu, E., Yu, H., Ray, P.C., 2008. Selective detection of mercury (II) ion using nonlinear optical properties of gold nanoparticles. J. Am. Chem. Soc. 130, 8038–8043.

Darroudi, M., Zak, A.K., Muhamad, M.R., Huang, N.M., Hakimi, M., 2012. Green synthesis of colloidal silver nanoparticles by sonochemical method. Mater. Lett. 66, 117–120.

Deme, P., Upadhyayula, V.V.R., 2015. Ultra–performance liquid chromatography atmospheric pressure photoionization high resolution mass spectrometric method for determination of multiclass pesticide residues in grape and mango juices. Food Chem. 173, 1142–1149.

Deme, P., Azmeera, T., Prabhavathi Devi, B.L.A., Jonnalagadda, P.R., Prasad, R.B.N., Vijaya Sarathi, U.V.R., 2014. An improved dispersive solid–phase extraction clean–up method for the gas chromatography–negative chemical ionisation tandem mass spectrometric determination of multiclass pesticide residues in edible oils. Food Chem. 142, 144–151.

Divsar, F., Habibzadeh, K., Shariati, S., Shahriarinour, M., 2015. Aptamer conjugated silver nanoparticles for the colorimetric detection of arsenic ions using response surface methodology. Anal. Methods 7, 4568–4576.

Dong, Y., Ding, L., Jin, X., Zhu, N., 2017. Silver nanoparticles capped with chalcon carboxylic acid as a probe for colorimetric determination of cadmium(II). Microchim. Acta 184, 3357–3362.

Du, J., Yin, S., Jiang, L., Ma, B., Chen, X., 2013. A colorimetric logic gate based on free gold nanoparticles and the coordination strategy between melamine and mercury ions. Chem. Commun. 49, 4196–4198.

Eksin, E., Erdem, A., Fafal, T., Kıvçak, B., 2019. Eco-friendly sensors developed by herbal based silver nanoparticles for electrochemical detection of mercury (II) ion. Electroanalysis 31, 1075–1082.

Elavarasi, M., Paul, M.L., Rajeshwari, A., Chandrasekaran, N., Mandal, A.B., Mukherjee, A., 2012. Studies on fluorescence determination of nanomolar Cr(iii) in aqueous solutions using unmodified silver nanoparticles. Anal. Methods 4, 3407–3412.

El-Sayed, W.N., Elwakeel, K.Z., Shahat, A., Awual, R., 2019. Investigation of novel nanomaterial for the removal of toxic substances from contaminated water. RSC Adv. 9, 14167.

Farhadi, K., Forougha, M., Molaei, R., Hajizadeha, S., Rafipour, A., 2012. Highly selective Hg^{2+} colorimetric sensor using green synthesized and unmodified silver nanoparticles. Sens. Actuators B 161, 880–885.

Ghaedi, M., Asadpour, E., Vafaie, A., 2006. Simultaneous preconcentration and determination of copper, nickel, cobalt, lead, and iron content using a surfactant-coated alumina. Bull. Chem. Soc. Jpn. 79, 432.

Ghoto, S.A., Khuhawar, M.Y., Jahangir, T.M., 2019. Silver nanoparticles with sodium dodecyl sulphate as colorimetric probe for detection of dithiocarbamate pesticides in environmental samples. Anal. Sci. 356, 631–637.

Godt, J., Scheidig, F., Grosse-Siestrup, C., Esche, V., Brandenburg, P., Reich, A., Groneberg, D.A., 2006. The toxicity of cadmium and resulting hazards for human health. J. Occup. Med. Toxicol. 1, 1–6.

Gracia, R.C., Snodgrass, W.R., 2007. Lead toxicity and chelation therapy. Am. J. Health–Syst. Pharm. 64, 45–53.

Gryczynski, I., Malicka, J., Shen, Y., Gryczynski, Z., Lakowicz, J.R., 2002. Multiphoton excitation of fluorescence near metallic particles: enhanced and localized excitation. Phys. Chem. B 106 (9), 2191–2195.

Guo, Y.M., Zhang, Y., Shao, H.W., Wang, Z., Wang, X.F., Jiang, X.Y., 2014. Label–free colorimetric detection of cadmium ions in rice samples using gold nanoparticles. Anal. Chem. 86, 8530–8534.

Harano, K., Hiraoka, S., Shionoya, M., 2007. 3 nm–scale molecular switching between fluorescent coordination capsule and nonfluorescent cage. J. Am. Chem. Soc. 129, 5300.

Hassan, K.M., Elhaddad, G.M., AbdelAzzem, M., 2019. Voltammetric determination of cadmium (II), lead(II) and copper(II) with a glassy carbon electrode modified with silver nanoparticles deposited on poly(1,8–diaminonaphthalene). Microchim Acta 186, 440.

He, Y., Xu, B., Li, W., Yu, H., 2015. Silver nanoparticle–based chemiluminescent sensor array for pesticide discrimination. J. Agric. Food Chem. 63, 2930–2934.

Huang, S.B., Stanton, J.S., Lin, Y., Yokley, R.A., 2003. Analytical method for the determination of atrazine and its dealkylated chlorotriazine metabolites in water using SPE sample preparation and GC–MSD analysis. J. Agric. Food Chem. 51, 7252–7258.

Huang, J., Li, Q., Sun, D., Lu, Y., Su, Y., Yang, X., Wang, H., Wang, Y., Shao, W., He, N., Hong, J., Chen, C., 2007. Biosynthesis of silver and gold nanoparticles by novel sundried Cinnamomum camphora leaf. Nanotechnology 18, 104.

Huang, L., Tao, H., Zhao, S., Yang, K., Cao, Q.-Y., Lan, M., 2020. A tetraphenylethylene–based aggregation–induced emission probe for fluorescence turn–on detection of lipopolysaccharide in injectable water with sensitivity down to picomolar. Ind. Eng. Chem. Res. 59, 8252–8258.

Iravani, S., 2011. Green synthesis of metal nanoparticles using plants. Green Chem. 13, 2638.

Ismail, M., Khan, M.I., Akhtar, K., Ali Khan, M., Asirib, A.M., Khan, S.B., 2018. Biosynthesis of silver nanoparticles: a colorimetric optical sensor for detection of hexavalent chromium and ammonia in aqueous solution. Physica E 103, 367–376.

Jena, B.K., Raj, C.R., 2008. Gold nanoelectrode ensembles for the simultaneous electrochemical detection of ultratrace arsenic, mercury, and copper. Anal. Chem. 80, 4836–4844.

Jiang, L.P., Xu, S., Zhu, J.M., Zhang, J.R., Zhu, J.J., Chen, H.Y., 2004. Ultrasonic–assisted synthesis of monodisperse single–crystalline silver nanoplates and gold nanorings. Inorg. Chem. 43, 5877.

Jiang, Y., Tian, J., Hu, K., Zhao, Y., Zhao, S., 2014. Sensitive aptamer–based fluorescence polarization assay for mercury(II) ions and cysteine using silver nanoparticles as a signal amplifier. Microchim. Acta 181, 1423–1430.

Jin, W., Huang, P., Wu, F., Ma, L.-H., 2015. Ultrasensitive colorimetric assay of cadmium ion based on silver nanoparticles functionalized with 5–sulfosalicylic acid for wide practical applications. Analyst 140, 3507–3513.

Kailas, S.K., Chandel, M., Park, T.K., 2018. Influence of ligand chemistry on silver nanoparticles for colorimetric detection of Cr^{3+} and Hg^{2+} ions. Spectrochim. Acta A 195, 120–127.

Kang, Y., Wu, T., Liu, B., Wang, X., Du, Y., 2014. Selective determination of mercury(II) by self–referenced surface–enhanced Raman scattering using dialkyne–modified silver nanoparticles. Microchim. Acta 181, 1333–1339.

Kant, T., Shrivas, K., Ganesan, V., Mahipal, Y.K., Devi, R., Deb, M.K., Shankar, R., 2020. Flexible printed paper electrode with silver nano–ink for electrochemical applications. Microchem. J. 155, 104687.

Kotas, J., Stasicka, J., 2000. Chromium occurrence in the environment and methods of its speciation. Environ. Pollut. 107, 263–283.

Krishna, N.S., Raju, K., Annapurna, N., Sreenivasulu, B., Kumar, H., 2015. Simple and reliable gas chromatography method for the trace level determination of oxalyl chloride content in zolpidem tartrate. Pharm. Lett. 7, 273–280.

Kruis, F., Fissan, H., Rellinghaus, B., 2000. Sintering and evaporation characteristics of gas–phase synthesis of size–selected PbS nanoparticles. Mater. Sci. Eng. B 69, 329–334.

Kumar, V.V., Anthony, S.P., 2014. Silver nanoparticles based selective colorimetric sensor for Cd^{2+}, Hg^{2+} and Pb^{2+} ions: tuning sensitivity and selectivity using co–stabilizing agents. Sens. Actuators B 191, 31–36.

Lakowicz, J.R., 2005. Radiative decay engineering 5: metal–enhanced fluorescence and plasmon emission, anal. Biochem. 337, 171–194.

Lee, Y.F., Huang, C.C., 2011. Colorimetric assay of lead ions in biological samples using a nanogold–based membrane. ACS Appl. Mater. Interfaces 3, 2747–2754.

Lee, S.H., Jun, B.H., 2019. Silver nanoparticles: synthesis and application for nanomedicine, Int. J. Mol. Sci 20, 865.

Li, J., Chen, L., Lou, T., Wang, Y., 2011. Highly sensitive SERS detection of As3+ ions in aqueous media using glutathione functionalized silver nanoparticles. ACS Appl. Mater. Interfaces 3, 3936–3941.

Li, P., Teng, Y., Nie, Y., Liu, W., 2018a. SERS detection of insecticide amitraz residue in milk based on Au@Ag core–shell nanoparticles. Food Anal. Methods 11, 69–76.

Li, Y., Chen, Y., Yu, H., Tian, L., Wang, Z., 2018b. Portable and smart devices for monitoring heavy metal ions integrated with nanomaterials. Trends Anal. Chem. 98, 190–200.

Liu, Y., Deng, Y., Dong, H., Liu, K., He, N., 2017. Progress on sensors based on nanomaterials for rapid detection of heavy metal ions. Sci. China Chem. 60, 329–337.

Liu, G., Yang, X., Li, T., Yu, H., Du, X., She, Y., Wang, J., Wang, S., Jin, F., Jin, M., Shao, H., Zheng, L., Zhang, Y., Zhou, P., 2015. Spectrophotometric and visual detection of the herbicide atrazine by exploiting hydrogen bond–induced aggregation of melamine modified gold nanoparticles. Microchim. Acta 182, 1983–1989.

Ly, S.Y., 2008. Assay of diazinon pesticides in cucumber juice and in the deep brain cells of a live carp. Microchim. Acta 163, 283–288.

Ly, N.H., Nguyen, T.H., Nghi, N., Kim, Y.H., Joo, S.W., 2019. Surface–enhanced raman scattering detection of fipronil pesticide adsorbed on silver nanoparticles. Sensors 19, 1355.

Ma, S., He, J., Guo, M., Sun, X., Zheng, M., Wang, Y., 2018. Ultrasensitive colorimetric detection of triazophos based on the aggregation of silver nanoparticles. Colloids Surf. A Physicochem. Eng. Asp. 538, 343–349.

Makwana, B.A., Vyas, D.J., Bhatt, K.D., Darji, S., Jain, V.K., 2016. Novel fluorescent silver nanoparticles: sensitive and selective turn off sensor for cadmium ions. Appl. Nanosci. 6, 555–566.

Moram, S.S.B., Byram, C., Shibu, S.N., Chilukamarri, B.M., Soma, V.R., 2018. Ag/au nanoparticle–loaded paper–based versatile surface–enhanced Raman spectroscopy substrates for multiple explosives detection. ACS Omega 3, 8190–8201.

Nantaphol, S., Chailapakul, O., Siangproh, W., 2015. Sensitive and selective electrochemical sensor using silver nanoparticles modified glassy carbon electrode for determination of cholesterol in bovine serum. Sens. Actuators B 207, 193–198.

Nriagu, J.O., Pacyna, M.J., 1988. Quantitative assessment of worldwide contamination of air, water and soils by trace metals. Nature 333, 134–139.

Obare, S.O., De, C., Guo, W., Haywood, T.L., Samuels, T.A., Adams, C.P., Masika, N.O., Murray, D.H., Anderson, G.A., Campbell, K., Fletcher, K., 2010. Fluorescent chemosensors for toxic organophosphorus pesticides: a review. Sensors 10, 7018–7043.

Ozcan, A., Hamid, F., Ozcan, A.A., 2020. Synthesizing of a nanocomposite based on the formation of silver nanoparticles on fumed silica to develop an electrochemical sensor for carbendazim detection. Talanta 222, 121591.

Palisoc, S., Lee, E.T., Natividad, M., Racines, L., 2018. Silver nanoparticle modified graphene paste electrode for the electrochemical detection of lead, cadmium and copper. Int. J. Electrochem. Sci. 13, 8854–8866.

Patel, G.M., Rohit, J.V., Singhal, R.K., Kailasa, S.K., 2015. Recognition of carbendazim fungicide in environmental samples by using 4–aminobenzenethiol functionalized silver nanoparticles as a colorimetric sensor. Sens. Actuators B 206, 684–691.

Pogacnik, L., Franko, M., 2003. Detection of organophosphate and carbamate pesticides in vegetable samples by a photothermal biosensor. Biosens. Bioelectron. 18, 1–9.

Poręba, R., Gać, P., Poręba, M., Andrzejak, R., 2011. Environmental and occupational exposure to lead as a potential risk factor for cardiovascular disease. Environ. Toxicol. Pharmacol. 31, 267–277.

Qi, L., Shang, Y., Wu, F., 2012. Colorimetric detection of lead (II) based on silver nanoparticles capped with iminodiacetic acid. Microchim. Acta 178, 221–227.

Qu, L.L., Geng, Y.Y., Bao, Z.N., Riaz, S., Li, H., 2016. Silver nanoparticles on cotton swabs for improved surface–enhanced Raman scattering, and its application to the detection of carbaryl. Microchim. Acta 183, 1307–1313.

Rafols, C.P., Serrano, N., Díaz-Cruz, J.M., Ariño, C., Esteban, M., 2016. Glutathione modified screen–printed carbon nanofiber electrode for the voltammetric determination of metal ions in natural samples. Talanta 155, 8–13.

Ravindran, A., Elavarasi, M., Prathna, T.C., Raichurb, A.M., Chandrasekaran, N., Mukherjee, A., 2012. Selective colorimetric detection of nanomolar Cr (VI) in aqueous solutions using unmodified silver nanoparticles. Sens. Actuators B 166–167, 365–371.

Riemenschneider, C., Seiwert, B., Goldstein, M., Al-Raggad, M., Salameh, E., Chefetz, B., Reemtsma, T., 2017. An LC–MS/MS method for the determination of 28 polar environmental contaminants and metabolites in vegetables irrigated with treated municipal wastewater. Anal. Methods 9, 1273–1281.

Rodríguez-Sánchez, L., Blanco, M.C., López-Quintela, M.A., 2000. Electrochemical synthesis of silver nanoparticles. J. Phys. Chem. B 104, 9683–9688.

Rohit, J.V., Solanki, J.N., Kailas, S.K., 2014a. Surface modification of silver nanoparticles with dopamine–dithiocarbamate for selective colorimetric sensing of mancozeb inenvironmental samples. Sens. Actuators B 200, 219–226.

Rohit, J.V., Kailasa, S.K., 2014b. 5-Sulfo anthranilic acid dithiocarbamate functionalized silver nanoparticles as a colorimetric probe for the simple and selective detection of tricyclazole fungicide in rice samples. Anal. Methods 6, 5934–5941.

Sahu, S., Sharma, S., Ghosh, K.K., 2020. Novel formation of Au/Ag bimetallic nanoparticles by a mixture of monometallic nanoparticles and their application for rapid detection of Lead in onion sample. New J. Chem. 2020 (44), 15010–15017.

Salkar, R.A., Jeevanandam, P., Aruna, S.T., Koltypin, Y., Gedanken, A., 1999. The sonochemical preparation of amorphous silver nanoparticles. J. Mater. Chem. 9, 1333–1335.

Sargazi, M., Kaykhaii, M., 2020. Application of a smartphone based spectrophotometer for rapid infield determination of nitrite and chlorine in environmental water samples. Spectrochim. Acta A 227, 117672.

Sharma, Y.C., 2003. Cr(VI) removal from industrial effluents by adsorption on an indigenous low–cost material. Colloids Surf., A 215, 155–162.

Sharma, P., Mourya, M., Goswami, M., Kundu, I., Prasad, M., Chandra, D., Tripathi, S.P., Guina, D., 2018. Thiol terminated chitosan capped silver nanoparticles for sensitive and selective detection of mercury (II) ions in water. Sens. Actuators B 268, 310–318.

Shen, J., Shi, M., Yan, B., Ma, H., Li, N., Ye, M., 2011. One–pot hydrothermal synthesis of Ag–reduced graphene oxide composite with ionic liquid. J. Mater. Chem. 21, 7795.

Shenashen, A., El-Safty, S.A., Elshehy, E.A., 2013. Synthesis, morphological control, and properties of silver nanoparticles in potential applications. Part. Part. Syst. Charact. 31, 293–316.

Shrivas, K., Nirmalkar, N., Ghosale, A., Thakur, S.S., 2016a. Application of silver nanoparticles for a highly selective colorimetric assay of endrin in water and food samples based on stereoselective endorecognition. RSC Adv. 6, 29855–29862.

Shrivas, K., Nirmalkar, N., Ghosale, A., Thakur, S.S., Shankar, R., 2016b. Enhancement of plasmonic resonance through an exchange reaction on the surface of silver nanoparticles: application to the highly selective detection of triazophos pesticide in food and vegetable samples. RSC Adv. 6, 80739–80747.

Shrivas, K., Nirmalkar, N., Ghosale, A., Thakur, S.S., 2016c. Application of silver nanoparticles for a highly selective colorimetric assay of endrin in water and food samples based on stereoselective endo–recognition. RSC Adv. 6, 29855–29862.

Shrivas, K., Ghosale, A., Nirmalkar, N., Srivastava, A., Singh, S.K., Shinde, S.S., 2017. Removal of endrin and dieldrin isomeric pesticides through stereoselective adsorption behavior on the graphene oxide–magnetic nanoparticles. Environ. Sci. Pollut. Res. 24, 24980–24988.

Shrivas, K., Sahu, B., Deb, M.K., Thakur, S.S., Sahu, S., Kurrey, R., Kant, T., Patle, T.K., Jangde, R., 2019a. Colorimetric and paper–based detection of lead using PVA capped silver nanoparticles: experimental and theoretical approach. Microchem. J. 150, 104156.

Shrivas, K., Sahu, S., Sahu, B., Kurrey, R., Patle, T.K., Kant, T., Karbhal, I., Satnami, M., Deb, M.K., Ghosh, K.K., 2019b. Silver nanoparticles for selective detection of phosphorus pesticide containing π–conjugated pyrimidine nitrogen and sulfur moieties through non–covalent interactions. J. Mol. Liq. 275, 297–303.

Singh, A.K., Kanchanapally, R., Fan, Z., Senapati, D., Ray, P.C., 2012. Synthesis of highly fluorescent water–soluble silver nanoparticles for selective detection of Pb(ii) at the parts per quadrillion (PPQ) level. Chem. Commun. 48, 9047–9049.

Sonkoue, B.M., Seumo Tchekwagep, P.M., Nanseu-Njiki, C.P., Ngameni, E., 2018. Electrochemical determination of arsenic using silver nanoparticles. Electroanalysis 30, 1–7.

Suherman, A.L., Ngamchuea, K., Tanner, E.E.L., Sokolov, S.V., Holter, J., Young, N.P., Compton, R.G., 2017. Electrochemical detection of ultratrace (picomolar) levels of Hg^{2+} using a silver nanoparticle–modified glassy carbon electrode. Anal. Chem. 89 (13), 7166–7173.

Sun, J., Guo, F., Shi, Q., Wu, H., Sun, Y., Chen, M., 2019. Electrochemical detection of paraquat based on silver nanoparticles/water–soluble pillar[5]arene functionalized graphene oxide modified glassy carbon electrode. J. Electroanal. Chem. 847, 113221.

Sung, Y., Wu, S., 2014. Colorimetric detection of Cd(II) ions based on di–(1H–pyrrol–2–yl) methanethione functionalized gold nanoparticles. Sens. Actuators B 201, 86–91.

Svorc, L., Rievaj, M., Bustin, D., 2013. Green electrochemical sensor for environmental monitoring of pesticides: determination of atrazine in river waters using a boron–doped diamond electrode. Sens. Actuators B 181, 294–300.

Sylvestre, J.P., Kabashin, A.V., Sacher, E., Meunier, M., Luong, J.H.T., 2004. Stabilization and size control of gold nanoparticles during laser ablation in aqueous cyclodextrins. J. Am. Chem. Soc. 126, 7176.

Tang, S., Tong, P., You, X., Lu, W., Chen, J., Li, G., 2016. Label free electrochemical sensor for Pb^{2+} based on graphene oxide mediated deposition of silver nanoparticles. Electrochim. Acta 187, 286–292.

Tang, J., Chen, W., Ju, H., 2019a. Rapid detection of pesticide residues using a silver nanoparticle coated glass bead as nonplanar substrate for SERS sensing. Sens. Actuators B 287, 576–583.

Tang, J., Chen, W., Ju, H., 2019b. Sensitive surface–enhanced Raman scattering detection of atrazine based on aggregation of silver nanoparticles modified carbon dots. Talanta 201, 46–51.

Thapa, R.K., Choi, J.Y., Poudel, B.K., Hiep, T.T., Pathak, S., Gupta, B., Choi, H.G., Yong, C.S., Kim, J.O., 2015. Multilayer–coated liquid crystalline nanoparticles for effective sorafenib delivery to hepatocellular carcinoma. ACS Appl. Mater. Interfaces 7, 20360–20368.

Tian, C., Mao, B., Wang, E., Kang, Z., Song, Y., Wang, C., Li, S., Xu, L., 2007. One–step, size–controllable synthesis of stable Ag nanoparticles. Nanotechnology 18, 285607.

Tsai, T.H., Thiagarajan, S., Chen, S.M., 2010. Green synthesis of silver nanoparticles using ionic liquid and application for the detection of dissolved oxygen. Electroanalysis 22, 680.

Vasimalai, N., Sheeba, G., John, S.A., 2012. Ultrasensitive fluorescence–quenched chemosensor for Hg(II) in aqueous solution based on mercaptothiadiazole capped silver nanoparticles. J. Hazard. Mater. 213–214, 193–199.

Wang, Z., Lee, J.H., Lu, Y., 2008. Label free colorimetric detection of metal ions using gold nanoparticles and DNAzyme with 3 nM detection limit and tunable dynamic range. Adv. Mater. 20, 3263–3267.

Wang, C., Cheng, H., Sun, Y., Xu, Z., Lin, H., Lin, Q., Zhang, C., 2015. Nanoclusters prepared from a silver/gold alloy as a fluorescent probe for selective and sensitive determination of lead(II). Microchim. Acta 182, 695–701.

WHO, Geneva, 2004. The WHO Recommended Classification of Pesticides by Hazards and Guideline to Classification.

Wu, X., Xu, Y., Dong, Y., Jiang, X., Zhu, N., 2013. Colorimetric determination of hexavalent chromium with ascorbic acid capped silver nanoparticles. Anal. Methods 5, 560–565.

Wu, J., Pan, J., Ye, Z., Zeng, L., Su, D., 2018a. A smart fluorescent probe for discriminative detection of hydrazine and bisulfite from different emission channels. Sens. Actuators B 274, 274–284.

Wu, F., Wang, H., Xu, J., Yuan, H.-Q., Zeng, L., Bao, G.-M., 2018b. A new fluorescent chemodosimeter for ultra–sensitive determination of toxic thiophenols in environmental water samples and cancer cells. Sens. Actuators B 254, 21–29.

Xia, H.-C., Xu, X.-H., Song, Q.-H., 2017. Fluorescent chemosensor for selective detection of phosgene in solutions and in gas phase. ACS Sens. 2, 178–182.

Xing, S., Xu, H., Chen, J., Shi, G., Jin, L., 2011. Nafion stabilized silver nanoparticles modified electrode and its application to Cr(VI) detection. J. Electroanal. Chem. 652, 60–65.

Yantasee, W., Lin, Y., Hongsirikarn, K., Fryxell, G.E., Addleman, R., Timchalk, C., 2007. Electrochemical sensors for the detection of lead and other toxic heavy metals: the next generation of personal exposure biomonitors. Environ. Health Perspect. 115.

Yao, Y., Wu, H., Ping, J., 2019. Simultaneous determination of Cd(II) and Pb(II) ions in honey and milk samples using a single–walled carbon nanohorns modified screen–printed electrochemical sensor. Food Chem. 2019 (274), 8–15.

Ye, Y., Liu, H., Yang, L., Liu, J., 2012. Sensitive and selective SERS probe for trivalent chromium detection using citrate attached gold nanoparticles. Nanoscale 4, 6442–6448.

Yin, B., Ma, H., Wang, S., Chen, S., 2003. Electrochemical synthesis of silver nanoparticles under protection of poly(N–vinylpyrrolidone). J. Phys. Chem. B 107, 8898–8904.

Yola, M.L., Eren, T., İlkimen, H., Atar, N., Yenikaya, C., 2014. A sensitive voltammetric sensor for determination of Cd(II) in human plasma. J. Mol. Liq. 197, 58–64.

Zacco, E., Pividori, M.I., Alegret, S., Galve, R., Marco, M.P., 2006. Electrochemical magneto–immunosensing strategy for the detection of pesticides residues. Anal. Chem. 78, 1780–1788.

Zehani, N., Dzyadevych, S.V., Kherrat, R., Renault, N.J.J., 2014. Sensitive impedimetric biosensor for direct detection of diazinon based on lipases. Front. Chem. 2, 1–7.

Zeng, L., Zeng, H., Jiang, L., Wang, S., Hou, J.-T., Yoon, J., 2019. A single fluorescent chemosensor for simultaneous discriminative detection of gaseous phosgene and a nerve agent mimic. Anal. Chem. 91, 12070–12076.

Zheng, M., Wang, C., Wang, Y., Wei, W., Ma, S., Sun, X., He, J., 2018. Green synthesis of carbon dots functionalized silver nanoparticles for the colorimetric detection of phoxim. Talanta 185, 309–315.

Zhou, Q., Lin, Y., Lin, Y., Wei, Q., Chen, G., Tang, D., 2016. Highly sensitive electrochemical sensing platform for lead ion based on synergetic catalysis of DNAzyme and Au–Pd porous bimetallic nanostructures. Biosens. Bioelectron. 78, 236–243.

CHAPTER

Applications of silver nanomaterial in agricultural pest control

19

Sharmin Yousuf Rikta[a] and P. Rajiv[b]

[a]*Department of Environmental Sciences, Jahangirnagar University, Dhaka, Bangladesh*
[b]*Department of Biotechnology, Karpagam Academy of Higher Education, Coimbatore, Tamil Nadu, India*

19.1 Introduction

Despite all industrial revolution and development, agricultural activity plays a vital role in human civilization by providing the most essential basic need of human beings—the food. Agricultural activity is more important for low- and middle-income countries for sustaining/achieving economic growth, development trends, and crucial for poverty reduction (The World Bank, 2011). Recently, agriculture has been considered as one of the factors of social, environmental, and economic crises, as well as the potential solution tool of those problems (Janvry, 2010).

Excessive pressure on increasing food demand, lower agricultural land availability, soil degradation, and pollution coupled with climate change impacts have made agricultural production a challenging task. The current world Application of silver nanomaterials in population is more than 7.5 billion and increasing day by day. With this ever-increasing population, the developing regions of the world are facing food shortages. Along with the impacts of floods, droughts on agricultural productivity (Joseph and Morrison, 2006) insect pest infestation have threatened the production rate (Rai and Ingle, 2012). Worldwide insect pests, plant pathogens, and weeds caused 14%, 13%, and 13% losses, respectively, which accounted for a value of US $2000 billion per year (Pimentel, 2009). Conventional pest management systems such as integrated pest management are unsatisfactory to meet the demand (Duhan et al., 2017). Despite various alternative techniques, agricultural pest control is still largely dependent on organic chemical-based elements and pesticides, which are neurotoxic to other nontargeted insects and mammals, and it has serious environmental implications (Athanassiou et al., 2018).

Recent application of nanotechnology in the agricultural sector has offered new methods of nanosized novel active ingredient designing, their formulation and targeted delivery (Athanassiou et al., 2018). The modern tools and techniques based on nanotechnology possess a great prospect to overcome the numerous difficulties of traditional agricultural practices (Duhan et al., 2017). It is expected that nanotechnology will help in promoting ecofriendly agriculture and reduce the cost of production through reduced chemical pesticides and fertilizers use; smart and targeted nanoparticle-based chemical pesticides and fertilizers delivery; and monitoring of insect pest infestation, soil condition, as well as plants and crop growth regulation using nanodevices and nanosensors (Athanassiou et al., 2018). Although there is a promising future of nanotechnology in the agricultural sector, the progress is still sluggish than other sectors. Among various nanomaterials, silver nanomaterials have been widely investigated because of their antimicrobial properties. Silver nanomaterials/particles show significant effectiveness against wide ranges of human pathogens. Hence, there is a growing interest to utilize the antimicrobial properties of silver nanoparticles for agricultural pest control (Duhan et al., 2017). This review article is a simple attempt to discuss the recent development in the applications of silver nanomaterials in the agricultural pest control/management sector.

19.2 Conventional pest control strategies

From the beginning of agricultural activities in human history, several traditional methods are being used to manage/control agricultural pests. Conventional techniques like crop rotation, alteration of seed sowing dates, introducing healthy crop varieties, and integrated pest management are commonly being used by the farmers. Before 20th-century crop rotation, changing sowing dates of seeds and using healthy crop varieties were the main tactics of pest management. With the discovery of various chemical pesticides, farmers became more reliant on those due to their higher activity against pests. Although at the beginning none could realize the detrimental impacts of chemical pesticides, later on, it was understood that recovering nontargeted insect species was very difficult. Moreover, targeted species were gradually developing resistance to chemical pesticides requiring repeated applications of pesticides in increased concentration (Duhan et al., 2017).

Among all the techniques, integrated pest management is the most popular form of management tactic. Integrated pest management involves the combination of various strategies such as cultural control, biological and chemical control, introduction of resistant varieties to avoid the negative effects of chemical pesticides on the environment. Implementation of this technique is knowledge-intensive as it requires an understanding of pest dynamics and skill for pest monitoring (Peshin et al., 2009; Duhan et al., 2017). Additionally, reduction in pesticides utilization may have the chance of reduced crop yields (Dhawan and Peshin, 2009). Thus, this technique becomes insufficient for agricultural pest management.

19.3 Nanotechnology in pest control

Nowadays, nanotechnology is a prominent sector for interdisciplinary research. The applications of nanotechnology range from pharmaceutical industries, medical science, cosmetics, and electronics industries to agricultural sector. This involves the use of nanomaterials, which have some novel characteristics that make them to be used in enormous fields. Nanomaterials often exhibit high electric conductivity, chemical reactivity, and high specificity. Recently, nanomaterials are becoming promising in the field of agricultural pest management. Nanomaterial-based pest management offers more benefits than conventional pest management in the agricultural sector (Fig. 19.1). Nanotechnology has provided the solutions of various agricultural problems by developing nanomaterial-based nanopesticides and polymers. Preparation of nanopesticides involves the use of nanosized particles, which bear some novel characteristics that provide great effectiveness against insect pests through their capacity of increasing solubility of poorly soluble ingredients in the insecticides. Nanopesticides provide enhanced affinity to targets as they contain large specific surface areas (Yan et al., 2005; Khot et al., 2012). It is also expected that the nanoinsecticides will release the active/essential ingredients in a controlled manner, which will reduce the overall wastage of the nanoinsecticides (Rikta et al., 2020; Ragaei and Sabry, 2014). Repeated use and higher concentration are required in conventional pesticides application to increase the efficiency and tackle the unexpected loss of chemical pesticides due to degradation, leaching, and evaporation (Sarlak et al., 2014). The application of nanopesticides can potentially reduce the unwanted movement of pesticides because of their improved dispersion and wettability properties (Bergeson, 2010).

Nanomaterials used in agricultural pest control can be in different forms through the common forms are as active elements in pesticides and pesticidal properties bearing nanostructured compounds. Some nanoparticles in aqueous solutions and filled in inorganic base (e.g., aluminosilicate) are being used for agricultural pest control (Ragaei and Sabry, 2014). Various amphiphilic polymers assembled as

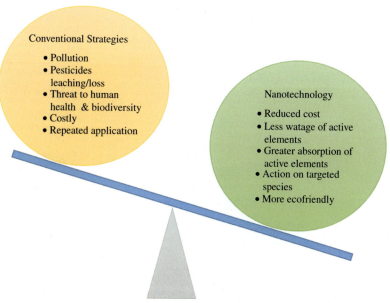

FIG. 19.1

Conventional pest control strategies vs nanotechnology in pest control.

nanoaggregates in aqueous solution can be used to control pests as controlled release formulation. Nanoencapsulations are used for the slow release of chemicals in an efficient way to control pests. Many nanopesticides, nanofungicides, and nanoherbicides are considered as nanoencapsulation due to their slow chemical release properties (Ragaei and Sabry, 2014). Pesticides encapsulated in water-soluble polymerized nanoparticles are more effective and stable compared to the conventional bulk pesticides (Sarlak et al., 2014). Additionally, polymeric nanoparticle encapsulated pesticides offer increased contact area for fungi resulting in more efficacy of the applied pesticides. Nanomaterials can act as both protectants and carriers. They are the protectants of plants in the forms of insecticides, fungicides, and RNA-interference molecules, which protect the plants from a wide range of pests and pathogens (Worrall et al., 2018). Moreover, they also carry active ingredients to the targeted sites and ensure controlled release. Fig. 19.2 demonstrates the roles of nanomaterials in plant protection as well as the benefits of using nanomaterials in agricultural pest control.

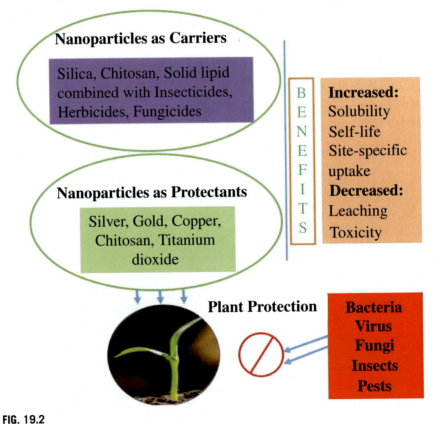

FIG. 19.2

Nanomaterials as protectants or carriers to provide crop protection.

Illustration modified after Worrall, E.A., Hamid, A., Mody, K.T., Mitter, N. and Pappu, H.R., 2018. Nanotechnology for plant disease management. Agronomy 8, 285. https://doi.org/10.3390/agronomy8120285.

19.3.1 Ag NPs and Ag-nanocomposites for insects' control

Manimegalai et al. (2020) found the effectiveness of biosynthesized silver nanoparticles from *Leonotis nepetifolia* against *Spodoptera litura* and *Helicoverpa armigera*. Synthesized particles exhibited strong antifeedant, larvicidal, and pupicidal activities against those pests. Sometimes the modification of bulk chemical pesticides with nanomaterials can significantly enhance the absorption of that chemical pesticides. Such a study was conducted by Yan et al. (2005), and it was found that TiO_2/Ag modified and conjugated with sodium dodecyl sulfate and dimethomorph remarkably increase the absorption of dimethomorph, which is a fungicide used to treat mildew and root rot caused by *Pythium* and *Phytophthora* species. Insecticidal effect of silica and silver nanoparticles on the cowpea seed beetle, *Callosobruchus maculatus* F. was reported by Rouhani et al. (2012). Alif Alisha and Thangapandiyan (2019) evaluated the effectiveness of silver nanoparticle for the mortality rate, ovipositional deterrent, repellent activity, and antifeedant activity of *Tribolium castaneum* (red flour beetle). The study found high insecticidal efficacy of silver nanoparticles while applied in combination with malathion. The effectiveness of silver nanoparticles against *Sitophilus oryzae* and baculovirus BmNPV was observed by Goswami et al. (2010). *Sitophilus oryzae* and baculovirus BmNPV are responsible for rice weevil and grasserie diseases, respectively, in silkworm. A study showed that hydrophilic silver nanoparticles were most effective on first-day exposure. More than 90% and 95% mortality were perceived after 2 and 7 days of exposure of hydrophilic silver nanoparticles. Biosynthesized silver nanoparticles from *Raphanus sativus* var. *aegyptiacus* (white radish) can limit the viability and activity of *Eobania vermiculata* (land snail) and reduce the frequency of the population of fungi in surrounding soil (Ali et al., 2015).

The nematicidal activity of *Euphorbia tirucalli* mediated silver nanoparticles was carried out at in vitro by Kalaiselvi et al. (2019). The different concentration of biogenic silver nanoparticles was inhibiting the second-stage juveniles of *M. incognita*. Harmala alkaloids mediated silver nanomaterials were exhibited insecticidal activity (Almadiy et al., 2018). Few investigators report the insecticide activity of *Ficus religiosa* and *Ficus benghalensis* assisted silver nanoparticles and concludes that silver nanoparticles were disturbing the Ha-gut protease activity (Kantrao et al., 2017). Larvicidal activity of green synthesized silver nanoparticles against *Tribolium castaneum* and *Callosobruchus maculatus* was determined by Jafer and Annon (2018). Moreover, they confirmed that the green synthesized silver nanoparticles exhibited maximum larvicidal activity.

19.3.2 Ag NPs and Ag-nanocomposites for control fungal plant diseases

Silver (Ag) nanomaterials are widely known for their antibacterial, antifungal, and insecticidal properties (Table 19.1). Antifungal efficacy of silver nanoparticles was studied by Jo et al. (2009). They have found the effectivity of silver nanoparticles against *Bipolaris sorokiniana,* which causes seedling diseases, common root rot, and

Table 19.1 Application of silver nanomaterials in agricultural pest control.

Silver nanomaterials	Affected pathogens/pest	Affected plant/experimental setting	Effective dose	Mortality/inhibition rate/effects	References
Green synthesized silver nanoparticle	*Spodoptera litura* and *Helicoverpa armigera*	Cotton, tomato, maize, tobacco, potato	150 ppm	Maximum larval mortality 78.49% and 72.70%, respectively. Maximum pupal mortality 84.66% and 77.44%, respectively.	Manimegalai et al. (2020)
Silver nanocolloid	*Tinea pellionella* L.	Wool fibers	20 ppm, 14 days after exposure	Almost complete mortality	Ki et al. (2007)
Silver nanoparticle	*Macrosiphum rosae* (Hemiptera: Aphididae)	Rose plants	500 ppm, 3 days after exposure	Complete mortality	Bhattacharyya et al. (2016)
Silver nanoparticle	*Pythium spinosum* & *Cylindrocarpon destructans*	In vitro experiment (using corn meal agar media)	25 ppm & 50 ppm, 14 days exposure	100% mortality	Kim et al. (2012)
Silver nanoparticle (containing stearic acid isolated from *Catharanthus roseus*)	*Earias vittella*	Cotton	200 µg/mL	maximum ovicidal activity 95.94% and oviposition deterrent activity 100%	Pavunraj et al. (2020)
Biosynthesized silver nanoparticle	*Lepidiota mansueta* & *Burmeister*	In vitro experiment	4.529 mg/mL & 9.580 mg/mL, 15 days exposure	LC_{50} and LC_{90}	Babu et al. (2014)
Silver nanoparticle	*Plutella xylostella*	Cabbage pest	24.5–38.23 ppm	LC_{50}	Roni et al. (2015)

Silver nanoparticle	*Sitophilus oryzae* L.	Rice pest	213.32, 247.90, 44.69 mg/kg and 1648.08, 2675.13, 168.28 mg/kg	LD_{50} and LD_{90}	Zahir et al. (2012)
Silver nanoparticle	*Helicoverpa armigera*	Cotton pest	10 ppm, after 4 days exposure	High mortality level (80%)	Durga Devi et al. (2014)
Silver nanoparticle	*Ariadne merione*	Pest of castor plant	–	Physiological and anatomical abnormalities	Moorthi et al. (2014)
Synthesized silver nanoparticles from *Euphorbia tirucalli*	*Meloidogyne incognita*	In vitro (mortality and hatching) and in-planta trials (infectivity and crop protection)	100–1000 ng/mL	Mortality rates 96%–100%, hatch inhibition rates 75.34%–100% and plant infestation significantly reduced (46.95%–100%)	Kalaiselvi et al. (2019)
Silver nanoparticles (50–60 nm)	*Spodoptera littoralis*	Cotton	10 mg/mL, 6 days	Reduced larval weight gain and pupal weight	Ibrahim and Ali (2018)
Silver nanoparticles (20–30 nm)	*Magnaporthe grisea*	In vitro and in vivo experiments	100 ppm	Detrimental effect on mycelial growth	Elamawi and El-shafey (2013)

spot blotch of barley and wheat, and *Magnaporthe grisea* liable for rice rotten neck, a blast of rice, rice seedling blight, and pitting diseases of rice. Phytopathogenic fungi like *Fusarium* and *Phoma* are also susceptible to silver nanoparticles (Gajbhiye et al., 2009). Phytopathogenic fungi *Raffaelea* sp. causing the oak tree mortality can be effectively controlled by silver nanoparticles. Kim et al. (2009) revealed that silver nanoparticles can control the spread of *Raffaelea* sp. by inhibiting hyphal growth and conidial germination. Effectiveness of sodium dodecyl sulfate-modified Ag/TiO$_2$ (titanium dioxide) imidacloprid nanoformulation was revealed for pest control (Guan et al., 2010). The application of silver nanoparticles has reduced the usage of chemical fungicides like azoxystrobin and isoprothiolane (Rabab and El-Shafey, 2013). Tuteja and Gill (2013) suggested silver–silica nanocomposites for the long-term control of pests and phytopathogens, which show resistance to antibiotics. Antifungal effectiveness of silver nanoparticles against *Botrytis cinerea* and *Alternaria alternate* was investigated by Ouda (2014). *Botrytis cinerea* is a fungus species that affects many plants, especially wine grapes, and *Alternaria alternate* causes leaf spot and other diseases. The study found the damage of fungal hyphae and conidia when silver nanoparticles were applied. Maximum inhibition of fungal hyphal growth was observed at 15 mg/L concentration of nanoparticle exposure. Antifungal activity of biosynthesized silver nanoparticles against *Rhizopus* sp. and *Aspergillus* sp. was reported by Medda et al. (2015) through the inhibition of hyphae growth, normal budding process, and conidial germination. Antifungal property of silver nanoparticles synthesized from *Trichoderma longibrachiatum* against *Fusarium verticillioides*, *Fusarium moniliforme*, *Penicillium brevicompactum*, *Helminthosporium oryzae*, and *Pyricularia grisea* phytopathogens was observed by Elamawi et al. (2018). It was reported that the application of synthesized silver nanoparticles can significantly reduce the number of fungal colonies, and the observed efficiency was up 90%.

Xiang et al. (2019) synthesized alginate-coated silver nanoparticles that had antifungal characteristics. The study found that the alteration of cell permeability, destruction of DNA structure, inhibition of DNA replication, and disturbance of insoluble protein synthesis are the main mechanisms of the antifungal activity of alginate-coated silver nanoparticles against plant pathogenic *Colletotrichum gloeosporioides* Penz.

Tomah et al. (2020) demonstrated that inhibition of *Sclerotinia sclerotiorum* using the fungal-assisted silver nanoparticles. They have used the cell-free aqueous filtrate of *T.virens* HZA14 for the synthesis of AgNPs. Phytopathogenic fungi (*Alternaria* sp.) was used for the fabrication of Ag nanoparticles, which kill the fungal pathogens (*Fusarium* sp. and *Alternaria* sp.) (Win et al., 2020). A few of the researchers concluded that mycogenic-mediated silver nanoparticles show the significant biological activity and act as potential bacterial agents to prevent the infections and wound healing activity. Malandrakis et al. (2020) were synthesized the silver nanoparticles using the Seed-borne fungus *Penicillium duclauxii* by mycosynthesis and it is inhibited the growth of *Bipolaris sorghicola* at in vitro. *As-synthesized silver nanoparticles (40 μg/mL) via extract of endophytic bacteria were killing the rice pathogens (Magnaporthe oryzae).* Sahayaraj et al. (2020) stated that *Pongamia glabra* mediated Ag nanoparticles exhibited antiphytopathogenic activity against *Rhizopus nigricans* which may be used as nanofungicide. Carbendazim-conjugated silver nanoparticles were produced by Nagaraju et al. (2020), which suppress the growth of anthracnose disease-causing agent *Colletotrichum gloeosporioides.* Basurto et al. (2020) explained the usage of silver nanoparticles coated with chitosan nanoparticles for controlling the vascular wilt on tomato. Silver nanoparticles coated with chitosan nanoparticles decreased the severity of vascular wilt on tomato seedling. Based on the above research reports gives the idea of Ag and Ag composites for control the fungal diseases in plants (Table 19.2).

19.3.3 Ag NPs and Ag-nanocomposites for control bacterial plant diseases

Antibacterial activity of silver nanomaterials against phytopathogens was investigated by limited researchers though many types of research have been done on antibacterial activity against human pathogenic bacteria (Baker et al., 2017). Antibacterial effectiveness of *Nephrolepis exaltata* L. synthesized silver nanoparticles against *Xanthomonas axonopodis* pv. *punicae* was reported by Bhor et al. (2014), which causes bacterial blight in pomegranate (*Punica granatum*). Biosynthesized silver nanoparticles from the leaf and stem extracts of *Piper nigrum* demonstrated antibacterial activity against *Erwinia cacticida* and *Citrobacter freundii*, which are phytopathogenic bacteria (Paulkumar et al., 2014). A similar result was also observed by a research conducted by Chowdhury et al. (2014). That research revealed potential antibacterial activity of biosynthesized silver nanoparticles (from *Macrophomina phaseolina*) against phytopathogenic *Agrobacterium tumefaciens.* Many citrus plant species are affected by bacteria. *Xanthomonas axonopodis* causes citrus canker, which is a serious disease affecting citrus species. Aravinthan et al. (2015) reported significant growth reduction of that bacteria while exposed to phytosynthesized silver nanoparticles. This study also found the same result against *Ralstonia solanacearum*, which causes bacterial wilt in a very wide range of potential host plants including tobacco.

Ibrahim et al. (2020) suggested that silver nanoparticles exhibited the antibacterial activity, and it suppresses and inhabited the growth of rice pathogenic bacteria. Silver nanoparticles were improving plant growth. The biofabricated silver nanomaterials (*Mentha arvensis* mediated) reduced the growth of different bacterial strains at dose-dependent on the manner (Javed et al., 2020). The biopolymer-assisted silver nanoparticles to control the *Ralstonia solanacearum* inducing disease (bacterial wilt) in tomato. The control of plant bacterial infections using silver nanoparticles was described by Shang et al. (2020) and stated that carboxymethylcellulose sodium-stabilized silver nanoparticles (13.53 ± 4.72 nm) have good antibacterial properties against *Xanthomonas oryzae* pv. *Oryzae*. An effective agriculture application of green-synthesized silver nanoparticles by *Stenotrophomonas*

Table 19.2 Some application of silver nanomaterials in plant diseases management.

Silver nanomaterials	Affected pathogens/pest	Affected plant/ experimental setting	Effective dose	Mortality/ inhibition rate/ effects	References
Silver nanoparticle	*Fusarium verticillioides, Fusarium moniliforme, Penicillium brevicompactum, Helminthosporium oryzae, & Pyricularia grisea*	Plant fungi	At a concentration of 105 spores/mL were incubated at 1 mM AgNO$_3$ solution, 48 h exposure	Efficiencies up to 90%	Elamawi et al. (2018)
Silver nanoparticle	*Sclerotinia sclerotiorum*	Plant fungi	–	Inhibition of sclerotia germination and mycelial growth	Guilger et al. (2017)

sp. was explained by the Mishra et al. (2017), and they demonstrated the AgNPs applications for controlling the collar rot of chickpea caused by *S. rolfsii* under greenhouse conditions. Vanti et al. (2019) synthesized *Gossypium hirsutum* mediated silver nanoparticles and evaluated the antibacterial efficiency using the phytopathogens (*Xanthomonas campestris* pv. Campestris and *Xanthomonas axonopodis* pv. *malvacearum*).

19.3.4 Ag NPs and Ag-nanocomposites for control viral plant diseases

The controlling of viral infection in crops is most difficult challenges because there is no effective control strategy. Hence, the researchers have proposed and investigated various methods to control the viral infections in plants for improving the crop production by nanotechnology approach. Past few decades, the researchers are concentrating and giving more attention to the antiviral activity of nanoparticles. A microbial-mediated silver nanoparticle shows the antiviral activity, and it inhibits the tobacco mosaic virus. Silver nanoparticles synthesized using the fermented broth by *Pseudomonas fluorescens* (Ahsan, 2020). Banana bunchy top virus (BBTV) can be killed by silver nanoparticles (Mahfouze et al., 2020). Few investigators were biosynthesized the silver nanoparticles and assessed their antiviral activity against the Mung bean yellow mosaic virus (MYMV). Pleurotus species (Mushroom) was used for the production of biogenic silver nanoparticles (Ansari, 2020). *Spirulina platensis* extract was employed for synthesis of silver nanoparticles and was performed by Tamilselvan et al. (2017). The interaction of *Bombyx mori* nuclear polyhedrosis virus (BmNPV) was characterized and observed by various microscopic techniques. Ag NPs were accumulated in the viral membrane and reached in capsid, which was confirmed by the TEM images. Till date, there is no study for determining the mechanisms of silver nanoparticles' action, dosage level, application intervals, and its effects.

19.3.5 Anti-plant pathogenic mechanisms

Several metals and metal oxide nanoparticles have been recognized for their bactericidal and fungicidal properties (Kikuchi et al., 1997; Lu et al., 2006). Silver nanomaterials are being increasingly used in the agricultural sector due to their insecticidal and antimicrobial properties (Zhang et al., 2012). Despite the availability of the huge volume of information about the toxicity against pests, pathogens, and vectors, any precise information on the possible mode of action of nanoparticles against pests and insects are stringently limited (Benelli, 2018). Benelli (2018) has done a detailed study on the mood of action of nanoparticles against insects. The mechanisms by which silver nanomaterials/particles kill insect pests vary but mainly through the destruction of cell membranes. Guilger et al. (2017) found that necrosis and apoptosis are responsible for substantial cell death. It was revealed that necrosis and apoptosis are caused by the disruption of cell membrane integrity and damage of DNA through chromatin condensation, respectively. Both necrosis and apoptosis can cause the death of pests though the severity depends on the size, concentration, and exposure time of silver nanoparticles (Kumar et al., 2015). Various studies showed that the production of reactive oxygen species (ROS) such as hydrogen peroxide, hydroxyl radical by silver, and other nanoparticles leads to cytotoxicity, which ultimately result in cellular destruction (Rikta, 2019; Guilger et al., 2017; Kumar et al., 2015; Zhang et al., 2015). A generalized concept of the antimicrobial mode of action of nanomaterials is depicted in Fig. 19.3.

Dalrymple et al. (2010) addressed intracellular and extracellular target sites that can be targeted by nanoparticles to destroy cells. Extracellular target sites are primarily cell wall and cell membrane, which consist of phospholipid bilayer, peptidoglycan, and liposaccharide layer. All those layers are susceptible to free radical attack produced by ROS. Nucleic acids, enzymes, and coenzymes are important intracellular target sites. Exposure to nanoparticles can inactivate enzymes and destroy nucleic acid through the production of superoxide (Dalrymple et al., 2010; Blum and Fridovich, 1985). The oxidation of coenzymes (Matsunaga et al., 1985) and breakage of DNA strand by ROS attack can damage biomolecules, leading to cell death (Dalrymple et al., 2010).

Although several kinds of research studies have been done on nanomaterials' use in agricultural pest control and many are in progress, it should be kept in mind that safeguarding human health and environment from long-term impacts of nanomaterials is one of the major concerns. Agriproducts and foodstuffs containing nanomaterials bearing pesticides residues could be a potential cause of public health impacts (Chaudhry and Castle, 2011). Further investigation is necessary to address the toxicity and ultimate fate of the specific nanomaterial in the environment.

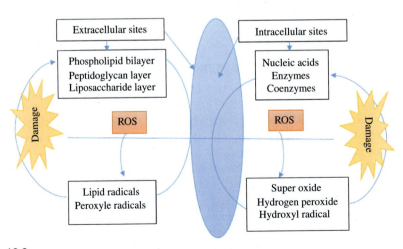

FIG. 19.3

Antimicrobial mode of action of nanomaterials.

19.4 Conclusion

This chapter highlights the applications of silver nanomaterials for controlling agricultural pests and pathogens. This also exposes the prospects of extensive use of silver nanomaterials in agricultural pest management. It may offer relatively green, ecofriendly, and efficient techniques of pest control in the agricultural sector. Recent advancement in nanotechnology has a promising future in the agricultural system that can help to meet the ever-growing food demand of the increasing world population at reduced costs. It has opened a new window of opportunities but not free from constraints. There is a risk of nanomaterials toxicity in the environment. Unused and residues of nanopesticides can affect the nontargeted species. Nanopesticides can act as a smart delivery system for releasing pesticides. A collaborative effort of researcher and expert groups is required to evaluate the transfer, transport, and transformation process of nanomaterials in the environment and to develop various biosynthesis techniques of nanomaterials to minimize toxicity and ensure biosafety. Biosynthesis of silver nanoparticles via an extract of plant and microbes is a perfect ecofriendly method for the management of insect pests. This chapter is authoritative for forthcoming registration and labeling of the silver nanoparticles if they are to be employed as fungicides for protection of crops. The bionanotechnology for the production of nanopesticides are high-grade, energy-efficient, low-cost method, which reduces the generated waste and usage of toxic chemicals. An introduction and implementation of usage of biogenic silver nanoparticles as an innovative powerful nanopesticide agent can change the significant impact of crop production. Bionanotechnology for production of nanopesticides using the biogenic silver is gradually being incorporated into the field of the agricultural industry. The biogenic silver nanoparticles will help to develop the environment-friendly nanomaterials for the disease management of the crops. This technology will prevent the excess usage of pesticides and prevent soil and water pollution and associated its hazardous problems, poor targeted delivery, and off-target waste accumulation in soil and water. Innovative technology is employed for the preparation of nanopesticides using the silver engineered nanomaterials. The conjugated silver nanomaterials or engineered nanomaterials are more responsive against plant pathogens and pest. The unique features of nanomaterials are exploited to control and manage the plant-pest chain. Biogenic silver nanoparticles did not show any negative effects on the germination and growth of various crops while applying the nanopesticides and other nanoformulations. More research should be investigated to determine the toxic effects of biogenic silver nanomaterials before use in field condition.

References

Ahsan, T., 2020. Biofabrication of silver nanoparticles from *Pseudomonas fluorescens* to control tobacco mosaic virus. Egypt J. Biol. Pest Control 30 (1), 1–4.

Ali, S.M., Yousef, N.M.H., Nafady, N.A., 2015. Application of biosynthesized silver nanoparticles for the control of land snail *Eobania vermiculata* and some plant pathogenic fungi. J. Nanomater. 2015, 10.

Alif Alisha, A.S., Thangapandiyan, S., 2019. Comparative bioassay of silver nanoparticles and malathion on infestation of red flour beetle, *Tribolium castaneum*. J. Basic Appl. Zool 80 (1), 55.

Almadiy, A.A., Nenaah, G.E., Shawer, D.M., 2018. Facile synthesis of silver nanoparticles using harmala alkaloids and their insecticidal and growth inhibitory activities against the khapra beetle. J. Pestic. Sci. 91 (2), 727–737.

Ansari, M., 2020. Biosynthesis of Silver Nanoparticles AgNPs and Sustainable Management of Mung Bean Yellow Mosaic Virus MYMV Disease. PhD Report. Department of Botany, Faculty of Science, Dayalbagh educational, institute Dayalbagh. Agra-282005.

Aravinthan, A., Govarthanan, M., Selvam, K., Praburaman, L., Selvankumar, T., Balamurugan, R., Kamala-Kannan, S., Kim, J.H., 2015. Sunroot mediated synthesis and characterization of silver nanoparticles and evaluation of its antibacterial and rat splenocyte cytotoxic effects. Int. J. Nanomedicine 10, 1977–1983.

Athanassiou, C.G., Kavallieratos, N.G., Benelli, G., Losic, D., Usha Rani, P., Desneux, N., 2018. Nanoparticles for pest control: current status and future perspectives. J. Pestic. Sci. 91, 1–15.

Babu, M.Y., Devi, V.J., Ramakritinan, C.M., Umarani, R., Taredahalli, N., Kumaraguru, A.K., 2014. Application of biosynthesized silver nanoparticles in agricultural and marine pest control. Curr. Nanosci 10 (3).

Baker, S., Volova, T., Prudnikova, S.V., Satish, S., Prasad, N., 2017. Nanoagroparticles emerging trends and future prospect in modern agriculture system. Environ. Toxicol. Pharmacol. 53, 10–17.

Basurto, D.A.E., Carvajal, F.A., Calderon, A.A., Soto, T.E.G., Miranda, E.E., Onofre, J.J., Ibarra, R.M., 2020. Silver nanoparticles coated with chitosan against *Fusarium oxysporum* causing the tomato wilt. Biotecnia 22 (3), 73–80.

Benelli, G., 2018. Mode of action of nanoparticles against insects. Environ. Sci. Pollut. Res. 25 (13), 12329–12341.

Bergeson, L.L., 2010. Nanosilver: US EPA's pesticide office considers how best to proceed. Environ. Qual. Manag. 19, 79–85.

Bhattacharyya, A., Prasad, R., Buhroo, A.A., Duraisamy, P., Yousuf, I., Umadevi, M., Bindhu, M.R., Govindarajan, M., Khanday, A.L., 2016. One-pot fabrication and characterization of silver nanoparticles using *Solanum lycopersicum*: an eco-friendly and potent control tool against rose aphid, Macrosiphum rosae. J. Nanosci 7, 4679410. https://doi.org/10.1155/2016/4679410.

Bhor, G., Maskare, S., Hinge, S., Singh, L., Nalwade, A., 2014. Synthesis of silver nanoparticles by using leaflet extract of *Nephrolepis exaltata* L. and evaluation of antibacterial activity against human and plant pathogenic bacteria. Asian J. Pharm. Tech. 02 (07). 2014.

Blum, J., Fridovich, I., 1985. Inactivation of glutathione peroxidase by superoxide radical. Arch. Biochem. Biophys. 240, 500–508.

Chaudhry, Q., Castle, L., 2011. Food applications of nanotechnologies: an overview of opportunities and challenges for developing countries. Trends Food Sci. Technol. 22, 595–603.

Chowdhury, S., Basu, A., Kundu, S., 2014. Green synthesis of protein capped silver nanoparticles from phytopathogenic fungus *Macrophomina phaseolina* (Tassi) Goid with antimicrobial properties against multidrug-resistant bacteria. Nanoscale Res. Lett. 9, 365.

Dalrymple, O.K., Stefanakos, E., Trotz, M.A., Goswami, D.Y., 2010. A review of the mechanisms and modeling of photocatalytic disinfection. Appl Catal B 98, 27–38.

Dhawan, A.K., Peshin, R., 2009. Integrated pest management: concept, opportunities and challenges. In: Peshin, P., Dhawan, A.K. (Eds.), Integrated Pest Management: Innovation-Development Process. Springer, Dordrecht, Netherlands, pp. 51–81.

Duhan, J.S., Kumar, R., Kumar, N., Kaur, P., Nehra, K., Duhan, S., 2017. Nanotechnology: the new perspective in precision agriculture. Biotechnol. Rep. 15, 11–23.

Durga Devi, G., Murugan, K., Selvam, C.P., 2014. Green synthesis of silver nanoparticles using *Euphorbia hirta* (Euphorbiaceae) leaf extract against crop pest of cotton bollworm, *Helicoverpa armigera* (Lepidoptera: Noctuidae). J. Biopest. 7 (Supp), 54–66.

Elamawi, R.M.A., El-shafey, R.A.S., 2013. Inhibition effects of silver nanoparticles against rice blast disease caused by *magnaporthe grisea*. Egypt J. Agric Res. 91 (4), 1271–1283.

Elamawi, R.M.A., Al-Harbi, R.E., Hendi, A.A., 2018. Biosynthesis and characterization of silver nanoparticles using *Trichoderma longibrachiatum* and their effect on phytopathogenic fungi. Egypt J. Biol. Pest Control 28, 28.

Gajbhiye, M., Kesharwani, J., Ingle, A., Gade, A., Rai, M., 2009. Fungus mediated synthesis of silver nanoparticles and its activity against pathogenic fungi in combination of fluconazole. Nanomedicine 5 (4), 282–286.

Goswami, A., Roy, I., Sengupta, S., Debnath, N., 2010. Novel applications of solid and liquid formulations of nanoparticles against insect pests and pathogens. Thin Solid Films 519, 1252–1257.

Guan, H., Chi, D., Yu, J., Li, H., 2010. Dynamics of residues from a novel nano-imidacloprid formulation in soyabean fields. Crop Prot. 29, 942–946.

Guilger, M., Pasquoto-Stigliani, T., Bilesky-Jose, N., Grillo, R., Abhilash, P.C., Fraceto, L.F., Renata de Lima, R.D., 2017. Biogenic silver nanoparticles based on *Trichoderma harzianum*: synthesis, characterization, toxicity evaluation and biological activity. Nat. Sci. Rep. 7, 44421.

Ibrahim, A.M., Ali, A.M., 2018. Silver and zinc oxide nanoparticles induce developmental and physiological changes in the larval and pupal stages of *Spodoptera littoralis* (Lepidoptera: Noctuidae). J. Asia Pac. Entomol. 21 (4), 1373–1378.

Ibrahim, E., Luo, J., Ahmed, T., Wu, W., Yan, C., Li, B., 2020. Biosynthesis of silver nanoparticles using onion endophytic bacterium and its antifungal activity against Rice pathogen *Magnaporthe oryzae*. J. Fungi 6 (4).

Jafer, F.S., Annon, M.R., 2018. Larvicidal effect of pure and green-synthesized silver nanoparticles against *Tribolium castaneum* (herb.) and *Callosobruchus maculatus* (fab.). J. Global Pharma. Technol. 10 (3), 448–454.

Janvry, A.d., 2010. Agriculture for development: new paradigm and options for success. Agric. Econ. 41, 17–36.

Javed, B., Raja, N.I., Nadhman, A., Mashwani, Z.U.R., 2020. Understanding the potential of bio-fabricated non-oxidative silver nanoparticles to eradicate Leishmania and plant bacterial pathogens. Appl. Nanosci, 123456789.

Jo, Y.K., Kim, B.H., Jung, G., 2009. Antifungal activity of silver ions and nanoparticles on phytopathogenic fungi. Plant Dis. 93, 1037–1043.

Joseph, T., Morrison, M., 2006. Nanotechnology in agriculture and food: a nanoforum report. Available at https://www.nanowerk.com/nanotechnology/reports/reportpdf/report61.pdf. (accessed on 02.04.2020).

Kalaiselvi, D., Mohankumar, A., Shanmugam, G., Nivitha, S., Sundararaj, P., 2019. Green synthesis of silver nanoparticles using latex extract of *Euphorbia tirucalli*: a novel approach for the management of root-knot nematode *Meloidogyne incognita*. Crop Prot. 117, 108–114.

Kantrao, S., Ravindra, M.A., Akbar, S.M.D., Jayanthi, P.K., Venkataraman, A., 2017. Effect of biosynthesized silver nanoparticles on growth and development of *Helicoverpa armigera* (Lepidoptera: Noctuidae): interaction with midgut protease. J. Asia Pac. Entomol. 20 (2), 583–589.

Khot, L.R., Sankaran, S., Maja, J.M., Ehsani, R., Schuster, E.W., 2012. Applications of nanomaterials in agricultural production and crop protection: a review. Crop Prot. 35, 64–70.

Ki, H.Y., Kim, J.H., Kwon, S.C., Jeong, S.H., 2007. A study on multifunctional wool textiles treated with nano-sized silver. J. Mater. Sci. 42, 8020–8024.

Kikuchi, Y., Sunada, K., Iyoda, T., Hashimoto, K., Fujishima, A., 1997. Photocatalytic bactericidal effect of TiO_2 thin films: dynamic view of the active oxygen species responsible for the effect. J. Photochem. Photobiol. A Chem. 106, 51–56.

Kim, S.W., Kim, K.S., Lamsal, K., Kim, Y.J., Kim, S.B., Jung, M., Sim, S.J., Kim, H.S., Chang, S.J., Kim, J.K., Lee, Y.S., 2009. An in vitro study of the antifungal effect of silver nanoparticles on oak wilt pathogen *Raffaelea* sp. J. Microbiol. Biotechnol. 19, 760–764.

Kim, S.W., Jung, J.H., Lamsal, K., Kim, Y.S., Min, J.S., Lee, Y.S., 2012. Antifungal effects of silver nanoparticles (AgNPs) against various plant pathogenic fungi. Mycobiology 40 (1), 53–58.

Kumar, G., Degheidy, H., Casey, B.J., Goering, P.L., 2015. Flow cytometry evaluation of in vitro cellular necrosis and apoptosis induced by silver nanoparticles. Food Chem. Toxicol. 85, 45–51.

Lu, J.W., Li, F., Guo, T., Lin, L., Hou, M., Liu, T., 2006. TiO_2 photocatalytic antifungal technique for crops diseases control. J. Environ. Sci. 18, 397–401.

Mahfouze, H.A., El-Dougdoug, N.K., Mahfouze, S.A., 2020. Virucidal activity of silver nanoparticles against Banana bunchy top virus (BBTV) in banana plants. Bull. Nat. Res. Cent. 44 (1), 1–11.

Malandrakis, A.A., Kavroulakis, N., Chrysikopoulos, C.V., 2020. Use of silver nanoparticles to counter fungicide-resistance in *Monilinia fructicola*. Sci. Total Environ. 747, 141287.

Manimegalai, T., Raguvaran, K., Kalpana, M., Maheswaran, R., 2020. Green synthesis of silver nanoparticle using *Leonotis nepetifolia* and their toxicity against vector mosquitoes of *Aedes aegypti* and *Culex quinquefasciatus* and agricultural pests of *Spodoptera litura* and *Helicoverpa armigera*. Environ. Sci. Pollut. Res. https://doi.org/10.1007/s11356-020-10127-1.

Matsunaga, T., Tomoda, R., Nakajima, T., Wake, H., 1985. Photoelectrochemical sterilization of microbial cells by semiconductor powders. FEMS Microbiol. Lett. 29, 211–214.

Medda, S., Hajra, A., Dey, U., et al., 2015. Biosynthesis of silver nanoparticles from *Aloe vera* leaf extract and antifungal activity against *Rhizopus* sp. and *Aspergillus* sp. Appl. Nanosci. 5, 875–880.

Mishra, S., Singh, B.R., Naqvi, A.H., Singh, H.B., 2017. Potential of biosynthesized silver nanoparticles using *Stenotrophomonas* sp. BHU-S7 (MTCC 5978) for management of soil-borne and foliar phytopathogens. Sci. Rep. 7, 45154.

Moorthi, P.V., Balasubramanian, C., Mohan, S., 2014. An improved insecticidal activity of silver nanoparticle synthesized by using *Sargassum muticum*. Appl. Biochem. Biotechnol. https://doi.org/10.1007/s12010-014-1264-9.

Nagaraju, R.S., Sriram, R.H., Achur, R., 2020. Antifungal activity of Carbendazim-conjugated silver nanoparticles against anthracnose disease caused by *Colletotrichum gloeosporioides* in mango. J. Plant Pathol. 102 (1), 39–46.

Ouda, S.M., 2014. Antifungal activity of silver and copper nanoparticles on two plant pathogens, *Alternaria alternata* and *Botrytis cinerea*. Res. J. Micbiol. 9, 34–42.

Paulkumar, K., Gnanajobitha, G., Vanaja, M., Rajeshkumar, S., Malarkodi, C., Pandian, K., Annadurai, G., 2014. *Piper nigrum* leaf and stem assisted green synthesis of silver nanoparticles and evaluation of its antibacterial activity against agricultural plant pathogens. Sci. World J. 2014, 829894. (2014), 9 pages.

Pavunraj, M., Baskar, K., Arokiyaraj, S., Rajapandiyan, K., Alqarawi, A.A., Allah, E.F.A., 2020. Silver nanoparticles containing stearic acid isolated from *Catharanthus roseus*: Ovicidal and oviposition-deterrent activities on *Earias vittella* and ecotoxicological studies. Pestic. Biochem. Physiol. 168, 104640.

Peshin, R., Bandral, R.S., Zhang, W.J., Wilson, L., Dhawan, A.K., 2009. Integrated pest management: a global overview of history, programs and adoption. In: Peshin, P., Dhawan, A.K. (Eds.), Integrated Pest Management: Innovation-Development Process. Springer, Dordrecht, Netherlands, pp. 1–50.

Pimentel, D., 2009. Pesticide and pest control. In: Peshin, P., Dhawan, A.K. (Eds.), Integrated Pest Management: Innovation-Development Process. Springer, Dordrecht, Netherlands, pp. 83–87.

Ragaei, M., Sabry, A.H., 2014. Nanotechnology for insect pest control. Int. J. Environ. Sci. Technol. 3 (2), 528–545. ISSN 2278–3687 (O).

Rai, M., Ingle, A., 2012. Role of nanotechnology in agriculture with special reference to management of insect pests. Appl. Microbiol. Biotechnol. 94, 287–293.

Rikta, S.Y., 2019. Application of nanoparticles for disinfection and microbial control of water and wastewater. In: Ahsan, A., Ismail, A.F. (Eds.), Nanotechnology in Water and Wastewater Treatment. Elsevier, pp. 159–176.

Rikta, S.Y., Parvin, F., Tareq, S.M., 2020. Prospects of hybrid nanomaterials in plant growth promotion. In: Abd-Elsalam, K.A. (Ed.), Multifunctional hybrid nanomaterials for sustainable Agri-food and ecosystems. Elsevier, pp. 319–333.

Roni, M., Murugan, K., Panneerselvam, C., JSubramaniam, J., Nicolett, M., Madhiyazhagan, P., Dinesh, D., Suresh, U., Khater, H.F., Wei, H., Canale, A., Alarfaj, A.A., Munusamy, M.A., Higuchi, A., Benell, G., 2015. Characterization and biotoxicity of *Hypnea musciformis*-synthesized silver nanoparticles as potential eco-friendly control tool against *Aedes aegypti* and *Plutella xylostella*. Ecotoxicol. Environ. Saf. 121, 31–38.

Rouhani, M., Samih, M.A., Kalantri, S., 2012. Insecticidal effect of silica and silver nanoparticles on the cowpea seed beetle, *Callosobruchus maculatus* F. (Col.: Bruchidae). J. Entomol. Res 4, 297–305.

Sahayaraj, K., Balasubramanyam, G., Chavali, M., 2020. Green synthesis of silver nanoparticles using dry leaf aqueous extract of *Pongamia glabra* vent (fab.), characterization and phytofungicidal activity. Environ. Nanotechnol. Monit. Manag 14, 100349.

Sarlak, N., Taherifar, A., Salehi, F., 2014. Synthesis of nanopesticides by encapsulating pesticide nanoparticles using functionalized carbon nanotubes and application of new nanocomposite for plant disease treatment. J. Agric. Food Chem. 62, 4833–4838.

Shang, H., Zhou, Z., Wu, X., Li, X., Xu, Y., 2020. Sunlight-induced synthesis of non-target biosafety silver nanoparticles for the control of Rice bacterial diseases. Nanomaterials 10 (10), 2007.

Tamilselvan, S., Ashokkumar, T., Govindaraju, K., 2017. Microscopy based studies on the interaction of bio-based silver nanoparticles with *Bombyx mori* nuclear polyhedrosis virus. J. Virol. Methods 242, 58–66.

The World Bank, 2011. Agriculture and Development: A Brief Review of the Literature. Available at https://documents.worldbank.org/en/publication/documents-reports/documentdetail/389411468330915034/agriculture-and-development-a-brief-review-of-the-literature. (Accessed on 05.04.2020).

Tomah, A.A., Alamer, I.S.A., Li, B., Zhang, J.Z., 2020. Mycosynthesis of silver nanoparticles using screened *Trichoderma* isolates and their antifungal activity against *Sclerotinia sclerotiorum*. Nanomaterials 10 (10), 1955.

Tuteja, N., Gill, S.S., 2013. Crop Improvement Under Adverse Conditions. Springer-Verlag, New York. https://doi.org/10.1007/978-1-4614-4633-0.

Vanti, G.L., Nargund, V.B., Vanarchi, R., Kurjogi, M., Mulla, S.I., Tubaki, S., Patil, R.R., 2019. Synthesis of *Gossypium hirsutum*-derived silver nanoparticles and their antibacterial efficacy against plant pathogens. Appl. Organomet. Chem. 33 (1), e4630.

Win, T.T., Khan, S., Fu, P., 2020. Fungus- (*Alternaria* sp.) mediated silver nanoparticles synthesis, characterization, and screening of antifungal activity against some Phytopathogens. J. Nanotechnol. 2020.

Worrall, E.A., Hamid, A., Mody, K.T., Mitter, N., Pappu, H.R., 2018. Nanotechnology for plant disease management. Agronomy 8, 285. https://doi.org/10.3390/agronomy8120285.

Xiang, S., Ma, X., Huan Shi, H., Ma, T., Tian, C., Chen, Y., Chen, H., Chen, X., Luo, K., Cai, L., Wang, D., Xue, Y., Huang, J., Sun, X., 2019. Green synthesis of an alginate-coated silver nanoparticle shows high antifungal activity by enhancing its cell membrane penetrating ability. ACS Appl Bio Mater 2 (9), 4087–4096.

Yan, J., Kelong, H., Yuelong, W., Suqin, L., 2005. Study on anti-pollution nano-preparation of dimethomorph and its performance. Chin. Sci. Bull. 50 (2), 108–112.

Zahir, A.A., Bagavan, A., Kamaraj, C., Elango, G., Rahuman, A.A., 2012. Efficacy of plant-mediated synthesized silver nanoparticles against *Sitophilus oryzae*. J. Biopest. 5 (supplementary), 95–102.

Zhang, Z., Kong, F., Vardhanabhuti, B., Mustapha, A., Lin, M., 2012. Detection of engineered silver nanoparticle contamination in pears. J. Agric. Food Chem. 60, 10762–10767.

Zhang, X.F., Choi, Y.J., Han, J.W., Kim, E., Park, J.H., Gurunathan, S., Kim, J.H., 2015. Differential nanoreprotoxicity of silver nanoparticles in male somatic cells and spermatogonial stem cells. Int. J. Nanomedicine 10, 1335–1357.

CHAPTER 20

Silver nanoparticles for insect control: Bioassays and mechanisms

Usha Rani Pathipati and Prasanna Laxmi Kanuparthi
Biology and Biotechnology Division, Indian Institute of Chemical Technology, Hyderabad, India

20.1 Introduction

The discovery of nanoparticles (NPs) and their utilization in insect pest control is a promising area of research. Nanoparticles exhibit completely new or improved properties as compared to the larger particles of the bulk material that they are composed. Due to these improved properties, NPs possess several bioactivities including antibacterial, antifungal, and antiviral, depending on their size, morphology, distribution, and other specific characteristics. Over the past several years, production of NPs has taken a diversification into the utilization of energy-efficient, low-cost, and nontoxic products such as plants, microbes, and algae. These biosynthesized NPs are more effective as pest and disease control agents and are assessed by standard methods for various biological activities. The high surface-to-volume ratio and unique optical properties combined with other specific characters NPs hold greater promise regarding their application in plant protection. Recently, biosynthetic strategies employing either bio-organisms or vegetative extracts have emerged as easy and feasible prospects compared to complex chemical synthesis methods (Fig. 20.1). The biogenic synthesized NPs are considered environmentally acceptable and show greater advantages than the chemically synthesized NPs.

Silver has been used in various forms and shapes since ancient times. Novel use of silver that has garnered high importance in recent years is its use in nanotechnology. Nanomaterials are nanoscale structures that while retaining the chemical structure of the parent product tend to have distinct physicochemical properties such as electric charge, reactivity, quantum effects, and specific area. These new properties can have great promise for developing new products and tools for crop protection. Excessive work is being done on nanosilver use in producing new pesticides, insecticides, and insect repellents, given the high accessibility of NPs across insect cell membranes owing to their high surface-area-to-volume ratio (Zahir and Rahuman, 2012; Zahir et al., 2012; Thamer et al., 2017; Sahayaraj, 2017) (Fig. 20.2).

The chemical and biological synthesis of nanosilver from its crude form has been reported extensively in recent times (Usha Rani and Rajasekhar Reddy, 2011; Rajasekharreddy et al., 2010; Hochella et al., 2015; Krishna and Singaracharya,

CHAPTER 20 Silver nanoparticles for insect control

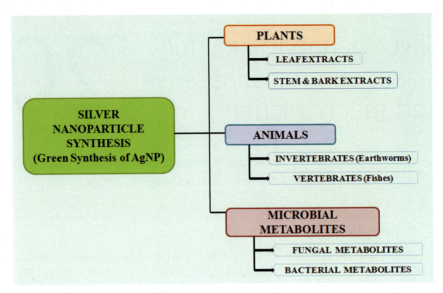

FIG. 20.1
Schematic diagram showing the green synthesis of nanosilver and the biosources.

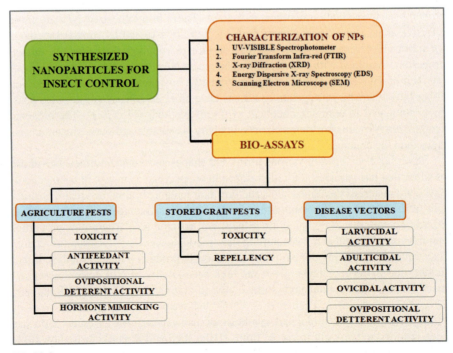

FIG. 20.2
Nanosilver synthesis and insect bioassays.

2016). Progressively advanced methods of synthesis have it made possible to manufacture and establish effective antibacterial and antifungal properties of silver nanoparticles (clusters of silver ions or atoms, AgNP) of sizes ranging from 1 to 100 nm (Gurunathan et al., 2009; Zhang et al., 2016). Interestingly, nanosilver has remarkably diverse and distinct properties than conventional silver, and it can provide new advantages and opportunities to multiple industries. Moreover, nanomaterials can be incorporated into any consumer product other than food products, enabling nanosilver to be widely impregnated into a broad range of consumer products (about more than 250 known till now) (Stark et al., 2015; Verma and Maheshwari, 2019). Among the most useful applications of nanosilver, recent development and most promising utilization is in insect and other pest management techniques, with a focus on crop protection and crop production and is relatively very recent and unexplored. Conventional methods of crop protection include the application of insecticides, fungicides, and herbicides often in larger quantities, which result not only in an increase of the crop production expenses but also in adverse environmental issues. As more than 90% of applied pesticides are unable to reach the target site, their efficacy is been reduced (Nuruzzaman et al., 2016); hence, the presence of active ingredients in minimum effective concentration in the form of a formulation is essential at the target sites. For preventing these losses, the development of nanoformulation or encapsulation of pesticides has been recognized. The controlled delivery techniques allow a measured amount of necessary/sufficient quantities of pesticides or herbicides to be released over some time. The large surface area of nanoparticles is advantageous for the effective delivery of agrochemicals; therefore, several mechanisms, such as absorption, capsulation, and entrapment into the nanomatrix of active ingredients were developed, which later became very popular for the efficient delivery of required crop protection chemical (Panpatte et al., 2016; Nuruzzaman et al., 2016; Pandey, 2018). This technique, known as nanoencapsulation, is a process through which a plant protection chemical is slowly released by the mechanism like dissolution, biodegradation, diffusion, or osmotic pressure with specific pH, and it is currently considered the most promising technology for crop protection and pest control. The potential application and prospects of nanotechnology are enormous, and there are great opportunities for the research to exploit nanotechnology in pest control in the future (Routray et al., 2016). In this chapter, we investigated nanosilver bioassays and mechanisms for insect control in the agricultural commodity.

20.2 Nanosilver in agricultural insect control

While nanomaterials have been successfully employed in various fields, their high potential for use in pest control and crop protection has remained largely unexplored. However, the utility of nanomaterials, particularly of biological origins with ecofriendly advantages, as pest control agents have been realized and are receiving considerable interest (Bhattacharyya et al., 2010; Athanassiou et al., 2018). In our laboratory, we studied the effects of AgNPs on the growth, development, and physiology of a few important agricultural pests; castor semilooper, *Achaea janata* L., and tobacco hornworm, *Spodoptera litura* F. Larval diets comprising of PVP coated-AgNPs treated castor leaf at different dilutions produced varied and positive effects on the insect growth, physiology, and development. The comparisons of the activities were made with that of silver nitrate ($AgNO_3$) treated leaf disks. The decrease in the larval and pupal body weights was observed along with the decrease in the concentrations of AgNPs and $AgNO_3$ in both the test insects. While small amounts of silver were noted to have accumulated in the larval guts, a majority of it was eliminated through the feces (Jyothsna and Rani, 2015).

Suaeda maritima, a flowering plant of the family Amaranthaceae, was known by the common name herbaceous seepweed. Synthesized AgNP was examined for its toxic properties to the tobacco cutworm, *S. litura*, using a no choice leaf disk method and was found to be effective against the larval pest (Suresh et al., 2018). Suresh et al. (2018) found that the AgNP treatment produced lethal effects in first to fourth instar larvae and pupae, and extended their life stage duration as well as reduction of female fecundity and male longevity. These long-term toxic effects evoked on *S. litura* larvae are similar to the ones presumed by Naresh Kumar et al. (2013) who described it to be due to the active compounds entering into the larval body causing reduction of ecdysone activity, so that failure of larval molt occurs and finally leads to the disruption of normal physiological and metabolic processes. *Helicoverpa armigera* Hubner is an important agricultural pest in India, which causes significant losses to cotton, cereals, legumes, and vegetable crops. This polyphagous insect pest has a worldwide distribution, hence, making it as one of the notorious pests. AgNPs synthesized from *Euphorbia hirta* have demonstrated multiple activities against *H. armigera*. A 24 h exposure of larvae to the AgNP treatments resulted in high larval fatality, i.e., 95% at 10 ppm and a remarkable deterrence in larval feeding. The treatment also affected the digestive enzyme physiology, caused a huge reduction in food consumption, effecting larval and pupal duration as well as weight, and decreased adult emergence and longevity, significant fall in growth rate, fecundity, and egg hatchability of *H. armigera* (Durga Devi et al., 2014). We obtained similar kind of results, in our laboratory.

While comparing the activities of AgNPs and SINPs (silica nanoparticles) on the biology of *H. armigera*, we discovered that the long-term exposure to these nanoparticles can modulate multiple biological processes, including reduced female fertility, increased larval durations, decreased pupal weights, and altered antioxidative and gut enzyme activities (Usha Rani et al., unpublished data).

The larvicidal activity of AgNPs against yet another pest, the oriental armyworm *Mythimna separata* (Walker), was demonstrated by Buhroo et al. (2017). In this study, the aqueous leaf extracts of *Trichodesma indicum* (Boraginaceae), a native Indian flowering plant, when used as a reducing agent, formed spherical shaped nanosilver particles. These biofabricated Ag NPs, when assayed by leaf surface treatment, produced larvicidal activity to the earlier stages of the oriental armyworm, *M. separata* (Walker), at a dosage of 500 ppm (Buhroo et al., 2017). Seaweeds were not exempted from nanomaterial synthesis. Longevity and fecundity of the cabbage pest, *Plutella xylostella* (the diamondback moth), has been strongly reduced due to the treatment of AgNP synthesized from the seaweed, *Hypnea musciformis* (Wulfen) (Roni et al., 2015), establishing both the diversity of existence and consistent efficacy of AgNPs in nature.

Studies against aphid pests have also yielded similar benefits. AgNPs synthesized from the leaves of *Solanum lycopersicum*, when tested for the toxicity by food treatment method against the rose aphid, *Macrosiphum rosae* (Hemiptera: Aphididae, a key pest on rose plants in Kashmir Valley, India), caused mortality in a dose-dependent manner (Bhattacharyya et al., 2016). Rouhani et al. (2012) reported entomotoxic effects of Ag and Zn nanoparticles on oleander aphid, *Aphis nerii*, a polyphagous pest on a wide variety of crops. It appears that aqueous extract of jujube leaf (*Ziziphus jujuba*) contain natural reducing and stabilizing compounds involved in the synthesis of AgNP, and these NP reduced the nymphal population of Whitefly, *Bemisia tabaci* on infested Al-Mustakbal eggplant hybrid grown in greenhouse conditions (Shammari et al., 2018).

Another interesting work is on the pests on textiles. The treated wool textiles with a sulfur nanosilver colloidal solution (SNSE) having Ag/S complex protected the fabric from moth pests (Ki et al., 2007). The pest activity was to the treatment was measured both by calculation of the fiber loss weight and visible assessment of attacked by larvae after 14 days. *S. litura* is one of the major agricultural pests, distributed all over the world; hence, several studies included this pest for toxicity assessments of AgNP. In interesting research, the impact of AgNP treatment on *S. litura* was explored by Chandrashekharaiah et al. (2015) who demonstrated a dose-dependent increase in mortality with AgNPs. They claimed that the exposure to AgNPs ceased active movements of *S. litura* larvae with the skin and entire larval body becoming rigid and discharge of body contents (lysis). Melanization occurred in these swollen larval bodies, demonstrating premature molting by larvae attaining pupal shape and eventually followed by larval death. Murugan and Jeyabalan (1999) described this kind of phenomenon as the compounds acting as inhibitors of neurosecretory cells, leading to the shrinking of internal cuticle and acting directly on epidermal cells that are responsible for the production of enzymes and finally leading to tanning or cuticular oxidation process. Insect growth and metabolism immensely depend on the digestive enzymes, which convert complex molecules into simple components. An interesting and a detailed study was reported on biogenic silver nanoparticles' impact on developmental and gut physiology of *S. litura* (Bharani and Rajanamasivayam, 2017). They noted that the biosynthesis of AgNP using the peel extract of the fruit, pomegranate (*Punica granatum*), has yielded highly stable, uniform, monodispersive, and very minute-sized silver nanoparticles, which, on treatment, arrested and restricted the developmental process and the gut physiology of *S. litura*. The laboratory analysis revealed that the gut physiology, nutritional index parameters such as gut enzymes, approximate digestibility, and efficiency of conversion of ingested and digested food; gut microbiota were affected in a dose-dependent manner. The biogenic AgNP inhibited larval and pupal mortality, reduced larval, pupal growth period, adult emergence, and adult longevity, which are important parameters for the survival of the pest (Bharani and Rajanamasivayam, 2017). Among different plant materials utilized for biosynthesis of AgNPs, the use of essential oils has also been explored as an ecofriendly option. Sahayaraj et al. (2016) evaluated the effects of pungamoil-based gold nanoparticles (PO-AuNPs), pungamoil-based silver nanoparticles (PO-AgNPs), and neem against *Pericallia ricini* for the larval development, mortality, and fecundity. Interestingly, the treatments drastically enhanced food consumption but reduced the larval growth by inhibiting food conversion, leading to growth retardation. Overall, the treatment resulted with the differences in larval, pupal, adult developmental periods, fecundity, and hatchability.

Mangroves are also being utilized for nanosilver production and explored for their effects on a few important crop pests. An economic and ecofriendly synthesis of silver nanoparticles using mangrove associate, *Hibiscus tiliaceus* leaf extract and further their evaluation for their insectistatic potential toward two key agricultural pests, tobacco cutworm, *S. litura*, and the cotton bollworm, *H. armigera*, and the comparisons with chemically synthesized nanoparticles, disclosed excellent antifeedant activity of the tested samples. The antifeedant activity was calculated by comparing the leaf area consumed in the treated leaves and the normal untreated or solvent treated leaf disks (controls). Decreased intake of food was observed in both the larvae following the exposure to synthetic silver and biogenic AgNPs. The highest percent of the antifeedant activity of $94.1 \pm 2.0\%$ was recorded for biogenic AgNPs at $200\,\mu g/21\,cm^2$ on *S. litura* (Usha Rani et al., 2016). Recently, we also studied the developmental rates of *H. armigera* larvae reared on chemically synthesized AgNPs incorporated artificial diets till pupation and compared to that of the larvae grown on their normal diets. The larval growth differed significantly among the treatment concentrations. Treated larval weights were higher than that of the untreated larvae, which increased gradually with the increase in consumption rate and age. Larval growth index decreased following treatments than in controls (Usha Rani et al., unpublished data). Silver nanoparticles increased the toxicity of pyrethroid insecticide bifenthrin to the western tarnished plant bug, *Lygus hesperus*, when combined. Interestingly, bifenthrin-only was more toxic than the bifenthrin mixed with nanosilver against the house cricket, *Acheta domesticus* (Orthoptera: Gryllidae), under field conditions (Louder, 2015).

20.3 The effects of AgNPs on insect vector management

A number of crucial challenges are occurring in parasitology, and the control of parasites and vectors, in particular, has become an important challenge. Several major diseases in tropical regions are vector-borne and mosquitoes act as vectors for many life-threatening diseases. There is currently an urgent need for effective preventive and/or curative tools against dreadful diseases such as malaria and arboviruses, chikungunya, dengue, filariasis, encephalitis, West Nile, and Zika virus (Benelli, 2015). Consequently, eradication of mosquitoes and other important vector species is critical. Over the past decade, there has been increased interest in the use of botanicals and compounds of natural origins in vector control as well as in agricultural and other pest management due to pressing environmental issues. The latest revelation of their novel use as stabilizing, reducing, and capping agents for the green synthesis of nanoparticles has now paved new avenues for the development of economical and ecofriendly methods for nanoparticle synthesis, suggesting scope for utilizing these molecules for vector control as well.

A growing number of plants have been successfully used for efficient and rapid extracellular synthesis of metal nanoparticles (Benelli, 2016a; Rajasekharreddy et al., 2010). Different parts of quiet a few plant varieties have been employed as reducing and stabilizing agents that resulted in the production of metal nanoparticles with various sizes, shapes, and different activities in insect vectors. The use of AgNPs, in particular, has increased the scope for utilizing these molecules for the vector control too. Many attempts were made to control major vector species using nanosilver with considerably success rate. Toxicity of silver nanoparticles (AgNPs) synthesized using the broom weed, *Sida acuta* Burm (Malvaceae), plant leaf extract was examined against late third instar larvae of three important mosquito vectors, namely, *Culex quinquefasciatus*, *Anopheles stephensi*, and *Aedes aegypti*. The synthesized AgNPs exhibited considerably high toxicity to all three mosquito species tested than the crude leaf extracts alone (Veerakumar et al., 2013), and the treatment affected the larval development too. AgNPs yielded from the aqueous root extract of *Parthenium hysterophorus* showed a dose-dependent larvicidal effect on the larvae of *Cx. quinquefasciatus* and were superior in toxicity when compared to the aqueous root extract or pure silver nitrate solution (Mondal et al., 2014), while AgNPs synthesized from the aqueous extract of red algae, *H. musciformis* (Wulfen), produced larval and pupal mortalities in dengue vector, *Aedes aegypti* (Roni et al., 2015).

In another study, the dry leaves, fresh leaves, and green berries of *Solanum nigrum* plant-mediated synthesis of nanosilver were assessed for the larvicidal activity against second and third instar larvae of the southern house mosquito, *Cx. quinquefasciatus*, and the Asian malaria vector, *A. stephensi*, using standard procedures by Rawani et al. (2013). Priyadarshini et al. (2012) synthesized AgNPs utilizing *E. hirta* leaf extract and bioassayed them against *A. stephensi* and found that the AgNPs caused highest larval mortality in all the larval instars tested, i.e., first to fourth. The green synthesis of silver nanoparticles from the fruit pulp of golden tree, *Cassia fistula*, and their evaluation against vector mosquitoes proved that these AgNPs are highly competent against first and fourth instar larvae of vector mosquitoes, *Ae. albopictus* and *Culex pipiens pallens* (Fouad et al., 2018). More recently, the toxicity bioassays with silver bio-nanoparticles (AgNPs) synthesized from nematode-symbiotic bacterial toxin complexes, namely, *Xenorhabdus indica*, *Xenorhabdus* spp., *Photorhabdus luminescens laumondii* HP88, *Photorhabdus luminescens akhurstii* HRM1, and *Photorhabdus luminescens akhurstii* HS1 against the larval stages of *Cx. pipiens* resulted in mortality in the treated mosquito larvae, which was concentration- and exposure time-dependent. The highest larval toxicity was observed when larvae were treated with bio-AgNPs synthesized from *P. luminescens akhurstii* HRM1 and *X. indica*, followed by *P. luminescens laumondii* HP88 and *Xenorhabdus* sp. (El-Sadawy et al., 2018). Green synthesis of NP is considered as one of the rapid, reliable, and cost-effective routes for the synthesis of silver and other metallic nanoparticles (NPs). In a study reported by Arumugam and Gopinath (2013), aqueous bark filtrate of *Terminalia arjuna*, a medicinal tree that consists of active compounds in high concentrations, notably, phytosterols, lactones, flavonoids, phenolic compounds, tannins, and glycosides, acted as a good reducing agent and produced spherical shaped AgNPs. The efficacy of the synthesized AgNPs was tested against third and fourth instar larvae of malarial vector, *A. stephensi*. The treated mosquito larvae showed high susceptibility toward the exposure to NP treatment. Another interesting study is with that of the coconut palm, *Cocos nucifera*, coir mediated AgNP synthesis and their outstanding bioefficacy against all the larval instars and pupae of *A. stephensi* and *C. quinquefasciatus* as claimed by Roopan et al. (2013). It is observed that plants with greater chemical profile aid in manufacturing superior quality nanoparticles having better activity. *Rhazya stricta* Decne (Apocynaceae) is a medicinal plant utilized for the preparation of herbal drugs to cure various diseases in South Asia (Pakistan, India, and Afghanistan) and in the Middle East (e.g., Saudi Arabia, Qatar, United Arab Emirates (UAE)). This plant is reported to contain several secondary metabolites (Bukhari et al., 2017). Biofabrication of silver nanomaterials using the plant, *R. stricta*, extract and the assessment of the AgNPs produced has shown excellent bioactivity to a few deadly mosquito vectors and microbial pathogens. *R. stricta*-mediated AgNPs were highly toxic to several mosquito strains and showed acute toxicity on key mosquito vectors from two different countries (India and Kingdom of Saudi Arabia, KSA) strains (Aziz et al., 2020). We presume that the presence of these secondary metabolite chemicals, particularly alkaloids, may be responsible for the production of highly bioactive AgNPs.

We at Indian Institute of Chemical Technology, Hyderabad, India, proposed an ecofriendly process for the synthesis of silver-(protein-lipid) nanoparticles (Ag-PL NPs) (core-shell) using the seed extract from wild Indian almond tree, *Sterculia foetida*. Silver nanoparticles (AgNPs) produced with the seed extract of this plant showed excellent mosquitocidal activity against IV instar larvae of *Aedes aegypti*, *A. stephensi*, and *Cx. quinquefasciatus*.

Mangrove plants are unique vegetation as they grow in different conditions than the normal terrestrial plants and consists valuable bioactive compounds in their leaves, roots, and stems (Usha Rani et al., 2016; Ali et al., 2014). These natural compounds also possess excellent insecticidal activity apart from being good reducing agents for AgNP production. It is evident from recent work that mangrove plants show good mosquitocidal properties toward various species of mosquitoes and are highly prospective in use as an alternate and ecofriendly source for insect management methods. A mangrove annual herb, *S. maritima* (L.), Dumort (Chenopodiaceae), grows in the salt marsh as well as very alkaline and moist saline soils (Raju and Kumar, 2016). Silver nanoparticles synthesized by using aqueous extract of this plant as a reducing and stabilizing agent disclosed the larvicidal and pupicidal toxicity toward the dengue vector, *Aedes aegypti* (Suresh et al., 2018). The treatment induced 100% reduction of the larval population after 72h of post-treatment and acted as adulticide in smoke toxicity tests. A study by Murugan et al. (2015a) highlighted the concrete potential of green-synthesized AgNPs in the fight against dengue virus. Aqueous extract of another mangrove plant, *Bruguiera cylindrica,* mediated synthesis of AgNPs, produced highly efficient antiviral and mosquitocidal silver nanoparticles. About 30 μg/mL of AgNPs inhibited the production of dengue viral envelope (E) protein in Vero cells and downregulated the expression of dengue viral E gene in in vitro experiments. The effects of the employed compound on nontarget organisms are another important area to be considered while applying/developing a pest control agent. Interestingly, the application of sub lethal doses of *B. cylindrica* synthesized AgNPs did not affect the predation rates of the fish, *Carassius auratus,* against *A. aegypti.* AgNPs synthesized using the leaf extract of *Avicennia marina* caused larvicidal activity to *Ae. aegypti* and *A. stephensi* at a significantly very low dose (Balakrishnan et al., 2016). Another mangrove, *R. mucronata,* fabricated AgNPs also exhibited mosquito larvicidal activity at a very low dose to *Ae. aegypti* and *Cx. quinquefasciatus,* indicating the high potentiality of the mangrove plant-mediated synthesis (Gnanadesigan et al., 2011). They presumed that the larvicidal activity of the AgNPs might be due to the denaturation of the sulfur-containing proteins or phosphorous-containing compounds like DNA and that leads to the denaturation of organelles and enzymes and thus reduces the cellular membrane permeability.

Although there are some studies on mosquito control using mangrove extract or extract synthesized AgNPs in a laboratory, very limited field bioassays investigated the bioefficacy of the green-synthesized AgNPs in field conditions. Complete elimination of mosquito larval population was recorded due to the application of *S. maritima* produced AgNP 72 h post-treatment (Suresh et al., 2018), which is a very good indication of their field use.

Controlling the insect pests at their early life stages, such as egg stage, is more beneficial than at the later stages. AgNPs have also been shown to be effective in reducing early instars of mosquito populations in the field, as well as to induce egg mortality and oviposition deterrence (Benelli, 2016a,b). Japanese wireweed, *Sargassum muticum,* mediated synthesis of AgNPs and further their application to the mosquito vectors, *A. stephensi, A. aegypti,* and *C. quinquefasciatus,* caused 100% inhibition of egg hatchability after a single exposure to the compound (Madhiyazhagan et al., 2015). Most of the available reports on AgNPs are on the mosquito controlling activity at the larval stages itself. However, it is also reported that a few plant species-mediated synthesis of silver nanoparticles also acted as good mosquito adulticides. The silver nanoparticles manufactured using dry leaf extracts of *Chomelia asiatica* L. (*Tarenna asiatica* (L.)) plant demonstrated good adulticide activity to female mosquitoes, *A. stephensi, Ae. aegypti,* and *Cx. quinquefasciatus.* The order of efficacy recorded was that of the adult of *A. stephensi, Ae. aegypti,* and *Cx. quinquefasciatus* (Muthukumaran et al., 2016).

Fascinatingly, in recent times, the field of nanosynthesis has extended beyond plants and plant extracts to invertebrates. Earthworm (*Eudrilus eugeniae*)-synthesized silver nanoparticles were evaluated for their acute toxicity against early instars of the malaria vector *A. stephensi* yielding excellent prospects (Jaganathan et al., 2016), suggesting exciting new directions that are waiting to be explored. Murugan et al. (2015b) in a novel method of AgNP biosynthesis used a low-cost seaweed extract of *Ulva lactuca* as a reducing and capping agent and tested them separately along with seaweed extract against larvae, pupae, and adults of the malaria vector *A. stephensi*. The biosynthesized silver nanoparticles were highly effective and produced toxicity at a very low dosage to first to fourth instar larvae. The mortality rates obtained in adult mosquitoes, by exposing them to treated smoke in the form of coils is comparable to that of the permethrin-based positive control. The bioassays on antiplasmodial activity of the samples, the leaf crude extracts and the extract mediated synthesized AgNPs of *U. lactuca* to CQ-resistant (CQ-r) and CQ-sensitive (CQ-s) strains of *Plasmodium falciparum* revealed that the bio fabricated AgNPs showed higher activity against *P. falciparum* when compared to chloroquine. Dengue, a mosquito-borne viral disease, has rapidly spread all over the world in recent years. The adult female mosquitoes *Ae. aegypti* are the causative for the transmission of this disease. A recent focus on the potential of green-synthesized nanoparticles from the extract of the marine alga *Centroceras clavulatum* revealed that these AgNPs act apart from producing toxicity in the tested mosquito species, inhibited dengue (serotype dengue virus type-2 (DEN-2) viral replication in Vero cells). *C. clavulatum*-synthesized AgNPs acted as inhibitors of the production of dengue viral envelope protein in Vero cells and down regulators of the expression of dengue viral E gene. Cellular internalization assays highlighted that untreated infected cells showed the high intensity of fluorescence emission, which indicates a high level of viral internalization. Also, AgNP-treated infected cells displayed lesser levels of fluorescence and failed to show substantial viral load (Murugan et al., 2015b).

Controlling mosquitoes in their early stages is advantageous than the control of more advanced stages. The sweet flag, *Acorus calamus* (Acoraceae), synthesized AgNPs have a good ovicidal activity to the malaria vector, *A. stephensi* (Chandramohan et al., 2015). In ovicidal and egg hatchability experiments, the application of 25 and 30 ppm of *S. maritima* synthesized AgNPs caused a 100% reduction of egg hatchability of *A. aegypti*, while there was a complete (100%) hatching in control eggs. Similar studies by Rajaganesh et al. (2016) pointed out the ovicidal activity of Uluhe fern; *Dicranopteris linearis* fabricated AgNPs against *Ae. aegypti*, while Madhiyazhagan et al. (2015) disclosed the effective ovipositional deterrent activity of *S. muticum*-synthesized AgNPs against *A. stephensi*, *Ae. aegypti*, and *Cx. quinquefasciatus* at considerably low dosages. Among the plant materials utilized as reducing agents for the green synthesis of AgNPs, the essential oils also proved to be efficient in all aspects. The essential oil, of *Aquilaria sinensis*, *Pogostemonis herba*, and *Pogostemon cablin* used in one-step production of nanosilver particles caused larval and pupal mortality in dengue and Zika virus vector *Ae. albopictus*. Compared to the tested essential oils, the biofabricated AgNPs showed the highest toxicity against larvae and pupae of *Ae. albopictus* (Ga'al et al., 2018). The pediculocidal and larvicidal activity of phytofabricated AgNPs using an aqueous leaf extract of *Tinospora cordifolia* indicated their efficacy against the head louse *Pediculus humanus* and fourth instar larvae of *An. subpictus*, as well as *Cx. quinquefasciatus*

(Jayaseelan et al., 2011). A major advantage of using nanoparticles is that they have little to no off-target toxicity or genotoxicity to aquatic organisms at the doses lethal to mosquito larvae. No evidence of toxicity was found in *Poecilia reticulata* (Poeciliidae) fish after being exposed to white frangipani, *Plumeria rubra,* and the trellis vine, *Pergularia daemia* (both belonging to Apocynaceae) synthesized AgNPs that were employed for the control of fourth instar larvae of *A. aegypti* and *A. stephensi* (Patil et al., 2012a,b). Similarly, *Vinca rosea* synthesized AgNPs failed to produce any toxicity to fish, *P. reticulata*, even after 72 h of exposure to doses toxic to *A. stephensi* and *C. quinquefasciatus* (Subarani et al., 2013), while the competence of *C. auratus,* a mosquito larval predator, was not affected by the exposure to sublethal doses of mangrove-fabricated AgNPs*,* on *A. aegypti* larvae (Murugan et al., 2015b). Murugan et al. (2015c) proved that the employment of ultralow quantities of biofabricated nanoparticles along with the biological control agents for the control of early larval instars was the better alternative as the presence of NP enhances their predation rates in the aquatic environment. Thus far, most of the available and published literature is predominantly focused on mosquitoes with comparatively little reported on the other disease-causing vectors. Ticks and hematophagous flies are also important parasitic species that cause considerable losses and efforts are being made to control them. *Hippobosca maculata* (Diptera: Hippoboscidae) is a serious pest of equines in a stud in India, whereas cattle tick, *Rhipicephalus* (Boophilus) *microplus* (Acari: Ixodidae), is a variety of livestock species that transmit infectious agents to vertebrate hosts, thus causing major constraints to livestock health. A first report on the antiparasitic activity of AgNPs synthesized from using aqueous extract of the stem of the popular medicinal plant, *Cissus quadrangularis* (Fam: Vitaceae), was reported against the adults of *H. maculata* and the larvae of *R. microplus* major tick species (Santhoshkumar et al., 2012). The potential of the plant-mediated synthesis of AgNPs, in producing lethal effects in both the parasite species, was evaluated using the contact toxicity method, and the results revealed that the biogenic AgNPs possess a good antiparasitic activity to the Arthropod species tested, which is superior to the crude plant extract alone. A noteworthy study tested the acaricidal and larvicidal activity of AgNPs synthesized using aqueous leaf extracts from *Musa paradisiaca* L. (Musaceae) against the larvae of *Haemaphysalis bispinosa* Neumann (Acarina: Ixodidae), hematophagous fly *H. maculata* Leach (Diptera: Hippoboscidae) and *A. stephensi*, Japanese encephalitis vector, *Cx. tritaeniorhynchus* Giles (Diptera: Culicidae), and the exposure to the treatment for 24 h produced excellent results (Jayaseelan et al., 2012). These findings suggest the efficacy of AgNPs to extend beyond specific species, providing exciting prospects to consider. The common house flies, *Musca domestica* (Diptera: Muscidae), the mechanical carriers and responsible for many bacterial, protozoan, helminthic diseases, and viral infections to humans and animals, are being the target of control since a long time. Many chemical and botanical insecticides have been used until now. A recent report on biosynthesized silver nanoparticle treatment effects on house flies, *M. domestica,* showed interesting and promising results. Phytochemical fabrication of silver nanoparticles by using leaf broth of *Manilkara zapota* (Sapotaceae) commonly known as sapodilla, an evergreen tree with edible fruits and leaves having medicinal value, yielded highly effective AgNPs. The exposure of adult house flies to different concentrations resulted with 100% mortality of treated flies within an exposure period of 3 h (Kamaraj et al., 2012).

20.4 AgNPs on postharvest insect management

Storage pests are as equally important as crop pests and cause severe damage. They infest the grain at the crop stage or during the storage. Storage pests mostly consist of beetles or moths and infest the stored products and grain stored for seed and other purposes. These infested pests cause physical damage, leading to deterioration in the weight, quality, vigor, and reduction in the germination of the grain, thus leading to the loss of market value. Their control is essential to save the production. After the chemical pesticides, botanical products took place of stored pest management, but the recent invention of the role of nanomaterials in production or enhancement of toxicity generated new interest. The unique characteristics of nanomaterials, particularly their extremely minute size, make them efficient agents in penetrating the cells, thus often making them responsible for the toxic properties or lethal effects.

Many studies have demonstrated the use of AgNPs in effectively reducing storage pest populations. The novel approach of synthesizing NPs using plants and other biomaterials has also been studied for acute toxicity, fumigant, antifeedant, repellent, and attractant properties, as well as reproductive inhibitors for many pest species (Mohammed, 2013). Repellent, toxic, antifeedant, and ovipositional deterrent activities of malathion, nanosilver, and both their combinations were tested against the major stored product pest, the red flour beetle, *Tribolium castaneum* Herbst. Through different bioassay methods such as food grain treatment (for mortality) and Petri dish assay (for repellency), they assessed the efficacy of the samples. The obtained data revealed that the nanosilver suspension with malathion was effective in producing lethal effects in the pests and the activities increased with the increase of dosage and the exposure time in the beetles. The tested samples also showed ovipositional deterrent activity (Alisha and Thangapandiyan, 2019).

AgNPs synthesized from the peel extract of sweet orange, *Citrus sinensis* (L.) (Rutaceae, Sapindales), were bioassayed in filter-paper residue and feeding methods against the adult tenebrionid beetle, *T. confusum* (Duval) (Sedighi et al., 2019). Interestingly, while the citrus extract itself lacked toxicity, the green-synthesized AgNPs triggered a significantly high mortality rate, suggesting exclusive toxicity of AgNPs superior to the parent extract. In our laboratory, we assessed the insecticidal activity of biologically as well as chemically synthesized AgNPs derived from the leaf extract of mangrove associate, *H. tiliaceus*, against three major stored product pests, flour beetle, *T. castaneum*, lesser grain borer, *Rhyzopertha dominica* F., and rice weevil, *Sitophilus oryzae* L., and discovered that the toxic properties of biologically synthesized AgNPs were higher than that of the chemically synthesized AgNPs (Usha Rani et al., 2016) (shown in Table 20.1).

Table 20.1 Comparative contact toxicity of biosynthesized AgNP using leaf extract of mangrove associate, Hibiscus tiliaceus and synthetic nanosilver against three major stored product pests.

	Dose (µg/cm²)	T. casutaneum	R. dominica	S. orzae
Biogenic	50	21.3±0.12[a]	22.8±0.32[b]	19.4±0.45[a]
	100	39.2±0.24[c]	41.2±0.24[c]	33.2±0.38[c]
	150	45.2±0.24[e]	42.8±0.24[e]	37.4±0.16[e]
Synthetic silver	50	17.6±0.12[a]	26.8±0.24[a]	19.2±0.24[a]
	100	31.3±0.13[c]	29.8±0.24[c]	30.0±0.21[d]
	150	32.8±0.17[e]	31.4±0.16[e]	33.8±0.24[e]
Control	—	0±0[f]	5.1±0.26[f]	9.4±0.17[f]

Values are mean ± standard error. Means within a column followed by the same letter are significantly different at $P < 0.001$.
From Usha Rani, P., Prasanna Laxmi, K., Vadlapudi, V., Sreedhar, B., 2016. Phytofabrication of silver nanoparticles using the mangrove associate, Hibiscus tiliaceus plant and its biological activity against certain insect and microbial pests. J. Biopestic. 9, 167–179.

Use of another mangrove, plant, *A. marina*, as a reducing agent has resulted with good yield as well as high potent AgNPs, and this is attributed to the presence of important secondary metabolites such as tannins, flavonoids, and polyphenols (Gnanadesigan et al., 2011). Sankar and Abideen (2015) discovered that the stored grain pest, *S. oryzae*, is highly susceptible to the exposure to silver nanoparticles biosynthesized by using mangrove plant, *A. marina*, and the treatment resulted with remarkably high toxicity of the beetles (100%). Other reports have confirmed the efficacy of biologically synthesized AgNPs. Nanosilver obtained through biological synthesis using aqueous leaf extracts of *Euphorbia prostrata* (Euphorbiaceae) caused lethal effects to the granary weevil, *S. oryzae* (Zahir et al., 2012). The exposure of beetle pests to AgNP reduced the movement of larvae and made the body wall strong and rigid and ultimately turned to dark brown. Nanoparticles loaded with essential oil of garlic are shown to be capable of producing toxicity to *T. castaneum* Herbst (Yang et al., 2009). In addition to botanical sources, AgNPs have also been successfully synthesized from the biomass of the fungus *Metarhizium anisopliae* Sorok in and shown to induce mortality and demelanization of the cuticle of adult khapra beetle, *Trogoderma granarium*, when treated during the early developmental stage (Thamer et al., 2017). Using a slightly different approach, Allahvaisi (2016) demonstrated the efficacy of AgNPs incorporated into the packaging polymer on the growth of an important storage pest, *Sitotroga cerealella* (Lep. Gelechiidae) and associated fungi. The silver nanoparticle application has influenced the eradication of the pathogen and produced toxic effects on the insect pests. Feeding damage by one of the major stored product pests, *Callosobruchus chinensis* (L.) (pulse beetle), on stored pulses results in weight loss of the infested grain and seed vigor. The curry leaf plant, *Murraya koenigii*, mediated synthesis of AgNPs' effect on this storage pest was studied using soya beans (Yerragopu et al., 2019), and the results obtained indicated the excellent toxic activity of the AgNPs.

20.5 Mode of action of AgNPs and their effect on insect biochemistry and physiology

The use of nanomaterials in various sectors and their increased utilization also attracted the grave environmental concerns and led to a rapid burst in studies to assess their toxicity. Recent revelation and progressively increasing number of reports on nanosilver toxicity to various organisms led to the exploration of their effects on the environment. Beyond certain concentration, silver nanoparticles (AgNPs) are potentially toxic and cause several pathophysiological abnormalities in the exposed organism, leading to compromised survival. Although considerable work has been done in nanoparticle research, precise information on the mechanisms of action of nanoparticles against insects and mites are very limited and the complete mechanism of AgNP toxicity remains undetermined. A few studies performed on a spectrum of organisms have been able to list a variety of toxic effects from organismal exposure to nanoparticles.

The increased surface area of NP makes them different from their counterparts, the bulk materials. This feature renders them distinct properties that affect their ability to penetrate skin, pass through cell membranes, cross the blood-brain barrier, be absorbed into lymphatic channels, and reach bone marrow, lymph nodes, heart, lungs, and the central nervous system (Oberdörster et al., 2005). Due to their size, it is likely that nanoparticles may enter cells through endocytosis, and the hydrophobic particles can penetrate the membranes. Studying the organism's response to endocytosed nanomaterials can provide information about their mode of action, enrich our understanding, and consequently lead to the development of enhanced measures of insect control. The application of nanosilver to insect species generates morphological, physiological, and biochemical variations in their bodies and causes changes in their lifecycle like development, growth reproduction, and often fecundity. The mechanism of toxicity of nanomaterials includes disruption of membranes, interruption of energy transduction, genotoxicity, oxidation of proteins, the formation of reactive oxygen species (ROS), and even the release of toxic constituents. AgNPs, when administered, adhere to the cellular surface, altering the membrane properties and affecting its permeability and cell respiration (Morones et al., 2005). The release of toxic silver ions due to DNA damage can also occur. Armstrong et al. (2013) stated that the AgNPs can interfere with copper transport into cells and create a deficiency of copper, thus becoming toxic to the exposed insects. There is a normal production of protein in the enzyme, but there is a decreased activity as the copper is not incorporated into the enzyme active site. The lack of melanin pigments resulting in *T. granarium* due to the exposure of nanosilver particles was presumed to be due to their interference in intracellular copper (Cu^+) transport. It is assumed that the presence of nanosilver in the extracellular environment inhibits the process of intracellular entry of Cu due to Ag and Cu competing for the same copper transporters. It is also reported that the metal nanoparticles can bind to S and P in proteins and nucleic acids, respectively, leading to a decrease in membrane permeability, thus causing organelle and enzyme denaturation, followed by cell death. It is known that AgNPs up- and down-regulates key insect genes and reduce protein synthesis and gonadotrophin release, causing developmental damages and reproductive failure (Benelli, 2018). Armstrong et al. (2013) published excellent research on the mode of action of silver nanoparticles' effects on *Drosophila melanogaster*. The dietary administration of silver nanoparticles at sublethal doses to early developmental stages of the wild strain of *D. melanogaster* over an extended period affected the cuticular formation. The exposure to AgNPs caused demelanization of adult cuticle, and therefore, all adult animals appeared discolored owing to the lack of melanin pigments (Fig. 20.3).

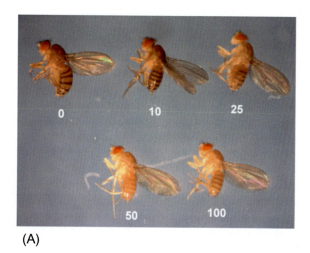

(A)

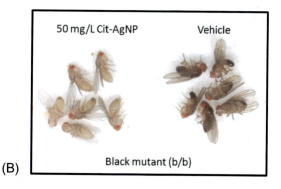

(B)

FIG. 20.3

(A) The lighter body color of flies grown on silver nanoparticles at varied concentrations. The eye color of *Drosophila* is not affected at 50 mg/L and above, indicating the selective interference of AgNPs with melanin pigmentation. Numbers denote the dosage employed.
(B) Accumulation of excessive melanin pigmentation in *Drosophila* black (b) mutants. (b) was effectively eliminated from their body, after exposure to 50 mg/L AgNP turning their bodies pale.

Picture courtesy: Armstrong, N., Ramamoorthy, M., Lyon, D., Jones, K., Duttaroy, A., 2013. Mechanism of silver nano particles action on insect pigmentation reveals intervention of copper homeostasis. PLoS One 8, 53186.

All adult flies raised in AgNP-doped food showed compromisation infertility and ability of vertical movement. Their biochemical analysis suggested that there is a significant decrease in the activity of copper-dependent enzymes, tyrosinase, and copper-zinc superoxide dismutase since tyrosinase activity is essential for melanin biosynthesis, and this explains the process of demelanization in the treated flies. The availability of extracellular silver causes sequestration of copper, without influencing the total copper uptake (Fig. 20.3).

AgNPs also seem to have an impact on the digestive system. In an in vivo toxicity study, Nouara et al. (2018) assessed the distribution and toxicity of AgNPs administered via food ingestion on the growth and development of silkworm, *Bombyx mori*. To investigate the transcriptomic responses in the silver nanoparticle treated insects, the RNA sequencing was performed. All these tests revealed interesting information that an increase of AgNPs in the larval bodies may cause several changes in histopathology of the midgut, alteration of the gene expressions, and also affect the carbohydrate and energy metabolisms. They concluded that the AgNPs may induce harmful effects on the alimentary system by deregulating the energy, digestion, and nutrition that lead to an imbalance in absorption. A decrease in food utilization efficiency measures, as well as the levels of midgut digestive enzyme profiles, has also been reported in response to AgNP exposure (Durga Devi et al., 2014). In a similar study made in our laboratory, we have demonstrated the significant changes in the feeding index and protease activity of larval *Achaea janta* and *S. litura* fed with diets treated with AgNP suspensions at various concentrations (Jyothsna and Usha Rani, 2014) (Fig. 20.4). These results together demonstrate that AgNPs induce harmful effects on the alimentary system by deregulating the energy, digestion and nutrition leading to imbalances in absorption.

In certain insect species, the AgNP treatment causes a reduction in the egg-laying, which indicates that the nanoparticles can control the population growth and the progeny, and this decline in oviposition at higher doses have been attributed to the interruption of vitellogenesis and damage to the egg chamber (Pandey and Khan, 1998). However, Vanmathi et al. (2010) predicted that the oviposition deterrence occurred after the exposure to AgNPs maybe because of the changes induced in physiology and behavior of the insect. It appears that the compounds which have oviposition preventive effect can also influence hormonal imbalance and reproductive physiology of insects (Rehman et al., 2009). The toxic properties of AgNP to young instars of mosquitoes are attributed to the ability of nanoparticles to penetrate through the exoskeleton of the larval body and is suggested that in the intracellular space, AgNP can bind to sulfur from proteins or to phosphorus from DNA, leading to the rapid denaturation of enzymes and organelles, which will further cause a decrease in membrane permeability and ultimately loss of cellular function and cell death (Benelli, 2016a,b). Stoimenov et al. (2002) has shown that silver nanoparticles can bind to proteins and enzymes in cells and can make adhesive interactions with cellular membranes and produce highly reactive and toxic radicals like ROS, which cause inflammation and destroy cells like mitochondria. They also subsequently produce apoptogenic factors that cause cell death and necrosis in the cellular environment. Thus, AgNPs can show serious toxic effects on the mitochondrial function and cell viability. Silver

20.5 Effect on insect biochemistry and physiology

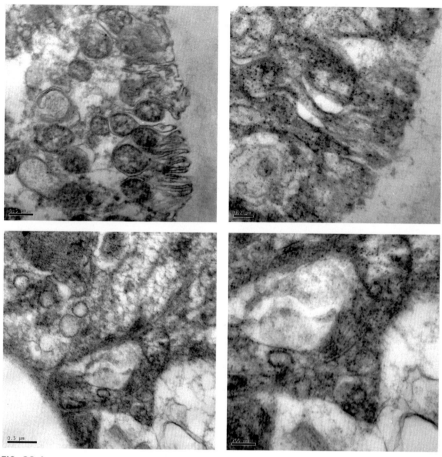

FIG. 20.4

Microscopical images of the normal and AgNP-treated *Achaea janata* larvae. (A) Ultrastructure of midgut in normal *A. janata* larvae showing microvilli, abundant mitochondria, vacuoles, and vesicles. (B) Accumulation of AgNPs in the gut cells and nucleolus (*dark spots*). (C) Midgut region of larvae with metal depositions in cell organelles. (D) Magnified view of rough endoplasmic reticulum with well-developed cisternae and Ag on surface region.

From Jyothsna, Y., Usha Rani, P., 2014. Silver nano particle induced changes in feeding index and protease activity of Achaea janata L. and Spodoptera litura. F. Hexapoda 19, 51–68.

nanoparticle application has been shown to have a substantial impact on insect biochemical parameters such as antioxidant and detoxifying enzymes that may lead to oxidative stress and cell death. Ag nanoparticles also reduced acetylcholinesterase activity. Ahamed et al. (2010) reported that in AgNPs exposed organisms malondialdehyde level, which is an end product of lipid peroxidation was significantly higher while antioxidant glutathione content was considerably lower.

Mortality and detrimental effects were recorded with AgNP-treated fruit fly, *D. melanogaster* and nanoparticles also triggered the accumulation of ROS in the fly tissues leading to ROS-mediated apoptosis, DNA damage, and autophagy at sublethal doses (Mao et al., 2018). The enzyme activities of protease, amylase, lipase, and invertase were decreased in *S. litura* treated with silver nanoparticles; besides this, there is a decrease in the gut microflora and extracellular enzyme production as well as weight, pH, and total heterotrophic bacterial population (Bharani and Rajanamasivayam, 2017). The exposure to AgNPs caused a decrease in total protein levels, AchE, α and β carboxylesterase activities in *A. albopictus* and *Cx. pipiens pallens* (Fouad et al., 2018). In an interesting study with the fruit fly, *D. melanogaster*, which was treated with coated and uncoated AgNPs and tested for heat shock proteins (Hsp70), demonstrated that Hsp70 protein expression was higher in AgNP exposed flies as compared to the negative control (Posgai et al., 2011).

20.6 Conclusion

AgNPs have been proven to be highly efficient as antibacterial and antifungal agents for a long time. Recent research suggests a beneficial role for AgNPs in insect pest management as well, and several studies have now established AgNPs as important components with promising prospects in vector management. Literature, thus far, indicates several biological agents including plants and microbes to be efficient and cost-effective candidates for the green synthesis of AgNPs, and despite presently limited information on their mode of action, it suggest tremendous potential for these particles in the agricultural sector. With a further detailed research, AgNPs and the field of nanotechnology have the potential to transform the agricultural landscape, providing convenient, affordable, and effective solutions for pest management.

Acknowledgments

I gratefully acknowledge the valuable support rendered by Jyothsna Yasur, Rajasekhar Reddy Pala, Peta Devanand, Pratyusha Sambangi, and Sandeepti Pathipati.

References

Ahamed, M., Posgai, R., Gorey, T.J., Nielse, M., Hussain, S.M., Rowe, J.J., 2010. Silver nanoparticles induced heat shock protein 70, oxidative stress and apoptosis in *Drosophila melanogaster*. Toxicol. Appl. Pharmacol. 1 (242), 263–269.

Ali, M.S., Ravikumar, S., Beula, J.M., Anuradha, V., Yogananth, N., 2014. Insecticidal compounds from Rhizophoraceae mangrove plants for the management of dengue vector *Aedes aegypti*. J. Vector Borne Dis. 51, 106–114.

Alisha, A.A.S., Thangapandiyan, S., 2019. Comparative bioassay of silver nanoparticles and malathion on infestation of red flour beetle, *Tribolium castaneum*. J. Basic Appl. Zool. 80, 55.

Allahvaisi, S., 2016. Effects of silver nanoparticles (AgNPs) and polymers on stored pests for improving the industry of packaging foodstuffs. J. Entomol. Zool. Stud. 4, 633–640.

Armstrong, N., Ramamoorthy, M., Lyon, D., Jones, K., Duttaroy, A., 2013. Mechanism of silver nano particles action on insect pigmentation reveals intervention of copper homeostasis. PLoS One 8, 53186.

Arumugam, A., Gopinath, K., 2013. Green synthesis, characterization of silver, gold and bimetallic nanoparticles using bark extract of *Terminalia arjuna* and their larvicidal activity against malaria vector, *Anopheles stephensi*. Int. J. Recent Sci. Res. 4, 904–910.

Athanassiou, C.G., Kavallieratos, N.G., Benelli, G., Losic, D., Usha Rani, P., Desneux, N., 2018. Nanoparticles for pest control: current status and future perspectives. J. Pest. Sci. 91, 1–15.

Aziz, A.T., Alshehri, M.A., Alanazi, N.A., Panneerselvam, C., Trivedi, S., Maggi, F., Sut, S., Dall'Acqua, S., 2020. Phytochemical analysis of *Rhazya stricta* extract and its use in fabrication of silver nanoparticles effective against mosquito vectors and microbial pathogens. Sci. Total Environ. 700 (2020), 134443.

Balakrishnan, S., Srinivasan, M., Mohanraj, J., 2016. Biosynthesis of silver nano-particles from mangrove plant *Avicennia marina* extract and their potential mosquito larvicidal property. J. Parasit. Dis. 40, 991–996.

Benelli, G., 2015. Research in mosquito control: current challenges for a brighter future. Parasitol. Res. 114, 2801–2805.

Benelli, G., 2016a. Plant-mediated biosynthesis of nanoparticles as an emerging tool against mosquitoes of medical and veterinary importance: a review. Parasitol. Res. 115, 23–34.

Benelli, G., 2016b. Plant-mediated synthesis of nanoparticles: a newer and safer tool against mosquito-borne diseases? Asian Pac. J. Trop. Biomed. 6, 353–354.

Benelli, G., 2018. Mode of action of nanoparticles against insects. Environ. Sci. Pollut. Res. Int. 25, 12329–12341.

Bharani, A.R.S., Rajanamasivayam, K.S., 2017. Biogenic silver nanoparticles mediated stress on developmental period and gut physiology of major lepidopteran pest *Spodoptera litura* (Fab.) (Lepidoptera: Noctuidae) an eco-friendly approach of insect pest control. J. Environ. Chem. Eng. 5, 453–467.

Bhattacharyya, A., Bhaumik, A., Usha Rani, P., Mandal, S., Epidi, T.T., 2010. Nanoparticles—a recent approach to insect pest control. Afr. J. Biotechnol. 9, 3489–3493.

Bhattacharyya, A., Prasad, R., Buhroo, A.A., Yousuf, I., Khanday, A.L., Duraisamy, P., Umadevi, M., Bindhu, M.R., Govindarajan, M., 2016. One-pot fabrication and characterization of silver nanoparticles using *Solanum lycopersicum*: an eco-friendly potent control tool against rose Aphid, *Macrosiphum rosae*. J. Nanosci. Res., vol 2016, Open Access Article ID 4679410. doi:https://doi.org/10.1155/2016/4679410.

Buhroo, A.A., Nisa, G., Asrafuzzaman, S., Prasad, R., Rasheed, R., Bhattacharyya, A., 2017. Biogenic silver nanoparticles from *Trichodesma indicum* aqueous leaf extract against *Mythimna separata* and evaluation of its larvicidal efficacy. J. Plant Prot. Res. 57, 194–200.

Bukhari, N.A., Al-Otaibi, R.A., Ibhrahim, M.M., 2017. Phytochemical and taxonomic evaluation of *Rhazya stricta* in Saudi Arabia. Saudi J. Biol. Sci. 24, 1513–1521.

Chandramohan, B., Murugan, K., Kovendan, K., Panneerselvam, C., Mahesh Kumar, P., Madhiyazhagan, P., Dinesh, D., Suresh, U., Subramaniam, J., Amaresan, D., Nataraj, T., Nataraj, D., Hwang, J.S., Alarfaj, A.A., Nicoletti, M., Canale, A., Mehlhorn, H., Benelli, G., Mehlhorn, H., 2015. Do nanomosquitocides impact predation of *Mesocyclops edax* against *Anopheles stephensi* larvae? In: Nanoparticles in the Fight Against Parasites. Parasitology Research Monographs, vol. 8. Springer, pp. 2192–3671.

Chandrashekharaiah, M., Kandakoor, S., Basana Gowda, G., Kammar, V., Chakravarthy, A., 2015. Nanomaterials: a review of their action and application in pest management and evaluation of DNA-tagged particles. In: Chakravarthy, A. (Ed.), New Horizons in Insect Science: Towards Sustainable Pest Management. Springer, New Delhi.

Durga Devi, G., Murugan, K., Panneerselvam, C., 2014. Green synthesis of silver nanoparticles using *Euphorbia hirta* (Euphorbiaceae) and leaf extract against crop pest of cotton ball, *Helicoverpa armigera*. J. Biopestic. 7 (Suppl), 54–66.

El-Sadawy, H.A., El-Namaky, A.H., Hafez, E.E., Baiome, B.A., Ahmed, A.M., Ashry, H.M., Ayaad, T.H., 2018. Silver nanoparticles enhance the larvicidal toxicity of *Photorhabdus* and Xenorhabdus bacterial toxins: an approach to control the filarial vector, *Culex pipiens*. Trop. Biomed. 35, 392–407.

Fouad, H., Hongjie, L., Hosni, D., Wei, J., Abbas, G., Gaal, H., Jianchu, M., 2018. Controlling *Aedes albopictus* and *Culex pipiens* pallens using silver nanoparticles synthesized from aqueous extract of *Cassia fistula* fruit pulp and its mode of action. J. Artif. Cells Nanomed. Biotechnol. 46 (3), 558–567.

Ga'al, H., Fouad, H., Mao, G., Tian, J., Jianchu, M., 2018. Larvicidal and pupicidal evaluation of nanoparticles synthesized using *Aquilaria sinensis* and *Pogostemon cablin* essential against dengue and zika viruses vector *Aedes albopictus* mosquito and its histopathological analysis. Artif. Cells Nanomed. Biotechnol. 46, 1171–1179.

Gnanadesigan, M., Anand, M., Ravikumar, S., Maruthupandy, M., Vijayakumar, V., Selvam, S., Dhineshkumar, M., Kumaraguru, A.K., 2011. Biosynthesis of silver nanoparticles by using mangrove plant extract and their potential mosquito property. Asian Pac. J. Trop. Med., 799–803.

Gurunathan, S., Kalishwaralal, K., Vaidyanathan, R., Venkataraman, D., Pandian, S.R., Muniyandi, J., Hariharan, N., Eom, S.H., 2009. Biosynthesis, purification and characterization of silver nanoparticles using *Escherichia coli*. Colloids Surf. B: Biointerfaces 74, 328–335.

Hochella, M.F., Spencer, M.G., Jones, K.L., 2015. Nanotechnology: nature's gift or scientists' brain child? Environ. Sci. Nano 2, 114–119.

Jaganathan, A., Murugan, K., Panneerselvam, C., Madhiyazhagan, P., Dinesh, D., Vadivalagan, C., Aziz, A.T., Chandramohan, B., Suresh, U., Rajaganesh, R., Subramaniam, J., Nicoletti, M., Higuchi, A., Alarfaj, A.A., Munusamy, M.A., Kumar, S., Benelli, G., 2016. Earthworm-mediated synthesis of silver nanoparticles: a potent tool against hepatocellular carcinoma, *Plasmodium falciparum* parasites and malaria mosquitoes. Parasitol. Int. 65, 276–284.

Jayaseelan, C., Rahuman, A.A., Rajakumar, G., Vishnu Kirthi, A., Santhosh kumar, T., Marimuthu, S., Bagavan, A., Kamaraj, C., Zahir, A.A., Elango, G., 2011. Synthesis of pediculocidal and larvicidal silver nanoparticles by leaf extract from heartleaf moonseed plant, *Tinospora cordifolia* Miers. Parasitol. Res. 109 (1), 185–194.

Jayaseelan, C., Rahuman, A.A., Rajakumar, G., Santhoshkumar, T., Kirthi, A.V., Marimuthu, S., Bagavan, A., Kamaraj, C., Zahir, A.A., Elango, G., Velayutham, K., Rao, K.V., Karthik, L., Raveendran, S., 2012. Efficacy of plant-mediated synthesized silver nanoparticles against hematophagous parasites. Parasitol. Res. 111, 921–933.

Jyothsna, Y., Rani, P.U., 2015. Lepidopteran insect susceptibility to silver nanoparticles and measurement of changes in their growth, development and physiology. Chemosphere 124, 92–102.

Jyothsna, Y., Usha Rani, P., 2014. Silver nano particle induced changes in feeding index and protease activity of *Achaea janata* L. and *Spodoptera litura*. F. Hexapoda 19, 51–68.

Kamaraj, C., Rajakumar, G., Rahuman, A.A., Velayutham, K., Bagavan, A., Zahir, A.A., Elango, G., 2012. Feeding deterrent activity of synthesized silver nanoparticles using *Manilkara zapota* leaf extract against the house fly, *Musca domestica* (Diptera: Muscidae). Parasitol. Res. 111, 2439–2448.

Ki, H.Y., Kim, J.H., Kwon, S.C., Jeong, S.H., 2007. A study on multifunctional wool textiles treated with nano sized silver. J. Mater. 42, 8020–8024.

Krishna, G., Singaracharya, M., 2016. Synthesis of silver nanoparticles by chemical and biological methods and their antimicrobial properties. J. Exp. Nanosci. 11 (9), 714–721.

Louder, J.K., 2015. Nanotechnology in Agriculture: Interactions Between Nanomaterials and Cotton Agrochemicals. Ph.D. Thesis, Texas Tech University, Texas, USA.

Madhiyazhagan, P., Murugan, K., Kumar, A.N., Nataraj, T., Dinesh, D., Panneerselvam, C., Subramaniam, J., Mahesh Kumar, P., Suresh, U., Roni, M., Nicoletti, M., Alarfaj, A.A., Higuchi, A., Munusamy, M.A., Benelli, G., 2015. *Sargassum muticum*-synthesized nanoparticles: an effective control tool against mosquito vectors and bacterial pathogens. Parasitol. Res. 114, 4305–4317.

Mao, B.H., Chen, Z.Y., Wang, Y.J., Yan, S.J., 2018. Silver nanoparticles have lethal and sublethal adverse effects on development and longevity by inducing ROS-mediated stress responses. Sci. Rep. 8, 2445.

Mohammed, H.H., 2013. Repellency of ethanolic extract of some indigenous plants against *Tribolium confusum* (Duval) (Coleoptera: Tenebrionidae). Agric. Vet. Sci. 2, 27–31.

Mondal, N.K., Chowdhury, A., Dey, U., Mukhopadhya, P., Chatterjee, S., Das, K., Datta, J.K., 2014. Green synthesis of silver nanoparticles and its application for mosquito control. Asian Pac. J. Trop. Med. Dis. 4, 204–210.

Morones, J.R., Elechiguerra, J.L., Camacho, A., Ramirez, J.T., 2005. The bactericidal effect of silver nanoparticles. Nanotechnology 16, 2346–2353.

Murugan, K., Jeyabalan, D., 1999. Effect of certain plant extracts against the mosquito, *Anopheles stephensi* Liston. Curr. Sci. 76, 631–633.

Murugan, K., Benelli, G., Ayyappan, S., Dinesh, D., Panneerselvam, C., Nicoletti, M., Hwang, J.S., Kumar, P.M., Subramaniam, J., Suresh, U., 2015a. Toxicity of seaweed-synthesized silver nanoparticles against the filariasis vector *Culex quinquefasciatus* and its impact on predation efficiency of the cyclopoid crustacean *Mesocyclops longisetus*. Parasitol. Res. 114, 2243–2253.

Murugan, K., Samidoss, C.M., Panneerselvam, C., Higuchi, A., Roni, M., Suresh, U., Chandramohan, B., Subramaniam, J., Madhiyazhagan, P., Dinesh, D., Rajaganesh, R., Alarfaj, A.A., Nicoletti, M., Kumar, S., Wei, H., Canale, A., Mehlhorn, H., Benelli, G., 2015b. Seaweed-synthesized silver nanoparticles: an eco-friendly tool in the fight against *Plasmodium falciparum* and its vector *Anopheles stephensi*? Parasitol. Res. 114, 4087–4097.

Murugan, K., Dinesh, D., Paulpandi, M.A., Althbyani, D.M., Subramaniam, J., Madhiyazhagan, P., Wang, L., Suresh, U., Kumar, P.M., Mohan, J., Rajaganesh, R., Wei, H., Kalimuthu, K., Parajulee, M.N., Mehlhorn, H., Benelli, G., 2015c. Nanoparticles in the fight against mosquito-borne diseases: bioactivity of Bruguiera cylindrica-synthesized nanoparticles against dengue virus DEN-2 (in vitro) and its mosquito vector *Aedes aegypti* (Diptera: Culicidae). Parasitol. Res. 114, 4349–4361.

Muthukumaran, U., Govindarajan, M., Rajeswary, M., Veerakumar, K.A., Amsath, A., Muthukumaravel, K., 2016. Adulticidal activity of synthesized silver nanoparticles using *Chomelia asiatica* (family: Rubiaceae) against *Anopheles stephensi*, *Aedes aegypti*, and *Culex quinquefasciatus* (Diptera: Culicidae). Int. J. Zool. Appl. Biosci. 1, 118–129.

Naresh Kumar, A., Murugan, K., Madhiyazhagan, P., 2013. Integration of botanicals and microbials for management of crop and human pests. Parasitol. Res. 112, 313–325.

Nouara, A., Peng, L., Keping, C., 2018. Silkworm, *Bombyx mori*, as an alternative model organism in toxicological research. Environ. Sci. Pollut. Res. Int. 25, 35048–35054.

Nuruzzaman, M., Rahman, M.M., Liu, Y., Naidu, R., 2016. Nanoencapsulation, nanoguard for pesticides: a new window for safe application. J. Agric. Food Chem. 64, 1447–1483.

Oberdörster, G., Oberdörster, E., Oberdörster, J., 2005. Nanotoxicology: an emerging discipline evolving from studies of ultrafine particles. Environ. Health Perspect. 113, 823–839.

Pandey, G., 2018. Challenges and future prospects of agri-nanotechnology for sustainable agriculture in India. Environ. Technol. Innov. 11, 299–307.

Pandey, S.K., Khan, M.B., 1998. Screening and isolation of leaf extract of *Clerodendrum siphonanthus* and their effects of *Callosobruchus chinensis* through injection method. Indian J. Toxicol. 6, 57–65.

Panpatte, D.G., Jhala, Y.K., Shelat, H.N., Vyas, R.V., 2016. Nanoparticles: the next generation technology for sustainable agriculture. In: Microbial Inoculants in Sustainable Agricultural Productivity. Springer, New Delhi, India, pp. 289–300.

Patil, C.D., Borase, H.P., Patil, S.V., Salunkhe, R.B., Salunke, B.K., 2012a. Larvicidal activity of silver nanoparticles synthesized using *Pergularia daemia* plant latex against *Aedes aegypti* and *Anopheles stephensi* and non-target fish *Poecilia reticulate*. Parasitol. Res. 111, 555–562.

Patil, C.D., Patil, S.V., Borase, H.P., Salunke, B.K., Salunkhe, R.B., 2012b. Larvicidal activity of silver nanoparticles synthesized using *Plumeria rubra* plant latex against *Aedes aegypti* and *Anopheles stephensi*. Parasitol. Res. 110, 1815–1822.

Posgai, R., Cipolla-McCulloch, C.B., Murphy, K.R., Hussain, S.M., Rowe, J.J., Nielsen, M.G., 2011. Differential toxicity of silver and titanium dioxide nanoparticles on *Drosophila melanogaster* development, reproductive effort, and viability: size, coatings and antioxidants matter. Chemosphere 85, 34–42.

Priyadarshini, K.A., Murugan, K., Panneerselvam, C., Ponarulselvam, S., Hwang, J.S., Nicoletti, M., 2012. Biolarvicidal and pupicidal potential of silver nanoparticles synthesized using *Euphorbia hirta* against *Anopheles stephensi* Liston (Diptera: Culicidae). Parasitol. Res. 111, 997–1006.

Rajaganesh, R., Murugan, K., Panneerselvam, C., Jayashanthini, S., Aziz, A.T., Roni, M., Suresh, U., Trivedi, S., Rehman, H., Higuchi, A., Nicoletti, M., Benelli, G., 2016. Fern-synthesized silver nanocrystals: towards a new class of mosquito oviposition deterrents? Res. Vet. Sci. 109, 40–51.

Rajasekharreddy, P., Usha Rani, P., Sreedhar, B., 2010. Qualitative assessment of silver and gold nanoparticle synthesis in various plants: a photobiological approach. J. Nanopart. Res. 12, 1711–1721.

Raju, A.J.S., Kumar, R., 2016. On the reproductive ecology of *Suaeda maritima*, *S. monoica*, and *S. nudiflora* (Chenopodiaceae). J. Threat. Taxa 8, 8860–8876.

Rawani, A., Ghosh, A., Chandra, G., 2013. Mosquito larvicidal and antimicrobial activity of synthesized nano-crystalline silver particles using leaves and green berry extract of *Solanum nigrum* L. (Solanaceae: Solanales). Acta Trop. 128, 613–622.

Rehman, J., Jilani, G., Ajab Khan, M., Masih, R., Kanvil, S., 2009. Repellent and oviposition deterrent effect of indigenous plant extracts to peach fruit fly, *Bactrocera zonata* Saunders (Diptera: Tephridae). Pak. J. Zool. 41, 101–108.

Roni, M., Murugan, K., Panneerselvam, C., Subramaniam, J., Nicoletti, M., Madhiyazhagan, P., Dinesh, D., Suresh, U., Khater, H.F., Wei, H., Canale, A., Alarfaj, A.A., Munusamy, M.A., Higuchi, A., Benelli, G., 2015. Characterization and biotoxicity of *Hypnea musciformis*-synthesized silver nanoparticles as potential eco-friendly control tool against *Aedes aegypti* and *Plutella xylostella*. Ecotoxicol. Environ. Saf. 121, 31–38.

Roopan, S.M., Rohit, G., Madhumitha, G., Abdul Rahuman, A.C., Kamaraj, C., Bharathi, A., Surendra, T.V., 2013. Low-cost and eco-friendly phyto-synthesis of silver nanoparticles using *Cocos nucifera* coir extract and its larvicidal activity. Ind. Crop. Prod. 43, 631–635.

Rouhani, M., Samih, M.A., Kalantari, S., 2012. Insecticidal effect of silica and silver nanoparticles on the cowpea seed beetle, *Callosobruchus maculatus*, F. (Col.: Bruchidae). J. Entomol. Res. 4, 297–305.

Routray, S., Damayanti, D., Baral, S., Patil, V., 2016. Potential of nanotechnology in insect pest control. Prog. Res. Int. J. 0973-6417. 11, 903–906. Online ISSN: 2454-6003 in Society for Scientific Development Print Agriculture and Technology.

Sahayaraj, K., 2017. Nano and bio-nanoparticles for insect control. Res. J. Nanosci. Nanotechnol. 7, 1–9.

Sahayaraj, K., Madasamy, M., Anbu Radhika, S., 2016. Insecticidal activity of bio-silver and gold nanoparticles against *Pericallia ricini* Fab. (Lepidaptera: Archidae). J. Biopestic. 9, 63–72.

Sankar, M.V., Abideen, S., 2015. Pesticidal effect of green synthesized silver and lead nanoparticles using *Avicennia marina* against grain storage pest *Sitophilus oryzae*. Int. J. Nanomater. Biostruct. 5, 32–39.

Santhoshkumar, T., Rahuman, A.A., Bagavan, A., Marimuthu, S., Jayaseelan, C., Kirthi, A.V., Kamaraj, C., Rajakumar, G., Zahir, A.A., Elango, G., Velayutham, K., Iyappan, M., Siva, C., Karthik, L., Rao, K.V., 2012. Evaluation of stem aqueous extract and synthesized silver nanoparticles using *Cissus quadrangularis* against *Hippobosca maculata* and *Rhipicephalus (Boophilus) microplus*. Exp. Parasitol. 132, 156–165.

Sedighi, S., Imani, G.R., Kashanian, M., Najafi, H., Fathipour, Y., 2019. Efficiency of green synthesized silver nanoparticles with sweet orange, *Citrus sinensis* (L.) (Rutaceae, Sapindales) against *Tribolium confusum* Duval. (Coleoptera, Tenebrionidae). J. Agric. Sci. Technol. 21, 1485–1494.

Shammari, H.I.A.I., Khazraji, H.I.A.L., Falih, S.K., 2018. The effectivity of silver nano particles prepared by jujube Ziziphus sp. extract against whitefly, *Bemisia tabaci* nymphs. Res. J. Pharm. Biol. Chem. Sci. 9, 551–558.

Stark, W.J., Stoessel, P.R., Wohlleben, W., Hafner, A., 2015. Industrial applications of nanoparticles. Chem. Soc. Rev. 44, 5793–5805.

Stoimenov, P.K., Klinger, R.L., Marchin, G.L., Klabunde, K.J., 2002. Metal oxide nanoparticles as bactericidal agents. Langmuir 18, 6679–6686.

Subarani, S., Sabhanayakam, S., Kamaraj, C., 2013. Studies on the impact of biosynthesized silver nanoparticles (AgNPs) in relation to malaria and filariasis vector control against *Anopheles stephensi* Liston and *Culex quinquefasciatus* Say (Diptera: Culicidae). Parasitol. Res. 112, 487–499.

Suresh, U., Murugan, K., Murugan, K., Panneerselvam, C., Rajaganesh, R., Roni, M., Aziz, A., Hatem, A., Al-Aoh, Trivedi, S., Rehman, H., Kumar, S., Higuchi, A., Canale, A., Benelli, G., 2018. *Suaeda maritima*-based herbal coils and green nanoparticles as potential bio pesticides against the dengue vector *Aedes aegypti* and the tobacco cutworm *Spodoptera litura*. Physiol. Mol. Plant Pathol. 101, 225–235.

Thamer, M.N., Mahmood, E.A., Hussam, E., 2017. The effect of silver nanoparticles on second larval instar of *Trogoderma granarium* everts (Insecta: Coleoptera: Dermestidae). Int. J. Sci. Nat. 8, 303–307.

Usha Rani, P., Rajasekhar Reddy, P., 2011. Green synthesis of silver-protein (core–shell) nanoparticles using *Piper betle* L. leaf extract and its ecotoxicological studies on *Daphnia magna*. Colloids Surf. A Physicochem. Eng. Asp. 389, 188–194.

Usha Rani, P., Prasanna Laxmi, K., Vadlapudi, V., Sreedhar, B., 2016. Phytofabrication of silver nanoparticles using the mangrove associate, *Hibiscus tiliaceus* plant and its biological activity against certain insect and microbial pests. J. Biopestic. 9, 167–179.

Vanmathi, J.S., Padmalatha, C., Ranjith Singh, A.J.A., Sudhakar, I.S., 2010. Efficacy of selected plant extracts on the oviposition deterrent and adult emergence activity of *Callosobruchus chinensi* F. Global J. Sci. Front. Res. 10, 12–14.

Veerakumar, K., Govindarajan, M., Rajeswary, M., 2013. Green synthesis of silver nanoparticles *Sida acuta* (Malvaceae) leaf extract against *Culex quinquefasciatus*, *Anopheles stephensi*, and *Aedes aegypti* (Diptera: Culicidae). Parasitol. Res. 112, 4073–4085.

Verma, P., Maheshwari, S.K., 2019. Applications of silver nanoparticles in diverse sectors. Int. J. Nano Dimens. 10 (1), 18–36.

Yang, F.L., Li, X.G., Zhu, F., Lei, C.L., 2009. Structural characterization of nanoparticles loaded with garlic essential oil and their insecticidal activity against *Tribolium castaneum* (Herbst) (Coleoptera: Tenebrionidae). J. Agric. Food Chem. 57, 10156–10162.

Yerragopu, P.S., Hiregoudar, S., Nidoni, U., Ramappa, K.T., Sreenivas, A.G., Doddagoudar, S.R., 2019. Effect of plant-mediated synthesized silver nanoparticles on pulse beetle, *Callosobruchus chinensis* (L). Int. J. Curr. Microbiol. Appl. Sci. 8 (9), 1965–1972.

Zahir, A.A., Rahuman, A.A., 2012. Evaluation of different extracts and synthesised silver. Nanoparticles from leaves of *Euphorbia prostrata* against *Haemaphysalis bispinosa* and *Hippobosca maculata*. Vet. Parasitol. 187, 511–520.

Zahir, A.A., Bagavan, A., Kamaraj, C., Elango, G., Rahuman, A.A., 2012. Efficacy of plant mediated synthesized silver nanoparticles against *Sitophilus oryzae*. J. Biopestic. 288, 95–102.

Zhang, X.F., Liu, Z.G., Shen, W., Gurunathan, S., 2016. Silver nanoparticles: synthesis, characterization, properties, applications, and therapeutic approaches. Int. J. Mol. Sci. 17, 1534.

CHAPTER 21

Silver-based nanomaterials for plant diseases management: Today and future perspectives

Heba I. Mohamed[a], Kamel A. Abd-Elsalam[b], Asmaa M.M. Tmam[a], and Mahmoud R. Sofy[c]

[a]Biological and Geological Sciences Department, Faculty of Education, Ain Shams University, Cairo, Egypt
[b]Plant Pathology Research Institute, Agricultural Research Center (ARC), Giza, Egypt
[c]Botany and Microbiology Department, Faculty of Science, Al-Azhar University, Cairo, Egypt

21.1 Introduction

Crop production declines by plant pests and pathogens infection with global losses estimated at 20%–40% per year (Allinne et al., 2016; Cerda et al., 2017; Cromwell et al., 2014). Traditional control of pests relies heavily on pesticides, such as insecticides, fungicides, and herbicides. Given many benefits, such as high quality, quick action, and efficacy, pesticides have harmful side effects on nontarget species, the revival of the pest population, and resistance (Mohamed and Akladious, 2017; Mohamed et al., 2018). Also, 90% of the pesticides used are projected to be lost during or after application (Ghormade et al., 2011). As a result, there is increased support to produce cost-effective, high-performance pesticides, and ecofriendly compounds that are less environmentally harmful (Aly et al., 2012, 2013, 2017; Abd El-Rahman and Mohamed, 2014; Asran and Mohamed, 2014; Mohamed et al., 2012, 2018; Sofy et al., 2020).

Nanotechnology has contributed to the production of new technologies and agricultural products with tremendous potential for solving the above problems. Nanotechnology in medicine has advanced greatly, although its significance in agricultural applications is comparatively less (Balaure et al., 2017; Sinha et al., 2017). In the sense of hormonal transmission, germination, management of the water and gene transfer targets, tracking nanosensors and control of releases of agricultural products, agriculture nanotechnology use is currently being explored (Hayles et al., 2017). Nanoparticles were engineered by material scientists with specific characteristics such as shape, pore size, and surface properties, so that they can then be used as defensive devices or for accurate and targeted distribution through the adsorption, encapsulation, and/or conjugation of active substances like pesticides (Khandelwal et al., 2016). When agricultural nanotechnology develops, there will be a major increase in the ability to provide a new generation of pesticides and other plant disease management tools. Use of nanoparticles for plant defense can take place through two separate mechanisms: (a) nanoparticles which provide protection for crops themselves and (b) nanoparticles as carriers of known pesticides and other active substances, like double-stranded RNA (dsRNA), which may be applied to plants as foliar spray to tissue or drilling/soaking of seeds. As carriers, nanoparticles have many benefits, such as (i) improved shelf-life, (ii) increased solubility of pesticides with low water-solubility, (iii) decreased toxicity, and (iv) improved site-specific target pest uptake (Hayles et al., 2017). Another potential nanocarrier gain involves an improvement in the effectiveness of nanopesticide under environmental pressures (UV and rain) operation and stability and minimizing the number of applications dramatically, thus reducing risk and reducing their costs (Fig. 21.1).

This schematic shows different characterization techniques of nanomaterials and the role of insecticides, fungicides, herbicides, and antibacterial.

Nanoparticles are materials between 10 and 100 nm, which can be built to distinguish between their molecular and bulk counterparts by certain chemical, biological, and physical characteristics (Bakshi et al., 2014). Nanoparticles may be used directly by soaking in plant seeds, foliar spray on leaves or injection in soils to protect against pests and pathogens, including nematodes, bacteria, fungi, or viruses. The studied metals nanoparticles such as platinum, copper, zinc oxide, silver, and titanium dioxide have demonstrated antimicrobial and antiviral properties (Kah and Hofmann, 2014; Kim et al., 2018).

Recently, the widely synthesized form of silver nanoparticles from plants or microorganisms have increased their popularity (Rafique et al., 2017). Well diffusion assay was used to detect the antifungal effect of silver nanoparticles against

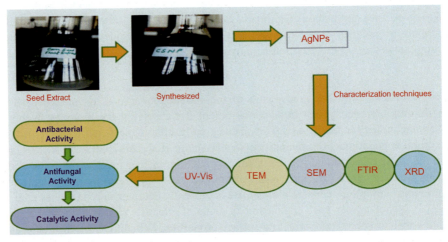

FIG. 21.1

Schematic presentation of synthesized AgNPs, characterization, and their applications.

Macrophomina phaseolina, Rhizoctonia solani, Botrytis cinerea, and *Curvularia lunata* (Worrall et al., 2018). Also, silver nanoparticles have an antiviral effect against sun-hemp rosette virus (Jain and Kothari, 2014) and bean yellow mosaic virus (Elbeshehy et al., 2015). While plant disease controls against bacterial and fungal pathogens, silver nanoparticles have shown huge potential, they are linked to important obstacles such as growth, toxicity, and soil interaction (Kah and Hofmann, 2014; Mishra and Singh, 2015).

21.1.1 Antimicrobial effect of silver nanoparticles

Because of their antimicrobial capacity, silver nanoparticles are popular and are more harmful to microorganisms than other metallic nanoparticles. Owing to their emergence, these particles are now called antibiotics of the next generation as an alternative to antibiotic therapy. Besides, these particles demonstrated efficacy against different bacteria (both gram-positive and gram-negative) with multidrug resistance (MDR). Despite extensive research and revolutionary research, the exact mechanism of their antimicrobial action is not yet completely understood (Dakal et al., 2016).

The antimicrobial activity of silver nanoparticles is mainly related to the formation of silver ions, and several evidences indicate that Ag^+ silver ions are related to the surface of particles. The high surface area of particles has been shown to release greater Ag^+ concentrations, leading to greater antimicrobial activity, while lower surface area releases depressed levels and gradually reduces antimicrobial activity (Möhler et al., 2018).

21.1.2 Mechanism of antimicrobial effect of silver nanoparticles

AgNPs control the growth of pathogens by several mechanisms. The exact mechanism of AgNPs by which antimicrobial activity is induced is not well known, which is one of the topics covered. AgNPs inhibit growth and prevent different activities of the cellular and molecular systems, and particles can be said to modify the profile of phosphorylation bacterial peptides. Therefore, analyzing this provides an important way of evaluating AgNP toxicity or mechanical behavior (Dakal et al., 2016).

Silver nanoparticles as an antimicrobial mode of action (Patil and Kim, 2017; Tang and Zheng, 2018) such as:

1. Adhesive to the cell membrane surface
2. Alter the structure of cell wall and caused damage to it.
3. Pit formation occurs on the cell surface at the site of disruption in the cell membrane and deposition of AgNPs.
4. Only short exposure to AgNPs allows the precursor envelope to accumulate, conducting proton motive power degeneration.
5. Minimizing the development of cell walls.
6. Inhibit phosphate ion absorption and exchange resulting in membrane leakage for sugar (mannitol), proteins, and amino acid (glutamine and proline).

7. Generation of reactive oxygen species (ROS) and free radical.
8. Inhibit the translation process by binding with ribosomal subunit 30S.
9. The intercalation of DNA bases contributes to block the transcription cycle of DNA.
10. Suppression of DNA replication, blocking of the required electrical potential differences in cytoplasmic membranes and suppression of the respiratory chain.
11. Interact with thiol group of enzymes and protein.
12. Inactivate enzymes which are involved in ATP production.
13. Shift in phosphorylation profile that inhibits the pathway of signal transduction.
14. Prevent the formation of biofilms by halting synthesis of exopolysaccharides

21.1.3 Antibacterial effect of silver nanoparticles

Some NPs influence bacterial growth and stress tolerance, bacterial infection resistance of plants, and processes of contact between plant and associated bacteria (Chen et al., 2016). It has been shown that silver nanoparticles demonstrate effective antibacterial activity against a variety of bacteria that cause diseases (Table 21.1). Nitrification may be the most susceptible microbial phase of the nitrogen cycle, though mild stimulatory impacts can also result from exposure to a limited range of sublethal concentrations of AgNP. Reduced nitrification activity takes place before such nitrogen cycling activities if AgNP levels reach inhibitory levels (Yang et al., 2013).

An increase in silver nanoparticles' (AgNP) antibacterial activity against phytopathogenic bacteria *Ralstonia solanacearum* following stabilizations with certain surfactants such as sodium dodecyl sulfate (SDS), sodium dodecyl benzenesulfonate (SDBS), octylphenol polyethoxylate (TX-100), and polysorbate 80 (Tween 80) as compared to silver ion has been identified by Chen et al. (2016). Owing to the beneficial synergistic effects of the AgNPs and surfactant, Tween 80 is the most preferred stabilizer of AgNPs. All the surfactants, however, almost did not affect Ag^+ antibacterial activity. In vitro, the highest bactericidal activity against *R solanacearum* was seen on Tween 80-stabilized AgNPs. Also, various concentrations of the green synthesized AgNPs (10, 20, 30, and 40 ppm) were exogenously added to *Citrus reticulata* before inoculums of *Xanthomonas axonopodis* pv. *citri* was used to determine canker disease incidence at different intervals of the day. With time, the infection index values were gradually increased in all the treatments applied. AgNPs with a concentration of 30 ppm have been found to be more beneficial for resistance to canker disease (Hussain et al., 2018).

At 20 μg mL^{-1}, the synthesized AgNPs displayed a strong antibacterial activity against the pathogen of rice bacterial leaf blight and bacterial brown stripe while *Xanthomonas oryzae* pv. *oryzae* strain LND0005 had an inhibition zone of 17.3 and 16.0 mm and *Acidvorax oryzae* strain RS-1, respectively. In addition,

Table 21.1 Silver nanomaterials employed against plant pathogenic bacteria.

Nanomaterials	Bacterial pathogens	References
Ag NPs	Pantoea agglomerans, Ralstonia solanacearum, Erwinia amylovora, and Pseudomonas lachrymans	Mohammad and Abd El-Rahman (2015)
	Escherichia coli, Salmonella typhi, and Pseudomonas aeruginosa	Abalaka et al. (2017)
	Pseudomonas aeruginosa, Escherichia coli, and Bacillus subtilis	Mohanta et al. (2017)
	Ralstonia solanacearum	Chen et al. (2016)
	Bacillus cereus, Listeria monocytogenes, Staphylococcus aureus, Escherichia coli, and Salmonella typhimurium	Patra and Baek (2017)
	Escherichia coli	Ukah et al. (2018)
	Escherichia coli	Shu et al. (2020)
	Pseudomonas syringae	Marpu et al. (2017)
	Xanthomonas axonopodis	Hussain et al. (2018)
	Clavibacter michiganensis	Rivas-Cáceres et al. (2018)
	Acidovorax oryzae strain RS-2	Masum et al. (2019)
	Bacillus sp. and Enterobacter cloacae	Singh et al. (2019)
	Acidovorax oryzae strain RS-2	Ali and Abdallah (2020)
	Staphylococcus aureus Pseudomonas aeruginosa	Bhuyar et al. (2020)
	Xanthomonas oryzae	Ahmed et al. (2020)
	Candidatus Liberibacter asiaticus	Stephano-Hornedo et al. (2020)
	Leishmania tropica	Javed et al. (2020)
Ag-Chitosan	Pseudomonas syringae	Shahryari et al. (2020)

the AgNPs synthesized significantly inhibited bacterial growth (Ibrahim et al., 2019). AgNPs reported lowest inhibitory concentrations of 6.25 and 12.5 $\mu g\ mL^{-1}$ against *Xanthomonas axonopodis* pv punicae and *Ralstonia solanacearum* bacterial plant pathogens, respectively. Disk-diffusion in vitro assay showed inhibition zones of 11.4 ± 1 mm and 18.1 ± 1 mm for *R. solanacearum* and *X. axonopodis* pv. *Punicae*, respectively, treated with AgNPs of $50\ \mu g\ mL^{-1}$ (Vanti et al., 2020). Possible function of biosynthesized AgNPs against *Xanthomonas oryzae* pv. *oryzae* was studied. At concentrations of 20, 30, and 50 $\mu g\ mL^{-1}$, we observed good antibacterial activity of biosynthesized AgNPs (size ~ 12 nm) against Xoo.

Even at the low dose of 5 μg/m, the major inhibitory effect of AgNPs on Xoo's biofilm formation was noted. Additionally, the suppression of disease by biosynthesized AgNPs under greenhouse conditions was authenticated. AgNP foliar spray significantly reduced rice sheath blight symptoms as shown by 9.25% DLA (%Diseased Leaf Area) compared to 33.91% DLA in Xoo inoculated rice plants (Mishra et al., 2020).

AgNPs antimicrobial effect was assessed using paper disk diffusion, colony growth, conidia germination, and in vitro inoculation methods. AgNPs' concentration of 50% inhibition (IC50) against *Setosphaeria turcica* was 170.20 μg mL^{-1}. It also showed a significant antifungal synergistic effect when AgNPs were combined with epoxiconazole in the ratios of 8:2 and 9:1 (Huang et al., 2020).

21.1.4 Mechanism of antibacterial effect of silver nanoparticles

The extensive research in this field has demonstrated that silver nanoparticles can interfere in several cellular and metabolic pathways of cells because of their mechanism of antibacterial action (Arya et al., 2019; Gupta et al., 2018; Aziz et al., 2016) (Fig. 21.2). It can be divided into two groups: (1) the nonoxidative and (2) the oxidative system.

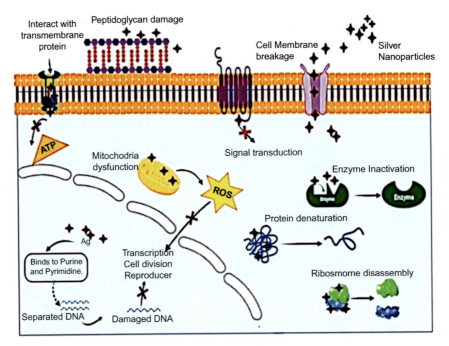

FIG. 21.2

The diagrammatic representation antibacterial mechanisms of silver nanoparticles.

21.2 Nonoxidative mechanism

Anchoring AgNPs to Ag^+ cellular wall formation, modifying membrane structure, destroy of cell walls, penetration in the cell membrane, leakage of the cell portion, and impaired transport activities are the nonoxidative mechanisms for the antibacterial action of AgNP.

21.2.1 Cell wall and cell membrane attachment in bacteria

As AgNPs are contacted with bacterial cells, silver nanoparticles impart the negative charge on the membrane, which makes it easy to attach those AgNPs to the cell wall due to the strong electrostatic interaction. Following attachment, some of these particles are separated to the biologically active Ag^+, and these ions are further connected with the cell wall and released into a cell by Trojan horse mechanism (Singh et al., 2015). These ions further combine with the sugar (β-1,4-conjoined N-acetyl glucosamine and N-acetyl muramic acid) and amino acid in the layer of peptidoglycan (Agnihotri et al., 2014). This interaction results in morphological changes involving cytoplasmic shrinking and removal of cell membrane (Anitha et al., 2018). This Ag also interacts in the cell wall with sulfur-containing protein, which affects the integrity of the lipid bilayer and causes permeability. This permeability influences the transportation process and impedes the phosphate and potassium ion pump. Excessive permeability eventually leads to cellular leakage in surrounding media causing cell deterioration and inhibition of the cell function, leading to necrosis and cell death (Anitha et al., 2018).

Additionally, this antibacterial activity of nonoxidative process often depends on the type of bacteria, whether gram-positive or gram-negative. The cell wall is composed of 30 nm thick negatively charged peptidoglycan layer in case of gram-positive bacteria, while in gram-negative bacteria the layer is between 3 and 4 nm thick. These changes in the structure and composition make the gram-negative bacteria more sensitive at the same concentration to the antibacterial action of AgNPs rather than gram-positive bacteria (Patil and Kim, 2017).

21.2.2 Damaging the intracellular structure in bacteria

This is more vulnerable to the release of silver ions by penetration following degradation of the bacteria cell wall. The use of surfactant along with these AgNPs has been shown to improve the penetration of AgNPs in the bacterial cells. Such particles interact with protein, lipid, DNA, and other biomolecules after cell membrane penetration (Hsueh et al., 2015; Kumar et al., 2016). AgNPs interact with DNA, leading to denaturation and damage to the DNA. These particles cause in the loss of DNA replicability because of a transition in relaxed to condensed state (Khalandi et al., 2017). The Ag^+ produced by AgNPs interacts separately with the purine and pyrimidine base and destabilizes the double-helical and β-sheet DNA structure, resulting in inhibition of gene transcription, so that, this contributes to cell division inhibition and their reproduction (Hsueh et al., 2015; Kumar et al., 2016). In addition, the AgNPs interact with the ribosomal subunit and help desaturate the translation cycle. These ions also inhibit proteins that take part in the generation of ATP transmembrane by reacting with the thiol group, interacting with the protein functional group and inhibiting its deactivation (Arya et al., 2019). Furthermore, these ions are interfering with synthesized protein by altering or blocking its active binding site and impairing its overall efficiency, resulting in a general functional defect in the microorganism (Arya et al., 2019).

21.3 Oxidative-stress mechanism

Oxidative stress is a disorder caused by the difference in the reactive oxygen species level (ROS) and free radical so that increases the free radicals can cause toxic effects (Quinteros et al., 2016). Increased oxidative cell stress of microbes is the responsibility of AgNPs, which is consistent with particulate antibacterial capacity. The increased Ag^+ concentration produced via AgNPs surface area after bacterial interaction is shown by the electron spin resonance spectroscopy studies, which are expected to cause higher ROS like hydrogen peroxide (H_2O_2), hydroxyl radical, superoxide anion (O_2^-), hydroxyl radical (OH), hypochlorous acid (HOCl), singlet oxygen, and free radical production (El-Beltagi and Mohamed, 2013). The large amounts of ROS and free radical caused breakage of the mitochondrial membrane and single-strand DNA, and the death of cells. As AgNPs interacts with bacterial cell, the Ag^+ released interacts with the thiol group of reduced glutathione and other enzymes that perturb the scavenging mechanisms. GSH is converted into oxidized forms of GSSH glutathione disulfide (Tang and Zheng, 2018). This Ag^+ also binds to the bacterial cell membrane and induces instability in the respiratory electron transporter chain and inhibits enzymes in the respiratory chain (Korshed et al., 2016; Gupta et al., 2018; Pareek et al., 2018; Prasad et al., 2017).

21.3.1 Antifungal effect of silver nanoparticles

Some researchers documented the antifungal activity of AgNPs on certain pathogenic fungi (Table 21.2). The mycelial growth and germination of incubated spores is observed to significant decrease with the use of AgNPs (Ahmed, 2017).

Many AgNPs have antifungal properties for other fungi, such as *Fusarium* species and other phytopathogenic fungi. Different studies have shown that antifungal activity against fungal pathogens can be caused by the fungal pathogens' suppression of enzymes and toxins (Dean et al., 2012). *Colletotrichum gloeosporioides*, which causes fruit anthracnose, has been used to test the antifungal activity of silver NPs. The mycelial growth of *Colletotrichum gloeosporioides* decreased dramatically in dose-dependent ways (Aguilar-Méndez et al., 2011).

The synthesized Ag NPs demonstrated excellent fungicidal activity against *Cladosporium fulvum*, which is the main cause of a serious plant disease called tomato

Table 21.2 Silver nanomaterials used as fungicides against plant pathogenic fungi.

Nanomaterials	Fungal pathogens	References
Ag NPs	*Penicillium digitatum*, *Alternaria citri*, and *Alternaria alternata*	Abdelmalek and Salaheldin (2016)
	Candida albicans	Ali and Abdallah (2020)
	Phomopsis vexans	Mahawar and Prasanna (2018)
	Alternaria alternata	Abbas et al. (2019)
	Alternaria alternata, *Penicillium digitatum*, and *Alternaria citri*	Abdelmalek and Salaheldin (2016)
	Aspergillus flvus and *Fusarium solani*	Villamizar-Gallardo et al. (2016)
	Macrophomina phaseolina	Mahdizadeh et al. (2015)
	Candida spp., *Aspergillus* spp., and *Fusarium* spp	Xue et al. (2016)
	Macrophomina phaseolina, *Alternaria Alternata*, and *Fusarium oxysporum*	Bahrami-Teimoori et al. (2017)
	Sclerotinia homoeocarpa	Li et al. (2017)
	Aspergillus niger, *Aspergillus flavus*, and *Aspergillus fumigates*	Menon et al. (2017)
	Alternaria solan	Abdel-Hafez et al. (2016)
	Fusarium oxysporum	Maroufpoor et al. (2019)
	Rhizoctonia solani, *Fusarium oxysporum*, and *Curvularia* sp.	Balakumaran et al. (2016)
	Alternaria solani	Kumari et al. (2017)
	Alternaria alternate, *Rhizoctonia solani*, *Botrytis cinereal*, and *Fusarium oxysporum*	Almadiy and Nenaah (2018)
	Fusarium moniliforme	Kalia et al. (2020)
	Setosphaeria turcica	Huang et al. (2020)
	Botrytis cinerea	Faghihi et al. (2020)
Starch stabilized AgNPs	*Candida* species	Prasher et al. (2018)
Ag core-DHPAC shell nanocluster	*Phytophthora capsici*, *Phytophthora nicotianae*, and *Phytophthora colocasiae*	Ho et al. (2015)
	Fusarium graminearum	Ibrahim et al. (2020)
AgNPs/chitosan	*Corynespora cassiicola*	Nhien et al. (2018)
AgNPs-Ce	*Alternaria alternata*	Mahawar et al. (2020)
Essential oils (EOs) of thyme (*Thymus daenensis* L.) and dill (*Anethum graveolens* L.) with AgNPs	*Colletotrichum nymphaeae*	Weisany et al. (2019)

leaf mold. The increase in inhibition zone reduction activity is observed with the increase of Ag NPs concentration. Such promising results can be exploited further by using the AgNPs against different pathogenic fungi to assess their range of fungicidal activity (Elgorban et al., 2017). Kaman and Dutta (2017) studied the effect of silver nanoparticles at various concentrations (10, 30, 50, and 100 ppm) as an antifungal against four soil-borne plant pathogens. The results found that the growth of pathogenic mycelia at 100 ppm of the silver nanoparticles. Also, two concentrations of AgNPs (50 and 100 $\mu g\,mL^{-1}$) were used in in vitro experiments to research their role as an antifungal action against plant pathogens *Xanthomonas axonopodis* pv. *malvacearum* and *Xanthomonas campestris* pv. *campestris*. The results showed zone of inhibition 11.0 ± 1.0 and 12.3 ± 0.5 mm for *X. axonopodis* pv. *malvacearum* and 9.7 ± 0.6 and 15.33 ± 1.0 mm for *X. campestris* pv. campestris (Vanti et al., 2019). The silver nanoparticles' antimicrobial activity was tested against *Fusarium oxysporum* and *Colletotrichum gloeosporioides*. Total inhibition on *F. oxysporum* was detected at a silver nanoparticles concentration of 75 ppm. No growth at 100 ppm silver nanoparticles was observed in vitro for *C. gloeosporioids* (Singh et al., 2019). The inhibitory activity of silver nanoparticles (AgNP) at concentrations of 10, 20, 50, and 100 ppm was assessed against two phytopathogens: *Bipolaris sorokiniana* and *Alternaria brassicicola*. The results showed that 20 ppm of AgNP caused a greater reduction in germinating spores of *B. sorokiniana* and *A. brassicicola*. Alternatively, 100 ppm of AgNP may be preferred to restrict the mycelial development of these pathogens (Kriti et al., 2020). Talie et al. (2020) found that AgNPs at highest concentrations (20 $mg\,mL^{-1}$) caused the highest inhibition in spore germination and maximum zone of inhibition against *Penicillium chrysogenum* followed by *Aspergillus niger* and *Alternaria alternata*, respectively.

In addition, AgNPs synthesized by the leaf extract from *Ligustrum lucidum* exhibited prominent antifungal activity against *Setosphaeria* (Huang et al., 2020). Moreover, after exposure to AgNPs both in vivo and in vitro, effective inhibition of *Fusarium oxysporum*, the causal agent of tomato wilting, was achieved. In vitro studies showed repressed mycelial fungal growth with an inhibition of 79%–98% compared to control (Anjum and Ashraf, 2020). Biosynthesized AgNPs showing strong inhibition of mycelium formation, spore germination, germ tube length, and mycotoxin production of wheat *Fusarium* head blight pathogen *Fusarium graminearum* (Ibrahim et al., 2020).

21.4 Mechanism of antifungal effect of silver nanoparticles

Ion efflux dysfunction may result in a rapid accumulation of silver ions by reacting to the molecules and prevent cellular activity at lower levels, for example, metabolism and respiration. Silver ions can also generate reactive oxygen species that are dangerous to cells by interacting with oxygen and cause nucleic acid, lipid, and protein damage (Hwang et al., 2008).

Some hypotheses suggest the Ag NPs mechanism is antifungal (Fig. 21.3).

21.4 Mechanism of antifungal effect of silver nanoparticles

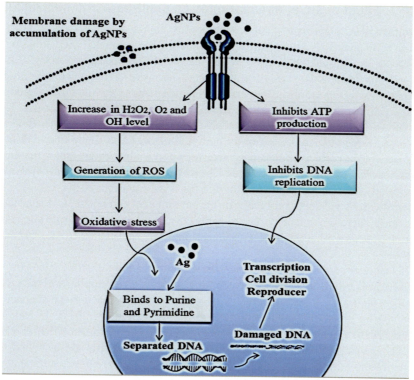

FIG. 21.3

The diagrammatic representation antifungal mechanism of silver nanoparticles.

1. The increase in Ag NPs could impregnate and adhere to fungal hyphae, thus inactivating pathogenic plant fungi (Lemire et al., 2013).
2. Ag NPs linked to the cell wall trigger cell lysis, causing structural damage and destroying the cell's proper activity and contributing to the end of cell life of microbes (Rai et al., 2009).
3. After Ag application, the DNA will not be able to replicate and ribosomal protein expression will be disabled. In addition, certain other cellular proteins and enzymes are required to produce ATP. Ag^+ is thought to primarily affect the role of membranous enzymes, such as those in the respiratory chain (Kim et al., 2012).
4. Cells die by using Ag NPs because of their effect on ribosome causing protein inhibition (Prabhu and Poulose, 2012).
5. In another study by Kumar et al. (2016), Ag NPs enter via phagosome process into the fungal cell membrane and nucleus, then bind to chromosomes, and cause chromosomal damage.

21.4.1 Effect of silver nanoparticles on plant-parasitic nematodes

More than 4000 plant pathogenic nematodes are known to cause significant harm to nearly all plants (Nicol et al., 2011). Annually, the estimated crop loss due to nematodes reaches about 14.5% worldwide (Abd-Elgawad and Askary, 2015). The phylum Nematoda plant-parasitic nematodes were divided into two orders: Tylenchida and Dorylaimida. Both the Dorylaimida genera (Xiphinema, Longidorus, Paralongidorus, Trichodorus, and Paratrichodorus) are migratory ectoparasites. They are known to target trees and herbaceous crops by transmitting phytopathogenic nepo- and tobra-viruses that seriously harm the plants (Abd-Elgawad and Askary, 2015). All plant parts are infected by nematode species like roots, bulbs, rhizomes, stems, leaves, buds, flowers, and seeds. The root-knot nematode (*Meloidogyne* spp.) is especially a severe pathogen for a lot of crops (Abd-Elgawad, 2014). The wide use of nematicides can lead to problems for the environment and health, and nematodes resistance. Therefore, it is important to apply effective, low-cost, and safe alternative control strategies to producers, consumers, and the environment (Abd-Elgawad, 2014).

In laboratory and field experiments, the nematicide effect of silver nanoparticles (AgNPs) was evaluated. They concluded that high doses of application of AgNPs (average 90.4 mg/m^2) were found to be effective in decreasing the number of *Meloidogyne graminis* juveniles in turfgrass (Cromwell et al., 2014). In addition, Nassar (2016) assessed the effectiveness of Ag-nano formulations of *Urtica urens* extracts against root-knot nematodes (*Meloidogyne incognita*), and he found that the petroleum ether extract effect and its Ag nanoparticles provide a sufficient and environmentally safe way to minimize *Meloidogyne incognita*. Moreover, Abdellatif et al. (2016) confirmed the substantial reduction of root galls due to root-knot nematode (*Meloidogyne javanica*) infection after application of AgNPs. Moreover, Taha and Abo-Shady (2016) reported that treatment with AgNP concentration (1500 ppm) caused a reduction in *Meloidogyne incognita* nematode populations associated with tomatoes. Results revealed nematicidal activity of leaf extracts of *Conyza dioscoridis* that were prepared as silver nanoparticles (AgNPs) had great nematicidal activity against the 2nd stage juvenile (J2) and eggs of *Meloidogyne incognita*.

Also, Nour El-Deen and El-Deeb (2018) observed that the use of AgNP in a tomato field against the root-knot nematode was very successful when compared with the application of control or silver nitrate. A series of laboratory tests (water and sand screening) and glasshouse experiments (using a soilless method, autoclaved soil, and naturally infested soil) were conducted to investigate AgNP nematicide effects on *Meloidogyne gramnicola*. Results from laboratory assays showed 0.1 µg mL^{-1} in the water screening test as the minimum concentration for 100% irreversible nematode mortality after 12 h. However, after 24 h of incubation, tests from the sand screening test showed 100% nematicidal effect of AgNP at 2 µg mL^{-1}. In glasshouse assays in soilless rice cultivation method, 1 µg mL^{-1} AgNP concentration applied directly to the trays achieved substantial suppression of the formation of root gall.

In addition, Barbosa et al. (2019) found that AgNP (*Duddingtonia flagrans*) demonstrated nematicide effectiveness, being the only treatment able to penetrate the larvae cuticles and cause their subsequent death. Hamed et al. (2019) found that in vitro assay against the nematode of the root-knot *Meloidogyne javanica* showed that AgNPs significantly reduced eggs hatching with *M. javanica* applied at various concentrations (3%, 6%, 12%, 25%, and 50%, v/v). The highest reduction rate (94.66%) was caused by 50% of AgNPs. Also, the AgNPs and AgNO$_3$ substantially improved larval percentages mortality of second-stage juvenile (J2). AgNPs or AgNO$_3$ at 2, 4 mL L^{-1}, 24 h completely inhibited J2 growth compared to 23% inhibition with aqueous cyanobacterial extract. Nazir et al. (2019) showed that silver nanoparticles' (AgNPs) nematicidal behavior was investigated against nematode with the most damaging root-knot (*Meloidogyne incognita*). Maximum juvenile mortality was reported at a concentration of 100 mg mL^{-1}, followed by AgNP at 75 mg mL^{-1}. AgNP at 25 mg mL^{-1} showed minimum mortality, the rise in concentration, a related increase in juvenile mortality revealed a clear relationship between mortality and concentration of nanoparticles. Also, the effective dose to kill nematodes in field soil assays was 3 μg mL^{-1}, which is less than the 150 μg mL^{-1} value (Baronia et al., 2020).

21.4.2 Mechanism of nematocidal effect of silver nanoparticles

AgNP has demonstrated high potential as a strong nematicide, whose impact is associated with oxidative stress induction in PPN cells and the disruption of membrane permeability and ATP synthesis by the cellular mechanism (Lim et al., 2012). The nematicide function of AgNPs for microorganisms may be associated with development silver surface free radicals, which lead to increased oxidative stress and damage to the membrane (Khalil and Badawy, 2012).

21.4.3 Effect of silver nanoparticles on antiviral

Apart from being an antibacterial and antifungal agent, AgNPs have also been recognized as antiviral agents. Plant viruses are occurred in nature and cause serious diseases in different plants. They were causing significant crop quality and yield losses. Virus diseases account for 47% of this decline, out of 15% of world food production is lost due to plant diseases (Boualem et al., 2016; Sofy et al., 2020).

Postinfection application with AgNPs resulted in a significant decrease in virus concentration, infection percentage, and severity of the disease. Accordingly, the AgNPs were confirmed to be effective against a prototype arenavirus when administered early after initial exposure to the virus (Speshock et al., 2010). Also, Jain and Kothari (2014) found that spraying of 50 ppm aqueous solution silver nanoparticles on cluster bean leaves inoculated with sunhemp rosette virus (SHRV) showed complete disease severity reduction, suggesting that silver nanoparticles are antiviral-effective. In addition, treatment with AgNPs led to a decrease in the percentage of BYMV infection and disease frequency (Elbeshehy et al., 2015). El-Dougdoug et al. (2018)

studied the efficiency of silver nanoparticles (AgNPs) as an antiviral reagent inducing systemically acquired resistance (SAR) to tomato mosaic virus (ToMV) and potato virus Y (PVY) and suppressing tomato plant infection. The findings showed that the severity of the disease and the relative concentration of both viruses were significantly reduced by treating AgNPs at 50 ppm vs the other treatments. Examination of the clarified sap of both viruses by transmission electron microscope (TEM) showed morphological evidence that AgNPs bind to coat particles of the protein virus.

21.4.4 Mechanism of antifungal effect of silver nanoparticles

Ag inactivates viruses in various ways by denaturing enzymes by reactions with sulfhydryl, amino, carboxyl, phosphate and imidazole groups (Rai et al., 2009). AgNPs may inhibit the replication of viral nucleic acid and its antiviral activity may be due to several factors such as: particle size, distribution of ligand/receptor molecules that interact (Papp et al., 2010). The silver nanoparticles can communicate with microbial cells directly. Silver ions can disrupt the transmembrane electron transfer, oxidize cell components and perpetration in the cell chain and make the cells susceptible to reactive oxygen species (ROS), or dissolve heavy metal ions causing harm in particular, the enzymes in the respiratory chain and the permeability of phosphates and protons (Rajeshkumar and Malarkodi, 2014). In addition, more damage to bacterial cells can be caused by cell permeation, where they interact with proteins, DNA, and other cell-containing constituents of sulfur and phosphorus (Nayak et al., 2015). AgNPs have also shown to resist viral membrane fusion, thereby preventing the virus from entering the host cell (Mohamed and Abd-Elsalam, 2018) (Fig. 21.4).

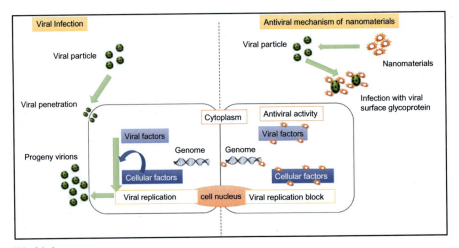

FIG. 21.4

Schematic model of viruses that infect live cells and the metal nanomaterial antiviral mechanism.

21.4.5 Postharvest diseases

It is estimated that about 20%–25% of the fruits and vegetables harvested are contaminated by pathogens even in developing countries during postharvest handling. Postharvest losses are often more severe in developing countries due to insufficient storage and transport facilities (Yahaya and Mardiyya, 2019). The Food and Agriculture Organization of the United Nations announced the world's food supply loss of 35% after losses in the production system have accumulated because of the disease pathogen (Gastavsson et al., 2011).

Several pathogenic agents such as fungi and bacteria are responsible for causing plant diseases. It is nevertheless well known that the big postharvest losses are caused by fungi such as *Alternaria*, *Aspergillus*, *Botrytis*, *Colletotrichum*, *Diplodia*, *Monilinia*, *Penicillium*, *Phomopsis*, *Rhizopus*, *Mucor*, and *Sclerotinia* and bacteria such as *Erwinia* and *Pseudomonas* (Sheikh, 2017). The highest risk of mycotoxin contamination arises when contaminated plants are used for food or animal feed production (Greeff-Laubscher et al., 2020). At present, controlling degradation of fruit and vegetables after harvest is currently achieved by applying chemical fungicides such as imazalil, thiabendazole, pyrimethanil, and fludioxonil to citrus or boscalid and iprodione to grapes (Palou, 2018). When used as sanitary agents, chloride-based chemicals can formulate organic chlorinated substances, including chloramines, dichloramines, and trichloromethane, known to be respiratory irritants (Fallanaj et al., 2013). On the other hand, the widespread use of chemical fungicides pre- and postharvest has caused resistant strains of pathogens that break down the efficacy of the fungicide (Hao et al., 2011). Although these techniques are expensive and time-consuming, the technological existence of fungicides after a period of time is essential (Vitale et al., 2016). In addition, customers are concerned about the use of chemical fungicides, as their active ingredients and coformulants are linked to health problems and environmental pollution (Nicolopoulou-Stamati et al., 2016). Nanomaterial therapies have recently studied and shown promising results to minimize the use of synthetic fungicides in fruit and vegetable rot after harvest (Ruffo et al., 2019).

In the section of postharvest fruit and vegetable industry, nanotechnology approach is helpful for postharvest disease control, the launch of innovation for packaging films, to prevent the impact of gases and unsafe rays, to improve the appearance of packaging, and help with multiple chips (nanobiosensors) to mark fresh products (Ruffo et al., 2019).

The antifungal effects of nanocomposites SiO_2/Ag_2S (300 nm) were first recorded in *Aspergillus niger*, the most common postharvest fungus causing many diseases of the fresh fruit like apples and oranges (Plascencia-Jatomea et al., 2014). The results indicate that the sporulation and growth of fungi are decreased by SiO_2/Ag_2S nanocomposites. This effect was observed because Ag_2S nanophases and the synergistic effect of Ag_2S and SiO_2 surfaces in support of fungal adsorption were present (Fateixa et al., 2009). It has been suggested that silver species associate in the plasma membranes of susceptible pathogens with the sulfhydryl groups of respiratory enzymes, causing changes in the membrane permeability (Ruffo et al., 2019). The binding of Ag^+ to microbial genetic material was another theory to clarify this mechanism (Han et al., 2011).

AgNPs (38 nm), CuNPs (20 nm), and Ag/CuNP were examined in vitro studies at various levels (0, 1, 5, 10, and 15 mg L^{-1}) against *Alternaria alternata* and *Botrytis cinerea*. The optimal level is 15 mg L^{-1} for inhibiting both pathogens. The nanoparticles weakened the hyphae and conidia of *A. alternata* using a scanning electron microscope. In *B. cinerea*, fungal hyphae were found to have weakened their surface, leading to the release of inner cell materials with shrinking pathogen hyphae. Also, total culture filtrate protein and total cell wall protein and *N*-acetyl glucoseamine decreased, while total culture filtrate and cell wall lipids increased after treatments with silver nanoparticles (Ouda, 2014).

In another in vitro study, silver nanoparticles (50 nm) were prepared and mixed with the fungicide tolclofosmethyl at various concentrations (25–100 mg L^{-1}). The mixture worked better against *B. cinerea* than individual compounds (Derbalah et al., 2011).

The composite of chitosan–silver NPs (diameter 495–616 nm) was prepared, and the silver NPs were distributed in 10–15 nm size composite. The composite exhibited greater antifungal effect against *Colletotrichum gloeosporioides* conidial germination. A complete spore germination inhibition was obtained when a composite of chitosan-silver NPs was applied at 100 μg mL^{-1}. Furthermore, in vivo studies showed a decrease in anthracnose disease of about 45.7% and 71.3% by composite chitosan-silver nanoparticles of 0.5% and 1%, respectively (Chowdappa et al., 2014).

The copper oxychloride-conjugated silver NPs were tested by Raghavendra et al. (2019) for their action against the mango anthracnose causative agent (*C. gloeosporioides*). The nanomaterials with an average particle size of 21–25 nm displayed the pathogen maximum growth inhibition (~187%) compared with copper oxychloride.

Ag NPs caused the release of Ag ions at the desired rate and position to naturally affect the microbe cell. The presence of ligands in the bacteria's vicinity is considered a problem using Ag ions directly as a toxic to bacteria, as it can bind to Ag ions, blocking them from reaching their target (Xiu et al., 2012). The AgNPs, therefore, serve as a highly efficient source for Ag ions to the cytoplasm and membrane of bacterial cells, where natural ligands are less associated with exposure or reduction. Consequently, several efforts were made to modify the NP surfaces. AgNPs modified with glucose, lactose, oligonucleotides, and these ligands' combinations have been tested for their cytotoxicity (Sur et al., 2010). Silver nanoparticles demonstrated excellent bactericidal properties against several microorganisms (Siddiqi and Husen, 2016; Wei et al., 2015). Owing to its inherent characteristic, it displayed an antimicrobial characteristic and a significant role even in solid state (Siddiqi et al., 2018). Several studies revealed the inhibitory nature of Ag NPs particles linked to sterilization (Kim et al., 2011). Inhibitory effect of silver nanoparticles was recorded on plant postharvested disease pathogens, *Botrytis cinereal* (Salem et al., 2019), *Penicillium verrucosum* (Mukherjee et al., 2016), *Fusarium oxysporum* (Abkhoo and Panjehkeh, 2016; Bahrami-Teimoori et al., 2017), *Alternaria alternata* (Bahrami-Teimoori et al., 2017), *Colletotrichum musae* (Jagana et al., 2017), *Sclerotinia sclerotiorum* (Guilger-Casagrande et al., 2019), *Penicillium digitatum* and *Alternaria citri* (Abdelmalek and Salaheldin, 2016), and *Aspergillus niger* (Al-Zubaidi et al., 2019).

Due to its nanometric size, which facilitates interaction with microbial cells, silver nanoparticles have a biocide effect, small AgNPs interact with cell membranes, modify the lipid bilayer, increase membrane permeability, and finally, cause cell death (Li et al., 2013). Ag NPs have been found to be detrimental to DNA (Han et al., 2011). It can form complexes with DNA nitrogen bases, thus condensing and reducing the process of replication and reacting with proteins, thus affecting the respiratory and cell division (Ruffo et al., 2019).

21.4.6 Pathogens detection and diagnosis

A global food shortage will continue in the next 40 years, as steady growth in the human population is anticipated. To meet the needs, a 70% increase in food production is required by 2050. Unsafe food is a concern to human health and the economies of countries, which affects people primarily. There are believed to be 600 million cases of food-borne diseases and 420,000 deaths worldwide annually (Zorraquin-Pena et al., 2020).

In-plant disease diagnosis, various direct and indirect methods are used, which are laboratory-based techniques such as nucleic acid detection based on polymerase chain reaction (PCR), in situ fluorescence hybridization, enzyme-linked immunosorbent assay (ELISA), immunofluorescence, flow cytometry, thermography, fluorescence imaging, hyperspectral techniques, and gas chromatography (Fang and Ramasamy, 2015). The key downside of these approaches is long analytical time ranging from several hours to days, usually with different phases of pretreatment (Omanović-Mikličanina and Maksimović, 2016).

The lack of traditional methods for solving new food safety challenges leads to the development of new techniques with miniaturized and quick analytical power and low detection limits. Key solutions in developing new nanotechnology-based devices integrated with analytical tools (Omanović-Mikličanina and Maksimović, 2016). Nanoparticles can be synthesized using various types of materials for electronics and sensing applications because they exhibit fascinating electronic and optical properties (Sun et al., 2014, 2017). Recently, various types of phytopathogens such as bacteria, viruses, and fungi were detected through amperometric biosensors based on nanoparticles (Chartuprayoon et al., 2010). The use of nanosensors as a detector of pathogens, toxins, pollutants, and food freshness emerges as a promising method in agriculture and food protection (Joyner and Kumar, 2015). It also uses physiological signals in detection and converts them into standardized, often electrical signals which can be quantified from analog to digital signals.

Due to the low cost, simplicity, and convenience of these methods, metal nanoparticules colorimetric assays have received considerable attention (Laksanasopin et al., 2015). Metal nanoparticles have been used as marker tags in several biosensor formats to replace labeling enzymes or other chromophores such as gold (AuNPs) and silver (AgNPs) nanoparticles, which have been tremendously used to detect analytes. AgNP is a broad-spectrum antimicrobial agent, which can affect plant pathogens and bacteria (Wu et al., 2011). These results may be due to its mode of action because it demonstrated strong adhesion on the surface of bacterial and fungal cells (Seong and Lee, 2017). For its simplicity and practicality, colorimetric detection has been widely applied (Song et al., 2011).

For the detection of multiplex mycotoxins with ultrahigh sensitivity, high throughput, and simplicity, a novel surface-plasmon-coupled chemiluminescent immunosensor was developed. The immunosensor was built by immobilizing carboxyl-modified silver nanoparticles (AgNPs) and bovine serum albumin combined antigens sequentially on the modified amino-grouped glass chip. The optical properties of AgNPs could amplify the chemiluminescence (CL) generated on the chip by the resonance phenomenon of surface plasmons (Jiang et al., 2020).

The SERS-sensor holds promise for the rapid quantification of Ochratoxin-A (OTA) and aflatoxin-B1 (AFT-B1) in cocoa beans at pg/mL level to enable safety assurance in the cocoa bean industry. AgNP@pH-11 was selected to manufacture SERS-sensor with the highest SERS-EF (1.45/108) and coupled with two chemometric algorithms to predict OTA and AFT-B1 in standard prepared solutions (SS) and spiked-cocoa-beans samples (SCBS) (Kutsanedzie et al., 2020).

21.5 Silver nanoparticles for bacterial detection

1. Surface-enhanced Raman Spectroscopy (SERS)

Surface-enhanced Raman spectroscopy is a method of amplifying Raman signals using nanoscale metal surfaces. Such metallic nanosystems are called plasmonic nanoparticles, and when exposed to laser light, they exhibit different optical properties. AgNPs is used in amplifying (SERS) for detection of living bacteria such as *Salmonella* and *Staphylococcus aureus* (Liu et al., 2018). Monoclonal antibodies with high recognition capacity are used to target a specific surface protein of the bacteria (Lin et al., 2014).

2. Electrochemical biosensors

Electrochemical detection depends on the electrical current produced by oxidation or reduction of reactions to biological processes for the detection of molecular biomarkers from various pathogenic infections, a very sensitive method showed an accuracy (Sismaet and Goluch, 2018). In particular within the range of 1–10nm, silver nanoparticles may interact with the surface of some gram-negative bacteria, disturbing its functioning and inducing a bactericidal effect (Sepunaru et al., 2015). Because of the high affinity of silver nanoparticles to the bacteria surface, this concept was used to decorate bacteria with silver nanoparticles about 10nm in diameters (Wu et al., 2018). A carbon fiber electrode was used to measure the chronoamperometric response of the silver nanoparticles.

3. Fluorescence methods

Fluorescence immunoassay is a highly sensitive technique for measuring mainly proteins and for quantifying antigens from viruses or bacteria as well as many other substances (Colino et al., 2018). The monoclonal antibody of *Pseudomonas aeruginosa* was appended to labeled AgNP with a pyrimidine fluorescent derivative. Due to the features of the metallic nanoparticle, the fluorescence response was able to enhance the metal surface when the fluorophore was located near it. This nanosystem is used for the identification of bacteria in water, crop samples, soil, and various food types (Li et al., 2011).

21.5.1 Mycotoxin detection

Nanoscale material displayed physicochemical properties differing from its larger counterpart due to a high area-to-volume ratio leading to increased reactivity (Borase et al., 2015). A single portable nanosensor can be fitted with thousands of nanoparticles to detect correctly harmful spores of the fungus, trace pollutants, and food toxins in a short period (Handford et al., 2014).

More than 300 mycotoxins were detected, and the serious effects of these toxins on animal and human health have been recorded (Lizárraga-Paulín et al., 2011). Analysis of mycotoxins is important and essential to ensure food safety because of the serious health risks associated with the presence of mycotoxins in food and its impact on domestic and international trade (Rhouati et al., 2017).

Due to the low cost, simplicity, and convenience of these methods, metal nanoparticules' colorimetric assays have received considerable attention. Furthermore, the manufacture of aptasensors depends on nanomaterials that play a role in the detection of toxins, amplifying the signal, mediating, and labeling the artificial enzyme (Rhouati et al., 2017).

The AgNPs have been widely used to detect analytes of mycotoxins (Velu and DeRosa, 2018). AgNPs demonstrated an efficient way of testing mycotoxin ochratoxin (OAT) concentration depending on the optical method of color detection (Velu and DeRosa, 2018). Nanosilver has found aptasensing applications as a signal amplifier to detect a wide range of molecules (Bahrami et al., 2016). AgNPs were also used to detect ochratoxin mycotoxin as aptamer carriers or as signal producing probes for aptasensing (Rhouati et al., 2017). Also, Chen et al. (2014) used a DNA-scaffolded-silver-nanocluster (AgNCs) as a fluorophore for the detection of OTA fluorescent. Also, the importance of silver nanoparticles in electro-oxidation and determination of aflatoxin M1 in milk samples was asserted (Shadjou et al., 2018).

21.5.2 Mycotoxin management

A quarter of cereal-based crops worldwide are contaminated with filamentous fungi and their mycotoxins and must be rejected at the expense of food supplies to a steadily increasing world population (WHO, 2018). Climate change contributes to worsening the situation further (Van de Perre et al., 2015). Fungi that produce mycotoxins are of paramount importance. Aflatoxins (*Aspergillus*), trichothecenes and fumonisins (*Fusarium*), alternariol and tenuazonic acid (*Alternaria*), and *Penicillium ochratoxins*, patulin and citrinin, are among the most important mycotoxins (Mukherjee et al., 2016). The most important mycotoxins are aflatoxins (AFs), ochratoxin A (OTA), fumonisins, trichothecenes, and zearalenones (JECFA, 2017). The widely distributed mycotoxigenic fungi, which cause risk on human health, are *Aspergillus ochraceus*, *Aspergillus niger*, *Aspergillus flavus*, *Aspergillus carbonarius*, and *Aspergillus parasiticus* (JECFA, 2018). The application of nanoparticles, e.g., as spray-disinfectant, could be an effective opportunity to avoid food and feed contamination due to mycotoxigenic fungi (Mittal et al., 2015).

Nutritional supplements may suppress mycotoxin toxicity, reduce tissue damage caused by oxidative stress, and enable the body to maintain a functional immune system that can eliminate the pathogens (Stroka and Maragos, 2016). Nanotechnology methods seem to be a promising, efficient, and low-cost way of reducing mycotoxins health effects (Horky et al., 2018). Recently, Mousavi and Pourtalebi (2015) showed that silver nanoparticles could severely inhibit the production of AFB1 aflatoxins at a dose of $135\,\mu g\,mL^{-1}$. Also, honey-derived silver nanoparticles showed a decrease in aflatoxin production to 66.5% and an inhibitory effect of ochratoxin production to 79.8% (El-Desouky and Ammar, 2016). Also, silver nanoparticles applied to *Aspergillus flavus* and *Aspergillus ochraceus* showed a reduction of mycotoxins aflatoxins (AFs) and ochratoxin A (OTA) production (Khalil et al., 2019).

AgNPs demonstrated superior antibacterial against *Staphylococcus aureus*, antifungal against *Aspergillus flavus* and *A. parasiticus* activities, and decreased the production of aflatoxins as opposed to Fe-NPs and Cu-NPs (Asghar et al., 2018). Three different sizes of citrated silver-coated nanoparticles inhibit aflatoxin biosynthesis at various effective doses of *Aspergillus parasiticus*, the pathogenic filamentous fungus of the plant (Mitra et al., 2019). Also, the AgNPs demonstrated notable antifungal activity against *Aspergillus flavus* and ability to thwart the production of mycotoxin (Khalil et al., 2019). AgNPs alone or as an active ingredient could be a good strategy for managing the main aflatoxigenic and ocratoxigenic species that affect contamination of food and aflatoxins (AFs) and ochratoxin A (OTA) (Gómez et al., 2019). Also, *Juniperus procera* leaf extract at $100\,mg\,mL^{-1}$ showed a growth inhibition of 35.83% and 44.09% but increased to 50.55% and 59.06% after 50 ppm silver nanoparticles (AgNPs) were added to *Aspergillus fumigatus* and *Fusarium chlamydosporum*, respectively (Bakri et al., 2020).

21.6 Conclusion and future perspectives

The application of large quantities of microbial pesticides and the presence of new microbial-resistant strains are the major critical concerns for food safety. This study depends on the use of NPs as an antifungal factor and antibacterial against the pathogens of plants as a new and successful tool for future prevention and control of pathogens, since bacteria and fungus are highly resistant to conventional control methods. We are therefore close to implementing modern agricultural techniques and new emerging technologies to more carefully and precisely manage these threats because traditional agricultural practices cannot adequately regulate these threats without jeopardizing human health. Effective NP biocidal can influence beneficial and deleterious bacteria, fungi, and other microorganisms under laboratory and environmental conditions, in particular, in soils and plants. AgNPs were found to regulate a wide variety of bacterial, fungal pathogens, and viral infections based on these findings. AgNPs may interfere with the replication, growth, and development of pathogens, thereby stopping or dying.

References

Abalaka, M., Akpor, O., Osemwegie, O., 2017. Green synthesis and antibacterial activities of silver nanoparticles against *Escherichia coli*, *Salmonella typhi*, *Pseudomonas aeruginosa* and *Staphylococcus aureus*. Adv. Life Sci. 4, 60–65.

Abbas, A., Naz, S.S., Syed, S.A., 2019. Antimicrobial activity of silver nanoparticles (agnps) against *Erwinia carotovora* subsp. atroseptica & *Alternaria alternata*. Pak. J. Agric. Sci. 56, 113–117.

Abd-Elgawad, M., 2014. Plant-parasitic nematode threats to global food security. J. Nematol. 46, 130.

Abd-Elgawad, M.M.M., Askary, T.H., 2015. Impact of phytonematodes on agriculture economy. In: Askary, T.H., Martinelli, P.R.P. (Eds.), Biocontrol Agents of Phytonematodes. CABI, Wallingford, UK, pp. 3–49.

Abd El-Rahman, S.S., Mohamed, H.I., 2014. Application of benzothiadiazole and *Trichoderma harzianum* to control faba bean chocolate spot disease and their effect on some physiological and biochemical traits. Acta Physiol. Plant. 36 (2), 343–354.

Abdel-Hafez, S.I., Nafady, N.A., Abdel-Rahim, I.R., Shaltout, A.M., Daròs, J.-A., Mohamed, M.A., 2016. Assessment of protein silver nanoparticles toxicity against pathogenic *Alternaria solani*. 3 Biotech 6, 199.

Abdellatif, K.F., Abdelfattah, R.H., El-Ansary, M.S.M., 2016. Green nanoparticles engineering on root-knot nematode infecting eggplants and their effect on plant DNA modification. Iran. J. Biotechnol. 14, 250–259.

Abdelmalek, G., Salaheldin, T., 2016. Silver nanoparticles as a potent fungicide for citrus phytopathogenic fungi. J. Nanomed. Res. 3, 00065.

Abkhoo, J., Panjehkeh, N., 2016. Evaluation of antifungal activity of silver nanoparticles on *Fusarium oxysporum*. Int. J. Inf. Secur. 4 (2), e41126.

Agnihotri, S., Mukherji, S., Mukherji, S., 2014. Size-controlled silver nanoparticles synthesized over the range 5–100 nm using the same protocol and their antibacterial efficacy. RSC Adv. 4, 3974–3983.

Aguilar-Méndez, M.A., San Martín-Martínez, E., Ortega-Arroyo, L., Cobián-Portillo, G., Sánchez-Espíndola, E., 2011. Synthesis and characterization of silver nanoparticles: effect on phytopathogen *Colletotrichum gloesporioides*. J. Nanopart. Res. 13, 2525–2532.

Ahmed, A., 2017. Chitosan and silver nanoparticles as control agents of some *Faba bean* spot diseases. J. Plant Pathol. Microbiol. 8 (9), 1000421.

Ahmed, T., Shahid, M., Noman, M., Niazi, M.B.K., Mahmood, F., Manzoor, I., Zhang, Y., Li, B., Yang, Y., Yan, C., 2020. Silver nanoparticles synthesized by using *Bacillus cereus* szt1 ameliorated the damage of bacterial leaf blight pathogen in rice. Pathogens 9, 160.

Ali, E.M., Abdallah, B.M., 2020. Effective inhibition of candidiasis using an eco-friendly leaf extract of *Calotropis gigantean*-mediated silver nanoparticles. Nano 10, 422.

Allinne, C., Savary, S., Avelino, J., 2016. Delicate balance between pest and disease injuries, yield performance, and other ecosystem services in the complex coffee-based systems of Costa Rica. Agric. Ecosyst. Environ. 222, 1–12.

Almadiy, A.A., Nenaah, G.E., 2018. Ecofriendly synthesis of silver nanoparticles using potato steroidal alkaloids and their activity against phytopathogenic fungi. Braz. Arch. Biol. Technol. 61, e18180013.

Aly, A.A., Mansour, M., Mohamed, H.I., Abd-Elsalam, K.A., 2012. Examination of correlations between several biochemical components and powdery mildew resistance of flax cultivars. Plant Pathol. J. 28, 149–155.

Aly, A.A., Mohamed, H.I., Mansour, M.T., Omar, M.R., 2013. Suppression of powdery mildew on flax by foliar application of essential oils. J. Phytopathol. 161, 376–381.

Aly, A., Mansour, M., Mohamed, H., 2017. Association of increase in some biochemical components with flax resistance to powdery mildew. Gesunde Pflanzen 69, 47–52.

Al-Zubaidi, S., Al-Ayafi, A., Abdelkader, H., 2019. Biosynthesis, characterization and antifungal activity of silver nanoparticles by *Aspergillus niger* isolate. J. Nanotechnol. Res. 1 (1), 023–036.

Anitha, R., Ramesh, K., Ravishankar, T., Kumar, K.S., Ramakrishnappa, T., 2018. Cytotoxicity, antibacterial and antifungal activities of ZnO nanoparticles prepared by the *Artocarpus gomezianus* fruit mediated facile green combustion method. J. Sci. Adv. Mater. Devices 3, 440–451.

Anjum, T., Ashraf, H., 2020. Microwave-assisted green synthesis and characterization of silver nanoparticles using *Melia azedarach* for the management of fusarium wilt in tomato. Front. Microbiol. 11, 238.

Arya, G., Sharma, N., Mankamna, R., Nimesh, S., 2019. Antimicrobial silver nanoparticles: future of nanomaterials. In: Microbial Nanobionics. Springer, pp. 89–119.

Asghar, M.A., Zahir, E., Shahid, S.M., Khan, M.N., Asghar, M.A., Iqbal, J., Walker, G., 2018. Iron, copper and silver nanoparticles: green synthesis using green and black tea leaves extracts and evaluation of antibacterial, antifungal and aflatoxin B1 adsorption activity. LWT Food Sci. Technol. 90, 98–107.

Asran, A.A., Mohamed, H.I., 2014. Use of phenols, peroxidase, and polyphenoloxidase of seed to quantify resistance of cotton genotypes to *Fusarium wilt* disease. Bangladesh J. Bot. 43, 353–357.

Aziz, N., Pandey, R., Barman, I., Prasad, R., 2016. Leveraging the attributes of *Mucor hiemalis*-derived silver nanoparticles for a synergistic broad-spectrum antimicrobial platform. Front. Microbiol. 7, 1984.

Bahrami, S., Abbasi, A.R., Roushani, M., Derikvand, Z., Azadbakht, A., 2016. An electrochemical dopamine aptasensor incorporating silver nanoparticle, functionalized carbon nanotubes and graphene oxide for signal amplification. Talanta 159, 307–316.

Bahrami-Teimoori, B., Nikparast, Y., Hojatianfar, M., Akhlaghi, M., Ghorbani, R., Pourianfar, H.R., 2017. Characterisation and antifungal activity of silver nanoparticles biologically synthesised by *Amaranthus retroflexus leaf* extract. J. Exp. Nanosci. 12, 129–139.

Bakri, M.M., El-Naggar, M.A., Helmy, E.A., Ashoor, M.S., Ghany, T., 2020. Efficacy of Juniperus procera constituents with silver nanoparticles against *Aspergillus fumigatus* and *Fusarium chlamydosporum*. BioNanoScience 10, 62–72.

Bakshi, M., Singh, H., Abhilash, P., 2014. The unseen impact of nanoparticles: more or less? Curr. Sci. 106, 350–352.

References

Balakumaran, M., Ramachandran, R., Balashanmugam, P., Mukeshkumar, D., Kalaichelvan, P., 2016. Mycosynthesis of silver and gold nanoparticles: optimization, characterization and antimicrobial activity against human pathogens. Microbiol. Res. 182, 8–20.

Balaure, P.C., Gudovan, D., Gudovan, I., 2017. Nanopesticides: A New Paradigm in Crop Protection, New Pesticides and Soil Sensors. Elsevier, pp. 129–192.

Barbosa, A., Silva, L.P.C., Ferraz, C.M., Tobias, F.L., de Araujo, J.V., Loureiro, B., Braga, G., Veloso, F.B.R., Soares, F.E.F., Fronza, M., Braga, F.R., 2019. Nematicidal activity of silver nanoparticles from the fungus *Duddingtonia flagrans*. Int. J. Nanomedicine 14, 2341–2348.

Baronia, R., Kumar, P., Singh, S.P., Walia, R.K., 2020. Silver nanoparticles as a potential nematicide against *Meloidogyne graminicola*. J. Nematol. 52, 1–9.

Bhuyar, P., Rahim, M.H.A., Sundararaju, S., Ramaraj, R., Maniam, G.P., Govindan, N., 2020. Synthesis of silver nanoparticles using marine macroalgae Padina sp. and its antibacterial activity towards pathogenic bacteria. Beni-Suef Univ. J. Basic Appl. Sci. 9, 1–15.

Borase, H.P., Salunkhe, R.B., Patil, C.D., Suryawanshi, R.K., Salunke, B.K., Wagh, N.D., Patil, S.V., 2015. Innovative approach for urease inhibition by Ficus carica extract-fabricated silver nanoparticles: an in vitro study. Biotechnol. Appl. Biochem. 62, 780–784.

Boualem, A., Dogimont, C., Bendahmane, A., 2016. The battle for survival between viruses and their host plants. Curr. Opin. Virol. 17, 32–38.

Cerda, R., Avelino, J., Gary, C., Tixier, P., Lechevallier, E., Allinne, C., 2017. Primary and secondary yield losses caused by pests and diseases: assessment and modelling in coffee. PLoS One 12, e0169133. https://doi.org/10.1371/journal.pone.0169133.

Chartuprayoon, N., Rheem, Y., Chen, W., Myung, N., 2010. Detection of plant-pathogen using LPNE grown single conducting polymer nanoribbon. In: Meeting Abstracts. The Electrochemical Society, Pennington, NJ, USA.

Chen, J., Zhang, X., Cai, S., Wu, D., Chen, M., Wang, S., Zhang, J., 2014. A fluorescent aptasensor based on DNA-scaffolded silver-nanocluster for ochratoxin A detection. Biosens. Bioelectron. 57, 226–231.

Chen, J., Li, S., Luo, J., Wang, R., Ding, W., 2016. Enhancement of the antibacterial activity of silver nanoparticles against phytopathogenic bacterium *Ralstonia solanacearum* by stabilization. J. Nanomater. 2016. https://doi.org/10.1155/2016/7135852, 7135852.

Chowdappa, P., Shivakumar, G., Chethana, C.S., Madhura, S., 2014. Antifungal activity of chitosan-silver nanoparticle composite against *Colletotrichum gloeosporioides* associated with *Mango anthracnose*. Afr. J. Microbiol. Res. 8, 1803–1812.

Colino, C.I., Millan, C.G., Lanao, J.M., 2018. Nanoparticles for signaling in biodiagnosis and treatment of infectious diseases. Int. J. Mol. Sci. 19, 1627. https://doi.org/10.3390/ijms19061627.

Cromwell, W.A., Yang, J., Starr, J.L., Jo, Y.K., 2014. Nematicidal effects of silver nanoparticles on root-knot nematode in Bermudagrass. J. Nematol. 46, 261–266.

Dakal, T., Kumar, A., Majumdar, R., Yadav, V., 2016. Mechanistic basis of antimicrobial actions of silver nanoparticles. Front. Microbiol. 7, 1831.

Dean, R., Van Kan, J.A., Pretorius, Z.A., Hammond-Kosack, K.E., Di Pietro, A., Spanu, P.D., Rudd, J.J., Dickman, M., Kahmann, R., Ellis, J., Foster, G.D., 2012. The top 10 fungal pathogens in molecular plant pathology. Mol. Plant Pathol. 13, 414–430.

Derbalah, A.S., Elkot, G.A., Hamza, A.M., 2011. Laboratory evaluation of botanical extracts, microbial culture filtrates and silver nanoparticles against *Botrytis cinerea*. Ann. Microbiol. 62, 1331–1337.

El-Beltagi, H.S., Mohamed, H.I., 2013. Reactive oxygen species, lipid peroxidation and antioxidative defense mechanism. Notulae Bot. Horti Agrobot. Cluj-Napoca 41, 44–57.

Elbeshehy, E.K., Elazzazy, A.M., Aggelis, G., 2015. Silver nanoparticles synthesis mediated by new isolates of *Bacillus* spp., nanoparticle characterization and their activity against bean yellow mosaic virus and human pathogens. Front. Microbiol. 6, 453.

El-Desouky, T., Ammar, H., 2016. Honey mediated silver nanoparticles and their inhibitory effect on aflatoxins and ochratoxin A. J. Appl. Pharm. Sci., 83–90.

El-Dougdoug, N., Bondok, A.M., El-Dougdoug, K.A., 2018. Evaluation of silver nanoparticles as antiviral agent against ToMV and PVY in tomato plants. Middle-East J. Appl. Sci. 8 (1), 100–111.

Elgorban, A.M., El-Samawaty, A.E.-R.M., Abd-Elkader, O.H., Yassin, M.A., Sayed, S.R., Khan, M., Adil, S.F., 2017. Bioengineered silver nanoparticles using *Curvularia pallescens* and its fungicidal activity against *Cladosporium fulvum*. Saudi J. Biol. Sci. 24, 1522–1528.

Faghihi, R., Larijani, K., Abdossi, V., Moradi, P., 2020. Silver nanoparticles produced by green synthesis using *Citrus paradise* peel inhibits *Botrytis cinerea* in vitro. J. Hortic. Postharvest Res. 3, 151–160.

Fallanaj, F., Sanzani, S.M., Zavanella, C., Ip Polito, A., 2013. Salt addition improves the control of citrus post-harvest diseases using electrolysis with conductive diamond electrodes. J. Plant Pathol. 95, 373–383.

Fang, Y., Ramasamy, R.P., 2015. Current and prospective methods for plant disease detection. Biosensors (Basel) 5, 537–561.

Fateixa, S., Neves, M.C., Almeida, A., Oliveira, J., Trindade, T., 2009. Anti-fungal activity of SiO_2/Ag_2S nanocomposites against *Aspergillus niger*. Colloids Surf. B: Biointerfaces 74, 304–308.

Gastavsson, J., Cederberg, C., Sonesson, U., 2011. Global Food Losses and Food Waste. Food and Agriculture Organization (FAO) of the United Nations, Rome, Italy.

Ghormade, V., Deshpande, M.V., Paknikar, K.M., 2011. Perspectives for nano-biotechnology enabled protection and nutrition of plants. Biotechnol. Adv. 29, 792–803.

Gómez, J.V., Tarazona, A., Mateo, F., Jiménez, M., Mateo, E.M., 2019. Potential impact of engineered silver nanoparticles in the control of aflatoxins, ochratoxin A and the main aflatoxigenic and ochratoxigenic species affecting foods. Food Control 101, 58–68.

Greeff-Laubscher, M.R., Beukes, I., Marais, G.J., Jacobs, K., 2020. Mycotoxin production by three different toxigenic fungi genera on formulated abalone feed and the effect of an aquatic environment on fumonisins. Mycology 11 (2), 105–117.

Guilger-Casagrande, M., Germano-Costa, T., Pasquoto-Stigliani, T., Fraceto, L.F., Lima, R.E., 2019. Biosynthesis of silver nanoparticles employing *Trichoderma harzianum* with enzymatic stimulation for the control of *Sclerotinia sclerotiorum*. Sci. Rep. 9, 14351.

Gupta, N., Upadhyaya, C.P., Singh, A., Abd-Elsalam, K.A., Prasad, R., 2018. Applications of silver nanoparticles in plant protection. In: Nanobiotechnology Applications in Plant Protection. Springer, pp. 247–265.

Hamed, S.M., Hagag, E.S., El-Raouf, N.A., 2019. Green production of silver nanoparticles, evaluation of their nematicidal activity against *Meloidogyne javanica* and their impact on growth of *faba bean*. Beni-Suef Univ. J. Basic Appl. Sci. 8 (9), 1–12.

Han, X., Gelein, R., Corson, N., Wade-Mercer, P., Jiang, J., Biswas, P., Finkelstein, J.N., Elder, A., Oberdorster, G., 2011. Validation of an LDH assay for assessing nanoparticle toxicity. Toxicology 287, 99–104.

Handford, C.E., Dean, M., Henchion, M., Spence, M., Elliott, C.T., Campbell, K., 2014. Implications of nanotechnology for the Agri-food industry: opportunities, benefits and risks. Trends Food Sci. Technol. 40, 226–241.

Hao, W., Li, H., Hu, M., Yang, L., Rizwan-ul-Haq, M., 2011. Integrated control of citrus green and blue mold and sour rot by *Bacillus amyloliquefaciens* in combination with tea saponin. Postharvest Biol. Technol. 59, 316–323.

Hayles, J., Johnson, L., Worthley, C., Losic, D., 2017. Nanopesticides: A Review of Current Research and Perspectives, New Pesticides and Soil Sensors. Elsevier, pp. 193–225.

Ho, V.A., Le, P.T., Nguyen, T.P., Nguyen, C.K., Nguyen, V.T., Tran, N.Q., 2015. Silver core-shell nanoclusters exhibiting strong growth inhibition of plant-pathogenic fungi. J. Nanomater. 2015, 241614.

Horky, P., Skalickova, S., Baholet, D., Skladanka, J., 2018. Nanoparticles as a solution for eliminating the risk of mycotoxins. Nanomaterials (Basel) 8 (9), 727. https://doi.org/10.3390/nano8090727.

Hsueh, Y.-H., Lin, K.-S., Ke, W.-J., Hsieh, C.-T., Chiang, C.-L., Tzou, D.-Y., Liu, S.-T., 2015. The antimicrobial properties of silver nanoparticles in *Bacillus subtilis* are mediated by released Ag+ ions. PLoS One 10, e0144306.

Huang, W., Yan, M., Duan, H., Bi, Y., Cheng, X., Yu, H., 2020. Synergistic antifungal activity of green synthesized silver nanoparticles and epoxiconazole against *Setosphaeria turcica*. J. Nanomater. 2020, 9535432.

Hussain, M., Raja, N.I., Mashwani, Z.-U.-R., Iqbal, M., Chaudhari, S.K., Ejaz, M., Aslam, S., Yasmeen, F., 2018. Green synthesis and characterization of silver nanoparticles and their effects on disease incidence against canker and biochemical profile in *Citrus reticulata* L. Nanosci. Nanotechnol. Lett. 10, 1348–1355.

Hwang, E.T., Lee, J.H., Chae, Y.J., Kim, Y.S., Kim, B.C., Sang, B.I., Gu, M.B., 2008. Analysis of the toxic mode of action of silver nanoparticles using stress-specific bioluminescent bacteria. Small 4, 746–750.

Ibrahim, E., Fouad, H., Zhang, M., Zhang, Y., Qiu, W., Yan, C., Li, B., Mo, J., Chen, J., 2019. Biosynthesis of silver nanoparticles using endophytic bacteria and their role in inhibition of rice pathogenic bacteria and plant growth promotion. RSC Adv. 9, 29293–29299.

Ibrahim, E., Zhang, M., Zhang, Y., Hossain, A., Qiu, W., Chen, Y., Wang, Y., Wu, W., Sun, G., Li, B., 2020. Green-synthesization of silver nanoparticles using endophytic bacteria isolated from garlic and its antifungal activity against wheat fusarium head blight pathogen *Fusarium graminearum*. Nanomaterials 10, 219.

Jagana, D., Hegde, Y., Lella, R., 2017. Green nanoparticles—a novel approach for the management of banana anthracnose caused by *Colletotrichum musae*. Int. J. Curr. Microbiol. Appl. Sci. 6, 1749–1756.

Jain, D., Kothari, S., 2014. Green synthesis of silver nanoparticles and their application in plant virus inhibition. J. Mycol. Plant Pathol. 44, 21–24.

Javed, B., Raja, N.I., Nadhman, A., Mashwani, Z.U.R., 2020. Understanding the potential of bio-fabricated non-oxidative silver nanoparticles to eradicate Leishmania and plant bacterial pathogens. Appl. Nanosci. 10, 2057–2067.

JECFA, 2017. Evaluation of Certain Contaminants in Food. Eighty-Third Report of the Joint FAO/WHO Expert Committee on Food Additives (JECFA). Food and Agriculture Organization of the United Nations, World Health Organization, WHO Technical Report Series 1002. World Health Organization, Geneva.

JECFA, 2018. Safety evaluation of certain contaminants in food. In: Prepared by the Eighty-Third Meeting of the Joint FAO/WHO Expert Committee on Food Additives (JECFA). WHO Food Additives Series No. 74, FAO JECFA Monographs 19 Bis. World Health Organization/Food and Agriculture Organization of the United Nations, Geneva/Rome.

Jiang, F., Li, P., Zong, C., Yang, H., 2020. Surface-plasmon-coupled chemiluminescence amplification of silver nanoparticles modified immunosensor for high-throughput ultrasensitive detection of multiple mycotoxins. Anal. Chim. Acta 1114, 58–65.

Joyner, J.R., Kumar, D.V., 2015. Nanosensors and their applications in food analysis: a review Int. J. Sci. Technol. 1 (4), 80–90.

Kah, M., Hofmann, T., 2014. Nanopesticide research: current trends and future priorities. Environ. Int. 63, 224–235.

Kalia, A., Kaur, J., Kaur, A., Singh, N., 2020. Antimycotic activity of biogenically synthesised metal and metal oxide nanoparticles against plant pathogenic fungus *Fusarium moniliforme* (*F. fujikuroi*). Indian J. Exp. Biol. 58 (4), 263–270.

Kaman, P.K., Dutta, P., 2017. In vitro evaluation of biosynthesized silver nanoparticles (Ag NPs) against soil borne plant pathogens. Int. J. Nanotechnol. Appl. 11, 261–264.

Khalandi, B., Asadi, N., Milani, M., Davaran, S., Abadi, A.J.N., Abasi, E., Akbarzadeh, A., 2017. A review on potential role of silver nanoparticles and possible mechanisms of their actions on bacteria. Drug Res. 11, 70–76.

Khalil, M.S., Badawy, M.E.I., 2012. Nematicidal activity of a biopolymer chitosan at different molecular weights against root-knot nematode, *Meloidogyne incognita*. Plant Prot. Sci. 48, 170–178.

Khalil, N.M., Abd El-Ghany, M.N., Rodriguez-Couto, S., 2019. Antifungal and anti-mycotoxin efficacy of biogenic silver nanoparticles produced by *Fusarium chlamydosporum* and *Penicillium chrysogenum* at non-cytotoxic doses. Chemosphere 218, 477–486.

Khandelwal, N., Barbole, R.S., Banerjee, S.S., Chate, G.P., Biradar, A.V., Khandare, J.J., Giri, A.P., 2016. Budding trends in integrated pest management using advanced micro-and nanomaterials: challenges and perspectives. J. Environ. Manag. 184, 157–169.

Kim, M.-J., Kim, S., Park, H., Huh, Y.-D., 2011. Morphological evolution of Ag_2O microstructures from cubes to octapods and their antibacterial activities. Bull. Kor. Chem. Soc. 32, 3793.

Kim, S.W., Jung, J.H., Lamsal, K., Kim, Y.S., Min, J.S., Lee, Y.S., 2012. Antifungal effects of silver nanoparticles (agnps) against various plant pathogenic fungi. Mycobiology 40, 53–58.

Kim, D.Y., Kadam, A., Shinde, S., Saratale, R.G., Patra, J., Ghodake, G., 2018. Recent developments in nanotechnology transforming the agricultural sector: a transition replete with opportunities. J. Sci. Food Agric. 98, 849–864.

Korshed, P., Li, L., Liu, Z., Wang, T., 2016. The molecular mechanisms of the antibacterial effect of picosecond laser-generated silver nanoparticles and their toxicity to human cells. PLoS One 11 (8), e0160078. https://doi.org/10.1371/journal.pone.0160078.

Kriti, A., Ghatak, A., Mandal, N., 2020. Inhibitory potential assessment of silver nanoparticle on phytopathogenic spores and mycelial growth of *Bipolaris sorokiniana* and *Alternaria brassicicola*. Int. J. Curr. Microbiol. App. Sci. 9, 692–699.

Kumar, N., Das, S., Jyoti, A., Kaushik, S., 2016. Synergistic effect of silver nanoparticles with doxycycline against *Klebsiella pneumoniae*. Int. J. Pharm. Pharm. Sci. 8, 183–186.

Kumari, M., Pandey, S., Bhattacharya, A., Mishra, A., Nautiyal, C., 2017. Protective role of biosynthesized silver nanoparticles against early blight disease in *Solanum lycopersicum*. Plant Physiol. Biochem. 121, 216–225.

Kutsanedzie, F.Y.H., Agyekum, A.A., Annavaram, V., Chen, Q., 2020. Signal-enhanced SERS-sensors of CAR-PLS and GA-PLS coupled AgNPs for ochratoxin A and aflatoxin B1 detection. Food Chem. 315, 126231.

Laksanasopin, T., Guo, T.W., Nayak, S., Sridhara, A.A., Xie, S., Olowookere, O.O., Cadinu, P., Meng, F., Chee, N.H., Kim, J., Chin, C.D., Munyazesa, E., Mugwaneza, P., Rai, A.J., Mugisha, V., Castro, A.R., Steinmiller, D., Linder, V., Justman, J.E., Nsanzimana, S., Sia,

S.K., 2015. A smartphone dongle for diagnosis of infectious diseases at the point of care. Sci. Transl. Med. 7. 273re271.

Lemire, J.A., Harrison, J.J., Turner, R.J., 2013. Antimicrobial activity of metals: mechanisms, molecular targets and applications. Nat. Rev. Microbiol. 11, 371–384.

Li, H., Qiang, W., Vuki, M., Xu, D., Chen, H.Y., 2011. Fluorescence enhancement of silver nanoparticle hybrid probes and ultrasensitive detection of IgE. Anal. Chem. 83, 8945–8952.

Li, J., Rong, K., Zhao, H., Li, F., Lu, Z., Chen, R., 2013. Highly selective antibacterial activities of silver nanoparticles against Bacillus subtilis. J. Nanosci. Nanotechnol. 13, 6806–6813.

Li, J., Sang, H., Guo, H., Popko, J.T., He, L., White, J.C., Dhankher, O.P., Jung, G., Xing, B., 2017. Antifungal mechanisms of ZnO and Ag nanoparticles to *Sclerotinia homoeocarpa*. Nanotechnology 28, 155101.

Lim, D., Roh, J.Y., Eom, H.J., Choi, J.Y., Hyun, J., Choi, J., 2012. Oxidative stress-related PMK-1 P38 MAPK activation as a mechanism for toxicity of silver nanoparticles to reproduction in the nematode *Caenorhabditis elegans*. Environ. Toxicol. Chem. 31, 585–592.

Lin, H.Y., Huang, C.H., Hsieh, W.H., Liu, L.H., Lin, Y.C., Chu, C.C., Wang, S.T., Kuo, I.T., Chau, L.K., Yang, C.Y., 2014. On-line SERS detection of single bacterium using novel SERS nanoprobes and a micro fluidicdi electrophoresis device. Small 10, 4700–4710.

Liu, T., Gao, L., Zhao, J., Cao, Y., Tang, Y., Miao, P., 2018. A polymyxin B-silver nanoparticle colloidal system and the application of lipopolysaccharide analysis. Analyst 143, 1053–1058.

Lizárraga-Paulín, E.G., Moreno-Martínez, E., Miranda-Castro, S.P., 2011. Aflatoxins and their impact on humanand animal health: An emerging problem. In: Aflatoxins-Biochemistry and Molecular Biology. Rijeka, Croatia, InTech.

Mahawar, H., Prasanna, R., 2018. Prospecting the interactions of nanoparticles with beneficial microorganisms for developing green technologies for agriculture. Environ. Nanotechnol. Monit. Manag. 10, 477–485.

Mahawar, H., Prasanna, R., Gogoi, R., Singh, S.B., Chawla, G., Kumar, A., 2020. Synergistic effects of silver nanoparticles augmented *Calothrix elenkinii* for enhanced biocontrol efficacy against Alternaria blight challenged tomato plants. 3 Biotech 10. https://doi.org/10.1007/s13205-020-2074-0, 102.

Mahdizadeh, V., Safaie, N., Khelghatibana, F., 2015. Evaluation of antifungal activity of silver nanoparticles against some phytopathogenic fungi and *Trichoderma harzianum*. J. Crop Prot. 4, 291–300.

Maroufpoor, N., Alizadeh, M., Hamishehkar, H., Lajayer, B.A., Hatami, M., 2019. Engineered nanoparticle-based approaches to the protection of plants against pathogenic microorganisms. In: Nanobiotechnology Applications in Plant Protection. Springer, pp. 267–283.

Marpu, S., Kolailat, S.S., Korir, D., Kamras, B.L., Chaturvedi, R., Joseph, A., Smith, C.M., Palma, M.C., Shah, J., Omary, M.A., 2017. Photochemical formation of chitosan-stabilized near-infrared-absorbing silver nanoworms: a "green" synthetic strategy and activity on gram-negative pathogenic bacteria. J. Colloid Interface Sci. 507, 437–452.

Masum, M.M.I., Siddiqa, M.M., Ali, K.A., Zhang, Y., Abdallah, Y., Ibrahim, E., Qiu, W., Yan, C., Li, B., 2019. Biogenic synthesis of silver nanoparticles using *Phyllanthus emblica* fruit extract and its inhibitory action against the pathogen *Acidovorax oryzae* strain RS-2 of rice bacterial brown stripe. Front. Microbiol. 10. https://doi.org/10.3389/fmicb.2019.00820, 820.

Menon, S., Agarwal, H., Kumar, S.R., Kumar, S.V., 2017. Green synthesis of silver nanoparticles using medicinal plant *Acalypha indica* leaf extracts and its application as an antioxidant and antimicrobial agent against foodborne pathogens. Int. J. Appl. Pharm. 9, 42–50.

Mishra, S., Singh, H., 2015. Biosynthesized silver nanoparticles as a nanoweapon against phytopathogens: exploring their scope and potential in agriculture. Appl. Microbiol. Biotechnol. 99, 1097–1107.

Mishra, S., Yang, X., Ray, S., Fraceto, L.F., Singh, H., 2020. Antibacterial and biofilm inhibition activity of biofabricated silver nanoparticles against *Xanthomonas oryzae* pv. oryzae causing blight disease of rice instigates disease suppression. World J. Microbiol. Biotechnol. 36, 1–10.

Mitra, C., Gummadidala, P.M., Merrifield, R., Omebeyinje, M.H., Jesmin, R., Lead, J.R., Chanda, A., 2019. Size and coating of engineered silver nanoparticles determine their ability to growth-independently inhibit aflatoxin biosynthesis in *Aspergillus parasiticus*. Appl. Microbiol. Biotechnol. 103, 4623–4632.

Mittal, A., Schulze, K., Ebensen, T., Weissmann, S., Hansen, S., Lehr, C.M., Guzman, C.A., 2015. Efficient nanoparticle-mediated needle-free transcutaneous vaccination via hair follicles requires adjuvant action. Nanomedicine 11, 147–154.

Mohamed, M.A., Abd-Elsalam, K.A., 2018. Nanoantimicrobials for Plant Pathogens. Springer International Publishing AG, p. 2087. Part of Springer Nature 2018.

Mohamed, H.I., Akladious, S.A., 2017. Changes in antioxidants potential, secondary metabolites and plant hormones induced by different fungicides treatment in cotton plants. Pestic. Biochem. Physiol. 142, 117–122.

Mohamed, H., El-Hady, A.A., Mansour, M., El-Samawaty, A.E.-R., 2012. Association of oxidative stress components with resistance to flax powdery mildew. Trop. Plant Pathol. 37, 386–392.

Mohamed, H.I., El-Beltagi, H.S., Aly, A.A., Latif, H.H., 2018. The role of systemic and non-systemic fungicides on the physiological and biochemical parameters in *Gossypium hirsutum* plant, implications for defense responses. Fresenius Environ. Bull. 27, 8585–8593.

Mohammad, T.G., Abd El-Rahman, A., 2015. Environmentally friendly synthesis of silver nanoparticles using *Moringa oleifera* (Lam) leaf extract and their antibacterial activity against some important pathogenic bacteria. Mycopathologia 13 (1), 1–6.

Mohanta, Y.K., Panda, S.K., Bastia, A.K., Mohanta, T.K., 2017. Biosynthesis of silver nanoparticles from *Protium serratum a*nd investigation of their potential impacts on food safety and control. Front. Microbiol. 8, 626.

Möhler, J.S., Sim, W., Blaskovich, M.A., Cooper, M.A., Ziora, Z.M., 2018. Silver bullets: a new lustre on an old antimicrobial agent. Biotechnol. Adv. 36, 1391–1411.

Mousavi, S.A.P., Pourtalebi, S.M., 2015. Inhibitory effects of silver nanoparticles on growth and aflatoxin B1 production by *Aspergillus parasiticus*. Iran. J. Med. Sci. 40, 501–506.

Mukherjee, A., Kotzybik, K., Gräf, V., Kugler, L., Stoll, D.A., Greiner, R., Geisen, R., Schmidt-Heydt, M., 2016. Influence of different nanomaterials on growth and mycotoxin production of *Penicillium verrucosum*. PLoS One 11, e0150855.

Nassar, A.M.K., 2016. Effectiveness of silver nano-particles of extracts of *Urtica urens* (Urticaceae) against root-knot nematode *Meloidogyne incognita*. Asian J. Nematol. 5, 14–19.

Nayak, B.K., Chitra, N., Nanda, A., 2015. Comparative antibiogram analysis of AgNPs synthesized from two *Alternaria* spp. with amoxicillin antibiotics. J. Chem. Pharm. Res. 7, 727–731.

Nazir, K., Mukhtar, T., Javed, H., 2019. In vitro effectiveness of silver nanoparticles against root-knot nematode (*Meloidogyne incognita*). Pak. J. Zool. 51 (6), 2077–2083.

Nhien, L.T.A., Luong, N.D., Tien, L.T.T., Luan, L.Q., 2018. Radiation synthesis of silver nanoparticles/chitosan for controlling leaf fall disease on rubber trees causing by *Corynespora cassiicola*. J. Nanomater. 2018. https://doi.org/10.1155/2018/7121549, 7121549.

Nicol, J.M., Turner, S.J., Coyne, D.L., Nijs, L.D., Hockland, S., Maafi, Z.T., 2011. Current nematode threats to world agriculture. In: Jones, J., et al. (Eds.), Genomics and Molecular Genetics of Plant-Nematode Interactions. Springer, Heidelberg, Germany, pp. 21–43.

Nicolopoulou-Stamati, P., Maipas, S., Kotampasi, C., Stamatis, P., Hens, L., 2016. Chemical pesticides and human health: the urgent need for a new concept in agriculture. Front. Public Health 4, 148.

Nour El-Deen, A., El-Deeb, B., 2018. Effectiveness of silver nanoparticles against root-knot nematode, *Meloidogyne incognita* infecting tomato under greenhouse conditions. J. Agric. Sci. 10 (12), 148–156.

Omanović-Mikličanina, E., Maksimović, M., 2016. Nanosensors applications in agriculture and food industry. Bull. Chem. Technol. Bosnia Herzegovina 47, 59–70.

Ouda, S.M., 2014. Antifungal activity of silver and copper nanoparticles on two plant pathogens, *Alternaria alternata* and *Botrytis cinerea*. Res. J. Microbiol. 9, 34–42.

Palou, L., 2018. Postharvest treatments with GRAS salts to control fresh fruit decay. Horticulture 4, 46.

Papp, I., Sieben, C., Ludwig, K., Roskamp, M., Bottcher, C., Schlecht, S., Herrmann, A., Haag, R., 2010. Inhibition of influenza virus infection by multivalent sialic-acid-functionalized gold nanoparticles. Small 6, 2900–2906.

Pareek, V., Gupta, R., Panwar, J., 2018. Do physico-chemical properties of silver nanoparticles decide their interaction with biological media and bactericidal action? A review. Mater. Sci. Eng. C 90, 739–749.

Patil, M.P., Kim, G.-D., 2017. Eco-friendly approach for nanoparticles synthesis and mechanism behind antibacterial activity of silver and anticancer activity of gold nanoparticles. Appl. Microbiol. Biotechnol. 101, 79–92.

Patra, J.K., Baek, K.H., 2017. Antibacterial activity and synergistic antibacterial potential of biosynthesized silver nanoparticles against foodborne pathogenic bacteria along with its anticandidal and antioxidant effects. Front. Microbiol. 8, 167.

Plascencia-Jatomea, M., Yépiz-Gómez, M.S., Velez-Haro, J.M., 2014. Aspergillus spp. (Black Mold). In: Post-Harvest Decay, Control Strategies. Academic Press, New York, NY, USA, pp. 267–286.

Prabhu, S., Poulose, E.K., 2012. Silver nanoparticles: mechanism of antimicrobial action, synthesis, medical applications, and toxicity effects. Int. Nano Lett. 2, 32.

Prasad, R., Gupta, N., Kumar, M., Kumar, V., Wang, S., Abd-Elsalam, K.A., 2017. Nanomaterials act as plant defense mechanism. In: Nanotechnology. Springer, pp. 253–269.

Prasher, P., Singh, M., Mudila, H., 2018. Green synthesis of silver nanoparticles and their antifungal properties. BioNanoScience 8, 254–263.

Quinteros, M., Aristizábal, V.C., Dalmasso, P.R., Paraje, M.G., Páez, P.L., 2016. Oxidative stress generation of silver nanoparticles in three bacterial genera and its relationship with the antimicrobial activity. Toxicol. In Vitro 36, 216–223.

Rafique, M., Sadaf, I., Rafique, M.S., Tahir, M.B., 2017. A review on green synthesis of silver nanoparticles and their applications. Artif. Cells Nanomed. Biotechnol. 45, 1272–1291.

Raghavendra, S.N., Raghu, H.S., Divyashree, K., Rajeshwara, A.N., 2019. Antifungal efficiency of copper oxychloride-conjugated silver nanoparticles against *Colletotrichum gloeosporioides* which causes anthracnose disease. Asian J. Pharm. Clin. Res., 230–233.

Rai, M., Yadav, A., Gade, A., 2009. Silver nanoparticles as a new generation of antimicrobials. Biotechnol. Adv. 27, 76–83.

Rajeshkumar, S., Malarkodi, C., 2014. In vitro antibacterial activity and mechanism of silver nanoparticles against foodborne pathogens. Bioinorg. Chem. Appl. 2014, 1–10.

Rhouati, A., Bulbul, G., Latif, U., Hayat, A., Li, Z.H., Marty, J.L., 2017. Nano-aptasensing in mycotoxin analysis: recent updates and progress. Toxins (Basel) 9. https://doi.org/10.3390/toxins9110349, 349.

Rivas-Cáceres, R.R., Stephano-Hornedo, J.L., Lugo, J., Vaca, R., Del Aguila, P., Yañez-Ocampo, G., Mora-Herrera, M.E., Díaz, L.M.C., Cipriano-Salazar, M., Alaba, P.A., 2018. Bactericidal effect of silver nanoparticles against propagation of *Clavibacter michiganensis* infection in *Lycopersicon esculentum* Mill. Microb. Pathog. 115, 358–362.

Ruffo, R.S., Youssef, K., Hashim, A.F., Ippolito, A., 2019. Nanomaterials as alternative control means against postharvest diseases in fruit crops. Nanomaterials (Basel) 9. https://doi.org/10.3390/nano9121752, 1752.

Salem, E.H., Nawito, M.A.S., El-Raouf Ahmed, A.E.-R.A., 2019. Effect of silver nanoparticles on gray mold of tomato fruits. J. Nanotechnol. Res. 1, 108–118.

Seong, M., Lee, D.G., 2017. Silver nanoparticles against *Salmonella enterica* serotype typhimurium: role of inner membrane dysfunction. Curr. Microbiol. 74, 661–670.

Sepunaru, L., Tschulik, K., Batchelor-McAuley, C., Gavish, R., Compton, R.G., 2015. Electrochemical detection of single *E. coli* bacteria labeled with silver nanoparticles. Biomater. Sci. 3, 816–820.

Shadjou, R., Hasanzadeh, M., Heidar-Poor, M., Shadjou, N., 2018. Electrochemical monitoring of aflatoxin M1 in milk samples using silver nanoparticles dispersed on alpha-cyclodextrin-GQDs nanocomposite. J. Mol. Recognit. 31, e2699.

Shahryari, F., Rabiei, Z., Sadighian, S., 2020. Antibacterial activity of synthesized silver nanoparticles by sumac aqueous extract and silver-chitosan nanocomposite against *Pseudomonas syringae* pv. *syringae*. J. Plant Pathol., 1–7.

Sheikh, M.A., 2017. Post harvest diseases of temperate fruits and their management strategies—a review. Int. J. Pure Appl. Biosci. 5, 885–898.

Shu, M., He, F., Li, Z., Zhu, X., Ma, Y., Zhou, Z., Yang, Z., Gao, F., Zeng, M., 2020. Biosynthesis and antibacterial activity of silver nanoparticles using yeast extract as reducing and capping agents. Nanoscale Res. Lett. 15, 14.

Siddiqi, K.S., Husen, A., 2016. Fabrication of metal and metal oxide nanoparticles by algae and their toxic effects. Nanoscale Res. Lett. 11, 363.

Siddiqi, K.S., Husen, A., Rao, R.A.K., 2018. A review on biosynthesis of silver nanoparticles and their biocidal properties. J. Nanobiotechnol. 16, 14.

Singh, S., Park, I.S., Shin, Y., Lee, Y.S., 2015. Comparative study on antimicrobial efficiency of $AgSiO_2$, ZnAg and Ag-zeolite for the application of fishery plastic container. J. Mater. Sci. Eng. 4 (4), 1–5.

Singh, R., Gupta, A., Patade, V., Balakrishna, G., Pandey, H., Singh, A., 2019. Synthesis of silver nanoparticles using extract of *Ocimum kilimandscharicum* and its antimicrobial activity against plant pathogens. SN Appl. Sci. 1, 1652.

Sinha, K., Ghosh, J., Sil, P.C., 2017. New pesticides: a cutting-edge view of contributions from nanotechnology for the development of sustainable agricultural pest control. In: New Pesticides and Soil Sensors. Elsevier, pp. 47–79.

Sismaet, H.J., Goluch, E.D., 2018. Electrochemical probes of microbial community behavior. Annu. Rev. Anal. Chem. (Palo Alto, Calif) 11, 441–461.

Sofy, A.R., Dawoud, R.A., Sofy, M.R., Mohamed, H.I., Hmed, A.A., El-Dougdoug, N.K., 2020. Improving regulation of enzymatic and non-enzymatic antioxidants and stress-related gene stimulation in *Cucumber mosaic cucumovirus*-infected cucumber plants treated with glycine betaine, chitosan and combination. Molecules 25 (10). https://doi.org/10.3390/molecules25102341, 2341.

Song, Y., Wei, W., Qu, X., 2011. Colorimetric biosensing using smart materials. Adv. Mater. 23, 4215–4236.

Speshock, J.L., Murdock, R.C., Braydich-Stolle, L.K., Schrand, A.M., Hussain, S.M., 2010. Interaction of silver nanoparticles with *Tacaribe virus*. J. Nanobiotechnol. 8, 19.

Stephano-Hornedo, J.L., Torres-Gutiérrez, O., Toledano-Magaña, Y., Gradilla-Martínez, I., Pestryakov, A., Sánchez-González, A., García-Ramos, J.C., Bogdanchikova, N., 2020. Argovit™ silver nanoparticles to fight Huanglongbing disease in Mexican limes (*Citrus aurantifolia* Swingle). RSC Adv. 10, 6146–6155.

Stroka, J., Maragos, C.M., 2016. Challenges in the analysis of multiple mycotoxins. World Mycotoxin J. 9, 847–861.

Sun, L., Yi, S., Wang, Y., Pan, K., Zhong, Q., Zhang, M., 2014. A bio-inspired approach for in situ synthesis of tunable adhesive. Bioinspir. Biomim. 9, 016005.

Sun, L., Zheng, C., Webster, T.J., 2017. Self-assembled peptide nanomaterials for biomedical applications: promises and pitfalls. Int. J. Nanomedicine 12, 73–86.

Sur, I., Cam, D., Kahraman, M., Baysal, A., Culha, M., 2010. Interaction of multi-functional silver nanoparticles with living cells. Nanotechnology 21, 175104–175114.

Taha, E.H., Abo-Shady, N.M., 2016. Effect of silver nanoparticles on the mortality pathogenicity and reproductivity of entomopathogenic nematodes. Int. J. Zool. Res. 12, 47–50.

Talie, M.D., Wani, A.H., Ahmad, N., Bhat, M.Y., War, J.M., 2020. Green synthesis of silver nanoparticles (AgNPs) using *Helvella leucopus* pers. and their antimycotic activity against fungi causing fungal rot of apple. Asian J. Pharm. Clin. 13, 161–165.

Tang, S., Zheng, J., 2018. Antibacterial activity of silver nanoparticles: structural effects. Adv. Healthc. Mater. 7, 1701503.

Ukah, U., Glass, M., Avery, B., Daignault, D., Mulvey, M., Reid-Smith, R., Parmley, E., Portt, A., Boerlin, P., Manges, A., 2018. Risk factors for acquisition of multidrug-resistant Escherichia coli and development of community-acquired urinary tract infections. Epidemiol. Infect. 146, 46–57.

Van de Perre, E., Jacxsens, L., Liu, C., Devlieghere, F., De Meulenaer, B., 2015. Climate impact on Alternaria moulds and their mycotoxins in fresh produce: the case of the tomato chain. Food Res. Int. 68, 41–46.

Vanti, G.L., Nargund, V.B., Vanarchi, R., Kurjogi, M., Mulla, S.I., Tubaki, S., Patil, R.R., 2019. Synthesis of *Gossypium hirsutum*-derived silver nanoparticles and their antibacterial efficacy against plant pathogens. Appl. Organomet. Chem. 33, e4630.

Vanti, G.L., Kurjogi, M., Basavesha, K., Teradal, N.L., Masaphy, S., Nargund, V.B., 2020. Synthesis and antibacterial activity of solanum torvum mediated silver nanoparticle against *Xxanthomonas axonopodis* pv. *punicae* and *Ralstonia solanacearum*. J. Biotechnol. 309, 20–28.

Velu, R., DeRosa, M.C., 2018. Lateral flow assays for Ochratoxin a using metal nanoparticles: comparison of "adsorption-desorption" approach to linkage inversion assembled nano-aptasensors (LIANA). Analyst 143, 4566–4574.

Villamizar-Gallardo, R., Cruz, J.F.O., Ortíz-Rodriguez, O.O., 2016. Fungicidal effect of silver nanoparticles on toxigenic fungi in cocoa. Pesq. Agrop. Brasileira 51, 1929–1936.

Vitale, A., Panebianco, A., Polizzi, G., 2016. Baseline sensitivity and efficacy of fluopyram against B*otrytis cinereafrom* table grape in Italy. Ann. Appl. Biol. 169, 36–45.

Wei, L., Lu, J., Xu, H., Patel, A., Chen, Z.S., Chen, G., 2015. Silver nanoparticles: synthesis, properties, and therapeutic applications. Drug Discov. Today 20, 595–601.

Weisany, W., Amini, J., Samadi, S., Hossaini, S., Yousefi, S., Struik, P.C., 2019. Nano silver-encapsulation of *Thymus daenensis* and *Anethum graveolens* essential oils enhances antifungal potential against *Strawberry anthracnose*. Ind. Crop. Prod. 141, 111808.

WHO, 2018. FOSCOLLAB Database. In: Chemical Overview Dashboard—Integrated Summary Elements From JECFA Evaluations Database, GEMS/Food Contaminants Database and the WHO Collaborating Centres Database. WHO.

Worrall, E.A., Hamid, A., Mody, K.T., Mitter, N., Pappu, H.R., 2018. Nanotechnology for plant disease management. Agronomy 8, 285.

Wu, B., Kuang, Y., Zhang, X., Chen, J., 2011. Noble metal nanoparticles/carbon nanotubes nanohybrids: synthesis and applications. Nano Today 6, 75–90.

Wu, R., Ma, Y., Pan, J., Lee, S.H., Liu, J., Zhu, H., Gu, R., Shea, K.J., Pan, G., 2018. Efficient capture, rapid killing and ultrasensitive detection of bacteria by a nano-decorated multifunctional electrode sensor. Biosens. Bioelectron. 101, 52–59.

Xiu, Z., Zhang, Q., Puppala, H.L., Colvin, V.L., Alvarez, P.J.J., 2012. Negligible particle-specific antibacterial activity of silver nanoparticles. Nano Lett. 12, 4271–4275.

Xue, B., He, D., Gao, S., Wang, D., Yokoyama, K., Wang, L., 2016. Biosynthesis of silver nanoparticles by the fungus *Arthroderma fulvum* and its antifungal activity against genera of Candida aspergillus and fusarium. Int. J. Nanomedicine 11, 1899.

Yahaya, S.M., Mardiyya, A.Y., 2019. Review of post-harvest losses of fruits and vegetables. Biomed. J. Sci. Technol. Res. 13 (14).

Yang, Y., Wang, J., Xiu, Z., Alvarez, P.J., 2013. Impacts of silver nanoparticles on cellular and transcriptional activity of nitrogen-cycling bacteria. Environ. Toxicol. Chem. 32, 1488–1494.

Zorraquin-Pena, I., Cueva, C., Bartolome, B., Moreno-Arribas, M.V., 2020. Silver nanoparticles against foodborne bacteria. Effects at intestinal level and health limitations. Microorganisms 8. https://doi.org/10.3390/microorganisms8010132, 132.

CHAPTER 22

Nematicidal activity of silver nanomaterials against plant-parasitic nematodes

Benay Tuncsoy

Adana Alparslan Turkes Science and Technology University, Faculty of Engineering, Department of Bioengineering, Adana, Turkey

22.1 Introduction

Nematodes are a fascinating, biologically diverse group of organisms. Their ability to adapt to a wide variety of habitats, including marine, soil, and aquatic, provides an evolutionary advantage for species longevity (Bernard et al., 2017). Plant-parasitic nematodes within the phylum of nematode, which forms 80% of multicellular organisms, are important pests that cause 10%–20% product loss in the agricultural area. They infect all parts of the plant: roots, bulbs, rhizomes, stems, leaves, buds, flowers, and seeds. Plant-pathogenic nematode life cycles consist of the egg, four juvenile stages, and the adults. The duration of the life cycle is dependent on host suitability and environmental conditions, and varies greatly among genera, ranging from a few days to several months.

Nematicides used to manage plant-parasitic nematode species are high molecular soil fumigants and carbamate or organic phosphorus compounds (Whitehead, 1998; Bakker et al., 1993). The extensive use of nematicides might lead to environmental and health problems on nontarget organisms and also may cause to develop the nematode resistance. Therefore, it is important to use alternative control strategies, which are effective, cheap, and safe to farmers, consumers, and the environment (Fernandez et al., 2001).

The implication of the nanotechnology research in the agricultural sector has become a necessary key factor for sustainable developments (Prasad et al., 2017). Synthesis of metal nanoparticles can be done by a wide variety of chemical, physical, and biological methods. Physical and chemical methods are the most used for the synthesis of nanoparticles, but these production methods are expensive and are potentially hazardous to the environment and living organisms because of the use of toxic compounds during the synthesis. Otherwise, the biological method for the synthesis of metal nanoparticles, called "green synthesis," is the most advantageous due to being cost-effective, ecologically correct, reproducible, and energy efficient (Ovais et al., 2018). During the past decade, it has been demonstrated that many biological systems, including plants and algae (Govindaraju et al., 2008), diatoms (Scarano and Morelli, 2002, 2003), bacteria (Lengke et al., 2007), yeast (Kowshik et al., 2002), fungi (Rautaray et al., 2003), and human cells (Anshup et al., 2005), can transform inorganic metal ions into metal nanoparticles via the reductive capacities of the proteins and metabolites present in these organisms (Mukundan and Vasanthakumari, 2017). Among these nanoparticles, silver nanoparticles are frequently used in medicine, pharmacy, and agriculture due to their chemical stability, and catalytic and antibacterial activity (Zhang et al., 2018; Fouda et al., 2020). Several studies have also revealed that Ag nanoparticles synthesized from plants are a potentially effective nematicide and used in agriculture (Baronia et al., 2020; Nazir et al., 2019; Kalaiselvi et al., 2017; Dura et al., 2019b; Nour El-Deen and El-Deen, 2018).

The working principle of the plant extracts with silver nanoparticles are based on the principle of adjusting the particle size of the active phenol compounds in the plant extracts between 1 and 100 nm, allowing harmful microorganisms to pass through the cell walls more actively, and by destroying the protein structures and DNA in the cell walls of the pathogen and blocking the enzyme activities of the cells. Silver nanoparticles can be used as a fertilizer as well as antimicrobial and nematicide, in addition to many important plant diseases. In addition, it produces toxic effects by causing oxidative stress on target nematode cells (Lim et al., 2012).

Nanoparticles (NP) have a high capacity surface volume, which increases their activity and biochemical properties. Likewise, silver nanoparticles are used as sensors, catalysts, anticancer agents, antimicrobial agents, and antioxidants due to being one of the advanced material types. In this chapter, it was mentioned that the recent advances in using silver nanoparticles as a nanopesticide and nanofertilizer, antioxidant and antimicrobial properties of silver nanoparticle, affects plant-parasitic nematodes as well as the types and management of plant-parasitic nematodes.

22.2 Types of plant-parasitic nematodes

There are more than 4100 species of plant-parasitic nematode described till date (Decraemer and Hunt, 2006), and, collectively, they represent an important constraint on agriculture. These nematodes are usually small, soil-borne pathogens, and the symptoms they cause are often nonspecific (Jones et al., 2013).

22.2.1 Root-knot nematodes (*Meloidogyne* spp.)

Root-knot nematodes are endoparasite nematodes. Yield losses caused by root-knot nematodes differ according to the population density and plant species, and this rate is 15%–85% in the vegetables. Although it varies according to the host and nematode species, there is an average of 400–500 eggs in a package, and even this number can reach up to 2000. They are easily recognized by generating large and small damages at the root of the host plant. They cause swelling in the root system of the host plant and disrupts the transmission tissues of the host and restricts the exchange of water and nutrients from the soil. As a result, the growth of the plant slows down and stops, and stunting appears. Due to the ripening of leaves, the flowers and fruits drop. If the infection is severe, the plant can dry completely. The size and shape of the swelling formed at the root vary according to the species and age of the host plant. Also, nematode can infect only in the second instar larval period. If the soil temperature is below 10 °C, it cannot develop, and its damage starts at 15 °C.

The second instar larvae and males are in the form of thread, whereas the females are in the form of pears and lemons. The female dies after laying her eggs in a gelatinous substance, some of which are buried in the root or on the root surface, just behind her body.

22.2.2 Cyst nematodes (*Heterodera* and *Globodera* spp.)

Cyst nematodes (*Heterodera* and *Globodera* spp.) are the most economically important types of parasitic nematodes due to their constant endoparasite nutrition, damage to vascular tissues, specialized parasitism mechanisms in the host plant and a high number of eggs per female. They appear on the roots of the host plant. It causes discoloration, yellowing, collapse, side root formation, and excessive bifurcation, especially in the above-ground accent. It is found in cold climates. The genus *Heterodera* prefers plants in the *Cruciferae* group and the *Globodera* (Solanaceae). Cereal cyst nematodes are endoparasitic nematodes and are fixed to the root after the 2nd larval stage and continue their next period by creating a feeding area (Syncytium) here. While females feed on the root, taking the shape of a pear or lemon, males roam freely in the soil in filamentous form during their maturity. After the females die, they turn into cysts and store their eggs in the body. Cysts hatch after the need for cooling, and the larvae hatch. As a result of the feeding of grain cyst nematode species in the plant, growth of the plant is declined due to bifurcation and swelling in the roots, as well as massing. Moreover, the deterioration of the water food intake pattern of the roots, wilting, and stunting occurs in the plant (Agrios, 1997).

22.2.3 *Ditylenchus* species

The genus *Ditylenchus* comprises many species, of which four are currently known to be important pests of crop plants, *Ditylenchus dipsaci* (Kuhn) Filipjev, *D. destructor*, *D. angustus*, and *D. africanus* (Wendt et al., 1995).

D. dipsaci, which is a stem nematode, can cause serious damage to Allium species and especially garlic in cool mountainous tropical and subtropical regions in winter season. The female and male of *D. dipsaci* are in the form of a roundworm and 1–1.3 mm in length. It is an endoparasitic nematode of alliaceous plant. When the living conditions of the host plants become unsuitable (when the plant rots), they leave the plant and move into the soil.

D. destructor is the second ranking potato cyst nematode. They can live on a very wide range of fungi and plant (Bridge and Starr, 2007; Zheng et al., 2016).

D. angustus is a plant-parasitic nematode and leads to ufra disease in rice and is responsible for substantial yield losses in a most agricultural area in the world. All life stages of *D. angustus* can infest the plant by entering the collar region, feeding on young foliar tissues, and completing its life cycle approximately in 10–20 days at 27–30 °C (Khanam et al., 2018).

Also, there are more species that include plant-parasitic nematodes such as root lesion nematodes (*Pratylenchus* spp.), the burrowing nematodes (*Radopholus similis*, *Ditylenchus dipsaci*), the pine wilt nematode (*Bursaphelenchus xylophilus*), and the reniform nematode (*Rotylenchulus reniformis*, *Xiphinema index*, *Nacobbus aberrans*, and *Aphelenchoides besseyi*).

22.3 Management of plant-parasitic nematodes

Plant-parasitic nematodes are among the most important harmful agricultural organisms. The annual crop loss caused by plant-parasitic nematodes in the world varies between 125 and 173 billion dollars depending on the tropical and subtropical regions (Elling, 2013). Nematodes cause cell death and growth inhibition through movement and penetration within the plant or damage the host plant by interrupting water uptake from the roots and the flow of nutrients from the leaves to the roots. Difficulties in management with plant-parasitic nematodes include their invisible to the naked eye, the need to take samples from the soil and root for detection, the adverse effects of other organisms, and the symptoms caused by environmental conditions.

Among the management methods include cultural measures, solarization, biofumigation, chemical applications, the use of resistant plant species, the use of hanging seedlings, and biological control.

Cultural measures include removing contaminated plants from the environment, keeping irrigation water clean, and using clean production materials.

Solarization is a physical method of soil disinfection. Solarization application, which is cheap, easy to apply, and safe to use, is the heating of the soil with solar energy. However, solarization is not always effective against some diseases, especially nematodes, when applied alone. For this reason, it is necessary to include integrated management against nematodes to increase the effect and protect the soil and the environment. One of these methods is biofumigation. The nematicides used for the control of plant-parasitic nematodes are generally high molecular soil fumigants and carbamate or organic phosphorus compounds. Fumigation is a method of combating with chemicals that become gas when mixed with air. The main disadvantages of fumigant nematicides are that they are broad-spectrum, highly toxic, carcinogenic, and cause residue problems in both soil and plants. For example, methyl bromide, one of the most effective fumigants, used extensively for the control of nematodes, weeds, and underground pathogens, has been banned due to damage to the ozone layer of the atmosphere (Abawi and Widmer, 2000). Thus, biofumigation studies, which can be defined as the application of fumigation by natural ways, not by chemicals, have been extensively investigated in recent years not only for nematode control but also for combating weed and soil-borne diseases. Biofumigation is defined as the control of harmful and disease factors and weeds in the soil by using biocidal sulfides and mainly isothiocyanates, which are contained in some natural materials such as *Brassica* sp., as green fertilizers or rotation plants. In recent years, Allium species containing sulfur amino acids have been added to this term in addition to *Brassica* (Matthiessen and Kirkegaard, 2006). Patalano et al. (2008) described biofumigation as an extremely natural technique that increases the fertility of the soil, keeps the plants healthier, and significantly reduces the use of pesticides. The use of solarization together with biofumigation increases the microbiological activity in the soil, improves the chemical and physical properties of the soil, and increases the effect of solarization.

Although chemical nematicides are easy to apply and can be seen quickly, they are banned in many countries due to human and environmental health concerns. For this reason, biological control is one of the alternative methods that have been studied extensively in recent years. Today, many microorganisms are known as antagonists of plant-parasitic nematodes. *Pasteuria penetrans*, *Pseudomonas* spp., *Bacillus* spp., *Paecilomyces lilacinus*, *Arthrobotrys oligosporave*, and *Burkholderia cepaciaise* can be qualified as antagonists in biological control of nematodes (Uysal and Göze Özdemir, 2020). Moreover, one of the biological control methods used to manage with plant-parasitic nematodes is using plant extracts. Extracts obtained from many plant species such as *Azadirachta indica*, *Withania somnifera*, *Eucalyptus citriodora*, and *Hunteria umbellate* are applied as nematicides. Many studies have reported that extracts obtained from *H. umbellate* and *Mallotus oppsitifolius* leaves significantly affect egg hatching and larval development in root-knot nematodes (Okeniyi et al., 2013). In another study, it was found that ethanol extracts from *A. indica* leaves, *Copsicum annuum* fruits, and *Parkia biglobosa* seeds significantly increase plant length and fruit production (Bawa et al., 2014; Bernard et al., 2017). In addition, in recent years, effective results against plant-parasitic nematodes have been obtained as a result of obtaining metal nanoparticles from plants by green synthesis method (El-Batal et al., 2019; Al Banna et al., 2020; Akhter et al., 2020).

Plants develop different defence mechanisms to protect themselves from diseases and pests. Resistance, one of these mechanisms, is defined as the plant's ability to prevent, eliminate, or reduce disease factors and attacks of pests. In recent years, nematode-resistant transgenic bacteria have been developed with the application of protein inhibitors and RNA interference that prevent nematodes feeding. Transgenic constructs, which specifically target genes for functional characterizations, have been developed since the discovery of RNA-interference. To express double-stranded RNA, which are silence genes in plant-parasitic nematodes, plants have been engineered (Huang et al., 2006; Yadav et al., 2006) (Fig. 22.1).

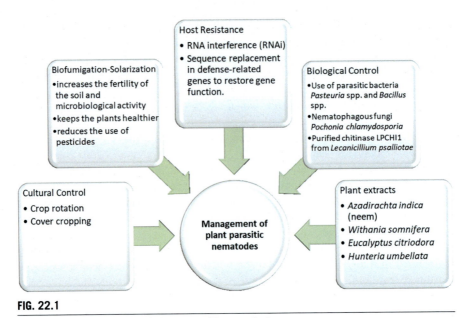

FIG. 22.1

Management mechanism of plant-parasitic nematodes.

22.4 Nanotechnology on agriculture

Nanotechnology is used in many fields, including physics, chemistry, pharmacy, material science, medicine, and agriculture. Recently, nanotechnology is mostly used in precision agricultural applications in fields. Precision agriculture is the use of the latest agricultural technologies to obtain maximum efficiency, taking into account the principles of economy and environmental protection. Any material having structures or components on their structure that are 100 nm or less in the least one dimension are examples of nanotechnology (Dowling et al., 2004; Nowack and Bucheli, 2007). Nanoparticles are frequently utilized in nutrition and agricultural control methods in plant protection, due to their small size, high surface volume, and optical properties (Ghormade et al., 2011). Chitosan, silver, zinc, and titanium nanoparticles are mostly used in agricultural areas in seed treatment and as biopesticide due to fighting potential with fungal infections. Moreover, nanoencapsulation plays an important role to protect environment via reducing leaching and evaporation of harmful substances (Duhan et al., 2017).

The use of chemical insecticides in agricultural areas has frequently seen adverse effects such as being beneficial against microorganisms in the soil or reducing the fertility of the soil, increasing the use of more effective and controlled release pesticides (Ragaei and Sabry, 2014). Also, quantum dots, which constitute an important group in nanotechnology, are among the methods frequently used in plant pathology. Furthermore, nanotechnology is a method used in the management of pests in agricultural areas. Studies have reported that silica, silver, aluminum, zinc, and titanium nanoparticles are effective as insecticides in species such as *Bombyx mori*, *Sitophilus oryzae*, and *Rhyzopertha dominica*.

One of the application methods of nanoparticles in agriculture is using as a fertilizer. The application nanofertilizers in the soil increase the efficiency of the elements, minimize their toxicity, and on the nontarget organisms in the soil. Consequently, nanofertilizer applications lead to their controlled release in the soil and reduce the frequency of the application.

22.4.1 Types of nanoparticles

Nowadays, with the increasing number of nanotechnology research studies and usage areas, revolutionary studies are emerging in this field. Nanoparticles have different effects than microparticles with the same chemical sequence. This different effect is due to two main factors. One of them is the surface effect, and the other is the quantum effect. These two factors cause nanoparticles to behave differently from large-scale materials, as these two factors change the chemical reactivity of the substance as well as its mechanical, optical, and magnetic properties (Amelia et al., 2012; Tang et al., 2012).

22.4.2 Metal nanoparticles

Having large surface areas and high reactivity, nanomaterials' activities have attracted considerable attention in recent years. As a result of the rapid development of nanotechnology, these nanomaterials of various sizes and diameters are frequently used in commercial and industrial areas (Amelia et al., 2012; Tang et al., 2012).

Metal-based nanomaterials are frequently used in the glass and paint industry, polishing, electronics production, pharmaceuticals and food additives, cosmetics and personal care products, biotechnology, and medical imaging, and they pollute the environment. Nanoparticles (NPs) are defined as small-sized and high surface area particles ranging 1–100 nm in size (Gomes et al., 2013).

Considering their chemical composition, the most important of the nanoparticles are carbon-derived nanoparticles and metal-containing metal oxide nanoparticles (Klaine et al., 2008; Bhatt and Tripathi, 2011).

In recent years, toxicological research has focused on metal oxide nanoparticles from these nanoparticle types (Ringwood et al., 2010; Buffet et al., 2011). Examples of these them are zinc, titanium, copper, and silver nanoparticles (Klaine et al., 2008).

22.5 Silver nanoparticles

Ag NPs have distinctive physicochemical properties such as Raman scattering catalytic activity and nonlinear optical properties developed with a high electrical and thermal conductivity surface. Also, these kinds of nanoparticles have strong antibacterial and antifungal properties and are mostly used in various fields (Le Ouay and Stellacci, 2015). Ag NPs do not have antibacterial or antifungal properties on their own, and they release silver ions by binding to electron donor groups of biological molecules containing sulfur, oxygen, or nitrogen. Two different mechanisms are responsible for the antimicrobial activity of Ag NPs, the first is the deterioration of lipopolysaccharides by adhering to the cell surface and the increase in permeability and the second is that it causes DNA damage by entering the bacterial cell (Sondi and Salopek-Sondi, 2004; Chowdhury et al., 2019). Ag NPs are widely used as antimicrobial agents in the healthcare industry, food packaging, water treatment, agriculture, and textile coatings. It is also used in absorbers, photocatalysts, and sensors to detect and remove environmental contaminants. In addition, Ag NPs have a wide range of uses in cosmetics, personal care products, and electronic devices (Chowdhury et al., 2019). Recently, Ag nanoparticles (AgNPs) have shown evidence of being a potentially effective nematicide (Cromwell et al., 2014; Abdellatif et al., 2016; Hassan et al., 2016; Nassar, 2016; Taha, 2016).

Green synthesis of silver nanoparticles is a good alternative route of nanoparticle synthesis from natural sources such as plant extracts microorganisms and enzymes. Green synthesis has many advantages, including low energy consumption and moderate operating conditions (e.g., pressure and temperature) without using harmful chemicals (Mie et al., 2014). Therefore, green synthesis techniques have been developed using various biological materials such as yeast, mold, algae, and bacteria and plant extracts for nanoparticle synthesis (Fig. 22.2) (Singh et al., 2017; Niknejad et al., 2015; Das et al., 2014; Kaviya et al., 2011; Balashanmugam and Kalaichelvan, 2015).

Although nanoparticles synthesized using biological sources such as plants and microorganisms are easier to bind to active ingredients, there are many metabolites in medicinal plants that have pharmacological activity. Many studies have indicated that these metabolites are more effective by binding to the synthesized nanoparticles and give more properties to the nanoparticles (Ge et al., 2014; Sintubin et al., 2012).

22.5.1 Silver nanoparticles as nanopesticides

Pesticides are chemicals that are frequently used in agriculture to protect plants from pests and diseases. Although the desired results are obtained as a result of the use of pesticides in these areas, in recent years, it has been understood that they lead to adverse effects on the environment and nontarget organisms. The main reason for these effects is the unnecessary and excessive use of pesticides. Many studies have shown that pesticides are safer against nature and other living species as a result of the use of lower concentrations. In recent years, nanotechnological methods have been used to control the use of pesticides and to ensure that they are applied at lower concentrations.

Because the surface/volume ratio of nanoparticles is much higher than microparticles, the reactivity and biochemical activities of the nanoparticles increase. Metal nanoparticles have a suppression effect on the pests and pathogens (Khan and Rizvi, 2014). Thus, they can be used as nanopesticides in agriculture. The simplest application method is that nanoparticles can be applied directly to plant parts such as soil,

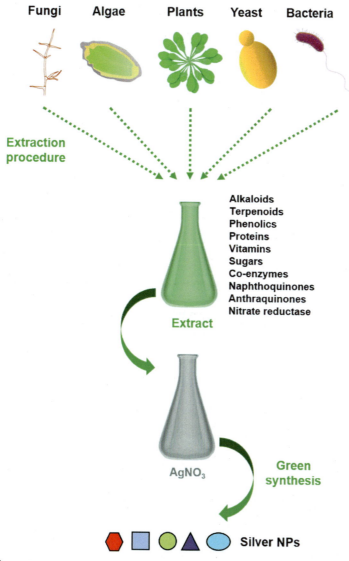

FIG. 22.2

Mechanism of green synthesized silver nanoparticles using various biological entities.

From Roy et al., 2019.

seeds, or leaves to protect plants from pathogens. Nanomaterials called nanopesticides are much more effective plant protection agents designed using the active ingredient of the classical pesticide (Kookana et al., 2014). Nanopesticides may offer a way to control delivery of pesticide and to achieve greater effectiveness of much smaller dose of a chemical.

Researchers are continuing to develop systems that detect pathogens and release the required dose only to the relevant regions in order to restrict pesticide applications that are more than the required dose and applied to the whole area (Khot et al., 2012). Nanomaterials are of a potential for pests and disease management in agricultural applications. Nanotechnology can be used in these kind of applications in two ways: (i) using nanoparticles due to the toxic effects to pests and pathogens and (ii) nanomaterials as a carrier of pesticides.

Silver nanoparticles have received significant attention as a pesticide for agricultural applications (Afrasiabi et al., 2016). The reason of using silver nanoparticles is having inhibitory effects on microorganisms. Hence, it is used for controlling various plant pathogens in a relatively safer way compared to synthetic fungicides (Park et al., 2006). In various studies, it was demonstrated that silver nanoparticles have adverse effects to fungicides. Jo et al. (2009) showed that silver nanoparticles inhibited the colony formation of *Bipolaris sorokiniana* and *Magnaparthe grisea*. Another study with silver nanoparticles also demonstrated that silver nanoparticles inhibited the growth of *Spaheratheca fusca* and *Fusarium culmorum* (Lamsal et al., 2010; Kasprowicz et al., 2010). Moreover, silver nanoparticles have potential insecticidal effects of pests. Rouhani et al. (2012) presented that silver nanoparticles inhibited the density of *Aphis nerri*. Sahayaraj et al. (2016) also reported that Ag NPs enhanced the food consumption and drastically reduced the bodyweight of *Pericallia ricini* Fab. larvae. In another study, it was stated that green synthesized Ag NPs obtained from *Periplaneta americana* wing's extract caused high mortality in *Aphis gossypii* according to the control under laboratory conditions.

Using synthetic insecticides for controlling pests leads to adverts effects on nontarget organisms and cause development of resistance. Ag NPs are also utilized for pest management to overcome these issues. The insecticidal activity of Ag NPs against two major insect pests *Helicoverpa armigera* and *Aedes aegypti* has been recorded (Kantrao et al., 2017; Parthiban et al., 2018). It was demonstrated that Ag NPs has an impact effect on insecticide resistance.

22.5.2 Silver nanoparticles as nanofertilizers

Chemical fertilizers, such as urea, diammonium phosphate (DAP), and single superphosphate (SSP), are used to maintain the shortage of nitrogen, phosphorus, and potassium in the soil. However, most of these fertilizers are lost as they flow or evaporate. It has been determined in studies that 40%–70% of the nitrogen, 80%–90% of the phosphorus, and 50%–70% of the potassium of the fertilizers applied are lost in this way in the environment and consequently cannot be absorbed by the applied plant. In addition, it has been reported to cause financial losses as well as environmental pollution (Trenkel, 1997; Ombodi and Saigusa, 2000; Duhan et al., 2017).

Nanomaterials have an important place in fertilizer applications in agriculture and feed the plant slowly in a controlled manner (Sohrab et al., 2016). In recent years, the use of slow-release fertilizers has become one of the important innovative technologies in which fertilizers are entrapped with nanoparticles to save fertilizer consumption and to minimize environmental pollution. Nanofertilizers have a positive effect on plants and environment by reducing environmental risk factors arising from soil pollution caused by chemical fertilizers (Naderi et al., 2011). One of the biggest advantages is that it is used less than other fertilizers (Selivanov and Zorin, 2001; Reynolds, 2002; Raikova et al., 2006; Batsmanova et al., 2013; Subramanian et al., 2015). Besides their large surface areas, in many studies on nanofertilizers, it has been demonstrated that plants have smaller sizes than their leaf and root pore sizes, and they increase the efficiency of use by making the intake of nutrients easier (Singh et al., 2017). With nanoparticle applications in the soil, a suitable and sufficient microelement content can be found, thus increasing the resistance of plants against pathogens.

The small size of nanofertilizers allows the plant to easily benefit from minerals by allowing the mineral to pass through stomata easily. Thus, maximum efficiency can be obtained from the applied fertilizers. Nanofertilizers have many advantages over conventional fertilizers (Liu and Lal, 2016; Singh et al., 2017; Daghan, 2017, Ahmad et al., 2019). Some of these advantages are as follows: the fact that the highest yield can be obtained at the lowest cost by using a very little amount of fertilizers, increasing the efficiency of using the current nutrients in the plant, increasing the efficiency of fertilizer use, avoiding the continuous use of fertilizers, and minimizing the possible negative effects on the environment by reducing the losses of nutrients useful for plants and soil. It can be listed as reducing the risk of possible toxicity and increasing soil fertility and product quality. Nanofertilizers increase product yield and nutritional value by allowing the plant to grow and develop healthily throughout the growing period. Thus, nanofertilizers are important for the healthy plant to gain more resistance against diseases and adverse environmental conditions. In most of the field applications, chitosan, nitrogen, phosphorus, titanium, and zinc nanoparticles are used as fertilizers.

Silver nanoparticles provide an improved uptake of nutrients from the soil than the bulk one due to having many unique properties and being a nutrient for the plants. It has been reported that silver nanoparticles could be used to enhance seed germination potential and can be used to enhance the nutrient uptake in the plants (Duhan et al., 2017; Anand and Bhagat, 2019). Therefore, it can be applied as nanofertilizers in crop protection.

22.5.3 Antioxidant and antimicrobial properties of silver nanoparticles

Nanoparticles can be synthesized by physical, chemical, or biological methods. Biological method is preferred more due to being environmentally friendly and economical. Biological method is to obtain nanoparticles as a result of extraction of microorganisms or plants. According to the Krithiga et al. (2015), plants are better options for nanoparticle synthesis due to having nontoxic properties, providing natural capping agents and reducing the cost of microorganism extraction. The compounds of the plants provide antioxidant properties and provide the plants additional antioxidant and antibacterial property to the nanoparticles because biomolecules such as protein, phenols, and flavonoids in plants play a role in reducing the ions to nanosize and play an important role in capping of nanoparticles (Pasupuleti et al., 2013; Kuppusamy et al., 2015). It is known that silver has been used in many areas such as water disinfection and burn treatment for many years. Hence, silver has many important advantages as an antimicrobial agent. These advantages are that silver is a very broad-spectrum antibiotic, and there is almost no bacterial resistance in silver and has no toxic effects at low concentrations (Rai et al., 2009). It has been reported that silver binds to the bacterial cell wall, and cell membrane deactivates proteins in thiol (–SH) groups by interacting with them and replaces the hydrogen cation by lowering membrane permeability, thus causing the death of bacterial cells (Duncan, 2011). The antimicrobial mechanism of action of silver is explained by the absorption of ions by the microorganism cell, their accumulation in the cell, the shrinkage of the cytoplasm membrane, or the attraction of the cytoplasm to itself by the cell wall. DNA molecules are damaged, and cells lose their ability to reproduce due to the infiltration of Ag^+ and it affects the -SH bonds of proteins and causes them to be inactivated (Feng et al., 2000). Moreover, silver acts as a catalyst for the oxidation of microorganisms in oxygen-added solutions (Cho et al., 2005). Keshari et al. (2020) reported that Ag NPs synthesized by *Cestrum nocturnum* have more antioxidant activity as compared to vitamin C, and it was also stated that Ag NPs have strong antibacterial activity due to the presence of bioactive molecules on the surface of Ag NPs. Moreover, Otunola et al. (2017) demonstrated that Ag NPs synthesized from aqueous extract of *Allium sativum*, *Zingiber officinale*, and *Copsicum frutescens* showed strong antibacterial activity. Also, it was reported that the synthesized Ag NPs exhibited potential free-radical scavenging activity against radicals. In another study with synthesized Ag NPs, it was demonstrated that Ag NPs from *Chenopodium murale* and *Piper longum* also enhanced the antioxidant activity (Ojha et al., 2012; Vivekanandhan et al., 2012). Mohanta et al. (2017) studied antimicrobial, antioxidant, and cytotoxic activity of Ag NPs synthesized by leaf extract of *Erythrina suberosa* (Roxb.) and found that Ag NPs extracts can be used for health preservation against different oxidative stress and Ag NPs also have the strong antimicrobial potential for microorganisms such as *Staphylococcus aureus* and *Pseudomonas aeruginosa*. In addition to these results, Ag NPs showed excellent anticancer activity against A431 osteosarcoma cell line. Salari et al. (2019) reported that biosynthesized Ag NPs from apple extract showed a higher antioxidant and antibacterial activity compared to *Prosopis farcta* fruit extract alone. Furthermore, the comparison studies with synthesized Ag NPs from plants such as *Chenopodium murale* leaf extract (Abdel-Aziz et al., 2014), spruce bark extract (Tanase et al., 2019), *Nothapodytes foetida* leaf extract (Datkhile et al., 2020), *Lysilom acapulcensis* (Garibo et al., 2020), *Camellia sinensis* tea leaf extract (Loo et al., 2012), *Berberis vulgaris* leaf and root aqueous extract (Behravan et al., 2019), and *Salvia spinosa* (Pirtarighat et al., 2019) reveal that Ag NPs have a potential antioxidant and antimicrobial activity.

22.6 Effects of silver nanoparticles on plant-parasitic nematodes

The use of chemical nematicides in the fight against plant-parasitic nematodes is not only expensive but also causes resistance development, negative effects on nontarget organisms and the environment. Therefore, more effective, safe, and inexpensive methods are preferred. Green synthesis, which is a safe method in agricultural fields, has been used in recent years with the developing nanotechnological methods. In particular, nanoparticles obtained from plants by this method are used in applications such as pesticides, fertilizers, and nutrition in agricultural areas (Fernandez et al., 2001).

The first report on the bioefficacy of AgNP on a plant-parasitic nematode was provided by Cromwell et al. (2014) against root-knot nematode on Bermuda grass with mixed results, the lab assays revealed promising results, but field experiments were not conclusive. In the studies with plant-parasitic nematodes carried out in later years, Hassan et al. (2016) compared with silver nanoparticles and two nematicides and found that the results demonstrated the positive effect of silver nanoparticles on *M. incognita* alone and in mixture with the nematicides and also obtained enhancement on growth parameters to the host plant. In another study with *M. incognita*, Nassar (2016) stated that LD_{50} value of silver nanoparticles extracted from *Urtica urens* were effective in the management of *M. incognita*. Moreover, Kalaiselvi et al. (2019) demonstrated that green synthesis Ag NPs using latex extract of *Euphorbia tirucall* has nematicidal effects on *M. incognita*. It was found that the toxicity of Ag NPs leads to reduce nematode infection without causing any phytotoxicity in tomato plants. In another study with *M. incognita*, it was reported that due to the Ag NPs applications, 100% of nematodes becomes inactive within 24 and 48h (Bernard et al., 2019). Soliman et al. (2017) presented that Ag NPs synthesized from *Artemisia judaica* leave extracts has potential adverse effects on mortality of *M. incognita*. It was evaluated in another study with *M. incognita* that silver nanoparticle synthesized by utilizing a naturally occurring biopolymer, which is called chitosan, can be used as a potential biopesticide (Bernard et al., 2019). Abbassy et al. (2017) investigated that the extractives of *Conyza dioscoridis*, *Melia azedarach*, and *Moringa oleifera* were synthesized in the silver nanoparticle form and found that *C. dioscoridis* extractives have great nematicidal activity against the second stage juvenile and eggs of *M. incognita*. Fouda et al. (2020) also showed that the microcrystalline cellulose embedded silver nanoparticles are an alternative econematicide on *M. incognita* owing to causing high mortality ratio. Taha (2016) established that Ag NPs inhibited the nematode growth and egg hatchability. Otherwise, Ag NPs did not show toxic effect on the plant growth of *Lycopersicon esculentum* L. var. Castel rock. Baronia et al. (2020) also demonstrated that silver nanoparticle has effective nematicidal activity against *M. graminicola* even at low concentrations. In another study with *Meloidogyne* spp., it was presented that green synthesized Ag NPs from two different algae *Ulva lactuca* and *Turbinaria turbinate* have adverse effects on the second-stage juveniles of *M. javanica* and also lead to changes in DNA profile (Abdellatif et al., 2016) (Table 22.1).

Table 22.1 Mode of action of green synthesized silver nanoparticles on plant-parasitic nematodes.

Plant species	Part	Characterization	Plant-parasitic nematode species	Mode of action	References
Zingiber officinale	Rhizomes	UV-Vis spectra analysis, TEM	Meloidogyne incognita	Improving plant growth and reduced nematode infection, lowest numbers of galls and egg masses	Nour El-Deen and El-Deen (2018)
Lantana camara L.	Leaves	UV-Vis spectra analysis	Anguina tritici Thorne, 1949	Mortality	Dura et al. (2019b)
Azadirachta indica, Curcuma longa	Fresh leaves	UV-Vis spectra analysis, XRD, FT-IR, DLS, zeta potential	Meloidogyne incognita	Mortality	Kalaiselvi et al. (2017)
Moringa oleifera L.	Leaves	UV-Vis spectra analysis	Meloidogyne incognita	Mortality	Dura et al. (2019a,b)
Conyza dioscoridis, Melia azedarach, Moringa oelifera	Leaves	UV-Vis spectra analysis, SEM	Meloidogyne incognita	Nematicidal activity, inhibit the two stages: egg and J2	Abbassy et al. (2017)
Azadirachta indica, Vernonia amygdalina L., Ocimum gratissimum, Morinda oleifera Lam.	Leaves	UV-Vis spectra analysis	Meloidogyne incognita	Inhibit egg hatchability and development of nematode	Youssef et al. (2014)
Urtica urens	Leaves	UV-Vis spectra analysis, SEM	Meloidogyne incognita	Suppressive effect to egg hatchability, toxic effect on second larval stage of the nematode	Nassar (2016)
Cyanobacterium Nostoc sp.	Aqueous extract	UV-Vis spectra analysis, TEM, XRD, FT-IR, zeta potential	Meloidogyne javanica	Completely inhibited the growth of J2 with remarkable enhancement of faba bean growth	Hamed et al. (2019)

22.7 Conclusion

The use of nematicides in the fight against plant-parasitic nematodes is a very expensive method and has negative effects on environmental health. The use of nematicides tends to be increasingly banned around the world due to their negative effects on the environment and human health. Plant extracts are considered as an alternative to chemical nematicides in the fight against nematodes due to their natural nature and ease of use and preparation. The number of studies on the usage possibilities of plant extracts against nematodes is quite high. Nowadays, with the technological advances in the 21st century, the number of scientific studies on the use of nanosilver-added plant extracts in agriculture tends to increase.

References

Abawi, G.S., Widmer, T.L., 2000. Impact of soil health management practices on soilborne pathogens, nematodes and root diseases of vegetable crops. Appl. Soil Ecol. 15, 37–47.

Abbassy, M.A., Abdel-Rasoul, M.A., Nassar, A.M.K., Soliman, B.S.M., 2017. Nematicidal activity of silver nanoparticles of botanical products against rootknot nematode, *Meloidogyne incognita*. Arch. Phytopathol. Plant Protect. 50 (17–18), 909–926.

Abdel-Aziz, M.S., Shaheen, M.S., El-Nekeety, A.A., Abdel-Wahhab, M.A., 2014. Antioxidant and antibacterial activity of silver nanoparticles biosynthesized using *Chenopodium murale* leaf extract. J. Saudi Chem. Soc. 18 (4), 356–363.

Abdellatif, K.F., Hamouda, R.A., El-Ansary, M.S.M., 2016. Green nanoparticles engineering on root-knot nematode infecting eggplants and their effect on plant DNA modification. Iran. J. Biotechnol. 14, 250–259.

Afrasiabi, Z., Popham, H.J., Stanley, D., Suresh, D., Finley, K., Campbell, J., Kannan, R., Upendran, A., 2016. Dietary silver nanoparticles reduce fitness in a beneficial, nut not pest, insect species. Arch. Insect Biochem. Physiol. 93, 190–201.

Agrios, G.N., 1997. Plant Pathology, fourth ed. Academic Press Inc., London, p. 250.

Ahmad, S., Munir, S., Zeb, N., Ullah, A., Khan, B., Ali, J., Bilal, M., Omer, M., Alamzeb, M., Salman, S.M., Ali, S., 2019. Green nanotechnology: a review on green synthesis of silver nanoparticles—an ecofriendly approach. Int. J. Nanomedicine. 14, 5087–5107.

Akhter, G., Khan, A., Ali, S.G., Khan, T.H., Siddiqi, K.S., Khan, H.M., 2020. Antibacterial and nematicidal properties of biosynthesized Cu nanoparticles using extract of holoparasitic plant. SN Appl. Sci. 2, 1268.

Al Banna, L.S., Salem, N.M., Awwad, A.M., 2020. Green synthesis of sulfur nanoparticles using *Rosmarinus officinalis* leaves aqueous extract and nematicidal activity against *Meloidogyne javanica*. Chem. Int. 6 (3), 137–143.

Amelia, M., Lincheneau, C., Silvi, S., Credi, A., 2012. Electrochemical properties of CdSe and CdTe quantum dots. Chem. Soc. Rev. 41, 5728–5743.

Anand, R., Bhagat, M., 2019. Silver nanoparticles (AgNPs): as nanopesticides and nanofertilizers. MOJ Biol. Med. 4 (1), 19–20.

Anshup, A., Venkataraman, J.S., Subramaniam, C., Kumar, R.R., Priya, S., Kumar, T.R., Omkumar, R.V., John, A., Pradeep, T., 2005. Growth of gold nanoparticles in human cells. Langmuir 21 (25), 11562–11567.

Bakker, J., Folkertsma, R.T., Rouppe van der Voort, J.N.A.M., de Boer, J.M., Gommers, F.J., 1993. Changing concepts and molecular approaches in the management of virulence genes in potato cyst nematodes. Annu. Rev. Phytopathol. 31, 169–190.

Balashanmugam, P., Kalaichelvan, P.T., 2015. Biosynthesis characterization of silver nanoparticles using *Cassia roxburghii* DC. aqueous extract, and coated on cotton cloth for effective antibacterial activity. Int. J. Nanomedicine 10, 87–97.

Baronia, R., Kumar, P., Singh, S.P., Walia, R.K., 2020. Silver nanoparticles as a potential nematicide against *Meloidogyne graminicola*. J. Nematol. 52, 1–9.

Batsmanova, L.M., Gonchar, L.M., Taran, N.Y., Okanenko, A.A., 2013. Using a colloidal solution of metal nanoparticles as micronutrient fertilizer for cereals. Proc. Int. Conf. Nanomater. 2 (4), 04NABM14.

Bawa, J.A., Mohammed, I., Liadi, S., 2014. Nematicidal effect of some plants extracts on root-knot nematodes (*Meloidogyne incognita*) of tomato (*Lycopersicon esculentum*). World J. Life Sci. Med. Res. 3 (3), 81–87.

Behravan, M., Panahi, A.H., Naghizadeh, A., Ziaee, M., Mahdavi, R., Mirzapour, A., 2019. Facile green synthesis of silver nanoparticles using *Berberis vulgaris* leaf and root aqueous extract and its antibacterial activity. Int. J. Biol. Macromol. 124, 148–154.

Bernard, G.C., Egnin, M., Bonsi, C., 2017. The impact of plant-parasitic nematodes on agriculture and methods of control. In: Shah, M.M. (Ed.), Nematology: Concepts, Diagnosis and Control. IntechOpen, London, pp. 121–151.

Bernard, G.C., Fitch, J., Min, B., Shahi, N., Egnin, M., Ritte, I., Collier, W.E., Bonsi, C., 2019. Potential nematicidal activity of silver nanoparticles against the root-knot nematode (*Meloidogyne incognita*). J. Altern. Complement. Med. 2 (2), 1–4.

Bhatt, I., Tripathi, B.N., 2011. Interaction of engineered nanoparticles with various components of the environment and possible strategies for their risk assessment. Chemosphere 82 (3), 308–317.

Bridge, J., Starr, J.L., 2007. Cereals: rice (*Oryza sativa*). In: Bridge, J., Starr, J.L. (Eds.), Plant Nematodes of Agricultural Importance: A Colour Handbook. Manson Publishing, London, pp. 52–60.

Buffet, P.E., Tankoua, O.F., Pan, J.F., Berhanu, D., Herrenknecht, C., Poirier, L., Amiard-Triquet, C., Amiard, J.C., Bérard, J.B., Risso, C., Guibbolini, M., Roméo, M., Reip, P., Valsami-Jones, E., Mouneyrac, C., 2011. Behavioural and biochemical responses of two marine invertebrates *Scrobicularia plana* and *Hediste diversicolor* to copper oxide nanoparticles. Chemosphere 84, 166–174.

Cho, N.K., Seo, D.S., Lee, J.K., 2005. Preparation and stabilization of silver colloids protected by surfactant. Mater. Forum 29, 394–396.

Chowdhury, A.H., Debnath, R., Manirul Islam, S., Saha, T., 2019. In: Inamuddin, Thomas, S., Kumar Mishra, R., Asiri, A.M. (Eds.), Impact of Nanoparticle Shape, Size, and Properties of Silver Nanocomposites and Their Applications. Springer, pp. 1067–1091.

Cromwell, W.A., Yang, J., Starr, J.L., Jo, Y.K., 2014. Nematicidal effects of silver nanoparticles on root-knot nematode in Bermuda grass. J. Nematol. 46, 261–266.

Daghan, H., 2017. Nano fertilizers. Turk. J. Agric. Res. 4 (2), 197–203.

Das, V.L., Thomas, R., Varghese, R.T., Soniya, E.V., Mathew, J., Radhakrishnan, E.K., 2014. Extracellular synthesis of silver nanoparticles by the *Bacillus strain* CS11 isolated from industrialized area. 3 Biotech 4, 121–126.

Datkhile, K.D., Patil, S.R., Durgavale, P.P., Patil, M.N., Jagdale, N.J., Deshmukh, V.N., 2020. Studies on antioxidant and antimicrobial potential of biogenic silver nanoparticles synthesized using *Nothapodytes foetida* leaf extract (wight) sleumer. Biomed. Pharmacol. J. 13 (1), 441–448.

Decraemer, W., Hunt, D.J., 2006. Structure and classification. In: Perry, R.N., Moens, M. (Eds.), Plant Nematology. CAB International, Wallingford, UK, pp. 3–32.

Dowling, A., Clift, R., Grobert, N., Hutton, D.R., Oliver, R., 2004. Nanoscience and Nanotechnologies: Opportunities and Uncertainties. The Royal Society, The Royal Academy of Engineering, UK.

Duhan, J.S., Kumar, R., Kumar, N., Kaur, P., Nehra, K., Duhan, S., 2017. Nanotechnology: the new perspective in precision agriculture. Biotechnol. Rep. (Amst.) 24 (15), 11–23.

Duncan, T.V., 2011. Applications of nanotechnology in food packaging and food safety: barrier materials, antimicrobials and sensors. J. Colloid Interface Sci. 363 (1), 1–24.

Dura, O., Sarı, Y., Tınmaz, A.B., Sönemaz, I., Yesilayer, A., Kepenekci, I., 2019a. Determination of the effectiveness of nano silver additive aqueous extract of *Moringo oliefera* L. (Brassica: Moringaceae) against root-knot nematode (*Meloidogyne incognita* (Kofoid & White, 1919) Chitwood, 1949 (nematode: Meloidogynidae)) under laboratory conditions. Bahce 48 (1), 19–25.

Dura, O., Tülek, A., Sönmez, I., Erdoğuş, F.D., Yeşilayer, A., Kepenekci, I., 2019b. Effects of silver nanoparticle (AgNPs) applications prepared using *Lantana camara* L. (Lamiales: Verbenaceae)'s aqueous extract on wheat gal nematode [*Anguina tritici* Thorne, 1949 (Nematoda: Anguinidae)]. Plant Prot. Bull. 59 (2), 49–53.

El-Batal, A.I., Attia, M.S., Nofel, M.M., El-Sayyad, G., 2019. Potential nematicidal properties of silver boron nanoparticles: synthesis, characterization, in vitro and in vivo root-knot nematode (*Meloidogyne incognita*) treatments. J. Clust. Sci. 30, 687–705.

Elling, A.A., 2013. Major emerging problems with minor *Meloidogyne* species. Phytopathology 103 (11), 1092–1102.

Feng, Q.L., Wu, J., Chen, G.Q., Cui, F.Z., Kim, T.N., Kim, J.O., 2000. A mechanistic study of the antibacterial effect of silver ions on *Escherichia coli* and *Staphylococcus aureus*. J. Biomed. Mater. Res. 52, 662–668.

Fernandez, C., Kabana, R.R., Warrior, P., Kloepper, J.W., 2001. Induced soil suppressiveness to a root-knot nematode species by a nematicide. Biol. Control 22, 103–114.

Fouda, M.M.G., Abdelsalam, N.R., Gohar, I.M.A., Hanfy, A.E.M., Othman, S.I., Zaitoun, A.F., Allam, A.A., Morsy, O.M., El-Naggar, M., 2020. Utilization of high throughput microcrystalline cellulose decorated silver nanoparticles as an eco-nematicide on root-knot nematodes. Colloids Surf. B: Biointerfaces 188, 110805.

Garibo, D., Borbón-Nuñez, H.A., de León, J.N.D., et al., 2020. Green synthesis of silver nanoparticles using *Lysiloma acapulcensis* exhibit high-antimicrobial activity. Sci. Rep. 10, 12805.

Ge, L., Li, Q., Wang, M., Ouyang, J., Li, X., Xing, M.M.Q., 2014. Nanosilver particles in medical applications: synthesis, performance, and toxicity. Int. J. Nanomedicine 9, 2399–2407.

Ghormade, V., Deshpande, M.V., Paknikar, K.M., 2011. Perspectives for nano-biotechnology enabled protection and nutrition of plants. Biotechnol. Adv. 29, 792–803.

Gomes, T., Pereira, C.G., Cardoso, C., Bebianno, M.J., 2013. Differential protein expression in mussels *Mytilus galloprovincialis* exposed to nano and ionic Ag. Aquat. Toxicol. 136 (137), 79–90.

Govindaraju, K., Khaleel Basha, S., Ganesh Kumar, V., Singaravelu, G., 2008. Silver, gold and bimetallic nanoparticles production using single cell protein (*Spirulina platensis*) Geitler. J. Mater. Sci. 43, 5115–5122.

Hamed, S.M., Hagag, E.S., El-Raouf, N.A., 2019. Green production of silver nanoparticles, evaluation of their nematicidal activity against *Meloidogyne javanica* and their impact on growth of faba bean. Beni-Suef Univ. J. Basic. Appl. Sci. 8 (2), 9.

Hassan, M.E., Zawam, M., El-Nahas, H.S., Desoukey, A.F., 2016. Comparative study between silver nanoparticles and two nematicides against *Meloidogyne incognita* on tomato seedlings. Plant Pathol. J. 15, 144–151.

Huang, G., Allen, R., Davis, E., Baum, T., Hussey, R., 2006. Engineering broad root-knot resistance in transgenic plants by RNAi silencing of a conserved and essential root-knot nematode parasitism gene. PNAS 103, 14302–14306.

Jo, Y.K., Kim, B.H., Jung, G., 2009. Antifungal activity of silver ions and nanoparticles on phytopathogenic fungi. Plant Dis. 93, 1037–1043.

Jones, J.T., Haegeman, A., Danchin, E.G.J., Gaur, H.S., Helder, J., Jones, M.G.K., Kikuchi, T., Manzanilla-López, R., Palomares-Rius, J.E., Wesemael, W.M.L., Perry, R.N., 2013. Top 10 plant-parasitic nematodes in molecular plant pathology. Mol. Plant Pathol. 14 (9), 946–961.

Kalaiselvi, D., Sundararaj, P., Premasudaha, P., Hafez, S.L., 2017. Nematicidal acitivity of green syhthesized silver nanoparticles using plant extracts against root-knoy nematode *Meloidogyne incognita*. Int. J. Nematol. 27 (1), 81–94.

Kalaiselvi, D., Mohankumar, A., Shanmugam, G., Nivitha, S., Sundararaj, P., 2019. Green synthesis of silver nanoparticles using latex extract of *Euphorbia tirucalli*: a novel approach for the management of root knot nematode, *Meloidogyne incognita*. Crop Prot. 117, 108–114.

Kantrao, S., Ravindra, M.A., Akbar, S.M.D., Jayanthi, P.D.K., Venkataraman, A., 2017. Effect of biosynthesized silver nanoparticles on growth and development of *Helicoverpa armigera* (Lepidoptera: Noctuidae): interaction with midgut protease. J. Asia Pac. Entomol. 20 (2), 583–589.

Kasprowicz, M.J., Kozioł, M., Gorczyca, A., 2010. The effect of silver nanoparticles on phytopathogenic spores of *Fusarium culmorum*. Can. J. Microbiol. 56 (3), 247–253.

Kaviya, S., Santhanalakshmi, J., Viswanathan, B., Muthumary, J., Srinivasan, K., 2011. Biosynthesis of silver nanoparticles using *Citrus sinensis* peel extract and its antibacterial activity. Spectrochim. Acta A 79 (3), 594–598.

Keshari, A.K., Srivastava, R., Singh, P., Yadav, V.B., Nath, G., 2020. Antioxidant and antibacterial activity of silver nanoparticles synthesized by *Cestrum nocturnum*. J. Ayurveda Integr. Med. 11 (1), 37–44.

Khan, M.R., Rizvi, T.F., 2014. Nanotechnology: scope and application in plant disease management. Plant Pathol. J. 13, 214–231.

Khanam, S., Bauters, L., Singh, R.R., Verbeek, R., Haeck, A., Sultan, S.M.D., Demeestere, K., Kyndt, T., Gheysen, G., 2018. Mechanisms of resistance in the rice cultivar Manikpukha to the rice stem nematode *Ditylenchus angustus*. Mol. Plant Pathol. 19 (6), 1391–1402.

Khot, L.R., Sankaran, S., Maja, J.M., Ehsani, R., Schuster, E.W., 2012. Applications of nanomaterials in agricultural production and crop protection: a review. Crop Prot. 35, 64–70.

Klaine, S.J., Alvarez, P.J.J., Batley, G.E., Fernandes, T.F., Handy, R.D., Lyon, D.Y., Mahendra, S., Mclaughlin, M.J., Lead, J.R., 2008. Nanomaterials in the environment: behavior, fate, bioavailability and effects. Environ. Toxicol. Chem. 27 (9), 1825–1851.

Kookana, R.S., Boxall, A.B., Reeves, P.T., Ashauer, R., Beulke, S., Chaudhry, Q., Cornelis, G., Fernandes, T.F., Gan, J., Kah, M., Lynch, I., Ranville, J., Sinclair, C., Spurgeon, D., Tiede, K., Van den Brink, P.J., 2014. Nanopesticides: guiding principles for regulatory evaluation of environmental risks. J. Agric. Food Chem. 62 (19), 4227–4240.

Kowshik, M., Deshmukh, N., Vogel, W., Urban, J., Kulkarni, S.K., Paknikar, K.M., 2002. Microbial synthesis of semiconductor CdS nanoparticles; their characterization, and their use in the fabrication of an ideal diode. Biotechnol. Bioeng. 78, 583–588.

Krithiga, N., Rajalakshmi, A., Jayachitra, A., 2015. Green synthesis of silver nanoparticles using leaf extracts of *Clitoria ternatea* and *Solanum nigrum* and study of its antibacterial effect against common nosocomial pathogens. J. Nanosci. 2015, 1–8.

Kuppusamy, P., Yusoff, M.M., Govindan, N., 2015. Biosynthesis of metallic nanoparticles using plant derivatives and their new avenues in pharmacological applications—an updated report. Saudi Pharm. J. 24 (4), 473–484.

Lamsal, K., Sang-Woo, K., Jung, J.H., Kim, Y.S., Kim, K.U., Lee, Y.S., 2010. Inhibition effects of silver nanoparticles against powdery mildews on cucumber and pumpkin. Mycobiology 39 (1), 26–32.

Le Ouay, B., Stellacci, F., 2015. Antibacterial activity of silver nanoparticles: a surface science insight. Nano Today 10 (3), 339–354.

Lengke, M.F., Fleet, M.E., Southam, G., 2007. Biosynthesis of silver nanoparticles by filamentous cyanobacteria from a silver(I) nitrate complex. Langmuir 23 (5), 2694–2699.

Lim, D., Roh, J.Y., Eom, H.J., Choi, J.Y., Hyun, J., Choi, J., 2012. Oxidative stress-related PMK-1 P38 MAPK activation as a mechanism for toxicity of silver nanoparticles to reproduction in the nematode *Caenorhabditis elegans*. Environ. Toxicol. Chem. 31 (3), 585–592.

Liu, R., Lal, R., 2016. Nanofertilizers. In: Lal, R. (Ed.), Encyclopedia of Soil Science, third ed. CRC Press, pp. 1511–1515.

Loo, Y.Y., Chieng, B.W., Nishibuchi, M., Radu, S., 2012. Synthesis of silver nanoparticles by using tea leaf extract from Camellia sinensis. Int. J. Nanomedicine 7, 4263–4267.

Matthiessen, J., Kirkegaard, J., 2006. Biofumigation and enhanced biodegradation: opportunity and challenge in soilborne pest and disease management. Crit. Rev. Plant Sci. 25, 235–265.

Mie, R., Samsudin, M.W., Din, L.B., Ahmad, A., Ibrahim, N., Adnan, S.N.A., 2014. Synthesis of silver nanoparticles with antibacterial activity using the lichen *Parmotrema praesorediosum*. Int. J. Nanomedicine 9, 121.

Mohanta, Y.K., Panda, S.K., Jayabalan, R., Sharma, N., Bastia, A.K., Mohanta, T.K., 2017. Antimicrobial, antioxidant and cytotoxic activity of silver nanoparticles synthesized by leaf extract of *Erythrina suberosa* (Roxb.). Front. Mol. Biosci. 4 (14).

Mukundan, D., Vasanthakumari, R., 2017. Phytoengineered Nanomaterials and their applications. In: Prasad, R., Kumar, V., Kumar, M. (Eds.), Nanotechnology. Springer, Singapore.

Naderi, M., Danesh Shahraki, A.A., Naderi, R., 2011. Application of nanotechnology in the optimization of formulation of chemical fertilizers. Iran. J. Nanotechnol. 12, 16–23.

Nassar, A., 2016. Effectiveness of silver nanoparticles of extracts of *Urtic aurens* (Urticaceae) against root-knot nematode, *Meloidogyne incognita*. Asian Nematol. 5, 14–19.

Nazir, K., Mukhtar, T., Javed, H., 2019. In vitro effectiveness of silver nanoparticles against root-knot nematode (*Meloidogyne incognita*). Pak. J. Zool. 51 (6), 2077–2083.

Niknejad, F., Nabili, M., Daie Ghazvini, R., Moazeni, M., 2015. Green synthesis of silver nanoparticles: advantages of the yeast *Saccharomyces cerevisiae* model. Curr. Med. Mycol. 1 (3), 17–24.

Nour El-Deen, A., El-Deen, A.B., 2018. Effectiveness of silver nanoparticles against root knot nematode, *Meloidogyne incognita* infecting tomato under greenhouse conditions. J. Agric. Sci. 10 (2), 148–156.

Nowack, B., Bucheli, T.D., 2007. Occurrence, behaviour and effects of nanoparticles in the environment. Environ. Pollut. 150, 5–22.

Ojha, A.K., Behera, S., Rout, J., Dash, M.P., Nayak, P.L., 2012. Green synthesis of silver nanoparticles from syzygium aromaticum and their antibacterial efficacy. Int. J. Pharm. Biol. Sci. 1, 335–341.

Okeniyi, M.O., Afolami, S.O., Fademi, O., Oduwaye, O., 2013. Effect of botanical extracts on root-knot nematode (*Meloidogyne incognita*) infection and growth of cashew (*Anacardium occidentale*) seedlings. Acad. J. Biotechnol. 1 (6), 081–086.

Ombodi, A., Saigusa, M., 2000. Broadcast application versus band application of polyolefin coated fertilizer on green peppers grown on andisol. J. Plant Nutr. 23, 1485–1493.

Otunola, G.A., Afolayan, A.J., Ajayi, E.O., Odeyemi, S.W., 2017. Characterization, antibacterial and antioxidant properties of silver nanoparticles synthesized from aqueous extracts of *Allium sativum, Zingiber officinale*, and *Capsicum frutescens*. Pharmacogn. Mag. 13 (2), 201–208.

Ovais, M., Khalil, A.T., Islam, N.U., et al., 2018. Role of plant phytochemicals and microbial enzymes in biosynthesis of metallic nanoparticles. Appl. Microbiol. Biotechnol. 102 (16), 6799–6814.

Park, H.J., Kim, S.H., Kim, H.J., Choi, S.H., 2006. A new composition of nanosized silica-silver for control of various plant diseases. Plant Pathol. J. 22, 2295–2302.

Parthiban, E., Manivannan, N., Ramanibai, R., Mathivanan, N., 2018. Green synthesis of silver-nanoparticles from *Annona reticulata* leaves aqueous extract and its mosquito larvicidal and anti-microbial activity on human pathogens. Biotechnol. Rep. (Amst.) 21, e00297.

Pasupuleti, V.R., Prasad, T.N., Shiekh, R.A., Balam, S.K., Narasimhulu, G., Reddy, C.S., Ab Rahman, I., Gan, S.H., 2013. Biogenic silver nanoparticles using *Rhinacanthus nasutus* leaf extract: synthesis, spectral analysis, and antimicrobial studies. Int. J. Nanomedicine 8, 3355–3364.

Patalano, G., Lazeri, L., D'Avino, L., Mazzoncini, M., 2008. Innovative approach for producing high value products in non food agro-industrial chains. In: Third Internetional Biofumigation Symposium, 21–25 July. CSIRO Discovery Centre, Canberra, Australia, p. 27.

Pirtarighat, S., Ghannadnia, M., Baghshahi, S., 2019. Green synthesis of silver nanoparticles using the plant extract of *Salvia spinosa* grown in vitro and their antibacterial activity assessment. J. Nanostruct. Chem. 9, 1–9.

Prasad, R., Bhattacharyya, A., Nguyen, Q.D., 2017. Nanotechnology in sustainable agriculture: recent developments, challenges, and perspectives. Front. Microbiol. 8, 1014.

Ragaei, M., Sabry, A.H., 2014. Nanotechnology for insect pest control. Int. J. Environ. Sci. Technol. 3, 528–545.

Rai, M.K., Yadav, A.P., Gade, A.K., 2009. Silver nanoparticles as a new generation of antimicrobials. Biotechnol. Adv. 27 (1), 76–82.

Raikova, O.P., Panichkin, L.A., Raikova, N.N., 2006. Studies on the effect of ultrafine metal powders produced by different methods on plant growth and development. Nanotechnologies and information technologies in the 21st century. In: Proceedings of the International Scientific and Practical Conference, pp. 108–111.

Rautaray, D., Ahmad, A., Sastry, M., 2003. Biosynthesis of $CaCO_3$ crystals of complex morphology using a fungus and an actinomycete. J. Am. Chem. Soc. 125 (48), 14656–14657.

Reynolds, G.H., 2002. Forward to the future nanotechnology and regulatory policy. Pac. Res. Inst. 24, 1–23.

Ringwood, A.H., McCarthy, M., Bates, T.C., Carroll, D.L., 2010. The effects of silver nanoparticles on oyster embryos. Mar. Environ. Res. 69 (1), 549–551.

Rouhani, M., Samih, M.A., Kalantari, S., 2012. Insecticied effect of silver and zinc nanoparticles against *Aphis nerii* Boyer of fonscolombe (Hemiptera: Aphididae). Chil. J. Agric. Res. 72, 590–594.

Roy, A., Bulut, O., Some, S., Mandal, A.K., Yilmaz, M.D., 2019. Green synthesis of silver nanoparticles: biomolecule-nanoparticle organizations targeting antimicrobial activity. RSC Adv. 9, 2673.

Sahayaraj, K., Madasamy, M., Anbu Radhika, S., 2016. Insecticidal activity of bio-silver and gold nanoparticles against *Pericallia ricini* fab. (Lepidaptera: Archidae). J. Biopestic. 9 (1), 63–72.

Salari, S., Esmaeilzadeh Bahabadi, S., Samzadeh-Kermani, A., Yosefzaei, F., 2019. In-vitro evaluation of antioxidant and antibacterial potential of green synthesized silver nanoparticles using *Prosopis farcta* fruit extract. Iran. J. Pharm. Sci. 18 (1), 430–455.

Scarano, G., Morelli, E., 2002. Characterization of cadmium- and lead-phytochelatin complexes formed in a marine microalga in response to metal exposure. Biometals 15 (2), 145–151.

Scarano, G., Morelli, E., 2003. Properties of phytochelatin-coated CdS nanocrystallites formed in a marine phytoplanktonic alga (*Phaeodactylum tricornutum*, Bohlin) in response to Cd. Plant Sci. 165, 803–810.

Selivanov, V.N., Zorin, E.V., 2001. Sustained action of ultrafine metal powders on seeds of grain crops. Perspekt. Mater. 4, 66–69.

Singh, T., Jyoti, K., Patnaik, A., Singh, A., Chauhan, R., Chandel, S.S., 2017. Biosynthesis, characterization and antibacterial activity of silver nanoparticles using an endophytic fungal supernatant of *Raphanus sativus*. J. Genet. Eng. Biotechnol. 15, 31–39.

Sintubin, L., Verstraete, W., Boon, N., 2012. Biologically produced nanosilver: current state and future perspectives. Biotechnol. Bioeng. 109, 2422–2436.

Sohrab, D., Tehranifara, A., Davarynejada, G., Abadía, J., Khorasani, R., 2016. Effects of foliar applications of zinc and boron nano-fertilizers on pomegranate (*Punica granatum* cv. Ardestani) fruit yield and quality. Sci. Hortic. 210, 57–64.

Soliman, B.S.M., Abbassy, M.A., Abdel-Rasoul, M.A., Nassar, A.M.K., 2017. Efficacy of silver nanoparticles of extractives of *Artemisia judiaca* against root knot nematode. J. Environ. Stud. Res. 7 (2), 1–13.

Sondi, I., Salopek-Sondi, B., 2004. Silver nanoparticles as antimicrobial agent: a case study on *E. coli* as a model for gram-negative bacteria. J. Colloid Interface Sci. 275 (1), 177–182.

Subramanian, K.S., Manikandan, A., Thirunavukkarasu, M., Sharmila Rahale, C., 2015. Nano-fertilizers for balanced crop nutrition. In: Rai, M., Ribeiro, C., Mattoso, L., Duran, N. (Eds.), Nanotechnologies in Food and Agriculture. Springer International Publishing, Switzerland, pp. 69–80.

Taha, E.H., 2016. Nematicidal effects of silver nanoparticles on root-knot nematodes (*Meloidogyne incognita*) in laboratory and screenhouse. Mansoura J. Plant Protect. Pathol. 7, 333–337.

Tanase, C., Berta, L., Coman, N.A., Roşca, I., Man, A., Toma, F., Mocan, A., Nicolescu, A., Jakab-Farkas, L., Biró, D., Mare, A., 2019. Antibacterial and antioxidant potential of silver nanoparticles biosynthesized using the spruce bark extract. Nanomaterials 9 (11), 1541.

Tang, F., Li, L., Chen, D., 2012. Mesoporous silica nanoparticles: synthesis, biocompatibility and drug delivery. Adv. Mater. 24, 1504–1534.

Trenkel, M.E., 1997. Controlled-Release and Stabilized Fertilizers in Agriculture. International Fertilizer Industry Association, Paris, pp. 234–318.

Uysal, G., Göze Özdemir, F.G., 2020. Bacteria in the management of plant parasitic nematodes. Turk. J. Sci. Rev. 13 (1), 53–72.

Vivekanandhan, S., Christensen, L., Misra, M., Mohanty, A.K., 2012. Green process for impregnation of silver nanoparticles into microcrystalline cellulose and thin antimicrobial bionanocomposite films. J. Biomater. Nanobiotechnol. 3, 371–376.

Wendt, K.R., Swart, A., Vrain, T.C., Webste, J.M., 1995. *Ditylenchus africanus* sp. n. from South Mrica; a morphological and molecular characterization. Fundam. Appl. Nemawl. 18 (3), 241–250.

Whitehead, A.G., 1998. Semi-endoparasitic nematodes of roots (Helicotylenchus, Rotylenchulus and Tylenchulus). In: Plant Nematode Control. CAB International, Wallingford, UK, pp. 90–137.

Yadav, B., Veluthambi, K., Subramaniam, K., 2006. Host-generated double stranded RNA induces RNAi in plant-parasitic nematodes and protects the host from infection. Mol. Biochem. Parasitol. 148, 219–222.

Youssef, M.M.A., El-Nagdi, W.M.A., Eissa, M.F.M., 2014. Population density of root knot nematode *Meloidogyne incognita* infecting date palm under stress of aqueous extracts of some botanicals and a commercial bacterial by product. Middle East J. Appl. Sci. 4 (4), 802–805.

Zhang, Z., Shen, W., Xue, J., Liu, Y., Liu, Y., Yan, P., Liu, J., Tang, J., 2018. Recent advances in synthetic methods and applications of silver nanostructures. Nanoscale Res. Lett. 13 (1), 54.

Zheng, J., Peng, D., Chen, L., Liu, H., Chen, F., Xu, M., Ju, S., Ruan, L., Sun, M., 2016. The *Ditylenchus destructor* genome provides new insights into the evolution of plant parasitic nematodes. Proc. Biol. Sci. 283 (1835), 20160942.

Silver nanoparticles applications and ecotoxicology for controlling mycotoxins

CHAPTER 23

Velaphi C. Thipe[a], Caroline S.A. Lima[a], Kamila M. Nogueira[a], Jorge G.S. Batista[a], Aryel H. Ferreira[a], Kattesh V. Katti[b], and Ademar B. Lugão[a]

[a]Laboratório de Ecotoxicologia, Centro de Química e Meio Ambiente, Instituto de Pesquisas Energéticas e Nucleares (IPEN), Comissão Nacional de Energia Nuclear, IPEN/CNEN-SP, São Paulo, Brazil
[b]Institute of Green Nanotechnology, Department of Radiology, University of Missouri, Columbia, Columbia, MO, United States

23.1 Introduction

The agricultural and food systems sector suffers from a vast array of contaminations resulting in losses in agricultural productivity, which accounts to approx. 1 billion metric tons of foods (i.e., crops, cereals, and food derivatives/products) destined for human and animal consumption amid global food demands and evolutional climate change patterns (Yin et al., 2018). Of these contaminations, mycotoxins account for 25% and the most frequent contaminant according to the Food and Agriculture Organization (FAO). Food insecurity is still associated with high mortality rates, which are exacerbated by mycotoxins, especially in developing countries. Increased global food demands are inevitable according to projected population growth of 9.7 billion by 2050 and 10.9 billion by 2100 (United Nations, 2019). Fungi ubiquitously occur in different environmental conditions and produce toxigenic mycotoxins that may have detrimental toxic effects to both human and animal health. Fungi can invade crops with subsequent production of mycotoxins at different stages of the food chain, during preharvesting, harvesting, processing, transport, and storage (Abd-Elsalam et al., 2017; Thipe et al., 2018). The biological activities of mycotoxins are specifically to induce toxicity, phytotoxicity, and antibiosis in food and feed, yielding a cascade of ecotoxicological ramifications to human and animal health (El-Waseif et al., 2019).

The agricultural sector is manifested by problems caused by aflatoxins (AFs), ochratoxins (OT), fumonisins (FUM), trichothecenes, and zearalenone (ZEN), with aflatoxin B_1 (AFB$_1$) known to be the most dangerous mycotoxin to human health because of its high hepatocarcinogenicity (Thipe et al., 2018). Exposure through acute or chronic to mycotoxins via consumption of contaminated foods/feeds can result in mycotoxicosis, provoking problems such as cancers and other generally irreversible effects. Over time, chronic exposure to mycotoxin can result in the deterioration of target organs such as the liver or lungs in animals and human beings (Cinar and Onbasi, 2019). Despite scientific advances, mycotoxins are endocrine and economic disruptors as they impose a tremendous burden on the healthcare system and economic losses from fungal infections with subsequent mycotoxins contamination. The staggering incidences of mycotoxins prompt consistent mandatory measures to monitor and control mycotoxins in the food segment. Prevention of mycotoxin manifestation can be achieved by mitigating the fungal growth using chemical compounds such as fungicides to protect crops. Nevertheless, these substances are not entirely efficient and present secondary toxicity due to 90% of fungicides is susceptible to runoff, thus leaching to the water supply that can increase costs and greater damage to the environment (Zubrod et al., 2019). Moreover, pesticides also affect human/animal health and plants, and this is due to their high toxicity and non-biodegradable properties (Abd-Elsalam et al., 2017; Abd-Elsalam and Alghuthaymi, 2015).

As we foster innovative solutions against mycotoxin, new technologies such as nanotechnology could be the solution to eliminating mycotoxins with no to minimal ecotoxicology. To circumvent the abovementioned challenges, nanotechnology and material sciences have received considerable attention due to the properties and feasibility of engineered nanomaterials in various applications (e.g., antiseptics and disinfectants, cosmetics, electronics, textiles, diagnostic imaging, and targeted drug delivery). This provides innovative, creative, and safe solutions for the detection, control, and prevention of mycotoxins. The high surface-to-volume ratio exhibited by nanomaterials exposes surface atoms, thus allowing multifunctionalization with active compounds at low concentration for increased efficacy (Yin et al., 2018). The minimal utilization of nanoparticles to obtain good efficiency in plant protection permits low toxic effects on the environment and human/animal health. Among the nanomaterials developed, metallic nanoparticles are known as the most promising nanomaterials because of their antimicrobial characteristics and ability to target plant pathogens, sparing good microorganisms and insects within the soil and aquatic microbiota (Abd-Elsalam et al., 2017).

Nanoparticles can originate from anthropogenic or natural sources or intentionally produce for different purposes. The popularity of nanoparticles, especially silver nanoparticles (AgNPs), which are one of the most prevalent nanoparticles, stems from their antimicrobial properties (controlling diverse and multidrug-resistant microorganisms), which makes them ideal substitutes for fungicides in agriculture as a new class of agrochemicals, antimicrobials, disinfectants, and food protectant (i.e., food packaging) acting over 650 different microorganisms, including mycotoxigenic fungi (Siddiqi and Husen, 2016; Tortella et al., 2020). Antimicrobial additives in various products consisting of Ag complexes, mainly AgNPs due to their high reactivity.

AgNPs have been suggested to be very effective antifungal agents against mycotoxigenic fungi, thereby ensuring agricultural sustainability and food safety toward achieving the food security in the 2030 Agenda for a sustainable development program for ending poverty and hunger to responding to climate change and sustaining natural resources, food and agriculture (FAO, 2020). Nanoparticles have offered enormous benefits in a myriad of applications due to their physicochemical properties; however, we must not be naïve in fully understanding their ecotoxicology profile before using them in agriculture and other applications (Yan and Chen, 2019). The surge in the application of AgNPs as antifungal agents in the agricultural sector has been under great concerns due to their safety and ecotoxicology and is intensified by the limited knowledge of AgNPs lifecycle analysis studies. It is paramount to fully understand the interaction between AgNPs and eukaryotic cells (e.g., animals, plants, fungus, and human) to best engineer materials effective at low doses with minimal to no toxicity on environmental and human health for ecological and ecotoxicological insight.

Ideally, AgNPs at much lower doses than those typically used in toxicological and ecotoxicological studies for application as mycotoxins inhibitors is important for food/feed and human/animal safety. This is advantageous because the low dose substantially reduces the concern of human toxicity and provides inhibition of the mycotoxins and virulence factors (Jesmin and Chanda, 2020). It is important to acknowledge that the efficacy and safety of nanoparticles including AgNPs are dependent on their size and capping/stabilizing ligand. The AgNPs-bio interaction dictates the efficacy and safety of AgNPs in agriculture as mycotoxins inhibitors. While the AgNPs interaction with fungal cell via endocytosis (clathrin-dependent and receptor-mediated) or cell surface lipid raft-associated domain (caveolae) often results in cellular and membrane damage facilitated by the dissolution of the AgNPs into Ag^+ ions producing reactive oxygen species (ROS) causing membrane rupture and leakage (Vergallo et al., 2020). This is attributed to AgNPs exhibiting several modes of action that guarantees their high efficiency as biocides. These nanoparticles can inhibit fungal growth and also interfere with mycotoxins biosynthetic pathways (Lara et al., 2015; Pietrzak et al., 2015; Thipe et al., 2020). When acting as a mycotoxin inhibitor in the toxin-producing environment (endosome), AgNPs do not necessarily inhibit the growth of the fungus. The fungicidal activity or mycotoxin inhibitory effect of AgNPs against fungi is dependent on the several factors (e.g., the concentration of AgNPs, release of Ag^+ ions, surface chemistry, size, coating ligand, shape, and polydispersity) (Jesmin and Chanda, 2020; Jogee et al., 2017). In this context, there are concerns with them leaching/migrating into the environment and entering the food ecosystem through water/soil bodies (Montes de Oca-Vásquez et al., 2020). Yan and Chen (2019) further elaborated on the importance of phytotoxicity of AgNPs, as plants are the primary constituents in the ecosystem that serves as important sources of food for both animals and humans.

Moreover, the effectiveness of AgNPs is dependent on their physicochemical properties such as their stability. This was corroborated by Mitra et al. (2019), where they demonstrated that PVP-coated AgNPs having high stability than citrated-coated AgNPs, which are unstable, exhibited inhibition of aflatoxin biosynthesis over a broader concentration range than citrated-coated AgNPs even at higher concentrations. They showed that nanoparticles with 15 nm are more easily captured by the endosome than the bigger ones (20 and 30 nm) and that the concentrations of inhibitors are directly linked to the size of the nanoparticles—the smaller the size of the AgNPs, the lower the minimum inhibitory concentration (MIC). The material coating also influences AgNPs action, and those encapsulated with PVP (36 nm) showed a higher concentration range for inhibition (from 40 to 75 ppm), while AgNPs coated with citrate presented punctual concentrations for inhibitory action: 25 ppm for 15 nm nanoparticles, 40–60 ppm for 20 nm nanoparticles, and 50 ppm for 30 nm nanoparticles (Mitra et al., 2019).

23.2 Silver nanoparticles in agriculture as antifungicides
23.2.1 Antifungal activity of commercial AgNPs

The global application of AgNPs has increased during the past few years; this is fueled by the desirable properties that AgNPs exhibit. Approximately 2000 products with nanomaterials are produced daily, with about 1/5 (435) of nanoproducts are Ag-based (Pulit-Prociak and Banach, 2016). Applications of AgNP or AgNPs formulations have seen a boom within the agricultural sector as alternative antimicrobial agents and packaging material constituents for protection against fungal growth, mycotoxins, and increasing shelf-life of various products (Liu et al., 2020). This warrants the application of AgNPs in agriculture as an alternative to toxic chemical fungicides that result in the development of resistance. New generation packing materials made of nanocomposites protect food by their antimicrobial properties. Polymers containing metal nanoparticles, including AgNPs, are used to produce these packaging materials that are also capable of reducing mycotoxin contamination by interacting with the outer cell membrane of the fungi and cause changes in their permeability and functionality (Thipe et al., 2018). Tarus et al. (2019) developed electrospun cellulose acetate and poly(vinyl chloride) nanofiber mats with AgNPs aiming to produce antifungal packaging material. Their results demonstrated that the presence of AgNPs decreased the development of mold on fruit surfaces.

Work by El-Waseif et al. (2019) evaluated the commercial prospects of AgNPs synthesized using *Streptomyces clavuligerus* against mycotoxigenic *F. oxysporum*, which produces fumonisins, zearalenone, trichothecenes, and nivalenol. Their investigation was on 4-week old tomato seedlings with pot experiment at an experimental farm station (using different methods of treatment such as foliar shoot, root immersion (RI), and foliar shoot+root immersion) for evaluating the effects of AgNPs on the percentage of disease incidence and protection of tomato plants infected with *F. oxysporum*. Results demonstrated that AgNPs have significant antifungal activity against *F. oxysporum*, in addition to the reduced percent disease index (PDI) at 8.00% and increased protection at 90.40% against infection compared to control infected tomato plants with PDI at 83.33% and no protection. Treatment with AgNPs at 60 ppm revealed optimum protection with increased phenolic level, proline amount as inhibitors of preinfection, thereby providing tomatoes with a specific level of essential resistance against *F. oxysporum*.

Villamizar-Gallardo et al. (2016) evaluated the fungicidal activity of commercial Biopure-AgNPs from NanoComposix against *A. flavus* and *F. solani* on cocoa (*Theobroma cacao*) crops. Their results revealed that Biopure-AgNPs did not significantly affect the growth of *A. flavus* and *F. solani* even at 100 ppm in liquid and solid synthetic culture media. However, in plant tissues (especially in the cortex), AgNPs inhibited the fungal growth of *A. flavus* at 80 ppm and it was noted that once fungi penetrated through the cocoa pods, their growth is unavoidable, and the effect of AgNP was reduced, while in *F. solani*, AgNPs only induced some texture and pigmentation changes. This demonstrated that AgNPs used in this study are more effective in plant tissues than in culture media, providing an opportunity for their utilization in-field application to control the growth of *A. flavus* and the production of aflatoxins.

Kotzybik et al. (2016) also investigated the use of commercial AgNPs (Citrate BioPure™ Silver) with different sizes (5, 50, and 200 nm) at varying concentrations (1, 5, 25, 50 and 100 ppm) from NanoComposix, in addition to MesoSilver® with size ≤ 0.65 nm at different concentrations (1, 2, 4, 6, and 8 ppm) from Purest Colloids. These AgNPs were tested for their influence on *P. verrucosum* growth and mycotoxin biosynthesis. Before studies were conducted on *P. verrucosum*, the authors evaluated the stability of AgNPs in the Malt Extract Agar (MEA) medium (with Tween/NaCl/Agar) used in the experiment. They noticed that Citrate BioPure™ Silver were not stable in MEA medium while MesoSilver® exhibited high stability over 7 days. Moreover, their antifungal activity studies revealed that smaller sized AgNPs (0.65 and 5 nm) application preinoculation of spore germination strongly inhibited germination and fungal growth of *P. verrucosum* with a subsequent decrease in mycotoxins biosynthesis at 2 and 5 ppm, respectively. The larger sized AgNPs were not effective, and this can be attributed to that smaller AgNPs have a high rate of Ag^+ dissolution, which is also dependent some factors (size, shape, concentration, capping agent, and the colloidal state of AgNPs) due to their high surface area to volume ratio, increased bioavailability, enhanced distribution, and toxicity of Ag compared with larger AgNPs (Ferdous and Nemmar, 2020). These AgNPs accumulate the fungal cell mycelial filaments, the Ag^+-ions/particles interfere with electron-transport-systems with subsequent production of ROS and damage to proteins associated with ATP production, lipids, and nucleic acids, ultimately modulating the intracellular signaling pathways resulting in apoptosis (Ferdous and Nemmar, 2020).

Kwak et al. (2012) evaluated the use of commercial product Pyto-patch®, which contains ≤ 5 nm AgNPs antifungal activity against *Botrytis cinerea*, *Colletotrichum gloeosporioides*, and *Sclerotinia sclerotiorum* for controlling anthracnose. Their results demonstrated that pepper fruits sprayed with 10 ppm Pyto-patch® treatment reported 75.8% survival, whereas in the untreated pepper fruits, no survival was observed. The results obtained by Kotzybik et al. (2016) corroborate those reported by Jo et al. (2009) where they used different forms of Ag [$AgNO_3$, AgCl, AgNPs (20–30 nm), where AgNPs are denoted as Ag(p), and electrochemical Ag are denoted as Ag(e)] to evaluate their effectiveness against plant-pathogenic fungi, *Bipolaris sorokiniana* and *Magnaporthe grisea*, and their disease severity in plants for phytotoxicity. Their results revealed that plants sprayed with AgNO$_3$, Ag(e), and Ag(p) before inoculation with spores inhibited colony formation. The half-maximal effective concentration (EC$_{50}$) values for *M. grisea* and *B. sorokiniana* ranged from 0.8 to 1.0 ppm AgNO$_3$, 3.9 to 4.7 ppm Ag(p), and 5.6 to 6.8 ppm Ag(e) at 1 to 6 h after treatment.

The utilization of AgNPs incorporated into packaging matrices to serve as antimicrobial agents to provide enhanced material strength, and inactivation of fungi and mycotoxins is desirable. Kim et al. (2012) compared three different types of AgNPs provided by BioPlus Co. Ltd. (Korea) applied for antifungal treatment in plant pathogens. The nanoparticles were WA-CV-WA13B, WA-AT-WB13R, and WA-PR-WB13R, and they were tested on 18 plant-pathogenic fungi of agricultural interest. The fungi were grown in vitro on three types of agar PDA (potato dextrose), MEA (malt extract), and CMA (corn flour). AgNPs size varied from 7 to 25 nm and concentrations from 10 to 100 ppm, the most significant inhibition was in PDA with AgNPs at 100 ppm. The results of this work indicated that AgNPs have antifungal properties against these plant pathogens at various levels depending on the type of AgNPs, fungi, culture medium, and concentration.

23.2.2 AgNPs commercial applications in controlling mycotoxins

The European Union has endorsed the use of commercial products fabrication and/or formulated with AgNPs for the mitigation of mycotoxins (Kotzybik et al., 2016). Nanotechnology has been beneficial at protecting crops from contaminates and extending food shelf life with immense security through in-field stage (i.e., enhanced agricultural crop cultivation and protection practices) and cross-stage (i.e., food packaging material and nanoenabled sensors for fungal and mycotoxin detection) applications to protect during all subsequent stages (i.e., harvest, storage, processing, and transportation) to the endpoint—consumer (Yin et al., 2018). The current use of AgNPs in a variety of products for increased product shelf-life and protecting from fungal and mycotoxins contamination has spiked due to catalytic and antimicrobial properties exhibited by AgNP through multimodal activities (Ferdous and Nemmar, 2020; Jo et al., 2009). This provides added functionality and enhanced efficiency for advanced agricultural sustainability.

The application of AgNPs supplements such as spray-impregnation for foods, which are normally washed, shelled, or peeled (e.g., melons, apples, or citrus fruits) in the field or applied via dehumidifier during the storage and transport, would be ideal to provide a protective coating against fungal manifestation and mycotoxins contamination (Kotzybik et al., 2016). AgNPs are already in products of daily use; this includes Ag stabilized hydrogen peroxide for soil fumigation and sterilization, as fungicides in agriculture and agricultural spray adjuvants, and in food context, especially in packaging material and container, as shown in Table 23.1 (Bumbudsanpharoke and Ko, 2015).

The use of AgNPs and/or incorporation polymeric matrices in packaging material provides stronger packaging barriers against fungal growth and mycotoxins contamination. According to a new Polaris Market Research report released on August 17, 2020, for "Nano-Enabled Packaging Market Share, Size, Trends, and Industry Analysis Report, By

Table 23.1 Commercial AgNP nanofungicides.

Company	AgNP formulation product	Application	Accreditation/certification
Chemtex Specialty Ltd., Vedic Orgo LLP, Anand Agro Care, Future India Chemicals, Acuro Organics Limited, India	Alstasan Silvox: Ag stabilized H_2O_2 Nano Shield fungicide Siddhi Nano: Ag + H_2O_2 Acurosil Nano+	Agriculture: – Open field cultivation – Protective cultivation – Animal husbandry – Aquaculture – Floriculture – Soil disinfectant Food & Beverages[a]	Certified by FDA, approved by National Toxicology Center up to 25 ppm in portable water. Approved in several countries as regarded as safe according to the EEC, WHO, and US EPA
Sistema Plastics, New Zealand	Fresher Longer™: AgNPs (25 nm) in polypropylene (PP) matrix	Food packaging: – Adhesives – Antifungal ingredients Food contact materials: – Storage containers	Phthalate and BPA free, products are not recommended for microwave
Huangshan Yongrui Biotechnology Co., Ltd., China	5–10 nm AgNP solution (20 ppm)	Household appliance, clothes, tableware & cookware, hospital appliance	Certified ISO9001, Ce, FDA, MSDS, Coa, Product Safety Test R
Nano Koloid, Poland	Nanosilver	Agriculture: – Animal husbandry – Food processing, storage and transport – Plant care and protection – Antifungal activity against *Aspergillus* spp., *Candida* spp., and *Saccharomyces*	Not mentioned
Libra Biotech, India	Nanosil Antifungal: AgNPs and peroxy acid	Agriculture via Foliar Application through a Knap Sack Sprayer and soil drenching, product is recommended for seed, soil, and plant disinfection of fungal and other related contaminations	Dermal LD_{50} > 2000 mg/kg Oral LD_{50} = 2896 mg/kg Ag rated as nontoxic according to EPA/FDA. Drinking water safe level: 180 ppb/day USA, Canada, Russia, Japan 0.05 ppm European Union 0.08 ppm Switzerland 0.10 ppm

[a] Packaging materials; LD50, mean lethal dose of AgNPs/AgNPs formulation that is lethal for 50% of tested animal model used within the dose group.

Type (Active, Intelligent, Others); By End-User (Food and Beverages, Pharmaceutical, Personal Care, Others); By Regions: Segment Forecast, 2018–2026." The global nanoenabled packaging market for food and beverages is estimated to grow at a compound annual growth rate (CAGR) of 12.7% to reach about $89.0 billion by 2026 (Polaris Market Research, 2020). Polyolefins are among the most widely used polymeric material for food packaging material. A polypropylene (PP) film, which is a type of polyolefin, is widely used due to its chemical inertness, brilliance, low specific weight, and transparency. In addition to PP, various grades of polyethylene and polymer blends are used as support matrices for AgNPs to create nanocomposites as *active* packaging materials for controlling food spoilage by fungi and mycotoxins.

23.2.3 Mycotoxin detection using AgNPs

AgNPs may also can be used to detect mycotoxins, and as previously mentioned, there is some difficulty and limitations in detecting some mycotoxins once they can be masked and possess some nonextractable compounds. Utilization of AgNPs for increasing the surface-enhanced Raman scattering can improve the signal amplification (Thipe et al., 2018). A majority of mycotoxins are present in trace levels (parts per billion or nanograms) and/or coexist (e.g., more than one type of mycotoxin can exist in a single food matrix), which makes their detection challenging. Biosensors can be applied in several ways in the food industry, from storage to packaging, guaranteeing safety of consumer products with the advantage of simplicity through point-of-care testing (POCT) and low cost. For mycotoxins identification, different methods may be used such as optical (surface plasmon resonance and fluorescence), piezoelectric (quartz crystal microbalance), and electrochemical (impedimetric, potentiometric, and amperometric) spectroscopies. Normally, nanomaterials are applied in these biosensors for signal amplification, improving their sensitivity. Moreover, the strategy of uniting nanoparticles and fluorescence techniques is especially interesting for mycotoxin detection (Adeyeye, 2020; Chauhan et al., 2016; Ng et al., 2016; Santana Oliveira et al., 2019). Khan et al. (2018) developed a functionalized silver nanoclusters assembled on a molybdenum disulfide thin layer for T-2 toxin detection. Anfossi et al. (2018) used gold and Ag nanoparticles to quench Quantum Dots emission by Foster resonance energy transfer and inner filter effect. This caused the decrease of the background noise caused by the Quantum Dots fluorescence used on an analytical platform such as lateral flow immunoassay with fluorescence. Biosensor development that uses AgNPs for mycotoxin detection is still at its infancy but postulated to grow as the knowledge about their ecotoxicological profile increases.

23.2.4 Green synthetic AgNP against mycotoxigenic fungi

The green production of AgNPs may also present advantages to the environment because of the absence of toxic waste byproducts. This is possible because AgNPs may be synthesized by green methods using microorganisms or plant extracts through green nanotechnology (Siddiqi and Husen, 2016). Antifungal action of Ag has been previously demonstrated against several strains such as *F. solani*, *Alternaria alternata*, *A. flavus*, and *A. ochraceus* (Abd-Elsalam et al., 2017). Jogee et al. (2017) synthesized AgNPs using 10 different plants to inhibit fungi that contaminate peanut, which includes *Aspergillus* spp., *Penicillium* spp., and *Macrophomina phaseolina*, the synthesis of AgNPs generated polydisperse particles from spherical to irregular in the range of 31–40 nm. The authors found that AgNPs mediated using *Cymbopogon citratus* leaf showed greater antifungal potential and their MIC was 2 ppm. In this case, the fungal potential of AgNPs was associated with the type of plant used in the synthesis.

Xue et al. (2016) performed optimization studies of AgNPs biosynthesis using a strain of *Arthroderma fulvum* isolated from soil. The produced bio-AgNPs were tested against *Candida* spp., *Aspergillus* spp., and *Fusarium* spp., in all species; results revealed considerable antifungal activity. The use of *A. fulvum* provides a green-sustainable, robust, and large-scalability of AgNPs production with antifungal capabilities. A study by Ashraf et al. (2020) used *Melia azedarach* leaf extract to produce AgNPs (MLE-AgNPs) against *F. oxysporum*, which produces trichothecenes mycotoxins (T-2 toxin, HT-2 toxin, diacetoxyscirpenol, and 3′-OH T-2 (TC-1)). Their results revealed that the AgNPs repressed fungal mycelia growth by 98% through the cell wall and cellular membrane damage attributed by elevated levels of ROS production and enhanced tomato seedlings without affecting plants grown, as shown in Fig. 23.1.

Abdel-Hafez et al. (2016) investigated the use of an endophytic nonpathogenic strain of *Alternaria solani* F10 (KT72914) as a reducing agent for the synthesis of AgNPs due to the vast repertoire of constituents and in turn evaluated the antifungal activity of the produced AgNPs against toxigenic *A. solani*. Their results demonstrated that AgNPs at a lower concentration (10 ppm) exhibited a higher antifungal activity when compared to standard treatment with Ridomil Gold® plus (2002.28 ppm). This justifies the use of AgNPs as antifungal agents at low concentration can become a safe, effective, and economical strategy in agriculture for controlling fungi and mycotoxin contamination. Furthermore, this would limit the onset of resistance due to the multimodality of AgNP's mechanism of action via disrupting cell walls and destroying the membrane lipid bilayer, as shown in Fig. 23.2. Moreover, AgNPs bind to nucleic acid and proteins containing sulfur to cause direct damage and inhibition of DNA replication through the release of Ag^+ ions that generates ROS and oxidative stress (Guilger-Casagrande and de Lima, 2019).

Elamawi et al. (2018) investigated *Trichoderma longibrachiatum* synthesized AgNPs and their effects against phytopathological fungi (*F. verticillioides*, *F. moniliforme*, *Helminthosporium oryzae*, *P. brevicompactum*, and *Pyricularia grisea*). Their results revealed that the produced AgNPs as an antifungal agent significantly reduced the growth of all the tested fungi by 90%. Also, Khalil et al. (2019) synthesized AgNPs through *Fusarium chlamydosporum* NG30 and *Penicillium chrysogenum* NG85 fungi to inhibit *A. flavus* and *A. ochraceus* with subsequent inhibition of aflatoxin and ochratoxin biosynthesis, respectively. The synthesis generated spherical particles with size varying between 6 and 26 nm for *F. chlamydosporum*

FIG. 23.1

Culture plates after 7 days of postincubation at 28°C and effect of treatment on the development of tomato plants after 45 days under greenhouse conditions infected with *F. oxysporum*: (A) control, (B) Nativo® fungicide treatment, and (C–K) antifungal activity of MLE-AgNPs at different concentrations (5, 10, 20, 60, 80, 100, 120, and 140 ppm).

Modified and reprinted by permission from Ashraf, H., Anjum, T., Riaz, S., Naseem, S., 2020. Microwave-assisted green synthesis and characterization of silver nanoparticles using Melia azedarach for the management of fusarium wilt in tomato. Front. Microbiol. 11, 1–22 distributed under the terms of the Creative Commons Attribution License (CC BY).

and 9 and 17.5 nm for *P. chrysogenum*. The MIC of *F. chlamydosporum* produced AgNPs was 45 ppm for *A. flavus* and MIC of *P. chrysogenum* produced AgNP was 48 ppm. On the other hand, for the total inhibition of aflatoxin production, the concentrations were 5.6 and 6.1 ppm for *F. chlamydosporum* AgNPs and *P. chrysogenum* AgNPs, respectively; even at these concentrations, no cytotoxicity against normal human melanocyte cells (HFB4) was observed. This study demonstrated the possibility to determine the concentration of AgNPs as fungal or mycotoxin inhibitors. Le et al. (2020) utilized phytoconstituents in leaf extracts of *Achyranthes aspera* and *Scoparia dulcis* to produce AA-AgNPs and SD-AgNPs, respectively. These AgNPs were tested for their fungicidal activity against *A. niger*, *A. flavus*, and *F. oxysporum*; they exhibited high antifungal activity against all tested fungi, with *F. oxysporum*

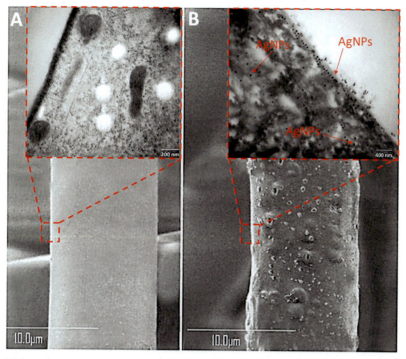

FIG. 23.2

Field emission scanning electron micrographs and corresponding high-resolution electron microscopy image insert of pathogenic *Alternaria solani* F11 (KT721909) hyphae: (A) Fungal hyphae before treatment with AgNPs which showing regular smooth surface and well-distinguished cell components and (B) fungal hyphae after treatment with AgNPs showing pores and cavities formed on the surface and AgNPs accumulation in the cytoplasm and membrane.

Modified and reprinted by permission from Abdel-Hafez, S.I.I., Nafady, N.A., Abdel-Rahim, I.R., Shaltout, A.M., Daròs, J.A., Mohamed, M.A., 2016. Assessment of protein silver nanoparticles toxicity against pathogenic Alternaria solani. 3 Biotech 6 (2), 199 distributed under the terms of the Creative Commons Attribution License (CC BY).

being the most susceptible. In a similar study, Ibrahim et al. (2020) demonstrated biosynthesized AgNPs using endophytic bacterium *Pseudomonas poae* strain CO isolated from *Allium sativum* against *F. graminearum*. The AgNPs inhibited mycelium growth, spore germination, and mycotoxin production.

Toxicity profile of AgNPs on human health was investigated by Panáček et al. (2009); this study compared the toxicity of ionic Ag and AgNPs used as a fungicide in the inhibition of *Candida* spp. Their results showed that the AgNPs were cytotoxic for human fibroblasts in concentrations higher than 30 ppm, while ionic Ag showed a lethal concentration (LC_{100}) = 1 ppm. This corroborates the results previously reported in the literature that at low concentrations, AgNPs do not present health risks. Yassin et al. (2017) evaluated the use of biosynthetic AgNPs using *P. citrinum* against aflatoxigenic *A. flavus* var. *columnaris*. Fungal growth decreased as a function of AgNPs concentration; the median effective dose (ED_{50}) and the effective dose required for 95% of the exposed population (ED_{95}) was 224.5 and 4001.8 ppm, respectively. The use of biogenic AgNPs production using fungi can be a feasible route for controlling fungal growth and mycotoxin production (Guilger-Casagrande and de Lima, 2019).

23.3 Toxicology profile of AgNPs
23.3.1 Ecotoxicology and phytotoxicology of AgNPs

Albeit, AgNPs applications are advantageous in controlling fungal growth and mycotoxin contaminations in the agricultural sector. Their potential toxic effects cannot be overlooked; thus, rigorous nanotoxicological studies are consistently required, especially AgNPs products deemed for agricultural application to mandate their overall safety to ecology (Ferdous and Nemmar, 2020). The growing utilization of AgNPs and Ag formulations in food packaging and antifungal agents already in the market has prompted concerns on the integrity, possible migration, and toxicity of AgNPs (Bumbudsanpharoke et al., 2018; Bumbudsanpharoke and Ko, 2015; Iavicoli et al., 2017; Tortella et al., 2020). Commercial NanoAg low-density polyethylene (LDPE) bags from Sunriver Industrial Co., Ltd., was evaluated in migration studies according to the Chinese standard GB/T 5009.60-2003 using ultrapure water, 4% acetic acid, 95% ethanol, and hexane as simulating solutions at different temperatures for 15 days. The fresh NanoAg bag contained 100 μg Ag per 1 g LDPE, and atomic absorption spectroscopy (AAS) analysis revealed that the amount of Ag migration increased as a function of time and temperature. This was attributed to the AgNPs on the surface layer leaching out and diffusion of simulants and change the overall crystalline state with simultaneous oxidation of AgNPs.

To extensively investigate the ecotoxicology and phytotoxicology of AgNPs concerning the vast majority of increased fabrications of these nanoparticles in the agriculture sector for controlling microbial manifestation such as fungal contamination with subsequent production of mycotoxins. It is of paramount to carry out such ecotoxicological studies in multispecies model organisms (e.g., live terrestrial or aquatic organisms: vertebrates, invertebrates, plants, algae, and microorganisms as biological markers) and also include phytotoxicity investigated, especially for in-field application of AgNPs exposure to crops (Tortella et al., 2020; Yin et al., 2018). Rodrigues et al. (2020) studied the effects of silver nanomaterials (AgNM) against standard ecotoxicological model organism—*Enchytraeus crypticus*, via water and soil exposure to four different AgNPs [polyvinylpyrrolidone (PVP) coated (PVP-AgNM), noncoated (NC-AgNM), JRC reference Ag NM300K, and silver nitrate ($AgNO_3$)]. They tested all life stages (cocoons, juveniles, and adults) evaluating five endpoints, including hatching success, survival, reproduction, avoidance, and gene expression of the animals upon the exposure to the four tested materials. Acute and chronic exposure from 1 to 21 days was investigated, and the main impact observed was that acute exposure via water to cocoons caused longer-term effects on survival and reproduction. On the other hand, chronic exposure to cocoons from 11 to 17 days, hatch delay, and impairment were observed. Also, juveniles were more sensitive than adults concerning survival. Generally, the toxicity profile rankings of the tested materials were as follows: $AgNO_3 \geq Ag\ NM300K \gg NC\text{-}AgNM \geq PVP\text{-}AgNM$. Table 23.2 shows some of the AgNPs formulations investigated for their antifungal and ecotoxicological imprint against fungi and their mycotoxins/phytotoxins.

Nanoencapsulation of fungicides presents less cytotoxicity and genotoxicity to plants than the conventional pristine formulation, suggestive of the benefit of nanomaterials in providing crop protection with no crop damage and improved crop production (Nuruzzaman et al., 2016; Shang et al., 2019). The stability of the AgNPs is very important as it also contributes to the overall efficiency and toxicity of the AgNPs. Most importantly, a study by Jo et al. (2009) revealed that the treatment of *Oryza sativa* L. leaves with electrochemical Ag and AgNPs was effective against plant-pathogenic fungi, *B. sorokiniana* and *M. grisea*, and did not present any phytotoxicity to *O. sativa* L. leaves, which is ideal for practical agricultural use. Likewise, any use of AgNPs and/or formulation and products containing AgNPs must adhere to local, national, and international standards. In the United States, the US Environmental Protection Agency (EPA) in coalition with the US Conference of Governmental Industrial Hygienists established metallic Ag limit at 0.1 ppm, soluble compounds of Ag at 0.01 ppm. Besides, the US National Institute for Occupational Safety and Health (NIOSH) set limits of all forms of Ag at 0.01 ppm (EPA, 2012). In the European Union, AgNPs are regulated through the Registration, Evaluation, Authorization, and Restriction of Chemicals (REACH) program.

The US Center for Disease Control (CDC) with the Department of Health and Human Services and NIOSH has suggested recommended subchronic exposure limits for AgNPs sized 15–20 nm. The bulletin contains a physiologically based pharmacokinetic (PBPK) model for estimating exposures to AgNPs for occupational monitoring that would not induce adverse lung or liver effects. Estimates ranged from 0.19 to 195 $\mu g/m^3$ exposed for 8 h time-weighted average (TWA) airborne concentration, assuming biologically relevant tissue dose (soluble or total silver), particle size, shape and charge, PBPK model parameters, and laboratory rat effect level estimate (NIOSH, 2015). Human beings who would have developed argyria were estimated to be exposed to 47–253 $\mu g/m^3$ of AgNPs for 45 years. The current NIOSH limit for metallic dust of Ag and soluble compounds containing AgNPs is 10 $\mu g/m^3$ as an 8 h TWA concentration (European Chemicals Agency, 2020; NIOSH, 2015). WHO has advised that 10 g of Ag lifetime should present No Observed Adverse Effect Levels (NOAELS), and on the other hand, the US EPA derived an oral reference dose (RfD), which represents an ingested daily dose of Ag for a lifetime NOAELS in human beings to be approximately 0.0056 mg/kg/day based on the minimal dose that can induce the onset of argyria (i.e., 1 g of metallic Ag) in 7.8% of individuals

Table 23.2 Antifungal effects and ecotoxicology profile of AgNPs formulations against fungi.

AgNP formulation product	Capping/ stabilizing ligand	Fungi	Mycotoxins/phytotoxins	Ecotoxicity profile	Ref
AgNPs (18.9 nm)	Biosynthesis by Serratia spp.	Bipolaris sorokiniana	Prehelminthosporol, dihydroprehelminthosporol, victoxinine, and prehelminthosporolactone	**Soil invertebrates:** Concentration/dose: 0.05, 0.1 and 0.5 ppm Exposure time: 24–72 h Main effects: Reproduction potential and expression of the sod-3 and daf-12 genes decreased at 0.1 and 0.5 ppm	Mishra et al. (2014)
AgNPs (20–30 nm)	Not coated	B. sorokiniana and Magnaporthe grisea	Prehelminthosporol, dihydroprehelminthosporol, victoxinine, prehelminthosporolactone, and tenuazonic acid	**Soil invertebrates:** Concentration/dose: 0.1, 0.5 and 1 ppm Exposure time: 24 h Main effects: AgNPs caused decreases in ROS formation, expression of PMK-1, p38 MAPK, and hypoxia-inducible factor, GST enzyme activity, and reproductive potential in wild type, unlike the pmk-1 mutant 48 h LC_{50}: Adult zebrafish: 7.07 (6.04–8.28) ppm Juvenile zebrafish: 7.20 (5.9–8.6) ppm 48 h LC_{50}: **Aquatic invertebrates:** Daphnia pulex adults: 0.040 (0.030–0.050) ppm Ceriodaphnia dubia neonates: 0.067 ppm Pseudokirchneriella subcapitata: 0.19 ppm	Yun et al. (2015); Courtois et al. (2019)

AgNPs (10 nm)	Trisodium citrate dehydrate, polyvinyl pyrrolidone, sodium tetrahydridoborate	Alternaria alternata, P. digitatum, and Alternaria citri	Tenuazonic acid, alternariol, alternariol monomethyl ether, altenuene, altertoxin I, citrinin, cyclopiazonic acid, ochratoxin A, patulin, penicillic acid, penitrem A, frequentin, palitantin, mycophenolic acid, viomellein, gliotoxin, citreoviridin, and rubratoxin B	**Soil invertebrates:** Concentration/dose: 500 ppm Exposure time: 5 weeks Main effects: Accumulation of Ag in organisms (more with AgNPs than with AgNO$_3$) and a decrease in the unsaturated degree of fatty acids LC50$_{8-h}$ in *Daphnia manga* 5.04 ± 0.84 ppm	Courtois et al. (2019)
AgNPs (15.5 nm)	Biosynthesis by the fungus *A. fulvum*	*Candida* spp., *Aspergillus* spp., *Fusarium* spp.	Deoxynivalenol, 3-acetyl deoxynivalenol, 15-acetyl deoxynivalenol, nivalenol, fusarenon X, T-2 toxin, HT-2 toxin, neosolaniol, diacetoxyscirpenol, zearalenone, fumonisin B$_1$, fumonisin B$_2$, and fusaric acid	**Soil invertebrates:** Concentration/dose: 0.05, 0.1 and 0.5 ppm Exposure time: 24–72 h Main effects: Reproduction potential and expression of the *sod-3* and *daf-12* genes decreased at 0.1 and 0.5 ppm	Xue et al. (2016); Courtois et al. (2019)

daf-12, abnormal dauer formation protein 12; GST, glutathione S-transferase; LC$_{50}$, mean lethal dose of AgNPs/AgNPs formulation that is lethal for 50% of tested animal model used within the dose group; p38 MAPK, p38 mitogen-activated protein kinases; PMK-1, mitogen-activated protein kinase pmk-1; ROS, reactive oxygen species; sod-3, superoxide dismutase 3.

after intravenous medical therapy for 2–9.75 year. When this dosage is converted into oral administration of 10 g advised by WHO accounting for oral absorption (i.e., 1 g ÷ 0.1 = 10 g) translates to a lifetime dose as follows:

$$10 \text{ g over a lifetime} = 10{,}000 \text{ mg} \div 70 \text{ years} \div 365 \text{ days/year} \div 70 \text{ kg body weight}$$
$$= 0.0056 \text{ mg/kg/day}$$

Although short- or long-term dietary investigations with AgNPs are still not available, a 28–90 day gavage dosing studies with AgNPs showed NOAELS for traditional experimental animal model toxicological endpoints range from 0.5 to ~500 mg Ag/kg/day (EFSA ANS Panel, 2016; European Chemicals Agency, 2020). Wang et al. (2020) investigated the anti-biofilm and fungicidal activities of positively charged AgNPs synthesized and stabilized with trimethylchitosan nitrate (TMCN) against *C. albicans* SC5314-virulent strain with strong biofilm activity, as well as other reference strains: *C. albicans* ATCC 76615, *C. tropicalis* ATCC 750, and *C. glabrata* ATCC 15545. Cytotoxicity was monitored using L929 fibroblast cell lines and embryotoxicity evaluated using *Danio rerio* (zebrafish). Results revealed that TMCN-AgNPs exhibited the best antifungal potential at 6.2 and 49.4 ppm for *C. tropicalis* ATCC 750 and both *C. albicans* ATCC 76615 and *C. glabrata* ATCC 15545, respectively. Furthermore, cytotoxicity data showed no significant toxicity toward L929 cells even at 31.7 ppm; embryotoxicity assay revealed that zebrafish eggs preinfected with *C. albicans* SC5314 and then treated with TMCN-AgNPs (24.8 ppm) greatly diminished the amount of biofilm formation on eggs and no damage was observed, thereby rescuing embryos survival by 70%. Therefore, dosage >24.8 ppm induced some deaths affecting heart and organ development. This work established toxicity limits for TMCN-AgNP which is paramount for mandating ecotoxicology and phytotoxicology limits to regulate the use of AgNPs in agriculture for controlling fungal growth, mycotoxin contamination, and other related pathogens/diseases.

A study by Yin et al. (2011) monitored AgNP uptake and their effect on grass (*Lolium multiflorum*), and it was reported that AgNPs <6 nm strongly affected *L. multiflorum* growth and induced morphological changes; this was not observed with similar concentrations of AgNO$_3$ and AgNPs >25 nm, signifying the role that size contributes to toxicity. There is an alarming concern regarding the interaction and effects of AgNPs on plants and soil microorganism's homeostasis (Montes de Oca-Vásquez et al., 2020). AgNPs can influence the growth of plants and rhizospheric microbial communities with symbiotism, such as arbuscular mycorrhizal fungi, which play an important role in plant development. It is also imperative to evaluate the environmental impact of biosynthetic AgNPs; AgNPs produced using *A. tubingensis* were investigated against aerobic heterotrophs soil microorganisms, rice seeds (*O. sativa*), and zebrafish (*D. rerio*). AgNPs had a low effect on soil microbiota compared to AgNO$_3$, however; they influenced the germination of rice seeds and subsequent growth development had a dose-dependent inhibitory effect, and AgNPs at 7.1 ppm did not induce mortality of the *D. rerio*.

A study by Cao et al. (2017) investigated growth responses of maize (*Zea mays* L.) and rhizospheric soils with mycorrhizal fungal colonies treated with different

AgNPs concentrations (0.025, 0.25, and 2.5 ppm) and Ag. The results demonstrated that 2.5 ppm of AgNPs significantly reduced plant root biomass attributed by Ag accumulation in plant tissues, resulting in increased antioxidant enzyme activity that causes oxidative stress and decreased photosynthetic carbon for fungi, as well as affecting the diversity and composition of the soil colonies. Furthermore, this simultaneously changed ecological functions, thereby weakening the phosphorus (P) cycle in the soil as a result of decreased soil alkaline phosphatase activity, thus limiting P content availability in plants for development. The authors also emphasized the importance of evaluating the effects of AgNPs on agricultural ecosystems, especially signaling responses between fungi and symbiotic microorganism.

According to food safety regulations, AgNPs in polymeric matrices utilized for active packaging material to increase the shelf-life of food and provide protection against fungal growth with subsequent production of mycotoxins warrant rigorous in vivo cytotoxicity investigations in multispecies animal models to reflect true real-life simulations of their ecotoxicity (Bahrami et al., 2020; Liu et al., 2020). A recent study by Maziero et al. (2020) highlighted the importance of multispecies in vitro and in vivo ecotoxicological analysis of tri-alanine-phosphine peptide ("Katti Peptide") stabilized gum arabic protein AgNPs (AgNP-GP) in different experimental models as follows: (i) *Daphnia similis* for the acute ecotoxicity (EC_{50}) tests; (ii) *D. rerio* for the evaluation of acute embryotoxicity (LC_{50}) tests; and (iii) Sprague Dawley rats for toxicity behavioral investigations. Their results revealed that AgNP-GP exhibited EC_{50} in *D. similis* was 4.40 ppm, LC_{50} in *D. rerio* was 177 ppm, and oral administration of AgNP-GP in Sprague Dawley rats in both male and female animals for 28 days showed no adverse effects in doses of up to 10.0 mg/kg body weight. Such investigations are appropriately tailored to assess the overall real-life systemic toxicity of AgNPs through utility of relevant animal species, for testing the in vivo toxicity of AgNPs, which assumes a paramount role in gaining long-term toxicity information. Toxicity studies in animals with similar receptor/epitope distribution as human beings provide vital information on in vivo toxicity, ultimately translating the data obtained in assessing tissue cross-reactivity profile in human tissues. This will help in defining AgNPs toxicity limits for various formulations in a myriad of applications such as antifungal agents in agriculture.

23.4 Life-cycle analysis of AgNPs for controlling mycotoxins in the agricultural sector

Among the nanomaterials with commercial applications, AgNPs are one of the most used forms of metallic nanoparticles. Europe Union, in 2014, there was a total production of approximately 50 tons of Ag, which eventually led to the accumulation of AgNPs released into the environment and contaminates the soil. Full life cycle assessments of AgNPs have sparked considerable interest in many researchers considering their applications in several sectors, including agriculture and food. Pourzahedi et al. (2017) and Elamawi et al. (2018) emphasized the importance of establishing rigorous full life cycle assessments to extensively observe and quantify all environmental burdens from upstream to beyond ecotoxic effects, thereby considering the benefits provided by these products during and after utilization (Iavicoli et al., 2017). Also, design detailed ecotoxicological studies that are much longer and carried out in multispecies experimental models to determine the toxic effect of AgNPs to potential toxicity in human beings before mass production and use of agricultural applications are imperative.

Pourzahedi et al. (2017) evaluated all these aspects of several commercial products containing AgNPs, including a food container and a plastic bag for food storage, and estimate cradle-to-gate environmental impacts. They concluded that the main environmental impact caused by AgNPs was related to the electricity use and emissions from Ag mining. The concentration of AgNPs found in these products was between 0.005% and 3%, which means that the contribution of Ag was very much dependent on the composition of the product. Therefore, impacts related to nonnano emissions upstream, like the production of the plastic matrix, may offer much more burden to the environment than the AgNPs themselves. In addition, studying AgNPs release profiles are imperative to define the minimum concentration of AgNPs, thus balancing product performance and reducing the impacts on the environment.

Another important point to consider on full life cycle assessment of AgNP is their route of synthesis which can dictate their overall toxicity profile in addition to their capping/stabilizing agents used during synthesis (Iavicoli et al., 2017). Pourzahedi et al. (2017) compared seven different routes observing the magnitude and patterns of impacts considering different environmental categories and presented a cradle-to-gate life cycle inventory. The differences in antimicrobial activity of the nanoparticles according to their route of synthesis were considered during the assessment. The seven methods were as follows: chemical reductions using trisodium citrate, sodium borohydride ethylene glycol and soluble starch from potatoes, and physical techniques, flame sprays pyrolysis, arc plasma, and reactive magnetron sputtering. The authors observed that, albeit the chemical reduction using starch is considered a bio-based reduction method, it presented the highest impacts for ozone depletion, acidification, eutrophication, noncarcinogenic, and ecotoxicity. This showed the importance of analyzing the whole life cycle impacts before adopting a synthetic route. Moreover, it is also important to mention that the researchers emphasize that for all routes, the greatest contribution of the process is related to Ag extraction because it involves mining, refining, and transport. Hence, recovering Ag from spent solutions may be a way to minimize these effects (Pourzahedi and Eckelman, 2015). This further emphasizes the consistent need for continued innovation for raw material acquisition (upstream processing) and end-of-life (downstream) stages, which is ubiquitous across several industrial sectors, as shown in Fig. 23.3.

Furthermore, it is of paramount to study the effects of the accumulation and major interactions of AgNPs with soil, soil biota, sediment, water, biota in water bodies, plants, and animals/human beings to fully comprehend the ecotoxicology in multispecies associated with their use in agriculture to control fungal growth and mycotoxin contamination (Maziero et al., 2020; Mishra and Singh, 2015). These

23.4 Life-cycle analysis of AgNPs for controlling mycotoxins

FIG. 23.3

Life-cycle assessments of AgNPs for quality assurance and quality control measures from various synthetic routes to possible migration to the soil/water microbiota. A cascade of pathways that include agricultural application in controlling fungal growth and mycotoxin contamination, product integrity maintenance through antifungal storage containment against fungal manifestation, end-product to consumer mycotoxins-free, active antifungal packaging material, end-of-life can facilitate a feedback loop of migration to the environment.

nanoparticles may enter agriculture through plant sludges that are applied to fields during crop cultivation. AgNPs tend to oxidize or dissolve releasing Ag^+ ions when they are in the soil after application. Eventually, Ag^+ may get subsidized and become less toxic to microorganisms. The half-life of this new substance varies from some years to over a century depending on its redox conditions (Dale et al., 2013; Riebeling and Kneuer, 2014). When exposed to organisms, the contact to body fluids containing proteins, lipids, and polysaccharides may form a corona around the nanoparticle, and this corona will influence the cellular uptake (Riebeling and Kneuer, 2014). As there is a growing number of AgNP incorporated packaging materials, it is imperative to consider the added complexity and potential exposure routes in minimizing AgNPs concentration per realized benefit from upstream embedded energy, downstream handling, and economic perspective to comply with regulatory entity standards (Bahrami et al., 2020). Moreover, the rate of recycling has increased over the years; therefore, it is important to consider packaging recycling when incorporating

AgNPs as this generates complexities in the recycling process, which changes and influences the physical, chemical, and biological reactivity of the adjacent environment, including plants, animals, and human beings through the discharge of AgNPs and/or Ag^+ ions into terrestrial or aquatic environments, which may increase the probability of ecotoxicity (European Union, 2018; Tortella et al., 2020).

The disposal and migrating of AgNPs into wastewater leads to AgNPs and its derivatives (e.g., AgCl, Ag^+ ions, silver sulfide nanoparticles (Ag_2S-NPs), etc.), resulting from their chemical biotransformation to the predominant Ag species—Ag_2S, which has a longer shelf-life over several years/decades, in sewage sludge. Part of the sewage sludge is rediverted to utilization in agricultural irrigation, the soil being a recipient of contamination by AgNPs and derivatives (Montes de Oca-Vásquez et al., 2020). In the first review published by Courtois et al. (2019), a primary study was carried out on the potential impact of Ag species introduced to the soil via sewage sludge, from ecologically important microorganisms, fungi, algae, terrestrial plants, and soil invertebrates, all-important in soil ecosystem stability and sustainability.

AgNPs and their chemical biotransformation products can enter cells through biological membranes causing physiological, biochemical, and molecular effects of organisms and microorganisms. In soils, exposure to AgNPs can modify biomass and microbial diversity, retard the growth of plants, and inhibit the reproduction of invertebrates. Generally, Ag_2S has been reported to have low bioavailability, therefore less toxic to plants as compared to Ag^+ ions. Some studies have demonstrated that the ecotoxicological profile hierarchy of Ag as follows: $AgNO_3$ > free Ag^+ ions > AgNPs > Ag_2S (Yan and Chen, 2019) and smaller size particles exhibit high toxicity than larger particles. However, the evaluation of the toxicity of AgNPs and Ag-derived species is complex, and there is much to be studied about the ecotoxicology of Ag species in soils, such as the possibility of translocation along the trophic chain due to the bioaccumulation in plant and animal tissues for better risk assessment. A recent study by Montes de Oca-Vásquez et al. (2020) investigated the effects of environmental realistic AgNPs concentration on soil microbial diversity exposed for 7–60 days. Their results revealed that after 7 days postincubation (pi), microbial biomass increased, and after 60 days pi, a decrease in biomass (mainly Gram + and actinobacteria) was observed, including relative abundance of the fungal community. However, little effects were observed on the overall microbial diversity and soil biogeochemical properties and cycles mediated by extracellular enzyme activities.

The tracking analysis of AgNPs exposure and bioaccumulation in the environment is difficult and remains unknown as to the levels that could affect the overall food chain. This is exacerbated by the limited knowledge of a comprehensive understanding of the ecotoxicological behavior of AgNPs, possible interactions with agrosystem coformulants, regulatory and legislative frameworks for AgNP nanotoxicity that govern occupational safety practices and policies in food contact materials that can yield first-in-human (FIH) translational ecotoxicological interpretation (Iavicoli et al., 2017; Maziero et al., 2020). Aid in the development of regulatory consensus, and nanospecific risk assessment analysis on the application of AgNPs in agriculture with emphasis to the long-standing principles of *"prevention through design"* and *"safety by design,"* this can mitigate unintended consequences. It is important to acknowledge that various factors such as organic matter content, pH, presence of HA, ionic strength, AgNPs concentration, size, or the presence and nature of capping/stabilizing ligands in combination, which can influence the fate, transformation/translocation (such as aggregation, chlorination, dissolution, oxidation, and sulfidation) and migration of AgNPs, which ultimately dictates the realistic evaluation of their ecotoxicological profile (Tortella et al., 2020).

23.5 Prospective ecotoxicological framework

An alternative strategic framework for the development of Green nanoenabled agricultural products is of paramount for evaluating both ecotoxicity and phytotoxicity dynamics through sustainable ecological practices (SEP) model for smart agricultural use of AgNPs. This SEP model would enhance agricultural efficiency and sustainability by elucidating ways to maximize the safety of AgNPs utilization in both up- and downstream of food production, processing, and use in controlling fungal and mycotoxin contamination, thereby ensuring a safe sustainable food supply chain. It is critical to fully understand the mechanistic insight of the ecotoxicology, phytotoxicity, nanospecific risk assessment, management, and governance of AgNPs designs and applications in agriculture for controlling fungal growth and mycotoxin contamination. This contributes to the realistic continued innovation in product design for food/feed protection from fungal and mycotoxins contamination, thus allowing for rapid advancements that promote agricultural sustainability through the SEP model, as shown in Fig. 23.4.

Detail information on the following is mandated to be disclosed:

(i) AgNPs concentration range and crop type(s) used for exposure studies
(ii) Metrics for benefit quantification (e.g., mycotoxins monitoring) for application studies
(iii) Ecotoxicology and phytotoxicology results between studies must be compared

All these critical information is necessary to quantify trade-offs of nanoenabling agriculture early in the design process to preclude future unintended consequences. The SEP model, in addition to the abovementioned criteria, aims to provide a platform for a synergistic next-generation AgNPs design that protects plants from fungal manifestation and mycotoxins contamination with the added armament of enhancing plant growth through improving nitrogen-fixation capability and photosynthesis in roots and leaves to encourage conversion efficiency and the energy utilization for a sustainable agricultural standard, which will be the driving force in modern agriculture and agrifood nanobiotechnology. The SEP model would help not only create but monitor ongoing nano-bio industrial products toward a concerted understanding of the complex interplay between agroecosystems, nanoparticles, and their level of exposure to human/animal health.

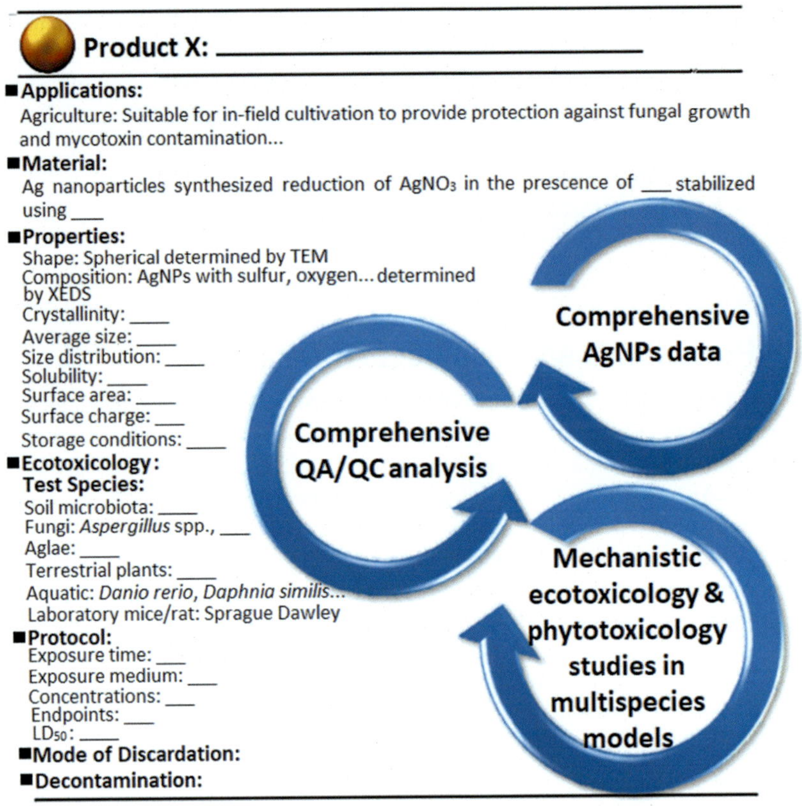

Product X: _____

- **Applications:**
  Agriculture: Suitable for in-field cultivation to provide protection against fungal growth and mycotoxin contamination...
- **Material:**
  Ag nanoparticles synthesized reduction of AgNO$_3$ in the presence of ___ stabilized using ___
- **Properties:**
  Shape: Spherical determined by TEM
  Composition: AgNPs with sulfur, oxygen... determined by XEDS
  Crystallinity: ___
  Average size: ___
  Size distribution: ___
  Solubility: ___
  Surface area: ___
  Surface charge: ___
  Storage conditions: ___
- **Ecotoxicology:**
  Test Species:
  Soil microbiota: ___
  Fungi: *Aspergillus* spp., ___
  Aglae: ___
  Terrestrial plants: ___
  Aquatic: *Danio rerio*, *Daphnia similis*...
  Laboratory mice/rat: Sprague Dawley
- **Protocol:**
  Exposure time: ___
  Exposure medium: ___
  Concentrations: ___
  Endpoints: ___
  LD$_{50}$: ___
- **Mode of Discardation:**
- **Decontamination:**

Comprehensive QA/QC analysis → Comprehensive AgNPs data → Mechanistic ecotoxicology & phytotoxicology studies in multispecies models

FIG. 23.4

Sustainable ecological practices (SEP) model framework for AgNPs products destined for application in agriculture to provide comprehensive data reporting, specifically as it relates to the active ingredient used. The mechanism (proposed and when possible, empirically supported) underlying ecotoxicology and phytotoxicology reports which govern the desired application and outcome, thus informing the continued sustainable design of promising solutions in accordance to set environmental limits.

23.6 Conclusion

The increased utilization of AgNPs and their derivatives in agriculture has prompted concerns on the scarcity of ecotoxicology and phytotoxicology multispecies studies on AgNPs already in the market because they demonstrate/provide optimal production yields and protection from fungi and mycotoxin contamination. These sidelines, the potential environmental and human/animal/plant health secondary collateral consequences, are often neglected from the initial AgNPs-enabled product design objectives. The green nanotechnological synthetic routes of AgNPs as new fungicides are explored with the aid to limit and generate no toxic byproducts to the environment.

However, further investigations into life cycle assessments of AgNPs with their ecotoxicology and phytotoxicology profiles in multispecies experimental models to yield realistic FIH translational ecotoxicological interpretation to AgNPs exposure must be mandatory to establish regulatory limit consensus on AgNPs. Finally, the suggested SEP model framework for AgNPs products/formulations would be the driving force in modern agriculture and agrifood nanobiotechnology governing safety and sustainability.

Acknowledgments

The authors would like to thank the Fundação de Amparo à Pesquisa do Estado de São Paulo (FAPESP), Grant No. 2019/15154-0, for support to Velaphi Clement Thipe.

References

Abdel-Hafez, S.I.I., Nafady, N.A., Abdel-Rahim, I.R., Shaltout, A.M., Daròs, J.A., Mohamed, M.A., 2016. Assessment of protein silver nanoparticles toxicity against pathogenic *Alternaria solani*. 3 Biotech 6 (2), 199.

Abd-Elsalam, K., Alghuthaymi, M., 2015. Nanobiofungicides: are they the next-generation of fungicides? J. Nanotechnol. Mater. Sci. 2, 1–3.

Abd-Elsalam, K.A., Hashim, A.F., Alghuthaymi, M.A., Said-Galiev, E., 2017. Chapter 10: Nanobiotechnological strategies for toxigenic fungi and mycotoxin control. In: Grumezescu, A.M. (Ed.), Nanotechnology in the Agri-Food Industry. Academic Press, pp. 337–364.

Adeyeye, S.A.O., 2020. Aflatoxigenic fungi and mycotoxins in food: a review. Crit. Rev. Food Sci. Nutr. 605, 709–721.

Anfossi, L., Di Nardo, F., Cavalera, S., Giovannoli, C., Spano, G., Speranskaya, E.S., Goryacheva, I.Y., Baggiani, C., 2018. A lateral flow immunoassay for straightforward determination of fumonisin mycotoxins based on the quenching of the fluorescence of CdSe/ZnS quantum dots by gold and silver nanoparticles. Microchim. Acta 1852, 94.

Ashraf, H., Anjum, T., Riaz, S., Naseem, S., 2020. Microwave-assisted green synthesis and characterization of silver nanoparticles using *Melia azedarach* for the management of fusarium wilt in tomato. Front. Microbiol. 11, 1–22.

Bahrami, A., Delshadi, R., Assadpour, E., Jafari, S.M., Williams, L., 2020. Antimicrobial-loaded nanocarriers for food packaging applications. J. Colloid Interface Sci. 278, 102140.

Bumbudsanpharoke, N., Ko, S., 2015. Nano-food packaging: an overview of market, migration research, and safety regulations. J. Food Sci. 805, R910–R923.

Bumbudsanpharoke, N., Lee, W., Ko, S., 2018. A comprehensive feasibility study on the properties of LDPE-Ag nanocomposites for food packaging applications. Polym. Compos. 399, 3178–3186.

Cao, J., Feng, Y., He, S., Lin, X., 2017. Silver nanoparticles deteriorate the mutual interaction between maize *Zea mays L.* and arbuscular mycorrhizal fungi: a soil microcosm study. Appl. Soil Ecol. 119, 307–316.

Chauhan, R., Singh, J., Sachdev, T., Basu, T., Malhotra, B.D., 2016. Recent advances in mycotoxins detection. Biosens. Bioelectron. 81, 532–545.

Cinar, A., Onbasi, E., 2019. Mycotoxins: the hidden danger in foods. In: Sabuncuoğlu, S. (Ed.), Mycotoxins and Food Safety. IntechOpen, https://doi.org/10.5772/intechopen.89001.

Courtois, P., Rorat, A., Lemiere, S., Guyoneaud, R., Attard, E., Levard, C., Vandenbulcke, F., 2019. Ecotoxicology of silver nanoparticles and their derivatives introduced in soil with or without sewage sludge: a review of effects on microorganisms, plants and animals. Environ. Pollut. 253, 578–598.

Dale, A.L., Lowry, G.V., Casman, E.A., 2013. Modeling nanosilver transformations in freshwater sediments. Environ. Sci. Technol. 4722, 12920–12928.

EFSA ANS Panel, 2016. Scientific opinion on the re-evaluation of silver E 174 as food additive. EFSA J. 141, 1–64.

Elamawi, R.M., Al-Harbi, R.E., Hendi, A.A., 2018. Biosynthesis and characterization of silver nanoparticles using *Trichoderma longibrachiatum* and their effect on phytopathogenic fungi. Egypt. J. Biol. Pest Control 281, 1–11.

El-Waseif, A.A., Attia, M.S., El-Ghwas, D.E., 2019. Influence of two extraction methods on essential oils of some Apiaceae family plants. Egypt. Pharm. J. 18, 160–164.

EPA, 2012. Nanomaterial Case Study: Nanoscale Silver in Disinfectant Spray Final Report. U.S. Environmental Protection Agency, Washington, DC. EPA/600/R-10/081F.

European Chemicals Agency, 2020. Silver EC Number: 231-131-3. [Online]. Available: https://echa.europa.eu/registration-dossier/-/registered-dossier/16155/7/6/1. (Accessed 23 August 2020).

European Union, 2018. Colloidal Silver Nano. [Online]. Available: https://ec.europa.eu/health/sites/health/files/scientific_committees/consumer_safety/docs/sccs_o_219.pdf. (Accessed 23 August 2020).

FAO, 2020. Sustainable Development Goals. [Online]. Avaialble: http://www.fao.org/sustainable-development-goals/en/. (Accessed 23 August 2020).

Ferdous, Z., Nemmar, A., 2020. Health impact of silver nanoparticles: a review of the biodistribution and toxicity following various routes of exposure. Int. J. Mol. Sci. 217, 2375.

Guilger-Casagrande, M., de Lima, R., 2019. Synthesis of silver nanoparticles mediated by fungi: a review. Front. Bioeng. Biotechnol. 7, 1–16.

Iavicoli, I., Leso, V., Beezhold, D.H., Shvedova, A.A., 2017. Nanotechnology in agriculture: opportunities, toxicological implications, and occupational risks. Toxicol. Appl. Pharmacol. 329, 96–111.

Ibrahim, E., Zhang, M., Zhang, Y., Hossain, A., Qiu, W., Chen, Y., Wang, Y., Wu, W., Sun, G., Li, B., 2020. Green-synthesization of silver nanoparticles using endophytic bacteria isolated from garlic and its antifungal activity against wheat fusarium head blight pathogen *Fusarium graminearum*. Nanomaterials 102, 219.

Jesmin, R., Chanda, A., 2020. Restricting mycotoxins without killing the producers: a new paradigm in nano-fungal interactions. Appl. Microbiol. Biotechnol. 1047, 2803–2813.

Jo, Y.-K., Kim, B.H., Jung, G., 2009. Antifungal activity of silver ions and nanoparticles on phytopathogenic fungi. Plant Dis. 9310, 1037–1043.

Jogee, P.S., Ingle, A.P., Rai, M., 2017. Isolation and identification of toxigenic fungi from infected peanuts and efficacy of silver nanoparticles against them. Food Control 71, 143–151.

Khalil, N.M., Abd El-Ghany, M.N., Rodríguez-Couto, S., 2019. Antifungal and anti-mycotoxin efficacy of biogenic silver nanoparticles produced by *Fusarium chlamydosporum* and *Penicillium chrysogenum* at non-cytotoxic doses. Chemosphere 218, 477–486.

Khan, I.M., Zhao, S., Niazi, S., Mohsin, A., Shoaib, M., Duan, N., Wu, S., Wang, Z., 2018. Silver nanoclusters based FRET aptasensor for sensitive and selective fluorescent detection of T-2 toxin. Sens. Actuators B Chem. 277, 328–335.

Kim, S.W., Jung, J.H., Lamsal, K., Kim, Y.S., Min, J.S., Lee, Y.S., 2012. Antifungal effects of silver nanoparticles AgNPs against various plant pathogenic fungi. Mycobiology 401, 53–58.

Kotzybik, K., Gräf, V., Kugler, L., Stoll, D.A., Greiner, R., Geisen, R., Schmidt-Heydt, M., 2016. Influence of different nanomaterials on growth and mycotoxin production of Penicillium verrucosum. PLoS One 113, 1–16.

Kwak, Y., Kim, S., Lee, J., Kim, I., 2012. Synthesis of Pyto-patch as silver nanoparticle product and antimicrobial activity. J. Bio-Environ. Control 212, 140–146.

Lara, H.H., Romero-Urbina, D.G., Pierce, C., Lopez-Ribot, J.L., Arellano-Jiménez, M.J., Jose-Yacaman, M., 2015. Effect of silver nanoparticles on *Candida albicans* biofilms: an ultrastructural study. J. Nanobiotechnol. 13, 91.

Le, T., Nguyen, D.H., Ching, Y., Nguyen, N., Nguyen, D., Ngo, C., Nguyen, H., Hoang Thi, T.T., 2020. Silver nanoparticles ecofriendly synthesized by *Achyranthes aspera* and *Scoparia dulcis* leaf broth as an effective fungicide. Appl. Sci. 10, 2505.

Liu, W., Zhang, M., Bhandari, B., 2020. Nanotechnology—a shelf life extension strategy for fruits and vegetables. Crit. Rev. Food Sci. Nutr. 6010, 1706–1721.

Maziero, J., Thipe, V., Rogero, S., Cavalcante, A., Damasceno, K., Ormenio, M., Martini, G., Batista, J., Viveiros, W., Katti, K., Raphael Karikachery, A., Mohandoss, D., Dhurvas, R., Nappinnai, M., Rogero, J., Lugão, A.B., Katti, K., 2020. Species specific in vitro and in vivo evaluation of toxicity of silver nanoparticles stabilized with arabic gum protein. Int. J. Nanomedicine 15, 7359–7376.

Mishra, S., Singh, H.B., 2015. Biosynthesized silver nanoparticles as a nanoweapon against phytopathogens: exploring their scope and potential in agriculture. Appl. Microbiol. Biotechnol. 993, 1097–1107.

Mishra, S., Singh, B.R., Singh, A., Keswani, C., Naqvi, A.H., Singh, H.B., 2014. Biofabricated silver nanoparticles act as a strong fungicide against *Bipolaris sorokiniana* causing spot blotch disease in wheat. PLoS One 95, e97881.

Mitra, C., Gummadidala, P.M., Merrifield, R., Omebeyinje, M.H., Jesmin, R., Lead, J.R., Chanda, A., 2019. Size and coating of engineered silver nanoparticles determine their ability to growth-independently inhibit aflatoxin biosynthesis in *Aspergillus parasiticus*. Appl. Microbiol. Biotechnol. 10311, 4623–4632.

Montes de Oca-Vásquez, G., Solano-Campos, F., Vega-Baudrit, J.R., López-Mondéjar, R., Odriozola, I., Vera, A., Moreno, J.L., Bastida, F., 2020. Environmentally relevant concentrations of silver nanoparticles diminish soil microbial biomass but do not alter enzyme activities or microbial diversity. J. Hazard. Mater. 391, 122224.

Ng, S.M., Koneswaran, M., Narayanaswamy, R., 2016. A review on fluorescent inorganic nanoparticles for optical sensing applications. RSC Adv. 626, 21624–21661.

NIOSH, 2015. In: Zumwalde, R.D., Kuempel, E.D., Holdsworth, G. (Eds.), External Review Draft—Current Intelligence Bulletin: Health Effects of Occupational Exposure to Silver Nanomaterials. U.S. Department of Health and Human Services, Centers for Disease Control and Prevention, National Institute for Occupational Safety and Health, Cincinnati, OH.

Nuruzzaman, M., Rahman, M.M., Liu, Y., Naidu, R., 2016. Nanoencapsulation, nano-guard for pesticides: a new window for safe application. J. Agric. Food Chem. 647, 1447–1483.

Panáček, A., Kolář, M., Večeřová, R., Prucek, R., Soukupová, J., Kryštof, V., Hamal, P., Zbořil, R., Kvítek, L., 2009. Antifungal activity of silver nanoparticles against Candida spp. Biomaterials 3031, 6333–6340.

Pietrzak, K., Twaruzek, M., Czyzowska, A., Kosicki, R., Gutarowska, B., 2015. Influence of silver nanoparticles on metabolism and toxicity of moulds. Acta Biochim. Pol. 624, 851–857.

Polaris Market Research, 2020. Nano-Enabled Packaging Market Share, Size, Trends, and Industry Analysis Report, by Type (Active, Intelligent, Others); by End-User (Food and Beverages, Pharmaceutical, Personal Care, Others); by Regions: Segment Forecast, 2018–2026. [Online]. Available: https://bulletinline.com/2020/08/17/nano-enabled-packaging-market-global-industry-analysis-size-share-growth-trends-and-forecasts-2020-2026/. (Accessed 12 August 2020).

Pourzahedi, L., Eckelman, M.J., 2015. Comparative life cycle assessment of silver nanoparticle synthesis routes. Environ. Sci. Nano 24, 361–369.

Pourzahedi, L., Vance, M., Eckelman, M.J., 2017. Life cycle assessment and release studies for 15 nanosilver-enabled consumer products: investigating hotspots and patterns of contribution. Environ. Sci. Technol. 5112, 7148–7158.

Pulit-Prociak, J., Banach, M., 2016. Silver nanoparticles—a material of the future…? Open Chem. J. 141, 76–91.

Riebeling, C., Kneuer, C., 2014. Case study: challenges in human health hazard and risk assessment of nanoscale silver. In: Wohlleben, W., Kuhlbusch, T.A.J., Schnekenburger, J., Lehr, C.-M. (Eds.), Safety of Nanomaterials Along Their Lifecycle: Release, Exposure, and Human Hazards. CRC Press, pp. 417–436.

Rodrigues, N.P., Scott-Fordsmand, J.J., Amorim, M.J.B., 2020. Novel understanding of toxicity in a life cycle perspective—the mechanisms that lead to population effect—the case of Ag nanomaterials. Environ. Pollut. 262, 114277.

Santana Oliveira, I., da Silva Junior, A.G., de Andrade, C.A.S., Lima Oliveira, M.D., 2019. Biosensors for early detection of fungi spoilage and toxigenic and mycotoxins in food. Curr. Opin. Food Sci. 29, 64–79.

Shang, Y., Hasan, M.K., Ahammed, G.J., Li, M., Yin, H., Zhou, J., 2019. Applications of nanotechnology in plant growth and crop protection: a review. Molecules 2414, 2558.

Siddiqi, K.S., Husen, A., 2016. Fabrication of metal nanoparticles from fungi and metal salts: scope and application. Nanoscale Res. Lett. 111, 1–15.

Tarus, B.K., Mwasiagi, J.I., Fadel, N., Al-Oufy, A., Elmessiry, M., 2019. Electrospun cellulose acetate and polyvinyl chloride nanofiber mats containing silver nanoparticles for antifungi packaging. SN Appl. Sci. 13, 245.

Thipe, V.C., Keyster, M., Katti, K.V., 2018. Sustainable nanotechnology: mycotoxin detection and protection. In: Abd-Elsalam, K.A., Prasad, R. (Eds.), Nanobiotechnology Applications in Plant Protection. Springer International Publishing, pp. 323–349.

Thipe, V.C., Bloebaum, P., Khoobchandani, M., Karikachery, A.R., Katti, K.K., Katti, K.V., 2020. Chapter 7: Green nanotechnology: nanoformulations against toxigenic fungi to limit mycotoxin production. In: Rai, M., Abd-Elsalam, K.A.B.T.-N. (Eds.), Nanomycotoxicology: Treating Mycotoxins in Nanoway. Academic Press, pp. 155–188.

Tortella, G.R., Rubilar, O., Durán, N., Diez, M.C., Martínez, M., Parada, J., Seabra, A.B., 2020. Silver nanoparticles: toxicity in model organisms as an overview of its hazard for human health and the environment. J. Hazard. Mater. 390, 121974.

United Nations, 2019. World Population Prospects 2019. [Online]. Available: https://population.un.org/wpp/. (Accessed 5 July 2020).

Vergallo, C., Panzarini, E., Tenuzzo, B.A., Mariano, S., Tata, A.M., Dini, L., 2020. Moderate static magnetic field 6 mT-induced lipid rafts rearrangement increases silver NPs uptake in human lymphocytes. Molecules 256, 1398.

Villamizar-Gallardo, R., Cruz, J.F.O., Ortíz, O.O., 2016. Fungicidal effect of silver nanoparticles on toxigenic fungi in cocoa. Pesq. Agrop. 5112, 1929–1936.

Wang, S.H., Chen, C.C., Lee, C.H., Chen, X.A., Chang, T.Y., Cheng, Y.C., Young, J.J., Lu, J.J., 2020. Fungicidal and anti-biofilm activities of trimethylchitosan-stabilized silver nanoparticles against Candida species in zebrafish embryos. Int. J. Biol. Macromol. 143, 724–731.

Xue, B., He, D., Gao, S., Wang, D., Yokoyama, K., Wang, L., 2016. Biosynthesis of silver nanoparticles by the fungus Arthroderma fulvum and its antifungal activity against genera of Candida, Aspergillus and Fusarium. Int. J. Nanomedicine 11, 1899–1906.

Yan, A., Chen, Z., 2019. Impacts of silver nanoparticles on plants: a focus on the phytotoxicity and underlying mechanism. Int. J. Mol. Sci. 205, 23–25.

Yassin, M.A., El-Samawaty, A.E.R.M.A., Dawoud, T.M., Abd-Elkader, O.H., Al Maary, K.S., Hatamleh, A.A., Elgorban, A.M., 2017. Characterization and anti-Aspergillus flavus impact of nanoparticles synthesized by *Penicillium citrinum*. Saudi J. Biol. Sci. 246, 1243–1248.

Yin, L., Cheng, Y., Espinasse, B., Colman, B.P., Auffan, M., Wiesner, M., Rose, J., Liu, J., Bernhardt, E.S., 2011. More than the ions: the effects of silver nanoparticles on *Lolium multiflorum*. Environ. Sci. Technol. 456, 2360–2367.

Yin, J., Wang, Y., Gilbertson, L.M., 2018. Opportunities to advance sustainable design of nano-enabled agriculture identified through a literature review. Environ. Sci. Nano 51, 11–26.

Yun, C.-S., Motoyama, T., Osada, H., 2015. Biosynthesis of the mycotoxin tenuazonic acid by a fungal NRPS-PKS hybrid enzyme. Nat. Commun. 6, 8758.

Zubrod, J.P., Bundschuh, M., Arts, G., Brühl, C.A., Imfeld, G., Knäbel, A., Payraudeau, S., Rasmussen, J.J., Rohr, J., Scharmüller, A., Smalling, K., Stehle, S., Schulz, R., Schäfer, R.B., 2019. Fungicides: an overlooked pesticide class? Environ. Sci. Technol. 537, 3347–3365.

PART 5

Water treatment and purification

CHAPTER 24

Comparing the biosorption of ZnO and Ag nanomaterials by consortia of protozoan and bacterial species

Anza-vhudziki Mboyi[a], Ilunga Kamika[b], and Maggy N.B. Momba[a]

[a]*Department of Environmental, Water and Earth Sciences, Faculty of Science, Tshwane University of Technology, Pretoria, South Africa*
[b]*Institute for Nanotechnology and Water Sustainability; School of Science; College of Science, Engineering and Technology; University of South Africa, Florida Campus, Johannesburg, South Africa*

24.1 Introduction

A plethora of research have announced various beneficial applications of nanomaterials (NMs), such as silver (Ag) and zinc oxide (ZnO), due to their toxicity effects (toward microbes, algae, and fungi) and photocatalytic properties when intended for environmental remediation and improved livelihoods (Bondarenko et al., 2013; Huang et al., 2020; Du et al., 2020). Therefore, continuous applications of nanomaterials (NMs) such as silver (nAg) and zinc oxide (nZnO) into numerous commodities will results in unintended emission into wastewater treatment plants (WWTPs). Nonetheless, toxicity reactivity effects of NMs on living organisms is still under debate and unclear when compared to their bulk counterparts, especially in dynamic and stochastic systems (Bundschuh et al., 2016). According to a growing body of literature, toxicity effects of nZnO and nAg have been reported toward bacteria (Simon-Deckers et al., 2009; Ivask et al., 2014; Mboyi et al., 2017; de Souza et al., 2019: Huang et al., 2020), algae (Franklin et al., 2007; Navarro et al., 2008; Ivask et al., 2014;), fungi and nematodes (Wang et al., 2009; Ivask et al., 2014; Yamindago et al., 2019), mammalian cells (Brunner et al., 2006; Horie et al., 2009; Sharma et al., 2009; Zheng et al., 2009; De Berardis et al., 2010; Ivask et al., 2014), protozoa (Lashkenari et al., 2019), and zebrafish (Bai et al., 2010; Xiong et al., 2011; Ivask et al., 2014) due to bioaccumulation at varying concentrations of nanomaterials.

In WWTP systems, the microbial biomass in the reactor is known as activated sludge (Sun et al., 2013; Johnston et al., 2019; Yang et al., 2020). An assortment of microscopical wide-ranging organisms is contained in these systems

(Salvadó et al., 1995; Yang et al., 2020), which ciliated protozoa plays an important role in ensuring efficient wastewater purification treatment processes (Johnston et al., 2019). The presence of protozoan isolates in activated sludge not only is beneficial as these organisms act as disinfection agents and can be used as an indicator to assess sludge health (Salvadó et al., 1995; Papadimitriou et al., 2010; Johnston et al., 2019)). Although ciliated protozoa maybe resist certain concentrations of heavy metals (Martín-González et al., 2006; Fernandez-Leborans and Herrero, 2000), their resistance toward nanomaterials is still unknown as these are emerging pollutants and their presence is posing an alarming concern.

However, to date, a wide variety of microorganisms like *Bacillus licheniformis*, *Pseudomonas putida*, and *Brevibacillus laterosporus* have been reported as efficient sorption materials in the removal of NMs as well as heavy metals (Krell et al., 2012; Sun et al., 2013; Kulkarni et al., 2014; Todorova et al., 2019; Biswas et al., 2020). The resistance of these microbes, especially *Pseudomonas* spp., toward high concentrations of heavy metals suggests their suitability for heavy metal removal (Kamika and Momba, 2011, 2012; Kulkarni et al., 2014; Al-Dhabi et al., 2019). Previous research recorded that *Bacillus licheniformis* and *Brevibacillus laterosporus* could remove cadmium (Cd (II)), nickel (Ni^{2+}), and chromium (Cr (VI)) (Zouboulis et al., 2004; Kamika and Momba (2011, 2012)). *Bacillus*, *Pseudomonas*, *Klebsiella*, and *Escherichia* were reported to remove 96% and 54% of zinc and nickel, respectively (Arjomandzadegan et al., 2014). The efficient removal of zinc oxide NMs from wastewater has been reported to be mainly from sorption to the sludge (Puay et al., 2015).

Thus far, no other study has explicitly investigated the ability of the consortium protozoa to remove NMs, or compared this ability with those of the bacterial consortium. Therefore, the current study aims to investigate the adsorption and removal of nZnO and nAg by consortia of protozoan species (*Trachelophyllum* sp., *Peranema* sp., and *Aspidisca* sp.) and of bacterial species (*Pseudomonas putida*, *Bacillus licheniformis*, and *Brevibacillus laterosporus*). The laboratory batch equilibrium experiments were conducted with consortia as adsorbent and NMs as the adsorbate. The following objectives were pursued: (i) to determine and compare the biosorption potential of protozoan and microbial consortia by kinetic modeling of nAg and nZnO biosorption; (ii) to assess the effect of contact time, pH, and initial concentration of NMs concentrations on the biomass and their induced different levels of cytotoxicity; and (iii) to quantify the interaction of NMs with the cellular structures.

24.2 Materials and methods

24.2.1 Wastewater solution and adsorbent (NMs) preparation

Wastewater samples from WWTPs were filtered and profiled in terms of COD, DO, and pH. Nevertheless, samples were further detected for the presence of zinc and silver ions and quantify their concentrations using inductively coupled plasma optical emission spectrometry (ICP-OES) (Spectro 145 Ciros CCD, Spectro Analytical

Instruments, Kleve, Germany) prior to the initial experiment. The filtered wastewater was modified by adding of D-glucose, anhydrous, magnesium sulphate monohydrate (MgSO$_4$.H$_2$O) and potassium nitrate (KNO$_3$) as nutrients and carbon (Kamika and Momba, 2013). The pH of the modified wastewater mixed liquor as working culture media was adjusted using HCl (1.0 M) and NaOH (1.0 M) (Merck, South Africa) and autoclaved.

24.2.2 Nanomaterial characterization and solution preparation

Zinc oxide and Ag nanopowder were purchased from Sigma-Aldrich (Johannesburg, South Africa) (544906, Lot # MKBD9523V and 576832, Lot # MKBN3581V), respectively. Test NMs (with the particle size between 100 and 70 nm) were diluted to make up a working solution with a concentration of 200 g/L for nZnO and 20 g/L for nAg. Fourier transform infrared spectroscopy (FT-IR) and transmission electron microscopy (TEM, JEOL-JEM 2100) absorption spectra at a frequency range of 3500–500 cm^{-1} using Smart FT-IR connected to a component accessory with a Nicolet 380 FT-IR Spectrometer (Thermo Scientific, USA) were used to characterise the particles. Zetasizer instrument (Malvern Nano S90) was used to analyze zeta potential for nanomaterials to evaluate the surface charges.

24.2.3 Consortia of protozoan isolates and bacterial isolates

Three protozoans (*Peranema* sp., *Trachelopyllum* sp., and *Aspidisca* sp.) were isolated from wastewater, and three bacterial isolates (*Brevibacillus laterosporus* ATCC 64, *Pseudomonas putida* ATCC 31483, and *Bacillus licheniformis* ATCC 12759), obtained from Quantum Biotechnologies (Randburg, South Africa), were used in consortia per their domains. Before the experimental study, the ability of bacterial and protozoan species to adsorb NMs from the environment at several concentrations [(1.25, 2.5, 5, and 10) µg/L and (1, 5, 10, and 15) mg/L] was investigated. Inverted microscope (100 × to 400 × magnification) (Axiovert 40CFL, Zeiss, Germany) was used to monitor protozoan growth by measuring their wavelength at 600 nm (OD$_{600}$) using a spectrophotometer (Spekol 1300, Analytikjena, Germany). The aseptically prepared bacterial consortium growth was measured using UV-spectrometer at 600 nm wavelength, and the results were interpreted as optical density (OD). The measured optical density values were converted to biomass equivalant to 100 cells/mL and CFU/mL from protozoan and bacterial isolates respectively, using the following equation as given by Kachieng'a and Momba (2015):

$$y = mx + C \tag{24.1}$$

where:

- y = is the absorbance at OD$_{600\,nm}$.
- m = is the constant value of the gradient of the curve (0.005).
- x = is the actual biomass of the protozoa (cells/mL)/ bacteria (CFU/mL).
- C y-intercept of the curve (0.001).

24.2.4 Batch biosorption kinetics studies

To evaluate the biosorption kinetics, modified mixed liquor medium (MMLM) was spiked with nZnO (0.015–40 g/L) and nAg (0.015–2 g/L) by adjusting pH to 2, 7, or 10 in 100 mL Erlenmeyer flasks. These concentration ranges were considered because the target microbial species were able to grow in the set concentrations considered to be higher than the environmentally acceptable concentrations for both nZnO and nAg. The freshly grown consortium of protozoan species or bacterial species with a known biomass concentration (10^2 CFU/mL or 10^2 cells), respectively, was added in MMLM. Inoculated media were agitated for a specific contact time (5 days) at a constant temperature (30°C). Positive control (biomass) and negative control (NMs) were treated the same. The effect of contact time, the initial pH, and NMs-concentration was evaluated at 30°C. Experimental samples were then analyzed on the initial day and final day using ICP/OES.

24.2.5 Uptake of nanomaterials and percentage sorption

The uptake was determined by considering the concentration of Zn and Ag present in the MMLM, and the added initial concentration of nZnO and nAg was quantified using ICP/OES. In addition, to calculate the total removal efficiency (%), the following formula was used:

$$NMs_{removal}(\%) = (Conc_{initial} - Conc_{final} / Conc_{initial}) \times 100 \qquad (24.2)$$

where Conc.$_{initial}$ is the NMs concentration before the contact time with biomass and Conc.$_{final}$ is the NMs concentration after the contact time with biomass (Park et al. (2013)). Upon further quantification on the NMs removal efficiency, the uptake was calculated in percentage by a simple concentration difference method. The uptake of nZn and nAg was determined by a mass balance equation (Ajaelu et al. (2011)):

$$Q_e = (Conc_{initial} - Conc_{final} / m)V \times 100 \qquad (24.3)$$

where uptake/sorption (Q_e (%)) of nZnO and nAg; initial concentration (mg/L) as Conc$_{initial}$; final concentration (mg/L) as Conc$_{final}$; volume in mL as V; and mass in g as m. The experiment samples were conducted in duplicates.

24.2.6 pH effect on the biosorption

A series of 100 mL bottles containing MMLM material in varying concentrations of nZnO (0.015–40 g/L) and Ag (0.015–2 g/L) together with (biomass concentrations) were agitated in the incubator over 5 days at a temperature of 30 °C. The pH variations were assessed using a pH probe (Model). The extent of the pH effect on the biosorption of ions was carried out accordingly. The pH variation of each experiment was adjusted using the 1.0 M (NaOH) or 1.0 M (HCl) to the desired values of pH 2, 7, and 10. The biomass was then filtered, and the ion concentration residuals were analyzed.

24.2.7 Concentration effect on the biosorption

This study evaluated the effect of the concentration range of 0.015–40 g/L for nZnO and 0.015 – 2 g/L for nAg on biosorption. Approximately 1 mL of protozoan species (cells/mL) and bacterial species (CFU/mL) (Eq. 24.1) were each inoculated in Erlenmeyer flask (100 mL) containing 80 mL of modified mixed liquor and different NMs concentrations of nZnO (0.015–40 g/L) and Ag (0.015–2 g/L) at different pHs: 2, 7, and 10. These flasks were incubated at 30 °C and agitated at a speed of 100 rpm for 5 days. The samples were filtered, and the residual ions in the solution were determined. Two flasks were used for a particular concentration. The results obtained were further analyzed based on the isotherm equation of Langmuir, isotherm linearized:

$$C_e / q_e = (1/q_{max}) + (1/K_L)C_e \tag{24.4}$$

where q_{max} is a constant adsorption capacity and K_L is related to constant adsorption energy,

Freundlich isotherm linearized model with adsorption capacity (K_f as constant value) and adsorption intensity ($1/n$ as constant parameter)

$$\log q_e = \log K_f + (1/n \log C_e) \tag{24.5}$$

24.2.8 Passive accumulation on cellular structure

The interaction of NMs with consortia microbial populations was observed using the high-resolution transmission electron microscope. The interaction was further analyzed using the chemical tracking composition (NDS) for the present metal on the microbial surface a structure.

24.2.9 Statistical analysis

Data were analyzed statistically to compare removal efficiency capacity, the capabilities of the consortia of test organisms to remove COD and the uptake of DO in the presence of NMs using Kruskal-Wallis test. In addition, the correlation between COD, growth survival, and varying pH over various NMs concentrations was compared at 95% confidence using Pearson correlation test.

24.3 Results

24.3.1 Characterization of NMs

FT-IR absorption peaks signals of the wurtzite nZnO, as shown in Fig. 24.1, revealed the presence of several vibration bands observed at 500, 600, 700, 850, 990, 1000, and 1050 cm^{-1} assigned to strong carbonyl compounds (C = O), and nitrile (C = N) and alkyne (C = C) functional groups. Absorption spectra were observed between 1490 and 1600 cm^{-1}, which are vibration attributed to C = C stretch signals.

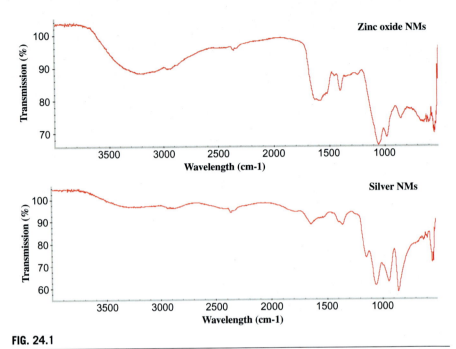

FIG. 24.1

FT-IR absorption micrographs revealing functional groups present in nZnO and nAg samples.

Similarly, the nAg particles have functional groups identified at 3000–2500 cm^{-1}, which were assigned to the stretching vibrations of amines, respectively, corresponding to the bending vibrations of a weak double bond at the wavelength of 1600 cm^{-1}. There were also single bonds of –C=O, H–C–H, –C–O, and –C–N stretching seen at 900 and 1500 cm^{-1}, respectively, which confirms amide I regions. Furthermore, there was a minor visible vibration peak at 2400 cm^{-1}, representative of C=O functional group.

24.3.2 Kinetic modeling of nZnO and nAg biosorption by protozoan and bacterial consortia

The biosorption of nAg and nZnO by protozoan and bacterial consortia was assessed and their performance was compared. Two common adsorption isotherms, the Langmuir and Freundlich models, were used to describe adsorption. Adsorption data were fitted to the linear forms using Freundlich equation (Eq. 24.5) and Langmuir equation (Eq. 24.4). The adsorption isotherms were fitting well with Langmuir isotherm model. Results showed that the effective biosorption of nZnO and nAg by the consortium of protozoan species, and that of bacterial species, decreased with the increasing concentrations of both NMs in the aqueous solution (Fig. 24.2). However, an increase in biosorption capacity at pH 7 was observed significantly by both target

24.3 Results

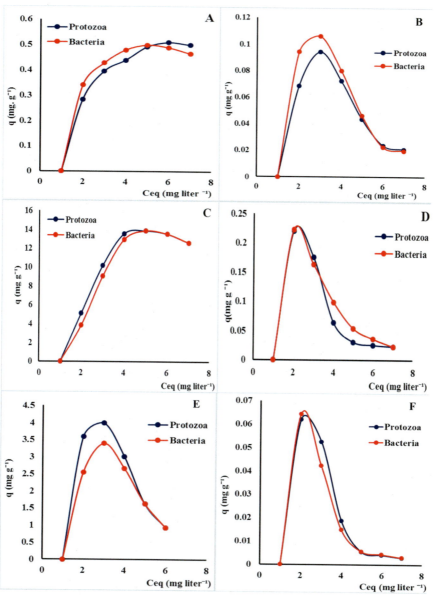

FIG. 24.2

Comparison of nZnO biosorption capacity (A, C, E) and nAg biosorption capacity (B, D, F) of the protozoan and bacterial consortia applying the Langmuir model at pH 2 (A and B), pH 7 (C and D), and pH 10 (E and F). (C_{eq} measures the concentration equilibrium (mg/L) of metal in solution, and q is the amount adsorbed (mg/g)).

organisms in the presence of nZnO. The consortium of protozoan and bacterial species exhibited the best average biosorption capacity (11.48 and 10.98), respectively, at pH 7. The bacterial consortium revealed a moderate coefficient correlation of $r = 0.3$ to $r = 0.7$ when compared to the consortium of protozoan species.

24.3.3 Reaction kinetics

From the results of the isotherm models depicted in Fig. 24.2, it was possible to establish the correlation coefficient parameters (Table 24.1). The biosorption capacities of the target microbial wastewater consortia were dependent on the microbial species and metal type. Based on the isotherm results, maximum nZnO biosorption capacities of the protozoan consortium were observed at 0.42, 4.47, and 1.17 at pH 2, 7, and 10, respectively, while bacteria consortium had maximum nZnO biosorption capacities exhibited at 3.00, 4.97, and 2.97 at pH 2, 7, and 10, respectively. Noticeably, based on the Langmuir model, the consortium of protozoan species significantly exhibited the highest absorption affinity or energy in the presence of nZnO (with K_L values of 58.69 at pH 2) and nAg (with K_L of 15.90 at pH 7) when compared to the bacterial consortium (Table 24.1). Using the Freundlich model, the highest absorption capacity was also recorded by the protozoan consortium in the presence of nZnO (with K_F values of 10. 46 at pH 2) and nAg (with K_F of 4.49 at pH 7). Overall, biosorption capacities for both target consortia were greater ($P < 0.05$) in the presence of nZnO than nAg. Additionally, Langmuir model analysis illustrated a suitable fit for the adsorption data obtained for the protozoan consortium, based on the regression (R^2) coefficients (0.99 and 0.97 for nZnO and nAg, respectively). In contrast, most correlation coefficients calculated based on the Freundlich model yielded better R^2 values ($R^2 = 1$) for the bacterial consortium in the presence of nZnO at pH 2 and 10. Statistically, there was a significant correlation between adsorption capacity of nZnO and nAg by the consortium of bacterial species with a coefficient value of $r = 0.478$. Additionally, there was a moderate correlation coefficient ($r = 0.3$ and $r = 0.7$) when compared with the consortia in the presence of NMs.

24.3.4 The removal of NMs by the microbial consortia

The effect of pH 2, 7, and 10 on the nZnO and nAg biosorption capacity of protozoan and bacterial consortia was evaluated (Fig. 24.3). In the presence of nZnO, the optimum biosorption was observed in this ascendant order: pH 7 > pH 2 > pH 10 for both target consortia. Evidently, maximum sorption capacity of nZnO was significantly higher with protozoan species (78.13%) when collated to those of bacterial species (34.89%) at pH 7. The lowest removal efficiency of nZnO by both microbial consortia was noted at pH 10 (18.41% and 6.26% by the consortium of protozoan species and bacterial species, respectively). Nevertheless, the biosorption of nAg by the microbial consortia was noted to be significant ($P < 0.05$), irrespective of the consortium species (14.55% and 15.75% by the consortium of protozoan species and bacterial species, respectively). Statistics revealed that there was a coefficient correlation of $r = 0.501$ between the removal of nZnO by both consortia. Furthermore, there was a

Table 24.1 Langmuir and Freundlich isotherm equations showing the kinetics isotherm obtained from biosorption of nZnO and nAg by bacterial and protozoan consortia.

pH	NMs	Species	Q_e	Langmuir equation			Freundlich equation		
				Q_{max} (mg/L)	K_L	R^2	K_F	n	R^2
2	ZnO	Protozoa	0.24	0.42	58.69	0.35	10.46	7.43	0.35
		Bacteria	0.53	3.00	1.40	0.00	1.00	0.00	1
	Ag	Protozoa	1.67	1.73	0.06	0.99	0.78	0.06	0.59
		Bacteria	2.10	1.16	0.54	0.96	2.00	0.08	0.89
7	ZnO	Protozoa	4.02	4.47	0.05	0.30	0.63	0.06	0.83
		Bacteria	4.09	4.97	0.91	0.99	0.01	0.40	0.86
	Ag	Protozoa	0.67	0.51	15.90	0.90	4.49	0.07	0.88
		Bacteria	1.17	2.00	0.19	0.97	3.72	0.15	0.95
10	ZnO	Protozoa	1.07	1.17	1.20	0.91	2.58	0.10	0.82
		Bacteria	0.62	2.75	0.45	0.47	1.00	0.00	1
	Ag	Protozoa	1.42	1.03	0.99	0.93	1.95	0.00	0.00
		Bacteria	0.85	1.01	2.00	0.91	1.88	0.01	0.00

R^2: Regression coefficient of kinetic curves resulting from Freundlich and Langmuir model equation.

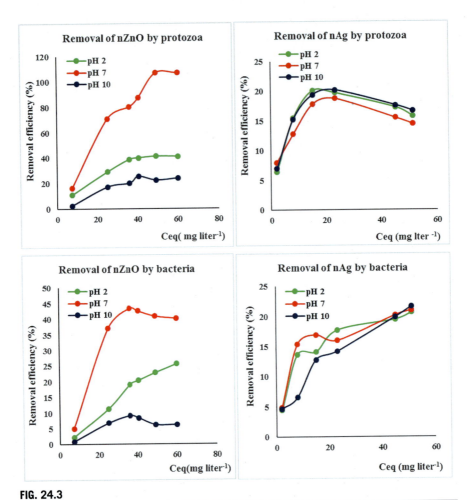

FIG. 24.3

Removal effeciency of nZnO and nAg by protozoan and bacterial consortia in modified wastewater under varying pH conditions.

correlation between protozoan ($r = 0.232$) and bacterial consortia ($r = -0.478$) due to change in pH, which illustrated higher removal efficiency by the consortium of protozoan species compared to the consortium of bacterial species.

24.3.5 The removal of COD by the microbial consortia

Ability of microbial consortia to remove organic compounds was evaluated based on the present of NMs concentration in media. It was observed that the positive controls for both domains (protozoa and bacteria) were able to remove a high level of COD, with 80.32% and 77.25%, respectively, corresponding to pH 2 and 10 (Fig. 24.4). A gradual decrease in COD removal was noted with increasing concentrations of nZnO, irrespective of the pH conditions (Figs. 24.4 and 24.5).

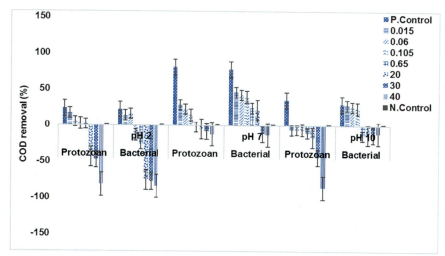

FIG. 24.4

The removal of COD by protozoan and bacterial consortia in modified wastewater mixed liquor added with nZnO under varying pH conditions.

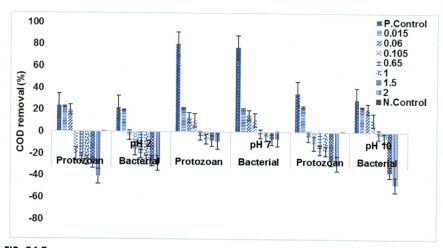

FIG. 24.5

The removal of COD by the protozoan and bacterial consortia in the modified wastewater mixed liquor added with nAg under varying pH conditions.

Despite the lower COD removal efficiency observed in both microbial consortia, there was a significant increase in COD release with approximately -80% at pH 2 and 10 (by a consortium of protozoan species) as nZnO concentration increases (40 g/L) in the solution. Thus, a negative moderate correlation coefficient of $r = -0.593$ between the consortium of protozoan species and the consortium of bacterial species was noted. Similarly, the protozoan consortium significantly removed net COD, relative to the consortium of bacterial species in the absence of nAg (Fig. 24.5). Increasing concentration of resulted in the significant decrease of COD removal, and hence, statistical analysis revealed negative deviation values of COD. It was observed that bacterial consortium was more efficient in removing COD significantly, across the pH range, unlike the consortium of protozoan species, and needed concentration of 0.105 g/L to induce COD release (3.03%), while the consortium of protozoan species released COD when exposed to a lower concentration of 0.06 g/L (26.49%) at pH 2. Furthermore, moderate correlation ($r = -0.104$) was obtained when comparing the consortium of protozoan species and the consortium of bacterial species within pH variations. Nonetheless, COD removal in the presence of NMs was moderately correlated ($r = 0.255$ to $r = 0.407$), with nAg significantly inducing the lethal effects that impacted the removal of COD.

24.3.6 Effects of passive accumulation of contaminants on the cellular structure of the organisms

The interaction between nZnO and nAg and microbial species was observed and identified using electron microscopic instruments coupled with EDS (Figs. 24.6 and 24.7). A significant cluster of microbial species was evident (Fig. 24.6A and B), but there was obvious formation of nAg observed to be accumulating on the cell surfaces.

The elemental mapping did confirm that the high concentration of aggregation detected was indeed dispersed nAg particles on the cell surface (Fig. 24.6B), while the chemical composition spectrum revealed a high presence of chemical constituents depicted to be Ag on the cell surface of microbial species (Fig. 24.6E and F). Furthermore, nZnO was noted to interact with the respective microbial isolates; however, such phenomenon increased around the cells as aggregates increased to relatively larger than 0.5 µm (Fig. 24.7A).

Some nZnO particles were observed to be dispersed on the cell membrane (Fig. 24.7C, D, and G) as shown by the SEI-STEM and elemental mapping images. The chemical composition showed high constituents spectrum assigned to ZnO on the cell membranes (Fig. 24.6H).

24.4 Discussion

For decades, anthropogenic activities and industrialization have led to drastic environmental pollution. As a result, several techniques, such as adsorption, have been developed to deal with this concern and are now regarded as being of paramount worldwide in industrial and environmental protection (Tansel, 2008). However, due to the growing environmental distress and undesired waste generation, the use of biosorption (adsorption principals and biological remediation techniques) is currently considered as one of the most reliable processes for the removal of pollutants. Even though toxicity mechanisms of several pollutants, such as NMs, have not yet been completely elucidated, zinc oxide and silver NMs are known to present public and environmental risks (Nogueiraa et al., 2015; Starner et al., 2015; Zhang et al., 2016).

24.4 Discussion

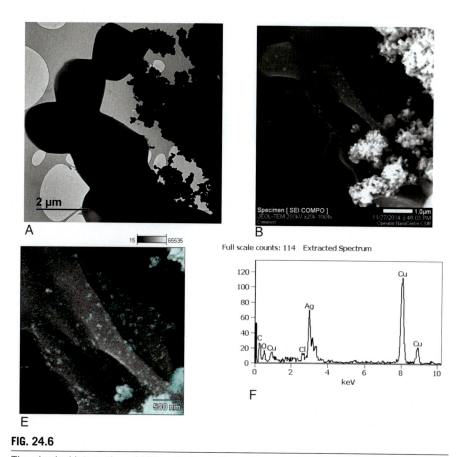

FIG. 24.6

The physical interaction of NMs with the cell membrane of the microbial species in the modified wastewater mixed liquor.

The present study aimed to investigate the adsorption or removal efficiency of nZnO and nAg using a consortium of bacterial and protozoan species. Notably, microbial species were initially acclimatized in the media containing a minimum environmental concentration of 1 μg/L to a maximum of 150 μg/L (Bundschuh et al., 2016), and their removal was also assessed before the experimental study. The study revealed that microbial species were able to grow and remove up to 150-fold (100%) of the environmental concentration of both tested NMs (data not reported). However, in the concentration range of 0.015–40 g/L for nZnO and 0.015–2 g/L for nAg, microbial biosorption capacity was noted to decrease as the concentration of nZnO and nAg increased in the aqueous solution. However, it was demonstrated that the protozoan consortium was more effective in adsorbing NMs with correlation coefficients that were moderate ranging from $r = 0.3$ to $r = 0.7$ when compared to the consortium of bacterial isolates (Fig. 24.2). Furthermore, it was also apparent that the consortium of protozoan species significantly exhibited the highest absorption affinity or energy

FIG. 24.7

The interaction of NMs with microbial cell membranes in the modified wastewater mixed liquor.

in the presence of nZnO (with K_L values of 58.69 at pH 2) and nAg (with K_L of 15.90 at pH 7) when compared to the bacterial isolates (Table 24.1). Thus, organisms tend to require higher energy to cope in toxic environments to be able to grow to utilize either intrinsic or induced mechanisms coupled with other environmental factors, for instance, pH, speciation, EPS, and redox reaction (Krell et al., 2012; Decho and Gutierrez, 2017; Biswas et al., 2020; Song et al., 2020). This study further revealed a higher removal of nZnO than nAg and this could be due to stronger antibacterial properties exhibited by nAg toward microbial cells which results in their inability to remove nAg. Similarly, it was observed when bacterial and protozoan consortia were only able to remove 15.75% and 14.55% of nAg, respectively. This confirms previous study results which reported high removal of nanoparticles could be associated with physical and chemical protperties of particles, zeta potential, size, and surface funtinalization. However, another factor that plays a role is the disintergrated cellular membrane transforms and stabilize acting as a protective barrier to reduce their toxic effects (Kiser et al., 2012). According to these authors, microbiome in activated sludge was able to remove 39%–62% of nAg when coated with citrate (CIT-Ag), polyvinylpyrrolidone (PVP-Ag), and carboxylate (Car-Ag). Despite the high removal ability portrayed by the microbiota, the removal of this unusual high concentration during the study could have also been due to the size of particle investigated NMs with varying sizes between 100 and 70 nm (Song et al., 2020).

The maximum adsorption capacity of both bacterial and protozoan consortia was assessed, and reaction kinetics parameters were determined utilizing Langmuir and Freundlich models. These results revealed higher R^2 values for Langmuir reaction than Freundlich model (Table 24.1). It was evident that the biosorption capacity was dependent on the microbial domain and metal type, as the bacterial consortium removed exceptionally more nZnO than nAg compared to the protozoan. Thus, there was a moderate correlation ($r = 0.478$) between the absorption capacity of nZnO and nAg by the consortium of bacterial species. Likewise, Babarinde et al. (2008) observed an increase of the biosorption efficiency with the increasing Zn^{2+} concentration, but reached a plateau at 50 mg/L but followed by a decrease in biosorption efficiency as the metal/biomass ratio increased. Nonetheless, several researchers have shown that *Pseudomonas* sp. (*P. aeruginosa* AT18, *P. veronii* 2E, and *P. putida* isolates), *Bacillus laterosporus*, *Bacillus licheniformis*, *Trachelophyllum* sp., *Aspidisca* sp., and *Peranema* sp. have a high affinity for a wide variety of heavy metals in polluted waters and can act as biosorbents of heavy metal ions, thus immobilizing and removing metals (Zouboulis et al., 2004; Choi et al., 2009; Silva et al., 2009; Li et al., 2014). Hence, the adsorption capacity by a consortium of protozoan species was observed to be pronounced compared to that of the bacterial species. The current experimental findings also showed that sorption capacity was affected by pH changes, especially bacterial consortium than protozoan consortium (Fig. 24.3). Furthermore, positive moderate correlation of $r = 0.2318$ was observed for protozoan consortium than the bacterial consortium, which had a negative moderate correlation coefficient $r = -0.4783$ under varying pH conditions. As a result, increasing EPS also increased pH, which affected the sorption capacity (Biswas et al., 2020). Furthermore, these NMs had negative zeta surface charge of -17.6 mV and -12.0 for nZnO and nAg, respectively, in the modified wastewater without microbial species (results not shown in this study). Previously, pH has been reported to significantly influence the solubility of the metallic ions, solution chemistry, and adsorbent zeta potential (surface charge), but most importantly, the ionization state of the cell wall and concentration of counterions and on functional groups such as imidazole, carboxylate, amino groups, and phosphate and speciation of adsorbent (Say et al., 2003; Nadeem et al., 2008; Özdemir et al., 2009; Ajaelu et al., 2011; Hsiao and Huang, 2011; Li et al., 2014; Yamindago et al., 2019). Li et al. (2014) postulated that cell surface charge was altered by low pH and the microbe becomes positively charged, and hence repels the positively charged metal cations. Furthermore, protons could compete with metal ions for binding sites, resulting in a decrease of metal ions interaction with the microbial cells. However, the opposite occurs under higher pH conditions, where precipitation of metals as insoluble hydroxides to the extent that it disrupts the sorption process is achieved (Leung et al., 2000; Elmaci et al., 2007; Li et al., 2014; Nong et al., 2019; Song et al., 2020).

Although the removal mechanism of ZnO and nAg NMs is still unknown, yet the present study revealed close contact of nZnO and nAg with the surface of the microbial cell membrane (Javed et al., 2020); these contacts resulted in possible toxicity to occur as interactions between NMs and organisms (Figs. 24.6 and 24.7). FT-IR absorption spectra of the nZnO revealed absorption bands in the 500 to 1600 cm^{-1} regions, associated with $C=O$ and $C=N$ and $C=C$ functional groups (amide I) (Fig. 24.1) can be assigned to the symmetrical and asymmetrical stretching modes; these signals were also observed by Esparza et al. (2011). According to Becheri et al. (2008), the peak intensity at 3450 and 2350 cm^{-1} are depicted to belong to –OH and $C=O$ residues. On the other hand, for nAg, the amide 1 bands were also visible, represented by single bonds of $C=O$, H–C–H, –C–O, and –C–N chemical functional classes at wavelengths of between 900 and 1500 cm^{-1} with a minor shift at a wavelength of 2400 cm^{-1} representing the $O=C=O$ functional group. As reported by Khan et al. (2014), however, there were minor shifts on the vibration as compared to those obtained by Theivasanthi and Alagar (2011). Therefore, it can be postulated that these NMs are prone to enter into the cell membrane (Vazquez-Muñoz et al., 2019; Zorraquin-Peña et al., 2020).

The toxicity of nZnO and nAg was also assessed by monitoring their impact on COD removal in the media. A decrease in COD removal efficiency was observed when the concentration of nZnO and nAg increases (Figs. 24.4 and 24.5). The effect of NMs on the microbial consortia resulted in a negative correlation of $r = -0.593$, which implies that the functionality of the microbial population was significantly impaired. Contrary to these findings, although functional bacterial species changed with high dosage, a slight reduction of COD and NH_4-N removal efficiencies was reported when ZnO-NP and Ag-NP were added (Chen et al., 2014).

24.5 Conclusion

In this study, the biosorption capacity of the protozoan and bacterial consortia was evaluated. Results obtained revealed higher removal efficiency of nZnO than nAg by both microbial consortia, corroborated by the Langmuir adsorption isotherm model. However, they were more susceptible to growth in the presence of nZnO than nAg NMs. Furthermore, pH variation and toxic effects of NMs (particularly for nAg) are likely to affect biosorption capacity by the microbial consortia, as higher removal of NMs was exhibited at pH 7. However, the lowest removal efficiency (nZnO) was noted at pH 10. The capacity of the consortia to remove the test NMs diminishes with the increasing concentrations of NMs, which ultimately reduces the COD removal, and the COD release is detected. Microbial species could remove the environmental concentration of target NMs up to 150-fold (approximately 100%). At extreme concentration (> 0.015 g/L), maximum sorption capacity for the nZnO was higher for the protozoan consortium (78.13%) when compared to the bacterial consortium. Nevertheless, the biosorption of nAg was noted to be ($P < 0.05$) irrespective of the consortium species (14.55% and 15.75% by protozoan and bacterial consortia, respectively) significantly. This is the first work investigating biosorption on/by the protozoan consortium.

References

Ajaelu, C., Ibironke, O., Adedeji, V., Olafisoye, O., 2011. Equilibrium and kinetic studies of the biosorption of heavy metal (cadmium) on *Cassia siamea* bark. AEJSR 6, 123–130.

Al-Dhabi, N.A., Esmail, G.A., Ghilan, A.M., Arasu, M.V., 2019. Optimizing the management of cadmium bioremediation capacity of metal-resistant *Pseudomonas* sp. strain Al-Dhabi-126. Isolated from industrial city of Saudi Arabian environment. Int. J. Environ. Res. Public Health 16, 1–12.

Arjomandzadegan, M., Rafiee, P., Moraveji, M., Tayeboon, M., 2014. Efficacy evaluation and kinetic study of biosorption of nickel and zinc by bacteria isolated from stressed conditions in a bubble column. Asian Pac. J. Trop. Biomed. 4, 285–632.

Babarinde, N., Babalola, J., Adebisi, O., 2008. Kinetic, isotherm and thermodynamic studies of the biosorption of zinc (II) from solution by maize wrapper. Int. J. Phys. Sci 3, 050–055.

Bai, W., Zhang, Z., Tian, W., He, X., Ma, Y., Zhao, Y., Chai, Z., 2010. Toxicity of zinc oxide nanoparticles to zebrafish embryo: a physicochemical study of toxicity mechanism. J. Nanopart. Res. 12, 1645–1654.

Becheri, A., Durr, M., Nostro, P., Baglioni, P., 2008. Synthesis and characterization of zinc oxide nanoparticles: application to textiles as UV-absorbers. J. Nanopart. Res. 10, 679–689.

Biswas, J.K., Banerjee, A., Sarkar, B., Sakar, D., Sarkar, S.K., Rai, M., Vithanage, M., 2020. Exploration of an extracellular polymeric substance from earthworm gut bacterium (Bacillus licheniformis) for bioflocculation and heavy metal removal potential. Appl. Sci. 10, 349.

Bondarenko, O., Juganson, K., Ivask, A., Kasemets, K., Mortimer, M., Kahru, A., 2013. Toxicity of Ag, CuO and ZnO nanoparticles to selected environmentally relevant test organisms and mammalian cells in vitro: a critical review. Arch. Toxicol. 87, 1181–1200.

Brunner, T., Wick, P., Manser, P., Spohn, P., Grass, R., Limbach, L., Stark, W., 2006. In vivo cytotoxicity of oxide nanoparticles: comparison to asbestos, silica, and the effects of particles solubility. Environ. Sci. Technol. 40, 4374–4381.

Bundschuh, M., Seitz, F., Rosenfeldt, R.R., Schulz, R., 2016. Effects of nanoparticles in fresh waters: risks, mechanisms and interactions. Freshw. Biol. 61 (12), 2185–2196.

Chen, J., Tang, Y., Li, Y., Nie, Y., Hou, L., Li, X., Wu, X., 2014. Impacts of different nanoparticles on functional bacterial community in activated sludge. Chemosphere 104, 141–148.

Choi, J., Lee, J., Yang, J., 2009. Biosorption of heavy metals and uranium by starfish and *Pseudomonas putida*. J. Hazard. Mater. 161, 157–162.

De Berardis, B., Civitelli, G., Condello, M., Lista, P., Pozzi, R., Arancia, G., Meschini, S., 2010. Exposure to ZnO nanoparticles induces oxidative stress and cytotoxicity in human colon carcinoma cells. Toxicol. Appl. Pharmacol. 246, 116–127.

de Souza, R.C., Haberbeck, L.U., Riella, H.G., Ribeiro, D.H.B., Carciofi, B.A.M., 2019. Antibacterial activity of zinc oxide nanoparticles synthesized by solochemical process. Braz. J. Chem. Eng. 36, 885–893.

Decho, A.W., Gutierrez, T., 2017. Microbial extracellular polymeric substance (EPSs) in ocean system. Front. Microbiol. 8, 922. https://doi.org/10.3389/fmicb.2017.00922.

Du, J., Zhang, Y., Yin, Y., Zhang, J., Ma, H., Li, K., Wan, N., 2020. Do environmental concentrations of zinc oxide nanoporicles pose ecotoxicological risk to aquatic fungi associated with leaflitter decomposition? Water Res. 178, 115840. https://doi.org/10.1016/j.watres.2020.115840.

Elmaci, A., Yonar, T., Ozengin, N., 2007. Biosorption characteristics of copper (II), chromium (III), nickel (II), and lead (II) from aqueous solutions by *Chara sp.* and *Cladophora sp.* Water Environ. Res. 79, 1000–1005.

Esparza, I., Paredes, M., Martinez, R., Gaona-Couto, A., Sanchez-Loredo, G., Flores-Velez, L., Dominguezo, O., 2011. Solid state reactions in Cr_2O_3-ZnO nanoparticles synthesized by triethanolamine chemical precipitation. Mater. Sci. Appl. 2, 1584–1592.

Fernandez-Leborans, G., Herrero, Y.O., 2000. Toxicity and bioaccumulation of lead and cadmium in marine protozoan communities. Exotoxicol. Environ. Saf. 47, 266–276.

Franklin, N., Rogers, N., Apter, S., Batley, G., Gadd, G., Casey, P., 2007. Comparative toxicity of nanoparticulates ZnO, bulk ZnO, and $ZnCl_2$ to a freshwater microalga (*Pseudokirchneriella subcapitata*). Environ. Sci. Technol. 41, 8484–8490.

Horie, M., Nishio, K., Fujita, K., Endoh, S., Miyauchi, A., Saito, Y., Yoshida, Y., 2009. Protein adsorption of ultrafine metal oxide and its influence on cytotoxicity toward cultured cells. Chem. Res. Toxicol. 22, 543–553. https://doi.org/10.1021/tx800289z.

Hsiao, I., Huang, Y., 2011. Effects of various physicochemical characteristics on the toxicities of ZnO and TiO_2 nanoparticles toward human lung epithelial cells. Sci. Total Environ. 409, 1219–1228.

Huang, J., Huang, G., An, C., Xin, X., Chen, X., Zhao, Y., Feng, R., Xiong, W., 2020. Exploring the use of ceramic disk filter coated with Ag/ZnO nanocomposites as an innovative approach for removing *Escherichia coli* from household drinking water. Chemosphere 245, 125545.

Ivask, A., Kurvet, I., Kasemets, K., Blinova, I., Auroja, V., Suppi, S., Kahru, A., 2014. Size-dependent toxicity of silver nanoparticles to bacteria, yeast, algae, crustaceans and mammalian cells in vitro. PLoS One 9, 102–108. https://doi.org/10.1371/journal.pone.0102108.

Javed, B., Raja, N.I., Nadhman, A., Mashwani, Z., 2020. Understanding the potential of bio-fabricated non-oxidate silver nanoparticles to eradicate Leihmania and plant bacterial pathogens. Appl. Nanosci. 1-11.

Johnston, J., LaPara, T., Behrens, S., 2019. Composition and dynamics of the activated sludge microbiome during seasonal nitrification failure. Sci. Rep. 9, 4565. https://doi.org/10.1038/s41598-019-40872-4.

Kachieng'a, L.O., Momba, M.N.B., 2015. Biodegradation of fats and oils in domestic wastewater by selected protozoan isolates. Water Air Soil Pollut. 226, 1–15.

Kamika, I., Momba, M., 2011. Comparison of the tolerance limits of selected bacterial and protozoan species to vanadium in wastewater systems. Water Air Soil Pollut. 223, 2525–2539.

Kamika, I., Momba, M., 2012. Comparison the tolerance limit of selected bacterial and protozoan species to nickel in wastewater systems. Sci. Total Environ. 410–411, 172–181.

Kamika, I., Momba, M., 2013. Assessing the resistance and bioremediation ability of selected bacterial and protozoan species of heavy metals in metal rich industrial wastewater. BMC Microbiol. 13, 1–14.

Khan, S., Ahmed, B., Raghuvanshi, S., Wahab, M., 2014. Structural, morphological and optical properties of silver doped polyvinylpyrrolidone composites. Indian J. Pure ApPhy. 52, 192–197.

Kiser, M., Ladner, D., Hristovski, K., Westerhoff, P., 2012. Nanomaterial transformation and association with fresh and freeze-dried wastewater activated sludge: implications for testing protocol and environmental fate. Environ. Sci. Technol. 46, 7046–7053.

Krell, T., Lacal, J., Guazzaroni, M.E., Busch, A., Silva-Jiménez, H., Fillet, S., Reyes-Darías, J.A., Muñoz-Martínez, F., Rico-Jiménez, M., García-Fontana, C., Duque, E., Segura, A., Ramos, J., 2012. Responses of *Pseudomonas putida* to toxic aromatic carbon sources. J. Biotechnol. 160, 25–32.

Kulkarni, R., Shetty, K., Srinikethan, G., 2014. Cadmium (II) and nickel (II) biosorption by *Bacillus laterosporus* (MTCC 1628). J. Taiwan Inst. Chem. E. 45, 1628–1635.

Lashkenari, S.M., Nikpay, A., Soltani, M., Gerayeli, A., 2019. In vitro antiprotozoal activity of poly(rhodanine)-coated zinc oxide nanopricles against *Trichomonas gallinae*. J. Dispers. Sci. Technol. 41, 495–502. https://doi.org/10.1080/01932691.2019.1591972.

Leung, W., Wong, M., Chua, H., Lo, W., Yu, P., Leung, C., 2000. Removal and recovery of heavy metals by bacteria isolated from activated sludge treating industrial effluents and municipal wastewater. Water Sci. Technol. 41, 233–240.

Li, Y., Wu, Y., Wang, C., Wang, P., 2014. Biosorption of copper, manganese, cadmium, and zinc by *Pseudomonas putida* isolated from contaminated sediments. Desalin. Water Treat. 52, 37–39. https://doi.org/10.1080/19443994.2013.823567.

Martín-González, A., Díaz, S., Borniquel, S., Gallego, A., Gutiérrez, J.C., 2006. Cytotoxicity and bioaccumulation of heavy metals by ciliated protozoa isolated from urban wastewater treatment plants. Res. Microbiol. 157, 108–118.

Mboyi, A., Kamika, I., Momba, M.N.B., 2017. Detrimental effects of commercial zinc oxide and silver nanomaterials on bacterial populations and performance of wastewater systems. Phys. Chem. Earth 100, 158–169.

Nadeem, R., Hanif, M., Shaheen, F., Perveen, S., Zafar, M., Iqbal, T., 2008. Physical and chemical modification of distillery sludge for Pb (II) biosorption. J. Hazard. Mater. 150, 335–342.

Navarro, E., Piccapietra, F., Wagner, B., Marconi, F., Kaegi, R., Odzak, N., Behra, R., 2008. Toxicity of silver nanoparticles to *Chlamydomonas reihardtii*. Environ. Sci. Technol. 42, 8959–8964.

Nogueiraa, V., Lopesa, I., Rocha-Santosb, T., Goncalvesa, F., Pereirad, R., 2015. Toxicity of solid residues resulting from wastewater treatment with nanomaterials. Aquat. Toxicol. 165, 172–178.

Nong, Q., Yuan, K., Li, Z., Chen, P., Huang, Y., Hu, L., Jiang, J., Luan, T., Chen, B., 2019. Bacterial resistance to lead: chemical basis and environmental relevance. Int. J. Environ. Sci., 46–55.

Özdemir, S., Kilinc, E., Poli, A., Nicolaus, B., Guven, K., 2009. Biosorption of Cd, Cu, Ni, Mn and Zn from aqueous solutions by thermophilic bacteria, *Geobacillus toebii sub.sp. decanicus Geobacillus thermoleovorans sub.sp. stromboliensis*: equilibrium kinetic and thermodynamic studies. Chem. Eng. J., 195–206.

Papadimitriou, C.A., Papatheodoulou, A., Takavakoglou, V., Zdragas, A., Samaras, P., Sakellaropoulos, G.P., Lazaridou, M., Zalidis, G., 2010. Investigation of protozoa as indicators of wastewater treatment efficiency in constructed wetlands. Desalination 250, 378–382.

Park, H., Kim, H.Y., Cha, S., Ahn, C.H., Roh, J., Park, S., Choi, K., Yi, J., Kim, Y., Yoon, J., 2013. Removal characteristics of engineered nanoparticles by activated sludge. Chemosphere 92, 524–528.

Puay, N., Qiu, G., Ting, Y., 2015. Effect of zinc oxide nanoparticles on biological wastewater treatment in a sequencing batch reactor. J. Clean. Prod. 88, 139–145.

Salvadó, H., Gracia, M.P., Amigó, J.M., 1995. Capability of ciliated protozoa as indicators of effluent quality in activated sludge plants. Water Res. 29, 1041–1050.

Say, R., Yilmaz, N., Denizli, A., 2003. Biosorption of cadmium, lead, mercury and arsenic ions by the fungus *Penicillium purpurogenum*. Sep. Sci. Technol. 38, 2039–2053.

Sharma, V., Shukla, R., Saxena, N., Parmar, D., Das, M., Dhawan, A., 2009. DNA damaging potential of zinc oxide nanoparticles in human epidermal cells. Toxicol. Lett. 185, 211–218.

Silva, R., Rodrıguez, A., De Oca, J., Moreno, D., 2009. Biosorption of chromium, copper, manganese and zinc by *Pseudomonas aeruginosa* AT18 isolated from a site contaminated with petroleum. Bioresour. Technol. 100, 1533–1538.

Simon-Deckers, A., Loo, S., Mayne-L'hermite, M., Herlin-Boime, N., Menhuy, N., Reynaud, C., Carriere, M., 2009. Size, composition and shape dependent toxicological impact of metal oxide nanoparticles and carbon nanotubes towards bacteria. Environ. Sci. Technol. 43, 8423–8429.

Song, K., Zhang, W., Sun, G., Hu, X., Wang, J., Yao, L., 2020. Dynamics cytoxicity of ZnO nanoparticles and bulk particles to Escherichia coli: a view from unfixed ZnO paricles: Zn^{2+} ratio. Aquat. Toxicol. 220, 105407.

Starner, D.L., Unrine, J.M., Starnes, C.P., Collins, B.E., Oostveen, E.K., Ma, R., Lowry, G.V., Bertsch, P.M., Tsyusko, O.V., 2015. Impacts of sulfidation on the bioavailability and toxicity of silver nanoparticles of *Cenorhabditis elegans*. Environ. Pollut. 196, 239–245.

Sun, X., Sheng, Z., Liu, Y., 2013. Effects of silver nanoparticles on microbial community structure in activated sludge. Sci. Total Environ. 443, 828–835.

Tansel, B., 2008. New technologies for water and wastewater treatment: a survey of recent patents. Recent Patents on Chemical Engineering 1, 17–26.

Theivasanthi, T., Alagar, M., 2011. Electrolytic synthesis and characterizations of silver nanopowder. Nano Biomed. Engin. 4, 58.

Todorova, K., Velkova, Z., Stoytcheva, M., Kirova, G., Kostadinova, S., Gochev, V., 2019. Novel composite biosornet from Bacillus cereus for heavy metals removal from aqueous solutions. Biotechnol. Biotechnol. Equip. 33, 730–738.

Vazquez-Muñoz, R., Meza-Villezcas, A., Fournier, P.G.J., Soria-Castro, E., Juarez-Moreno, K., Gallego-Hernández, A.L., Bogdanchikova, N., Vazquez-Duhalt, R., Huerta-Saquero, A., 2019. Enhancement of antibiotics antimicrobial activity due to silver nanoparticles impact on the on the cell membrane. PLoS One 14, 1–18.

Wang, H., Wick, R., Xing, B., 2009. Toxicity of nanoparticulate and bulk ZnO, Al_2O_3 and TiO_2 to the nemaotode *Caenorhadbitis elegans*. Environ. Pollut. 157, 1171–1177.

Xiong, D., Fang, T., Yu, L., Sima, X., Zhu, W., 2011. Effects of nano-scale TiO_2, ZnO and their bulk counterparts on zebrafish: acute toxicity, oxidative stress and oxidative damage. Sci. Total Environ. 409, 1444–1452.

Yamindago, A., Lee, N., Woo, S., Yum, S., 2019. Impact of zinc oxide nanopricles on the bacterial community of *Hyda magnipapillata*. Mol. Cell. Toxicol. 16, 63–72.

Yang, Y., Wang, L., Xiang, F., Zhao, L., Qiao, Z., 2020. Activated sludge microbial community and treatment performace of wastewater treatment plants in industrial and municipal zonesInt. J. Environ. Res. Pub. Health. 17, 436.

Zhang, J., Dong, Q., Liu, Y., Zhou, X., Shi, H., 2016. Response to shock load of engineered nanoparticles in an activated sludge treatment system: insight into microbial community succession. Chemosphere 144, 1837–1844.

Zheng, Y., Li, R., Wang, Y., 2009. In vitro and in vivo biocompatibility of ZnO nanoparticles. Int. J. Mod. Phys. B 23, 1566–1571.

Zorraquin-Peña, I., Cueva, C., Bartolomé, B., Moreno-Arribas, M.V., 2020. Silver nanoparticles against food borne bacteria. Effcets at intestinal level and helath limitations. Microorganisms 8, 132.

Zouboulis, A., Loukidou, M., Matis, K., 2004. Biosorption of toxic metals from aqueous solutions by bacteria strains isolated from metal-polluted soils. Process Biochem. 39, 909–916.

Silver-doped metal ferrites for wastewater treatment

CHAPTER 25

Nimra Nadeem[a], Muhammad Zahid[a], Muhammad Asif Hanif[a], Ijaz Ahmad Bhatti[a], Imran Shahid[b], Zulfiqar Ahmad Rehan[c], Tajamal Hussain[d], and Qamar Abbas[e]

[a]Department of Chemistry, University of Agriculture, Faisalabad, Pakistan
[b]Environmental Science Centre, Qatar University, Doha, Qatar
[c]Department of Polymer Engineering, National Textile University Faisalabad, Faisalabad, Pakistan
[d]Institute of Chemistry, University of the Punjab, Lahore, Pakistan
[e]Institute for Chemistry and Technology of Materials, Graz University of Technology, Graz, Austria

Abbreviations

Ag-BFO	silver-doped bismuth ferrite
Ag-MFNPs	silver-doped metal ferrite nanoparticles
AgNPs	silver nanoparticles
BFO	bismuth ferrite
BQ	benzoquinone
CFU	colony-forming units
DBPs	disinfectants by-products
FCC	face centered cubic
FNPs	ferrite nanoparticles
IPA	isopropanol
KI	potassium iodide
LB	Lysogeny broth
ROS	reactive oxygen species
SQUID	superconducting quantum interference device
T_c	curie temperature
ZOI	zone of inhibition

25.1 Introduction

Continuous supply of clean and safe water is a crucial challenge to human society. Owing to the fast development in industrialization and the rapid increase in population growth, the demand for freshwater resources is increasing day by day. It is expected that the problem associated with accessing freshwater will be worse in the future. According to an estimation, about 50% population of developing countries do not have access to fresh water (Figoli et al., 2017). Although diseases associated with contaminated water exist in both developed and developing countries, the deficiency in producing freshwater system is more problematic in developing countries. Interestingly, the countries containing conventional wastewater remediation technologies and distribution systems spread effectively, and new challenges are emerging due to waterborne problems. The anthropogenic or natural activities supply several pollutants in water, making it nonportable and unfit for human, animal, and plant consumption. The consumption of such water may pose health consequences that vary from mild effects to hazardous and even consumers' death depending upon the pollutant nature.

Different water purification methods are used, which either aims at removing the toxic and poisonous chemical from water in a process called decontamination or removing parasites and pathogens from water under the disinfection process. Adsorption, coagulation, and chelation are the methods used for the treatment of chemical pollutants from infected water, whereas boiling, UV treatment, ozonation are the methods used for disinfection. The effectiveness of UV treatment in degrading organic pollutants is highly recognized (Devi and Kavitha, 2013), but the treatment cost is the real bottleneck in accessing economical solutions to wastewater, particularly for its pilot plant applications. The conventional treatment technologies may not be economical, particularly when scaled up. These may result in secondary pollution. Some technologies may only kill the vegetative pathogenic cells and not effective to kill resistant pathogens. The secondary pollutants from chemical treatment maybe even carcinogenic. Therefore, improved water treatment technologies aimed at making the processes effective, safe, accessible, and economical are being developed.

Ferrite nanoparticles (FNPs) have been studied for their extensive applications in wastewater treatment (Zahid et al., 2019; Mushtaq et al., 2020; Nadeem et al., 2020; Tabasum et al., 2020). The role of FNPs in photocatalytic degradation of various organic pollutants as an adsorbent for contaminants, pathogen disinfection, analyte extraction, membrane modification is well recognized. The broad range applications of FNPs are attributed to their fascinating structural, chemical, thermal, and magnetic properties. Moreover, the decoration of FNPs with noble metals not only improves the structural properties but also the optical properties of FNPs. Studies revealed that substituting the small percentage of ferric ions (Fe^{3+}) with the metal ions of rare-earth induces strain results in structural distortion and, therefore, modifies the electrical and magnetic properties of FNPs. Nanoparticles with antibacterial applications are well recognized for their applications in water disinfection. Silver is the widely used noble metal as a capping or doping agent for a variety of NPs, including FNPs.

This chapter focuses on the effective use of ferrite NPs with special reference to their silver-doped nanocomposites, their properties, and applications in wastewater treatment. The possible reaction mechanism during the synthesis of Ag-doped metal ferrite nanoparticles (Ag-MFNPs), and the mechanism of photocatalytic and antibacterial activity is also presented.

25.2 Chapter objectives

The proposed chapter aims at exploring the potential of Ag-doped metal ferrites for their applications in wastewater treatment. The environmental perspectives of water pollutants, their health hazards, and some remediation technologies have also been presented. Further, the focus has been laid on the metal ferrites due to their superior properties if compared with conventional methods. The role of silver in promoting the wastewater treatment applications of metal ferrite is also discussed. This chapter also holds the physicochemical properties of Ag-doped metal ferrites. In the end, the antibacterial and photocatalytic mode of action against water pollutants is also presented.

25.3 Disinfection of wastewater

The chemical disinfection technology and UV irradiation technology can potentially control microbial pathogens such as viruses, algae, fungi, bacteria, mold spores, and yeast (Nangmenyi and Economy, 2009). The chemical disinfection agents include ozone, chlorine, chlorine dioxide to treat the waterborne microbiological contamination from wastewater. However, disinfection treatment technologies using chemical disinfection revealed challenges due to the production of disinfectants by-products (DBPs) (Matilainen and Sillanpää, 2010; Grellier et al., 2015). The chlorinated disinfection agents are supposed to react with iodide ions, bromide ions, and organic matter to generate a variety of DBPs. Some DBPs reported in the literature include trihalomethane (THMs), chlorite, haloacetic acid, haloacetonitriles, halonitromethanes, halofuranones, haloamides, iodo-THMs, halofuranones, nitrosamines, and many others (Singer, 2006; Hua and Reckhow, 2007; Matilainen and Sillanpää, 2010). The formaldehyde and bromate are the DBPs formed during ozonation (Matilainen and Sillanpää, 2010) and the nitrogenous DBPs formed during UV irradiation (Kolkman et al., 2015). Moreover, prolonged UV exposure for effective disinfection is another drawback. Furthermore, some pathogens are resilient to the conventional disinfectants and required high disinfection dose, resulting in more DBPs generation (Li et al., 2008b). Although the use of advanced wastewater treatment technologies by developed countries reported disease epidemics from waterborne infections (Reynolds, 2007), advanced disinfections treatment technologies are required not only to enhance their reliability but also to avoid the production of DBPs. Recently, several research papers have presented the microbicidal efficiency of various nanomaterials such as metallic nanoparticles, nanophotocatalysts, and carbon-supported nanomaterials (Li et al., 2008b). The described nanomaterials are weedy oxidizing agents. Consequently, the formation of DBPs is not likely in the disinfection process.

25.4 Silver nanoparticles

Nanometallic particles, particularly the silver (Ag), exhibit antimicrobial characteristics owing to the oligodynamic effect (Mahmoudi and Serpooshan, 2012). Wide study and commercialization of silver nanoparticle (AgNPs) is due to their following features:

- widespread microbicidal action,
- low toxicity as compared to other metallic nanoparticles,
- easy incorporation into a variety of cheaper materials for their applications in filters, and
- historic applications of Ag ions and Ag compounds as disinfectants.

Various bactericidal action of AgNPs have been reported, but the exact mechanism is still concerning. The bactericidal efficiency of AgNPs is attributed to the oligodynamic effect of Ag ions and distinguished physicochemical characteristics of AgNPs, which are responsible for the bacterial destruction (Li et al., 2008b). Three mechanisms are presented here (Morones et al., 2005; Kittler et al., 2010; Loo et al., 2015):

1. Ag^+ ions are released in small concentrations by AgNPs and taken by bacterial cells. Ag^+ ions instantly react with the thiol groups of enzymes and producing reactive oxygen species (ROS) like hydrogen peroxides and superoxide radicle species. At once, Ag^+ ions inhibit ATP synthesis and transport of e^- in the respiratory chain.
2. AgNPs adhere to the outer membrane surface and disturb the surface membrane properties. Furthermore, the reaction of AgNPs with hydrogen peroxide produces hydroxyl radicles, which abolish the cell membrane and lead to cell death.
3. AgNPs can degrade the membrane's lipopolysaccharide molecules and generate "pits." In this way, the membrane permeability increases, and AgNPs degrade the DNA molecule.

The AgNPs cannot be used directly for disinfection of water. For real applications, the AgNPs cannot be used without supports (Figoli et al., 2017). AgNPs have been used extensively as an antibacterial agent due to their stability, antibacterial activity under broad-spectrum range, and durability (Li et al., 2008a; Sharma et al., 2009). AgNPs are clusters of atomic silver (Ag^0), which range from 1 to 100 nm in diameter. Presently, the AgNPs have emerged as a new generation of antibacterial agents (Rai et al., 2009), being used in hygiene (Zhang et al., 2011), antibacterial water filter (Jain and Pradeep, 2005), and medical applications (Lok et al., 2006; Atiyeh et al., 2007; Roe et al., 2008). Among other metallic NPs such as gold, titanium, zinc, and copper, the silver NPs possess the highest antibacterial efficiency (Rai et al., 2009). Further, the research has shown that the AgNPs exhibit superior antibacterial performance as compared to other Ag compounds and bulk silver as well. The treatment of nosocomial water—infected frequently with the antibiotic-resistant bacteria—can be done potentially using AgNPs (Salvatorelli et al., 2005). A high concentration of chlorine compounds is required to treat nosocomial water, which may offer high cancer risk to human beings due to halomethane formation (Chang, 1982).

The antibacterial action of the silver ions is perhaps like that of silver nanoparticles (AgNPs) (Kong and Jang, 2008). The ultrasmall size is the principal characteristic of AgNPs, leading to the high surface area. However, the AgNPs having a diameter < 20 nm tend to aggregate when used and decrease their antibacterial activity.

To overcome the problem associated with the aggregation, a broad range of materials like ferrites (Amarjargal et al., 2013), zeolites (Inoue et al., 2002), SiO_2 (Oh et al., 2006), TiO_2 (Akhavan, 2009), graphitic carbon nitride (Saher et al., 2020), and activated carbon (AC) (Chen et al., 2005) have been used to immobilize the AgNPs so that the homogeneous formation of ultrafine AgNPs can be achieved without any aggregations, and such nanocomposites offer many their advantages like their improved band structure and high stability and reusability (Kooti et al., 2013).

25.5 Ferrite nanoparticles

Ferrites are ferrimagnetic particles and can be obtained as ceramic or powder. The main component of these chemical compounds is Fe_2O_3. The composition of these iron oxides (Fe_2O_3, FeO) can be partly different with other transition metal oxides (e.g., Mn, Zn, Co, and Ni) to have a variety of metal ferrites with distinguished properties. Among ferrites and iron oxides, the magnetite (Fe_2O_3) is of interest. Owing to the crystalline structure, these metal ferrites are classified into three groups: (a) spinel ferrites (MFe_2O_4), (b) hexagonal ferrites ($MFe_{12}O_{19}$), and (c) garnet ferrites ($M_3Fe_5O_{12}$), where M is a bivalent transition metal (Fe, Mn, Ni, Co, Cu, and Zn). Besides pristine metal ferrites, the doped metal ferrites are also attractive candidates toward various example applications, photocatalytic degradation of organic pollutants, adsorption of heavy metals from aqueous media, magnetic and electronic devices, production of sustainable hydrogen, and many more. These ferrites prevail the property of being magnetically separable from the reaction media after completion of catalytic reactions. So far, only zinc, nickel, cobalt, copper, manganese, core − shell, and many mixed metal ferrites have been used widely for the photocatalytic degradation of several pollutants. Zahid and co-workers have describes the commonly used synthesis techniques for metal ferrites. The magnetic ferrite nanoparticles have been recognized as potential material for the effective degradation of many pollutants due to their fascinating properties (Zahid et al., 2019).

25.6 Properties of ferrite NPs

Ferrites are well known due to their thermal and chemical stability in the aqueous system. These are stable in alkaline and near-neutral conditions in photochemical cells. However, in acidic media, ferrites suffer from corrosion. Besides the crystallographic structure, mono- or disubstituted cations in ferrites affect the physicochemical properties of ferrite such as the conductivity (resistivity), the variation in bandgap structure, optical properties, and the behavior as a p/n-type material. Hence, the selection of mono-, bi-, or multisubstitution of cations in ferrite network provides the way to tune surface as well as bulk physicochemical properties of ferrites as photocatalysts for environmental remediation.

The exposed surface area of ferrite materials depends strongly on the synthesis method. Although the ferrites exhibit low- to medium-specific surface area (< 100 $m^2 g^{-1}$) as compared to the specific surface area of commercial or lab synthesized TiO_2 photocatalyst (200–300 $m^2 g^{-1}$), the calcination of ferrites may be a suitable option for the improvement in the surface area of ferrite materials. However, the outcomes of moderate or low calcination may not necessarily be in the form of an increase in specific surface area of ferrites, instead, it may result in a decrease in photocatalytic activity due to low crystallinity and hence high recombination rate.

The ferrites are usually semiconductor materials and possess the advantage of being visible light-responsive with a low bandgap (about 2 eV) corresponding to wavelength below 621 nm. This contrasts the famous UV light-responsive TiO_2 with the bandgap of 3.2 eV corresponding to the wavelength below 388 nm. The ferrites usually have band structure (location of valance and conduction) suitable for either the reduction of protons or oxidation of water, the generation of highly reactive · OH radicals, and the generation of superoxide radicle by the reduction of dioxygen, all necessary for water splitting and environmental applications.

Likewise, the influence of substituted metals (mono-, bi-, or multisubstituted) on the reactivity of thermal catalysis, their ratios, and chemical nature also affects the photocatalytic properties such as optical (bandgap), conductivity, material reflectivity, and n/p type performance of the semiconductor. The position of valence and conduction bands are influenced by the n- or p-type nature of spinel ferrites and subsequently the energy bandgap. For example, the spinel ferrite $MgFe_2O_4$ has a bandgap of 1.8 eV (more cathodic) for p-type ferrite as compared to the bandgap of n-type spinel ferrite, which is 0.3 eV smaller than the p-type ferrite. The ABO_3 transition metal oxides (orthoferrite) are attractive candidates toward visible light photocatalysis due to their small bandgap (2.0–2.7 eV). The visible light receptive photocatalytic activity of these orthoferrites is due to the relative location of valance and conduction bands.

25.7 Ag-doped metal ferrite nanoparticles (Ag-MFNPs)

The silver-coated or silver substituted ferrite NPs exhibit improved chemical and biological properties (Fig. 25.1). The positive role of silver in developing electrical and thermal conductivities has been observed (Okasha, 2008). The decoration of ferrite NPs with Ag can open up the doors of its applications in a wide range like targeted drug delivery in biomedical applications, magnetic resonance imaging, and cancer treatment using hyperthermia (Kumar and Mohammad, 2011; Amiri and Shokrollahi, 2013; Pour et al., 2017). Furthermore, owing to the antibacterial activities of Ag-MFNPs, these can be used in packaging and food processing, water

25.7 Ag-doped metal ferrite nanoparticles (Ag-MFNPs)

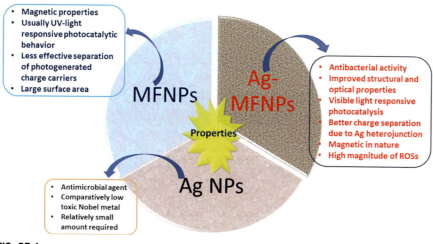

FIG. 25.1
Properties of silver nanoparticles, metal ferrite nanoparticles, and silver-doped metal ferrite nanoparticles.

disinfection, and biomedical devices (Husanu et al., 2018; Krishnakumar et al., 2018; Pande et al., 2018).

25.7.1 Structural properties

The Ag affects the structural properties of Ag-MFNPs considerably. It has been observed that no silver oxide formed even when the sintered sample (under air) was characterized, which suggests that the silver oxide is less stable than the metallic silver (Okasha, 2008). Okasha and colleagues also reported that the lattice constant (physical dimension of the unit in the crystal lattice) decreases with increasing the silver content up to critical concentration (i.e., X = 0.2) followed by a slight increase and then nearly stable values. The reason for the change may logically be due to the different ionic radius of Ag^+ (1.24 Å) and Fe^{3+} (0.64 Å) ions. The increase in the Ag^+ ions at the expense of Fe^{3+} ions causes a decrease in the volume of the unit cell.

25.7.2 Magnetic properties

Measurements of the magnetic properties of Ag-MFNPs are usually carried out using a SQUID magnetometer, and the magnetization hysteresis curves are obtained. The magnetization curves of both simple metal ferrites and silver-doped metal ferrites exhibit high magnetization values. However, the variations in the magnetic properties of Ag-MFNPs as compared to native MFNPs are attributed to the insertion of silver in Ag-MFNPs. For example, the saturation magnetization of Fe_3O_4 (magnetite) is reported as 66.24 emu/g with no coercivity or remanence magnetization detected, provided that the magnetite is superparamagnetic, whereas the significant increase in the saturation magnetization value of Ag/Fe-Fe$_2$O$_4$ was recorded at 76.02 emu/g. A weak ferromagnetic behavior was observed in Ag/Fe-Fe$_2$O$_4$, coercivity (hc = 57.03 Oe), and saturation remanence (Mrs = 1.54 emu/g). Such high-value magnetization of Ag-MFNPs reflects that doping of metal ferrite with silver improves the magnetic behavior of nanocomposites.

25.7.2.1 Effect of particle size on the magnetization

It has been observed that high magnetic moment is related to the decrease in the size of metal ferrites. For example, Thapa et al. (2004) suggested that the increase in the magnetization values can be understood based on the crystal structure of magnetite—a ferrimagnetic iron oxide with inverse spinel cubic structure having oxygen anions with face-centered cubic (FCC) close packing and iron cations sited at interstitial octahedral and tetrahedral sites. At room temperature, the electrons may trip between ferrous (Fe^{2+}) and ferric (Fe^{3+}) ions in the octahedral sites, imparting partial magnetic properties to Fe$_3$O$_4$. The unit cell magnetic moment arises from Fe^{2+} ions. Also, with the decrease in particle size, the relative decrease in oxygen content is observed, which might lower the valance state of cations. The increase in the volume of the unit cell with the decrease in the particle size of magnetite may imply the increase in Fe^{2+} in the sample (i.e., Fe^{2+} ionic radius (0.74 Å) greater than the ionic radius of Fe^{3+} (0.64 Å)). The change in the magnetization of Fe$_3$O$_4$ nanocomposites was due to two factors:

- First, the paramagnetic moment associated with the strong exchangeable interactions between ions located in tetrahedral (A-sites) space and in octahedral (B-sites) space in unit crystal with FCC inverse spinel structure.
- Second, the decrease in domain size with the increase in dopant concentration.

By considering the above two factors, the magnetic moment of silver-doped M-ferrites can be described. More interestingly, the variations in the magnetic moment of both metal ferrites and silver-doped metal ferrites are due to the variation in Ag content present in their structure and improved magnetization is observed in Ag/Fe$_3$O$_4$.

The change in curie temperature (temperature where a sharp change in the magnetic properties of magnetic material takes place) with the change in Ag content is also reported (Okasha, 2008). Imparting Ag (with an ionic radius of 1.26 Å) results with an increase in the Fe^{2+}/Fe^{3+} ratio on B sublattice sites of silver-doped ferrites, resultantly few Fe^{2+} ions can migrate from B site to A site; therefore, a direct decrease in the magnetization of the system takes place. The exchange in the interaction between magnetic ions located on both A and B sites increases with the magnetic moment and density of magnetic ions. High thermal energy is mandatory to balance the exchange interaction effect. The high Ag^+ ion concentration (above critical) decreases the Fe^{3+} ion concentration on B sites. Subsequently, the rearrangement of metal cations occurs in the spinel matrix and increases curie temperature (T_C).

25.7.3 Optical properties

Optical properties are associated with the behavior of a material under light sources. The role of the light source in detraining the photocatalytic response of Ag-MFNPs is crucial for evaluating their commercial applications. The photocatalytic degradation potential of materials is associated with the generation and the stabilization of photogenerated charge carriers (e^-/h^+). Bandgap energy is the amount of energy required to successfully transfer an electron from valance band (VB) to the conduction band (CB) of the material. The as-generated electron and proton play important role in the water purification system either by disinfection, degradation, or mineralization processes. Several studies reported narrowing the bandgap of FNPs and other material doping them with Ag. The reduction in bandgap energy with the insertion of Ag concentration may be due to the strong adhesion between the host substrate and deposited material that results in bigger grain size and therefore reducing the bandgap energies. According to a study conducted by Zeeshan and coworkers (Zeehan et al., 2019), the reduced bandgap of silver-doped cobalt – chromium ferrite thin films is possibly due to the formation of unsaturated bonds during deposition of silver on thin-film, which causes inadequate numbers of atoms. The defects generated in thin films due to these bonds create localized states and the thickness of film increases. The increase in film thickness is responsible for the increase in the width of the localized states in a bandgap. Consequently, the optical bandgap decreases, and the bandgap energy reduces. Furthermore, the addition of silver in reducing bandgaps may also be attributed to the electron dependence behavior on the band shift. The bandgap is associated with the electron transition from VB to the fermi levels of CB for semiconductor degeneration. According to the Moss Burstein effect, the increase in silver concentration should increase the band values, but the bandgap energy values decrease with silver doping. This is due to an increase in defects like stresses, impurities, oxygen vacancies, the extent of the disorder, and partial replacement of dopant with host atom (Chen et al., 2009).

25.8 The formation mechanism of Ag-MFNPs using hydrothermal method

The synthesis of typical silver-doped metal ferrites usually follows the chemical precipitation method. The detailed formation mechanism is presented here by selecting Ag-doped Fe_3O_4 core – shell as a model Ag-MFNPs (Fig. 25.2).

The formation route of Fe_3O_4 NPs with ferrous and ferric salts at mole ratio 1:2 is commonly referred to as "coprecipitation" synthesis. The stoichiometric ratio of Fe^{2+} (ferrous) and Fe^{3+} (ferric) ion react with the basic solution ($NaOH/NH_3OH$) to form Fe_3O_4 (Kim et al., 2001; Iida et al., 2007). The precursor used for the synthesis of Ag-doped Fe_3O_4 includes ferrous chloride tetrahydrate ($FeCl_2 \cdot 4H_2O$) as a ferrous ion (Fe^{2+}) source and silver nitrate ($AgNO_3$) as a silver ion (Ag^+) source using a one-pot hydrothermal method using ammonium hydroxide (NH_3OH). The synthesis

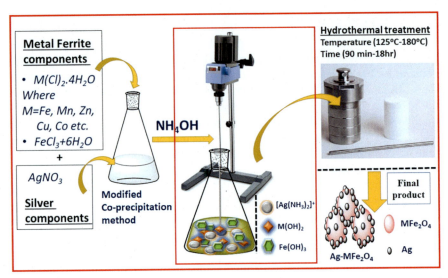

FIG. 25.2

Hydrothermal synthesis route for Ag-doped metal ferrite nanoparticles.

using ferrous salt only in the absence of ferric precursor calls for the diverse reaction mechanism for Fe_3O_4 NPs synthesis. The mixing of two separate aqueous solutions results in the intermediate form, which is responsible for the large NPs of Fe_3O_4 (Iida et al., 2007; Zheng et al., 2010). An interesting mechanism proposed by Genin et al. (Olowe and Génin, 1991; Refait and Génin, 1993) follows the alkalization reaction of Fe^{2+} ion for the formation of iron oxide and iron hydroxide. They projected the following reaction mechanism for Fe_3O_4 (Eqs. (25.1)–(25.3)):

$$Fe^{2+} + 2OH^- \rightarrow Fe(OH)_2 \tag{25.1}$$

$$2Fe(OH)_2 + \frac{1}{2}O_2 \rightarrow Fe(OH)_2 + 2FeOOH + H_2O \tag{25.2}$$

$$Fe(OH)_2 + 2FeOOH \rightarrow Fe_3O_4 + 2H_2O \tag{25.3}$$

whereas the reduction of silver diamine complex $[Ag(NH_3)_2^+]$ into AgNPs may occur due to the involvement of sonochemical reaction, microwave irradiations, and the addition of reducing agents. The reduction is carried out using ammonium hydroxide (Eqs. (25.4)–(25.6)).

$$2Ag^+ + 2OH^- \rightarrow Ag_2O + H_2O \tag{25.4}$$

$$Ag_2O + 2NH_4^+ \rightarrow Ag(NH_3)_2^+ + H_2O \tag{25.5}$$

$$Ag(NH_3)_2^+ \rightarrow Ag^0 \tag{25.6}$$

When a solution holding Ag^+ and Fe^{2+} ions are entertained with ammonium solution, the dark brown precipitates produced immediately and gradually changed into black precipitates, indicating the formation of intermediate before the formation of Fe_3O_4 NPs.

Moreover, at elevated temperature, the hydrolysis reaction may promote the dehydration of Fe^{3+} precursor whereas at low temperature (room temperature), the oxides of Fe^{3+} were important in inhibiting the Fe_3O_4 formation. At elevated temperature, the increased hydrolysis and dehydration increases Fe_3O_4 formation and provide fine crystal structure too (Wang et al., 2005). Therefore, an effective hydrothermal treatment increases the rate of reaction for the synthesis of Fe_3O_4.

Amarjargal and colleagues proposed the detail of the synthesis mechanism of Ag-doped Fe_3O_4 (Amarjargal et al., 2013). First, the ferrous hydroxide [$Fe^{II}(OH)_2$] and ferric oxyhydroxide ($Fe^{III}OOH$) (Eq. 25.3) intermediates were formed, and then, the partial oxidation of ferrous hydroxide takes place (Eq. 25.2). The partial oxidation reaction of $Fe^{II}(OH)_2$ occurs by the addition of ammonia into ferrous chloride, and it depends upon the ratio (r) of the initial concentration of reactants (r = Cl^-/OH^-) (Refait and Génin, 1993). The pot-treatment as a hydrothermal reaction facilitates the formation of the Fe_3O_4 component on AgNPs which further reduces by the residual complexions. The stability constant of AG2HL is lower (1 g $K_f(Ag_2H)$ = 7.1) as compared to the stability constant of FeL (1 g $K_f(FeL)$ = 25.0), the concentration of free Ag^+ ion exceeds the concentration of Fe^{2+} ions, and hence, the formation of Ag^+ also increases as compared to the formation of the Fe^{2+} ions, resulting in the large size of Ag than Fe_3O_4. Finally, the composite formed with the core of a single crystal of Ag and shell of Fe_3O_4 polycrystal (Amarjargal et al., 2013).

25.9 Antibacterial activity test of Ag-MFNPs

The antibacterial action of Ag-MFNPs has been described in various research studies. As usual, the determination of antibacterial performance is carried out using ZOI and the bacterial inactivation measurements. Gram-negative *E. coli* top 10 strain is used as a model microorganism. The growth of *E. coli* is carried out in lysogeny broth (LB) having distilled water, sodium chloride, tryptone, and yeast extract in proper proportion. Before bacterial inoculation, sterilization of LB is mandatory, and it is done by autoclaving it at 121°C for 20 min and 15 psi. One *E. coli* colony is cultured in 500 μL LB medium and incubated at 35°C ± 0.1°C and placed in an incubator shaker with the shaking speed of 200 rpm for 24 h.

To evaluate the antibacterial potential of a typical Ag-MFNPs, a stock solution is prepared to contain 100 μL LB medium, which is then further diluted up to 50 mL distilled water with the final *E. coli* concentration of about ~ 105 CFU/mL (Amarjargal et al., 2012). To measure bacterial inactivation, 3 mg Ag-MFNPs nanocomposite is to be dispersed in the stock solution (300 μL) of bacterial suspension and distilled water (27 mL) followed by incubation in dark at room temperature for

2 h. An agar plate is used to place the small amount of each sample in triplicate for the record of their media and then incubated at 37°C overnight.

For the ZOI, first, the agar nutrient plates were incubated using 1 mL suspension of bacterial culture with 105 CFU (colony forming units). The same amount of Ag-MFNPs nanocomposite is then allowed to place gently on incubation plates and place in incubation at 37°C for 24 h. To determine the ZOI, the area (mm) is measured where no bacterial growth is observed around every antibacterial agent.

25.10 Applications of Ag-MFNPs in wastewater treatment

Extraordinary features of Ag-MFNPs, such as high surface area, antibacterial activity, visible light-responsive photocatalytic activity, Fenton-like mechanism for oxidative degradation of organic pollutants, and magnetic nature, make them efficient materials for wastewater treatment. Table 25.1 contains the application review of various Ag-MFNPs.

25.10.1 Photodegradation of dyes using Ag-MFNPs

The role of silver in the photodegradation of dyes can be elucidated by considering the research conducted by Fernandes et al. (2019). No photodegradation of Rhodamine B was observed after 120 min of light irradiation ($\lambda > 400$ nm) using $Zn_{0.5}Ca_{0.5}Fe_2O_4$ NPs. However, when silver-decorated $Zn_{0.5}Ca_{0.5}Fe_2O_4$ NPs were used, a significant enhancement in the photodegradation of RhB was observed under visible light irradiation. The boosted photodegradation ability of Ag–$Zn_{0.5}Ca_{0.5}Fe_2O_4$ NPs was attributed to the better charge separation due to the transport of e^- to silver clusters attached to the surface of magnetic NPs (Chidambaram et al., 2016); therefore, the recombination susceptibility of photogenerated charge carriers reduces. The results showed that most of the dye degradation with the first 30 min of irradiation. The 2 mg/mL concentration of calcined NPs (Ag–$Zn_{0.5}Ca_{0.5}Fe_2O_4$) was chosen, and first five minutes of reaction time under visible light irradiation were recorded very fast followed by stabilization. A plateau was observed for other samples but decreases with the increase in particle concentration.

Another thing to be considered is that the degradation rate highly depend upon the structure of Azo dyes (Guillard et al., 2003). For example, the degradation of Congo Red (CR) was reported 0.112 min^{-1} with ZnO photocatalysts under UV irradiation (365 nm) emitted from 1000 W lamp (Chen et al., 2017). A comparable photodegradation rate was observed for the related class of azo dyes (CI Reactive Red 195) containing two naphthalene sulfonated groups associated with azo moiety (like CR dye) under visible light irradiation ($\lambda > 400$ nm) with low lamp power (200 W). Another textile dye Hispamin Black CA with a comparable structure (two naphthalene-sulfonated functionalities associated with azo group) showed a very low degradation rate (0.05 min^{-1}) using oxygen peroxide oxidation under UV irradiation (Cisneros et al., 2002).

Table 25.1 Application review of various Ag-MFNPs.

Catalyst	Ag (wt%)	Mode of action	Pollutant	Photocatalytic degradation					Estimated particle size	Degradation efficiency	References
				Effective light source	Experimental conditions	Bandgap				
AgNWs@ ZnFe$_2$O$_4$	16.66	Biofilm Inhibition	*Candida albicans*	Solar simulator	Catalyst does = 100 μg/mL Time = 80 min	–		~70 nm	55.7%	Thakur et al. (2019)
Ag@ Zn$_{0.5}$Ca$_{0.5}$Fe$_2$O$_4$	33.6	Photodegradation	Rhodamine B	200 W Xenone Arc lamp	2 mg/mL Time = 30 min	–		3.97 nm		Fernandes et al. (2019)
AgFeO$_2$	50	Persulfate degradation (PDS)	Rhodamine B	–	PDS conc. = 0.2 g/L Time = 100 min. Cat. Dose = 1 g/L pH = 4.65	1.6 eV		10–20 nm	99.4%	Li et al. (2019)
(AgFeO$_2$-G)	–	Photocatalytic degradation	Methyl orange Methylene blue	50 W xenon lamp	Time = 40 min	–		–	96% for MO 100% for MB	Hosseini et al. (2017)
AgFeO$_2$/PANI	–	photocatalytic degradation	Tetracycline	300 W xenon lamp	Cat. Dose = 30 mg/L Time = 60 min	2.54 eV		–	Up to 91.8%	Chen et al. (2019)
ZnFe$_2$O$_4$/AgI	–	Catalytic disinfection of E.coli Photocatalytic degradation of RhB	*E. coli* and RhB	300 W Xe lamp filtered through a UV cut off (k < 420 nm)	Time = 80 min for E. coli removal, 40 min for RhB removal	1.92 eV for ZnFe2O4 2.8 eV for AgI		–	100% removal of E.coli (98.5% removal of RhB	Xu et al. (2018a,b)

Continued

Table 25.1 Application review of various Ag-MFNPs—cont'd

Photocatalytic degradation

Catalyst	Ag (wt%)	Mode of action	Pollutant	Effective light source	Experimental conditions	Bandgap	Estimated particle size	Degradation efficiency	References
Ag-doped BiFeO3	–	Photocatalytic degradation	MG	105 W compact fluorescent lamp	Cat.dose = 0.1 g/100 mL Time = 250 min pH = 5.4	2.53 eV	–	~100%	Jaffari et al. (2019)
Ag@MgFe$_2$O$_4$	0.692	Photocatalytic degradation	MG	UV lamp (345–400nm)	Cat.dose = 1 g/L pH = 5.8–5.9	2.0 eV	–	70%	Tsvetkov et al. (2019)
Ag@ZnFe$_2$O$_4$	0.692	Photocatalytic degradation	MG	Visible light	Cat.dose = 1 g/L pH = 5.8–5.9	1.83 eV	22 nm	72%	Tsvetkov et al. (2019)
Ag@Co$_2$Fe$_2$O$_4$	0.32	Photocatalytic degradation	MG	UV lamp (345–400nm)	Cat.dose = 1 g/L pH = 5.8–5.9	1.26 eV	31 nm	27%	Tsvetkov et al. (2019)
Zn$_{0.85}$Fe$_{0.05}$Ag$_{0.1}$O	10	Photocatalytic activity	MB	High-pressure mercury lamp (UV-365 nm)	Cat. dose = 10 mg/L Time = 4 h	3.14 eV	34.57 nm	87%	Mahmood et al. (2014)

Antimicrobial activities

Catalyst	Ag (wt%)	Mode of action	Crystallite size	Synthesis method	Culture used	Saturation magnetization (emu/g)	Zone of inhibition	References
Ag@CoFe$_2$O$_4$	0.02	Antibacterial activity	21.6 nm	Sol–gel auto combustion	Escherichia coli, and Listeria monocytogenes	49.95	8 mm for E. coli 9 mm for L. monocytogenes	Mahajan et al. (2019)
Ag@CoFe$_2$O$_4$	0.02	Antibacterial activity	19.05 nm	Green synthesis using Tulsi extract	Escherichia coli, and Listeria monocytogenes	49.71	9 mm for E. coli 9 mm for L. monocytogenes	Mahajan et al. (2019)

Material		Application	Size	Method	Microorganism		Result	Reference
Ag@ CoFe$_2$O$_4$	0.02	Antibacterial activity	18 nm	Green synthesis using Garlic extract	Escherichia coli, and Listeria monocytogenes	28.89	7 mm for E. coli 8 mm for L. monocytogenes	Mahajan et al. (2019)
Ag–MgFe$_2$O$_4$	–	–	100 nm	Solid state combustion	Escherichia coli Enterococcus faecal Pseudomonas aeruginosa	–	22 mm for E. coli 18 mm for Enterococcus faecal 16 mm for Pseudomonas aeruginosa	Lagashetty et al. (2019)
Ag-doped CuFe$_2$O$_4$	3.0	Disk diffusion assay for antimicrobial activity	21 nm	Sol – gel	Candida albicans, Aspergillus, Aspergillus flavus	–	23, 21, and 20 mm for Candida albicans, Aspergillus, Aspergillus flavus, respectively	Thakare et al. (2018)
MnFe$_2$O$_4$/Ag	–	Antibacterial activity	–	Hydrothermal method	Escherichia coli, S. aureus	44.48	Radius 1.78 cm for E. coli and 2.14 cm for S. aureus	He et al. (2014)
NiFe$_2$O$_4$@Ag	–	Antibacterial activity, catalytic properties	22.62 nm	Thermal decomposition	Alternaria solani and Fusarium oxysporum		Complete inhibition of fungi	Golkhatmi et al. (2017)

The high concentration of TiO$_2$ photocatalyst, i.e., 3 g/L, was used to decolorize 98% of Reactive Red 198 (Remazol Red 133, with naphthalene-bisulfonated functionalities associated with azo group) within 120 min of UV irradiation, whereas the Ag-decorated metal ferrite (Ag/Zn$_{0.5}$Ca$_{0.5}$Fe$_2$O$_4$) with comparatively lower NPs concentration (2 g/L) was reported for Reactive Red 195 under visible light irradiation.

Beside silver insertion, the role of calcium incorporation in forming mixed metal ferrite is also appreciable, as the decrease of 40% degradation efficiency was observed using 22.7% silver-decorated ZnFe$_2$O$_4$ (at zero Ca) for 25 min under visible light irradiation using 500 W Xenon lamp (Cao et al., 2007).

25.10.1.1 Possible photocatalytic degradation mechanism and the role of Ag dopant

The schematic representation of the possible charge carrier mechanism in boosting the photocatalytic degradation is presented in Fig. 25.3. Under the specific light source (depending upon the energy band gap of Ag-MFNPs), valence band (VB) electrons of MFNPs may be photoexcited and then transferred to the conduction band (CB). Simultaneously, the large number of photogenerated charge carriers may recombine, making them less available for photodegradation processes. Doping the MFNPs with Ag could provide the trapping site for e$^-$, therefore, extending the lifetime of photogenerated charge carriers (h$^+$/e$^-$). Then, the stored electrons in Ag dopant can capture dissolved oxygen to produce hydrogen peroxide and endured series of chemical reactions to provide active $^{\bullet}$OH radicals (Niu et al., 2015; Zhang et al., 2015;

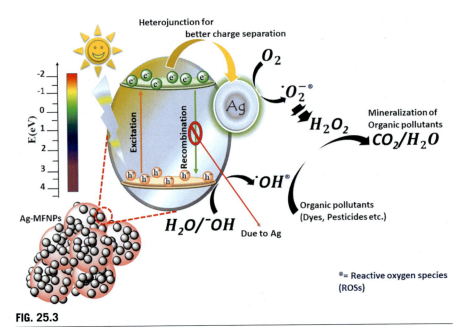

FIG. 25.3

Photocatalytic degradation mechanism for Ag-doped metal ferrite nanoparticles.

Lam et al., 2018). Meanwhile, the accumulated h^+ in the VB of MFNPs can oxidize the organic pollutants directly into CO_2 and H_2O. Moreover, the scavenging experiment (Jaffari et al., 2019) has witnessed the pivotal role of ROS in the photodegradation of organic pollutants.

Therefore, it is concluded that the doping of Ag not only improves the inhibition of charge carrier's recombination but also provides additional active site reaction.

25.10.2 An antibacterial activity using Ag-doped BFO nanocomposite

Unfortunately, wastewater may also contain a diverse variety of bacteria and others posing a threat to human health and ecological stability. The *E. coli* (gram-negative bacterium) and *M. luteuc* (gram-positive bacterium) are the prime causatives for the food and waterborne diseases like food poising, cholera, diarrhea, blood infection, urinary tract infections, and others (Karthik et al., 2018; Quek et al., 2018). Therefore, the study of Ag-MFNPs in terms of their antibacterial activity is indispensable. Jaffari and colleagues performed a test using bismuth ferrite (BFO) and silver-doped bismuth ferrite (Ag-BFO). In photolysis (without catalysts) experiments, no considerable inhibition of bacteria (both *E. coli* and *M. luteuc*) was observed under visible light, which shows that both strains were resistant to inhibition under visible light irradiations. In contrast, BFO and Ag-doped BFO exhibit antibacterial activities to some extent in dark conditions. However, the catalysts Ag-BFO performed very well under visible light with strong antibacterial activities, i.e., ZOI of 32 ± 1 mm and 28 ± 1 mm for *M. luteuc* and *E. coli* while BFO exhibited bactericidal activities with ZOI of 23 ± 2 mm and 22 ± 3 mm for *M. luteuc* and *E. coli*, respectively. So, the nanocomposite (Ag-BFO) manifested effective and excellent wide spectral antibacterial properties. The enhanced antibacterial performance of Ag-doped BFO may be attributed to the electrostatic forces between the negatively charged bacterial cell and positively charged catalyst surface (Najma et al., 2018; Xu et al., 2018b) responsible for the growth inhibition of bacterial culture. Another probable reason may be the generation of an interface between Ag and BFO to achieve efficient separation of charge carriers responsible to produce ample ROS (h^+, H_2O_2, and $^{\bullet}OH$ radicals) (Adhikari et al., 2015). Stimulatingly, the antibacterial performance of BFO and Ag-BFO against *M. luteus* bacterium was more prominent as compared to the antibacterial performance against *E. coli*. The resistant behavior of *E. coli* could be due to the outer membrane, which acts as a preventive shield against antibacterial activity or may slow down the entrance of antibacterial agents (Dai et al., 2017; Ahmed et al., 2018).

25.10.3 The ROS antibacterial activity mechanism of Ag-doped catalysts

The role of metal oxide NPs in antibacterial activity may be pacified by the H_2O_2 or superoxide anion radicals. The working action of these active free radicals known as the reactive oxygen species (ROS) mechanism (Song et al., 2018;

Wang et al., 2018). ROS is a comprehensive term associated with the assembly of partly reduced oxygen-containing species like peroxide, superoxide, and free radicles. The ferrous ions (Fe^{2+})—from ferrite NPs—produces H_2O_2 after reacting with dissolved oxygen. The produced H_2O_2 can react with Fe^{2+} ions to produce OH radicles result in intercellular stress, depolymerization of polysaccharides, damaging the DNA strand ultimately cellular necrosis (Wang et al., 2014, 2018). The mechanism includes the disturbance of ionic and electronic transportation acROS cell membrane due to the generation of ROS (by the reaction of metal oxides with oxygen species) attributed to the strong attraction of NPs for the cell membrane (Wang et al., 2014). In the first stage, the NPs react with cell membrane disturbing the fluid flow acROS bacterial cell, which in turn triggers other processes. After the disruption of the cell membrane, three main processes describe the antibacterial activity. ROS generation increases the concentration of hydrogen peroxide and/or superoxide anions, which in turn increases the intercellular stress and ultimately bacterial death. The diffusion of OH radicles into DNA strands inhibits the replication of DNA and stops or inhibits the bacterial growth. The radicles disrupt the enzyme functioning, which retard the bacterial growth and even the death of bacteria (Fig. 25.4).

Another important factor that affects the antibacterial efficiency is the concentration of dopant used. Palak et al. (Mahajan et al., 2019) observed the increase in the ZOI, which is the increase in antibacterial activity by increasing the concentration of Ag (dopant). The dopant persuades the fast generation rate of the surface located oxygen species, which causes bacterial death (Mahajan et al., 2019).

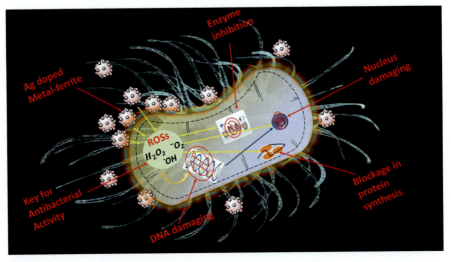

FIG. 25.4

Role of Ag-doped metal ferrite nanoparticle in antibacterial activity.

25.10.4 Role of ROS in degradation

The role of ROS species in promoting photocatalytic degradation of various organic pollutants has been investigated extensively. For example, Jaffarai and co-workers performed trapping experiments for ROS using Ag-BFO nanocomposites for the degradation of MG dyes (Jaffari et al., 2019). Potassium iodide (KI), catalase, benzoquinone (BQ), isopropanol (IPA) were used as trapping agents for h^+, H_2O_2, $^•O_2^-$, and, $^•OH$ ROS species respectively. The scavenger, which causes a maximum reduction in photodegradation of MG, was graded as the most active species. The degradation of MG reduces from 89.8% to 33.7%, 36.5%, and 55.8% using KI, H_2O_2, and IPA, respectively, while the addition of BQ was found to pose no negative impact on the MG degradation. These results suggest that the degradation mechanism of MG using silver-doped bismuth ferrite nanocomposite follows the combined effect of ROS, including h^+, H_2O_2, and $^•OH$ radicles, whereas $^•O_2^-$ showed a nominal effect on the degradation process.

25.11 Conclusion

In summary, the silver-doped metal ferrite NPs are the capitative agents being used for wastewater treatment. Considering the effectiveness of Ag in activity enhancement, prevention in charge recombination, and efficient production of ROS, the Ag-MFNPs can be used on a large scale. The hydrothermal synthesis is among the facile strategy to decor MFNPs with Ag. The Ag-MFNPs may be successfully used against the photocatalytic degradation of various organic pollutants. Additionally, the improved antibacterial activity of these materials ensures their application in biological research and biomedical applications. The magnetic properties of these NPs offer easy separation from the reaction mixture and better reusability.

References

Adhikari, S.P., Pant, H.R., Kim, J.H., Kim, H.J., Park, C.H., Kim, C.S., 2015. One pot synthesis and characterization of Ag-ZnO/g-C3N4 photocatalyst with improved photoactivity and antibacterial properties. Colloids Surf., A 482, 477–484.

Ahmed, B., Ojha, A.K., Singh, A., Hirsch, F., Fischer, I., Patrice, D., Materny, A., 2018. Well-controlled in-situ growth of 2D WO3 rectangular sheets on reduced graphene oxide with strong photocatalytic and antibacterial properties. J. Hazard. Mater. 347, 266–278.

Akhavan, O., 2009. Lasting antibacterial activities of Ag–TiO2/Ag/a-TiO2 nanocomposite thin film photocatalysts under solar light irradiation. J. Colloid Interface Sci. 336 (1), 117–124.

Amarjargal, A., Tijing, L.D., Ruelo, M.T.G., Lee, D.H., Kim, C.S., 2012. Facile synthesis and immobilization of Ag–TiO2 nanoparticles on electrospun PU nanofibers by polyol technique and simple immersion. Mater. Chem. Phys. 135 (2), 277–281.

Amarjargal, A., Tijing, L.D., Im, I.-T., Kim, C.S., 2013. Simultaneous preparation of Ag/Fe3O4 core–shell nanocomposites with enhanced magnetic moment and strong antibacterial and catalytic properties. Int. J. Chem. Eng. 226, 243–254.

Amiri, S., Shokrollahi, H., 2013. The role of cobalt ferrite magnetic nanoparticles in medical science. Mater. Sci. Eng. C. 33 (1), 1–8.

Atiyeh, B.S., Costagliola, M., Hayek, S.N., Dibo, S.A., 2007. Effect of silver on burn wound infection control and healing: review of the literature. Burns 33 (2), 139–148.

Cao, X., Gu, L., Lan, X., Zhao, C., Yao, D., Sheng, W., 2007. Spinel ZnFe2O4 nanoplates embedded with Ag clusters: preparation, characterization, and photocatalytic application. Mater. Chem. Phys. 106 (2–3), 175–180.

Chang, S.-L., 1982. The safety of water disinfection. Annu. Rev. Public Health 3 (1), 393–418.

Chen, S., Liu, J., Zeng, H., 2005. Structure and antibacterial activity of silver-supporting activated carbon fibers. J. Mater. Sci. 40 (23), 6223–6231.

Chen, K.J., Hung, F.Y., Chang, S.J., Hu, Z.S., 2009. Microstructures, optical and electrical properties of in-doped ZnO thin films prepared by sol–gel method. Appl. Surf. Sci. 255 (12), 6308–6312.

Chen, X., Wu, Z., Liu, D., Gao, Z., 2017. Preparation of ZnO photocatalyst for the efficient and rapid photocatalytic degradation of azo dyes. Nanoscale Res. Lett. 12 (1), 143.

Chen, S., Huang, D., Zeng, G., Gong, X., Xue, W., Li, J., Yang, Y., Zhou, C., Li, Z., Yan, X., 2019. Modifying delafossite silver ferrite with polyaniline: visible-light-response Z-scheme heterojunction with charge transfer driven by internal electric field. Int. J. Chem. Eng. 370, 1087–1100.

Chidambaram, S., Pari, B., Kasi, N., Muthusamy, S., 2016. ZnO/Ag heterostructures embedded in Fe3O4 nanoparticles for magnetically recoverable photocatalysis. J. Alloys Compd. 665, 404–410.

Cisneros, R.L., Espinoza, A.G., Litter, M.I., 2002. Photodegradation of an azo dye of the textile industry. Chemosphere 48 (4), 393–399.

Dai, L., Liu, R., Hu, L.-Q., Si, C.-L., 2017. Simple and green fabrication of AgCl/Ag-cellulose paper with antibacterial and photocatalytic activity. Carbohydr. Polym. 174, 450–455.

Devi, L.G., Kavitha, R., 2013. A review on non metal ion doped titania for the photocatalytic degradation of organic pollutants under UV/solar light: role of photogenerated charge carrier dynamics in enhancing the activity. Appl. Catal. 140, 559–587.

Fernandes, R.J., Magalhães, C.A., Amorim, C.O., Amaral, V.S., Almeida, B.G., Castanheira, E., Coutinho, P.J., 2019. Magnetic nanoparticles of zinc/calcium ferrite decorated with silver for photodegradation of dyes. Materials. 12 (21), 3582.

Figoli, A., Dorraji, M.S.S., Amani-Ghadim, A.R., 2017. 4 - application of nanotechnology in drinking water purification. In: Grumezescu, A.M. (Ed.), Water Purification. Academic Press, pp. 119–167.

Golkhatmi, F.M., Bahramian, B., Mamarabadi, M., 2017. Application of surface modified nano ferrite nickel in catalytic reaction (epoxidation of alkenes) and investigation on its antibacterial and antifungal activities. Mater. Sci. Eng. C. 78, 1–11.

Grellier, J., Rushton, L., Briggs, D.J., Nieuwenhuijsen, M.J., 2015. Assessing the human health impacts of exposure to disinfection by-products — a critical review of concepts and methods. Environ. Int. 78, 61–81.

Guillard, C., Lachheb, H., Houas, A., Ksibi, M., Elaloui, E., Herrmann, J.-M., 2003. Influence of chemical structure of dyes, of pH and of inorganic salts on their photocatalytic degradation by TiO2 comparison of the efficiency of powder and supported TiO2. J. Photochem. Photobiol. A. 158 (1), 27–36.

He, Q., Liu, J., Liang, J., Huang, C., Li, W., 2014. Synthesis and antibacterial activity of magnetic MnFe2O4/Ag composite particles. Nanosci. Nanotechnol. 6 (5), 385–391.

Hosseini, S.M., Hosseini-Monfared, H., Abbasi, V., 2017. Silver ferrite–graphene nanocomposite and its good photocatalytic performance in air and visible light for organic dye removal. Appl. Organomet. Chem. 31 (4), e3589.

Hua, G., Reckhow, D.A., 2007. Characterization of disinfection byproduct precursors based on hydrophobicity and molecular size. Environ. Sci. Technol. 41 (9), 3309–3315.

Husanu, E., Chiappe, C., Bernardini, A., Cappello, V., Gemmi, M., 2018. Synthesis of colloidal Ag nanoparticles with citrate based ionic liquids as reducing and capping agents. Colloid Surf. A. 538, 506–512.

Iida, H., Takayanagi, K., Nakanishi, T., Osaka, T., 2007. Synthesis of Fe3O4 nanoparticles with various sizes and magnetic properties by controlled hydrolysis. J. Colloid Interface Sci. 314 (1), 274–280.

Inoue, Y., Hoshino, M., Takahashi, H., Noguchi, T., Murata, T., Kanzaki, Y., Hamashima, H., Sasatsu, M., 2002. Bactericidal activity of ag–zeolite mediated by reactive oxygen species under aerated conditions. J. Inorg. Biochem. 92 (1), 37–42.

Jaffari, Z.H., Lam, S.-M., Sin, J.-C., Zeng, H., 2019. Boosting visible light photocatalytic and antibacterial performance by decoration of silver on magnetic spindle-like bismuth ferrite. Mater. Sci. Semicond. Process. 101, 103–115.

Jain, P., Pradeep, T., 2005. Potential of silver nanoparticle-coated polyurethane foam as an antibacterial water filter. Biotechnol. Bioeng. 90 (1), 59–63.

Karthik, K., Dhanuskodi, S., Gobinath, C., Prabukumar, S., Sivaramakrishnan, S., 2018. Multifunctional properties of microwave assisted CdO–NiO–ZnO mixed metal oxide nanocomposite: enhanced photocatalytic and antibacterial activities. J. Mater. Sci. Mater. 29 (7), 5459–5471.

Kim, D., Zhang, Y., Voit, W., Rao, K., Muhammed, M., 2001. Synthesis and characterization of surfactant-coated superparamagnetic monodispersed iron oxide nanoparticles. J. Magn. Magn. Mater. 225 (1–2), 30–36.

Kittler, S., Greulich, C., Diendorf, J., Koller, M., Epple, M., 2010. Toxicity of silver nanoparticles increases during storage because of slow dissolution under release of silver ions. Chem. Mater. 22 (16), 4548–4554.

Kolkman, A., Martijn, B.J., Vughs, D., Baken, K.A., van Wezel, A.P., 2015. Tracing nitrogenous disinfection byproducts after medium pressure UV water treatment by stable isotope labeling and high resolution mass spectrometry. Environ. Sci. Technol. 49 (7), 4458–4465.

Kong, H., Jang, J., 2008. Antibacterial properties of novel poly (methyl methacrylate) nanofiber containing silver nanoparticles. Langmuir 24 (5), 2051–2056.

Kooti, M., Saiahi, S., Motamedi, H., 2013. Fabrication of silver-coated cobalt ferrite nanocomposite and the study of its antibacterial activity. J. Magn. Magn. Mater. 333, 138–143.

Krishnakumar, S., Janani, P., Mugilarasi, S., Kumari, G., Janney, J.B., 2018. Chemical induced fabrication of silver nanoparticles (Ag-NPs) as nanocatalyst with alpha amylase enzyme for enhanced breakdown of starch. Biocatal. Agric. Biotechnol. 15, 377–383.

Kumar, C.S., Mohammad, F., 2011. Magnetic nanomaterials for hyperthermia-based therapy and controlled drug delivery. Adv. Drug Deliv. Rev. 63 (9), 789–808.

Lagashetty, A., Pattar, A., Ganiger, S.K., 2019. Synthesis, characterization and antibacterial study of Ag doped magnesium ferrite nanocomposite. Heliyon. 5 (5), e01760.

Lam, S.-M., Jaffari, Z.H., Sin, J.-C., 2018. Hydrothermal synthesis of coral-like palladium-doped BiFeO3 nanocomposites with enhanced photocatalytic and magnetic properties. Mater. Lett. 224, 1–4.

Li, Q., Mahaendra, S., Lyon, D.Y., Brunet, l., Liga, M.V., Li, D., Alvarez, P.J.J., 2008a. Antimicrobial nanomaterials for water disinfection and microbial control: potential applications and implications. Water Res. 42, 4591–4602.

Li, Q., Mahendra, S., Lyon, D.Y., Brunet, L., Liga, M.V., Li, D., Alvarez, P.J.J., 2008b. Antimicrobial nanomaterials for water disinfection and microbial control: potential applications and implications. Water Res. 42 (18), 4591–4602.

Li, C., Yang, W., Guo, S., Yang, Z., Fida, H., Liang, H., 2019. Efficient degradation of refractory contaminants with silver ferrite for persulfate activation. FML. 12 (6), 1950083.

Lok, C.-N., Ho, C.-M., Chen, R., He, Q.-Y., Yu, W.-Y., Sun, H., Tam, P.K.-H., Chiu, J.-F., Che, C.-M., 2006. Proteomic analysis of the mode of antibacterial action of silver nanoparticles. J. Proteome Res. 5 (4), 916–924.

Loo, S.-L., Krantz, W.B., Fane, A.G., Gao, Y., Lim, T.-T., Hu, X., 2015. Bactericidal mechanisms revealed for rapid water disinfection by superabsorbent cryogels decorated with silver nanoparticles. Environ. Sci. Technol. 49 (4), 2310–2318.

Mahajan, P., Sharma, A., Kaur, B., Goyal, N., Gautam, S., 2019. Green synthesized (Ocimum sanctum and Allium sativum) Ag-doped cobalt ferrite nanoparticles for antibacterial application. Vacuum 161, 389–397.

Mahmood, A., Ramay, S.M., Al-Zaghayer, Y.S., Imran, M., Atiq, S., Al-Johani, M.S., 2014. Magnetic and photocatalytic response of Ag-doped ZnFeO nano-composites for photocatalytic degradation of reactive dyes in aqueous solution. J. Alloys Compd. 614, 436–442.

Mahmoudi, M., Serpooshan, V., 2012. Silver-coated engineered magnetic nanoparticles are promising for the success in the fight against antibacterial resistance threat. ACS Nano 6 (3), 2656–2664.

Matilainen, A., Sillanpää, M., 2010. Removal of natural organic matter from drinking water by advanced oxidation processes. Chemosphere 80 (4), 351–365.

Morones, J.R., Elechiguerra, J.L., Camacho, A., Holt, K., Kouri, J.B., Ramírez, J.T., Yacaman, M.J., 2005. The bactericidal effect of silver nanoparticles. Nanotechnology 16 (10), 2346–2353.

Mushtaq, F., Zahid, M., Mansha, A., Bhatti, I., Mustafa, G., Nasir, S., Yaseen, M., 2020. MnFe2O4/coal fly ash nanocomposite: a novel sunlight-active magnetic photocatalyst for dye degradation. Int. J. Environ. Sci. Technol. 17, 4233–4248.

Nadeem, N., Zahid, M., Tabasum, A., Mansha, A., Jilani, A., Bhatti, I.A., Bhatti, H.N., 2020. Degradation of reactive dye using heterogeneous photo-Fenton catalysts: ZnFe2O4 and GO-ZnFe2O4 composite. Mater. Res. Express. 7 (1), 015519.

Najma, B., Kasi, A.K., Khan, K.J., Akbar, A., Bokhari, S.M.A., Stroe, I.R.C., 2018. ZnO/AAO photocatalytic membranes for efficient water disinfection: synthesis, characterization and antibacterial assay. Appl. Surf. Sci. 448, 104–114.

Nangmenyi, G., Economy, J., 2009. Chapter 1–Nanometallic particles for Oligodynamic microbial disinfection. In: Diallo, M., Duncan, J., Street, A., Sustich, R. (Eds.), Nanotechnology Applications for Clean Water Savage N. William Andrew Publishing, Boston, pp. 3–15.

Niu, F., Chen, D., Qin, L., Gao, T., Zhang, N., Wang, S., Chen, Z., Wang, J., Sun, X., Huang, Y., 2015. Synthesis of Pt/BiFeO3 heterostructured photocatalysts for highly efficient visible-light photocatalytic performances. Sol. Energy Mater. Sol. 143, 386–396.

Oh, S.-D., Lee, S., Choi, S.-H., Lee, I.-S., Lee, Y.-M., Chun, J.-H., Park, H.-J., 2006. Synthesis of Ag and Ag–SiO2 nanoparticles by γ-irradiation and their antibacterial and antifungal efficiency against salmonella enterica serovar typhimurium and Botrytis cinerea. Colloid Surf. A. 275 (1–3), 228–233.

Okasha, N., 2008. Influence of silver doping on the physical properties of Mg ferrites. J. Mater. Sci. 43 (12), 4192–4197.

Olowe, A.A., Génin, J.M.R., 1991. The mechanism of oxidation of ferrous hydroxide in sulphated aqueous media: importance of the initial ratio of the reactants. Corros. Sci. 32 (9), 965–984.

Pande, J.V., Bindwal, A.B., Pakade, Y.B., Biniwale, R.B., 2018. Application of microwave synthesized Ag-Rh nanoparticles in cyclohexane dehydrogenation for enhanced H2 delivery. Int. J. Hydrog. Energy 43 (15), 7411–7423.

Pour, S.A., Shaterian, H.R., Afradi, M., Yazdani-Elah-Abadi, A., 2017. Carboxymethyl cellulose (CMC)-loaded Co-Cu doped manganese ferrite nanorods as a new dual-modal simultaneous contrast agent for magnetic resonance imaging and nanocarrier for drug delivery system. J. Magn. Magn. Mater. 438, 85–94.

Quek, J.-A., Lam, S.-M., Sin, J.-C., Mohamed, A.R., 2018. Visible light responsive flower-like ZnO in photocatalytic antibacterial mechanism towards Enterococcus faecalis and Micrococcus luteus. J. Photochem. Photobiol. B Biol. 187, 66–75.

Rai, M., Yadav, A., Gade, A., 2009. Silver nanoparticles as a new generation of antimicrobials. Biotechnol. Adv. 27 (1), 76–83.

Refait, P., Génin, J.M.R., 1993. The oxidation of ferrous hydroxide in chloride-containing aqueous media and pourbaix diagrams of green rust one. Corros. Sci. 34 (5), 797–819.

Reynolds, K., 2007. Water quality monitoring: lessons from the developing world. Water Condition. Purif. 39, 66–68.

Roe, D., Karandikar, B., Bonn-Savage, N., Gibbins, B., Roullet, J.-B., 2008. Antimicrobial surface functionalization of plastic catheters by silver nanoparticles. J. Antimicrob. 61 (4), 869–876.

Saher, R., Hanif, M., Mansha, A., Javed, H., Zahid, M., Nadeem, N., Mustafa, G., Shaheen, A., Riaz, O., 2020. Sunlight-driven photocatalytic degradation of rhodamine B dye by Ag/FeWO 4/gC 3 N 4 composites. Int. J. Environ. Sci. Technol. 18, 927–938.

Salvatorelli, G., Medici, S., Finzi, G., De Lorenzi, S., Quarti, C., 2005. Effectiveness of installing an antibacterial filter at water taps to prevent Legionella infections. J. Hosp. Infect. 61 (3), 270–271.

Sharma, V.K., Yngard, R.A., Lin, Y., 2009. Silver nanoparticles: green synthesis and their antimicrobial activities. Adv. Colloid Interf. Sci. 145 (1–2), 83–96.

Singer, P.C., 2006. DBPs in drinking water: additional scientific and policy considerations for public health protection. J Am Water Works Ass. 98 (10), 73–80.

Song, Y., Qi, J., Tian, J., Gao, S., Cui, F., 2018. Construction of Ag/g-C3N4 photocatalysts with visible-light photocatalytic activity for sulfamethoxazole degradation. Int. J. Chem. Eng. 341, 547–555.

Tabasum, A., Bhatti, I.A., Nadeem, N., Zahid, M., Rehan, Z.A., Hussain, T., Jilani, A., 2020. Degradation of acetamiprid using graphene-oxide-based metal (Mn and Ni) ferrites as Fenton-like photocatalysts. Water Sci. Technol. 81 (1), 178–189.

Thakare, P., Padole, P., Bodade, A., Chaudhari, G., 2018. Microstructural and antifungal properties of silver substituted copper ferrite Nanopowder synthesized by sol-gel method. Indo Am. j. pharm. 5 (1), 52–63.

Thakur, D., Govindaraju, S., Yun, K., Noh, J.-S., 2019. The synergistic effect of zinc ferrite nanoparticles uniformly deposited on silver nanowires for the biofilm inhibition of Candida albicans. Nano 9 (10), 1431.

Thapa, D., Palkar, V., Kurup, M., Malik, S., 2004. Properties of magnetite nanoparticles synthesized through a novel chemical route. Mater. Lett. 58 (21), 2692–2694.

Tsvetkov, M., Zaharieva, J., Milanova, M., 2019. Ferrites, modified with silver nanoparticles, for photocatalytic degradation of malachite green in aqueous solutions. Catal. Today 357, 453–459.

Wang, J., Deng, T., Dai, Y., 2005. Study on the processes and mechanism of the formation of Fe3O4 at low temperature. J. Alloys Compd. 390 (1), 127–132.

Wang, J., Wei, Y., Zhang, J., Ji, L., Huang, Y., Chen, Z., 2014. Synthesis of pure-phase BiFeO3 nanopowder by nitric acid-assisted gel. Mater. Lett. 124, 242–244.

Wang, H., Wu, D., Liu, C., Guan, J., Li, J., Huo, P., Liu, X., Wang, Q., Yan, Y., 2018. Fabrication of Ag/In$_2$O$_3$/TiO$_2$/HNTs hybrid-structured and plasma effect photocatalysts for enhanced charges transfer and photocatalytic activity. J. Ind. Eng. Chem. 67, 164–174.

Xu, Y., Liu, Q., Xie, M., Huang, S., He, M., Huang, L., Xu, H., Li, H., 2018a. Synthesis of zinc ferrite/silver iodide composite with enhanced photocatalytic antibacterial and pollutant degradation ability. J. Colloid Interface Sci. 528, 70–81.

Xu, Y., Wen, W., Wu, J.-M., 2018b. Titania nanowires functionalized polyester fabrics with enhanced photocatalytic and antibacterial performances. J. Hazard. Mater. 343, 285–297.

Zahid, M., Nadeem, N., Hanif, M.A., Bhatti, I.A., Bhatti, H.N., Mustafa, G., 2019. Metal Ferrites and their Graphene-Based Nanocomposites: Synthesis, Characterization, and Applications in Wastewater Treatment. In: Magnetic Nanostructures. Springer, pp. 181–212.

Zeehan, T., Anjum, S., Waseem, S., Riaz, M., Zia, R., 2019. Tuning of structural, magnetic and optical properties of silver doped cobalt chromium ferrite ferrites thin film by PLD technique. Dig. J. Nanomater. Biostructures. 14 (4), 855–866.

Zhang, X., Niu, H., Yan, J., Cai, Y., 2011. Immobilizing silver nanoparticles onto the surface of magnetic silica composite to prepare magnetic disinfectant with enhanced stability and antibacterial activity. Colloid Surface A. 375 (1), 186–192.

Zhang, X., Wang, B., Wang, X., Xiao, X., Dai, Z., Wu, W., Zheng, J., Ren, F., Jiang, C., 2015. Preparation of M@ BiFeO3 nanocomposites (M = Ag, Au) bowl arrays with enhanced visible light photocatalytic activity. J. Am. Ceram. Soc. 98 (7), 2255–2263.

Zheng, B., Zhang, M., Xiao, D., Jin, Y., Choi, M.M., 2010. Fast microwave synthesis of Fe$_3$O$_4$ and Fe$_3$O$_4$/Ag magnetic nanoparticles using Fe 2+ as precursor. Inorg. Mater. 46 (10), 1106–1111.

CHAPTER 26

Silver-doped ternary compounds for wastewater remediation

Noor Tahir[a], Muhammad Zahid[a], Haq Nawaz Bhatti[a], Asim Mansha[b], Khalid Mahmood Zia[b], Ghulam Mustafa[c], Muhammad Tahir Soomro[d], and Umair Yaqub Qazi[e]

[a]*Department of Chemistry, University of Agriculture, Faisalabad, Pakistan*
[b]*Department of Chemistry, Govt. College University Faisalabad, Faisalabad, Pakistan*
[c]*Department of Chemistry, University of Okara, Okara, Pakistan*
[d]*Center of Excellence in Environmental Studies (CEES), King Abdulaziz University, Jeddah, Saudi Arabia*
[e]*Department of Chemistry, University of Hafar Al Batin, Hafar Al Batin, Saudi Arabia*

26.1 Introduction

Water is certainly an essential natural resource for the survival of life on earth. Almost 70% of Earth's surface is covered with water, but only 3% of this is available as fresh and clean water. The increase in overall climate change, population growth, subsequent utilization of water resources, agricultural and industrial activities, and generation of waste has resulted in grave problems of water scarcity and water pollution (Fardood et al., 2020). The concentration of recalcitrant hazardous pollutants in water bodies has increased as a result of the continuous release of effluents and sludge from various industrial evacuation, untreated wastewater, and frequent disposal of solid materials into water reservoirs (Rasheed et al., 2020). As a basic prerequisite for all living beings, the concerns related to the accessibility of freshwater continue to attract the attention of researchers to develop new technologies for the treatment and reuse of wastewater (Anjum et al., 2017). Wastewater treatment demands a high cost for the removal of contaminants so that water can be reused again. Hence, it necessitates advanced technologies with lesser requirements and less cost (Noreen et al., 2017). For this purpose, researchers have been working on novel sustainable strategies for the treatment of wastewater (Nadeem et al., 2020).

Numerous techniques have been developed over the years for wastewater treatment. The main objective of all these technologies was the development of ecofriendly nanomaterials (Yahya et al., 2018). Photocatalysis is a potential substitute for conventional technologies for wastewater treatment. Wastewater treatment is the most studied aspect of photocatalysis. It is a promising advance oxidation process (AOPs), owing to its ability to utilize semiconductor materials, which easily get excited by solar light and degrade hazardous inorganic and organic pollutants (Kubacka et al., 2012). Nanotechnology utilizing semiconductor materials has tremendous potential in water and wastewater treatment. Therefore, the perfect synthesis approach utilizes the integration of principles of nanotechnology, materials chemistry, and green chemistry (Soo et al., 2018). Semiconductor materials have great technological importance for environmental remediation owing to their capability to produce charge carriers when triggered by solar light. The favorable electronic structure, light absorption characteristics, appropriate bandgap and charge transport abilities of semiconductor materials make them effective photocatalysts (Friedmann et al., 2010).

Ternary metal oxides and sulfides are used as photocatalysts in wastewater treatment owing to their ability to produce highly oxidizing charge carriers, chemical stability, and photostability. Semiconductor photocatalysts, when used in slurry and bulk, cause nanoparticle agglomeration, which causes hindrance in photocatalytic performance. Transition (Cr, Co, Ni, Mn, Fe, and Zn) and noble metals (silver, gold, platinum, and palladium) are usually used as dopants to transform the structure of ternary compounds to make them visible light-responsive. Among the noble metals, silver is the most attractive metal owing to its natural antimicrobial properties, photoactivity, the lesser material cost in comparison to other noble metals, and ease of preparation. It is used as a good adsorbent, nanofiller, and cocatalyst in wastewater treatment. It is also a very good disinfectant and is a potential and safe alternative to chemical disinfectants like chlorine, iodine, and ozone. Silver has also found extensive applications in the development of optical and electronic devices (Koe et al., 2019).

Doping noble metals into ternary compounds (metal oxides and sulfides) has become an effective way to improve the movement of electrons from photocatalyst surface to surface of noble metal. It enhances separation efficiency of light-induced charge carriers and improves the visible light absorption range (Ge, 2008). Extensive research on the doping of silver metals has shown that metallic levels of the noble metals are lower than conduction band of ternary metal oxides or sulfides, so they act as a sink for electrons generated after light irradiation (Zahid et al., 2019). Hence, studies have confirmed that silver doping in ternary compounds enhances the solar light response by increased e^- h^+ pairs and suppression of their recombination efficiency (Niu et al., 2015). The surface plasmon response of silver nanoparticles enhances the efficiency of ternary compounds by acting as a sink of electrons by trapping electrons coming from the conduction band of ternary compounds and prevents their recombination. The silver modified ternary compounds present a novel class of visible light active compounds (McEvoy and Zhang, 2014).

26.2 Objectives

The objective of this chapter is to illustrate comprehensively the effects due to silver doping on the photocatalytic behavior and properties of ternary compounds, overcoming their limitations of bandgap as visible light active photocatalysts. The chapter includes a study of all major classifications of ternary compounds and their silver-doped compounds for wastewater treatment and the comparisons of their photocatalytic activity. The chapter also aims to give insights into the doping mechanism for the degradation of dye and antimicrobial activity of silver-doped compounds.

26.3 Semiconductor photocatalytic ternary compounds

The ternary compounds show the enhanced photocatalytic response for the development of hybrid nanocomposite materials through coupling or doping with transition metals and plasmonic metals for the wastewater treatment (Zahid et al., 2020). Numerous ferrites, tungstates, vanadates, indicates, oxyhalides, and oxysulfides are categorized as photocatalytic ternary compounds. Several of the ternary compounds are principally being investigated for the splitting of water and destruction of lethal waste products of organic and inorganic nature under solar light. These metal oxide-based ternary compounds are found suitable to exploit the visible light. The photoactive metal oxides of titanium, vanadium, chromium, and tungstate distributed on the high surface area supports represent new photocatalytic systems of binary and ternary nanocomposites (Di Paola et al., 2012).

26.3.1 Metal tungstates

A new class of metal oxide-based ternary semiconductors, which have attained considerable attention, is metal tungstates. Nanostructured photocatalytic tungstate materials have been investigated for their widespread use in photocatalysis, synthesis of optical fibers, and supercapacitors, as well as in photoluminescence. The photocatalytic activity and optical properties of these metal oxide-based tungstates depend strongly on the nature of the metal (Mahapatra et al., 2008). Bismuth tungstate (Bi_2WO_6) is a promising ternary photocatalyst for environmental remediation owing to its photostability, exceptional charge separation ability, and the simple production methods (Khojeh et al., 2017). Porous zinc tungstate films synthesized as heteronuclear complex exhibit high photocatalytic activity. The $ZnWO_4$ nanorods are usually prepared via hydrothermal methods using surfactants, while $ZnWO_4$ powders having various morphologies are prepared using the coprecipitation method. Cadmium tungstate ($CdWO_4$) is another ternary tungstate, which shows high photocatalytic activity and is generally prepared by the hydrothermal process (Zhao and Zhu, 2006).

26.3.2 Metal vanadates

Vanadates that are prominent ternary metal oxide compounds, exhibiting layered and tetragonal structures, have gained substantial attention as competent photocatalyst for the water splitting and deterioration of pollutants. Bismuth vanadate is a typical example of mixed metal oxide for photocatalysis and subsequently for water splitting. Bismuth vanadate ($BiVO_4$) classified as a ternary compound shows a typical layered structure that usually occurs in diverse phases, including tetragonal zircon, monoclinic, and a sheelite type structure. Silver vanadates are powders synthesized in the form of nanoflowers, nanowires, and powder forms, and their catalytic activity has been checked for the removal of various pollutants. These monoclinic and perovskite structures are usually made by sol − gel, hydrothermal, solvothermal, co-precipitation, and solid-state reaction procedures, involving the mixing of high purity metal oxides precursors. The silver vanadate particles produced by a facile coprecipitation method showed excellent visible light degradation of dye acid red B. The samples with an excess of vanadium oxide and silver showed more photoefficiency than binary TiO_2 and ZnO (Zhang et al., 2008). Many other silver vanadates have been made employing the simple hydrothermal process. The experimentation showed that by altering the time for hydrothermal treatment and the addition of surfactants could change the morphology of prepared catalyst samples. The samples consisted of Ag_3VO_4 or $Ag_4V_2O_7$ or their mixed phases, which showed maximum photocatalytic activity (Huang et al., 2009). $CeVO_4$ and $PrVO_4$ are other ternary compounds that have been produced by the solid-state and microwave-assisted methods, respectively. The efficiency of these ternary compounds is inspected for the degradation of various phenols and dyes. Those vanadates, which were synthesized by microwave method, showed higher degradation rates than those prepared through solid-state technique (Mahapatra et al., 2008).

26.3.3 Ternary metal ferrites

Ternary ferrites like $ZnFe_2O_4$ and $BiFeO_3$ having spinel and perovskite assemblies, respectively, have been excessively explored for wastewater treatment. The $ZnFe_2O_4$ and $BiFeO_3$ ternary photocatalysts show better degradation of dyes and other pollutants in both ultraviolet and visible light (Anchieta et al., 2014). $BiFeO_3$ is of particular interest owing to its magnetic properties, photostability, and bandgap energy in a narrow range (2.1–2.5 eV). The $BiFeO_3$ nanoparticles having around 150–200 nm diameter may be prepared via different methods. The hydrothermal treatment can be used to synthesize phase-controlled bismuth ferrites by regulating the concentration of oxidants. The prepared photocatalysts sample shows different morphologies like nanoflakes and nanoparticles self-assembly depending upon the concentration of oxidant like KOH (Yang et al., 2013) $BiFeO_3$ synthesized via the ultrasonication method is prepared in comparatively less time and at less temperature as compared to hydrothermal treatment (Soltani and Entezari, 2013). During recent years, a variety of silver-doped ternary compounds have been synthesized effectively for the wastewater treatment. Fig. 26.1 shows the main groups of silver-doped ternary compounds for wastewater treatment.

26.4 Possible mechanisms of silver nanoparticles doping on ternary compounds

26.4.1 Schottky barrier and surface plasmon response

The plasmonic photocatalysts comprise noble metal nanoparticles (Ag, Au, Pd, and Pt) along with other semiconductor materials. Among all noble metal nanomaterials,

26.4 Possible mechanisms of silver nanoparticles

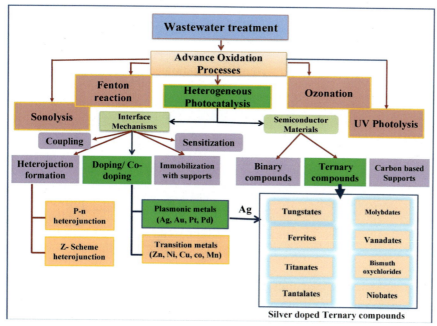

FIG. 26.1

Silver doped ternary compound groups for wastewater treatment.

silver nanoparticles (Ag NPs) are mostly selected owing to the formation Schottky barrier and largely because of their strong surface plasmon response (SPR). At the photocatalyst and metal interface, the silver metal nanoparticles form a Schottky barrier, which helps in electron-hole separation (Hsieh et al. 2012). Studies have shown that surface plasmon response helps in enhancing the visible light absorption, eventually increasing the photodegradation efficiency. Doping of silver nanoparticles with ternary metal oxides, chalcogenides, and support materials like graphene oxide and graphitic carbon nitride has shown improved optical absorption capability of these conventional semiconductor photocatalysts (Saher et al., 2020). Doping of silver nanoparticles introduces additional energy bands or states in the forbidden bandgap of ternary semiconductors. Consequently, because of an interaction between the photocatalyst and the silver nanoparticles, an internal electric field is created at the interface of the semiconductor material and Ag nanoparticles. This phenomenon is shown in Fig. 26.2. The light-induced charge carriers produced in the ternary compounds because of irradiation move in various directions as a result of electric field phenomena. These dopants also serve to move the absorption edge of semiconductor materials toward higher wavelengths. This improves the overall charge separation efficacy of semiconductor photocatalysts (Singh and Dhaliwal, 2020). When doped with silver metal particles, the electrons diffuse into the doping metals. The silver ions (Ag^+) act as electron scavengers and trap the electrons. The electrons are excited from the valence band of the ternary compound to its conduction band. These

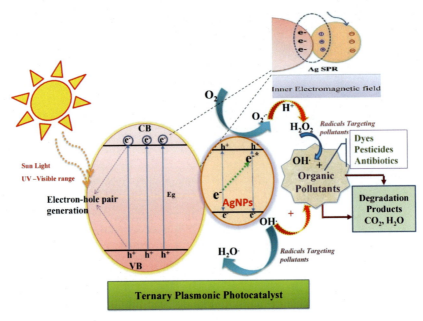

FIG. 26.2

Degradation mechanism of pollutant molecules by silver-doped ternary compound.

electrons could easily recombine in the absence of silver dopant. The energy level of silver dopant is positioned lower than the conduction band of photocatalyst so that the silver dopants act as trapping sites for electrons and consequently increase the lifecycle of charge carriers by repressing their recombination (Jaffari et al., 2019). Hence, the photocatalytic activity of silver-doped ternary compounds is primarily due to increased production of e^-/h^+ pairs eventually enhancing the lifetime of the charge carriers by hindering their recombination (Ekthammathat et al., 2016). Hence, when a silver-doped ternary compound is irradiated by a photon having energy higher than bandgap energy of ternary compound, the light absorption excites electrons from valence band to conduction band, causing the formation of electron-hole pairs. The electrons on surface of ternary compounds get trapped by silver metal nanoparticles owing to the lower Fermi level of silver. These electrons react readily with oxygen adsorbed on the surface of silver and yield highly reactive oxygen radicals. Hence, the photoexcited e^-/h^+ pairs are separated efficiently, increasing the overall quantum efficiency (Shen et al., 2015).

26.4.2 Dye removal mechanism for wastewater treatment using silver-doped ternary compounds

Silver is a valuable plasmonic metal. In nature, it is found in the surface and ground waters. In pure form, it is found in the form of alloys with gold or other minerals. Silver and its nanoparticles have been widely used because of their antimicrobial properties. Many different types of dyes and pigments are utilized daily on an industrial basis and the majority of these dyes are released into water streams after the dyeing process (Noreen et al., 2014). Hence, the role of silver nanoparticles has been extensively investigated for wastewater treatment, which includes the degradation of recalcitrant dyes, removal of heavy metals, bacteria, and fungi. Silver nanoparticles upon their release into the water are chemically transferred into various species, which are not detrimental to the environment. The surface coating of silver nanoparticles is lost, forming aggregates and latter getting oxidized to silver oxide. This then finally dissolves and releases as silver ions. The silver nanoparticles are effectively used for the removal of the above contaminants. Silver nanoparticles exhibit excellent antimicrobial properties due to their exceptional physicochemical properties. The general mechanism of dye removal by silver nanoparticles (Ag NPs) includes adsorption of dye particles on to Ag NPs and subsequent photocatalytic degradation (Jiang et al., 2005). Silver because of its exceptional stability and light absorption capability is usually doped with ternary metal oxides, sulfides, and oxyhalides. The resulting nanostructures exhibit excellent photocatalytic activity owing to SPR effect of silver nanoparticles. The collective oscillations of electrons because of SPR effect, radiating from the outermost band toward the higher energy state eventually results in contact with the molecular oxygen present in water. This causes the production of free radicals. The resultant holes formed because of the excitation of electrons to higher energy states, accept an electron from the dye adsorbed finally reducing dye molecules. Furthermore, additional free radicals produced as a result of H^+ ions oxidation, produced by water splitting causes oxidation of the dye molecule (Marimuthu et al., 2020). Fig. 26.2 illustrates photocatalytic degradation of the organic pollutants like a dye by silver-doped ternary compounds.

26.4.3 Antimicrobial mechanism

The antimicrobial mechanism displayed by silver nanoparticles is still ambiguous. The pathogens like bacteria, viruses, and other microbes are agents of infections in living beings, causing serious infectious diseases worldwide (Naqvi et al., 2019). Silver is a well-known antibacterial agent, and owing to its antimicrobial properties, it is being used for wastewater treatment (Rehan et al., 2020). The silver nanoparticles get attached to cell wall of most bacteria, eventually changing the permeability of cell membrane and resulting in formation of "pits." The silver nanoparticles accumulate on bacterial surface forming free radicals that penetrate inside the bacterial cell ultimately causing cell death. Apart from releasing free radicals, silver ions cause inhibition of respiratory enzyme, which produces reactive oxygen species, hence destroying the cell. The silver ions are also supposed to cause changes in the bacterial DNA by combining with sulfur and phosphorous present in DNA and hence inactivating the microbe (Prabhu and Poulose, 2012). Particle size and shape are important parameters for determining antimicrobial activity. If the AgNPs have a size of less than 20 nm, it gets strongly attached to DNA protein having sulfur, increasing permeability through the bacterial membrane and finally killing the microbe (Morones et al., 2005).

26.5 Influence of various synthesis approaches on doping behavior and morphology of silver-doped compounds

The difference in the photocatalytic activity and morphology of silver-doped ternary compounds synthesized by different methods is based on the oxidation state of silver on the photocatalyst surface. In some synthesis methods, Ag^+ ions are reduced to silver, while in other methods, silver is deposited at the metal oxide surface as silver ions. Some studies suggest that during the synthesis of silver-doped bismuth tungstate prepared by mild hydrothermal method, silver ions consume electrons and are reduced to metallic silver, ultimately lessening the recombination of electron-hole pairs. Here, electron scavenging because of O_2 at excited semiconductor particle surface cannot compete well with the electron transfer to silver ion. Consequently, electrons are scavenged by molecular oxygen at the photocatalyst surface producing reactive oxygen radical species. The valence band holes do not react with H_2O molecules or OH^- to form hydroxyl radicals. This is on account of the fact that the valence band redox potential is more negative as compared to that of OH^-/OH. Hence, the photocatalytic activity of Ag-doped Bi_2WO_6 made through a hydrothermal process is credited to light-induced holes and reactive O_2 radical oxygen species (Li et al., 2014). Other studies suggest that in some methods, the loading of silver metal on the metal oxide surface can speed up the transport of photoinduced electrons to outer systems. Such transfer of electrons to outer metal deposits results in deposits becoming negatively charged. For example, in silver-doped tungstate prepared by photoreduction approach, the doped silver nanoparticles mainly act as acceptors of electrons on the bismuth tungstate surface, thus hindering charge carriers from recombining. Therefore, significantly, enhancing photocatalytic efficiency (Guo et al., 2012).

The photocatalytic behavior of silver-doped ternary compounds could be well understood through a dye-sensitized mechanism in connection with the formation of adsorbents and newly formed recombination centers. In such a mechanism, a major role is played by key active species (h^+ $\cdot O_2^-$), while hydroxyl radical plays a minor role but, during Ultraviolet irradiation all oxidizing species $\cdot O_2^-$, h^+, and $\cdot OH$ have role in the degradation process of dyes (Kim et al., 2014). Hence, various studies have suggested that synthesis methods have a notable effect on the doping behavior and morphology of doped photocatalysts. Silver-doped molybdates prepared by different methods exhibited diverse morphologies and photocatalytic behavior for the degradation of same pollutant Rhodamine B dye. Silver-doped $PbMoO_4$, prepared by refluxing method, showed single phase spherical nanoparticles. The light-induced charge carriers were transferred to silver-doped lead molybdate surface, where they reacted with oxygen adsorbed on the photocatalyst surface. Superoxide radical is created as a result of molecular oxygen reduction, and hydroxyl radicals are created as a result of the reaction of water with positively charged holes. These radical species react with dye molecules and convert them into less harmful species like CO_2 and H_2O (Phuruangrat et al., 2015). In silver-doped molybdates prepared from a hydrothermal method, studies show that the electrons upon irradiation are excited from VB of molybdates to CB forming charge carriers. These charge carriers are transferred to the silver nanoparticles from the CB of photocatalyst as a result of Schottky energy barrier formed at the interface. These electrons react with the oxygen molecules adsorbed on silver nanoparticles surface and finally transfer oxygen molecules into the superoxide anion radicals. The hydroxyl radical species, formed as an outcome of the interaction of the holes with water molecules and the superoxide radical, convert the dye molecule into water and carbon dioxide (Phuruangrat et al., 2020).

26.6 Silver-doped ternary compounds for wastewater remediation

Recently, Ag-doped ternary compounds have been broadly studied owing to their visible light absorption capability and improved photocatalytic activity owing to their narrow bandgap energy. The design and development of these novel silver-doped ternary compounds have resulted in heterostructures with the improved photocatalytic performance for wastewater treatment. Various synthesis methods are employed for the formation of silver-doped ternary compounds as shown in Fig. 26.3.

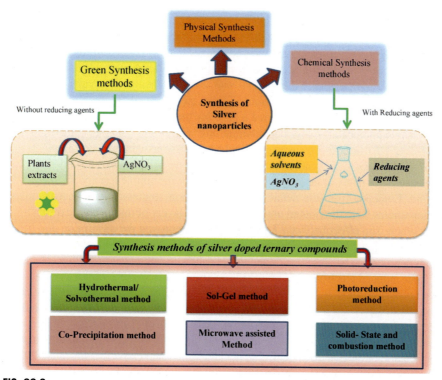

FIG. 26.3

General method for synthesis of silver-doped ternary compounds.

26.6.1 Silver-doped tungstates for wastewater treatment

Bismuth tungstate belongs to the category of ternary aurivillius oxides having a layered slab structure and a bandgap of 2.7–2.8 eV. Wang et al. prepared bismuth tungstate nanoparticles employing a hydrothermal method. The silver nanoparticles were deposited on the spherical shaped bismuth tungstate compound using a simple photoreduction approach. Various characterization studies showed that the metallic silver nanoparticles possessing standard particle size ranging from 12 to 15 nm were uniformly dispersed over the surface of bismuth tungstate. The photocatalytic potential of the individual and doped compounds was determined by desulfurization of thiophene and degradation of Rhodamine B dye by sunlight. Studies revealed that doping of silver nanoparticles greatly improved the catalytic activity of bismuth tungstate (Wang et al., 2012). In another study, three-dimensional flowers like and spherical $Ag-Bi_2WO_6$ nanoparticles were made via hydrothermal process. The as-synthesized catalyst displayed excellent photocatalytic activity for the removal of dye Rhodamine B (Li et al., 2014). The photocatalytic properties of silver-doped bismuth tungstate were also evaluated by Chen et al. for Rhodamine B degradation under visible light. The doped samples displayed about a 26% enhancement in photocatalytic activity in comparison to the undoped samples (Chen et al., 2013a,b). The Raman spectra and Fourier transform infrared spectroscopy of $Ag-Bi_2WO_6$ showed WO_6 octahedrons symmetric and asymmetric stretching and bending modes and the translation of bismuth and tungsten ions. In this study, 3% of the doped samples showed nearly 100% contaminant degradation efficiency (Dumrongrojthanath et al., 2013). Aslam et al. described the performance and efficiency of silver-doped tungsten oxide for removing 2-nitrophenol and 2-chlorophenol. The vital information about the behavior of prepared photocatalyst was revealed by the electrochemical characterization. The study provided a base for the synthesis of silver-based visible light active photocatalysts for decontaminating wastewater (Aslam et al., 2019).

Silver NPs loading on microspherical Sb_2WO_6 was achieved via simple solvothermal process. The synthesized Ag-doped Sb_2WO_6 samples showed enhanced degradation of many organic dyes in comparison to the simple ternary Sb_2WO_6 compound. Evidence from the detailed characterization studies showed that the silver nanoparticles were distributed uniformly over the layers of microspherical substrate. The synergistic effect between Ag and Sb_2WO_6 improved the adsorption and hence photocatalytic activity. Samples with less than 20% doping of silver showed maximum degradation because this doped sample showed optimized migration efficiency of light-induced electrons that inhibited electron and holes recombination produced as a result of irradiation (Chen et al., 2018). Codoping of another metal ion like Zn^{2+} along with silver ion in tungstates has reported shifting bandgap to higher energies due to the formation of defects that subsequently prevent electron-hole pairs recombination. Silver-doped and codoped samples in calcium tungstate showed utmost degradation of the dye methylene blue than the individual $CaWO_4$ (Neto et al., 2020).

Strontium tungstate is a ternary tungstate material having the central and tetragonal configuration, where WO_4^{2-} molecular ions are freely attached to Sr^{2+} cations. It is usually active under UV light. Flower and star-shaped microcrystals of

Ag-SrWO$_4$ were made using a coprecipitation method at a temperature of 70 °C by a group of researchers. Under visible light, silver-doped SrWO$_4$ exhibited 92% degradation than pristine SrWO$_4$, which showed 50% degradation (Sedighi et al., 2019). ZnWO$_4$ has been studied for pollutant mineralization under UV light. Hence, it remained a challenge to make it active under visible light. Nanorods of ZnWO$_4$ fabricated through ultrasonic irradiation showed excellent degradation of dye RhB. The highly dispersed silver particles served as electron traps encouraging the separation of charge carriers and subsequently promoting photoactivity (Yu and Jimmy, 2009). Similar Ag$^+$ doped ZnWO$_4$ samples, prepared using a hydrothermal method, showed higher photocatalytic degradation for dye Eosin B under UV light (Wang et al., 2009).

26.6.2 Silver-doped molybdates for wastewater treatment

Molybdates are another important class of ternary compounds that have been of immense importance in diverse technological applications. Their unique structure and extraordinary physical and chemical properties make them promising photocatalysts (Ahmad et al., 2018). The structure of molybdates usually supports the decrease of the excitation energy and enhances luminescence efficiency, but they usually have wide bandgaps, which make them ineffective in visible light (Di Paola et al., 2012). Doping of noble metals in molybdates has shown higher visible light absorption due to improved and lesser recombination rate of e$^-$/h$^+$ pairs as a result of the surface plasmonic resonance (SPR) effect. Doping of various molybdates with silver metal ion has been studied for dye and drug degradation.

Ag nanoparticles finely dispersed on Bi$_2$MoO$_6$ microspheres via an in situ reduction approach showed improved photoefficiency for pollutant degradation. Various characterization studies revealed that the doped molybdate had enhanced surface area as a result of a fine dispersion of silver nanoparticles. Furthermore, the well-dispersed Ag nanoparticles avoid aggregation on Bi$_2$MoO$_6$ nanosheets, consequently resulting in inefficient charge separation and transfer (Yuan et al., 2013). As a very important part of the molybdate family, silver molybdate (Ag$_2$MoO$_4$) has been broadly studied for its use in photoluminescence and photocatalytic degradation of pollutants. The deposition of Ag nanoparticles on silver molybdate has shown enhanced absorption of visible light. In a study, octahedral shaped silver-doped silver molybdate composites were prepared hydrothermally, which showed the enhanced ability for Rhodamine B degradation (RhB) in sunlight. This higher catalytic activity was credited to the fact that the light-induced electrons in silver nanoparticles were transferred to the conduction band of Ag$_2$MoO$_4$, forming a massive quantity of active radical species due to the increased SPR effect (Yang et al., 2017). Ag-doped PbMoO$_4$ prepared by a simple sonochemical method showed increased degradation of dye Indigo Carmine when irradiated by sunlight (Gyawali et al., 2013). Uniform dispersal of the silver nanoparticles on PbMoO$_4$ surface considerably enhanced photocatalytic activity due to SPR effect caused by Ag nanoparticles (Namvar et al., 2019). Novel tetragonal shaped silver-doped cadmium molybdate nanoparticles, synthesized through a sonochemical method, have shown that the silver dopant acts as an effective trap for the charge carriers and retards recombination. The silver-doped $CdMoO_4$ showed enhanced decolorization of pollutant methyl orange from wastewater up to 98% under solar light irradiation (Hosseinpour-Mashkani et al., 2016).

Barium molybdate ($BaMoO_4$) is a typical ternary compound having a large bandgap, which makes it ultraviolet light active only. A fine dispersion of Ag NPs over octahedral shaped microcrystals of $BaMoO_4$ has shown to remarkable increase in response of doped nanomaterial in visible light range. The experiments showed a rapid increase in the degradation of dye Rhodamine B and antibiotic Ibuprofen by the addition of silver nanoparticles (Ray et al., 2017). Nickel molybdate is a visible light-responsive ternary compound with a narrow bandgap of (2.7–2.9 eV) and exhibits a monoclinic structure. The octahedral α-phase $NiMoO_4$ is the most ecofriendly and chemical stable compound. Ray et al. utilized a microwave hydrothermal process for the deposition of silver nanoparticles on α-$NiMoO_4$. The photocatalyst showed excellent mineralization of drug tetracycline up to 98% and good photostability even after the fifth cycle of drug degradation (Ray et al., 2019). In another study, Bi_2MoO_6 doped with Ag nanoparticles in various weight ratios were effectively synthesized using a sonochemical method, followed by hydrothermal method. The catalytic activities of Ag/Bi_2MoO_6 and Bi_2MoO_6 were assessed by Rhodamine B degradation under solar light. Under sunlight, 10 wt% Ag-doped Bi_2MoO_6 catalyst showed highest catalytic activity for RhB degradation up to 94–98% and hence proved as a promising and novel photocatalyst for wastewater remediation (Phuruangrat et al., 2015).

26.6.3 Silver-doped ferrites for wastewater treatment

The wastewater treatment through conventional technologies like adsorption, biodegradation, and membrane filtration is ineffective. The photofenton process utilizing ferrites and oxidants is the most effective process among all advanced oxidation processes (Tabasum et al., 2020). Metal ferrites are ternary compounds with ferromagnetic properties that are generally ascribed to the presence of iron oxides (FeO, Fe_2O_3) as main components, which are partially replaced by other transition metal oxides. Ferrites, when combined with dopants, oxidants, or with other semiconductor materials, have shown significance in wastewater treatment due to their physical and magnetic properties, adsorption ability, reusability, and ease of recovery (Mushtaq et al., 2020). These ternary metal oxides show excellent photocatalytic and antibacterial activities for eliminating hazardous products and minimizing bacterial pollution from water. Metal ferrites, because of their magnetization, mechanical stability, and easy recovery, have been used for the degradation of a large number of poisonous pesticides and dyes (Tabasum et al., 2018).

Magnesium ferrite ($MgFe_2O_4$) is a well-designed magnetic ternary compound with a normal cubic spinel structure. In nanoscience, it has found exceptional applications in magnetic and electric devices, as a photocatalyst in wastewater treatment, sensor, and biomedicines. In a research study, silver-doped magnesium ferrite was synthesized by a solid-state combustion method. The Ag-$MgFe_2O_4$ showed enhanced antibacterial activities due to the inhibition of bacteria that is resistant to multidrugs. This was due to the bigger surface area because the condensed particle size leads to improved surface reactivity (Lagashetty et al., 2019). Another vital member of the ternary spinel ferrites family is the cobalt ferrite ($CoFe_2O_4$), which shows enhanced electric and magnetic properties (Gulzar et al., 2020) Cobalt ferrite attains enhanced properties when it is combined with noble metals like Ag and Au. Considering the intrinsic antimicrobial properties of silver, it is anticipated that the doping of Ag and $CoFe_2O_4$ will enhance the antimicrobial activity (Gingasu et al., 2016). During a recent study done by Satheeshkumar et al., silver-doped $CoFe_2O_4$ was prepared using a novel honey-assisted combustion method. The silver-doped sample showed enhanced antibacterial activity against bacteria (*Escherichia coli* and *Staphylococcus aureus*) and fungus, *Candida ablicans* (Satheeshkumar et al., 2019).

Magnetic silver-doped bismuth ferrite nanocomposite was made using a simplistic hydrothermal method with the loading of silver in various compositions. Detailed characterization revealed that silver particle loading in different weight percentages produced no significant changes in bismuth ferrite morphology. All nanocomposites showed spindle shapes with surfaces getting rougher with the increase in loading of silver nanoparticles percentage, but the composite with 3% silver loading showed excellent photocatalytic dye degradation under ultraviolet light and invisible light too (Jaffari et al., 2019). Similarly, silver-doped magnetite (Fe_3O_4) was used for the degradation of eosin Y. The magnetic nanoparticles of magnetite prepared through the coprecipitation method were further doped with Ag nanoparticles to get higher photocatalytic efficiency. Under UV light irradiation, the Ag-Fe_3O_4 system showed excellent degradation of azo dyes (Alzahrani, 2015). A sol – gel method followed by a chemical reduction method was utilized for the synthesis of silver-doped bismuth ferrite nanocomposite. Silver particles having a fine diameter of 20–40 nm were effectively anchored on the ternary bismuth ferrite surface as were revealed by various characterization studies. The Ag-doped $BiFeO_3$ exhibited excellent degradation efficiency for dye methyl orange in sunlight irradiation in comparison to pristine $BiFeO_3$. The doped catalyst showed excellent photoactivity, photostability, and recoverability, all credited to strong SPR effect of silver nanoparticles (Lu et al., 2015).

26.6.3.1 Ozonation technique using silver-doped ferrites for wastewater treatment

The ozonation technique is among those advanced oxidation processes that match well with other facilities available for water treatment, without requiring further heat or photon sources. Ozone is extensively used for pollutant decomposition and disinfection of water. The efficiency of ozonation process for wastewater treatment can be enhanced by coupling with heterogeneous photocatalysis. This type of metal oxide-based photocatalytic ozonation reduces the rate of adverse side reactions and increases the rate of pollutant degradation by supporting formation of hydroxyl radicals from ozone decomposition. These types of photocatalysts work well in acidic and basic conditions and do not require extra input of energy. A ternary compound manganese ferrite ($MnFe_2O_4$) is frequently used for the degradation of hazardous pollutants

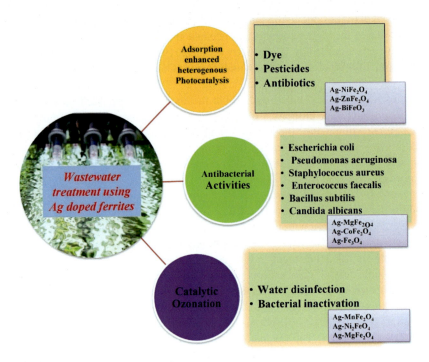

FIG. 26.4

Various applications of silver-doped ferrites for wastewater treatment.

through ozonation treatment. An effective way to enhance ozonation activity of $MnFe_2O_4$ regarding wastewater treatment is Ag doping. Silver doping improves the process of cycling and electron transfer across Ag and Mn several valences, converting molecular ozone into highly reactive radical species. It also leads to the formation of more hydroxyl groups on catalyst surface that are crucial for ozone deterioration and hydroxyl radical formation, subsequently leading to enhanced pollutant degradation (Wang et al., 2018). A hybrid ternary silver-doped nickel ferrite photocatalyst was made through a sol-gel method and effectively utilized for catalytic ozonation for wastewater coming from the paper industry. The addition of silver to nickel ferrite was found to enhance the chemical oxygen demand. The hydroxyl radicals happened to be the core reactive species for the recalcitrant wastewater by Ag-doped $NiFe_2O_4$ based ozonation process (Zhao et al., 2020). Silver-doped ternary ferrites have found multiple applications in wastewater treatment as depicted in Fig. 26.4.

26.6.4 Silver-doped vanadates for wastewater treatment

Xu et al. prepared a highly efficient visible light active silver-doped bismuth vanadate photocatalyst with a simple hydrothermal approach. Several characterization studies revealed that either bulk or monoclinic $BiVO_4$ or silver nanoparticles formed microspheres of uniform size distribution that showed increased degradation of methylene blue dye than individual $BiVO_4$ sample (Xu et al., 2016). Series of hybrid silver-doped vanadates were synthesized using a heteronuclear and amorphous complexing method. The X-ray diffraction method (XRD) showed monoclinic scheelites structure of doped vanadates. The silver-doped $BiVO_4$ showed enhanced photoactivity for degradation of 2,4-dichlorophenol and dye methyl blue in visible light. The UV–visible spectra exhibited that bandgap of $BiVO_4$ was narrowed down after doping with silver nanoparticles (Zhou et al., 2011). Likewise, silver-doped iron vanadate photocatalyst was prepared by Zhou et al. using a coprecipitation method as a visible light catalyst for wastewater treatment. Characterization studies revealed that after doping, there were more Fe^{2+} ions in the silver-doped iron vanadate sample as compared to the undoped $FeVO_4$ sample. It shows that doping of Ag^+ enhances the density of surface oxygen vacancies of catalysts, retarding electron-hole recombination and increasing photoefficiency. The Ag^+-doped samples showed 20% more degradation against methyl orange than pure $FeVO_4$ (Wang and Liu, 2011).

26.6.5 Silver-doped Titanates for wastewater remediation

Metal titanates are ternary inorganic compounds having the common formula $MTiO_3$ (where M is Ni, Cr, Fe, Zn, Sr, Ba, and Cu) and have been of immense research interest for over a decade because of their remarkable optical, magnetic, electrical, and ferroelectric properties. Such a combination of various functional properties makes them exceptional materials for developing microelectronic devices, sensors and for photocatalytic applications in the degradation of dyes and pesticides. Functional metal titanate materials are obtained in the form of nanopowder or thin nanosheets (Hosseinpour-Mashkani et al., 2017). It is a known fact that $NiTiO_3$ is the only ternary metal compound exhibiting ilmenite structure. In this structure, Ti and Ni both favor octahedral coordination. In silver-doped $NiTiO_3$, silver ions replace Ni ions. The photocatalytic activity of silver-doped $NiTiO_3$ photocatalysts, made by the modified Pechini method, is assessed by the reduction of dye MB under UV and visible light. The silver nanoparticles serve as charge carrier separation centers; hence, increasing the silver content in Ag-$NiTiO_3$ showed enhanced catalytic activity (Lin et al., 2009). The layered nanosheets show more photocatalytic activity than other forms. Moreover, various studies have shown that the incorporation or doping of silver nanoparticles into the titanate nanostructure renders photocatalytic activity of photocatalysts toward visible light range. Sodium titanate is a well-known photocatalyst of the metal titanate family that has captured attention owing to its low-cost and high surface area. However, its wide bandgap (3.4 to 3.7 eV) limits its applications in the visible region of the sunlight. Doping of silver into layered nanostructure of sodium titanate has shown significant changes in the structure and electronic and optical properties. In a research study, sodium titanate and nanosheets of Ag-doped sodium titanate were synthesized through a solvothermal method. Ti atoms were partially replaced by Ag^{3+} ions, enhancing the thermal stability of the titanate cluster structure. Moreover, the interference of silver atoms in titanate structure and the electronic recombination of bandgap showed enhanced degradation of dye methyl orange and phenol in the visible region up to 90% degradation (Rahut et al., 2017). In a similar research study, silver nanoparticles-based hydrogen tri-titanate nanotubes were fabricated using a simple microwave supported hydrothermal method. The Ag nanoparticles aided in fabrication of $H_2Ti_3O_7$ nanotubes more comprehensively. The silver-doped $H_2Ti_3O_7$ showed enhanced methyl orange dye degradation in sunlight (Rodríguez-González et al., 2012). A novel Ag-doped $K_2Ti_4O_9$, prepared by a sol-gel method showed higher degradation of dye methylene blue up to nearly 80% under sunlight. Doping altered the microstructure and electronic structure of $K_2Ti_4O_9$, effectively reducing the grain size and minimizing the possibility of electron and holes recombination. Hence, doping with Ag enhanced the visible light response (Xiaoli et al., 2015).

26.6.6 Silver-doped bismuth oxyhalides for wastewater remediation

Bismuth oxyhalides (BiOX, where X = Cl, Br, and I) belong to a class of ternary compounds having appropriate band structure and stability for photocatalysis. Bismuth oxyhalides have recently emerged as a class of most investigated potential photocatalysts for wastewater treatment. These multicomponent metal oxyhalides include members of the chief group family and exhibit matlockite tetragonal arrangement. These bismuth-based photocatalysts are less detrimental, corrosion-resistant, and chemically inert. These ternary compounds exhibit excellent performance for oxygen and hydrogen removal, pollutant degradation, and water disinfection. The redox potential is the main parameter for any photocatalyst for the conversion of oxygen into the superoxide radicals and for converting hydrogen into h^+. BiOX alone has relatively weak reducing potential owing to its bandgap. Doping or loading of these bismuth oxyhalides with the noble metals like Pt, Au, and Ag have shown to augment their photocatalytic activity (Dutta et al., 2019). Bismuth oxyiodide (BiOI) has a layered structure with a bandgap of 1.77–1.92 eV, making it a promising catalyst for the conversion of pollutants into less harmful materials. Doping with silver further enhances its photocatalytic potential. A silver-doped BiOI catalyst was made through a unique hydrothermal method with various percentages of silver content. The photocatalytic activity of silver-doped BiOI samples was found by optimizing and evaluating the degradation of various dyes. The degradation efficiency was determined by the production and separation of light-induced charge carriers. While doping with silver, the photogenerated electrons in the photocatalyst diffuse into the doping metals. The silver ions act as electron scavengers by decreasing recombination rate of $e-/h+$ pairs, enhancing the catalytic effect of photocatalysts for the removal of unwanted pollutants. The Ag-doped BiOI (1.0 mol%) showed 73.16 and 85.25% decolorization efficiency of dyes MO and Mb, respectively, respectively (Ekthammathat et al., 2016). In another study, the flower-like silver-doped BiOI photocatalyst prepared through a solvothermal method was used for the degradation of various dyes MO, MB, and Rhb. The 5 mol % Ag-BiOI showed enhanced degradation efficiency for all three dyes under daylight (Park et al., 2014). The effect of solvent on silver-doped compounds was studied by Kim et al. Silver-doped bismuth oxychloride prepared to employ a solvothermal method was assessed for the degradation of dyes methyl orange and Rhodamine B. Doping was done utilizing two solvents, ethylene glycol and water. The SEM images showed that Ag-BiOCl samples prepared in the water had flower-like morphology while those prepared in water had plate-like configuration. Moreover, the photocatalytic performance of doped sample enhanced upon increasing silver content (Kim et al., 2014).

26.6.7 Silver-doped Tantalates for wastewater treatment

Tantalates are a representative class of ternary compounds with broad potential in wastewater treatment, air purification, and water purification. The crystal structure of these compounds is of perovskite-type. Tantalates have conduction bands with Ta5d orbital located at higher negative potential as compared to titanates (Ti3d) having a conduction band located at a higher potential.

The photocatalytic activity of Ag–$K_2Ta_2O_6$ nanocomposite with different concentrations of silver content fabricated by a photochemical reduction approach is measured by tetracycline degradation under visible light. Various characterization techniques reveal uniform dispersal of the AgNPs on the surface of octahedral $K_2Ta_2O_6$. The excited electrons gather on silver nanoparticles surface. The probable mechanism includes a transfer of visible-light-induced electrons from silver nanoparticles to $K_2Ta_2O_6$ surface and subsequently producing · OH radicals by reacting with O_2. This shows that silver contributes to the improved photocatalytic activity of $K_2Ta_2O_6$ in sunlight (Xu et al., 2015). Similarly, Ag-doped Bi_3TaO_7 was also prepared by a simple reduction method. The doped sample showed excellent degradation of tetracycline. The improved activity is due to the SPR effect of Ag nanoparticles (Luo et al., 2015). Silver-doped potassium tantalate ($KTaO_3$), prepared via solvothermal process and later by a photodeposition method, was studied for photocatalytic hydrogen evolution in ultraviolet light (Kalaiselvi et al., 2020). Ag-$Na_2Ta_2O_6$ was studied for the removal of Rhodamine B. The incorporation of silver nanoparticles into $Na_2Ta_2O_6$ host matrix shows lattice expansion and narrowing of bandgap via the introduction of an interband, resulting in the formation of a well-controlled electronic structure. Silver ions not only enhance the absorption efficiency of $Na_2Ta_2O_6$ in the visible range but also cause the modulation of defective oxygen species that were causing hindrance in photocatalytic performance (Wang et al., 2015).

26.6.8 Silver-doped niobates for wastewater treatment

Metal niobates are key members of inorganic semiconductor materials and have found extensive applications in the field of optoelectronics, photoluminescence, and photocatalysis. During the previous years, extensive research has been done to elaborate on the role of niobates as photocatalysts. Coupling with other semiconductor materials and doping with plasmonic metal nanoparticles has resulted in the synthesis of heterostructures that are visible light-responsive. Zinc niobate is a prominent member of the family, exhibiting two-dimensional columbite structures having NbO_6 octahedral units, which are attached to oxygen atoms bridging with NbO_6 units. The silver doping of zinc niobate was done by using a solid-state method after which photoreduction of silver was done. The homogeneous doping of silver nanoparticles in $ZnNb_2O_6$ matrix resulted in enhanced visible light response, which was attributed to formation of heterojunctions along with strong surface plasmonic resonance effect. The doped ternary niobate with optimal 6 weight percentage loading of silver content showed the highest degradation for various dyes (Bi et al., 2018). Ag-doped Bi_3NbO_7, synthesized by a combined coprecipitation and a photodeposition method, was effectively studied for the removal of dye pollutants. The sample with a 2 wt% loading of silver nanoparticles showed excellent degradation efficiency. The improvement in photocatalytic activity was highly credited to combined electron exciting ability, electron trapping ability, and the SPR effect of silver nanoparticles (Shen et al., 2015). Similarly, heterostructured $Ag-Bi_5Nb_3O_{15}$ was made through a simplistic hydrothermal method, followed by a photodeposition process. This octahedral sheet-like silver-doped $Bi_5Nb_3O_{15}$ revealed enhanced degradation for tetrabromobisphenol A under both UV and visible light, which is attributed to the layered structure of $Bi_5Nb_3O_{15}$, SPR effect, and the electron trapping ability of silver nanoparticles (Guo et al., 2011). Potassium niobate ($K_4Nb_6O_{17}$) is a ternary compound having a typical orthorhombic structure having four NbO_6 layers stacked along two axes.

Silver-doped potassium niobate, synthesized with the help of a solid-state method, showed photocatalytic disinfection in the visible light. Under visible light, the photocatalytic activity was evaluated for inactivation of bacteria *E. coli* (Lin and Lin, 2012). Doping of silver in three different forms of potassium niobate (perovskite, pyrochlore, and layered) was effectively attained through a hydrothermal method. The perovskite and pyrochlore types of silver-doped potassium niobate showed enhanced photocatalytic activity than the doped layered form (Withanage et al., 2018).

26.6.9 Other silver-doped ternary compounds for wastewater treatment

Ternary indicates a new type of semiconductor photocatalysts having a narrow bandgap showing strong absorption in visible light region of solar spectrum. Silver indium sulfide ($AgIn_5S_8$) is a ternary compound that gets excited easily by visible light, hence producing photogenerated pairs of charge carriers. These type of chalcogenides exhibit excellent potential in the field of photovoltaics and wastewater treatment. Doping of these ternary compounds with plasmonic metals results in enhanced degradation of recalcitrant contaminants. In a detailed study done by Deng et al., doping of silver particles with silver indium sulfide was done by a solvothermal method followed by a photoreduction method. The deposited silver on the ternary in-date showed an evident effect on the separation of photoinduced charges. Nearly 95% degradation of tetracycline hydrochloride was achieved with the sample having 2.5% loading of silver nanoparticles. This is credited to the synergetic effect between the SPR effect of silver nanoparticles and the proper visible light active bandgap of $AgIn_5S_8$ (Deng et al., 2018). Cobalt chromite ($CoCr_2O_4$) is a ternary compound having a spinel cubic structure, in which the Co^{2+} ions occupy positions at tetrahedral sites while the Cr^{3+} ions at octahedral sites. The photocatalytic behavior of silver-doped $CoCr_2O_4$ nanoparticles, prepared by the photodeposition process, was investigated by the degradation of various dyes (Abbasi et al., 2017). Various other silver-doped ternary compounds are listed in Table 26.1.

26.7 Effects of silver doping on ternary compounds

26.7.1 Effect of silver doping on the morphology of ternary compounds

Doping of silver nanoparticles into the lattice structure of semiconductor photocatalysts may lead to changes in the morphology, photocatalytic activity, and optical and electrical properties. Surface and interface effects are very important as they are directly linked to physical properties (Abbasi et al., 2017). The contents of dopants in the ternary heterostructures also lead to controllable optical properties and the photocatalytic activity (Ekthammathat et al., 2016). Characterization techniques are used to study the morphology and composition of silver-doped and undoped ternary compounds. Effective and simple methods are being employed for decades to fabricate these photocatalytic materials with controlled morphology and size. Detailed experimentation on silver doping in bismuth ferrite by different synthesis methods showed different morphologies. The Ag-doped $BiFeO_3$ produced by an easy sol-gel method showed spherical shape and enhanced photodegradation than pure bismuth ferrite nanocomposite for the degradation of dye methyl orange. The photocatalytic activity showed an increase with an increasing amount of silver nanoparticles (Lu et al., 2015). The Ag-$BiFeO_3$ prepared with hydrothermal method showed spindle-like shapes and showed 95% degradation of malachite green (MG) dye (Jaffari et al., 2019). The Ag-doped bismuth ferrite composite made by a sol-gel scheme showed perovskite structure and was used as gas sensors (Bagwaiya et al., 2018). Silver-doped bismuth ferrite prepared using flash auto combustion technique had hexagonal and rhombohedral platelet shape and was used in fuel cell applications (Ahmed et al., 2014).

26.7.2 Effect of silver doping on physical properties of ternary compounds

The chemical and physical properties of nanomaterials are usually influenced by the morphology of nanoparticles. Any variation in size or morphology may also alter the physical properties of these compounds (Kudo et al., 2007). The modification in morphology and properties of these compounds by the introduction of a small amount of impurities depends upon dopants and the preparation conditions. The irregular shaped particles of magnesium ferrite become compact sheet-like after the introduction of silver atoms as dopants. The scanning electron microscope (SEM) images illustrate that the doped sheet-like spherical particles display micro-self-doping phenomena forming micronetting. This type of micro-self-doped morphology is of immense significance in catalysis (Lagashetty et al., 2019).

Table 26.1 Other Silver-doped Ternary compounds for wastewater treatment.

Silver-doped Ternary compound	Synthesis method	Morphology of doped photocatalyst	Model pollutant	Degradation efficiency	Light sources	References
Ag-Ag$_8$W$_4$O$_{16}$	Microwave method	Roasted rice beads	Methylene blue, Rhodamine B, *Escherichia coli*, *Staphylococcus aureus*	75.8% MB, 80%, 50 min (80% Rhodamine B)	Visible	Selvamani et al. (2016)
Ag-BiFeO$_3$	Hydrothermal	Spindle shape	Malachite green, *Escherichia coli*, and *Micrococcus luteus*	90.5%	Visible	Jaffari et al. (2019)
Ag-doped Bi$_2$MoO$_6$	Hydrothermal and Sonochemical	Two-dimensional plates	RhB	98% within 80 min	UV, Visible	Phuruangrat et al. (2015)
Ag/BiOI	Hydrothermal and photodeposition method	Flower-shaped	Methylorange, Rhodamine B, acid orange II	92.1% for AR 80.0% MO, 50% for Rhb	UV, Visible	Liu et al. (2012)
Ag-H$_2$Ti$_3$O$_7$	Microwave assisted method	Cylinderical multiwalled nanotubes	Methyl orange	90%	UV	Rodriguez-González et al. (2012)
Ag-PbMoO$_4$	Sonochemical and photoreduction method	Oval shaped structure	Indocyanine green	85%	UV/Visible	Gyawali et al. (2013)
Ag/Bi$_2$WO$_6$	Photoreduction approach	Flower-like microspheres	Rhodamine B	99.2%	UV/Visible	Wu et al. (2018)
Ag-doped PbMoO$_4$	Reflux method	Spherical shape	Rhodamine B	97.33%	UV	Phuruangrat et al. (2020)
Ag-doped SnZrMoP	Green synthetic method using mulberry leaves	Spherical/globular shape	*E. coli* and *Bacillus*	88% inactivation	UV	Kaur et al. (2020)
Ag-Bi$_4$NbO$_8$Cl	Liquid-Combustion Approach	Ellipsoidal shape particles	Rhodamine B	100%	UV	Qu et al. (2019)
Ag-Bi$_4$Ti$_3$O$_{12}$	Sonochemical method	Orthorhombic	Rhodamine B	100%	UV	Dutta and Tyagi (2016)
Ag-LaTiO$_3$	Hydrothermal method	Orthorhombic crystalline structure	Atrazine herbicide	96%	Visible	Shawky et al. (2020)
Ag-Na$_2$Ti$_3$O$_7$	Ion-Exchange method	Layered stacking	Methylene blue	90%	Visible	Vithal et al. (2013)
Ag-ZnTiO$_3$	Microwave assisted method	Elongated Nanorods	Rhodamine B, *Escherichia coli*, and *Bacillus subtilis*	100% under UV, 93% and visible light 90% inactivation	UV, Visible	Dutta et al. (2014)
Ag-Ag$_2$WO$_4$	Precipitation and calcination process		Rhodamine B	95%	UV, Visible	Song et al. (2020)
Ag-Ag$_2$MoO$_4$	Microwave-assisted hydrothermal process	Cube like	Rhodamine B	100%	Visible	Li et al. (2013)
Ag-Mg$_4$Ta$_2$O$_9$	Hydrothermal method	Flakes like structure	Atrazine	100%	Visible	Alkayal and Hussein (2019)

Doping of silver nanoparticles in ternary compounds like ferrites, molybdates, and vanadates has shown to significantly increase the magnetic properties and mechanical strength, as well as electrical conductivity of doped compounds. Doping with silver nanoparticles also improves density by reducing porosity and decreases the sintering temperature (Okasha, 2008). All these changes are directly linked with the changes in the crystal structure and morphology. Silver nanoparticles are homogeneously dispersed in the nanocomposite matrix where they act as nucleation centers, which increase crystallinity and thermal stability of nanocomposites (Miranda et al., 2009). Also, the silver-doped compounds show reduced compactness and uniform lattice fringes indicating high crystallinity. Silver-doped $Li_4Ti_5O_{12}$ prepared by spray pyrolysis method shows an enhanced crystallite size and crystallinity than the pristine $Li_4Ti_5O_{12}$. The silver particles on the surface of lithium titanium oxide have been shown to enhance its electrochemical properties (Karhunen et al., 2016).

26.7.3 Effect of silver doping on enhancement of visible light response and photocatalytic activity

The silver-doped ternary compounds show better photocatalytic activity as compared with their undoped samples. In-band structure of most ternary semiconductor compounds, the bottom of a conduction band, is at a negative potential than the redox potential of H^+/H_2, and the top of the valence band is more positive as compared to the redox potential of O_2/H_2O, for water splitting by photocatalysis. Bandgaps of most metal oxides and sulfides for photocatalysis are wider than 3 eV, which makes them responsive under ultraviolet light only. The creation of new bands above valence band is indispensable for construction of visible light active photocatalysts. Hence, doping of silver introduces additional electron donor levels making these compounds visible light active. In a research study, Rahut et al. shows that the presence of silver on the layers of titanate compound favors the rearrangement of light-induced electrons to silver atoms, which subsequently improves separation of charge carriers. The introduction of silver atoms in the titanate layers increased the photo-oxidation of pollutants on the titanate surface. Pristine titanate is UV light sensitive and degrades dyes up to 90% under UV light only. Changes in the titanate nanostructure as a result of silver doping and electronic reconstruction of the valence band and conduction bands prove to be evidence of photocatalytic activity in visible region of solar spectrum (Rahut et al., 2017). According to the UV–visible spectra, the silver-doped Bi_2WO_6 shows a redshift and increased photoabsorption in the visible light region. This is credited to SPR effect caused by the silver nanoparticles (Li et al., 2014).

26.7.4 Effect of doping content of silver

A suitable content of silver nanoparticles is crucial for the efficient separation of charges. Usually, it is considered that increasing the dopant concentration will enhance the photoactivity by providing more interfaces. However, studies have shown that silver content beyond a certain limit will form clusters acting as recombination centers and hinder the photocatalytic activity (Gyawali et al., 2013). By increasing the concentration of the AgNPs in silver-doped bismuth tantalate beyond the optimal limit which was 1%, the unnecessary silver nanoparticles started covering most of active sites on the surface of ternary compound and suppress the light absorption. This led to a decline in the photocatalytic activity (Luo et al., 2015).

26.8 Other applications of silver-doped ternary compounds

The efficiency Ag-doped nanoparticles used in different applications depends upon various factors. This includes types of precursors for silver doping, reducing agents, and various solvents used for synthesis (Sajjadi et al., 2017). Besides the wastewater treatment and antimicrobial properties of silver-doped ternary compounds, it is also used for the inactivation of viruses (Liga et al., 2011). These doped compounds are used in numerous day to day applications like in electronics, as sensors, as antireflection coatings, as bone cements for artificial joint replacement (Morley et al., 2007). They are also used as photoanodes, gas sensors, for hydrogen production, and in preparation of ceramics. A list of various silver-doped ternary compounds for various applications other than wastewater treatment is described in Table 26.2.

Table 26.2 Various other applications of silver-doped ternary compounds for.

Silver-doped ternary compound	Synthesis method	Application	References
Ag-LaFeO$_3$	Sol-gel/microwave	Gas sensors	Zhang et al. (2014)
Ag-doped NaTaO$_3$	Sol-gel method	Perovskite semiconductors	Sfirloaga et al. (2015)
Ag-NiFe$_2$O$_4$	Citrate method	Enhanced structural and ferromagnetic character	Ahmed et al. (2011)
Ag-Ag$_2$V$_2$O$_5$	melt quenching technique	Ternary ionic–electronic conducting glass system	Masoud and Mousa (2015)
Ag-LiTaO$_3$	Ag$^+$ ion implantation	Ultrafast laser generation	Pang et al. (2019)
Ag/BiFeO$_3$	Sol-gel	Thin film composites	Cheng et al. (2017)
Ag -CoCr$_2$O$_4$	Photodeposition	Thermal resistivity of epoxy acrylate	Abbasi et al. (2017)
Ag-PbBi$_4$Te$_7$	Czochralski (CZ) technique based seed growth method	Enhanced thermoelectric properties	Shelimova et al. (2008)
Ag-ZnInSe	Reflux method	Bioimaging Optical coding	Wang et al. (2014)

Table 26.2 Various other applications of silver-doped ternary compounds for—cont'd

Silver-doped ternary compound	Synthesis method	Application	References
Ag-BiVO$_4$	Photoreduction technique	Photoanode; Hydrogen production	Zhang et al. (2009)
Ag-KTaO$_3$	Solvothermal method	Hydrogen evolution activity under visible light	Kalaiselvi et al. (2020)
Ag-CaHPO$_4 \cdot$2H$_2$O	Sintering and ball mill Method	Production of Antimicrobial cement	Ewald et al. (2011)
Ag-Ca$_3$(PO$_4$)$_2$	Coprecipitation route	Study of Antimicrobial and toxic effects toward prokaryotic cells	Peetsch et al. (2013)
BiOBr-Ag	Hydrothermal method	Photocatalysis	Paudel et al. (2018)
BaTiO$_3$–Ag	Hydrothermal sol gel method	Ceramics with enhanced dielectric properties	Du et al. (2009)
Ag-CuMoO4	Extraction–pyrolysis process	Oxidation of carbon black	Makarevich et al. (2010)
Ag-BiFeO$_3$	Flash auto combustion technique	Promotion of multiferroic in solid oxide fuel cell applications.	Ahmed et al. (2014)
Ag-LaCrO$_3$	Microwave combustion method	Enhanced magnetization	Athawale and Desai (2011)
Ag-Cu$_2$SnS$_3$	Solar cell	Monoclinic heterostructures	De Wild et al. (2017)

Conclusion

Silver-doped ternary compounds are nowadays being commonly used as visible light active photocatalytic materials. The photoefficiency of these silver-doped compounds is largely based on their morphology, properties, and doping mechanism. These heterostructures show enhanced photoactivity and photostability due to strong surface plasmonic response. These doped compounds are being used for wastewater treatment. Apart from that silver-doped ternary compounds are being used as antibacterial and antifungal agents. Ternary compounds alone usually have less photocatalytic activity in visible range owing to rapid recombination of charge carriers. Doping with silver enhanced their photocatalytic, environmental, and industrial importance.

References

Abbasi, A., Hamadanian, M., Salavati-Niasari, M., Mazhari, M.P., 2017. Hydrothermal synthesis, characterization and photodegradation of organic pollutants of $CoCr_2O_4$/Ag nanostructure and thermal stability of epoxy acrylate nanocomposite. Adv. Powder Technol. 28 (10), 2756–2765.

Ahmad, M.Z., Qureshi, K., Bhatti, I.A., Zahid, M., Nisar, J., Iqbal, M., Abbas, M., 2018. Hydrothermal synthesis of molybdenum trioxide, characterization and photocatalytic activity. Mater. Res. Bull. 100, 120–130.

Ahmed, M., El-Dek, S., El-Kashef, I., Helmy, N., 2011. Structural and magnetic properties of nano-crystalline Ag^+ doped $NiFe_2O_4$. Solid State Sci. 13 (5), 1176–1179.

Ahmed, M., Mansour, S., El-Dek, S., Abu-Abdeen, M., 2014. Conduction and magnetization improvement of $BiFeO_3$ multiferroic nanoparticles by Ag^+ doping. Mater. Res. Bull. 49, 352–359.

Alkayal, N.S., Hussein, M.A., 2019. Photocatalytic degradation of atrazine under visible light using novel Ag@ $Mg_4Ta_2O_9$ nanocomposites. Sci. Rep. 9 (1), 1–10.

Alzahrani, E., 2015. Photodegradation of eosin Y using silver-doped magnetic nanoparticles. Int. J. Anal. Chem 15, 1–11.

Anchieta, C.G., Sallet, D., Foletto, E.L., da Silva, S.S., Chiavone-Filho, O., do Nascimento, C.A., 2014. Synthesis of ternary zinc spinel oxides and their application in the photodegradation of organic pollutant. Ceram. Int. 40 (3), 4173–4178.

Anjum, M., Oves, M., Kumar, R., Barakat, M., 2017. Fabrication of ZnO-ZnS@ polyaniline nanohybrid for enhanced photocatalytic degradation of 2-chlorophenol and microbial contaminants in wastewater. Int. Biodeterior. Biodegradation 119, 66–77.

Aslam, M., Tahir Soomro, M., Ismail, I.M.I., Salah, N., Waqar Ashraf, M., Qari, H.A., Hameed, A., 2019. The performance of silver modified tungsten oxide for the removal of 2-CP and 2-NP in sunlight exposure: optical, electrochemical and photocatalytic properties. Arab. J. Chem. 12 (8), 2632–2643.

Athawale, A., Desai, P., 2011. Silver doped lanthanum chromites by microwave combustion method. Ceram. Int. 37 (8), 3037–3043.

Bagwaiya, T., Khade, P., Reshi, H.A., Bhattacharya, S., Shelke, V., Kaur, M., Debnath, A., Muthe, K., Gadkari, S., 2018. Investigation on gas sensing properties of Ag doped $BiFeO_3$. In: Paper presented at the AIP Conf. Proc., p. 080076.

Bi, X., Wu, N., Zhang, C., Bai, P., Chai, Z., Wang, X., 2018. Synergetic effect of heterojunction and doping of silver on $ZnNb_2O_6$ for superior visible-light photocatalytic activity and recyclability. Solid State Sci. 84, 86–94.

Chen, R., Hu, C.H., Wei, S., Xia, J.H., Cui, J., Zhou, H.Y., 2013a. Synthesis and activity of Ag-doped Bi_2WO_6 photocatalysts. J. Mater. Sci. 743, 560–566.

Chen, R., Hu, C.H., Wei, S., Xia, J.H., Cui, J., Zhou, H.Y., 2013b. In: Synthesis and Activity of Ag-DopED BI2WO6 Photocatalysts. Paper Presented at The Mater. Sci. Forum, pp. 560–566.

Chen, S., Zhou, M., Li, T., Cao, W., 2018. Synthesis of Ag-loaded Sb_2WO_6 microsphere with enhanced photocatalytic ability for organic dyes degradations under different light irradiations. J. Mol. Liq. 272, 27–36.

Cheng, S., Xu, Q., Hao, X., Wang, Z., Ma, N., Du, P., 2017. Formation of nano-Ag/$BiFeO_3$ composite thin film with extraordinary high dielectric and effective ferromagnetic properties. J. Mater. Sci.: Mater. Electron. 28 (7), 5652–5662.

De Wild, J., Babbe, F., Robert, E.V., Redinger, A., Dale, P.J., 2017. Silver-doped Cu_2SnS_3 absorber layers for solar cells application. IEEE J. Photovoltaics 8 (1), 299–304.

Deng, F., Zhao, L., Luo, X., Luo, S., Dionysiou, D.D., 2018. Highly efficient visible-light photocatalytic performance of Ag/AgIn$_5$S$_8$ for degradation of tetracycline hydrochloride and treatment of real pharmaceutical industry wastewater. Chem. Eng. J. (Lausanne) 333, 423–433.

Di Paola, A., García-López, E., Marcì, G., Palmisano, L., 2012. A survey of photocatalytic materials for environmental remediation. J. Hazard. Mater. 211, 3–29.

Du, F., Yu, P., Cui, B., Cheng, H., Chang, Z., 2009. Preparation and characterization of monodisperse Ag nanoparticles doped barium titanate ceramics. J. Alloys Compd. 478 (1–2), 620–623.

Dumrongrojthanath, P., Thongtem, T., Phuruangrat, A., Thongtem, S., 2013. Synthesis and characterization of hierarchical multilayered flower-like assemblies of Ag doped Bi$_2$WO$_6$ and their photocatalytic activities. Superlattice. Microst. 64, 196–203.

Dutta, D.P., Tyagi, A., 2016. Facile sonochemical synthesis of Ag modified Bi$_4$Ti$_3$O$_{12}$ nanoparticles with enhanced photocatalytic activity under visible light. Mater. Res. Bull. 74, 397–407.

Dutta, D.P., Singh, A., Tyagi, A., 2014. Ag doped and Ag dispersed nano ZnTiO$_3$: improved photocatalytic organic pollutant degradation under solar irradiation and antibacterial activity. J. Environ. Chem. Eng. 2 (4), 2177–2187.

Ekthammathat, N., Kidarn, S., Phuruangrat, A., Thongtem, S., Thongtem, T., 2016. Hydrothermal synthesis of Ag-doped BiOI nanostructure used for photocatalysis. Res. Chem. Intermed. 42 (6), 5559–5572.

Ewald, A., Hösel, D., Patel, S., Grover, L.M., Barralet, J.E., Gbureck, U., 2011. Silver-doped calcium phosphate cements with antimicrobial activity. Acta Biomater. 7 (11), 4064–4070.

Fardood, S.T., Forootan, R., Moradnia, F., Afshari, Z., Ramazani, A., 2020. Green synthesis, characterization, and photocatalytic activity of cobalt chromite spinel nanoparticles. Mater. Res. Express 7 (1), 015086.

Friedmann, D., Mendive, C., Bahnemann, D., 2010. TiO$_2$ for water treatment: parameters affecting the kinetics and mechanisms of photocatalysis. Appl. Catal., B 99 (3–4), 398–406.

Ge, L., 2008. Novel Pd/BiVO$_4$ composite photocatalysts for efficient degradation of methyl orange under visible light irradiation. Mater. Chem. Phys. 107 (2–3), 465–470.

Gingasu, D., Mindru, I., Patron, L., Calderon-Moreno, J.M., Mocioiu, O.C., Preda, S., Stanica, N., Nita, S., Dobre, N., Popa, M., 2016. Green synthesis methods of CoFe$_2$O$_4$ and Ag-CoFe$_2$O$_4$ nanoparticles using hibiscus extracts and their antimicrobial potential. J. Nanomater. 2016, 75–86.

Gulzar, N., Zubair, K., Shakir, M.F., Zahid, M., Nawab, Y., Rehan, Z., 2020. Effect on the EMI shielding properties of cobalt ferrites and coal-Fly-ash based polymer nanocomposites. J. Supercond. Novel Magn., 1–6.

Guo, Y., Chen, L., Ma, F., Zhang, S., Yang, Y., Yuan, X., Guo, Y., 2011. Efficient degradation of tetrabromobisphenol A by heterostructured Ag/Bi$_5$Nb$_3$O$_{15}$ material under the simulated sunlight irradiation. J. Hazard. Mater. 189 (1–2), 614–618.

Guo, L., Wang, D.J., Fu, F., Qiang, X.D., Xu, T., 2012. Synthesis of Ag, Pd-loaded Bi$_2$WO$_6$ and its photocatalytic activities. Adv. Mater. Res. 518, 833–836.

Gyawali, G., Adhikari, R., Joshi, B., Kim, T.H., Rodríguez-González, V., Lee, S.W., 2013. Sonochemical synthesis of solar-light-driven Ag̊-PbMoO$_4$ photocatalyst. J. Hazard. Mater. 263, 45–51.

Hosseinpour-Mashkani, S.M., Maddahfar, M., Sobhani-Nasab, A., 2016. Novel silver-doped CdMoO$_4$: synthesis, characterization, and its photocatalytic performance for methyl orange degradation through the sonochemical method. J. Mater. Sci.: Mater. Electron. 27 (1), 474–480.

Hosseinpour-Mashkani, S.M., Maddahfar, M., Sobhani-Nasab, A., 2017. Novel silver-doped $NiTiO_3$: auto-combustion synthesis, characterization and photovoltaic measurements. S. Afr. J. Chem. 70, 44–48.

Hsieh, J., Yu, R., Chang, Y., Li, C., 2012. Structural analysis of TiO_2 and TiO_2-Ag thin films and their antibacterial behaviors. J. Phys. 339, 11–24.

Huang, C.-M., Pan, G.-T., Li, Y.-C.M., Li, M.-H., Yang, T.C.-K., 2009. Crystalline phases and photocatalytic activities of hydrothermal synthesis Ag_3VO_4 and $Ag_4V_2O_7$ under visible light irradiation. Appl. Catal., A 358 (2), 164–172.

Jaffari, Z.H., Lam, S.-M., Sin, J.-C., Zeng, H., 2019. Boosting visible light photocatalytic and antibacterial performance by decoration of silver on magnetic spindle-like bismuth ferrite. Mater. Sci. Semicond. Process. 101, 103–115.

Jiang, Z.-J., Liu, C.-Y., Sun, L.-W., 2005. Catalytic properties of silver nanoparticles supported on silica spheres. J. Phys. Chem. B 109 (5), 1730–1735.

Kalaiselvi, C., Ravi, P., Senthil, T., Sathish, M., Kang, M., 2020. Synthesis of Ag and N doped potassium tantalate perovskite nanocubes for enhanced photocatalytic hydrogen evolution. Mater. Lett. 128166.

Karhunen, T., Välikangas, J., Torvela, T., Lähde, A., Lassi, U., Jokiniemi, J., 2016. Effect of doping and crystallite size on the electrochemical performance of $Li_4Ti_5O_{12}$. J. Alloys Compd. 659, 132–137.

Kaur, R., Kaushal, S., Singh, P., 2020. Biogenic synthesis of a silver nanoparticle–SnZrMoP nanocomposite and its application for the disinfection and detoxification of water. Mater. Adv, 1–10.

Khojeh, B., Zanjanchi, M., Golmojdeh, H., 2017. Preparation of catalytically active bismuth tungstate: effects of organic additives and dopants. Mater. Res. Innov. 21 (5), 341–349.

Kim, W.J., Pradhan, D., Min, B.-K., Sohn, Y., 2014. Adsorption/photocatalytic activity and fundamental natures of BiOCl and $BiOCl_xI1 - x$ prepared in water and ethylene glycol environments, and Ag and Au-doping effects. Appl. Catal., B 147, 711–725.

Koe, W.S., Lee, J.W., Chong, W.C., Pang, Y.L., Sim, L.C., 2019. An overview of photocatalytic degradation: photocatalysts, mechanisms, and development of photocatalytic membrane. Environ. Sci. Pollut. R., 1–44.

Kubacka, A., Fernandez-Garcia, M., Colon, G., 2012. Advanced nanoarchitectures for solar photocatalytic applications. Chem. Rev. (Washington, DC, U. S.) 112 (3), 1555–1614.

Kudo, A., Niishiro, R., Iwase, A., Kato, H., 2007. Effects of doping of metal cations on morphology, activity, and visible light response of photocatalysts. Chem. Phys. 339 (1–3), 104–110.

Lagashetty, A., Pattar, A., Ganiger, S.K., 2019. Synthesis, characterization and antibacterial study of Ag doped magnesium ferrite nanocomposite. Heliyon 5 (5), e01760.

Li, Z., Chen, X., Xue, Z.-L., 2013. Microwave-assisted hydrothermal synthesis of cube-like $Ag-Ag_2MoO_4$ with visible-light photocatalytic activity. Science China Chem. 56 (4), 443–450.

Li, J., Guo, Z., Zhu, Z., 2014. Ag/Bi_2WO_6 plasmonic composites with enhanced visible photocatalytic activity. Ceram. Int. 40 (5), 6495–6501.

Liga, M.V., Bryant, E.L., Colvin, V.L., Li, Q., 2011. Virus inactivation by silver doped titanium dioxide nanoparticles for drinking water treatment. Water Res. 45 (2), 535–544.

Lin, H.-Y., Lin, H.-M., 2012. Visible-light photocatalytic inactivation of Escherichia coli by $K_4Nb_6O_{17}$ and Ag/Cu modified $K_4Nb_6O_{17}$. J. Hazard. Mater. 217, 231–237.

Lin, Y.-J., Chang, Y.-H., Chen, G.-J., Chang, Y.-S., Chang, Y.-C., 2009. Effects of Ag-doped $NiTiO_3$ on photoreduction of methylene blue under UV and visible light irradiation. J. Alloys Compd. 479 (1–2), 785–790.

Liu, H., Cao, W., Su, Y., Wang, Y., Wang, X., 2012. Synthesis, characterization and photocatalytic performance of novel visible-light-induced Ag/BiOI. Appl. Catal., B 111, 271–279.

Lu, H., Du, Z., Wang, J., Liu, Y., 2015. Enhanced photocatalytic performance of Ag-decorated $BiFeO_3$ in visible light region. J. Sol-Gel Sci. Technol. 76 (1), 50–57.

Luo, B., Xu, D., Li, D., Wu, G., Wu, M., Shi, W., Chen, M., 2015. Fabrication of a Ag/Bi_3TaO_7 plasmonic photocatalyst with enhanced photocatalytic activity for degradation of tetracycline. ACS Appl. Mater. Interfaces 7 (31), 17061–17069.

Mahapatra, S., Nayak, S.K., Madras, G., Guru Row, T., 2008. Microwave synthesis and photocatalytic activity of nano lanthanide (Ce, Pr, and Nd) orthovanadates. Ind. Eng. Chem. Res. 47 (17), 6509–6516.

Makarevich, K., Lebukhova, N., Chigrin, P., Karpovich, N., 2010. Catalytic properties of $CuMoO_4$ doped with Co, Ni, and Ag. Inorg. Mater. 46 (12), 1359–1364.

Marimuthu, S., Antonisamy, A.J., Malayandi, S., Rajendran, K., Tsai, P.-C., Pugazhendhi, A., Ponnusamy, V.K., 2020. Silver nanoparticles in dye effluent treatment: A review on synthesis, treatment methods, mechanisms, photocatalytic degradation, toxic effects and mitigation of toxicity. J. Photochem. Photobiol. B 205, 111823.

Masoud, E.M., Mousa, M., 2015. Silver-doped silver vanadate glass composite electrolyte: structure and an investigation of electrical properties. Ionics 21 (4), 1095–1103.

McEvoy, J.G., Zhang, Z., 2014. Antimicrobial and photocatalytic disinfection mechanisms in silver-modified photocatalysts under dark and light conditions. J. Photochem. Photobiol. C 19, 62–75.

Miranda, D., Sencadas, V., Sánchez-Iglesias, A., Pastoriza-Santos, I., Liz-Marzán, L., Ribelles, J., Lanceros-Méndez, S., 2009. Influence of silver nanoparticles concentration on the α-to β-phase transformation and the physical properties of silver nanoparticles doped poly (vinylidene fluoride) nanocomposites. J. Nanosci. Nanotechnol. 9 (5), 2910–2916.

Morley, K., Webb, P., Tokareva, N., Krasnov, A., Popov, V., Zhang, J., Roberts, C., Howdle, S., 2007. Synthesis and characterisation of advanced UHMWPE/silver nanocomposites for biomedical applications. Eur. Polym. J. 43 (2), 307–314.

Morones, J.R., Elechiguerra, J.L., Camacho, A., Holt, K., Kouri, J.B., Ramírez, J.T., Yacaman, M.J., 2005. The bactericidal effect of silver nanoparticles. Nanotechnology 16 (10), 2346.

Mushtaq, F., Zahid, M., Mansha, A., Bhatti, I., Mustafa, G., Nasir, S., Yaseen, M., 2020. $MnFe_2O_4$/coal fly ash nanocomposite: a novel sunlight-active magnetic photocatalyst for dye degradation. Int. J. Environ. Sci. Technol. 17, 4233–4248.

Nadeem, N., Zahid, M., Tabasum, A., Mansha, A., Jilani, A., Bhatti, I.A., Bhatti, H.N., 2020. Degradation of reactive dye using heterogeneous photo-Fenton catalysts: $ZnFe_2O_4$ and GO-$ZnFe_2O_4$ composite. Mater. Res. Express 7 (1), 15–19.

Namvar, F., Abass, S.K., Soofivand, F., Salavati-Niasari, M., Moayedi, H., 2019. Sonochemical synthesis of Pr_6MoO_{12} nanostructures as an effective photocatalyst for waste-water treatment. Ultrason. Sonochem. 58, 104687.

Naqvi, S.A.R., Nadeem, S., Komal, S., Naqvi, S.A.A., Mubarik, M.S., Qureshi, S.Y., Ahmad, S., Abbas, A., Zahid, M., Raza, S.S., 2019. Antioxidants: Natural Antibiotics Antioxidants. Vol. 10 IntechOpen, pp. 55–90.

Neto, N.A., Dias, B., Tranquilin, R., Longo, E., Li, M., Bomio, M., Motta, F., 2020. Synthesis and characterization of Ag^+ and Zn^{2+} co-doped $CaWO_4$ nanoparticles by a fast and facile sonochemical method. J. Alloys Compd. 823, 153617.

Niu, F., Chen, D., Qin, L., Gao, T., Zhang, N., Wang, S., Chen, Z., Wang, J., Sun, X., Huang, Y., 2015. Synthesis of Pt/$BiFeO_3$ heterostructured photocatalysts for highly efficient visible-light photocatalytic performances. Sol. Energy Mater. Sol. Cells 143, 386–396.

Noreen, S., Bhatti, H.N., Nausheen, S., Zahid, M., Asim, S., 2014. Biosorption of Drimarine blue HF-RL using raw, pretreated, and immobilized peanut hulls. Desalin. Water Treat. 52 (37–39), 7339–7353.

Noreen, S., Bhatti, H.N., Zuber, M., Zahid, M., Asgher, M., 2017. Removal of Actacid Orange-RL dye using biocomposites: modeling studies. Pol. J. Environ. Stud. 26 (5).

Okasha, N., 2008. Influence of silver doping on the physical properties of Mg ferrites. J. Mater. Sci. 43 (12), 4192–4197.

Pang, C., Li, R., Li, Z., Dong, N., Wang, J., Ren, F., Chen, F., 2019. Plasmonic Ag nanoparticles embedded in lithium tantalate crystal for ultrafast laser generation. Nanotechnology 30 (33), 334001.

Park, Y., Na, Y., Pradhan, D., Min, B.-K., Sohn, Y., 2014. Adsorption and UV/visible photocatalytic performance of BiOI for methyl orange, rhodamine B and methylene blue: Ag and Ti-loading effects. CrystEngComm 16 (15), 3155–3167.

Paudel, S., Adhikari, P.R., Upadhyay, O.P., Kaphle, G.C., Srivastava, A., 2018. First principles study of electronic and optical properties of pristine and X (Cu, Ag and Au) doped BiOBr. J. Instr. Sci. Technol. 22 (2), 63–69.

Peetsch, A., Greulich, C., Braun, D., Stroetges, C., Rehage, H., Siebers, B., Köller, M., Epple, M., 2013. Silver-doped calcium phosphate nanoparticles: synthesis, characterization, and toxic effects toward mammalian and prokaryotic cells. Colloids Surf., B 102, 724–729.

Phuruangrat, A., Putdum, S., Dumrongrojthanath, P., Ekthammathat, N., Thongtem, S., Thongtem, T., 2015. Enhanced properties for visible-light-driven photocatalysis of Ag nanoparticle modified Bi_2MoO_6 nanoplates. Mater. Sci. Semicond. Process. 34, 175–181.

Phuruangrat, A., Thongtem, S., Thongtem, T., 2020. Refluxing synthesis and characterization of UV-light-driven Ag-doped $PbMoO_4$ for Photodegradation of rhodamine B. J. Electron. Mater. 49, 4212–4220.

Prabhu, S., Poulose, E.K., 2012. Silver nanoparticles: mechanism of antimicrobial action, synthesis, medical applications, and toxicity effects. International nano letters 2 (1), 32.

Qu, X., Liu, M., Zhai, H., Zhao, X., Shi, L., Du, F., 2019. Plasmonic Ag-promoted layered perovskite oxyhalide Bi_4NbO_8Cl for enhanced photocatalytic performance towards water decontamination. J. Alloys Compd. 810, 151919.

Rahut, S., Panda, R., Basu, J.K., 2017. Solvothermal synthesis of a layered titanate nanosheets and its photocatalytic activity: effect of Ag doping. J. Photochem. Photobio. A 341, 12–19.

Rasheed, I.A., Rehan, Z.A., Khalid, T., Zahid, M., Ahmad, H., 2020. Prospects of Nanocomposite Membranes in Commercial Scale Nanocomposite Membranes for Water and Gas Separation. Elsevier, pp. 457–473.

Ray, S.K., Kshetri, Y.K., Dhakal, D., Regmi, C., Lee, S.W., 2017. Photocatalytic degradation of rhodamine B and ibuprofen with upconversion luminescence in Ag-$BaMoO_4$: Er^{3+}/Yb^{3+}/K^+ microcrystals. J. Photochem. Photobio. A 339, 36–48.

Ray, S.K., Dhakal, D., Gyawali, G., Joshi, B., Koirala, A.R., Lee, S.W., 2019. Transformation of tetracycline in water during degradation by visible light driven Ag nanoparticles decorated α-$NiMoO_4$ nanorods: mechanism and pathways. Chem. Eng. J. (Lausanne) 373, 259–274.

Rehan, Z.A., Zahid, M., Akram, S., Rashid, A., Rehman, A., 2020. Prospects of Nanocomposite Membranes for Water Treatment by Pressure-Driven Membrane Processes Nanocomposite Membranes for Water and Gas Separation. Elsevier, pp. 237–256.

Rodríguez-González, V., Obregón-Alfaro, S., Lozano-Sánchez, L., Lee, S.-W., 2012. Rapid microwave-assisted synthesis of one-dimensional silver–$H_2Ti_3O_7$ nanotubes. J. Mol. Catal. A Chem. 353, 163–170.

Saher, R., Hanif, M., Mansha, A., Javed, H., Zahid, M., Nadeem, N., Mustafa, G., Shaheen, A., Riaz, O., 2020. Sunlight-driven photocatalytic degradation of rhodamine B dye by Ag/FeWO$_4$/gC$_3$N$_4$ composites. Int. J. Environ. Sci. Technol., 1–12.

Sajjadi, M., Nasrollahzadeh, M., Sajadi, S.M., 2017. Green synthesis of Ag/Fe$_3$O$_4$ nanocomposite using Euphorbia peplus Linn leaf extract and evaluation of its catalytic activity. J. Colloid Interface Sci. 497, 1–13.

Satheeshkumar, M., Kumar, E.R., Srinivas, C., Suriyanarayanan, N., Deepty, M., Prajapat, C., Rao, T.C., Sastry, D., 2019. Study of structural, morphological and magnetic properties of Ag substituted cobalt ferrite nanoparticles prepared by honey assisted combustion method and evaluation of their antibacterial activity. J. Magn. Magn. Mater. 469, 691–697.

Sedighi, F., Esmaeili-Zare, M., Behpour, M., 2019. Fabricant and characterization of SrWO$_4$ and novel silver-doped SrWO$_4$ using co-precipitation method: their photocatalytic performances for methyl orange degradation. Journal of Nanostructures 9 (2), 331–339.

Selvamani, M., Krishnamoorthy, G., Ramadoss, M., Sivakumar, P.K., Settu, M., Ranganathan, S., Vengidusamy, N., 2016. Ag@ Ag$_8$W$_4$O$_{16}$ nanoroasted rice beads with photocatalytic, antibacterial and anticancer activity. Mater. Sci. Eng. C 60, 109–118.

Sfirloaga, P., Miron, I., Malaescu, I., Marin, C., Ianasi, C., Vlazan, P., 2015. Structural and physical properties of undoped and Ag-doped NaTaO$_3$ synthesized at low temperature. Mater. Sci. Semicond. Process. 39, 721–725.

Sharma, K., Dutta, V., Sharma, S., Raizada, P., Hosseini-Bandegharaei, A., Thakur, P., Singh, P., 2019. Recent advances in enhanced photocatalytic activity of bismuth oxyhalides for efficient photocatalysis of organic pollutants in water: A review. J. Ind. Eng. Chem.

Shawky, A., Mohamed, R., Mkhalid, I., Youssef, M., Awwad, N., 2020. Visible light-responsive Ag/LaTiO$_3$ nanowire photocatalysts for efficient elimination of atrazine herbicide in water. J. Mol. Liq. 299, 112163.

Shelimova, L., Karpinskii, O., Svechnikova, T., Nikhezina, I.Y., Avilov, E., Kretova, M., Zemskov, V., 2008. Effect of cadmium, silver, and tellurium doping on the properties of single crystals of the layered compounds PbBi$_4$Te$_7$ and PbSb$_2$Te$_4$. Inorg. Mater. 44 (4), 371–376.

Shen, Y., Wei, Q., Guo, W., Fan, L., Liu, D., Li, S., 2015. Fabrication of Ag loaded Bi3NbO7 nanoparticles and its photocatalytic activity under visible light irradiation. J. Alloys Compd. 618, 311–317.

Singh, J., Dhaliwal, A., 2020. Plasmon-induced photocatalytic degradation of methylene blue dye using biosynthesized silver nanoparticles as photocatalyst. Environ. Technol. 41 (12), 1520–1534.

Soltani, T., Entezari, M.H., 2013. Sono-synthesis of bismuth ferrite nanoparticles with high photocatalytic activity in degradation of rhodamine B under solar light irradiation. Chem. Eng. J. (Lausanne) 223, 145–154.

Song, Y., Xie, W., Yang, C., Wei, D., Su, X., Li, L., Wang, L., Wang, J., 2020. Humic acid-assisted synthesis of Ag/Ag$_2$MoO$_4$ and Ag/Ag$_2$WO$_4$ and their highly catalytic reduction of nitro-and azo-aromatics. J. Mat. Res. Tech. 9 (3), 5774–5783.

Soo, J.Z., Ang, B.C., Ong, B.H., 2018. Influence of calcination on the morphology and crystallinity of titanium dioxide nanofibers towards enhancing photocatalytic dye degradation. Mater. Res. Express 6 (2), 025039.

Tabasum, A., Zahid, M., Bhatti, H.N., Asghar, M., 2018. Fe$_3$O$_4$-GO composite as efficient heterogeneous photo-Fenton's catalyst to degrade pesticides. Mater. Res. Express 6 (1), 015608.

Tabasum, A., Bhatti, I.A., Nadeem, N., Zahid, M., Rehan, Z.A., Hussain, T., Jilani, A., 2020. Degradation of acetamiprid using graphene-oxide-based metal (Mn and Ni) ferrites as Fenton-like photocatalysts. Water Sci. Technol. 81 (1), 178–189.

Vithal, M., Krishna, S.R., Ravi, G., Palla, S., Velchuri, R., Pola, S., 2013. Synthesis of Cu2+ and Ag+ doped Na2Ti3O7 by a facile ion-exchange method as visible-light-driven photocatalysts. Ceram. Int. 39 (7), 8429–8439.

Wang, M., Liu, Q., 2011. Influence of Ag doping on the crystal structure and photocatalytic activity of FeVO$_4$. Adv. Mater. Res. 197, 919–925.

Wang, J.-Y., Huang, M.-L., Zhong, Q.-Q., Lin, J.-M., Wu, J.-H., 2009. Photocatalysis and preparation of nano-sized ZnWO4/Ag by hydrothermal process. Journal of Functional Materials 40 (9), 1442–1444.

Wang, D., Xue, G., Zhen, Y., Fu, F., Li, D., 2012. Monodispersed Ag nanoparticles loaded on the surface of spherical Bi$_2$WO$_6$ nanoarchitectures with enhanced photocatalytic activities. J. Mater. Chem. 22 (11), 4751–4758.

Wang, C., Xu, S., Shao, Y., Wang, Z., Xu, Q., Cui, Y., 2014. Synthesis of Ag doped ZnInSe ternary quantum dots with tunable emission. J. Mater. Chem. C 2 (26), 5111–5115.

Wang, T., Lang, J., Zhao, Y., Su, Y., Zhao, Y., Wang, X., 2015. Simultaneous doping and heterojunction of silver on Na$_2$Ta$_2$O$_6$ nanoparticles for visible light driven photocatalysis: the relationship between tunable optical absorption, defect chemistry and photocatalytic activity. CrystEngComm 17 (35), 6651–6660.

Wang, Z., Ma, H., Zhang, C., Feng, J., Pu, S., Ren, Y., Wang, Y., 2018. Enhanced catalytic ozonation treatment of dibutyl phthalate enabled by porous magnetic Ag-doped ferrospinel MnFe$_2$O$_4$ materials: performance and mechanism. Chem. Eng. J. (Lausanne) 354, 42–52.

Withanage, W., Yanagida, S., Takei, T., Kumada, N., 2018. Hydrothermal doping of Ag into three types of potassium niobates. J. Ceram. Soc. Jpn. 126 (10), 784–788.

Wu, S., Nianyuan, T., Donghui, L., Bing, Y., 2018. Photoinduced synthesis of hierarchical flower-like Ag/Bi$_2$WO$_6$ microspheres as an efficient visible light photocatalyst. Int. J. Photoenergy 18, 38–64.

Xiaoli, J., Shijiang, W., Jie, S., Xiujian, Z., 2015. Sol-gel process synthesis and visible-light photocatalytic degradation performance of Ag doped K$_2$Ti$_4$O$_9$. Integr. Ferroelectr. 161 (1), 62–69.

Xu, D., Liu, K., Shi, W., Chen, M., Luo, B., Xiao, L., Gu, W., 2015. Ag-decorated K2Ta2O6 nanocomposite photocatalysts with enhanced visible-light-driven degradation activities of tetracycline (TC). Ceram. Int. 41 (3), 4444–4451.

Xu, X., Du, M., Chen, T., Xiong, S., Wu, T., Zhao, D., Fan, Z., 2016. New insights into Ag-doped BiVO$_4$ microspheres as visible light photocatalysts. RSC Adv. 6 (101), 98788–98796.

Yahya, N., Aziz, F., Jamaludin, N., Mutalib, M., Ismail, A., Salleh, W., Jaafar, J., Yusof, N., Ludin, N., 2018. A review of integrated photocatalyst adsorbents for wastewater treatment. J. Environ. Chem. Eng. 6 (6), 7411–7425.

Yang, X., Zhang, Y., Xu, G., Wei, X., Ren, Z., Shen, G., Han, G., 2013. Phase and morphology evolution of bismuth ferrites via hydrothermal reaction route. Mater. Res. Bull. 48 (4), 1694–1699.

Yang, X., Wang, Y., Xu, X., Qu, Y., Ding, X., Chen, H., 2017. Surface plasmon resonance-induced visible-light photocatalytic performance of silver/silver molybdate composites. Chinese J. Catal. 38 (2), 260–269.

Yu, C., Jimmy, C.Y., 2009. Sonochemical fabrication, characterization and photocatalytic properties of Ag/ZnWO4 nanorod catalyst. Mater. Sci. Eng. B 164 (1), 16–22.

Yuan, B., Wang, C., Qi, Y., Song, X., Mu, K., Guo, P., Meng, L., Xi, H., 2013. Decorating hierarchical Bi$_2$MoO$_6$ microspheres with uniformly dispersed ultrafine Ag nanoparticles by an in situ reduction process for enhanced visible light-induced photocatalysis. Colloids Surf. A Physicochem. Eng. Asp. 425, 99–107.

Zahid, M., Nadeem, N., Hanif, M.A., Bhatti, I.A., Bhatti, H.N., Mustafa, G., 2019. Metal Ferrites and Their Graphene-Based Nanocomposites: Synthesis, Characterization, and Applications in Wastewater Treatment Magnetic Nanostructures. Springer, pp. 181–212.

Zahid, M., Nadeem, N., Tahir, N., Majeed, M.I., Naqvi, S.A.R., Hussain, T., 2020. Hybrid Nanomaterials for Water Purification Multifunctional Hybrid Nanomaterials for Sustainable Agri-Food and Ecosystems. Elsevier, pp. 155–188.

Zhang, Y., Ding, H., Liu, F., Wei, S., Liu, S., He, D., Xiao, F., 2008. Synthesis of nanosized Ag_3VO_4 particles and their photocatalytic activity for degradation of rhodamine B under visible light. Chinese J. Catal. 29 (8), 783–787.

Zhang, X., Zhang, Y., Quan, X., Chen, S., 2009. Preparation of Ag doped $BiVO_4$ film and its enhanced photoelectrocatalytic (PEC) ability of phenol degradation under visible light. J. Hazard. Mater. 167 (1–3), 911–914.

Zhang, Y., Lin, Y., Chen, J.L., Zhang, J., Zhu, Z., Liu, Q., 2014. A high sensitivity gas sensor for formaldehyde based on silver doped lanthanum ferrite. Sensors Actuators B Chem. 190, 171–176.

Zhao, X., Zhu, Y., 2006. Synergetic degradation of rhodamine B at a porous $ZnWO_4$ film electrode by combined electro-oxidation and photocatalysis. Environ. Sci. Technol. 40 (10), 3367–3372.

Zhao, J., Cao, J., Zhao, Y., Zhang, T., Zheng, D., Li, C., 2020. Catalytic ozonation treatment of papermaking wastewater by Ag-doped $NiFe_2O_4$: performance and mechanism. J. Environ. Sci. 97, 75–84.

Zhou, B., Zhao, X., Liu, H., Qu, J., Huang, C., 2011. Synthesis of visible-light sensitive M–$BiVO_4$ (M = Ag, Co, and Ni) for the photocatalytic degradation of organic pollutants. Sep. Purif. Technol. 77 (3), 275–282.

PART 6

Silver nanoparticles in veterinary science

CHAPTER 27

Potential of silver nanoparticles for veterinary applications in livestock performance and health

Moyosore Joseph Adegbeye[a], Mona M.M.Y. Elghandour[b], P. Ravi Kanth Reddy[c], Othman Alqaisi[d], Sandra Oloketuyi[e], Abdelfattah Z.M. Salem[b], and Emmanuel K. Asaniyan[f]

[a]Department of Agriculture, College of Agriculture and Natural Sciences, Joseph Ayo Babalola University, Ilesha, Osun State, Nigeria
[b]Facultad de Medicina Veterinaria y Zootecnia, Universidad Autónoma Del Estado de México, Mexico
[c]Department of Livestock Farm Complex, College of Veterinary Science, Sri Venkateswara Veterinary University, Proddatur, Andhra Pradesh, India
[d]Animal and Veterinary Sciences Department, College of Agricultural & Marine Sciences, Sultan Qaboos University, Al-Khoud, Oman
[e]Laboratory for Environmental and Life Sciences, University of Nova Gorcia, Nova Gorcia, Slovenia
[f]Department of Animal Production and Health, Olusegun Agagu University of Science and Technology, Okitipupa, Ondo State, Nigeria

27.1 Introduction

The use of nanoparticles in human lives and activities is increasing each day. The concept of nanoparticles has been applied in food packaging, equipment manufacturing, and health care and could extend their applications in diverse fields as more discoveries unfold. Agriculture is currently facing challenges and backlashes on its present way of practice and the effect of its previous activities. Many practices have been banned or limited because of their relative effects on human health due to resistance, environmental pollution, and residues in food production. Because of these complications, farmers, academics, and researchers are dealing with implications to combat various health-related issues, thereby reducing morbidity and mortality of livestock. Several options have been explored, especially on the use of materials of phytogenic origins for animal health. Although anecdotal, several successful cases were reported as evidenced by the ethnoveterinary practices among rural and nomadic farmers. However, the modern-day intensive system of livestock farming faces increased pressure of meeting the demand for animal products. To meet this demand, the antimicrobial feed additives or drugs are essential for either curative or preventive purposes. Epidemic outbreaks of zoonotic or animal-related diseases usually have devastating effects on livestock production activities in nations, with many farmers being unable to recover economically from the shock. For instance, the outbreak of Avian influenza in Nigeria and African swine fever in Asian countries in 2006 and 2019 caused the loss of millions of birds and pigs, which has economic and protein security implications. Other challenges in animal agriculture include the prevention of environmental pollution, disease outbreak, re-emergence of infectious diseases, antimicrobial resistance of strains to established drugs, and the emergence of resistance against newly developed antibiotics. The problems demand the use of potential alternatives in veterinary medicine and animal health.

Nanotechnology holds some potential for use in the activities related to animal production, health, and veterinary medicine (Meena et al., 2018). This technique uses biogenic, organic, and inorganic minute-sized (usually between 1 and 100 nm) materials for various applications (Adegbeye et al., 2019). The term nanoparticles refers to their small sizes. These nanoparticles have a high surface area, charges, catalytic activity, and adsorption activity (Khurana et al., 2019). Furthermore, they could provide alternatives for developing new drugs, delivery of vaccines, adjuvants to improve immune responses, antigen stability, and immunogenicity (Sekhon, 2014; Zhao et al., 2014; Hill and Li, 2017; El-Sabry et al., 2018) and has high potential for use in veterinary medicine. As the field of nanotechnology continues to gather attention, its use in animal agriculture will be more expansive (Hill and Li, 2017) and could contribute to the development of nanovaccines, nanoantibiotics, and nano-antibiotics-hybrids with various diagnostic and therapeutic applications.

Several nanoparticles such as silver, gold, calcium, iron, selenium, silicon, titanium, and zinc-based particles have been used in various agricultural and environmental applications. Of these nanoparticles, silver nanoparticles have distinguished themselves as strong antimicrobial agents by causing death and inhibiting pathogenic organisms (bacterial, fungal, and viral origin). The objective of this chapter is to explore various ways by which silver nanoparticles could have veterinary applications in livestock farming. Also, this chapter explored some negative impacts of silver nanoparticles on livestock performance and health.

27.2 Brief on silver nanoparticle synthesis

The increased applications of silver nanoparticles in health, cosmetics, electrochemistry, material science, food, and agriculture result from their distinctive physicochemical properties and varying methods of synthesis. These methods include chemical, physical, and biological (microbial and plant-based); however, due to concerns about the hazardous byproducts, expensive, low yield, and labor-intensive from chemical (requires additional reducing and stabilizing/capping agent) and physical method, green synthesis of silver nanoparticles methods are being used (Ledwith et al., 2007). The green (biological) method of synthesis is a sustainable approach that involves controllable design and processes of cost-effective and less/no toxic nanoparticle substances (Ahmad et al., 2019; Tripathi et al., 2019). For a sustainable livestock agricultural management, the use of green synthesis approach is an interesting area because of its economic, biocompatible, and eco-friendly benefits over chemical and physical methods. These methods include pest and disease control, disinfection of livestock, home, and utensils, and as feasible alternative to antibiotics (Huang et al., 2014).

Silver nanoparticles have been synthesized from plant extracts such as *Cleome viscose*, Alfalfa sprouts, *Elaeagnus latifolia*, Geranium, *Ganoderma neojaponicum*, *Glycyrrhiza glabra*, *Feronia elephantum*, *Amphipterygium adstringens*, *Aloe vera*, neem, and bamboo (Gardea-Torresdey et al., 2003; Yasin et al., 2013; Rodríguez-Luis et al., 2016; Lakshmanan et al., 2018; Eze et al., 2019). Bacteria and fungi are major sources for microbial-based synthesis of silver nanoparticles. The bacteria (*Pseudomonas stutzeri*, *Bacillus* spp., *Escherichia coli*, *Xanthomonas* spp., *Staphylococcus* spp., *Deinococcus radiodurans*, and lactic acid bacteria) and fungi (*Aspergillus* spp., *Arthroderma fulvum*, *Trametes ljubarskyi*, *Fusarium oxysporum*, and *Ganoderma enigmaticum*) are known for their great potential in the biosynthesis of silver nanoparticles with controllable uniformity and stability (Bhainsa and D'Souza, 2006; Gudikandula et al., 2017; Javaid et al., 2017). Subsequent studies demonstrated other cell-mediated silver nanoparticles synthesis with macro- and microalgae such as *Caulerpa racemosa*, *Sargassum muticum*, *Chlorella vulgaris*, *Spirulina platensis*, *Chaetoceros calcitrans*, *Padina pavonia*, *Isochrysis galbana*, and *Tetraselmis gracilis* because of their high silver metal uptake potential (Azizi et al., 2013; Kathiraven et al., 2015; Annamalai and Nallamuthu, 2016; Abdel-Raouf et al., 2019; Khanna et al., 2019). Biosynthesis of silver nanoparticles involves either intracellular or extracellular reduction of Ag^+ to Ag^0 (Fig. 27.1) facilitated by active compounds such as alkaloids, terpenes, fatty acids, and amino acids in biological extracts for stabilization (Khanna et al., 2019). These processes including dimension and morphology of the nanoparticles are influenced by factors such as temperature, pH, extract concentration, exposure/reaction time, interactions, and biochemical activities (Pathak et al., 2019). The potential delivery methods of silver nanoparticles have been extensively discussed by Hill and Li (2017) and Fahimirad et al. (2019).

27.3 Potential routes of administration

Silver nanoparticles are efficient materials for therapeutics and drug delivery in veterinary practices. The minute-sized particles provide potentiality in bypassing many-body barriers such as placenta and blood–brain barriers. Silver nanoparticles could be applied through several means such as topical, oral, intranasal, intravenous,

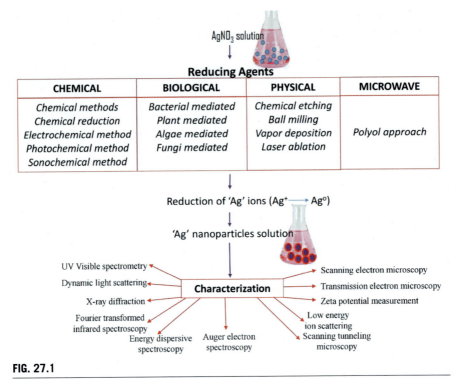

FIG. 27.1

Synthesis and characterization techniques of silver nanoparticles.

muscular, and transdermal nanodelivery systems (Sekhon, 2014). Other means include *in ovo*, intravenous (Lee et al., 2018; Mathur et al., 2018), intragastric (Melnik et al., 2013), intraperitoneal (Doudi and Sertoki, 2014), and subcutaneous (Tang et al., 2007; Mathur et al., 2018) injections. Administering silver nanoparticles through inhalation for 6 h/day for 90 days revealed no genotoxicity with a reduced lung burden of 24 h postexposure (Kim et al., 2011).

27.4 Potential for nanoveterinary application of silver nanoparticles

Nanomedicine refers to the procedure of applying nanoparticles in diagnosis, prognosis, and treatment, while nanopharmacy and nanotherapy relate to their utilization in drug-making-related applications and animal rehabilitation, respectively. It involves the "smart" delivery of a drug to the target tissue with a profile that drugs are delivered as needed (Scott, 2005). These drugs could be packed with biodegradable nanoparticles so that they would be delivered to the intestine for absorption (Simon et al., 2016) and reach other target sites within the body. Further, the use of silver nanoparticles could be extended to treat the ailments caused by various diseases, parasites, open injuries, and other microbial infections. This section is meant to discuss the various applications of silver nanoparticles in livestock health and well-being.

27.5 Endoparasites (helminths)

Helminthiasis is a major challenge in extensive livestock management systems involving grazing animals for a significant period. The mortality and morbidity rates will be higher with a parasitic disease burden, which ultimately decreases the flock's production performance. To overcome the helminths, conventional anthelminthic like albendazole and some herbs such as neem and pineapple leaves are generally used. However, the chemical antihelminth is expensive and could be unaffordable to many farmers in developing countries. Often, the standard dosage of active ingredients on the label differs from the actual dosage of the active ingredient. In regions where standards are upheld, these helminthic parasites have evolved resistance to various anthelmintic, such as benzimidazole, imidazothiazole, and ivermectin (Waller, 2003), which is a major concern worldwide. Also, despite the presence of many herbs with potent anthelminthic activity, the nonuniversality of these herbs and possibility of low financial gain from the economy led to low or limited adoption by global farming companies. Among these helminths, *Haemonchus contortus* is a gastrointestinal nematode that affects small ruminants in tropical, subtropical, and temperate regions. Tomar and Preet (2017) and Avinash et al. (2017a) have shown that neem-mediated nanoparticle is more effective than the individual neem or albendazole drug against helminth. Furthermore, the IC_{50} and IC_{90} values of neem mediated AgNp were 99.5% and 97.2% lower than albendazole (Avinash et al., 2017a). This implies that neem-mediated AgNp is more lethal than albendazole. Another study showed that 1–25 μg/mL of neem-mediated AgNp resulted in 15%–85% motility of adult *H. contortus*, whereas 200–1200 μg/mL of neem was required to do same, and the IC_{50} for the adult mortality was 7.89 μg/mL (Tomar and Preet, 2017). In another study by Preet and Tomar (2017), the LC_{50} of *Ziziphus jujuba* leaf extract biofabricated AgNp was 98.37% lower than the individual leaf extract in raw form. The authors reported that the nanoparticle worked by altering the egg morphology and depleting the nutrients (glycogen, lipid, and protein) of adult worms in a range of 5.69%–21.81%. The low concentration required by herbal conjugated AgNp suggests the importance of silver nanoparticles in potentiating the lethal effects of herbal extracts against pathogens and the development of "nanoherbal medicines."

Fascioliasis, an infectious parasitic disease caused by liver fluke, tops all zoonotic helminths globally. The infestation affects various ruminants and pseudoruminants, including sheep, goats, cattle, buffalo, horses, donkeys, camels, and rabbits. Its prevalence could be up to 90% in some zones causing huge animal losses (Farag, 1998). Triclabendazole is a safe and effective drug of choice against the fluke. However, there are reports from Australia and Ireland about the inconsistencies in the obtained results, presumably due to antiparasitic drug resistance (Alvarez-Sanchez et al., 2006). Despite the resistance, developing a new drug or breaking the resistance mechanism of the fluke is essential. In this regard, the efficiency of the established drug could be improved by employing nanotechnology. Due to the minute-size conjugation of silver nanoparticle with triclabendazole, this could allow the particle to serve as a carrier of the drug into the cell membrane, thereby increasing the efficiency. An in vitro report (Gherbawy et al., 2013) showed that the conjugation of triclabendazole drug with *Trichoderma harzianum* biosynthesized silver nanoparticle inhibited the egg hatchability by 89.67% compared to the 69.67% of those hatched under triclabendazole drug control group. The increased inhibitory activity was due to a pit-like perforation on the egg surface, which was not observed in the drug alone and untreated group. This study shows that the silver nanoparticle could be combined with a drug to aid its quick delivery and enhance efficiency at the target site.

Cystic hydatid disease is a helminth infection and a major neglected cyclozoonotic disease caused by *Echinococcus granulosus* in many countries globally (Adibpour et al., 1998). It is an infection that causes economic and animal protein losses such as decreased meat, milk, fiber, and mortality of offspring. An invasive method is practiced to treat this disease, but there are setbacks such as anaphylactic shock, mortality, and even potential for reoccurrence (Rouhani et al., 2013). Other noninvasive methods such as hypertonic saline, mannitol, chlorhexidine gluconate, *Allium sativum*, *Sambucus ebulus*, and fungal chitosan have been used, but their usage is not recommended because of low efficacy, high toxicity, and undesirable effects (Rahimi et al., 2015). Discovering a noninvasive and nontoxic treatment for hydatid cysts is essential. An in vitro exposure of protoscolices eggs to 0.1–0.15 mg/mL Ag-Nps for 60 and 120 min caused a 79%–80% and 83%–90% mortality, respectively (Rahimi et al., 2015). The liver contains a high residue of nanoparticles after administration of nanotechnology-based medication. Since liver is the organ that the disease attack, silver nanoparticles has a great possibility of treating the liver-related ailment. This suggests that silver nanoparticle could be used against the disease without resorting to an invasive method. Indeed, the Ag-NPs decreased protoscolices by 40% in 10 min, even at a lower concentration. Because of the Ag-NPs' strong scolicidal activity, they could be projected as an ideal and economical scolicidal agents against the disease without resorting to an invasive method.

27.6 Ectoparasites (ticks)

Grazing animals, both under semi-intensive and nomadic systems, are infested by various ectoparasites, which affect their productive efficiency by competing for nutrients with the host (Adegbeye et al., 2018). Ticks serve as vectors to various infections such as anaplasmosis, babesiosis, borreliosis, and ehrlichiosis. Because of the resistance phenomenon, the effectiveness of acaricidal products against several tick species of tropical and subtropical countries is declining. The residues in meat and milk are another major concern and discovering new acaricidal product is cost-intensive (National Research Centre, 1986; Perez-Cogollo et al., 2010; Benelli et al., 2017). Hence, there is a need to find acaricidal and repellent products to mitigate tick resistance. Silver nanoparticles could help in reducing difficulties and expenses related to manufacturing new acaricidal against ticks (such as *Rhipicephalus* (Boophilus) *microplus*, *Haemaphysalis bispinosa*, and *Hyalomma anatolicum*). Further, silver nanoparticles could be projected as a novel strategy against acaricide resistance. Another study proved that the neem-coated silver nanoparticles are toxic to larvae and adult *Rhipicephalus microplus* ticks (Avinash et al., 2017b). Additionally, 10 and 25 µg/mL of silver nanoparticles had 100% mortality against the larvae of *Rhipicephalus* (Boophilus) *microplus* and *Haemaphysalis bispinosa* adults, respectively (Santhoshkumar et al., 2012; Rajakumar and Rahuman, 2012; Zahir and Rahuman, 2012), whereas the silver nanoparticles at 50 mg/L of silver nanoparticle killed 40% (*Rhipicephalus* (Boophilus) *microplus*) adults (Johari, 2016).

The resistance against deltamethrin, the most common chemical agent used against ticks, is widespread and making vector control programs vulnerable. Hybridizing this drug with silver nanoparticles could provide a synergistic effect on tick. In this view, Avinash et al. (2017a, b) reported that 50 ppm of deltamethrin neem-coated silver nanoparticles killed *Rhiphicelus microplus* larvae, while 360 ppm of deltamethrin was required for 100% ticks' mortality. Furthermore, the deltamethrin neem-coated silver nanoparticles killed 93.33% of the adults and had 99.16% oviposition inhibitory activity. In addition, the LC_{99} and IC_{99} against both larvae and adults of *R. microplus* and for oviposition inhibitory activity were lower than deltamethrin alone. The tick activity could be controlled using conjugated AgNP-coated deltamethrin as topical agents by immersion in dip or pour-on sprays.

27.7 Potential application in poultry and hatcheries

The major constraints in livestock production are due to the use of antibiotic feed additives, mortality, morbidity, environmental challenges, and vaccine failures. Because of the lesser contribution of poultry to greenhouse gases, environmentalists endorse chicken as nonvegetarian protein source compared to beef, carabeef, and pork. Since the recent past, global warming potential of diet is an alarming concept, and hence, the poultry market has a great potential for expansion soon. Vaccination is done in poultry hatcheries and farms to prevent the devastating effects of pathogenic diseases caused by both bacteria and viruses. Failure of immunization programs due to improper vaccinations may cause huge losses to poultry farmers. Commercial poultry birds are periodically vaccinated *in ovo*, orally, or through the wing web against diseases such as infectious bursal disease, fowl pox, Newcastle, Marek's disease, avian influenzas, and infectious bronchitis.

Most recently, researchers are developing nanoparticles to challenge viruses by delivering enzymes that prevent the replication of the virus in the blood system of humans or livestock (Meena et al., 2018). Infectious bursa disease or gumboro is caused by a virus and can spread by contact, feces, or contaminated feed. Silver nanoparticles at a concentration of 20 ppm were found to act as both preventive and therapeutic agents by decreasing the growth of IBD virus in embryonated eggs (Pangestika et al., 2017). The preventive technique was developed by mixing silver nanoparticle and IBD virus 2h before inoculation, while therapeutic techniques were developed by inoculating virus first and then injecting silver nanoparticles 48h postinfection. In the preventive methods, the silver nanoparticle prevented the penetration of the virus into the host cell, while the therapeutic methods inhibited the interaction between liver nanoparticle and the DNA, consequently hindering the replication of the virus (Galdiero et al., 2011). Silver nanoparticles could be employed with these methods to build the immunity in chicks against IBD virus before hatching. Thus, the preventive and therapeutic application of silver nanoparticles on IBDV may be a novel strategy to prevent virus replication at an early stage. In addition, Kordestani et al. (2015) reported SilvoSept® at 4ppm had anti-H1N1 influenza A virus activity reducing optical absorption by 99%. Hence, silver nanoparticles could be applied to improve the biosecurity of poultry ventures (farms and hatchery) to prevent the spread of diseases. In addition, silver nanoparticles could be used to develop vaccines as adjuvant or in other capacities for these deadly poultry diseases.

Embryonic development is important in the poultry industry as the finishing age of the commercial broiler is reducing (Goel et al., 2017) due to improved nutrition, genetic, and consumers demand tender meat. *In ovo* injection with silver nanoparticles could improve the bird's growth and immunocompetence of the late-term embryo or post-hatch chicks. Toll-like receptors (TLR-2) and TLR-4 play a key role in innate adaptive immune systems and recognize the invading pathogens by a series of pathogen-associated molecular patterns (Beutler, 2004; He et al., 2006). Silver nanoparticles enhance the vulnerability of macrophages to inflammatory stimulation by activating the specific ligands on the toll-like receptor (Castillo et al., 2008). Injecting the AgNPs at 12.5, 25, or 50 µg into egg increased the bursal weight, spleen weight, and hatchability, along with immune parameters such as foot web index and expression of TLR2 and TLR 4 genes (Goel et al., 2017). Bursa and spleen play important roles in imparting immunity and elicit cellular and humoral immune response in chicks. The increased response indicates a better immunological health status *in ovo*-injected birds. This improvement in immunity could be through enhanced early maturation of the immune system and higher phagocytic activity producingmore antibodies against invading pathogens (Goel et al., 2017). Therefore, if the *in ovo* silver nanoparticle application can boost the embryo's immune system, it can enhance the body's first line of defense such that there is a reduction in the use of antibiotics for preventive or therapeutic purposes.

Because of the faster growth rate, leg paralysis is one of the main challenges in broiler production. Adding nano-Ag to the chicken embryo at 0.25 µg Ag/g egg improved the calcium, iron, and copper content in the embryo skeleton by 3%, 12%, and 9%, respectively (Sikorska et al., 2010). Furthermore, about 8% of silver nano was settled in the thigh bone. This shows that silver nanoparticles improve mineralization in chicken. Moreover, the ability of the silver nanoparticle in penetrating and settling in the embryo reveals that the silver nanoparticle could be a potential carrier of drugs and minerals. Further, the increased copper, iron, and calcium after AgNano use suggest that the silver nanoparticle can mitigate rickets and brittleness of chickens' bone by stimulating the hydroxylapatite formation. A silver nanoparticle can also improve the hatchability of eggs. Another study reported that injecting silver nano at 10, 20, or 30 μL/mL into allontoic cavity of eggs increased hatchability by 69%, 75%, and 81%, respectively (Kathiresan et al., 2019).

In hatcheries, formaldehyde is used to fumigate the hatchable eggs before incubation to eliminate pathogenic microbes. However, Chmielowiec-Korzeniowska et al. (2015) revealed the toxic and carcinogenic nature of formaldehyde. The same authors attempted to find an alternative to formaldehyde for fumigation and revealed that silver nanoparticles in spray form decreased the pathogenic load and lowered the residues in visceral organs such as the liver, GIT, and eggshell. Because of the established antimicrobial activity, silver nanoparticles could be useful to disinfect hatchers and eggshell instead of formaldehyde. Silver nanoparticles could protect the chicken embryo from pathogenic infection as well as support the growth of a healthy embryo.

27.8 Immunity

The immune system comprises innate (first line of defense) and adaptive responses. The former is present and mobilized rapidly during infections (Marquardt and Li, 2018). The nuclear factor κB (NF-κB) is a transcriptional factor that plays a key role in the defense of the organism. The defense mechanisms inlcude proinflammatory pathways and could be activated or stimulated by pathogenic bacteria or their products (LPS and endotoxins), viruses, and reactive oxygen species (Sawosz et al., 2010). The phosphorylation of IκB activates NF-κB and releases of P50 and P65 subunit, which binds to genes involved in immune defense activities (D'Acquisto et al., 2002). Preinjection of 0.3 mL colloidal Ag nano *in ovo* at 50 ppm concentration in eggs challenged with *Escherichia coli* strain 0111:B4 LPS (0.4 mg/egg) showed improvement in chicks body weight by 6.5% besides limiting the expression of proinflammatory NF-κB mRNA (Sawosz et al., 2010). The silver nanoparticles, in colloidal form, could be used to improve the immune system of chicks so that the chicks could manage the intestinal microbial imbalance and immune disorders. In stressed animals, the oxidation stress-led ROS can trigger the translocation of Nrf2, leading to the production of various antioxidant genes (e.g., *sod1*, *sod2*, *cat*, *gclc*, *gstD*, and *gstE*) (Nguyen et al., 2009). A study by Mao et al. (2018) found that the use of AgNPs at lethal and sublethal doses (50 μg/mL) caused damages to the DNA of the brain and salivary gland and gut apart from the activation of Nrf2-dependent antioxidant pathway, consequently triggering autophagy. Usually, autophagy is induced by activating Nrf2/antioxidant response element-dependent antioxidant system during cellular stress or homeostasis or removing misfolded protein and damaged organelles) (Kraft et al., 2010). As such, AgNPs may trigger the body Nrf2-dependent antioxidant pathway in a time-dependent manner to help an animal when they are stressed cellularly. Furthermore, caution is advised to limit the high doses, leading to chronic cumulative effects on the host.

Injecting Ag-NPs (5 and 10 mg/L) at 2.87 and 12.25 mg/bird stimulated the production of B lymphocytes, ultimately producing IgA and Ig A immunoglobulins at 95% and 37%, respectively. The study showed exertion of proinflammatory effects by elevating IL-6 by 125% and increased ESR by 97% (Kulak et al., 2017). This suggests that AgNPs exert an immunotropic effect on livestock if applied appropriately at right doses and appropriate administration methods. The importance of neutrophils in regulating immune networks and innate defense stresses the impact of silver nanoparticles on immunoregulation (Fraser et al., 2018). Besides, the phagocytic activity of circulating neutrophils is an indispensable defense mechanism of the immune system. They also play an important role in releasing cytokines and chemokines, which contribute to modulation of the immune network and responses (Scapini et al., 2000; Pelletier et al., 2010). Exposing to 20 µg AgNPs/10^5 cells for 20 h triggered the activation and maturation of circulating neutrophils, thereby increasing the key cytokine release including, IL-8, IL-16, and IL-27. However, the increased cytokine levels did not cause proinflammatory or damaging effects (Fraser et al., 2018). The above-mentioned effects of silver nanoparticles on immune system suggest their usage in the veterinary sector to strengthen the immune system and avoid invading pathogens. However, appropriate dosage and period of contact have to be established to prevent inflammatory pathogenesis because of inappropriately recruited or activation of neutrophils (Jorch and Kubes, 2017) (Fig. 27.2).

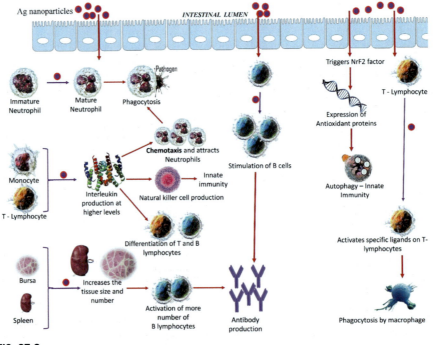

FIG. 27.2

Immunomodulatory effects of silver nanoparticles.

27.9 Wound and burn healing

In nations with a transition economy, the usage of animals for carting or ploughing purpose is common, causing open injuries, sunburns, and other stress-related problems. Although farmers use herbal extracts or gentian violet to improve the healing process, those topical medicines could not heal the wounds at a faster rate. Several swine breeds such as large white, Yorkshire, and Poland China are susceptible to sunburn, causing an open wound, which may take a long time to heal. Kitsyuk and Zvyagintseva (2018) found a quick restoration of the normal structure of the epidermis, density index fibroblasts, and reduction in the thickness of the epidermis in guinea pigs exposed to ultraviolet irradiation and Ag Np-tiotriazolin ointment. In contrast, untreated groups had a thick epidermal layer, dystrophic changes in epidermocytes and dyskeratosis, increased thickness of fibroblasts and dermis collagenization, changes in the content and structure of elastic fibers, and uneven derma fibrosis along with sclerotic changes. In the future, climate change could increase the pronocity for sunburn, as such, silver nanoparticle could be applied for quick healing of the wounds. Salih et al. (2016) showed that olive leaves extract-based silver nanoparticles promoted the healing of burned wounds. The silver nanoparticles improved in the formation of a thin epithelial layer in 14 days and reduced the burn diameter by 94.23%. Similarly, silver nanoparticle was shown to prevent and treat burn infection wounds faster than established drug-silver sulfadiazine, which is used globally. Ag Np-aloevera combination (containing 7 cc nanosilver, 0.2 g *Aloe vera*, and other materials) increased rate of epidermis re-epithelialization and decreased the total body surface area affected by burns faster than silver sulfadiazine (Mousavi et al., 2019). The superiority of silver nanoparticles in healing process compared implies that the herbal–nanoparticle formulation could be promoted as an alternative to the conventional treatments in livestock. The silver nanoparticles could also be used to treat certain surgical infectious diseases such as caseous lymphadenitis, a chronic and potentially zoonotic disease caused by *Corynebacterium pseudotuberculosis* bacterium in ruminants small ruminants (Santos et al., 2019).

Surgery is often carried out on animals depending on the disease and discomfort. Draining and cauterizing the lesions with 10% iodine solution could consume more time to heal. Hence, other alternatives to hasten the healing process are necessary. In this regard, Santos et al. (2019) used an ointment formulation based on biogenic AgNPs mixed with natural waxes and oils. The healing rates of surgical wounds of goat and sheep treated with silver nanoparticles were 5 and 8 days faster than those treated with iodine. Further, the wounds treated with silver nanoparticles had less purulent discharge and lower leukocyte counts and anti-C pseudotuberculosis antibody titers. Therefore, it could be concluded that postsurgical treatment of wounds with AgNP-based ointment may be a promising tool to enhance the healing rate of surgical wounds. Similarly, Kordestani et al. (2015) showed that rinsing the wound with SilvoSept® was effective against a wide spectrum of microorganisms and could be used to rinse wounds as an alternative to iodine-based ointments. The wound healing activity of AgNPs could be related to the antimicrobial properties and autoinflammatory effects, which are implicated in wound-healing responses. Furthermore, silver could modulate the cytokines involved in tissue repair (Tian et al., 2007; Vasile et al., 2020). These nanoparticles could be applied as ointment, spraying, cream, and powder.

27.10 Antimicrobial activity and synergy of silver nanoparticles

Antimicrobial resistance to medical and veterinary drugs worldwide is a cause for concern and one of the greatest threats to human health (Marquardt and Li, 2018). Also, resistance poses a threat to animal-derived protein security. Affordability, excessive use, and adulteration of antibiotics has led to increased resistance of pathogenic gram-positive and gram-negative bacteria to drugs. In addition, accumulation of antibiotic residues in animal products has resulted in the resistance of transmissible food pathogens to antibiotics in humans (World Health Organisation (WHO), 2017). Moreover, the rate at which these microbes generate resistance outcompetes new antibiotics (Vazquez-Muñoz et al., 2019) and biofilm is one of the modes whereby microorganisms build resistance to antibiotics. Therefore, there is a need for developing new antibiotics and finding a way to break the resistance of the microbes. Kathiresan et al. (2019) reported that silver nanoparticles (10–30 µL/mL) were able to inhibit the biofilm formation of pathogenic microbes.

Nonjudicious utilization of antimicrobial agents causes the spread of resistance, consequently reducing their efficacy. According to the collected literature, several pathogens such as *Escherichia coli*, *Salmonella* spp., *Pasteurella multocida*, and *Actinobacillus* spp. have shown antimicrobial resistance at different levels. As the threat of antimicrobial resistance continues to grow globally, the impact of drug resistance could be mitigated by employing combination therapy of different antibiotics or antibiotics with nonantibiotic agents such as silver nanoparticles. Smekalova et al. (2015) revealed that silver nanoparticles are able to act in synergy with amoxycillin, Penicillin G, gentamicin, and colistin against some resistant microbes like *Staphylococcus aureus*, *Actinobacillus pleuropneumoniae*, *Streptococcus uberis*, and *Pasteurella multocida*. Besides, Tetracycline-conjugated *Oscillatoria limnetica* synthesized-silver nanoparticles and cefaxone-conjugated *Oscillatoria limnetica*-silver nanoparticles showed higher inhibition zone diameter (26 and 24 mm, respectively) than 19 and 18 mm for cefaxone and tetracycline against *E. coli* and *B. aureus* (Hamouda et al., 2019).

In Nigeria, antibiotics such as tetracycline, penicillin, and gentamycin are used as either feed additives or injectables in livestock. However, few authors have reported resistance of pathogenic microbes to antibiotics used in Nigerian livestock sector (Oluwasile et al., 2014; Nsofor et al., 2013). Despite the antimicrobial properties of silver nanoparticle, few reports of resistance were observed in some gram-positive and gram-negative bacteria (Panacek et al., 2018; Jose et al., 2019; Mohammed and Aziz, 2019). As such, formulating antibiotics to overcome the resistance challenge is essential. The gram-negative resistance was due to the production of flagellin, a flagellum protein, while the same from gram-positive was due to the role of efflux pump on the cell wall. In a recent study, Khatoon et al. (2019) tested the efficacy of a nanoformulation involving ampicillin antibiotic (AMP), silver nanoparticles (Ag-NPs), and a combination of silver nanoparticles and ampicillin antibiotic (AMP-Ag-NPs). They revealed that the MIC_{90} of AMP-silver nanoparticles against the bacterial strains was 3–28 µg/mL lower than the AMP (12–720 µg/mL) and synthesized silver nanoparticles (280–640 µg/mL). Further, the authors revealed no evidence of resistance mechanism on testing the AMP-silver nanoparticles against bacterial strains in 15 repeated cycles. Hence, hybridizing silver nanoparticle antibiotics could be done in livestock industries to mitigate the resistance of pathogenic microbes, thus reducing the morbidity and mortality of livestock suffering from antibiotic-resistant infections.

Besides, the localized surface plasmon resonance properties of silver nanoparticles make them attractive against antimicrobial-resistant bacterial strains with excellent antimicrobial activities at lower concentrations (Jose et al., 2019). The peptidoglycans on the cell wall of gram-positive microbes may play a pivotal role in preventing the cytoplasmic entry of nanoparticles protecting them from cell death. The coevolution of microbes against antibiotics is of greatest concern over the excessive use of silver nanoparticles. As such, the overuse of silver nanoparticles could also lead to an evolution in bacteria to make themselves resistant to it. More recently, Jose et al. (2019) found high toxicity levels of silver-silica nanoparticles to *Bacillus subtilis* and *Escherichia coli* at low concentration (20 µg/mL); however, *S. aureus*, a gram-positive bacterium was resistant with only 20% mortality even at 100 µg/mL concentration. It was observed that gram-positive *Bacillus subtilis* and *S. aureus* maintained 60% and 80% of their respective cell walls within the exposure period. The authors found that the resistance was due to the role of efflux pump; hence, inhibiting the efflux pump with calcium channel blockers such as verapamil may facilitate the entry of silver–silica nanoparticles into the cell, ultimately causing DNA damage and cell death. Therefore, when treating gram-positive bacteria, an antimicrobial formulation of silver nanoparticle could be administered along with efflux pump inhibitors to breach the cell wall of gram-positive microbes. Furthermore, the sensible use of silver nanoparticles is essential to prevent the possibility of antimicrobial resistance. The combination of antimicrobial agents with silver nanoparticles could be a promising way to decrease the number of antibiotics used in livestock and enhance the efficiency of injectable antibiotics at a lower quantity.

27.11 Infectious diseases

27.11.1 Mastitis

In practice, many farmers and veterinarians administer antibiotics and drugs to overcome mastitis-related illness and death based on previous experience or recommendation rather than scientifically informed drug prescription. Bovine mastitis is an important economic disease that affects cost of milk production and cows performance, which greatly decreases milk production. The excessive use of antibiotics in cattle leads to antibiotic-resistance of mastitis-causing bacteria. Mastitis is primarily caused by *Staphylococcus aureus*, *Streptococcus agalactiae*, *Escherichia coli*, *Pseudomonas aeruginosa*, *Corynebacterium bovis*, and *Bacillus cereus* (Yuan et al., 2017). Mastitis is a critical threat to the dairy industry because of its devastating effects on animal health, milk production, and the cost of milk production. Indiscriminate antibiotic usage against mastitis is common in developing nations such as India; hence, the disease is heavily related to the antibiotic-resistance. As *Staphylococcus aureus* is a major pathogen of mastitis prevalence in dairy herds, the resistance of the bacteria against antimicrobial agents is well-documented. Silver nanoparticles had MIC values ranging from 1.25 to 10 μg/mL, which inhibited 50% and 90% of *S. aureus* by 7 min contact time at a concentration of 5 and 10 μg/mL of silver nanoparticle, respectively (Dehkordi et al., 2011). Furthermore, 11 nm-sized spherical silver nanoparticles had a MIC of 1 and 2 μg/mL against *Pseudomonas aeruginosa* and *Staphylococcus aureus*, respectively (Yuan et al., 2017).

27.11.2 Tuberculosis

Mycobacterium bovis, the causative agent of bovine tuberculosis, can be responsible for human tuberculosis, thus having zoonotic importance (Allix-Be'guec et al., 2009). Both *M. tuberculosis* and *M. bovis* cause serious tuberculosis infection in both human beings and animals and have high mortality than any infectious disease. Large cattle population reported to be infected with bovine tuberculosis worldwide (WHO, 2010; Selim et al., 2018). Often *M. bovis* infected cattle are sold even in local markets, which could be purchased unnoticeably. Consumption of beef with white sphere-like structures predisposes humans to tuberculosis infection. The antimycobacterial activity of silver nanoparticles against *M. bovis*, *M. tuberculosis* H37Rv, and multidrug-resistant *M. tuberculosis* inhibited them at MIC of 1, 4, and 16 μg/mL, respectively (Selim et al., 2018).

27.12 Contaminated/infected water

Contaminated drinking water is another concern for livestock disease burden. For instance, *Cryptosporidium parvum*, a major coccidian in contaminated drinking water, causes a significant losses in farms due to the higher mortality rate of preruminant calves, especially those below one month age (Thomson, 2015). Surprisingly, the oocysts have low infection doses and even resist chlorinated water treatment (Rose et al., 2002). Silver nanoparticles at 500 μg/mL resulted in 33% excystation far lower compared to 83% in control, while 5–500 μg/mL of silver nanoparticles caused 60%–93% decrease in sporozoite or shell ratio. The excystation process includes the rupture of oocyst releasing sporozoites that initiate infection in the host cell. Therefore, it could be assumed that silver nanoparticle prevents infection in host cells. The impact of silver nanoparticles was due to the ability of Ag ions in breaking the cell wall and entering into the oocyst wall, ultimately destroying the sporozoites (Cameron et al., 2016; Bravo-Guerra et al., 2020). In typical farms of developing nations, the freshwater sources and the disposed of wastewater from agriculture and human activities get mixed up through percolation to the groundwater, resulting in a high pathogenic bacterial population, which indirectly affects livestock production. A study conducted in India found that 15 nm-sized silver nanoparticles had a MIC for *E. coli* from farm water at 50 mg/L. Adding 15 nm-sized silver nanoparticles to the poultry diets improved feed intake and body weight, decreased mortality (4.92% vs 14.13% in the control group), and the meat was fit for consumption (Kumar and Bhattacharya, 2019). Therefore, AgNPs can be used as water disinfectant, surface disinfectant, and therapeutic material in livestock; and aquatic industry (Deshmukh et al., 2019; Prosposito et al., 2020).

27.13 Biosecurity/disinfection

The ability of silver nanoparticles against many bacteria shows that it could be used as an antimicrobial agent against many aerosol microbes such that it could be sprayed as foam and hung between buildings to reduce the exchange of infected air. These particles could also be used on-farm to minimize the spread of disease as an aid of biosecurity measures such as spraying and dipping. Combining antimicrobial agents with silver nanoparticles is a promising way to decrease antibiotic usage in the extensive livestock production systems. Furthermore, silver nanoparticles can play an essential role in agriculture and animal production by using sterilization tools and equipment in animal buildings.

The endospores of *Bacillus* and *Clostridium* species are means of transmitting spore-mediated diseases like anthrax, gas gangrene, botulism, tetanus, food poisoning, and pseudomembranous colitis, which are resistant to heat, chemical, and UV radiation treatment (Nicholson et al., 2000; Aminianfar et al., 2019). A study showed that 90% of the *Bacillus* and *Clostridium* endospores were inhibited with chemicals such as glutaraldehyde (20 mg/mL), sodium hypochlorite (0.25 mg/mL), and formaldehyde (5 mg/mL) in 25, 20.6, and 11.8 min, respectively (Gopinath et al., 2016). However, biogenic nanosilver (75 µg/mL) inhibited more than 90% of the *Streptomyces* sp. in 20 min. This inhibitory effect of nanosilver at lower concentration compared to chemical methods could be useful in disinfection of farms during hazardous spores' epidemic attack or outbreak. Furthermore, silver nanoparticles proved to be an effective disinfectant by sterilizing cages co-contaminated with opportunistic pathogens—*B. cereus* and *C. difficile* at 1×10^6 spores. The animals in cages void of AgNPs had infected lungs, inflammation, submucosa edema, ulceration, and hyperplasia of the GIT, deformation of hepatic parenchyma, and lympho-monocytic infiltration around portal vein (Gopinath et al., 2016), whereas no signs of pathological lesions were observed in the rats maintained in nanosilver sterilized cages. This suggests that sterilizing the cages and livestock houses regularly with nanosilver could improve their biosecurity and enhance animal safety against endospore infections. Therefore, nanosilver could be applied as a surface disinfectant against environmental spores as well as for several theragnostic applications. The nanosilver adhere to the spore's coat, leading to pitting formation by denaturing the proteins and glycosidic bonds of the peptidoglycan N-acetylglucosamine and N-acetylmuramic acid (Mirzajani et al., 2011; Gopinath et al., 2016; Ismail et al., 2019).

27.14 Mechanism of action

Silver nanoparticles exhibit an array of mechanisms of action involving antimicrobial activity against bacteria, fungi, and viruses, as antioxidants, nutraceuticals, drug delivery systems and immunological responses (Hill and Li, 2017). The silver nanoparticle has various means of action against microbes. The antimicrobial mechanisms include inactivation of enzymes, change of protein expression, damaged respiratory chain, production of reactive oxygen species, and increasing the membrane permeability resulting in cytoplasmic leakages and disruption of the cell membrane (Choi and Hu, 2008; Jin et al., 2010; Hartemann et al., 2015; Wu et al., 2016; Bondarenko et al., 2018). Besides, there are reports of a compromise of notable organelles in bacteria causing interruption of transmembrane electron transport (Potbhare et al., 2019; Eze et al., 2019). Due to the hydrophobic nature of silver nanoparticles, the nanoparticles interact and alter membrane permeability through cell penetration. They inactivate and inhibit the lactate dehydrogenase activity leading to increased leakage of proteins, sugars, and DNA, structural damage, severe disturbance to cell function, and cell death (Prabhu and Poulose, 2012; Yuan et al., 2017). This mechanism can also be attributed to silver cations, which specifically bind to thiol (–SH) groups of bacterial proteins by displacing the hydrogen atom to form –S–Ag, thereby suppressing the enzymatic function of affected protein leading to cell death (Kim et al., 2011). Most antibiotics are ineffective to inhibit multidrug-resistant (MDR) bacterial strains. However, silver nanoparticles can eliminate MDR bacteria such as *Staphylococcus aureus*, *Pseudomonas aeruginosa*, *Enterococcus faecalis*, and *Arcanobacterium pyogenes* responsible for mastitis, metritis, gastrointestinal, and respiratory infections in livestock (cattle, sheep, goats, pigs, and horses) by inhibiting the respiratory chain dehydrogenases and generating reactive oxygen species, thereby affecting ATP synthesis and cellular metabolic process (Gurunathan et al., 2018). Other means include the variation in the zeta potential on the surface of AgNP and microbes, synergistic ability of silver nanoparticles with β-lactam antibiotics in inhibiting hydrolytic β-lactamases, and biofilm disruption by inhibiting exopolysaccharide production (Kalishwaralal et al., 2010; Hwang et al., 2012).

In other words, there are electrostatic forces between the positively charged surface of NP and negatively charged surface of parasites allowing closer attraction and interaction for its scolicidal activity (Franci et al., 2015; Rahimi et al., 2015). Silver nanoparticles act in synergy with β-lactam antibiotics by inhibiting the hydrolytic β-lactamases produced by bacteria, disrupting the biofilm, consequently penetrating the cell and altering the cellular function (Hwang et al., 2012). The hydrophobic nature of AgNPs makes it easier for them to pass through cellular membranes and act as a

27.14 Mechanism of action

carrier for hydrophilic antibiotics (Jamaran and Zarif, 2016). Reports have shown the ability of silver nanoparticles to exert bactericidal activity through a Trojan-horse mechanism, a mechanism involved in the cellular uptake of nanoparticles leading to cellular respiration impairment and release of intracellular metallic toxic silver ion (Hsiao et al., 2015). The antifasciolasis activity of AgNp when working in synergy with the established drug is by causing a pit-like structure on egg surface, leading to penetration of drug and cytoplasmic leakage causing death (Gherbawy et al., 2013). The antifungal activity of silver nanoparticles occurs by transcriptional inhibition of many aflatoxin genes, especially the two key regulatory genes for secondary metabolism, viz., laeA and veA and an associated decrease in total reactive oxygen species (ROS) (Mitra et al., 2017). The probable mechanism during healing is the increased blood flow to the wound area and decreased inflammatory response caused by silver nanoparticles (Li et al., 2006).

Drug solubility and bioavailability are major challenges in medical sector those need to be solved. Because of the properties, such as small size and the ability to withstand gastric enzyme and pH, nanoparticles have been used for encapsulation and targeted delivery of drugs and bioactive (hydrophobic and hydrophilic) compounds. Depending on silver nanoparticles composition and surface modification, these nanoparticles have been evaluated for their ability to enhance cellular and humoral immunity by increasing the production of lymphocytes, monocytes and neutrophils (Al-Rhman et al., 2016). The mechanism of silver nanoparticles in initiation and regulation of the immune response is not clearly understood, though it is suggested to be attributed to its interaction with macrophages which stimulates upregulation of proinflammatory genes (interleukin-IL1 and IL6), cytokine release, and leukocyte recruitment (Shin et al., 2007; Greulich et al., 2009) (Fig. 27.3).

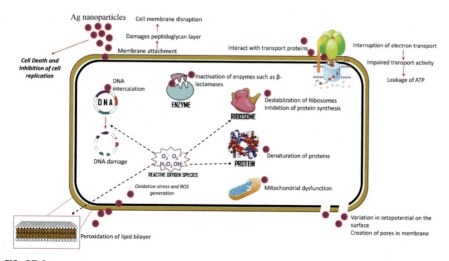

FIG. 27.3

Mechanism of antimicrobial activity of silver nanoparticles.

27.14.1 Nutrient deliveries for fetuses/neonates

Diarrhea is a major cause of loss in neonatal and preweaned livestock. This loss has economic implications. Administering nanoparticles either orally or intragastrically in pregnant and lactating mothers, respectively, can lead to accumulation in fetuses (Lee et al., 2012, 2018; Melnik et al., 2013). These reports indicate the possibility of nanoparticles to cross natural biological barriers like placenta or blood–brain barrier and pass from mother to offspring. During the transition from weaning to solid food in calves or lambs, certain gut microbial profile changes cause an influx of some pathogenic bacteria leading to diarrhea. The correct application of silver nanoparticles as a carrier of drugs or additives in the dam could be used as a medium of stabilizing the gut microbial community, thereby preventing disruption or gut microbial imbalance through breast milk. Pregnancy is important to all livestock operations as it represents the next generation of milk and meat-producing animals. At this stage, the cellular and hormonal system of offspring will be established and maternal nutrition can have an indelible influence on the lifetime productivity and health of progeny, which could enhance or limit productivity and efficiency (Greenwood et al., 2017). During mid to late pregnancy, the dam undergoes physiological changes to help maintain the increase in metabolic demand (Lemley, 2017). Abnormality at this stage can increase the risk of morbidity and mortality during the early neonatal stage (Sawalha et al., 2007) and other lifelong complications and developmental disabilities (Reynolds et al., 2013). The placenta helps in transporting nutrients between the mother and fetus, and this transfer is vital specifically in the growth and development of the fetus during the last half of gestation (Redmer et al., 2004). Size and nutrient transporter abundance are important factors affecting placental nutrient transfer capacity (Fowden et al., 2006). Due to its size and ability to cross the barrier, nanosilver can be designed to transport therapeutic supplements that could improve fetal growth, increase the average body weight at birth, and improve the chance of survival of offspring. Producers can use nanoparticle features as a specific strategy to improve reproductive efficiency of livestock (Lemley, 2017). Particularly, nanoparticles could be used to deliver drugs or vaccines to a fetus during growing epidemic regions.

27.14.2 Potential side effects

Many drugs used in both human beings and animals show some side effects, either mild or severe. For example, application of 50 ppm AgNP through drinking water reduced the growth and modulated the immune functions (Vadalasetty et al., 2018). Injecting nanosilver particles could cause lesions in liver and lungs by damaging the tissue (Doudi and Sertoki, 2014; Loghman et al., 2012). Application of nanosilver through drinking water reduced the yolk weight and hen-day egg production of Japanese quail layers (Farzinpour and Karashi, 2013).

Furthermore, oral subchronic exposure to silver nanoparticles for 13 days decreased the expression of immunomodulatory genes and altered microflora of ileal mucosa shifting the population towards gram-negative microbes, i.e., lowered *Firmicutes* phyla and *Lactobacillus* genus (Williams et al., 2015). Oral administration of silver nanoparticles at 0.5, 1.0, and 1.5 mg/kg BW showed a dose-dependent reduction in absorption of minerals such as K and Fe (Ognik et al., 2017). Intravenous administration of silver nanoparticles at 0.6 mg/kg BW reduced the sperm concentration in the first 21 days and had a similar effect to the control for the later 126 days period. However, the nanoparticles reduced sperm motility and sperm speed while increasing the sperm anomalies (Castellini et al., 2014). The aforementioned phenomena imply that silver nanoparticle has several negative effects on functions related to reproduction, mineral absorption, and metabolism.

The antimicrobial resistance phenomenon necessitates the precise use of nanoparticles according to the purpose. Further, we suggest investigation into other route of administration of nanoparticles that ensure minimal negative effects.

27.15 Conclusion

The use of silver nanoparticles as antimicrobial compounds possesses great potentiality in the animal husbandry sector. Despite the beneficial antimicrobial activity, silver nanoparticles pose negative effects on the environment due to the chemicals involved in the synthesis. Hence, green synthesis methods are most commonly encouraged nowadays. A silver nanoparticle can help overcome the resistance of disease-causing organisms by working in synergy with these established drugs and reducing the cost of developing a new drug. Nanoparticles could be used in fetal programing as delivery agents to immunity enhancement or delivery of mineral to a growing fetus. The wound healing properties of AgNp is outstanding as it aids quick healing of surgery wound and could be applied for healing of sunburn. The immunomodulatory function through the maturation of neutrophils and increased cytokine production is well evidenced. Overall, the diverse roles of silver nanoparticles in animal husbandry sector include, but not limited to, wound healing agents, health promoters, vaccine carriers, growth promoters, immunostimulants, synergic agents, and microbial-resistance preventive agents. However, projecting the silver nanoparticles as complete feed additives for livestock need extensive research, which is lacking at present. Similarly, the selecting appropriate dosages needs to be further studied by conducting a meta-analysis of all the data available on the usage of silver nanoparticles. Thus, the silver nanoparticle has excellent potentials for various veterinary applications.

References

Abdel-Raouf, N., Al-Enazi, N.M., Ibraheem, I.B.M., Alharbi, R.M., Alkhulaifi, M.M., 2019. Biosynthesis of silver nanoparticles by using of the marine brown alga *Padina pavonia* and their characterization. Saudi J. Biol. Sci. 26 (6), 1207–1215.

Adegbeye, M.J., Elghandour, M.M.Y., Faniyi, T.O., Perez, N.R., Barbabosa-Pilego, A., Zaragoza-Bastida, A., Salem, A.Z.M., 2018. Antimicrobial and antihelminthic impacts of black cumin, pawpaw and mustard seeds in livestock production and health. Agrofor. Syst. https://doi.org/10.1007/s10457-018-0337-0.

Adegbeye, M.J., Elghandour, M.M.Y., Barbabosa-Pliego, A., Monroy, J.C., Mellado, M., Ravi Kanth Reddy, P., Salem, A.Z.M., 2019. Nanoparticles in equine nutrition: mechanism of action and application as feed additives. J. Equine Vet. Sci. 78, 29–37.

Adibpour, A., Jamaly, R., Kazami, A., 1998. Seroepidemiological study of hydatid cyst prevalence in north west of Iran (1995-96). Parasitol. Int. 47, 320.

Ahmad, S., Munir, S., Zeb, N., Ullah, A., Khan, B., Ali, J., Bilal, M., Omer, M., Alamzeb, M., Salman, S.M., Ali, S., 2019. Green nanotechnology: a review on green synthesis of silver nanoparticles—an ecofriendly approach. Int. J. Nanomedicine 14, 5087–5107.

Allix-Be'guec, C., Fauville-Dufaux, M., Stoffels, K., Ommeslag, D., Walravens, K., Saegermane, C., Supply, P., 2009. Eur. Respir. J. 35, 692–694.

Al-Rhman, R.M., Ibraheem, S.R., Al-Ogaidi, I., 2016. The effect of silver nanoparticles on cellular and humoral immunity of mice in vivo and in vitro. Iraqi J. Biotechnol. 15 (2), 21–29.

Alvarez-Sanchez, M.A., Mainar-Jaime, R.C., Perez-Garcia, J., Rojo Vazquez, F.A., 2006. Resistance of *Fasciola hepatica* to triclabendazole and albendazole in sheep in Spain. Vet. Rec. 159, 424–425.

Aminianfar, M., Parvardeh, S., Soleimani, M., 2019. In vitro and in vivo assessment of silver nanoparticles against *Clostridium botulinum* Type A Botulinum. Curr. Drug Discov. Technol. 16 (1), 113–119.

Annamalai, J., Nallamuthu, T., 2016. Green synthesis of silver nanoparticles: characterization and determination of antibacterial potency. Appl. Nanosci. 6, 259–265.

Avinash, B., Supraja, N., Charitha, V.G., Adeppa, J., Prasad, T.N., 2017a. Evaluation of the anthelmintic activity (in- vitro) of neem leaf extract-mediated silver nanoparticles against *Haemonchus contortus*. Int. J. Pure Appl. Biosci. 5, 118–128.

Avinash, B., Venu, R., Alpha, R.M., Srinivasa Rao, K., Srilatha, C., Prasad, T.N., 2017b. In vitro evaluation of acaricidal activity of novel green silver nanoparticles against deltamethrin resistance Rhipicephalus (Boophilus) microplus. Vet. Parasitol. 237, 130–136.

Azizi, S., Namvar, F., Mahdavi, M., Ahmad, M.B., Mohamad, R., 2013. Biosynthesis of silver nanoparticles using brown marine macroalga *Sargassum muticum* aqueous extract. Materials (Basel) 6 (12), 5942–5950.

Benelli, G., Maggi, F., Romano, D., Stefanini, C., Vaseeharan, B., Kumar, S., Higuchi, A., Alarfaj, A., Mehlhorn, H., Canale, A., 2017. Nanoparticles as effective acaricides against ticks—a review. Ticks Tick Borne Dis. 8, 821–826.

Beutler, B., 2004. Innate immunity: an overview. Mol. Immunol. 40, 845–859.

Bhainsa, K.C., D'Souza, S.F., 2006. Extracellular biosynthesis of silver nanoparticles using the fungus *Aspergillus fumigatus*. Colloids Surf. B: Biointerfaces 47 (2), 160–164.

Bondarenko, O.M., Sihtmae, M., Kuzmiciova, J., Rageliene, L., Kahru, A., Daugelavicius, R., 2018. Plasma membrane is the target of rapid antibacterial action of silver nanoparticles in *Escherichia coli* and *Pseudomonas aeruginosa*. Int. J. Nanomedicine 13, 6779–6790.

Bravo-Guerra, C., Cáceres-Martínez, J., Vásquez-Yeomans, R., Pestryakov, A., Bogdanchikova, N., 2020. Lethal effects of silver nanoparticles on *Perkinsus marinus*, a protozoan oyster parasite. J. Invertebr. Pathol. 169, 107304.

Cameron, P., Gaiser, B.K., Bhandari, B., Bartley, P.M., Katzer, F., Bridle, H., 2016. Silver nanoparticles decrease the viability of *Cryptosporidium parvum* oocysts. Appl. Environ. Microbiol. 82, 431–437.

Castellini, C., Ruggeri, S., Mattioli, S., Bernardini, G., Macchioni, L., Moretti, E., Collodel, G., 2014. Long-term effects of silver nanoparticles on reproductive activity of rabbit buck. Syst. Biol. Reprod. Med. 60, 143–150. https://doi.org/10.3109/19396368.2014.891163.

Castillo, P.M., Herrera, J.L., Fernandez-Montesinos, R., Caro, C., Zaderenko, A.P., Mejias, J.A., Pozo, D., 2008. Tiopronin monolayer-protected silver nanoparticles modulate IL-6 secretion mediated by toll-like receptor ligands. Nanomedicine 3, 627–635.

Chmielowiec-Korzeniowska, A., Leszek, T., Magdalena, D., Marcin, B., Nowakowicz-Dębek, B., Bryl, M., Drabik, A., Tymczyna-Sobotka, M., Kolejko, M., 2015. Silver (Ag) in tissues and eggshells, biochemical parameters and oxidative stress in chickens. Open Chem. 13, 1269–1274.

Choi, O., Hu, Z., 2008. Size-dependent and reactive oxygen species-related nanosilver toxicity to nitrifying bacteria. Environ. Sci. Technol. 42, 4583–4588.

D'Acquisto, F., May, M.J., Ghosh, S., 2002. Inhibition of nuclear factor kappa B (NF-κB): an emerging theme in anti-inflammatory therapies. Mol. Interv. 2, 235.

Dehkordi, S.H., Hosseinpour, F., Kahrizangi, A.E., 2011. An in vitro evaluation of antibacterial effect of silver nanoparticles on *Staphylococcus aureus* isolated from bovine subclinical mastitis. Afr. J. Biotechnol. 10, 10795–10797.

Deshmukh, P., Patila, S.M., Mullania, S.B., Delekara, S.D., 2019. Silver nanoparticles as an effective disinfectant: a review. Mater. Sci. Eng. C 97, 954–965.

Doudi, M., Sertoki, M., 2014. Acute effect of nanosilver to function and tissue liver of rat after intraperitoneal injection. J. Biol. Sci. 14, 213–219.

El-Sabry, M.I., McMillin, K.W., Sabliov, C.M., 2018. Nanotechnology considerations for poultry and livestock production systems—a review. Ann. Anim. Sci. 18, 319–334.

Eze, F.N., Tola, A.J., Nwabor, O.F., Jayeoye, T.J., 2019. *Centella asiatica* phenolic extract-mediated biofabrication of silver nanoparticles: characterization, reduction of industrially relevant dyes in water and antimicrobial activities against foodborne pathogens. RSC Adv. 9, 37957.

Fahimirad, S., Ajalloueian, F., Ghorbanpour, M., 2019. Synthesis and therapeutic potential of silver nanomaterials derived from plant extracts. Ecotoxicol. Environ. Saf. 168, 260–278.

Farag, H.F., 1998. Human fascioliasis in some countries of the eastern Mediterranean region. East Mediterr. Health J. 4, 156–160.

Farzinpour, A., Karashi, N., 2013. The effects of nanosilver on egg quality traits in laying Japanese quail. Appl. Nanosci. 3, 95–99.

Fowden, A.L., Ward, J.W., Wooding, F.P.B., Forhead, A.J., Constancia, M., 2006. Programming placental nutrient transport capacity. J. Physiol. 572, 5–15.

Franci, G., Falanga, A., Galdiero, S., Palomba, L., Rai, M., Morelli, G., Galdiero, M., 2015. Silver nanoparticles as potential antibacterial agents. Molecules 20, 8856–8874.

Fraser, J.A., Kemp, S., Young, L., Ross, M., Prach, M., Hutchison, G.R., Malone, E., 2018. Silver nanoparticles promote the emergence of heterogeneic human neutrophil subpopulations. Sci. Rep. 8, 7506.

Galdiero, S., Falanga, A., Vitiello, M., Cantisani, M., Marra, V., Galdiero, M., 2011. Silver nanoparticles as potential antiviral agents. Molecules 16, 8894–8918.

Gardea-Torresdey, J.L., Gomez, E., Peralta-Videa, J.R., Parsons, J.G., Troiani, H., Jose-Yacaman, M., 2003. Alfalfa sprouts: a natural source for the synthesis of silver nanoparticles. Langmuir 19, 1357–1361. https://doi.org/10.1021/la020835i.

Gherbawy, Y.A., Shalaby, I.M., Abd El-sadek, M.S., Elhariry, H.M., Banaja, A.A., 2013. The antifasciolasis properties of silver nanoparticles produced by *Trichoderma harzianum* and their improvement of the anti-fasciolasis drug triclabendazole. Int. J. Mol. Sci. 14, 21887–21898.

Goel, A., Bhanja, S.K., Mehra, M., Majumdar, S., Mandal, A., 2017. In ovo silver nanoparticle supplementation for improving the post-hatch immunity status of broiler chickens. Arch. Anim. Nutr. https://doi.org/10.1080/1745039X.2017.1349637.

Gopinath, P.M., Ranjani, A., Dhanasekaran, D., Thajuddin, N., Archunan, G., Akbarsha, M.A., Gulyás, B., Padmanabhan, P., 2016. Multi-functional nanosilver: a novel disruptive and theranostic agent for pathogenic organisms in real-time. Sci. Rep. 6, 34058.

Greenwood, P., Clayton, E., Bell, A., 2017. Developmental programming and beef production. Anim. Front. 7, 38–47.

Greulich, C., Kittler, S., Epple, M., Muhr, G., Koller, M., 2009. Studies on the biocompatibility and the interaction of silver nanoparticles with human mesenchymal stem cells (hMSCs). Langenbeck's Arch. Surg. 394, 495–502.

Gudikandula, K., Vadapally, P., Singara Charya, M.A., 2017. Biogenic synthesis of silver nanoparticles from white rot fungi: their characterization and antibacterial studies. Open Nano 2, 64–78. https://doi.org/10.1016/j.onano.2017.07.002.

Gurunathan, S., Choi, Y.J., Kim, J.H., 2018. Antibacterial efficacy of silver nanoparticles on endometritis caused by *Prevotella melaninogenica* and *Arcanobacterum pyogenes* in dairy cattle. Int. J. Mol. Sci. 19, 1210.

Hamouda, R.A., Hussein, M.H., Abo-Elmagd, R.A., Bawazir, S.S., 2019. Synthesis and biological characterization of silver nanoparticles derived from the cyanobacterium *Oscillatoria limnetica*. Sci. Rep. 9, 13071.

Hartemann, P., Hoet, P., Proykova, A., Fernandes, T., Baun, A., De Jong, W., Filser, J., Hensten, A., Kneuer, C., Maillard, J.-Y., 2015. Nanosilver: safety, health and environmental effects and role in antimicrobial resistance. Mater. Today 18, 122–123.

He, H., Genovese, K.J., Nisbet, D.J., Kogut, M.H., 2006. Profile of toll-like receptor expressions and induction of nitric oxide synthesis by toll-like receptor agonists in chicken monocytes. Mol. Immunol. 43, 783–789.

Hill, E.K., Li, J., 2017. Current and future prospects for nanotechnology in animal production. J. Anim. Sci. Biotechnol. 8, 1–13.

Hsiao, I.-L., Hsieh, Y.-K., Wang, C.-F., Chen, I.-C., Huang, Y.-J., 2015. Trojan-horse mechanism in the cellular uptake of silver nanoparticles verified by direct intra- and extracellular silver speciation analysis. Environ. Sci. Technol. 49, 3813–3821.

Huang, S., Wang, L., Liu, L., Hou, Y., Li, L., 2014. Nanotechnology in agriculture, livestock, and aquaculture in China. A review. Agron. Sustain. Dev. 35 (2), 369–400.

Hwang, I., Hwang, J.H., Choi, H., Kim, K.-J., Lee, D.G., 2012. Synergistic effects between silver nanoparticles and antibiotics and the mechanisms involved. J. Med. Microbiol. 61, 1719–1726.

Ismail, A.E.M.A., Kotb, S.A., Mohamed, I.M., Abdel-Mohsein, H.S., 2019. Silver nanoparticles and sodium hypochlorite inhibitory effects on biofilm produced by *Pseudomonas aeruginosa* from poultry farms. J. Adv. Vet. Res. 9 (4), 178–186.

Jamaran, S., Zarif, B.H., 2016. Synergistic effect of silver nanoparticles with neomycin or gentamicin antibiotics on mastitis-causing *Staphylococcus aureus*. Open J. Ecol. 6, 452–459.

Javaid, A., Oloketuyi, S.F., Khan, M.M., Khan, F., 2017. Diversity of bacterial synthesis of silver nanoparticles. BioNanoScience 8, 43–59.

Jin, X., Li, M., Wang, J., Marambio-Jones, C., Peng, F., Huang, X., 2010. High-throughput screening of silver nanoparticle stability and bacterial inactivation in aquatic media: influence of specifications. Environ. Sci. Technol. 44, 7321–7328.

Johari, P., 2016. http://shodhganga.inflibnet.ac.in/handle/10603/2363. (Accessed November 2016).

Jorch, S.K., Kubes, P., 2017. An emerging role for neutrophil extracellular traps in noninfectious disease. Nat. Med. 23, 279–287.

Jose, J., Anas, A., Jose, B., Puthirath, A., Athiyanathil, S., Jasmin, C., Anantharaman, M.R., Nair, S., Subrahmanyam, C., Biju, V., 2019. Extinction of antimicrobial resistant pathogens using silver embedded silica nanoparticles and an efflux pump blocker. ACS Appl. Bio Mater. https://doi.org/10.1021/acsabm.9b00614.

Kalishwaralal, K., Barath Mani Kanth, S., Pandian, S.R.K., Deepak, V., Gurunathan, S., 2010. Silver nanoparticles impede the biofilm formation by *Pseudomonas aeruginosa* and *Staphylococcus epidermidis*. Colloids Surf. B: Biointerfaces 79 (2), 340–344.

Kathiraven, T., Sundaramanickam, A., Shanmugam, N., Balasubramanian, T., 2015. Green synthesis of silver nanoparticles using marine algae *Caulerpa racemosa* and the antibacterial activity against some human pathogens. Appl. Nanosci. 5, 499–504.

Kathiresan, G., Kanimozhi, Arulnathan, N., 2019. Silver nanoparticle: a bactericidal agent for pathogenic poultry bacteria. Int. J. Recent Technol. Eng. 7, S2.

Khanna, P., Kaur, A., Goyal, D., 2019. Algae-based metallic nanoparticles: synthesis, characterization and applications. J. Microbiol. Methods 163, 105656.

Khatoon, N., Alam, H., Khan, A., Raza, K., Sardar, M., 2019. Ampicillin silver nanoformulations against multidrug resistant bacteria. Sci. Rep. 9, 6848.

Khurana, A., Tekula, S., Saifi, M.A., Venkatesh, P., Godugu, C., 2019. Therapeutic applications of selenium nanoparticles. Biomed. Pharmacother. 111, 802–812.

Kim, J.S., Sung, J.H., Ji, J.H., Song, K.S., Lee, J.H., Kang, C.S., Yu, L.I., 2011. In vivo genotoxicity of silver nanoparticles after 90-day silver nanoparticle inhalation exposure. Saf. Health Work 2, 34–38.

Kitsyuk, N.I., Zvyagintseva, T.V., 2018. The influence of the thiotriazoline ointment with silver nanoparticles on morphological lesions of Guinea pigs' skin due to the local effects of ultraviolet rays at the remote terms after irradiation. J. Educ. Health Sport 8, 274–279.

Kordestani, S., Nayeb Habib, F., Saadatjo, M.H., 2015. A novel wound rinsing solution based on nano colloidal silver. Nanomed. J. 1, 315–323.

Kraft, C., Peter, M., Hofmann, K., 2010. Selective autophagy: ubiquitin-mediated recognition and beyond. Nat. Cell Biol. 12, 836–841.

Kulak, E., Sembratowicz, I., Stępniowska, A., Ognik, K., 2017. The effect of administration of silver nanoparticles on the immune status of chickens. Ann. Anim. Sci. https://doi.org/10.1515/aoas-2017-0043.

Kumar, I., Bhattacharya, J., 2019. Assessment of the role of silver nanoparticles in reducing poultry mortality, risk and economic benefits. Appl. Nanosci. https://doi.org/10.1007/s13204-018-00942-x.

Lakshmanan, G., Sathiyaseelan, A., Kalaichelvan, P.T., Murugesan, K., 2018. Plant-mediated synthesis of silver nanoparticles using fruit extract of *Cleome viscosa* L.: assessment of their antibacterial and anticancer activity. Karbala Int. J Mod. Sci. 4 (1), 61–68.

Ledwith, D.M., Whelan, A.M., Kelly, J.M., 2007. A rapid, straight-forward method for controlling the morphology of stable silver nanoparticles. J. Mater. Chem. 17 (23), 2459.

Lee, Y., Choi, J., Kim, P., Choi, K., Kim, S., Shon, W., Park, K., 2012. A transfer of silver nanoparticles from pregnant rat to offspring. Toxicol. Res. 28, 139–141.

Lee, H.J., Gulumian, M., Faustman, E.M., Workman, T., Jeon, K., Yu, I., 2018. Blood biochemical and hematological study after subacute intravenous injection of gold and silver nanoparticles and coadministered gold and silver nanoparticles of similar sizes. Biomed. Res. Int. 2018, 8460910.

Lemley, C.O., 2017. Investigating reproductive organ blood flow and blood perfusion to ensure healthy offspring. Anim. Front. 7, 18–24.

Li, Y., Leung, P., Yao, L., Song, Q.W., Newton, E., 2006. Antimicrobial effect of surgical masks coated with nanoparticles. J. Hosp. Infect. 62, 58–63.

Loghman, A., Iraj, S.H., Naghi, D.A., Pejman, M., 2012. Histopathologic and apoptotic effect of nanosilver in liver of broiler chickens. Afr. J. Biotechnol. 11, 6207–6211.

Mao, B.H., Chen, Z.-Y., Wang, Y.-J., Yan, S.-J., 2018. Silver nanoparticles have lethal and sublethal adverse effects on development and longevity by inducing ROS-mediated stress responses. Sci. Rep. 8, 2445.

Marquardt, R.R., Li, S., 2018. Antimicrobial resistance in livestock: advances and alternatives to antibiotics. Anim. Front. 8, 30–37.

Mathur, P., Jha, S., Ramteke, S., Jain, N.K., 2018. Pharmaceutical aspects of silver nanoparticles. Artif. Cells Nanomed. Biotechnol. 46, 115–126.

Meena, N.S., Sahni, Y.P., Thakur, D., Singh, R.P., 2018. Applications of nanotechnology in veterinary therapeutics. J. Entomol. Zool. Stud. 6, 167–175.

Melnik, E.A., Buzulukov, Y.P., Demin, V.F., Demin, V., Gmoshinski, I.V., Tyshko, N.V., Tutelyan, V.A., 2013. Transfer of silver nanoparticles through the placenta and breast milk during in vivo experiments on rats. Acta Nat. 5, 107–115.

Mirzajani, F., Ghassempour, A., Aliahmadi, A., Esmaeili, M.A., 2011. Antibacterial effect of silver nanoparticles on *Staphylococcus aureus*. Res. Microbiol. 162, 542–549.

Mitra, C., Gummadidala, P.M., Afshinnia, K., Merrifield, R.C., Baalousha, M., Lead, J.R., Chanda, A., 2017. Citrate-coated silver nanoparticles growth-independently inhibit aflatoxin synthesis in *Aspergillus parasiticus*. Environ. Sci. Technol. 51, 8085–8093.

Mohammed, A.N., Aziz, S.A.A.A., 2019. Novel approach for controlling resistant *Listeria monocytogenes* to antimicrobials using different disinfectants types loaded on silver nanoparticles (Ag NPs). Environ. Sci. Pollut. Res. 26 (2), 1954–1961.

Mousavi, S.A., Mousavi, S.J., Zamani, A., Nourani, S.R., Abbasi, A., Nasiri, E., Aramideh, J.A., 2019. Comparison of burn treatment with nano silver-aloe vera combination and silver sulfadiazine in animal models. Trauma Mon. 24, e79365.

National Research Centre, 1986. Pesticide Resistance: Strategies and Tactics for Management. National Academy Press, Washington, p. 471.

Nguyen, T., Nioi, P., Pickett, C.B., 2009. The Nrf2-antioxidant response element signaling pathway and its activation by oxidative stress. J. Biol. Chem. 284, 13291–13295.

Nicholson, W.L., Munakata, N., Horneck, G., Melosh, H.J., Setlow, P., 2000. Resistance of Bacillus endospores to extreme terrestrial and extraterrestrial environments. Microbiol. Mol. Biol. Rev. 64, 548–572.

Nsofor, C.A., Olatoye, I.O., Amosun, E.A., Iroegbu, C.U., Davis, M.A., Orfe, L.H., Call, D.R., 2013. *Escherichia coli* from Nigeria exhibit a high prevalence of antibiotic resistance where reliance on antibiotics in poultry production is a potential contributing factor. Afr. J. Microbiol. Res. 7, 4646–4654.

Ognik, K., Stępniowska, A., Kozłowski, K., 2017. The effect of administration of silver nanoparticles to broiler chickens on estimated intestinal absorption of iron, calcium, and potassium. Livest. Sci. 200, 40–45.

Oluwasile, B.B., Agbaje, M., Ojo, O.E., Dipeolu, M.A., 2014. Antibiotic usage pattern in selected poultry farms in Ogun state. J. Vet. Sci. 12, 45–50.

Panacek, A., Kvitek, L., Smekalova, M., Vecerova, R., Kolar, M., Roderova, M., Dycka, F., Sebela, M., Prucek, R., Tomanaec, O., Raek, Z., 2018. Bacterial resistance to silver nanoparticles and how to overcome it. Nat. Nanotechnol. 13, 65–71.

Pangestika, R., Ernawati, R., Suwarno, S., 2017. Antiviral activity effect of silver nanoparticles (AgNps) solution against the growth of infectious bursal disease virus on Embryonated chicken eggs with Elisa test. In: The Vet. Med. Int. Conf. KnE Life Sciences, pp. 536–548.

Pathak, J., Rajneesh, Ahmed, H., Singh, D.K., Pandey, A., Singh, S.P., Sinha, R.P., 2019. Recent developments in green synthesis of metal nanoparticles utilizing cyanobacterial cell factories. In: Nanomaterials in Plants, Algae and microorganisms. Academic Press, pp. 237–265.

Pelletier, M., Maggi, L., Micheletti, A., Lazzeri, E., Tamassia, N., Costantini, C., Cosmi, L., Lunardi, C., Annunziato, F., Romagnani, S., Cassatella, M.A., 2010. Evidence for a crosstalk between human neutrophils and Th17 cells. Blood 115, 335–343.

Perez-Cogollo, L.C., Rodriguez-Vivas, R.I., Ramirez-Cruz, G.T., Miller, R.J., 2010. First report of the cattle tick *Rhipicephalus microplus* resistant to ivermectin in Mexico. Vet. Parasitol. 168, 165–169.

Potbhare, A.K., Chaudhary, R.G., Chouke, P.B., Yerpude, S., Mondal, A., Sonkusare, V.N., Rai, A.R., Juneja, H.D., 2019. Phytosynthesis of nearly monodisperse CuO nanospheres using *Phyllanthus reticulatus/Conyza bonariensis* and its antioxidant/antibacterial assays. Mater. Sci. Eng. C 99, 783–793.

Prabhu, S., Poulose, E.K., 2012. Silver nanoparticles: mechanism of antimicrobial action, synthesis, medical applications, and toxicity effects. Int. Nano Lett. 2, 32.

Preet, S., Tomar, R.S., 2017. Anthelmintic effect of biofabricated silver nanoparticles using *Ziziphus jujuba* leaf extract on nutritional status of *Haemonchus contortus*. Small Rumin. Res. 154, 45–51.

Prosposito, P., Burratti, L., Venditti, I., 2020. Silver nanoparticles as colorimetric sensors for water pollutants. Chem. Aust. 8 (2), 26.

Rahimi, M.T., Ahmadpour, E., Esboei, B.R., Spotin, A., Koshki, M.H., Alizadeh, A., Honary, S., Barabadi, H., Mohammadi, M.A., 2015. Scolicidal activity of biosynthesized silver nanoparticles against *Echinococcus granulosus* protoscolices. Int. J. Surg. 19, 128–133.

Rajakumar, G., Rahuman, A.A., 2012. Acaricidal activity of aqueous extract and synthesized silver nanoparticles from Manilkara zapota against *Rhipicephalus* (Boophilus) *microplus*. Res. Vet. Sci. 93, 303–309.

Redmer, D.A., Wallace, J.M., Reynolds, L.P., 2004. Effect of nutrient intake during pregnancy on fetal and placental growth and vascular development. Domest. Anim. Endocrinol. 2, 99–217.

Reynolds, L.P., Vonnahme, K.A., Lemley, C.O., Redmer, D.A., GrazulBilska, A.T., Borowicz, P.P., 2013. Maternal stress and placental vascular function and remodeling. Curr. Vasc. Pharmacol. 11, 564–593. https://doi.org/10.2174/1570161111311050003.

Rodríguez-Luis, O.E., Hernandez-Delgadillo, R., Sánchez-Nájera, R.I., Martínez-Castañón, G.A., Niño-Martínez, N., Sánchez Navarro, M.D.C., Cabral-Romero, C., 2016. Green synthesis of silver nanoparticles and their bactericidal and antimycotic activities against oral microbes. J. Nanomater. 2018, 1–10. 9204573.

Rose, J.B., Huffman, D.E., Gennacaro, A., 2002. Risk and control of waterborne cryptosporidiosis. FEMS Microbiol. Rev. 26, 113–123.

Rouhani, A., Parvizi, P., Spotin, A., 2013. Using specific synthetic peptide (p176) derived AgB 8/1-kDa accompanied by modified patients sera: a novel hypothesis to follow-up of cystic echinococcosis after surgery. Med. Hypotheses 81, 557–560.

Salih, N.A., Ibrahim, O.M.S., Eesa, M.J., 2016. Biosynthesis of silver nanoparticles and evaluate its activity in promoting burns healing in rabbits. Al-Anbar J. Vet. Sci. 9, 47–58.

Santhoshkumar, T., Rahuman, A.A., Bagavan, A., Marimuthu, S., Jayaseelan, C., Kirthi, A.V., Kamaraj, C., Rajakumar, G., Zahir, A.A., Elango, G., Velayutham, K., Iyappan, M., Siva, C., Karthik, L., Rao, K.V.B., 2012. Evaluation of stem aqueous extract and synthesized silver nanoparticles *using Cissus quadrangularis* against *Hippobosca maculate* and *Rhipicephalus* (Boophilus) *microplus*. Exp. Parasitol. 132, 156–165.

Santos, L.M., Stanisic, D., Menezes, U.J., Mendonça, M.A., Barral, T.D., Seyffert, N., Azevedo, V., Durán, N., Meyer, R., Tasic, L., Portela, R.W., 2019. Biogenic silver nanoparticles as a post-surgical treatment for *Corynebacterium pseudotuberculosis* infection in small ruminants. Front. Microbiol. 10, 824.

Sawalha, R.M., Conington, J., Brotherstone, S., Villanueva, B., 2007. Analyses of lamb survival of Scottish blackface sheep. Animal 1, 151–157.

Sawosz, E., Grodzik, M., Lisowski, P., Zwierzchowski, L., Niemiec, T., Zielinska, M., Szmidt, M., Chwalibog, A., 2010. Influence of hydrocolloids of Ag, Au, and Ag/Cu alloy nanoparticles on the inflammatory state at transcriptional level. Bull. Vet. Inst. Pulawy 54, 81–85.

Scapini, P., Lapinet-Vera, J.A., Gasperini, S., Calzetti, F., Bazzoni, F., Cassatella, M.A., 2000. The neutrophil as a cellular source of chemokines. Immunol. Rev. 177, 195–203.

Scott, N.R., 2005. Nanotechnology and animal health. Rev. Sci. Technol. 24, 425–432.

Sekhon, B.S., 2014. Nanotechnology in agri-food production: an overview. Nanotechnol. Sci. Appl. 7, 31–53.

Selim, A., Elhaig, M.M., Taha, S.A., Nasr, E.A., 2018. Antibacterial activity of silver nanoparticles against field and reference strain of mycobacterium tuberculosis, mycobacterium bovis and multiple-drug-resistant tuberculosis strain. Rev. Sci. Tech. 37, 1–16.

Shin, S.H., Ye, M.K., Kim, H.S., Kang, H.S., 2007. The effects of nano-silver on the proliferation and cytokine expression by peripheral blood mononuclear cells. Int. Immunopharmacol. 7, 1813–1818.

Sikorska, J., Szmidt, M., Sawosz, E., Niemiec, T., Grodzik, M., Chwalibog, A., 2010. Can silver nanoparticles affect the mineral content, structure and mechanical properties of chicken embryo bones? J. Anim. Feed Sci. 19, 286–291.

Simon, L.C., Sabliov, C.M., Stout, R.W., 2016. Bioavailability of orally delivered alpha-tocopherol by poly (lactic-co-glycolic) acid (PLGA) nanoparticles and chitosan covered PLGA nanoparticles in F344 rats. Nano 3. https://doi.org/10.5772/63305.

Smekalova, M., Aragon, V., Panacek, A., Prucek, R., Zboril, R., Kvitek, L., 2015. Enhanced antibacterial effect of antibiotics in combination with silver nanoparticles against animal pathogens. Vet. J. 209, 174–179.

Tang, J., Xiong, L., Wang, S., Wang, J., Liu, L., Li, J., Wan, Z., Xi, T., 2007. Influence of silver nanoparticles on neurons and blood-brain barrier via subcutaneous injection in rats. Appl. Surf. Sci. 225, 502–504.

Thomson, S., 2015. Cryptosporidiosis in Farmed Livestock (Ph.D. Thesis). University of Glasgow, Glasgow, United Kingdom.

Tian, J., Wong, K.K.Y., Ho, C.-M., Lok, C.-N., Yu, W.-Y., Che, C.-M., 2007. Topical delivery of silver nanoparticles promotes wound healing. ChemMedChem 2, 129–136.

Tomar, R.S., Preet, S., 2017. Evaluation of anthelmintic activity of biologically synthesized silver nanoparticles against the gastrointestinal nematode, *Haemonchus contortus*. J. Helminthol. 91 (454), 461.

Tripathi, D., Modi, A., Narayan, G., Rai, S.P., 2019. Green and cost-effective synthesis of silver nanoparticles from endangered medicinal plant *Withania coagulans* and their potential biomedical properties. Mater. Sci. Eng. C 100, 152–164.

Vadalasetty, K.P., Lauridsen, C., Engberg, R.M., Vadalasetty, R., Kutwin, M., Chwalibog, A., Sawosz, E., 2018. Influence of silver nanoparticles on growth and health of broiler chickens after infection with *Campylobacter jejuni*. BMC Vet. Res. 14, 1–11.

Vasile, B.S., Birca, A.C., Musat, M.C., Holban, A.M., 2020. Wound dressings coated with silver nanoparticles and essential oils for the management of wound infections. Materials 13 (7), 1682.

Vazquez-Muñoz, R., Meza-Villezcas, A., Fournier, P.G.J., Soria-Castro, E., Juarez-Moreno, K., Gallego-Hernandez, A.L., Bogdanchikova, N., Vazquez-Duhalt, R., Huerta-Saquero, R., 2019. Enhancement of antibiotics antimicrobial activity due to the silver nanoparticles impact on the cell membrane. PLoS One 14, e0224904.

Waller, P.J., 2003. The future of anthelmintics in sustainable parasite control programs for livestock. Helminthologia 40, 97–102.

WHO, 2010. Global tuberculosis control: WHO report 2010. WHO, Geneva, Switzerland, p. 218. http://apps.who.int/iris/bitstream/10665/44425/1/9789241564069_eng.pdf. (Accessed 15 April 2010).

Williams, K., Milner, J., Boudreau, M.D., Gokulan, K., Cerniglia, C.E., Khare, S., 2015. Effect of subchronic exposure of silver nanoparticles on intestinal microbiota and gut-associated immune responses in the ileum of Sprague-Dewley rats. Nanotoxicology 9, 279–289.

World Health Organisation (WHO), 2017. Guidelines on Use of Medically Important Antimicrobials in Food-Producing Animals. World Health Organization, Geneva (Licence: CC BY-NC-SA 3.0 IGO).

Wu, H., Lin, J., Liu, P., Huang, Z., Zhao, P., Jin, H., Ma, J., Wen, L., Gu, N., 2016. Reactive oxygen species acts as executor in radiation enhancement and autophagy inducing by AgNPs. Biomaterials 101, 1–9.

Yasin, S., Liu, L., Yao, J., 2013. Biosynthesis of silver nanoparticles by bamboo leaves extract and their antimicrobial activity. J. Fiber Bioeng. Inform. 6, 77–84.

Yuan, Y.G., Peng, Q.L., Gurunathan, S., 2017. Effects of silver nanoparticles on multiple drug-resistant strains of Staphylococcus aureus and *Pseudomonas aeruginosa* from mastitis-infected goats: an alternative approach for antimicrobial therapy. Int. J. Mol. Sci. 18, 569.

Zahir, A.A., Rahuman, A.A., 2012. Evaluation of different extracts and synthesised silver nanoparticles from leaves of *Euphorbia prostrata* against *Haemaphysalis bispinosa* and *Hippobosca maculata*. Vet. Parasitol. 187, 511–520.

Zhao, L., Seth, A., Wibowo, N., Zhao, C.X., Mitter, N., Yu, C., Middelberg, A.P.J., 2014. Nanoparticle vaccines. Vaccine 32, 327–337.

CHAPTER 28

Silver nanoparticles in poultry health: Applications and toxicokinetic effects

Vinay Kumar[a], Neha Sharma[b], Sivarama Krishna Lakkaboyana[c], and Subhrangsu Sundar Maitra[b]

[a]Department of Biotechnology, Indian Institute of Technology Roorkee, Uttarakhand, India
[b]School of Biotechnology, Jawaharlal Nehru University, New Delhi, India
[c]Department of Chemical Technology, Faculty of Science, Chulalongkorn University, Bangkok, Thailand

28.1 Introduction

In the recent years, silver nanoparticles have emerged as a multiuse material with wide applications range. The advantage of using silver nanoparticles is that it has a great surface to volume ratio, which makes it useful material. They are used extensively in various applications such as food, healthcare, electrical, and textile products (Calderón-Jiménez et al., 2017; Rai et al., 2014a,b). Due to their better performance, the demand for silver nanoparticles will rise in 2022 in the electrical appliances and electronics (Syafiuddin et al., 2017). It is surprising that one fifth of the produced nanomaterials are based on silver, which account for approximately 320–420 tons (Pulit-Prociak and Banach, 2016; Wilson, 2007). Silver nanoparticles are used in many environmental applications (Mukherjee et al., 2017; Yadav et al., 2017) and wastewater treatments (Kühr et al., 2018; Moustafa, 2017). Silver nanoparticles have various potential uses in commercial applications. Due to this reason, they are much popular among all the nanoparticles characterized so far (Natsuki et al., 2015; Zhang et al., 2016). The silver nanoparticles are characterized by various special properties such as greater surface area, small size, agglomeration, and ion release efficiency (Zhang et al., 2016; Zivic et al., 2018). They have electrical, optical, catalytic, and antimicrobial properties (Syafiuddin et al., 2017). Due to the above-mentioned properties, they are used in preservation of food during storage, household utensils, health care industries, and others environmental applications (Zivic et al., 2018).

Various methods are used to synthesize silver nanomaterials. These methods include electrochemical (Nasretdinova et al., 2015), physicochemical (Flores-Rojas et al., 2020), and chemical reduction (Khan et al., 2017). These methods generate toxic residues, which are released to the environment ultimately. It has been observed that the waste products accumulate from the silver nanoparticles manufacture and their disposal (Durán et al., 2017; Khan et al., 2017; Maurer-Jones et al., 2013).

The toxicity of the nanoparticles is affected by various parameters such as pH, ionic strength, humic acid, fulvic acid, and organic matter (Akaighe et al., 2012; Chambers et al., 2014; Delay et al., 2011; Yang et al., 2018). Although industrial production of silver nanoparticles contributes a major portion of the toxic materials to the environment, artificial synthesis is not the only means for interaction with environment. It has been observed that nanoparticles can be produced naturally also (Wimmer et al., 2018). It has been observed that metal nanoparticles (silver, gold, zinc, copper, nickel, and iron), oxide nanomaterials (titanium oxide, ferrous oxide, silicon dioxide, selenium oxide, aluminum oxide, and manganese dioxide), and metal sulfides (ferrous, silver, copper, and zinc) (Sharma et al., 2015) are formed during microbial activity, industrial processes, mining, precipitation, bioaccumulation, and biosorption (Durán et al., 2017; Griffin et al., 2018; Yu et al., 2013).

Broiler industry is a major source for providing the food to the overgrowing population. Development and growth of flocks is a major concern to farmers associated with the broiler industry (Stanley et al., 2012). In poultry, the chickens are resistant to the antibiotics due to their drugs involving routine antibiotics (van den Bogaard and Stobberingh, 2000). There is always a possibility for transfer of antibiotic resistance zoonotic pathogens to the humans. There are two major parameters to determine the growth performance in poultry. These are body weight gain and feed conversion ratio (FCR). Diet is one of the major contributors to the FCR, which can include fat sources (Skrivan et al., 2000), and proteins (Fancher and Jensen, 1989). Assimilation of the feed is one of the major factor that affects digestive tract physiology as well as the overall performance (Amerah et al., 2007). In the recent years, the inclusion of the nanoparticles has gained much popularity due to their great overall performance (Samanta et al., 2019; Vijayakumar and Balakrishnan, 2015). They have been known to improve the growth, including FCR in the poultry (Ahmadi et al., 2009). Silver nanoparticles are much for popular in poultry and have more applications as compared to other nanoparticles. This chapter is focused on the evaluation the application of silver nanoparticles in poultry. The chapter is also aimed at analyzing the detrimental effect of silver nanoparticles on poultry animals.

28.2 Silver nanoparticles application in poultry

28.2.1 As antibacterial agents

In the recent years, silver nanoparticles have been used as a carrier for nutrients and drugs. They are popular due to the properties such as reducing drug doses and improvement in FCR (Seil and Webster, 2012). Studies have reported about the use of silver nanoparticles as an antibacterial agent for circulating avian microflora. The introduction of nanoparticles is responsible for improvement of feed intake, nutrient absorption, and weight gain (Andi et al., 2011). In a study, feeding of nanoparticles supplements leads to the decrease in lymphoid organs in a dose-dependent manner (Ahmadi and Branch, 2012). This reduction in lymphoid organs may be due to silver nanoparticles' antimicrobial properties. In another study, silver nanoparticles caused the reduction in the number of harmful bacteria such as *Salmonella* in the litter of broilers. The antibacterial activity of silver nanoparticles was investigated on chickens (Lane Pineda et al., 2012). When broilers were fed with nanoparticles in the range of 2–8 ppm/kg, a reduction in the pathogenic bacteria such as *E. coli* was observed (Elkloub et al., 2015). While surprisingly, it did not affect the population of *Lactobacilli*.

28.2.2 In poultry feed

Nowadays, nanoparticles find an immense use in poultry production (Abd El-Ghany, 2019; Fisinin et al., 2018). The nanoparticles used in poultry include silver (Chauke and Siebrits, 2012; Elkloub et al., 2015; Sawosz et al., 2012b), copper (Mroczek-Sosnowska et al., 2016; Pineda et al., 2013; Sawosz et al., 2018; Scott et al., 2016; Sizova et al., 2020; Wang et al., 2011), selenium (Cai et al., 2012; Fuxiang et al., 2008; Mohapatra et al., 2014; Selim et al., 2015; Shirsat et al., 2016; Wang, 2009; Zhou and Wang, 2011), and combined with some other nutrients (Yausheva et al., 2016). The advantage of using nanoparticles is that as compared to the regularly used sources of minerals such as P and Ca, the nanoparticles are effective even at low doses (Swain et al., 2015). There are several types of nanoparticle feed available for poultry (Table 28.1).

It has been known that macroelements play an important role in the development of the birds, including size. Therefore, nanotechnology has been focused on production of effective birds feed supplements. There are various examples where nanoparticles have been applied as a feed in the birds' diet. Some of the examples include nanoparticles containing Ca-P (Samanta et al., 2019; Vijayakumar and Balakrishnan, 2014, 2015),

Table 28.1 Type of nanoparticles feed and the parameters studied on broiler chickens.

Feed type	Organism used	Parameters studied	References
Ca-P	Broiler chicken	Tibial bone morphometry, bone and serum mineral	Vijayakumar and Balakrishnan (2014, 2015)
Ca-P	Broiler chicken	Body weight	Samanta et al. (2019)
Ca carbonate	Broiler chicken	Body weight, spleen weight, and bone Ca concentration	Salary et al. (2017)
Dicalcium phosphate	Broiler chicken	Carcass characteristics and bone measurements	Hassan et al. (2016) and Mohamed et al. (2016)
Hydroxyapatite	Broiler chicken	Body weight, feed intake	Ali (2017)

Ca carbonate (Salary et al., 2017), dicalcium phosphate (Hassan et al., 2016; Mohamed et al., 2016), and hydroxyapatite (Ali, 2017). The approach used here was to utilize much more effective bioavailable elements for effective absorption rather using a high amount of the selected element (Gonzales-Eguia et al., 2009). Studies were conducted to analyze the adverse effects for Ca phosphate nanoparticles. It was observed that when broiler chickens were fed with Ca phosphate nanoparticles, no harmful effects were observed as compared to the conventional dicalcium phosphate diet (Vijayakumar and Balakrishnan, 2014). The studies were conducted considering effects on biochemical markers such as albumins, glucose, cholesterol, and hematological parameters (Vijayakumar and Balakrishnan, 2014, 2015). Another study conducted on broilers diet demonstrated that compared to conventional dicalcium phosphate, the feed with nanoparticle dicalcium phosphate improved measured bone parameters (Mohamed et al., 2016). The study demonstrated that with nanoparticle dicalcium phosphate, only 25% P is sufficient for broilers as compared to the 100% conventional dicalcium phosphate. In a study, it was observed that nanoparticles feed resulted in 50% less excretion of P and Ca (Hassan et al., 2016). The inclusion of nanoparticles in the birds feed led to about 200% increase in the bioavailability of Ca and P as compared to the conventional diets (Vijayakumar and Balakrishnan, 2015).

It has been observed that there were no changes in the biochemical parameters, carcass characteristics, feed conversion ratio, and feed consumption ratio on application of Ca phosphate nanoparticles when used either in combination or as only source (Samanta et al., 2019). However, a 50% increase in the growth performance was observed with nanoparticles feed. Hydroxyapatite nanoparticles feed was used in poultry to determine its effect on broilers. It was observed that this feed leads to an increase of 6%–8% in intake and improved the body weight (Ali, 2017). This is due to the better assimilation of P and Ca in nanoparticle feed, which causes lesser release in excreta. It has been demonstrated that introduction of Ca carbonate to the broiler chicken hatchlings resulted in increased body as well as organ weight (Salary et al., 2017). Most of the studies conducted so far are focused on the nanoparticles administration in the food and only a few of the works have mentioned use of P and Ca nanoparticles on the skeletal systems. In a study, dicalcium phosphate nanoparticles effects were observed on the long bones quality of broiler chickens (Mohamed et al., 2016). It was observed that on a 26-day old had optimistic effects on morphometric measurements.

28.2.3 In poultry production

It is considered that silver nanoparticles have the potential to target a specific cell type and they can act as antibacterial agent (Mahendra Rai et al., 2009). When silver nanoparticles are coupled with adenosine triphosphate (ATP), they act as ATP carrier, thereby causing its transport and distribution in cells. This caused increased density and maturation of muscles and muscle fibers (Sawosz et al., 2012b, 2013). Silver nanoparticles can promote development and growth of muscle cells, which can be the reason to enlarged breast muscles, thereby leading to body weight increase

(Sawosz et al., 2012a,b, 2013). If silver nanoparticles are combined with hydroxyproline and injected to embryos, it causes the increased cartilage collagen fiber and blood vessel dimensions (Beck et al., 2015). It was demonstrated that if 50 ppm of combined silver and palladium nanoparticles was exposed to the embryos, it can reverse the hypertrophy effect (Studnicka et al., 2009). It was observed that when silver nanoparticles were combined with amino acids such as cysteine and threonine, it can improve adaptive as well as innate immunity in chickens as well embryos. It can also lead to improved oxygen consumption in eggs as well as reduction in liver weight (Bhanja et al., 2015). In a study, silver nanoparticles in a concentration range of 0–900 ppm were supplemented to broilers. At 900 ppm, broiler weight improvement was observed at 600 ppm gives best FCR values (Ahmadi, 2009). The introduction of silver nanoparticles demonstrated improved absorption of nutrients in liver and intestine (Ahmadi et al., 2009). It was observed that when silver nanoparticles were included in feed water, it leads to increased nitrogen intake (Pineda et al., 2012). Silver nanoparticles have the potential to control the protozoan disease such as coccidiosis (Chauke and Siebrits, 2012; Dalloul and Lillehoj, 2006).

28.3 Nanoparticles uptake by cells

Surface interaction of silver nanoparticles with cells causes induction of various processes. Internalization of the silver nanoparticles is the first step to the cell toxicity. To determine toxicity mediated by silver nanoparticles, it is quite essential to determine the uptake mechanism. Some of the methods for uptake are endocytosis, phagocytosis, and diffusion (Murugan et al., 2015). Silver nanoparticles enter human macrophages by either phagocytic or nonphagocytic pathway (Haase et al., 2012). In mammalian cells, silver nanoparticles are engulfed by the cells. The particle size and types are the major factor determining the engulfment of silver nanoparticles (AshaRani et al., 2009; Jiang et al., 2013; Sahu et al., 2014). Other modes of nanoparticles' entry include ion-mediated or flip-flop mechanism (Haase et al., 2011).

28.4 Silver nanoparticles distribution

Distribution of silver nanoparticles takes place through various routes. A few means of the transfer are discussed below.

28.4.1 Distribution through placenta

Study of placenta transfer of silver nanoparticles was conducted on mice and rats using three modes: oral administration, intraperitoneal injection, and intravenous injection. In a study, transmission electron microscopy was used to identify silver nanoparticles with an oral dose ranging from 10 to 1000 mg/kg in mouse dams. The results identified nanoparticles in fetal kidneys and livers (Philbrook et al., 2011).

Citrate-capped silver nanoparticles were administered orally to determine silver levels in the offspring at 250 mg/kg/day at different stages of sexual and reproductive stages (Lee et al., 2012). The level of silver was in increasing order from brain, liver, lungs, and it was highest in kidneys. In another study, female rats were administered with silver labeled nanoparticles at 2.11 mg/kg (Melnik et al., 2013). The results demonstrated that the labeled silver nanoparticles were accumulated in fetus. The determined level of the silver nanoparticles was higher as compared to that in the maternal brain.

Silver nanoparticles with an average size of 50 nm were injected intravenously at gestation days from seven to nine, and autopsy were done on gestation day 10 (Austin et al., 2012). The results demonstrated that the silver level accumulation was higher in visceral yolk and endometrium. Localization of silver nanoparticles with size ranging from 30 to 50 nm was confirmed in the endodermal cells. In another study, silver nanoparticles with average size of 9 nm were injected to a pregnant mice in between day 7 and 9 (Austin et al., 2016). In maternal spleen and liver, the level of silver accumulation was higher as compared to observed in the fetus. Polyvinylpyrrolidone-coated silver nanoparticles were injected intraperitoneally in female and male mice with average size range of 25 nm. After injection of the nanoparticles, both the mice were mated, and embryos were collected on gestation day 14.5 (Wang et al., 2013). Male rat dams offspring was injected with silver nanoparticles in the size range of 20–50 nm (Wu et al., 2015). The study confirmed enhanced accumulation of silver in the hippocampus. The study also confirmed that toxicity observed was mainly due to silver ions. The above results demonstrated that silver nanoparticles were present during early gestation.

28.4.2 Distribution through blood barrier

Testis-blood barrier transfer is another mode for silver nanoparticles transfer in the animals. Oral administration of silver nanoparticles at 30–1000 mg/kg/day in rat for 28 days resulted in elevated silver concentrations in testis (Kim et al., 2008). It was observed that the level of silver was higher in testis as compared to the blood and brain. In another study, a dose-dependent increment in the silver level was observed when silver nanoparticles were administered orally for 13 weeks with a dose range from 30 to 500 mg/kg/day (Kim et al., 2010). A similar trend was observed in this study where the silver levels detected in testis were higher as compared to those in the lungs, brain, liver, and blood. In a study conducted on mice, silver nanoparticles were administered orally with a diameter range from 22 to 323 nm for 2 weeks (Park et al., 2010). The study reported that the size of accumulated nanoparticles was less than 71 nm. The accumulated nanoparticles were detected in all the tissues, but their concentration was higher in the testis. The study demonstrated that small size nanoparticles were easily absorbed in the gastrointestinal tract and thereby transported to the blood-testicular barrier.

A study was conducted on male rats where silver nanoparticles were administered orally for 28 days (van der Zande et al., 2012). The results from the study demonstrated that after 29 days, silver levels were in decreasing order following gastrointestinal tract, liver, testes, brain, and lungs. In another study, male rats were administered orally with silver nanoparticles with size range of 100 to 500 mg/kg/day for 4 weeks. The study demonstrated that the size of the nanoparticle does not have any effect on nanoparticle uptake. The level of the nanoparticles decreased gradually during a 4 week recovery period.

Silver nanoparticles' transport via means of blood — testis barrier was evidenced in rats (Lankveld et al., 2010; Wang et al., 2013), male mice, and rabbits (Castellini et al., 2014). In these studies, higher level of silver level was detected in spleen, liver, and kidney in male mice. In testis, a significant amount of silver was detected after 4 months of silver nanoparticle injection (Wang et al., 2013). In rats, intravenous injection of silver nanoparticles for five consecutive days was administrated. The results demonstrated that the silver level was considerably lower in testis as compared to kidney and spleen. It was also observed that nanoparticles size do not have a major role to play in the distribution of silver nanoparticles in testis (Lankveld et al., 2010).

28.5 Silver nanoparticles toxicity

Silver nanoparticles can enter the cells by passing through biological membranes. Therefore, they cause toxicity in organisms. The level of toxicity caused depends on the types of organism affected and various factors. These factors include nanoparticles size, shape, aggregation, concentration, surface charge, chemical coating, and their route of synthesis (Akter et al., 2018; Durán et al., 2017). There are various organs associated with the silver nanoparticles toxicity (Fig. 28.1).

28.5.1 Nanoparticles size and toxicity

The cytotoxicity performed on different cell lines suggests that the size of the silver nanoparticles is an important parameter on toxicity assessment (Gliga et al., 2014). When different cell lines were exposed to the silver nanoparticles with varying particles size, it affected lactate dehydrogenase activity (Gliga et al., 2014), cell viability, and reactive oxygen species production (Carlson et al., 2008). It was observed that smaller nanoparticles impart greater toxicity as compared to the larger nanoparticles size (Akter et al., 2018). A study demonstrated that silver nanoparticles with a size of 15 nm imparts greater toxicity as compared to 55 nm nanoparticles (Carlson et al., 2008). In another study, 5 nm silver nanoparticles were proved to be more toxic as compared to 20 and 50 nm size nanoparticles (Liu et al., 2010). This study demonstrated that the smaller nanoparticles have greater penetration power as compared to the larger nanoparticles. In a similar study, it was observed that smaller silver nanoparticles cause great toxicity in mice lungs with neutrophilic inflammation (Wang et al., 2014).

The shape of nanoparticles also affects the cellular uptake by the cells, which ultimately affects the cytotoxicity. In an instance, it was observed that wire-shaped silver nanoparticles show cytotoxic effects, but spherical nanoparticles do not have

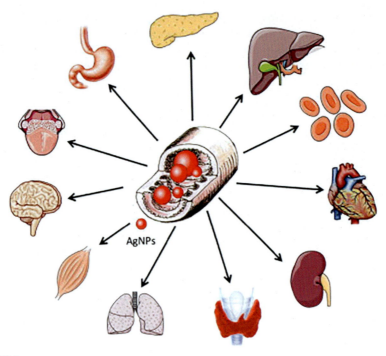

FIG. 28.1

Target organs associated with silver nanoparticle toxicity.

Reprinted from with permission from Gaillet, S., Rouanet, J.-M. 2015. Silver nanoparticles: Their potential toxic effects after oral exposure and underlying mechanisms–A review. Food Chem. Toxicol. 77, 58–63. https://doi.org/10.1016/j.fct.2014.12.019 Elsevier.

any cytotoxic effects (Stoehr et al., 2011). The toxicity of silver nanoparticles on different cell lines varies. Silver nanoparticles with size ranging from 5 to 43 nm were applied at 2.0 mg/L on cell lines such as RAW 264.7, A498, Neuro 2A, and A 549 (Singh and Ramarao, 2012). The results demonstrated that macrophages were more sensitive as compared to other cell lines.

Various parameters such as cell survival, cell proliferation, and migration were affected when bovine retinal endothelial cells were exposed to silver nanoparticles with a particle size of 50 nm at 500 nm concentration. The mechanism involves the suppression of Akt phosphorylation and activation of caspase-3 (Gurunathan et al., 2009; Kalishwaralal et al., 2009). The study on rat brain microvessel endothelial cells in the size range of 25–80 nm resulted in permeability and blood – brain barrier inflammation (Grosse et al., 2013). The results from the study suggested that silver nanoparticles toxicity is affected by particle size, dose, surface area, and exposure time for a specific cell line (Grosse et al., 2013). Stem cells exposed to the silver nanoparticles caused enhanced toxicity. In murine spermatogonia stem cells, silver nanoparticles caused prolonged apoptosis, LDH leakage, and reduced cell viability

(Braydich-Stolle et al., 2005). Dose-dependent cell cytotoxicity was observed when human mesenchymal stem cells were treated with silver nanoparticles with a size range of 100 nm (Greulich et al., 2009).

28.5.2 Nanoparticles toxicity mechanisms

Silver nanoparticles are highly bactericidal and have strong inhibitory effects (Firdhouse and Lalitha, 2015; Nam et al., 2016). Therefore, they are used on wounds and catheters for protection from infection (Catauro et al., 2004; Crabtree et al., 2003; Roe et al., 2008; Tian et al., 2007). It has been suggested that these nanoparticles interact with the bacterial membrane, thereby providing the antimicrobial effects (Le Ouay and Stellacci, 2015; Yun'an Qing et al., 2018).

The mechanism of silver nanoparticles toxicity is investigated by various researchers (Durán et al., 2017; Liao et al., 2019; Tripathi et al., 2017) (Fig. 28.2). It has been observed that silver nanoparticles and silver ions are known to cause toxicity in the organisms (Durán et al., 2017). The toxicity is caused by mechanisms such as reactive oxygen species production, membrane damage, protein oxidation, mitochondrial dysfunction, and protein denaturation, inhibiting cell proliferation and DNA damage (Liao et al., 2019; Zhang et al., 2014). It was reported that the interaction of silver nanoparticles with sulfur containing macromolecules or protein is one of the important toxicity mechanisms. This shows the high affinity of silver nanoparticles toward sulfur and silver (McShan et al., 2014).

Silver nanoparticles are highly toxic in the size range of 10–100 nm at a concentration of 5–10 $\mu g \, mL^{-1}$ (Ahamed et al., 2008; Arora et al., 2009). It is known that they disrupt mitochondrial function, but it is not clear whether silver nanomaterials interact with mitochondria directly or indirectly (Foley et al., 2002). The cytotoxicity of silver nanoparticles is induced by mitochondrial pathway, including high lipid peroxidation, reducing glutathione, DNA damage, necrosis, and apoptosis (Haase et al., 2012). Studies have shown that silver nanoparticles are responsible for morphological deformities in mammalian animal models and have adverse effects on reproduction (Zhang et al., 2015). Studies have confirmed that a minimal release of silver ions causes silver nanoparticle-mediated toxicity (Beer et al., 2012; Gorth et al., 2011; Kruszewski et al., 2011; van der Zande et al., 2012).

Silver nanoparticles' toxicity causes production of reactive oxygen species (ROS) after entering the cells (van Aerle et al., 2013). The level of GSH increases with an increase in ROS production, which ultimately causes apoptosis (He et al., 2012). In neuronal cells, ROS-mediated oxidative stress is responsible for calcium dysregulation (Wang et al., 2009; Ziemińska et al., 2014). In human cell lines, oxidative stress leads to DNA damage, damage to antioxidant defense capacity, and ultimately to apoptosis (Awasthi et al., 2013; Cheng et al., 2013; Haase et al., 2012; He et al., 2012; van Aerle et al., 2013). Apoptosis is promoted in cells due to the reduction in the levels of glutathione, protein bound sulfhydryl group, and antioxidant enzyme activity (Miethling-Graff et al., 2014). In this way, the silver nanoparticles' cell cytotoxicity is mediated by apoptosis (De Gusseme et al., 2011).

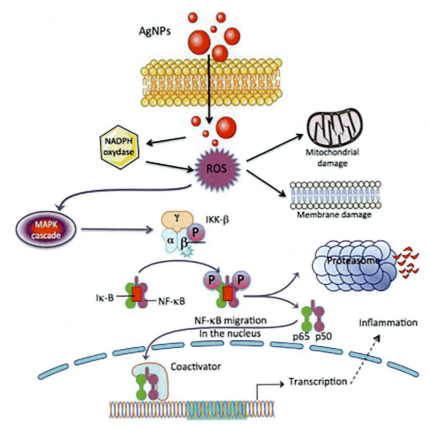

FIG. 28.2

Toxicity of silver nanoparticles including inflammatory responses.

Reprinted from with permission from Gaillet, S., Rouanet, J.-M. 2015. Silver nanoparticles: Their potential toxic effects after oral exposure and underlying mechanisms–A review. Food Chem. Toxicol. 77, 58–63. https://doi.org/10.1016/j.fct.2014.12.019 Elsevier.

28.5.3 ROS-mediated toxicity

In cells, in vitro studies have confirmed that ROS-mediated toxicity is responsible for alterations in cellular processes (van Aerle et al., 2013). The possible mechanism of silver nanoparticles toxicity is oxidative stress (Fig. 28.3). Reactive oxygen species is required for the maintenance of normal process in the cells. However, higher ROS production can lead to damage to proteins, lipids, and DNA by affecting the antioxidant defense system (Sriram et al., 2012). ROS is mainly released by mitochondria, which is responsible for DNA damage, ATP synthesis disruption, and apoptosis (Gurunathan et al., 2009). When the silver nanoparticles enter the cells, they generate ROS (van Aerle et al., 2013). The increase in ROS level ultimately leads to apoptosis (Ahmadi and Branch, 2012; Cheng et al., 2013; Haase et al., 2011; He et al., 2012; van Aerle et al., 2013).

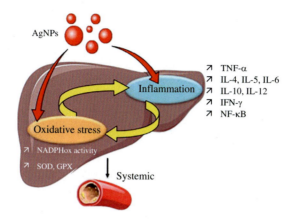

FIG. 28.3

Mechanism of silver nanoparticles toxicity.

Reprinted from with permission from Gaillet, S., Rouanet, J.-M. 2015. Silver nanoparticles: Their potential toxic effects after oral exposure and underlying mechanisms–A review. Food Chem. Toxicol. 77, 58–63. https://doi.org/10.1016/j.fct.2014.12.019 Elsevier.

28.5.4 Reproductive toxicity of silver nanoparticles

In rats, silver nanoparticles with an average size of 70 nm were administered orally at 25–200 mg/kg/day (Baki et al., 2014; Miresmaeili et al., 2013). The rats were administered every 12 h for 8 weeks (Miresmaeili et al., 2013). Silver nanoparticles affected the spermatogonia cells at 200 mg/kg/day. While it affected spermatid, spermatocytes and spermatozoa number at 50 mg/kg/day. Silver nanoparticles did not affected Sertoli cell number. When rats were administered with silver nanoparticles every 12 h for 45 days, it resulted in low-grade sperm quality and reduced Leydig cells number. The silver nanoparticles affected the levels of serum luteinizing hormone and testosterone. The findings from the study resulted in the harmful effect on spermatogenesis.

Prepubertal exposure of silver nanoparticles (60 nm) to rat male reproductive system at varying postnatal days resulted in alteration in epididymal sperm parameters (Mathias et al., 2015; Sleiman et al., 2013). In rats treated with silver nanoparticles, various histopathological changes were observed. In spermatic cells, silver nanoparticles resulted in decreased plasma membrane integrities, mitochondrial activity, and reduced acrosome (Sleiman et al., 2013). The studies demonstrated that prepubertal silver nanoparticle exposure cause obstruction in onset of puberty.

Rats orally administered with silver nanoparticles for 90 days at 20 μg/kg/day show accumulation of nanoparticles near basement membrane in testes (Thakur et al., 2014). In a study, PVP silver nanoparticles were administered orally to male rats at 50–200 mg/kg/day for 90 days (Lafuente et al., 2016). The study was conducted with Organization for Economic Co-operation and Development (OECD) guidelines 408, and epididymal sperm parameters were evaluated (Co-operation & Development, 2008). In the study, abnormal sperm count was observed, which was not dose-dependent. In the histopathology of epididymis and testis, no significant changes were observed. The oral administration of silver nanoparticles resulted only in a mild toxicity on sperm parameters.

In a study conducted on mice, silver nanoparticles were injected subcutaneously at 1 and 5 mg/kg/day for 8–21 postnatal days with five times a day (Austin et al., 2016; Zhang et al., 2015). The study concluded that at high doses of silver nanoparticles, an increase in the abnormal sperms while decreased sperm count in the epididymis were observed. Toxicity of silver nanoparticles was studied on rats, mice, and rabbits (Garcia et al., 2014). In the study, it was observed that silver nanoparticles caused the increased levels of testicular and serum testosterone. The researchers suggested that increased testicular and serum testosterone was due to alterations of Leydig cell function by silver nanoparticles. In another study, male mice were administered intravenously with silver nanoparticles at 0.5 and 1 mg/kg/day for 4 weeks (Han et al., 2016). The study evidenced a significant toxicity on testis. The toxicity includes atrophy of seminiferous tubules, lack of spermatids, and down regulation of biomarker genes. In another study, male rats were administered with 5 mg/kg of silver nanoparticles (Gromadzka-Ostrowska et al., 2012). The study demonstrated that silver nanoparticles do not have adverse effects on relative or absolute weights of the epididymis or testis. The study demonstrated that at 20 nm silver nanoparticles reduced epididymal sperm count while 200 nm silver nanoparticle caused histopathological changes in the testis of rats.

28.6 Conclusion

With the advancement of technologies, nanoparticles have attained a considerable attention in daily applications. As compared to various nanoparticles in use today, silver nanoparticles have enormous applications. With large-scale production and industrialization of silver nanoparticles, their abundance has been noticed in the environment recently. They have been detected in various environments and living organisms. Although they have useful applications, but they also produce toxic effects to the animals. Silver nanoparticles uptake is one of the first step for their entry in the organisms. After uptake, they are distributed in various organs of the organisms. After distribution to the organs, they cause toxicity to the organism. There are various parameters that affect nanoparticles' toxicity to the organisms. Silver nanoparticles cause cell toxicity by various mechanisms. The results from the toxicity studies conducted conclude that silver nanoparticles caused toxicity to animal models. However, to arrive at a conclusion, more comprehensive and organ-specific toxicity data are required.

References

Abd El-Ghany, W., 2019. Nanotechnology and its considerations in poultry field: an overview. J. Hell. Vet. Med. Soc. 70 (3), 1611–1616.

Ahamed, M., Karns, M., Goodson, M., Rowe, J., Hussain, S.M., Schlager, J.J., Hong, Y., 2008. DNA damage response to different surface chemistry of silver nanoparticles in mammalian cells. Toxicol. Appl. Pharmacol. 233 (3), 404–410.

Ahmadi, J., 2009. Application of different levels of silver nanoparticles in food on the performance and some blood parameters of broiler chickens. World Appl. Sci. J. 7 (1), 24–27.

Ahmadi, F., Branch, S., 2012. Impact of different levels of silver nanoparticles (Ag-NPs) on performance, oxidative enzymes and blood parameters in broiler chicks. Pakistan Vet. J. 32 (3), 325–328.

Ahmadi, J., Irani, M., Choobchian, M., 2009. Pathological study of intestine and liver in broiler chickens after treatment with different levels of silver nanoparticles. World Appl. Sci. J. 7 (S1), 28–32.

Akaighe, N., Depner, S.W., Banerjee, S., Sharma, V.K., Sohn, M., 2012. The effects of monovalent and divalent cations on the stability of silver nanoparticles formed from direct reduction of silver ions by Suwannee River humic acid/natural organic matter. Sci. Total Environ. 441, 277–289. https://doi.org/10.1016/j.scitotenv.2012.09.055.

Akter, M., Sikder, M.T., Rahman, M.M., Ullah, A.K.M.A., Hossain, K.F.B., Banik, S., et al., 2018. A systematic review on silver nanoparticles-induced cytotoxicity: physicochemical properties and perspectives. J. Adv. Res. 9, 1–16. https://doi.org/10.1016/j.jare.2017.10.008.

Ali, A.A., 2017. Use of nano-calciumand phosphors in broiler feeding. Egypt. Poult. Sci. J. 37 (2), 637–650.

Amerah, A., Ravindran, V., Lentle, R., Thomas, D., 2007. Influence of feed particle size and feed form on the performance, energy utilization, digestive tract development, and digesta parameters of broiler starters. Poult. Sci. 86 (12), 2615–2623.

Andi, M.A., Hashemi, M., Ahmadi, F., 2011. Effects of feed type with/without nanosil on cumulative performance, relative organ weight and some blood parameters of broilers. Global Veterinaria 7 (6), 605–609.

Arora, S., Jain, J., Rajwade, J., Paknikar, K., 2009. Interactions of silver nanoparticles with primary mouse fibroblasts and liver cells. Toxicol. Appl. Pharmacol. 236 (3), 310–318.

AshaRani, P.V., Mun, G.L.K., Hande, M.P., Valiyaveettil, S., 2009. Cytotoxicity and genotoxicity of silver nanoparticles in human cells. ACS Nano 3 (2), 279–290. https://doi.org/10.1021/nn800596w.

Austin, C.A., Umbreit, T.H., Brown, K.M., Barber, D.S., Dair, B.J., Francke-Carroll, S., et al., 2012. Distribution of silver nanoparticles in pregnant mice and developing embryos. Nanotoxicology 6 (8), 912–922.

Austin, C.A., Hinkley, G.K., Mishra, A.R., Zhang, Q., Umbreit, T.H., Betz, M.W., et al., 2016. Distribution and accumulation of 10 nm silver nanoparticles in maternal tissues and visceral yolk sac of pregnant mice, and a potential effect on embryo growth. Nanotoxicology 10 (6), 654–661. https://doi.org/10.3109/17435390.2015.1107143.

Awasthi, K.K., Awasthi, A., Kumar, N., Roy, P., Awasthi, K., John, P., 2013. Silver nanoparticle induced cytotoxicity, oxidative stress, and DNA damage in CHO cells. J. Nanopart. Res. 15 (9), 1898.

Baki, M.E., Miresmaili, S.M., Pourentezari, M., Amraii, E., Yousefi, V., Spenani, H.R., et al., 2014. Effects of silver nano-particles on sperm parameters, number of Leydig cells and sex hormones in rats. Iran. J. Reprod. Med. 12 (2), 139.

Beck, I., Hotowy, A., Sawosz, E., Grodzik, M., Wierzbicki, M., Kutwin, M., et al., 2015. Effect of silver nanoparticles and hydroxyproline, administered in ovo, on the development of blood vessels and cartilage collagen structure in chicken embryos. Arch. Anim. Nutr. 69 (1), 57–68.

Beer, C., Foldbjerg, R., Hayashi, Y., Sutherland, D.S., Autrup, H., 2012. Toxicity of silver nanoparticles-nanoparticle or silver ion? Toxicol. Lett. 208 (3), 286–292. https://doi.org/10.1016/j.toxlet.2011.11.002.

Bhanja, S.K., Hotowy, A., Mehra, M., Sawosz, E., Pineda, L., Vadalasetty, K.P., et al., 2015. In ovo administration of silver nanoparticles and/or amino acids influence metabolism and immune gene expression in chicken embryos. Int. J. Mol. Sci. 16 (5), 9484–9503.

Braydich-Stolle, L., Hussain, S., Schlager, J.J., Hofmann, M.-C., 2005. In vitro cytotoxicity of nanoparticles in mammalian germline stem cells. Toxicol. Sci. 88 (2), 412–419.

Cai, S., Wu, C., Gong, L., Song, T., Wu, H., Zhang, L., 2012. Effects of nano-selenium on performance, meat quality, immune function, oxidation resistance, and tissue selenium content in broilers. Poult. Sci. 91 (10), 2532–2539.

Calderón-Jiménez, B., Johnson, M.E., Montoro Bustos, A.R., Murphy, K.E., Winchester, M.R., Vega Baudrit, J.R., 2017. Silver nanoparticles: technological advances, societal impacts, and metrological challenges. Front. Chem. 5, 6.

Carlson, C., Hussain, S.M., Schrand, A.M., Braydich-Stolle, K., Hess, K.L., Jones, R.L., Schlager, J.J., 2008. Unique cellular interaction of silver nanoparticles: size-dependent generation of reactive oxygen species. J. Phys. Chem. B 112 (43), 13608–13619.

Castellini, C., Ruggeri, S., Mattioli, S., Bernardini, G., Macchioni, L., Moretti, E., Collodel, G., 2014. Long-term effects of silver nanoparticles on reproductive activity of rabbit buck. Syst. Biol. Reprod. Med. 60 (3), 143–150.

Catauro, M., Raucci, M., De Gaetano, F., Marotta, A., 2004. Antibacterial and bioactive silver-containing $Na_2O \cdot CaO \cdot 2SiO_2$ glass prepared by sol–gel method. J. Mater. Sci. Mater. Med. 15 (7), 831–837.

Chambers, B.A., Afrooz, A.N., Bae, S., Aich, N., Katz, L., Saleh, N.B., Kirisits, M.J., 2014. Effects of chloride and ionic strength on physical morphology, dissolution, and bacterial toxicity of silver nanoparticles. Environ. Sci. Technol. 48 (1), 761–769.

Chauke, N., Siebrits, F., 2012. Evaluation of silver nanoparticles as a possible coccidiostat in broiler production. S. Afri. J. Anim. Sci. 42 (5), 493–497.

Cheng, X., Zhang, W., Ji, Y., Meng, J., Guo, H., Liu, J., et al., 2013. Revealing silver cytotoxicity using Au nanorods/Ag shell nanostructures: disrupting cell membrane and causing apoptosis through oxidative damage. RSC Adv. 3 (7), 2296–2305.

Co-operation & Development, 2008. Repeated dose 28-day oral toxicity study in rodents. In: OECD Guideline for the Testing of Chemicals. Section 4: Health Effects. Co-operation & Development.

Crabtree, J.H., Burchette, R.J., Siddiqi, R.A., Huen, I.T., Hadnott, L.L., Fishman, A., 2003. The efficacy of silver-ion implanted catheters in reducing peritoneal dialysis-related infections. Perit. Dial. Int. 23 (4), 368–374.

Dalloul, R.A., Lillehoj, H.S., 2006. Poultry coccidiosis: recent advancements in control measures and vaccine development. Expert Rev. Vaccines 5 (1), 143–163.

De Gusseme, B., Hennebel, T., Christiaens, E., Saveyn, H., Verbeken, K., Fitts, J.P., et al., 2011. Virus disinfection in water by biogenic silver immobilized in polyvinylidene fluoride membranes. Water Res. 45 (4), 1856–1864.

Delay, M., Dolt, T., Woellhaf, A., Sembritzki, R., Frimmel, F.H., 2011. Interactions and stability of silver nanoparticles in the aqueous phase: influence of natural organic matter (NOM) and ionic strength. J. Chromatogr. A 1218 (27), 4206–4212. https://doi.org/10.1016/j.chroma.2011.02.074.

Durán, N., Durán, M., Souza, C.E.d., 2017. Silver and silver chloride nanoparticles and their anti-tick activity: a mini review. J. Braz. Chem. Soc. 28 (6), 927–932.

Elkloub, K., El Moustafa, M., Ghazalah, A., Rehan, A., 2015. Effect of dietary nanosilver on broiler performance. Int. J. Poult. Sci. 14 (3), 177.

Fancher, B.I., Jensen, L.S., 1989. Influence on performance of three to six-week-old broilers of varying dietary protein contents with supplementation of essential amino acid requirements. Poult. Sci. 68 (1), 113–123.

Firdhouse, M.J., Lalitha, P., 2015. Biosynthesis of silver nanoparticles and its applications. J. Nanotechnol. 2015, 1–19.

Fisinin, V., Miroshnikov, S., Sizova, E., Ushakov, A., Miroshnikova, E., 2018. Metal particles as trace-element sources: current state and future prospects. Worlds Poult. Sci. J. 74 (3), 523–540.

Flores-Rojas, G., López-Saucedo, F., Bucio, E., 2020. Gamma-irradiation applied in the synthesis of metallic and organic nanoparticles: A short review. Radiat. Phys. Chem. 169.

Foley, S., Crowley, C., Smaihi, M., Bonfils, C., Erlanger, B.F., Seta, P., Larroque, C., 2002. Cellular localisation of a water-soluble fullerene derivative. Biochem. Biophys. Res. Commun. 294 (1), 116–119.

Fuxiang, W., Huiying, R., Fenghua, Z., Jinquan, S., Jianyang, J., Wenli, L., 2008. Effects of nano-selenium on the immune functions and antioxidant abilities of broiler chickens. Chin. Agric. Sci. Bull. 2, 831–835.

Garcia, T.X., Costa, G.M., França, L.R., Hofmann, M.-C., 2014. Sub-acute intravenous administration of silver nanoparticles in male mice alters Leydig cell function and testosterone levels. Reprod. Toxicol. 45, 59–70.

Gliga, A.R., Skoglund, S., Wallinder, I.O., Fadeel, B., Karlsson, H.L., 2014. Size-dependent cytotoxicity of silver nanoparticles in human lung cells: the role of cellular uptake, agglomeration and Ag release. Part. Fibre Toxicol. 11 (1), 1–17.

Gonzales-Eguia, A., Fu, C.-M., Lu, F.-Y., Lien, T.-F., 2009. Effects of nanocopper on copper availability and nutrients digestibility, growth performance and serum traits of piglets. Livest. Sci. 126 (1–3), 122–129.

Gorth, D.J., Rand, D.M., Webster, T.J., 2011. Silver nanoparticle toxicity in Drosophila: size does matter. Int. J. Nanomedicine 6, 343.

Greulich, C., Kittler, S., Epple, M., Muhr, G., Köller, M., 2009. Studies on the biocompatibility and the interaction of silver nanoparticles with human mesenchymal stem cells (hMSCs). Langenbecks Arch. Surg. 394 (3), 495–502.

Griffin, S., Sarfraz, M., Farida, V., Nasim, M.J., Ebokaiwe, A.P., Keck, C.M., Jacob, C., 2018. No time to waste organic waste: Nanosizing converts remains of food processing into refined materials. J. Environ. Manage. 210, 114–121. https://doi.org/10.1016/j.jenvman.2017.12.084.

Gromadzka-Ostrowska, J., Dziendzikowska, K., Lankoff, A., Dobrzyńska, M., Instanes, C., Brunborg, G., et al., 2012. Silver nanoparticles effects on epididymal sperm in rats. Toxicol. Lett. 214 (3), 251–258. https://doi.org/10.1016/j.toxlet.2012.08.028.

Grosse, S., Evje, L., Syversen, T., 2013. Silver nanoparticle-induced cytotoxicity in rat brain endothelial cell culture. Toxicol. In Vitro 27 (1), 305–313.

Gurunathan, S., Lee, K.-J., Kalishwaralal, K., Sheikpranbabu, S., Vaidyanathan, R., Eom, S.H., 2009. Antiangiogenic properties of silver nanoparticles. Biomaterials 30 (31), 6341–6350.

Haase, A., Tentschert, J., Jungnickel, H., Graf, P., Mantion, A., Draude, F., et al., 2011. In: Toxicity of Silver Nanoparticles in Human Macrophages: Uptake, Intracellular Distribution and Cellular Responses. Paper presented at the Journal of Physics-Conference Series.

Haase, A., Rott, S., Mantion, A., Graf, P., Plendl, J., Thünemann, A.F., et al., 2012. Effects of silver nanoparticles on primary mixed neural cell cultures: uptake, oxidative stress and acute calcium responses. Toxicol. Sci. 126 (2), 457–468. https://doi.org/10.1093/toxsci/kfs003.

Han, J.W., Jeong, J.-K., Gurunathan, S., Choi, Y.-J., Das, J., Kwon, D.-N., et al., 2016. Male- and female-derived somatic and germ cell-specific toxicity of silver nanoparticles in mouse. Nanotoxicology 10 (3), 361–373.

Hassan, H., Samy, A., El-Sherbiny, A., Mohamed, M., Abd-Elsamee, M., 2016. Application of nano-dicalcium phosphate in broiler nutrition: performance and excreted calcium and phosphorus. Asian J. Anim. Vet. Adv. 11 (8), 477–483.

He, D., Dorantes-Aranda, J.J., Waite, T.D., 2012. Silver nanoparticle algae interactions: oxidative dissolution, reactive oxygen species generation and synergistic toxic effects. Environ. Sci. Technol. 46 (16), 8731–8738.

Jiang, X., Foldbjerg, R., Miclaus, T., Wang, L., Singh, R., Hayashi, Y., et al., 2013. Multi-platform genotoxicity analysis of silver nanoparticles in the model cell line CHO-K1. Toxicol. Lett. 222 (1), 55–63.

Kalishwaralal, K., Banumathi, E., Pandian, S.R.K., Deepak, V., Muniyandi, J., Eom, S.H., Gurunathan, S., 2009. Silver nanoparticles inhibit VEGF induced cell proliferation and migration in bovine retinal endothelial cells. Colloids Surf. B Biointerfaces 73 (1), 51–57.

Khan, Z., Hussain, J.I., Hashmi, A.A., Al-Thabaiti, S.A., 2017. Preparation and characterization of silver nanoparticles using aniline. Arab. J. Chem. 10, S1506–S1511. https://doi.org/10.1016/j.arabjc.2013.05.001.

Kim, Y.S., Kim, J.S., Cho, H.S., Rha, D.S., Kim, J.M., Park, J.D., et al., 2008. Twenty-eight-day oral toxicity, genotoxicity, and gender-related tissue distribution of silver nanoparticles in Sprague-Dawley rats. Inhal. Toxicol. 20 (6), 575–583. https://doi.org/10.1080/08958370701874663.

Kim, Y.S., Song, M.Y., Park, J.D., Song, K.S., Ryu, H.R., Chung, Y.H., et al., 2010. Subchronic oral toxicity of silver nanoparticles. Part. Fibre Toxicol. 7 (1), 20.

Kruszewski, M., Brzoska, K., Brunborg, G., Asare, N., Dobrzyńska, M., Dušinská, M., et al., 2011. Toxicity of silver nanomaterials in higher eukaryotes. In: Advances in Molecular Toxicology. Vol. 5. Elsevier, pp. 179–218.

Kühr, S., Schneider, S., Meisterjahn, B., Schlich, K., Hund-Rinke, K., Schlechtriem, C., 2018. Silver nanoparticles in sewage treatment plant effluents: chronic effects and accumulation of silver in the freshwater amphipod Hyalella azteca. Environ. Sci. Eur. 30 (1), 1–11.

Lafuente, D., Garcia, T., Blanco, J., Sánchez, D., Sirvent, J., Domingo, J., Gómez, M., 2016. Effects of oral exposure to silver nanoparticles on the sperm of rats. Reprod. Toxicol. 60, 133–139.

Lankveld, D.P., Oomen, A.G., Krystek, P., Neigh, A., Troost-de Jong, A., Noorlander, C., et al., 2010. The kinetics of the tissue distribution of silver nanoparticles of different sizes. Biomaterials 31 (32), 8350–8361.

Le Ouay, B., Stellacci, F., 2015. Antibacterial activity of silver nanoparticles: a surface science insight. Nano Today 10 (3), 339–354.

Lee, Y., Choi, J., Kim, P., Choi, K., Kim, S., Shon, W., Park, K., 2012. A transfer of silver nanoparticles from pregnant rat to offspring. Toxicol. Res. 28 (3), 139–141.

Liao, C., Li, Y., Tjong, S.C., 2019. Bactericidal and cytotoxic properties of silver nanoparticles. Int. J. Mol. Sci. 20 (2), 449.

Liu, W., Wu, Y., Wang, C., Li, H.C., Wang, T., Liao, C.Y., et al., 2010. Impact of silver nanoparticles on human cells: effect of particle size. Nanotoxicology 4 (3), 319–330.

Mathias, F.T., Romano, R.M., Kizys, M.M., Kasamatsu, T., Giannocco, G., Chiamolera, M.I., et al., 2015. Daily exposure to silver nanoparticles during prepubertal development decreases adult sperm and reproductive parameters. Nanotoxicology 9 (1), 64–70.

Maurer-Jones, M.A., Gunsolus, I.L., Murphy, C.J., Haynes, C.L., 2013. Toxicity of engineered nanoparticles in the environment. Anal. Chem. 85 (6), 3036–3049. https://doi.org/10.1021/ac303636s.

McShan, D., Ray, P.C., Yu, H., 2014. Molecular toxicity mechanism of nanosilver. J. Food Drug Anal. 22 (1), 116–127. https://doi.org/10.1016/j.jfda.2014.01.010.

Melnik, E., Demin, V., Demin, V., Gmoshinski, I., Tyshko, N., Tutelyan, V., 2013. Transfer of silver nanoparticles through the placenta and breast milk during in vivo experiments on rats. Acta Naturae 5 (3), 107–115.

Miethling-Graff, R., Rumpker, R., Richter, M., Verano-Braga, T., Kjeldsen, F., Brewer, J., et al., 2014. Exposure to silver nanoparticles induces size-and dose-dependent oxidative stress and cytotoxicity in human colon carcinoma cells. Toxicol. In Vitro 28 (7), 1280–1289.

Miresmaeili, S.M., Halvaei, I., Fesahat, F., Fallah, A., Nikonahad, N., Taherinejad, M., 2013. Evaluating the role of silver nanoparticles on acrosomal reaction and spermatogenic cells in rat. Iran. J. Reprod. Med. 11 (5), 423.

Mohamed, M., Hassan, H., Samy, A., Abd-Elsamee, M., El-Sherbiny, A., 2016. Carcass characteristics and bone measurements of broilers fed nano dicalcium phosphate containing diets. Asian J. Anim. Vet. Adv. 11 (8), 484–490.

Mohapatra, P., Swain, R., Mishra, S., Behera, T., Swain, P., Mishra, S., et al., 2014. Effects of dietary nano-selenium on tissue selenium deposition, antioxidant status and immune functions in layer chicks. Int. J. Pharm. 10 (3), 160–167.

Moustafa, M.T., 2017. Removal of pathogenic bacteria from wastewater using silver nanoparticles synthesized by two fungal species. Water Sci. 31 (2), 164–176.

Mroczek-Sosnowska, N., Łukasiewicz, M., Wnuk, A., Sawosz, E., Niemiec, J., Skot, A., et al., 2016. In ovo administration of copper nanoparticles and copper sulfate positively influences chicken performance. J. Sci. Food Agric. 96 (9), 3058–3062.

Mukherjee, T., Chakraborty, S., Biswas, A.A., Das, T.K., 2017. Bioremediation potential of arsenic by non-enzymatically biofabricated silver nanoparticles adhered to the mesoporous carbonized fungal cell surface of Aspergillus foetidus MTCC8876. J. Environ. Manage. 201, 435–446. https://doi.org/10.1016/j.jenvman.2017.06.030.

Murugan, K., Choonara, Y.E., Kumar, P., Bijukumar, D., du Toit, L.C., Pillay, V., 2015. Parameters and characteristics governing cellular internalization and trans-barrier trafficking of nanostructures. Int. J. Nanomedicine 10, 2191.

Nam, G., Purushothaman, B., Rangasamy, S., Song, J.M., 2016. Investigating the versatility of multifunctional silver nanoparticles: preparation and inspection of their potential as wound treatment agents. Int. Nano Lett. 6 (1), 51–63.

Nasretdinova, G.R., Fazleeva, R.R., Mukhitova, R.K., Nizameev, I.R., Kadirov, M.K., Ziganshina, A.Y., Yanilkin, V.V., 2015. Electrochemical synthesis of silver nanoparticles in solution. Electrochem. Commun. 50, 69–72. https://doi.org/10.1016/j.elecom.2014.11.016.

Natsuki, J., Natsuki, T., Hashimoto, Y., 2015. A review of silver nanoparticles: synthesis methods, properties and applications. Int. J. Mater. Sci. Appl 4 (5), 325–332.

Park, E.-J., Bae, E., Yi, J., Kim, Y., Choi, K., Lee, S.H., et al., 2010. Repeated-dose toxicity and inflammatory responses in mice by oral administration of silver nanoparticles. Environ. Toxicol. Pharmacol. 30 (2), 162–168.

Philbrook, N.A., Winn, L.M., Afrooz, A.N., Saleh, N.B., Walker, V.K., 2011. The effect of TiO2 and Ag nanoparticles on reproduction and development of Drosophila melanogaster and CD-1 mice. Toxicol. Appl. Pharmacol. 257 (3), 429–436.

Pineda, L., Chwalibog, A., Sawosz, E., Lauridsen, C., Engberg, R., Elnif, J., et al., 2012. Effect of silver nanoparticles on growth performance, metabolism and microbial profile of broiler chickens. Arch. Anim. Nutr. 66 (5), 416–429.

Pineda, L., Sawosz, E., Vadalasetty, K., Chwalibog, A., 2013. Effect of copper nanoparticles on metabolic rate and development of chicken embryos. Anim. Feed Sci. Technol. 186 (1–2), 125–129.

Pulit-Prociak, J., Banach, M., 2016. Silver nanoparticles–a material of the future…? Open Chem. 14 (1), 76–91.

Rai, M., Yadav, A., Gade, A., 2009. Silver nanoparticles as a new generation of antimicrobials. Biotechnol. Adv. 27 (1), 76–83.

Rai, M., Birla, S., Ingle, A.P., Gupta, I., Gade, A., Abd-Elsalam, K., et al., 2014a. Nanosilver: an inorganic nanoparticle with myriad potential applications. Nanotechnol. Rev. 3 (3), 281–309. https://doi.org/10.1515/ntrev-2014-0001.

Rai, M., Kon, K., Ingle, A., Duran, N., Galdiero, S., Galdiero, M., 2014b. Broad-spectrum bioactivities of silver nanoparticles: the emerging trends and future prospects. Appl. Microbiol. Biotechnol. 98 (5), 1951–1961. https://doi.org/10.1007/s00253-013-5473-x.

Roe, D., Karandikar, B., Bonn-Savage, N., Gibbins, B., Roullet, J.-B., 2008. Antimicrobial surface functionalization of plastic catheters by silver nanoparticles. J. Antimicrob. Chemother. 61 (4), 869–876.

Sahu, S.C., Zheng, J., Graham, L., Chen, L., Ihrie, J., Yourick, J.J., Sprando, R.L., 2014. Comparative cytotoxicity of nanosilver in human liver HepG2 and colon Caco2 cells in culture. J. Appl. Toxicol. 34 (11), 1155–1166.

Salary, J., Hemati Matin, H., Ghafari, K., Hajati, H., 2017. Effect of in ovo injection of calcium carbonate nanoparticles on bone post hatched characteristics and broiler chicken performance. Iran. J. Appl. Anim. Sci. 7 (4), 663–667.

Samanta, G., Mishra, S., Behura, N., Sahoo, G., Behera, K., Swain, R., et al., 2019. Studies on utilization of calcium phosphate Nano particles as source of phosphorus in broilers. Anim. Nutr. Feed. Technol. 19 (1), 77–88.

Sawosz, F., Pineda, L., Hotowy, A., Hyttel, P., Sawosz, E., Szmidt, M., et al., 2012a. Nano-nutrition of chicken embryos. The effect of silver nanoparticles and glutamine on molecular responses, and the morphology of pectoral muscle. Baltic J. Comp. Clin. Syst. Biol 2, 29–45.

Sawosz, F., Pineda, L., Hotowy, A., Jaworski, S., Prasek, M., 2012b. Influence of Ag nanoparticles, ATP and biocomplex of Ag nanoparticles with ATP on morphology of chicken embryo pectoral muscles. M. Brzozowski, R. Głogowski, A. Grzeszczak-Pytlak. Reproductive efficiency of mink females, selected for weaned litter size after first season of reproduction 5, 127.

Sawosz, F., Pineda, L., Hotowy, A., Jaworski, S., Prasek, M., Sawosz, E., Chwalibog, A., 2013. Nano-nutrition of chicken embryos. The effect of silver nanoparticles and ATP on expression of chosen genes involved in myogenesis. Arch. Anim. Nutr. 67 (5), 347–355.

Sawosz, E., Łukasiewicz, M., Łozicki, A., Sosnowska, M., Jaworski, S., Niemiec, J., et al., 2018. Effect of copper nanoparticles on the mineral content of tissues and droppings, and growth of chickens. Arch. Anim. Nutr. 72 (5), 396–406.

Scott, A., Vadalasetty, K.P., Sawosz, E., Łukasiewicz, M., Vadalasetty, R.K.P., Jaworski, S., Chwalibog, A., 2016. Effect of copper nanoparticles and copper sulphate on metabolic rate and development of broiler embryos. Anim. Feed Sci. Technol. 220, 151–158.

Seil, J.T., Webster, T.J., 2012. Antimicrobial applications of nanotechnology: methods and literature. Int. J. Nanomedicine 7, 2767–2781. https://doi.org/10.2147/IJN.S24805.

Selim, N., Radwan, N., Youssef, S., Eldin, T.S., Elwafa, S.A., 2015. Effect of inclusion inorganic, organic or nano selenium forms in broiler diets on: 2-physiological, immunological and toxicity statuses of broiler chicks. Int. J. Poult. Sci. 14 (3), 144.

Sharma, V.K., Filip, J., Zboril, R., Varma, R.S., 2015. Natural inorganic nanoparticles–formation, fate, and toxicity in the environment. Chem. Soc. Rev. 44 (23), 8410–8423.

Shirsat, S., Kadam, A., Mane, R.S., Jadhav, V.V., Zate, M.K., Naushad, M., Kim, K.H., 2016. Protective role of biogenic selenium nanoparticles in immunological and oxidative stress generated by enrofloxacin in broiler chicken. Dalton Trans. 45 (21), 8845–8853.

Singh, R.P., Ramarao, P., 2012. Cellular uptake, intracellular trafficking and cytotoxicity of silver nanoparticles. Toxicol. Lett. 213 (2), 249–259.

Sizova, E., Miroshnikov, S., Lebedev, S., Usha, B., Shabunin, S., 2020. Use of nanoscale metals in poultry diet as a mineral feed additive. Anim. Nutr. 6 (2), 185–191.

Skrivan, M., Skrivanova, V., Marounek, M., Tumova, E., Wolf, J., 2000. Influence of dietary fat source and copper supplementation on broiler performance, fatty acid profile of meat and depot fat, and on cholesterol content in meat. Br. Poultry Sci. 41 (5), 608–614.

Sleiman, H.K., Romano, R.M., Oliveira, C.A.D., Romano, M.A., 2013. Effects of prepubertal exposure to silver nanoparticles on reproductive parameters in adult male Wistar rats. J. Toxic. Environ. Health A 76 (17), 1023–1032.

Sriram, M.I., Kalishwaralal, K., Barathmanikanth, S., Gurunathani, S., 2012. Size-based cytotoxicity of silver nanoparticles in bovine retinal endothelial cells. Nanoscience Methods 1 (1), 56–77.

Stanley, D., Denman, S.E., Hughes, R.J., Geier, M.S., Crowley, T.M., Chen, H., et al., 2012. Intestinal microbiota associated with differential feed conversion efficiency in chickens. Appl. Microbiol. Biotechnol. 96 (5), 1361–1369.

Stoehr, L.C., Gonzalez, E., Stampfl, A., Casals, E., Duschl, A., Puntes, V., Oostingh, G.J., 2011. Shape matters: effects of silver nanospheres and wires on human alveolar epithelial cells. Part. Fibre Toxicol. 8 (1), 36.

Studnicka, A., Sawosz, E., Grodzik, M., Chwalibog, A., Balcerak, M., 2009. Influence of nanoparticles of silver/palladium alloy on chicken embryos' development. Ann. Warsaw. Agricult. Univ–SGGW, Anim. Sci. 46, 237–242.

Swain, P.S., Rajendran, D., Rao, S., Dominic, G., 2015. Preparation and effects of nano mineral particle feeding in livestock: A review. Vet. World 8 (7), 888.

Syafiuddin, A., Salmiati, Salim, M.R., Beng Hong Kueh, A., Hadibarata, T., Nur, H., 2017. A review of silver nanoparticles: research trends, global consumption, synthesis, properties, and future challenges. J. Chin. Chem. Soc. 64 (7), 732–756. https://doi.org/10.1002/jccs.201700067.

Thakur, M., Gupta, H., Singh, D., Mohanty, I.R., Maheswari, U., Vanage, G., Joshi, D., 2014. Histopathological and ultra structural effects of nanoparticles on rat testis following 90 days (chronic study) of repeated oral administration. J. Nanobiotechnol. 12 (1), 1–13.

Tian, J., Wong, K.K., Ho, C.M., Lok, C.N., Yu, W.Y., Che, C.M., et al., 2007. Topical delivery of silver nanoparticles promotes wound healing. ChemMedChem 2 (1), 129–136.

Tripathi, K., Tripathi, R., Seraji-Bozorgzad, N., 2017. Diplopia and Sjogren's disease: A rare case report. J. Neuroimmunol. 302, 7–9. https://doi.org/10.1016/j.jneuroim.2016.11.013.

van Aerle, R., Lange, A., Moorhouse, A., Paszkiewicz, K., Ball, K., Johnston, B.D., et al., 2013. Molecular mechanisms of toxicity of silver nanoparticles in zebrafish embryos. Environ. Sci. Technol. 47 (14), 8005–8014.

van den Bogaard, A.E., Stobberingh, E.E., 2000. Epidemiology of resistance to antibiotics: links between animals and humans. Int. J. Antimicrob. Agents 14 (4), 327–335.

van der Zande, M., Vandebriel, R.J., Van Doren, E., Kramer, E., Herrera Rivera, Z., Serrano-Rojero, C.S., et al., 2012. Distribution, elimination, and toxicity of silver nanoparticles and silver ions in rats after 28-day oral exposure. ACS Nano 6 (8), 7427–7442.

Vijayakumar, M., Balakrishnan, V., 2014. Evaluating the bioavailability of calcium phosphate nanoparticles as mineral supplement in broiler chicken. Indian J. Sci. Technol. 7, 1475–1480.

Vijayakumar, M., Balakrishnan, V., 2015. Assessment of calcium phosphate nanoparticles as safe mineral supplement for broiler chicken. Indian J. Sci. Technol. 8 (7), 608.

Wang, Y., 2009. Differential effects of sodium selenite and nano-se on growth performance, tissue se distribution, and glutathione peroxidase activity of avian broiler. Biol. Trace Elem. Res. 128 (2), 184–190.

Wang, J., Rahman, M.F., Duhart, H.M., Newport, G.D., Patterson, T.A., Murdock, R.C., et al., 2009. Expression changes of dopaminergic system-related genes in PC12 cells induced by manganese, silver, or copper nanoparticles. Neurotoxicology 30 (6), 926–933.

Wang, C., Wang, M., Ye, S., Tao, W., Du, Y., 2011. Effects of copper-loaded chitosan nanoparticles on growth and immunity in broilers. Poult. Sci. 90 (10), 2223–2228.

Wang, Z., Qu, G., Su, L., Wang, L., Yang, Z., Jiang, J., et al., 2013. Evaluation of the biological fate and the transport through biological barriers of nanosilver in mice. Curr. Pharm. Des. 19 (37), 6691–6697. https://doi.org/10.2174/13816128113199370012.

Wang, X., Ji, Z., Chang, C.H., Zhang, H., Wang, M., Liao, Y.P., et al., 2014. Use of coated silver nanoparticles to understand the relationship of particle dissolution and bioavailability to cell and lung toxicological potential. Small 10 (2), 385–398.

Wilson, W., 2007. A Nanotechnology Consumer Product Inventory. International Center for Scholars.

Wimmer, A., Kalinnik, A., Schuster, M., 2018. New insights into the formation of silver-based nanoparticles under natural and semi-natural conditions. Water Res. 141, 227–234.

Wu, J., Yu, C., Tan, Y., Hou, Z., Li, M., Shao, F., Lu, X., 2015. Effects of prenatal exposure to silver nanoparticles on spatial cognition and hippocampal neurodevelopment in rats. Environ. Res. 138, 67–73. https://doi.org/10.1016/j.envres.2015.01.022.

Yadav, K.K., Singh, J.K., Gupta, N., Kumar, V., 2017. A review of nanobioremediation technologies for environmental cleanup: A novel biological approach. J. Mater. Environ. Sci. 8 (2), 740–757.

Yang, Y., Li, H.Y., Lin, M.M., Xie, B., 2018. Batch studies for removing vanadium(V) and chromium(VI) from aqueous solution using anion exchange resin. *Vol. Part F5*. In: Minerals, Metals and Materials Series, pp. 291–298.

Yausheva, E., Miroshnikov, S., Kosyan, D., Sizova, E., 2016. Nanoparticles in combination with amino acids change productive and immunological indicators of broiler chicken. Agric. Biol. 51 (6), 912–920.

Yu, S.J., Yin, Y.G., Liu, J.F., 2013. Silver nanoparticles in the environment. Environ Sci Process Impacts 15 (1), 78–92. https://doi.org/10.1039/c2em30595j.

Yun'an Qing, L.C., Li, R., Liu, G., Zhang, Y., Tang, X., Wang, J., et al., 2018. Potential antibacterial mechanism of silver nanoparticles and the optimization of orthopedic implants by advanced modification technologies. Int. J. Nanomedicine 13, 3311.

Zhang, T., Wang, L., Chen, Q., Chen, C., 2014. Cytotoxic potential of silver nanoparticles. Yonsei Med. J. 55 (2), 283–291. https://doi.org/10.3349/ymj.2014.55.2.283.

Zhang, X.-F., Gurunathan, S., Kim, J.-H., 2015. Effects of silver nanoparticles on neonatal testis development in mice. Int. J. Nanomedicine 10, 6243.

Zhang, X.-F., Liu, Z.-G., Shen, W., Gurunathan, S., 2016. Silver nanoparticles: synthesis, characterization, properties, applications, and therapeutic approaches. Int. J. Mol. Sci. 17 (9), 1534.

Zhou, X., Wang, Y., 2011. Influence of dietary nano elemental selenium on growth performance, tissue selenium distribution, meat quality, and glutathione peroxidase activity in Guangxi yellow chicken. Poult. Sci. 90 (3), 680–686.

Ziemińska, E., Stafiej, A., Strużyńska, L., 2014. The role of the glutamatergic NMDA receptor in nanosilver-evoked neurotoxicity in primary cultures of cerebellar granule cells. Toxicology 315, 38–48.

Zivic, F., Grujovic, N., Mitrovic, S., Ahad, I.U., Brabazon, D., 2018. Characteristics and applications of silver nanoparticles. In: Commercialization of Nanotechnologies–A Case Study Approach. Springer, pp. 227–273.

CHAPTER 29

Nanosilver-based strategy to control zoonotic viral pathogens

Yasemin Budama-Kilinc[a], Burak Ozdemir[b], Tolga Zorlu[c], Bahar Gok[b], Ozan Baris Kurtur[b], and Zafer Ceylan[d]

[a]Department of Bioengineering, Yildiz Technical University, Istanbul, Turkey
[b]Yildiz Technical University, Graduate School of Natural and Applied Science, Department of Bioengineering, Istanbul, Turkey
[c]Department of Physical and Inorganic Chemistry and EMaS, Universitat Rovira I Virgili, Tarragona, Spain
[d]Van Yuzuncu Yıl University, Faculty of Tourism, Department of Gastronomy and Culinary Arts, Van, Turkey

29.1 Introduction

Zoonotic diseases are defined as infectious diseases that are transmitted from vertebrate animals to human beings and vice versa. These diseases can be of any pathogen origin, including bacteria, parasites, fungi, viruses, and prions (Wang and Crameri, 2014). Rabies, anthrax, tuberculosis, plague, yellow fever, and influenza can be counted as zoonotic parasitic diseases among the best-known zoonosis before the 20th century (İnci et al., 2018). Sixty percentage of the 300 infectious agents identified between 1940 and 2004 were classified as a zoonosis, but most of these infections belong to ignored diseases (Budke et al., 2006). More than 200 zoonoses have been identified up till now, and the epidemiological appearance and distribution of these diseases have occurred in a sporadic, endemic, epidemic, and pandemic forms that cause death among human beings, livestock, and wildlife (Narrod et al., 2012).

Zoonotic infections are also accepted as one of the most important causes of poverty because they cause great economic losses (Pendell et al., 2016). In the second half of the last century, many industrialized countries needed expensive infrastructure investments to control or eliminate zoonotic diseases. Additionally, the control of the diseases mentioned above has been accomplished through rigorous and coordinated interventions such as animal testing and segregation, feed bans, pets and mass vaccination in wildlife, health education, and milk pasteurization. Besides, the development of new antiviral agents has become a necessity, as various pathogenic viruses, including those that are the source of zoonotic disease, cause outbreaks and the mentioned viruses develop resistance to the known antiviral drugs.

Today, nanoscale materials emerged as new "antimicrobial agents" due to their high surface area/volume ratio and unique chemical and physical properties (Kim et al., 2007; Morones et al., 2005). In this respect, nanotechnology is defined as a field of applied science and advanced technology that uses the physicochemical properties of nanomaterials as a tool to produce nanoscale materials and control their sizes, surface areas, and shapes. Among these nanomaterials, metal-based ones have attracted great interest in medicine as they allow the development of antivirals that stop infections, especially by affecting viral pathogens during cell entry. Some studies reported that silver nanoparticles (Ag NPs) are much more effective than the other metal-based NPs due to their antiviral properties (Dung et al., 2019). In one of these studies, Ag NPs inhibited replication of human immunodeficiency virus 1 (HIV-1) and also did not indicate acute cytotoxicity on Hut/CCR5 cell line and human peripheral blood monocular cells (Alghrair et al., 2019). Apart from HIV-1, Ag NPs can also interact with other human viral diseases such as herpes simplex virus type 1 and type 2 (HVS-1 and HSV-2) (Baram-Pinto et al., 2009; Orłowski et al., 2018), hepatitis B (HBV) (Lu et al., 2008), respiratory syncytial virus (RSV) (Morris et al., 2019), H1N1 influenza (Li et al., 2016), H3N2 influenza (Xiang et al., 2013), human parainfluenza type 3 (Gaikwad et al., 2013), poliovirus (Huy et al., 2017), and adenovirus type 3 (Chen et al., 2013).

Zoonotic diseases are evaluated to be among the most major problems, which cause huge economic losses worldwide and seriously threaten both human and animal health. Therefore, in this chapter, we focused on the antiviral effects of Ag NPs, which have an important place among metal-based nanomaterials.

29.2 Production methods of silver nanoparticles

Various methods are used for the synthesis of Ag NPs. Some of them are laser ablation, gamma irradiation, electron irradiation, chemical reduction, photochemical methods, and microwave processing (Iravani et al., 2014). However, these methods are ideal for the use of Ag NPs in physicochemical and advanced nanotechnological studies. When it comes to fighting various infections or eliminating viral pathogens, there are basically two different synthesis methods: (i) chemical synthesis and (ii) green synthesis. For this reason, these two synthesis methods and the studies that were carried out in recent years are mentioned in the following sections in accordance with the subject of this chapter.

29.2.1 Chemical synthesis

Chemical synthesis is the method used to obtain not only for Ag NPs but also for producing all other metal-based NP sorts. In this method, different metal salts and reducing agents are reacted under certain synthesis conditions. In this way, the continuity of the colloidal stability of the NPs is ensured. One of the most commonly used metal salts to produce of Ag NPs is $AgNO_3$. The reason for this is that NO_3^- easily separates from the compound, and Ag^+ ions are released. The released Ag^+ ions become Ag^0, and as following the reaction, the nucleation phase and the growth of the particles occur. In this reaction, a reducing agent catalyses the releasing of Ag^+ ion from NO_3^-. The agent used for the reaction can be ascorbic acid, citric acid, or sodium borohydride. In general, although the chemical process of Ag NPs takes place in this way, the number of chemicals that are used or their combinations may vary in the reaction depending on the desired sizes or morphological shapes. For instance, even though the same chemicals are used in the synthesis of Ag NPs with different shapes such as spherical, rod-shaped, or triangular morphology; quantities, temperature, and even the order in which the chemicals are added to the reaction medium may vary. In some cases, procedures can be changed even to obtain particles of different sizes but with the same morphology. In this regard, Xing et al. (2019) synthesized spherical Ag NPs in different sizes (from 40 to 300 nm) (Xing et al., 2019). For this, the authors synthesized Ag seed (~23 nm) from $AgNO_3$ metal salt with the help of citric acid and NaCl, and then prepared a growth solution containing $AgNO_3$ and ammonia. In this sense, they revealed that spherical Ag NPs with different sizes could be obtained when they added different amounts of seed solution to growth solution.

Recently, even though there are different spherical Ag NP synthesis methods, the procedure is generally based on the same logic. By boiling, the water, whose kinetic energy is maximized, catalyses the separation of Ag^+ ion from NO_3 more easily, and with the help of various agents, these ions easily transform into NPs with spherical morphology in the reaction mixture. However, in some cases, more complex and exothermic methods are needed for the synthesis of Ag NPs with morphologies such as rod and triangle. For instance, Liu et al. (2020) synthesized rod-shaped Ag NPs in their study (Liu et al., 2020). For this, they took assistance from certain temperatures and different metals and compounds such as gold and $CuCl_2$. In addition, methods such as ultrasonication and centrifugation were used to obtain monodisperse rod-shaped Ag NPs. In the study, first, the gold seeds were mixed with cetyltrimethylammonium chloride (CTAC), and then, a certain amount of $CuCl_2$ was added to the solution at 60°C. The authors suggested that CTAC plays a key role in synthesis and controls both the growth and shape of the particles. Then, $AgNO_3$ solution was added drop by drop to this solution, and the synthesis was continued for another 10 min. Finally, ascorbic acid was also added drop by drop to the synthesis medium, and the reaction was continued for 6 h under the same conditions. The authors obtained monodisperse and pure rod-shaped Ag NPs after centrifugation and ultrasonication of the product. In another study, Hu et al. (2019) synthesized triangle-shaped Ag NPs (Hu et al., 2019). For this, the seed solution consisting of $AgNO_3$, citric acid, and sodium borohydride was added to the growth solution consisting of $AgNO_3$ and cetyltrimethylammonium bromide (CTAB). Sodium hydroxide (NaOH) was then added to the final solution, and the reaction was continued at 25°C. At the end of the study, the authors reported that the amounts of the chemicals mentioned above are very important to obtain monodisperse triangle-shaped Ag NPs.

Obtaining of Ag NPs by chemical synthesis method is mostly developed for some specific applications such as sensing and detection (Deng et al., 2019; Li et al., 2020;

Roh et al., 2019). The primary reason for this is that the particles obtained by this method often require advanced surface modifications for their use in antiviral and antibacterial applications. In one of these studies, Ag NPs were conjugated with the "FluPep" peptide, which included CVVVTAAA sequence, and the effects of the materials against influenza virus were examined (Alghrair et al., 2019). The results showed that Ag NP-FluPep conjugates had higher antiviral activity than free-peptides. In another study, successful results were obtained even though no surface modification was made to Ag NPs synthesized by chemical synthesis method (Dung et al., 2019). The authors investigated the antiviral effects of Ag NPs against African swine fever virus (ASFV) and reported that very low doses of Ag NPs (0.78 ppm) did not cause any cytotoxic effects on alveolar macrophage cells; moreover, they completely inhibited ASFV.

29.2.2 Green synthesis

Today, one of the other most used methods in the synthesis of Ag NPs is defined as green synthesis (Asgary et al., 2016). In this technique, similarly to chemical synthesis methods, using a precursor is mandatory. $AgNO_3$ precursor is usually selected for this process. Green synthesis has some advantages as compared to other synthesis methods: (i) it only needs precursor, (ii) it is environmentally friendly due to the use of natural compounds as a reducing agent, and (iii) the materials synthesized by these natural compounds also exhibit the action paths of the compounds.

While many plants and plant parts such as roots, leaves, fruits and even stems are used during the synthesis, some viruses and bacteria strains can also be selected for this process. Thus, green synthesis can also be seen as a part of a more complex system called biofabrication. Due to reducing agents such as ascorbic acid and citric acid that are commonly used in the chemical synthesis of Ag NPs are naturally found in most plants, the reaction mixture does not require any other different reducing agent. However, uniform sizes and shapes obtained by changing the quantitative parameters of the chemicals used during chemical synthesis are more difficult to present in green synthesis conditions. On the other hand, some medicinal properties that are added from the herbal ingredients to the surface of Ag NPs, which are ideal for antiviral and antibacterial applications and turn the particles into more powerful weapons in viral applications. In one study, Sharma et al. (2019) tested Ag NPs synthesized from three different plant sources such as *Andrographis paniculata*, *Phyllanthus niruri*, and *Tinospora cordifolia* against chikungunya virus (CHIKV) (Sharma et al., 2019). The authors synthesized the particles in dark conditions to prevent the photoactivation dynamics of $AgNO_3$, and equal amounts of $AgNO_3$ were added to the extracts obtained from each plant species. Then, the reaction was continued for 24h at room temperature. Thus, Ag NPs with different antiviral and physicochemical properties were obtained. Although all three plant species have antiviral properties, these properties are different from each other. The results revealed this situation, and the authors noted that Ag NPs synthesized with *A. paniculata* extract had a stronger effect against CHIKV than those synthesized with other extracts. In another study, the antiviral effects of Ag NPs synthesized via sea sponge (*Amphimedon* spp.) against Hepatitis C virus (HCV) were examined (Shady et al., 2020). Some organisms in the marine ecosystem such as *Amphimedon* spp. have very high bioactive natural compound contents. For this reason, the authors tried to obtain a novel material against HCV by combining these compounds of the sea sponge with nanosized silvers. For this aim, *Amphimedon* spp. extract was prepared, and this solution reacted with a certain amount of $AgNO_3$. The results showed that Ag NPs obtained by this method are promising materials against HCV.

Generally, the particle morphology that is obtained from the green synthesis methods using different biological sources, is spherical. On the other hand, not only size but also shape is one of the most important parameters to rely on a successful viral application with Ag NPs. The disadvantage that occurs in green synthesis methods is one of the biggest problems. Besides, the particle sizes which are obtained are generally far from uniform. Therefore, these problems pose an obstacle to the use of green synthesis methods. Nevertheless, recent studies suggested some approaches that allow controlling not only the particle size but also the particle shape. In one of these approaches, Logaranjan et al. (2016) reported that the size and shape of Ag NPs can be controlled using the extract of the *Aloe vera* plant (Logaranjan et al., 2016). In the study, the authors put certain amounts of $AgNO_3$ into the extract of *Aloe vera* and then applied microwave irradiation to this solution. The results showed that *Aloe vera* had a great effect on the shape and size of the particles, and Ag NPs with 5–50 nm sizes and octahedron shapes can be obtained. Microwave irradiation is preferred in many green synthesis methods because it gives high temperatures at short intervals (Jahan et al., 2019; Karthik et al., 2018; Mellinas et al., 2019). Therefore, the particles can be produced in more uniform sizes and different shapes.

29.3 Biological activities of silver nanoparticles

It is known that Ag NPs also have many important biological activities due to their unique chemical and physical properties (Zhang et al., 2016). One of the most impressive biological activities of these particles is antimicrobial effects (Chen et al., 2014; Rai et al., 2014). It is reported that Ag NPs can also be used in medical tools such as intravenous catheters, endotracheal tubes, dressings, bone cements, and dental fillings to prevent microbial infections (Prabhu and Poulose, 2012). For this reason, in the next subsections, we will discuss and emphasize their different biological properties such as antibacterial, antifungal, and antiviral effects.

29.3.1 Antibacterial effects

Ag NPs are one of the most used metallic NPs in the modern antimicrobial applications due to their broad bactericidal effects against both gram-negative and gram-positive bacteria strains (Shao et al., 2018). Previous studies indicated that there are several mechanisms for the bactericidal effects of Ag NPs. In one of them, the particles with a positive surface charge can adhere to the bacterial cell membrane, which is negatively charged by electrostatic interaction. In this way, the adhesion causes to change of the permeability of the bacterial cell membrane (Dakal et al., 2016). For instance, Abbaszadegan et al. (2015) synthesized positive, neutral, and negatively charged Ag NPs to evaluate the bactericidal role of the electric charge on the surface of Ag NPs against various gram-positive and gram-negative bacteria strains (Abbaszadegan et al., 2015). The results showed that positively charged Ag NPs had the highest bactericidal activity against all bacteria strains that were tested. On the other hand, negatively charged Ag NPs were reported as the least antibacterial activity.

The interaction of Ag NPs in the cell could be associated with DNA (Yang et al., 2009). The mentioned interaction can occur with sulfur and phosphorus in the DNA backbone, which causes problems in the DNA replication of bacteria. Therefore, the particles can cause harmful effects to various bacteria strains (Prabhu and Poulose, 2012). Ag NPs can also overcome bacterial resistance to antibiotics (Sondi and Salopek-Sondi, 2004). Qais et al. (2019) synthesized Ag NPs using aqueous extract of *Murraya koenigii* leaves (Qais et al., 2019). The bactericidal activities of MK-Ag NPs on *Escherichia coli* and methicillin-resistant *Staphylococcus aureus* (MRSA) producing extended-spectrum β-lactamase (ESβL-), which are multidrug resistant pathogens, were evaluated. Synthesized MK-Ag NPs were reported to effectively inhibit the growth of the pathogens. In another study, Sivakumar et al. (2020) synthesized Ag NPs using *Parthenium hysterophorus* leaf extract and examined the antibacterial activities of these NPs on *E. coli*, *Pseudomonas aeruginosa*, *Bacillus subtilis*, *S. aureus,* and *Enterococcus faecalis* (Sivakumar et al., 2020). The authors suggested that Ag NPs are effective materials against these bacteria strains.

29.3.2 Antifungal effects

Fungal infections are common in patients who have suppressed their immune system because of cancer chemotherapy and HIV infection. It is known that Ag NPs could play an important role as antifungal agents against different diseases caused by fungi. For instance, Mallmann et al. (2015) evaluated the antifungal activity of Ag NPs that were synthesized with a green approach against *Candida albicans* and *Candida tropicalis* (Mallmann et al., 2015). The results indicated that Ag NPs show similar antifungal activities with amphotericin B against *C. albicans* and *C. tropicalis*, and for this reason, the authors reported that Ag NPs may be an alternative to treat fungal infections.

Shah et al. (2020) synthesized Ag NPs from *Silybum* sp. and coded these particles according to different parts of the plant (Shah et al., 2020). The particles that were coded as NP1 (wild Ag NPs), NP2 (wild extract), NP3 (seed Ag NPs), and NP4 (seed extract) showed various biological activities such as antibacterial, antioxidant, antifungal, anti-inflammatory effects and also be reported that Ag NPs coded as NP1 were more effective on *C. albicans*, which is a multidrug-resistant human pathogen, than others. Bocate et al. (2019) evaluated antifungal activity of biogenic Ag NPs that were synthesized from fungi, and simvastatin (SIM), a semisynthetic drug, onto five *Aspergillus* species such as *A. flavus*, *A. nomius*, *A. parasiticus*, *A. ochraceus*, and *A. ochraceus* (Bocate et al., 2019). The authors reported that Ag NPs were more effective than SIM, and the combination showed an even more effective result on fungi. Khatoon et al. (2018) synthesized Ag and Au NPs by chemical method and investigated their antifungal activities (Khatoon et al., 2018). The authors reported that Ag NPs were more effective on *C. albicans* and *Saccharomyces cerevisiae* strains than Au NPs.

Various mechanisms related to antifungal effects of Ag NPs were investigated in different ways. It was reported that the small sizes of Ag NPs can cause to failure of fungal cell walls (Khatami et al., 2018). Another mechanism is serving of Ag NPs as a reservoir for releasing of Ag^+ ion (Holt and Bard, 2005). This mechanism causes to stop of the production of ATP and inhibition of DNA replication by ROS and hydroxy radicals. In one study, Musa et al. (2018) evaluated the antifungal activity of Ag NPs against *C. albicans* (Musa et al., 2018). The authors reported that *C. albicans* ICL1 (CaICL1) transcription and isocitrate lyase (ICL) protein expression were suppressed in the cells treated with Ag NPs compared to control.

29.3.3 Antiviral effects

A pathogenic virus that has emerged and developed resistance to known antiviral drugs causes epidemics and pandemics. Therefore, many studies have been carried out to develop new antiviral agents (Gurunathan et al., 2020). The occurrence of viral diseases depends on the binding of the virus with surface components of the cell membrane and its internalization to the host cells (Deshmukh et al., 2019). This uptake mechanism is shown in Fig. 29.1. Blocking the binding and internalization processes of the virus is the best strategy to develop new antiviral agents. Considering the mechanism of action of metal-based NPs on microorganisms, one of the strongest candidates among new antivirals is Ag NPs. Since Ag NPs have wide target-oriented attack ranges, they negate the resistance of microorganisms against these particles (Salleh et al., 2020).

29.4 Viral applications of silver nanoparticles
29.4.1 Paramyxoviruses

Paramyxoviridae that includes negative-sense single-stranded RNA viruses is one of the important virus families. The viruses belonging to the family, paramyxoviruses, can infect vertebrates like human. These viruses, as with all enveloped viruses, must fuse with the plasma membrane of a host cell for viral entry and infection. Paramyxoviruses contain two different glycoproteins in their membranes to infect and entry to the host cells. One of them, the attachment glycoprotein, is required for the binding of virion and the other one, fusion protein, is directly related to bind of viral and host cell membranes (Bossart et al., 2013). *Paramyxoviridae* has one of the

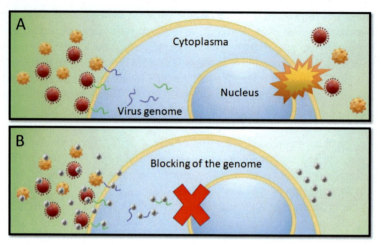

FIG. 29.1

Illustration of action mechanism of Ag NPs on viral materials. In normal circumstances, viral proteins that are located in virus surface bind to a receptor of a specific cell line (A). After this process, virus genomes that are released from virus capsid bind to related regions of DNA in the nucleus and make the copies of virus' itself. These copies, then, leave the host cell for the new infections. On the other hand, Ag NPs can block to be made of the copies of viruses by either binding to viral proteins in the surface of the virus or viral genome in the cytoplasma of the host cell (B).

widest biological diversity among the other virus families, and for this reason, it has many members. Newcastle virus, canine distemper virus (CDV), human parainfluenza virus type 3 (HPIV-3), and respiratory syncytial virus (RSV) are some of the important viruses that belong to this family.

In the past years, various studies were carried out with Ag NPs to control and eliminate the viruses in question. In one of them, Mehmood et al. (2020) examined to control of Newcastle virus (Mehmood et al., 2020). For this purpose, Ag NPs, which were synthesized via an aqueous extract of the buds of the *Syzygium aromaticum* plant, were investigated for their antiviral activity on the virus. As a result of this study, the authors reported that high doses of Ag NPs have antiviral activity against Newcastle virus as in vivo and in vitro. In another study, the use of Ag NPs in the treatment of distemper disease in dogs was investigated (Bogdanchikova et al., 2016). In the study, polyvinylpyrrolidone (PVP) was used as an agent to protect the stabilization of Ag NPs, and it was found that Ag@PVP NPs are very effective against acute CDV. Besides, it was observed that Ag@PVP NPs did not show any side effects. Gaikwad et al. (2013) aimed to determine the antiviral potential of Ag NPs which were synthesized via using Dryopteris fungus, on HPIV-3 and it was reported that the particles reduced viral infectivity, and block the interaction between virus and the host cell (Gaikwad et al., 2013). Furthermore, it was suggested that this blocking mechanism occurs depending on the size and surface charge of Ag NPs, and the particles that have smaller sizes was found to cause inhibition of infectivity of the viruses. In another study, Morris et al. (2019) investigated the antiviral and immunomodulatory effects of Ag NPs on RSV (Morris et al., 2019). The results showed that Ag NPs exhibited antiviral effects on RSV and produced proinflammatory cytokines in mouse lungs and epithelial cell lines. In another study related to RSV, the authors modified Ag NPs with curcumin and examined the antiviral effects of them (Yang et al., 2016). The results pointed out that Ag NPs modified with curcumin can be useful tools for combating with RSV and they have high antiviral activity against the RSV.

29.4.2 Coronaviruses

Coronaviruses (CoVs) cause mild or severe infections in vertebrates, like the paramyxoviruses we described in the previous section. However, the diseases caused by the virus are observed in a wide range from respiratory to systemic disorders. The virus also causes cold or pneumonia in human beings (Wang and Crameri, 2014b). Four species of CoVs, which are known as HCoV 229E, HCoV NL63, HCoV HKU1 and HCoV OC43, cause the symptoms like a cold in immunocompetent individuals (Ksiazek et al., 2003; Peiris et al., 2003). The two diseases that are caused by CoVs known as coronavirus associated with the severe acute respiratory syndrome (SARS-CoV) and coronavirus associated with the middle east respiratory syndrome (MERS-CoV) are zoonotic, and they can be deathful in human. As a very close example, the presence of pneumonia patients with unknown etiology in China was reported to The World Health Organization (WHO) by national authorities in 2019. The virus caused the symptoms that were officially identified as SARS-CoV-2, and the current outbreak of the coronavirus-associated acute respiratory disease was labeled as coronavirus disease 19 (COVID-19) (Tan et al., 2020).

SARS was identified first in Asian, North American, and European countries, and the cause of the disease still is not fully known. It is an infectious disease that progresses rapidly and can cause acute respiratory failure (ARDS). The findings of fever and lower respiratory tract infection, the contact with the diseased people and traveling to the specified geographical regions are defined as a suspect SARS case by WHO. As of 27 March 2003, 367 cases in Hong Kong and 1400 cases worldwide were identified as SARS, and 53 deaths in total were reported (Ge et al., 2013). The studies that were investigated by Ag NPs on SARS are quite less. Because of this scarcity, a study carried out in the past years can shed light on us for the future. In 2003, Rentz investigated the effects of oligodynamic Ag ions on the SARS virus, and as a result of the study, the author determined that oligodynamic Ag ions had antiviral potential and might be useful for the viral infections (Rentz et al., 2003).

MERS is also a human disease caused by coronavirus species. The virus that caused MERS was first described in a male who died of severe lung disease in June 2012 (Al-Osail and Al-Wazzah, 2017). The using of Ag NPs for diagnosis or treatment of MERS is a quite noval approach. Nevertheless, one study showed that detection of oligonucleotides of MERS-CoV virus using paper-based colorimetric DNA sensor by aggregation of Ag NPs is possible (Teengam et al., 2017). On the other hand, it was determined that the diagnostic system created has also some limitation for diagnosis of MERS-CoV.

29.4.3 Filoviruses

Filoviruses are viruses that belong to the *Filoviridae* family and have an elongated fibrous morphology. *Paramyxoviridae*, *Rhabdoviridae*, and *Bornaviridae* are the other main families of this class that are genetically closely related to filoviruses. Filoviruses are divided into two types: Marburg virus (MARV) and Ebola virus (EBOV). MARV contains a single species (Lake Victoria Marburgvirus). EBOV is divided into five types: Zaire, Ivory Coast, Sudan, Reston, and Bundibugyo (recently discovered in Uganda) (Towner et al., 2008). Some studies were carried out with Ag NPs on filoviruses. In one of these studies, Hentrich (2015) investigated the role of using nanosilver solution as a cure with immunization buffers in the fight against EBOV (Hentrich, 2015). As a result of the study, it was determined that the cure prepared with nanosilver solution was effective to combat with EBOV. In another study, Yen et al. (2015) aimed to develop lateral flow test based on Ag NPs for diagnosis of EBOV, dengue, and yellow fever diseases (Yen et al., 2015). As a result of the study, when the positive patient samples applied to the lateral flow test, it was observed that the positive and control lines in the relevant regions became clear.

29.4.4 West Nile virus

West Nile Virus (WNV), an endemic and neurotropic virus, is a type of virus belonging to the *Flaviviridae* family (Suthar et al., 2013). The transmission of the virus between organisms is provided by mosquitoes. It is known that more than a hundred mammal species, including many bat species, are susceptible to infections caused by the WNV virus. It has been determined that the risk of occurrence of this infection increases due to the fact that the habitats of animal and human populations are intertwined today (Jeffrey Root, 2013).

WNV was first seen in Uganda in 1937 in a patient with a high fever. Then, WNV, which was seen in North America about 62 years after this case, spreads to all of North America in a short time. Approximately 80% of the infections seen in human beings are subclinical. In addition, fever, neurological disorders, and deaths are seen in symptomatic infections in human beings (Suthar et al., 2013). The second major epidemic of WNV occurred in the United States in 2012, the number of cases in this epidemic is the second-highest number of cases seen in history (Arnold, 2012). In the same year, large numbers of WNV cases were reported in Europe. In this WNV epidemic within the borders of the European Union, the number of cases was determined as 224. In countries not within the borders of the European Union, the number of cases was determined to be 538 (Arnold, 2012). As a result of epidemiological studies, it is suspected the reproduction of the vectors that are effective in the spread of this disease in the recurrence of the WNV outbreak.

In the studies that cope with the WNV, substances that affect virus vectors rather than viruses have been used as targets. For instance, in one study, the larvicidal effect of Ag NPs synthesized using aqueous extract obtained from the leaves of the *Bauhinia variegata* plant was tested on *Anopheles subpictus*, *Aedes albopictus*, and *Culex tritaeniorhynchus* species, which are of WNV vectors (Govindarajan et al., 2016). As a result of the study, it was reported that Ag NPs have larvicidal effects. In another study, the larvicidal effect of Ag NPs, which were synthesized using essential oil of *Curcuma zedoaria*, on the larvae of *Culex quinquefasciatus* was investigated (Sutthanont et al., 2019). The results showed that the larvae of *C. quinquefasciatus*, which are sensitive and resistant to larvicidal effect, were killed by Ag NPs. Parthiban et al. (2019) synthesized Ag NPs with the aqueous extract obtained from the leaves of *Annona reticulata* plant (Parthiban et al., 2019). In the study, it was investigated that the larvicidal effect of the synthesized these Ag NPs on the larvae of the *Aedes aegypti* species. For this, the authors applied the different concentrations of Ag NPs to the larvae, and it was determined that 4.43 µg/mL dose of Ag NPs killed 50% of the larvae. Veerakumar et al. (2013) synthesized Ag NPs by using a green synthesis method (Veerakumar et al., 2013). The authors aimed to determine the effects of the Ag NPs on *C. quinquefasciatus*, *Anopheles stephensi* and *A. aegypti* larvae. As a result of the study, it was concluded that Ag NPs obtained by green synthesis could be an effective biopesticide that does not have any pollute on the environment.

29.4.5 Chikungunya virus

CHIKV, a member of the *Togaviridae* family and genus *Alphavirus*, was first detected in a patient with high fever in Tanzania in 1952 (Burt et al., 2012). This enzootic virus has been seen in Africa, parts of Asia, and islands in the Indian Ocean. CHIKV was reported to be protected among monkeys, small mammals, *Aedes* species on the African Continent. The research showed that monkeys are hosts in CHIKV transmission, while *Aedes* sp. is the major vector (Wang and Crameri, 2014). CHIKV outbreaks began to appear from the beginning of the 20th century, while it was not common in the 19th century. After decades, the reappearance of CHIKV is considered quite dramatic. CHIKV was seen for the first time in the Democratic Republic of Congo. Later, it was also seen in Indonesia, Kenya, Comoros Islands, India, and Maldives (Burt et al., 2012).

Jinu et al. (2018) synthesized Ag nanohybrids by using the aqueous extract of the leaves of *Cleistanthus collinus* and *Strychnos nux* and investigated the effects of these nanohybrids on *A. stephensi* and *A. aegypti* larvae (Jinu et al., 2018). As a result of the study, it was reported that the Ag nanohybrids had larvicidal properties. Kaushik et al. synthesized Ag NPs by using the extract from the *Carica papaya* (Kaushik et al., 2020). Antiviral characteristics of Ag NPs were investigated on CHIKV. As a result of the study, the authors revealed that Ag NPs increased the survival of CHIKV-infected cells by 14%. Another study related with CHIKV, Sharma et al. studied the antiviral effects of Ag NPs, which were synthesized by *A. paniculata*, *Phyllanthus nirurive*, and

T. cordifolia extracts (Sharma et al., 2019). As a result of the study, it was determined that Ag NPs synthesized by using *A. paniculata* increased the survival rate of infected cells by 25.69%.

29.4.6 Crimean-Congo hemorrhagic fever

Crimean-Congo hemorrhagic fever virus (CCHFV), a member of the *Bunyaviridae* family, is transmitted by ticks belonging to the *Hyalomma* genus (Bente et al., 2013). These ticks can infect many mammals by biting. In most infected animals, the disease can be asymptomatic. CCHFV can be transmitted to humans in two ways: (i) the first of which must be bitten by the virus that carries the disease and (ii) the second way is direct contact with the body fluid of infected animals. Human CCHFV infections are reported in Asia, Southeast Europe, the Middle East, and Africa since its first recognition in 1944. CCHFV infections generally cause a mild and nonspecific febrile illness. However, the serious hemorrhagic disease can be observed in some patients. The death rate of infection caused by the CCHFV was determined as the lowest 5% and the highest 30%. This change depends on the region where the epidemic occurred, the health system in that region, and the strain of the virus (Mertens et al., 2013).

Hyalomma marginatum and *Hyalomma anatolicum* tick species are known as the vector of CCHFV, which cause CCHFV. Acaricide or acaricidal substances are used to combat these tick species. For instance, the acaricidal effect of Ag NPs synthesized by using the extract of *Ocimum canum* plant on *H. marginatum* and *H. anatolicum* was investigated (Jayaseelan and Rahuman, 2012). As a result of the study, the authors reported that Ag NPs had acaricidal properties and caused a decrease in the number of ticks.

29.4.7 Zika virus

Zika virus (ZIKV) is a virus belonging to the flavivirus family and transmitted through mosquitoes (Kazmi et al., 2020). The infection progresses asymptomatically in approximately 80% of adults (Gorshkov et al., 2019). However, skin symptoms such as rash and itching are prominent in symptomatic cases (He et al., 2017).

The substances with larvicidal properties are also used to cope with the ZIKV just like WNV and CHIKV. In one study, Mahyoub et al. (2017) was investigated the larvicidal effect of Ag NPs synthesized by using the extract of the *Halodule uninervis* and tested on *A. aegypti* larvae (Mahyoub et al., 2017). The results showed that Ag NPs had a lethal potential on larvae.

29.5 Conclusion and future perspectives

Long-term use of drugs in the treatment of zoonotic viral pathogens causes increased side effects and the emergence of resistant viral species. Therefore, researchers are interested in alternative methods for treating infections caused by zoonotic viral pathogens and vector of these pathogens. One of the most important of alternatives is Ag NPs obtained as a result of nanotechnological methods. Ag NPs have caused to increase the lifetime of infected cells in many studies because of their antiviral effects. Because Ag NPs can be used as a biomedical weapon for zoonotic viral pathogens. Besides, they can be used in the treatment and control of future pandemics such as COVID-19 with the developing technology. However, the in vivo toxic potential of Ag NPs should be determined depending on the exposure time. Also, it is necessary to determine how to use Ag NPs without harming the environment and human beings, and the methods of their whose shelf life are exhausted in the future.

References

Abbaszadegan, A., Ghahramani, Y., Gholami, A., Hemmateenejad, B., Dorostkar, S., Nabavizadeh, M., Sharghi, H., 2015. The effect of charge at the surface of silver nanoparticles on antimicrobial activity against gram-positive and gram-negative bacteria: a preliminary study. J. Nanomater. 2015. https://doi.org/10.1155/2015/720654.

Alghrair, Z.K., Fernig, D.G., Ebrahimi, B., 2019. Enhanced inhibition of influenza virus infection by peptide-noble-metal nanoparticle conjugates. Beilstein J. Nanotechnol. 10, 1038–1047.

Al-Osail, A.M., Al-Wazzah, M.J., 2017. The history and epidemiology of Middle East respiratory syndrome coronavirus. Multidiscip. Respir. Med. https://doi.org/10.1186/s40248-017-0101-8.

Arnold, C., 2012. West Nile virus bites back. Lancet Neurol. https://doi.org/10.1016/S1474-4422(12)70278-8.

Asgary, V., Shoari, A., Baghbani-Arani, F., Shandiz, S.A.S., Khosravy, M.S., Janani, A., Bigdeli, R., Bashar, R., Cohan, R.A., 2016. Green synthesis and evaluation of silver nanoparticles as adjuvant in rabies veterinary vaccine. Int. J. Nanomedicine 11, 3597–3605.

Baram-Pinto, D., Shukla, S., Perkas, N., Gedanken, A., Sarid, R., 2009. Inhibition of herpes simplex virus type 1 infection by silver nanoparticles capped with mercaptoethane sulfonate. Bioconjug. Chem. 20, 1497–1502.

Bente, D.A., Forrester, N.L., Watts, D.M., McAuley, A.J., Whitehouse, C.A., Bray, M., 2013. Crimean-Congo hemorrhagic fever: history, epidemiology, pathogenesis, clinical syndrome and genetic diversity. Antivir. Res. https://doi.org/10.1016/j.antiviral.2013.07.006.

Bocate, K.P., Reis, G.F., de Souza, P.C., Oliveira Junior, A.G., Durán, N., Nakazato, G., Furlaneto, M.C., de Almeida, R.S., Panagio, L.A., 2019. Antifungal activity of silver nanoparticles and simvastatin against toxigenic species of Aspergillus. Int. J. Food Microbiol. 291, 79–86.

Bogdanchikova, N., Vázquez-Muñoz, R., Huerta-Saquero, A., Pena-Jasso, A., Aguilar-Uzcanga, G., Picos-Díaz, P.L., Pestryakov, A., Burmistrov, V., Martynyuk, O., Luna-Vazquez-Gomez, R., Almanza, H., 2016. Silver nanoparticles composition for treatment of distemper in dogs. Int. J. Nanotechnol. 13, 227–237.

Bossart, K.N., Fusco, D.L., Broder, C.C., 2013. Paramyxovirus entry. Adv. Exp. Med. Biol. 790, 95–127. https://doi.org/10.1007/978-1-4614-7651-1_6.

Budke, C.M., Deplazes, P., Torgerson, P.R., 2006. Global socioeconomic impact of cystic echinococcosis. Emerg. Infect. Dis. 12, 296–303.

Burt, F.J., Rolph, M.S., Rulli, N.E., Mahalingam, S., Heise, M.T., 2012. Chikungunya: a re-emerging virus. Lancet 379, 662–671.

Chen, N., Zheng, Y., Yin, J., Li, X., Zheng, C., 2013. Inhibitory effects of silver nanoparticles against adenovirus type 3 in vitro. J. Virol. Methods 193, 470–477.

Chen, G., Lu, J., Lam, C., Yu, Y., 2014. A novel green synthesis approach for polymer nanocomposites decorated with silver nanoparticles and their antibacterial activity. Analyst 139, 5793–5799.

Dakal, T.C., Kumar, A., Majumdar, R.S., Yadav, V., 2016. Mechanistic basis of antimicrobial actions of silver nanoparticles. Front. Microbiol. 7. https://doi.org/10.3389/fmicb.2016.01831.

Deng, H., Wang, B., Wu, M., Deng, B., Xie, L., Guo, Y., 2019. Rapidly colorimetric detection of caffeine in beverages by silver nanoparticle sensors coupled with magnetic molecularly imprinted polymeric microspheres. Int. J. Food Sci. Technol. 54, 202–211.

Deshmukh, S.P., Patil, S.M., Mullani, S.B., Delekar, S.D., 2019. Silver nanoparticles as an effective disinfectant: a review. Mater. Sci. Eng. C. https://doi.org/10.1016/j.msec.2018.12.102.

Dung, T.T.N., Nam, V.N., Nhan, T.T., Ngoc, T.T.B., Minh, L.Q., Nga, B.T.T., Phan Le, V., Quang, D.V., 2019. Silver nanoparticles as potential antiviral agents against African swine fever virus. Mater. Res. Express 6. https://doi.org/10.1088/2053-1591/ab6ad8.

Gaikwad, S., Ingle, A., Gade, A., Rai, M., Falanga, A., Incoronato, N., Russo, L., Galdiero, S., Galdiero, M., 2013. Antiviral activity of mycosynthesized silver nanoparticles against herpes simplex virus and human parainfluenza virus type 3. Int. J. Nanomedicine 8, 4303–4314.

Ge, X.Y., Li, J.L., Yang, X.L., Chmura, A.A., Zhu, G., Epstein, J.H., Mazet, J.K., Hu, B., Zhang, W., Peng, C., Zhang, Y.J., Luo, C.M., Tan, B., Wang, N., Zhu, Y., Crameri, G., Zhang, S.Y., Wang, L.F., Daszak, P., Shi, Z.L., 2013. Isolation and characterization of a bat SARS-like coronavirus that uses the ACE2 receptor. Nature 503, 535–538.

Gorshkov, K., Shiryaev, S.A., Fertel, S., Lin, Y.W., Huang, C.T., Pinto, A., Farhy, C., Strongin, A.Y., Zheng, W., Terskikh, A.V., 2019. Zika virus: origins, pathological action, and treatment strategies. Front. Microbiol. https://doi.org/10.3389/fmicb.2018.03252.

Govindarajan, M., Rajeswary, M., Veerakumar, K., Muthukumaran, U., Hoti, S.L., Mehlhorn, H., Barnard, D.R., Benelli, G., 2016. Novel synthesis of silver nanoparticles using *Bauhinia variegata*: a recent eco-friendly approach for mosquito control. Parasitol. Res. 115, 723–733.

Gurunathan, S., Qasim, M., Choi, Y., Do, J.T., Park, C., Hong, K., Kim, J.-H., Song, H., 2020. Antiviral potential of nanoparticles—can nanoparticles fight against coronaviruses? Nano 10, 1645. https://doi.org/10.3390/nano10091645.

He, A., Brasil, P., Siqueira, A.M., Calvet, G.A., Kwatra, S.G., 2017. The emerging zika virus threat: a guide for dermatologists. Am. J. Clin. Dermatol. https://doi.org/10.1007/s40257-016-0243-z.

Hentrich, M., 2015. Curing Ebola: resistant strain injections with nano silver. SSRN Electron. J. https://doi.org/10.2139/ssrn.2549306.

Holt, K.B., Bard, A.J., 2005. Interaction of silver(I) ions with the respiratory chain of *Escherichia coli*: an electrochemical and scanning electrochemical microscopy study of the antimicrobial mechanism of micromolar Ag. Biochemistry 44, 13214–13223.

Hu, G., Zhang, W., Zhong, Y., Liang, G., Chen, Q., Zhang, W., 2019. The morphology control on the preparation of silver nano-triangles. Curr. Appl. Phys. 19, 1187–1194.

Huy, T.Q., Hien Thanh, N.T., Thuy, N.T., Chung, P.V., Hung, P.N., Le, A.T., Hong Hanh, N.T., 2017. Cytotoxicity and antiviral activity of electrochemical—synthesized silver nanoparticles against poliovirus. J. Virol. Methods 241, 52–57.

İnci, A., Doğanay, M., Özdarendeli, A., Düzlü, Ö., Yıldırım, A., 2018. Overview of zoonotic diseases in Turkey: the one health concept and future threats. Turkiye Parazitol. Derg. https://doi.org/10.5152/tpd.2018.5701.

Iravani, S., Korbekandi, H., Mirmohammadi, S.V., Zolfaghari, B., 2014. Synthesis of silver nanoparticles: chemical, physical and biological methods. Res. Pharm. Sci. 9 (6), 385.

Jahan, I., Erci, F., Isildak, I., 2019. Microwave-assisted green synthesis of non-cytotoxic silver nanoparticles using the aqueous extract of *Rosa santana* (rose) petals and their antimicrobial activity. Anal. Lett. 52, 1860–1873.

Jayaseelan, C., Rahuman, A.A., 2012. Acaricidal efficacy of synthesized silver nanoparticles using aqueous leaf extract of *Ocimum canum* against *Hyalomma anatolicum* anatolicum and *Hyalomma marginatum* isaaci (Acari: Ixodidae). Parasitol. Res. 111, 1369–1378.

Jeffrey Root, J., 2013. West Nile virus associations in wild mammals: a synthesis. Arch. Virol. https://doi.org/10.1007/s00705-012-1516-3.

Jinu, U., Rajakumaran, S., Senthil-Nathan, S., Geetha, N., Venkatachalam, P., 2018. Potential larvicidal activity of silver nanohybrids synthesized using leaf extracts of *Cleistanthus collinus* (Roxb.) Benth. Ex Hook. F. and Strychnos nux-vomica L. nux-vomica against dengue, Chikungunya and Zika vectors. Physiol. Mol. Plant Pathol. 101, 163–171.

Karthik, K., Shashank, M., Revathi, V., Tatarchuk, T., 2018. Facile microwave-assisted green synthesis of NiO nanoparticles from *Andrographis paniculata* leaf extract and evaluation of their photocatalytic and anticancer activities. Mol. Cryst. Liq. Cryst. 673, 70–80. https://doi.org/10.1080/15421406.2019.1578495.

Kaushik, S., Jangra, G., Kundu, V., Yadav, J.P., Kaushik, S., 2020. Anti-viral activity of Zingiber officinale (ginger) ingredients against the Chikungunya virus. Virus Dis. 31, 270.

Kazmi, S.S., Ali, W., Bibi, N., Nouroz, F., 2020. A review on Zika virus outbreak, epidemiology, transmission and infection dynamics. J. Biol. Res. https://doi.org/10.1186/s40709-020-00115-4.

Khatami, M., Sharifi, I., Nobre, M.A.L., Zafarnia, N., Aflatoonian, M.R., 2018. Waste-grass-mediated green synthesis of silver nanoparticles and evaluation of their anticancer, antifungal and antibacterial activity. Green Chem. Lett. Rev. 11, 125–134.

Khatoon, U.T., Rao, G.V.S.N., Mohan, M.K., Ramanaviciene, A., Ramanavicius, A., 2018. Comparative study of antifungal activity of silver and gold nanoparticles synthesized by facile chemical approach. J. Environ. Chem. Eng. 6, 5837–5844.

Kim, J.S., Kuk, E., Yu, K.N., Kim, J.H., Park, S.J., Lee, H.J., Kim, S.H., Park, Y.K., Park, Y.H., Hwang, C.Y., Kim, Y.K., Lee, Y.S., Jeong, D.H., Cho, M.H., 2007. Antimicrobial effects of silver nanoparticles. Nanomed. Nanotechnol. Biol. Med. 3, 95–101.

Ksiazek, T.G., Erdman, D., Goldsmith, C.S., Zaki, S.R., Peret, T., Emery, S., Tong, S., Urbani, C., Comer, J.A., Lim, W., Rollin, P.E., Dowell, S.F., Ling, A.-E., Humphrey, C.D., Shieh, W.-J., Guarner, J., Paddock, C.D., Rota, P., Fields, B., DeRisi, J., Yang, J.-Y., Cox, N., Hughes, J.M., LeDuc, J.W., Bellini, W.J., Anderson, L.J., 2003. A novel coronavirus associated with severe acute respiratory syndrome. N. Engl. J. Med. 348, 1953–1966.

Li, Y., Lin, Z., Zhao, M., Xu, T., Wang, C., Hua, L., Wang, H., Xia, H., Zhu, B., 2016. Silver nanoparticle-based codelivery of oseltamivir to inhibit the activity of the h1n1 influenza virus through ROS-mediated signaling pathways. ACS Appl. Mater. Interfaces 8, 24385–24393.

Li, Z., Wang, Z., Khan, J., LaGasse, M.K., Suslick, K.S., 2020. Ultrasensitive monitoring of museum airborne pollutants using a silver nanoparticle sensor array. ACS Sensors. https://doi.org/10.1021/acssensors.0c00583.

Liu, S., Liang, L., Meng, L., Tian, X., Zhang, Z., Yu, Y., Lan, Z., Wu, J., Zhang, J., Gao, P., 2020. Synergy of plasmonic silver nanorod and water for enhanced planar perovskite photovoltaic devices. Sol. RRL 4, 1900231.

Logaranjan, K., Raiza, A.J., Gopinath, S.C.B., Chen, Y., Pandian, K., 2016. Shape- and size-controlled synthesis of silver nanoparticles using *Aloe vera* plant extract and their antimicrobial activity. Nanoscale Res. Lett. 11, 520.

Lu, L., Sun, R.W.Y., Chen, R., Hui, C.K., Ho, C.M., Luk, J.M., Lau, G.K.K., Che, C.M., 2008. Silver nanoparticles inhibit hepatitis B virus replication. Antivir. Ther. 13, 252–262.

Mahyoub, J.A., Aziz, A.T., Panneerselvam, C., Murugan, K., Roni, M., Trivedi, S., Nicoletti, M., Hawas, U.W., Shaher, F.M., Bamakhrama, M.A., Canale, A., Benelli, G., 2017. Seagrasses as sources of mosquito nano-larvicides? Toxicity and uptake of halodule uninervis-biofabricated silver nanoparticles in dengue and zika virus vector *Aedes aegypti*. J. Clust. Sci. 28, 565–580.

Mallmann, E.J.J., Cunha, F.A., Castro, B.N.M.F., Maciel, A.M., Menezes, E.A., Fechine, P.B.A., 2015. Atividade antifúngica de nanopartículas de prata obtidas por síntese verde. Rev. Inst. Med. Trop. Sao Paulo 57, 165–167.

Mehmood, Y., Farooq, U., Youusaf, H., 2020. Antiviral activity of green silver nanoparticles produced using aqueous buds extract of *Syzygium aromaticum*. Pak. J. Pharm. Sci. 33, 839–845.

Mellinas, C., Jiménez, A., Del Carmen Garrigós, M., 2019. Microwave-assisted green synthesis and antioxidant activity of selenium nanoparticles using *Theobroma cacao* l. bean shell extract. Molecules 24.

Mertens, M., Schmidt, K., Ozkul, A., Groschup, M.H., 2013. The impact of Crimean-Congo hemorrhagic fever virus on public health. Antivir. Res. https://doi.org/10.1016/j.antiviral.2013.02.007.

Morones, J.R., Elechiguerra, J.L., Camacho, A., Holt, K., Kouri, J.B., Ramírez, J.T., Yacaman, M.J., 2005. The bactericidal effect of silver nanoparticles. Nanotechnology 16, 2346–2353.

Morris, D., Ansar, M., Speshock, J., Ivanciuc, T., Qu, Y., Casola, A., Garofalo, R., 2019. Antiviral and immunomodulatory activity of silver nanoparticles in experimental rsv infection. Viruses 11.

Musa, S.F., Yeat, T.S., Kamal, L.Z.M., Tabana, Y.M., Ahmed, M.A., El Ouweini, A., Lim, V., Keong, L.C., Sandai, D., 2018. *Pleurotus sajorcaju* can be used to synthesize silver nanoparticles with antifungal activity against *Candida albicans*. J. Sci. Food Agric. 98, 1197–1207.

Narrod, C., Zinsstag, J., Tiongco, M., 2012. One health framework for estimating the economic costs of zoonotic diseases on society. EcoHealth 9, 150–162. https://doi.org/10.1007/s10393-012-0747-9.

Orłowski, P., Kowalczyk, A., Tomaszewska, E., Ranoszek-Soliwoda, K., Węgrzyn, A., Grzesiak, J., Celichowski, G., Grobelny, J., Eriksson, K., Krzyzowska, M., 2018. Antiviral activity of tannic acid modified silver nanoparticles: potential to activate immune response in herpes genitalis. Viruses 10. https://doi.org/10.3390/v10100524.

Parthiban, E., Manivannan, N., Ramanibai, R., Mathivanan, N., 2019. Green synthesis of silver-nanoparticles from *Annona reticulata* leaves aqueous extract and its mosquito larvicidal and anti-microbial activity on human pathogens. Biotechnol. Rep. 21, e00297.

Peiris, J.S.M., Lai, S.T., Poon, L.L.M., Guan, Y., Yam, L.Y.C., Lim, W., Nicholls, J., Yee, W.K.S., Yan, W.W., Cheung, M.T., Cheng, V.C.C., Chan, K.H., Tsang, D.N.C., Yung, R.W.H., Ng, T.K., Yuen, K.Y., 2003. Coronavirus as a possible cause of severe acute respiratory syndrome. Lancet 361, 1319–1325.

Pendell, D.L., Lusk, J.L., Marsh, T.L., Coble, K.H., Szmania, S.C., 2016. Economic assessment of zoonotic diseases: an illustrative study of Rift Valley fever in the United States. Transbound. Emerg. Dis. 63, 203–214.

Prabhu, S., Poulose, E.K., 2012. Silver nanoparticles: mechanism of antimicrobial action, synthesis, medical applications, and toxicity effects. Int. Nano Lett. 2, 1–10. https://doi.org/10.1186/2228-5326-2-32.

Qais, F.A., Shafiq, A., Khan, H.M., Husain, F.M., Khan, R.A., Alenazi, B., Alsalme, A., Ahmad, I., 2019. Antibacterial effect of silver nanoparticles synthesized using *Murraya koenigii* (L.) against multidrug-resistant pathogens. Bioinorg. Chem. Appl. 2019.

Rai, M., Kon, K., Ingle, A., Duran, N., Galdiero, S., Galdiero, M., 2014. Broad-spectrum bioactivities of silver nanoparticles: the emerging trends and future prospects. Appl. Microbiol. Biotechnol. https://doi.org/10.1007/s00253-013-5473-x.

Rentz, E., Cnmo, D.C., Rentz, E.J., 2003. Viral pathogens and severe acute respiratory syndrome: oligodynamic Ag + for direct immune intervention. J. Nutr. Environ. Med. 13, 109–118.

Roh, S.G., Robby, A.I., Phuong, P.T.M., In, I., Park, S.Y., 2019. Photoluminescence-tunable fluorescent carbon dots-deposited silver nanoparticle for detection and killing of bacteria. Mater. Sci. Eng. C 97, 613–623.

Salleh, A., Naomi, R., Utami, N.D., Mohammad, A.W., Mahmoudi, E., Mustafa, N., Fauzi, M.B., 2020. The potential of silver nanoparticles for antiviral and antibacterial applications: a mechanism of action. Nano 10, 1566.

Shady, N.H., Khattab, A.R., Ahmed, S., Liu, M., Quinn, R.J., Fouad, M.A., Kamel, M.S., Muhsinah, A.B., Krischke, M., Mueller, M.J., Abdelmohsen, U.R., 2020. Hepatitis c virus ns3 protease and helicase inhibitors from red sea sponge (Amphimedon) species in green synthesized silver nanoparticles assisted by in silico modeling and metabolic profiling. Int. J. Nanomedicine 15, 3377–3389.

Shah, M., Nawaz, S., Jan, H., Uddin, N., Ali, A., Anjum, S., Giglioli-Guivarc'h, N., Hano, C., Abbasi, B.H., 2020. Synthesis of bio-mediated silver nanoparticles from *Silybum marianum* and their biological and clinical activities. Mater. Sci. Eng. C 112, 110889.

Shao, Y., Wu, C., Wu, T., Yuan, C., Chen, S., Ding, T., Ye, X., Hu, Y., 2018. Green synthesis of sodium alginate-silver nanoparticles and their antibacterial activity. Int. J. Biol. Macromol. 111, 1281–1292.

Sharma, V., Kaushik, S., Pandit, P., Dhull, D., Yadav, J.P., Kaushik, S., 2019. Green synthesis of silver nanoparticles from medicinal plants and evaluation of their antiviral potential against chikungunya virus. Appl. Microbiol. Biotechnol. 103, 881–891.

Sivakumar, M., Surendar, S., Jayakumar, M., Seedevi, P., Sivasankar, P., Ravikumar, M., Anbazhagan, M., Murugan, T., Siddiqui, S.S., Loganathan, S., 2020. Parthenium hysterophorus mediated synthesis of silver nanoparticles and its evaluation of antibacterial and antineoplastic activity to combat liver cancer cells. J. Clust. Sci. 32, 1–11.

Sondi, I., Salopek-Sondi, B., 2004. Silver nanoparticles as antimicrobial agent: a case study on *E. coli* as a model for gram-negative bacteria. J. Colloid Interface Sci. 275, 177–182.

Suthar, M.S., Diamond, M.S., Gale, M., 2013. West Nile virus infection and immunity. Nat. Rev. Microbiol. https://doi.org/10.1038/nrmicro2950.

Sutthanont, N., Attrapadung, S., Nuchprayoon, S., 2019. Larvicidal activity of synthesized silver nanoparticles from *Curcuma zedoaria* essential oil against *Culex quinquefasciatus*. Insects 10. https://doi.org/10.3390/insects10010027.

Tan, W., Zhao, X., Ma, X., Wang, W., Niu, P., Xu, W., Gao, F., Wu, G., 2020. A novel coronavirus genome identified in a cluster of pneumonia cases—Wuhan, China 2019–2020. China CDC Wkly 2, 61–62.

Teengam, P., Siangproh, W., Tuantranont, A., Vilaivan, T., Chailapakul, O., Henry, C.S., 2017. Multiplex paper-based colorimetric DNA sensor using pyrrolidinyl peptide nucleic acid-induced AgNPs aggregation for detecting MERS-CoV, MTB, and HPV oligonucleotides. Anal. Chem. 89, 5428–5435.

Towner, J.S., Sealy, T.K., Khristova, M.L., Albariño, C.G., Conlan, S., Reeder, S.A., Quan, P.-L., Lipkin, W.I., Downing, R., Tappero, J.W., Okware, S., Lutwama, J., Bakamutumaho,

B., Kayiwa, J., Comer, J.A., Rollin, P.E., Ksiazek, T.G., Nichol, S.T., 2008. Newly discovered Ebola virus associated with hemorrhagic fever outbreak in Uganda. PLoS Pathog. 4, e1000212.

Veerakumar, K., Govindarajan, M., Rajeswary, M., 2013. Green synthesis of silver nanoparticles using *Sida acuta* (Malvaceae) leaf extract against Culex quinquefasciatus, *Anopheles stephensi*, and *Aedes aegypti* (Diptera: Culicidae). Parasitol. Res. 112, 4073–4085.

Wang, L.F., Crameri, G., 2014. Emerging zoonotic viral diseases. Rev. Sci. Technol. https://doi.org/10.20506/rst.33.2.2311.

Xiang, D., Zheng, C., Zheng, Y., Li, X., Yin, J., O'Conner, M., Marappan, M., Miao, Y., Xiang, B., Duan, W., Shigdar, S., Zhao, X., 2013. Inhibition of A/Human/Hubei/3/2005 (H3N2) influenza virus infection by silver nanoparticles in vitro and in vivo. Int. J. Nanomedicine 8, 4103.

Xing, L., Xiahou, Y., Zhang, P., Du, W., Xia, H., 2019. Size control synthesis of monodisperse, quasi-spherical silver nanoparticles to realize surface-enhanced raman scattering uniformity and reproducibility. ACS Appl. Mater. Interfaces 11, 17637–17646.

Yang, W., Shen, C., Ji, Q., An, H., Wang, J., Liu, Q., Zhang, Z., 2009. Food storage material silver nanoparticles interfere with DNA replication fidelity and bind with DNA. Nanotechnology 20. https://doi.org/10.1088/0957-4484/20/8/085102.

Yang, X.X., Li, C.M., Huang, C.Z., 2016. Curcumin modified silver nanoparticles for highly efficient inhibition of respiratory syncytial virus infection. Nanoscale 8, 3040–3048.

Yen, C.W., De Puig, H., Tam, J.O., Gómez-Márquez, J., Bosch, I., Hamad-Schifferli, K., Gehrke, L., 2015. Multicolored silver nanoparticles for multiplexed disease diagnostics: distinguishing dengue, yellow fever, and Ebola viruses. Lab Chip 15, 1638–1641.

Zhang, X.F., Liu, Z.G., Shen, W., Gurunathan, S., 2016. Silver nanoparticles: synthesis, characterization, properties, applications, and therapeutic approaches. Int. J. Mol. Sci. https://doi.org/10.3390/ijms17091534.

Index

Note: Page numbers followed by *f* indicate figures and *t* indicate tables.

A

Abscisic acid (ABA), 343–344
Acetamiprid, electrochemical detection of, 396–397
Acetamiprid-specific aptamer, 396–397
Acetylcholinesterase (AChE) inhibition, 383–385
Achaea janata larvae, AgNP-treated, 486, 487*f*
Activated sludge, 579–580
Activity-based protein profiling (ABPP), 208
Adenosine triphosphate (ATP), 331–334
Administration routes, of AgNPs, 659–660
Adsorption
    elimination of toxic pollutants, 413–414
        onto high surface area nanostructures, pesticides, 401
Adsorption NMs, 580–581
Advance oxidation process (AOPs), 623–624
Aedes aegypti, 477–482
Aflatoxin B1 (AFB1), 115–116
Ag-based nanohybrid fungicides, 192–193
Ag-doped metal ferrite nanoparticles (Ag-MFNPs), 600–601, 605*f*
    antibacterial activity test, 609–610
    application, 611–613*t*
    hydrothermal method, 607–609
    magnetic properties, 605–606
    optical properties, 607
    properties of, 603–604
    silver nanoparticles, 601–603
    structural properties, 605
    wastewater disinfection, 601
    wastewater treatment, 611–613*t*
        photodegradation of dyes, 610–615
Aggregation behavior, of AgNPs, 362–363
Ag-nanocomposites and Ag NPs
    for bacterial plant disease control, 461–463
    for fungal plant disease control, 457–461
    for insects' control, 457
    for viral plant disease control, 463
Ag-NanoZymes, 396–397
AgNP-based biocides, toxic effects, 193
AgNP-FluPep conjugates, 707–708
AgNPs. *See* Silver nanoparticles (AgNPs)
AgNPs/histidine-functionalized graphene quantum dots/graphene hybrid (Ag/His-GQD/G), 396–397
Ag@PVP NPs, 712–713
Agricultural activity, 453
Agricultural insect control. *See* Insects, control
Agricultural pest control, 453
Agri-food sectors, 3–4, 5*f*, 104
    contaminant, prevention and mitigation, 105–107
    food contaminant detection, 114–116
    nanosensors, 104
    silver nanosensors
        in agriculture, 107
        heavy metal contamination, 107–109
        heavy metal detection, 111–112
        metal ions, 109–111
        pesticide detection, 112–114
        significance of, 105
Agrotech, 201–202
Air decontamination, AgNPs, 80
Alternaria solani F10 (KT72914), 557
American Society for Testing and Materials (ASTM), 206
4-aminobenzenethiol (ABT) functionalized AgNPs, 387–388
    for carbendazim detection, 444
Amphimedon spp. extract, 708–709
Amplified fragment length polymorphism (AFLP), 340
Andrographis paniculata extract, 708–709
Anopheles stephensi, 477–482
Antagonists, of plant-parasitic nematodes, 531
Antibacterial activity
    Ag-MFNPs, 609–610, 616*f*
    AgNPs, 68–72
Antibacterial effect, of AgNPs, 709–710
    mechanism of action, 500
    nitrification activity, 498
    nonoxidative mechanism
        cell wall and membrane attachment, in bacteria, 501
        intracellular structural damage, 501–502
    oxidative-stress mechanism, 502–504
    against plant pathogenic bacteria, 499*t*
    against *Ralstonia solanacearum*, 498
    against *Xanthomonas oryzae* pv. *oryzae*, 498–500
Antibiofilm activity, AgNPs, 70–72
Antifeedant activity, for biogenic AgNPs, 476
Antifungal activity
    AgNPs, 72–73
    and mycotoxin control, 82–84, 83*t*
    in phytopathogens, 195–196, 196–197*t*
Antifungal agents
    mechanisms of, 18–22, 19*f*

723

Antifungal agents *(Continued)*
  nanoparticles, 17–18
  silver nanoparticles, 18
    biosynthesis, 25–26
    human fungal infections, 26–29
    mechanistic approach of, 18, 23f
    plant fungal infections, 29–32
Antifungal effect, of AgNPs, 710–711
  of AgNPs formulations against fungi, 562–563t
  diagrammatic representation, 505f
  inhibition zone reduction activity, 502–504
  mechanism of, 504–512
  on plant-parasitic nematodes, 506–507
  against plant pathogenic fungi, 502, 503t
Antimicrobial action
  food packaging, 129–130
Antimicrobial agents, 41–42, 157–165t, 272–273, 272f
Antimicrobial effect, of AgNPs, 497
  assessment methods, 500
  mechanism of, 673f
  mode of action, 497
  and synergy, 668–669
Antimicrobial food packaging, 128
Antimicrobial mechanism, 629
Antimicrobial packaging materials
  biodegradable, 166–168
  commercial marketplace, 169–173
  food quality maintenance, 168–169
  nonbiodegradable food packaging, 156–166
Antimicrobial peptides (AMPs), 21–22
Antimicrobial properties, of AgNPs, 537–538
Antimicrobial resistance (AMR), 39, 147
  to medical and veterinary drugs, 668
Antioxidant properties, of AgNPs, 537–538
Antioxidants, AgNPs, 271
Anti-plant pathogenic mechanisms, 463–464
Antiviral activity, AgNPs, 73–74
Antiviral effect, of AgNPs, 508f, 711
  efficiency, 507–508
  mechanism of, 508
Aphid pests, AgNP effects on, 475
Apoptosis, 463–464
Aptamer-based assays, 396–397
Aptamer-based nanoprobe, for malathion pesticide detection, 443–444
Aptamer conjugated AgNPs, for As(III) determination, 439–440
Aptasensor, 396–397
Arsenic (As) detection, 439–440
Ascorbate peroxidase (APX), 341–342
Ascorbic acid (AA) capped AgNPs, for Cr(VI) determination, 441–442

Atrazine determination, 444
Au@AgNPs based colorimetric sensor, for organophosphate pesticides, 392, 393f
Autocatalytic sensor, for Hg(II) detection, 438–439
Azinphosmethyl (AM) pesticides, 389–390
Azoles, 20

B

Bacterial AgNP synthesis, 267–268
Bacterial cells, 71f
Bacterial consortia, 584–586
Bacterial detection, AgNPs for
  electrochemical biosensors, 512
  fluorescence immunoassay, 513
  mycotoxin
    detection, 513
    management, 513–514
  surface-enhanced Raman spectroscopy, 512
Bacterial isolates, 581
Bacterial plant disease control, Ag NPs and Ag-nanocomposites for, 461–463
Barium molybdate (BaMoO$_4$), 634
Barley, germination of
  copper oxide nanoparticles, 291–292
  multiwalled carbon nanotubes, 289–290
  nickel oxide nanoparticles, 293–294
  silver nanoparticles, 285–289
  titanium dioxide nanoparticles, 290–291
Beneficial microbes, AgNPs
  negative impacts on
    abundance, 310
    activity of, 311–312
    structure of, 310–311
β-1,3-glucan, 20
Bioaerosols, 189
Biochemical processes, plant, 308–310
Biodegradable food packaging, 166–168
Biodegradable polymers, 133
Biodeterioration, 84–85
Biofumigation, 530–531
Biogenic silver nanoparticles, 41–42, 465
Biogenic synthesized NPs, 471
Biological activity, of AgNPs
  antibacterial effects, 709–710
  antifungal effects, 710–711
  antiviral effects, 711
Biological control, of plant-parasitic nematodes, 531
Biological synthesis method, 537–538
Bio-nanocomposite film, food packaging, 133
Bionanotechnology, 465
Biopesticides, 199

Biopolymers, in food packaging, 133
Biosafety regulations, 7–9
Biosecurity/disinfection, of farms, 671–672
Biosorption
  concentration effect on, 583
  kinetics studies, 582
  microbial consortia, 586–588
  nZnO and nAg, 584–586, 587t
  pH effect on, 582
Biosynthesized AgNPs (bAgNPs), 25–26
Bismuth ferrite (BFO) nanocomposite, 615
Bismuth oxyhalides, wastewater treatment, 638–639
β-lactam antibiotics, 672–673
Botanic fungicides, 199
Broiler chickens, nanoparticle feeds and parameters, 687t
Broiler industry, 686
Bronchus-associated lymphoid tissue (BALT), 73
Bruguiera cylindrica mediated AgNP synthesis, 478–479
Burn healing activity, of AgNPs, 667–668

C

Cadmium (Cd) detection, 441
Candida albicans, 26
Carbendazim detection, 387–388
Carbon dots-assisted dual readout sensor, for carbaryl, 391, 392f
Carbon fiber microelectrode (CFME), 115
Carboxymethylcellulose (CMC), 167
Catalytic degradation, AgNPs, 245
Cefaxone-conjugated *Oscillatoria limnetica* synthesized-AgNPs, 668
Centroceras clavulatum-synthesized silver nanoparticles, 479–480
Cetyltrimethylammonium bromide (CTAB), 318–320
Chemical-based assays, 385–390
Chemical fertilizers, 269
Chemically active species (CAS), 240
Chemical sensor, for pesticides and metal ions detection
  colorimetric-based, 435–436
  electrochemical-based, 437
  fluorescence sensing, 436–437
  surface-enhanced Raman spectroscopy based, 437–438
Chemical speciation, of AgNPs, 361–362
Chemical synthesis, 706–708
Chemiluminescent (CL) sensor, 113–114
Chlorinated pesticides, 442–443
Chlorination, of AgNPs, 366–367

Chloropyrifos (CP) degradation, 414f
Cholinergic syndrome, 383–385
Chromium (Cr) detection, 441–442
Citrate BioPure™ Silver, 553
Citrated-coated *vs.* PVP-coated AgNPs, 551–552
Cl⁻ ion mediated shape transformation, of AgNPs, 366–367
Coating, AgNP, 330–334, 332–333t
COD removal, by microbial consortia, 588–590
Colloidal stabilization, AgNPs
  aqueous phase transfer of, 42–44
  synthetic paradigm for, 40–41
Colorimetric-based chemical sensor, for pesticides and metal ions detection, 435–436
Colorimetric detection method, 431
Comet assay, 339
Commercial AgNP nanofungicides, 555t
Conduction band (CB), 607, 614–615
Constructive strategy, NPs, 266
Contaminated/infected water, 670–671
Controlled release, AgNPs
  as antimicrobial agents, 272–273, 272f
  as antioxidant, 271
  as nanofertilizers, 269–270
  as nanopesticides, 274–275
Conventional pest control strategies, 454
  (*vs.* nanotechnology based pest control, 454–455, 455f, *see also* Pest control, in nanotechnology)
Copper oxide (CuO)
  barley, germination of, 291–292
  nanopreparations, 284–285
Coprecipitation synthesis, 607–608
Coronavirus associated with the middle east respiratory syndrome (MERS-CoV), 713
Coronavirus associated with the severe acute respiratory syndrome (SARS-CoV), 713
Coronavirus disease 19 (COVID-19) pandemic, 713, 716–717
Coronaviruses (CoVs), 713–714
Crimean-Congo hemorrhagic fever virus (CCHFV) infection, 716
Crop plants, AgNPs
  beneficial microbes
    abundance, 310
    activity of, 311–312
    structure of, 310–311
  growth processes
    biochemical characteristics, 308–310
    physiological characteristics effect, 303–308
    phytotoxic effects of, 305–307t
    plant morphological characteristics, 303
    seed germination, 302

Index

Crop production, 495
Culex quinquefasciatus, 477–482
Cultural measures, of plant-parasitic nematodes, 530
Cyclen dithiocarbamate-functionalized silver nanoparticles (CN-DTC-AgNPs), 386–387, 387f
Cyclodextrin stabilized AgNPs/GO/GCE, for As(III) determination, 439–440
Cypermethrin colorimetric sensor, 388
Cystic hydatid disease, 662
Cyst nematodes, 529
Cytotoxicity, AgNPs, 334–340

D

Deltamethrin neem-coated silver nanoparticles, 663
Destructive strategy, NPs, 266
Detection of pesticides, 398–399t
  aptamer-based assays, 396–397
  chemical-based assays, 385–390
  enzyme-based assays, 390–392
  surface-enhanced Raman spectroscopy based sensing, 392–395
Diazinon detection, 443–444
2,7-Di chloroflourescin-di acetate (DCFH2-DA), 174f
Diclofopmethyl (DM), 303–308
Didecyldimethylammonium bromide (DDAB), 318–320
Dihydroethidium (DHE), 337
Dimethyldidodecylammonium bromide, 331–334
Disinfectants by-products (DBPs), 601
Ditylenchus angustus, 530
Ditylenchus destructor, 530
Ditylenchus dipsaci, 529
Dopamine dithiocarbamate functionalized silver nanoparticles (DDTC-AgNPs), 386–387
Drosophila melanogaster, AgNPs mode of action on, 484–486, 485f, 488
Dynamic light scattering (DLS), 320–324
Dynamic SERS (D-SERS), 393–394

E

Earthworm (*Eudrilus eugeniae*)-synthesized silver nanoparticles, 479–480
Ebola virus (EBOV), 714
Ecotoxicology, of AgNPs, 560–565
Ectoparasites (ticks), 662–663
Electrochemical-based chemical sensor, for pesticides and metal ions detection, 437
Electrochemical biosensors, for bacterial detection, 512
Electrochemiluminescence (ECL) aptasensor, 396–397

EN 1275, 202
EN 1650, 202
EN 12353, 202
Endocrine-disrupting chemicals (EDCs), 416–417
Endoparasites (helminths), 661–662
Endoplasmic reticulum (ER), 346
Endrin detection, 442–443
Environmental concern, of pesticides, 416–417
Environmental exposure, of AgNPs, 358–360, 359f
Environmental sectors, AgNPs, 6
Enzyme-based assays, 390–392
Ergosterol, 20
Essential minerals, NPs, 236–237
Ethylene (ET), 343–344
European Committee for Standardization (CEN), 202
European norms (EN), 202
Excystation process, 670–671
External polymeric substances (EPS), 72
Extracellular target sites, 464

F

Face-centered cubic (FCC), 606
Fascioliasis, 661–662
Feed conversion ratio (FCR), 686
Fe_3O_4 nanocomposites, 606
Ferredoxin-NADP reductase (FNR), 344–346
Ferrite nanoparticles (FNPs), 600, 603
Ferrites, wastewater treatment, 634–636
Filoviruses, 714
First-in-human (FIH) translational ecotoxicological interpretation, 568–569
Fluorescence based chemical sensor, for pesticides and metal ions detection, 436–437
Fluorescence-based sensor, 388
Fluorescence immunoassay, for bacterial detection, 513
Fluorescence probe, for Cr(VI) determination, 441–442
Fluorescence-quenching method, for Hg(II) detection, 439
FNPs. *See* Ferrite nanoparticles (FNPs)
Foodborne diseases (FBDs), 147
  for food spoilage, 148–150
Food contaminant detection, 114–116
Food insecurity, 549
Food packaging
  antimicrobial action, 129–130
  biodegradable, 166–168
  commercial marketplace, 169–173
  food quality maintenance, 168–169
  food safety, importance, 126–127

food systems, 136, 136–137t
mechanism of action, 173–175
nanotechnology, 127–128
nonbiodegradable, 156–166
overview, 125–126
safety aspects and regulations, 175–177
safety assessments, 138
silver as nanofiller, benefits, 128–129
silver-based nanomaterials
  antimicrobial effect of, 135t
  bio-nanocomposite film, 133
  edible films/nanocoatings, 133–134
  nondegradable nanocomposite film, 132
  polymer blend nanocomposites, 134
  preparation of, 131
silver nanomaterials, 156–173, 157–165t
Food processing, contaminant mycobiota in, 199–200
Food production, 201–202
Food quality, maintenance, 168–169
Food sectors, 104
Food systems, 136, 136–137t
Fourier-transform infrared spectroscopy (FITR), 220
F. oxysporum, AgNPs antifungal activity, 552
Freundlich equation, 584–586, 587t
Fumigant nematicides, disadvantages of, 530–531
Fumigation, 530–531
Fungal infections, 17
Fungal plant disease control, Ag NPs and Ag-nanocomposites for, 457–461
Fungal resistance to biocides, 197t
Fungicidal activity, of commercial Biopure-AgNPs, 553

G

Gene expression, 340–341
Genotoxicity, AgNPs, 334–340
Glassy carbon electrode (GCE)-based nanosensors, 110
Glutathione (GSH) functionalized AgNPs, for As(III) determination, 439–440
Glutathione peroxidase (GPX), 341–342
Glutathione reductase (GR), 342
Glyceraldehyde-3-phosphate dehydrogenase (GAPDH), 344–346
Graphene oxide sheet (RGO)–AgNPs composite synthesis, 433
Green malt extract, 285–287, 288f
Green nanotechnology, 265
Green synthesis, 265, 527–528, 534, 539, 658–659, 708–709
  of nanosilver, 471, 472f
  using bacteria, 267–268
  using plants, 268–269
Green-synthesized Ag-based compounds, for pesticide degradation, 418–420
Green synthetic AgNP, 556–560

H

Hatcheries, AgNP role in, 663–665
Hazard analysis and critical control point (HACCP), 150, 154
Hazardous effects of AgNPs, 312
Heavy metal
  contamination, 107–109
  detection, 111–112
Helicoverpa armigera, AgNPs
  larval growth, 476
  vs. silica nanoparticles activity, 474
Helminthiasis, 661
Helminths, 661–662
Hsp70 protein expression, 488
Human fungal infections, 26–29
Hydrothermal synthesis, Ag-MFNPs, 607–609
Hydroxyapatite nanoparticles feed, 688
Hyperbranched polyethyleneimine-protected silver nanocluster (hbPEI-Ag NC), 114

I

Immobilization efficiency of AgNPs, 241f
  catalytic degradation, 245
  chemical immobilization and adsorption, 243–244
  phytoremediation, 241–243, 242f
  pollutant-sensing properties, 245
Immune system, 665–666
Immunomodulatory effects, of AgNPs, 666, 666f
Inductively coupled plasma atomic emission spectroscopy (ICPAES), 320–324
Inductively coupled plasma mass spectrometry (ICP-MS), 320–324
Inductively coupled plasma optical emission spectrometry (ICP-OES), 580–581
Infectious bursa disease virus (IBDV) penetration, prevention of, 663–664
Infectious diseases
  mastitis, 669–670
  tuberculosis, 670
In ovo injection, with AgNPs, 664
Insect
  bioassays and nanosilver synthesis, 471, 472f
  biochemistry and physiology, AgNPs effect on, 484–488
  control, AgNPs

Insect *(Continued)*
    effects on insect vector management, 476–482
    mode of action, 484–488
    on postharvest insect management, 482–483
    nanosilver in insect control, 473–476
Insecticides, classes of, 383–385
Insects' control, Ag NPs and Ag-nanocomposites for, 457
Insect vector management, 476–482
Integrated pest management, 454
Inter simple sequence repeat (ISSR), 340
Intracellular target sites, 464
Invasive treatment method, for cystic hydatid disease, 662
Iodonitrotetrazolium chloride (INT), 272–273
Iodonitrotetrazolium formazan (INF), 272–273

K

Kinetic modeling, nZnO, 584–586

L

Label-free electrochemical aptasensor, 396–397
Langmuir equation, 584–586, 587t
Lead (Pb) detection, 440–441
Life-cycle AgNPs analysis, for mycotoxin control, 565–569
Lindane removal, 400–401
Linear low-density polyethylene (LLDPE), 166
Lipopolysaccharides (LPS), 69–70
Livestocks
    contaminated/infected water, 670–671
    drug side effects, 674–675
    nutrient deliveries, for fetuses/neonates, 674
    production, constraints in, 663
    sterilizing cages and houses, 671–672
Localized surface plasmon resonance (LSPR), 104
Lolium multiflorum, AgNPs uptake and effect on, 564
Lunar caustic, 67–68

M

Magnesium ferrite (MgFe$_2$O$_4$), 634–635
Magnetization Ag-MFNPs, 605–606
Malassezia furfur, 28
Malate dehydrogenase (MDH), 344–346
Mammalian cells, AgNPs in, 689
Manganese oxide nanoparticle-based nanofiber catalyst, 417
Marburg virus (MARV), 714
Mastitis-causing bacteria, antibiotic-resistance of, 669–670

Medium-density fiberboard (MDF), 220
Meloidogyne gramnicola, AgNP nematicide effects on, 506
Meloidogyne incognita
    AgNPs effect on, 539
    nematode populations, 506
Meloidogyne javanica, AgNP effects on, 507
2-Mercapto-5-nitrobenzimidazole capped silver nanoparticles (MNBZ-AgNPs), 388
3-Mercapto-1,2-propanediol (MP), 109–110
4-Mercaptopyridine (4-MPY) functionalized AgNPs, for As(III) determination, 439–440
Mercaptoundecanoic acid (MUA), 275
Mercury (Hg) detection, 438–439
MesoSilver®, 553
Metal enhanced fluorescence (MEF), 436–437
Metal ions detection, AgNPs based chemical sensors
    arsenic (As), 439–440
    cadmium (Cd), 441
    chromium (Cr), 441–442
    lead (Pb), 440–441
    mercury (Hg), 438–439
Metal ions, silver nanosensors, 109–111
Metallothionein (MT), 342
Metal nanoparticles, 1
Metal nanoparticles colorimetric assays, 511–512
Metal niobates, 639–640
Metal titanates, 637–638
Metal tungstates, 625
Metal vanadates, 625–626
Microbial biomass, 579–580
Microbial consortia
    COD, 588–590
    nanomaterials, 586–588
Micropollutants, handling of, 420
Middle east respiratory syndrome (MERS), 713–714
Mineralization
    of organic and inorganic pollutants, 408–409
    of pesticides, 397–401
Minimum inhibitory concentration (MIC), 27, 167, 551–552
Mitotic index (MI), 338–339
Mode of entry, 246–248, 247f
Modified mixed liquor medium (MMLM), 582
Molybdates, wastewater treatment, 633–634
Morphology, silver-doped compounds
    doping behavior, 630–631
    effect of, 641
Mosquito control, 480–482. *See also* Insect
Multidrug resistance (MDR), 148

Multidrug-resistant (MDR) microbes, 46, 47–50t
Multidrug-resistant (MDR) microorganisms, 82
Multiwalled carbon nanotubes (MWCNTs)
  barley, germination of, 289–290
  nanopreparations, 284
Musca domestica, AgNP treatment effects on, 480–482
Mycobacterium bovis, 670
Mycobacterium tuberculosis, 670
Mycobiota, food process, 199–200
Mycogenic-mediated silver nano-fungicides, 190–191
Mycotoxins, 4
  biological activities of, 549
  chronic exposure, 549–550
  detection, 513
  management, 513–514
  prevention of, 549–550
  silver nanoparticles application in
    control of, 554–556, 565–569
    detection using, 556
Mythimna separata, AgNPs larvicidal activity, 474–475

N

Nafion stabilized AgNPs, for Cr(VI) determination, 441–442
nAg biosorption, 584–586
Nanocomposite fabrics, AgNPs, 79–80
Nanocomposite film, in food packaging, 132
Nanoencapsulations, 455–456, 471–473
Nanofertilizers, 269–270, 357–358, 361, 533, 536–537
Nanogold particles, 417
Nanoherbal medicines, 661
Nanomaterial-based apta sensing, 115
Nanomaterials
  characterization, 581, 583–584
  food packaging, 154
  and percentage sorption, 582
Nanomedicine, 660–661
Nano Mission, 176
Nanoparticles, 657–658
  application in poultry and hatcheries, 663–665
  defined, 533
  description, 496
  fabrication by plant leaf extract
    activation phase, 419–420
    concluding phase, 419–420
    flavonoids, 418–419
    progressive phase, 419–420
    with reduction process, 418–419, 419f
  ferrite properties, 603–604
  metal, 533
  for plant defense, 495–496
  silver
    antioxidant and antimicrobial properties, 537–538
    biological synthesis method, 537–538
    green synthesis, 527–528, 534, 539
    as nanofertilizers, 536–537
    as nanopesticides, 534–536
    properties, 533–534
    synthesis methods, 658–659, 706–709
      and characterization techniques, 660f
      from plant extracts, 659
  synthesis
    constructive strategy, 266
    destructive strategy, 266
  types of, 533
Nanopesticides, 274–275, 361, 454–455, 534–536
Nanopreparations
  copper oxide (CuO), 284–285
  multiwalled carbon nanotubes, 284
  nickel oxide (NiO), 285
  silver nanopreparation (Ag), 285
  titanium dioxide (TiO_2), 284
Nanosensors, 104. *See also* Silver nanosensors
Nanosilver
  impregnated textiles, 358–360
  impregnated wound dressings and plasters, 360–361
  loaded medical bandages, 360–361
  synthesis and insect bioassays, 471, 472f
Nanosilver in agricultural insect control, 473–476
Nanotechnology, 265, 408–409
  agricultural applications, 495–496, 532–533
  defined, 706
  food packaging, 127–128
  mycotoxins management, 514
  in pest control, 458–459t
    vs. conventional pest control strategies, 454–455, 455f
    nanoencapsulations, 455–456
    nanopesticides, 454–455
Nanoveterinary application, 660–661
Natural AgNPs sources, 358
Necrosis, 463–464
Neem-mediated AgNP, 661
Nematicide effect, of AgNPs, 506
Nematicides, 527
Nematodes, 527
  plant-parasitic, 506–507 (*see also* Plant-parasitic nematodes)
Newcastle virus control, 712–713

Nickel oxide (NiO)
  barley, germination of, 293–294
  nanopreparations, 285
Niobates, wastewater treatment, 639–640
Nitrogen containing pesticides, 444
Noble metallic nanoparticles, 188
Nonbiodegradable food packaging, 156–166
Nonenzymatic assays. *See* Chemical-based assays
Noninvasive treatment method, for cystic hydatid disease, 662
Nonoxidative mechanisms, for AgNPs antibacterial action
  cell wall and membrane attachment, in bacteria, 501
  intracellular structural damage, 501–502
Nonwoven (NW)@polydopamine@AgNPs fabric, 393–394
No Observed Adverse Effect Levels (NOAELS), 561–564
Novel silver-based nanomaterials
  Ag-based nanohybrid fungicides, 192–193
  biocide disinfectant effectivity test, 202–207
  mycobiota in food processing, 199–200
  mycogenic-mediated silver nano-fungicides, 190–191
  phytopathogens
    antifungal activities, 195–196, 196–197t
    mycobiota in agriculture, 197–199
  proteomic-based fungicides, 207–208
  silver nanoparticle biocides, 188–190
  toxic effects, 193
nZnO biosorption, 584–586

O

Octamethoxy resorcin (4) arene tetrahydrazide (OMRTH)-functionalized AgNPs, 441
Ointments, AgNP-based, 667–668
Oligodynamic antimicrobial mechanism, AgNPs, 44–45, 45f
Olive leaves extract-based AgNPs, 667
Operational taxonomic units (OTUs), 310–311
Organization for Economic Cooperation and Development (OECD), 206–207
Organs associated, with AgNP toxicity, 691, 692f
Oriented strand board (OSB), 220
Oxidative dissolution, of AgNPs, 363–365
Ozonation technique, wastewater treatment, 635–636

P

Packaging food, AgNPs, 78–79
Paramyxoviridae, 711–712
Paraquat pesticide, electrochemical-based sensing of, 442–443
Particleboard (PB), 220
Pathogens detection and diagnosis, 511–512
PDOPA functionalized AgNPs, for Pb(II) and Cu(II) detection, 440–441
Peptide-based nanoprobe, for malathion pesticide detection, 443–444
Pericallia ricini, pungamoil-based Ag and Au NPs effects, 475–476
Peroxidase (POD), 331–334
Peroxidase-mimic tyrosine capped silver NanoZymes (Ag-NanoZyme), 396–397
Pest control, in nanotechnology, 458–459t
  Ag NPs and Ag-nanocomposites
    for bacterial plant disease control, 461–463
    for fungal plant disease control, 457–461
    for insects' control, 457
    for viral plant disease control, 463
  vs. conventional pest control strategies, 454–455, 455f
  nanoencapsulations, 455–456
  nanopesticides, 454–455
Pesticides
  adsorption onto high surface area nanostructures, 401
  AgNPs, 80, 81t
  banned pesticides, in India, 413
  classification of, 383, 384f, 429–430, 430f
  degradation
    with AgNPs, 413–414
    green-synthesized Ag-based compounds, 418–420
  detection of, 398–399t
    aptamer-based assays, 396–397
    chemical-based assays, 385–390
    enzyme-based assays, 390–392
    surface-enhanced Raman spectroscopy based sensing, 392–395
  detection, silver nanosensors, 112–114
  determination using AgNPs chemical sensors
    chlorinated pesticides, 442–443
    nitrogen containing pesticides, 444
    phosphorus pesticides, 443–444
  environmental concern of, 416–417
  exposure to, 383
  harmful effects of, 429–430
  migration, stages of, 408f
  mineralization of, 397–401
  organic pollutant-related incidents, in India, 409t
  photocatalytic degradation mechanism, 415, 416f
  physical and chemical assets of, 410–411t
  remediation by AgNPs, 412t
  sources of, 407

and water pollution, 407
Phosalone (PS) pesticides, 389–390
Phosphorus pesticides, 443–444
Photocatalytic activity, silver-doped ternary compounds, 643
Photocatalytic degradation mechanism, 614–615
Photocatalytic efficiency, of AgNPs, 414–415
Photochemical transformations, of AgNPs, 367–368
Photodegradation, Ag-MFNPs, 610–615
Photosystem (PS) II proteins, 344–346
Phoxim detection, 443–444
Physicochemical transformations, of AgNPs, 369f
    aggregation, 362–363
    chlorination, 366–367
    effect on agroecosystems, 368, 370–372t
    oxidative dissolution, 363–365
    photochemical transformation, 367–368
    speciation, 361–362
    sulfidation for toxicity mitigation, 365–366
Physiologically based pharmacokinetic (PBPK) model, 561–564
Phytodegradation, 242
Phytoextraction, 242
Phytohormone modulators, 248–249
Phytoimmobilization, 242
Phytopathogens
    antifungal activities, 195–196, 196–197t
    mycobiota in agriculture, 197–199
Phytoremediation, 241–243, 242f
    of pollutants, 241f
Phytostabilization, 242
Phytotoxic effects, plants, 305–307t
Phytotoxicity mechanisms
    biological media for plant growth, 318–329
    coating-dependent effects, 330–334, 332–333t
    cytotoxic and genotoxic effects, 334–340
    gene and protein expression, 340–346
    synthesis and stabilization, 318, 319f
Phytotoxicity of AgNPs, 301–308
Phytotoxicology, of AgNPs, 560–565
Phytovolatilization, 242
Placenta transfer, of AgNPs, 689–690
Plant biochemical, AgNPs, 245–252
    disease-suppression, 249–252
    mode of entry, 246–248, 247f
    physiology and phytohormone modulators, 248–249
Plant defense, nanoparticles for, 495–496
Plant disease management, 462t
    Ag NPs and Ag-nanocomposites
        for bacterial plant disease control, 461–463
        for fungal plant disease control, 457–461
        for viral plant disease control, 463
    pest control (*see* Pest control, in nanotechnology)
Plant diseases, 191f
Plant exposure, AgNP, 321–323t, 325f
Plant fungal infections, 29–32
Plant-mediated AgNPs synthesis, 480–482
Plant morphology, 303
Plant-parasitic nematodes
    AgNPs effect on, 506–507, 539, 540t
    cyst nematodes (*Heterodera* and *Globodera* spp.), 529
    Ditylenchus species, 529–530
    management methods, 532f
        biofumigation, 530–531
        biological control, 531
        cultural measures, 530
        resistance, 531
        solarization, 530–531
    root-knot nematodes (*Meloidogyne* spp.), 528–529
Plant pathogenic fungi, 208–209
Plant-pathogenic nematodes life cycles, 527
Plant physiology, 240, 248
    biochemical parameter, 239, 246
    phytohormones in, 247f
Plant protection, nanomaterials role in, 455–456, 456f
Plants synthesis, AgNPs, 268–269
Plasmonic silver nanoparticle bacterial nanocellulose paper (AgNP-BNCP), 394–395
Pollutant-sensing properties, AgNPs, 245
Polyethylene glycol (PEG), 318–320
Polyethylene glycol functionalized AgNPs, for As(III) determination, 439–440
Polyhexamethylene biguanide (PHMB), 318–320
Polymerase chain reaction (PCR), 340
Polymer bio-nanocomposites, 152
Polymer blend nanocomposites, 134
Poly methacrylic acid (PMAA), 167
Polypropylene (PP) film, 554–556
Polystyrene-*block*-poly(2-vinylpyridine) (PS-b-P2VP), 389–390
Polyvinyl alcohol (PVA), 318–320
Polyvinylpyrrolidone (PVP), 318–320
Polyvinylpyrrolidone-coated silver nanoparticles, 690
Postharvest disease, 509–511
    control
        Ag/CuNPs, 510
        chitosan–silver NPs, 510
        copper oxychloride-conjugated silver NPs, 510
        SiO_2/Ag_2S nanocomposites, 509
Postharvest insect management, 482–483

Poultry, AgNP role, 663–665
  as antibacterial agents, 686–687
  poultry feed, 687–688
  poultry production, 688–689
Protein-targeted mutants (PTMs), 208
Proteomic-based fungicides, 207–208
Proteomics, 340–342, 344–346
Protozoan, 584–586, 587t
Protozoan isolates, 581
Pseudomonas aeruginosa, 41–42
p-Sulfonatocalix[n]arene-modified AgNPs (pSC$_n$-Ag NPs), 385–386
Purpureocillium lilacinum, 204, 205f
PVP-coated AgNPs, 551–552
Pyrethroids, 416–417
Pyto-patch®, 553–554

Q

Quality analysis critical control point QACCP, 154

R

Reaction kinetics, 586
Reactive oxygen species (ROS), 3–4, 21, 250, 271, 308–310, 317–318
  Ag-doped catalysts, 615–616
  degradation role, 617
  production, 693
Redox transformations, in paddy soils, 363–365
Reduced graphene oxide-silver nanocomposites (RGO@Ag), 400–401
Reproductive toxicity, of AgNPs, 695–696
Resistance, 531
rGO-PEI-AgNPs nanosensor, 387–388
Rhizodegradation, 242
Rhizosphere microbes, 301–302
Rhodamine B coated silver nanoparticles (RB-AgNPs), 391
Root-knot nematodes, 528–529
ROS-mediated toxicity, of AgNPs, 694

S

Sargassum muticum-synthesized AgNPs, 480–482
SARS-CoV-2, 713
Scanning electron microscope (SEM), 641
Schottky barrier, 626–628
Sclerotinia sclerotiorum, 30
Seed germination, crops, 302
Semiconductor photocatalytic ternary compounds
  metal tungstates, 625
  metal vanadates, 625–626
  ternary metal ferrites, 626
Severe acute respiratory syndrome (SARS), 713

SFP. *See* Sustainable food packaging (SFP)
Silver-based nanomaterials, in food packaging
  antimicrobial effect of, 135t
  bio-nanocomposite film, 133
  edible films/nanocoatings, 133–134
  nondegradable nanocomposite film, 132
  polymer blend nanocomposites, 134
  preparation of, 131
Silver-coated gold nanoparticles (Au@AgNPs), 393–394
Silver-doped ternary compounds
  antimicrobial mechanism, 629
  applications of, 644, 644–645t
  effect of, 643–644
  morphology
    doping behavior, 630–631
    effect of, 641
  numerous techniques, 623–624
  objectives, 624–625
  photocatalytic activity, 643
  physical properties of, 641–643
  Schottky barrier and surface plasmon response, 626–628
  semiconductor photocatalytic ternary compounds
    metal tungstates, 625
    metal vanadates, 625–626
    ternary metal ferrites, 626
  visible light response, 643
  wastewater treatment, 642t
    bismuth oxyhalides, 638–639
    dye removal mechanism, 628–629
    ferrites, 634–636
    molybdates, 633–634
    niobates, 639–640
    remediation, 631–641
    tantalates, 639
    ternary compounds, 640–641
    titanates, 637–638
    tungstates, 632–633
    vanadates, 636–637
Silver-doped zinc oxide nanoparticles, 415
Silver nanomaterials
  antimicrobial potential of, 175
  bacterial isolates, 581
  biosorption
    concentration effect on, 583
    kinetics studies, 582
    microbial consortia, 586–588
    nZnO and nAg, 584–586, 587t
    pH effect on, 582
  cellular structure, organisms, 590, 591–592f

food packaging, 156–173, 157–165t, (see also Food packaging)
nanomaterials
  characterization, 581, 583–584
  and percentage sorption, 582
  passive accumulation on cellular structure, 583
  protozoan isolates, 581
  reaction kinetics, 586
  statistical analysis, 583
  wastewater solution and adsorbent, 580–581
Silver nanoparticles (AgNPs), 1
  administration routes, 659–660
  advantages, 7
  Ag-doped metal ferrite nanoparticles, 601–603
  in agriculture, 237–240, 238f
    as antifungicides, 552–560
  agri-food applications, 3–4, 5f
  antibacterial activity of, 68–72
  antibiofilm activity of, 70–72
  antifungal activity of, 72–73
  antifungal agents, 18, 550–551
    biosynthesis, 25–26
    human fungal infections, 26–29
    mechanistic approach of, 18, 23f
    plant fungal infections, 29–32
  antimicrobial activity and synergy, 668–669
  antimicrobial agents, 41–42
  antioxidant and antimicrobial properties, 537–538
  antiviral activity of, 73–74
  applications
    air decontamination, 80
    antifungal activity and mycotoxin control, 82–84, 83t
    monuments, 84–85
    multidrug-resistant microorganisms, 82
    nanocomposite fabrics, 79–80
    packaging food, 78–79
    pesticide, 80, 81t
    water sanitization, 80–82
  based chemical sensors, for metal ions detection
    arsenic (As), 439–440
    cadmium (Cd), 441
    chromium (Cr), 441–442
    lead (Pb), 440–441
    mercury (Hg), 438–439
  biocides, 188–190
  biological activities of, 709–711
  biological synthesis method, 537–538
  biosafety and regulations, 7–9
  brewing barley, germination of
    copper oxide nanoparticles, 291–292
    germination of, 285–289
    multiwalled carbon nanotubes, 289–290
    nickel oxide nanoparticles, 293–294
    titanium dioxide nanoparticles, 290–291
  characterization techniques, 435, 435t
  colloidal stabilization
    aqueous phase transfer of, 42–44
    synthetic paradigm for, 40–41
  controlled release
    as antimicrobial agents, 272–273
    as antioxidant, 271
    as nanofertilizers, 269–270
    as nanopesticides, 274–275
  distribution through blood barrier, 690–691
  distribution through placenta, 689–690
  effect on insect biochemistry and physiology, 484–488
  environmental applications, 6
  food packaging, 129f (see also Food packaging)
  global future trends, 9–10, 10f
  green synthesis, 527–528, 534, 539
    using bacteria, 267–268
    using plants, 268–269
  immobilization efficiency of, 241f
    catalytic degradation, 245
    chemical immobilization and adsorption, 243–244
    phytoremediation, 241–243, 242f
    pollutant-sensing properties, 245
  and immune system, 665–666
  infectious bursa disease virus penetration, prevention of, 663–664
  MDR microbes, 46, 47–50t
  mechanism of action, 672–675
  as nanofertilizers, 536–537
  as nanopesticides, 534–536
  nanopreparations
    copper oxide (CuO), 284–285
    multiwalled carbon nanotubes, 284
    nickel oxide (NiO), 285
    silver nanopreparation (Ag), 285
    titanium dioxide (TiO$_2$), 284
  nanoveterinary application, 660–661
  nutritional qualities, 245–252, 250f
  oligodynamic antimicrobial mechanism, 44–45, 45f
  perspective and future trends, 85
  plant biochemical, 245–252
    disease-suppression, 249–252
    mode of entry, 246–248, 247f
    physiology and phytohormone modulators, 248–249
  positive effects of, 276
  potential applications of, 2, 2f

Silver nanoparticles (AgNPs) *(Continued)*
　properties, 533–534
　synergism antibiotics, 46, 51–56*t*
　synthesis, 432*f*, 658–659
　　biological-assisted method, 434
　　and characterization techniques, 660*f*
　　constructive strategy, 266
　　destructive strategy, 266
　　electrochemical deposition, 434
　　hydrothermal approach, 433
　　laser ablation technique, 433
　　physical approach, 433
　　from plant extracts, 659
　　sonochemical synthesis, 434
　　using *Durenta erecta* aqueous extract, 441–442
　　wet-chemical method, 431–433, 432*f*
　synthesis methods
　　chemical synthesis, 706–708
　　green synthesis, 708–709
　synthetic approaches, 43*f*
　toxicity, 74–78
　　mechanisms, 693
　　nanoparticles size and, 691–693
　　reducing nanoparticles, 76–78
　　reproductive toxicity, 695–696
　　ROS-mediated toxicity, 694
　uptake by cells, 689
　veterinary applications, 4–6
　wood-based panels
　　physical and mechanical characteristics, 222–225
　　textiles, 225–229
　　water purification filter, 225
　　wood and wood products, 220–222
Silver nanopreparation, 285
Silver nanosensors
　in agriculture, 107
　heavy metal contamination, 107–109
　heavy metal detection, 111–112
　metal ions, 109–111
　pesticide detection, 112–114
　significance of, 105
SilvoSept®, 667–668
Single-cell gel electrophoresis (SCGE), 339
Slovak Technical Norm (STN), 202–203
Sodium dodecyl sulfate capped silver nanoparticles (SDS-AgNPs), 389–390
Sodium 2-mercapto-ethanesulfonate-functionalized AgNPs, for Hg(II) detection, 439
Soil functionality, 311–312
Solanum nigrum plant-mediated nanosilver synthesis, 477–478
Solarization, 530–531

Speciation, of AgNPs, 361–362
STN EN 1650, 203
Suaeda litura
　AgNP treatment on, 475–476
　long-term toxic effects on, 474
Sucrose-capped AgNPs, for endrin detection, 442–443
Sulfidation, for toxicity mitigation, 365–366
5-Sulfo anthranilic acid dithiocarbamate functionalized silver nanoparticles (SAADTC-AgNPs), 386–387
　for tricyclazole analysis, 444
Superoxide dismutase (SOD), 331–334
Support-based mineralization process, 400*f*
Surface-enhanced Raman scattering (SERS), 104
　bacterial detection, 512
Surface-enhanced Raman spectroscopy based chemical sensor, for pesticides and metal ions detection, 437–438
Surface-enhanced Raman spectroscopy based sensor, 392–395
Surface interaction, of AgNPs, 689
Surface-plasmon-coupled chemiluminescent immunosensor, 512
Surface plasmon response, 626–628
Sustainable agriculture, 235–236
Sustainable ecological practices (SEP) model, 569, 570*f*
Sustainable food packaging (SFP)
　active packaging, 154
　demands, 150–152
　edible packaging, 154
　improved packaging, 154
　intelligent packaging, 154
　nanomaterials, 155
　nanotechnological aspects, 152–155
Synergism antibiotics, AgNPs, 46, 51–56*t*
Synergistic activity of Ag-NPs, 194*f*
Systemic acquired resistance (SAR), 343

T

Tantalates, wastewater treatment, 639
Ternary compounds. *See* silver-doped ternary compounds
Ternary metal ferrites, 626
Ternary metal oxides, 624
Testis-blood barrier transfer, 690
Tetracycline-conjugated *Oscillatoria limnetica* synthesized-AgNPs, 668
Textiles, pest control in, 475–476
Thiol terminated chitosan-AgNPs, for Hg(II) determination, 438–439
Ticks, 662–663

TiO$_2$/AgNPs hybrid materials, 414–415
Titanates, wastewater treatment, 637–638
Titanium dioxide (TiO$_2$)
    barley, germination of, 290–291
    nanopreparations, 284
Toxicity, AgNPs, 74–78
    mechanisms, 693
    nanoparticles size and, 691–693
    reproductive toxicity, 695–696
    ROS-mediated toxicity, 694
Toxicology profile, of AgNPs, 560–565
Toxic properties, of biologically synthesized AgNPs, 482, 483t
Transcriptomic responses in silver nanoparticle treated insects, 486
Transmission electron microscopy (TEM), 320–324
Tri-alanine-phosphine peptide stabilized gum arabic protein AgNPs, 565
Triazophos-Cit-AgNP complex, 388–389
Triazophos detection, 443–444
Tricarboxylic acid cycle (TCA), 344–346
Trichoderma viride, 30
Trichodesma indicum, aqueous leaf extracts of, 474–475
Triclabendazole, 661–662
Trihalomethane (THMs), 601
Trimethylchitosan nitrate (TMCN)-AgNPs, 564
Triosephosphate isomerase (TPI), 344–346
Trojan-horse mechanism, 672–673
Tuberculosis, 670
Tungstates, wastewater treatment, 633–634

U

UV-mediated photochemical transformations, of citrate fabricated AgNPs, 367–368

V

Vaccination, in poultry hatcheries and farms, 663
Valence band (VB), 607, 614–615
Vanadates, wastewater treatment, 636–637
Veterinary sectors, AgNPs, 4–6
Viral applications, of AgNPs
    Chikungunya virus, 715–716
    coronaviruses, 713–714
    Crimean-Congo hemorrhagic fever virus, 716
    filoviruses, 714
    paramyxoviruses, 711–713
    West Nile virus, 714–715
    Zika virus, 716
Viral plant disease control, Ag NPs and Ag-nanocomposites for, 463

W

Wastewater
    disinfection of, 601
    solution, 580–581
Wastewater treatment, 611–613t
    antibacterial activity
        bismuth ferrite nanocomposite, 615
        reactive oxygen species, 615–616
    photodegradation of dyes, 610–615
    silver-doped ternary compounds, 642t
        bismuth oxyhalides, 638–639
        dye removal mechanism, 628–629
        ferrites, 634–636
        molybdates, 633–634
        niobates, 639–640
        ozonation technique, 635–636
        remediation, 631–641
        tantalates, 639
        ternary compounds, 640–641
        titanates, 637–638
        tungstates, 632–633
        vanadates, 636–637
    ternary metal oxides, 624
Wastewater treatment plants (WWTPs), 579–581
Water barrier properties (WBPs), 167
Water-based AgNPs synthesis, 434
Water purification filter, AgNP, 225
Water purification methods, 600
Water sanitization, AgNPs, 80–82
Water vapor transmission rate (WVTR), 152
West Nile Virus (WNV), 714–715
Wood-based panels, AgNPs
    physical and mechanical characteristics, 222–225
    textiles, 225–229
    water purification filter, 225
    wood and wood products, 220–222
Wood products
    classification of, 223f
    durability properties, 220–222
    physical and mechanical characteristics, 222–225
    preservation *versus* conventional treatments, 227t
    textiles, 225–229
    water purification filter, 225
Wound healing activity, of AgNPs, 667–668

Z

Zika virus (ZIKV), 716
Zoonotic diseases, 705–706

Printed in the United States
by Baker & Taylor Publisher Services